Étienne Danchin, Luc-Alain Giraldeau, Frank Cézilly

**Behavioural Ecology**

Étienne Danchin, Luc-Alain Giraldeau, Frank Cézilly

# Behavioural Ecology

OXFORD
UNIVERSITY PRESS

OXFORD
UNIVERSITY PRESS

Great Clarendon Street, Oxford OX2 6DP

Oxford University Press is a department of the University of Oxford.
It furthers the University's objective of excellence in research, scholarship,
and education by publishing worldwide in

Oxford New York

Auckland Cape Town Dar es Salaam Hong Kong Karachi
Kuala Lumpur Madrid Melbourne Mexico City Nairobi
New Delhi Shanghai Taipei Toronto

With offices in

Argentina Austria Brazil Chile Czech Republic France Greece
Guatemala Hungary Italy Japan Poland Portugal Singapore
South Korea Switzerland Thailand Turkey Ukraine Viernam

Oxford is a registered trade mark of Oxford University Press
in the UK and in certain other countries

Published in the United States
by Oxford University Press Inc., New York



First published 2008



British Library Cataloguing in Publication Data
Data available

Library of Congress Cataloging in Publication Data
Danchin, Etienne.
[*Ecologie comportmentale*. English]
Behavioural ecology / Etienne Danchin,
Luc-Alain Giraldeau, Frank Cézilly.
p. cm.
Includes bibliographical references and indexes.
ISBN 978–0–19–920629–2
1. Animal behavior. 2. Animal ecology. 3. Animal behavior—Evolution.
I. Giraldeau, Luc-Alain, 1955– II. Cézilly, Frank. III. Title.
QL751.D34513 2008
591.5—dc22
2007037569

Typeset by Graphicraft Limited, Hong Kong
Printed in China by Asia Pacific Offset

ISBN 978–0–19–920629–2

1 3 5 7 9 10 8 6 4 2

*We dedicate this book to the memory of the late Chris Barnard, whose research in many areas of behavioural ecology and whose impressive textbook on animal behaviour have contributed to the success of this exciting field.*

*To our students, to whom we wish to convey our passion for evolution in general and animal behaviour in particular.*

# Preface

If there is single discipline in which students find significance and frequently even a calling, animal behaviour is the one. Many student generations, including ourselves, have been motivated by dreams of observing fierce carnivore hunts or extravagant avian sexual displays sitting in the middle of African savannahs or Amazonian jungles, binoculars hanging from our neck. However, beyond its obvious romantic beauty, the scene inevitably raises questions about the logic of life itself, a logic that lies in the workings of evolution. This book is about the evolutionary origin of behaviour. Behaviour is indeed part of the phenotype and as such it is under selection just as any other trait. Intuitively, behaviour can be pictured as a fantastic form of plasticity, and thus as one of the main processes of adaptation. To understand it, we must think in terms of evolution.

For millennia, humans believed that the world had remained unchanged in the past and would remain as such forever. Many cultures and religions are rooted in the vision of a fixed, unchanging universe that avoids scientific questions about why the world is as it is. Such a view may seem archaic today, but dynamic views of the universe are recent, emerging as late as a mere two centuries ago. At the beginning of the 20th century, for instance, Albert Einstein, one of the brightest minds of modern times, felt he needed to add a parameter to his equations to reconcile Hubble's expanding universe to his own fixed view of the universe. Similarly, Wegener's theory of continental drift in the early 20th century was universally rejected and it was not until the 1960s with the emergence of theories of plate tectonics that a dynamic view of our own planet became widely accepted. Now the view that nothing in the universe is fixed and that everything changes inescapably seems more readily accepted by all. Quantum theory even teaches us that atoms have finite lives.

Life is above all the ability to self-replicate. Replication constitutes a property that bears enormous potential for change over time. Indeed, the evolutionary process itself is inescapably triggered by the capacity for self-replication. No self-replicating system can occur without information duplication and transfer across generations. However, even the most complex duplication mechanisms are bound to either duplicate or transfer with error, errors that are called mutations. The consequence of these mutations is the origin of different types of individuals possessing different combinations of mutations. If these differences among individuals affect their ability to duplicate and transfer, that is to produce offspring, then inescapably, the frequency of traits that allow individuals to produce more offspring will increase in the population. Eventually, very many generations later, there should be a time when only descendants of that lineage will be present in the population. All the other lineages will have become extinct as a result of competition and selection. The lineage that produced more descendants was selected. This unavoidable process implies that populations change, we say evolve, over evolutionary time.

Historically, Jean-Baptiste Lamark (1744–1829) was the first to come to the conclusion that species gradually change over time as a result of some force; a theory he called 'transformism'. However, it took more than half a century after Lamark's *Zoological Philosophy* for Charles Darwin and Alfred Russel Wallace to publish their view of transformism and propose, not a force, but a material process they called 'natural selection' as the cause of evolution. Since then, the evolutionary perspective has cast a new light on a number of unresolved questions, raised many new ones and lead to the restructuring of biology as a whole. The evolutionary perspective has particularly rejuvenated natural history. The importance of evolution in the study of biology is perfectly summarized in Dobshansky's statement that 'Nothing in biology make sense but in the light of evolution'.

The study of behaviour was obviously also deeply affected by the evolutionary perspective, but its full impact awaited the emergence of behavioural ecology in the 1960s and 1970s. Behavioural ecology considers animal behaviour as an historical process, at the scale of species (macroevolution), populations (microevolution), and the development of individual animals (ontogeny). The approach is rooted in our current knowledge of biological evolution, population genetics, and mechanisms of development. The success of behavioural ecology can be assessed in the number of scientific books and journals that are entirely dedicated to that discipline, as well as in its influence on other disciplines such as experimental psychology, anthropology, and even medicine.

When behavioural ecology emerged, ethology was the sole naturalistic approach to behaviour. Ethology, developed by Nikko Tinbergen and Konrad Lorenz, initially set the stage for the coming of behavioural ecology. Today, it is difficult to distinguish ethology from behavioural ecology unless one wishes to point out that ethology studies mainly the proximate causes of behaviour involving physiology, psychology, and the neurosciences whereas behavioural ecology focuses more on the survival value or adaptive function of behaviour. However, though the dichotomy between these two approaches is often overemphasized, the frontier between them has become fuzzy ever since proximate causation questions have become part of the general framework of behavioural ecology. Consequently, several authors have stressed the fact that these approaches are intimately intertwined (see, for instance, Avital and Jambloka 2000), and it is likely that the future lies more in the collaboration or even the merging of these approaches than in the stressing of their differences.

## *Behavioural ecology today*

There are several textbooks about behavioural ecology. The classic series of *Behavioural Ecology: An Evolutionary Approach*, co-edited by Sir John R. Krebs and Nicholas B. Davies, and their famous co-authored *Introduction to Behavioural Ecology* have more than anything else in the field served to establish behavioural ecology as a coherent research area. The first edition of the co-edited book was published in 1978 and then regularly updated, but not deeply restructured, three times up to 1997 which was more than 10 years ago. The last version of the Introduction is even older, dating back to 1993.

Since then the field has greatly changed and among other things has become increasingly interdisciplinary, incorporating approaches from ethology, population biology, genetics, cognition, physiology, anthropology, and neuroscience. Behavioural ecology is thus a dynamic field that continues to integrate new fields while its definition remains the same. Behavioural ecology can be defined in relation to the hierarchical organization of life, from molecules, to genes, cells, organisms, populations, species, and communities, as the part of biology that is centred at the level of individual organisms. Because behaviour is a property of individual organisms, behavioural ecology is a part of evolutionary sciences that mainly views the individual as a unit of selection. Throughout this book, we will see that behavioural ecology envisages behaviour as a decision-making process that involves information gathering and processing. Thus, behavioural ecology naturally adopts an information-driven approach to behavioural evolution.

Thirty years after the first edition of the Krebs and Davies multi-authored book, and 10 years after its last edition, the field has evolved sufficiently to envisage a new, more up-to-date book on the topic. This is our goal. Being behavioural ecologists, we examined the strategies available to us to accomplish this objective. Two options emerged: either select a few researchers as co-authors with whom to write the book; or, instead, ask a larger number of specialists in each field to write specific chapters. The former has the advantage of homogeneity of style and consistency in terminological usage. However, the downside is that inescapably some fields would have been treated by non-specialists and coverage of the material could be uneven. The alternative solution solves that problem but at the expense of the homogeneity provided by the first option. Given that neither strategy seemed entirely satisfactory, we allowed for a third strategy, a compromise between the two. We asked each specialist from each field to write a first draft of their chapter with the agreement that we, as editors, could make major and substantial changes to ensure consistency of style and vocabulary throughout the book. This third strategy, we feel, provides the best returns and we hope you will agree.

This book was first published in French and was meant to stimulate the development of behavioural ecology among many French-speaking countries. We were quite naturally flattered when English-speaking reviewers contacted by Oxford University Press commented that our French book was *a propos* and addressed many of the fundamental issues they felt their own students should be exposed to. This more recent and updated English version incorporates three new chapters that were not in the original French edition (Chapters 4, 5, and 20).

### *The general principles of the outline of the book*

The book can be read at different levels. Readers without a deep knowledge of evolutionary biology should start reading Chapters 2 and 3, which explain the basic concepts and methods that will be used throughout the rest of the book. Readers who have a good knowledge of evolutionary principles may skip these chapters and go directly to Chapter 4.

For each chapter, we start with an overview of the ideas on the topic that can be found in the literature. However, whenever possible, we also incorporated recent results and ideas to provide the current state of the art view of the field. For each topic, we tried to present the main approaches and schools of thought, so readers may sometimes have the impression that approaches within a field are rather contradictory. They are. However, we preferred an approach that reflects the current state of a field rather than forging an artificial consensus to make things appear coherent when they are not. Debating is part of science.

When possible, we also introduce concepts with the help of case studies, most often taken from animals. This is mainly due to the preponderance of animal examples in behavioural ecology and because most of the contributors mainly (if not only) study animals. However, we regularly use plant examples to recall that the adaptive definition of **behaviour**[1] we adopt applies to plants as well as animals.

### *Book outline*

The book is divided into six parts. **Part one** 'The Behavioural Ecology Approach' introduces behavioural ecology as a scientific field by giving its history, main concepts, as well as general principles and methods. **Chapter 1** 'A History of Behavioural Ecology' discusses the origin of the discipline and specifies its relationships and its originality to other behaviour centred approaches. **Chapter 2** 'Fundamental Concepts of Behavioural Ecology' presents the major concepts necessary for the study of the adaptive function of behaviour. **Chapter 3** 'Research Methods in Behavioural Ecology' explains the main methods and approaches used in behavioural ecology, the hypothetico-deductive approach, theoretical approaches and **optimization**, the advantages of the experimental approach over the correlative approach, the measurement of **fitness**, the **comparative approach**, etc. **Chapter 4** advocates the necessity of adopting 'An Information-Driven Approach to Behaviour'. It first provides the taxonomy of biological information that is used in this book. It then reviews recent studies that demonstrate that behavioural patterns that were thought to be essentially innate are also under significant environmental (and particularly social) influences.

**Part two** 'The Development of the Behavioural Phenotype' deals with a field that has developed particularly rapidly since the early 1990s. It comprises two chapters. **Chapter 5** about 'Life History Strategies, Multidimensional Trade-Offs, and behavioural syndromes' develops an integrated view of the link between genes and behaviour. It covers biotic sources of trade-offs including genes, physiology, mating systems, and social systems. It links genetic approaches and behavioural studies. **Chapter 6** about 'Hormones

1 We use specific conventions throughout the book: key terms in bold set in grey boxes have a specific entry in the Glossary at the end of the book; italics are for foreign language; we have emboldened terms or phrases when we want to stress something.

and Behaviour' continues some of the themes of the preceding chapter in dealing with the specific role of hormones, in the development of the phenotype and particularly that of behaviour.

**Part three** 'Exploiting the Environment' deals with questions of that range from foraging to breeding habitat selection to dispersal. **Chapter 7** 'Solitary Foraging Strategies' and **Chapter 8** 'Social Foraging' present the whole spectrum of issues about foraging separated into two chapters to highlight the distinctive methods used to model non-social and social foraging. **Chapter 9** 'Choosing Where to Reproduce: Breeding Habitat Selection' deals with issues of breeding habitat choice. In fact, the spatial and temporal scales of breeding habitat choice may differ so profoundly from those of foraging that the strategies that are likely to be selected may differ sharply from those involved in foraging. The evolution of breeding habitat choice strategies is directly linked to that of dispersal and gene flow among populations. **Chapter 10** deals with the 'Evolution of Dispersal' and its consequences in terms of population structuring and even speciation.

**Part four** 'Sex and Behaviour' deals with behavioural aspects of reproduction. **Chapter 11** 'Sexual Selection: Another Evolutionary Process' develops the fundamental evolutionary consequences of sexual reproduction. It is one of the fields of behavioural ecology that has developed the most in the past two decades. **Chapter 12** 'Mating Systems and Parental Care' presents the major types and ecological determinants of mating systems. **Chapter 13** 'Sex Allocation' deals with the questions of sexually dimorphic parental investment in offspring of either sex. It tackles questions of why parents may have an interest in investing more in the production of offspring of one sex or the other according to their own condition as well as environmental and social situations.

**Part five** 'Social Interactions among Individuals' deals with the obvious fact that in the vast majority of species, every individual is bound to interact with others, whether they be conspecifics or not. **Chapter 14** 'Animal Aggregation: Hypotheses and Controversies' is about the question of the evolution of group living through the parasocial pathway, which is as a consequence of individuals choosing to live together at some stage in their life. **Chapter 15** 'The Adaptive Evolution of Social Traits' deals with the general dilemma of cooperation. How is it that individuals selected to behave selfishly can nonetheless cooperate at some stage or in some circumstances? Cooperation has been raised as an obvious case against evolution by natural selection. We will see that cooperation no longer raises intractable issues to evolutionary biologists. **Chapter 16** 'Communication, Sensory Ecology, and Signal Evolution' places a particular emphasis on the environmentally induced physical constraints that influence the evolution of communication using different sensory channels. **Chapter 17** 'Interspecific Parasitism and Mutualism' develops behavioural questions in the context of long lasting interspecific interactions such as parasitism and mutualism. The former type of interaction is a relatively well-studied form of long-lasting interaction, but its behavioural aspects are sometimes neglected. The latter type of interaction appears to be particularly unstable and thus poses interesting evolutionary questions.

**Part six** 'Humans and Animals' compensates for the paucity of human examples presented in all previous parts that could give rise to the impression that behavioural ecology does not have anything important to say about human behaviour. The goal of this part, therefore, is to focus on evolutionary issues that are particularly relevant to human behaviour. Applying evolutionary reasoning to humans does not constitute a simple approach, and may easily be questioned. However, it is important to tackle such questions as readers will certainly feel concerned while reading several parts of this book. Furthermore, humans are biological organisms and must in one way or another be under similar selective forces as any other living organism. **Chapter 18** 'Behavioural Ecology and Conservation' treats the question of the specific input that behavioural ecology can bring to conservation biology. **Chapter 19** 'Behavioural Ecology of Humans' deals with the question of the role of evolutionary processes in shaping human behaviour. This will raise interesting issues about whether behavioural ecology may help understanding our biological nature. **Chapter 20** 'Cultural Evolution' deals with an emerging topic in behavioural ecology, that of the potential existence of cultural inheritance in a wide variety of animals. This chapter is linked to the review presented in Chapter 4. One of the main goals of this chapter is to provide a general non-human-centred and testable definition of culture to allow the study of animal culture. Another important goal is to show how important cultural processes can be in evolution in general, not only for humans but also for a wide array of animals.

Chapter 20 ends the book by underlining the major, and often underestimated, role of behaviour in evolutionary processes. Authors often claim that behaviour is an important part of adaptation, but few accept that it is a major **actor** of information inheritance, and thus of evolution (Avital and Jambloka 2000). It is surprising to observe that it is mainly students of behaviour that seek genetic variance at the origin of behavioural variance, rather than incorporate social influences in their reasoning. With cultural transmission, behaviour and learning become as important as DNA duplication for the study of biological evolution in general. Behaviour is thus at the origin of another system of information transmission across generations. Our prediction is that the study of animal culture is likely to develop into a major field of behavioural ecology in the near future.

### *Citation*

Chapters from this book should be cited as, for example:

Cézilly, F. 2007. A history of behavioural ecology. In É. Danchin, L.-A. Giraldeau and F. Cézilly (eds), *Behavioural Ecology: An Evolutionary Perspective on Behaviour*, pp. XXX–YYY. Oxford: Oxford University Press.

### *Further Reading*

Avital, E. & Jablonka, E. 2000. *Animal Traditions. Behavioural Inheritance in Evolution.* Cambridge: Cambridge University Press.

# Learning from this Book

We have adopted several strategies to make this book as easy to learn from as possible.

Extensive **sub-headings** break the text down into short sections, and allow for easy navigation within and between chapters.

ganizing new life cycles. Simple changes in thyroxine regulation can generate spectacular rearrangements of life history cycles (Box 5.3 and Figure 5.5a).

5.3.3 **The miniaturization of plethodontid salamanders**

*The loss of larval stages frees up evolution to refine novelty . . .*

Plethodontid salamanders lost metamorphosis and acquired direct development early in their evolutionary history (Figure 5.5). Direct development constitutes a novel life history adaptation for terrestrial environments. In salamanders with direct develop-

do with making room for an elaborate tongue and supporting skeleton that is used in feeding.

*. . . to rearrangements in bones originally evolved for gills into a tongue . . .*

In the larval form of the ancestral salamander with metamorphosis the hyoid bones served a dual role in gas exchange and feeding. Hyoid bones are derived from vertebrate gill arches. In larval salamanders, the bones are used to support and ventilate the gills (i.e. gas exchange). The bones are also used to expand the buccal cavity and suck prey into the mouth. When these salamanders metamorphose, the hyoid bones are completely reorganized into a tongue projection mechanism used to capture prey in a

Emboldened text is used to draw out **points of emphasis**, or key terms and phrases.

14.2.3.3 *An ongoing debate*

The debate about the ICH continues. In 1996, Marzluff, Heinrich, and Marzluff published experimental results obtained for raven (*Corvus corax*) roosts. **Their study was very impressive because it was one of the only real experimental approaches applied to this hypothesis**. They found compelling evidence that some information transfer was involved in foraging ravens. This article led to a forum in the same journal (*Animal Behaviour*) a few years later (Danchin and Richner 2001; Marzluff and Heinrich 2001; Mock 2001; Richner and Danchin 2001). The forum emphasized the importance of Marzluff *et al.*'s (1996) results, but insisted, again, cass. Indeed, 'local territorial pairs can successfully defend any carcass from one or two juveniles, but give way only once six or more juveniles gather together (Heinrich 1990, Marzluff and Heinrich 1991, Heinrich *et al.* 1993)' (Wright *et al.* 2003). This clearly suggests that by recruiting foraging mates individuals are likely to exploit food patches better, which suggests that raven roosts likely function more as recruitment than information centres.

However, Marzluff and Heinrich (2001) continue to believe that raven roosts correspond to an information centre mechanism, as do other authors. For instance, Wright *et al.* (2003) published another very interesting study of raven roosts in which they placed carcasses baited with colour coded plastic

The book ends with a **Glossary** that provides the definitions of more than 320 general and fundamental terms in behavioural ecology.

When such terms are used in the chapters, they are usually emboldened in grey boxes to remind the reader that there is a corresponding entry in the Glossary.

Can more generally be part of sexual dimorphism versus sexual monomorphism.

**Dilution**: Reduction of the probability of falling victim to a predator by virtue of the presence of alternative potential victims. The probability of being the victim within a group of $n$ individuals is $1/n$. Chapter 14.

**Direct or indirect benefit**: **In sexual selection**. Direct: **benefit** in terms of viability of the progeny that comes from the quality of the mate or parent. Indirect: benefit that comes from the genetic quality of the mate that is **heritable** and will be effective only in the next generation. **In kin**

E

**Eavesdropping**: In the context of **communication**, the **behaviour** of a **receiver** that extracts **information** from the **signals** or **cues** resulting from an interaction in which it does not take part. Chapter 4.

**Economic approach**: Approach that consists of analysing the adaptive value of a trait through a **cost benefit** analysis in terms of its fitness. Most commonly in behavioural ecology these involve **optimization** and **game theory**.

**Effective population size**: The number of individuals in a **population** with efficient reproduction.

**Emancipation**: Developmental phase of the young

Finally, use of ISI may lead to the transmission of behavioural patterns among individuals (i.e. **transmittability**). The possibility of among-individual transmission of behavioural patterns through learning may be called **culture** as it leads to certain forms of inheritance of variation of learned behaviour. This is the subject of Chapter 20.

Another interesting line of research concerns the implications of information use for the evolution of **cooperation** (see Chapter 15). Several recent hypotheses underlined the prominent role of **social prestige** in the evolution of cooperation (Zahavi 1995; Nowak and Sigmund 1998; Wedekind and Milinski 2000; Milinski *et al.* 2002; and see Chapter 15). According to such models, the capacity to mon- information is much more common and influential than previously recognized, and there is now growing evidence that ISI is used in many fitness-affecting decisions. Although studies of true communication have stressed the importance of intentionally produced information (i.e. signals) in the evolution of behaviour, studies reviewed here strongly suggest that the inadvertent component of performance information also plays a prominent role (Danchin *et al.* 2004). Furthermore, it is likely that ISI may be the platform from which many forms of true communication have evolved (Lotem et al. 1999, Danchin *et al.* 2004, 2005). All these results and considerations stress the importance of adopting an information-driven approach to studying the evolution of behaviour.

**Enhanced explanations** of key concepts or points augment the main text, and are denoted as follows:

**of multidimensional trade-offs, and phylogenetic constraints**

At a macro-evolutionary scale involving millions of years and speciation, trade-offs evolve over stretches of time, which if sufficiently vast, allow for novelty to evolve. As innovations are refined and functional integration proceeds, this process constrains subsequent innovation, generating a contingency to trade-offs that makes them taxon specific (e.g., dichotomy between avian and mammalian life histories; Box 5.2). Phylogenetic constraints are idiosyncratic with respect to lineage-specific suites of evolved traits, which is illustrated with an example of salamander endocrine axis, hormones govern gene transcription and regulation through an **endocrine cascade**.

For the moment, we can define a supergene to be a key regulatory gene of development that controls the expression of an **endocrine cascade**. Later, we will broaden our definition to include salient genes and gene interactions that include all aspects of behaviour.

In all vertebrates, metabolism is regulated by thyroxine, a small chain of eight amino acids called an octopeptide. In all vertebrates, a diverse class of protein hormones collectively called a growth hor-

**Boxes** usually take the reader one step further into the subject, by giving further experimental evidence for the concepts or theories being presented, for example.

Box 8.3 **A deterministic model of the producer–scrounger game**

A model is constructed according to a plausible scenario which determines its initial assumptions. Here, we assume that a group of $G$ individuals are searching for patches each containing $F$ indivisible prey items. Among these individuals, $pG$ adopt the producer strategy and $qG = (G - pG)$ adopt the scrounger strategy. A producer discovers a patch at frequency $\lambda$. The $pG$ producers uncover patches at a total frequency of $pG\lambda$. When a producer discovers a patch, it consumes $a$ prey items before the arrival of the scroungers: this is the **finder's** into account the alternative's frequency in the population. We can calculate the equilibrium frequency of producers $\hat{p}$ algebraically given the prediction of equal payoffs at equilibrium.

$$W_p = W_s$$

$$\lambda\left(a + \frac{A}{qG}\right) = pG\lambda\left(\frac{A}{1+qG}\right)$$

$$\hat{p} = \frac{a}{F} + \frac{1}{G}$$

The **further reading** lists at the end of each chapter guide the reader towards further published resources that they might like to consult to take their studies further.

» Further reading

> *The following books and articles provide a general review on the topic of solitary foraging:*
**Begon, M., Harper, J.L. & Townsend, C.R.** 1990. *Ecology: Individuals, Populations and Communities*, 2nd edn. Blackwell Scientific Publications, Boston, MA.
**Charnov, E.** 1976. Optimal foraging, the marginal value theorem. *Theoretical Population Biology*, 9: 129–136.
**MacArthur, R.H. & Pianka, E.R.** 1966. On optimal use of a patch environment. *American Naturalist* 100: 603–609.
**Stephens, D.N., Brown, J.S. & Ydenberg, R.C.** 2007. *Foraging Behavior and Ecology*. The University of Chicago Press, Chicago.
**Stephens, D.W. & Krebs, J.R.** 1986. *Foraging Theory*. Princeton University Press.

**Questions** at the end of each chapter encourage the reader to think more deeply about the topics covered in the chapter. There is often no right or wrong answer to these questions. Instead, they encourage the reader to critically assess the information put before them.

» Questions

1. In the prey model, try to explain why the abundance of a given prey item has nothing to do with its inclusion in the optimal diet. One way to do this would be to redo by yourself the algebra presented in the prey model section.

2. Can you imagine what would happen to the rate of exploitation of a patch if the prey, instead of being distributed randomly, were spread in a regular and predictable fashion, allowing the predator to search for them systematically such that exploitation functions would become linear up to a horizontal plateau? Hint: you will need to use the tangent method to solve this one.

3. You have no doubt noticed that the currency used in the models presented was essentially the gross energy intake rate. Try to think out and draw what the exploitation functions presented in

An **Index of Species Names** precedes the **General Index**, which has been developed to be as useful to the reader as possible.

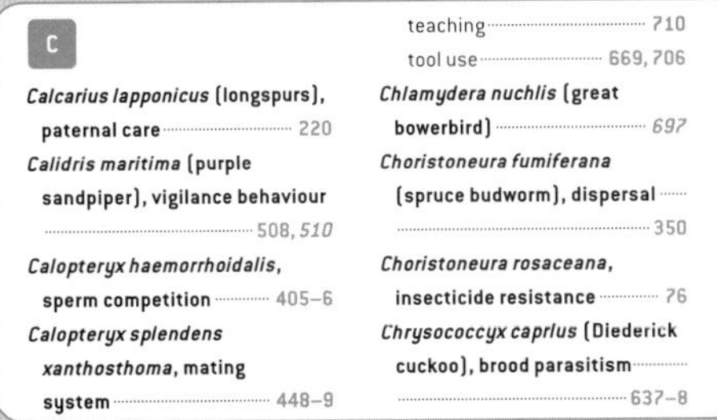
C
*Calcarius lapponicus* (longspurs), paternal care 220
*Calidris maritima* (purple sandpiper), vigilance behaviour 508, *510*
*Calopteryx haemorrhoidalis*, sperm competition 405–6
*Calopteryx splendens xanthosthoma*, mating system 448–9
teaching 710
tool use 669, 706
*Chlamydera nuchlis* (great bowerbird) *697*
*Choristoneura fumiferana* (spruce budworm), dispersal 350
*Choristoneura rosaceana*, insecticide resistance *76*
*Chrysococcyx caprius* (Diederick cuckoo), brood parasitism 637–8

Arnold, S.J. 78
Arnqvist, G. *et al.* 422, 423–5
aromatase 192
role in sex determination 198, 199
Arroyo, B.E. *et al.* 462
artificial selection 33–4
artificial selection experiments 79–80, 95
action of pleiotropy 158–9
associationism 4
associative learning 564, 565

B
babies' cries 685–6
bacteria, aggregation 538
badges 371, **728**
Baird, R.N. and Dill, L.M. *271*, *272*, *273*
Baker, R.H. *et al.* 383
Baker, R.R. and Bellis, M.A. 405
Barnard, C. and Brown, C.A.J. *253*
Barnard, C. and Sibly, R. *274*

# Online Support

This book is augmented by an Online Resource Centre at www.oxfordtextbooks.co.uk/orc/danchin/which includes many figures from the book in electronic format, ready for downloading by registered adopters of the book for use in lectures, etc.

# Acknowledgements

This book results from a collective endeavour. Thus, the role of all the contributors has been invaluable. We are particularly grateful to all of them, and especially for their universal acceptance of the sometimes substantial changes we the editors made to their text in order to adapt it to the general concept of the book. Many of the contributors from France are members of the *Groupement de Recherche Écologie Comportementale* (GDR-CNRS 2155), which emerged in the late 1990s and had the making of the French version of this book as one of its goals. In parallel with the contributors, many others helped in the writing of the book and we are very grateful for their input. In alphabetical order these are: Carlos Bernstein, Caroline Bouteiller, Jacques Bovet, Anne Chapuisat, Philippe Christe, Mike N. Clout, Anne-Marie Combarieu, Isabelle Coolen, Marc Girondot, Bernard Godel, Gérard Lacroix, Laurent Lehmann, Don Merton, Marie-Jeanne Perrot-Minnot, Thierry Rigaud, François Sarrazin, Susana Varela, Éric Wajnberg.

We offer our special thanks and gratitude to Graeme Ruxton who provided particularly helpful and enthusiastic comments to us about the chapters all through the long process of translation and editing. We are particularly grateful to him for his enthusiasm and support.

The French Publisher, Edition Dunod, and particularly Ms Anne Bourguignon and her collaborators, provided invaluable help from the very beginning of this project.

We were helped in the translations by Adrienne Boon, Isabelle Coolen, Madeleine Doiron, France Landry, Kimberley Mathot, Phoenix Bouchard-Kerr, and Stephanie Surveyer.

Other people have more or less indirectly participated to this book, but this does not mean that their role was not fundamental. We are particularly thankful to the sadly missed François Bourlière who has been in France a precursor of behavioural ecology. Several of us were once influenced by his bright mind. More recently, Robert Barbault and Pierre-Henri Gouyon have been keen promoters of ecological and evolutionary studies in France.

Finally many colleagues and students by their discussions have been fantastic friends during this long writing process. Monique Avnaim greatly helped with figures and references all along the writing process. We are particularly thankful to (in alphabetic order): Paul Alibert, Jean-Christophe Auffray, Jesús Avilés, Andy Bennett, Manuel Berdoy, Angéline Bertin, Maryse Barrette, Mathilde Baude, Keith Bildstein, Loïc Bollache, Vincent Boy, François Bretagnolle, Vincent Bretagnolle, Charles R. and Mary Brown, Bernard Brun, Emmanuelle Cam, the sadly missed Jean-Pierre Desportes, Claire Doutrelant, Amélie Dreiss, Frédérique Dubois, Patrick Duncan, Bruno Faivre, Mauro Fasola, Claudia Feh, the sadly missed Heinz Hafner, Philipp Heeb, Fabrice Helfenstein, Philippe Jarne, Alan Johnson, Jim Kushlan, Jean-Dominique Lebreton, Sarah Leclaire, Louis Lefebvre, Sandrine Maurice, Karen McCoy, Agnès Mignot, Hervé Mulard, Ruedi Nager, Isabelle Olivieri, Mark Pagel, Deseada Parejo, Cécile Rolland, Mike Siva-Jothy, Nicola Saino, Nadia Silva, Anne Thibaudeau, Frédéric Thomas, Joël White, Jacques Zafran, and René Zayan.

**Étienne Danchin, Luc-Alain Giraldeau, and Frank Cézilly**
**Toulouse, France; Montréal, Canada; and Dijon, France**
**May 2007**

# Brief Contents

# Full Contents

# Contributing Authors

*Editors*

**Étienne Danchin**, Directeur de Recherche CNRS, Évolution et Diversité Biologique (EDB), UMR CNRS-UPS 5174, Université Paul Sabatier – Toulouse III, France. E-mail, edanchin@cict.fr

**Luc-Alain Giraldeau**, Professeur et Directeur du Département des Sciences Biologiques. Université du Québec à Montréal. Canada. E-mail, giraldeau.luc-alain@uqam.ca

**Frank Cézilly**, Professeur à l'Université de Bourgogne, Dijon, France. E-mail, frank.cezilly@u-bourgogne.fr

*Contributors*

**Thierry Boulinier**, Directeur de Recherche CNRS, Centre d'Écologie Fonctionnelle et Évolutive (CEFE), Montpellier, France. E-mail, Thierry.boulinier@cefe.cnrs.fr

**Frank Cézilly**, Professeur, UMR CNRS 5561 Biogéosciences, Université de Bourgogne, Dijon, France. E-mail, frank.cezilly@u-bourgogne.fr

**Michel Chapuisat**, Senior Scientist, Department of Ecology and Evolution, Biophore, University of Lausanne, Lausanne, Switzerland. E-mail, Michel.Chapuisat@unil.ch

**Jean Clobert**, Directeur de Recherche CNRS, Station d'Écologie Expérimentale du CNRS à Moulis USR 2936, Saint-Girons, France. E-mail, jean.clobert@lsm.cnrs.fr

**Étienne Danchin**, Directeur de Recherche CNRS, Évolution et Diversité Biologique (EDB), UMR CNRS-UPS 5174, Université Paul Sabatier – Toulouse III, France. E-mail, edanchin@cict.fr

**Michelle de Fraipont**, Ingénieur de Recherche CNRS, Station d'Écologie Expérimentale du CNRS à Moulis USR 2936, Saint-Girons, France. E-mail, michelle.defraipont@lsm.cnrs.fr

**Blandine Doligez**, Chargé de Recherche CNRS, Laboratoire de Biométrie et Biologie Evolutive, Université de Lyon; Villeurbanne, France. E-mail, doligez@biomserv.univ-lyon1.fr

**Alfred M. Dufty Jr.**, Professor of Biology and Associate Dean of the Graduate College, Boise State University, ID, USA. E-mail, ADUFTY@boisestate.edu

**Régis Ferrière**, Professeur, École normale supérieure, Membre de l'Institut universitaire de France, Laboratoire Écologie & Évolution, Paris, France. E-mail, regis.ferriere@ens.fr

**Luc-Alain Giraldeau**, Professeur et Directeur du Département des Sciences Biologiques. Université du Québec à Montréal. Montréal, Canada. E-mail, giraldeau.luc-alain@uqam.ca

**Philipp Heeb**, Chargé de Recherche CNRS. Évolution et Diversité Biologique (EDB), UMR CNRS-UPS 5174, Paul Sabatier – Toulouse III, France. E-mail, heeb@cict.fr

**Jean-François Le Galliard**, Chargé de Recherche CNRS, Laboratoire d'Écologie, École Normale Supérieure, Paris, France. E-mail, galliard@wotan.ens.fr

**Mylène Mariette**, PhD student, School of Biological, Earth and Environmental Sciences, University of New South Wales, Sydney, Australia. E-mail, mmariette7@hotmail.com

**Anders P. Møller**, Directeur de Recherche CNRS, Laboratoire de Parasitologie Evolutive. CNRS UMR 7103, Université Pierre et Marie Curie, Paris, France. E-mail, amoller@snv.jussieu.fr

**Barry Sinervo**, Professor of Ecology and Evolutionary Biology, Department of Ecology and Evolutionary Biology, Earth and Marine Sciences Building, University of California, Santa Cruz, CA, USA. E-mail, sinervo@biology.ucsc.edu

**Gabriele Sorci**, Directeur de Recherche, UMR CNRS 5561 Biogéosciences, Université de Bourgogne, Dijon, France. E-mail, Gabriele.Sorci@u-bourgogne.fr

**Marc Théry**, Chargé de Recherche CNRS, CNRS UMR 7179 – GDR 2155 Écologie Comportementale, Muséum National d'Histoire Naturelle, Brunoy. France. E-mail, thery@mnhn.fr

**Richard H. Wagner**, Senior Scientist. Konrad Lorenz Institute for Ethology, Austrian Academy of Sciences. Vienna, Austria. E-mail, r.wagner@klivv.oeaw.ac.at

one

Part One

# The Behavioural Ecology Approach

Any attempt to characterize a scientific approach implies the description of how it emerged historically, how it builds its theories, and how it tests them (Soler 2000). Each of these themes is developed in a separate chapter. Despite the relatively large audience for evolutionary thinking in our society, we are often surprised by how superficial is the common knowledge of graduating students about the fundamentals of evolution. Part one thus provides a brief but comprehensive overview of the basic knowledge that readers need to master before reading the rest of the book.

Chapter 1 presents a history of behavioural sciences. Experienced readers will detect how, at its beginning, this history is intertwined with that of the study of evolution in general and how behavioural ecology emerged as a fully identified scientific domain only 30–40 years ago.

Chapter 2 develops the major concepts used in behavioural ecology. Most of them are not unique to behavioural ecology but in fact pertain to any evolutionary approach.

Chapter 3 presents some general principles and methodologies used in behavioural ecology. Again, most of them are common to any scientific approach.

Chapter 4 stresses the fact that behavioural ecology is essentially an information-driven approach to behaviour. It first provides a taxonomy of information; then it reviews recent evidence that many fitness-enhancing decisions that used to be considered as essentially under genetic control may in fact be deeply affected by social influences.

### Reference

Soler, L. 2000. *Introduction à l'Epistémologie*. Éditions Ellipses, Paris.

# 1

1

# A History of Behavioural Ecology

Frank Cézilly

## 1.1 Introduction

More than 30 years after Wilson's *Sociobiology* (1975) and Krebs and Davies' *Behavioural Ecology: An Evolutionary Approach* (1978) were first published, the term 'behavioural ecology' remains somewhat unfamiliar to the general public. The position is barely any different in academic circles, especially outside the Anglo-Saxon world, where there seems even to be some difficulty in identifying for sure what sets behavioural ecology apart from other disciplines. And the fact remains that the study of animal behaviour is customarily associated more readily with the terms 'animal psychology' or 'ethology'. This state of affairs may be attributed to nothing more than teething troubles or it may be regarded more ominously as symptomatic of a pseudo-discipline with fuzzy demarcations and flimsy theoretical underpinnings. It is essential therefore to characterize behavioural ecology from the outset. This is a particularly pressing requirement in a competitive academic world where the research grants awarded to a discipline and its place in the curriculum are largely dependent upon how it is perceived both by the scientific community as a whole and by decision makers.

Historically, it may be asked whether behavioural ecology is just a continuation of preceding disciplines within what are commonly called the **behavioural sciences** or whether it stands irreclaimably apart from them through a wholesale rearrangement of the theoretical content. This chapter sets out to answer these questions without claiming to make any in-depth epistemological analysis, which lies beyond the scope of a textbook designed primarily for teaching behavioural ecology. We shall confine ourselves initially, then, to tracing the major stages in the history of the behavioural sciences so as to gain a better understanding of how behavioural ecology emerged. We shall then try to clarify how behavioural ecology truly differs from the disciplines that preceded it. Lastly, we shall delimit its field of enquiry, which shall be illustrated in detail by the chapters that follow.

## 1.2 A short history of the behavioural sciences

### 1.2.1 The forerunners

#### 1.2.1.1 *The remote origins*

The scientific analysis of behaviour is relatively recent, first emerging in the late 19th century. However, the antecedents to its study are much older. The observation of animal behaviour must have originated at the dawn of humanity, when the first humans were both prey and predators, and for their own survival had to be attentive to how the animal species around them lived. This special attention was often manifest in spiritual practices of which just a few pictorial representations or rock paintings now remain, such as those adorning the walls of the caves of Lascaux or Tautavel.

The earliest recorded enquiries into behaviour we have are by the philosophers of Ancient Greece, and foremost among them Plato (427–347 Before the Common Era (BCE)) and Aristotle (384–322 BCE). The divergence between the two philosophers over the status of human knowledge foreshadowed the theoretical schisms that were to mark the study of learning many centuries later. Whereas Plato separated mind from body and minimized the role of sensory experience in knowledge, which he deemed could only be attained through reason, Aristotle united the two entities and related knowledge to learning the laws governing nature. The mind cannot conceive of these laws independently of sensory experience, which, with Aristotle, became the medium of cognition. Yet Aristotle's influence on the behavioural sciences did not end there (Dewsbury 1999). For the founder of peripateticism, a distinction had to be made among the various kinds of causes accounting for a phenomenon. Aristotle's classification of causes is commonly illustrated by the example of making a statue. The **material cause** is the matter from which the statue is made, such as clay, marble or bronze. The **formal cause** corresponds to the specific shape imparted to the material, such as human or animal form. The **efficient cause** is defined as the agent responsible for making the statue, the sculptor. Lastly, the **final cause** is the object's purpose, the statue having been created out of some aesthetic intent or to immortalize a famous figure. This emphasis on recognizing that different logical antecedents may contribute to producing a given effect prefigures the discussion about different levels of behavioural analysis (Dewsbury 1999) that continued until the advent of behavioural ecology.

The Greek philosophers exerted a substantial and lasting influence. Intellectual activity in the following centuries came down largely to exegesis of the writings of the ancients. Only in the 17th century was new impetus imparted. This era was highly anthropocentric and intent on demarcating the workings of the human mind as separate from those of animals. This outlook found its fullest expression in the theory of **automata** professed by the Frenchman René Descartes du Perron (1596–1650). This theory holds that humans share certain characteristics with animals but humans alone have souls and are endowed with reason. Animals are merely machines whose movements can be reduced entirely to mechanical principles (the famous Cartesian mechanisms) which must be easily explainable. This conception, as radical as it was, still played a leading part centuries later in the advent of physiological reductionism as a way of studying behaviour. Across the English Channel, the empiricist movement was developed by John Locke (1632–1704) and David Hume (1711–1776). Like Descartes, the empiricists likened the mind's characteristics to a machine working on simple principles. One important strand in British empiricism was **associationism**, which was believed to be the basis of mental activity. By this mechanistic theory, ideas, or sensations were associated when they arose simultaneously. This concept was to be widely taken up in the early theories of learning. However, neither Descartes nor empiricists really went down the experimental road. Their reasoning continued to be based on anecdotal evidence, and their thinking remained both speculative and subjective.

#### 1.2.1.2 *The early developments of sensory physiology: vitalists versus mechanists*

The quest for necessary objectivity was to come in the late 18th and early 19th centuries with the development of sensory physiology and the divergence between the **vitalist** and **mechanist** approaches to behaviour. In the 18th century, biology had not yet taken off in the same way as mathematics, physics, or chemistry. Medical practitioners of the time tended to reduce biology to mechanics and hydraulics. At the same time, chemists and physicists attempted to subjugate biology to their own disciplines. Vitalism developed in opposition to this imperialism (Mayr 1982). Vitalists held that physiological truths are of a higher order than those of physics. This stance was defended notably by the French anatomist Xavier Bichat (1771–1802) for whom life was constantly in opposition to the laws of physics. Consequently, medicine and biology could be based on observation only and so were not amenable to experimentation.

The mechanist approach arose in reaction to vitalism, arguing instead for out-and-out empiricism and giving credit to experiment alone. This strand of thought began with the studies of two physiologists, the Englishman Charles Bell (1774–1842) and the Frenchman François Magendie (1783–1855), who illustrated that nerves carried both sensory and motor signals and showed by experiment the respective pathways of sensory and motor influxes in the spinal nerves. These studies were considerably enhanced by the Frenchman Pierre Flourens (1794–1867), who devised experiments to prove the direct connection between nervous structures and behaviour (Flourens 1842). His experiments, consisting of removing pigeons' cerebral lobes, made a big impact at the time. The subject birds seemed to have lost all their mental 'faculties' and in the absence of any mechanical stimulus remained motionless to the point of starving to death. Flourens was notable too for thinking of the nervous system as being divided into large units each with its own function. These discoveries fostered study of the nerve structures and nervous influxes and completed the physiological foundations of Cartesian mechanisms in the early 20th century with the publication of the studies of the English neurophysiologist Charles Scott Sherrington (1857–1952).

Alongside this, the early mechanical models of behaviour were developing with the studies of the German-American biologist Jacques Loeb (1859–1924) on **tropism**. These are directional growth phenomena influenced by some external stimulus which Loeb first studied in plants. He then transposed the idea to animals to describe the directional movements he studied in invertebrates. He then showed that some directional animal responses result from the stimulation of various intensity of their different receptors. Phototaxis (i.e. orientation movements to the light), thermotaxis (movements according to temperature), and even rheotaxis (orientation movements by aquatic organisms to an oncoming current) became the basic components of activity of living organisms. According to this theory, tropism is akin to sums of reflexes and should be able to account for the behaviour of all life forms. While Loeb's theory involved the serious drawback of reducing behaviour to forced, automatic motion, to its credit it nevertheless clarified the ideas of **stimulus** and **response**. And it prompted further research into directional mechanisms, particularly those of Herbert Spencer Jennings (1868–1947) on protozoans.

The mechanistic strand made further headway with the works of Russian physiologist Ivan Pavlov (1849–1936). His research into the workings of the nervous system brought to light one form of learning, **classical conditioning**. From a series of observations of the behaviour of dogs to which food was shown, Pavlov noticed that gastric secretions or salivation could be induced by stimuli that usually preceded the arrival of food such as the sight of the bowl or of the keeper. Stimuli without any direct relation to the response (**conditioned** stimuli) could trigger responses if they consistently preceded the occurrence of the natural trigger of the reaction (the **unconditioned** stimulus). In this way Pavlov demonstrated classical conditioning. Whereas the idea of reflex, which was already current in Pavlov's time, referred to an automatic and involuntary mechanical manifestation, the Russian physiologist and his co-workers defined the **conditioned response** as a law-governed response to an environmentally determined factor. They made this the fundamental unit of all animal and human learning. Behaviour acquired through training or upbringing could supposedly be reduced to a string of conditioned reflexes (Plotkin 2004).

#### 1.2.1.3 *The naturalists*

Another approach to behaviour developed in parallel to sensorial physiology in the 18th and 19th centuries was that of the **naturalists**. Their detailed descriptions of **animal habits** contrasted with the reductionist conceptions of the mechanists. The early naturalists were not scientists and more often than not simply catalogued and described animal species, much like Thomas Morton (1579–1647), a lawyer, trader, and adventurer who settled in Massachusetts, to whom we owe, among other things, a detailed description of the behaviour of beavers (Dewsbury 1989). The naturalists' approach then became directed increasingly at an ever more thorough description of patterns of animal activity. This concern for detail taken to extremes was typical, for instance, of the work of French physicist and entomologist René-Antoine Ferchault de Réaumur (1683–1757), who engaged in the precise and meticulous observation of insects of agricultural interest and more especially social insects. A few years later, the French naturalist Charles-Georges Leroy (1723–1789) emphasized the importance of field observations and argued about animal's ability to modify their behaviour by experience (Leroy 1802).

The naturalist approach truly took off with Georges-Louis Leclerc, Comte de Buffon (1707–1788), and the publication of the first volumes of his *Histoire Naturelle* in 1749. Buffon objected to the classification of species on essentially morphological and anatomical criteria; he recommended including various ecological and behavioural variables such as their social organization, their use of habitat, or their exploitation of food resources. Animal behaviour became an essential feature of taxonomy. Across the Atlantic, the naturalist tradition met with some success particularly with the American John James Audubon (1785–1851). He was famous for his engravings showing animals in their natural settings and he made countless notes reporting a wealth of observations about the behaviour of North American fauna (Dewsbury 1989). It seems even that the great American naturalist did not settle for merely observing and describing nature. Audubon's assertion that American vultures used sight and not smell for locating their prey is said to have been based on experiments conducted in Louisiana (Chatelin 2001).

The vitalist movement was to find a marked extension among naturalists with the **instinctivist** strand whose most famous representative was undoubtedly the French entomologist Jean-Henri Fabre (1823–1915; Box 1.1). For Fabre and the naturalists, instinct, a sort of unconscious **motivation** which leads animals inexorably towards a goal that they are unaware of, is the basis of the organism's relational life and ensures the preservation of both the individual and the species. However, although Fabre's published descriptions of behaviour are thorough and detailed, they are based on a limited number of often chance observations of each species, sometimes made decades apart. They lack that systematic character that is the hallmark of any truly scientific investigation.

Thus two major avenues of research into behaviour were opened up with sensory physiology and the naturalist movement. The first, which was experimental and focused exclusively on the study of mechanisms, sought to be reductionist. It limited its field of investigation to just a few types of organism. The second was more descriptive and involved cross-referencing and extrapolation from often anecdotal evidence. However, it dealt with a much larger number of species. These two approaches contributed to advances throughout the 19th century in research into behaviour, but the evidence they marshalled was still interpreted within a creationist framework. Despite some questioning by first-rank scholars like Buffon and Pierre-Louis Moreau de Maupertuis (1698–1759), the vast majority of scientists remained convinced of the fixity of species which, it was

Box 1.1 **Jean-Henri Fabre, the forerunner of behavioural ecology**

Only the study of animals in their biotopes found favour with this tireless observer of insects, born at Sant-Léons du Lévezou in southwest France. He collected his countless observations in the ten volumes of his major work *Souvenirs Entomologiques*, subtitled *Études sur l'Instinct et les Mœurs des Insectes* (1879–1908; republished in 1989). In it, Fabre claimed that 'nothing is impossible for the instinct' and supported his claim with the example of the bee, which is able to make perfectly hexagonal cells without any 'understanding of algebra'. He also observed that many parasitoid Hymenoptera are capable, from the time they are first captured and in the absence of any learning, of clinically inserting their sting into a nerve centre of their prey. Moreover, some simple manipulations confirmed Fabre in his opinion that insects, being prisoners of their instinct, have very limited scope for adjusting their behaviour to unpredictable changes in their environment.

Fabre listed the essential features of instinct from many varied observations: it is inborn, pre-formed, fixed, and specific. In this respect Fabre may be seen as a forerunner of ethology. But in some respects he was also a forerunner of behavioural ecology. He consistently emphasized the importance of reasoning by analogy and recommended comparison with similar species to understand insect behaviour. He also suggested that animal behaviour complies with a *loi d'économie de la force* (somehow equivalent to the rule of sparing forces), which he likened to economic principles prevailing in industrial society. Notably, Fabre was the first to invoke this principle in analysing the pathways travelled by various insects. These two aspects of Fabre's conception foreshadow the two methodological pillars of behavioural ecology, the comparative method and optimization (cf. Chapter 3).

assumed, had not evolved since they were first created.

### 1.2.2 The contribution of transformism: from Lamarck to Darwin

Observations by physiologists and anatomists, coupled with those of naturalists, revealed that organisms are complex and well adapted to the surroundings in which they live. This complexity is not some ragged overlap of independent parts. On the contrary, the parts are arranged and interconnected to form a coherent whole. This coherence could only be interpreted in one of two ways at the time (Sober 1993). Either organisms were designed and created by some intelligent entity, or they owed their existence merely to the action of physical forces that transformed inert matter into living forms (Strick 2000). For William Paley (1743–1805), the first interpretation was the correct one. To support his argument, the English archdeacon proposed at the beginning of his *Natural Theology – Or Evidences of the Existence and Attributes of the Deity Collected from the Appearance of Nature* (1802) an analogy that has remained famous. Suppose a walker crossing the heath trips on a simple object like a stone. He will not be particularly surprised and will continue on his way sure that the stone on his path had always been there and that there was nothing exceptional about its presence. But what if the same walker should stumble upon some more complex object such as a watch? For Paley finding such an object with such fine and well-adjusted mechanisms in the middle of the heath begged some explanation. For him only one could be given: a watchmaker had made it. If the argument held for a watch, it held for all other organisms and for the complex organs we can observe in nature. And Paley developed his argument by drawing a parallel between the technical perfection of the eye and that of the telescope, each of which must have a designer. For the English theologian, the perfection of organisms was direct proof of the existence of God. This argument from design won support from the ablest scientists and philosophers of the 19th century (Thomson 2005). The naturalists Réaumur and Fabre saw signs of divine understanding everywhere in the adapted behaviour of organisms. This, then, was the seemingly unpropitious context in which evolutionist views were to develop and irremediably transform the study of behaviour.

#### 1.2.2.1 *Lamarck and transformism*

Between 1788, the date of Buffon's death, and 1800, the date the first transformist hypotheses were formulated, France, the cradle of evolutionism, traversed troubled times. The Revolution entailed profound social and institutional upheavals. These changes also affected the organization of research in the natural sciences and sparked debate about the need for a reform of natural history (Corsi 2001). The advances in the natural sciences and the rigour introduced prompted some scientists of the time to construct a theory capable of combining a unitary vision of nature with the requirement for precision that asserted itself. The prime mover behind this new theory was to be Jean-Baptiste Pierre-Antoine Monet de Lamarck (1744–1829).

Lamarck was a disciple of Buffon and already a reputed botanist when he was appointed professor of invertebrates at the Muséum National d'Histoire Naturelle in 1793. He proved first to be a fixist but then embarked belatedly upon a radical change of attitude (Mayr 1982; Buican 1989). In his *Discours d'Ouverture* delivered in 1800 he set out his new transformist conceptions which he was to develop later in his main work *Philosophie Zoologique* (1809). Lamarck claimed that nature produced in succession different animal species in different lineages which tended inexorably to become more complex over the course of time. Lamarck argued that the more complex species were the present-day representatives of the older lineages while simpler species belonged to lineages that had appeared recently and had little time to reach a high level of complexity (Bowler 2003). The trend towards increasing complexity within lineages was, for Lamarck, a law of nature needing no explanation.

For the founder of transformism, species spread over the course of geological time to different regions of the planet where they developed particular characteristics under the influence of the local environment. Such successive transformations within lineages are explained by a fundamental process, **the law of use and disuse**. In any animal that has not reached its final stage of growth, the repeated use of any organ increases its size but, if unused, the organ becomes atrophied. Behaviour was central to Lamarck's theory as it implied the use of an organ to satisfy some need. In attempting to reach high leaves to satisfy its appetite, the young giraffe strives to lengthen its neck, which consequently grows. Changes induced by more or less intensive use of the organ during development were then passed on to offspring, if they were common to both parents. This principle was renowned as **inheritance of acquired characters**. Clearly Lamarck can be considered the first and a leading evolutionist. His theory broke with the prevailing view of his day of a static world. He argued for a form of gradual evolution that was also adaptive evolution. However, the mechanisms underlying **adaptation** as postulated by Lamarck proved to be mistaken.

#### 1.2.2.2 *Darwin's work*

The dissemination of transformist idea throughout Europe after 1830 contributed to the emergence of a pro-evolutionist culture. Moreover, evidence for evolution accumulated from biogeography, systematics, and comparative anatomy. However, scholars were still reluctant to shift paradigms (Mayr 1982; Bowler 2003). It was at this juncture that the publication on 24 November 1859 of *The Origin of Species by Means of Natural Selection* by Charles Darwin (1809–1882;

box 1.2) completely overthrew the philosophical, religious, and scientific bases of the study of nature. The English naturalist claimed that organisms had not been created unchanging and independently of each other. They all derived from some remote common ancestor and had become transformed and differentiated over millions of years. This differentiation had occurred by a process known as **natural selection**.

Darwin's ideas developed slowly. First his voyage aboard HMS *Beagle* allowed him to notice on many occasions the extent of biological diversity both in terms of the number of species and in terms of variation among individuals of the same species. Then consideration of the mechanism involved in the domestication of animals and plants inspired the naturalist. He recognized in artificial selection practised by man a process of cumulative selection capable of retaining or eliminating slight variations and eventually producing clearly differentiated breeds. Some analogous selection process in nature might account for differences among species. It was through reading the British economist Thomas Robert Malthus (1766–1834) that Darwin hit upon the final component of his theory, the **struggle for life**. Species had demographic potential that far outstripped the rate of renewal of the resources on which they depended. Overpopulation inevitably meant **competition** for resources among individuals. The outcome of such competition was the survival of those individuals best adapted to the environment and the differential transmission of

### Box 1.2 Charles Darwin, naturalist, geologist, and theorist

Charles Darwin was born near Shrewsbury in England on the eve of the Industrial Revolution. From the age of eight he indulged his interest in nature by becoming a keen collector of shells, eggs, and minerals. He began studying medicine at Edinburgh in 1825 but broke off his studies in 1828 at his father's bidding to study theology at Cambridge. A mediocre student, his university years allowed him above all to perfect his naturalist knowledge through studying geology, botany, and entomology. His acceptance as naturalist aboard HMS *Beagle* was the turning point in his life. He embarked on 27 December 1831 aged 22 and did not return to England until 20 October 1836, after visiting South America, the Pacific islands, Australia and New Zealand, the islands of the Indian Ocean, and Africa. On returning, he gathered together a team of brilliant naturalists to classify and study the collections he had brought back from his travels. He was soon acknowledged as an eminent naturalist and rubbed shoulders with the great scholars of Victorian times. While coordinating the work of his assistants, he set himself to writing several books including two on geology, four on the classification of cirripede crustaceans, and the account of his voyage on board the *Beagle*. From 1839 to 1844 he worked on the manuscript that was to become *The Origin of Species* but which was not actually published until 1859. His now poor health left him but a few hours a day to study. However, he did not stop publishing new works including (apart from his books on sexual selection and behaviour) *The Variation of Animals and Plants under Domestication* (1868), *Insectivorous Plants* (1875), and *The Formation of Vegetable Mould, through the Action of Worms* (1881). The entire work of Charles Darwin was marked by the keen observations of this insatiable naturalist, the originality of his theories and the richness of his written style (Browne 2002; Stott 2003). Featuring in the history of science as the most important scientist of the 19th century, his influence is still decisive today, particularly in behavioural ecology.

their characteristics to the next generation[1]. All that was missing from Darwin's theory was a mechanism for inheritance. Darwin considered heredity as a given, but did not account for it within his theory. The rediscovery, a few decades later, of the laws of heredity identified during Darwin's lifetime by the Austrian monk and botanist Gregor Mendel (1822–1884) consolidated the Darwinian edifice after a few trials and tribulations (cf. Gayon 1998).

Darwin's ideas ran counter to Lamarck's on several points. Lamarck claimed it was the direct influence of the environment that engendered variability. Darwin, however, argued that variability came before the influence of the environment, which merely acted to select among what existed. The innate tendency towards increasing complexity that Lamarck held dear was superseded in Darwin's work by the role of chance. With Darwin the deterministic view of nature gave way to a probabilistic view. The adaptation of organisms to their environment no longer conformed to any grand scheme of things, the watchmaker had become blind (Dawkins 1989b).

The repercussions for the study of behaviour of Darwin's theories as set out in *The Origin of Species* are obvious enough. In asserting continuity between animals and humans, Darwin implicitly postulated continuity in mental processes, opening up the way to **comparative psychology** (Plotkin 2004). Behaviour, like any other of an organism's characteristics, was liable to evolve by natural selection and the seeds of the behavioural characteristics of a species had to be apparent in its ancestral species. Darwin gave over the entire seventh chapter of *The Origin of Species* to discussing instinct and the way it evolved through the accumulation of gradual changes. Darwin's influence on the study of behaviour was further asserted with the publication of two other books. In *The Descent of Man and Selection in Relation to Sex* (1871) Darwin picked up on and expanded the idea of **sexual selection**, a compartment of his theory already introduced in *The Origin of Species*. Darwin invoked sexual selection to explain the evolution of certain dimorphic features occurring in one sex only, usually males, and which seem at first sight to have adverse consequences in terms of survival. Such traits include singing, bright colouring, extravagant **ornament**, or certain offensive or defensive features. For example, the calls and decorative plumage of males make them more prominent than (the usually better camouflaged) females and so more readily identifiable to predators. The horns of some Coleoptera or the lion's mane are used as weapons or shields, but do not seem to have evolved to counter predators. If they had, females would have them too. For Darwin such dimorphic features were selected for because of the advantage they procured one sex in **competition for access to mates**. Darwin drew a distinction between traits involved in the direct confrontation between males and those involved in a form of indirect competition arbitrated by the choice of females. Weapons and shields were part of the first category whereas bright colouring and ornament supposedly stimulated the females' aesthetic senses. Although this interpretation of how sexual dimorphism evolved was not greeted enthusiastically when first published, it is now one of the leading fields of investigation in behavioural ecology (cf. Chapter 11).

In *The Expression of Emotions in Man and Animals* (1872), Darwin explained how emotions are expressed on the basis of three main principles which he termed serviceable associated habits, the principle of antithesis and the direct action of the nervous system. The first principle is a sort of preparation for action associated with an emotion and which takes on an adaptive character. Hitting out, for example is a form of behaviour closely associated with anger. Darwin interpreted the movement and tension of nerves and muscles as a habit associated with a serviceable feature which might express anger by revealing the intention to strike out. The second principle posits that opposing emotions have

1 For a time, Darwin was reluctant to publicize his new theory of natural selection. He was however forced to do it when he realized that a younger scientist, Alfred Russel Wallace (1823–1913), had independently come to the same conclusions and was prepared to air his views on the evolution of species. Finally, the contributions of Darwin and Wallace were presented jointly during a meeting of the Linnean Society of London on 1 July 1858.

opposing attitudes or postures. For example, the relaxation of muscles is a sign of calm. The third principle was not as clearly defined as the first two and it seems Darwin used it to explain expressions that could not be accounted for by the first two concepts such as the habit of trembling when frightened. Darwin's influence was to have some impact on the study of communication, particularly the study of signals as movements of intent. However, the various arguments Darwin developed about behaviour were usually illustrated only by anecdote and observation recounted by explorers, naturalists, or zoo-keepers. There is no trace in Darwin's work of any truly experimental approach in support of his theories of behaviour.

#### 1.2.2.3 *Initial outlines of an evolutionist approach to behaviour*

Darwin's hypotheses were to meet with varying fortunes (Mayr 1982; Gayon 1998; Bowler 2003). Generally, it can be said that the idea of evolution by descent with modification was adopted quite rapidly by biologists in Britain, Germany, and the United States and with some delay in France where there was no chair of evolutionism until 1888 (Mayr 1982). Natural selection, however, was long held to be unacceptable.

The advancement of Darwin's ideas was slowed by the development of **philosophical evolutionism** advocated by the Englishman Herbert Spencer (1820–1903). An engineer by trade, Spencer was a prolific writer and published many books setting out his theories which helped to popularize the term 'evolution' (a term Darwin himself cared little for) throughout Europe and the United States (Gould 1974). Spencer's evolutionist apparatus ran counter to Darwin's, though, both in its logical structures and in its dependence on Lamarckism (Bowler 2003; Plotkin 2004). Spencer's system was more metaphysical and based on an analogy between evolution and ontogenetic development implying, like Lamarckism, a deterministic progression of living organisms towards greater complexity. Spencer saw natural selection as merely a principle of preservation of the type species allowing deviants to be removed. This principle does not account for the occurrence of new characteristics, a role Spencer attributed entirely to the inheritance of acquired characters.

Spencer came up with a sociological theory which has most unfortunately been remembered as 'social Darwinism'. It consisted in thinking of human and animal societies as organisms whose characteristics are conditioned by the instincts of the individuals of which they are composed. According to this theory, human evolution should be viewed as a moral and not an organic process. Ill-adapted social conduct does not entail death but moral suffering which progressively brings the individual to change his behaviour and eventually to temper his selfishness and act more altruistically (Kaye 1986). Such 'morally proper' behaviour then spreads through society by inheritance of acquired characters. This view of social regulation based on environmental determinism of human conduct was then taken up by others to legitimize highly objectionable educational programmes and social policies, which were often tinged with racism (Valade 1996). This excess explains the negative connotations that social Darwinism has since acquired and its influence rapidly dissipated after the German biologist August Weissmann (1834–1914) finally refuted the hypothesis of inheritance of acquired characters.

The spread of Spencer's evolutionist ideas did, however, have a direct impact on the development of studies of animal societies. The French sociologist Alfred Espinas (1844–1922) published a book in 1876 on the different forms of association, from groups of cells through to human societies, in which he proposed a classification of monospecific animal societies by their function: nutrition, reproduction. He argued that animal societies implied highly structured relations among individuals of the same species (Espinas 1876). These forms of relations were independent of taxonomic relations among species and should be considered as the expression of direct influence of the environment on organisms' characteristics. Some of Espinas' considerations on the contrast between monogamy and polygamy or on the distribution of sea-bird colonies relative to food resources are evidence of a very relevant acknowledgement of the influence of ecological factors on social organization. From this standpoint Espinas can be seen as a forerunner of socio-ecology. In the same vein, the works of Emile Waxweiler (1867–1916) in Belgium aimed to bring together biology, behaviour, and sociology. This approach to animal societies, strongly coloured by Lamarckism, continued to exert great influence especially in France (Hachet-Souplet 1928) until the early 1930s.

Despite Spencer's influence and Flourens' resistance, Darwin's work on behaviour also had an impact on the development of psychology. This influence was particularly marked with George John Romanes (1848–1894) who sought to prove a continuity of mental states between humans and animals. Romanes (1882) was the first to propose methods for developing true comparative psychology as an extension of Darwin's ideas. His main method was that of **subjective inference**, consisting of considering that animal behaviour was analogous to human behaviour and that mental states associated with such behaviour in humans could also be found in animals. Romanes also considered that human emotions arose in other animals depending on how complex they were. He argued that fish could be jealous or angry, birds could be proud and apes feel shame or remorse. However, the empirical basis of Romanes' work remained limited, like Darwin's, to anecdotes or facts related by amateur naturalists. His method was therefore largely inductive, with the converging impressions of each observer supposedly underpinning the generalization of reasoning based on a few individual cases.

By the late 19th century the results of the influence of transformist ideas on the study of behaviour therefore remained mixed. On one side, many naturalists were not won over by transformist arguments and dismissed any influence of the process of natural selection on animal behaviour. This was particularly the position of Fabre, who, despite the esteem in which he held Darwin (which was mutual), consistently declined to side with the evolutionists. On the other side, studies of animal societies were driven merely by the hope of finding general organizational principles which would be valid for the human species. Lastly, proclaiming the continuity of mental states between humans and animals had the unfortunate consequence of promoting exaggerated anthropomorphism which did not baulk, on flimsy empirical foundations, at attributing elaborate cognitive abilities to animals. The scientific bases for the study of animal behaviour still lay ahead.

### 1.2.3 The behaviourist approach

The excessive anthropomorphism of the early evolutionists soon prompted reactions. In 1894 Conwy Lloyd Morgan (1852–1936) published *An Introduction to Comparative Psychology*. In this fundamental text for the study of behaviour he proposed his famous 'canon', a sort of rule of parsimony. Where there were alternative explanations, Morgan advised using the one which required the fewer assumptions. As applied to behaviour, Morgan's canon meant scientists should avoid construing animal behaviour in terms of feelings and emotions felt by humans. He claimed it was unwarranted to refer to higher psychological structures (such as intention or will) when simple systems of the reflex or tropism type could adequately account for the behaviour observed. Morgan's canon gradually gained support from many scientists especially in North America where the **behaviourist** movement formed under the leadership of Edward Lee Thorndike (1874–1949), John Broadus Watson (1878–1958), Clark Leonard Hull (1884–1952), and Burrhus Frederic Skinner (1904–1990). Although heirs to the mechanistic school of thought, behaviourists ignored the internal cogwheels of behaviour. They broke away from a psychology of states of consciousness to concentrate exclusively on the objectively observable behaviour in which organisms engaged in response to stimuli. Behaviourism placed great emphasis on acquired conduct and tended to reduce the explanation of any behaviour to the exposure of learning mechanisms.

Thorndike argued that animal conduct was dictated neither by instinct nor by some form of reasoning analogous to human reasoning but simply by learning 'through trial and error with chance success'. The experimental paradigm he developed was quite straightforward. A hungry animal, say a cat, was placed in a **puzzle box** and food was set down outside the box. To escape from the box and reach the food the animal had to activate some mechanism. In the first few attempts not much happened except for the animal spending a lot of time trying to bite and scratch its way out. Thorndike measured learning by the time it took the animal to escape. At first learning was a slow business. But with repetition of the experiment Thorndike noticed that the animal made fewer inappropriate responses and was ever faster in displaying the behaviour allowing it to open the gate. Thorndike claimed these results suggested a learning process in which the animal progressively eliminated any inappropriate responses and eventually retained only the right response. Learning the solution to the problem involved making connections between the environmental stimuli and the organism's response. Several laws were suggested to explain how the connections were made, the most important being the 'law of effect'. This law introduced the ideas of positive and negative **reinforcement**. The first part of the law states that when a situation entails several responses, those which lead on to a 'satisfactory' state for the animal will be the most closely connected with the situation so that when the same situation recurs, there will be a higher probability of the same responses appearing. The second part of the law suggests that responses entailing a state of discomfort for the animal will have weaker connections and will be less likely to recur when the animal is confronted with the same situation again. The law of effect is important in that it establishes a direct linkage between an animal's behaviour and its effect on the environment.

Skinner, for his part, set out the laws of **operant** conditioning; he opposed these to Pavlov's classical conditioning, which was renamed **respondent** conditioning. The latter concerned conducts whose occurrence could be related to one or more earlier events to which the behaviour was a response. Operant conditioning corresponded to spontaneous behaviour, the occurrence of which could not be related to any prior stimulus. To study this type of conditioning, an animal is placed in a cage where there is a lever, for example. Activating the lever triggers the distribution of food. The animal discovers this relation by chance. The animal then makes the connection between its activity (pressing the lever) and the consequences of that activity on the environment (the arrival of food). The experimenters' degree of control, being free to reinforce any particular action by the animal, enables them to choose the conduct to be studied.

With behaviourism, learning was reduced in some sense to making connections between stimuli and responses. This movement was thus the complete opposite of the naturalist approach. Its purpose was not to describe the diversity of behaviour of animal species but rather to find their invariants. The ultimate end was to reach a level of prediction of learning where behaviour could be controlled. Animals were simply models worth studying for the lessons that could be derived for understanding learning in humans. The choice of biological model was dictated for behaviourists by practical considerations and only the few species which were easy to breed and keep in the laboratory (rats, mice, pigeons, cats) were regularly confronted with the perfunctory environment of the puzzle box. Despite this naturalistic 'deficit', the behaviourist movement did have its implications for behavioural ecology. First, it imposed the requirement of resorting to parsimonious explanations. This mandatory reference to Morgan's canon is paramount in arbitrating among various explanations and has maintained all its heuristic value in behavioural ecology, whose adaptationist discourse remains vulnerable to anthropomorphic excess (Kennedy 1992). Moreover, the behaviourist movement prioritized the experimental approach and to this end developed automated appliances for controlling situations in which to evaluate animal performance. These experimental devices are still used today in behavioural ecology especially in experimental verification of predictions of optimal foraging models (cf. Chapter 7).

### 1.2.4 The cognitivist approach

The behaviourists' reductionist radicalism came in for fierce criticism in its turn. The systematic refusal to invoke any intermediate variable was challenged in the early 20th century by **gestalt psychology**, one of whose leading proponents was Wolfgang Köhler (1887–1967). Unlike the behaviourists, Köhler considered that the requirement for rigour and parsimony need not necessarily be reflected by a strictly quantitative conception of the study of behaviour. Quite the contrary, he emphasized that preliminary qualitative and more global observations were required to bring out the problems which should then be examined in detail. The gestalt approach stood apart from the connectionist behaviourist approach in the type of organisms it studied. Köhler's (1925) work included a meticulous description of how chimpanzees managed to resolve a problem whose solution is related to the animal's understanding of spatial and causal relations, for example using a stick to drag an object within reach. Köhler noticed that two stages occurred before the solution was reached. In the first stage, the animal made disorderly movements. Then came a clearly marked stoppage stage before the resumption of activity which was now coordinated and flowing and which quickly led to the solution. For proponents of gestalt psychology, the stoppage stage was a phase of rearranging the spatial information and restructuring of the perceptual field providing a grasp of the significant relations of the environment and leading to a direct apprehension of the situation, a process known as **insight**.

Edward Tolman (1886–1959) was another pioneer of cognitivism. He emphasized the flexibility of animal behaviour, which he refused to reduce to a string of connections of stimuli and responses. He thought animal behaviour fulfilled a purpose which made the activity observed by the experimenter meaningful. Thus, the animal's first movements in the problem box were not irrational given the situation. They were not haphazard movements. The starving cat was trapped and crashed into the cage walls in an attempt to force its way out to reach the food outside. Its activity was directed from the outset, proof that the animal had grasped the various relevant components of the situation. The works of Tolman and his assistants on the behaviour of the rat in a maze led in particular to the idea of a **cognitive map**. This was a mental picture the animal made of its path in successive passages through the maze and Tolman demonstrated it was not the same as simple connections between spatial stimuli and motor responses.

Cognitivism clearly opposed behaviourism in claiming that there were unobservable internal processes allowing information extracted from the environment to be integrated. The opposition between cognitivism and connectionism was to deeply mark the course of comparative psychology. The input from cognitivists was decisive for the study of behaviour which shook off the straitjacket of an overly restrictive interpretation of Morgan's canon. This did not actually state that the simplest explanation was necessarily the best. A simple explanation may be abandoned for a more complex interpretation as long as it better accounts for the situation. In a sense, the cost of an assumption, in terms of the more or less complex mechanisms it implies, is to be gauged against the benefits derived from it in terms of predictive power.

## 1.3 Ethology

Whether affiliated to connectionism or to cognitivism, comparative psychology in the early 20th century remained focused on laboratory study. This situation could not satisfy the zoologists of the day. As heirs of the naturalist movement, they refused to subscribe to a science of behaviour that ignored the diversity and complexity of animal conduct as observable in nature. Behaviourist studies concentrated exclusively on animals whose behaviour zoologists suspected was deformed by domestication and captivity. Moreover, the situations animals were confronted with (problem box, maze, Skinner cage) were to their minds only over-simplified schematizations of the natural environment. It was the very relevance of the comparative psychology approach that was called into question with the development of **ethology, a biological study of animal behaviour that was resolutely naturalistic and evolutionist**.

The term 'ethology' was introduced back in 1854 by the French zoologist Isidore Geoffroy Saint-Hilaire (1805–1861) to designate the study of the ways of being of animals.[2] Initially this branch of zoology with its still hazy outlines developed in reaction to the preponderance of another branch, comparative anatomy. This had thrived in France and in Europe under the influence of the French zoologist and palaeontologist Georges Cuvier (1769–1832) and field observations had been progressively abandoned for dissecting dishes (Jaynes 1969). The return to the ethological approach was advocated in particular by the French zoologist and evolutionist Alfred Giard (1846–1908) whose interest in shoreline organisms could hardly be catered for by the investigative conditions offered by laboratories of the time. A convinced evolutionist, Giard did not dismiss natural selection but remained attached to Lamarck's ideas and considered that environmental factors were the main driving forces behind evolution through their direct influence on animal behaviour. He called for the development of an 'external physiology', that is a study of animal habits and their interactions with the environment, which he explicitly distinguished from comparative psychology (Giard 1904). Subsequent to Giard, the proponents of the ethological method were to maintain the distinction between ethology and comparative psychology (Jaynes 1969).

### 1.3.1 The early development of ethology: 1900–1935

The late 19th century was marked by the ultimate rejection of Lamarckism (Weismann 1892). The growing adhesion of naturalists to the theory of evolution by natural selection meant that, at the turn of the 20th century, the value of behaviour for systematics became an inescapable question. Answering it implied studying species that were phylogenetically close, easily observable and which displayed remarkable behavioural characteristics. Birds were a favourite biological model for the forerunners of ethology. Among the leading figures were Charles Otis Whitman (1842–1910) in the United States, Oskar Heinroth (1887–1945) in Germany, and Julian Huxley (1887–1975) in England. Whitman did substantial work on the behaviour of pigeons (published posthumously, Whitman 1919), whereas Heinroth concentrated on the behaviour of the Anatidae (ducks and geese; Heinroth 1911). Impressed one day by the observation of a young bird of prey which, although it had never seen a pheasant, reacted instantly at the sight of a prey and seized it in a flash, Heinroth became convinced that instinctive behaviour is quite perfect and fully functional from the outset (Lorenz 1977). Such a phenomenon assumes the existence of innate information. Heinroth consequently proposed that **specific instinctive behaviour** was the fundamental component of behaviour. Such behaviour could not be reduced to a sequence of unconditioned reflexes. Triggering it required some endogenous mechanism whose structure contained exclusively the information acquired in the course of phylogenesis. Whitman (1898) reached the same conclusions. Similarities and differences between bodily movements of species of the same group were ordered by taxonomic proximity, and behaviour might therefore be considered an infallible sign of kinship among species in the same way as morpho-anatomic characteristics. In their wake, Huxley advanced a conceptual interpretation of the phenomenon he called **ritualization** (Huxley 1914) based on observation of the courtship behaviour of the great crested grebe (*Podiceps cristatus*). The complex and spectacular courtship displays of different bird species were much described at the time, although understanding of them was limited. In grebes, the courtship display is characterized by stereotyped behaviour. Male and female face each other, dive and return to the surface several times, each time holding pieces of water

2 The term was first used in its modern sense as the study of the behaviour of animals in their natural environment, in North America in 1902 by William Morton Wheeler (1865–1937), an expert on arthropod behaviour, in a paper published by the review journal *Science* (Wheeler 1902).

| | Ethology | Comparative psychology |
|---|---|---|
| **Main geographical origin** | Europe | North America |
| **Main subject of study** | Innate behaviour (instinct) | Acquired behaviour (learning) |
| **Study location** | Natural or accurately reconstructed environment | Purposefully simplified experimental arrangement |
| **Level of control of subjects** | Minimal | Stringent |
| **Preferred biological models** | Species favoured by naturalists: insects, birds, fish | Rodents, pigeons, humans |
| **Type of measurements** | Detailed recording of behavioural components | Simple objective responses (pressing a lever) |

Adapted from Brain 1989.

Table 1.1 **Key differences between ethology and comparative psychology**

plants in their bills. Huxley suggested this component of grebe courtship was derived from the activity of nest construction and, in the course of evolution of the species, had acquired a role in coordinating the social behaviour of both sexes. In considering behaviour as hereditary motor coordination built up over the course of evolutionary history of the species and in asserting its relevance for phylogenetic reconstruction, in one fell swoop Whitman, Heinroth, and Huxley explicitly related understanding of behaviour to its evolutionary history and so laid the foundations for modern ethology (Hess 1962; Lorenz 1977).

### 1.3.2 The heyday of ethology: 1935–1975

It is generally agreed that the birth of modern ethology coincided with the advent of the **objectivist school** which arose in Europe in the 1930s in reaction to behaviourism. At this time, the idea of instinctive behaviour was harshly criticised by lab-coated North American behaviourists (Dunlap 1919; Kuo 1924), who viewed it as a vague and pointless concept. Naturalists, with their boots and binoculars, were still perceived as amateurs, albeit enlightened ones, but lacking the experimental rigour that necessarily characterizes any scientific discipline. The objectivist school then set out to wrong-foot the behaviourist movement (Table 1.1). It developed a zoocentric approach based on a methodology and a theoretical analysis that was totally removed from behaviourism. Its two historical leaders were the Austrian Konrad Lorenz (1903–1989) and the Dutch-born Niko Tinbergen (1907–1988; Box 1.3).

#### 1.3.2.1 *Tinbergen's four questions*

With them ethology was defined by its subject matter, the study of the **natural behaviour** of animals. Priority was given to observing animals in their habitual surroundings. Laboratory study was justified out of practical concern only (better control of observations) and implied satisfactorily reconstituting natural conditions. Four levels of causation were clearly identified by ethologists in analysing behaviour (Huxley 1942; Kortlandt 1940; Tinbergen 1963): immediate causation, ontogeny, adaptive value, and evolution. **Immediate causation** related to the physiological mechanisms directly involved in performing behaviour at a given moment depending on the animal's internal state. Thus a bird sings in springtime because the longer period of daylight

### Box 1.3 Lorenz and Tinbergen, the founding fathers of ethology

**Konrad Lorenz**. Austrian by birth, Konrad Lorenz developed a lively interest in observing the living world from an early age and intended to study zoology and palaeontology. However, after secondary school, obeying his father's wishes, he resigned himself to going through medical school. And yet it was in those years of study, marked by a real enthusiasm for comparative anatomy and embryology, that he learned strict scientific reasoning and became convinced of the importance of the comparative approach. An amateur naturalist initially, Lorenz remained a passionate student of the behaviour of birds, especially ducks and crows. Without giving up his medical studies, he trained in ethology under the influence of Heinroth and then of Craig and Whitman. He took up his first appointment in 1937 and began to develop his views on instinct and imprinting. From 1951 onwards, in view of how successful his theories were, the Max Planck Gesellschaft granted him substantial resources with which to develop an ethology research institute based in Seewiesen. He then concentrated his research on the study of aggression and its regulating mechanisms. Late in life, Lorenz took an interest in the evolution of culture and the relations modern humans maintain with their environment. He wrote several books including *Evolution and Modification of Behavior* (1965), *On Aggression* (1966), *Studies in Animal and Human Behaviour* (1970), and *Behind the Mirror* (1977).

**Niko Tinbergen**. A nature lover, Niko Tinbergen first took an interest in studying animal behaviour in the Netherlands of the early 20th century, under the influence of the writings of Fabre and the works of Von Frisch. After writing a thesis on the behaviour of digger wasps, he began his academic career at Leyden University where he developed the teaching of comparative anatomy and ethology. In 1936 he met Konrad Lorenz. It was the beginning of a long friendship between the two whose complementary nature (Lorenz was more of a theorist, Tinbergen more of an experimenter) was to be decisive for the development of ethology. After the Second World War, Tinbergen made many contacts with scientists in the United States and Britain, where he went to live in the 1950s. In his wake, research in ethology became preponderant at the Department of Zoology of Oxford University. Towards the end of his career, he concentrated on an ethological approach to the problem of autism in children. His best-known works are *The Study of Instinct* (1951) and *Social Behaviour in Animals* (1953).

stimulates the production of certain hormones which in turn induce the singing behaviour. In terms of **ontogeny**, behaviour is analysed in a historical dimension, mainly in relation to experience early in life. The characteristics of any given individual's song are related to certain events that occurred during its development (for example, the variable richness of the sonic environment in which the individual developed). The **adaptive value** of behaviour is seen by ethologists as its current utility in the animal's natural surroundings. This utility is defined by the consequences of behaviour for the individual and its environment. The bird sings 'in order to' defend its territory or 'in order to' attract a mate. This level of causation introduces apparent purposiveness into the interpretation of behaviour (cf. Chapter 3). Lastly, the question of the **evolution of behaviour** is part of another historical dimension, that of the evolutionary history of species. The bird sings because it belongs to species in whose ancestors singing behaviour appeared and was maintained and transformed over the course of evolution. Over time, under the influence of certain commentators (Dewsbury 1999), these four levels of analysis, often called '**Tinbergen's four questions**', were merged into two tiers: proximate causes (physiological causation and ontogeny) and ultimate causes (adaptive value and evolution).

Ethology began to develop as an institutional discipline in the 1930s with the formation of the first learned societies (Durant 1986). The Deutsche Gesellschaft für Tierpsychologie was founded in Germany in 1936 followed two years later by the Institute for the Study of Animal Behaviour (ISAB), which became the Association for the Study of Animal Behaviour (ASAB) in 1949 in Britain. In 1937 Lorenz and his colleagues launched the first specialist journal in the field, *Zeitschrift für Tierpsychologie* (recently renamed '*Ethology*'). The outbreak of the Second World War disrupted the development of ethology in Europe for a time and slowed its influence in the United States. However, the ties among ethologists on both sides of the Atlantic before the war (Lorenz having visited the University of Columbia at New York in 1922 and Tinbergen having visited various American institutions in 1938) soon re-formed at the end of the conflict (Dewsbury 1989; Kruuk 2003). In 1948 Tinbergen, with one of his Dutch students, Gerard Pieter Baerends (1916–1999), and the British ethologist William Homan Thorpe (1902–1986), founded the journal '*Behaviour*', with American scientists on its editorial board. In 1953 the bulletin published by the ASAB expanded and became the '*British Journal of Animal Behaviour*'. In 1958 the journal was renamed '*Animal Behaviour*' and published jointly by the ASAB and the Animal Behaviour and Sociobiology Section of the Ecological Society of America.

Exchange among ethologists of different geographical and disciplinary backgrounds further intensified in the 1950s with the organization of international conferences. As early as 1949 an important conference was organized at Cambridge at the joint initiative of the ISAB and the Society for Experimental Biology. It was coordinated by Tinbergen and Thorpe and brought together the leading European and American scientists, reforming ties that had been severed during wartime. The programme, initially given over to the study of physiological mechanisms of behaviour, addressed the major questions of the time, particularly questions about central versus peripheral control of behaviour (as behaviour was still thought to be the observable outcome of a time sequence of muscular contractions, this raised the problem of whether the timing of these contractions was predetermined by the central nervous system or induced by external stimuli before the movement was performed) and about the nature of instinct. Three years later the first international conference on ethology was held at Buldern in Germany, and the second the following year at Oxford in England. Ever since 1955 this conference has been held every two years.[3] Although initially it brought together mostly European ethologists, in the 1960s it attracted a growing number of North American scientists and since the 1970s it has regularly been organized by non-European countries. By this time ethology had become a recognized discipline firmly established in universities on both sides of the Atlantic.

This institutional success was crowned in 1973 by the joint award of the Nobel Prize for medicine and physiology to Lorenz, Tinbergen, and Karl von Frisch (1886–1982), the founder of modern comparative psychology and author of a study of the now famous 'dance' of bees (von Frisch 1955). This somewhat belated recognition, in fact, rewarded a long series of original works which, since the 1930s, had profoundly changed conceptions of animal behaviour. One of the early ambitions of objectivist ethologists had been to save the concept of instinct from the rather sterile debate between vitalist and mechanist viewpoints (Baerends 1976). Even at the 1949 Cambridge conference, Tinbergen had insisted on the undeniable variability and plasticity of so-called instinctive behaviour, which seemed an insuperable barrier for the mechanistic interpretation. Lorenz then proposed a solution that was supposed to satisfy mechanists and vitalists alike. He suggested confining the concept of instinct to **fixed action**

3 The 2007 conference took place in Halifax, Nova Scotia, Canada.

**patterns**, which were defined as simple, highly stereotyped and species-specific behavioural units (Lorenz 1950). These units were triggered, Lorenz claimed, in a near reflex manner by **innate releasing mechanisms**, inborn perceptual filters, as it were, involved in distinguishing among various stimuli. The highly specific fixed action patterns gave them some value as taxonomic characters, much like morphological structures. They also provided an approach to the study of evolution and genetic control of behaviour based on the study of the behaviour of hybrid forms (Lorenz 1958; Ramsay 1961). In addition, through his 'psychohydraulic' model of behaviour, Lorenz introduced the somewhat vitalist concept of **action specific energy**, which he claimed accounted for variability in the expression of behaviour through changes in organisms' motivational states (Box 1.4).

The essential contribution of objectivist ethologists in the first half of the 20th century was without contest the revelation of the phenomenon of behavioural imprinting (*Prägung*; Lorenz 1935). For most organisms, social contacts (kinship or mating) are generally between individuals of the same species, which assumes that individuals are able to 'recognize' their conspecifics. In the early 20th century, identifying fellows was held to be an instinctive phenomenon not involving any learning. But a few cursory observations by Heinroth on ducks and by the American Wallace Craig (1876–1954) on pigeons (Craig 1908) implied that the ability to recognize fellows was in fact an acquired trait. Lorenz, to his credit, undertook a systematic study of this phenomenon. He soon established that ducklings confronted upon hatching with a human being readily adopted humans as a maternal substitute and in adulthood directed their sexual behaviour towards people. Lorenz went on to validate this phenomenon with many other bird species. In his first conception of imprinting, whether it manifested itself in the filial attachment of the young animal or in its sexual orientation in adulthood, the German ethologist identified a few essential characteristics. The phenomenon was limited in time to a critical period, corresponding to a specific physiological state of maturity. Its effects were deemed irreversible. The imprint was at a supra-individual level: the animal was impregnated by the general characteristics of the object and not its individual features. This dimension of the phenomenon explained that the animal recognized the species to which it belonged. Imprinting was therefore a highly original form of learning. Lorenz's work on imprinting provoked many strong reactions in comparative psychology circles because it called into question the behaviourist conception of learning which was limited to Pavlov's classical conditioning and to Thorndike's trial-and-error learning process.

### 1.3.3 The time of controversies

Ethology entered a period of controversies from the 1950s onwards. Criticisms initially came mostly from comparative psychology. Concerns were related to the classification of behaviour, the concept of instinct, the various levels of analysis used, and especially the physiological assumptions associated with energy models, and an over-simplified conception of motivation (Lehrman 1953; Hebb 1953; Kennedy 1954; Beach 1955; Schneirla 1956). The central concept of fixed action patterns was also challenged (Barlow 1968). Criticisms peaked at the 1973 international ethology conference which neither Lorenz nor Tinbergen attended.

As time went by, research in ethology had come to be organized around three separate explanatory systems (Gervet 1980). The first was a **reductive system** which explained behaviour on the basis of hypothetical elementary biological processes of the sensory, physiological, or neuro-endocrinal type. For such ethologists, recourse to this explanatory system did not necessarily imply the use of physiological techniques (Tinbergen 1951). The second explanatory system was a **structural system**, viewing behaviour as a set of simple actions related by implication. This explanatory system was tied in with a hierarchical conception of behaviour that supposedly reflected the hierarchical organization of nerve centres (Tinbergen 1950, 1952; Baerends *et al.* 1970;

## Box 1.4 **Lorenz's 'psychohydraulic' model**

To explain the relationship between the concept of motivation and the way behaviour was expressed, Konrad Lorenz (1950) developed a mechanistic model to which many ethologists subsequently made more or less explicit reference (Figure 1.1).

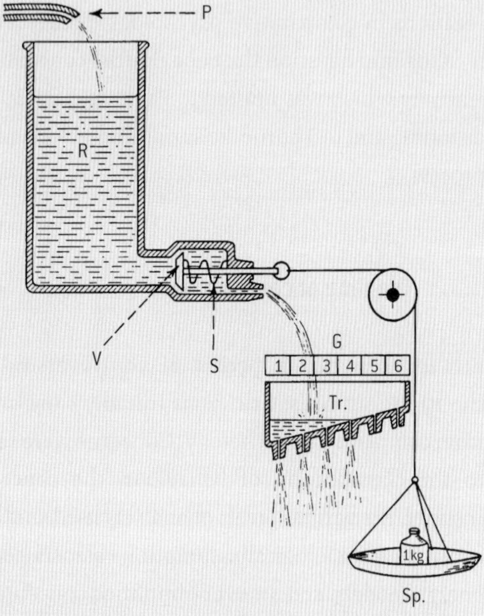


Figure 1.1 **The psychohydraulic model of behaviour by Konrad Lorenz (1950)**
This figure was published in *Animal Behaviour*, volume 4. Lorenz, K., The comparative method in studying innate behaviour patterns, pp. 221–268 © Elsevier (1950).

The model seeks to explain the connection between internal motivation and external stimulus in triggering a given behaviour, generally of the consumption type. The general idea is of an internal mechanism which allows an endogenous force to build up (action specific energy) rather like an electronic capacitor does. As the stored energy increases, the level of stimulation required to trigger the activity falls. Lorenz suggested representing the interaction between internal and external factors by a simple diagram. Action-specific energy supposedly builds up in the organism when the behaviour is not expressed. This phenomenon is represented in the model by a constant flow of water from a pipe (P) which gradually builds up in a reservoir (R). The quantity of action specific energy available is represented by the amount of water that has accumulated. The tank has an outlet closed by a valve (V) held by the pressure of a spring (S). The spring is connected by a pulley to a tray holding a weight (Sp) representing the strength of the stimulus. So it is the combined and variable action of the push of the fluid and the pull exerted by the weight that releases the valve. The level of the fluid then determines the level of the response. If only a small amount of water is stored then a heavy weight is required to trigger the behaviour. When motivation has run out (which corresponds to an empty reservoir in the diagram) the motor response cannot be triggered however strong the stimulus. Similarly, if a very large amount of water has built up, the valve may be released without there being any weight on the tray, explaining the observation of 'vacuous' behaviour (that is, behaviour expressed in the absence of any relevant stimulus). The strength of the stream of water spurting from the tank into a trough (Tr) represents the intensity of the behavioural response. It can be measured on a graduated scale (G). If very little water flows through, only the first graduations of the trough are filled. This represents the lowest level of motor activity and supposedly corresponds to what is called appetitive behaviour. When the flow is stronger, the maximum level of motor response is achieved. On the basis of this model Lorenz emphasized the need to know both the level of motivation for judging the strength of a triggering stimulus and vice versa.

Baerends 1976). In practice, it was more often than not a matter of establishing whether the occurrence of a given action could be predicted from knowledge of previous actions (Nelson 1964; Delius 1969) and whether relations of precedence connecting the different actions could be described concisely either by statistical models (Chatfield and Lemon 1970; Dawkins and Dawkins 1973; Morgan *et al.* 1976) or by syntactic models (Fentress and Stilwell 1973). The third explanatory system was the **adaptationist**

**approach**. It construed behaviour from its **adaptive significance**, considered at the time as the suitability of behaviour in view of the characteristics of the natural environment (Gervet 1980). This perspective closely associated the interpretation of behaviour with the ecological context in which it occurred and was directly related to the problem of the phylogenetic evolution of behaviour.

Ultimately, none of the three systems appeared to be really satisfactory; but not for the same reasons. The first two systems remained entirely dependent on intermediate variables that lacked any validation. Over time, the reductive approach to behaviour did not really produce concepts separate from those of neurophysiology, the discipline that had succeeded sensory physiology. Thus the development of the neurosciences inexorably brought about the 'substantialization' (Parot 2000) of capacities inferred from the observation of behaviour. Specifically behaviour-related concepts, such as the concept of motivation, proved to be of little practical relevance to physiology and were abandoned (Gervet 1980). The structural model, despite some highly refined mathematical developments, remained dependent on causal models borrowed from other disciplines (cybernetics, systems theory). Thus, the reductive and hierarchical approaches were destined to move closer to the behavioural neurosciences.

The fate of the adaptationist approach was markedly different. Its ability to form a coherent explanatory system depended, in the mid 1970s, on ethologists' ability to integrate the conceptual advances of evolutionary biology which at the time formed a powerful theoretical body, **neo-Darwinism**, able to account for the anatomical, physiological and behavioural characteristics of organisms in terms of causation (Futuyma 1998; Bowler 2003). Yet, 'classical' ethologists, while claiming to be Darwin's heirs, seemed unable to fully subscribe to neo-Darwinism (Barlow 1968). Indeed, their conception of an adaptationist approach to behaviour hardly went beyond the stage of reporting that behaviour was consistent with environmental factors; this goodness of fit was supposed to have evolved on the basis of the benefit it brought for the 'survival of the species'. For instance, Lorenz (1966) clearly believed in a 'good for the species' or 'survival of the group'. We now know that selection acts primarily acts on individual organisms not on species, groups or populations. A new conception of the adaptationist approach to behaviour was to emerge in the shape of behavioural ecology.

## 1.4 The advent of behavioural ecology

### 1.4.1 Sociobiology and its origins

Despite ethologists' insistence on the need to study animals in their natural environments, their concepts initially had little relation with ecology. In the 1960s a few rare attempts to bring closer together psychological and ethological approaches to animal ecology remained limited to the analysis of relations between species (Klopfer 1962). This situation can be explained in part by the particular interest of ecologists of the time for more systemic approaches concerning energy flows in ecosystems and the composition of plant and animal communities. In particular, the role of social behaviour in relation to population biology was ignored. Within classical ethology, social behaviour was mostly interpreted in terms of interactions between organisms within which the fixed action patterns of one individual entailed reciprocal behaviour by another. In ecology the organisms' behaviour was treated like a black box. The trend reversed markedly in 1975 with the publication of *Sociobiology, The New Synthesis* (Wilson 1975), a seminal book for the advent of behavioural ecology. Authored by the already famous Harvard entomologist Edward O. Wilson (Box 1.5), the book presented **sociobiology** as *'the systematic study of the biological basis of all social behavior'* and set out its objective as reaching *'an ability to predict features of social organization from a knowledge of... population parameters combined with information on the behavioral constraints imposed by the genetic constitution of the species'* (Wilson 1975, p. 5). This 'new synthesis' was based on the study of behaviour but also on work done since 1930 in ecology and population genetics

**Box 1.5 Edward O. Wilson, the founder of sociobiology**

Born in Birmingham, Alabama, in 1929 Edward O. Wilson is above all an expert on social insects. After graduating from Harvard in 1955 he taught zoology there from 1964 and then directed the entomology department of its museum of comparative zoology from 1973. He is one of the foremost experts on ant societies, which he has studied in the laboratory and in the field, especially in the South Pacific and New Guinea. As a naturalist, population biologist, and theorist, he has contributed in particular to developing the theoretical foundations of island biogeography. He is considered the founding father of sociobiology and is interested in conservation biology and the preservation of biodiversity. He has written several books including *Biophilia* (1984), *Consilience: The Unity of Knowledge* (1998) and has twice won the prestigious Pulitzer Prize.

and its avowed intent was to establish sociobiology as a branch of evolutionary biology and in particular of population biology. From the outset, Wilson (1975, p. 5) set the study of animal behaviour against other biological disciplines. He argued that comparative psychology and ethology were no longer the unifying disciplines of the study of animal behaviour. Back in 1975, Wilson announced their future integration into both neurophysiology and behavioural ecology.

These views were not strictly revolutionary at the time. Four years earlier, Wilson (1971) himself had already called for the unification of the concepts used in interpreting the social organization of vertebrates and invertebrates by forging connections between ecology, sociobiology, and phylogenetic studies. *Sociobiology* purports, then, to be a long plea for such unification based on a masterful synthesis of the main works of the previous 40 years in ethology, ecology, population genetics, and evolution. The connection between behaviour and population biology was progressively forged from observational works by ecologists and zoologists in the 1930s, especially in North America, where the naturalist tradition remained solidly anchored (Collias 1991). One of the figures at the forefront of this domain at the time was the ecologist Warder Clyde Allee (1885–1955), who wrote three books on social behaviour (Allee 1931, 1933, 1938). Based at Chicago University, he contributed through his research to connecting social behaviour and community ecology. From studies of organisms as varied as protozoans, rotifers, insects, crustaceans, fish, reptiles, birds, and mammals, he emphasized the role of cooperation in regulating social interactions. He exerted considerable influence on the development of studies of social behaviour (Barlow 1989; Collias 1991). Three other scientists of the same period deserve a mention. G. Kingsley Noble (1894–1940) was a forerunner of studies on sexual selection (Chapter 11), especially in his attempts to distinguish the respective influences of female choice and competition among males in the mating process (Noble 1936, 1938). A specialist on the social behaviour of birds, Margaret Morse Nice (1883–1974) opened up the way to modern work on the behavioural ecology of territoriality by observing that male song buntings (*Melospiza melodia*) are virtually invincible on their own territories. A. F. Skutch (1904–2004) was the first to consider the conditions under which cooperative behaviour in rearing young could develop in birds (Skutch 1935). These early works were followed in the 1960s and 1970s by more in-depth empirical studies showing the importance of allowing for ecological variables in interpreting social structures (e.g. Orians 1961; Ashmole 1963; Brown 1964; Crook 1964; Jarman 1974). They were supplemented by theoretical work on the advantages of gregariousness against the risk of predation

(Hamilton 1971; Pulliam 1973; Vine 1973) and the adaptive value of coloniality (Ward and Zahavi 1973; see Chapter 14).

At the same time, the theoretical aspects of evolution were more fully mastered. Population genetics was advancing by leaps and bounds. Ronald A. Fisher (1890–1962) and John B.S. Haldane (1892–1964) in Britain and Sewal Wright (1889–1988) in the United States developed a mathematical theory of population genetics showing that adaptive evolution depended on the combined action of the phenomenon of mutation and the process of selection. These studies were extended to the study of social organization with the work of William D. Hamilton (1936–2000; Box 1.6). The British evolutionist was the first (Hamilton 1964a, b) to offer a solution to the problem of reproductive altruism in hymenopteran insect societies (the fact that workers sacrifice their reproduction to raise their fellows) taking account of the genetic proximity of individuals (Chapter 2). The importance of Hamilton's work was not recognized immediately in Europe but soon caught the attention of Wilson, who understood that the repercussions of the work extended far beyond hymenopteran insects. The theoretical analysis of social behaviour was pushed a little further with the work of Trivers (Box 1.6). He showed, among other things, how altruistic cooperative behaviour can, under certain conditions of reciprocity, develop outside of any kinship structure (Trivers 1971). Moreover, he profoundly changed the conceptions of the day about interactions binding parents and their offspring (Trivers 1972, 1974). Parental behaviour was, from then on, no longer perceived as pacific cooperation between parents and offspring, but rather as a conflict of interests (Chapter 12). In turn, this work prompted resort to game theory in analysing social behaviour. Under the impetus of John Maynard Smith (1920–2004; Box 1.6), behaviour was analysed in terms of strategies whose benefits depended on how frequent they were in the population and how frequent alternative strategies were. This approach soon found multiple applications in the analysis of the social organization of breeding or the social exploitation of resources (Chapters 8, 9, 10, and 12).

The rapid success of the sociobiological approach was also due to the publication in 1976 of Richard Dawkins' highly educational book *The Selfish Gene*. Illustrated by various problems related to the adaptive value of behaviour, the book aims essentially at familiarizing readers with an informational logic of natural selection and adaptation of organisms. Behaviour is a phenotypic trait partly determined on a genetic basis. Any **genetic information** whose phenotypic manifestations (expressed in individuals) promote its own duplication (through survival and reproduction of individuals) is positively selected for (Chapter 2). The behaviour of individuals (like other phenotypic traits) can therefore be considered in terms of a strategy put in place by genes to perpetuate themselves. Here again, the content of the book is not really novel. Apart from the works of Hamilton, it picks up on the main lines of the book published earlier by evolutionist George C. Williams (1966), *Adaptation and Natural Selection*.

### 1.4.2 Behavioural ecology

Without really marking the birth of behavioural ecology, the publication of Wilson's and Dawkins' books coincided with a phase of unprecedented interest for the evolutionist approach to behaviour. In the 1960s and 1970s various **optimal foraging** methods had already been proposed by several North American ecologists (Emlen 1966; MacArthur and Pianka 1966; Schoener 1971; Charnov 1976). Within these models, animal behaviour is analysed as a decision-making process (where to forage? what type of diet to adopt?) allowing selection among a range of options (Chapter 7). As a result of the optimizing action of natural selection on **decision-making** processes, the choice made by the animal is supposed to maximize its phenotypic fitness. The logic underlying the use of optimization models is masterfully expounded in the work of David McFarland and Alasdair Houston published in 1981, *Quantitative Ethology: The State Space Approach*. The models, which are initially simple, can be used to make predictions that may be easily tested in the field or in the

### Box 1.6 **The great theorists of behavioural ecology**

**William D. Hamilton**. Bill Hamilton was born in Cairo, Egypt, of New Zealander parents and grew up in Britain mostly. Today he is considered one of the leading figures of modern Darwinism. His education had a decisive influence on his career. His parents encouraged him from an early age to develop his talents as a naturalist. His mother took him to visit Darwin's home, Down House, which made an impression on the young Bill. His father, an engineer, fostered his interest in mathematics. At the age of 14, Hamilton won a copy of *The Origin of Species* as a school prize. Reading it brought on his interest in the study of evolution. This precociousness probably explains why two of his most important works were published before he had completed his PhD. He began his studies at Cambridge University and then (after a few difficulties because of his teachers' failure to understand) obtained his PhD from University College London. He began his academic career at Imperial College (part of London University) and in 1977 became professor of evolutionary biology at Michigan University in the United States before becoming a Fellow of the Royal Society in 1980 and Research Professor at Oxford University's Department of Zoology from 1984. His most important work is certainly the resolution of the problem of reproductive altruism by weighing of costs and benefits by the degree of genetic proximity binding individuals (Chapters 2, 13, and 15). His other works relate to the evolution of secondary sexual characters in relation to resistance to parasites, optimization of the **sex ratio** and senescence. Although a theorist and model maker, Hamilton was also a field worker who enjoyed collecting insects in the tropical forests of Brazil. It was in the field, in the jungle of Congo, where he had gone to collect data to test his theories on the origin of the AIDS virus, that Hamilton contracted malaria in summer 2000. He was flown back to Britain where he died six weeks later at the age of 63, leaving a substantial, compelling scientific legacy containing several of the major theoretical foundations of behavioural ecology.

**John Maynard Smith**. Born in 1920, John Maynard Smith trained initially as an engineer, which led him to serve in the Royal Air Force during the Second World War. After the war, he turned towards biology and followed the teaching of the famous evolutionist J.B.S. Haldane. He began his career at University College London before moving to the University of Sussex at Brighton, where he became a professor. An outstanding naturalist, he is known above all for his contribution to the development of modelling in behavioural ecology, especially for the application of game theory to the study of behaviour, which was the origin of the concept of the **evolutionarily stable strategy** (Chapter 3). His other major contributions concerned the evolution of sexual reproduction and the evolutionary biology of the fruit fly. He wrote several books including *Models in Ecology* (1974), *The Evolution of Sex* (1978), *Evolution and the Theory of Games* (1982), and *Evolutionary Genetics* (1989).

**Robert L. Trivers.** Born in Washington, DC in 1944, this sometimes iconoclastic evolutionist first studied history before taking an interest in evolution. His research, conducted mainly at the University of California at Santa Cruz, allowed him to address different problems related to the evolution of behaviour such as **reciprocity**, **parental investment**, control of the sex ratio and kinship selection. He is notably the author of *Social Evolution* (1975).

laboratory. The approach proved successful and experimental tests were multiplied with various organisms (Schoener 1987). The synthesis between the evolutionary analysis of social behaviour and the **economic** analysis of exploitation of resources by animals, two approaches that share the same theoretical foundations (Krebs 1985), came about quickly and gave birth to behavioural ecology. This coming together was hailed in several works, including *Behavioral Mechanisms in Ecology* (Morse 1980) and *Sociobiology and Behavior* (Barash 1982). However, the greatest contribution undoubtedly came from the four editions of *Behavioural Ecology: An Evolutionary Approach* (Krebs and Davies 1978, 1984, 1991, 1997). For about 20 years, this series of multi-author books has regularly provided an update of work in behavioural ecology.

In the 1980s and 1990s it became clear that behavioural ecology occupied a dominant position (Gross 1994). The first specialist journal in the field appeared in 1976, *Behavioral Ecology and Sociobiology*. In 1985, at the international ethology conference in Toulouse (France) a delegation of students and scientists from New York State University at Albany approached their colleagues with a view to founding a behavioural ecology association with a new journal aimed at an international readership. By the following year, a first international congress on behavioural ecology was organized at Albany. In 1988, the second congress was held in Vancouver. It was there that the International Society for Behavioural Ecology (ISBE) was officially founded. Since then the ISBE congress has been held every two years (alternating with the international ethology conference) and brings together 600–900 scientists from around the world.[4] In 1990 the ISBE launched the journal *Behavioral Ecology*, which soon proved itself one of the leading journals in behavioural sciences.

However, the development of behavioural ecology was not all plain sailing. It has been criticized in particular for what is judged to be excessive or even grotesque adaptationism (Gould and Lewontin 1979, cf. Chapter 3). This antagonism was catalysed by sharp scientific, political, and emotional reactions to the application of the sociobiological approach to human behaviour, initiated by Wilson himself in the final chapter of *Sociobiology* (Wilson 2000; Laland and Brown 2002). His proposal to extend evolutionist concepts to the human species, always a delicate issue and a source of polemic, has now been extended in the emergence of evolutionary psychology (Cartwight 2000; Plotkin 2004). These criticisms, even if sometimes farfetched, have not been pointless and over time evolutionists have become more cautious in their interpretations and more open to alternative interpretations (Pigliucci and Kaplan 2000). Behavioural ecology has come through this consolidated, with a more pluralistic and diversified approach to behaviour.

### 1.4.3 The current standing of the discipline and changes in its themes

At the present time, behavioural ecology differs from classical ethology in its explicit theoretical rooting in neo-Darwinism, in having a single explanatory system and in the way it formulates its questions (Table 1.2 and Chapter 3). The generalized use of genetic markers has made it possible to show that the analysis of social structures and reproductive strategies could not be fully comprehended by relying solely on observable behaviour (Hughes 1998; Zeh and Zeh 2001; Birkhead and Møller 1992, 1998). In many cases, for example, paternity cannot be deduced simply from observed copulations or rank orders established among males but must be established through the use of molecular tools (Chapter 11). However, the great conceptual advance has been the move beyond Tinbergen's (1963) four questions. In fact, these four levels of analysis are not on the same footing and it is pointless describing them as alternative approaches. The study of proximate causes cannot in itself suffice to claim any true understanding of behaviour. Ultimately, inter-specific (or even interpopulational) diversity of physiological mechanisms and modes of development of behaviour can be understood only in a comparative and

4 The latest was held in Tours, France, in 2006.

| | Classical ethology | Behavioural ecology |
|---|---|---|
| **Formulation of questions** | Essentially inductive: hypotheses often follow observation | Essentially hypothetico-deductive: hypotheses often precede observation and sometime generated by models |
| **Methodology** | Detailed description and quantification of behaviour | Performance measurement (fitness)<br>Experimental approach<br>Molecular tools |
| **Explanatory systems** | Reductive<br>Structural<br>Adaptationist | Adaptationist: attributing a central role to the notion of fitness |
| **Level of selection** | Mostly: species, group | Individual, gene |
| **Influence of genetics** | Weak (hybrid crossings) | Weak (theoretical models, molecular tools) |
| **Influence of development** | Strong | Weak |

Table 1.2 **Contrasts between ethology and behavioural ecology**

evolutionary perspective. We can therefore predict the future emergence of comparative neurophysiology, which will be able to link up the diversity of development and of the working of the nervous system with species ecology.

The multiplicity of the questions now addressed by behavioural ecology is reflected in the chapters that follow. This is in itself evidence of the vitality of a discipline that, although only recent, has already managed to evolve and renew itself. After an initial development marked by a great interest for the study of exploitation of resources, the discipline has seen over the past decade or so unprecedented enthusiasm for the study of sexual selection processes (Chapter 11). At the same time, the study of mechanisms and of development has seen a clear revival of interest (Chapters 5 and 6), reflecting the vital need to clearly understand the constraints on the expression of behaviour before making an adaptationist interpretation. An effort is still required, however, to better integrate behavioural genetics within behavioural ecology. The future of behavioural ecology, although still unpredictable, should be fruitful and yet richer. We can wager that new investigative techniques combined with constant advances in computing, electronics and biotechnologies will long allow it to explore new and fertile horizons.

## 1.5 Conclusion

The aim of this chapter was to describe the historical stages that have progressively led to the emergence of behavioural ecology as an autonomous discipline. Behavioural ecology has been able to diversify its fields of enquiry, although devoting much of its work to the study of the sexual selection process and its repercussions. More recently, behavioural ecologists have developed an interest in the study of physiological mechanisms, in particular in relation to the emerging field of immunoecology. Even though studies so far seem to have raised more questions that they have answered (Roberts *et al.* 2004), they have played an important part in restoring the phenomenological approach in a world of modern biology dominated by reductionism and molecular approaches (Little *et al.* 2005). Indeed, the long-lasting success of behavioural ecology may lie in its ability to contribute to advances at the mechanistic level from the consideration of important adaptive phenomena.

## » Further reading

**Corsi, P.** 2001. *Lamarck. Genèse et Enjeux du Transformisme 1770–1830*. CNRS Editions, Paris.
**Darwin, C.** 1859. *On the Origin of Species by Means of Natural Selection*. Murray, London.
**Darwin, C.** 1871. *The Descent of Man and Selection in Relation to Sex*. John Murray, London.
**Darwin, C.** 1872. *The Expression of Emotions in Man and Animals*. John Murray, London.
**Dawkins, R.** 1976. *The Shelfish Gene*. Oxford University Press, Oxford.
**Dawkins, R.** 1989. *The Blind Watchmaker*. Oxford University Press, Oxford.
**Dewsbury, D.A.** 1999. The proximate and ultimate: past, present, and future. *Behavioural Processes* 46: 189–199. Futuyma, D.J. 1998. *Evolutionary Biology* 3rd edition. Sinauer, Sunderland, Massachussets.
**Gayon, J.** 1998. *Darwinism's Struggle for Survival*. Cambridge University Press, Cambridge.
**Gould, S.J. & Lewontin, R.** 1979. The spandrels of San Marco and the Panglossian paradigm: a critique of the adaptationist programme. *Proceedings of the Royal Society of London Series B* 205: 581–598.
**Sober, E.** 1993. *Philosophy of Biology*. Oxford University Press, Oxford.
**Tinbergen, N.** 1963. On aims and methods of ethology. *Zietschrift für Tierpsychologie* 20: 410–433.
**Wilson, E.O.** 2000. *Sociobiology. The New Synthesis* 25th anniversary edition. Belknap Press of Harvard University Press, Cambridge, Massachussets.

## » Questions

1. What influence do you think political and socio-economic changes may have had on the study of animal behaviour?

2. The philosopher Gaston Bachelard asserted that science 'was as old as its measuring instruments'. Is his assertion relevant to behavioural ecology?

3. What, if anything, remains of the vitalist and mechanist approaches in behavioural ecology?

2

2

# Fundamental Concepts in Behavioural Ecology

Étienne Danchin, Frank Cézilly, and Luc-Alain Giraldeau

## 2.1 Introduction: what is 'behavioural ecology'?

Throughout history, the study of living creatures in their natural environment has been a source of fascination for people. From Aristotle to Darwin, the diversity, complexity, and sheer exuberance of animal behaviour has never ceased to astonish and to defy complete understanding. Indeed, the reasoning and logic behind certain behaviours seem at first to elude us. For instance, how does one rationally explain the sexual cannibalism of the praying mantis, or the sterility of the working caste within ant societies, or even the parasitic reproductive strategy of the common cuckoo (*Cuculus canorus*)? Individuals of the last species do not construct a nest, but rather lay their eggs in the nests of other bird species, often of smaller size. The newly hatched cuckoo rapidly excludes the eggs and young of its 'adoptive parents', monopolizing their attention and care. We cannot help but wonder what drives individuals of the parasitized species to accept the heavy cost of the young cuckoo rather than seek to raise their own offspring properly?

Certain behaviours are so remarkable that they appear *unnatural* to some observers. The infanticide in lions is one such behaviour. In this social species, groups of up to a dozen individuals are composed of adult females, their cubs, their immature daughters, and one to six males. The females within a group are closely related, having usually grown up within a pride that has existed for multiple generations. Females typically remain within the same group for a long period of time. However, this is not so for male lions. They are excluded from their natal group as soon as they are independent of their mother, and subsequently form coalitions with their brothers or with unrelated young males. If a coalition recruits a sufficient number of young males, they can eventually take over a pride of females after forcefully evicting the adult males of the group. These power reversals lead to many spontaneous abortions in females, undoubtedly caused by the stress involved with such situations. Females for which pregnancy is too advanced for abortion will give birth, but the new dominant males quickly kill the cubs, particularly the males. Because it reduces the reproductive output of females, infanticide may at first appear to be an unusual behaviour.

Behaviours such as these seem paradoxical unless considered within their evolutionary framework which, as we shall see, can reveal the logic behind them. The interpretation of behaviour within this framework characterizes the approach of behavioural ecology.

### 2.1.1 An evolutionary approach to behaviour

Behavioural ecology seeks to explore the relationships between behaviour, ecology, and evolution. Here, behaviour is considered as the collection of decisive processes by which individuals adjust their state and situation according to variations in their environment, both abiotic and biotic. Decision making does not necessarily refer to elaborate cognitive processes but simply to the fact that an animal is regularly confronted with multiple alternatives, each differing in their consequences in terms of individual survival and reproduction. Other more restrictive or mechanistic definitions of behaviour exist (see, for example, Manning 1979). The advantage of the above functional definition is that it emphasizes the crucial role of behaviour in adaptation and therefore in evolution.

The goal of behavioural ecology is to understand how behaviour results **from a combination** of the evolutionary history of a species, recent or current events occurring at the population level, and characteristics particular to individuals and to the conditions in which they developed. It also involves using our knowledge of biological evolution to construct an analytical framework of behaviour and to identify the various factors, both internal and external, that induce or constrain the expression of behaviour. Behavioural ecology belongs to neo-Darwinism, an evolutionary movement based on a modification of Darwin's ideas in light of discoveries on the nature of heredity. The field of sociobiology also uses this evolutionary approach to behaviour. However, the differences between sociobiology and behavioural ecology are subtle. They both use the same hypothetico-deductive approach, and in fact differ only in their focus of study. Sociobiologists are primarily concerned with the interactions between individuals within animal groups or societies, whereas behavioural ecologists are interested in all aspects of behaviour. Sociobiology is thus a subset of behavioural ecology (Krebs and Davies 1981; Krebs 1985), and accordingly there is no strict advantage to regarding it as a separate field.

The questions typically asked in behavioural ecology are numerous and diverse. Why does a given predator focus its energy on a particular type of prey while another shows much greater diversity? Why do males provide parental care in one species and not in another that is closely related? Why do certain individuals within an animal society have the ability to reproduce whereas others are limited to sterile auxiliary roles? Why does the song of a certain passerine species differ from one individual to the next? The questions we ask can be even more specific, and expressed in quantitative terms. Why does the European starling (*Sturnus vulgaris*) bring its young a maximum of six food items per visit to the nest? Why does mating last on average 7 hours 40 minutes in a particular species of mite?

To answer these questions, behavioural ecologists often favour a utilitarian approach to behaviour. The underlying question, whether explicit or implicit, is then the value of a behavioural trait for the survival and/or reproduction of individuals, and for the replication of their genotype above others, as per the process of natural selection (Grafen 1984; Dawkins 1989). The number of food items brought back to the nest by a starling is then understood as a balance between the maximization of food brought to the nest and the costs in energy and travel time involving the distance between the nest and the food source. The presence of sterile individuals is interpreted with respect to the help they bring to the reproductive individuals to which they are closely related. However, if evolution by natural selection is indeed an optimizing process (Endler 1986; Dawkins 1986), it follows that the possibilities are never infinite (Jacob 1981). Throughout evolutionary time, various random events have demarcated areas within which species can evolve towards new forms and functions. With respect to this, the goal of behavioural ecology is also to establish the extent to which an observed behavioural trait results from the historical limitations that have guided its evolution. Behavioural ecologists must generally keep themselves from hastily drawing the conclusion that the behaviour they study is the **direct** result of a selective process. This situation is illustrated in the following example.

### 2.1.2 Size homogamy in gammarids

Gammarids (genus *Gammarus*) are aquatic amphipod crustaceans distributed across the globe. An important characteristic of their reproductive behaviour is precopulatory **mate guarding**, or **amplexus**, in which males compete for the possession of females by guarding them before copulation. As in all crustaceans, amphipods regularly undergo moult, and it is only at that moment that females can mate. The time lapse between two consecutive moults can spread over several weeks depending on water temperature, an interval during which females cannot be inseminated. By contrast, males are always able to reproduce, apart from the time they are themselves moulting. Although there are equal proportions of males and females within natural populations, the **operational sex ratio** (i.e. the relative proportions of males and females available for mating at any moment) is strongly male-biased, resulting in strong competition among males for access to the few moulting females capable of being inseminated. Male gammarids have the ability to detect the hormonal status of females, and when a male encounters a female nearing her moult, he may choose to attach himself to her back, holding on to her with his gnathopods, two pairs of hypertrophied appendages, thus forming an amplexus. This nuptial ride persists until the female moults, after which the male that was guarding her fertilizes her eggs.

When collecting gammarids in amplexus, it is generally evident that individuals do not pair randomly. There exists a positive relationship between the size of the male and female within a pair (Figure 2.1). Larger males tend to attach themselves to larger females, while smaller males typically hold on to smaller females. This size-based association between males and females is referred to as size **homogamy**.

How do we interpret such a phenomenon? First, a selective, utilitarian interpretation. We know that

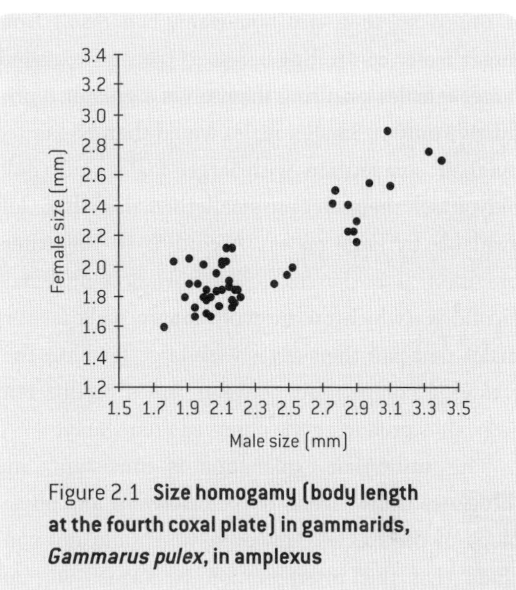


Figure 2.1 **Size homogamy (body length at the fourth coxal plate) in gammarids, *Gammarus pulex*, in amplexus**

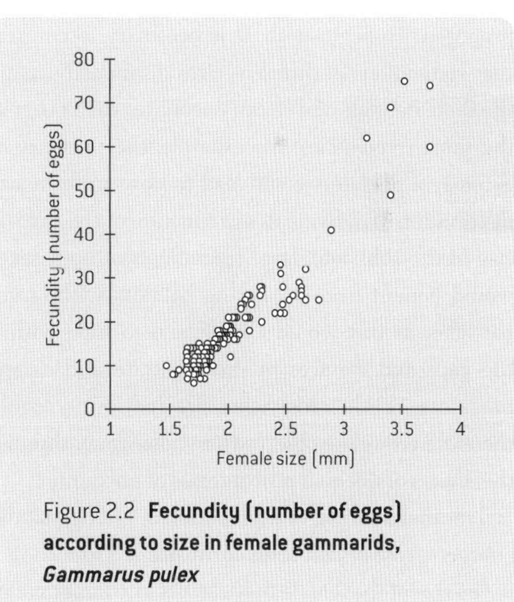


Figure 2.2 **Fecundity (number of eggs) according to size in female gammarids, *Gammarus pulex***

the number of eggs a female can lay increases exponentially with her size (Figure 2.2). For a male, therefore, a small size difference between two females can mean a large difference in the number of eggs fertilized. On an evolutionary scale, males that choose to pair with larger females will have a greater number of descendants. Moreover, larger males can also dominate smaller ones and can thus dislodge them from females onto which they have amplexed.

If this is so, then size homogamy can result from larger males controlling access to the most coveted, larger females, ensuring themselves a greater reproductive output. Smaller males would then be limited by their competition with larger males and could only attach themselves to smaller females (those left available by large males). According to this interpretation, it is the size-linked variation in female fecundity (which on an evolutionary scale favours males that pair themselves with large females) that creates a competition between males of different sizes, thus producing the observed homogamy.

This utilitarian explanation is consistent, and integrates many aspects of the behaviour and physiology of the species. However, we should not conclude that this interpretation is correct without considering some alternatives. Various alternative explanations, seemingly just as consistent, can be proposed. To begin with, it is necessary to confirm that individuals of different sizes do indeed occupy the same habitats and so can encounter each other at the same frequencies. For example, the presence of gammarid predators could lead to non-uniform size distribution of individuals as a function of the particle size of the substrate: large individuals of both sexes would benefit from occupying habitats with larger particles, as this would allow them to hide within the gaps and crevices in the substrate and escape predators. This heterogeneous size distribution could therefore create the observed size homogamy through the effects of unequal probabilities of encounter.

Another alternative explanation might be that the energetic cost of amplexus for males is proportional to the size of the female they carry. Only larger males could bear the energetic costs of pairing up with larger females. Smaller males would therefore have no interest in pairing up with larger females because they are literally too expensive for them. A final possibility is that morphological limitations prevent large males from attaching themselves effectively to small females, and conversely prevent small males from attaching to large females.

It is important to note that these various alternatives hypotheses are not mutually exclusive and it is possible that they have a cumulative effect to produce the observed pattern. Before assuming the primary importance of competition between males for large females, we must consider the significance of all the above explanations (and any others that we might imagine; Bollache *et al.* 2000). A valuable approach to deciding which interpretation carries the most weight is to search for situations in which the many alternatives make very different predictions, which would allow us to differentiate between them. This simple example illustrates the precautionary measures that must always be taken when interpreting behavioural phenomena: **it is essential to rule out alternative explanations before raising our confidence that a given interpretation is true**. This is a fundamental principle behind any scientific process.

## 2.2 Behavioural ecology: an evolutionary perspective

### 2.2.1 What is evolution?

Behavioural ecology provides an evolutionary perspective on behaviour. It is appropriate then to define exactly and explicitly what we mean by evolution. The word evolution invokes, in its literal sense, a series of progressive transformations. In its common usage, the word has at least four different meanings: (1) a fact, (2) a transformation, (3) a particular course, and (4) an improvement. Although the first three meanings are valid, the fourth is not to be used in the context of evolutionary biology. Accordingly, the word evolution refers to the now well-established fact (Box 2.1) that species have progressively transformed over time. Taken in this sense, the word evolution incorporates the first three meanings seen in current use. It represents a history, that of life on earth.

Today, evolution is no longer just a theory, but a fact (see Box 2.1). Apart from some creationist religious movements, very few people challenge the reality of evolution. When we speak of the 'theory of evolution', we use 'theory' in its scientific sense to mean an organized body of knowledge and not its everyday

### Box 2.1 **Empirical evidence for evolution**

Since the time of Jean-Baptiste de Lamarck (1809), the empirical evidence for evolution is varied and stems from diverse domains of biology. The following is a summary of certain important arguments, each independently consistent with the evolution of species.

(1) Fossils show that ancient life forms are radically different from contemporary species. It can even be difficult at times to find present-day representatives for many fossil groups that lived tens or hundreds of millions of years ago. This is a result of two key processes: extinction and radiation. Most organisms that existed in the past are extinct today, whereas there also exists at present a great number of different forms, some of which did not exist in the distant past.

(2) There is a strong resemblance between the early stages of embryonic development within different members of a taxonomic group. For instance, vertebrate embryos pass through initial stages that resemble the adult forms of ancestral groups. In this manner, the embryos of mammals (considered derived forms) possess vestigial gill arches during early developmental stages, similar to those of fish (considered ancestral forms). In these ancestral species, development is complete at stages that for mammals correspond to the very beginnings of development. This is evidence that mammals arose from fish, because their development retains the traces of their ancestral history; evolutionary processes can only build on what already exists.

(3) The field of comparative anatomy has revealed remarkable similarities between the anatomical structures of related groups. This notable likeness can be seen when comparing the bones of our arms with those of a bird's wing, the forelimbs of a horse, a lizard or a frog, or the anterior fins of certain fish. The similarities between these homologous organs strongly suggest that they are related through common descent, and that they gradually transformed over evolutionary time.

(4) Functionally complex organs are no longer used in some groups, indicating their abandonment over the course of evolution. For example, in bed bugs (heteropterans), females have well-developed copulatory organs, yet fertilization occurs directly across the abdominal wall (Stutt and Siva-Jothy 2001). Moreover, whales and snakes possess a vestigial pelvic girdle as well as rudimentary hind limbs. Such vestigial structures have no apparent function, and are just evolutionary leftovers inherited from their quadruped ancestors. How else can we explain their existence if not by evolution?

(5) The discipline of molecular genetics has shown that there lies an intriguing unity behind the enormous diversity of life. This dualism between external diversity and structural and functional unity is found at various levels. The same processes regulate the metabolism of creatures as diverse as snails, frogs and bacteria. The same genetic code allows the transmission of information from generation to generation in plants as well as in animals. The incredible diversity in form and colour seen in birds rests upon the same internal organization for all members of this group. The explanations behind this unity become obvious when we realize that these forms are all derived from more or less ancient common ancestors.

(6) For thousands of years, man has practised artificial selection on domesticated species. By the simple process of artificial selection for more suitable forms, man has, for example, created incredibly diverse dog breeds. This

demonstrates the enormous potential for the diversification of life from a common root. This type of evidence has been demonstrated in the laboratory an uncountable number of times.

(7) Biogeography has shown that portions of the earth's crust that have long been isolated tend to have life forms that resemble ancient fossilized forms found on neighbouring continents. For instance, Australia is home to mammal groups that are closely phylogenetically related to mammals that existed in Eurasia at the beginning of mammal radiation, about 100 million years ago. Equally, New Caledonia is today the only place where one can find gymnosperm groups that have been extinct in the rest of the world for a long time.

All those phenomena, and many more, offer factual evidence for the reality of evolution.

meaning of an uncertain hypothesis. Evolution as such comprises several co-existing theories that give various degrees of significance to the different processes affecting biological evolution. The recognition of evolution as a fact then creates a framework in which the assorted theories explaining it can be placed. Scientific progress regularly allows for the subtle modification of evolutionists' viewpoints about the prevalence of each of the known evolutionary processes. A thorough understanding of biological evolution can only be attained by integrating the findings of such diverse fields as palaeontology, ecology, taxonomy, behavioural ecology, population dynamics, evolutionary physiology, population and quantitative genetics, and even developmental biology.

### 2.2.2 The logic of evolution: information, replicators, and vehicles

*Evolution*

In the absence of any knowledge in the field of genetics, the study of evolution was for a long time based on the study of morphological changes in organisms over time, principally by comparing fossilized forms among themselves and with extant forms.

> Today, our concept of evolution integrates observed transformations with changes that occur over time in the frequency of different genes or **replicators**, which have the ability to self-replicate and hence ensure their existence over time. In other words, evolution is the process by which the frequencies of genes change over time.

*Genotype and phenotype*

Before proceeding, it is essential to define two fundamental concepts: genotype and **phenotype**.

> The **genome** corresponds to all of the genetic information carried by an individual (Mader 2004), although the same word is sometimes used to refer to a particular collection of genes forming a functional unit (genome of species, mitochondrial genome, genome inherited from an ancestral species in a polyploid individual, etc.). Because of the infinite combinations of genes and alleles that are possible, two individuals cannot possess exactly the same genome, apart from identical twins or clones. It is particularly important to understand the difference that exists between the two related words genome and genotype. In its strict sense, as defined by population geneticists, the genotype corresponds to the allelic composition of the gene locus or loci studied in an individual. In that sense, the genotype refers only to a class of individuals and then two different individuals can

belong to the same genotype or not depending on which locus (or loci) is (are) considered. The phenotype, on the other hand, is the collection of an organism's characteristics that result from the interaction between its genome and the environment in which it developed. In population genetics, however, the term phenotype refers in the narrow sense to only a subset of the characteristics of an organism, i.e. those dependent on a particular locus or on any other specific portion of the genotype. Depending on the gene or genes considered, two individuals can in that sense belong to the same phenotype or to two different phenotypes. We will consider the links between these two concepts in further detail in Chapter 5 and at the start of Chapter 6.

In behavioural ecology, we generally consider categories of individuals that share a particular characteristic (referred to as a strategy). Thus we tend to use a narrow definition of the term phenotype. We speak for example of the 'dispersing' phenotype to denote individuals that have moved between territories during their life, as opposed to the 'resident' phenotype that denotes individuals that remained on the same territory throughout their life.

*From genes . . .*

At this point, it is appropriate to emphasize that the word gene has a double meaning (Dawkins 1989, Haig 1997). In one sense, a gene refers to a material structure, a group of atoms organized into a particular sequence of deoxyribonucleic acid (DNA). In another sense, however, a gene refers to a more abstract concept, that of the information that is carried by that particular DNA sequence and is used to produce a protein. We therefore speak of **material** genes to refer to the many material copies of the same **informational gene** that exist in the same or different organisms. Ultimately, it is the information carried by the structure of DNA that persists across generations, and not the structure itself (Dawkins 1989). A gene is consequently a unit of information whose material support is a DNA sequence. We will come back to the question of information in biology in more detail in Chapter 4.

*. . . to avatars*

We use the term 'avatar' to designate the material forms produced by information (Gilddon and Gouyon 1989). Individuals are but 'avatars', that is to say by-products of genes (Dawkins 1982, Gilddon and Gouyon 1989). The word 'avatar' is taken from the Hindu religion, where it refers to the material forms taken by the god Vishnu during his visits to earth. Information is the sole target of selection, and avatars are merely its vehicle. Hence, an avatar is the entity under selection, but it is the self-replicating information it contains that is the absolute target of selection.

The characteristics of organisms can be seen as strategies put in place by genes to survive and replicate, while individuals are their **vehicles**. An individual exists only as a temporary vehicle, or a machine created by genes to help the genes survive and replicate (Dawkins 1976). Another way of putting it is that individuals are genes dressed in an elaborate external phenotype (Wittenberger 1981). When viewed in this manner, the gene reveals itself as the true unit of selection. As the famous evolutionist from the United States Georges Williams said in his now classic 1966 book on selection, 'genotypes are mortal', whereas 'genes are potentially immortal'. Indeed, a genotype disappears upon the death of its associated organism, but also during the process of meiosis, which allows only the genes to endure not their combination. Because of recombination during meiosis, the lifespan of genes is outstandingly longer than that of genotypes. The complex structures (cells, organs, organisms, societies) and behaviour that have evolved over the course of natural selection can be considered as adaptations for the benefit of genetic information, that is for the good of genes (replicators) rather than individuals (vehicles). This notion remains compatible with the fact that every individual is an organized whole, within which the fate of a given gene is linked to the degree of cooperation or coordination between itself and other genes (Haig 1997).

### 2.2.3 Phenotype, genotype, and reaction norm

Natural selection can be analysed at two levels. The first is the level at which the individuals of the same species within a population are sorted. This sorting takes place as a function of the characteristics that vary from one individual to the next and that differentially affect their survival and reproduction. The second is the level at which the frequency of different alleles changes between two consecutive generations as a consequence of selection. This is the **response to selection**. There is a strong dividing line between these two levels which can be made further explicit through the notion of the **uncoupling of genotype and phenotype**. This notion uses several different concepts, many of them developed in quantitative genetics. We will limit ourselves here to the fundamental principles (but see Roff (1992, 1997), Stearns (1992), and Futuyma (1998) for a more complete discussion).

#### 2.2.3.1 *Heredity and heritability*

The phenotypic differences between the individuals of a population can be of genetic and/or environmental origin. Specifically, phenotypic variation can result from differences in the genetic information contained in the fertilized eggs from which these individuals originated, or from differences between the environments in which they developed (Cockburn 1991). In most cases, individual variation in a given trait results concurrently from genetic and environmental effects. We can thus divide the phenotypic variance of a trait ($V_P$) according to the following formula:

$$V_P = V_G + V_E + V_{G \times E} \quad (2.1)$$

where $V_G$ and $V_E$ correspond to genetic variance and environmental variance, respectively, and $V_{G \times E}$ corresponds to the variance that is due to the interaction between the genotype and the environment.

The total genetic variance of a character controlled by multiple genes can also be separated into numerous components. A part of genetic variance is from the specific interactions that occur within a given individual, including variance from the dominance of alleles at the same locus ($V_D$) and from **epistatic** interactions between alleles from different loci ($V_I$). The resemblance between parents and their offspring is determined by the additive effects of the alleles at each locus and at the different loci involved (called additive genetic variance, $V_{AG}$).

$$V_G = V_{AG} + V_D + V_I \quad (2.2)$$

When $V_{AG}$ approaches a value of zero, $V_D$ and $V_I$ also approach zero. If there is no additive variance, there cannot be variance due to dominance because it is the same for all individuals; nor, for the same reason, can there be variance due to epistatic interactions between loci.

> It is the additive variance alone that allows for a response to selection. The **heritability** of a trait is therefore defined as the ratio of the additive variance and the phenotypic variance ($V_{AG}/V_P$), usually represented by the term $h^2$ (Falconer 1981). Heritability is then the portion ($V_{AG}$) of the differences between individuals ($V_P$) that is transmitted to descendants. Stated otherwise, heritability is the 'heredity of differences'. The problems stemming from measuring heritability are discussed in Box 2.2 and later in Chapter 20.

Because heritability is the proportion of additive genetic variance contributing to phenotypic variance, it indicates whether or not there is any genetic variation in a specific sample on which natural selection can act. However, even when there is appreciable genetic variation among individuals, heritability can be low because of substantial environmental variation. In a symmetric way, heritability can be high

Box 2.2 **How is heritability measured?**

In practice, completely eliminating the environmental component of phenotypic variance is not an easy task. To do so, all environmental factors that could potentially act upon phenotypic variance must be controlled. One manner of solving this problem is to conduct experiments in which selection can be simulated and observe how it affects variation in a trait in the next generation. Suppose that we are studying the heritability of a trait that shows continuous variation in size across a population (Figure 2.3a, solid line). We can simulate selection by having, for instance, only individuals showing a larger size of the trait in question reproduce (Figure 2.3a, dotted line). The difference between the average value of the entire population ($T_{mt}$ in Figure 2.3a) and the average value of the reproducing sample ($T_{ms}$ in Figure 2.3a) is the selection pressure $S$ exerted in our experiment. We can then raise the descendants under standard conditions to remove environmental effects on phenotypic variance as much as possible. Upon measuring the average value of the trait in the descendent generation ($T_{mF1}$), we can calculate the difference $R = T_{mt} - T_{mF1}$, which represents the selective response that we have created (Figure 2.3b). The ratio ($R/S$) measures the proportion of the selection pressure that is transmitted to descendants, the heritability of the trait in question.

However, there are numerous situations in which such controlled experiments are not possible. In these cases, we can use the resemblance between parents and their offspring. Unfortunately, such measures can be misleading as this resemblance may be due to

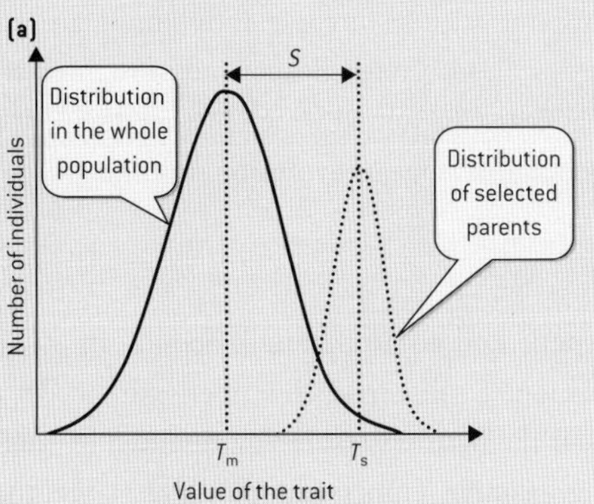


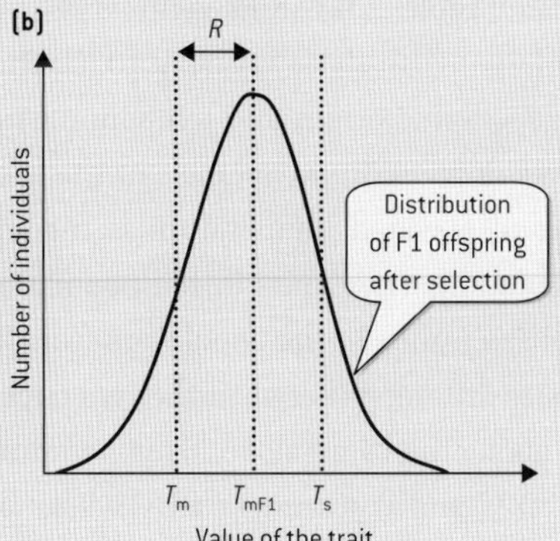


Figure 2.3 **Selection experiment to measure the heritability of a character**

**a.** We are interested in the heritability of a trait showing continuous variation (*x* axis). The distribution of this trait in the population prior to selection is represented by the solid line. The trait has an average value of $T_{mt}$ in the total population. In the parental generation, the experimenter only allows individuals with a large size of the trait (on average $T_{ms}$) to reproduce. By doing so, the experimenter exerts a selection pressure that can be measured by the difference $S$ between the average value of the trait in the whole population and the average value in the selected population.

**b.** In the F1 descendant generation, the average value of the trait ($T_{mF1}$) allows us to estimate the response to selection, $R$. $R$ can vary between zero and $S$. The ratio $R/S$ provides a measure of heritability, the portion of the differences between parents that is transmitted to the next generation.

**maternal effects**, environmental differences that cannot be controlled because they are linked to the parents, or to social (cultural) influences. This is especially the case when parents help raise and care for their young: for example, larger mothers might provide more food to their young (perhaps because they are dominant and can thus access better resources), which would lead to descendants having a larger average size, regardless of genetic variance. In such a situation one would find significant heritability of body size that would in fact result from parental effects, not necessarily from genetic variation among families. One way to avoid this problem is to make reciprocal transfers by exchanging half the descendants between two families. Such cross-fostering experiments allow us to distinguish, at least in part, maternal effects (being raised within a same family) from effects due to genetic variance (being born from the same parents but raised by different parents). However, the resemblance between siblings from a same litter can be due to dominance ($V_D$) or **epistatic** interactions ($V_I$), which are likely to be similar between siblings. The best way to avoid these problems is to work with half-siblings that share the same father but different mothers. The resemblance between half-siblings would then be due to genes from the father (see Falconer (1981) and Roff (1997) for a discussion on heritability estimation methods).

Another possible complication is when the effects of the environment and genotype are not additive, but rather there exist interactions between the phenotype and the environment. This is the case for myopia in humans. For a century in several societies, males with myopia have preserved their capacity to survive and reproduce successfully because their poor sight was compensated for by the use of glasses. In addition, strongly myopic individuals have benefited from exemptions from military duty during recent conflicts, which further increases this capacity relative to other individuals exposed to the dangers of war. This situation is unique to the modern world. In a hunter–gatherer society, myopic individuals are at a strong disadvantage. Readers who are interested may read Chapter 20 which further develops on heritability a major evolutionary concept.

despite moderate genetic variation, if there is little variation in the environment. It is therefore important to remember that heritability is always defined with regard to a specific sample with a specific genetic composition and environmental context. Therefore, any heritability estimate is specific to a given sample, and heritability estimates cannot be directly compared between samples not having the same genetic or environmental composition (Vitzhum 2003). This may explain why heritability may change over time within a population (Charmantier *et al.* 2006).

#### 2.2.3.2 *Phenotypic variance and reaction norm*

The division of phenotypic variance into separate components assumes that the two types of effects, genetic and environmental, operate in an **additive** fashion. To illustrate this phenomenon, let us consider the fictitious example presented in Table 2.1. In this example, French gammarids are larger than Irish gammarids (Table 2.1a), and gammarids raised at 18 °C are larger than those raised at 10 °C. These two effects (geographic origin and temperature) express themselves in an additive manner, in the sense that the effect of elevated temperature is the same for both populations (11.9 units in the population from France and 12.1 units in the population from Ireland) and the size difference between the two populations remains about the same at both temperatures (10.1 units at 10 °C and 9.9 units at 18 °C). The fact that temperature and geographic origin are additive indicates that the joint effect of the genotype and the

| a. Size at sexual maturity of the amphipod *Gammarus pulex* (arbitrary units) | | |
|---|---|---|
| **Origin of the population** | **Temperature during development** | |
| | **10 °C** | **18 °C** |
| France | 125.1 | 137.0 |
| Ireland | 115.0 | 127.1 |
| **b. Growth between 2 and 5 months of age in two lineages of rat as a function of dietary quality (arbitrary units)** | | |
| | **Enriched diet** | **Poor diet** |
| Lineage A | 55.2 | 36.9 |
| Lineage B | 48.6 | 42.3 |

Table 2.1 **Genotype–environment interaction**

environment is equal to the sum of each of these effects taken separately. In such condition, the term $V_{G \times E}$ in Equation (2.1) is zero.

Let us now consider the case of the growth of two rat lineages as a function of environmental quality (Table 2.1b). The growth of lineage A is better than that of lineage B when nutritional conditions are good, but the opposite is observed when nutritional conditions are poor. The effects of the genotype and the environment are no longer additive in this case: there is a **genotype–environment interaction**. In such a case, the term $V_{G \times E}$ in Equation (2.1) is not equal to zero.

> Depending on the environment in which it develops, a same genotype can produce different phenotypes. The term **norm of reaction** is used to designate all the phenotypes that can originate from a given initial genotype.

It is possible to depict graphically the additive and interaction forms of genetic and environmental effects in relation to the concept of reaction norms. When the two effects are additive, the reaction norms of the two genotypes (which correspond to the variation of the phenotypes expressed across an environmental gradient) follow parallel trajectories (Figure 2.4a). In the opposite situation, when an interaction exists, the two trajectories are no longer parallel (Figure 2.4b).

#### 2.2.3.3 *Phenotypic variance and phenotypic plasticity*

If $R$ is not different from zero (Figure 2.3), the trait in question is not heritable. However, this does not imply that the trait is not hereditary, that it is not genetically coded. It only means that differences between individuals are not caused by differences in the genes coding for this trait ($V_{AG}$ is not different from zero). Stated otherwise, there is no longer any genetic variation in the genes responsible for the expression of the trait. A simple example to illustrate the difference between heritability and heredity is the number of eyes in vertebrates. This character is not heritable because it has lost all genetic variation. Of course, this does not mean that eye number is not genetically controlled.

If a trait is not heritable, what causes the observable variation in the trait within the population? According to Equation (2.1), when additive variance is zero, all phenotypic variance can be attributed to environmental variance. Differences between individuals are then caused solely by the differential effects of the environment during development.

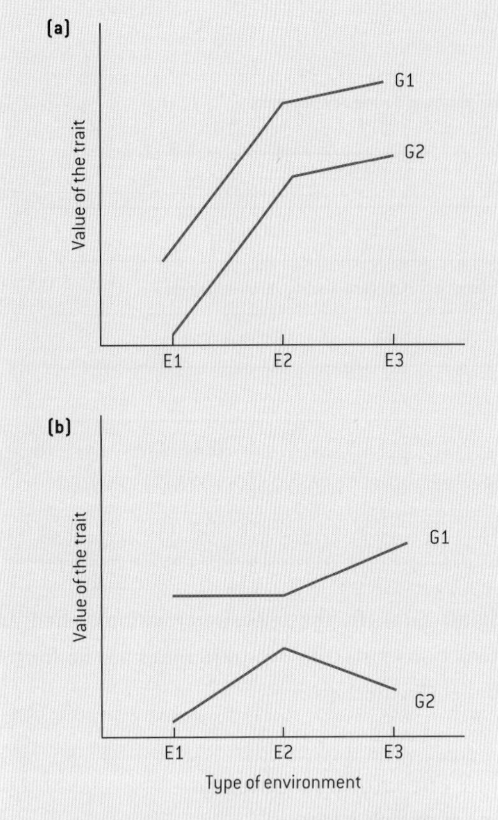


Figure 2.4 **Reaction norms for two genotypes (G1 and G2) in three environments (E1, E2, and E3)**

The curves joining the values of the trait along the gradient of environments represent the reaction norm of the study trait.

**a)** When the environmental and genetic components of phenotypic variance are additive, the reactions norms are more or less parallel.
**b)** When there are interactions between the genotype and the environment, the reaction norms are no longer parallel.

The environmental variance $V_E$ brings about what is called **phenotypic plasticity**, the capacity of a given genotype to produce different phenotypes according to the environment in which an individual develops. Equation (2.1) can now be approximately reformulated by stating that phenotypic variance results from heritability and phenotypic plasticity. In Chapter 20 we will decompose the term $V_E$ into its various components. This will allow us to better understand the various components of phenotypic plasticity.

Phenotypic plasticity can either be adaptive or can simply reflect physiological effects that have no adaptive significance. It is favoured by natural selection when it allows individuals from a population to adapt efficiently to different environments that they might encounter. However, there are certain costs linked to the maintenance of the sensory and regulatory machinery necessary to ensure this flexibility (DeWitt *et al.* 1998), such that phenotypic plasticity remains limited.

#### 2.2.3.4 *Phenotypic plasticity and adaptation*

The adaptive significance of phenotypic plasticity can be clearly demonstrated by looking at organisms that produce clones. For instance, water fleas, freshwater cladoceran crustaceans, have asexual reproductive phases during which many individuals possessing the same genotype are produced. However, the members of a given clone can develop an extension over the head (called rostrum) of variable sizes (Figure 2.5) depending on the surrounding conditions.

In a detailed study on the water flea *Daphnia cucullata*, Anurag A. Agrawal and colleagues (1999) analysed the environmental factors influencing the development of the rostrum as well as its adaptive significance. First, they showed that if development took place in the presence of predators placed in enclosures such that they could not interact with the water fleas, adult water fleas had rostrums almost twice as long as individuals of the same clone raised in the absence of predators. These results were similar for two different predator types, the cladoceran *Leptodora kindtii* (average rostrum length of controls $15.53 \pm 0.35$ versus $29.71 \pm 0.49$ for individuals raised in the presence of predators, $t = 23.72$, degrees of freedom (df) $= 303$, $P < 0.001$), and the dipteran *Chaoborus flavicans* (average rostrum length of controls

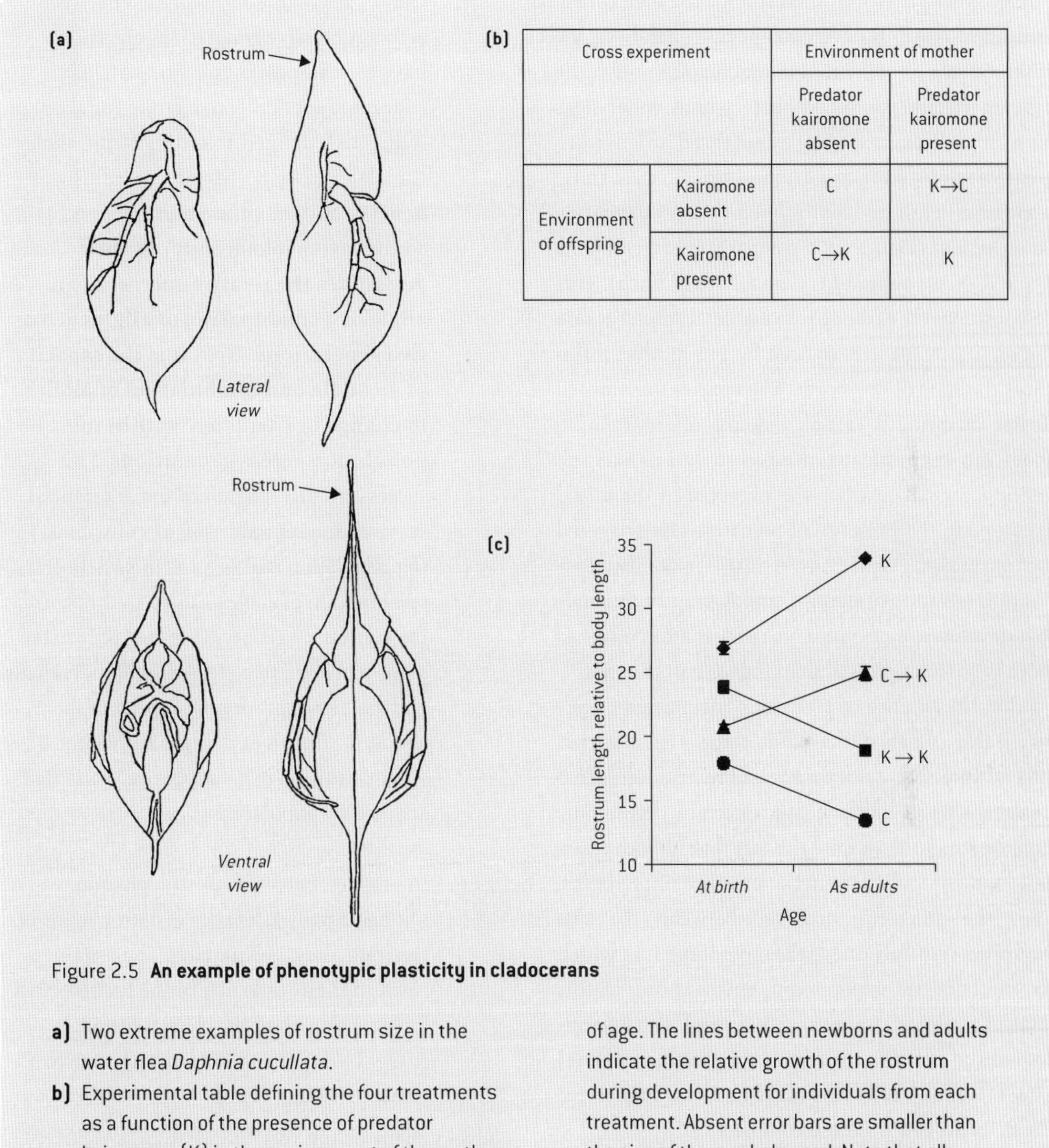


| Cross experiment | | Environment of mother | |
|---|---|---|---|
| | | Predator kairomone absent | Predator kairomone present |
| Environment of offspring | Kairomone absent | C | K→C |
| | Kairomone present | C→K | K |

Figure 2.5 **An example of phenotypic plasticity in cladocerans**

**a)** Two extreme examples of rostrum size in the water flea *Daphnia cucullata*.

**b)** Experimental table defining the four treatments as a function of the presence of predator kairomone (K) in the environment of the mother and her offspring.

**c)** Rostrum length relative to body length (average ± standard error) of individuals from the four treatments (C, K, C→K, and K→C) as a function of age. The lines between newborns and adults indicate the relative growth of the rostrum during development for individuals from each treatment. Absent error bars are smaller than the size of the symbols used. Note that all individuals have the same genotype, having been asexually produced. Adapted from Agrawal *et al.* (1999).

13.93 ± 0.15 versus 27.88 ± 0.28 for individuals raised in the presence of predators, $t = 44.29$, df = 470, $P < 0.001$). This first experiment demonstrates that it is indeed predators that influence rostrum development, and that water fleas recognize the presence of predators through chemical substances that the latter release in the environment. Subsequently, when water fleas with and without rostrums were placed in contact with predators, it was shown that possessing a rostrum dramatically reduced the chances of being captured by a predator, indicating that the rostrum undeniably has an adaptive function.

Agrawal *et al.* (1999) then studied the mechanism leading to rostrum development by controlling the environment in which the mother and offspring develop (Figure 2.5b). At birth, young water fleas born from mothers having experienced an environment with predators had a larger rostrum than those born from mothers having lived in a predator-free environment (Figure 2.5.c: comparison of treatments K and K→C to treatments C and C→K at birth). This demonstrates the existence of a maternal effect, because the environment experienced by the mother influences the state of newborns. The mother transmitted to her offspring a certain amount of information about the environment in which they are likely to develop. Similar to what was observed in the initial experiment, the presence of predators also triggered rostrum growth (Figure 2.5.c: comparison of treatments K and C→K to treatments C and K→C). Individuals that developed in the presence of predators invested more in rostrum growth than individuals that developed in the absence of predators. Furthermore, once water fleas reached the adult stage, the maternal effects were still evident. For example, adults from treatment C→K had a smaller rostrum than adults from treatment K, despite the fact that in both cases, development took place in the presence of predators. Thus the difference observed between the two groups was entirely due to the environment in which the mothers had experienced. Furthermore, in this same experiment, Agrawal *et al.* (1999) also demonstrated the possible existence of a **grand maternal effect**. Hence, the causes and effects of phenotypic plasticity are multiple and complex.

### 2.2.4 Genotypic and individual fitness

Another central concept in evolutionary biology is that of **genotypic fitness**, which can represent both an estimation and a prediction of the rate of natural selection. Let us consider the process of natural selection when it implies genetic heredity. Certain genotypes could be more successful than others, and it is possible to demonstrate this by studying a natural population (Endler 1986; Bell 1997).

Genotypes that replicate more effectively across successive generations are said to have a greater genotypic fitness. This concept is applicable to a group (or class) of individuals defined according to the allele (or alleles) they possess at a specific locus under consideration (or multiple loci considered simultaneously). **It thus designates the relative success of an allele or a combination of alleles across two generations.** Thus, **it is the capacity of an allele or a combination of alleles to change in frequency within the population across generations.** This success is a direct result of the differential survival and reproduction of individuals in the population that possess this allele or combination of alleles. Taken in this sense, the concept of genotypic fitness only makes sense when referring to classes of individuals, and it is not appropriate to speak of the genotypic fitness of a single individual. This strict definition of genotypic fitness is that adopted by population geneticists.

In practice, behavioural ecologists do not necessarily know the genetic origins of the traits they are studying. Their main concern is rather to determine the selective forces acting upon the character under study. To do so, it is more appropriate to measure at the individual scale the survival and reproductive consequences of variation (natural and experimentally induced) in the trait. Individual **fitness is the ability of a phenotype to produce mature descendants relative to other phenotypes in the same population at the same time.**

Measures of individual fitness (or simply fitness) are implicitly substituted for measures of genotypic fitness in most studies. It corresponds then to the average **demographic success** of a phenotype relative to the success of other phenotypes present in the population. The quantification of individual fitness can be, depending on the objectives of the study, limited to a short period in the life of an individual (winter survival, the number of offspring produced in a single reproductive event) or, ideally, according to the total reproductive success of an individual calculated over its entire lifetime. Individual fitness designates the **success of a trait within one generation**. The fundamental questions of how to measure individual fitness will be developed in Chapter 5.

The concepts of genotypic and individual fitness are only meaningful when considered within a single population. It makes no sense to compare these values between genotypes or between individuals belonging to different populations or living at different time periods.

### 2.2.5 Evolution, natural selection, and adaptation

#### 2.2.5.1 *What is natural selection?*

Behavioural traits, in the same manner as all other types of characters observed in living organisms, are the result of the history of species and populations, just like the organization of galaxies or the features of today's mountain ranges stem from past events. However, there is an important distinction between the organic and inorganic worlds. Living organisms have evolved over time because of, in large part, the process of selection (Endler 1986; Dawkins 1986, 1989). When the characteristics of organisms are beneficial to the survival and reproduction of individuals within a stable environment, copies of the allelic forms of the genes that are responsible for these characteristics multiply and spread throughout the population, which has the effect of increasing the frequency of these allelic forms in the population over time. This differential multiplication of genes according to their beneficial effects on the survival and reproduction of their vehicles within a given environment is the process of natural selection.

Since Darwin (1859), natural selection has played a pivotal role in the development of the modern synthesis of biological evolution (Fisher 1930; Williams 1966; Dawkins 1982; Endler 1986; Bell 1997). Despite this, its definition is often vague and imprecise, and reference to it still raises controversy. It is therefore necessary to define carefully and accurately what natural selection is, as well as explain its applications in the study of animal behaviour.

*Definition*

Natural selection is a process (Endler 1986), a group of related events linked through a chain of causality. There are certain requirements for natural selection, each independent of the other, and when these requirements are simultaneously met, the effects of selection ensue. There are three conditions that must necessarily be fulfilled to trigger natural selection.

(a) There is **variation** among individuals in a particular trait.
(b) There is a **consistent relationship** between this trait and the ability of individuals possessing the trait to survive (for example the ability to evade predators) and/or reproduce (for example the ability to produce viable offspring). In other words, there exists a consistent relationship between this trait and individual fitness. This is what is known as the **selection pressure**.
(c) Variation in the trait can be passed on to the next generation, independent of effects related to the fact that successive generations may develop within the same environment. In other words, the trait must be **heritable** (or more generally transmittable; see **transmittability** in Chapter 20).

When these three conditions are met simultaneously, foreseeable effects will automatically be produced within one generation (1) and between two successive generations (2):

1. The frequency distribution of the trait will vary **predictably** among age classes or among different life cycle stages, beyond differences linked to the ontogeny of the trait (with the provision that environmental conditions remain stable throughout the life cycle); at the genetic level, the frequencies of the alleles coding for the trait will vary **predictably** over time within the same cohort.
2. If the population is not at **equilibrium**[1], the distribution of the trait within a generation will differ **predictably** from the distribution within the parental generation, beyond the effects produced by conditions (a) and (c) alone; at the genetic level, the frequencies of the alleles coding for the trait will change **predictably** from one generation to the next.

This is one of the major points of evolution by natural selection. If conditions (a), (b), and (c) are fulfilled and if the environment remains stable between successive generations, it is possible to **predict** the direction in which evolution will take place between two generations. This does not mean that selection has a predetermined, organized goal, but rather that it is always the phenotype that produces the greatest number of descendants that increases within the population, until eventually replacing representatives of all other phenotypes. It is possible to distinguish between different modes of selection according to their effects on the mean and variance of a particular character (Box 2.3).

In sum, natural selection is a process in which individuals are sorted out according to their ability to survive and reproduce. The immediate response to selection is observable only at the level of changes in allelic frequencies. The gene therefore represents the **unit of selection**. Natural selection has been demonstrated within a wide variety of organisms (see Endler 1986; Bell 1997), and numerous studies have shown that its impact can be rapid and large (see, for example, Malhotra and Thorpe 1991).

1 By definition, when a population is at equilibrium, the same frequency distribution for a trait persists in each generation. The population is then not evolving. This can be the result of natural selection, or of its association with other antagonistic forces. If a population is not at equilibrium, evolution of the trait can occur. It is important to note that equilibrium is not an intrinsic property of a population and cannot be judged unless in relation to a given trait.

#### 2.2.5.2 *Evolution, natural selection, and genetic drift*

*Selection does not always lead to evolution*

If only conditions (a) and (b) of those described above are fulfilled, a selection pressure still exists in that only individuals in which the trait takes a certain value are better able to survive and reproduce. But in the absence of heritability of the trait, there is no selective response and the gene frequencies within the population will not change from one generation to the next. A selection pressure working on a trait is consequently not sufficient to cause evolution of the trait. For that matter, **evolution does not always imply natural selection**.

*Genetic drift: evolution does not necessarily imply selection*

Evolution can take place in a random manner by genetic drift, or by chance events such as large natural disasters, or the displacement of certain individuals. Thus, **evolution and selection are not interchangeable terms** (Sober 1984). Natural selection is one of the possible causes of biological evolution, but it is not the only one.

Genetic drift **corresponds to the random sampling of alleles between generations.** Genetic drift is a stochastic process. It emphasizes the random nature of transmitting alleles from one generation to the next given that only a fraction of all possible zygotes become mature adults. Because a pair of diploid sexually reproducing parents only have a limited number of offspring, not all of the parent's alleles will be passed on to their progeny because of chance assortment of chromosomes at meiosis. In a large population this will not have much effect in each generation because the random nature of the process will tend to average out. However, in a small

## Box 2.3 **Effects of selection on the mean and variance of a trait**

Selection can occur according to different modes. To keep matters simple, we will consider only two essential modes: **stabilizing** and **directional selection** (there is a third mode, **diversifying selection**, which leads to bimodality in the distribution of a trait by symmetrically favouring its extreme values). Let us consider a continuous trait that is normally distributed within a population and whose variance is equal to the mean (this type of distribution is typical of morphological characters: size, weight, number of segments, etc.). Under stabilizing selection (Figure 2.6), the genotype representing the average value of the character has the greatest fitness. Genotypes responsible for the other more extreme values of the character have a lower fitness, and more so the further they are from the mean value. This mode of selection results in a narrowing of the distribution about the mean, which remains unchanged across two generations, although the variance is reduced. Under directional selection (Figure 2.6), the genotypes that represent one of the two extremes of the distribution are favoured. The mean will then change in this direction between two generations, and the variance may also be reduced.

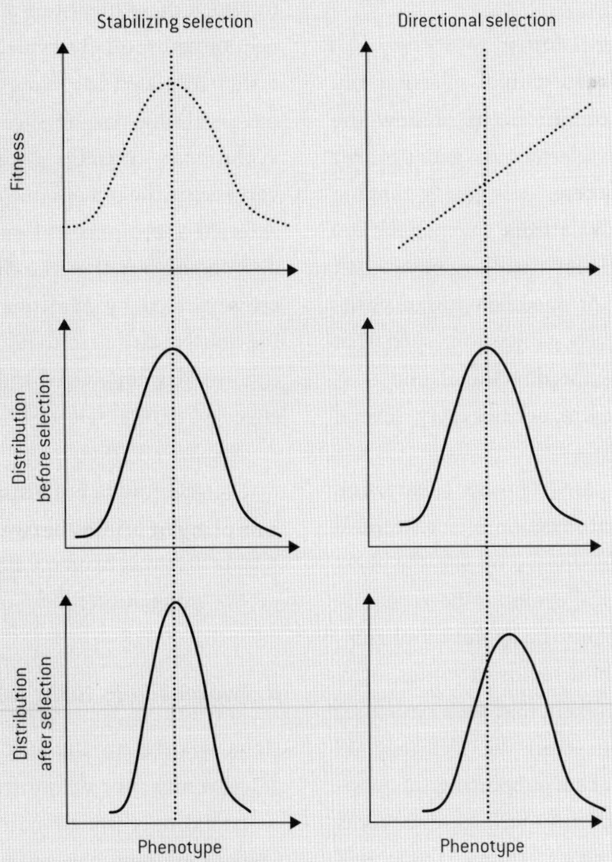


Figure 2.6 **Stabilizing and directional selection**

population the stochastic effect can be significant and lead to more lasting changes in allele frequencies between generations. The importance of genetic drift therefore depends critically on the size of the population. Actually, the exact probabilities that determine the duplication of alleles during sexual reproduction are only seen within populations of infinite size. Because all natural populations are of finite size, genetic drift always has a more or less moderate effect depending on the size of the population and the extent of other operating evolutionary forces (selection, dispersal). Hence, genetic drift and selection work concurrently within small populations. The essential difference between natural selection and genetic drift is found however in the conditions in which they take place. Variation in a trait does not have to have survival and/or reproductive consequences (condition (b) of selection) for genetic drift to occur.

*Natural selection versus genetic drift*

The only difference between natural selection and genetic drift is in the condition (b) of how the various phenotypes are selected. In genetic drift there is no link with fitness. It is just a random drawing of some individuals from the population. Thus, because of its random nature, genetic drift can affect the frequency of one allele either positively or negatively at each generation, such that it cannot be responsible for directional and consistent changes in allele frequencies over several generations.

If this is indeed the case, why do behavioural ecologists consider natural selection of such central importance? Natural selection is by far the most important force in biological evolution because only natural selection can explain the evolution of complex structures such as bird or insect wings, eyes, etc. It is indeed highly unlikely that such structures evolved for other reasons than the fact they do increase fitness. The argument supporting this position has been included in many well-known works (Dawkins 1982, 1986, 1989; Dupré 1987; Bell 1997), and we encourage the curious reader to explore them. We will confine ourselves here to the essential aspects. The privileged status that is bestowed upon natural selection simply results from its central role in understanding the logic behind the hierarchical organization of life. The complex traits that are observed today did not appear 'as if' following a simple mutation. They are more likely the result of a series of cumulative changes, where each intermediate step was better adapted to the environment than the preceding one (see Dawkins 1986). This phenomenon cannot be explained in terms of genetic drift, only through natural selection. Natural selection is at the basis of how organisms **adapt** to their surroundings.

#### 2.2.5.3 *Selection and adaptation*

The adaptive state of organisms is a narrow connection between the form taken by the organs that fulfil various functions and the environmental conditions in which these functions must be fulfilled. This connection is especially evident when species that are distantly related but living in the same environment have striking resemblances. This is referred to as evolutionary **convergence**. A typical example of convergence is the remarkable resemblance between the external appearances of marine mammals and fish, despite the fact that the origins of the two groups are separated by over one hundred million years. Previously attributed to divine intervention, adaptation is today considered the result of the cumulative effects of natural selection in the past.

> **Adaptation** is a historical concept (Sober 1984). A trait is considered an adaptation if it was fixed (or stabilized) within a population following a selection event.

##### a *Two meanings to the word 'adaptation'*

However, certain authors use the same word to describe any trait whose frequency in a population is increasing through natural selection. The term adaptation then has a certain ambiguity to it. It can simultaneously describe two different aspects of evolution, which need to be clearly distinguished:

**adaptation can refer to the final result of a completed process** or **a process currently underway**. The study of adaptive states or the degree of adaptation applies to the 'finished product' of a selection event. At this stage, the trait is fixed within the population or its frequency is stable. When a trait is fixed, there is no longer any genetic variation in this trait and its heritability is zero (hence the expectation that traits that contribute much to fitness should have low heritability (Falconer 1981). The study of allelic frequency changes is consequently not applicable to describe the degree of adaptation. Conversely, the study of selection in progress concerns the process itself. The continuation of the process necessarily implies that the trait maintains a certain degree of heritability over time. It then becomes possible to detect changes in allelic frequencies as long as the genetic origins of the trait have been identified.

*Four mechanisms of adaptation*

Overall, **the adaptive state of an organism can be produced by four important mechanisms**, all quite different (Gould and Lewontin 1979; Laland *et al.* 2000). First, the degree of adaptation can be the result of the Darwinian processes of mutation/selection as described above. Next, the processes of physiological and behavioural development allow organisms to shape themselves to the conditions prevalent during their development (through phenotypic plasticity; see Chapters 5 and 6 for further details). Of course, the adaptations themselves are not passed on, but the ability of organisms to display such plasticity is genetically transmitted. The third mechanism relates to cultural adaptation, which has become transmittable through the learning process (see Section 2.2.7 and Chapter 20). The fourth mechanism is rarely accounted for despite the fact that it plays a major role in human adaptation. It is 'niche construction' (Laland *et al.* 2000), the fact that organisms constantly modify their own environment, which in turns affect their fitness. In species where offspring often inherit their parents' territory, niche construction may strongly affect the fitness of descendants over many generations and such variation in environment becomes transmitted across many generations. In other words it becomes transmittable and thus open to selection. We will discuss such issues in Chapter 20.

### 2.2.6 Inclusive fitness

The use of individual fitness to characterize genotypic fitness, while justifiable within a limited framework, can lead under certain circumstances to erroneous conclusions. This is especially the case in the field of social behaviour. Individual fitness considers the consequences of an organism's behaviour only in light of the propagation of its genes **as a result of the survival and reproduction of that individual**. We have seen, however, that the individual is not the unit of selection (see Section 2.2.2). Therefore, it is appropriate to measure the success of a behavioural trait by estimating the consequences of this behaviour for the genes that are involved in its expression. This means considering not only the consequences related to the survival and reproduction of the individual exhibiting the behaviour under study but also to consequences for other individuals that are affected by this trait. This leads us to the introduction of the concept of **inclusive fitness** (Hamilton 1964a, b). This central concept in behavioural ecology originates from sociobiology where it served to replace group selection as a better account for the evolution of altruistic behaviour among related individuals (see Box 2.4).

#### 2.2.6.1 *A major but subtle concept*

Despite its importance the concept was not always well understood (see Wilson (1975) and Barash (1982), two 'stars' of sociobiology!), and so it is imperative to explain the logic behind it clearly (Maynard-Smith 1982; Grafen 1984; Creel 1990).

When formulating his ideas on inclusive fitness, Hamilton drew attention to the fact that a gene responsible for a particular social behaviour could be selected for or against depending on the effects of the

### Box 2.4 **Group selection: fact or fiction?**

In the early 1960s, many ethologists thought that certain phenotypic traits that reduced the survival and reproductive success of individuals that bore them could still be selected for if these traits otherwise increased the long-term stability and survival of the individual's group or the species. Selection it was therefore assumed could also operate **at the group level**. This idea was most clearly articulated by a Scottish ecologist Vero Wynne-Edwards (1962) in his work entitled *Animal Dispersion in Relation to Social Behaviour* (see also Wynne-Edwards 1986).

In the absence of any regulatory force, the growth rate of a population is exponential. The expansion of a population is however limited by the maximum number of individuals that the environment can sustain (called the **carrying capacity**). According to Wynne-Edwards, animals naturally tend to avoid overexploiting their environment, especially their food resources. This was achieved through altruistic behaviour in which certain individuals postponed or sacrificed their own reproduction in order to avoid overpopulation that would be fatal to all. Wynne-Edwards also proposed that certain social displays of species served to allow individuals to evaluate their population size and hence adjust their reproduction according to their perception of resource availability. In support of his theory, Wynne-Edwards cited numerous examples suggesting that animals do not always realise their full reproductive potential, and even actively regulate it. For example, in certain species, individuals do not immediately reproduce upon reaching sexual maturity, but rather defer reproduction for no apparent benefit. When a species has a distinct social structure, it is common for subordinate individuals to not reproduce at all. In certain cases, infanticide even occurs. According to Wynne-Edwards, these trends and events supported the existence of population autoregulation by individual behaviour.

This idea was strongly criticized and refuted by Williams (1966) in his work *Adaptation and Natural Selection*. Williams gave the example of a population in which each individual was genetically predisposed to limit its own reproduction. If a single individual that was less genetically inclined to sacrifice its own reproduction entered the picture, it would undoubtedly leave more descendants than the other members of the population. This individual's 'abnormal' behaviour would spread across successive generations, until it was the only strategy within the entire population, even if in the end the whole population crashed because of it. Natural selection is a **blind** cumulative process that simply cannot think ahead. The examples of reduced individual reproduction interpreted by Wynne-Edwards as the self-restraint are now interpreted as the result of social constraints related to competition between individuals. Today, despite some efforts to revitalize group selection (see Borrello 2005), the individual and gene level selection point of view defended by Williams and Dawkins still dominates scientific thought. The reader interested in thinking about different levels of selection is recommended to consult Keller (1999). For those interested in reading about selection at levels above that of the individual we also suggest the papers of this theory's strongest adherent David S. Wilson (1997) as a good place to start.

behaviour on individuals other than the immediate progeny of its vehicle, as long as there is a certain amount of genetic similarity by descent linking the vehicle to the other individuals. So for Hamilton the fitness of a trait should **include** its bearer's fitness but also the fitness of all individuals likely to be bearing the same gene and to have been affected by the behaviour of the bearer. As a result, an early but erroneous way of thinking of inclusive fitness has been to consider it as the total number of direct descendants plus the number of direct descendants of genetically similar individuals, where the latter value is corrected for the coefficient of genetic similarity between the vehicle and the other individuals. If we ponder this definition for a moment, it becomes clear that inclusive fitness would then always tend towards infinity because each individual has a certain degree of genetic relatedness by descent, albeit minimal, with a great number of individuals in the population. Reduced to this definition, the concept of inclusive fitness would be of very little interest and of no utility.

*Hamilton's first definition . . .*

The correct interpretation of inclusive fitness is both more complex and more subtle (Hamilton 1964a, b; Grafen 1984; Creel 1990; Bourke and Franks 1995). In his initial article, Hamilton (1964a, b) defines it as:

> 'The personal fitness which an individual actually expresses in its production of adult offspring... stripped of all components which can be considered as due to the individual's social environment, leaving the fitness which he would express if not exposed to any of the harms or benefits of that environment,... then augmented by certain fractions of the quantities of harm and benefit which the individual himself causes to the fitnesses of his neighbours. The fractions in question are simply the coefficients of relationship appropriate to the neighbours whom he affects.'

What Hamilton calls the 'social environment' corresponds to the portion of the environment comprising an individual's interactions with its neighbours. Neighbours can have a positive or a negative influence on a given individual's progeny, by either facilitating the individual's reproduction and/or the survival of its descendants, or conversely by hindering and limiting the reproductive potential of that individual through their behaviour. Likewise, an individual can have a positive or a negative effect on the reproduction of its neighbours by its actions. Within a given population, it is possible to calculate the total number of additional **direct** descendants imparted by the help of neighbours, divided by the total number of individuals (reproducers and helpers) in the population. This quantity corresponds to the average amount of help *per capita* within the population. Equally, it is possible to calculate the average amount of hindrance in the population, which would correspond to the total number of **direct** descendants lost due to the behaviour of neighbours divided by the total number of individuals.

*. . . was later improved*

However, Creel (1990) pointed out the necessity to replace a portion of Hamilton's original definition: '*all components which can be considered as due to the individual's social environment*' should be replaced by the term '***the per capita average effect (average amount of help – average amount of hindrance) due to the social environment of the individual***'. The importance of the adjustment brought by Creel can be illustrated using a simple example (Bourke and Franks 1995). Let us consider a population of animals in which there are two types of individual: reproducers and helpers. Individuals in the first category can only reproduce if they obtain help from conspecific helpers from the second category. Although the genetic basis of helping behaviour can be found in individuals from both categories, it is only expressed in certain individuals, the helpers (a similar situation is seen in certain species of social Hymenoptera, analysed in detail in Chapter 15). Then, according to Hamilton's original definition, the inclusive fitness of all reproducers would be zero. In fact, all the progeny of a reproductive individual is due entirely to the assistance of helpers, and that same reproducer does not contribute whatsoever to the reproductive success of others. This unreasonable result is corrected in the modified definition proposed by Creel, where only the **average** effect of one individual on the reproductive success of others is removed. It follows that in such a population, the inclusive fitness of reproducers ultimately depends on the help they manage to obtain. An individual that obtains more help than average would have a positive inclusive fitness once the average amount of help per reproductive individual was removed. The same reasoning can be applied to helpers. Because they are sterile by definition, their inclusive fitness depends directly on the amount of help they provide to genetically similar individuals. If this amount is great enough (or if help is predominantly given to the most genetically similar individuals), the genetic origin of helping behaviour will spread throughout the population. If helpers do not favour reproducers according to their degree of genetic similarity, the gene will not spread.

#### 2.2.6.2 *Kin selection*

How exactly do individuals evaluate the degree of genetic similarity between themselves and their neighbours? This has been a notorious problem since Dawkins (1976) proposed his 'green beard effect' metaphor. Let us suppose that a gene appears in a population that causes the two following effects (we know that a gene can have more than one effect; see Chapter 5): its owners all have a green beard and behave in an altruistic manner towards every other individual with a green beard. Let us also assume that it is impossible for a mutant to cheat by having a green beard but not displaying the altruistic behaviour. This situation is highly improbable, but if by chance it did occur the consequences would be clear: the gene would inevitably spread throughout the population.

Of course, nobody believes (not even Richard Dawkins) that the green beard effect could be commonly observed in its literal form. However, some evidence for green-beard gene have actually been found in nature, most specifically in the red fire ant (*Solenopsis invicta*, Keller and Ross 1998) and in the social amoeba *Dictyostelium discoideum* (Queller *et al.* 2003). Most of the time, however, individuals distinguish genetically similar individuals through less specific signs than a 'green beard'. Kinship and familiarity are examples of such signs. A brother, a sister, or more generally 'those that were raised with me' represent an indicator of genetic similarity. Furthermore, two recent independent theoretical studies have shown that social influences can play the role of a green beard effect. First, Hochberg *et al.* (2003) with a simple model showed that the evolution of social discrimination causes the congealing of phenotypically similar individuals into different, spatially distinct tribes (or **cultures**, see Chapter 20). They also showed that such a result was only obtained with altruistic and selfish behaviour, not with spiteful and mutualistic behaviour. Second, Jansen and van Baalen (2006) showed that if the green beard and altruism effects are caused by loosely coupled separate genes, altruism is facilitated through beard chromodynamics in which many beard colours cooccur. In fact, both of these models showed that 'culture' may play the role of such a green beard effect (we will come back to the question of the evolution of culture in Chapter 20).

> Natural selection preserving altruistic behaviour directed towards kin is aptly called **kin selection**.

#### 2.2.6.3 *Hamilton's rule*

The ideas proposed by Hamilton (1964a, b) simplified the calculations necessary to establish the conditions under which certain alleles spread within populations. Let us consider some social behaviour that involves two individuals, the **actor** (that performs the social behaviour) and the **recipient** (towards whom the social act is directed). **Hamilton's rule** involves three terms: $c$, the degree of modification to the actor's fitness (considered a cost to the actor); $b$, the degree of modification to the recipient's fitness (considered a benefit to the recipient); and $r$, the degree of genetic similarity between the actor and the recipient. This degree of genetic similarity is mathematically defined as a regression coefficient (see Bourke and Franks (1995, pp. 14–17) for a detailed explanation of the mathematical definition of the degree of genetic similarity; see also Chapters 13 and 15). Hamilton's rule is expressed by the formula:

$$br - c > 0$$

This rule is only valid under certain conditions. It is especially important to verify that costs and benefits are additive. An individual that is helped $x$ times and that helps $y$ times should in total experience a change of $xb - yc$ to its number of descendants.

The interest in Hamilton's rule lies in a large part in its simplicity. Indeed, the rule is easier to apply than the concept of inclusive fitness. In practice, it is convenient to use the coefficient of relatedness between the interacting individuals for the value of $r$. However, applying the rule to real situations requires following several well-determined steps. It is especially imperative to specify which behavioural alternatives are under study and to evaluate properly all of their consequences. For instance, one could study the following alternatives: a, to reproduce *versus* b, not to reproduce in order to help one or many relatives reproduce themselves. If the animal chooses option b, then the cost suffered is the number of offspring it would have produced had it chosen option a. The benefits are more difficult to calculate because they require estimating the difference between the recipient's reproductive success when the actor plays a and b. Various examples applying Hamilton's rule are presented in Chapter 15 in relation to the study of the evolution of **cooperation**.

## 2.3 Conclusion

When evaluating the purpose or function of a behaviour it is crucial to keep in mind several of the basic concepts discussed in this chapter. Although animal behaviour may appear to defy logic at first, the application of the evolutionary framework often makes the biological reasons behind the behaviour quite clear. An evolutionary perspective applied at the correct level of selection can often turn an apparently damaging behaviour into one that is evolutionarily profitable because it helps transmit the vehicle's genes to the next generation. We will see many such examples in this book. As the great geneticist and evolutionary biologist Dobzhansky (1973) said, '*nothing in biology makes sense except in the light of evolution*'. It is therefore important to keep in mind that the most obvious explanation is not always the correct one.

However, it is also important to emphasize that variation in behaviour may also involve learning that contributes to phenotypic plasticity. Furthermore, learning from others may lead to the transmission of behaviour across generations. Although the transmission of the variation in such learned behaviours may not necessarily involve genetic variation, the genetic basis of an individual's flexibility and capacity to learn is likely to involve genetic variation. The transmission across generation of non-genetic, yet Transmittable, phenotypic variation that involves learning is usually called 'cultural transmission'. The importance of such culture in the evolution of animal behaviour is addressed in Chapter 20.

## » Further reading

> *For a complete discussion of the genetics behind evolution:*

**Sober, E.** 1984. *The Nature of Selection. Evolutionary Theory in Philosophical Focus.* MIT Press, Harvard.

> *For a discussion of how genes drive evolution:*

**Dawkins, R.** 1989. *The Selfish Gene*, 2nd edn. Oxford University Press, Oxford.

> *For a complete overview of natural selection and how it can be detected in natural populations:*

**Endler, J.A.** 1986. *Natural Selection in the Wild.* Princeton University Press, Princeton.

> *For a case study about the genetic component of phenotypic variance and adaptation to the local environment:*

**Tracy, C.R.** 1999. Differences in body size among chuckwalla (*Sauromalus obesus*) populations. *Ecology* 80: 259–271.

> *For a case study of natural selection in response to environmental conditions:*

**Grant, B.R. & Grant, P.R.** 1989. Natural selection in a population of Darwin's finches. *The American Naturalist* 133: 377–393.

> *For a case study demonstrating heritability measurements of morphological traits:*

**Boag, P.T.** 1983. The heritability of external morphology in Darwin's ground finches (*Geospiza*) on Isla Daphne Major, Galapagos. *Evolution* 37: 877–894.

> *For a review of reaction norms and phenotypic plasticity:*

**Stearns, S.C.** 1989. The evolutionary significance of phenotypic plasticity. *BioScience* 39: 436–445.

**West-Eberhard, M.J.** 1989. Phenotypic plasticity and the origins of diversity. *Annual Review of Ecology and Systematics* 20: 249–278.

> *For a review of cooperative breeding in mammals integrating inclusive fitness and kin selection:*

**Jennions, M.D. & Macdonald, D.W.** 1994. Cooperative breeding in mammals. *Trends in Ecology and Evolution* 9: 89–93.

> *For a comprehensive look at kin selection in insects:*

**Queller, D.C. & Strassmann, J.E.** 1998. Kin selection and social insects. *BioScience* 48: 165–175.

## » Questions

1. Explain why the terms 'heredity' and 'heritability' are not synonymous.

2. How do genotype–environment interactions affect the phenotypic variance of a trait?

3. In what type of environment is phenotypic plasticity most likely to be favoured by selection?

4. What is the difference between genotypic fitness and individual fitness? Which concept is more useful to the field of behavioural ecology, and why?

5. What exactly is natural selection, and what conditions are necessary for it to occur?

6. Why can genetic drift alone not explain the adaptations of organisms to their environment?

7. How do the concepts of inclusive fitness and group selection differentially explain altruistic behaviour? Which concept has more validity?

8. What was the major problem with the initial interpretation of Hamilton's (1964a, b) definition of inclusive fitness? What did Creel (1990) propose to improve this definition?

9. How are the concepts of inclusive fitness and kin selection related?

10. Meerkats (*Suricata suricatta*) are cooperatively breeding mammals that live in packs consisting of two to three family units, each with a single breeding pair. Other pack members forego their own reproduction in order to assist the breeders. Design an experiment that tests if meerkat group behaviour follows Hamilton's rule.

11. You are studying a population of a granivorous passerine bird species that shows variation in beak size. Propose a detailed experimental design that would allow you to measure the heritability of beak size in this population.

12. Using the concepts you learned in this chapter, construct a logical argument defending that it is genes, and not individuals, that are the units of selection.

3

# 3

# Research Methods in Behavioural Ecology

Frank Cézilly, Étienne Danchin, and Luc-Alain Giraldeau

## 3.1 Introduction

Behavioural ecology differs from other behavioural disciplines in terms of both theoretical framework and methodologies. Historically, early progress in ethology was accomplished through inductivism that was dear to some of Darwin's followers and based on the generalization of anecdotes (cf. Chapter 1). This approach suffers from numerous limitations. For instance, it does not allow for any statistical considerations because it ignores the fundamental notion of sampling, leaving conclusions to the subjective appreciation of the observer. One of the major advances in the school of objectivists is to have anchored the study of behaviour into a **hypothetico-deductive approach**. This approach consists of formulating hypotheses and subsequently devising a research plan whose ultimate objective is not to confirm the hypotheses, but rather to disprove their predictive power. We can therefore summarize this approach as a tendency to refute or invalidate hypotheses, which was adopted by ethologists and then behavioural ecologists. What distinguishes ethology from behavioural ecology is the way in which the hypotheses are formulated. Behavioural ecology places itself explicitly within an adaptationist framework, the study of the evolutionary significance of behaviours in relation to how they increase the **fitness** of the organisms that display them, without necessarily making any assumptions concerning the mode of transmission of the behaviour across generations (Avital and Jablonka 2000). All observed behaviours are therefore understood as being the result of a historical process that plays out at different temporal scales, that of the ontogeny of individuals, the structuring of populations, and species differentiation (see Chapter 2 for more details). In contrast, today, ethology focuses more on the short-term mechanisms that produce behaviour. Ethology thus mainly studies processes occurring within individuals. Studying the link between behaviour and **adaptation** to the environment is broad and requires a combination of different lines of investigation that correspond to the study of differences between these diverse entities: individuals, populations, and species (Krebs and Davies 1984). The objective of this chapter is to present these different approaches, to which the subsequent chapters will regularly make reference. It opens with a general introduction to the application of the hypothetico-deductive approach in behavioural ecology. We then present three major approaches in behavioural ecology: the phenotypic approach, the genetic approach, and the comparative approach, before concluding with how they complement each other.

## 3.2 Theories, principles, models, and experiments

The distinction between the so-called 'hard' and 'soft' sciences refers generally to their stated objective. In hard sciences, the objective is to discover general laws whereas in soft sciences it is often to explain a particular sequence of events within a historical process. From this point of view, behavioural ecology lies somewhere between the two. On the one hand, behavioural ecology attempts to identify how behaviour adapts an organism to its environment. The question is most often elucidated by logical models, more or less formalized from the mathematical point of view. The models can serve to define the possible modes of evolution of phenotypic characters, or predict the **performance** of an organism in a given situation. However, behavioural ecology also uses real organisms to study the results of a historical process and therefore proposes scenarios that explain *a posteriori* the evolution of the organisms' traits.

### 3.2.1 The distinction between theory, principle, and model in behavioural ecology

Through the course of their work, researchers in behavioural ecology refer to a certain number of theories and principles, construct models, and test **predictions**. However, it is not always easy to understand the distinction between theory, principle, and model. In a general sense, theories are broader than principles. Principles are logical constructions, that are neither false nor true, but which must be coherent. The logical coherence of verbally set out principles is most often verified by a mathematical formalization. The principles define what would happen **if** a certain suite of conditions, called **assumptions**, were satisfied by the system. It is important to keep in mind that principles do not say **when** or **at what frequency** these conditions are upheld or were upheld in the past. In this sense, they resemble scientific laws. Models, however, correspond more to a representative context, approximate and schematic, but judged useful on the basis of their predictive capacity. They translate the application of a principle to a more or less realistic situation. One theory is therefore meant to correspond to a more global scientific system, which includes several principles, which in turn are based on distinct models.

In behavioural ecology, the word 'theory' is especially employed to designate a research programme and therefore corresponds to an ensemble of principles and models derived from a limited number of axioms. For example, optimal foraging theory, presented in Chapters 7 and 8, rests on the axiomatic proposal that the foraging behaviour of animals can be studied as a process of **choice**. One of the principles that may guide these choices is that of energy maximization. On this basis, different models predicting how animals should forage under various circumstances can be developed. Contradictory principles can often be included in a single theory. **Sexual selection** theory (discussed in Chapter 11) is based upon a simple axiom that stipulates that the evolution of certain secondary sexual traits results from the benefits linked to differential access to reproductive partners. Within this theory, different potentially incompatible principles (the **handicap principle**, Fisher's principle, **sensory exploitation** principle, and sexual conflict principle), coexist. It is possible that some of these principles are not realistic, or, more likely, that their pertinence varies from one biological model to another.

By definition, a model represents an approximation of reality. However, the degree of sophistication of models can vary. At one extreme, the simple laying out of a hypothesis constitutes a verbal model. At the other extreme, models can incorporate numerous equations relating a multitude of variables. Different types of model are constructed based on the question being asked, the complexity of the situation, and the objective to be attained. The scientific approach usually consists of setting out several simplifying assumptions in order to reduce a complex phenomenon to the interaction between a limited number of factors, thereby facilitating the understanding of the phenomenon under study. The advantage of this simplification resides in the generality (Box 3.1)

### Box 3.1 The concept of generalization in sciences

Scientific theories, whether in physics, chemistry, biology, or behavioural ecology have the objective of being general, that is of applying to very many classes of objects or situations. For instance, Newton's Universal Law of Gravitation applies to all point masses whether rocks, small planets, apples, birds, or feathers.

Behavioural ecological theory strives towards the same type of generality, which in our case implies correctly predicting the behaviour of a large number of animal species. However, in almost every theory there will be a trade-off between generality and the precision of predictions. For instance, gravity accounts for a large part of a pendulum's motion. However, to be precise in predicting the movement of a specific pendulum at a given time requires consideration of data that are not included in the general law. We would need to account for friction of the pendulum against the air, perhaps include wind speed, the width of the string linking the mass to its rotation pivot, friction at the pivot, the surface of the pendulum's bob, etc. Adding all these variables makes predicting the pendulum's behaviour more precise but at the expense of generality, that is of being able to predict the behaviour of other types of pendulum that oscillate under different sets of conditions. Similarly, Newton's theory is sufficient to predict the movement of most planets, but it fails at predicting some details in Mercury's orbit around the sun. This is because Mercury being very close to the sun is placed in a much stronger field of gravity, a condition for which Newton's theory is no longer valid. Although Newton's theory is sufficient to send rockets to the Moon or Mars in conditions of relatively low gravity, it would not be enough if we were to send a spaceship to Mercury or in areas where gravity becomes much more influential in the shaping of space. Then we would need to use Einstein's generalized theory of relativity which, as its name indicates, describes the cosmos under a much wider array of conditions. However, despite the fact that Einstein's theory is more general than that of Newton's, it would be unnecessarily complicated to use the former to place a telecommunication satellite around our planet.

The same is true for behavioural ecological theories. The best example is that of fitness. As long as we are dealing with non-social interactions, i.e. situations where the action of a given individual does not really influence the fitness of others, we can safely ignore the indirect component of inclusive fitness (see Chapter 2) because in such situations that component is negligible. However, if we are to explain ant societies (see Chapter 15) or some aspects of dispersal (see Chapter 10), then we need to use the more general concept of inclusive fitness, because the indirect component of inclusive fitness is no longer negligible in such conditions.

However, by adding special variables it may become possible to predict the foraging behaviour of a given individual of a particular species, but this precision would be achieved at the expense of generality; that is, these variables would not necessarily make for better predictions for other species foraging under different circumstances, or even another individual of the same species having a different knowledge of its environment. What we attempt to do in behavioural ecology is to provide the simplest general theory that applies to the largest number of species possible, knowing that by doing this we are discarding a large number of probably important variables that play some role in fine tuning the behaviour to each special circumstance in which it is used.

An alternative approach to prediction is known as forecasting models. These models are complicated computer simulations of the real world in which very many variables and their interactions are included to forecast the state of some phenomenon in time. Forecasting models often require very large computers to analyse the numbers required to predict. They are used in areas such as meteorological, economic, and atmospheric sciences and are increasingly important in applied ecology such as forest ecology and conservation. However, despite the fact they are extremely useful tools, they have only little value in terms of generality.

of the predictions that are generated from the models that are constructed in this way, that is to say, that these models can then be applied to a wide range of different organisms and situations. On the other hand, we can try to predict very precisely the value taken by a particular variable amid a complex phenomenon and try to describe very accurately the natural situation within the model. These are local application models, whose usefulness is usually limited to a single type of organism or population. This approach is more often used in conservation biology, where the model must serve to aid in decisions for specific cases. Generally, high realism in a model limits its generality and diminishes its simplicity, thereby reducing its didactic value. At the extreme, a model that is too complex, incorporating too many parameters, will not provide any analytical solutions and will only be able to provide predictions through computer simulation (Box 3.1). In practice, whichever model type is used, it always acts as a hypothesis. It generates certain predictions whose accuracy can only be predicted using an experimental approach. From this regard, the advantage of a simple, and therefore general, model is that the limited number of factors that it includes allows for them to be controlled later by an experimenter.

### 3.2.2 The experimental approach

It is fundamental to remember that **the real objective of experimental tests is to attempt to disprove a hypothesis, not to confirm it**. In fact, our understanding of the question being studied can only progress if results of these tests disprove our hypothesis, that is, that they are not in accordance with the predictions that are derived from that hypothesis. In such a situation we are certain the hypothesis is false. On the other hand, if the results fail to reject the hypothesis, the only conclusion that we can draw is that the hypothesis, as it is formulated, has not been disproved, not that it is true. There can always be other processes that we have failed to consider that could have generated the same results. We therefore have no certainty concerning the correctness of a hypothesis. However, when the results disprove our hypothesis, we must reconsider it and construct a new model, devise a new hypothesis that must *a priori* be compatible with the results of the previous test and finally generate new predictions that can then be tested. To illustrate the experimental approach we will use the example of the adaptive significance of the long tail ornaments of the barn swallow (*Hirundo rustica*).

#### 3.2.2.1 *The tail ornaments of barn swallows*

In barn swallows, males and females have two outer feathers on their tails (called streamers) that are dramatically longer than the other feathers (Figure 3.1). However, the streamers are significantly longer in males than in females. This raises the question of what selective pressures may have resulted in this difference. We will discuss in further detail questions related to differences in morphology between males and females of the same species in Chapter 11. One hypothesis that has been put forth is that females prefer to mate with males that have longer tails. If tail

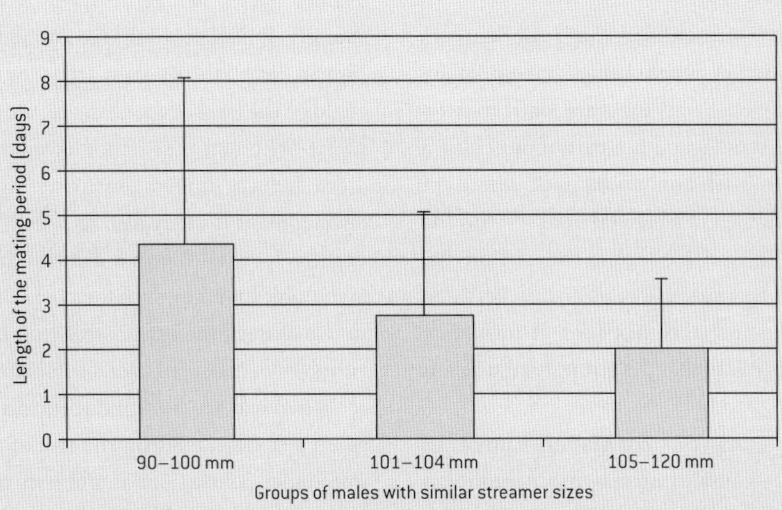


Figure 3.1 **The length of tail feathers and the latency to mating of male barn swallows (*Hirundo rustica*)**

Males were captured and marked at the moment of their arrival on the breeding grounds. They were then observed until they were clearly pair-bonded with a female, that is, until the day that a female arrived in the territory of the male and stayed to reproduce with him. The correlation is significant ($P < 0.05$, $r^2 = 0.08$) and negative, showing that the longer the tail the shorter the length of the mating period. Data extracted from Møller (1990).

length is **heritable**, over time, such a preference could lead to a lengthening of the tails of males as compared with those of females. To test this hypothesis we can put forth the following prediction: if females prefer to mate with males that have longer tail feathers, then we expect that there will be a negative relationship between male tail length and the latency to acquire a mate because the longer a male's tail the more quickly it will acquire a mate.

*A correlative approach*

Having formulated this prediction, we can now go to the field, record the arrival date of males, measure the lengths of their tails, observe the date of pair formation, and calculate the latency to pair formation. We can then plot latency to pair formation against tail length and analyse the relationship between the two with a regression analysis. All this has been done, and the relationship was statistically significant (Figure 3.1).

*Can we conclude from this that females actually prefer males with longer tails?*

No, because many other mechanisms could explain such a relationship. For example, males with longer tails may be better flyers, and are therefore better able to force females into pairing with them. Or, females may chose males based on some other criteria, which is itself correlated with tail length. To say that Figure 3.1 proves that females prefer males with longer tails would be mistaken in two ways. First, as we mentioned above, failing to reject a hypothesis does not mean it is true. Second, such a conclusion would infer a cause and effect relationship from a simple correlation analysis. Implicitly, it would amount to saying that tail length is the cause of the shortened pairing period. The following example of storks and the birth of baby humans illustrates the dangers of such a causal interpretation of correlational associations.

*Babies are brought by storks*

When you were a young child, perhaps your parents avoided the question of 'where babies come from' by telling you stories such as they are delivered by storks. This persistent legend is a part of our culture and is still used in birth announcements, where we see babies being carried in a blanket hooked onto the beak of a stork. To test this legend, we can predict that there should be a positive relationship between human birth rates and stork densities. It turns out, however, that this relationship is strongly significant (Figure 3.2).

Naturally we know this correlation cannot be telling us that the cause of increased birth rates in humans is the increased population density of storks. However, the reason for the correlation remains unclear. One possible explanation is that there exists a third factor which is itself independently but causally linked to both the number of storks and human birth rates (Figure 3.3). One candidate factor could be the level of economic development of a country. It is known that strong economic development often reduces human birth rate. Strong economic development is likely also to affect the environment negatively, resulting in a decrease in the number of favourable breeding sites for storks. If this interpretation is correct, then the correlation between the two variables of interest may be completely fortuitous.

Whenever we interpret a correlation in terms of cause and effect, we run the risk of drawing equally unfounded conclusions as when we tell our child that babies are delivered by storks. However, we cannot deny the value of a correlational approach. An absence of a relationship in the case of Figure 3.3

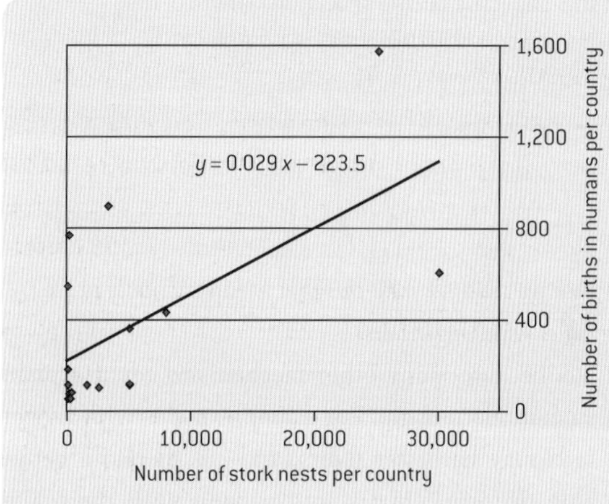


Figure 3.2 **Babies are delivered by storks**

Human birth rates by country (total number of births annually) as a function of the number of stork pairs in 17 European countries. The relationship is positive and strongly significant: $F_{1,15} = 9.40$; $p = 0.0079$; $r^2 = 0.385$. Adapted from Matthews (2000).

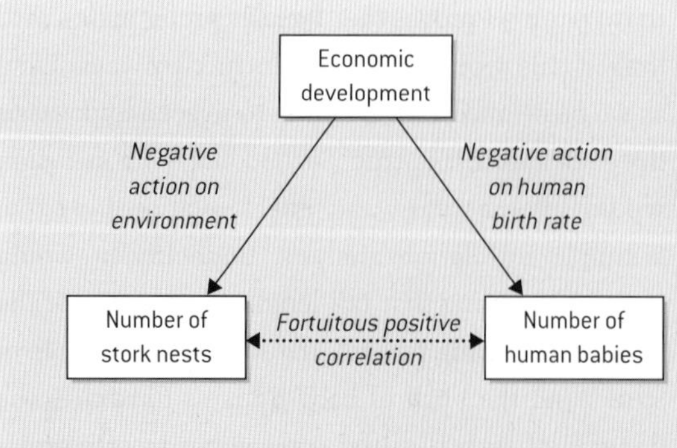


Figure 3.3 **Possible causal explanation for the correlation between the number of stork nests and human birth rates**

The existence of a third factor, itself simultaneously responsible for variation in the first two variables, could generate the correlation between human birth rate and the number of stork nests. According to the potential mechanism, only the solid thick arrows represent cause and effect relationships, the dotted thin arrow represents a fortuitous statistical correlation.

would have clearly invalidated the hypothesis. Furthermore, the accumulation of coherent correlations can, in certain cases, constitute an accumulation of evidence in favour of a given mechanism. However, it is important to keep in mind that the experimental approach remains the only means of testing a cause and effect relationship.

*Back to barn swallows*

For the hypothesis that female barn swallows prefer males with long tails, the critical test consists of artificially modifying the length of the male tails as they arrive in the spring, and then observing whether or not the modification has an effect on the latency to pair formation of that male. This experiment has been done by Anders P. Møller (1988), a Danish researcher now at Paris VI University. He trapped males as they arrived in spring, and marked them in order to be able to identify them individually. Each captured individual was subjected to one of four treatments, assigned at random: the first group had their tail feathers shortened by 2 cm; the second group had their tail feathers lengthened by 2 cm (the clippings obtained from the first group were glued to their tails with super glue); the third and fourth groups consisted of controls, one with individuals having their tail feathers clipped and re-glued without changing their length, the other comprising individuals that were simply handled and marked during capture.

Next, Møller observed the dates of each individual's pair-formation. The results are very informative (Figure 3.4): individuals whose tails were elongated paired with females significantly sooner than the controls (the two control types did not differ significantly from one another), who themselves took less time to pair with females than those males whose tails were experimentally shortened. The four groups differed only with respect to the length of their tails. Therefore, the experiment demonstrates clearly that female mate choice is directly influenced by this trait. This experiment has since been repeated in Canada and has yielded the same results (Smith and Montgomerie 1991). Such an experiment allows us to conclude that events occur as predicted by the hypothesis: the manipulation of the trait believed to influence mate choice in females did in fact have the predicted effect.

#### 3.2.2.2 *The virtues and methods of experimentation*

During an experiment, the objective is to test for a cause and effect association between two variables. To do so it is necessary to change a single variable at a time to observe its effect on the other. To be convincing an experiment must control for the effects of all other variables on the variable of interest. These other variables are called **confounding variables** when they themselves change with the factor that is being tested and so it is not always straightforward to design a strong experiment that controls for such confounding variables.

##### a *Accounting for individual heterogeneity*

A common problem in practically all experiments is the heterogeneity that exists among individuals. As seen in Chapter 2 with measures of heritability, it is very difficult to control the effects of this heterogeneity. There are several solutions. First, in the experimental protocol itself, it is fundamental to assign individuals to each experimental treatment randomly. By chance, it is possible that the individuals assigned to a particular treatment will all be of the same type. Randomization allows us to avoid the situation where the results obtained are due to the heterogeneity of individuals. For example, in the barn swallow study, all of the individuals assigned to the tail elongation treatment could have been males that already had long tails. The results would then be impossible to interpret. The opposite may also have occurred (where the individuals assigned to the tail lengthening treatment were all individuals with naturally short tails), which could have completely masked the results.

Second, it is always possible to verify *a posteriori* the quality of our sampling design. In the case of Figure 3.4, it is specified that a test on the mean initial tail length yielded non-significant results: no

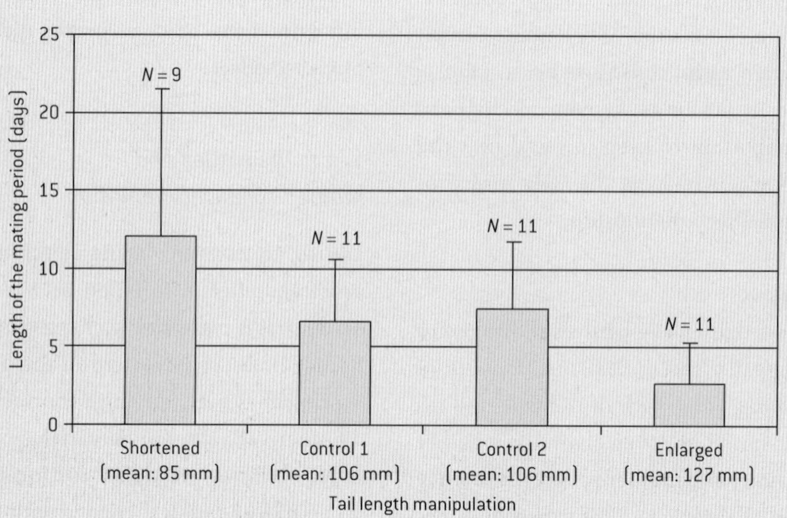


Figure 3.4 **Duration of the pairing period after manipulation of tail length in male barn swallows (*Hirundo rustica*)**

Duration of the paring period (measured in days between the migration arrival date and the date of pairing) for male barn swallows as a function of the treatment concerning tail length. The treatments were: Shortened: males whose tail feathers were shortened by 2 cm; Control 1: males whose tales were cut and re-glued without changing length; Control 2: males whose tail feathers were not manipulated; Elongated: males whose tail feathers were elongated by cutting the streamers and gluing a 2 cm piece in between.

Two males from the Shortened group never paired and were excluded (so sample size is 9 rather than 11).

The values are means (histograms) ± standard error (vertical lines). The sampling design was verified by testing whether significant differences in natural tail length existed between individuals of different groups before manipulation ($P > 0.10$, one factor analysis of variance). The effectiveness of the manipulation of tail length was tested by verifying that the length of tails after manipulation differed significantly between treatments ($P < 0.001$). The duration of the pairing period varied between groups ($P < 0.01$, analysis of covariance with tail length before manipulation as a covariable). All groups differed in paired comparisons ($P < 0.05$ in each case, Mann–Whitney *U*-tests) with the exception of the two controls ($P > 0.10$). Modified from Møller (1988).

significant differences between experimental groups existed for tail length before manipulation. In other words, this confirms that the sampling for each experimental treatment was done correctly, at least with respect to this trait. If this test had been significant, the interpretation of the results would have been more difficult. We must also verify that the manipulation had a detectable effect on the manipulated variable. In Figure 3.4, the author therefore verified that the length of the tails that did not differ among treatments before manipulation did differ significantly after the manipulation.

### b *The importance of confounding variables*

Finally, a multivariate approach can be used to attempt to take into account the effect of various potentially confounding factors that may influence the results of the experiment. This requires doing a statistical control *a posteriori*. In Figure 3.4, it is also specified that the length of the tails before manipulation was used as a covariable in the statistical analysis. This increases the power of the statistical test in the case where adding this covariable amounts to taking into account all the factors that are correlated with the length of the tail – factors which risk masking the real results of the experiment.

Another important methodological conclusion that is implied throughout this book, as illustrated in the above example, concerns the prominence of statistical analyses in any study of behaviour. Whether in correlational or experimental approaches, it is crucial to use the proper statistical model to test the reality of a relationship or of an artificial effect. Statistics is clearly beyond the scope of this book, but this is an important take home message. We thus often specify the statistics associated with a given result in the figure legends and sometimes provide information on the statistical methods used to obtain them.

#### C *The possibility to use natural experiments*

The nature of experiments can vary widely from one study to another. In certain experiments, we can use natural events as a type of manipulation. Natural catastrophes constitute an excellent means of testing hypotheses across large spatial scales, as long as measures on the state of the system exist from before the catastrophe. Also, many human actions have the effect of creating an experimental situation (Sarrazin and Barbault 1996). This is the case with species introductions and with reintroductions. These are situations that occur often, either accidentally or due to deliberate human activity, and are very rarely exploited as real experiments. At the other extreme, we can conduct experiments on organisms that are kept in the laboratory. This allows us to control more effectively the numerous other variables including confounding variables. It also allows for careful experimental manipulations to be performed that will allow alternative hypotheses to be distinguished. However, it will sometimes be necessary to check that laboratory results also apply to natural situations. All kinds of other experiments in nature lie between these two extremes. The barn swallows study described above represents a typical example of an experiment conducted in the field.

Theories, principles, and models allow us to establish predictions, which, depending on the adopted approach, can be directly subjected to experimental tests and whose coherence can be tested with formal models. The rest of this chapter is devoted to the presentation of these different approaches.

## 3.3 The phenotypic approach

We begin by introducing the phenotypic approach because it remains the most commonly used approach in behavioural ecology. It is based on the following assumption (Grafen 1984): '*whatever the nature of the genetic system involved, we assume that studies at the level of the phenotype are sufficient for identifying the selective pressures that exert themselves on the organism being studied*'. We start by presenting in detail the reasoning applied in this approach and discuss its formal implementation. Examples of its application will be introduced throughout most of the following chapters. It is up to the reader, in the following pages, to appreciate the daring nature of this assumption.

### 3.3.1 The optimization concept

Several studies in behavioural ecology are based on the following postulate: organisms, through their behaviour, maximize a certain value (we will return to this in Section 3.3.4) linked more or less directly to fitness. We can reasonably think that the efficiency of predators depends in large part on their capacity to maximize returns from hunting, whereas at the same time, that of prey must depend to some extent on their capacity to detect and prevent predator attacks. It is reasonable to believe that in the past, natural selection acted on the variants of the genes that influence such behaviours (searching for and capturing prey, anti-predator defence, etc.), or their development (learning itself can be considered as a process of optimization), retaining them on the basis of how efficiently they maximized the survival and reproduction of their vehicle organism. This simply involves considering that at some moment in the past there existed sufficient heritable variation for selection to act upon. This variation may often have been reduced through the course of evolution, and for many characters that strongly influence the survival and reproduction of organisms, it is quite possible that there no longer exists very much heritable variation. It follows then that at the moment when we are observing organisms, their phenotypic characteristics have already been optimized through the course of evolution.

The appeal of the optimization concept is widespread in biology (Baldwin and Krebs 1981; Dupré 1987; Weibel *et al.* 1998) and is not an innovation of behavioural ecology. Numerous phenotypic characteristics can be studied in terms of the relationship between form and function. For example, the structure of the arm or leg bones can be studied in animals in direct relation to various physical constraints (pressure, torsion) that exert themselves on those structures (Alexander 1996). Understanding the laws of physics permits us to demonstrate through formal calculations that the 'design' of bones is a perfect solution to the demands to which they are normally subjected to. In a similar manner, the efficiency of the sonar of bats can be appreciated in terms of the laws of physics that apply to echolocation (see Chapter 16). It is, however, more difficult to conceive of the 'forces' that exert themselves on the behaviour of organisms in a manner analogous to how the laws of physics exert themselves on bones or the propagation of sound. The solution to this problem came with the development of an 'economic' concept of behaviour.

#### 3.3.1.1 *Functional aspects of decision-making: an 'economic' approach*

Through the course of their activities, organisms are often faced with several alternatives: for example, different categories of prey are available, different patches can be prospected, different individuals of the opposite sex are all potential reproductive partners. Many options are therefore available, and from the point of view of the observer, a 'choice' must be made (see Chapter 9 for further comments on the concept of choice).

*Any trait involves costs and benefits*

The 'economic' concept of behaviour associates each option with a certain number of costs and benefits. For example, a bird can land on bare ground with no vegetative cover in the immediate proximity, to feed on seeds that are rich in energy and very abundant. But this option exposes it to predation risks because the bird is far from any refuge in the event of a predator attack. Clearly, the option holds a benefit (the food itself), and a cost (the risk of being captured by a predator). To estimate the value of each option, various parameters must be known, including the hunger state of the bird, the density of predators in the environment, and the probability of discovering another food source that exposes the bird to a lower risk of predation. If the animal is not particularly hungry, it may ignore food located in areas that are too dangerous. On the other hand, if its energy reserves are at their lowest and there are few feeding opportunities that carry a lower risk, rejecting the opportunity to feed could have a strong, negative effect on the bird's survival. Now imagine that the bird has entered a patch to feed. Pecking, with the head oriented downwards towards the ground, is incompatible with scanning for danger. Similarly, scanning the environment, with the head raised, is incompatible with feeding. Two options exist: the bird can peck rapidly at many seeds and return to a more sheltered location (which will minimize the time spent searching for food and exposed to predators); or it can alternate between episodes of pecking and scanning (which increases the time spent searching for food, but decreases the risk of a surprise predator attack). Again, the costs and benefits of each tactic must be evaluated in light of the characteristics of the animal (ability to detect an approaching predator, ability to ingest seeds rapidly) and the environment (distance to nearest refuge, visibility). It is easy to see that for all the daily activities that organisms must perform, several of options are available. The question is to know whether it is possible to quantify the costs and benefits of each of those options and whether there exists a general principle allowing us to understand the decision-making process of the animal.

*From the concept of utility in economics . . .*
One of the major advances of behavioural ecology was to view behaviour as choices through which some value (see Section 3.3.4) is maximized (McFarland and Houston 1981). This concept is often adopted by economists and sociologists to analyse consumer behaviour. It involves surmising that consumers conduct themselves as rational agents that exhibit consistent choices when repeatedly faced with the same alternatives. This assumption of rationality implies that different options can be ranked according to some order and that the choice that is made will be based upon some maximization principle. In economics, the value scale on which the options are categorized is called a 'utility scale' and we consider consumers as maximizing a certain value called **utility**. Utility often corresponds to the consumer's level of satisfaction, a subjective measure that can be variable depending on the consumer, the conditions and the questions asked. In practice, it is difficult to define utility in economics *a priori* because it is not immutable and can depend on trends that are unforeseeable and sometimes short-lived. Utility, therefore, can only be inferred based on observations of the consumer's behaviour (Samuelson 1965). In this sense, the concept of the utility in economics is a **descriptive** tool (Stephens and Krebs 1986). It does not allow us to predict differences without observing the choices made and therefore inevitably leads to circular reasoning (Eichner 1985).

*. . . to that of fitness in biology*
In biology, on the other hand, utility is not a totally arbitrary concept and its use is not tautological (Cézilly *et al.* 1991). If we accept that natural selection is optimizing (Oster and Wilson 1978; Maynard Smith 1982; Dupré 1987) we can expect that the choices made by animals tend to maximize their fitness. As such, the utility becomes a normative concept, allowing us to predict the choices that animals **should** make if they were **perfectly** adapted to their environments. Ideally, each option that offers itself to the animal should be able to be evaluated by its consequences in terms of survival and reproduction. The option that confers the greatest fitness is the option that the animal should take.

#### 3.3.1.2 *Optimization and perfection*

Fitness maximization is not free from constraints. For example, it is often difficult simultaneously to maximize the number of seeds eaten over some period (which implies no interruption of feeding) and the probability of detecting an approaching predator (which implies being vigilant at all times). Also, there are various intrinsic and extrinsic constraints operating on organisms (see Chapter 8 of Stephens and Krebs (1986), for a more detailed discussion of the subject). Intrinsic constraints are those linked to the sensory or cognitive capacities of organisms (for example, the human eye is unable to perceive ultraviolet light, and chickens are incapable of making a detour to reach a goal), and others are linked to physiology (for example, certain species of passerines cannot survive more than two hours of food deprivation). We will speak of limitations for the former, and tolerances for the latter (Stephens and Krebs 1986). Extrinsic constraints are those that are imposed by the environment. For example, for a diurnal predator, the time available each day for feeding depends on seasonal variations in day length. Intrinsic and extrinsic constraints are not necessarily mutually exclusive categories. In fact, the capabilities of organisms often interact with the characteristics of the environment. The running speed of a reptile, for example, depends both on its muscle physiology and the ambient temperature. Therefore, fitness maximization operates under numerous constraints. This idea is retained in the term **optimization**, which corresponds to maximization under constraints.

Although natural selection operates as a process of maximization, it would be naive to think that in reality animals are always maximizing their fitness. It helps to distinguish between the **proximate** motivational processes that direct the animal's choices when it encounters stimuli and decides to eat rather than drink, or fight rather than mate, or choose that mate because it is more attractive than the other. At a proximate level an animal is not comparing the fitness of alternative courses of action but choosing on the basis of sensory information that is integrated with its current internal state. The animal's decisions are set by some proximate goal, say to maximize sweetness when choosing among foods. The proximate objective or goal that is pursued by an animal is called the **objective function**. The fitness consequences that follow from such an objective function depend on the environment in which it operates. For example, humans when given a choice will generally prefer the sweeter-tasting food. In ancestral environments this objective function may have imposed low fitness costs (or large fitness benefits); however, today with such high availability of sweet foods in developed countries, the objective function may in fact impose large fitness costs. When the consequence of the objective function is expressed in terms of fitness costs it is called the **cost function** (McFarland and Houston 1981). We should keep in mind that a perfect correspondence between the objective and cost functions is unlikely if only because environments change in time and we expect that the evolution of objective functions will lag behind changes in cost functions.

**Optimization therefore does not imply that individuals are perfectly adapted to their environment**. As has been pointed out previously, the optimization approach to behaviour simply considers that the differences between the two functions should not be too great, and as a consequence, the use of the objective and cost functions is a heuristic starting point because it permits the identification of parameters that have the greatest influence on the choices made by animals. In fact, it is differences between the cost function (evaluated by the researcher) and the objective function (measured through observation of the animal) that allow progress in our understanding of behaviour. To determine the cost function, the researcher is led to evaluate, through mathematical formalizations, the consequences of various options. He or she then predicts the option that an animal should take. When the behaviour of the animal differs markedly from this prediction, the researcher will not be led to abandon the optimization approach, but will instead revise the variables and constraints that were incorporated in the formalization. Through successive adjustments and regular back-and-forth between the model and the empirical results, the researcher will eventually come to identify the relevant variables and constraints.

### 3.3.2 Static versus dynamic optimization

#### a *Time and energy, two limited resources . . .*

Economic constraints that exert themselves on behaviour are mainly related to two essential parameters: time and energy (Cuthill and Houston 1997). Each organism faces a certain number of needs that it must satisfy on a daily basis to survive, and consequently, to ensure its reproduction. However, the time available to organisms is not infinite. For some, life expectancy may be very short. For other, more long-lived species, the rhythm of night and day, or of the seasons, determines the time available to accomplish various activities. The time invested in any given activity limits the time available for other activities. Along the same lines, all activities require a certain amount of energy. Energy must be acquired by the organism, which often has a limited ability to store it. Energy is therefore available in limited quantities. As such, the allocation of time and energy to different, often conflicting activities, is important in terms of consequences for the survival and reproduction of the organism.

#### b *. . . that are the basis of trade-offs . . .*

In this context, the notion of a **trade-off**, as in all evolutionary approaches, is a key concept in behavioural ecology. (See Chapter 5 for a general description of trade-offs and their central roles in the evolution of behaviour.) Trade-offs arise due to the existence of conflicting needs (to feed/to avoid predators, to find a mate(s)/to defend a territory, etc.) that animals regularly face in natural situations. Even when two behaviours can be performed simultaneously, this generally results in a reduced efficiency compared with the situation where each behaviour is performed singly (Futuyma and Moreno 1988). The use of optimization models allows us to examine more closely the nature of the trade-offs to be made, and to explore their expected consequences for the behaviour of organisms.

### C ... *and can be studied using models*

One family of optimization models considers situations where the consequences of the behaviour of an organism are independent of the behaviour of its congeners. These include two categories of model: static optimization and dynamic optimization.

*Static optimization*

The first category is particularly appropriate for informing us on an action that is limited in time such that the internal state of the organism will not vary in a significant manner. The solution to the problem that the animal is faced with is therefore a single and unique decision. In this situation, we assume that the animal maintains its tactic choice for the entire period considered, this is known as a static optimization. Many examples of this approach are presented in Chapter 7.

*Dynamic optimization*

However, static optimization cannot inform us of all the situations which organisms face. For instance, there are cases where the consequences of an action taken by an individual alter the state of the organism and therefore modify the exact conditions of the problem. This complicates matters and it is no longer possible to summarize it with a simple decision. Imagine the example of a species with separate sexes and continuous growth, where reproductive success increases with body size. Each day, the organism has the choice to feed or to find a mate. Given that a relationship exists between the amount of food consumed and growth, and body size and reproduction, if the organism conducts itself in an optimal manner, it must use one or the other option over the course of successive days. This is a dynamic problem, because the decision (to feed or to find a mate) taken at a given time (on a given day in this example) affects the state of the individual (its size) in the future, and therefore has consequences on the optimal decision for the next time period. Dynamic optimization models allow us to determine the optimal sequence of decisions. In practice, dynamic optimization relies on a particular form of programming called stochastic dynamic programming (McFarland and Houston 1981; Mangel and Clark 1988). It is a numerical technique that allows us to determine the optimal decision of an animal at a given time and the state of the animal. The state may be characterized by an ensemble of variables (size, energy reserves, knowledge of the environment) and the consequences of an action are generally considered as stochastic phenomena (for example, an animal that chooses to feed will obtain a certain quantity of energy according to a probability distribution). To determine the optimal sequence of decisions that maximize fitness, the program works backwards from a terminal state at time $T$ for which the relationship between state and fitness is known. Then for each possible state at time $T - 1$, the optimal choice for that time period can be determined. We then obtain the expected fitness associated with each state at time $T - 1$. We can then proceed in an analogous manner to time $T - 2$. By repeating this procedure, we come to establish a matrix of decisions that constitute the optimal strategy enabling us to specify the best option for each state and at each time step. A simple example of this application is detailed in Box 3.2. A more detailed presentation with more complex examples is available in works by Mangel and Clark (1988) and Clark and Mangel (2000).

## 3.3.3 Optimization in situations with frequency dependence: game theory

In many situations, the consequences of an individual's choice are not determined solely by the interaction between its internal state and environmental factors, but depend largely on the choices made by other individuals in the same population. It is then impossible to calculate the return from a given

## Box 3.2 **A simple example of dynamic programming**

In continental zones where winter is harsh, night-time temperatures can be extremely low. Therefore, the survival of certain species of small passerine birds, unable to feed at night, depends on the energy reserves they hold at the end of the day. Alexander (1996) considers this situation to illustrate an example of the application of dynamic programming. Imagine a passerine that requires at least 10 units of energy to survive the night beginning at 18:00, the time when it begins to get too dark to be able to feed. The bird has the choice between two zones to feed in. In zone S, the availability of food **resources** is stable, and it is completely possible to predict the energy gain of the bird: it will gain reserves at the rate of one unit for each ▶

| Level of energy reserves | Patch type | 14 hours | 15 hours | 16 hours | 17 hours |
|---|---|---|---|---|---|
| 10 | S | 1 | 1 | 1 | 1 |
| | V | 1 | 1 | 1 | 1 |
| 9 | S | 1 | 1 | 1 | **1** |
| | V | 1 | 1 | 1 | 0.5 |
| 8 | S | 1 | 1 | **1** | 0 |
| | V | 1 | 1 | 0.75 | **0.5** |
| 7 | S | 1 | **1** | 0.5 | 0 |
| | V | 1 | 0.75 | 0.5 | 0 |
| 6 | **S** | **1** | 0,5 | 0 | 0 |
| | V | 0.81 | **0.63** | **0.25** | 0 |
| 5 | S | 0.63 | 0.25 | 0 | 0 |
| | V | 0.63 | 0.25 | 0 | 0 |
| 4 | S | 0.25 | 0 | 0 | 0 |
| | **V** | **0.38** | **0.13** | 0 | 0 |

Table 3.1 **Probability of survival as a function of strategy (S or V), time of day, and energy reserves already acquired**

This table presents, for each level of energy reserves and each time step, the probability of survival of a bird adopting strategy S or V for the hour that follows and subsequently adopts the best strategy. For each combination of energy reserve level and time step, the probability attached to the optimal strategy is noted in boldface. In the light zone the best strategy is S, in the grey zone the best strategy is V. To obtain this table, one must work backwards in time starting at the last hour, then the second-to-last hour, and so on. This explains the fact that for the energy reserve level at a given time step, the best strategy is independent of the state at the earlier time step and previous options taken.

hour spent in the zone. Zone V, however, is particularly variable and we cannot predict with certainty the energy gain of the bird: at any hour of the day, there is a 50% chance of gaining two units of energy and a 50% chance of gaining nothing. The mean rate of energy gain in the two zones is therefore equivalent. (A more detailed discussion of the problems of variance in expected gains is presented in Chapter 7.)

A bird that has nine units of energy stored at 17:00 (when there is only one hour left to feed), will attain the required energy stores of 10 units by 18:00 if it chooses to feed in patch S. On the other hand, a bird that held only eight units of energy at the same time in the day should prefer option V which gives him a 50% chance of survival, while option S would not allow him to reach the energy reserve level required to survive the night. A bird that has only seven units of energy at this time is condemned to die.

The column '17 hours' in Table 3.1 gives the probabilities of nocturnal survival as a function of the level of energy reserves given that the individual feeds in patch S or V. The options that allow it to maximize survival are noted in boldface.

Let's move backwards now and consider what happens at 16:00. A bird that has nine units of energy that feeds in S will already have attained 10 units of energy by 17:00 and assured its survival. If it feeds in V, it may gain two supplementary units of energy and survive, or it may gain nothing. It would then start the last hour with nine units of energy in reserves and could still assure its survival by feeding in zone S. A bird that had only eight units of energy at 16:00 could feed in S for the next two hours, in which case it would obtain two units of energy in the next two hours and would survive. It could also feed in V until 17:00 and would either have 10 units of energy and assure its survival, or eight units of energy. In the second case, the best option is noted in column 17:00: stay in V with a 50% chance of survival. The mean probability of survival is then 0.75 if the bird chooses option V at 16:00 and then makes the best choice at 17:00.

The example presented here is very simple, but illustrates nicely the two aspects of a situation when the use of dynamic programming is justified: the optimal strategy is conditional on the state of the animal which changes over the course of time as a direct consequence of the option taken at each time step.

behavioural strategy without taking into account the frequency of all strategies in the population. Take for example the decision to drive a car on the right or left side of the road. We can ask which of the sides is better; chauvinist types will tend to think that it is the side adopted by their country that is better than the other. However, the answer is not so simple. If you really had a free choice regarding the side of the road to drive on, the best strategy would be to choose the side chosen by the majority of other drivers. The best strategy in this case depends on the frequency of strategies in the population. The majority side is better because it minimizes the probability that you will be in a head-on collision. Therefore, two possible solutions to this game exist, either everyone drives on the left or everyone drives on the right. No other combination of strategies is profitable, only pure strategies (i.e. always driving on the right or never driving on the right) are possible and the two are equally valid. Which of the solutions is retained is due purely to historical chance.

The problem can be summarized as a game where two strategies are pitted against each other, right side versus left side. The analysis of the problem consists of finding the winning strategy that, in this case, consists of playing the same strategy as the opponent because the alternative strategy would be detrimental. Technically, we have just entered the world of **game theory**, a field stemming from mathematics and economics, originally developed for cooperation games by Oskar Morgenstern and John von Neumann, and later for selfish games by John Nash, in which the objective is to find the winning strategies for defined games that are often military or economic. It is thanks to John Maynard Smith (see Chapter 1) that behavioural ecology was able to take advantage of this mathematical technique for the analysis of various decisions, notably in the field of social behaviour. Maynard Smith adapted the approach to animal behaviour and demonstrated that searching for the wining strategies in an evolutionary context has a peculiar solution. He developed the basis for what is now known as evolutionary game theory.

#### 3.3.3.1 *The solution to evolutionary games: the evolutionarily stable strategy*

There are many criteria for selecting the best strategy in a purely economic or military game. For example, we can determine *a priori* that the winning strategy will, once adopted by all players, maximize the benefit to the group of players. However, in the world of biology, it is natural selection that decides between different strategies. In this section, we will see numerous examples where the best strategy is not necessarily the strategy that provides the greatest benefit to the population, but the strategy that, once adopted, cannot be invaded in an evolutionary sense by an alternative strategy (Maynard Smith 1984). The strategy will therefore be evolutionarily stable (ESS) because once established in a population all evolution ceases, no modification of the strategy can be favoured by natural selection. However, we will see in Chapter 15 that the situation can be much more complicated than this.

#### 3.3.3.2 *An example of an evolutionarily stable strategy: sex ratio*

We can illustrate the application of an evolutionarily stable strategy (ESS) with the problem of **sex ratio**. As we will see in detail in Chapters 11 and 13, in sexual animals, anisogamy (dimorphism of gametes) makes it such that males can produce more gametes than females. Because of this, the gametes produced by one male are sufficient to fertilize the gametes of more than one female. It follows then, from the point of view of the population, that population growth rate would be maximized by the production of many more females than males as long as there were enough males to fertilize all the females. It would therefore be more efficient for a species to have a female-biased sex ratio; all excess males represent a waste of resources into non-reproductive individuals.

However, natural selection does not operate for the good of the species or for the good of the population (see Chapter 2), it acts at the level of the individual. Imagine a population that produces the minimum number of males necessary for the fertilization of all the females. In this population, there would be no waste of males and all individuals would reproduce at their maximum capacity. However, a male would have a greater average number of offspring than a female by virtue of the fact that he fertilizes more than one female. Under these circumstances, a parent that possessed a mutation that allowed it to produce more sons than the other individuals in the population would have a higher fitness than the other parents (which produce more daughters). In practice, parents producing male-biased offspring would have, on average, more grandchildren, and would transmit more copies of their genes into the next generation. Natural selection would again favour a higher production of males as long as males held a higher share of the reproduction in the population compared with females. However, the success of males depends on the availability of females. When females are abundant, each male can fertilize several. When males are more abundant, they have access to relatively fewer females. Males are then subject to higher inter-sexual competition and their reproductive success decreases. Because of this frequency dependence that characterizes all such games, there comes a moment where the frequency of males in the population is such that they will have the same reproductive potential as females. Once this point is reached, selection is jammed, it no longer favours one sex over the other because both now have the same reproductive success. This constitutes an ESS, and it corresponds to the production of an equal number of sons and daughters: a 50% sex ratio. We will return to this problem in greater detail in Chapter 13.

This example illustrates a rather important point that can be drawn from ESSs: their solutions are often disadvantageous at the level of the population, and even sometimes at the level of the individual. The 50% sex ratio does not exist because it allows individuals to maximize productivity, on the contrary, it results in the waste of surplus males, but it exists because it is the only evolutionarily stable solution to the sex ratio game.

### 3.3.4 A recurrent problem: estimating fitness

Throughout all the previous examples, we have made reference to fitness as a measure of efficiency in terms of the evolution of various strategies. The concept of fitness, however, has no scientific value unless we are able to estimate its value under various circumstances (or at least to compare the values associated with different strategies). The question of how we can measure the effect of behaviour on fitness is therefore central to everything in evolutionary ecology and in particular, to behavioural ecology.

#### 3.3.4.1 *Behaviour, fitness, and demography*

As stated earlier, in behavioural ecology, we assume that a certain value is optimized through the course of evolution. This factor might be **genotypic fitness** or phenotypic (individual) fitness, depending on the type of question being asked (see Chapter 2). All studies in behavioural ecology involve measuring the impact of the studied behaviour on the fitness of individuals, that is to say, on the **capacity of the phenotype that expresses the behaviour in question to produce mature descendants relative to other phenotypes in the same population at the same time**.

To estimate this value, it is necessary to measure the consequences of each strategy, or more generally, of the variation of the behaviour (natural or experimental) on life-history traits (survival and/or reproductive potential). The idea is that the genes that underlie the behaviour of animals have been retained by natural selection by virtue of their ability to maximize the survival and reproduction of their vehicle. We therefore use a demographic measure of fitness. One way or another, there exists a link between demographic processes and evolution: differential demographic processes between different categories of individuals contribute to the variation in the frequency of genes, and therefore, eventually to evolution. As such, in most studies we can consider that this demographic measure of fitness can be substituted for genotypic fitness. This is the basis of the phenotypic approach of behaviour. This demographic measure of fitness corresponds to the mean demographic success of the phenotype being considered relative to the success of other phenotypes present in the population. It defines the success of a trait within a generation.

#### 3.3.4.2 *Common currency and fitness*

For self-evident practical reasons we do not always measure the consequences of a given strategy on every life history trait. Depending on the question being studied, we can sometimes indirectly quantify fitness by limiting ourselves to a short time period in the life of an individual (its survival over winter, the number of young produced in a single reproductive season), that is to say that we measure a component of fitness. In other cases, we can use even less direct measures by assuming that fitness is directly correlated with something that can be measured easily (which we call a **currency of fitness**, because it allows us to convert the effect of the strategy in terms of fitness). This common currency is assumed to be correlated with fitness such that the common currency associated with various strategies allows us qualitatively and quantitatively to rank the strategies according to their effect on fitness.

In a simple optimality approach, we consider the impact of one strategy in terms of fitness for a focal individual, independent of the strategy adopted by other members of the population. We will see many examples of this type of game, especially in Chapter 7. In a game theory approach, on the other hand, the reasoning lies on the principle that the evolutionary efficiency of a strategy depends on the strategies adopted by other individuals in the population. In this case, the selected strategy, usually called the ESS, is the one that will have the **greatest capacity to invade the other strategies and resist invasion from others once established**. We saw an example earlier with the sex ratio of a population.

#### 3.3.4.3 *Which currency of fitness to use?*

The nature of the currency varies greatly depending on the question being asked. For example, in a study on foraging strategies, we can assume that the relative success of one strategy can be measured in terms of the quantity of energy gained per unit time. For this, we implicitly assume that the amount of energy consumed per unit time is directly related to fitness (see example in Figure 7.1). In a mate choice experiment, the common currency may be, for example, the inverse of the time that it takes an individual to find a partner (as in Figure 3.4). In a sexual selection study, depending on the type of question asked, the common currency may be resistance to parasites, intensity of the immune response, intensity of body coloration, number of mates, intensity of sexual displays, etc. If we are studying the efficiency of predators, this depends on their ability to maximize the yield from hunting prey; that of prey depends on their capacity to detect and prevent attacks.

From these examples, it clearly follows that the nature of the common currency that we adopt depends principally on the question being studied. However, at some point, **it is necessary to study the exact nature of the relationship between the common currency used and fitness**.

#### 3.3.4.4 *Tools and methods for estimating fitness*

The methods used to estimate fitness depend strongly on the nature of the question being studied. These methods are numerous and would justify an entire book unto themselves. They are often borrowed from other domains in evolutionary ecology, principally demography and molecular biology.

##### a *Demographic tools*

The necessity of adopting a demographic approach to estimate fitness properly will be made more explicitly in Chapter 5. Here we just present the basic methods used in demographic studies that are also particularly useful to the study of behaviour. The first step in demographic and in behavioural ecological study requires finding a way to recognize individuals within a population, whether in a natural environment or the laboratory. For this, it is often necessary to place marks on the individuals and hence often to capture them. Marking individuals them allows us to identify behavioural polymorphisms. How can we know whether territorial and non-territorial (satellite) frogs exist if we cannot distinguish individuals? Moreover, if the individuals are not marked, we will have difficulties using statistical tests to verify our hypotheses, because we will never know if our data set involves multiple repeated measures of the same individual, which creates problems with lack of independence of the data. In practical terms, marking techniques vary from simply placing a metal or plastic ring on an animal's leg, neck, ear, or arm all the way to attaching or even implanting radio-transmitters. These latter devices have made it possible to study the spatial exploitation of marine environments by albatrosses, and the diving behaviour of penguins. More generally, radio transmitters allow animals to be located within their natural environment at any time.

The data obtained by following individuals allow us to extract **information** that can allow the estimation of demographic parameters. As such, following individuals allows the estimation of all the parameters that describe the **life history strategy** of various phenotypes such as age at maturity, fecundity, and survival as a function of age. Generally, capture–mark–recapture methods are valuable in terms of estimating demographic parameters. Once these parameters have been estimated, graphs of the life cycle corresponding to the study strategy can be used to build specific models, called Leslie matrix models, which allow estimation of the intrinsic growth rate of each phenotype, which provides a useful measure of fitness. This shows to what point demographic methods are fundamental if we plan to make precise estimates of fitness. We will discuss this important issue in more detail in Chapter 5.

### b *Molecular tools*

We will see in Chapter 12 that if we want to measure fitness, it is often necessary to verify that the young produced are in fact the genetic offspring of their putative parents. In socially monogamous birds, it turns out that in many species, including the tree swallow (*Tachycineta bicolor*), up to 80% of the chicks in a nest are not fathered by the male that raises them. In some species, there also exists intra-specific brood parasitism (cf. Chapter 17), such that the young in the nest may be the genetic offspring of neither putative parent. In these circumstances, it is difficult to imagine estimating fitness without taking into account the genetic paternity or maternity. Fitness measures must only take into account genetic descendents, that is, those among the 'legitimate' descendants that are genetic descendents, as well as those genetic 'illegitimate' descendents in neighbouring families.

To do this, beginning in the early 1990s, behavioural ecologists began assigning maternity and paternity using genetic fingerprints, based on comparisons of the DNA of parents with their offspring. DNA fingerprinting uses highly variable zones of DNA that are compared among the putative parents and their progeny to identify who the genetic parents of the individuals are. These methods play a critical role for testing certain hypotheses because they allow a more precise knowledge of the genetic reproductive system, which can be strongly disassociated from the social reproductive system. We will return to this in detail in Chapters 12 and 14.

## 3.3.5 Phenotypic engineering: a tool of the future?

The major goal of behavioural ecology is to understand the consequences of behaviour on the survival and reproductive success of organisms. However, not all traits that we consider to be 'behaviours' lend themselves well to being studied in terms of decisions. Let us take the example of grooming behaviour. Many species regularly groom or preen to free themselves of ectoparasites, to maintain the quality of their fur or feathers. The optimization approach can be useful for understanding how an animal allocates time to grooming/preening behaviour amidst other activities (feeding, scanning for predators, etc.). The ESS approach allows us to determine how reciprocal grooming/preening can arise between members of the same social group. However, the precise sequence of the acts during the course of inspection of the body escapes this analysis. Behaviours can often depend upon physiological mechanisms and cellular processes whose understanding requires an analysis at the molecular level, a level where it can be difficult to relate the consequences to the fitness of organisms. One possible solution is therefore to generate new variation in an experimental manner through manipulations of the phenotype.

Phenotypic manipulations allow us to evaluate the utility of traits by modifying them and comparing the performance of modified individuals with those individuals that were not manipulated (Sinervo and Basolo 1996). This approach, which consists of demonstrating the current utility of a trait, has been named **phenotypic engineering** (Ketterson *et al.* 1996; Ketterson and Nolan 1999).

Consider a phenotypic manipulation on a trait with a normal distribution. It will require producing phenotypes that possess a trait value that deviates significantly from the norm in each direction (greater and smaller). Three types of result may be obtained.

### a *Stabilizing selection*

In the first case, manipulated individuals may have a reduced fitness compared with non-manipulated individuals and this result would argue that the trait, under normal expression, is maintained by stabilizing selection. A classic example involves the bright red shoulder patches, underlined with yellow, that decorate the wings of male red-winged blackbirds, *Agelaius pheniceus*. This predominantly black-coloured bird lives in North America and normally establishes its territory in reed beds. Only the males have these shoulder patches, which can either be exposed or hidden under black scapular feathers. The shoulder patches function as a signal in the regulation of territorial behaviour. Individuals possessing a territory signal their intent to defend the territory by exposing their shoulder patches. Intruding individuals signal their desire to usurp the territory in the same manner. Smith (1972) painted the shoulder patches of certain territorial males in black, and covered the shoulder patches of a second group of territorial males using a clear solvent as a manipulated control group. The individuals with the reduced shoulder patches had a reduced fitness compared with non-manipulated birds: only a third of them were able to retain their territory after manipulation compared with 90% of males in the control group. These results were confirmed by other experiments manipulating the size of the shoulder patches (Peek 1972; Røskaft and Rohwer 1987). A decrease in the size of the shoulder patch seems to be selected against. On the other hand, enlarging the shoulder patch seems to be limited by the need to be able to cover them with the scapular feathers. Metz and Weatherhead (1992) showed that territorial individuals whose ability to cover their shoulder patches with their scapular feathers was compromised, were penalized because they faced stronger aggression from their neighbours.

### b *Neutral characters*

In the second case, the fitness of the manipulated individuals may not differ from the individuals whose trait was not modified. We can then conclude that the level of expression of the trait is actually a neutral character and does not have any current usefulness.

### c *Phenotypic manipulations that increase fitness*

The most intriguing results correspond to the third case, when the fitness of manipulated individuals is greater than that of control individuals. Such a result has been discussed earlier in this chapter with the example of the barn swallows (Figure 3.4). Such apparently paradoxical results lead us to ask why the level of expression of the trait that is most advantageous in terms of fitness is not achieved in nature. Two types of answer can be put forth. The first invokes selection pressures that were not directly measured through the course of the manipulation experiment. Such as, if the elongated tails of the barn swallows increases their sexual attractiveness, it may also come at the cost of attracting more attention from predators or diminishing manoeuvrability during flight (Møller 1994). The second type of answer draws on the notion of constraint. The expression of the studied trait may be correlated with the expression of other potentially disadvantageous traits, and it is selection against these other traits that limits the evolution of the trait being studied. In the case of the barn swallow reported in Figure 3.4, the manipulation affected the trait in isolation. To understand why male tail length does not increase more it would be necessary to increase the length of the tail feathers experimentally and to identify the suite of phenotypic traits that accompany their elongation. If, for example, the lengthening of the tail feathers resulted in a concomitant decrease in flying or foraging capacities such that the overall effect on fitness was negative, we could then conclude that the evolution of longer tail feathers is constrained.

Over the course of the past 15 years, numerous studies have experimentally modified the expression of various traits through manipulation of endocrine regulation in certain species and analysed the consequences of these manipulations on the expression of phenotypic traits and the fitness of organisms. For example, in the trout *Onchorhynchus mykiss*, manipulating the level of growth hormones resulted in an increase in the basal metabolic rate, food intake, and aggression (Johnsson *et al.* 1996; Jönssen *et al.* 1996).

Raising the hormone level also had the effect of increasing behaviours in young trout that increased their risk of being captured by a predator (Jönssen *et al.* 1996). We will see many examples of phenotypic manipulations throughout this book, and numerous examples of phenotypic engineering through hormonal manipulations in Chapter 6. These allow us to better understand the physiological mechanisms that underlie the development of behavioural phenotypes in general.

## 3.4 The genetic approach

The phenotypic approach tends to consider the genetic determinism of behaviours as a sort of black box (Grafen 1984). In fact, its principal goal is to provide evidence of the selective pressures that exert themselves on the characters being studied, and not to study the response to selection, i.e. changes in allele frequency over generations. The phenotypic approach often assumes that genes that cause individuals to survive and reproduce better in their environment have been selected over evolutionary time, and therefore assumes that the population being studied is at **equilibrium** and that allele frequencies are not changing at the loci influencing the trait being studied. In practice, the realism of this assumption is rarely tested. We can, however, resort to genetic models to describe the evolutionary trajectory of a population and evaluate its capacity to reach an equilibrium between selection and genetic variation. These indicate that the time required to converge on an optimum may correspond to several generations. During this time, the environment is susceptible to undergoing various perturbations, thereby modifying the selective pressures. It is therefore not certain that for all traits of interest, the optimum will always be stable. This suggests that natural populations are less often at equilibrium than we have generally thought in the past.

Furthermore, traits that greatly influence fitness are under strong selective pressures, which are in turn likely to diminish the variability of these traits. A consequence is that usually such traits are expected to have relatively low, or nearly zero, heritabilities. In fact, behavioural traits, even those that are significantly correlated with fitness, can have a non-negligible, even a strong, heritability (Mousseaux and Roff 1987). Such findings argue for a better understanding of the genetic mechanisms in behavioural ecology, at least to address some of the limitations of the phenotypic approach. Integrating genetic aspects into the set of methods available is one of the challenges of behavioural ecology (see Chapter 1). This has already begun to be taken into account over the past 10 years (Moore and Boake 1994) and may rapidly accentuate the preference for developing genetic techniques that open new perspectives for research on the evolution of behaviour (Tatar 2000). In the following sections, we will highlight some of the limitations of the phenotypic approach and describe how studies incorporating the relationship between genes and behaviour offer hope for better integration of the genetic aspects of behavioural ecology.

### 3.4.1 Some limitations of the phenotypic approach

The phenotypic approach often treats the genetic structure underlying the trait under investigation as being determined at a single locus within a haploid system. In the real world, few of the organisms studied are haploid, and few traits are dependent on a single locus. It is also possible that the phenotype that does best in a population is heterozygous. The most renowned example is that of sickle cell anaemia. In human populations that are affected by malaria, we find a polymorphism with respect to two alleles of a gene influencing, among other things, the shape of red blood cells: the N allele produces normal disc-shaped cells, the S allele produces sickle-shaped cells. These populations comprise individuals with each of the following genotypes: NN, NS, and SS. The red blood cells of SS homozygous individuals have the shape of a sickle. These individuals suffer from anaemia, which is generally fatal before reaching adulthood. The NN homozygote does not have sickle-shaped blood cells and therefore does not suffer from anaemia. The NS heterozygous individuals generally have a very low frequency of sickle shaped blood cells and are not anaemic. Despite the disadvantage that it presents, the S allele is not eliminated from the population. In fact, we observe that the NS heterozygous individuals have a greater resistance to malaria compared with NN homozygous individuals. This advantage is sufficient to maintain the S allele in populations exposed to malaria. Mendelian segregation mechanisms prevent the fittest phenotype from taking over the population because it is produced by a heterozygous genotype. It is clear that the coexistence of these three phenotypes cannot be understood by a strict phenotypic approach (Grafen 1984). In this case, understanding the distribution of phenotypes requires an understanding of their genetic determination.

On the other hand, in nature, selection often acts on a suite of traits that all contribute towards determining the fitness of individuals. Even when selection acts on a single trait, it may still have consequences on other traits through genetic or phenotypic correlation among the traits that are directly under selection and those that are not (Roff 1997). Some of these traits may be negatively correlated, in which case it becomes impossible to have positive selection on one trait without having a negative effect on the other. For example, insecticide resistance in the moth *Choristoneura rosaceana* was accompanied by concomitant modifications of certain life-history traits: increased incidence of diapauses and decreased larval mass (Carrière *et al.* 1994). We currently know very little about selection that, for example, favours an increase in each of two traits that are themselves negatively correlated. Rare examples of incompatible antagonistic artificial selection have provided results that are difficult to interpret (Roff 1997). We also do not yet understand the real importance of negative genetic correlations in nature or the regularity with which antagonistic selective forces turn out to be incompatible.

In face of the complexity of heredity mechanisms, should we doubt the validity of the phenotypic approach? Not necessarily. On the one hand, genetic systems like the one described for sickle cell anaemia appear not to be extremely common. Maynard Smith (1982) suggests that most genetic systems can be simplified into a haploid form without major consequences for the validity of the predictions formulated. As far as incompatible antagonistic selection is concerned, we can think of strong selective pressure as breaking a negative genetic correlation between the two traits whose expression both positively influence fitness. Without minimizing the value of the phenotypic approach (whose merits will be aptly demonstrated throughout the rest of this book), we emphasize that it is necessary to devote more attention to the genetic mechanisms assumed to support the behavioural adaptations that we study. The study of genetic factors that influence behaviour will certainly develop in the future, if only because of the increasing availability of the biotechnology required to study such genetic problems. Several chapters in this book will tackle these questions (Chapter 5 and 6 for instance, and Chapter 20 at the end of this book will come back to related questions). But to engage further in this field requires that we question the very nature of the relationship between genes and behaviour.

### 3.4.2 The relationship between genes and behaviour

The study of the relationship between genes and behaviour can take different forms. Let us begin by considering the status of the relationship between genes, the nervous system, and behaviour. There is some consensus that a relationship between genes and the nervous system exists. Genes specify proteins and other molecules. These determine the properties of cells, which in turn interact to promote development. The process of development, in interaction with environmental constraints, adjusts the phenotypic characteristics of individuals within whom the nervous system, neuro-endocrine regulation, and the ability to use past experience, operate.

#### 3.4.2.1 *What genetic determinism for behaviour?*

The relationship between genes and behaviour is less straightforward. Although our knowledge regarding the heredity and expression of genes is considerable, we are still far from clearly understanding how genes influence behaviour. Behaviour may be either controlled by a single gene or by a suit of genes.

*The existence of monogenic determinism . . .*

A simple direct correspondence between one gene and one behaviour is often difficult to establish. Monogenic determinism has, however, been accepted in the case of a limited number of behaviours. Rothenbuhler (1964) conducted pioneering work in this field. Certain lines of bees are called 'hygienic', because when a larva dies within the hive, the worker bees systematically remove the operculum that seals the cell that the larva is in and remove the dead body. The 'non-hygienic' lines do not display this behaviour and generally leave the dead larva in the cell. Crossing experiments between the two lines allowed us to determine that the 'non-hygienic' character is dominant (hybrids do not remove dead bodies). In fact, the results of further crossing experiments between the hybrids and hygienic lines were consistent with the existence of genetic determinism based on two pairs of alleles, the first controlling the tendency to uncap the cells containing dead larva, and the second controlling the tendency to remove the dead body (Rothenbuhler 1964). That behaviour would thus be under the control of two genes.

The fruit fly, *Drosophila melanogaster*, provides a rare example of a behavioural polymorphism expressed in nature that depends on a single major gene corresponding to the locus *for* (for 'foraging'; Sokolowski 1980; Sokolowski *et al.* 1984; De Belle and Sokolowski 1987). In the laboratory, the foraging behaviour of larvae is measured based on the length of total movement over a fixed time interval in a Petri dish with a yeast culture (the food for larvae). Larvae possessing the 'rover' allele at the *for* locus, have longer foraging trajectories than the individuals that are homozygous for the 'sitter' allele. The two types do not differ in their level of activity in the absence of food. Crossing experiments demonstrate that differences between the 'rover' and 'sitter' phenotypes have an autosomal basis (they depend on genes from non sex-linked chromosomes), with complete dominance of the 'rover' allele over the 'sitter' allele. The polymorphism conforms to a model of Mendelian heredity both in the laboratory and in nature.

*. . . Though polygenic determinism is more likely*

The existence of very few examples of a monogenetic determinism of behaviour suggest that most units of behaviour that we recognize are controlled by not one, but rather by a suite of genes. We therefore call their determination **polygenic**. As we will see in Chapter 4, behaviour can be seen as a decision-making process that encompasses a series of capacities to gather information about current environmental conditions and alternatives, memorize and process it and finally make a decision according to that information. Such a suite of capacities likely involve different neuronal processes that are unlikely to be controlled by a single gene. For instance, it is unlikely that learning capacities, as well as movement and orientation capacities involved in habitat selection and **dispersal** (see Chapters 9 and 10) are all under the control of a single gene. It is thus safe to imagine that most behaviours are under a polygenic control, with the various genes involved interacting to determine the behaviour that is performed in interaction with current environmental conditions. It is thus not so surprising that examples of a monogenetic determinism of behaviour are few.

Furthermore, genes can influence behaviour at various levels, but it would be wrong to think that they always firmly determine where, when, or why a given behaviour should be produced. In studies by Rothenbuhler (1964), the hygienic behaviour was not completely absent in the non-hygienic lines of bees. It appears that the workers from the latter lines require a particularly high stimulus threshold to show hygienic behaviour. From this point of view, it is useful to distinguish between performance and predisposition (Heisenberg 1997). In the case of hygienic behaviour, genes seem to regulate the predisposition of the workers to react to a certain level of a stimulus corresponding to the chemical signals that are present when there are dead larva (driving them to uncap the cell), and also their reaction to the dead larva body (driving them to remove the body). It does not necessarily specify the order of the acts that lead to the removal of the larva. In another example, performance can correspond to the flight taken by a migratory bird while predisposition would involve its increased level of agitation and preference for a southerly orientation as day length begins to shorten (Heisenberg 1997).

Different methods can be used to study how certain genes predispose individuals to manifest one behaviour or another. We have categorized them here into three main groups.

#### 3.4.2.2 *Methods to study the relationship between genes and behaviour*

##### a *The study of differences between populations*

Comparisons between populations of the same species that are geographically isolated from one another allow us to determine the genetic component of the adaptation of organisms to their environments. Studies published so far have shown that differences between populations can equally well exist for simple and complex behaviours, from the level of locomotion to choices in prey types and reactions to predators. Fleury *et al.* (1995) have shown that in the parasitoid insect *Leptopilina heterotoma*, the circadian rhythm of motor activity in females varies depending on their geographic origin. The results of crossing experiments between populations allowed us to determine the genetic basis for this variation and suggest an adaptation of the parasitoids to local prey populations.

*Garter snakes . . .*

Arnold (1981) studied two populations of garter snakes, *Thamnophis elegans*, in the southwestern United States. The coastal populations essentially feed on slugs, whereas the interior populations are aquatic and mainly capture frogs, fish, and leeches. Experiments conducted in the laboratory demonstrate that individuals from interior populations will not eat slugs that are readily eaten by individuals from the coastal populations. In tests with naive-newborn individuals, 73% of individuals from the coastal population captured and ate an offered slug, compared with 35% of individuals from interior populations. This difference, however, does not demonstrate a genetic component to the behaviour of the two populations. In this species, incubation takes place within the female and the preferences of young may therefore be influenced by the feeding regime of their mother. This potential **maternal effect** could be responsible for the differences between individuals from the two populations. Arnold (1981) noted that descendents from crossing experiments had intermediate preferences for slugs from either the coastal or interior populations, but which resembled their mothers and fathers equally. Arnold (1981) was therefore able to conclude that the differences observed between populations did in fact have a genetic basis.

*. . . and spiders*

Another example of a genetically determined trait uncovered by comparisons between populations concerns the anti-predator behaviour of the spider *Agelenopsis aperta* (Riechert and Hedrick 1990). This species is common in arid habitats in the south-western United States and Mexico. They weave webs that are horizontal, a few centimetres above the ground, which have a hole in the centre where a tunnel extends towards a small cavity or lump of grass. During periods of activity, the spider normally sits in the protected zone at the entrance of the tunnel. When a danger is present, the spider retreats into the interior of the tunnel. This behaviour protects it from predators because the tunnel is linked to a network of cracks in the ground. However, the spider remains exposed to predators, primarily birds, while it handles a prey, weaves its web, or defends its web against conspecifics.

Riechert and Hedrick (1990) studied the behaviour of spiders in two populations exposed to different levels of predation danger. In the first population, from a desert prairie in the south of New Mexico where there are few birds, predation events were incidental. In the other population, from a wooded habitat bordering a river in southwestern Arizona, birds could remove up to 50% of the spider population per week. The response of each population to a potential danger was measured by applying high amplitude vibrations to the spider webs to simulate a perturbation by a predator. The tests were first performed in the field for each population. In a second round of experiments, females were brought into the laboratory and isolated. Their progeny were then raised under standardized conditions to sexual maturity. Then, random crossings were performed within each population and the second generation was raised under the same conditions as previously. One member from each family was randomly chosen and subjected to the same test as in the field. This procedure again allowed possible maternal effects to be tested (for example, differences in prey abundance between the two populations could have had as an effect that the females from the better nourished population invested more nutrients in their eggs, which in turn may have had effects on the behaviour of their progeny). After the application of the vibration stimulus, the spiders displayed the characteristic retreat into the tunnel, which they did not leave to re-take their position on their web until a certain amount of time had passed. Both in the field and in the lab, differences in latency to return to the web were significant between the two populations. The population exposed to the higher predation danger in their natural environment had a longer latency period. These differences between populations again suggest a genetic component in the determinism of these behaviours.

### b *Artificial selection and the role of quantitative genetics*

Certain behavioural traits show discrete variation, such as the example of 'rover/sitter' described earlier. There are also numerous traits that show continuous variation. Such traits may include latency to respond to a stimulus or the intensity of begging behaviour in chicks. The study of the genetic determination of these traits relies on quantitative genetics (Roff 1997), of which some of its major principles (notions of heritability, genetic variation) have already been discussed in the previous chapter.

In practice, quantitative genetics apply most often to artificial selection experiments. This method allows us to determine if continuous behavioural variation has a genetic component. To do this, selected lines of individuals are created by keeping and breeding only the individuals showing the most extreme values of the trait being studied from each generation. This mode of artificial selection mimics the effects of a natural diversifying selection (see Box 2.3). A response to artificial selection in terms of changes in the mean value of a trait across generations proves that there is heritability of the trait being considered. If the feasibility of artificial selection experiments is limited by the generation time of organisms, an intense selection can still allow a response to be obtained rapidly.

Using this method, Wood-Gush (1960) demonstrated that it is possible in only three generations to substantially increase the frequency of mating behaviours in domestic chicks. The same procedure can be applied for samples taken from natural populations. In the cricket *Gryllus integer*, two strategies exist among males for finding partners. Some males call to attract females, while others remain silent and try to intercept the females as they move towards calling males. Using selected lines, Cade (1981) demonstrated that variation in the duration of sound emissions in males of this species had a genetic component. In the lepidopteran *Spodoptera exigua*, there is natural variation in the degree of **polyandry** of females. Torres-Villa *et al.* (2001) conducted a selection experiment by separating lines with high pairing levels (line H) from low pairing levels in females (line L). After six generation of selection, the initial average 1.57 pairing frequency of the population changed to 2.5 and 1.25 in lines H and L, respectively. This divergence between the selected lines became statistically significant only after the second generation. However, the percentage of polyandry, that is, females that pair with more than one male, stabilized over the course of the experiment with 90% and 25% of females being polyandrous in lines H and L respectively, such that it was impossible to obtain lines of purely polyandrous or purely monogamous females. The differences between the selected lines showed that the degree of polyandry in hybrids is proportional to the relative quantity of genes from line H. These studies therefore established that the degree of polyandry in *S. exigua* is an autosomal trait (the genes involved are not located on sex chromosomes), polygenic, and heritable. Demonstrating that additive genetic variation exists in wild populations indicates that selection can operate in nature and offers new means of investigation for understanding the processes that maintain levels of polyandry under natural conditions.

### C *The role of biotechnology*

Rapid developments in biotechnology offer new perspectives for understanding how natural selection shapes adaptation. Like in the case of the engineered phenotype, it is now possible to use genetic engineering to surpass the limits of natural variation. Genetic manipulation has various tools at its disposal that allow us to produce genetically identical individuals, introduce new genes into organisms, modify the number of copies of a gene, switch some genes on or off, or even generate mutations directed towards a specific locus (Tatar 2000; see Chapter 6). These different techniques allow us to uncover the points that are essential for our understanding of the relationships between genes, behaviour, and adaptation. Differences between individuals form the primary basis upon which natural selection can act. Behavioural **idiosyncrasies** (patterns of response that differ between individuals but are repeatable within individuals or clones), remain poorly understood in behavioural ecology and we tend to treat them as background noise. Inter-individual differences may arise from a certain adaptive flexibility, either through phenotypic variability or variability in the capacity to learn. In this context, the ability to clone individuals can prove a particularly powerful tool. Iguchi *et al.* (2001) recently demonstrated while studying two cloned lines of a salmon species, *Onchryhynchus masou macrotomus*, that differences between individuals in their basic behavioural responses such as distance travelled, space use and food searching technique, have a genetic basis and consequences for the growth of individuals, which suggests that inter-individual variations may have adaptive consequences.

The development of biotechnologies in the realm of behavioural ecology is still in its infancy if we compare it with the general usage of these techniques in behavioural neurosciences (Keverne 1997; Greespan and Ferveur 2000). Nonetheless, the use of these techniques may develop rapidly (Tatar 2000; Wolf 2001). We must stress that the use of these sophisticated techniques should not lead to a reductionist view of behaviour, systematically advocating genetic determinism. Keverne (1997), in a review of works that used the 'knock-out' technique (a technique that allows a gene to be made 'silent' in order to study the effects) to study the brains of mammals, concluded that if genes have their greatest influence during brain development, then the behaviour of mammals remains largely dependent on social interactions, context, and experience, three factors that are themselves capable of modifying the activity of genes. The important question of the links between genes and behaviour will be the main topic of Chapters 5 and 20 which will discuss the possibility of a non-genetic component in the variation that is transmitted across generations in animals.

## 3.5 The comparative approach

A final major approach in behavioural ecology is based upon the comparison of traits between species. Establishing comparisons between species to judge the adaptive character of a trait is a recurrent technique in evolutionary biology in general (Darwin 1872; Zangerl 1948; Felsenstein 1985; Cockburn 1991), and more specifically, in the behavioural sciences (Hinde and Tinbergen 1958; Greene and Burghardt 1978; Gittleman 1989; Martins 1996). However, the return to the comparative method in the study of adaptations intensified with the development of unprecedented methods (Felsenstein 1985; Brooks and MacLennan 1991; Harvey and Pagel 1991; Pagel 1997). The objective of this section is to present a succinct description of the main comparative methods used in behavioural ecology, to justify their use and to outline the conditions of their use. We will start by describing the qualitative methods initially used in ethology before discussing in greater detail the more recent quantitative methods that are currently used.

### 3.5.1 Qualitative methods

#### 3.5.1.1 *Lizards and mammals in the Chihuahuan desert*

At the most general level, the correspondence that can be observed between organisms and their environments leads us to ask about the adaptive nature of a given trait. A classic example of variation in animal colouration as a function of their environment will serve to illustrate the basis of the reasoning that is used in the comparative approach. The Tularosa Basin is a valley circled by mountains situated in the north of the Chihuahuan desert of New Mexico in the United States. At its centre, the White Sands National Monument extends over 715 km$^2$, the largest dune field of gypsum sand in the world. Gypsum, a mineral that is soluble in water, is not commonly found in sand form. However, the rain and snow that fall on the surrounding mountains dissolve the gypsum, separating it from the rocks and deposits it in the Tularosa Basin where it accumulates in the form of selenium crystals following evaporation of the water. Storms progressively erode the crystals into particles as fine as sand. Over geological time, an immense desert of white sand has formed that is so pure it sometimes resembles a vast expanse of snow. Various species of animals that live here have developed a white coloration that allows them to blend in with their environment. Three species of lizards, normally grey or brown, living in this desert with white sand, have a coloration that is much lighter than normal; some are even white. The pocket mouse, *Perognathus goldmani*, a small rodent that weighs less than 30 g, has a white fur with yellow nuances in the desert, whereas the same species that lives outside of the desert has a brownish-grey fur. Certain species of invertebrates, such as orthopterans and coleopterans, also have local light-coloured forms.

However, only a few kilometres to the northeast, still in the Tularosa Basin, the situation is quite different. In the volcanic region of Fire Valley, the ground, formed by solidified lava, is dark. Here, the same species of lizards and mice have an almost black colour, which allows them to be less conspicuous against the basalt rock formations. We can suppose that in each environment, natural selection has favoured the most cryptic variants because they were the least easily detected by visual predators. On the basis of this comparison between two sites and several species, the variation in body colour, light or dark depending on the environment, seems to constitute an adaptation to the immediate environment.

#### 3.5.1.2 *Predation danger and eggshell removal*

The same reasoning has been applied for a long time to assess in a more precise manner the adaptive value of certain behavioural traits (Gitleman 1989). A series of studies that related behavioural differences to certain ecological characteristic in various species of gulls (see Cullen 1957; Tinbergen *et al.* 1962a, b) pioneered this domain. The cryptic colour of the egg and plumage of chicks of the ground nesting black-headed gull provide camouflage in an environment made up of a mixture of sand and vegetation. However, after hatching, the white interior of the broken shell contrasts greatly with the substrate, ruining the camouflage. Niko Tinbergen and colleagues (Tinbergen *et al.* 1962a, b) observed that shortly after the hatching of a chick, parents systematically removed the eggshell debris and disposed of it at some distance from the nest. In the black-legged kittiwake (*Rissa tridactyla*), a closely related cliff-nesting species whose nests are inaccessible to ground predators, parents eventually remove the eggshell debris, but they deposit them immediately adjacent to the nest (Cullen 1957). This led Tinbergen to suggest that the distant removal of eggshells served to reduce the risk of predation for black-headed gulls. One series of experiments (Tinbergen *et al.* 1962a, b) allowed him to demonstrate that in the black-headed gull, the presence of eggshells close to the nest increases the chance that the nest would be discovered by a predator. If distant eggshell removal does have an adaptive value, we should expect to observe this behaviour in other ground-nesting birds, but not in cliff-nesting relatives. Complementary studies (Cullen 1960; Hailman 1965) demonstrated that this is in fact the case.

#### 3.5.1.3 *Solitary or colonial weaver-birds*

Other pioneering studies attempted to relate interspecific variability in social organization to various environmental characteristics. John Crook (1964) began listing the diversity of social organization in weaver birds (family Ploceidae), a group of passerines from Africa and Asia that construct elaborate hanging nests. Although very little morphological variation exists among species in this family, their social organization is highly varied. Some species are solitary and defend a territory, while others nest in colonies. Some exhibit **social monogamy**, whereas others exhibit **polygyny**. Crook (1964) established a link between these different aspects of social organization and different ecological factors, such as food availability and predation pressure. Solitary and monogamous species tend to occupy forested habitats and are essentially insectivores. Solitary foraging is the best strategy to exploit resources that are largely dispersed in the environment, such as insects. Forest species therefore defend territories and individuals of both sexes must cooperate to raise their young successfully. Colonial and polygamous species, on the other hand, live in savannahs and feed on seeds. In this habitat, food is aggregated and has high local availability. A social mode of exploitation is favoured in this habitat because it allows zones with food and areas of high food concentrations to be located more easily. The habitat being so open, it is difficult to camouflage nests and only a few acacias offer an area to build a nest. According to Crook, this 'housing crisis' constrains individuals to nest in close proximity to one another in the few available trees. The males that manage to control the best nesting sites obtain several females, whereas those that have access to the least favourable nesting sites often fail to attract even a single female. Also, because there is so much food available, females are able to care for the nests alone, and the male, freed from **parental care**, can invest more time in attracting additional females.

Peter Jarman (1974) also provided a pioneering study on African ungulates, relating the diversity of social organization to ecological conditions. Here again, the comparative method made it possible to link variations in social organization of species to the type of food resources they exploited, suggesting a causal relationship between these two variables.

### 3.5.2 Quantitative methods

The previous examples, all drawn from relatively old studies, are simple enough to give the impression that the comparative approach allows us to assess easily the adaptive value of a character. This is not really so. Without the necessary precautions, a simple direct comparison between species can often lead to erroneous conclusions. First, for reasons we have already explained a correlational association among behavioural traits and ecological characteristic does not necessarily demonstrate a cause and effect relationship. For example, in the case of the weaver birds described earlier, it is suggested that savannah species form colonies because of their seed feeding habit. It would be equally possible to envision that these birds adopt a colonial lifestyle due to predation, and that feeding on seeds, a divisible locally abundant resource is a necessary consequence for individuals living in groups (Krebs and Davies 1987).

Establishing the causal relationship between two phenotypic traits, or between one phenotypic trait and an ecological characteristic by way of species comparisons, requires us to be able to infer the ancestral state of the two variables and to determine an evolutionary scenario that would allow us to state the order in which the transitions in states appeared for each trait. Otherwise, we run the inherent risk of interpreting a causal relationship from a correlation (see Figure 3.3). Also, the simple qualitative method does not allow us to estimate rigorously the degree of association between the variables and remains a sort of 'verbal or descriptive model'. A more formal approach involves submitting a hypothesis of an association between two variables to a statistical test. For example, in the case of the gulls presented above, a first approach would consist of constructing a table that allowed species to be arranged according to whether they nested on the ground or on cliffs, and whether or not they removed eggshells after hatching. A chi-squared test would allow us to test the null hypothesis of an absence of association between the presence of the eggshell-removing behaviour and the nest site type.

However, it is not always possible to perform such statistical tests by comparing species with one another. It is necessary to first remove certain 'confounding' effects, which may mask the relationship between two variables, or create an erroneous correlation. The two principal perils of comparisons between species concern the effects of body size and phylogenetic dependence between species (Harvey and Pagel 1991).

#### 3.5.2.1 *The effect of body size: an allometry problem*

Different morphological, physiological, and behavioural traits of organisms tend to change in magnitude, form, or intensity in relation to body size. The relationship between the magnitude of a given trait and body size must be taken into account in comparative analyses. An excellent example of the confounding effect of body size in a comparative analysis concerns the study of variation in brain size among mammals (see Harvey and Bennett 1983; Cockburn 1991). By directly comparing species, we observe that aquatic mammals have a larger average brain size than terrestrial mammals. Based on this, is it possible to infer something regarding the influence of a marine habitat on the development of the nervous system? Can we conclude that marine habitats select for larger brain size? Certainly not. Let us start by considering the two groups. Among terrestrial mammals, we encounter many species of a very small size including shrews, which weigh only a few grams. On the other hand, we find the largest living animals among the aquatic mammals. To compare the relative brain size of ungulates and terrestrial carnivores with their marine homologues, it is necessary to take into account the issue of scale. Relative brain size does not show the same pattern of variation compared with body size. In small mammals, brain mass represents approximately 5% of the body mass, while this corresponds to about 0.05% of the body mass of large cetaceans. The relationship is therefore inverse to that of absolute brain size.

A simple way to resolve this problem of scale is to divide brain size by body size. However, this procedure is not fully satisfactory because it implicitly assumes that the relationship between brain and body size is linear. To describe changes in the size of a given organ (i.e. a measure of length) as a function of body size (usually measured as mass or volume), it is better to take into account the fact that mass is a function of length and use the allometric relationship. The allometric relationship between two variables X and Y is defined by the equation:

$$Y = aX^b$$

This exponential relationship can be made linear by using the log of each of the two variables. We then obtain the following relationship:

$$\log(Y) = b\log(X) + \log(a)$$

We can then estimate the value of the exponent $b$ in the allometric relationship from the slope of the log transformed regression between the two variables. This slope is a measure of the differential change in Y given X. When $b$ is greater than 1, Y increases in value more quickly than X, and we call this positive allometry. In the opposite case ($b < 1$),

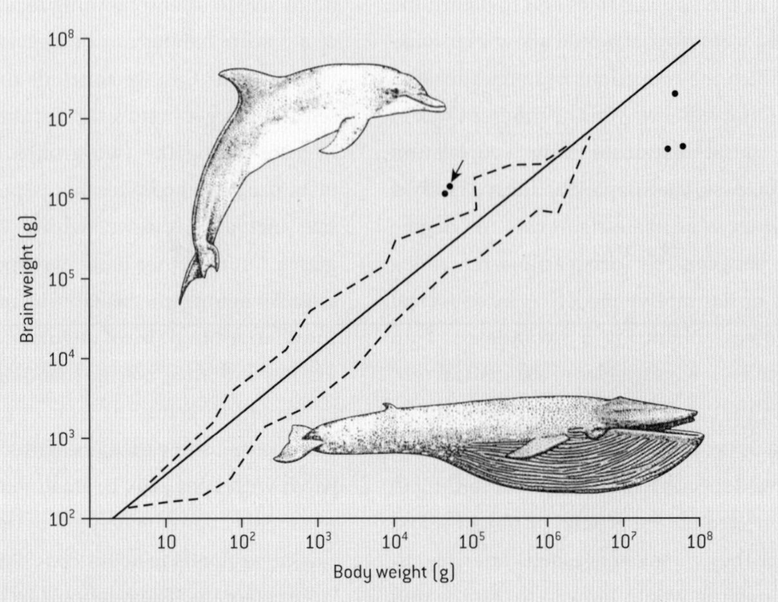


Figure 3.5 **Brain size and body size in mammals**
From Cockburn (1991), *An Introduction to Evolutionary Ecology*. © Blackwell (1991).

The regression line is calculated for all mammals. The dotted lines delimit the zone where the vast majority of mammal species lie. Only species that fall outside this range are shown. The arrow indicates the position of humans, which has the largest brain size relative to body size. The point situated immediately next to it is a dolphin, which has a similar relative brain size to that of our own species. This graph shows clearly that, once body size is taken into account, whales (the three dots to the right of the graph) have the smallest relative brain size among all mammals.

Y increases more slowly than X which corresponds to a negative allometry. When applied to the relationship between brain size and body size in mammals, we find that a coefficient value for $b$ of 0.75 describes the data well (Figure 3.5).

It appears that the greatest positive deviation from this straight regression corresponds to humans, closely followed by a species of dolphin. The greatest negative deviation, corresponding to a relatively reduced brain size for a given body size, is found in the whales, probably because these species have significant fat deposits. As such, the relationship is actually the inverse of what seemed evident with the raw data.

In several cases, it is therefore not possible to make direct comparisons between the values of a given trait between species without taking into account the effect of body size. This correction can be made by analysing not only the raw values, but also the residuals from the allometric relationship.

#### 3.5.2.2 *The effect of phylogeny*

Taking the allometric relationship into account does not remove all biases from a comparison between species. In a rigorous statistical analysis, all data points should be independent. In this case, they are not. If each point corresponds to one species, some species will belong to the same genus or family. Closely related species often have many characteristics in common (Felsenstein 1985). Harvey and Pagel (1991) analysed in detail the sources of resemblances between species and found that they cannot be considered independent points in a comparative analysis. To summarize, the characters being studied can be similar between different species for two

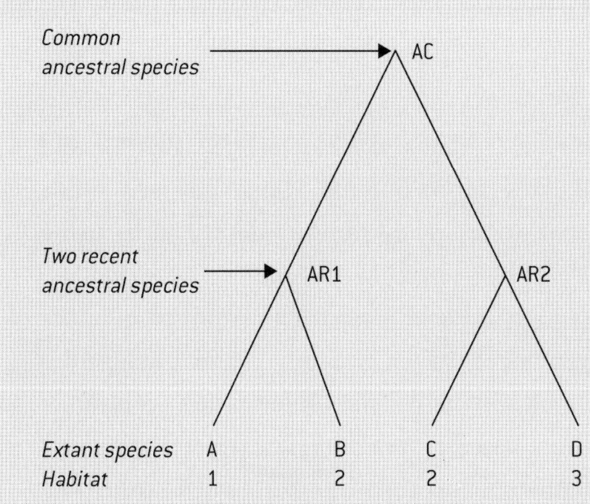


Figure 3.6 **Two reasons for resemblances between species**

(1) Sharing a relatively recent common ancestor results in closely related species resembling one another more so than distantly related species. This is because they had the same evolutionary history up until their common ancestor, which is most of their history. This is resemblance through **homology**. In this diagram, species A and B on one side and species C and D on the other side resemble one another: their traits are evolutionarily homologous since they have been inherited from a common ancestor (AR1 for species A and B, and AR2 for species C and D). Resemblances between the four extant species can also result from the fact that they share an ancestor (AC), less recently, but a common ancestor all the same.

(2) The characters of the species can also resemble one another because they are adapted to the same habitat. This is the case of species B and C in the diagram that both live in habitat 2. In this case, the characters are analogous, and if B and C resemble one another for certain traits, we speak of evolutionary convergence or homoplasy.

In a comparative analysis, only resemblances due to evolutionary convergence allow us to study the relationship between species traits in order to make inference on adaptation to given conditions. Comparative analysis methods require us to distinguish between characters that are homologous (due to evolutionary inertia) and those that really arise from adaptation.

distinct reasons (Figure 3.6): the existence of a common ancestor (the traits are therefore homologous), or the phenomenon of homoplasy (in which case the traits are analogous and arose through convergent or reverse evolution). Including non-independent data points in a statistical analysis is pseudo-replication, which has the effect of artificially increasing the sample size and the number of degrees of freedom in the analysis.

The fundamental point is that the comparative approach, as a means of studying adaptations, has the objective of evaluating empirical data in favour of the convergence phenomenon. The number of times a character evolved independently in response to the same environmental conditions must be identified. It is crucial to separate homology and analogy, to distinguish that which is similar from ancestry and that which is similar due to ecological convergence. Therefore, taking phylogeny into account is critical (Box 3.3).

As in all statistics, two types of error are possible while testing for a statistical association without taking phylogenetic dependency into account (Figure 3.7). In the case of a type I error, not taking phylogeny into account results in a rejection of the null hypothesis of the absence of a relationship between the two studied variables, when in fact, no relationship exists. In the case of a type II error, the null hypothesis is accepted when it should be rejected.

It is crucial to keep in mind that phylogenetic trees, for comparative analyses, are assumptions of relationships based on a limited set of data. Consequently, the results of a comparative analysis are directly linked to the accuracy of the phylogeny used. This depends on the quantity of information used to construct the genealogical relationships between species (Huelsenbeck *et al.* 1996).

It remains possible that in certain cases the variables being studied are little affected by phylogeny. In this case, direct comparisons between species as

### Box 3.3 **What is a phylogeny?**

Phylogenies allow us to describe the relationships believed to exist between species, by tracing a phylogenetic tree (broadly understandable as a genealogical tree). This tree illustrates, from an ancestral species, the ramifications that led to the species recognized today. When specified, the length of the branches of the tree that link the species two by two (in the case of a completely resolved phylogeny) indicates the time that has elapsed since their divergence. In the past, different characters (mainly morphological, but also physiological or behavioural) were used to establish phylogenies based on resemblances between species. However, when available, phylogenies based on molecular information are generally preferred over all others (Sibley and Ahlquist 1987; Bledsoe and Raikow 1990; but see Hillis 1987 and Hillis *et al.* 1994). They can be established by using DNA hybridization techniques; or better, by sequencing and comparing certain parts of the genome between species. When phylogenies based on molecular information contradict those based on phenotypic characters, it is recommended to trust the molecular information. This is because the existence of similarities between species in terms of phenotypic characters can be due to evolutionary convergence and therefore does not provide reliable information on the relationship between species (McCracken and Sheldon 1998).

well as comparisons that take phylogeny into account give generally the same results. However, several studies have shown that in the case where phylogenetic inertia strongly influences the state of the variables being studied, the two methods can yield very different results, perhaps even completely opposite results. Serge Morand and Robert Poulin (1998) studied the relationship between species richness in helminth parasites (cestodes, digenes, and trematodes), and the size and density of hosts in 79 species of terrestrial mammals. Larger host species were predicted *a priori* to host more parasite species (Gregory *et al.* 1996), because they are believed to offer a greater diversity of niches and can support a higher parasite load. However, the infestation level of individuals depends as much on the number of hosts available for colonization as the density of hosts, which are both factors determining the species richness of parasites (Bell and Burt 1991; Côté and Poulin 1995). The results obtained by Morand and Poulin (1998) are very telling (Figure 3.8). The simple comparison between species led to the conclusion that species richness in parasites load was positively correlated with host body size, and negatively correlated with host density. The comparative approach, removing the confounding effect of phylogeny, showed a lack of relationship between parasite species richness and host body size, and suggested a positive relationship between parasite species richness and host density (Figure 3.8). This example illustrates the importance of taking phylogenetic relationships between the species being studied into account if we want to extract from the comparison of species information on the process of adaptation.

#### 3.5.2.3 *The problem of ancestral characters*

To understand the evolution of a behavioural trait, it is not sufficient to know the different states of the character held by actual organisms, it is also necessary to determine the state of the trait in the putative ancestral species. This is not without problems because there is no fossil record of ancestral behaviour. It is therefore not possible, as with morphological traits, to calibrate a reconstruction based on fossil evidence. We can use phylogenetic

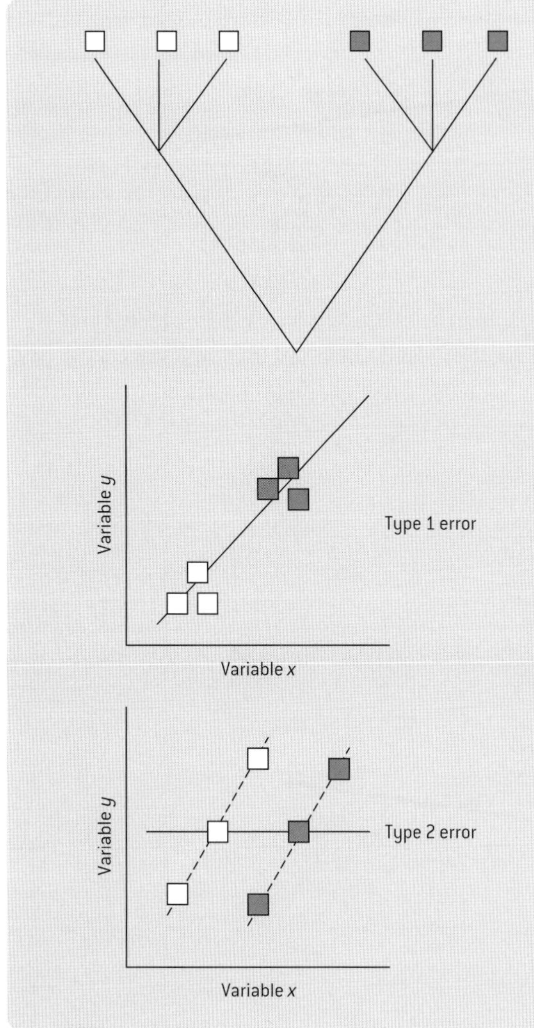


Figure 3.7 **The types of error possible in the absence of controlling for phylogenetic dependency between species**

Squares symbolize distinct species and their colour, white or black, indicates their belonging to the same taxonomic level (for example, species of the same colour all belong to the same genus). Closely related species (shown in the same colour), are liable to share more characters in common than more distant species (species shown in different colours) due to phylogenetic inertia.

**a.** In the first case, the correlation obtained between the variable *x* and the variable *y* is artificial because the species represented by the black square and those represented by the white square are not independent. We will therefore reject the null hypothesis of independence between the two variables when the null hypothesis is in fact true: this is a type I error.

**b.** In the second case, there is a real positive relationship between the two variables within each group of species, but the analysis, which does not correct for phylogenetic dependence, confirms the null hypothesis of an absence of a relationship when the null hypothesis is in fact false: this is called a type II error.

information to evaluate the state of different variables under study in ancestral species, given certain assumptions about the process of evolution. The reconstruction of ancestral characters can be done using different methods (Brooks and McLennan 1991; Harvey and Pagel 1991; Cunningham *et al.* 1999).

## a *Parsimony*

The method of maximum parsimony was for a long time the most commonly employed method for reconstructing ancestral characters. In involves finding the evolutionary scenario that minimizes the number of evolutionary transitions, that is, changes in the state of the character under study from ancestral to extant species. In the simplest example, if all the members of a monophyletic group (a group of species all descended from a single common ancestor) have a crest on the top of their head, it is parsimonious to assume that their ancestor also had this attribute.

Figure 3.9 illustrates the application of the principle of parsimony using a real example (cf Cummingham *et al.* (1998) for a more detailed presentation).

However, in practice, the cases are often much more complicated. Different parsimony algorithms exist that allow the reconstruction of ancestral traits, available with MacClade 3.0 software for example (Maddison and Maddison 1992) or Mesquite-version

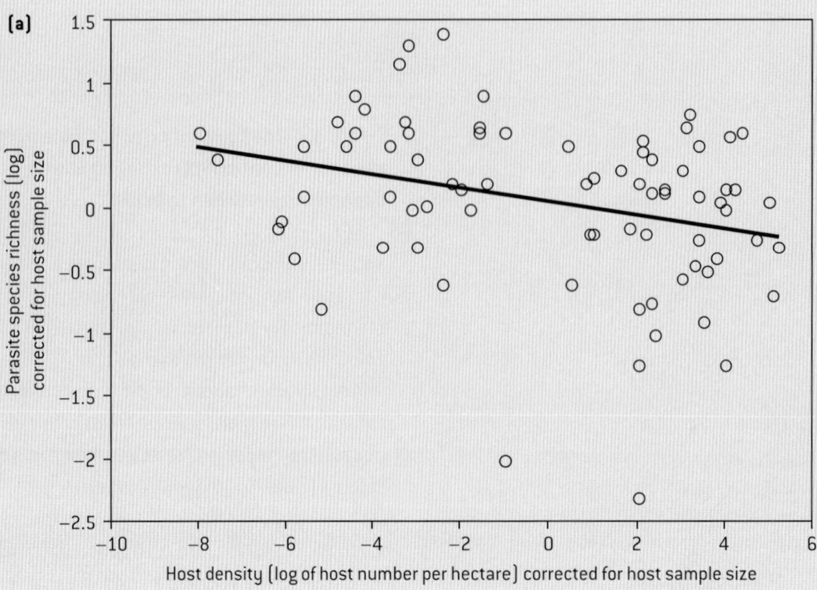


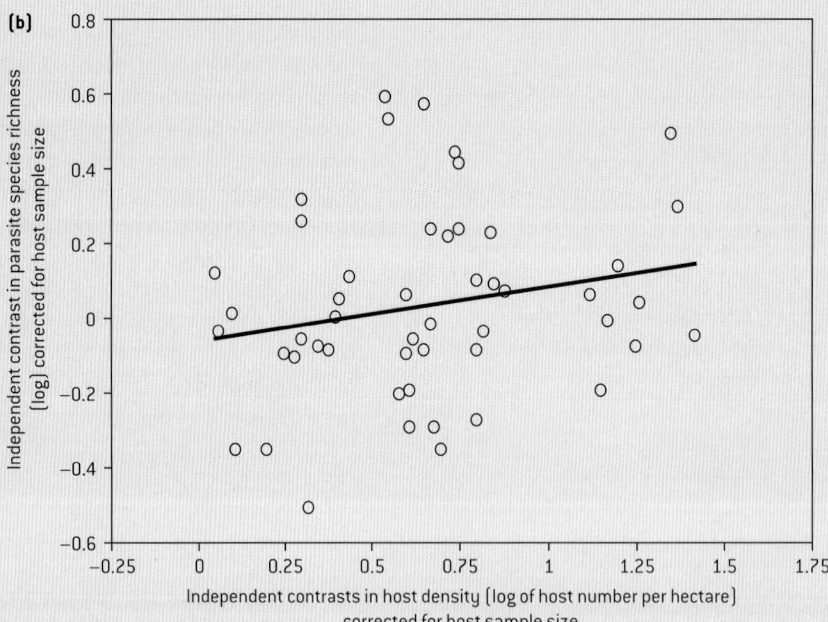


Figure 3.8 **The importance of phylogenetic inertia in comparative analyses**

Relationship between species richness of parasites in a given species of terrestrial mammal as a function of the density of the host species. The two axes are corrected for the effect of the number of hosts sampled because the number of species of parasites found on a given species increases as the number of hosts sampled increases. This is a classic sampling effect. The contrasts used in **b.** are independent and correct for the relationships between species (see Figure 3.10).

**a.** Results obtained without taking phylogenetic inertia into account. The species diversity of parasites diminishes as the density of hosts increases.

**b.** The same data, analysed while taking phylogenetic inertia into account through the method of contrasts. This time, species diversity increases with host species density. Modified from Morand and Poulin (1998).

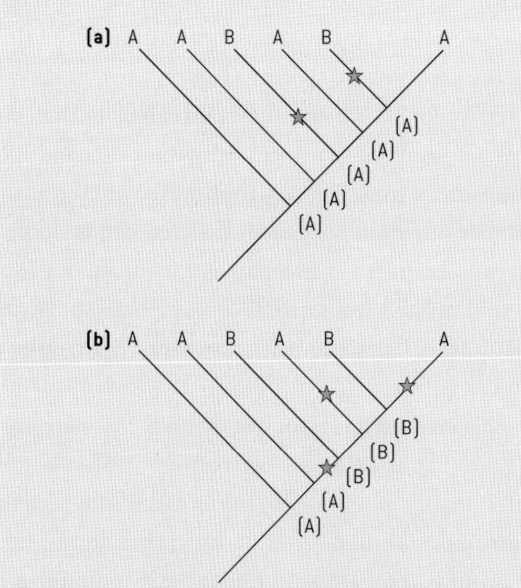


Figure 3.9 **Use of parsimony to reconstruct the ancestral state of a trait on a phylogenetic tree**

We consider a trait that can take on two distinct forms, A and B. The values indicated on the ends of the branches show the state that exists in six extant species. The values in parentheses at the nodes of branches indicate the reconstructed values of the trait. Two scenarios (among others) are possible.

a. Character A is judged to be ancestral, and only two independent transitions from A to B are necessary to obtain the distribution of traits A and B in extant species.
b. This alternative evolutionary scenario implies one independent transition from A to B and two transitions from B to A. This second scenario is less parsimonious than the first and therefore the first would be preferred.

4.5.2.0 (Maddison and Maddison 2003). However, the principle of parsimony in the reconstruction of characters on phylogenies suffers from limitations when we are comparing many species. For example, in a study on ecological characters in relation to the evolution of colonial breeding in birds, Cécile Rolland *et al.* (1998) obtained 750 different equally parsimonious reconstructions for 320 species. In other words, each of these reconstructions implicated the same number of evolutionary changes in total over the whole phylogenetic tree. Such a situation poses a problem for statistical treatment that will be resolved below.

## b *The major quantitative methods*

Different quantitative methods exist to test the adaptive character of a trait according to phylogenies (which we must remember, can vary in their level of precision). We can distinguish two major families of methods (see Harvey and Pagel 1991; Martins 2000).

Correlational methods verify the significance of the degree of association or dependence between two variables (two characters or one character and an environmental variable) while taking the genealogical relationships between species into account.

Directional methods retrace the evolution of specific characters along the phylogenetic tree. Based on the statistical reconstruction of ancestral states, it becomes possible to retrace the chronological order of successive changes in state and to attribute a probability associated with that chronology.

A complete description of the various methods used in comparative analyses is beyond the scope of the present textbook (detailed information can be found can be found in Harvey and Pagel (1991) and Martins (2000)). Therefore, we will limit ourselves to describing the most widely used methods.

*An example of the correlational method*

Historically, the first methods that allowed us to be freed from phylogenetic inertia attempted to analyse the degree of association between two variables of interest between extant species. For example, we can ask if carnivorous mammal species have larger territories than herbivores, or, as in Figure 3.8, if the number of parasite species that exploit a given host depends on the size of that host. These types of question amount to the study of correlations between various parameters while accounting for the phylogenetic relationships among species. In the case of the first question, it is a matter of knowing whether territory size is correlated with foraging mode. In the second, whether the number of parasite species is correlated with the size of the host.

We have seen earlier why it is necessary to take into account the phylogenetic relationship between neighbour species to answer such questions. Imagine that we are working with a phylogeny like the one shown in Figure 3.10a, and information on two continuous variables, *X* and *Y*, which we suspect of having evolved in correlation with one another. Feselstein (1985) was the first to propose a method to treat this type of question for continuous variables. His method is based on two major assumptions: (1) the evolution of characters along the branches of a phylogenetic tree occur according to a random process and, therefore, changes over time in the value of a trait can be modelled as Brownian movement; and (2) the changes that are produced in one branch of the tree are independent from those that occur in other branches. The general idea is that even though two species are not independent because they share a common ancestor, given these assumptions, we can consider that two species have evolved independently since their divergence. Therefore, if we are capable of quantifying the degree of differentiation between two species since their divergence, this degree of divergence constitutes information that is independent, from a statistical point of view, from the divergence that has taken place between two other taxa in the same phylogenetic tree. This degree of divergence between two closely related species can easily be represented by differences in the value of the trait of interest (*X* or *Y*) in the two species in question (Figure 3.10b). This difference quantifies a contrast between species. It is for this reason that the method is classically called the independent contrast method. If, for example, in the case of the relationship between variable *X* and variable *Y*, species A presented a *Y* value of 11 and an *X* value of 157, whereas species B presented a *Y* value of 15 and an *X* value of 160, the contrast between them in *X* would be 15 – 11 = 4 (in the units of measure for *X*), and for *Y*, 160 – 157 = 3 (in the units of measure for *Y*). In the case where the length of the branches between the ancestral species AR1 and species A and B is the same, we can also infer the value of the trait for the ancestral species AR1 as being the mean value of the trait for the two extant species (Figure 3.10), 158.5 for *X* and 13 for *Y*. We can therefore calculate the contrast between the two ancestral species AR1 and AR2 (contrast C3 from Figure 3.10b), and so on. If *N* is the number of extant species in the tree being studied, we can obtain *N* – 1 independent contrasts for each of the variables studied. We can then study the relationship between the two series of contrasts (Figure 3.10c).

This method has been widely used. It is necessary to note that associations between two variables can differ greatly when phylogeny is taken into account compared with direct comparisons that do not take phylogeny into account (Figure 3.8; Morand and Poulin 1998). It is important to emphasize, however, that transforming the data into contrasts in order to study the relationship between *X* and *Y* has the effect of modifying the question being asked. Our original question was 'Are variables *X* and *Y* correlated in extant species?' When studying the relationship between contrasts in *X* and contrasts in *Y*, the question becomes 'Is evolution in a given direction of the variable *X* correlated with evolution in a particular direction of the variable *Y*?'

*An example of directional methods*

The method of contrasts requires studies of continuous variables. Often however, behaviours are better classified as binomial variables: an animal does or does not do this, the species lives in mountains or elsewhere, animals of a given species forage in groups or alone. Various methods have been developed to allow comparative analyses of these kinds of data. One of these is the general method of comparative analyses for discrete variables developed by Mark Pagel (1994, 1997). We will illustrate this using a study of environmental factors associated with colonial reproduction in birds (Rolland *et al.* 1998). As we will see in Chapter 14, colonial reproduction is common in birds, and particularly in marine birds where more than 95% of species breed in colonies. Species can therefore reproduce solitarily (coded hereafter as state 0), or in colonies (coded as state 1). The

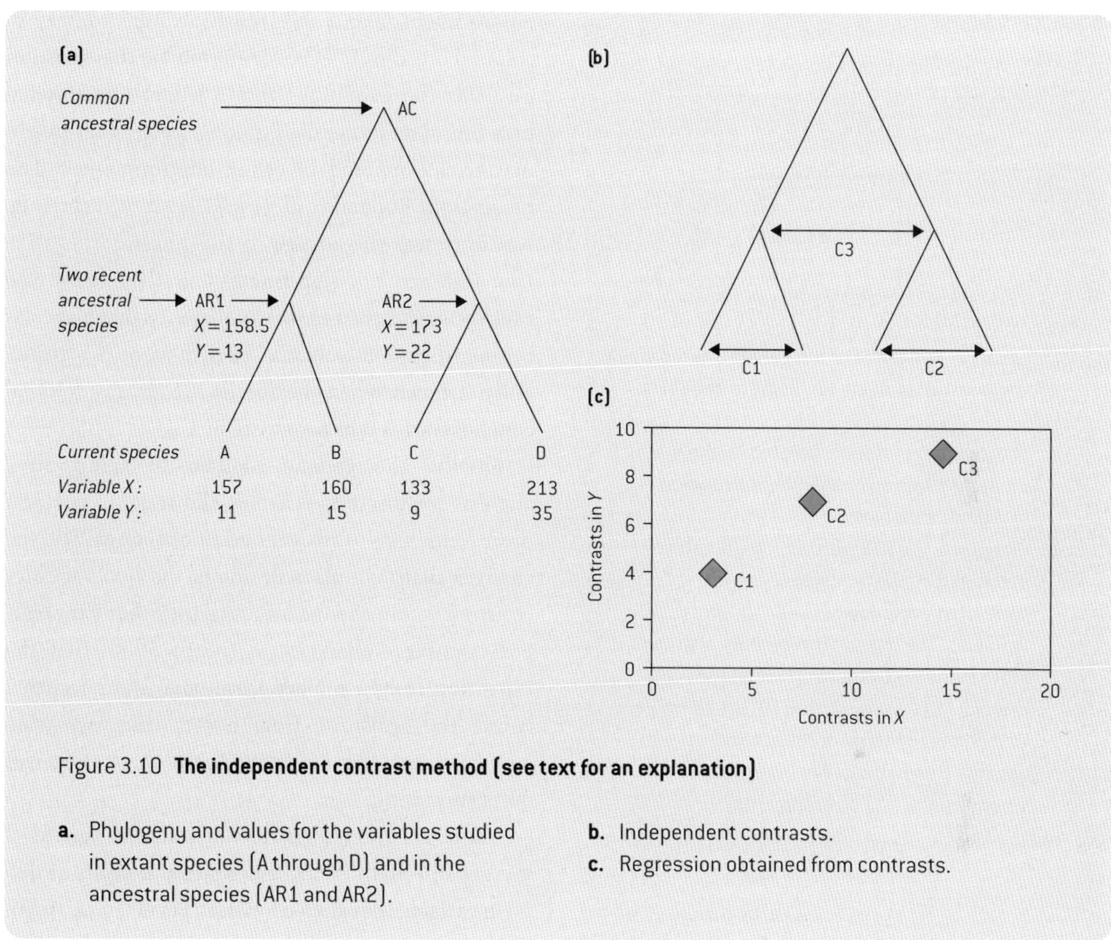


Figure 3.10 **The independent contrast method (see text for an explanation)**

**a.** Phylogeny and values for the variables studied in extant species (A through D) and in the ancestral species (AR1 and AR2).

**b.** Independent contrasts.

**c.** Regression obtained from contrasts.

question is to understand the ecological correlates of this behaviour. In particular, the prevalence of coloniality among marine birds has classically led to the belief that the marine environment required the animals that exploit it to breed in colonies. This amounts to an interpretation of causation based on a strong apparent correlation between marine environments and coloniality. The various equally parsimonious reconstructions of the character of reproductive mode indicate that in the portion of the phylogeny studied by Rolland and his collaborators, between 18 and 21 transitions to the marine environment have occurred. This allows for comparative tests with strong statistical power.

The general method of comparative analysis for discrete variables relies on the method of maximum likelihood to estimate the rate of change along the phylogenetic tree with respect to the state of the two variables of interest. It allows a certain number of subsequent tests to be conducted. A first model is fitted to the data under the hypothesis that the characters evolved independently from one another (Figure 3.11a). This model allows the estimation of the four rates of evolution that correspond to each of the transitions considered. A second model that allows for non-independent evolution of two variables, and thus includes eight transitions, is then fitted to the data (Figure 3.11b). A first test, called the omnibus test, is then performed by means of a likelihood ratio test to determine which model provided the best fit to the data. This test indicates whether evolution of the first variable (here for example, mode of reproduction) and that of the second variable (living in a marine environment or not) occurred in a correlated manner. If the second model with eight parameters provides the best fit, then the hypothesis

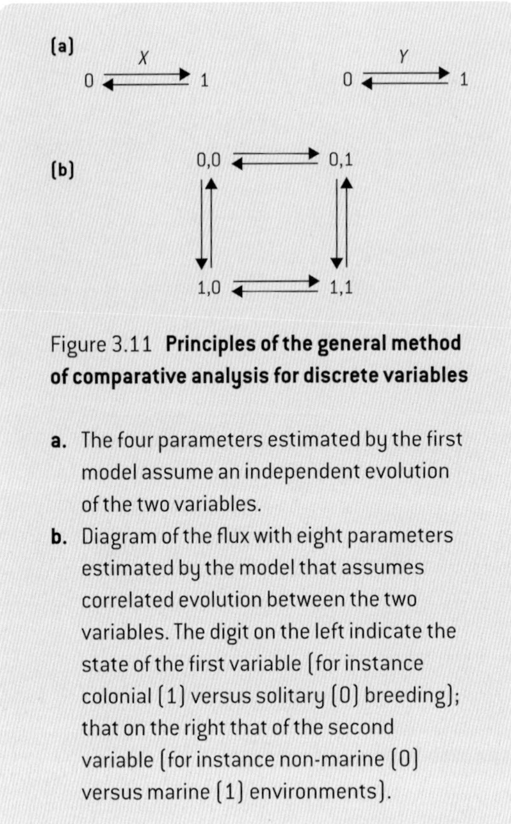


Figure 3.11 **Principles of the general method of comparative analysis for discrete variables**

**a.** The four parameters estimated by the first model assume an independent evolution of the two variables.
**b.** Diagram of the flux with eight parameters estimated by the model that assumes correlated evolution between the two variables. The digit on the left indicate the state of the first variable (for instance colonial (1) versus solitary (0) breeding); that on the right that of the second variable (for instance non-marine (0) versus marine (1) environments).

of non-independent evolution is statistically supported. Rolland and his collaborators suggest that in the case of the evolution of coloniality in relation to the marine environment in birds, this is the case: the two characteristics show a highly correlated evolution. This result may seem trivial to all people who study marine birds, but it shows that the obvious observable correlation in extant species is not the product of phylogenetic inertia. In other words, this association is not due to the fact that all seabirds are descended from the same ancestral species that lived in a marine environment and was a colonial breeder. This resemblance between species is therefore likely the result from selection and adaptation. The biological implications of this correlation remain to be interpreted, a goal that this method allows us to study, at least partly.

The general method of comparative analysis for discrete variables allows us to perform contingency tests, to ask questions such as 'Did living in a marine environment favour the transition from solitary to colonial breeding?' We can reformulate this question as: 'Was the evolution from a solitary to colonial breeding state more likely among species that live in marine, as opposed to other, environments?' The result from Rolland *et al.* (1998) is quite surprising. Not only was the answer to this question negative (the test was not significant), but they found the exact opposite pattern: the probability that a species begins to exploit a marine environment is significantly stronger when the species are already colonial compared with when they are not.

Finally, the general method of comparative analysis for discrete variables allows us to perform precedence tests to answer questions such as: 'Did the changes that take place in one of the two variables occurred before changes occur in the other variable?' For example, Rolland *et al.* (1998) found that the transition from a solitary to colonial state occurred before the transition from non-marine to marine environments (Figure 3.12). Again this result was quite surprising.

These unexpected results illustrate the benefit of the general method for comparative analysis of discrete variables developed by Mark Pagel (1994, 1997). This method allows for much more than the simple analysis of a correlation between two variables. The contingency and precedence tests allow us to interpret causation from these correlations. In the case of the evolution of coloniality, the precedence of coloniality over the transition towards marine environments seems to indicate that the past interpretations in terms of causality were inappropriate: the marine environment did not force species to become colonial, but living in colonies facilitated the transition towards marine environments whose exploitation seems particularly difficult for solitary individuals (questions about the evolution of group living are addressed in Chapter 14).

#### 3.5.2.4 *Strengths and weaknesses of the comparative approach*

We have only covered a few basic principles of the comparative approach. Numerous other methods

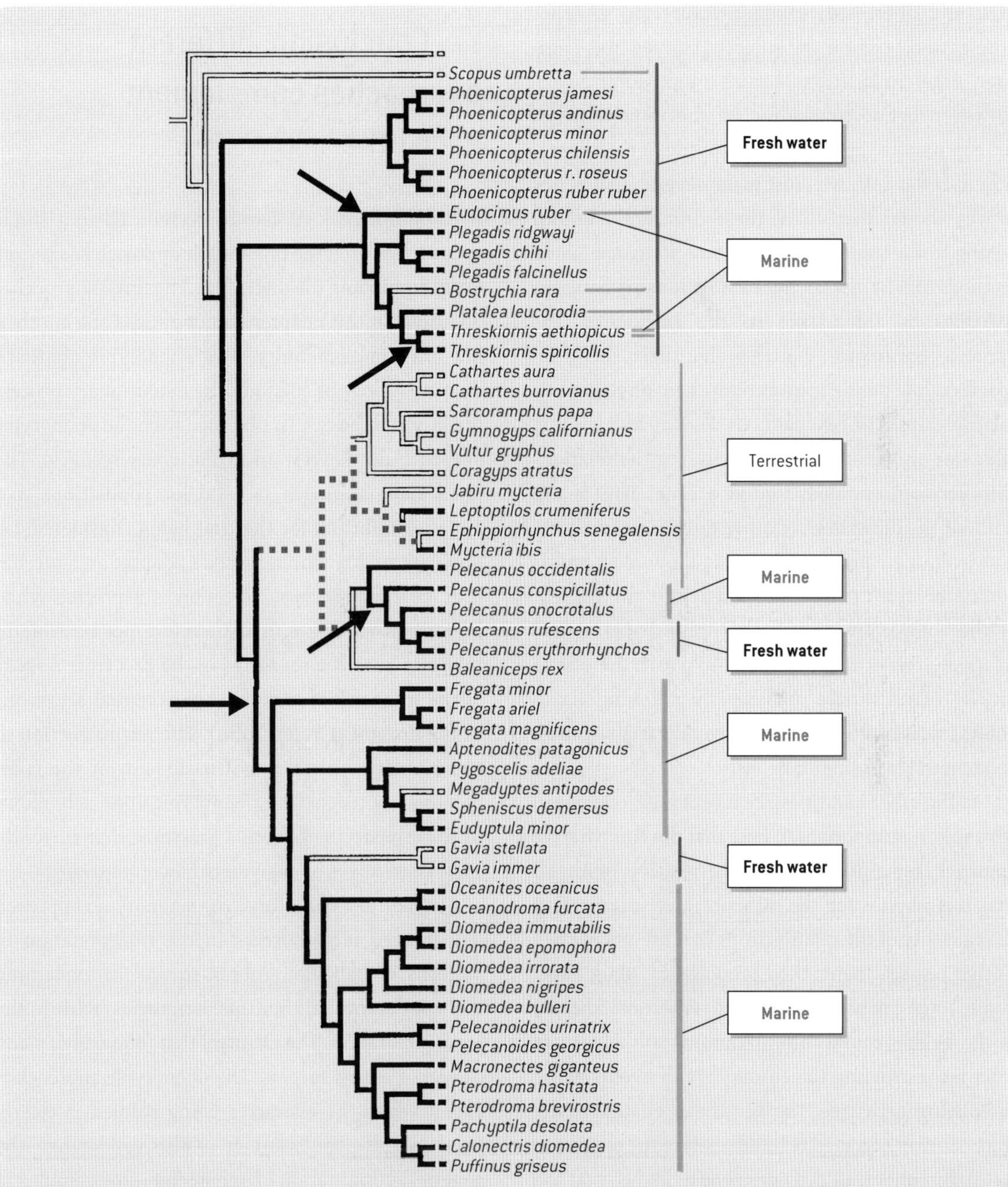


Figure 3.12 **Coloniality in birds evolved before transition to a marine environment**

Branches in black correspond to parts of the tree where species are reconstructed as having had colonial reproduction. The white parts correspond to portions of the tree where species are reconstructed as having had solitary reproduction. The dashed parts correspond to situations where the reconstruction of reproductive mode is ambiguous. The four arrows indicate the four branches where a transition from a non-marine to a marine environment is probable. It clearly shows that these four transitions all arose after species had already become colonial. The precedence of coloniality over transition to marine environments is significant ($\chi^2 = 7.0$; $P < 0.01$). This figure uses a portion of the phylogenetic tree studied by Rolland *et al.* (1998).

exist and interested readers are directed to other articles and texts on the subject (Harvey and Pagel 1991; Pagel 1994, 1997; Martins 1996; Rohlf 2001). However, before concluding on the importance of the comparative approach in the study of the adaptation of behaviour, we must make several comments. It was shown that all of these different methods belong to the same global family of statistics. These methods consist of studying the relationship between variables by integrating in the calculation the variance–covariance matrices that exist between species while taking into account their phylogenetic relationships. The two methods developed here only allow for univariate work by analysing the relationship between variables two at a time. This is not without the drawbacks of other univariate analyses. However, there are programs that allow most multivariate analyses to be transposed into the context of the comparative approach (Rohlf 2001). These relatively recent tools allow us to avoid the classic pitfalls of univariate analyses by allowing us to analyse relationships between several variables and a behavioural trait simultaneously.

On the other hand, despite recent methodological advances, it is important to note that the comparative approach remains essentially correlative, with all that this implies for the difficulties of interpreting the causality of the relationships that are uncovered. In that respect, the method proposed by Mark Pagel (1994, 1997) can be regarded as an important advance in the field as causal interpretations can be derived from reconstructing the chronology of events along the phylogenetic tree. The fact that a change in the state of a trait occurred regularly before the change in state of another trait provides good evidence that the state of the first trait may have influenced change in the second trait.

Finally, we must emphasize that, regardless of the power of the statistical analyses conducted, the value of comparative analyses depends in large part on the robustness of the phylogenies used and the ability of the researcher to define without ambiguity the characters whose correlated evolution he or she wishes to study.

## 3.6 Conclusion: different approaches complement each other

Behavioural ecology cannot be reduced to a single approach, even if the phenotypic approach has historically dominated the discipline. The three approaches described in this chapter do not oppose one another, but are complementary. The phenotypic approach does not allow us to study the evolutionary process directly and the results obtained from the phenotypic approach do not test an evolutionary scenario. The phenotypic approach is valid for populations that are at equilibrium for the trait being considered, i.e. populations for whom a stable adaptation state has been reached. Because selection is a process while adaptation is a state, it is therefore not possible to infer the selective pressures in the past from a study of the intensity of selection (or potential selection) in the present (Grafen 1984). The reconstruction of evolutionary scenarios is the domain of the comparative approach. The phenotypic approach allows us to estimate the selective pressures that currently exert themselves on a given behavioural trait.

Despite its limitations, the phenotypic approach has enjoyed large success in the interpretation of patterns of behaviour observed and in explaining the coexistence of multiple strategies within the same population. The chapters that follow give many examples of this success. The phenotypic approach is useful when the character being studied is clearly fixed in the population. It is also valid when the genetic determination of a character is too complex to take into account. Studies of evolutionary genetics, however, are necessary to understand cases where selection is still underway and certain complex equilibria.

Moreover, in certain cases it is crucial to be able to understand to what extent genetic constraints limit adaptive evolution. The primary interest of the genetic approach in behavioural ecology is to allow the formulation of new questions, complementary to those raised by the phenotypic approach. Experiments on the 'rover/sitter' polymorphism conducted by Sokolowski *et al.* (1997) in *Drosophila* have demonstrated that the advantage of each phenotype depends on the density of the larva populations. In three independent populations, the 'rover' type was favoured in high densities, whereas the 'sitter' type was selected for in low densities. This type of study offers an interesting perspective for testing the hypothesis that density-dependence in allele frequencies underlies the behavioural polymorphism that exists in natural populations.

Genetic studies can also precede the phenotypic approach. A good example is given by a study on territorial behaviour in *Drosophila melanogaster* (Hoffmann 1988; Hoffmann and Cacoyianni 1990). Males in this species can defend feeding territories against other males. The function of territory defence seems to be exclusively related to access to females that come to feed on the territory. However, territorial behaviour is not systematic. Artificial selection experiments have shown that genetic variation underlies variation in territorial success in different populations and that the character is heritable in nature. This information leads to two questions. First, we can ask why territorial behaviour is not often encountered in natural populations. We can also question the mechanisms that maintain genetic heterogeneity. Based on these questions, Hoffmann (1988) carries out several series of experiments whose results suggest that territorial behaviour in this species is a conditional strategy whose expression depends on the size of the food resource and the degree of competition between males. This study illustrates how a genetic approach, beyond simply identifying genetic determinism, can serve as a basis for a phenotypic approach. Because these three approaches complement each other, each one must receive as much consideration as the other. The objective of a study will determine which approach should be used to resolve a given problem at a given time. The rest of the chapters in this book will regularly make use of all three approaches to illustrate the different themes discussed.

## » Further reading

> *Different editions of the text* Behavioural Ecology: An Evolutionary Approach *by J.R. Krebs and N.B. Davies provide a theoretical framework for phenotypic and comparative approaches.*
**Alexander R.McN.** 1996. *Optima for Animals.* Princeton University Press, Princeton, constitutes an excellent introduction to the concept of optimization in biology.

> *A special issue of the review journal* BioEssays *(volume 19, number 12, December 1997) is dedicated to the study of the relationship between genes, molecules, and behaviour.*

> *Finally, a detailed account of the comparative approach can be found in:*
**Harvey P.H. & Pagel M.D.** 1991. *The Comparative Method in Evolutionary Biology.* Oxford University Press, Oxford.

4

4

# An Information-Driven Approach to Behaviour

Étienne Danchin, Luc-Alain Giraldeau, and Richard H. Wagner

## 4.1 Introduction

Replication is the essence of life. It is what differentiates living organisms from other entities. Life is thus essentially a matter self-replication, passing **information** about the replication process from one generation to the next. Many authors stress that selection targets not DNA molecules but the genetic information they carry (Gilddon and Gouyon 1989). Genes therefore are not merely pieces of DNA, but **bits of information** coded by DNA (Sections 2.2.2.3 and 2.2.2.4). The general impression that often emanates from such a view is that this self-replicated information is exclusively genetic. However, evidence is accumulating that non-genetically coded information can be transmitted from individual to individual. This will be the topic of the last chapter of this book.

In this chapter, which ends the first part of this book, our goal is to (1) define a lexicon of biological information, (2) review studies that provide evidence that non-genetically coded information can be transmitted from individual to individual, and (3) stress the importance of adopting an information driven approach to behaviour.

Behaviour is largely a matter of decision-making, which requires information, and hence information gathering, memorising, and processing. Consequently, behavioural ecology has increasingly adopted an information-driven approach to the study of behaviour and has just begun to consider the evolutionary consequences of individuals having the ability to use information extracted from the behaviour of others.

## 4.2 A lexicon of biological information

Information is everywhere in the study of behaviour. Consider its prevalence in this book. It first came up in Chapter 2 on the concepts required to study the evolution of behaviour. Then, in Part 3 'Exploiting the Environment', information will be an omnipresent theme in Chapters 7–10, which deal with processes of choice: foraging and breeding habitat selection, as well as dispersal. Then again, information will be a central though less explicitly stated recurring theme in Chapters 11–13 of Part 4 'Sex and Behaviour'. In Part 5, 'Social Interactions Between Individuals', information will structure the discussions of Chapters 14–17. So, clearly information is part of the study of most kinds of behaviour. This is not too surprising if one considers that behaviour is often defined as **the way by which organisms adjust to environmental variation**. Adjustment implies some reaction to cues in the environment hence making decisions based on information that must therefore be gathered and processed by individuals. **The study of behaviour can thus be viewed as the study of decision-making.**

### 4.2.1 When to transmit versus acquire information

In biology, variation in information creates phenotypic variance ($V_P$ in Chapter 2, Section 2.2.3.1). Thus **phenotypic variance is produced by variation in genetic and environmental information**. We have seen that there are two broad categories of biological information responsible for phenotypic variance. Most genetic information is the result of selection and essentially produces phenotypes that correspond to long-term characteristics of past environments. Variation in genetic information (i.e. non-silent variation in the coding part of the genome) contributes to phenotypic variation ($V_G$ in Chapter 2, Section 2.2.3.1). However, some phenotypic variation ($V_E$) does not result from genetic variation. Non-genetic information is at the origin of such phenotypic variation. Such information is transmitted through direct or indirect interaction with the environment and mainly involves some neuronal processing. The function of the genetic substratum is to transmit information across generations, not to acquire information de novo. The function of the neuronal substratum is both to transmit information and to acquire new information *de novo*.

Unless conditions are constant over many generations, information about the current state of the environment cannot be transmitted by genes alone (Figure 4.1). So, when conditions are changing even slightly individuals must acquire information by themselves about the current state of the world, which excludes the sole use of genetic information. Changing conditions in themselves, however, are not sufficient to favour the acquisition of information. Several theoretical approaches have shown that acquiring information is favoured only when the environmental cues that produce information predict the future conditions with some reliability (Stephens 1991 and Figure 4.1). When conditions change randomly, no cue can carry reliable information about the future such that selection produces behaviour that is independent of conditions. For instance, in randomly varying environments, models predict that natural selection favours animals that do not gather information but rather settle to breed at random (see Figure 9.4 for a detailed example). Interestingly however, at the other extreme, when the conditions are highly predictable between generations because the environment is constant, acquiring information is disfavoured because information can more effectively be transmitted by genes from one generation to the next. For example, constant environments across generations select for philopatry (Clobert *et al.* 2001; Doligez *et al.* 2003), a strategy

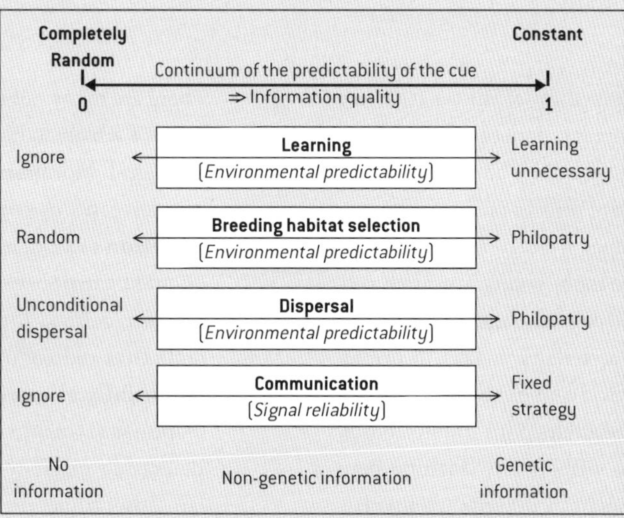


Figure 4.1 **Organisms are predicted to learn about their environment genetically or via trial and error according to the degree of predictability of the environment**

The importance of temporal autocorrelation in environmental variation for the evolution of condition-dependent strategies in four major evolutionary fields: animal learning, communication, breeding habitat selection and dispersal (Boyd and Richerson 1988; Stephens 1991; Feldman and Laland 1996; Bradbury and Vehrencamp 2000; Clobert *et al.* 2001; Doligez *et al.* 2003). All these fields in fact deal with processes that can be grouped under the general term of 'learning'.

The absolute value of predictability of the variation of any environmental component can vary from 0 when that variation is random to 1 when the environmental component is constant. In all these cases the question under study is how can animals acquire knowledge about the current state of the environment. For instance, in breeding habitat selection by prospecting different potential habitats, animals gather information about the current state of those habitats. In other words, they learn about their environment. The case of communication is particular here, as the components of the environment that are involved are **signals** provided by other individuals. The temporal autocorrelation involved here is linked to the reliability of the signal: a given signal usually carries the same information about the signal producer. Game theoretical models developed by Jack Bradbury and collaborators lead to the counter-intuitive result that communication is not selected for when the signal reliability is absolute.

Those four groups of models lead to the conclusion that information gathering strategies (i.e. learning either from others or directly from the environment) are only favoured when the concerned components of the environment are neither constant nor entirely randomly variable. This suggests that knowledge about components of the environment that are constant is likely to become transmitted across generations genetically. Genetic information describes the constant components of the environment, whereas learning allows organisms to react to short-term variations in the environment. At the other extreme, when the environment varies randomly, selection favours strategies that ignore the cues because in such conditions cues do not carry information. Thus, plastic strategies involving learning are only expected when the concerned components of the environment are neither totally random nor totally predictable in time. Potential cues only produce information when environmental predictability is intermediate.

This shows that environmental predictability is a major environmental parameter influencing most decision-making processes. However, environmental predictability is often difficult to measure. Nevertheless, several authors have claimed that this is a major parameter for understanding phenotypic plasticity. Because extremely constant or random environments are likely to be rare in nature, we should expect the acquisition of knowledge about the environment to be often based on learning. In other words, in a completely constant or random world, phenotypic plasticity, learning and human culture would have probably never developed.

that often makes the acquisition of information on the value of an environment unnecessary. This is because in such environments local breeding success directly reflects local future **fitness** prospects and so using a fixed unconditional strategy such as philopatry is effective because more young are produced in and return to the best patches. Another example involves birds that use a stellar compass to orient during migration in which juveniles migrate successfully independently of the adults. This innate ability to use the stellar compass without any learning suggests some form of information inheritance involving genetic information, no doubt because stellar maps are quasi-constant, changing extremely slowly.

However, learning capacities and the genetic encoding of information about the environment may interact. In a recent experiment of selection in the fruit fly (*Drosophila melanogaster*), Frédéric Méry and Tadeusz J. Kawecki (2004) showed that the evolution of innate and learned knowledge of the environment can interact. Under selection for preference for laying habitat in certain circumstances, an opportunity to learn about alternative habitats facilitated the evolution of an innate preference.

This suggests that knowledge about components of the environment that are constant are likely to become transmitted across generations. Such transmission is likely to be genetic, but might also be **cultural** (see Chapter 20), and no model has specifically studied the question of when such information should be transmitted culturally or genetically. However, models by Doligez *et al.* (2003) suggest that cultural transmission (here defined simplistically as learning from others) may be selected for over a wider range of temporal autocorrelation than genetic transmission.

### 4.2.2 What is non-genetic information?

There is much confusion about the meaning of non-genetic information. This section proposes an integrated view of the types of information relevant to biologists and focuses on non-genetic information, a fundamental concept in behavioural ecology.

> There are many definitions of information, some of which do not apply to biology. Biological information can be defined in its broadest sense as **any source of phenotypic variation** ($V_P$). In a more behaviourally oriented perspective, biological information has been defined as **genes and detectable facts that reduce uncertainty, potentially allowing a more adaptive response** (Danchin *et al.* 2005). The term 'detectable' means anything that organism can sense, which includes chemical recognition in plants and microorganisms, i.e. also organisms without brains. Non-genetic information can thus be defined as '**information that can be extracted from detectable facts**'.

Information changes the state of the receiver and can improve its fitness when it allows for a more effective response resulting from a reduction in uncertainty about current conditions. The **information value** resides in its power to identify current conditions that predict future conditions with improved accuracy.

#### a *Non-genetic information comes from detectable facts, or cues, and nothing else*

By definition, biological information can be derived from detectable facts (Figure 4.2) reflecting characteristics of the **environment**, i.e. physical **habitat** and organisms.

> We define **social cues** as detectable facts that are unintentionally produced by organisms. Although 'social' typically refers to interactions among individuals of the same species, we view social cues as also encompassing heterospecific interactions.

An appropriate lexicon of biological information should thus focus on cues that allow individuals to predict the future, potentially allowing a more adaptive response.

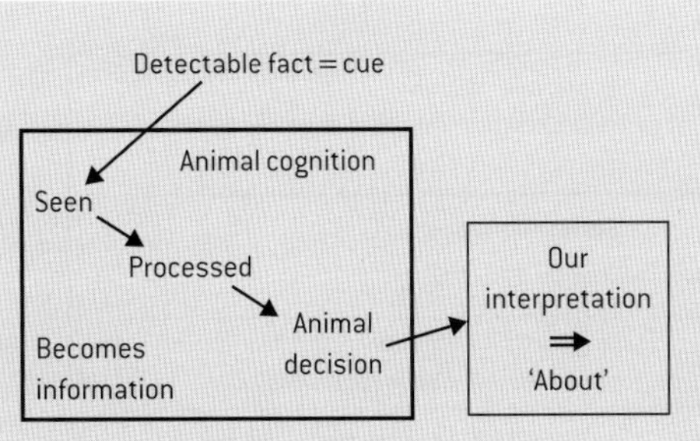


Figure 4.2 **Information results from detectable facts (i.e. cues)**

The cue may be detected and processed by a bystander, eventually leading to a decision by the bystander. The initial fact becomes information through the observation and processing by the bystander. In many animals such cognitive processing involves the brain, but plants can also gather information from the current state of the environment and make decisions even in the absence of a sophisticated neural system. Subsequently, the behaviour of the bystander may be witnessed by the researcher who provides an interpretation of the resulting decision. We usually report that interpretation in a statement starting with 'about' at the end of a definition of a category of information. For example, an early definition of 'public information' stated that it is information about the quality of a resource (see text). Adding such an 'about statement' in the definition of a category of information limits the interpretation of the fact from which information was extracted. Defining information should thus focus on the fact, or cue, that is used by individuals to make decisions.

### b *The cue becomes information only through an animal's cognition*

Many definitions of information also incorporate an interpretation, or a meaning, that is a statement starting with 'about'. Adding such a statement creates several levels of interpretation of the cue (Figure 4.2). The first level is the interpretation made by the animal that observes the cue. The second level is the one made by the researcher that interprets the animal's reaction and hence defines what the information was about. We can only infer the meaning that a focal individual extracted from a cue through its reaction to it. However, the same cue may provide different information to different individuals. It is therefore more efficient to study biological information by focussing on the fact from which it is extracted rather than assuming that it is about only one thing.

Throughout this book we illustrate how researchers study experimentally the use of information by animals. Because such experiments are placed in a specific context, researchers must assume that the fact they manipulate is about the parameter they are investigating. In the next section we offer a lexicon that focuses on the facts that may provide information, rather than on how it is used. That lexicon thus sticks to the facts from which information is extracted. This approach aims to build a broader conceptual framework for the study of biological information. However, this does not prevent us from providing an interpretation when studying a specific case, and we will regularly propose such interpretations in this book. But this will always be in a specific context rather than when defining the various categories of information, which is the goal of this section on the lexicon of information.

A gap between a cue and its potential meaning is illustrated by situations when different individuals under different ecological and social constraints interpret the same cue in different ways. For instance, Doligez *et al.* (2002, and Chapter 9, Section 9.5.3.3 for details on this study) showed in the collared flycatcher (*Ficedula albicollis*) that individuals use the average local reproductive success of conspecifics as information to choose their future breeding patch. However, in that population, males and females did not use that information in the same way. Females tended to remain faithful to the most successful patches while males did not (Doligez *et al.* 1999). Furthermore, different categories of males did not show the same use of former local reproductive success. Males with low competitive abilities tended to avoid the highly competitive patches, while highly competitive males showed no apparent effect of

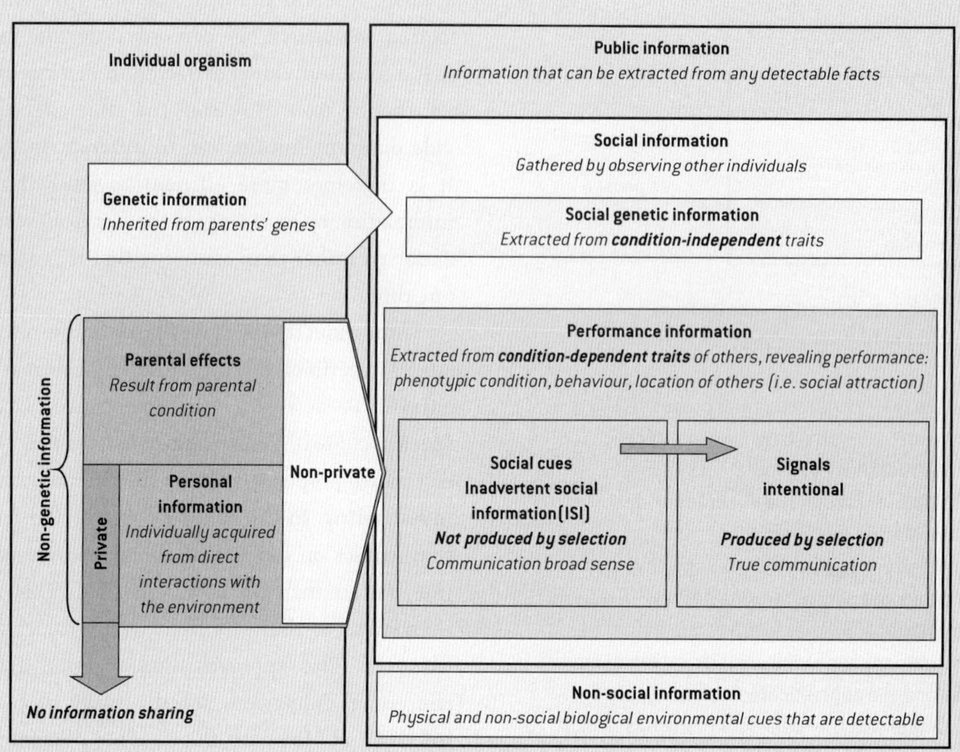


Figure 4.3 **A lexicon of non-genetic information that focuses on the facts from which information may be extracted rather than on how it is used**

**Individual information** encompasses genetic and non-genetic information (left side). Non-genetic information encompasses parental effects as well as personal information, which results from an individual's interaction with the environment. Parental effects involve the transfer of information from parent to offspring and moulds offspring phenotype to current environmental conditions. Part of personal information may remain **private** (i.e. non-accessible to other individuals). The part of personal information that is public (i.e. non-private) may also be used by others as social information (right side). All individuals thus have genetic, parental and personal as well as social information. From the information producer's perspective, information may be intentional (this is true **communication**) or unintentional (i.e. **inadvertent social information** (ISI)) which belongs to communication in the broad sense.

The concept of **intentionality** in an evolutionary context refers to the fact that individuals producing a signal attempt to extract a benefit from the reaction of others to the signal. The common thread of these two types of social information is that both signals and cues are produced by condition-dependent traits of other individuals that reveal their performance. Together they thus constitute **performance information**. Social information also encompasses **social genetic information** that results from condition independent traits. Animals can also use **non-social information** from relevant environmental cues (physical and non-social biological cues). Altogether, non-social and social information can be lumped within the general category of **public information**, a term that has previously been used more restrictively (see text). This figure presents the lexicon of information that is used throughout this book.

local reproductive success on future habitat choice (Doligez *et al.* 1999). The authors interpreted the lack of effect of patch reproductive success as due to two opposing effects that cancelled each other out: increased attraction and increased competition to recruit to highly successful patches.

### 4.2.3 Types of non-genetic information

There are three ways that non-genetic information can be acquired (Figure 4.3): (1) **parental effects** on the phenotype of their offspring; (2) **personal Information** is gathered by the individual's interaction with the environment; and (3) **social Information** is gathered by observing other individuals (Figure 4.3).

#### 4.2.3.1 *Parental effects . . .*

> The effects of the environment mediated via the parents on the growth and development of offspring are called parental effects (Cheverud and Moore 1994). Most of these effects are transmitted by mothers to offspring so that we often talk of **maternal effects**. However, fathers can transmit information to their offspring. For instance, in some human cultures the transmission of estates is from fathers to the first son.

Through parental effects, parents can thus shape the phenotype of their offspring according to the conditions that prevail in the environment. For instance, experimental **manipulations** of environmental conditions during pregnancy have shown that female common lizards (*Lacerta vivipara*) can adaptively modify the phenotype of their sons and daughters relative to their tendency to disperse from their natal territory (Massot and Clobert 2000). Furthermore, young female common lizards that are unlikely to die before their offspring reach adulthood tend to produce daughters that are more prone to disperse than the daughters of older females (see Figure 10.7). This diminishes the level of kin competition between mothers and daughters. Complementary experiments have shown the causality of that relationship as well as the role of kin recognition in the process of dispersal in that species. Several other examples in common lizards are given in Chapter 10. Through parental effects, individuals can thus transmit some information as they reduce the level of uncertainty about the environment with which their offspring will be confronted. Parental effects allow parents to tune the phenotype of their offspring to local conditions.

*Parental care and parental effects*

Usually, parental effects are mediated by physiological processes. For instance, kittiwakes (*Rissa tridactyla*) are often parasitized by ticks (*Ixodes uriae*) that transmit the Lyme disease spirochaete (*Borrelia burgdorferi*) to their hosts. Julien Gasparini and his collaborators (Gasparini *et al.* 2001) showed that female kittiwakes living in highly infested areas deposit more antibodies against the parasite in their eggs than females from sparsely infected areas. Such deposits have the potential to affect the development of chick immunity (Gasparini *et al.* 2006), and thus can have long-term consequences. Furthermore, the existence of grandparental effects has also been demonstrated (see Section 2.2.3.4), suggesting that these effects can be transferred across multiple generations. More generally, parental effects may involve many molecular and behavioural processes (see Chapter 4 in Avital and Jablonka 2000), some of which may act well before fertilization. However, as Avital and Jablonka (2000) underlined, parental care has almost never been considered as an efficient way of providing offspring with social information about the environment (see Chapter 20).

Such parental effects, whether physiological or behavioural, may play a prominent role in adaptation to local conditions, and hence in evolution. In particular, as parental effects may lead to parent-offspring resemblance if not properly controlled (which is particularly difficult to do) they may be incorrectly incorporated in the measure of **heritability**. We will return to this important issue in Chapter 20.

#### 4.2.3.2 . . . *Personal versus social information . . .*

The second form of non-genetic information is personal information (Figure 4.3). This is the information obtained by an individual's own interaction with the environment. One individual is considered as having more information than another if it has more independent interactions with the environment (Taylor 1998). Personal information can thus be obtained using trial-and-error tactics.

Part of personal information may remain private, in the sense that other individuals may not be able to access it. However, the outcome of an individual's interactions with the environment can often be observed by a bystander who may be able to extract social information (Figure 4.3). Here the term 'social' means obtained from other individuals.

#### 4.2.3.3 . . . *The various forms of social information*

Social information involves three categories that share fundamental characteristics (Figure 4.3). It can be based on **signals**: traits shaped by selection for organisms to modify the behaviour of other individuals to the benefit of the sender. Signals are involved in true communication which will be covered in Chapter 16. Alternatively, social information can be based on social cues produced **inadvertently** by individuals engaged in efficient performance of their activities. This is inadvertent social information (ISI; Danchin *et al.* 2004).

Both social cues and signals share the fundamental characteristic of being derived from **condition-dependent traits that reveal performance**. We thus lump them into performance information, that is **the information that can be derived from the performance of other individuals**. It may involve a wide variety of condition-dependent traits such as phenotypic condition, behaviour, and location of others (Figure 4.3).

Finally, social information also encompasses **social genetic information**: that is information that can be extracted from traits that are unconditionally (or quasi-unconditionally) determined by genes.

Although the question of signalling and animal decision-making is widely analysed, relatively few studies have focussed on the role of social cues in animal behaviour. In this chapter, we focus on the importance of inadvertent social information in animal decision-making, since true communication is the topic of Chapter 16.

#### a *Phenotypic condition and behaviour as inadvertent social information*

Phenotypic condition of other individuals may inadvertently carry information (Figure 4.3). In the common lizard for instance, Aragon *et al.* (2006) showed that a resident can use the body condition of immigrants to make decisions about space use. Furthermore, the behaviour of others may reveal important characteristics of the environment. This is the case of the detection of danger. If an individual in a group spots an imminent danger, his unavoidable reaction produces information to other group members. This is true even in the absence of any specific signalling from the individual that spotted the danger (no alarm call for instance). The simple reaction of fright carries valuable information and it is easy to envisage how flock members that can use such information may be favoured by **natural selection**. The same argument can be extended to the evolution of fright responses in fish exposed to alarm substances (Chivers and Smith 1998).

#### b *Location as inadvertent social information*

It has been proposed that information can be extracted from the **location** of other individuals that may indicate the presence of some resource (Figure 4.3). This has been termed **social** or **conspecific attraction** because using the presence of conspecifics in a patch leads bystanders to join the patch; another term often used is **local enhancement**, which is the same mechanism when the presence of conspecifics is detected from a distance. The classic example is that of seabirds that can detect foraging flocks from several nautical miles away and join the flock. This example is interesting as it shows how important it is to concentrate on the **cue** that actually produces information. Imagine a foraging flock of Peruvian boobies (*Sula variegata*) off the Pacific coast of South America. These seabirds spot fish schools from high in the air and dive straight into the schools, sometimes from more than 50 m above the sea surface. Because foraging booby flocks can involve thousands of diving individuals constantly impacting the sea, foraging flocks can look from a distance as if the sea was boiling below a dark cloud of birds in flight. In such a case, it is likely that what attracts new boobies from a distance is not the mere presence of other boobies, but rather the fact that these boobies are obviously catching fish. Then the information is more in the feeding performance of the observed boobies, than in their presence. This shows that most often, location and quality are inextricably mixed. It may prove difficult and hence often fruitless to try to separate these two components.

### c *Behavioural performance as inadvertent social information*

Further inadvertent social information may be extracted from the direct observation of the success and failure of other individuals (Danchin *et al.* 2001, 2004). As the performance of other individuals in a given activity is a component of fitness, its variation directly reflects environmental variation in quality. Performance may thus produce valuable information about the current state of the environment. Imagine a maturing individual just about to reproduce for the first time. That individual may have no clue of how good habitat looks and habitat quality may be very hard to assess. Evolutionarily, the question such an individual has to answer is 'where will I have the highest reproductive success?' An answer may be simply 'where conspecifics have the highest success.' Indeed, the reproductive performance of conspecifics shows that the environment was favourable throughout the breeding period. Similarly, a hungry individual searching for a foraging patch may be able to observe several conspecifics foraging for food with varying efficiency. It is clear that an individual animal capable of detecting and using such variation in the foraging success of others is likely to be favoured by natural selection. Thus, the performance of other individuals exploiting the same resource may be an excellent predictor of local conditions.

### d *The relative reliability of social cues and signals*

The outcome of an activity often affects fitness. Performance reveals the interaction between the **genotype** and the environment. Thus, it encompasses two intertwined and often correlated components: environmental quality and the performer's capacity to cope with that environment. Performance of others is thus a good predictor of environmental conditions, or a good social cue revealing the observed performer's quality.

In ISI, performance inadvertently produces information to a bystander that reveals either environmental conditions or the performer's quality. In **habitat copying**, for instance (Section 4.3.1.2aa and Chapter 9), variation among patches in the performance of breeders reveals among-patch variation in current environmental conditions. In **mate choice copying** or **eavesdropping**, within patch variation in male performance reveals within-population variation in male quality. From the bystander's point of view, the reliability of ISI resides in the fact that ISI is not intentionally produced; information producers are selected to perform as well as possible, rather than to manipulate the behaviour of other individuals to their own benefit. In other words, the reliability of ISI resides in the fact that it is based on fitness components of others. ISI thus provides rich and reliable information in that it usually predicts the fitness of choosing various alternatives. The benefits of using ISI are that it reduces the costs associated with trial-and-error learning and provides additional information that can lead to more accurate estimates of environmental parameters. For instance, we will see in Section 9.1.1 that in the case of breeding habitat choice the costs of personal information may be prohibitive relative to ISI.

In contrast to cues, signals are reliable because they are costly to produce (Zahavi 1975). Only individuals in good condition can produce them, so that signals also reveal performance. The difference between social cues and signals is that signals are the result of selection, whereas social cues are not (Figure 4.3). In signalling, individuals expect a benefit from the receptor's reaction to the information carried by the signal. Signals have thus been shaped by selection to reveal performance, while ISI is extracted from non-modified fitness components.

#### e *Social genetic information*

Some components of phenotypic variance that are rather robust to environmental variation may also produce valuable information about an individual, particularly in the context of mate choice (Figure 4.3). For example, eye colour in humans is relatively robust to variation in environmental conditions and so phenotypic differences are due almost exclusively to genetic differences among individuals. As a result, eye colour provides a form of social information we call social genetic information. Animals might also use such environmentally robust phenotypic information to assess the contents of another individual's genotype, something useful in mating and in estimating genetic kinship. For instance, in both mice and humans some body odours are closely related to an individual's major histocompatibility complex (MHC), a complex of genes that is involved in immunity. Both mice and humans are sensitive to these odours and use them to estimate their genetic similarity to others (see Figures 19.6 and 19.7, and Wedekind *et al.* 1995).

Some authors have recently suggested that mating strategies in songbirds and waders depend on the genetic similarity between social mates (Blomqvist *et al.* 2002; Foerster *et al.* 2003; Freeman-Gallant *et al.* 2003). However, relatively little is known about the role of social genetic information in animal decision-making. Social genetic information is the only form of social information (Figure 4.3) that is condition-independent and so does not reveal performance.

#### 4.2.3.4 *Public information and inadvertent social information*

In a seminal paper, Thomas J. Valone (1989) proposed that individuals could extract information about the quality of the resources exploited by others from their behaviour and called this 'public information'. Since then the expression has been used to describe several superficially similar but slightly different kinds of information. Giraldeau (1997) for instance used the term to describe all information that can be gained from observing others, as has Galef and Giraldeau (2001). However, Valone and Templeton (2002) view the term as restricted only to cases when the information concerns the quality of the resource, while Danchin *et al.* (2004) introduced the concept of inadvertent social information (ISI), which encompasses any information extracted from the location and the performance of others. The classification of the various types of social information has lacked consensus (Dall *et al.* 2005; Danchin *et al.* 2005) and the time is ripe to attempt some clarification of the terminology.

The term 'public' is used mostly to refer to anything that is non-private. We propose therefore to use the term 'public' to mean exactly that: the opposite of private (Figure 4.3). Public Information therefore encompasses all sources of information that are available to all. It thus can encompass several components, some that are non-social such as the height of a tree, the position of the sun in the sky, the physical characteristics of a marsh or of a cliff, etc. Others will be social as is the case of what we term performance information, information that can be extracted from the performance of other individuals (Figure 4.3) whether it is produced as a deliberate signal or is simply a cue emitted inadvertently. All the chapters in this book use definitions of information according to the terms provided in Figure 4.3 and in this section.

#### 4.2.3.5 *The performance of others carries the information*

When performance information involves the use of signals it is part of communication and we will not consider it further in this chapter. However when performance information is inadvertent, that is, it is not based on an evolved signal but merely an unavoidable by-product of resource exploitation and other behaviours, then it is inadvertent social information (ISI). Although signals and communication are well studied in animal decision-making, the role of the inadvertent component of information in animal decision-making has been largely ignored (Lotem *et al.* 1999). Section 4.3 reviews the recent evidence that ISI is an important component of animal decision-making in general.

#### 4.2.3.6 *Various situations of inadvertent social information use*

The use of ISI can vary from situations where the bystander parasitizes the performer's information (involving a cost to the performer), to commensalism (when the bystander's use of ISI is neutral to the performer), to mutualism (when both actors benefit). The use of ISI in cases when there are costs or benefits to the inadvertent information producer may create selective pressures on the information producer. When the producers of ISI suffer a cost, they are under selective pressure to conceal the information. For instance, in a foraging context, individuals capable of eating without revealing that they just found a food patch may avoid losing food to nearby watchful competitors. However, in many situations the inadvertent information producer simply cannot avoid producing information because doing so would reduce their fitness even more. This is the case, for instance, of inadvertent information given off by fright and flight responses in situations of danger.

Situations when ISI use leads to mutualism are also interesting as they may lead to the evolution of signalling. Over evolutionary time, the formerly inadvertent information producer may be selected to provide a signal that more effectively elicits the behaviour of bystanders to its own benefit: communication. For instance, an individual having found a large food patch that would be easier to exploit in a group than alone may be selected to inform others of its discovery. This may well be how food calls have evolved (Elgar 1986; Stoddard 1988). ISI may therefore under some circumstances set the stage for the evolution of true communication (Danchin *et al.* 2004, and Chapter 16, Section 16.3.3.2 in particular).

### 4.2.4 Summary: the four sources of individual variation

It follows from this section that organisms have four main sources of phenotypic variation, each resulting from variation in one of the four sources of information: genetic, parental, personal, and social information, of which the last three are non-genetically acquired.

Every classification has its pros and cons. A potential drawback of the classification proposed in Figure 4.3 is that classifying implies the definition of discrete boxes with discrete borders while the reality is more like a continuum. This is clearly the case with this lexicon of biological information. Here, for instance, we detailed the various forms of non-genetic information for the sake of completeness, but excessively detailed classification may be impracticable in specific situations. The classification of public information we provide here does away with some of the difficulties with earlier classifications. Take the example of the Peruvian booby developed above, in which an individual may spot two foraging flocks that appear equally promising from a distance. Probably the individual will join the foraging flock that is closest to save time and energy. This shows how location and quality may be indistinguishable, and the cues used as sources of information most often reveal location and quality simultaneously. Location of a food patch for instance is probably almost always associated with information on its potential quality and vice versa. Similarly, most animals often use simple strategies such as win–stay, lose–shift. Hence, the presence of individuals may often be synonymous with recent success.

However, theoretical approaches show that the situation may be more complex than that. For instance, in the context of breeding habitat choice, Doligez *et al.* (2003) found that strategies using different cues as sources of ISI did not show similar evolutionary stability (see Figure 9.4). In their game theoretical model, strategies using the breeding performance of conspecific breeders won over a strategy using the mere presence of conspecific breeders as a cue. This example suggests the importance of focussing on the specific cue that is used by animals to obtain information on the current state of the environment. What matters in the end for a classification of information is its value in predicting the future. The performance of currently behaving individuals may most often constitute a good predictor of the future. The classification we propose here does away with the difficulty of distinguishing quality and location information. These are subsets of performance information which themselves are a subset of public information (see Figure 4.3). This is achieved by focussing on the cues that are used as sources of information rather than on how it is used.

## 4.3 Inadvertent social information and decision-making: the evidence

Evidence has accumulated that not only animals but also plants use ISI. This section provides a brief overview of the various types of fitness-enhancing decision in which the use of ISI has been demonstrated.

### 4.3.1 Inadvertent social information and foraging

It is in the context of foraging that Valone (1989) proposed that animals can use the sampling of the environment by other individuals (called vicarious sampling) as a valuable source of information about the quality of the environment. This idea refines the reasoning of several previous authors. The information centre hypothesis (ICH) posited by Ward and Zahavi (1973) proposed that unsuccessful foragers can, by returning to the communal site, extract information from successful foragers, leading to group formation. This idea has been seriously challenged (Chapter 14), but may be involved in the maintenance of coloniality (Barta and Giraldeau 2001). Similarly, in his important series of experiments, Bennet J. Galef, the Canadian experimental psychologist from McMaster University, showed that some social information was involved in the transmission of a desire for novel food in Norway rats (*Rattus norvegicus*; see, for example, Galef *et al.* 1984). Similarly, as early as the early 1980s (Tinbergen and Drent 1980), it had been suggested that socially foraging animals may observe the sampling activities of other flock members while foraging.

#### 4.3.1.1 *Social foragers monitor conspecifics*

##### a *Foraging efficiency by starlings according to ISI availability*

The first experiments showing that wild animals can and do use ISI to make decisions about food patches were those of Jennifer Templeton, a Canadian scientist now at Knox College in the United States. Valone had performed some theoretical analyses of ISI use (Valone and Giraldeau 1993) and had failed to show experimentally that budgerigars (*Melopsitaccus undulatus*) could use ISI in their social foraging decisions. Then, using an elegant experimental approach, Templeton showed that European starlings (*Sturnus vulgaris*) left patches early or late depending on the quantity of ISI produced by a flock member (see Templeton and Giraldeau 1995). She manipulated the quantity of information produced by a companion by training it either to probe into many of potential patches or to probe in just three. All starlings were observed while the experimenter had provided them with totally empty foraging environments and noted the number of probes a focal individual would make before giving up the patch as empty. When the focal bird foraged with a companion that produced copious amounts of ISI, the focal bird gave up sooner as if the consistent lack of success of its companion produced information that the environment was empty. However, when the focal bird was placed with a low information companion that probed only three holes, it stayed longer and probed more holes before giving up. Finally, focal birds tested alone were the ones that stayed the longest and probed the most holes before giving up the empty environment. These results are consistent with focal birds using the lack of success of their companions as a means to assess the value of the environment they were currently exploiting.

### b *General characteristics of experiments on information*

In this first example, Templeton manipulated the amount of ISI available to focal birds. She achieved that goal by putting the focal bird in the experimental setting either (1) alone, or (2) with a companion trained to produce low information (only probing a few holes), or (3) with a companion that produced copious amounts of ISI (probing many holes). In other experiments, authors manipulated the accessibility of the cues. This was the case in another study on the starling in which the food was provided below an opaque cup that was either shallow (Figure 4.4a), or deep (Figure 4.4b). In the former case, other flock members could gather ISI from the companions, whereas in the latter case, they could not.

This shows one of the main characteristics of all the approaches reported in this section aiming at reviewing the many fitness enhancing decisions that have been shown to involve ISI use: such experiments always aim at manipulating the social cue that is suspected to be used by animals to extract information. This is because, as we have seen, **information is extracted from a cue**.

Since those experiments, others have been performed in the context of foraging that lead to similar conclusions. For instance, Julie W. Smith and collaborators (Smith *et al.* 1999) showed that red crossbills (*Loxia curvirostra*) also use ISI to assess the quality of a feeding patch. Red crossbills exploit seeds hidden within pine cones that individuals constantly probe while foraging. In an elegant experiment (Figure 4.5) they showed that individual foragers obviously monitor their foraging flock mates and are thus able to decide to leave the patch in a much more efficient way when in a group than when alone (Figure 4.5).

Figure 4.4 **How to manipulate ISI**

Starlings exploit prey that are hidden in the soil such that they must constantly probe to check whether or not there is food in a given patch. In the experiments, the food was hidden below a slit at the bottom of the two types of cups that the birds had to manipulate in order to detect if food was present.

a. Situations when flock members' feeding performances were detectable.
b. Because the cup was opaque, foragers could not extract ISI from their flock members. Using this technique, Templeton and Giraldeau were the first to evaluate the relative weight of personal and inadvertent social information. Birds that were in situation **b.** behaved as if they were foraging alone. Those in situation **a.** made faster decisions to depart empty patches.

From Templeton and Giraldeau (1995).

#### 4.3.1.2 *Pilfering jays that project their own behaviour on others*

Another fascinating series of experiments that implicitly used ISI is that of British scientist Nicola S.

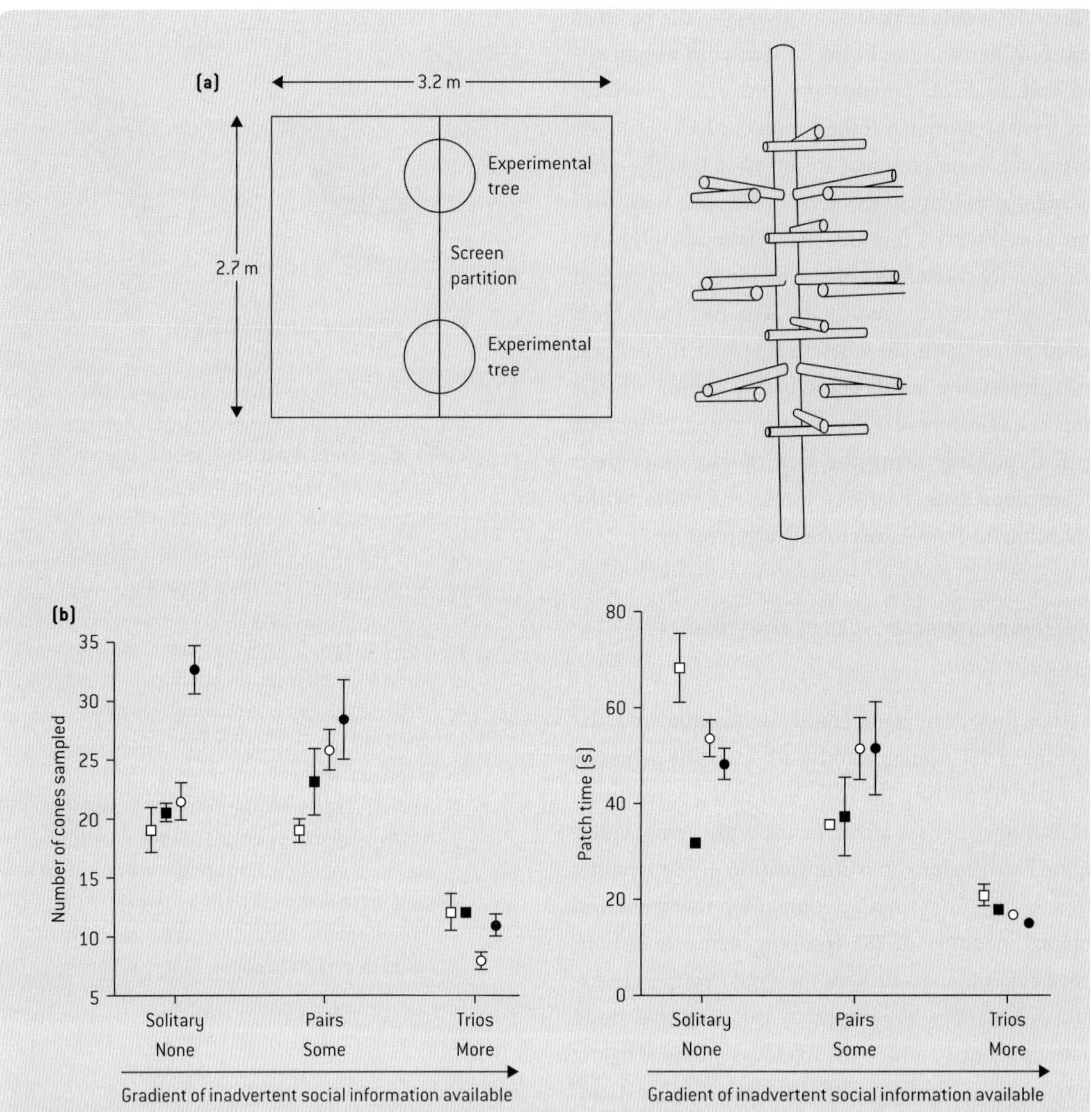


Figure 4.5 **Red crossbills use ISI to assess the quality of food patches**

**a.** The experimental setting used a two patch system. Each patch was made of an artificial tree on which cones were placed. Cones on one of the trees were empty, while every cone on the other tree contained seeds. Experimenters manipulated the amount of ISI by manipulating the size of the foraging flock: birds were either alone (no ISI), in pairs (some ISI) or in trios (more ISI).

**b.** When foraging with two flock mates, birds sampled approximately half the number of cones on the empty patch before departing as compared with solitary foragers. Consequently they departed more rapidly from the empty patch. Interestingly, as expected if ISI had been used, the variance in both the number of cones sampled and the time spent on the empty patch decreased in group foragers.

Modified from Smith *et al.* (1999).

Clayton's team in Cambridge, UK. During her research in the University of California at Davis, she noticed that food-storing scrub jays (*Aphelocoma coerulescens*) were often pilfered by others. She then brought jays to the laboratory to study the details of such behaviour. She and her students first showed that food-caching jays can remember what, where, and when they cached. She also observed that jays remembered where conspecifics had cached and pilfered these caches when the opportunity arose. Nathan J. Emery, working with Clayton, raised the question of whether pilferers could adjust their own caching strategy in order to minimize their own losses to pilfering.

The experiment comprised a total of 21 hand-raised scrub jays, all of which had the same experience of caching and recovering food. During one session, the birds could cache food items in a tray (Figure 4.6) and were offered the opportunity of re-caching these items in a subsequent session. The first caching session occurred either privately while the bird was alone or when another jay was seen observing the caching bird. The re-caching session was always done in private without other jays observing.

The results were amazing (Figure 4.7). Birds that cached in private did not tend to re-cache food items during the second session, and when they did so they often re-cached the item in the same place

Figure 4.6 **Scrub jays in the experimental cages**

Birds could cache food in the compartments of the plastic tray.

Picture provided by Nicky S. Clayton.

(Figure 4.7a). Birds that cached while being observed behaved quite differently during the re-caching session. Such birds re-cached the items much more often than those that cached in private, and when they did so they always re-cached in another place (Figure 4.7a). This first difference shows that individual jays can react to the potential threat of being pilfered by re-caching the food when they know that others have observed them caching.

The above experiment shows how individuals may parasitize ISI and how this leads to counter-adaptations. Clayton's group found another fascinating result that indicated how such defence was acquired. There were two groups of birds differing in their former experience as pilferers. Some birds had had an experience of pilfering other birds' caches. Others were naive relative to pilfering as they had never had the opportunity to pilfer others. Those two groups did not differ in the private situation (compare the right part of Figure 4.7b with that of a), but they differed significantly in the observed situation: only the experienced pilferers re-cached food during the second session of the experiment. The naive birds that cached in the observed situation did not re-cache food differently from those in the private situation (compare the left part of Figure 4.7b with that of a).

Emery and Clayton's experiments show that ISI can lead to subtle condition-dependent types of decisions. It also shows that ISI is a concept that may help link animal decision making and psychology. Mental time travel (episodic memory and future planning) and mental attribution were thought to be unique to humans. Although further work is needed to demonstrate this definitively, it seems that these skills may not be unique to humans. It could be that scrub jays are able to project their own experience onto other individuals. It seems as though they may have a reasoning process like: '*If I can pilfer others, others can pilfer me. I'd better protect myself against such a risk*'.

#### 4.3.1.3 *Especially competitors, including heterospecifics, may produce ISI*

Several studies extended the concept of ISI to cases involving two species exploiting the same kind of

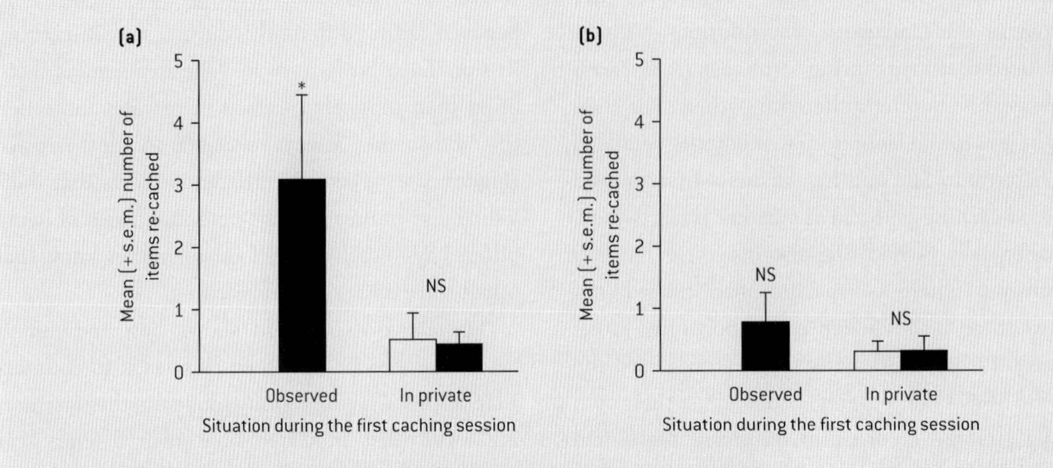


Figure 4.7 **The caching strategy of scrub jays is highly experience-dependent**

Experiments involved two categories of birds. Experienced pilferers had already pilfered others' caches, and inexperienced pilferers never had the opportunity to pilfer others. White bars: re-caches in the same compartment of the tray; black bars: re-caches in another compartment of the tray. * $N = 7, P < 0.05$. NS: non-significant.

**a.** Experienced pilferers that cached food in private did not re-cache food items when given the opportunity (right bars), while experienced pilferers that were observed while caching did re-cache in a new site when given the opportunity (left bar).

**b.** Relative to experienced pilferers, inexperienced pilferers did not show different behaviour after having cached in private (right bars), but they did not re-cache food items after having cached while being observed by another jay (left bar).

Modified after Emery and Clayton (2001).

resource. For instance, Carib grackles (*Quiscalus lugubris*) imitate Zenaida dove tutors (*Zenaida aurita*; Lefebvre *et al.* 1997). Furthermore, individuals of the same dove species learn better from Carib grackles than from conspecifics (Dolman *et al.* 1996). More recently, Isabelle Coolen, a French scientist then working with Kevin Laland in the UK, performed experiments with sticklebacks (Coolen *et al.* 2003). These authors showed that nine-spined sticklebacks (*Pungitius pungitius*) used the foraging performance of three-spined sticklebacks (*Gasterosteus aculeatus*) to choose their foraging patches (Figure 4.8). After observing two groups of three three-spined sticklebacks feeding at a poor and a rich patch for 10 minutes, nine-spined stickleback observers first moved towards the feeder that provided the highest amount of food and spent more time in its close vicinity over five minutes (Figure 4.9).

In another experiment, Coolen *et al.* (2005b) also ruled out the possibility that, in the same experimental setting and in the absence of any demonstrator, fish can cue on remains of bloodworm molecules or not. They did so by giving nine-spined sticklebacks the choice between one feeder that provided bloodworms and another that did not. In such a design, nine-spined sticklebacks did not show any preference for the side where bloodworms were delivered, thus showing that the results reported in Figure 4.9 did not result from the use of olfactory cues by observing nine-spined sticklebacks. Altogether, such results showed that individuals can monitor the behaviour of heterospecifics and make decisions according to the performance of heterospecific demonstrators.

Coolen's experiment raises the interesting idea that even competitors can produce valuable information precisely because they exploit the same resources, which is why they are competitors in the first place. Monitoring individuals exploiting resources that differ from the resources necessary

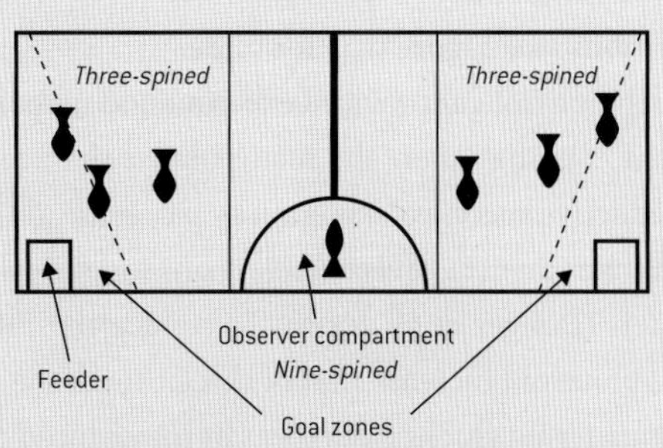


Figure 4.8 **Experimental design for the testing of heterospecific ISI use in sticklebacks**

The design involved a tank divided into compartments. The borders of the compartments were transparent, except the middle one (thick line). A nine-spined stickleback was placed in a central glass bulb compartment where it could observe the two other compartments, each with three foraging three-spined sticklebacks. The two feeders provided bloodworms at different rates. The rich one provided three times more food than the poor one. However, while the food was delivered on the rich side, a similar amount of water and bloodworm juice was delivered to the feeder on the poor side to control for any movement of the feeder and for possible residual chemical cues on the rich side. Feeders were designed so that the observing fish could not see the food but could only see the foragers feeding. Therefore, the only clue to the observer was the success of the foragers. The observation period lasted for 10 minutes. After that period, demonstrators were removed as well as the glass bulb that delimited the observer compartment. Researchers then measured the proportion of observer fish that entered the rich goal zone first and the proportion of time they spent in that goal zone over a test period of 5 minutes (see Figure 4.9 for results).

Modified after Coolen *et al.* (2003).

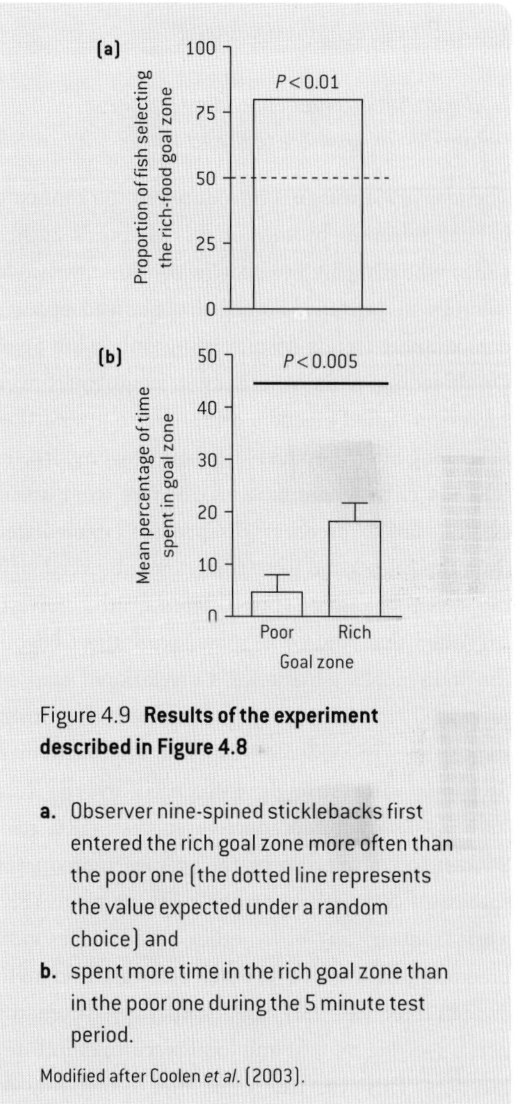


Figure 4.9 **Results of the experiment described in Figure 4.8**

**a.** Observer nine-spined sticklebacks first entered the rich goal zone more often than the poor one (the dotted line represents the value expected under a random choice) and
**b.** spent more time in the rich goal zone than in the poor one during the 5 minute test period.

Modified after Coolen *et al.* (2003).

for the observer would not produce useful information to the observer and we would not expect inter-specific use of ISI in those cases. The result of Coolen *et al.* (2003) thus leads to the extension of the definition of ISI to encompass inter-specific cues.

### 4.3.2 Inadvertent social information and breeding habitat selection

The choice of a breeding patch is a another important fitness-affecting decision in life history, perhaps even more so in spatially constrained species, such as sessile organisms, or birds and seals. As we will see in Chapter 9, the study of breeding habitat selection is made particularly difficult by the time and spatial scales involved. This is probably why evidence of ISI use in the breeding context came after evidence of its use in foraging.

#### 4.3.2.1 *Breeding habitat copying*

##### a *Birds that prospect breeding patches before deciding where to breed*

The idea of ISI was first introduced in the context of habitat selection by one of this chapter's authors (É.D.) who compared two kittiwake (*Rissa tridactyla*) colonies in Brittany, France that showed opposite demographic trends despite being only 2 km apart. On breeding cliffs where most neighbours were successful, failed breeders continued to attend their nests long after they had lost their progeny. In contrast, on cliffs where most neighbours were unsuccessful, failed breeders disappeared immediately from their nests and were sometimes observed visiting other colonies. This difference in behaviour suggested that there was some social effect beyond the individual's experience. Correlational analyses presented in Section 9.5 confirmed the initial impression. Individuals' dispersal and settlement decisions were strongly influenced by the focal individual's own reproductive success (IRS, *sensu* Switzer 1997), but also by the average reproductive success of conspecifics on that cliff, or patch (PRS, *sensu* Switzer 1997). In other words, the mean breeding performance of other kittiwakes on a cliff (which produces ISI) significantly predicted dispersal (see Figure 9.6) and local recruitment of new breeders. This was true even after controlling for the effect of individual reproductive success and other potentially confounding effects such as sex or year.

Juvenile kittiwakes, failed breeders, and non-breeders appear to prospect breeding colonies (Cadiou *et al.* 1994) to gather ISI in order to choose where to breed in the following year. We call this 'habitat copying' because it leads prospectors to copy the habitat choice of successful conspecifics. Thus, most prospecting in kittiwakes occurs when PRS can be estimated accurately in a single visit to the colony that occurs towards the end of a breeding season (Boulinier *et al.* 1996). Both simple optimality (Boulinier and Danchin 1997), and ESS models (Doligez *et al.* 2003) confirmed that habitat copying was evolutionarily expected under most natural conditions (see Figure 9.4) for birds.

Blandine Doligez provided experimental support for habitat copying when she manipulated the average reproductive success of a migratory passerine, the collared flycatcher (*Ficedula albicollis*) on Götland Island, Sweden (Doligez *et al.* 2002). To reduce the PRS of three plots and increase that of three others she simply transferred chicks from nests of the reduced plots to those of the increased plots. Her results confirm that individual flycatchers use ISI in their dispersal and settlement decisions (see Figure 9.7). Similar results were obtained in the blue tit (*Cyanistes caeruleus*), a non-migratory passerine (Parejo *et al.* 2007).

These results were extended recently to synchronously breeding birds that cannot obtain information about the number of young produced by neighbours but only the location of their nests. Nocera *et al.* (2006) experimentally deployed decoys and song playbacks of breeding males to produce ISI in either suitable or sub-optimal habitats during both pre- and post-breeding periods. They monitored territory establishment during the subsequent breeding season for a social bird, the bobolink (*Dolichonyx oryzivorus*), and a more solitary species, Nelson's sharp-tailed sparrow (*Ammodramus nelsoni*). The more solitary sparrows did not use ISI but the more social bobolinks responded strongly to post-breeding ISI irrespective of habitat quality. The following year, 17 of 20 sub-optimal plots to which bobolink males were recruited based on ISI use were defended for at least two weeks. Sixteen recruited males were natal dispersers, as expected if ISI was used when individuals have little opportunity to sample their natal habitat quality directly.

##### b *Lek-breeding antelopes that use ISI to select mating territories*

Evidence that animals use ISI in breeding habitat selection was also obtained for mammals: two antelope species in Uganda, the kob (*Kobus kob thomasi*) and kafue lechwe (*Kobu leche kafuensis*; Deutsch and Nefdt 1992). Although the studies do not explicitly place the results in the context of ISI their results nonetheless provide evidence that females use ISI to choose their breeding territory. Both species mate on **leks** where oestrus females congregate on just a few territories where the most matings occur. Although males that defend the most successful territories are replaced at one or two-day intervals, territories remain successful for months. The mating rate of different males holding the same territory are thus highly correlated and when a male changes territories, his success on the new territory is predicted by the success of his predecessor on that territory and not his own previous success. Hence females appear to select males based on their location rather than on their quality. The studies do not explicitly describe male–male competition for the best mating territories. However, because male ownership of such territories may change every day, it is likely that males fight intensely to possess them, as is true in many lekking species. Hence, by mating with males on such territories, females probably indirectly favour highly competitive males. After ruling out several alternative hypotheses, Deutsch and Nefdt (1992) concluded that females are attracted to territories on the basis of a territory's olfactory cues. Females urinate when sexually stimulated thereby producing olfactory cues about the visitation rate on a given male territory. Oestrus females that arrive at leks start off by repeatedly smelling the ground of male territories. The successful territories, which emit a strong smell, are dotted with yellow patches where females have deposited urine marks. The mean number of females on successful territories ends up being 35 times higher than in unsuccessful territories.

To test the idea that females are attracted by the olfactory cues of heavily used territories, they performed two separate experiments in the two species. After seven days of observation, each of the two most successful territories was paired with a neighbouring unsuccessful territory and experimental and control pairs were chosen at random. Deutsch and Nefdt (1992) then removed about 250 kg of soil and dead grass from each of the four territories, then exchanged the earth between two pairs of experimental territories, while they replaced the earth on the four controls. This procedure was repeated on four new territories each week for four weeks. In both species, the number of matings on the unsuccessful territories, which received the earth from formerly successful territories, increased more than tenfold (Figure 4.10), whereas mating success on control and unmanipulated territories remained unchanged.

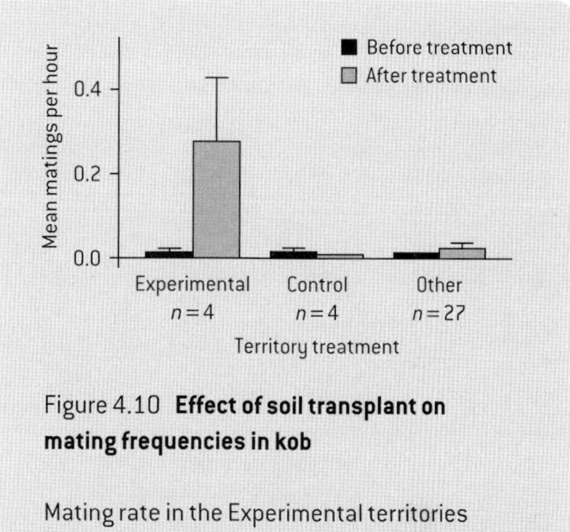

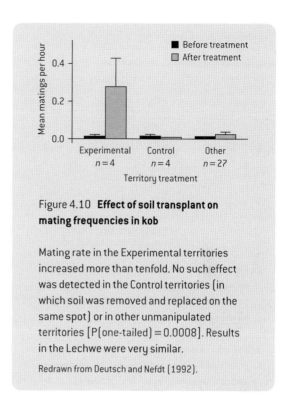

Figure 4.10 **Effect of soil transplant on mating frequencies in kob**

Mating rate in the Experimental territories increased more than tenfold. No such effect was detected in the Control territories (in which soil was removed and replaced on the same spot) or in other unmanipulated territories [P(one-tailed) = 0.0008]. Results in the Lechwe were very similar.

Redrawn from Deutsch and Nefdt (1992).

These results support the hypothesis that females of these two antelope species use olfactory cues to choose mates. This shows that according to the situation, social cues at the origin of ISI can be indirect. Those authors also report a result that suggests that the form of information involved is ISI. Indeed, in lechwe, the increase in mating success on territories to which earth had been transferred was significantly correlated with the former success of the territories from which the earth had been removed (Figure 4.11). In other words, olfactory cues carried information on past mating success on that territory.

#### 4.3.2.2 *Inadvertent social information use in brood parasites*

A recent development of the use of ISI in fitness-affecting decisions was proposed by Hannu Pöysä,

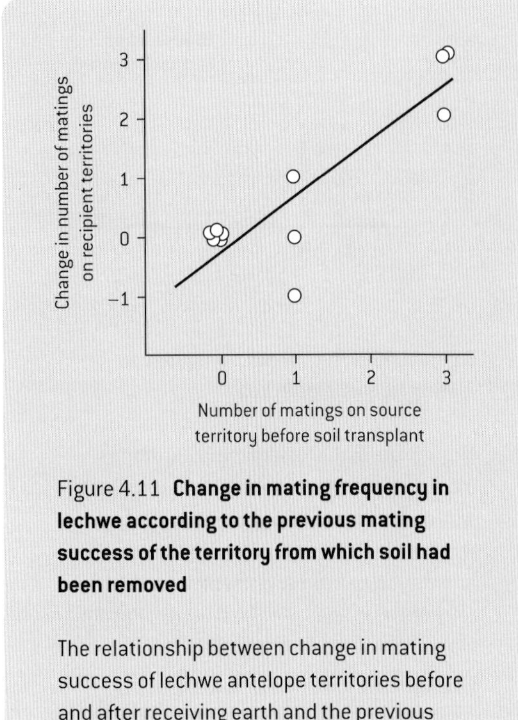

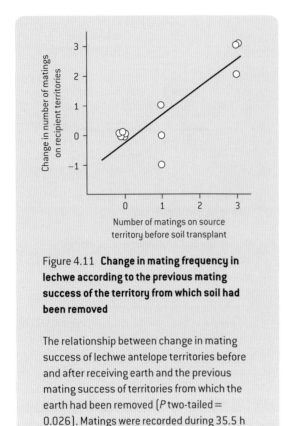

Figure 4.11 **Change in mating frequency in lechwe according to the previous mating success of the territory from which soil had been removed**

The relationship between change in mating success of lechwe antelope territories before and after receiving earth and the previous mating success of territories from which the earth had been removed ($P$ two-tailed = 0.026). Matings were recorded during 35.5 h before and 25.5 h after manipulation.

From Deutsch and Nefdt (1992).

a researcher at the Finnish Game and Fisheries Research Institute working on the common goldeneye (*Bucephala clangula*), a cavity-nesting duck that has a significant level of intra-specific brood parasitism (Pöysä 2006). In that species, some females follow a mixed strategy that consists of laying eggs in their own nest while also laying supplementary eggs in neighbouring nests of conspecifics. In previous experiments, Pöysä showed that parasitic females preferentially lay eggs in safe nest sites, implying that nest predation was an important ecological determinant of conspecific nest parasitism. In a subsequent experiment, he focussed on the nature of the information used by parasitic females to decide where to lay their eggs. At the end of the nesting season, female goldeneyes prospected active nest-sites more frequently than nest-sites that did not have a nest in the current season (Figure 4.12a). An amazing result was that nest-sites that had been prospected more frequently by females in one year had a higher probability of being parasitized in the following breeding season (Figure 4.12b). This suggests that parasitic females prospect active nests at the end of a breeding season and return to lay parasitic eggs in the formerly most successful ones. This enables brood parasites to target safe nests. These findings provide a new kind of fitness enhancing decision in which ISI may be involved. Furthermore, it may create a new dimension for the study of the evolution of conspecific brood parasitism. Such a mechanism has recently been extended to obligate brood parasites such as cuckoos by two Spanish researchers, Deseada Parejo and Jesús Ávilés (Parejo and Avilés 2007; see Section 1.3.3.2.d).

### 4.3.3 Inadvertent social information and mate choice

The choice of a mate is a crucial decision for all sexually reproducing organisms. This is the topic of Chapter 11. The literature on mate choice mainly focussed on intentional information carried by secondary sexual signals. However, in recent years, several approaches focussing on the nature of the information that is used in mate choice have underlined that ISI may also play an important role and involve mate choice copying and eavesdropping.

#### 4.3.3.1 *Mate choice copying*

Mate choice copying (or simply mate copying) is hypothesized to provide time constrained or inexperienced individuals with a means to increase the success of their mate choices by copying the mate choices of more experienced or less time constrained conspecifics. This idea emerged in the late 1980s to explain the strong skew in male mating success in avian lekking species. Convincing experiments were later developed in the 1990s and produced support at least in two animal taxa that prove more suitable for lab experiments: fish and birds.

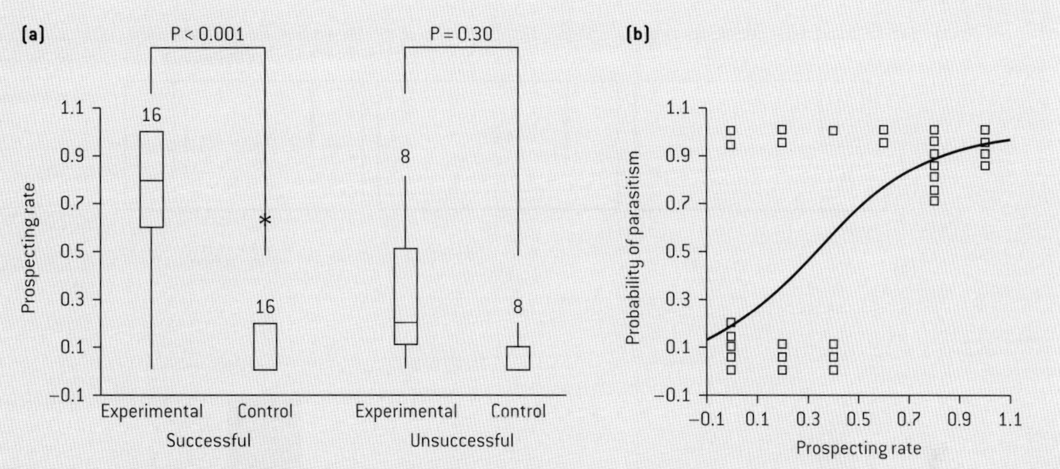


Figure 4.12 **Habitat copying in the facultative brood parasitic common goldeneye**

Each experimental replicate included a pair of close nest-boxes, one manipulated and one untouched control box. There were two treatments. In the successful treatment one of the nest-boxes (the experimental one) received an apparent cue of breeding success (egg shells). The other nest was not manipulated. In unsuccessful nest-boxes no such cues were provided. One was randomly selected as the experimental nest-box, the other becoming the control box.

**a.** Rate of prospecting in pairs of experimental nest-boxes (control and experimental). The statistical comparison between experimental and control successful nests was significant ($P < 0.001$). The same comparison in unsuccessful nests was non-significant ($P = 0.30$). Horizontal bar: median. Box: interquartile range. Whiskers: full range. Sample sizes of nests are given above the box plot. The star shows the position of an outlier in the successful control.

**b.** The probability of being parasitized clearly increased with the prospecting rate the year before (logistic regression: $P < 0.001$).

Redrawn from Pöysä (2006).

### a *Females that reverse their decisions after observing the choices of other females*

Mate choice copying experiments in fish usually involve compartmented tanks, opaque and transparent partitions, two males with contrasting phenotypes ($M_x$ and $M_y$) placed in opposite compartments and females that are willing to express their preference (Figure 4.13a). These experiments are conducted in three sequential steps.

1. Step 1, the natural preference of the observing female ($F_o$) relative to the two male phenotypes is measured and we label the male phenotype that is preferred by **most** females as $M_y$.
2. Step 2, the same observer female $F_o$ is placed in a transparent glass tub in the centre of the tank that prevents her from interacting with the males. Two demonstrator females $F_d$ are placed in compartments adjacent to each male but opaque partitions allow $F_o$ females to see both males but only one of the females (Figure 4.13a.), the one that is adjacent to $M_x$ the initially less preferred males. Seeing a female with this male is supposed to provide the observer female with the information that $M_x$ is more attractive to her than $M_y$ because another female apparently chose him over $M_y$. It is important to place a female next to the preferred male even though the observer female cannot see her to insure that both males

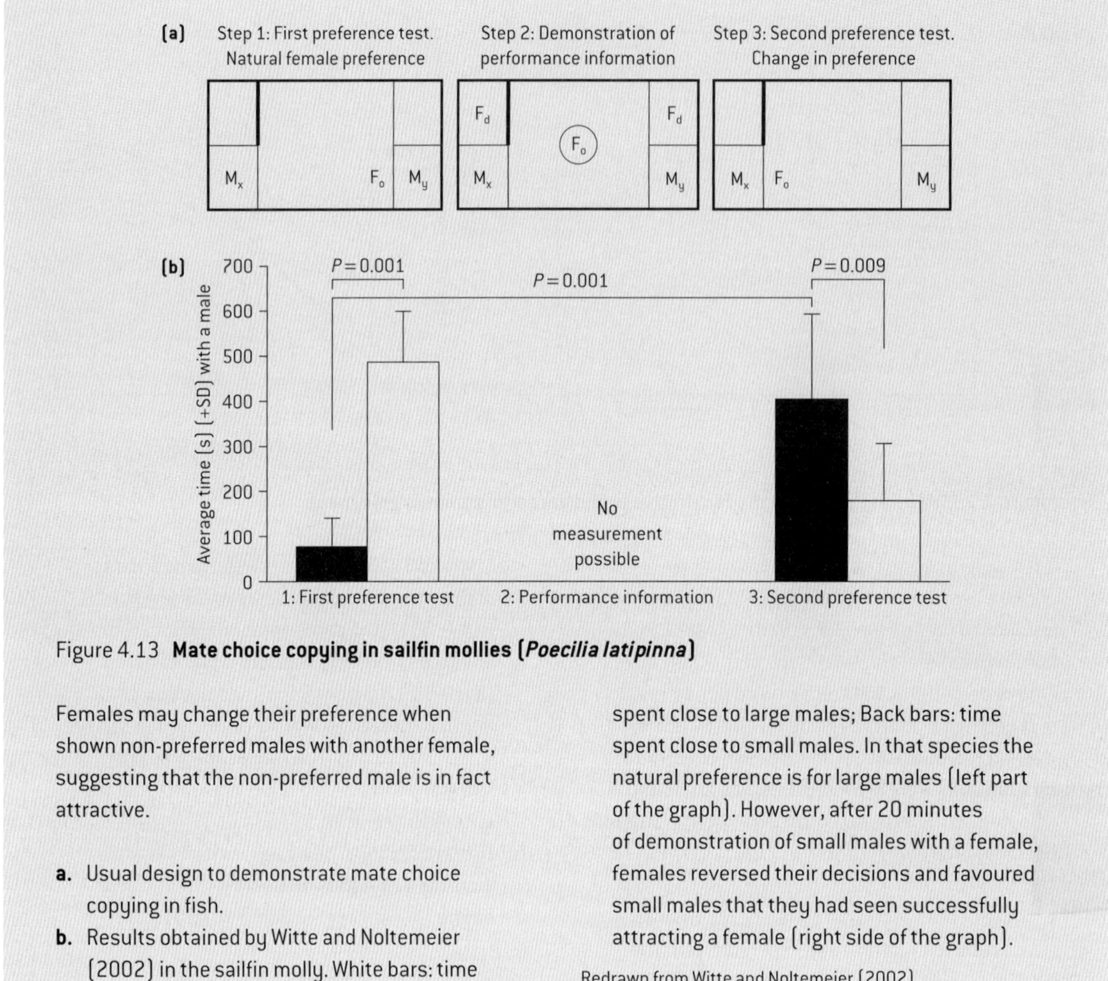


Figure 4.13 **Mate choice copying in sailfin mollies (*Poecilia latipinna*)**

Females may change their preference when shown non-preferred males with another female, suggesting that the non-preferred male is in fact attractive.

**a.** Usual design to demonstrate mate choice copying in fish.
**b.** Results obtained by Witte and Noltemeier (2002) in the sailfin molly. White bars: time spent close to large males; Back bars: time spent close to small males. In that species the natural preference is for large males (left part of the graph). However, after 20 minutes of demonstration of small males with a female, females reversed their decisions and favoured small males that they had seen successfully attracting a female (right side of the graph).

Redrawn from Witte and Noltemeier (2002).

are engaged in courtship keeping behavioural differences between the two at a minimum.

3. Step 3, the demonstrating females $F_d$ are removed, and $F_o$ is released so that her preference can be tested again in the same situation as in step 1 (Figure 4.13a).

Such an experiment in the sailfin molly (*Poecilia latipinna*) clearly showed that providing such ISI to females influences their future mate preferences (Figure 4.13b). Similar results were found in the Japanese quail (*Coturnix coturnix japonica*) by Galef and colleagues in a series of well-designed experiments. Galef's group even showed that one could use such a design to create female preference for artificially produced traits, such as a red or blue patch on the male's chest (Galef and White 2000). This was rather unexpected because most models of the evolution of male traits and female preference assume that variation in mating preferences is only due to variation in genes. Mate choice copying strongly suggests that genetic variation is not the only cause of variation in sexual preferences. We will develop this important issue in Chapter 20.

#### 4.3.3.2 *Eavesdropping*

The experiments on mate choice copying show that females use ISI produced by other females to select mates. Several studies reveal that females also observe male-male interactions to obtain information about mate quality. This has been termed eavesdropping by

Peter McGregor (McGregor and Peake 2000), a British scientist now at Cornwall College. He called it eavesdropping because **eavesdroppers extract information from observing interacting individuals without being involved in the interaction themselves.** By eavesdropping on the interactions of others, animals can thus gather valuable information on interacting conspecifics. Females, for instance, may learn that one male is a winner while the other male is a loser. Evidence of eavesdropping mainly comes from a few situations involving birds and fish but also plants (see section 4.3.3 for this last case).

### a *Eavesdropping influences female mating strategies*

Females have been shown to eavesdrop on male fights in the context of mate choice. For instance, female birds 'eavesdrop' (McGregor and Peake 2000) on song competitions between neighbouring males and obtain extra-pair fertilizations from the winning singer (Otter *et al.* 1999; Peake *et al.* 2001; Mennill *et al.* 2002). Interactive song broadcasting (i.e. song playback) to engage territorial male black-capped chickadees (*Podecile atricapilla*) in countersinging interactions (Figure 4.14) showed that high-ranking males that lost song contests with a simulated intruder lost paternity more often than those that received a playback simulating a submissive intruder (Figure 4.15). In other words, females paired with high-ranking males that lost the simulated playback intrusions somehow heard that their male lost and hence behaved as if they were paired with a low-ranking male. Only two short playback sessions were sufficient to alter females' perceptions of their mates' status in high-ranking, but not low-ranking, males. Information available through eavesdropping therefore may play an important role in female assessment of male quality.

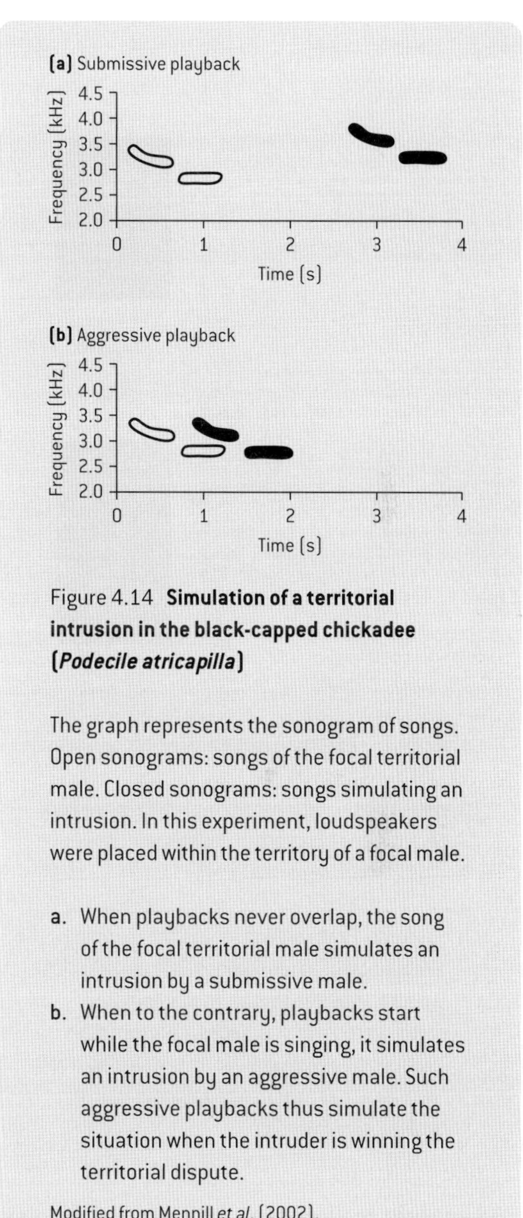


Figure 4.14 **Simulation of a territorial intrusion in the black-capped chickadee (*Podecile atricapilla*)**

The graph represents the sonogram of songs. Open sonograms: songs of the focal territorial male. Closed sonograms: songs simulating an intrusion. In this experiment, loudspeakers were placed within the territory of a focal male.

a. When playbacks never overlap, the song of the focal territorial male simulates an intrusion by a submissive male.
b. When to the contrary, playbacks start while the focal male is singing, it simulates an intrusion by an aggressive male. Such aggressive playbacks thus simulate the situation when the intruder is winning the territorial dispute.

Modified from Mennill *et al.* (2002).

### b *Implications of eavesdropping: inadvertent social information as a platform for the evolution of signalling . . .*

Eavesdropping is also suggested by the behaviour of female fighting fish (*Betta splendens*) that prefer to mate with winners of male-male contests (Johnstone 2001; Oliveira *et al.* 2001; Whitfield 2002). Females that had observed fights between two males preferred the winner whereas females that had not seen the fights showed no preference (Doutrelant and McGregor 2000). Interestingly, the presence of a female audience, but not that of a male audience, increased the fighting rates of males (Doutrelant *et al.* 2001 and see Figure 16.7).

Eavesdropping thus provides a peculiar situation in which males use signals to their competitors that

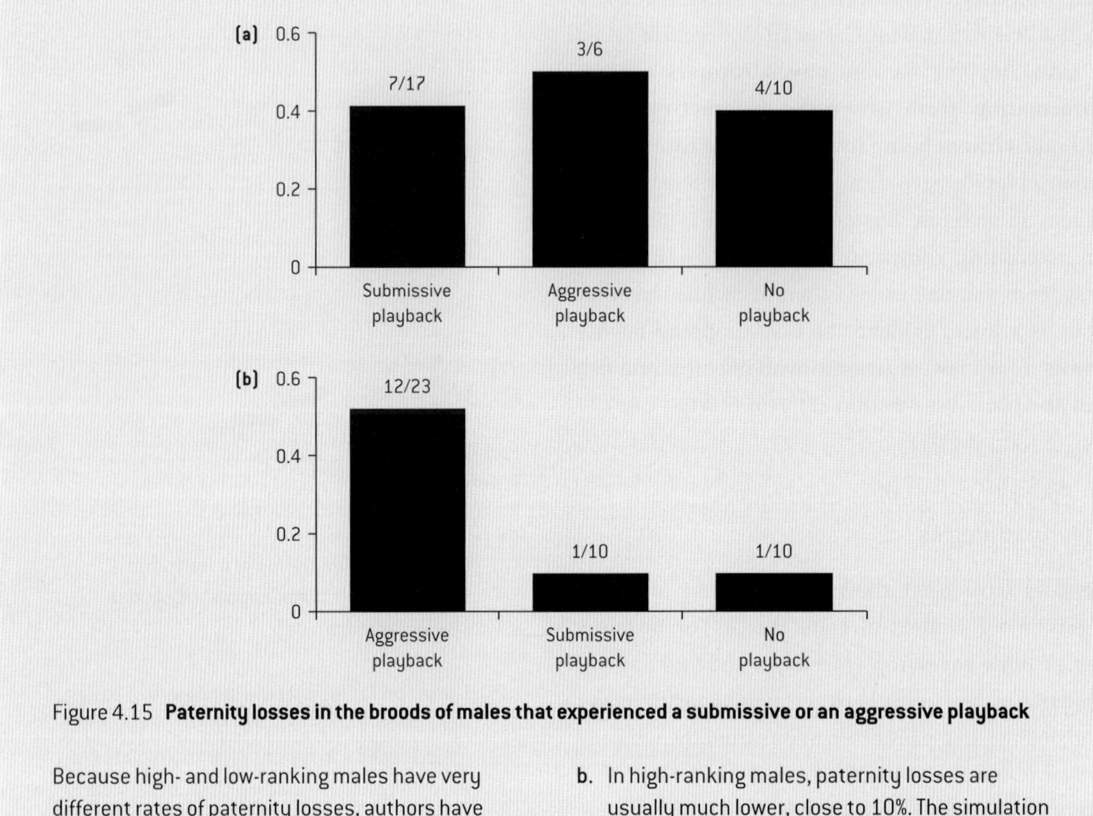


Figure 4.15 **Paternity losses in the broods of males that experienced a submissive or an aggressive playback**

Because high- and low-ranking males have very different rates of paternity losses, authors have analysed them separately.

**a.** In control low-ranking males (no playback) the frequency of paternity losses is usually close to 50%. In males that were manipulated with aggressive playbacks, paternity loss frequencies were similar. This was also true in males that experienced a simulation of a submissive intrusion.

**b.** In high-ranking males, paternity losses are usually much lower, close to 10%. The simulation of submissive intrusions did not change the frequency of paternity losses. However, broods of high-ranking males that experienced playback simulations of an aggressive intrusion had similar levels of paternity losses to that of low-ranking males.

Modified from Mennill *et al.* (2002).

inadvertently produce cues that may be exploited by observing females as cues of male competitive abilities. Two kinds of information therefore are produced by fighting males, signals to their competitor, and ISI cues (for instance, winning or losing). The signals to the receiver with whom it is most directly interacting thus become cues to the bystander. However, the audience effect (see Figure 16.7) further suggests that eavesdropping may entail a transition from social cues to signals because ordinary behaviours that did not evolve as signals may subsequently be modified by the audience effect (Lotem *et al.* 1999). Thus, ISI may be viewed in some contexts as the platform from which signals evolve.

### C ... *Communication in a network*

Another important implication of eavesdropping concerns the evolution of communication once it is established. The classical view of communication is that of a dyad involving a sender and a receiver. McGregor and collaborators pointed out that eavesdropping and the audience effect make this simplistic view of communication incorrect and one that comprises information exchange within a network of individuals more appropriate (McGregor and Peake 2000; Peake 2005). Obviously, the singing interaction between a territorial male and an intruder does not only involve the two males but a network of individuals (among which are the territorial male's mate and its neighbours) that witness interactions directly or indirectly. This has important implications for how communication evolves and is maintained.

#### d *Obligate brood parasites that may eavesdrop on male secondary sexual characteristics*

The most recent development in the topic of eavesdropping was proposed in a stimulating review article by two Spanish researchers (Parejo and Avilés 2007). The idea parallels the one developed in Section 4.3.2.2 above. These authors suggest that obligate brood parasites (which always lay their eggs in nests of other species) may also eavesdrop on secondary sexual signals to detect individual hosts that are likely to be the best parents. In many species where males participate in raising the brood, sexual signals are likely to reveal paternal quality, which may be an important criterion of female mate choice. On the other hand, brood parasites are essentially parasitizing parental care. It is thus easy to understand how a brood parasite that can exploit the within-host population variation in paternal ability would be favoured by natural selection. The stimulating idea is that brood parasites may exploit the information carried by sexual signals to choose a given host pair. If this is the case, it may change the balance of cost and benefits of sexual signalling in host populations.

In their review, Parejo and Avilés (2007) reinterpret previous results in view of their new idea and found several studies with results consistent with eavesdropping by brood parasites. For instance they found evidence for host choice among potential hosts on the basis of sexually selected traits, or at least revealing parental abilities. Furthermore, they found support for the existence of benefits linked to host selection by avian brood parasites. Finally, one of the reviewed studies reported on the attenuation of a sexual ornament in host populations under strong pressure by brood parasites. Such attenuation would be expected if brood parasites eavesdrop on sexual signals. Historically, all these findings had been interpreted as evidence for host selection by avian brood parasites among various species of potential hosts, based on the species' average conspicuousness of sexual signals, but Parejo and Avilés (2007) suggested that all these findings may also reveal eavesdropping on host signals by obligate brood parasites. In fact brood parasites not only tend to parasitize species with flashy sexual ornaments, but within such species they tend to parasitize those with the flashiest ornaments or behaviour because they are likely to be the best host parents for their offspring.

### 4.3.4 Inadvertent social information and danger

Acquiring information about danger may be risky. Not surprisingly, animals have evolved the ability to estimate levels of danger from the use of social cues that may unavoidably accompany it. For instance, the general tendency to copy flight-responses of an entire flock or herd and to respond to fright or stress signs of other animals, clearly demonstrate ISI use. When a predatory fish consumes a prey the damage inflicted to the prey's cells releases chemical substances that produce ISI about the ambient level of predation danger. These so-called alarm substances have been experimentally demonstrated in many fish (Chivers and Smith 1998).

#### 4.3.4.1 *Can plants use inadvertent social information?*

There is evidence that even plants may use ISI from an elegant experiment involving wild tobacco (*Nicotiana attenuata*) in California in the context of herbivory (Karban and Maron 2002). Herbivores can be considered as predators or parasites to the plants they feed on. To mimic herbivory Karban and Maron (2002) designed a simple experiment with two treatments. In both treatments the focal wild tobacco plant remained untouched. The manipulation of the experimental group consisted of clipping neighbouring sagebrush (*Artemisia tridentate*) bushes to mimic herbivory. In the control group no manipulation was performed. Karban and Maron (2002) observed the effect of surrounding sagebrush clipping on untouched wild tobacco, and they found an impressive effect (Figure 4.16). Wild tobacco plants surrounded

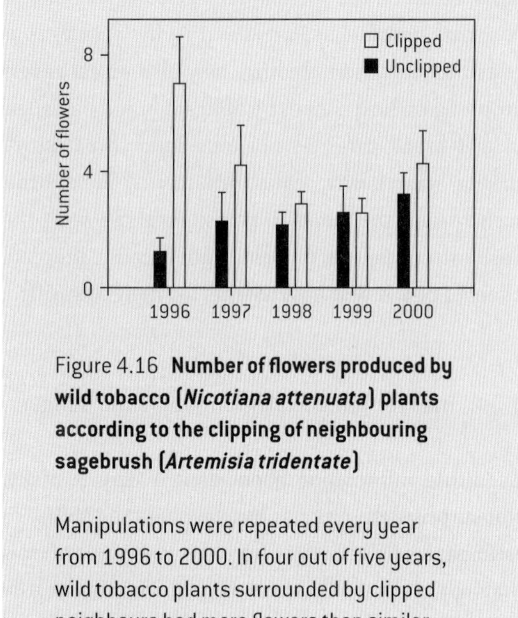

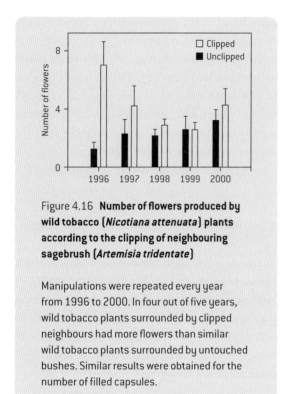

Figure 4.16 **Number of flowers produced by wild tobacco (*Nicotiana attenuata*) plants according to the clipping of neighbouring sagebrush (*Artemisia tridentate*)**

Manipulations were repeated every year from 1996 to 2000. In four out of five years, wild tobacco plants surrounded by clipped neighbours had more flowers than similar wild tobacco plants surrounded by untouched bushes. Similar results were obtained for the number of filled capsules.

Modified form Karban and Maron (2002).

by clipped bushes had more flowers and full capsules than similar plants surrounded by untouched bushes (Figure 4.16). The same team of scientists later showed that the information is carried by airborne molecules released by damaged heterospecific neighbours.

This example shows that social information use does not need to involve complex cognitive mechanisms. Here, information gathering probably only involves the recognition by a receptor molecule of the presence of the informative molecule in the air surrounding the plant. That molecule reveals predation risk as it is released at the site of the wounds made by herbivores.

The reaction of wild tobacco to the high risk of herbivory may be adaptive for two non-exclusive reasons. First, plants growing under high risk of herbivory have a shorter lifespan. Thus, individuals capable of detecting predation risk and prematurely diverting energy to reproduction may have higher fitness than individuals ignoring such information and continuing to grow while risking being eaten before reproducing. Second, intensive herbivory may decrease competition for local space between seeds of the remaining plants. Individuals that can divert more energy to making more seeds may ultimately have more numerous reproducing offspring.

#### 4.3.4.2 *Inadvertent social information in insects*

Until now, we have seen that different kinds of vertebrates such as fish, birds, and mammals, and even plants, can use ISI to make fitness-enhancing decisions. What about invertebrates such as insects? An elegant experiment by Isabelle Coolen (Coolen *et al.* 2005a), then a researcher in Tours (France), noticed that when in the presence of predatory spiders, juvenile wood crickets (*Nemobius sylvestris*) tended to be less visible to a human observer at the surface of the forest floor's leaf litter. She designed an experiment to test whether uninformed wood crickets would detect and react to the difference in behaviour of other wood crickets that had detected the presence of spiders (Figures 4.17 and 4.18).

Coolen's results showed that when spiders were present, wood crickets tended to remain hidden under leaves (Figure 4.18a) and that these behavioural changes were used by other crickets which then used similar antipredator behaviour (Figure 4.18c, d). This shows that some insects can also use ISI and opens the way for other studies with this enormous and diverse animal group.

### 4.3.5 Inadvertent social information use in humans: evidence and consequences

Humans also use ISI. We do know that most commercial advertising exploits our capacity to extract information from others. Many adverts carry a message like 'look how nice looking (performing) I am with this or that perfume or makeup'. 'Look, that socially performing person (e.g. a famous and sexy actor) is using this product'. The main principle of most commercial advertising is to link a product to performing people, whether famous or not.

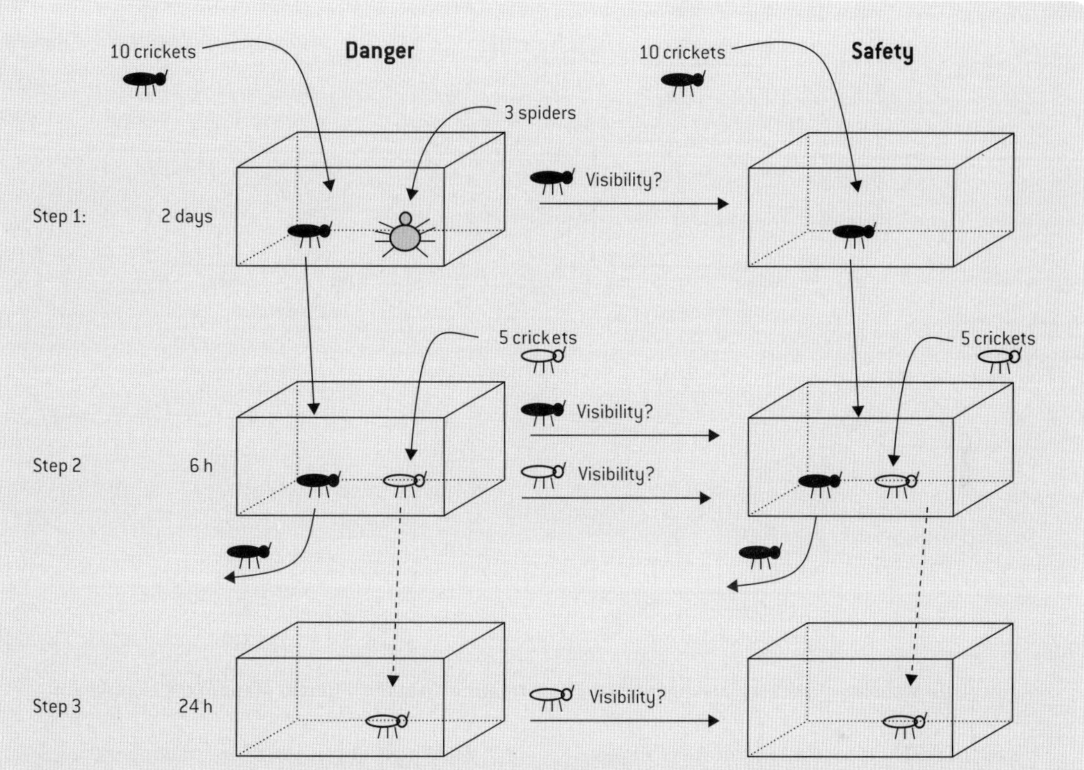


Figure 4.17 **Crickets can detect danger by observing the usual reaction of other crickets**

Demonstrating crickets are black. Observing crickets are white. Spiders are grey. Straight arrows indicate that subjects were transferred to new boxes, devoid of predators. Dashed arrows indicate that subjects remained in the box.

**Step 1.** Demonstrating crickets were set in boxes according to whether they were in the danger or safety treatment, which was created by the presence or absence of predatory spiders. The visibility of crickets was measured as the proportion of crickets seen at the surface of leaves during 30 seconds of observation (results in Figure 4.18a).

**Step 2.** Demonstrating crickets were transferred into new previously unused boxes where 5 new identifiable crickets were added. The visibility of the demonstrating crickets (results in Figure 4.18b) and that of the new uninformed observing crickets was measured after 6 hours (results in Figure 4.18c).

**Step 3.** Demonstrating crickets were removed and only the observing ones were left in the box. The visibility of the observing uninformed crickets was measured again after 24 h (results in Figure 4.18d).

Redrawn from Coolen *et al.* (2005a).

Commercial advertising is obviously intentional, but it exploits our natural tendency to cue on the behaviour and performance of others to extract information about our environment. Are we influenced by such advertising? The answer is in the billions of dollars invested every year in commercial advertising. Would firms spend so much money on advertising if we were oblivious to it? This is surely one example of humans using ISI.

The question of ISI use in humans is likely to be crucial to understanding human evolution of very close cooperation among unrelated individuals. As we will see in Chapter 15, human altruism has been difficult to explain by kin selection or reciprocity

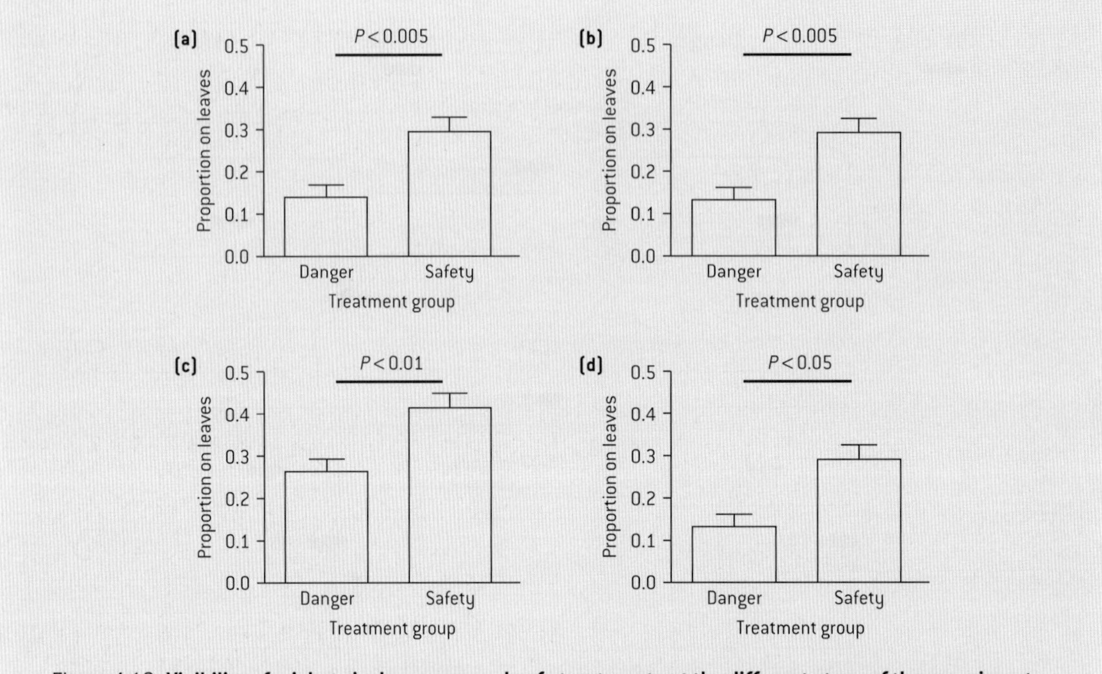


Figure 4.18 **Visibility of crickets in dangerous and safe treatments at the different steps of the experiment**

Visibility of crickets seen at the surface of leaves in the danger or safety treatments (i.e. with or without predatory spiders), measured as the proportion of individuals seen at the surface of leaves during 30 seconds of observation.

**Visibility of demonstrators** in the danger and safety treatments: **a.** During step 1, which is in the presence of spiders in the danger group, and **b.** during step 2, 6 hours after both treatment groups had been transferred in boxes devoid of spiders.

**Visibility of observers** in each treatment: **c.** during step 2 when demonstrators were still present and **d.** during step 3, 24 hours after the removal of demonstrators. Coolen *et al.* (2005a) carefully designed other experiments to rule out any other potential explanation to their results.

Redrawn from Coolen *et al.* (2005a).

alone (Semmann *et al.* 2003). Recently, the 'prestige hypothesis', an extension of the **handicap principle** (Zahavi and Zahavi 1997), has been proposed to account for its origin. The hypothesis proposes that, because apparent **altruism** is costly, only high-quality individuals can afford to exhibit it when faced with a choice: be altruistic or selfish. Performing altruistically thus provides an honest signal of wealth (social performance) that increases social reputation, which can enhance the likelihood to benefit from altruism in the future. This mechanism has been extended to situations when there is no variation in quality among members of the population. It is then called **indirect reciprocity**. The evolutionary logic of indirect reciprocity is apparently sound (Nowak and Sigmund 1998) suggesting that cooperation could emerge in a wide array of conditions, even when only a small fraction of the population witnesses altruistic actions. Reputation is clearly based on the acquisition and maintenance of long-term knowledge of the characteristics of other individuals. That knowledge is implemented continuously by the acquisition of performance information whether inadvertent or intentional. In humans, reputation appears to be used in our decisions to cooperate or not with specific individuals (Wedekind and Milinski 2000), and cooperation is perpetuated at a substantial level (Semmann *et al.* 2003). The need to maintain one's reputation may prevent individuals from abusing public goods and help avoid the '**tragedy of the commons**' (Milinski *et al.* 2002); that is the overexploitation of goods that are common to every group member (Hardin 1968).

### 4.3.6 Conclusion: inadvertent social information plays a prominent role in the evolution of animal decision-making

All the above examples share a common thread (Danchin *et al.* 2004). In all these experiments, the information is inadvertently produced and in most cases, it is derived from the performance of the information producer. For instance, in the mate choice copying experiments it is the capacity of males to attract females that is used by females to select among males. In the eavesdropping context, it is the fighting performance of males that is used by females. The experiment with the black-capped chickadees clearly manipulated the apparent fighting performance of high-ranking males showing that females constantly assess their male's performance and react accordingly.

Historically, studies have mainly stressed the importance of signals and true communication in the evolution of animal decision-making, that is in the evolution of animal behaviour in general. However, all of the above examples show that inadvertent social information also plays a prominent role in animal decision-making. This has been largely ignored until recently. We now conclude this chapter by developing in the final section how the study of animal behaviour can be centred around the concept of information.

## 4.4 An information-driven approach to animal behaviour

Understanding animal decision-making implies answering four complementary groups of questions about biological requirements, information, cognition, and the dynamic consequences of information use (Figure 4.19). We now detail these approaches while referring to the chapters that deal with some of these specific questions.

### 4.4.1 Biological requirements

To study animal decision-making one should first ask questions about the **fitness** components that are

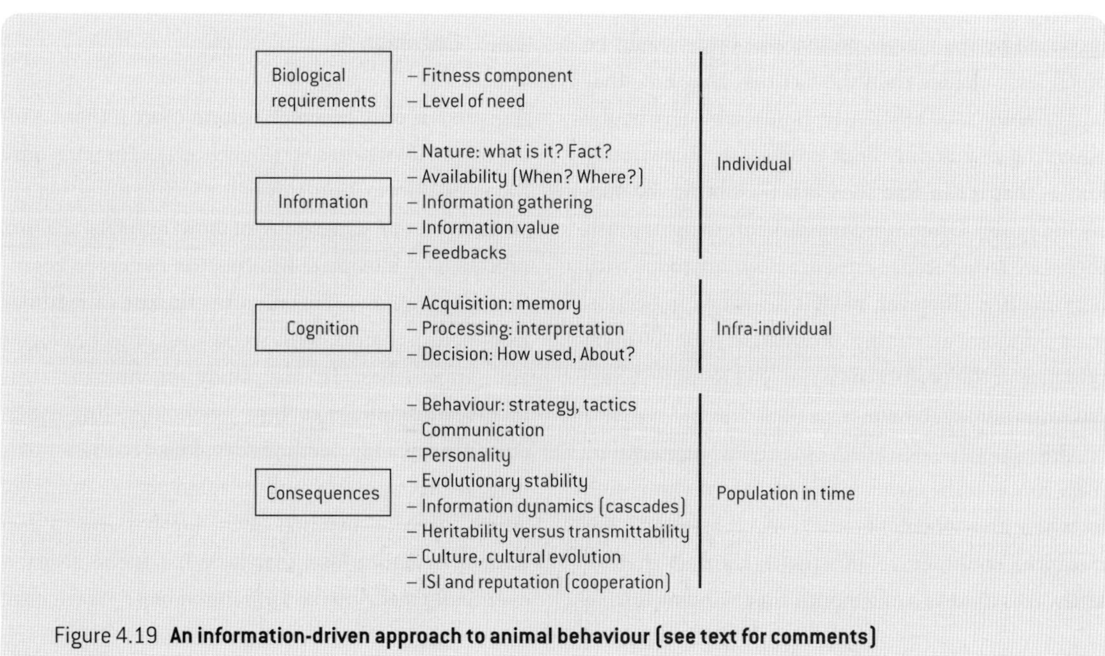


Figure 4.19 **An information-driven approach to animal behaviour (see text for comments)**

influenced by the decision under study. Is that decision about foraging for food, or choosing a mate, or finding shelter from predators or cold, etc.? It is obvious that cues that can inform about those various biological requirements are likely to differ strongly. Another important aspect concerns the current conditions of the focal individual. For instance, what is the level of need? A very hungry individual may react to the same information in quite a different way compared with a satiated one. Even if the information is that patch profitability may be low but sufficient to survive, a starving individual may decide to exploit it because it cannot afford to take the time to find another patch, while a satiated individual may ignore it and keep on seeking a better patch.

### 4.4.2 Information

The second group of questions concerns that of information *per se* (Figure 4.19). What is the nature of the source of information? What is it that informs? In other words, what is the fact that diminishes uncertainty? Despite the crucial importance of such questions, they are sometimes ignored in evolutionary sciences. The example of the feeding flocks of Peruvian boobies (Section 4.2.3.3) shows how subtle such questions can be. It is clear that the information produced by the simple presence of birds would be much less valuable than the fact that they are obviously finding food efficiently. Hence, a hungry booby observing a foraging flock from afar immediately knows that it can find food just by joining the successful group. Often it is very difficult to distinguish these two facts because presence is synonymous with performance. However, models described in Section 9.4.2 show that extracting information from the mere presence or the performance can lead to quite different evolutionary dynamics.

Once cues are identified, other relevant questions may concern the availability of information such as when and where the cue can be assessed? For instance, Boulinier *et al.* (1996) showed that in kittiwakes there is a one month time window during which individuals prospecting for breeding sites can accurately assess local reproductive success. Interestingly, these authors showed that the bulk of prospecting occurs exactly during that period (see Section 9.5.3).

Another key question is that of information value, which is the extent to which the observed fact predicts the future state of the environment. This mainly involves studying the temporal and spatial autocorrelation of the environment. Indeed, the time lags between information gathering and the decision may be long in some circumstances. In the meantime, conditions may have changed so that the information becomes obsolete. This is particularly true in habitat copying (Section 4.3.2.1 and Chapter 9) where individuals visit breeding grounds towards the end of the breeding season and settle in the following breeding season in the most successful patches.

Another important characteristic of information is that of the potential existence of ecological feedbacks that occur when information use changes the value of information. This is typically the case of resource exploitation. Information about resources may diminish the value of the resource and hence that of information. This may lead to particularly involved informational dynamics.

### 4.4.3 Cognition

A given cue only becomes 'information' if there is an organism to witness it (Figure 4.2). The cue only becomes information through the cognitive system of an observer. Questions of how animals acquire, process, and memorize information belong to cognition (Figure 4.19). This involves the use of methods from neurosciences, with its own set of technologies and approaches. In the study of humans, this involves disciplines such as psychology, linguistics, and anthropology. Methods developed in these fields may be extrapolated and adapted to the study of animal decision-making. Cognition has also to deal with information use, that is 'What is the decision eventually made by the focal individual?' As we have already seen, that decision may vary greatly among individuals. After observing the actual decision, we may make hypotheses to explain it.

We have seen (Section 4.3) that animals can account for subtle variations in the performance of conspecifics and of other species. This book is full of such examples. For instance, we will see in Chapter 13 that some animals can manipulate the **sex ratio** of their offspring according to subtle variations in environmental conditions. This suggests that organisms can sense subtle variations and react more adaptively. The plant example in Section 4.3.4.1 is particularly enlightening as it shows that even organisms without brains can sense and adjust to very subtle components of their environment. In the experiments on wild tobacco, the mechanism was apparently as simple as molecular recognition. Although that example shows that information gathering and processing does not necessarily involve sophisticated cognitive processes, cognitive questions remain a major issue for understanding the evolution of decision making in general.

More generally, the question of the cognitive implications of behaviour underlies every chapter of this book, although none of the chapters focus specifically on cognitive issues. It is probably one of the major challenges of behavioural ecology to incorporate cognition into the study of the evolution of behaviour. Although questions of cognition have been largely ignored by behavioural ecologists, they are at the heart of many approaches in ethology.

### 4.4.4 Consequences

Another important set of questions concern the consequences of information use at various temporal and spatial scales (Figure 4.19). The most basic consequence of decision-making is behaviour. We can study the various tactics and strategies of animals using information to their benefit. Another consequence of information use may be the evolution of communication (Chapter 16). Danchin *et al.* (2004) proposed that the use of ISI may provide the platform from which signals may evolve. As soon as individuals start using the information that another inadvertently produces it may entail costs or benefits for the information producer. Such costs and benefits create selective pressures that may modify behaviour so as to hide or show the information. From that moment on, it may no longer be ISI, but rather a signal that results from a history of selection.

A related question is that of the emergence of personalities, or **behavioural syndromes**. Recent studies in the great tit (*Parus major*) showed that various behavioural traits are not associated randomly (Dingemanse *et al.* 2003, 2004). For instance, boldness is associated with a high capacity of exploration and dispersal, whereas shyness is associated with low exploratory and dispersing behaviour. Bold individuals also appear to be more likely to become dominant (Dingemanse and de Goede 2004). Furthermore, such behavioural syndromes appear to be related to breeding success in a way that differs between the sexes (Both *et al.* 2005). This suggests that the same information may be used in very different ways by individuals of different behavioural syndromes.

Other questions deal with the evolutionary stability of strategies using ISI or other sources of information. For instance, Doligez *et al.* (2003) showed that habitat copying (i.e. using the breeding performance of conspecifics as a source of information to select breeding habitat) is evolutionarily stable if the environment does not vary randomly or is not totally constant in time. Habitat copying resists the invasion of other habitat selection strategies while being able to invade other strategies such as philopatry or **social attraction** when the environment varies with some level of temporal autocorrelation (see Figure 9.4).

We have seen that the use of ISI by others may lead to involved feedbacks that may diminish or enhance the value of information. Another potential consequence of the use of ISI concerns **informational cascades** (Bikhchandani *et al.* 1998; Giraldeau *et al.* 2002). An informational cascade occurs when

ISI overrides personal information such that all decisions are based entirely on the behaviour of others (Giraldeau *et al.* 2002). Although cascades are optimal solutions to obtaining information, they can sometimes lead to maladaptive responses.

Finally, use of ISI may lead to the transmission of behavioural patterns among individuals (i.e. **transmittability**). The possibility of among-individual transmission of behavioural patterns through learning may be called **culture** as it leads to certain forms of inheritance of variation of learned behaviour. This is the subject of Chapter 20.

Another interesting line of research concerns the implications of information use for the evolution of **cooperation** (see Chapter 15). Several recent hypotheses underlined the prominent role of **social prestige** in the evolution of cooperation (Zahavi 1995; Nowak and Sigmund 1998; Wedekind and Milinski 2000; Milinski *et al.* 2002; and see Chapter 15). According to such models, the capacity to monitor the behaviour of several group members, as well as to increase and memorize a score (or social reputation) of several companions, can be a powerful process favouring the evolution of cooperation in any animal species and particularly in humans (Nowak and Sigmund 1998; Wedekind and Milinski 2000; Milinski *et al.* 2002; Semmann *et al.* 2003). It is likely that the building of reputations of individuals within their social group involves any kind of performance information (communication and ISI). This implies that 'cognitive' abilities are important skills for the evolution of cooperation.

## 4.5 Conclusion: the necessity of an integrated approach

This chapter first showed how information in biology is linked to phenotypic variance, which is produced by variation in information. We then proposed a new lexicon of biological information that will be used throughout this book. One major conclusion is that most social information is derived from condition-dependent traits that produce performance information, whether intentionally (signals) or inadvertently (ISI). In turn, performance information reveals the interaction between individual and environmental quality. This chapter then reviewed evidence that animals can extract information by observing others in many contexts. Such inadvertently produced information is much more common and influential than previously recognized, and there is now growing evidence that ISI is used in many fitness-affecting decisions. Although studies of true communication have stressed the importance of intentionally produced information (i.e. signals) in the evolution of behaviour, studies reviewed here strongly suggest that the inadvertent component of performance information also plays a prominent role (Danchin *et al.* 2004). Furthermore, it is likely that ISI may be the platform from which many forms of true communication have evolved (Lotem *et al.* 1999, Danchin *et al.* 2004, 2005). All these results and considerations stress the importance of adopting an information-driven approach to studying the evolution of behaviour.

Figure 4.19 begs for two important comments. First, despite the importance of questions dealing with information *per se*, few approaches have really analysed the question of 'information in the evolution of behaviour'. Chapter 9 for instance will have the same conclusion in the case of breeding habitat choice. However, several approaches concentrate on the central question of information with experiments aiming at manipulating the facts suspected of being used as cues by animals. Some of these results were reviewed above (Section 4.5).

The second comment is that these approaches are complementary. Although Niko Tinbergen (1963) urged that only an integrated approach of animal decision-making can allow the understanding of the evolution of animal behaviour, we have not yet fully heeded his advice. We have seen in Chapter 1 that there are two major schools for the study of animal behaviour: ethology and behavioural ecology. Among ethologists, researchers have been particularly interested in questions of animal cognition, as in the case of the study of animal social learning. The time is ripe for behavioural ecologists to work jointly with ethologists to study the evolution of animal behaviour. Of particular interest is the question of **cultural evolution** because it gives behaviour a central role in evolution. Intuitively we know that **cultural transmission** is a mode of information transfer across generations whose transmission mechanism mainly involves behaviour. This will be the topic of the final chapter, Chapter 20, which will analyse the implications of the widespread use of performance information as a potential trigger of the inheritance of variation across generations.

## » Further reading

> *For a review on the ecology of information use:*

**Giraldeau, L.-A.** 1997. The ecology of information use. In *Behavioural Ecology: An Evolutionary Approach* (ed. J.R. Krebs & N.B. Davies), pp. 42–68. Blackwell Scientific Publications, Oxford.

> *About the potential disadvantages of using social information:*

**Giraldeau, L.A., Valone, T.J. & Templeton, J.J.** 2002. Potential disadvantages of using socially acquired information. *Philosophical Transactions of the Royal Society of London Series B* 357: 1559–1566.

> *About the links between environmental variation and animal learning:*

**Stephens, D.W.** 1991. Change, regularity and value in the evolution of animal learning. *Behavioral Ecology* 2: 77–89.

> *About the importance of considering genes as bits of information:*

**Gilddon, C.J. & Gouyon, P.H.** 1989. The units of selection. *Trends in Ecology and Evolution* 4: 204–208.

> *For the links between inadvertent and true communication:*

**Bradbury, J.W. & Vehrencamp, S.L.** 2000. Economic models of animal communication. *Animal Behaviour* 59: 259–268.

**Lotem, A., Wagner, R.H. & Balshine-Earn, S.** 1999. The overlooked signalling component of non-signalling behavior. *Behavioral Ecology* 10: 209–212.

> *About the interaction between learning capacity and the evolution of innate components in animal decisions:*

**Mery, F. & Kawecki, T.J.** 2004. The effect of learning on experimental evolution of resource preference in *Drosophila melanogaster*. *Evolution* 58: 757–767.

## » Questions

1. This chapter urges that the concept of information should be at the heart of the study of behaviour. It reviews several fitness-affecting decisions that were shown to be influenced by inadvertent forms of social information. Can you imagine other kinds of situation where ISI may prove valuable in decision-making?

2. To what extent is the concept of information linked to the notion of animal culture? Refer to Chapter 20 for further information on animal culture.

3. The theory of information was proposed by Shannon (1948) in the context of economy and human communication networks. This theory has been applied in various contexts since. For instance, in the context of thermodynamics where it appears to parallel the concept of entropy. Do you think this theory can be transposed easily to biological information?

4. Can you think of examples of humans using inadvertent performance information?

Part Two

# The Development of the Behavioural Phenotype

Having presented the history, general concepts, and methods of behavioural ecology, as well as having developed how it constitutes an information-driven approach to behaviour, we can now elaborate on the various domains of this discipline. This part is about the ontogeny of the behavioural phenotype. We have seen in the forewords that behavioural ecology constantly synthesizes new fields. In particular, from the beginning of the 1990s, behavioural ecology has participated in the emergence of evolutionary physiology, a discipline at the interface between pure physiology and evolution. This part of behavioural ecology thus naturally involves increasing interdisciplinary interactions with physiology, endocrinology, and medicine among other disciplines. This is undoubtedly the domain of behavioural ecology that has developed the most since then, and which is still growing in importance. The goal of this part is to present some of the major developments brought by the study of proximate mechanisms of behaviour to behavioural ecology in general.

Chapter 5 develops an integrated view of the link between genes and behaviour. It covers biotic sources of trade-offs including genes, physiology, mating systems, and social systems. It links genetic approaches and behavioural studies. In particular, it elaborates on the importance of the interaction between genetic and non-genetic information in the shaping of the adult phenotype.

Chapter 6 continues some of the themes of the preceding chapter in dealing with the specific role of hormones in the development of the phenotype and particularly that of behaviour.

5

# 5

# Life History Strategies, Multidimensional Trade-offs, and Behavioural Syndromes

Barry Sinervo and Jean Clobert

## 5.1 Introduction

We have seen in Chapter 2 that in its simplest form, the theory of **natural selection** predicts that individuals with maximum **fitness** will be selected. With such an expectation, natural selection should favour individuals that reproduce maximally and live forever. Yet, such creatures do not exist and the fundamental question is why? Evolutionists rapidly recognized that space and time impose limits on organisms. Travel time and food processing are constraints, which ultimately limit individual fitness. However, the main constraints arise from species interactions, such as prey limitation, predation, or parasitism, or from intra-specific interactions such as competition for resources. An obvious example would be the case of a predator or herbivore species that would consume its main living resource more than the latter can reproduce and grow. That consumer would rapidly go extinct.

Given that resources constrain the population growth rate of a species, it follows that survival and reproduction, which are the key determinants of population growth rate must be limited: for a given value of growth rate, only a set of combinations of survival and fecundity are possible, usually arranged along some curves named **trade-offs**. Lines of equal population growth rate track a given trade-off (Box 5.1).

*From life histories to life cycles . . .*

The **life cycle** of a species is the description of the way organisms are born, grow, reproduce, and die. An important characteristic is whether a species reproduces sexually or asexually, or both. Some remain unable to reproduce before a certain amount of time. Others reproduce almost immediately after birth. Some species grow to adult size and reproduce only once (e.g., salmon): a **semelparous** life history. Others reproduce more than once in a series of consecutive breeding periods: an **iteroparous** life history.

The life cycle of a species can be described with the set of parameters that compose the **life history strategy** of individuals (Stearns 1992). As some or all of these key parameters can vary among individuals within a population, we can conclude that various **phenotypes** within a population can have different life history strategies. For example, in some species males may be semelparous while females are iteroparous. Similarly the set of parameters can vary greatly among species.

**Life history traits** include demographic parameters such as the rate of reproduction, survival within reproductive seasons, survival between reproductive seasons, early or late fecundity, as well as any other traits including parameters such as dispersal, age at maturity, growth rate, adult size, etc. Competition for space and resources imply that not all combinations of these parameters are possible in organisms. An increase in one parameter will unavoidably imply a change in other parameters. This is the origin of the **primary demographic constraints** on the evolution of behaviours.

*. . . to trade-offs . . .*

The concept of trade-off describes the general fact that variations in one life history parameter necessarily imply co-variation(s) in one (or several) other life history parameter(s). The term **compromise** is sometimes used instead of trade-off. In this book, we use trade-off.

To illustrate trade-offs, let us imagine a continuously growing sexually reproducing organism. To transmit its genes before dying, it has to find food to grow and make eggs or sperm, and seek a mate, court, and copulate with the mate. Obviously, if more time is spent in foraging for growth or maintenance, less time will be available for courtship, and vice versa. In the same way, in species providing parental care, while an individual cares for previous offspring, it may not be able to forage for food, or potential sexual partners. These examples show that trade-offs are thus everywhere. In fact, life is mainly a matter of compromise. From animals to plants, bacteria to elephants, any organism is constantly facing trade-offs.

*. . . to behaviour . . .*

Trade-offs also underlie most decision-making that animals face during their entire life. In most of the examples above, living organisms can improve the way they solve these trade-offs by gathering information in one way or another about the current state of the environment. For instance, in the case of a species that provides parental care, the solution of the trade-off between caring for young or seeking new mates may differ greatly if there are many potential predators of their offspring in the environment, or if there are many individuals of the other sex that are currently free from parental duties. Similarly, the optimal solution to trade-offs that deal with the time spent in foraging, or the expense in other activities, may greatly vary according to the level of satiation of the individual, or to the probability of finding a significant amount of food in the next minute, etc.

All these examples show that the solution to trade-offs is highly likely to involve information gathering, processing and decision-making, which are the essence of behaviour. Thus, the concepts of trade-off and behaviour are intimately linked.

*. . . and behavioural ecology*
As we have seen in the three previous chapters, another fundamental component of the evolutionary study of behaviour is the impact of behaviour on fitness. We have seen in Chapter 1 that this is the fundamental question that led to the emergence of behavioural ecology within behavioural sciences. Now, this implies that we are able to estimate fitness in a quantitative way.

In Chapter 2, we defined **phenotypic fitness** as the average capacity of a phenotype to produce mature offspring relative to other phenotypes in the same population at the same time. This capacity depends on survival and reproductive performance of the phenotype being examined. Survival and reproduction are thus the two major components of fitness. Once we have estimated the relevant life history parameters for a given phenotype, we can combine the balance among their effects by using demographic models to compute the population growth rate of the phenotype (usually called $\lambda$, see Box 5.1). Indeed population growth rate can be used to quantify the actual capacity of a phenotype to grow in number under ecological conditions that prevailed in the population where and when the demographic parameters were estimated. Phenotype-dependent population growth rate is thus often used as a potential proxy for phenotype fitness.

This implies that, as in any other evolutionary approach, behavioural ecology has fundamental links with the study of life history strategies. Thus, if we are to study the fitness consequences of behaviour, we need to link behaviour to life history strategies, at the very least to be able to estimate fitness properly.

*The fundamental rationale for population growth*
An understanding of the link between trade-offs and behaviour begins with a mathematical description of the life history and interplay between demographic parameters. Box 5.1 illustrates compatible values for survival and reproduction that yield equal growth rate. From these figures it is easy to understand that when survival is high, reproduction should be low and vice versa, leading to what are called iso-fitness curves or trade-offs. Among the many possible trade-offs that are defined by the parameter space that comprises a species life cycle, natural selection will retain a few trade-offs, but at present there is no theory that predicts which trade-offs will act or even the way they are controlled (directly through the **pleiotropic** action of genes on two components of fitness or indirectly through physiology and/or behaviour, or both, etc.). The problem is compounded when we not only consider the trade-offs of natural selection, but also the trade-offs of sexual selection, mating systems, and social systems.

Space, food, or any resource, including access to conspecifics, can all be limiting. However, as animals evolve greater and greater behavioural complexity, new constraints are added to these evolving social systems, which not only constrain and interact with

### Box 5.1 **Species life history cycle, demographic parameters, trade-offs and iso-fitness curves**

To measure fitness, ones needs to describe the life cycle of the species considered (Figure 5.1), then estimate the demographic parameters associated with each behavioural (physiological, etc.) strategy, and finally compare population growth rate ($\lambda$) resulting from the different combinations of demographic parameters characterizing each strategy (Figure 5.2). When the growth is constrained to a given value, and strategies compete, then special techniques such as evolutionary game theory and concepts like the **evolutionarily stable strategies** (ESSs) are necessary to compare strategies (Maynard Smith 1982).

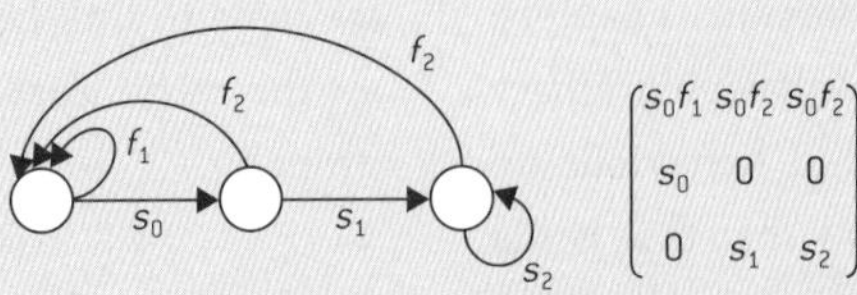


Figure 5.1 **Life cycle graph and the corresponding transition matrix**

This life cycle graph corresponds to a typical passerine where individuals reproduce in the spring after the year of birth. It corresponds to a post-breeding census (see Caswell 2001) for further explanations): $s_i$ is survival of different age classes and $f_i$ is the fecundity of different age classes. Fecundities subsume a quantity that describes proportion of breeders ($\alpha_i$) and clutch size ($cs_i$) such as $f_i = \Sigma \alpha_i \times cs_i$. All these parameters can be built into a corresponding Leslie matrix (provided on the right) which describes mathematically the life cycle of the phenotype in a very compact form.

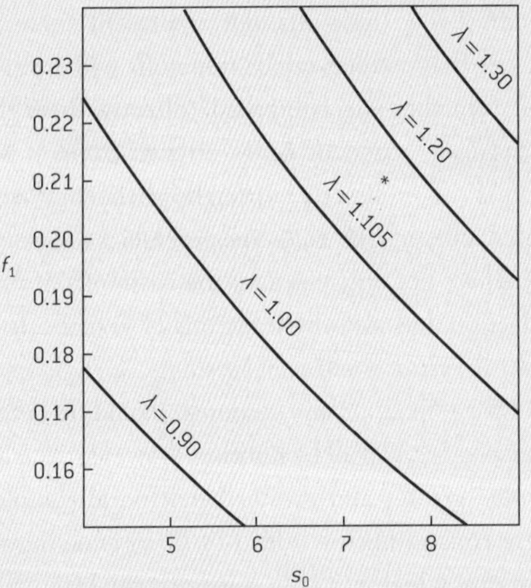


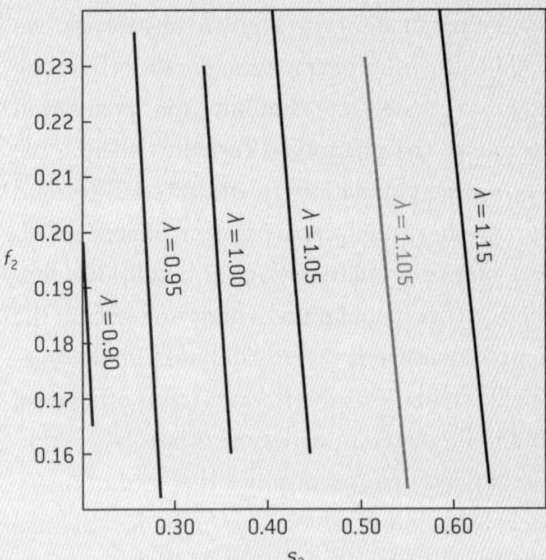


Figure 5.2 **Example of iso-fitness curves or trade-offs**

The points that lead to the same population growth rate (usually called $\lambda$) form curves that are called iso-fitness curves. In this example, values for $s_0$, $s_1$, and $s_2$ (defined in Figure 5.1) are respectively 0.2, 0.35, and 0.5, and $f_1 = f_2 = 3.5$. The computer program ULM (Legendre and Clobert 1995) was used to calculate iso-fitness curves. The corresponding value of $\lambda$ is provided on each iso-fitness curve. Two types of iso-fitness curve are illustrated here: the trade-off between fecundity at one year of age and survival from birth to first year of reproduction (panel on the left), and trade-off between adult fecundity and adult survival (panel on the right). As you can see, the shape and slope of trade-offs vary with the demographic parameters considered, as well as with the value of the growth rate. Here only bi-dimensional trade-offs are illustrated, but multi-dimensional ones are likely to exist. The light grey line indicates the actual value of growth rate in that theoretical population. The asterisk indicates the actual value of fecundity and survival of the species used as reference.

other trade-offs present in the ancestral condition, but which also create new opportunities for innovation and adaptation.

*Goals and structure of this chapter*

In this chapter we start with some historical perspectives on trade-offs and life history tactics that are defined by a specific (or a set of) trade-off(s). Then we cover biotic sources of trade-offs including genes, physiology, mating systems and social systems. We discuss the added levels of trade-offs that are present as mating systems and social systems evolve, and also highlight a few of the opportunities for alternative strategies that arise in the context of such social situations of increasing complexity. The kinds of adaptations that arise can be classified as either genetically fixed or phenotypically plastic. We outline trade-offs and constraints in both situations.

A fundamental thesis of this chapter is that all genetic trade-offs, which are expressed in mating and social systems, must involve interactions among many genes, not just the action of pleiotropy, which is the effects of one gene on many traits. Interactions among many genes are referred to as epistasis. Epistasis is fundamental to behaviour because behaviours are most often generated in a social context, involving a sender and a receiver. Genes controlling signals are likely to differ from those controlling signal reception. Pleiotropy is still involved in trade-offs because genes controlling signals, particularly honest signals, are often pleiotropically linked to life history trade-offs through physiology, the immune system, or the endocrine system. Thus, **life history trade-offs in a behavioural context must include a consideration of both pleiotropy and epistasis**. One important source of pleiotropy and epistasis lies in endocrine regulation, which by maintaining homeostasis generates trade-offs. We must therefore explain the proximate source of trade-offs in terms of endocrine regulation of behaviour and life history.

Whereas selection and fitness differences act on phenotypes, the actual source of trade-offs arises from the compromise that is expressed at the level of the genes, the fundamental units of inheritance. This genetic perspective is crucial to a comprehensive understanding of the evolution of behavioural systems. We must therefore provide sufficient background on genetics to generate a deeper understanding of pleiotropy and epistasis, which complements the perspective generated from a study of endocrine system trade-offs.

## 5.2 Body size, allometry, slow–fast continuum, and complex life history trade-offs

### 5.2.1 The *r*–*K* continuum

Variations in life style are considerable. The magnitude of inter-specific variation can be highlighted by comparing bacteria, which can reproduce every two hours and thus have several generations per day, to elephants or giant turtles, which reproduce hardly every second year and have a few generations per century. This led some researchers to propose that species were either *r*-selected for short generation time with high potential growth rate or *K*-selected for long generation time with low potential growth rate (MacArthur and Wilson 1967; Pianka 1970). Associated with *r*-selected life cycles were additional correlates such as small size, semelparity, short lifespan, and a lack of intra- or inter-specific interactions. *K*-selected forms exhibited opposite traits. According to that view, depending on the relative force of *r* versus *K* selection, species had a specific place on this so-called *r*–*K* continuum. These ideas had important impacts on research in the area. We discuss three ideas (density dependence, body size, and the trade-off dimensionality) that generated debate and are tightly linked to behavioural decision-making.

### 5.2.2 The prominent role of density dependence

The idea that the demographics of *r* species were mainly driven by **density independent** factors (Andrewartha and Birch 1954) led to the use of fitness maximization models based on **intrinsic rate of increase**, denoted as *r* (Roff 1992). However, this assumption is debatable. The current view is that most species are influenced by density-independent and **density-dependent** factors (Clobert *et al.* 1988, Juillard *et al.* 1999), which can act independently of **generation time**. Moreover, species with high *r* have an innate tendency to show complex population excursions (cycles or chaos, May 1976) when density-regulated, which might be confused with **stochasticity** of environmental origin, unless cycles are very regular (Sinervo *et al.* 2000b). Theoretical analysis of life histories now favours ESS modelling approaches (Maynard Smith 1982, and Chapter 3) where environmental stochasticity and density dependence are explicitly considered. Such models deploy more complex definitions of fitness, and feedbacks between ecology and evolution are taken into account (Dieckmann 1997).

### 5.2.3 Allometry, a classic assumption

A classic assumption is that species with a high potential *r* have small body size while those with small *r* have large body size. This assumption was indeed verified on a large species range of body size (when bacteria were compared with elephants), and **allometric** analyses, in which demographic parameters were scaled to body size with some magic exponent (0.25 for demographic traits), received considerable attention (Peters 1983; Calder 1984). Many reasons such as physiology and energetic constraints (West *et al.* 1997, Enquist *et al.* 1998) or mechanical constraints (McMahon 1973) were used to rationalize allometric scaling exponents.

A finer analysis, however, revealed considerable variation around the relationship between body size and demographic parameters. Important taxonomic variation was found to explain such scattering (Gaillard *et al.* 1989; Stearns 1983, 1984, 1992; an example of an allometric relationship is given in Figure 3.5). In other words, for the same unit of size, bats survived much better than mice, and giant turtles better than elephants or whales. Therefore, besides the strict effect of size, some species had longer lifespan, more delayed age at maturity, and reduced fecundity relative to others. Demographic parameters and life history adaptations (fitness components) such as age at sexual maturity, survival and fecundity are therefore the target of natural selection at least partly independent of allometric effects.

The notion that allometry cannot explain all evolutionary change among populations was demonstrated in a simple allometric engineering experiment (by manipulating egg yolk content; Sinervo 1990; Sinervo and Huey 1990). Lizard populations often vary in a physiological trait like stamina of hatchlings. Such variations may co-vary with a trait like egg size: larger hatchling lizards should have higher stamina. By experimentally reducing egg size of lizard hatchlings using yolk removal, Sinervo and Huey (1990) could only explain some of the inter-population variation in stamina of lizards. Unexplained variation among populations must therefore be due to evolved differences in stamina. In contrast, all variation in sprint speed, within and between populations, was explained by differences in yolk volume. The general message is that contrary to what was first assumed: we should expect that part of the variation among and within populations results from evolved size-independent effects, despite a general or near universal effects of size per se. The source of such evolved size-independent variation in physiological and behavioural systems is therefore of great interest to the field of life history evolution.

### 5.2.4 Are life histories bi- or multi-dimensional?

#### a *One-dimensional life histories*

According to the *r–K* continuum approach, life history strategies are supposed to vary along a single dimension, the *r–K* continuum. Thus, *r* species are expected to have low survival, early age at maturity, and high fecundity, whereas *K* species should display an opposite pattern. This implies that life history diversification should proceed mainly along an *r–K* continuum (Pianka 1970; Stearns 1992), also called slow–fast continuum (Gaillard *et al.* 1989). Assuming, among other things, stationarity of populations (i.e. no population excursions and stable equilibria), Charnov (1993) demonstrated theoretically that demographic parameters co-varied two-by-two along one direction with a slope of unity. In other words, the allometric scaling between, for example, age at maturity and fecundity was expected to be constant, a **dimensionless number**, regardless of species and body size (Charnov and Berrigan 1991; Charnov 1993).

### b *Life history diversification in a multidimensional space*

These predictions were tested at various taxonomic levels (Shine and Charnov 1992; Berrigan *et al.* 1993) and found to grossly match model predictions. These views were, however, challenged both theoretically and empirically. Ferrière and Clobert (1992) showed theoretically that removing the assumption of population stationarity revealed other directions of demographic parameter associations besides the classical slow–fast continuum (Box 5.2). Using a comparative approach, Gaillard *et al.* (1989) and later Clobert *et al.* (1998) found empirically that other demographic trait co-variations could occur besides the slow–fast dimension suggesting that life history strategies probably vary in a multidimensional way, of which the $r$–$K$ continuum was only one of the possible dimensions.

### c *Life history diversification likely also involves behavioural strategies*

Most of these studies however, only analysed a restricted set of possible demographic traits, and it is likely that if other traits had been incorporated, many other patterns of co-variation would have been revealed. Indeed, the inclusion of age/stage-specific demographic traits such as senescence (Charmantier

#### Box 5.2 **Life history co-variation with size and size-independent effects**

The role of body size is clearly taxon-dependent. In a comparative analysis, body size explains 60% of variation in demographic tactics in birds, but only 40% in mammals (possibly a role of flight constraints; Figure 5.3a). However, other parameter associations were visible besides size effects. When size effects were removed, another axis of variation appeared: the slow–fast continuum. This second axis (Figure 5.3a) explains slightly more variation in the association of life history traits than body size in mammals (45%), but less in birds (only 35%). Finally, a third-order strategy (*sensu* Western 1979), which is called lifetime reproductive effort (or semelparous-iteroparous gradient), explains a small part (especially for birds) of the remaining variation.

The existence of multiple trade-offs received some theoretical support when more demographic complexity was taken into account (Ferrière and Clobert 1992). In a theoretical approach, three axes of life history variation emerge (Figure 5.3b):

1. The classical **slow–fast continuum**, here represented by the x axis of adult survival rate (empty circles in Figure 5.3b).
2. A **semelparous intensity gradient**, here visualized as variation in the age at maturity, but in which adult survival rate after reproduction is always low (black stars in Figure 5.3b).
3. An **iteroparous gradient**: always implying early reproduction, but either only once or several times within life. Such strategies can be coupled with either low or high survival after the first reproduction (empty stars in Figure 5.3b).

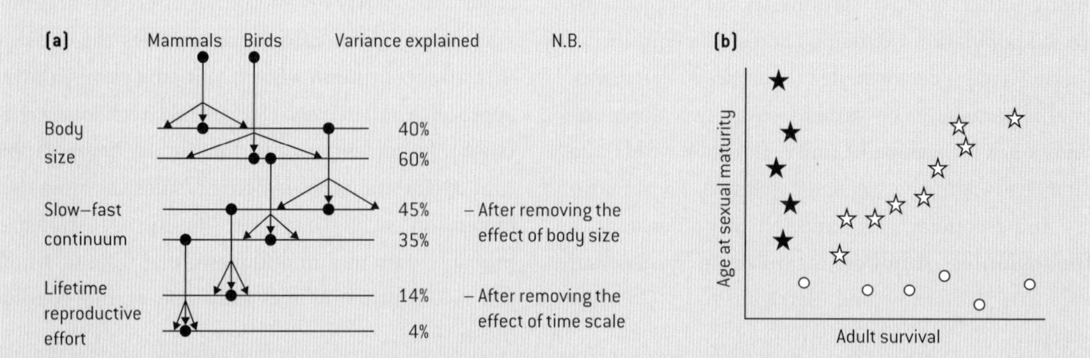


Figure 5.3 **Variations in life history trait associations**

**a.** In a comparative approach of mammals and birds: percentage of variation in life history traits explained by different levels of association between variables. The three effects (body size, slow fast, and lifetime reproductive effort) were analysed sequentially. The horizontal lines represent scales of either the time (top two lines) or the slow–fast continuum (middle two lines), or the lifetime reproductive effort (bottom two lines). Arrows pointing down from an axis illustrate the extent to which, for a given body size, variation in life history is left that can be explained by another life history axis of variation: for a given position along the body size axis or on the slow–fast continuum, the subsequent scale effect may explain part of the remaining variation in demographic tactics). The more widely spaced the arrows, the greater the portion of inter-specific variance explained by the relevant combination of parameters. In this comparative study the authors only considered three demographic parameters: yearly fecundity, age at maturity and yearly adult survival.

After Gaillard *et al.* (1989).

**b.** Results from a model of life history evolution. Each star or circle represents a combination of demographic parameters (here represented on the age at maturity versus adult survival rate plot). Each star or circle corresponds to a combination of demographic traits that is evolutionarily stable. A species was characterized by its inherent net reproductive rate $R_0$ (which is the product of survival and fecundities at different ages). $R_0$ grossly corresponds to the maximum growth rate that a given species can display. This model shows that many types of demographic trait association are theoretically possible, and most probably many more would be found if more life cycle complexity and demographic traits were incorporated into models.

Modified from Ferrière and Clobert (1992).

*et al.* 2006; Reznick *et al.* 2006; Williams *et al.* 2006) or inclusion of lower-level fitness components such as clutch size, number of clutches within a reproductive season, or even morphological, physiological, or behavioural traits, might deeply change our views about life history evolution. Other trait dimension, such as its variance, which is an indication of trait plasticity, might also co-vary with the trait mean or even with the mean or variance of other traits (Newman 1994; Ernande *et al.* 2004; Box 5.2).

Without going more deeply into the machinery that controls trade-offs and evolution, the general message is that we may still have a too simplistic view of life history diversification and life history trade-offs. Life history diversification is likely to occur within a multi-dimensional space incorporating many parameters, among which are distributed the behavioural strategies of a population.

In the following sections, we will examine other ways of looking at life history evolution by focusing more on the major selective pressures that act on species evolution, as well as the mechanisms that underpin such evolution, and constraints and trade-offs in a phylogenetic context. A complete genetic perspective begins with a phylogenetic perspective. We will purposely focus on behavioural traits.

## 5.3 Contingency, the evolution of multidimensional trade-offs, and phylogenetic constraints

At a macro-evolutionary scale involving millions of years and speciation, trade-offs evolve over stretches of time, which if sufficiently vast, allow for novelty to evolve. As innovations are refined and functional integration proceeds, this process constrains subsequent innovation, generating a contingency to trade-offs that makes them taxon specific (e.g., dichotomy between avian and mammalian life histories; Box 5.2). Phylogenetic constraints are idiosyncratic with respect to lineage-specific suites of evolved traits, which is illustrated with an example of salamander life cycles. Evolutionary transitions in salamander life cycles illustrate the contingency in trade-offs and interactions of life history trade-offs with behavioural, endocrine, neural, and morphological systems. We also use the amphibian example to illustrate how supergenes play a role in shaping super trade-offs, which can collapse as life history cycles are simplified. Endocrine systems, which regulate life cycles, comprise supergenes for physiology and behaviours.

### 5.3.1 The hierarchical structure of endocrine regulation arises from supergenes

Supergenes that control life cycles and behaviour are readily identified in the genes of endocrine regulation (Figure 5.4). The complex life cycle of *Ambystoma* is depicted in Figure 5.4. That species has an aquatic larva that transforms into a terrestrial adult. That transformation is referred to as **complex metamorphosis**. Examining the endocrine control of metamorphosis allows us to define hierarchical structure in endocrine systems. The endocrine systems involved in metamorphosis are hierarchically structured into **negative and positive regulatory networks** consisting of two main **endocrine axes**: the hypothalamic–pituitary (HP)–thyroid axis (a system regulating development) that interacts with the HP–prolactin axis (a system regulating growth). Thyroxine and prolactin form a third axis of regulation. Within each endocrine axis, hormones govern gene transcription and regulation through an **endocrine cascade**.

> For the moment, we can define a supergene to be a key regulatory gene of development that controls the expression of an **endocrine cascade**. Later, we will broaden our definition to include salient genes and gene interactions that include all aspects of behaviour.

In all vertebrates, metabolism is regulated by thyroxine, a small chain of eight amino acids called an octopeptide. In all vertebrates, a diverse class of protein hormones collectively called a growth hormone gene family regulates growth rate. In amphibians, a specific growth hormone called prolactin interacts with other growth hormones (Sawada *et al.* 2000) to regulate larval growth and inhibit metamorphosis (Denver 2000; Point 1 in Figure 5.4). High levels of prolactin inhibit metamorphosis, but enhance larval growth (Point 1 in Figure 5.4). At a critical larval size, growth slows (Point 2 in Figure 5.4) and prolactin secretion shuts down (Point 3 in Figure 5.4), thereby releasing inhibition of thyroxine-releasing hormones (TRH) by prolactin. Elevated TRH produced by the hypothalamus triggers release of thyroid-stimulating hormone (TSH) by the anterior pituitary (Point 4 in Figure 5.4). TSH simulates the thyroid to produce thyroxine, more specifically $T_4$. In specific tissues, an enzyme called monoiodinase converts $T_4$ to the more potent $T_3$ (ellipse on the left of Figure 5.4). In cells of these tissues, $T_3$ binds to two classes of thyroxine receptor, $\alpha$ and $\beta$ receptors, to trigger gene transcription, which reorganizes larval body form (cell death, resorption of gills and tail fin, melanization, reorganization of skull and hyoid bones). Thus, the HP–prolactin–thyroid systems achieves metamorphic climax from larval to adult form (Figure 5.4).

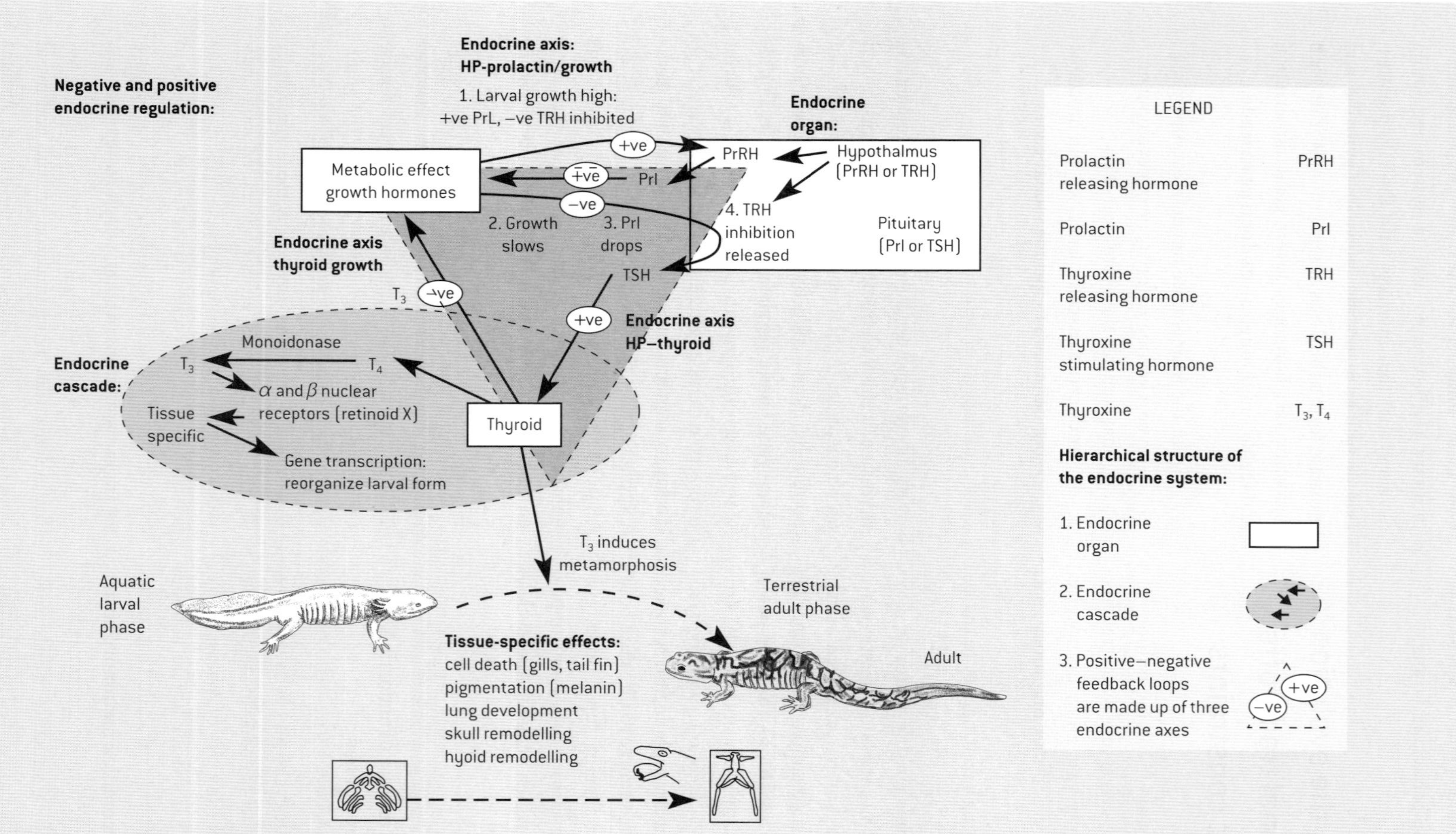


Figure 5.4 **Endocrine control of life cycle states in salamanders.**

Hormones that regulate metamorphosis and migration in a pond-breeding amphibian (*Ambystoma*). The hypothalamo–pituitary (HP–)–prolactin system (HP–prolactin system) interacts with the HP–thyroid system to generate larval growth and metamorphic climax (see text for a discussion of endocrine regulation and positive and negative regulation). The endocrine system is hierarchically organized into: (1) endocrine glands; (2) endocrine glands that trigger endocrine cascades of gene transcription; (3) two or more endocrine glands (and their endocrine cascades) that interact to form endocrine axes; and (4) three endocrine axes that compose the fundamental unit of positive and negative endocrine regulation that generate organismal homeostasis or that trigger behavioural and physiological changes in life cycles.

### 5.3.2 Rearrangements of life cycle diagrams are achieved by supergenes

In mammals, birds, and reptiles, life cycles only involve direct development in which the juvenile that hatches or is born already has the same form as the adult. This is not necessarily the case for most amphibians. Salamander taxa, for instance, have undergone fundamental changes in life cycle states from metamorphosis to paedomorphosis or to direct development, and then back to metamorphosis (Box 5.3). The loss of the adult stage from the ancestral **metamorphic** form, which had aquatic larvae and terrestrial adults, led to evolution of paedomorphic forms that mature

Box 5.3 **Phylogenetic origins of life history innovations and constraints**

The mapping of the various developmental strategies on the phylogeny of the salamanders. Amphibians provide evidence of multiple evolutions and reversals in the type of developmental strategies adopted by the various extant species Figure 5.5a. Paedomorphosis has repeatedly evolved in the plethodontid genus *Eurycea* (Figure 5.5a). In the most dramatic example of convergence, a cave salamander *Eurycea rathbuni* superficially resembles the more ancient and paedomorphic *Proteus* (in the Proteidae; Figure 5.5a), which constitutes another cave-dwelling taxon. Other *Eurycea* species have metamorphic life cycles (not shown in Figure 5.5a). Another branch leads to the genus *Thorius*, a direct developing salamander that exhibits the most extreme miniaturization seen in vertebrates with adults as small as 13 mm long, and heads as small as 3 mm (Figure 5.5a). Such miniaturization exerts profound allometric constraints on neurodevelopment and thus behaviour (see text for a discussion of cell size and genome size).

The evolution of direct development in plethodontid salamanders is not an irreversible trait, despite a dramatic rearrangement of life history trade-offs. *Desmognatus*, for instance, shows a reversal from an ancestral species with direct development to a species that shows more complex metamorphosis (Figure 5.5a). The presence of yolky eggs in desmognathine salamanders (e.g., *Desmognatus quadramaculatus* is depicted) reflects retention of life history attributes that are the hallmark of their direct-developing ancestors. A supergene associated with metamorphosis and thyroxine regulation allows for dramatic convergent evolution (paedomorphosis) and evolutionary reversals (direct-developing back to metamorphosis, see Figure 5.5a).

Interestingly, observations on the projectile tongue in salamanders shows that evolutionary reversal from direct development to metamorphosis is associated with reacquisition of ancestral features (four epibranchials in the hyoid bones, Figure 5.5b), which are referred to as atavisms.

By tracing up the branches to one twig in the phylogenetic tree (salamanders → Plethodontidae → *Thorius*), we can track a suite of multidimensional changes in life historical, morphological, and behavioural trade-offs associated with direct development. Other lineages exhibit reduced life cycle dimensionality (paedomorphosis: salamanders → Plethodontidae → *Eurycea*; Figure 5.5). Finally, evolution of **facultative paedomorphosis**, found in families with metamorphosis (salamanders → Ambystomatidae → *Ambystoma*; Figure 5.5), involves trade-offs faced by both paedomorphic and direct-developing lineages. However, at times these multi-dimensional trade-offs collapse, such as when *A. tigrinum* gave rise to the axolotl (Ambystomatidae; Figure 5.5a).

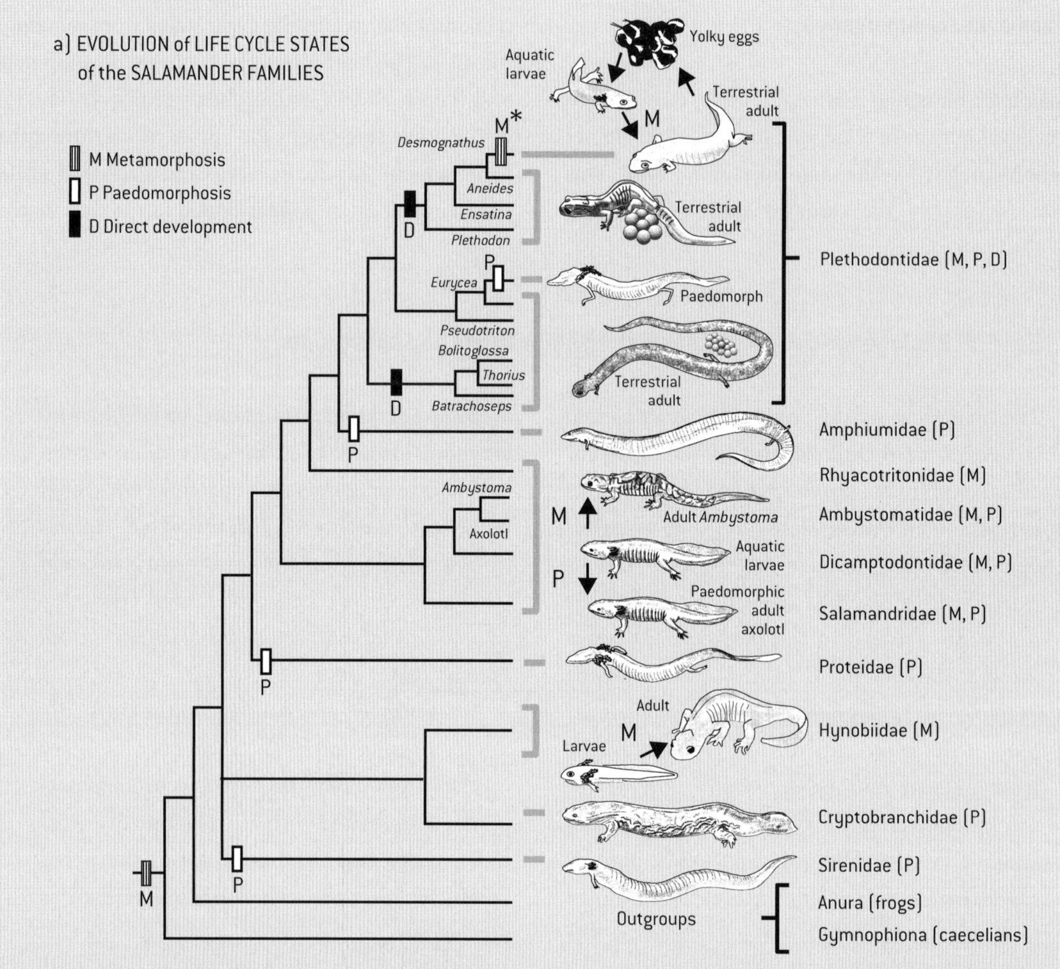


Figure 5.5 **The phylogeny of salamanders, and tongue and hyoid evolution in plethodontid salamanders**

**a.** Salamander diversity and evolution of life cycle states. Salamanders have evolved novel life cycles from ancestral **metamorphosis** (M), a condition also ancestral to other orders in the class amphibia (i.e. anura or frogs and Gymnophiona or caecelians). The evolution of paedomorphosis (P), where the terrestrial stage is lost, a convergent adaptation with multiple origins across salamander families (Amphiumidae: *Amphiuma means*, Proteidae: *Proteus anguis*, Cryptobranchidae: *Cryptobranchus allegheniensis*, and Sirenidae: *Siren intermedia* are depicted here). Paedomorphs are also recent in origin in Ambystomatidae, Dicamptontidae, Salamandridae. In these families, paedomorphs have evolved within species (facultative paedomorphosis) and between species. This is illustrated by the metamorphic *Ambystoma tigrinum* and paedomorphic axolotl, *A. mexicanum*, which is evolutionarily derived from *A. tigrinum*. The Plethodontidae exhibit the most diverse life cycles with metamorphosis, paedomorphosis, and direct development. Those with **direct development** have evolved either large yolky eggs as in *Batrachoseps attenuatus* and/or extended maternal care as in *Ensatina eschscholtzii*, where egg guarding is observed. Phylogeny adapted from Wiens *et al.* (2005). Placement of *Thorius* is from Mueller *et al.* (2004). Illustrations are by Sinervo.

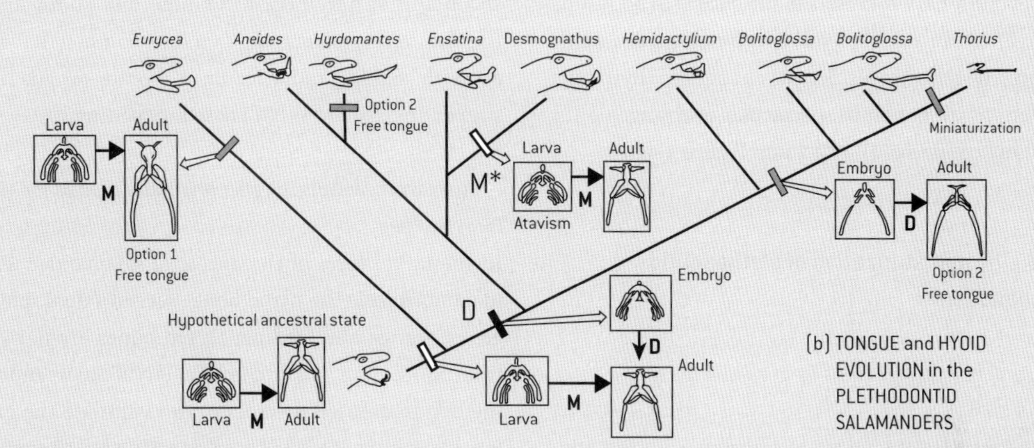


Figure 5.5 **(cont'd)**

**b.** Tongue and hyoid evolution in plethodontid salamanders. The drawings in the small boxes represent the shape of the hyoid bone structure in larvae and adults of the different taxa. Across the plethodontid clade evolution shows dramatic and parallel changes from a projectile tongue that is attached with a retractor muscle, to free tongues in which muscle attachment is lost. Evolutionary reversal from direct development to metamorphosis is associated with reacquisition of the ancestral features (the atavistic reacquisition of more epibranchials in the hyoid bones in *Desmognathus*). Three lineages have evolved a free tongue: metamorphic *Eurycea* (option 1) and direct-developing *Bolitoglossa* and *Hydromantes* (option 2). Option 2 allows more distance in tongue projection compared with option 1. The extreme miniaturization of *Thorius* salamanders allows for further refinements in option 2. Drawings, adapted by Sinervo, are from Wake and Larson (1987).

Phylogeny adapted from Mueller *et al.* (2004).

in an aquatic environment. Loss-of-function mutations in the genes of thyroxine regulation in an ancestral metamorphic form led to paedomorphs (Figure 5.5). In contrast, loss of aquatic larval forms led to evolution of **direct development** on land. Different trade-offs apply to organisms on land versus water, and higher dimension trade-offs apply to organisms that must be adapted to both environments due to an increase in life cycle complexity. Here we focus on the spectacular adaptations of one group, plethodontid salamanders, to terrestrial environments, but also describe the converse condition, paedomorphic salamanders that are strictly adapted to water.

In salamanders, evolutionary changes in life cycle arise from modification of a supergene that controls the hormone thyroxine (Voss and Shaffer 1997), which interacts with developmental systems that organize all of morphology and behaviour during metamorphosis. We will discuss the genomic structure of supergenes in the next section, but first we must describe all the functional systems that are modified when direct development evolves, and thus novel trade-offs that have been reshaped across these groups.

Direct development, which is the hallmark of plethodontids (a subset of the salamander radiation), evolved at least twice (Box 5.3). Amazingly, this is not an irreversible trait because members of the plethodontid salamanders in the genus *Desmognathus* re-evolved aquatic larvae (Mueller *et al.* 2004). Developmental interactions involving supergenes are thought to remain intact over long periods of evolutionary time (Wiens *et al.* 2005) allowing for such evolutionary reversals in life cycles. In addition, supergenes can rapidly evolve into new states, reorganizing new life cycles. Simple changes in thyroxine regulation can generate spectacular rearrangements of life history cycles (Box 5.3 and Figure 5.5a).

### 5.3.3 The miniaturization of plethodontid salamanders

*The loss of larval stages frees up evolution to refine novelty . . .*

Plethodontid salamanders lost metamorphosis and acquired direct development early in their evolutionary history (Figure 5.5). Direct development constitutes a novel life history adaptation for terrestrial environments. In salamanders with direct development, larval development is restricted to eggs and contrary to all other salamanders, juveniles hatch into an adult form. Larvae feed from a very yolky egg that allows for larger size at hatching. The developing embryo exchanges gas ($O_2$–$CO_2$) with the atmosphere through egg membranes and a circulatory system (like in chicken embryos, but lacking a calcified shell, a key innovation of amniote vertebrates). Besides yolky eggs, direct development is associated with extended egg guarding in some species (e.g., *Ensatina*, Box 5.3), a life history with high maternal investment and care compared with metamorphic ancestors with many small eggs and no care.

*. . . from the rearrangement of life cycles to the loss of lungs . . .*

In the ancestral condition of salamanders, aquatic larvae undertake a complete metamorphosis to produce a terrestrial adult and this adult breathes with lungs. Since the time that fish ventured onto land, terrestrial vertebrates have breathed with lungs. However, plethodontid salamanders have discarded lungs and instead use buccal ventilation, moving air rapidly across blood vessels in the throat. They also use skin gas exchange, moving gases across skin. Why do away with lungs, because allometric constraints of surface area to volume will surely limit the size of the salamander? One reason may have to do with making room for an elaborate tongue and supporting skeleton that is used in feeding.

*. . . to rearrangements in bones originally evolved for gills into a tongue . . .*

In the larval form of the ancestral salamander with metamorphosis the hyoid bones served a dual role in gas exchange and feeding. Hyoid bones are derived from vertebrate gill arches. In larval salamanders, the bones are used to support and ventilate the gills (i.e. gas exchange). The bones are also used to expand the buccal cavity and suck prey into the mouth. When these salamanders metamorphose, the hyoid bones are completely reorganized into a tongue projection mechanism used to capture prey in a terrestrial environment. The hyoid has three roles in salamanders with complex metamorphosis, each subject to a set of constraints and functional trade-offs: larval gas exchange and feeding, and adult tongue projection.

Thus, the many roles of the ancestral hyoid in larval gill ventilation and feeding were freed up in plethodontid salamanders, allowing morphology to specialize on feeding (Figure 5.5b). Consequently, plethodontids have highly refined projectile tongue, which involves exquisite modification of the ancestral hyoid. Loss of lungs in plethodontids further allowed the hyoid to extend into areas of the body normally required for lungs (Box 5.3). Long bones that support long tongues need lots of space for retraction. Lungs were probably reduced to house a long hyoid. Gas transport was transferred to the buccal cavity and skin, areas that ancestrally supported gas exchange, even in the first vertebrates that crawled onto land.

*. . . to miniaturization, cell size, and the scale of structures*

We can follow the branching phylogeny of plethodontid salamanders down to one twig, the genus *Thorius*. The case of the evolutionary miniaturization of plethodontid salamanders in the genus *Thorius*

(Wake and Larson 1987) illustrates the multilayered nature of behavioural and life history adaptation and it illustrates developmental constraints that are directly tied to allometry. Plethodontid salamanders that lack lungs must stay small compared with lunged salamanders because allometric surface to volume constraints limit gas exchange. Other allometric constraints arise from genome size and cell size.

In a final evolutionary step, salamanders of the genus *Thorius* underwent an evolutionary process of extreme miniaturization (Wake 1991; Hanken and Wake 1993). Normally, evolutionary miniaturization is not problematical, even if it does entail scaling constraints (Hanken and Wake 1993). However, salamanders have the largest genomes of vertebrates (Sessions and Larson 1987). Cell size of all metazoa scales allometrically with genome size (an example of a universal metazoan constraint). Thus, decreasing organismal cell size was not an option during miniaturization. Instead *Thorius* retained an elaborate nervous system, but cell number was reduced to a minimum, given the coordinated visual and nervous system that drives a projectile tongue (Wake and Larson 1987; Linke *et al.* 1993).

*A textbook example of the complex effects of allometry on behaviours and nervous system*

The salamander example serves to highlight three important points. First, design limitations or constraints often arise from complex genetic and developmental interactions among many functional systems. Secondly, allometry sets fundamental constraints on life history evolution, morphology, and even nervous systems. Thirdly, any general theory of life history trade-offs built upon fundamental principles of development and genetics, must have a framework for dealing with such complexity in functional design.

Viewing trade-offs from the framework of proximate mechanisms and phylogeny builds a whole-organism perspective of the role of life history in shaping behaviour (and vice versa). As new adaptations evolve in a given lineage, they close doors to certain options (i.e. constraint), but at the same time, they open new doors to further novelty. One cannot simply understand the multidimensional nature of life history evolution simply from the scaling effects of size. In salamanders, the evolution of life cycle involves major rearrangements of traits (e.g., contrast the convergent traits of aquatic paedomorphs with those of direct development: no aquatic larval stage, large, yolky eggs, maternal care, and egg guarding).

### 5.3.4 Expansion or collapse of multidimensional trade-offs due to supergenes

#### 5.3.4.1 *Multidimensional trade-offs collapse or expand with simplification or elaboration of life cycles*

As we have seen in Section 5.2, the dimension of a species life cycle (number of life cycle transitions) limits the number of potential demographic trade-offs, even if trade-offs at other levels (including morphological, physiological and/or behavioural traits) might be less constrained by life cycle complexity. Paedomorphs that only deal with aquatic environments, experience a lower dimensionality of trade-offs compared with metamorphs that are subject to both aquatic and terrestrial constraints. Evolution can act on the intensity and even geometry of trade-offs (Chevrud 1984; Phillips and Arnold 1989), increasing or decreasing trade-off dimensionality. Correlational selection that reorganizes trade-offs (Sinervo and Svensson 2002; Roff and Fairbairn 2006) acts on endocrine regulation (see Section 5.4.3).

#### 5.3.4.2 *Novel stages can be inserted into life cycles or deleted from life cycles*

Selection can impose new environmental conditions that favour the addition of new stages to a life cycle, resulting in the emergence of new transitions, that is new metamorphoses, and by consequence new sets of trade-offs become associated with the new metamorphic events (Shaffer *et al.* 1991). The reacquisition of metamorphosis in *Desmognathus* salamanders provides an example of the re-acquisition of ancestral trade-offs. However, these trade-offs are re-acquired in the context of more elaborate maternal care that is the hallmark of the plethodontid salamanders (e.g., more yolk and even egg guarding). This potentially imparts even more dimensionality for trade-offs exhibited by the re-evolved complex life cycle of *Desmognathus* compared with ancestral salamanders with complex metamorphosis.

An increase in life cycle complexity is exemplified by eastern newts, *Notophthalmus viridescens*, which have an aquatic larval form, a dispersive **red eft form**, and a stream breeding aquatic stage (Pope 1928, Chadwick 1950). Rather than just two body forms (larval, terrestrial adult), the eastern newt has three distinct body forms and behaviours, adding an entirely new phase to the life cycle typical of most amphibians. Thus, eastern newts exhibit a **bimetamorphic** life cycle. The red eft that has been inserted into the life cycle is protected by its striking aposematic red coloration, a behavioural signal to deter predation. These juveniles are not just colourful, a signal that deters predators, but also toxic, a physiological adaptation that reinforces the meaning of the signal. Evolution of aposomatism confers red efts with greater dispersal potential, a behavioural adaptation, than other metamorphic forms that are readily eaten by avian predators if they are in the open. Thus, anti-predator adaptations like aposematic signals further dimensionalize life history trade-offs by coupling morphology, physiology and behavioural adaptations into a single behavioural syndrome. When such aposematic forms evolve, they are invadable by mimicry strategies in other species that lack toxicity and any of the costs of toxicity (Sinervo and Calsbeek 2006), but benefit from expression of anti-predator signals. For example, *Pseudotriton ruber* is non-toxic but mimics toxic red efts. Sinervo and Calsbeek (2006) describe how such ramifying ecological interactions invoke even higher dimensionality to trade-offs, because aposematic models can experience costs from predators that learn to attack undefended mimics, which evolve to be common in frequency.

When life cycle phases disappear trade-offs are reorganized and may be simplified (e.g., paedomorphosis). Other behavioural innovations might evolve in the place of ancestral trade-offs. For example, the HP–prolactin axis regulates migratory behaviour in the vertebrate classes such as fish, birds, and amphibians (Rankin 1991). In terrestrial adult amphibians, an increase in prolactin (triggered in the HP by either photoperiod or rainfall) induces water drive behaviours that promote a migration to natal ponds. Notice that the deletion of the HP–thyroid function in paedomorphs collapses trade-offs associated with migration and any adult trade-offs (Box 5.4). Conversely, loss of the larval phase in direct developing salamanders eliminates the need for water drive, migratory behaviour, and their associated trade-offs.

We will see in Chapter 6 that HP–prolactin regulation has also been coopted by all the vertebrate classes into an endocrine system that regulates parental care. Migratory behaviour if present, and care-giving behaviour of all vertebrates has its origins in the water drive of amphibian migration (Rankin 1991; Schraden and Anzenburger 1999). **A phylogenetic analysis of the evolution of endocrine regulation in amphibians enriches our understanding of trade-offs and the origins of behavioural traits like migration and maternal care in other vertebrates.**

In summary, a study of endocrine mechanisms that generate trade-offs helps us to understand how likely they are to evolve, become organized, elaborated as new innovations, or collapse. Life history trade-offs are inextricably linked to all kinds of behavioural processes like anti-predator behaviours, social behaviours such as philopatry and dispersal, and mating system behaviours. In the face of such complexity, how can we begin to unravel the genetic sources of trade-offs?

Endocrine cascades are a potent class of interacting genes that comprise supergenes for behaviours. Presence or absence of trade-offs or elaboration of greater numbers of trade-offs are governed by these supergenes of the endocrine system. Thus, the number of trade-offs, which we refer to as **trade-off dimensionality**, has a direct mechanistic basis in the interacting components of endocrine regulation.

Box 5.4 **Endocrine supergenes, trade-off dimensionality, and life cycles**

Several axes of the endocrine system interact to achieve organismal homeostasis during reproduction. Energetic limits on organismal homeostasis in turn generate life history trade-offs. As we will see in Chapter 6, most of the physiological functions of vertebrates are controlled by the HP axis that involves a series of cascades of neurohormones and hormones that control growth, brain development, reproduction and adaptation to stress. The regulatory genes of hormone systems are present in all metazoa, but their mode of action on life history trade-offs have only been identified in a few cases (see Zera and Harshmann 2001). Endocrine regulation is particularly well studied in amphibians (see Denver 2000; Hayes 1997a, b).

As noted in Figure 5.4, endocrine regulation can be visualized as interacting endocrine axes (one side of each triangle) that complete a network of positive and negative regulation (one triangle) (Figure 5.6a). The HP organ is a master regulator of endocrine regulation in all vertebrates (triangle apex). Interplay between protein hormones produced by the HP, and

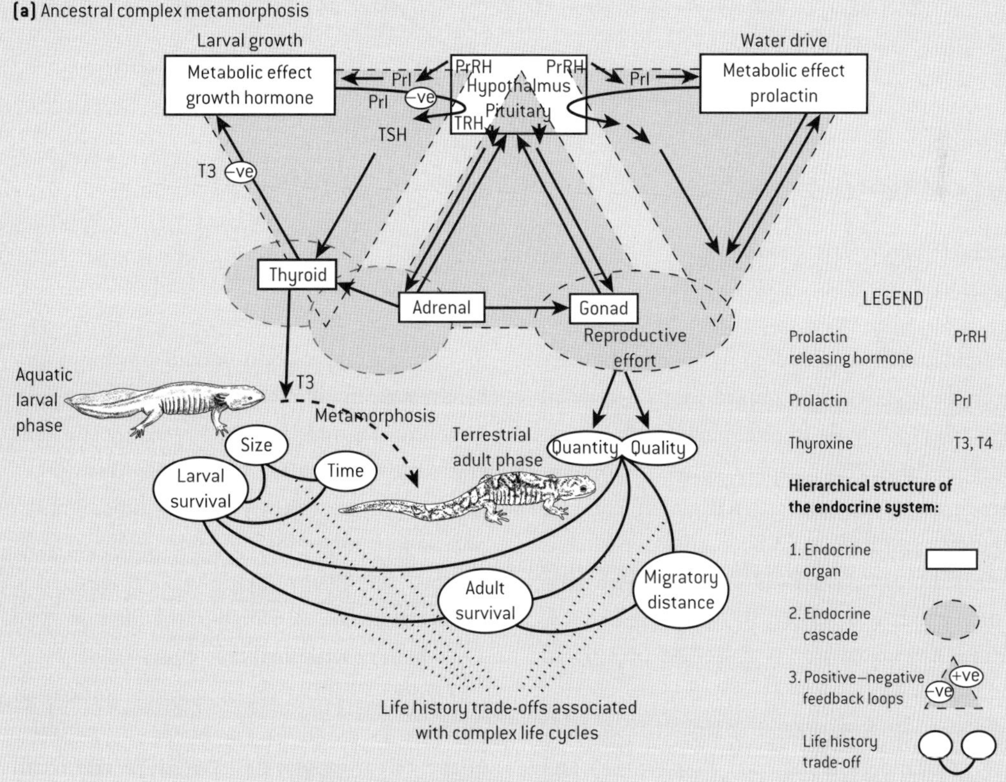


Figure 5.6 **The endocrine control of life cycle states in salamanders and the vertebrate endocrine system**

Three different life history cycles are depicted for salamanders:

**a.** Complex metamorphosis (*Ambystoma*);

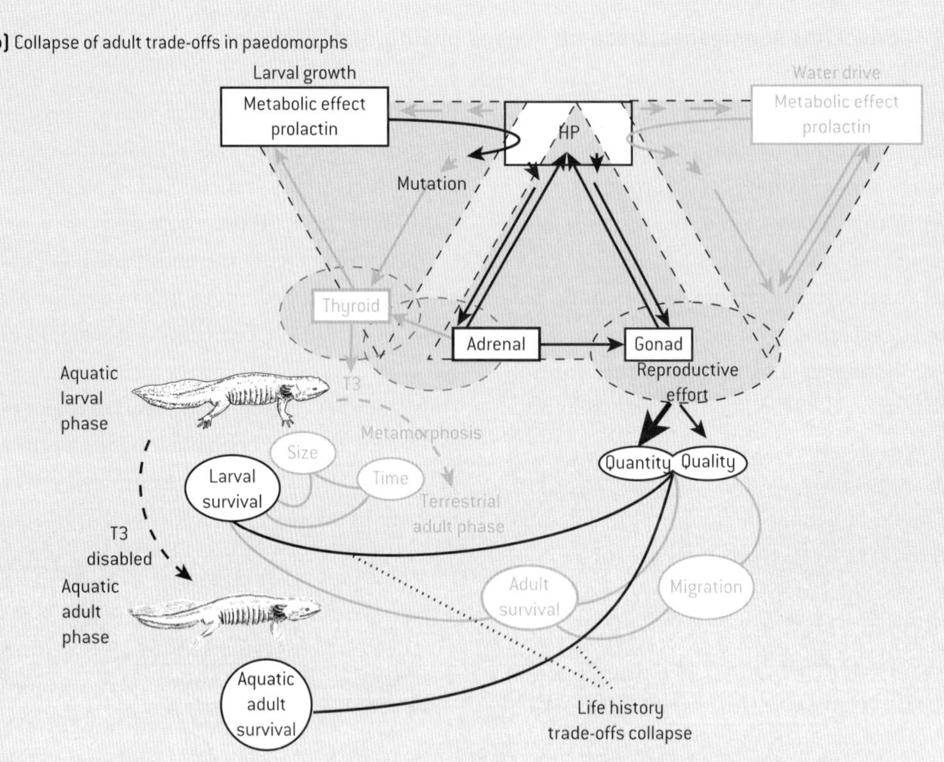


Figure 5.6 **(*cont'd*)**

**b.** Srictly paedomorphic (axolotl);

steroid hormones produced by gonads, regulate reproduction, which is termed the HP–gonadal axis. Reproduction is also modulated by the adrenal gland. The interplay between hormones produced by HP and adrenal glands controls energy mobilization into reproduction that is referred to as the HP–adrenal axis. The HP–adrenal axis is described in Boxes 5.5 and 5.6. The HP–gonadal axis is described in Sinervo and Calsbeek (2003).

Figure 5.6a. A generic complex life history for salamanders is depicted (*Ambystoma*) and the HP–thyroid–growth hormone axis regulates metamorphosis. In terrestrial adults, the HP–prolactin–growth hormone axis further regulates a behaviour called water drive that induces adults to migrate back to water and to reproduce. Prolactin modulates migratory behaviours in all vertebrates (Rankin 1991). Migration distance from the breeding site generates life history trade-offs. While the distance of migration alleviates local crowding around a pond, a long migration can lower adult survival, another life history trade-off. At least nine trade-offs can be identified in the life history of a salamander with metamorphosis.

Figure 5.6b. Loss-of-function mutations at different points along the regulatory pathway of the HP–thyroid axis have given rise to all of the families of paedomorphic salamanders depicted in Box 5.3. Changes in life history trade-offs from complex metamorphosis to paedomorphosis have been particularly well studied in the axolotl, which is derived from

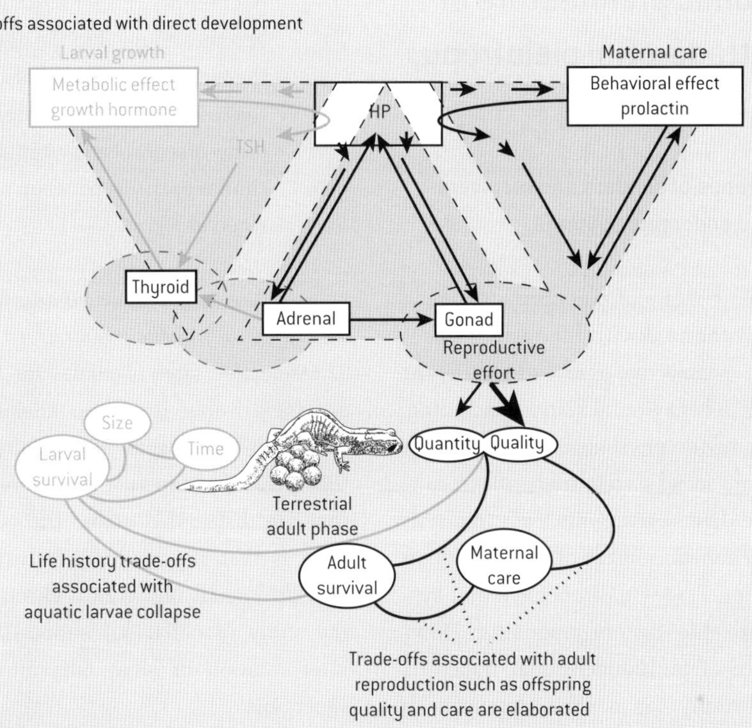


Figure 5.6 **(*cont'd*)**

**c.** Strictly terrestrial with direct development (*Ensatina escheschoItzii*).

*Ambystoma tigrinum*. Multi-dimensional trade-offs collapse when life cycle phases are deleted. The trade-off dimensionality in paedomorphic Axolotls is reduced to three from the ancestral dimensionality of nine trade-offs.

Figure 5.6c. Deletion of the larval phase has given rise to direct-developing plethodontid salamanders. In vertebrates, prolactin also regulates the expression of parental care (Schraden and Anzenburger 1999). Given the absence of water drive in terrestrial plethodontid species, prolactin machinery has likely been coopted for a novel function such as maternal care (e.g., observed in *Ensatina eschscholtzii* salamanders). Female *Ensatina* salamanders guard their eggs and secrete substances on their skin that is applied to eggs and inhibits fungal growth. It should be noted that in birds and mammals prolactin mainly plays a role in parental investment (see Chapter 6).

In summary, the number of different endocrine axes that are in operation across all of the life cycle phases is proportional to the number of interacting endocrine axes. Deletion of endocrine axes by loss-of-function mutations decreases the dimensionality of life history trade-offs. In other situations, loss of functional life cycle stages (e.g., aquatic larvae and pond-breeding adults in Plethodontidae) frees up the endocrine system for novel behaviours to evolve (e.g., maternal care through prolactin regulation, which was ancestrally associated with water drive and migration).

## 5.4 The genomic architecture of life history trade-offs: pleiotropy, epistasis, and plasticity

### 5.4.1 The supergenes of senders and receivers in behavioural and endocrine systems

The standard genetic views of pleiotropy and gene epistasis can be synthesized with life history trade-offs. This genetic perspective also links different views of trade-offs in behavioural systems including: genetics, development, physiology, **performance**, **mating systems**, and social system trade-offs. A consideration of these diverse trade-offs leads to a synthetic view of the genetic sources of constraints on behavioural systems. Trade-offs in behavioural systems interact with life history trade-offs. However, the existence of multidimensional trade-offs creates the opportunity for novel adaptations to arise, referred to as alternative behavioural strategies. Evolution of these strategies within species is controlled by supergenes of the endocrine system. New species can arise when supergenes collapse, or when supergenes evolve to be quite different, and elaborated to a higher degree. When new species evolve, ancestral trade-offs and endocrine systems are re-organized dramatically. When this occurs, individual genes of a supergene can rapidly evolve new functions (i.e., water drive in migration due to prolactin becomes coopted for a different and specialized role in maternal and/or paternal care-giving behaviours).

> The standard genetic view is that all trade-offs arise from pleiotropy (Reznick *et al.* 2000). This notion can be refuted for behavioural systems that must involve epistasis, the interaction of two or more genes. A simple example that illustrates why behavioural trade-offs arise from epistasis is found in the female preference for male exaggerated traits. In the Fisherian runaway process (see Chapters 11 and 20), female preference for extreme and exaggerated phenotypes (Fisher 1930; Lande 1981) becomes coupled to the genes that code for the males' signalling traits. Below, we show that other forms of social selection and the trade-offs of social systems arise from similar epistasis between sender and receiver loci.

*Other supergenes for behaviour arise from sender–receiver systems extrinsic to the organism*

The selective fates of alleles for female preference are inextricably linked to alleles for male traits, forming a 'supergene' for communication between the sexes. This supergene happens to involve genes that are distributed across the genome. The usual view is that supergenes are clusters of genes located at a small region of a chromosome. However, the study of behaviour supports a much broader interpretation of what falls into the category of supergene.

> A more general definition is thus that a supergene involves the interaction of two or more genes whose fates are inextricably linked by common physiological mechanism (i.e. sender and receiver molecules of the endocrine system) or by common behavioural mechanisms (i.e. communication between sender and receivers in the context of sexual selection or social selection).

The **physiological epistasis** of the endocrine system (sender and receiver molecules called hormones and hormone receptors) or **behavioural epistasis** of senders and receivers in mating systems can be teased apart experimentally (Sinervo and Basolo 1996). Experiments should be applied in tandem with genetic inference. Only genetic inference can resolve whether life history trade-offs are due to pleiotropy or gene epistasis and only experiments can elucidate endocrine effects that generate life history trade-offs. Below, we discuss experiments on the endocrine system and in sender–receiver communication.

Because physiological epistasis can arise from environmental causes (Boxes 4.5 and 4.6), we must explore the role of endocrine genes in plasticity. In addition to being plastic in the face of abiotic environmental cues, organisms may also use signals and/or presence of conspecifics to trigger life history plasticity. Thus behavioural epistasis that is generated by senders and receivers in a mating or social system, can also result in plastic changes in either behavioural or physiological traits. Such socially induced plasticity can alter life history trade-offs just like plasticity induced by the abiotic environment.

### 5.4.2 The control of life history plasticity arises from endocrine supergenes

The goal of this section is to elaborate on plasticity due to endocrine regulation, and the regulation of alternative behaviour patterns (and morphology) within a single species such as the alternative metamorph versus paedomorph strategies in *Ambystoma* (Semlitsch 1988; Semlitsch *et al.* 1988; Box 5.4). A portion of plasticity arises from the action of endocrine axes with feedback and information from the nervous system. In salamander species where individuals facultatively switch from paedomorph to metamorph (i.e., *Ambystoma* salamanders), or in species where other alternative larval morphs have evolved, plasticity is achieved through a synergism of the HP–adrenal and HP–thyroid axes (Figure 5.6). The source of most life history plasticity in development and reproduction is derived from the HP–adrenal axis.

For example, the hormone thyroxine, which governs amphibian metamorphosis, also governs the development of size and time to metamorphosis. The size versus time-to-metamorphosis trade-off is fundamental to amphibian life cycles. The cues used to induce early time to metamorphosis (but smaller size) arise from the HP–adrenal stress response as induced by social crowding (e.g., biotic cues) as well as by pond drying (e.g. abiotic cues; Denver 2000; Hayes 1997a, b; see examples in Chapter 6).

In addition to this trade-off of size versus time to metamorphosis, spade-foot toads exhibit plasticity with respect to a prey item, the fairy shrimp, which is either present or absent in their natal pond (Pfennig 1992). Two alternative larval strategies of omnivory versus carnivory are induced in spade-foot toads (Box 5.5), and the two strategies arise from plastic induction of the HP–thyroid axis. Fairy shrimp naturally contain high levels of thyroxine and thus, when larvae consume shrimp, they ingest exogenous thyroxine which triggers the development of a carnivorous morphology that is even more efficient at consuming shrimp.

The larval strategies of spade-foot toads are also associated with an antisocial behaviour of cannibalism versus social behaviour of schooling. Carnivorous tadpoles are more predisposed to exhibit cannibalistic behaviour, owing to large head size, stronger jaw muscles, and a highly keratinized beak. Omnivores are more social and school. Schooling behaviour increases the feeding efficiency of omnivores becaus a school of omnivores can more efficiently churn up the detritus on the bottom compared with solitary tadpoles.

Whereas carnivores eat conspecifics, kin altruism that is associated with the level of genetic relatedness can confer protection for larvae from their cannibalistic kin (Pfennig *et al.* 1993), thereby invoking social system trade-offs which we discuss below. Thus, plasticity and trade-offs in behaviour and life history that arise from endocrine regulation can be linked in very interesting ways to social selection.

In conclusion, pleiotropy in trade-offs can arise from an endocrine cascade. However, multidimensional trade-offs arise from physiological epistasis among multiple endocrine cascades, and behavioural epistasis of sender–receiver interactions among conspecifics. Diverse forms of life history plasticity in vertebrates are linked to the HP–adrenal axis and glucocorticoid steroid hormones like corticosterone. When life cycle complexity is reduced, one might expect that dimensionality reduction, which operated at a given level (genetic versus plastic, physiology, behaviour, demographic), generates more complexity at other levels by transferring trade-offs to these levels. We might also predict that plasticity and/or epistasis will be the favourite mechanisms by which transfers are facilitated (i.e. genetic to plastic control or vice versa). Furthermore, a reduction in true levels of genetic pleiotropy can then allow for fixation of trait combinations that are essential in the new life cycle configuration. One may even imagine that reduction always involves the lower levels and complexity is thus slowly transferred to higher levels (behaviour). Behavioural systems invoke a higher level of complexity because both endocrine and physiological effects, as well as the neural pathways of the central nervous system, govern the actualization of behaviours and coordinate the behaviours between senders and receivers.

### 5.4.3 Four genetic methods for resolving the genetic basis of trade-offs

> The standard view is that the trade-offs we observe among life history traits arise from a genetic trade-off (Reznick *et al.* 2000). A genetic correlation between traits that both positively affect fitness but are negatively related to each other, is termed a **negative genetic correlation**.

#### Box 5.5 **Endocrine regulation of life history plasticity in amphibian larvae**

The regulatory interactions of the endocrine system during larval development of the spade-foot toad (*Scaphiopus hammondii*) are shown in Figure 5.7 (see also Figures 6.1 and 6.8). As noted in Figure 5.4, control arises from the HP axis owing to gene regulation of peptide hormones such as PrRH and TRH which respond to metabolic inputs (growth, Point 1, Figure 5.7) and abiotic inputs (e.g., temperature, pond water level). However, the plasticity in the time or size of metamorphosis arises from the regulatory action of the HP–adrenal axis.

Life history plasticity in larval development of the spade-foot toad (Figure 5.7). Depending on either abiotic (point 2. in Figure 5.7) or biotic (point 3. in Figure 5.7) stressors (Denver 2000; Hayes 1997a, b), the hormone CRH triggers ACTH production by the pituitary, which in turns stimulates the adrenal to produce corticosterone. The steroid hormone corticosterone can induce earlier metamorphosis by acting on levels of PrRH, TRH, or the sensitivity of the thyroid to TSH. Under such stressful pond conditions, metamorphosis is accelerated and survival is enhanced, reflecting adaptive plasticity. Life history plasticity in larval development and the environmental cues that generate plasticity are shown: social crowding, pond drying, or the presence versus absence of fairy shrimp prey. Even though spade-foot toads exhibit pronounced plasticity, the changes in life history traits like size and time to maturity are still constrained by the basic life history trade-offs of amphibian larval development.

Social trade-offs and life history trade-offs in spade-foot toads (Figure 5.7). The basic trade-offs of amphibian development involve the time and size of metamorphosis, balanced against larval survival and survival of the juvenile after metamorphosis. Spade-foot toad larvae are subject to higher dimension trade-offs because of social interactions. Because carnivore morphs are more pre-disposed to cannibalism, the carnivores are selected to detect and avoid eating kin, but still eat unrelated tadpoles. Not eating kin constitutes a kin altruistic cost (of reduced growth and thus longer time to metamorphosis or smaller size) that benefits a genetically related tadpole. Thus, the alternative strategies of carnivore and omnivore invoke additional social trade-offs.

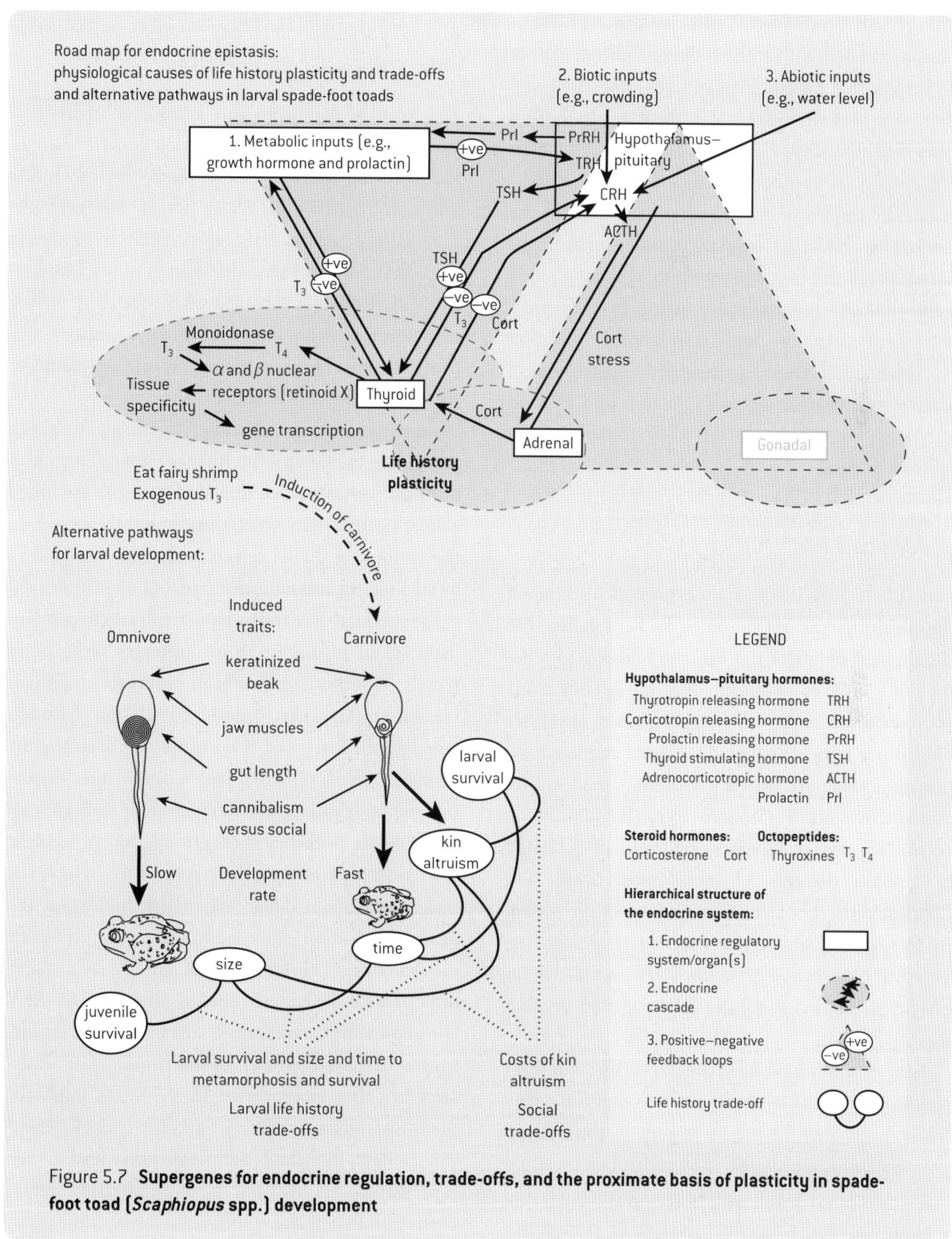


Figure 5.7 **Supergenes for endocrine regulation, trade-offs, and the proximate basis of plasticity in spade-foot toad (*Scaphiopus* spp.) development**

Genetic correlations and genetic trade-offs can have two very different causes. Trade-offs can arise from pleiotropy in which a single gene controls the expression of two or more traits, or from strong selection that generates linkage disequilibrium among physically unlinked genes, due to sender–receiver epistasis. Linkage disequilibrium, which is the non-random association of alleles at multiple loci (Lynch and Walsh 1998), can yield powerful genetic correlations. It is often assumed that because linkage will rapidly decay, negative genetic correlations associated with linkage disequilibrium cannot be responsible for life history trade-offs. However, there are several reasons why linkage disequilibrium will not always decay in the context of mating and social system dynamics. Here we discuss proximate origins of epistasis and pleiotropy in the context of endocrine regulation and genetic regulation of alternative strategies that can generate stable linkage disequilibrium owing to frequency-dependent selection and correlational selection on behavioural traits, particularly behaviours in sender and receiver systems (Sinervo and Clobert 2003).

#### 5.4.3.1 *Pleiotropy versus gene epistasis as a source of trade-offs*

Most life history theories are expressed in terms of the pleiotropic action of genes, such as genes that simultaneously affect the egg size and egg number trade-off (Sinervo 1999; Zera and Harshmann 2001) or other life history components (Box 5.4). Other life history trade-offs include costs of reproduction that is a trade-off between current versus future reproductive effort (Sinervo 1999; Reznick *et al.* 2000). The cost of reproduction trade-off is related to genetic theories of senescence (Rose and Charlesworth 1980a, b; Charmantier *et al.* 2006). In this genetic theory, senescence is thought to arise from a selective premium placed on alleles for early reproduction that have pleiotropic effects that shorten lifespan.

Four methods have been used to verify (or refute) pleiotropy in causing trade-offs: (1) artificial selection experiments; (2) negative genetic correlations estimated with a pedigree or by controlled breeding experiments; (3) manipulating gene expression and studying its ramifying effects on two or more traits; and (4) mapping multiple traits to single genes or refuting pleiotropy by mapping traits to many genes. The last two methods are similar in inference as both of these approaches directly address the genomic architecture of life history trade-offs that are associated with behavioural traits.

##### a *Artificial selection on one trait tugs on another, while sexual selection tugs on many traits*

The action of pleiotropy can be resolved in artificial selection experiments as associated genetic changes in other traits that are themselves not directly under selection (Lynch and Walsh 1998). For example, artificial selection on early reproduction in *Drosophila* can concomitantly reduce lifespan in only a few generations (Rose and Charlesworth 1980a, b). While such approaches are informative in a laboratory context, care must be taken in interpreting the results of such experiments in a natural context.

In nature, chronic selection (Sinervo and Calsbeek 2006) can create chronic linkage disequilibrium of unlinked genes (Lynch and Walsh 1998). As noted above, a situation in which this commonly arises is in the context of sexual selection that is predicted to build linkage disequilibrium among even unlinked sender and receiver traits. Thus artificial selection on male colour results in an evolutionary response of female preference, as has been observed in guppies (Houde 1994). The genetic correlation between female preference and male traits demonstrated by artificial selection on guppies must be due to gene interaction because the colour genes for male guppies reside on the Y chromosome (Houde 1994), whereas any genes that affect female preference, which the males pass on to daughters in the context of the artificial selection experiment, must logically reside on either an X chromosome or an autosome (sires only pass on the colour genes to sons through the Y chromosome). Complicating this analysis even further is the fact that each of the traits under sexual selection that is expressed in the sexes might be pleiotropically related to sex-specific traits in a natural context. In males, bright colour may lead to high reproductive success, but it may enhance rates of mortality. In females, preference for certain colour spectra may enhance a female's ability to perceive and feed on prey, but it pre-disposes females to prefer males with similar colours (Endler 1993; Ryan 1997). Thus, Fisherian runaway selection (see Chapter 11) that couples selective fates of male signals and female preference, likewise couples their disparate pleiotropic effects, thereby generating multi-dimensional trade-offs. Although artificial selection experiments are useful in uncovering pleiotropy, it is only in exceptional circumstances when the exact location of genes for given traits are known (e.g., colour resides on the Y chromosome in guppies; Houde 1994) that artificial selection can resolve an epistasis. This is why it is crucial to undertake gene-mapping studies (method d, below).

In summary, sexual or social selection in nature can continuously rebuild genetic correlations, even in the face of recombination and segregation. In Chapter 20, Section 20.3.7, we will see that this process remains valid even when female preference is not transmitted genetically but **culturally**. Constant sexual selection can generate 'linkage disequilibrium' between genes (or 'memes') that code for male traits and female preference, even if the latter is transmitted culturally (Sinervo and Calsbeek 2006). This remains valid between two cultural traits, but instead of genes, we talk about **'memes' that form the cognitive representations of the environment that are the units of cultural knowledge**.

### b *Pleiotropy assessed by genetic correlation and confirmed by endocrine experiments*

Another method to assess pleiotropy indirectly is to measure a negative genetic correlation between traits in the context of a pedigree (e.g., Svensson *et al.* 2001b) or through controlled breeding experiments (Sinervo *et al.* 2001; Ernande *et al.* 2004; Sinervo *et al.* 2006b). If two traits have a strong genetic correlation that promotes a negative relationship between fitness components (e.g., cost in one trait affords benefit in the other trait, behavioural traits included), the action of pleiotropy is suspected.

Definitive proof of pleiotropy does not arise from the sole measurement of negative genetic correlations per se. As noted above, **negative genetic correlations** can also arise from linkage disequilibrium generated by selection that correlates trait combinations (Sinervo and Svensson 2002; Sinervo and Calsbeek 2006). Therefore, experimental confirmation that a single endocrine pathway affects a pleiotropic trade-off on two or more traits is critically required. Data on the side-blotched lizard (*Uta stansburiana*; Figure 5.8) confirm that the simple trade-off between offspring size and offspring number that affects progeny survival does indeed arise from pleiotropy (Sinervo *et al.* 1992; Sinervo and Basolo 1996; Sinervo 2000). Follicle stimulating hormone (FSH) specifically controls both egg number and egg size in oviparous vertebrates like lizards, and thus it has pleiotropic effects on clutch size and egg mass. The genetic correlation for egg size and number (genetic correlation: $G = -0.92$) in a field pedigree of the side-blotched lizard is estimated in Figure 5.8.

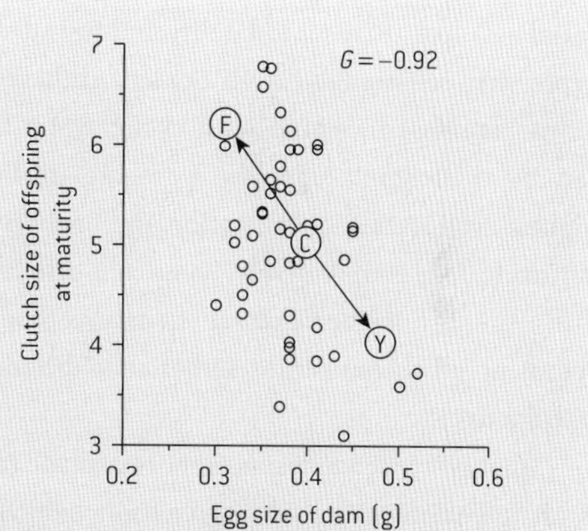


Figure 5.8 **Negative genetic correlations in egg production that impact survival of progeny of female side-blotched lizards (*Uta stansburiana*)**

**The negative genetic correlation of daughter's clutch size and dam's egg mass.** The sign and magnitude of the genetic correlation, $G = -0.92$, is indicative of a very strong negative genetic correlation (Sinervo *et al.* 2000b), which affects female fecundity and progeny survival, a classic offspring-size quality trade-off (see Sinervo *et al.* 1992). Endocrine and surgical experiments by injecting follicle stimulating hormone (F) or by ablating follicles on the ovary (Y), relative to controls (C), elucidate physiological bases of the negative genetic correlation observed between dam and daughter traits.

### c *Manipulating traits with gene manipulation and mapping traits to gene(s)*

A third and direct method to identify pleiotropy is by gene manipulation (deletion or augmentation) or by synthesizing transgenic breeds in laboratory experiments. All of these methods constitute genetic engineering, a subject discussed in Chapter 6. Consider the social amoeba, *Dictyostelium discoideum*, in which a single gene controls greenbeard (see Section 5.5.2.2) recognition (Queller *et al.* 2003). Greenbeard loci are key sender–receiver loci that mediate social behaviours. A greenbeard is thought to consist of a social supergene (Hamilton 1964; Dawkins 1976 with pleiotropic effects on a signal, recognition of that signal by conspecifics that allows social acts like altruism to be directed to individuals expressing self-signals. High density induces individual amoebas to signal with cyclic AMP. On receiving this signal, amoebas aggregate in areas of highest cyclic AMP concentration and begin to build a fruiting body cooperatively. Some cells differentiate into stalk, a non-reproductive structure that raises spores up off the substrate. Other cells differentiate into spores.

Although cyclic adenosine monophosphate (cAMP) is an important signal in aggregation behaviour, a second gene mediates cooperation. The *csA* gene, a cell adhesion protein, codes for a homophilic (self-loving) protein in the cell membrane. The *csA* gene is intimately involved in aggregation behaviour. In laboratory experiments, deletion of this single gene generates a mutant class of cheater cells that tend to become fruiting body, at the expense of non-mutant cells (wild type) that are pushed into stalk. This reflects a costly altruistic social act that arises in the context of mutant cheaters; a potential cost of carrying genes necessary for differentiation into either stalk or spores. Only spores get fitness. However, the cheater advantage is only realized in aggregations that are constructed of chimeras of mutant and wild-type amoebas. When allowed to self-aggregate on more natural substrates, homophilic (e.g., self-loving) properties of *csA* ensure that only self types (e.g., wild types) gain access to the fruiting body, thereby thwarting cheaters with a mutation at *csA* (e.g., non-self). Thus, the *csA* gene has pleiotropic effects on aggregation of cells into a multicellular slug, and it is involved in self-recognition behaviour that mediates social acts of altruism. Additional details on this example can be found in Chapter 15.

### d *Gene mapping many traits to single locus (pleiotropy) versus many loci (epistasis)*

A fourth and related genomic method is to use gene-mapping technology to map suites of traits to a single genetic locus. Paedomorphosis in salamanders has been mapped to a single locus of large effect (Voss and Shaffer 1997). Alternatively, pleiotropy is invalidated when multiple traits involved in a trade-off are mapped to multiple unlinked loci (Sinervo *et al.* 2006a). In this case, epistasis must be involved in the trade-off. This situation is discussed in greater detail below as the trade-off arises from the interactions among genes that are present in senders and receivers in a lizard social system that involves cooperation. This lizard mating system bears a striking resemblance to social dynamics of amoebas in the way self types self-organize into cooperative partnerships that thwarts the action of a class of cheater lizards.

In summary, traditional approaches to detecting life history trade-offs such as artificial selection, while effective in the laboratory, are ineffective at elucidating trade-offs expressed in natural contexts, where linkage disequilibrium can build owing to sexual and social selection. It is only in exceptional circumstances where gene maps of traits already exist (e.g., colour of male guppies resides on the Y chromosome, female preference on the X or autosomes) that artificial selection can resolve gene interactions. Genetic engineering methods are likewise useful in the laboratory, but their relevance to nature is dubious unless natural mutants are eventually identified that harbour similar traits (but see Foster *et al.* (2002 for a study using chimeras that shows promise). Although genetic engineering methods are extremely useful in uncovering pleiotropy, analogous to a targeted endocrine manipulation, gene manipulation will often ignore epistatic interactions. Of the four methods, gene mapping of behavioural traits in field pedigrees will become essential to uncover pleiotropy versus epistasis in trade-offs. When augmented with hormone experiments, which test for pleiotropic versus epistatic interactions among physiological pathways, we have a set of tools to study supergenes that generate behavioural trade-offs. However, the links between genetic and physiological (behavioural) pleiotropy and epistasis have yet to be elucidated.

#### 5.4.3.2 *Four kinds of epistasis: physiological, behavioural, genetic, and fitness epistasis*

A brief recap of the four types of epistasis is useful here before we proceed to our discussion of specific behavioural trade-offs. A way we can unify our perspective is to realize that **gene products** are used to build **physiological systems**, physiological systems interact to create sender and receiver molecules in **endocrine axes**, endocrine axes control the expression of **multiple genes**, sender and receivers interact in behavioural systems, and multiple genes (with genetic variants at each locus) interact to create **fitness surfaces** and **fitness epistasis**. Notice that physiological and behavioural epistasis are unified because both processes involve senders and receivers that act respectively intrinsically versus extrinsically to organisms.

*Genetic epistasis: nonlinear interactions among alleles at different loci*

**Genetic epistasis** is closely related to physiological epistasis (Sinervo and Calsbeek 2003; Sinervo and Svensson 2002). In the case of genetic epistasis, a specific gene interacts with another gene, perhaps shutting off that gene or alternatively, amplifying its gene products in a highly nonlinear way.

Indeed, the epistasis of multiple interacting endocrine systems can generate highly nonlinear physiological effects and nonlinear genetic effects. The product of one allele may shut off other gene functions or initiate a cascade of gene functions. For example, sex-determining factors induce testis development (rather than ovary). The testis then produces testosterone that turns on a series of genes for maleness. To some extent, genetic epistasis is opposite to the additive effects of genes (see Chapter 2).

When epistatic genetic variation goes to fixation (Goodnight 1995), such as when one of the loci contributing to epistasis fixes while others maintain segregating genetic variation, the genetic epistasis collapses and additive genetic variation is created. In such cases, the epistatic trade-offs, if they remain present, are converted to pleiotropic trade-offs. Thus, **there is a continuum between epistasis and pleiotropy** that depends on the relative contribution of additive genetic effects in an endocrine cascade versus epistatic effects. Epistasis versus additive genetic effects can be estimated with elaborate breeding experiments, but to our knowledge it has only been estimated in crosses between populations or between lines under artificial selection (Bradshaw *et al.* 2005). Partitioning of additive versus epistatic variation could be estimated within a single population in the case of discrete polymorphisms whose phenotypic expression clearly arises from two or more identifiable loci, which when crossed might reveal epistatic versus additive combinations between the two or more loci.

*Physiological and behavioural epistasis: nonlinear interactions among senders and receivers*

Sewall Wright (1969) considered physiological epistasis to be universal in genetic systems (Wade 2002; Sinervo and Svensson 2002). However, he theorized that genetic variation in epistatic networks destabilized organismal function, and he suggested that epistatic genetic variation is fixed in most species owing to such negative effects on fitness. Physiological epistasis is ubiquitous in the sender and receiver molecules of endocrine regulation, even if endocrine networks are fixed on a single type in the population.

Thus, endocrine networks of Figure 5.6 are largely fixed for genetic variants in most species. However, key regulatory loci may harbour alternative alleles in some species that generate alternative morphs (Figures 5.6d and 5.7). In such species, the expression of endocrine pathways is often altered by the morph loci. Morph loci are thought to consist of supergenes or key regulatory genes of the endocrine system that organize suites of behavioural, morphological, and life history traits into behavioural syndromes (e.g., amphibian HP–thyroid axis). It is precisely for this reason that alternative morphs are of great interest to life history theory in general, and the evolution of behaviour in particular.

*Fitness epistasis: nonlinear interactions of alleles and effects on fitness*

Genetic and physiological epistasis are closely related to fitness epistasis (Whitlock *et al.* 1995; Kelly 2000) in which nonlinearity among traits and fitness are so extreme that they create alternative optima on a fitness landscape (e.g., Figure 5.10). To the contrary, purely additive effects of traits (and underlying alleles) generate much simpler (e.g., single optimum) fitness landscapes (Sinervo and Svensson 2002).

Hybridization between species provides a useful model for visualizing fitness epistasis. Fitness epistasis, which is generated at a species contact zone, is resolved as hybrid unfitness relative to pure parental species. Hybrids consist of admixtures of alleles of the pure parental types. Speciation and hybrid unfitness provide a useful model for visualizing the action of key control loci for alternative morphs within species (Calsbeek and Sinervo 2003). Hybrid unfitness is thought to arise from physiological epistasis and the fact that each species harbours a **coadapted complex of genes** (Carson 1975) that work poorly when mixed.

Within species, a specific morph allele that codes for genetic determination of a specific morph may require interactions with 'other loci' to create ideal combinations of alleles that enhance fitness. A morph allele is a morph-determining factor. The 'other loci' can be considered strategic loci that enhance fitness depending on genetic combinations at the morph locus and at the strategic locus (Sinervo and Clobert 2003). The constant mixing during sexual reproduction of underlying multilocus **genotypes** that control alternative morphs generates fitness epistasis. In essence, morphs experience outbreeding depression when they mate with different morphs due to mixing of **coadapted morph complexes** (Sinervo and Clobert 2003). Species also contain different **coadpated gene complexes** (Carson 1975) that likewise mix poorly if hybrids are unfit. Within a species, alternative morphs select for mate choice by females for males of the same genotype to maintain intact these coadapted gene complexes of morphs (Bleay and Sinervo 2007). We can illustrate the concepts of fitness epistasis with a discussion of the two morphs that are fundamental to all sexual species, males and females. Fitness epistasis generates trade-offs in the design of the sexes.

### 5.4.4 Trade-offs in design of the two sexes that arise from epistasis: Ontogenetic conflict

Males and females reflect the fundamental morphs of sexual organisms. Recent advances in our understanding of life history trade-offs have identified different patterns of selection on the sexes as an important source of genetic variation (Rice and Chippindale 2001). Genetic trade-offs related to the functional design of the sexes are referred to as **intersexual ontogenetic conflict** or ontogenetic conflict (Rice and Chippindale 2001). Many alleles that are favoured under the force of sexual selection in the male morphology and physiology are of limited value during natural selection on the female morphology and physiology, and vice versa.

*The design of pelvic girdles in male and female primates*

Consider the human pelvic girdle, which is thought to be a strong allometric constraint on the evolution of large brain size in primates (Luetenegger 1979) and other vertebrates (Sinervo and Licht 1991). The presence of the birth canal means that the pelvis is under selection for large size in females so that the infant's head can pass through during birth. However, males are not subject to this functional trade-off. Alleles favoured in males should reach a pelvic girdle width that is optimal for foraging behaviour or perhaps in mate competition, under the optimizing force of selection on locomotor function. This might be the case were it not for the fact that males and females share nearly all of their genes in a common genome (aside from a few genes on the Y chromosome). The pelvis of females is under dual selection for locomotion and reproduction. Males are only optimised for locomotion.

Ontogenetic conflict can be ameliorated through sex-limiting steroid hormones that govern female and male traits (Sinervo and Calsbeek 2003). Gene promoters called estrogen response elements (EREs) control the differential gene transcription in the sexes (Freedman and Luisi 1993; Zajac and Chilco 1995; Sanchez *et al.* 2002). In females, oestrogen, through a gene cascade, loosens the pelvis around the time of parturition or egg laying. Vitellogenin is a key ERE controlled gene which contributes to yolk production. However, administering exogenous oestrogen to males induces vitellogenin production. Thus, males have, but do not express the vitellogenin gene (Sinervo and Calsbeek 2003). Females produce vitellogenin in abundance, when in oestrus.

*Endocrine systems have notoriously permeable expression*
Sex-limiting effects are, however, not perfect. Not every gene for male and female traits that are under ontogenetic conflict has an ERE. Thus, there is always selection on female traits, and counterbalancing selection on male traits. Any secondary sexual trait is likely to be under ontogenetic conflict, given that regulatory genes in male and female reproduction are shared between the sexes. The classic sex steroid hormones oestrogen and testosterone are expressed in both sexes, albeit at reduced levels in each sex. For example, testosterone secretion, which is required for female sex drive, might also trigger the action of gene expression for aggression, which may not be optimal for female life history (Sinervo and Calsbeek 2003). Similarly, variation in testosterone expression in males, which is associated with alternative male strategies in vertebrates (Brantley *et al.* 1993), might have permeable expression in females, thereby altering female behaviour and life history. By permeable we mean that the endocrine system and the receptors may allow for some expression of male function genes in females and vice versa.

Sex chromosomes that initiate the endocrine cascade of sex determination through single genes (e.g., *sry* is a gene identified to be a testis determining factor, see Chapter 6, Section 6.2.1.1), are clearly unlinked from autosomal genes where many EREs reside. Physiological epistasis is also involved given the many endocrine switches that are instantiated in response to most hormone secretion by the hypothalamus and pituitary (Boxes 5.4 and 5.5) (Sinervo and Calsbeek 2003). Thus, ontogenetic conflict, a fundamental life history trade-off between the sexes, is due to genetic epistasis between loci residing on sex chromosomes and those residing on autosomes, and the physiological epistasis of endocrine networks. The fitness effects of ontogenetic conflict on the sexes reflect the manifestation of fitness epistasis generated by genetic and physiological epistasis. When females choose males for these qualities, this again reflects fitness epistasis, but which involves behavioural epistasis of senders and receivers, and genetic and physiological epistasis governing quality of endocrine expression for regulatory networks that create maleness versus femaleness. These important issues are currently poorly studied.

### 5.4.5 Plasticity and trade-offs

*Plasticity generates ontogenetic conflict*
The kinds of trade-off that arise from ontogenetic conflict are closely related to trade-offs that arise from plastic control of morphology, physiology or behaviour. The analogy can be made explicit if one considers the steroid control of sexual determination to be a means by which the same autosomal genome is induced to give rise to male versus female

morphology. In chromosomal sex determination, the inducing cues are sex chromosomes. Inducing cues can also be external to organisms such as in temperature-dependent sex determination (see Chapter 6). Inducing cues can even be conspecifics. In some marine invertebrates, larvae are induced to become male if they settle on females. Similarly, coral reef fish may change sex under various circumstances (Wachtel *et al.* 1991). Ontogenetic conflict can thus also arise in plastic or environmental sex determination due to abiotic (temperature) or biotic (social) causes.

Trade-offs generated by ontogenetic conflict are general to all forms of plasticity, not just sex determination (see Boxes 5.4 and 5.5). All forms of plasticity are subject to a fundamental constraint that the same genome creates alternative morphologies (or behavioural types), or a broad spectrum of alternative morphologies (or behavioural types) arrayed along an environmental gradient. Given that **behaviours reflect the ultimate kind of plasticity** (Huey *et al.* 2003), variation in behaviour is subject to the fundamental constraint that the same genome creates alternative behavioural strategies.

More generally, although plastic responses appear to be advantageous, there are limits and costs to the benefits afforded by plasticity (DeWitt *et al.* 1998). Indeed, production of a new phenotype takes time, and the environment that induced a phenotype might have changed (environmental mismatch). Cues used to assess future environments may not be reliable, again leading to environmental mismatch. Plasticity itself may reduce the possibility of producing extreme well-adapted phenotypes or phenotypes produced might be of lower quality than a similar phenotype produced by genetically fixed mechanisms. For example, a shift from one anti-predator behaviour to another may take time compared with a genotype that exhibits one behaviour regardless of context. Time needed for shifts might limit benefits of plasticity.

*Other costs of plasticity*

There is a second type of constraint in the case of plasticity that is more elusive: the costs of plasticity per se. Costs of plasticity relate to additional genetic and physiological machinery that must be in operation to achieve alternative forms, relative to a hypothetical form lacking plasticity (i.e., genetically fixed). Many other potential costs of plasticity have been recognized (DeWitt *et al.* 1998) that can be categorized as five main costs (DeWitt *et al.* 1998; Ernande and Dieckmann 2004): (1) a **maintenance cost**, i.e. the cost for maintaining and forming necessary machinery to assess environmental fluctuations; (2) a **control cost**, i.e. avoiding developmental instability that can result from developmental routes that are too complex; (3) a **production cost**, i.e. those costs incurred when a given phenotype is produced for a given environment when compared with the same phenotype produced by a fixed genotype; (4) **information acquisition costs**, which are especially important for behaviour (exploration, etc.); and (5) **genetic costs** that arise from plastic genes or regulatory loci on the expression of other genes (pleiotropy and epistasis).

*Measuring the costs of plasticity is extremely difficult*

The actual costs of plasticity have not been measured (DeWitt 1988). Measuring costs and benefits of plasticity are complicated. First, a plastic trait has to be adaptive. Costly maladaptive plasticity is selected against. Secondly, plastic traits can often be less costly than constant responses. This may arise in cases where **plasticity is a passive response to environment** (i.e. passive plasticity), such as in the case of high growth and large size in good environments but poor growth and small size in poor environments. This corresponds to the term $V_{DE}$ of phenotypic variance defined in Section 20.2.2.2. Such growth effects that might impact the time allocated to foraging and/or predator avoidance (Clobert *et al.* 2000) are classically observed in metamorphic size of larval amphibians. The alternative, **active canalization** (Waddington 1966), may require elaborate developmental machinery to stabilize phenotype in the face of environmental perturbations. Targeted growth rate, in which juvenile growth does not cease until a constant size is reached (Riska *et al.* 1986), is a common feature of mammals and birds. Targeted growth requires elaborate genetic machinery to achieve homeostasis, which involves coordination of the genes for maternal care of the mother (e.g., lactation) and genes for growth homeostasis of the progeny (Cowley *et al.* 1989). Thirdly, one needs to compare plastic genotypes with a genotype that lacks plasticity, or there must exist polymorphism in the reaction norm (Ernande and Dieckmann 2004) such that genotypes can be compared across environments. This requires elaborate breeding experiments, which are difficult (Ernande *et al.* 2003), especially if the cost expressed by a genotype is environment dependent. Costs must then be measured across many environments. Because plasticity is present in most organisms, either such costs are not substantial or the benefits are enormous.

*Plasticity versus genetically fixed alternative forms*

Genetically fixed and plastic forms can be contrasted by the environmental conditions thought to give rise to each adaptive solution. Levins (1962a, b) formulated a solution when he described the role of environmental grain in the maintenance of plastic versus genetically fixed forms. In the case of a fine-grained environment that supplies predictable cues, adaptive plasticity will be favoured (Lively 1986). In the case of a coarse-grained environment with unreliable cues, genetically fixed strategies are favoured. Another way to view this process of selection is that if the size of the grains is bigger than the space over which an animal moves in its lifetime (e.g., a fixed territory) then genetic adaptation is favoured (see Figure 4.1 for a further discussion of this topic). A corollary to this idea is that there will also be a strong genetic basis for habitat selection in the case of genetically fixed types. Habitat selection behaviour will thus be under strong correlational selection with genetically fixed alternative strategies. The settlement behaviour of side-blotched lizard males (see Sinervo and Clobert 2003), which we discuss below, reflects such genetically based behaviours of dispersal and social habitat preference (Sinervo *et al.* 2006ab).

*Plasticity, behaviour and trade-offs*

Regardless of whether or not morphologies or behaviours are due to plasticity or genes, they are subject to identical life history trade-offs (e.g., see Box 5.5 and the trade-offs of larval amphibians). Plastic solutions cannot escape the same life history trade-offs that constrain genetic systems of determination even if plasticity does involve a set of costs that are different from genetically fixed types (noted above). However, plasticity may induce an unrecognized increase of the life cycle dimensionality, which might then offer more freedom and solution in trade-off organization (by trading trait variance against trait means; see Section 5.2.4). On the other hand, genetically fixed alternatives that are found in a single population generate much stronger costs related to recombination and segregation load among the genes that control morph determination and strategic loci (see Section 5.4.3.2c), unless genes can be consolidated into a tightly linked supergene. Behaviour, which is the ultimate form of plasticity, can thus serve to ameliorate these genetic costs because animals can arrive at a behavioural solution that completely ameliorates the effects of selection (Huey *et al.* 2003). For instance, Huey *et al.* (2003) discuss how the evolution of behavioural thermoregulation ameliorates selection on thermal physiology of lizards. Thus, behaviour can buffer the organism's physiology and selection on physiology from the effects of variable environments.

## 5.5 Mating and social systems and life history trade-offs

### 5.5.1 The synergism between mating system trade-offs and life history trade-offs

Although ontogenetic conflict of the sexes, noted above, is rarely measured in nature, an important step in this direction is **identifying the selection on males and females in different mating systems**. Mating system trade-offs (Wiley 2000) can be partitioned into those acting on females, those acting on males, and those impacting both sexes. Given that females are more often the care-giving sex, males are under sexual selection to increase mating number, referred to as Bateman's (1948) principle (see Chapter 11). Mating systems with sex-role reversal (Gwynne 1981; Gwynne and Simmons 1990; Chapter 12) provide a useful counterpoint to this generalization, which reflect exceptions that prove the rule. The optimal number of matings for a male is much higher than the optimal number for a female, owing to Bateman's (1948) principle. However, females in some mating systems are under selection to be promiscuous owing to factors like fertility assurance (Madsen *et al.* 1992). Other forces driving female promiscuity are related to more complex issues like ontogenetic conflict. A female might mate with more than a single male, to produce both sons and daughters of high genetic quality (Calsbeek and Sinervo 2002b, 2004).

Our goal is not to review all factors favouring polygyny or polyandry. These ideas are already deeply rooted in the tradition of behavioural ecology (Orians 1969; Emlen and Oring 1981) and will be developed in Chapter 12. Our goal is to point out that mating system trade-offs are not merely due to single gene effects, but also invoke behavioural epistasis, given that mating and social systems entail communication between senders and receivers.

The trade-offs of polyandry (and polygyny) are clear in birds, which exhibit bi-parental care (Reynolds *et al.* 2002). In such mating systems, polyandrous females might risk detection by a mate that subsequently abandons the female in a situation of compromised paternity. Conversely, highly polygamous males, owing to the action of testosterone, might be less effective parents than a monogamous mate. Such sexual strategies, via impacts on parental workload, will have cascading impacts on the offspring survival trade-off and on the cost of reproduction trade-off that affects survival of both parents.

Behavioural ecology has a rich tradition of naming new phenomena. Although such nomenclature is necessary for precisely ascribing genetic causes, the diversity of terms can obfuscate the commonality of all processes for the student of behavioural ecology. All forms of conflict can be circumscribed by the notion of life history trade-offs. Parent–offspring conflict (Trivers 1972) is a complex trade-off involving offspring quality and quantity, and costs of reproduction for parents. Asymmetries in relatedness among progeny, caused by to polyandry, can generate a genetic form of conflict referred to as genomic imprinting.

> **Genomic imprinting** is a form of parent–offspring conflict and mating system conflict between the sexes, which also invokes trade-offs of offspring quality versus quantity, and costs of reproduction (Haig and Westoby 1989). It involves paternal genes that evolve to silence maternal gene expression. The silencing of maternal alleles in females allows progeny to extract greater maternal reproductive effort through active copies transmitted by sire alleles.

Conflict only arises in the context of cooperation (and benefits). Otherwise individuals should abandon a relationship. Thus, conflict is the cost of cooperation in mating or social system interactions. All behavioural trade-offs have a life historical consequence and thus are collectively life history trade-offs. Costs versus benefits, or cooperation versus conflict, both fall under the umbrella of life history trade-offs.

Behavioural ecologists (Krebs and Davies 1987) also have a rich tradition of calling trade-offs a cost–benefit. However, here **we synonymize cost–benefit analysis with trade-off**. Below, we show that even in the case of the costs and benefits of social system interactions, these trade-offs have been linked to the pleiotropic action of genes. Nevertheless, the costs and benefits of social system interactions can also arise from epistasis, given that social system interactions are driven by sender–receiver interactions.

*Antagonistic sexual selection between the sexes*

The difference in number of mates that are optimal in the sexes (males generally with more mates) may promote **antagonistic sexual conflict**, which constitutes another mating system trade-off. Antagonistic sexual conflict arises in the context of polygamous situations where one mating partner is favoured to enhance its reproductive success (i.e. mate competition), even if it negatively impacts future reproductive success of its partners. This is because the partner subject to polygamy may only gain high fitness from the current mating, and has much less of a genetic stake (i.e., progeny) in future matings by its partner.

For example, seminal fluid proteins in *Drosophila* are hypothesized to reduce female remating rates and enhance paternity assurance (Rice 2000). This constitutes a form of male manipulation that enhances its current success, even if it has deleterious effects on future reproduction by females in subsequent reproductive episodes when the female mates with different males. Physiologically naïve females, which have been artificially selected under a monogamous mating system and thus not previously exposed to male strains that have evolved under polygyny, suffer higher rates of mortality when mated to polygamous strains compared with polyandrous females that have evolved under polygyny (Holland and Rice 1999).

Similar antagonistic selection has been identified as a driving force in water striders (see Rowe *et al.* 1994; Arnqvist and Rowe 2002), but the selective factors acting on females relate to the burdening effect of a male that maintains amplexus for extended periods of time, rather than physiological action of hormones that induce costs of reproduction. In extreme cases, remating pressure by males favours the evolution of alternative female morphology, such as in damselflies, in which females are differentially cryptic to males. A novel female morph of damselfly (called androchrome) evolves to resemble the male form, becoming more cryptic to males. Other rare female forms in the same damselfly species avoid detection through apostatic or rare advantage, relative to a common female form that is encountered by males at a much higher rate. Thus, trimorphisms comprising all three female types (andochrome, apostatic, and normal females) are common in damselflies of Europe and North America (Svensson *et al.* 2005). Female trimorphism is thought to arise from search image formation in males for common female morphs (Fincke 2004). Given intense sexual conflict arising from male harassment (Le Galliard *et al.* 2005), rare cryptic female morphs gain higher fecundity (Svensson *et al.* 2005), thus invoking previously noted trade-offs of costs of reproduction, fecundity, polyandry, and sexually antagonistic selection owing to polygyny as well as trade-offs in perceptual systems (Sinervo and Calsbeek 2006).

*Sexual dimorphism and sexually antagonistic selection*

Sexually antagonistic selection may also explain many sexually dimorphic traits, if dimorphism has evolved to limit the sex-specific costs of sexually antagonistic alleles (Rice 1984; Gibson *et al.* 2002). Mating system trade-offs of antagonistic sexual selection are closely related to trade-offs involving alternative optima in the sexes or ontogenetic conflict (see Section 5.4.4). This is because sexual conflict can operate on the same genes and the same traits in each sex (e.g., parental care in males versus females). This generates a vicious cycle of adaptation in which the traits that enhance male fitness conflict with female interests (e.g., lower levels of paternal care, higher polygyny). This may result in counter-adaptations in females (higher care, perhaps higher polyandry) that exacerbate costs of reproduction in females, or through sexually antagonistic selection and mate harming on the part of the male to enforce the male's reproductive strategy. Thus, counter-adaptations arising from sexually antagonistic selection (Shuster and Wade 2003; Gravilets and Hayashi 2005, 2006) can generate and integrate many primary demographic trade-offs and secondary trade-offs invoked by life history, physiology, and mating systems.

In summary, mating systems dimensionalize trade-offs to a high degree by invoking interactions of parent–offspring and cost-of-reproduction trade-offs that are asymmetric between the sexes. Trade-offs on the care-giving sex are intensified, but this intensifies sexually selected trade-offs on the other sex.

### 5.5.2 Social system trade-offs

**Social systems** comprise all interactions among juveniles, males, and females. Social system trade-offs can powerfully synergize with life history trade-offs. Mating system trade-offs are best viewed as a special case of social system trade-offs, because mating systems impact juveniles. Although juveniles of species may not interact with adults during the reproductive season, they may disperse to avoid adults or remain philopatric. A higher degree of symmetry in genetic similarity among social actors serves to structure and stabilize social systems, whereas genetic dissimilarity among actors generates social trade-offs.

#### 5.5.2.1 *Dispersal as a social system trade-off*

Dispersal has been studied in many species (see Chapter 10 for a more complete description of this behaviour), but some of the most complete studies involve two species of lizards. In the European common lizard (*Lacerta vivipara*), there are three dispersal strategies (Cote and Clobert 2007): one type of juvenile disperses to empty habitats, another type to low density habitats, and a third type to densely populated habitats. All types exhibit marked differences in social behaviour, ranking from asociality to cooperation and from neophily to neophoby. Most of the behavioural characteristics appear to be under maternal influence, although a direct genetic control cannot be excluded in some cases.

In the side-blotched lizard (*Uta stansburiana*), three dispersal strategies are likewise present. They are easy to spot because each strategy is associated with a different throat colour (see Figure 5.9). These strategies have been linked directly to a major gene, referred to as the OBY locus (*o*, *b*, *y* alleles), named for the alternative colours (orange, blue, and yellow) expressed on the throats of adults (see Sinervo and Clobert 2003; Sinervo *et al.* 2006a, b; Figure 5.9). Dispersal to a high-density habitat is associated with juveniles carrying one or more *o* allele, which also induces aggression when they mature as adult males (Calsbeek and Sinervo 2002a, b) through high plasma testosterone (Sinervo *et al.* 2000a). Philopatric behaviour is associated with one or more *y* alleles, which is also associated with cryptic morphology and behaviour in juveniles, and female mimicry in adult males. Dispersal and settlement beside genetically similar individuals is associated with two *b* alleles. The *b* allele is also associated with a cooperative strategy of territory defence (Sinervo *et al.* 2006a). In *Uta*, dispersal strategies were shown to be genetic because they are associated with sire OBY alleles in an extensive field pedigree and in controlled laboratory crosses with field release of progeny (Sinervo *et al.* 2006b). However, female

Figure 5.9 **Three of the six possible morphs of males side-blotched lizards, *Uta stansburiana***

Throat colour morphs of male side-blotched lizards, *Uta stansburiana*, illustrating three of the six possible genotypes. Depicted from left to right are *oo*, *bb*, and the heterozygote *by* male genotypes. These morphs are involved in a suite of social system trade-offs (see text). A colour version of this photo can be found on the Online Resource Centre.

Photo by Suzanne Mills and Sinervo.

egg-size strategies (Sinervo *et al.* 2000b) interact synergistically to enhance progeny dispersal (Sinervo *et al.* 2006b).

In both the North American side-blotched lizard and the European common lizard, dispersal trade-offs arise directly from life history trade-offs that are impacted by social system dynamics. In *Uta*, progeny dispersal interacts with trade-offs of offspring quantity and quality, and female costs of reproduction (Sinervo 1999), which are exacted during density-dependent competition (Sinervo *et al.* 2000b; Svensson *et al.* 2001a, b, 2002; Comendant *et al.* 2003). In the common lizard, progeny dispersal induction is intimately related to the survival of the female parent (see Figure 10.7). If common lizard females incur high costs of reproduction, which decrease their condition and elevate mortality risk, the females can induce their progeny to become philopatric, which may enhance the likelihood of them inheriting their territory.

Returning to our salamander example (see also the common lizard example, Chapter 10), we can see that dispersal invokes other levels of selection besides competition. Aquatic forms like paedomorphs are an evolutionary dead-end in the long term. The oldest lake, Lake Baikal in Siberia, is a few millions of years old, which is not ancient. Eventually all paedomorphic clades of salamanders will go extinct as the bodies of water that contain them wink out of existence. Thus, metamorphic or direct developing clades, with an intact dispersal potential have an advantage on long evolutionary time scales. Given that paedomorphosis can rapidly evolve, new forms will probably be derived either from clades with complex metamorphosis or clades with direct development. This represents a level of selection referred to as **species selection**.

Dispersal is clearly a phenomenon that requires an understanding of behaviour, genetics, and life history trade-offs, at diverse levels of selection (gene, individual, kin group, population, species). Forces that shape mating systems thus have multiple effects on juveniles. Trade-offs between traits beneficial for juvenile survival, dispersal and social behaviour promote different trait optima than those that enhance adult survival and reproduction (Sinervo *et al.* 1992; Calsbeek and Sinervo 2004). Thus, dispersal versus philopatry can be viewed as ontogenetic conflict that impacts juvenile and adult life history differently.

#### 5.5.2.2 *Two case studies of social trade-offs*

Cooperation can invoke novel social trade-offs. Cooperation can be broken down into two forms (Hamilton 1964): mutualism or altruism (see Chapters 15 and 17). Costs and benefits of cooperation can be discussed in terms of altruism and mutualism. Mutualisms evolve if two partners achieve high fitness through social acts, but both partners must participate to get high fitness. The reward benefits them both. Altruisms yield a fitness asymmetry from cooperation that is measured relative to a strategy that does not join in social acts (a going it alone strategy or loner strategy). Both partners in an altruistic relationship may incur a risk that ultimately benefits one party, a **recipient**. The altruistic partner obtains lower fitness than the loner strategy, and incurs a cost. Social systems of genetically related altruist-recipient pairs are **kin altruists**. Social systems of genetically unrelated altruist-recipient pairs are **true altruists** (Sinervo *et al.* 2006a). As noted above, genetic similarity and dissimilarity structures social system trade-offs.

In social situations, genetic similarity of cooperators enhances cooperation by stabilizing mutualism. Genetic dissimilarity destabilizes cooperation and generates costs either within a mutualistic relationship or from other antagonists that are external to the cooperative partnership. In the situation of altruism, the cooperative partner that pays the social costs is called the altruist, while the other partner that benefits is the **recipient**. When this altruistic cost arises purely in a social context, other social strategies that reside outside of the cooperative partnership exact the costs from altruists. Two basic social strategies exist by which fitness is extracted from altruists: costs can be exacted by individuals that adopt a cheater strategy, which is referred to as deceit, or costs can be exacted by an individual that adopts an aggressive strategy, which is referred to as **usurpation** (Sinervo *et al.* 2006a). Altruism in social systems must involve games between at least three players (altruists, recipients, and either usurpers and/or deceivers that exact costs).

Two situations with social trade-offs are developed here: (1) kin nepotism and queen killing behaviour in eusocial insects, and (2) greenbeard alleles for true altruism and mutualism in lizards. Asymmetry in genetic similarity is the root cause of social system conflicts, which are analogous to relatedness asymmetries that give rise to the mating system conflicts of genomic imprinting, noted above. The asymmetry in relatedness generates conflict in genetic interests, which invokes life history trade-offs in social systems that may be exacted by other members of the social group.

### a *Kin nepotism and greenbeards for queen assassination in the red fire ant*

The following example of queen assassination illustrates the Hamiltonian concept of greenbeard alleles, which was introduced above in the example of social amoebas. As we saw earlier in Chapter 2, greenbeard is a term coined by Richard Dawkins (1976) to explain models of genic selection first proposed by Hamilton (1964). Greenbeard loci exhibit three behaviours: (1) a signal, (2) self- or non-self recognition of the signal, and (3) signal recognition elicits social acts, which benefit others that share self signals or detriment others with nonself signals. The homophilic protein product of the *csA* gene in the social amoeba *Dictyostelium discodeum* harbours the three key greenbeard traits, albeit in a single pleiotropic mutation (see 4.4.3.1d). Such greenbeard loci are the source of many social system trade-offs.

#### *Variation in relatedness skew determines invasion of nepotism*

Eusocial systems with multiple queens are called polygyne, while ones with a single queen are called monogyne. Social systems with polygyne and monogyne generate social system trade-offs of kin nepotism (Alexander 1974). Consider a mutant allele that is present in only a subset of the queens and a subset of the workers. This will promote kin nepotism to evolve based on the shared similarity of this allele. Normally one thinks of helping kin as a kin nepotistic act. However, if you hurt non-kin, which has an indirect benefit to your kin, it also reflects a kin nepotistic act.

In the red fire ant, *Solenopsis invicta*, a nepotistic allele has evolved that has pleiotropic effects on non-self recognition and queen-killing behaviour in colonies that exhibit polygyny. Female workers that carry a single copy of the *B* allele at this nepotistic locus will kill any maturing queen that is homozygous for the *b* allele. Homozygous recessive *bb* workers lack such queen killing behaviour (Keller and Ross 1998). Experiments in which odours are extracted from *B* genotypes and applied to *bb* queens confirms that the lack of a cuticular compound (on *bb* queens) confers the queen killing behaviour by *BB* or *Bb* workers. Thus *BB* and *Bb* workers identify and kill newly maturing *bb* queens that lack this specific compound, whereas *BB* or *Bb* queens carry the compound in their cuticle and are protected. This reflects a greenbeard locus that indirectly enhances fitness of *B* workers by removing competition experienced by their closely related *B* queens, where relationship is due to the greenbeard locus, a form of greenbeard kin nepotism (Alexander 1974).

#### *The costs of kin nepotism will involve colony level life history traits*

This social system trade-off generates costs because the *BB* homozygous condition (but not *Bb*) has lethal effects on mature queens. Thus, only *Bb* queens are fertile. Moreover, queen-killing behaviour may impact colony reproductive output and fitness. If *bb* queens are removed from the colony and *BB* queens are not viable, colony output may drop, unless new *Bb* queens can be quickly produced by the colony to return the colony to the optimal number of queens. Costs may also arise due to conflict among worker genotypes (e.g., novel nepotistic alleles may arise that cause workers to kill any progeny lacking the odours, not just queens). Finally, *bb* queens, lacking the odour of *B* genotypes, pay the ultimate social cost at maturity.

The social system of polygyne may exact as yet unmeasured costs, when compared with the social system of monogyne. If a monogynous social system has higher colony output (because it has less relatedness asymmetry and thus less greenbeard nepotism) than a polygynous one, then the life history trade-offs are exacted at the level of colony reproductive output and survival, not merely among individuals. Such fitness relations that determine the full stability of the system have yet to be estimated but are central to the maintenance of alternative strategies of polygyne and monogyne and alternative greenbeard alleles.

### b *Altruism as a cost of mutualism in the alternative social behaviours of lizards*

*The genetics of the OBY locus*

One of the few social systems in which the fitness relations among all social actors have been estimated involves the genes controlling social behaviours of side-blotched lizards. Figure 5.9 shows the striking colour phenotypes of male side-blotched lizards (*Uta stansburiana*). Genetic crosses (Sinervo *et al.* 2001), gene mapping studies (Sinervo *et al.* 2006a) and theory (Sinervo 2001) confirm that colour transmission in *Uta* behaves like a single-locus factor with three alleles (*o*, *b*, *y*). This yields six colour phenotypes that reflect six genotypes (*oo*, *bo*, *yo*, *bb*, *by*, *yy*) of the OBY locus, named for the three colour strategies of males Figure 5.9). Alleles at the OBY locus have co-dominant effects on colour expression (two different alleles yields a throat with both colours), but dominant effects on male strategy and behaviour. In males, the *o* allele is genetically dominant to *b* and *y*, and *y* is genetically dominant to *b* (i.e., the O phenotype = *oo*, *bo*, or *yo*; B phenotype = *bb*; Y phenotype = *by* or *yy*). Male colour morphs exhibit physiologies correlated with mating behaviour, plasma testosterone, and territoriality (Sinervo *et al.* 2000a), and as noted above, dispersal and settlement (Sinervo and Clobert 2003; Sinervo *et al.* 2006b). O males have high stamina, low survival, and patrol large territories with large female harems (Calsbeek and Sinervo 2002a, b). O can be invaded by the sneaker strategy of Y males, which have low stamina, no territoriality, mimic female behaviour, and cuckold harems of O males at high rates (Zamudio and Sinervo 2000). Y males are beaten in turn by pairs of B males that cooperate. OBY appears to be a super-gene that pleiotropically controls many male traits.

*Male strategies generate a rock–paper–scissors dynamic due to a greenbeard*

The male mating system of *Uta* is referred to as a rock–paper–scissors system because each strategy beats one strategy, but is each is beaten in turn by another, leading to an evolutionary cycle among three players: rock beats scissors, paper beats rock, and scissors beats paper. The highly cooperative blue strategy (mutualism) is invadable by O. When the competitive strategy of O (usurpation) becomes common, Y beats O by crypsis and female mimicry (deception). When Y becomes common, the cooperation of two genetically similar B males is mutualistic as these B males have much higher fitness than loner blue males, which lack a genetically similar partner to help it to chase away Y males (e.g., we refer to these B males as loners, because they lack a genetically similar partner to help them defend their territories, but these B loners live in similar crowded conditions as all male types). The genetic similarity of B male cooperators does not arise from kin philopatry. A deep pedigree for *Uta* (20 generations) indicates that completely unrelated B males find one another using a gene complex for self-recognition of blue colour and of genetic similarity. Thus, cooperative B males satisfy three conditions for a greenbeard: they carry signal genes that recognize self, and self-recognition elicits social acts that enhance fitness.

To cooperate, pairs of B neighbours must also share alleles at many self-recognition loci, including *b* alleles of the OBY locus (Figure 5.10). When they find a genetically similar male partner their fitness is greatly increased. Thus, loci for self-recognition and settlement next to genetically similar males will become coupled to *b* alleles by the effects of correlational selection. In essence blue males have a social habitat preference for genetically similar blue neighbours, which enhances cooperation greatly and

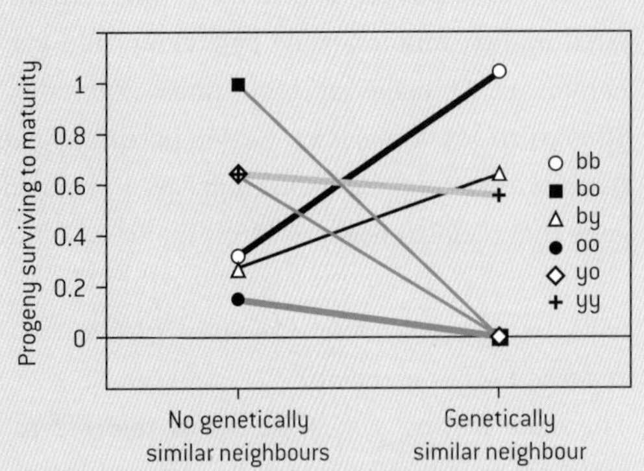


Figure 5.10 **Correlational selection on the OBY locus with settlement beside genetically similar males**
From Sinervo & Clobert (2003), Morphs, dispersal, genetic similarity and the evolution of cooperation, *Science* 300: 1949–1951. Reprinted with permission from AAAS.

Line colours are for the different strategies (O, B, and Y, thick lines indicate homozygotes, thin lines heterozygotes). Cooperation among genetically similar blue males is favoured because this clearly increases the average number of progeny recruits at maturity (two black lines going up from left to right). The reverse is true of O males: O males next to genetically similar males receive no fitness (three mid-grey lines going down from left to right). Genetic similarity seems to have no impact on the fitness of Y males. However, when genetically similar neighbouring B males are found in nature, pedigree data show that they are in fact completely unrelated. Furthermore, gene-mapping studies (Sinervo *et al.* 2006a) indicate that settlement behaviours are controlled by three unlinked loci involved in self- and colour-recognition. We refer to these loci and their hypothetical alleles as *Ss*, *Cc*, and *Pp*, for the putative self-recognition systems that may be involved including scent, colour, and push-ups, the three main behaviours that are used in male–male communication in lizard social systems.

enhances fitness. This social habitat preference is highly heritable in the wild ($h^2 = 0.89$). Moreover the genes have analogous effects on female mating preferences for genetically similar mates ($h^2 = 1.05$) (Sinervo *et al.* 2006a). Conversely, alleles for self-repulsion become coupled to *o* alleles given that settlement next to genetically similar individuals generates zero fitness for O males, but high fitness if they settle with genetically dissimilar neighbours (Figure 5.10). Finally, the *y* allele is relatively neutral for settlement next to genetically similar neighbours (Figure 5.10). Correlational selection on settlement behaviour and the OBY colour locus is an example of fitness epistasis. B males only obtain high fitness if they settle next to genetically similar partners. O males get no fitness in such situations.

Gene mapping studies in the deep field pedigree of *Uta* indicate that self-recognition genes are all unlinked (on different linkage groups) and are also unlinked from the OBY locus (Sinervo *et al.* 2006a), but yet under profoundly strong correlational selection. This correlational selection generates sender–receiver epistasis and very strong fitness epistasis (Figure 5.10). The fact that all these genes are shared between the sexes also generates a mating preference in females for self similarity (Sinervo *et al.* 2006; Bleay and Sinervo 2007), but this preference is not a pleiotropic consequence of OBY, rather it is caused by unlinked loci that govern social and mating partner preference (Sinervo *et al.* 2006a).

*B males experience social trade-offs from the kind of rock–paper–scissors neighbours*

Social system trade-offs arise because selection that favours cooperative behaviour in B males (Figure 5.11) generates social conditions for invasion of alternative social strategies (O beats B), which converts relations between B mutualists (B beats Y) into altruistic ones (Figures 5.10 and 5.11). These greenbeard loci ('blue-beard' loci really) also result in a cost of cooperation that generates altruism on the part of one B male in a partnership. The B male whose territory is next to more O males ends up receiving no fitness, but this male buffers his B partner from the aggressive O male strategy. When O is rare and Y is common, both B males enjoy high fitness and social relations are thus mutualistic between B males (Figure 5.12). Two cooperating and genetically similar B males can more efficiently defend females from Y sneakers.

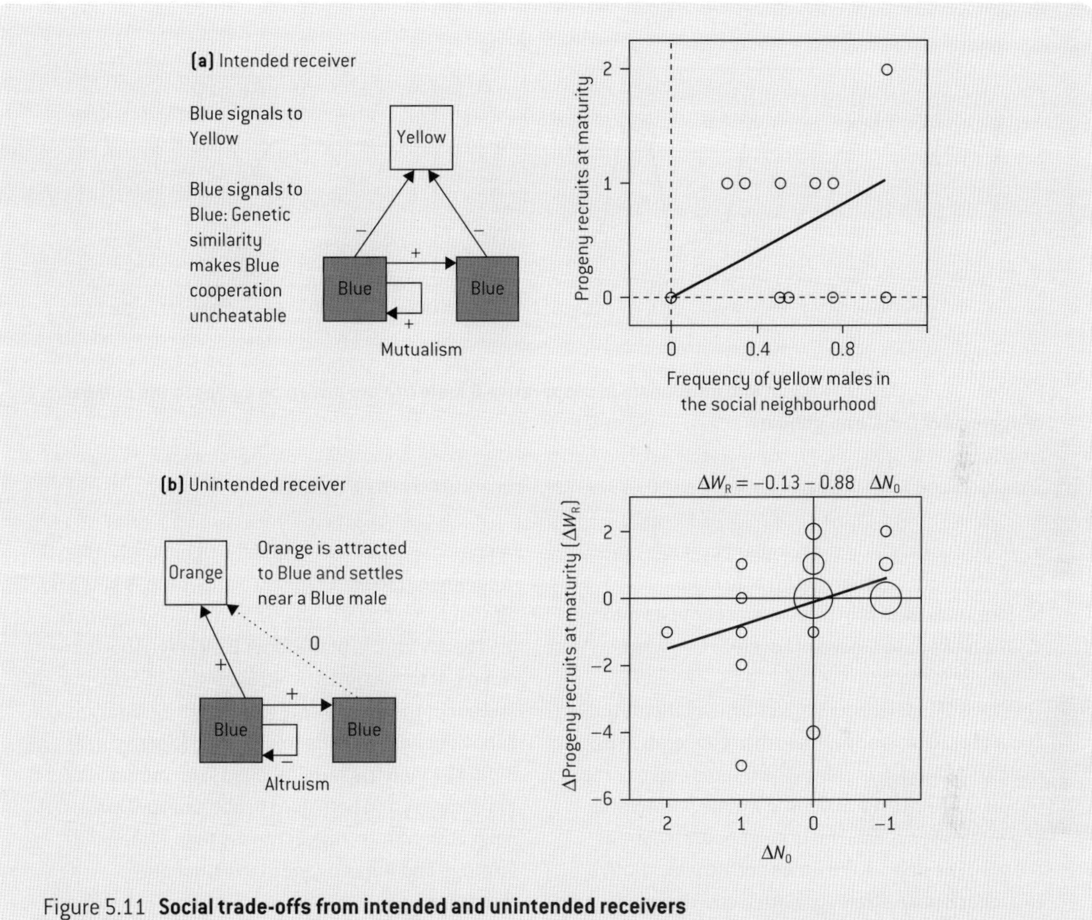


Figure 5.11 **Social trade-offs from intended and unintended receivers**

Situations with three strategies generate social trade-offs among intended and unintended receivers.

**a.** For example, B males are involved in honest signalling relations that are enforced by a requirement of genetic similarity at the OBY locus, as well as at self-recognition loci (of genetic similarity) distributed across the genome. Cooperation between B males thwarts the weaker Y strategy and a high frequency of Y males in their neighbourhood (*x*-axis) enhances the fitness of both B partners (*y*-axis), which is given by progeny recruits in the next generation.

**b.** However, aggression by O destabilizes B cooperation. Asymmetry in the number of O neighbours (*x*-axis) demotes one B male to altruism (*y*-axis is the difference in fitness between two genetically similar B neighbours). B males with more O male neighbours produce fewer progeny recruits compared with their genetically similar B partners with fewer O neighbours.

#### 5.5.2.3 *The causes of social system trade-offs: sender–receiver networks*

Other biological mutualisms may involve hidden costs that are reflected as transient or cryptic altruism. Social trade-offs need not merely reside within a species, but may be fundamental to co-evolutionary interactions between species, which may either be mutualistic, or involve costly parasitism (Thompson 2005). For example, the moth *Greya pollitella* pollinates flowers of the woodland star genus *Lithophragma*, as an incidental by-product of oviposition behaviour. *Greya* larvae feed on some seeds, imposing a reproductive cost. A geographic mosaic across both species'

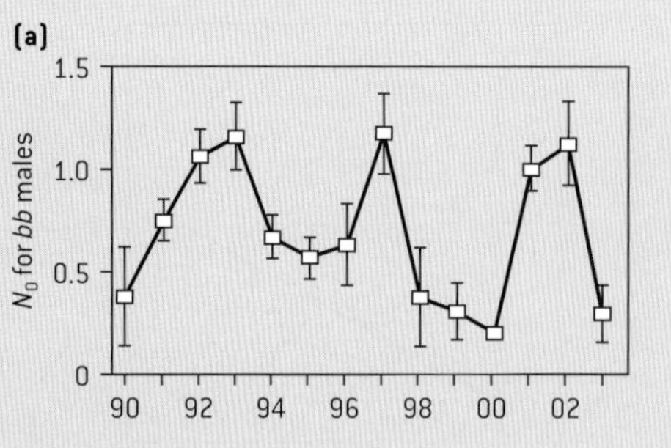


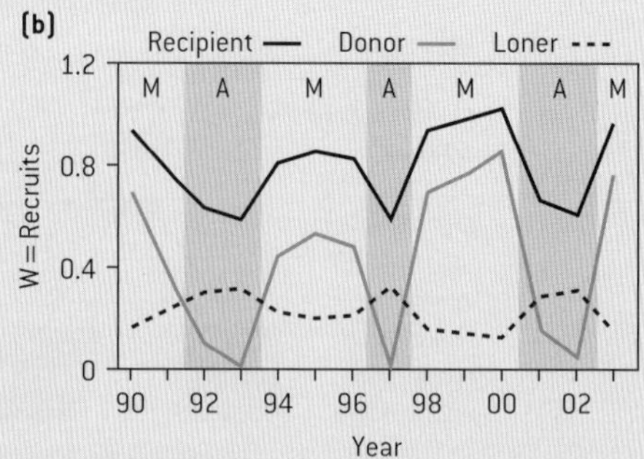


Figure 5.12 **Cycles of altruism and mutualism in cooperative B male *Uta* are driven by cycles in the number of aggressive O male neighbours**

From Sinervo *et al.* (2006), Self-recognition, colour signals and cycles of greenbeard mutualism and transient altruism, *Proceedings of the National Academy of Sciences* USA, volume 103, 7372–7377. © National Academy of Sciences.

**a.** Population cycle in the number of orange male neighbours experienced by blue males.
**b.** Blue male fitness in three social contexts during the rock–paper–scissors cycle.

The cycle of O neighbours drives an evolutionary cycle of altruism and mutualism in cooperative B males. The black line shows the fitness curve of the recipient of altruistic actions in a male partnership of genetically similar B male neighbours, whereas the grey line depicts the fitness of altruist, relative to a loner B male strategy (dotted curve), which does not engage in group behaviours (i.e. loner B males lacks genetically similar neighbours). When O is common (labelled A for altruism), fitness of the B altruist (grey curve), which has more O neighbours, dips below fitness of the loner B strategy (dotted curve). At these times, B altruists should defect, and adopt the loner strategy, but this egoistic behaviour has never been observed (Sinervo *et al.* 2006a). When O is rare (labelled M for mutualism), Y is common and both genetically similar B male neighbours obtain higher fitness than the loner B strategy and thus engage in evolutionary mutualism.

Figure from Sinervo *et al.* 2006a.

ranges arises from parasitic costs versus mutualistic benefits of *Greya* pollinating *Lithophragma*, relative to neutral pollinators like flies (see Thompson and Pellmyr 1992). Sinervo and Calsbeek (2006) review many other inter-specific examples. The example of the red eft of *Notophthalmus viridescens* and mimetic *Pseudotriton ruber* discussed in Section 5.3.4 reflects an example of the costs of aposomatism that is invoked by invasion of mimicry.

Formal discussion of social system trade-offs requires an explicit discussion of the root cause of behavioural trade-offs in social systems, which rests on sender and receiver networks (Section 5.5.2.2c). The extent to which mating or social system trade-offs act depends on information transfer during communication (Wiley 2000). Sender–receiver systems generate a trade-off with intended and unintended receivers (Figure 5.13). Unintended receivers can be conspecifics or predators, or workers of different genotype in eusocial colonies. In the case of male competition, a male that signals its resource holding potential to a conspecific male through a physiologically costly badge of status risks detection by alternative male strategies of crypsis that intercept such signals. This observation underlines the key importance of information in the evolution of trade-offs and of behaviour between intended and unintended receivers. See Chapter 10 for detailed developments on the role of information in evolution.

Honesty (see Chapter 16 for details about this concept) in sender-intended receiver interactions forms a core component of the theory of communication; cheats either pay a cost of retaliation (Enquist and Liemar 1983; Enquist 1985) or individual

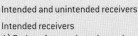

Intended and unintended receivers

Intended receivers
1) Badge of status in male–male conflicts: Orange beats Blue
2) Blue is a badge of cooperation
3) Both Orange and Blue badges are used in female choice

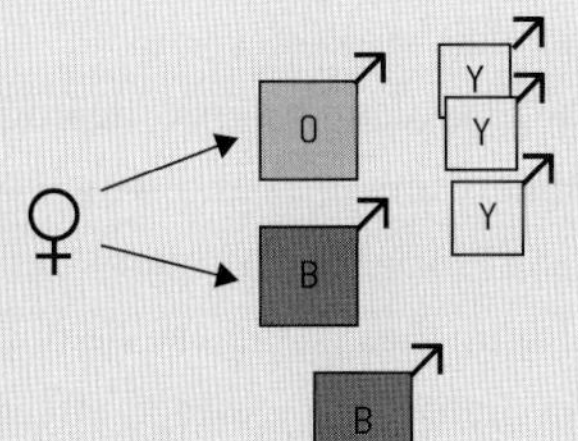


4) The unintended Yellow receivers sneaker males also cue in on the honest Orange badge to be able to cheat on Orange, but not Blue males which have a genetically similar partner

Figure 5.13 **The rock–paper–scissors dynamic results from a complex interplay between three senders and their intended and unintended receivers**

The male rock–paper–scissors dynamic also involves female mate choice for males as a function of male colour (Bleay and Sinervo 2007) and genetic similarity (Sinervo *et al.* 2006). 1. Badge of status in male–male conflicts: Orange beats Blue. 2. Blue is a badge of cooperation. 3. Both Orange and Blue badges are used in female choice. These are the intended receivers. 4. The Unintended Yellow receiver, the sneaker male strategy, also cue in on the honest Orange badge and are able to cheat on the aggressive Orange males, but not the cooperative B males.

recognition limits deceptive signals (van Rhijn and Vodegel 1980). As signals evolve to become more honest, which enhances reliability of information transfer between senders and receivers, they become more physiologically costly (this is the **handicap principle**). An illustration of such physiological costs is seen in the case of survival costs of testosterone production (Marler and Moore 1991; Sinervo *et al.* 2000a), which are implicit in expression of aggressive male strategies. Thus, signals become pleiotropically linked to life history trade-offs of survival selection or immune function. Colours on side-blotched lizard throats (Figure 5.13) reflect signals with pleiotropic effects on testosterone production (Sinervo *et al.* 2000a). Colours are used honestly (orange: usurp with badge of status; blue: cooperate with greenbeard badge of cooperation; female choice for good genes is also in play for both B and O males), or dishonestly in communication (yellow: female mimicry).

In addition to sender–receiver trade-offs and unintended receivers, other trade-offs arise from the cost versus benefit of alternative social behaviours in social networks. Usurpation and cooperation reflect alternative social solutions to resource acquisition and territory defence (Calsbeek and Sinervo 2002b; Calsbeek *et al.* 2002). Cooperation is also common in foraging systems; however, such cooperation is vulnerable to invasion by strategies that either take by force or as noted above acquire information by guile without the costs of information collection (Zoltán and Szép (1994); see Chapter 8 for details on the producer-scrounger game and Chapter 4 for inadvertent social information in a foraging context). Thus, cooperative strategies are invadable by strategies that usurp, which is a strategy that is vulnerable to deceptive strategies. Rock–paper–scissors social strategies are thus very common in social systems, and the social trade-offs of the rock–paper–scissors dynamic are likewise common (Sinervo and Calsbeek 2006).

## 5.6 A synthesis: multivariate trade-offs

A consideration of diverse trade-offs allows for a more synthetic view of the sources of constraints on behavioural systems and a modification of Zera and Harshmann's (2001) diagram for trade-offs. The putative organization of life history versus behavioural trade-offs can be predicted on allocation graphs that are inspired from Zera and Harshman (2001). An allocation graph depicts the optimal allocation of total resources into two pathways. These models are sometimes called Y-allocation graphs, in the case of a pair-wise allocation trade-off. However, dimensionality invoked by behavioural, mating, and social system trade-offs increases the number of branches enormously. The different components of multidimensional trade-offs are characterized in Figure 5.14 by the possible shape of each type of trade-off (a–c) then with respect to an integrated view of the multidimensional trade-offs of behaviour (d).

### 5.6.1 The primacy of demographic constraints in establishing trade-offs

First, demographic constraints set up limits on what and which type of trade-offs are possible among the primary demographic traits (Figure 5.14a). By primary demographic traits, we mean traits that have direct effects on fitness (age at maturity, stage-dependent survival, and fecundity), or their variance (trait plasticity). Under the hypothesis that constraints (either internal or external) set limits on population growth rate (population sizes are confined to some

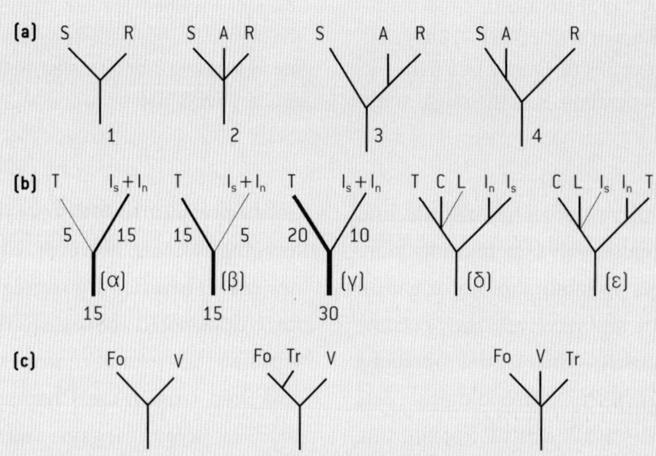


Figure 5.14 **Graphs for visualizing the multidimensionality of life history trade-offs**

**a.** Demographic trade-offs. Bi-dimensional or multi-dimensional trade-offs (1 versus 2). Star- or hierarchical trade-offs (3 versus 4). The basal line represents some environmental constraints, which are similar for all potential trade-off organizations. A change in the thickness of the basal line or stem mimics a change in the strength of the environmental constraint. (S: survival; R: reproduction; A: age at maturity).

**b.** Physiological trade-offs. Potential trade-offs between the hormonal and immune system. α and β are two genotypes investing differently between testosterone and immune function. The correlation between investment in testosterone and immunity across genotype is then negative. However, if individuals vary in their resource availability (thickness of root 15 versus 30), genotype γ can invest more in both testosterone and immunity when compared with β, leading to a positive correlation between testosterone and immunity across genotypes. However, the omnipotence of hormonal actions (Dufty *et al.* 2002) is likely to result in more complex trade-off organizations (for examples δ and ε involve hormones corticosterone and leptin). Trade-offs might even change during development since both immune and hormonal systems have organizational effects and might set up different trade-offs late versus early in life (e.g., role of prolactin in amphibian metamorphosis, migration, or care; see Box 5.4). (T: testosterone; Cort: corticosterone; L: leptin; $I_s$: specific immunity; $I_n$: non-specific immunity).

**c.** Behavioural trade-offs. Behavioural trade-offs involve foraging (Fo) patterns, travel time (Tr), and vigilance (V).

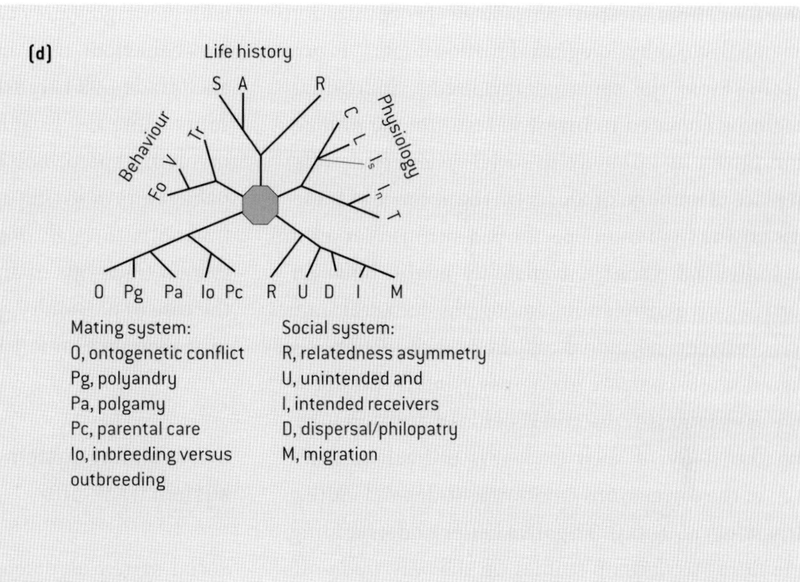


Figure 5.14 **(*cont'd*)**

**d.** Allocation among demography, physiology, behaviour, mating, and social systems. Thus, life history trade-offs really involve physiological and behavioural systems, as well as mating systems and social system dynamics. This figure thus regroups trade-offs presented in the previous graphs into a single multidimensional trade-off (same legend). The size of the circle in the centre (analogous to the width of the basal line or stem in **a** and **b**) of one possible system of trade-offs indicates the total amount of resources that is divided up into a complex pie, with conflicting needs of each system receiving optimal levels of investment.

density or frequency attractor), then all primary demographic parameters cannot vary independently from one another (because growth rate is fixed, see Section 5.1), with the consequence that they should co-vary. Usually, an increase in a trait, which positively influences fitness, negatively influences other traits, which positively influence fitness. However, positive co-variation between traits can also occur if they each negatively influence a third.

Thus multivariate trade-offs organize themselves among demographic traits, resulting in equal effects on fitness. Such trade-off cascades can then be identified by their relative influence on fitness. Even greater multi-dimensionality of trade-offs might be the rule if they arise from epistatic interactions with physiological and/or behavioural machinery (Figure 5.14). In some cases, however, demographic constraints might be relaxed and species might escape certain trade-offs.

For example, when a species expands its range (e.g., at northern margins of many species because of global warming), or when a species expands its niche (e.g., because of a new mutation or by modification of competition with other species), or when an alternative type replaces others in a population (e.g., mutants which replace residents, frequency-dependent polymorphism, etc.), one might find situations where the above predictions will fail and other types of constraint will prevail. This may reorganise the trade-offs or even result in positive trait correlations that are transient over evolutionary time (Reznick *et al.* 2000). When this happens the system may re-organise trade-offs to such a degree that speciation results.

### 5.6.2 Trade-offs due to endocrine and physiological systems

Physiological trade-offs (Costa and Sinervo 2004) express internal constraints such as those arising from energetic constraints (food processing, foraging time) or from deeper physiological organization (metabolism, physiological functions, etc.; Figure 5.14b). From the same energy intake, individuals might be favoured to invest in testosterone production (given its influence in social system trade-offs) in place of immune products (in the absence of trade-offs arising from inter-specific parasitism). Variation in individual resource availability might, however, turn negative correlations among physiological traits (i.e., between hormonal and immune function, see Figure 5.14b: compare genotypes γ and β) into positive associations (Zera and Harshman 2001), rendering the study of such trade-offs difficult and the sign of the correlations environment-dependent (e.g. Ernande *et al.* 2004). To gain a better understanding, one needs to develop studies on the way trade-offs control and are controlled by hormones (Sinervo 1999, 2000; Dufty *et al.* 2002; Sinervo and Calsbeek 2003; see also Boxes 5.4 and 5.5), as well as the interaction among important physiological functions such as immunity (Svensson *et al.* 2002), reserve storage and maintenance (Zera and Harshman 2001; Costa and Sinervo 2004). For example, steroid hormones associated with sexually or socially selected traits often generate a cascade of pleiotropic and/or epistatic effects, which impact many physiological pathways that affect life history trade-offs. Physiological and genetic epistasis can be so strong that it sustains fitness epistasis and alternative optima (Figure 5.10).

Physiological trade-offs may or may not result in demographic trade-offs. Behavioural trade-offs may or may not result in demographic or physiological trade-offs, and vice versa. For example, if a trade-off between vigilance and foraging results in trade-offs between survival and reproduction, a trade-off between sit-and-wait versus widely foraging strategies might not result in any differences in survival and reproduction because both incur survival risk: parasitism for sit-and-wait and predation for wide-ranging foragers (Clobert *et al.* 2000). Sit-and-wait and wide-ranging foraging tactics also invoke behavioural trade-offs of signal detection by their predators: cryptic patterns are often favoured in sit-and-wait prey, versus alternative escape patterns in widely foraging prey types. Selection on morphology and behaviour may amplify physiological performance trade-offs because wide-ranging foragers must adopt more costly anti-predator escape behaviours involving speed, not just those involving stamina. Behaviours also impart greater dimensionality to trade-offs through ontogenetic conflict and mating system trade-offs of antagonistic sexual selection in the case of polyandry, polygyny, parental care, paternity assurance, or fertility assurance.

### 5.6.3 Mating system and social system trade-offs

Social system dynamics further dimensionalize life history trade-offs (Figure 5.14c, Box 5.5). Although social trade-offs have usually been expressed in terms of cost–benefit equations (i.e., Hamilton's (1964) famous equation for kin altruism), these constraints can also be viewed as genetically based trade-offs that arise from pleiotropy (e.g., the action of the OBY locus of side-blotched lizards on life history traits, behaviour, endocrine system, and physiology) or from epistasis such as in sender–receiver interactions (e.g., self-recognition loci that interact with OBY generates conditions for cooperation; Figure 5.11). One way to visualize social trade-offs is to view benefits of mutualism as generating conditions for invasion of alternative strategies that convert mutualisms to altruisms (or parasitisms) (and vice versa; Figure 5.12).

In some mating systems, such as those with biparental care, males and females gain mutualistic benefits from jointly rearing progeny, but this mating system may be invaded by a non-care strategy in one partner (of either sex) that abandons their mate in favour of polygamy (see Chapter 12). The abandoned partner is subject to 'mating system' altruism and bears the costs of reproduction associated with uniparental care. In some social systems, highly aggressive strategies may be subject to invasion by highly deceptive social strategies. Highly aggressive strategies are incompatible with co-operation because of their egoist tendencies. More generally, social trade-offs arise because of the fundamental requirement of communication in social systems, and because benefits of communication with intended receivers come with attendant risks of exploitation by unintended receivers (conspecifics or predators), and ultimately because of asymmetry versus symmetry in genetic similarity among all of the social actors.

*The universality of epistasis in sender and receiver networks*

We have argued here that epistasis or gene interaction is universal in behavioural interactions because social behaviour requires senders and receivers, and loci governing signals will rarely be the same as those governing signal reception or the social acts. There may be exceptions to this generalization, such as greenbeard nepotism observed in red fire ants, where a simple cuticular compound seems to elicit the behaviours fundamental to greenbeard interactions (signal, reception, social acts). However, olfaction of the cuticular compounds must invariably involve many more receiver loci than a cuticular signal (Ross *et al.* 1996; Keller and Ross 1998), thus even this example is likely to involve epistatic interactions of other sender and receiver loci, which have yet to be identified, and which build by correlational selection. The same is true of the pleiotropic gene identified for social amoeba. The demonstration of a pleiotropy on the aggregation behaviour and altruism of the social amoeba, owing to the single *csA* gene, does not preclude the interaction of many other genes that might enhance cooperation or alternatively exclude cheaters from the system. Other sender or receiver genes may be under strong correlational selection with the *csA* gene (see Section 5.4.3.1), but genetic engineering experiments per se cannot resolve such effects. Only gene mapping technology (e.g., Sinervo *et al.* 2006a) allows us to resolve the single locus versus multilocus genetic bases of social behaviour and cascading effects on life history trade-offs.

Not all behaviours will generate such synergism. In the absence of sender–receiver interactions, behavioural traits will be controlled by many non-interacting loci that simply sum up in their genetic effects to generate the phenotype (e.g., additive genetic effects). However, behavioural traits with sender–receiver interactions generate gene epistasis, and thus, fitness epistasis. It is only by combining genetic/endocrine engineering and/or gene mapping that we can resolve the synergetic effects among genes (a mix of methods b, c, and d for resolving genetic epistasis: Section 5.4.3.1).

For example, three greenbeard traits (and loci) that comprise Hamiltonian greenbeards for cooperation are bundled together by correlational selection (Sinervo and Clobert 2003; Sinervo *et al.* 2006a; Figure 5.10) into social supergene for signal, signal recognition, and social acts. Evolutionary stability of cooperative and altruistic behaviours requires that inter-individual benefits be protected from competition, cheating, and defection. Without such safeguards, selfish strategies will eliminate altruistic strategies.

Recent theory suggests that genetic epistasis does not need to be strong to be effective in linking signal, signal recognition, and donation into a multiple greenbeard(s) system (Axelrod *et al.* 2004; Jansen and van Baalen 2006). Epistasis might even be advantageous whenever the association between cooperation and a particular greenbeard signal needs to be changed regularly to avoid invasion by cheaters (chromodynamics, Jansen and van Baalen 2006). Supergenes can thus theoretically be built up from weak epistasis, but it is likely that loose associations among unlinked genes are open to cheater mutations in other genes which either end up eliminating social acts or lead to stronger and stonger epistasis including a strengthening of social interactions and/or alternative strategies (such as between B males in *Uta*). This may ultimately lead to sympatric speciation due to social system dynamics (Hochberg *et al.* 2003; Sinervo and Calsbeek 2006), particularly when alleles for a social preference for self-neighbours are also expressed in females (Sinervo *et al.* 2006a), which generates assortative mating as a function of social signals (Bleay and Sinervo 2007).

Thus, Hamiltonian greenbeards need not arise from pleiotropic supergenes, but supergenes can be bundled by correlational selection on cooperative behaviour and assortative mating in females. In sexual social systems, female mate choice for cooperative behaviour in male mates is a logical consequence of loci for partner recognition, if loci have pleiotropic effects on female choice. Given that mating behaviour is ubiquitous in sexual organisms, **sexual selection will likewise generate strong fitness epistasis between loci for male signals and unlinked female preference loci**. These sender–receiver loci will become coupled by runaway sexual selection (Fisher 1931; Lande 1981). When female mate preferences or territory preferences exert ramifying trade-offs through costs of reproduction (e.g., by mating system, polygyny threshold), systems of trade-offs become multi-dimensional and are a property of genetic interactions, not just pleiotropy. In this regard, behaviours make the study of trade-offs far more difficult, but far more interesting, given the higher dimensionality that is invoked as mating or social system complexity evolves.

### 5.6.4 Using phylogenetic inference to unravel trade-offs

Returning to our original example of multidimensional trade-offs, salamander life cycles, we can identify methods for studying the sequence by which multidimensional trade-offs are added to lineages. The phylogenetic sequence (Box 5.3) provides a clear picture of how innovations are added and thus, how trade-offs are added, thereby increasing system complexity. Application of phylogenetic methods (see Chapter 3) to evolved mating and social system behaviours (Brooks and McLennan 1991; Ryan 1997; Rolland *et al.* 1998; Reynolds *et al.* 2002; Figure 3.12) provides a way to assess the evolutionary sequence by which trade-off dimensions are added to an evolving clade (Wake and Larsen 1987).

This historical perspective is, however, inherently *post hoc*. Our goal is to generate predictive power. In this regard, modelling approaches like the ESS, when incorporating explicit trade-offs, may eventually allow us to predict when certain trade-offs dominate a given mating or social system, when trade-offs remain suppressed, or when trade-off dimensionality collapses. The cost–benefit approach in the context of ESS provides a clear way to identify these trade-offs in pair-wise trait interactions. We will need to increase the dimensionality of cost–benefit ESS analysis to be able to predict the direction of and evolving constraints on the evolution of novelties and life history strategies (Ferrière and Clobert 1992). Combining phylogenetic analysis with ESS models that capture genetics in a multi-locus context may allow us to eventually predict which trade-offs will be present in a given social or ecological context.

## 5.7 Conclusion

We hope that the reader is convinced of the intimate links that bind a study of behaviour to that of life history cycles and trade-offs. To appreciate these links it is necessary to understand the interplay between genes and plasticity and the induction of behaviours that are due to either pleiotropy or epistasis. Pleiotropy can arise from a single gene (or endocrine cascade) that generates a life history trade-off, mating or social system trade-off. However, trade-offs of higher dimensions can also arise from epistasis among many endocrine cascades (axes). A useful homology between the physiological epistasis of endocrine systems and the behavioural epistasis of social systems is that both kinds of epistasis involve senders and receivers. In the case of physiological epistasis sender–receiver systems are hormones and hormone targets (e.g., tissue- or sex-specific gene promoters), which are intrinsic to the organism. Behavioural epistasis incorporates sender–receiver systems that are extrinsic to the organism and a property of the mating or social system. Trade-offs naturally emerge from the interplay between the genetic expression and regulation, hormonal regulation, as well as mating and social system regulation (mating, competition, communication, etc.) and environmental constraints. The take-home message is that behaviour is an essential dynamical component of these prominent processes of evolution.

Behaviour increases the dimensionality of life history trade-offs in profound and important ways. Moreover, behaviour has emergent and self-organizing effects on the evolution of life history trade-offs, but depending on environmental constraints such complexity can collapse into simpler states, or expand enormously into highly dimensionalized social systems like eusocial insects. Because behaviour can invoke both physiological epistasis and behavioural epistasis of sender–receiver systems, behaviour can help to build genetic correlations among unlinked traits. Moreover, the evolutionary dynamics of behavioural systems can generate stable ESS cycles (or states), which hold these unlinked sender and receiver loci as stable genetic correlations (Sinervo and Calsbeek 2006), until selection has enough time to functionally integrate the genetic correlations into more permanent and stable physiological epistasis. Thus, physiological epistasis of unlinked sender–receiver loci can be cemented into pleiotropic physiological states of a single integrated endocrine cascade. Life history trade-offs evolve by the process of correlational selection (Sinervo 2000; Sinervo and Svensson 2002; Roff and Fairbairn 2006), which shapes the negative genetic correlations that arise from endocrine regulation. Unravelling the multi-dimensional nature of behavioural complexity should be a central goal of future work. Life history theory has tended to address trade-offs in a pair wise fashion, but we should now begin to assemble these trade-offs into a more synthetic and multidimensional framework to understand the more general rules by which mating systems and social systems evolve. In particular, evaluating the pleiotropic or epistatic nature of the different kind of trade-offs might be better achieved by looking at several types of trade-off simultaneously, and by verifying at which levels of multidimensionality equal fitness among traits combinations are attained. Constraints are not necessarily the same at the genetic, physiological, behavioural, and demographic level, either because constraints may be specific to each level, and also because of the **bijective**, **injective**, and/or **surjective** nature of the functions linking the different levels of trade-off organization.

Bijective, injective, and surjective are terms that are used when qualifying mathematical functions that link points in two spaces. Bijective refers to a space in which each point of one space maps onto one point of a second space. Injective refers to multiple points of one space that map onto one point of the second space (e.g., a hierarchy). Surjective means the reverse situation. The analogy is useful here with behavioural traits in that the spaces comprising the genetic, physiological, and behavioural layers are due to complex ramifying causes across these levels. The multiple causes at a lower level (endocrine regulation) may collapse and form a single behavioural trade-off. Alternatively one endocrine system may ramify into multiple behavioural trade-offs in certain social contexts, especially in the case of sender–receiver interactions. The terms bijective, injective, and surjective therefore qualify the types of links that are possible between these layers. The links are not simple and linear, nor are they hierarchical (genetic within physiology, within behaviour). Rather the links between genes, physiology, and behaviour form a complex set of spaces with constraints that are specific within each level of organization, yet in total constrain organismal function.

## » Further reading

> *The links between behavioural ecology and life history evolution have deep roots, but surprisingly, behavioural ecologists tend to compartmentalize studies into either male or female behaviours. Integration of such studies is required to elucidate the relationship between life history trade-offs and theories in behavioural ecology. Hanna Kokko (2001) provides a review of data that are necessary for integrating theories of sexual selection with life history components of the male and female life history. Derek Roff and Daphne Fairbairn (2006) provide a review of the state of life history theory and directions that need to be pursued for a more comprehensive theory.*

*Our understanding of social evolution has focused on interactions between individuals, rather than a focus on the routes by which life history trade-offs are exacted. In this regard, theories of social evolution are well described by Alexander (1974), while more recent treatments of the meaning, and costs and benefits of cooperation can be found in West* et al. *(2007), Jansen and van Baalen (2006), and Lehman and Keller (2006).*

*Kingsolver* et al. *(2001) review patterns of selection in nature (see also Rousset 2004) and note the sparse information on correlational selection. Sinervo and Calsbeek (2006) review the role of frequency-dependent selection in the genesis of correlational selection between sender and receiver systems (across mating systems and ecosystems).*

**Alexander, R.** 1974. The evolution of social behavior. *Annual Review of Ecology and Systematics* 5: 325–383.

**Jansen, V.A. & van Baalen, M.** 2006 Altruism through beard chromodynamics. *Nature* 440: 663–666.

**Kingsolver, J.G., Hoekstra, H.E., Hoekstra, J.M., Berrigan, D., Vignieri, S.N., Hill, C.E., Hoang, A., Gibert, P. & Beerli, P.** 2001. The strength of phenotypic selection in natural populations. *American Naturalist* 157: 245–261.

**Kokko, H.** 2001 Fisherian and 'good genes' benefits of mate choice: how not to distinguish between them. *Ecology Letters* 4: 322–326.

**Lehmann, L. & Keller, L.** 2006. The evolution of cooperation and altruism. A general framework and classification of models. *Journal of Evolutionary Biology* 19: 1365–1725.

**Roff, D.A. & Fairbairn, D.J.** 2006 The evolution of trade-offs: where are we? *Journal of Evolutionary Biology* 20: 433–447.

**Rousset, F.** 2004. *Genetic Structure and Selection in Subdivided Populations.* Princeton University Press, Princeton, NJ.

**Sinervo, B. & Calsbeek, R.** 2006. The developmental and physiological causes and consequences of frequency dependent selection in the wild. *Annual Review of Ecology and Systematics* 37: 581–610.

**West, S.A., Griffin, A.S. & Gardner, A.** 2007 Social semantics: altruism, cooperation, mutualism, strong reciprocity and group selection. *Journal of Evolutionary Biology* 20: 415–432.

## » Questions

1. The trade-offs for thyroxine induction of metamorphosis are outlined in Boxes 5.4 and 5.5. The carnivorous tadpole morph is also involved in a social trade-off that involves levels of relatedness and true altruism balanced against the selfish act of cannibalism (Pfennig and Collins 1992). How might kin recognition come about? Posit a plausible mechanism for self- and non-self recognition based on the physiological systems outlined in Box 5.5 and another physiological system. Based on this answer, with what other physiological systems might the HP–thyroid axis and HP–adrenal axis interact epistatically?

2. Are the trade-offs for males and females the same? Does the process of sexual selection on male ornaments intensify the survival trade-offs for costs of reproduction in males? If yes, by what process of selection? What sorts of signal traits are involved in density-dependent female competition? Are the costs of reproduction in males thus greater than the costs of reproduction in females?

3. The links and analogies between language and epistasis could also be made. Phonemes and elements of language have analogous variations to alleles or mutation, but to constitute a language you need to link language elements (epistasis). Dialects (intonation, inflection) are just genotypes. Many languages are different, but common roots and elements, sometimes slightly modified, can be identified and just are organized in different ways creating different languages to the extent that they are no longer understandable by others (they become species). Trade-offs controlled by epistasis should evolve more rapidly (given that genes can be unlinked) than trade-offs arising through pleiotropy (because they affect one genes structure more deeply). Analogously, in language, the elements are more slowly evolving than the combination of these elements. Discuss these analogies by creating a table comparing and contrasting genetic and language evolution.

4. Why is dispersal versus philopatry a social system trade-off? Why is dispersal versus philopatry a life history trade-off?

5. Describe the three greenbeard traits proposed by Hamilton (1964). Outline a version of these traits that comprises a pleiotropic Hamiltonian supergene, and a version comprising an epistatic Hamiltonian supergene. Which of these scenarios is more likely from the point of view of the number of mutations required? Which is more likely from the point of view of a sequence of traits in an evolving social system? Describe the sequence by why altruism is converted into mutualism? Map it onto a phylogeny for a mating system that involves your organism of choice.

6

# 6

# Hormones and Behaviour

Alfred M. Dufty Jr. and Étienne Danchin

## 6.1 Introduction

The first steps in an individual's life put into play all the developmental processes by which a young organism becomes an adult. The **phenotype** of an individual is formed during his development, with input from the environment. Thus, important behavioural characteristics of an adult are shaped during development. Therefore, to analyse the adaptive value and evolution of behaviour, it is necessary to understand the forces that have directed phenotypic processes. Since the early 1990s, studies in behavioural ecology increasingly have attempted to incorporate an evolutionary approach into the study of phenotypic development. The goal of this chapter is to provide basic insights into some of the mechanisms concerned with the development and expression of behaviours. This chapter is deliberately biased in favour of vertebrates, although equally complex processes occur in invertebrates and plants. Furthermore, we have chosen to concentrate on the relationship between **hormones** and behaviour. We do not claim to provide exhaustive coverage of this huge question. Rather, we have attempted to present the diversity of research approaches used in this field.

In behavioural ecology the terms 'ontogeny' and 'development' are equivalent, describing processes that unfold during an individual's entire life. Therefore, for a behavioural ecologist, the ontogeny of an individual lasts its entire lifetime, and any event that may modify the phenotype participates in that individual's ontogeny.

### 6.1.1 Relation between phenotype and genotype

We have seen in Chapter 2 that the **phenotype** of an organism is its suite of observable traits, things that one might see, feel and measure. For example, the phenotype of a chimpanzee includes the colour of its eyes, the texture of its hair, its height, its sexual characteristics, its thyroid gland, the length of its toes, as well as any other characteristics in which one might be interested. Thus, an organism's phenotype includes all of its physical characteristics.

The phenotype is the product of the visible expression of the **genotype** (a collection of genetic information stored as genes), knowing that gene expression is affected by environmental conditions encountered during development. Genes are nucleotide sequences (made of deoxyribonucleic acid, DNA) that code for specific proteins that affect the properties of cells. And cells, whether acting singly (as in bacteria) or together (as in multi-cellular organisms), produce the attributes that collectively make up the phenotype.

Genes form an integral part of an organism and provide a framework for phenotypic development. They are vital to all parts of an organism, from its internal physiology to its external features. Thus, genes are integral to, and provide the framework for, phenotypic development. The transcription and translation of genes into ribonucleic acid (RNA) and proteins, respectively, are fundamental properties shared among all living organisms, regardless of their level of complexity. Genetic messages thus put into action, direct, and coordinate the development of limbs, antennae, cilia, feathers, kidneys, swim bladders, and all other physical properties that make up different phenotypes.

However, in addition to physical properties, the concept of phenotype also includes an organism's behaviour. An individual's behaviour is every bit as distinctive and measurable as any of its physical characteristics and is also involved in its survival and reproduction. Does it exhibit a diurnal or nocturnal activity rhythm? Is it a predator? Is it prey? Is it both? Does it seek out a new mate each year? Does it mate for life? In the case of the chimpanzee, for example, one may observe the extent to which an individual is sociable by the frequency with which it exhibits grooming behaviour or solicits such behaviours from others. One also may determine dominance rank by gauging its aggressiveness and its ability to control access to resources such as food or sexual partners. Finally, one can include in the phenotype all elements directly influenced by the organism in question. These elements, known as the extended phenotype (Dawkins 1982), include particular structures produced by the activity of an individual: a nest or a tool, for example.

### 6.1.2 Genes and behaviour

If genes direct phenotypic development and if the phenotype includes behavioural traits, then one way or another, genes must affect behaviour. Although few would doubt this simple exercise in deductive reasoning, it is difficult to link genes with specific behaviours, especially in vertebrates. Briefly stated, behaviour does not occur in a vacuum. That is, possessing an appropriate genotype does not guarantee that a given behaviour will be expressed. In general, animals exhibit behaviours in response to sensory input. For example, the appearance of a predator elicits mobbing behaviour in birds (see, for example, Clode *et al.* 2000). However, in the absence of this stimulus the behaviour does not occur, despite the presence of an appropriate genotype.

Another problem is the apparent lack of variability in many behaviours. For example, experienced male mice prefer to associate with sexually receptive females (see, for example, Huck and Banks 1984). There is strong selective pressure on males to behave in this way, because males that do not have this preference would find sexual partners less rapidly than those possessing the attribute. Preferences for unreceptive females would have consequences even more harmful in terms of reproduction: all else being equal, that behaviour would become increasingly rare or even disappear because such individuals would have no or very few offspring. Although a preference for receptive females has obvious reproductive advantages, the lack of variability in this male trait makes it very difficult to manipulate and study its underlying genetic basis.

As a consequence, non-vertebrate models have dominated the study of behavioural genetics (Dudai 1988) and have produced some fascinating discoveries about the importance of single genes in directing behaviour (Demir and Dickson 2005). However, the relatively recent technical advances leading to the production of knockout and transgenic mice have greatly facilitated the study of genes and behaviour, at least in mammals. In knockout mice a precise gene is removed (in fact, rendered inactive), whereas in transgenic mice a totally new gene is inserted into their genotype (Lee *et al.* 1996; Nizielski *et al.* 1996; Ryffel 1996). Such explicit, single-gene removals or additions make possible investigations where experimental and control groups differ genetically by only one gene. Although problems remain in the interpretation of results (Gingrich and Hen 2000), knockout and transgenic organisms have become important tools in the study of genes (and their products) and their links with behaviour (Nelson 1997; Götz *et al.* 2004).

### 6.1.3 Non-genomic factors

Although phenotypic development is orchestrated by genes, the genotype is by no means the only factor that affects the process. Indeed, even if one knew the entire genomic make-up of an organism, gene by gene and nucleotide by nucleotide, one could not predict the phenotype with certainty. This is because non-genomic factors stemming from the external environment affect developmental pathways and thus vary the phenotypic end-product. If this were not the case – that is, if genotype predicted phenotype with certainty – then the latter would simply be a complement, or a copy of the former, similar to the relationship between DNA and RNA.

What types of environmental factor affect phenotypic development? The answers to this question are numerous, and some of those answers are discussed in detail later in this chapter. They consist of things like endocrine disruptors that can irrevocably alter the developmental process (Choi *et al.* 2004), or dietary restrictions that can affect physical characteristics and/or physiological processes (Nowicki *et al.* 2002). Seasonal changes in day length also provide predictable cues used by animals to modify their phenotype according to the activities they engage in at different times of their annual cycle (Jacobs and Wingfield 2000). We will see that even characteristics as fundamental as the sex of individuals may be influenced by environmental factors in some species (Crews 2003).

Many of these environmental effects are the products of **maternal effects**. This has been demonstrated most effectively with mammals, where there is ample opportunity for trans-placental communication between mother and offspring, and for *in utero* interactions among the offspring themselves (Godfrey 2002, Ryan and Vandenbergh 2002). However, subtle maternal effects also have been demonstrated in egg-laying taxa (see, for example, Schwabl 1993; Gasparini *et al.* 2001; Groothuis *et al.* 2005), suggesting that this phenomenon is widespread. Furthermore, neonatal interactions between mother and offspring also can have developmental repercussions that affect, for example, the strength of the physiological response of offspring to stressful situations (van Oers *et al.* 1998).

It is important to recognize that the **genome** is the basis of, and sets constraints on, all aspects of an organism's phenotypic development, including behaviour. However, as van der Steen (1998) notes, the study of behavioural genetics focuses on events at the level of the population, and provides little information regarding the development of individual phenotypes. Insofar as our interest in the present chapter lies with phenotypic development as it relates to behaviour, we will concentrate on post-transcriptional processes that affect behaviour. We will examine proximate mechanisms that integrate genetic commands and environmental information at critical developmental stages, such as during the prenatal phase, the early postnatal phase, puberty, and adulthood. This integration of genetic and ecological factors produces phenotypic variation that may improve an organism's **fitness** by improving its chances of surviving and reproducing in a given environment. Therefore, to put phenotypic development in an ecological context, we also discuss the possible adaptive significance of variation in phenotypic development.

### 6.1.4 Hormones and their regulation

The integrative mechanisms that coordinate developmental changes typically involve hormonal messengers. **Hormones** are molecules produced and released into the blood by certain organs (called endocrine glands) and act on target cells located elsewhere in the organism. Hormones produced by the endocrine system are but one of several types of chemical messengers. The other types are neurotransmitters (produced by the nervous system) and cytokines (produced by the immune system). Once thought to be independent mechanisms, these three systems are now known to affect one another, and the investigation of their interactions represents one of the most active areas of biological research (Cardinali *et al.* 2000; Haddad *et al.* 2002).

Most endocrine systems are regulated by negative *feedback* mechanisms, as illustrated by the hypothalamic–pituitary–adrenal (HPA) axis (Figure 6.1; see Table 6.1 for terminology), which is involved in the regulation of energy stores and the response of animals to stressors. The hypothalamus produces corticotropin-releasing hormone (CRH), a peptide hormone that acts on the anterior pituitary to stimulate secretion of a protein, adrenocorticotropic hormone (ACTH). ACTH, in turn, stimulates the release of glucocorticoids from the adrenal cortex. In addition to facilitating a suite of physiological and behavioural responses in target cells, elevated plasma glucocorticoid levels interact with the hypothalamus and the anterior pituitary to inhibit additional secretion of CRH and ACTH, thereby regulating their own production.

#### 6.1.4.1 *Hormone transport and target cells*

*Circulating hormone levels and binding proteins*

Hormones are transported in the blood, either circulating freely in the plasma or in association with albumin or specific binding proteins. Consequently, hormones rapidly come into contact with most of an organism's cells. However, only the fraction of circulating hormone that is free (i.e., not attached to binding proteins) is biologically active. Changes in the concentration of binding proteins thus can alter the amount of biologically active hormones without changing the overall concentration of hormones in the blood (Figure 6.2a). This phenomenon is often

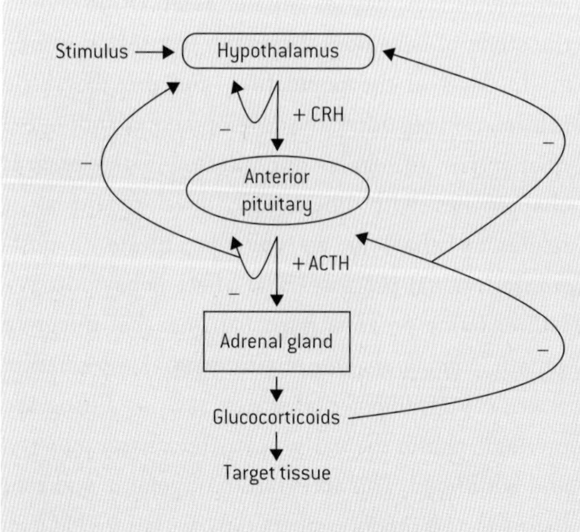


Figure 6.1 **The hypothalamic–pituitary–adrenal (HPA) axis of vertebrates**

In response to a stimulation, the hypothalamus releases corticotropin-releasing hormone (CRH), which, in turn, stimulates (action represented by the symbol +) the synthesis and release of adrenocorticotropic hormone (ACTH) by the anterior pituitary. ACTH then stimulates the release of glucocorticoids, such as corticosterone and cortisol, which modify the metabolic activity of target cells. The secretion of glucocorticoids is regulated by a negative feedback mechanism: hormones of the HPA inhibit the endocrine glands that are upstream in the axis, thus regulating their own secretion.

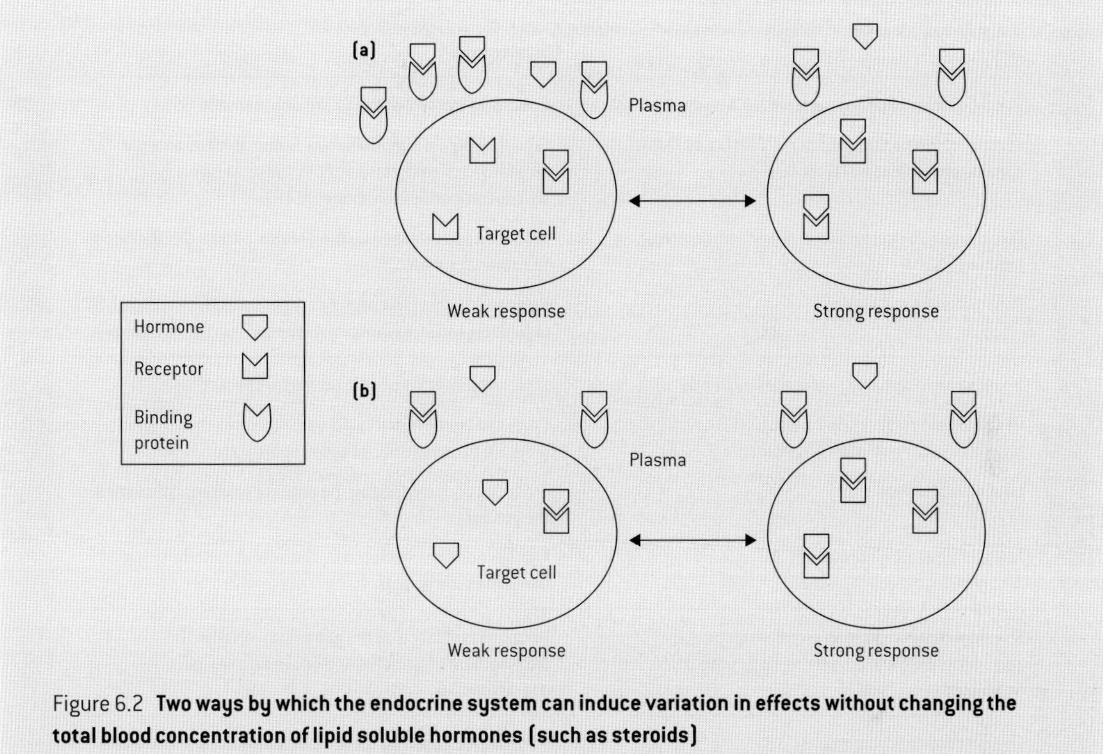


Figure 6.2 **Two ways by which the endocrine system can induce variation in effects without changing the total blood concentration of lipid soluble hormones (such as steroids)**

**a.** Changes in the concentration of transport proteins induce variation in the concentration of biologically active hormone. Only steroid hormones unattached to carrier molecules are biologically active.

**b.** Changes in target cell receptor numbers can produce variability in the response to the same blood hormone concentration. Simultaneous changes in a. and b. also can occur. Note that in the two cases, the total concentration of the hormone, represented by the number of symbols, is the same: there are six symbols in both cases.

neglected, despite the fact that it can play a major role in explaining certain behavioural differences (e.g., Jennings *et al.* 2000).

*Receptors*

For hormones to act, they first must be recognized by the organism's cells. This recognition is mediated through specific proteins, called hormone receptors, which have the capacity to bind dynamically to a specific hormone molecule. This association is based on a thermodynamic interaction between the hormone and the receptor. Only cells that possess the appropriate hormone receptors are capable of responding to a given hormone. These cells are called target cells. Receptors for peptide and protein hormones, which are not lipid soluble and therefore cannot penetrate the cell membrane, are located on cell surfaces. When a hormone combines with its receptor, the activated receptor induces a change in target cell activity, often creating intra-cellular second messengers (the hormone itself is the first messenger) that activate (or deactivate) pre-existing cytoplasmic enzymes (Hadley 1996). The result is a relatively quick cellular response. An unresolved concern with this model is that there are relatively few possible biochemical shapes that second messengers can assume, whereas the variety of messages that are transmitted is quite large.

| Terms | Significance |
|---|---|
| Hypothalamic–pituitary–adrenal (HPA) axis | Structures interlinked by a hormonal cascade |
| Corticotropin-releasing hormone (CRH); also known as corticotropin-releasing factor (CRF) | Produced by the hypothalamus. Stimulates the anterior pituitary to release corticotropin, better known as adrenocorticotropic hormone (ACTH). |
| Corticotropin; better known as adrenocorticotropic hormone (ACTH) | Produced and released by the anterior pituitary. Stimulates the adrenal glands. |
| Corticosterone and cortisol | Released by the adrenal glands; associated with energy regulation and stress responsiveness |
| Hypothalamic–pituitary–gonadal (HPG) axis | Structures interlinked by a hormonal cascade |
| Gonadotropin-releasing hormone (GNRH). Also called luteinizing hormone-releasing hormone (LHRH) | Produced by the hypothalamus. Stimulates the anterior pituitary to release the gonadotropins (LH and FSH) |
| Luteinizing hormone (LH) | Produced and released by the anterior pituitary. Stimulates the gonads. |
| Follicle-stimulating hormone (FSH) | Another hormone secreted by the anterior pituitary that stimulates the gonads. |
| Sex steroids: testosterone in males, œstrogens (e.g., œstradiol) and progestins (e.g., progesterone) in females | Hormones released by the gonads; involved in the development of gametes and secondary sexual characteristics. |
| Thyroid stimulating hormone (TSH) | Made and released by the anterior pituitary; stimulates the thyroid gland. |
| Thyroid hormones | Hormones produced by the thyroid; primarily involved in regulating metabolism. |
| Leptin | Protein hormone released by adipose cells; acts mainly on energy balance. |
| Arginine vasotocin (AVT) or simply vasotocin | Neurohormone. Affects reproductive behaviour. |
| Prolactin | Protein hormone produced by the anterior pituitary. The mammalian placenta produces lactogen, which is very similar to prolactin. Induces, among other things, a variety of maternal behaviour in different taxa. |
| Oxytocin | Produced by the posterior pituitary; among other things, plays a role in the induction of parental behaviour. |
| Gonadectomy | Surgical removal of the gonads. |
| Anti-Müllerian Hormone (AMH) | Peptide hormone produced by developing testes; inhibits development of female reproductive structures. |

Table 6.1 **Table of the principal specific terms cited in the chapter**

Receptors for steroid and thyroid hormones, which are lipid soluble, have traditionally been thought to be located not on the membrane, but in the cytoplasm or the cell nucleus. These types of hormone pass through the cell membrane and combine with a receptor, sometimes by displacing a chaperone molecule. If the resulting complex is not already in the nucleus, then it is taken there where it stimulates (or inhibits) gene transcription. Thus, steroid and thyroid hormones also alter the activity of target cells, but because they act directly at the level of the genome, their effects take longer (at least 30 minutes;

usually hours) to occur. This traditional view has been called into question, both by empirical results showing the existence of rapid, non-genomic steroid hormone effects and by the discovery of membrane steroid hormone receptors (Orchinik *et al.* 1991, Moore and Evans 1999, Dallman 2005). For example, mountain chickadees (*Parus gambeli*) store seeds in caches. They remember the location of the caches, which they use when environmental conditions deteriorate. Chickadees given corticosterone five minutes before a test of their cache recovery ability retrieved more seeds than did control birds (Saldanha *et al.* 2000). This behavioural change is too rapid to have been induced by genomic effects. In fact, the classification of membrane and intracellular receptors as fast-acting and non-genomic or slow-acting and genomic, respectively, may not be absolute (Orchinik and McEwen 1995).

Target cells can have multiple receptor types for a given hormone that differ in the behavioural responses they induce. For example, oestrogen receptors (ERs) α and β play distinct roles in regulating mating behaviour (Greco *et al.* 2003). Elimination of ER-α, but not ER-β, reduces aspects of male copulatory behaviour in mice (Ogawa *et al.* 2000). Receptors can vary in their affinity, or biochemical attraction, for a hormone and this may explain the observed variability in behavioural responses with different levels of circulating hormones. For example, corticosterone, an adrenal glucocorticoid, has two receptor types: type I receptors have a high affinity for corticosterone and, therefore, are activated at low corticosterone levels, whereas type II receptors have a low affinity for the hormone and are activated only at elevated corticosterone levels (McEwen *et al.* 1988). The relatively low corticosterone levels that occur during the diurnal cycle of animals are thought to mediate corticosterone's role in regulating daily energy use through its interactions with type I receptors. When elevated corticosterone levels are produced, such as during stressful events, type II receptors are activated, and these trigger additional physiological and behavioural responses that are not seen at the lower hormonal levels (Wingfield and Ramenofsky 1999).

*Receptor density*

The number of receptors can change in response to changes in hormonal secretion. In many cases, an increase in concentration of a hormone will decrease the number of its receptors in (or on) target cells, a process called down-regulation. This renders target cells less sensitive to the hormone, despite the latter's higher concentration in the blood (Figure 6.2b). Conversely, chronically low levels of a hormone can increase, or up-regulate, receptor numbers, making target cells more sensitive to stimulation by the hormone in question. In addition, under some conditions increased plasma levels of a hormone, such as prolactin or oxytocin, can up-regulate its receptor numbers, making target cells more sensitive to stimulation by the hormone. However, many hormones are released in a pulsatile manner that prevents large changes in receptor number. Furthermore, limiting the duration of maximum (or minimum) plasma hormone concentrations also reduces the likelihood of significant down- (or up-)regulation of receptor number. Finally, some hormones increase or decrease the number of receptors, and thus the efficacy, of other hormones, in what is termed a permissive effect. For example, oestrogens are known to increase receptor numbers for progesterone (Godwin and Crews 1999). Thus, although measuring absolute hormone levels in the blood often provides a sufficiently clear indication of a physiological or behavioural response, subtle changes in receptor number or receptor type can elicit behavioural changes in the absence of changes in absolute hormone level.

#### 6.1.4.2 *Other regulatory mechanisms*

Study of the brain mechanisms by which hormones elicit behavioural responses represents one of the most interesting and rapidly unfolding areas of scientific research. Long-accepted dogma is being revised in the face of new and exciting ideas. For example, studies suggest that steroid hormones can be synthesized *de novo* in the brain or may be created there from inactive precursors (Baulieu 1998; Soma *et al.* 2005). This would allow localized production and use of steroid hormones, changes that would not be reflected in the general circulation. Local brain production of testosterone would help to explain the results of studies in which hormonal bases for behaviour exist under some conditions but not others. For instance, the sex-steroid hormone testosterone (T) facilitates territorial behaviour in male song sparrows (*Melospiza melodia*) in spring, but autumnal territoriality occurs in the absence of hormonal correlates in the peripheral circulation (Wingfield and Hahn 1994). Local brain production of testosterone and use of testosterone in autumn could induce changes in territorial behaviour without exposing other parts of the body to testosterone (Soma *et al.* 2000). Testosterone can incur fitness costs through reduced survival (Dufty 1989), decreased body mass (Ketterson *et al.* 1991), and impaired immune function (Nelson and Demas 1996). Finally, testosterone might stimulate seasonally inappropriate behaviours, such as autumnal courtship singing.

Another way to regulate interactions between a hormone and its receptor is through molecular chaperones, which, by their connection with the receptor, could modulate the dynamics of the hormone–receptor complex (Li and Sánchez 2005). Finally, the behavioural and physiological effects of some hormones involve their conversion from biologically active precursors. For example, testosterone frequently is changed in the brain to oestradiol, which stimulates sexual differentiation of the brain and male sexual behaviour. This conversion is accomplished in a single step by an enzyme, cytochromes P450 aromatase. Interestingly, some effects of environmental cues on behaviour appear to be mediated by changes in brain aromatase activity, suggesting that this enzyme plays a prominent regulatory role beyond that of a simple catalyst (Balthazart *et al.* 2006).

### 6.1.5 Effects of behaviour on hormones

Although the focus of this chapter is on the effects of hormones on behaviour, it should not be forgotten that the converse also is true: behaviour affects hormone secretions (see Rissman 1996 for a review). This phenomenon has been described in detail in ring doves (*Streptopelia risoria*), where behavioural interactions between the two members of a pair enable them to coordinate their progression through the various stages of the breeding cycle (Lehrman 1965). Behavioural displays can stimulate both the recipient (i.e., the mate) and the individual producing the display. For example, nest coos are a vocalization produced by both sexes of *S. risoria*. Their production by the female helps to stimulate the development of her own ovaries (Cheng 1986). Female nest coos, in turn, are induced by male courtship, and both auditory and proprioceptive (i.e. mechanical information about the movements and position of the body parts) components of male courtship behaviour are necessary to stimulate full ovarian growth (Cheng *et al.* 1988). Thus, complete female reproductive development is a complex process involving cues from the mate as well as expression of specific behaviours by the female herself. A physiological link has been discovered between brain areas associated with the perception of nest coos and the release of luteinizing hormone (LH), a gonadotropin that stimulates ovarian growth (Cheng *et al.* 1998). Furthermore, these brain areas respond differently to nest coos produced by females and those produced by males, attesting to the high degree of specificity of the mechanism. However, such effects are not limited to females. For example, social interactions also affect endocrine secretions in male ring doves (O'Connell *et al.* 1981a, b). Furthermore, other environmental factors associated with reproduction, such as the number of young in a clutch (ten Cate *et al.* 1993), can affect both members of a pair. For example, hormonal differences have been induced following the manipulation of clutch size, with individuals raising two young exhibiting lower LH levels than individuals caring for one young.

Similar effects have been found in other animals. For example, exposure to vocal cues can stimulate steroid hormone production (Burmeister and Wilczynski 2000, Figure 6.3). In birds, the presence of conspecifics of either sex can affect androgen secretion in adult males (see Dufty and Wingfield 1984b; Dufty and Wingfield 1990; Wikelski *et al.*

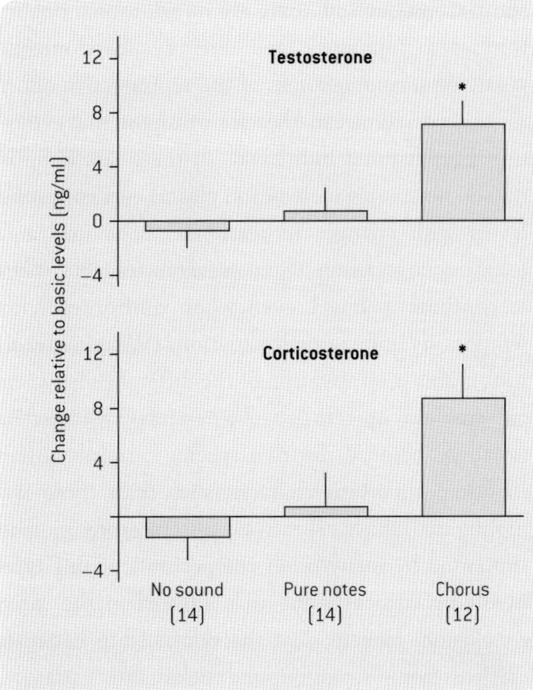


Figure 6.3 **Hearing songs can affect hormone levels**

Mean (+standard deviation) changes of steroid hormone levels in three groups of male green treefrog (*Hyla cinera*) that were exposed to the following stimuli for several nights: no sound (the controls), pure notes, and a chorus of the species' mating songs. The endocrine changes for each hormone were analysed separately by repeated-measures analysis of variance (ANOVA) (testosterone: $F_{2,37} = 6.76, P = 0.003$; corticosterone: $F_{2,37} = 5.98, P = 0.006$). Additional analyses using Pearson correlations show that exposure to choruses of mating songs significantly increases the circulating level of the two hormones (testosterone: $F_{1,11} = 12.02, P = 0.005$; corticosterone: $F_{1,11} = 19.79, P = 0.001$). The symbol * indicates a significant difference between the levels before and after the experimental treatment. Sample sizes are given in parentheses.

After Burmeister and Wilczynski (2000).

1999). Likewise, in mammals, the presence of adult males can accelerate the onset of puberty in immature females, although the overall level of sociality of a species influences this effect (Levin and Johnston 1986). Similarly, the act of winning or losing an aggressive encounter also can affect subsequent endocrine levels in a variety of taxa and may be related to changes in brain aromatase activity (Hsu *et al.* 2006).

Finally, **coloniality** (see Chapter 14) offers an interesting setting for the investigation of social influences on reproductive endocrinology. For example, in colonies of naked mole-rats (*Heterocephalus glaber*), only one female breeds and her aggression towards other females keeps them from ovulating (Faulkes *et al.* 1990). Upon removal of the breeding female, subordinate females exhibit hormonal changes, begin to ovulate, and increase their level of aggression (Margulis *et al.* 1995). Similarly, in *Cryptomys* mole rats, reproductive subordination results in reduced release of gonadotropin-releasing hormone, a reproductive hormone (Du Toit *et al.* 2006). Social suppression of breeding also occurs in primates, where subordinate females in marmoset colonies are prevented from breeding by a combination of behavioural, visual, and olfactory cues from the dominant, breeding female (Barrett *et al.* 1993). Similar effects have long been known to occur in social invertebrates. For example, in hymenoptera, the reproductive bias in favour of the single queen is maintained by the release of pheromones and aggressive behaviour towards other females in the colony (Butler 1954, Velthuis 1976). Thus, if one wants to understand the interactions between hormones and behaviour, then it is important to remember that the relationship is bi-directional. Not only do hormones have profound effects on behaviour, but the reverse is also true: the expression of a given behaviour can significantly modify patterns of hormone secretion.

### 6.1.6 The role of context in behavioural endocrinology and adaptation

When exploring the endocrine underpinnings of behaviour, it is important to know as much as possible about the animal model used and the context in which one is examining its behaviour. Hormonal patterns and behavioural responses become difficult to interpret if one does not understand how physical and social environments affect the organism's behaviour.

An example involving corticosterone will serve to illustrate this idea. Corticosterone is a hormone that is released in response to stressful stimuli; it induces physiological changes that regulate energy use, and behavioural changes that reduce energy expenditure and enhance energy intake (Sapolsky 1992; Wingfield and Ramenofsky 1999). Free-living male song sparrows given corticosterone-filled implants (to mimic a stressful situation) reduce their territorial behaviour (Wingfield and Silverin 1986), probably because it is an energetically costly activity that cannot be maintained during stressful periods. Thus, few corticosterone-implanted song sparrow males respond to simulated territorial intrusions. For those that do respond, the latency period increases compared with control birds, and their plasma levels of testosterone (a hormone involved in territorial behaviour) decline. In contrast, American tree sparrows (*Spizella arborea*), treated with corticosterone in exactly the same way, exhibit no behavioural change: Males continue to respond quickly and aggressively to simulated territorial intrusions, and plasma testosterone levels remain similar to those in controls (Astheimer *et al.* 2000). To understand the role of hormones in the development of behaviour, we must reconcile these two apparently contradictory results. The answer may, in fact, be fairly simple, but it requires knowledge of the behavioural and ecological contexts for each species. Song sparrows breed in a temperate climate where the summer is long and affords the birds multiple opportunities to breed (Figure 6.4). When faced with a stressor, either real (such as a late-spring snowstorm) or simulated (the corticosterone implants), these males abandon their territories, at least temporarily. The long duration of the breeding season makes it possible for them to return to their territories and re-nest after conditions have improved. The situation for American tree sparrows is very different: In an Arctic population of this species (Figure 6.4), the breeding season is very short. Consequently, there are no additional opportunities to reproduce, and territory abandonment results in a complete loss of the reproductive effort for that year. One can, therefore, propose that evolution has favoured individuals of these Arctic birds that do not couple increases in plasma corticosterone levels with changes in territorial behaviour and testosterone secretion. This perseverance enables them to continue to breed, even when confronted with temporarily unfavourable situations (Wingfield *et al.* 1995). This interpretation of the difference between two species is open to debate when it is considered in isolation, because the two species certainly differ in other important characteristics than those we highlighted above. However, this interpretation is supported by many other comparisons of this type involving corticosterone that all lead to the same conclusion; namely, that the relationship between corticosterone secretion and behaviour varies as a function of the conditions experienced by the species. We return to this idea in Section 6.5.3.

Thus, without a firm understanding of the context in which these hormonal manipulations are conducted, it would be difficult to reconcile such seemingly contradictory observations. The above example, in fact, reflects the general idea that many organisms must be equipped to face a wide variety of situations within a range of possible environments. Despite the conservative nature of evolution – many taxa have enzymes, hormones, and other substances that are identical or nearly identical in structure to those found in both closely and distantly related taxa – conserved substances are used in a wide variety of ways by organisms within and among taxa. By comparing the responses of species having different social systems and experiencing different environmental conditions, one can begin to develop a more general understanding of the underlying behavioural and physiological mechanisms, as well as their flexibility. As such comparisons accumulate, their interpretation can be refined, general features will become apparent, and the extent (and limits) of flexible responsiveness as a function of context will emerge. The strength of these generalizations can be tested by comparing species with similar types of

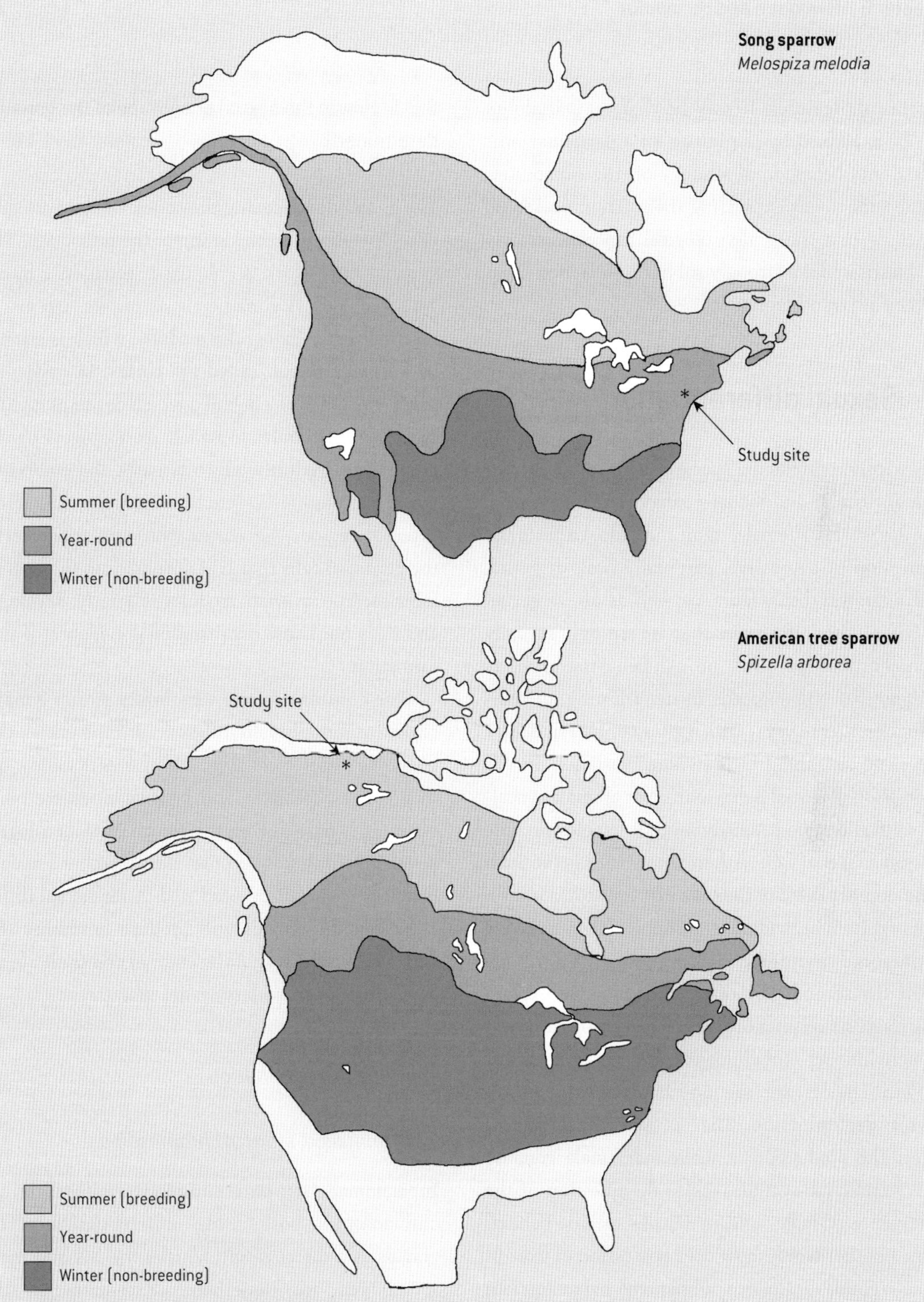


Figure 6.4 **Distribution of the song sparrow (*Melospiza melodia*) and the tree sparrow (*Spizella arborea*)**

Distribution map of two North American sparrows in which territorial behaviour has been tested in males that received corticosterone implants. Male song sparrows tend to abandon their territory whereas male tree sparrows maintain their territory, probably because of the short duration of the reproductive season of the latter favours individuals that persevere despite unfavourable conditions.

Map reproduced with permission from the Cornell Laboratory of Ornithology/NatureServe. The intensity of shading represents estimates of pair density in the zone in question (see Sauer *et al*. 1997).

contrasting features. It may be that the variability can be explained by the impact of a particular hormonal and behavioural strategy on the capabilities of the individuals expressing them. Thus, one could conclude that these physiological processes clearly are plastic and are strongly subject to selection. That is, they are true adaptations.

## 6.2 Sexual differentiation

In sexually reproducing species, the most important influence on an individual's behaviour is its gender. The sex, male or female, of an individual affects the expression of aggressive, parental, vocal, dispersal, and migratory behaviours, as well as all its sexual activity. Indeed, in species that are strongly sexually monomorphic, it is often through behavioural observations that one succeeds in determining an individual's sex. And yet within a group of individuals of the same sex, there is individual variation in the expression of sexual behaviours. Not all males vocalize with the same intensity, not all females provide the same level of care to their young, and so on. Such behavioural plasticity can take extreme forms in fish, where certain individuals can change sex in response to the social environment (Bass 1996).

In the second part of this chapter, we explore three aspects of phenotypic and behavioural development during embryological and early postnatal growth. The first sub-section examines the question of sex determination. That is, what mechanisms direct the production of male and female gonads? The second sub-section tackles the question of the emergence of male- and/or female-specific behaviours. In other words, how does possession of male or female gonads influence subsequent behaviour? In the third sub-section we will explore a question that recurs several times in this chapter: How do environmental factors influence phenotypic development? In the context of sexual development, we ask to what extent does the prenatal environment, and especially the maternal environment, affect the expression of sex-specific behaviour? Some aspects of these questions will be re-visited in Chapter 13.

### 6.2.1 How is the sexual phenotype of the gonad determined?

The gonads form during the course of embryological development by a thickening of the mesoderm. This gonadal framework is colonized secondarily by the germinal cells. A gonad consists of two regions: an internal region called the medulla, which is covered by an outer region called the cortex. In males, the testis consists of the medulla which develops into the seminiferous tubules in which spermatogenesis will occur during adulthood. In females, the embryonic gonad differentiates into an ovary in which the external region, the cortex, thickens and receives the germinal cells that will develop into the ovarian follicles. The ovarian medulla is poorly developed and does not show organizational evidence of tube formation.

One constant among vertebrates is that gonadal structure initially is indifferent, or bipotential; that is, simple histological observation does not allow one to predict whether it will become a testicle or an ovary. It is only during the course of embryological development that it becomes clear whether a male or female gonad will be produced. After sexual differentiation, the gonads will produce hormones that differentiate the rest of the embryo and facilitate the appearance of secondary sexual characteristics.

#### 6.2.1.1 *Sex determination*

##### a *Mammals: a story of chromosomes*

In mammals, sex determination originates in the chromosomes, with sexual differentiation of the gonads as a direct result: females are XX and males XY. There are nonetheless exceptions to this mamalian rule. For instance, lemmings have two different types of X chromosome, only one of which is masculinizing (Wiberg and Güntler 1985). Note that the XX/XY system enables fertilization to produce as many male (XY) embryos as females (XX), the primary **sex ratio** (used here as a measure of masculinity) being 0.5. We will reconsider this point again in Chapter 13.

The X and Y chromosomes have a common evolutionary origin and retain a trace of that history in the form of the pseudo-autosomal area, an area within which **crossing over** occurs. The other areas of X and Y chromosomes are specific to each and not subject to recombination; these are the heterosomal parts within which one nonetheless can find genes held in common, but that do not exhibit recombination. There are about 10 genes on the Y chromosome that do not have a counterpart on the X. Among them is the SRY (sex-specific region of the Y chromosome) gene, the major gene in sex determination in mammals (Gubbay *et al.* 1990). The protein resulting from this gene initiates a series of genetic responses that lead to the development of the testicles (Kanai *et al.* 2005). Gonads that develop in the absence of this testis-determining factor become ovaries. So, individuals that do not have the SRY gene (either because they carry an XX pair or because this area of their Y chromosome is not functional) develop into a female phenotype (Goodfellow and Lovell-Badge 1993). Conversely, the addition of this area of chromosome Y by transgenic methods to an XX individual stimulates the development of testicles (Koopman *et al.* 1991). One finds other specific genes on chromosome Y in mammals and they are, for the most part, involved in spermatogenesis, which explains why transgenic XX mice, to which only the SRY gene is added, do not undergo spermatogenesis. Furthermore, although the absence of the SRY gene results in a female phenotype, other genes are involved in this developmental process. Although the mechanism is not fully understood, a variety of genes and gene products also are known to be active during sexual differentiation of females (Kobayashi and Behringer 2003, Park and Jameson 2005).

### b *Sauropsids (birds and reptiles): chromosomes and/or temperature*

While birds and reptiles are biologically distinct, this separation is not clear cut in the context of sex determination. Within sauropsids one finds a broad range of mechanisms of sex determination (Crews 2003). In snakes and birds, sex determination is chromosomally based, but the chromosomes are named Z and W. Males have two copies of the Z chromosome in all their cells, and females have one copy of the Z and one copy of the W. This ZW/ZZ system, as in the XX/XY type, allows for the conception of equal numbers of males and females. One finds sex determination using a ZW/ZZ chromosomal mechanism in squamates (snakes, lizards) and chelonia (turtles, tortoises), but in these two groups one also finds chromosomal determination of sex using the XX/XY system. Strangely enough, in most tortoises, all the crocodilians, both sphenodons, and some squamates, sex determination can be influenced by the temperature at which the eggs are incubated, a mechanism called temperature-dependent sex determination, or TSD.

### c *Three important types of environmental determination*

Three types of sex determination sensitive to incubation temperature have been observed in reptiles (see Crews *et al.* 1988). The MF type is observed for the most part in chelonians, with males produced from eggs incubated at low temperatures and females at high temperatures (from which we get the abbreviation MF; Figure 6.5b). The FM type represents the reverse situation where females are obtained with low incubation temperatures and males at higher temperatures. The FM profile is observed in squamates (Figure 6.5a) and was initially described in crocodilians. We now know that in this group very low incubation temperatures generate additional females, and this profile is called FMF (Figure 6.5c). It has been proposed that all reptiles in which sex determination is sensitive to egg incubation temperature fall into the FMF category, but that the low (FM) or high (MF) parts of the profile are not observed because the temperatures required to produce them do not allow embryonic development. Interestingly, species that use temperature-dependent sex determination are vulnerable to extinction should there be a rapid, sustained environmental temperature change. Indeed, it has been argued that this was a factor in the extinction of dinosaurs, and that current

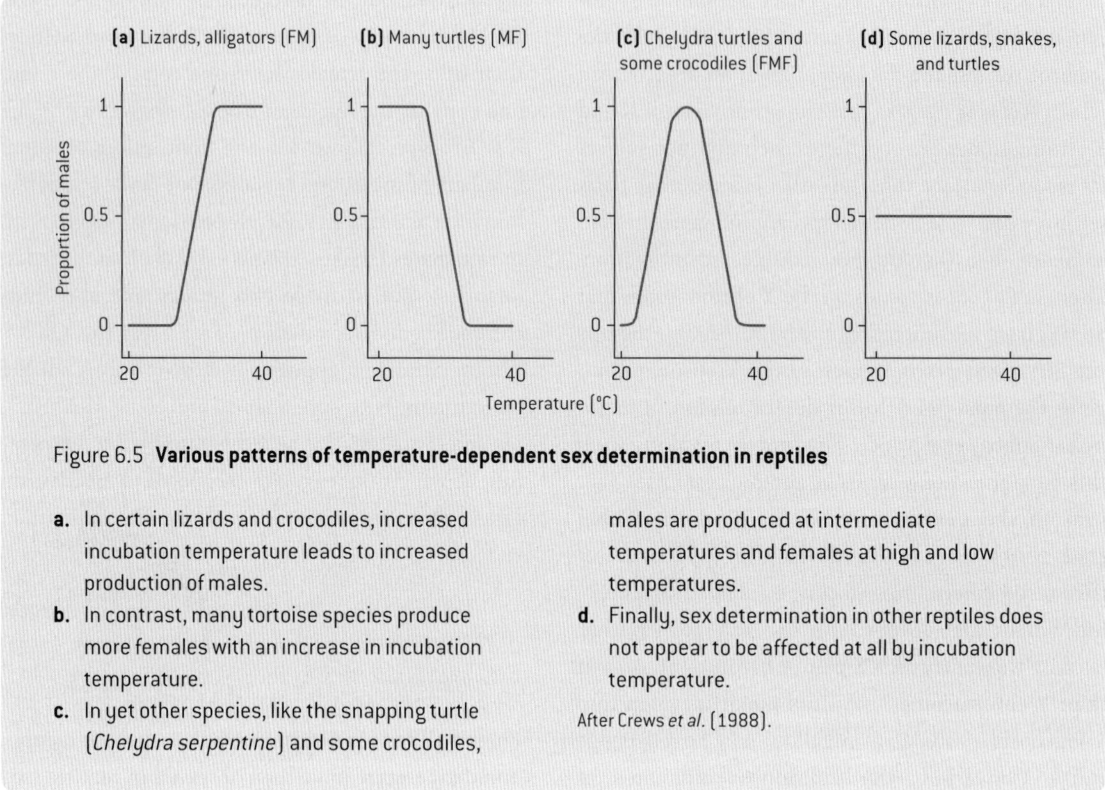


Figure 6.5 **Various patterns of temperature-dependent sex determination in reptiles**

**a.** In certain lizards and crocodiles, increased incubation temperature leads to increased production of males.
**b.** In contrast, many tortoise species produce more females with an increase in incubation temperature.
**c.** In yet other species, like the snapping turtle (*Chelydra serpentine*) and some crocodiles, males are produced at intermediate temperatures and females at high and low temperatures.
**d.** Finally, sex determination in other reptiles does not appear to be affected at all by incubation temperature.

After Crews *et al.* (1988).

global warming trends may have similar effects on extant species with TSD (Miller *et al.* 2004).

## d *Sensitive period*

The period of embryonic development during which temperature acts to determine sex corresponds to about 15 days, at the initial stages of gonadal formation (Raynaud and Pieau 1985). The biochemical mechanisms of action of temperature are also well established: the amount of oestrogen hormones (oestradiol and oestrone) in the gonad during its formation directly influences its differentiation into an ovary or a testicle. The amount of oestrogens in the embryonic gonad depends on the activity of cytochrome-P450 aromatase that converts testosterone into oestrogen and androstenedione into oestrone. The period of embryonic development during which the regulation of cytochrome-P450 aromatase is sensitive to egg incubation temperature (thermosensitive period; Desvages and Pieau 1992) is the same as that during which temperature influences sex determination (Desvages *et al.* 1993). Furthermore, aromatase inhibitors administered to eggs incubated at feminizing temperatures mimic the effects of masculinizing temperatures (Richard-Mercier *et al.* 1995).

## e *Molecular mechanism*

The biochemical factors regulating sex determination also have been elucidated, in part, in other sauropsids, especially in chickens, which are characterized by chromosomal sex determination. Once again, we find a major action of aromatase in this species (Burke and Henry 1999; Vaillant *et al.* 2001). The regulation of this enzyme seems to be the critical point in explaining sex determination in all sauropsids. If its gene is transcribed repeatedly in the putative gonad during development, and if the precursor (testosterone) and receptors for oestradiol-17β are present, then the gonad will differentiate into an ovary. If these conditions are not met, the gonad will develop into a testicle.

This model allows us to explain the many examples available in the literature. For instance, the right ovary of many birds does not develop because of the absence of oestrogen receptors. The model also enables us to explain the strong maternal component in sex determination based on incubation temperature; indeed, we know that the quantity of oestrogens deposited in the egg yolk is quite variable (Bowden *et al.* 2001) and changes over the course of the breeding season (Bowden *et al.* 2000). However, aromatase transcription in sauropsids could be activated by oestrogens (Pieau 1996), which explains the feminizing effect when additional maternal oestrogens are added to the egg yolk. Thanks to this model, we also can easily explain the existence of a mixed situation between genetic and environmental determinism observed for instance in the European pond turtle, *Emys orbicularis.* This species is characterized by sexual determinism typical of the MF type (Figure 6.5b). However, in natural conditions, sex will be determined by the individual's genotype approximately 90% of the time (Girondot *et al.* 1994). Such a situation can be explained by the presence of a genetic polymorphism in response to the transcription temperature of cytochrome-P450 aromatase.

#### 6.2.1.2 *Secondary sexual characters*

##### a *Wolffian and Müllerian ducts*

Vertebrate embryos initially have both male (Wolffian) and female (Müllerian) genital tracts. These tracts connect the gonads with the external environment. Once gonadal sex is determined, one set of these tracts develops while the other disappears. In males, secretions of an anti-Müllerian peptide hormone (of the TGF-β family) produced by the testicles is necessary for the regression of the Müllerian ducts (Visser and Themmen 2005). The hormone 5α-dihydrotestosterone (5α-DHT), the active form of testosterone catalyzed by 5α-reductase, also is synthesized by the testicles. It enables the growth and differentiation of the Wolffian ducts into the ejaculatory duct, epididymis, ductus deferens, and the accessory secretory glands that empty into the reproductive tract (i.e., the seminal vesicles, the prostate, and the bulbourethral glands). Furthermore, 5α-DHT stimulates the development of external genital organs (George *et al.* 1989). Thus, male vertebrates typically undergo simultaneous masculinizing and de-feminizing effects, both resulting from testicular secretions.

In female mammals, it seems that 17β-oestradiol is responsible for the regression of the Wolffian ducts and the development of the Müllerian ducts that develop into the oviducts, uterus, and part of the vagina (Rey and Picard 1998; Nelson 2000). In addition, tissues that, in the presence of androgens, form the penis and the scrotum in males, become, in the absence of androgens, the clitoris and the vaginal folds. The feminizing (development of the Müllerian ducts and external genital organs) and de-masculinizing (degeneration of the Wolffian ducts) effects that occur during sexual differentiation in females are analogous to the events that occur during male differentiation.

##### b *Establishing the hypothalamic–pituitary–gonadal axis during development*

In mammals, as we have seen above (Table 6.1), the hypothalamic–pituitary–gonadal (HPG) axis plays prominent roles in the regulation of most physiological functions of the organism such as growth, brain development, reaction to stress, general metabolism associated with reproduction, such as gonadal development, gender orientation, libido, and sexual behaviour. During fetal development, the HPG axis is set up according to a precise chronology. Work in humans, monkeys (the crab-eating (or long-tailed) monkey *Macaca fascicularis*), and feral pigs (*Sus scrofa*), have shown that differentiation of this axis begins from the bottom upwards (Danchin 1980; Danchin and Dang 1981; Danchin *et al.* 1981; Danchin and Dubois 1982; Figure 6.6). Indeed, in these two species the organization of the HPG axis is very similar, and it is possible that it is a general feature of mammalian development. The gonads differentiate first and release sex hormones very early in fetal life. As we have already seen, this differentiation is influenced by genes of the sex chromosomes. The anterior pituitary differentiates later in fetal life and

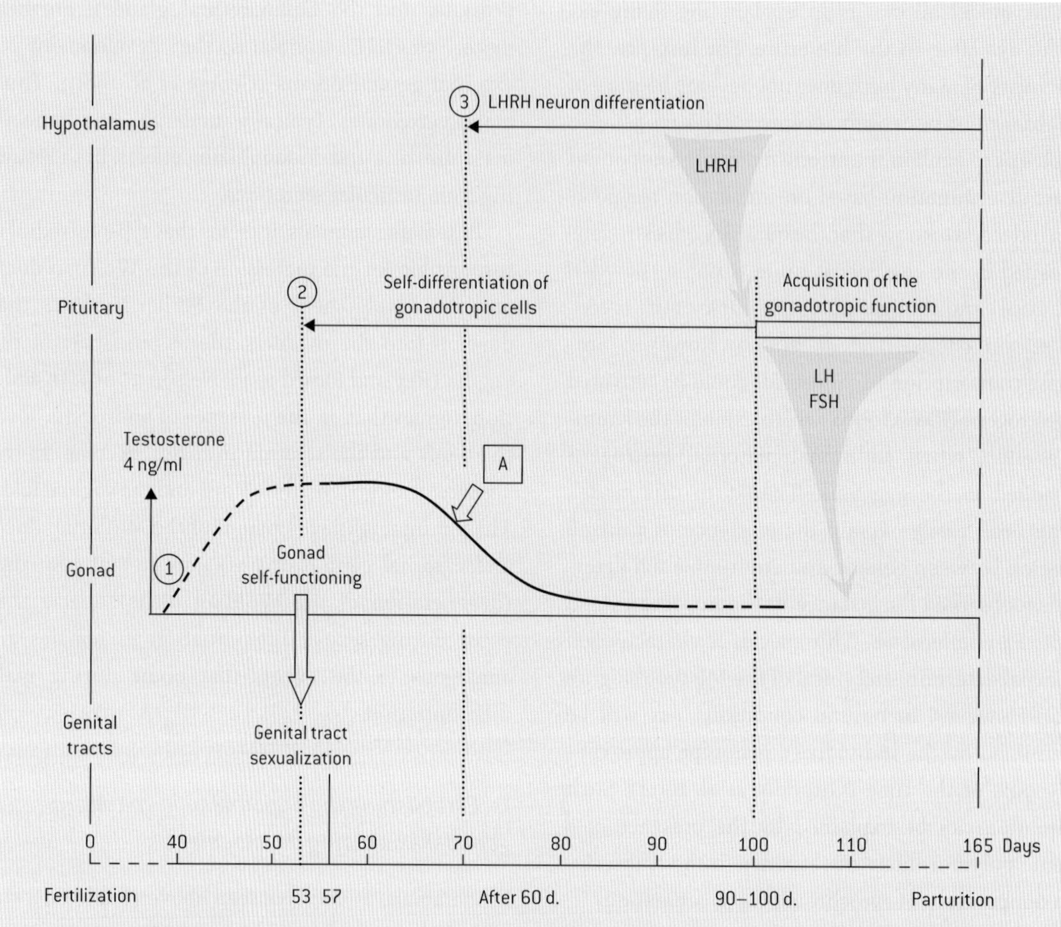


Figure 6.6 **Establishment of the hypothalamic–pituitary–gonadal axis in the crab-eating monkey (*Macaca fascicularis*)**

Gestation in this species lasts 165 days. Data on the differentiation of endocrine cells making pituitary and hypothalamic hormones were obtained by immunocytology. This immunofluorescence technique allows investigation of the presence of hormones in the structures in question by means of antibodies that interact specifically with those hormones. This facilitates documentation of the appearance of hormone molecules in the cytoplasm of endocrine cells.

(1) First of all, the gonads develop autonomously. Then they secrete sex steroid hormones (as evidenced by the levels of circulating hormones) that sexualize the genital tracts. Next (in A) gonadal activity begins to depend on gonadotropins. However, the lack of gonadotropic stimulation (caused by undeveloped higher brain levels) results in a reduction in circulating testosterone, which remains at low levels during the rest of gestation.

(2) The hypophyseal level only begins to differentiate later and does not seem to release gonadotropic hormones (LH and FSH) before 100 days of age. (3) Finally, cells of the hypothalamus do not seem to begin to differentiate until around 70 days of gestation, with the release of hormones stimulating the gonadotropic function of the hypothalamus occurring around 100 days of age. Beginning at this time, the hypothalamic–pituitary–gonadal axis seems to function from the top downwards. Differentiation of the HPG axis takes place from the bottom upwards, while in adults this axis functions from the top downwards. This diagram includes results of other studies on morphological differentiation of gonads and external genital organs and on the appearance of sex steroids in fetal blood.

After Danchin (1980) and Danchin and Dang (1981) and Danchin and Dubois (1982).

becomes capable of releasing the hormones that regulate gonadal function. Finally, the hypothalamus differentiates still later in fetal development, at a time when the gonads and the pituitary have been functional for a relatively long time. Cells of the preoptic area of the hypothalamus begin to synthesize hormones that control the function of the pituitary and, by extension, the gonads. Thus, each level of the HPG axis differentiates on its own, without control by higher levels. In fact, during development the HPG axis seems to function in the opposite manner from what occurs in adults: secretions from the lower levels contribute to the differentiation of the higher levels of the axis. So, the establishment of the axis during fetal life proceeds from the bottom to the top of the axis (Figure 6.6), whereas the axis functions from top to bottom in adults. A very similar scenario appears to be true for the other components of the hypothalamic–pituitary–somatic axis. For example, the hypothalamic–pituitary–adrenal (HPA) axis controls the secretion of corticosteroids involved in stress responsiveness (Figure 6.1). This phase corresponds to the organizational phase of the organizational/activational hypothesis detailed in Section 6.2.2.1.

The developmental sequence was the object of a rather significant controversy involving reptiles and sex determination as it relates to egg incubation temperature. Based on experiments using gonadal tissue cultures, it was proposed that the HPG axis was reversed in embryos and that it was the brain that was the temperature action zone, thus controlling sex determination (Salame-Mendez *et al.* 1998). These results are contrary to data obtained in the slow worm (*Anguis fragilis*; Raynaud and Raynaud 1961) and the mouse (Vigier *et al.* 1989) in which sex determination of a decapitated embryo *in ovo* (slow worm) or of a gonadal culture (mouse) conformed to the genotype of the individual from which it came. This point was resolved recently thanks to isolated gonadal cultures of marine turtles that showed sexual differentiation conforming to the temperature of the culture, thus excluding the HPG axis as a factor in sexual determination (Moreno-Mendoza *et al.* 2001). However, note that the presence of nerve endings at the level of the gonad (Gutiérrez-Ospina *et al.* 1999) indicates that the brain, nonetheless, could have an effect on the gonads: not on sexual determination itself, but rather on gonadal growth.

Once fully active, the HPG axis functions in the fetus in a manner similar to a sexually mature adult, with relatively high circulating hormone levels. After birth, at times that vary among species, higher brain areas seem to silence the HPG by inhibiting the hypothalamus and, as a consequence, the pituitary and the gonads. Thus the beginning of infancy marks a period during which circulating sex hormone levels are very low. Infancy is terminated at puberty, the time when the HPG axis again becomes functional and induces the appearance of all adult sexual behaviours. This phase corresponds to the activational phase of the organizational/activational hypothesis presented in the following sections.

### 6.2.2 How do typical male and female behaviours emerge?

We have seen how embryological events involving hormones participate in gonadal differentiation. But this poses the question of 'how do the gonads themselves subsequently affect behaviour?' That is, after having directed the bipotential gonad to become either a testis or an ovary, how does the brain express appropriate adult male or female behavioural patterns, patterns that do not emerge until much later in life? As was the case with sexual differentiation, the answer involves hormones.

#### 6.2.2.1 *The organizational/activational hypothesis*

A fundamental tenet of brain sexual differentiation is the organizational/activational hypothesis (Phoenix *et al.* 1959). Briefly, the hypothesis states that the presence or absence of hormones at a specific stage of early development (either prenatally or perinatally) modifies neural structures associated with sexually dimorphic behaviour to produce either male-like or female-like behaviours. This organizational effect is followed at sexual maturation by additional exposure to sex steroid hormones that facilitate the expression of those gender-specific behaviours. Thus, sexually appropriate behaviours are activated as the individual becomes sexually competent. Although these processes and the neural basis of sex differences in behaviour have been found to be more complex than originally thought (Romero 2003; Sisk *et al.* 2003; and see below), the organizational/activational hypothesis remains at the heart of much of the work on sex differences in behaviour. Below we give examples of sexually dimorphic brain structures that, when activated by exposure to the appropriate hormones, are associated with sex-specific behaviours.

### a *Mammals*

Above, we discussed some of the early endocrine events that serve to organize the brain. In rats, for example, testosterone and its metabolites, experienced before birth and through the first 20 days after birth, masculinize the brain, the HPG axis, and sexual behaviour (see Kelley *et al.* (1999) for a review). These behavioural and phenotypic differences develop at sexual maturity, when the HPG axis is activated (Sisk and Foster 2004).

The best-studied region of the adult mammalian brain to exhibit sexual dimorphism in structure is the medial preoptic area (mPOA) of the hypothalamus (Raisman and Field 1973). This region includes an area of remarkable sexual dimorphism, the sexually dimorphic nucleus of the preoptic area (SDN–POA), which is up to five times larger in males than females (Gorski *et al.* 1980). The mPOA is involved in copulatory behaviour in males (Sachs and Meisel 1988), and lesions of the mPOA impair copulatory behaviour in adult male rats (Liu *et al.* 1997a, b). These effects are only partly reversed by exposure to androgens (Christensen *et al.* 1977; Arendash and Gorski 1983). This contrasts with the response of males to castration, which also compromises copulatory activity, where testosterone therapy can maintain or restore normal sexual behaviour (Davidson 1966). Furthermore, adult male rats treated with female sex hormones (estrogens and progesterone) normally do not exhibit female-like sexual behaviour, but males treated with these same hormones after receiving lesions in the mPOA display lordosis, the stereotypical posture assumed by female rats during copulation (Hennessey *et al.* 1986). Finally, the mPOA in both sexes has extensive neural connections with other areas of the brain that also exhibit sexual dimorphism and, thus, may affect the expression of other sexually distinct physiological and behavioural characteristics (De Vries and Simerly 2002).

The mPOA is also involved with sexual behaviour in females, where it has been implicated in lordosis, parental behaviour, and in the regulation of the oestrus cycle (Gray and Brooks 1984; Jakubowski and Terkel 1986). In addition, females exhibit male-like synaptic patterns in the brain if they are injected with testosterone before four days of age (Raisman and Field 1973).

Other areas of the brain not closely involved with reproductive behaviour also develop sexual dimorphism following early exposure to hormones (see Beatty (1979) for a review). For example, the hippocampus, a structure associated with learning and memory, exhibits sexual dimorphism. Male rats castrated as newborns exhibit female-like learning patterns as adults, whereas treatment of female neonates with oestradiol produces male-like patterns (Williams and Meck 1991). Differences in spatial discrimination are thought to be mediated by organizational effects of steroids on the oestrogen receptor, ER-$\alpha$ (Fugger *et al.* 1998).

### b *Birds*

Avian brains also show sex differences in the preoptic area (POA) (reviewed by Schlinger 1998). The medial pre-optic nucleus (POM) of the POA is involved in copulatory behaviour and is larger in males than in females. Early hormonal exposure organizes this behavioural dimorphism, producing sexual differences in neural connections (Panzica *et al.* 1998), and subsequent exposure to steroid hormones activates copulatory behaviour. For example, oestrogens produced by female embryos during development demasculinize copulatory behaviour. If eggs containing female embryos are treated early in incubation with a substance that prevents oestrogens from interacting with their receptors, then the adult reproductive behaviour of the females is masculinized (Adkins 1976). On the other hand, if eggs containing male embryos are treated with testosterone or oestrogens, then they are demasculinized (Adkins-Regan 1987). In unmanipulated embryos, female (but not male) embryos normally are exposed to elevated plasma steroid levels before hatch; this suggests that exposure of female embryos to oestrogen is responsible for the demasculinization of copulatory behaviour, while male embryos produce quantities of sex steroids that are too low to demasculinize male copulatory behaviour (Balthazart and Foidart 1993).

Interestingly, in quail, although copulatory behaviour is organized by early hormones, the size of the POM is not affected by these hormones. Instead, the size of the POM responds directly to the activational effects of testosterone. Castration reduces adult male POM volume, and testosterone therapy restores it. Similarly, ovariectomized females treated with testosterone have POM volumes as large as those of intact adult males (Panzica *et al.* 1987).

Finally, not all avian brain sexual differentiation necessarily is regulated by endocrine secretions. For example, while treatment with sex steroid hormones affects development of the avian neural song system, it is unclear if steroid hormones exert these effects normally (Schlinger 1998). In particular, because females of many species sing in some contexts, singing is not as strictly sex-dependent as is copulatory behaviour. This has led some authors to suggest direct genetic control of sexual differentiation of avian song systems (Arnold 1996; Schlinger 1998). Thus, while the basic mechanism of the organizational/activational hypothesis is similar in mammals and birds, there also are interesting taxonomic differences.

### C *Reptiles*

*The case of lizards with temperature-dependent sex determination*

Incubation temperature and hormones interact to determine gonadal sex in lizards with temperature-dependent sex determination. However, the question remains whether or not temperature and/or hormones have organizational or activational effects on gender-specific behaviour in such species. Rhen and Crews (1999, 2000) have shown that incubation temperature, early hormone exposure, and adult hormone production are important in leopard geckos (*Eublepharis macularis*). Gonadectomized males treated with testosterone and incubated at temperatures that produce mostly females perform less scent-marking (a typically male behaviour) than do similarly treated males hatched at temperatures that produce mostly males (Figure 6.7a). In addition, a developmental effect of gonadal sex is shown by an increase in courtship behaviour of adult, gonadectomized males in response to testosterone treatment, a response which is lacking in females (Figure 6.7b). Thus, the organizational/activational hypothesis applies here, too, but with the added effect of incubation temperature during development.

*The case of unisexual lizards*

Some of the most interesting species, in terms of the development of their sexual behaviour (or, more accurately, their 'pseudosexual' behaviour), are the parthenogenic lizards of the genus *Cnemidophorus*. In some *Cnemidophorus* species, all individuals are female and their unfertilized eggs all produce female offspring. These species are thought to be hybrids of two sexually reproducing ancestral species (Cole 1984). Interestingly enough, individuals in many of these species engage in pseudosexual copulatory behaviour; that is, some exhibit male-like mounting behaviour whereas others display female-like receptive behaviour (Crews 1987). Furthermore, these behaviours are mediated by changes in hormone secretion. Female-like behaviour occurs just before ovulation, when plasma oestradiol levels are elevated, whereas male-like behaviour is expressed after an individual has ovulated; that is, under the influence of progesterone (Crews 1987). Differences in brain metabolism (i.e., in the use of energy substrates) also are evident when the lizards are expressing male-like or female-like behaviour (Rand and Crews 1994). This is not simply some sort of behavioural

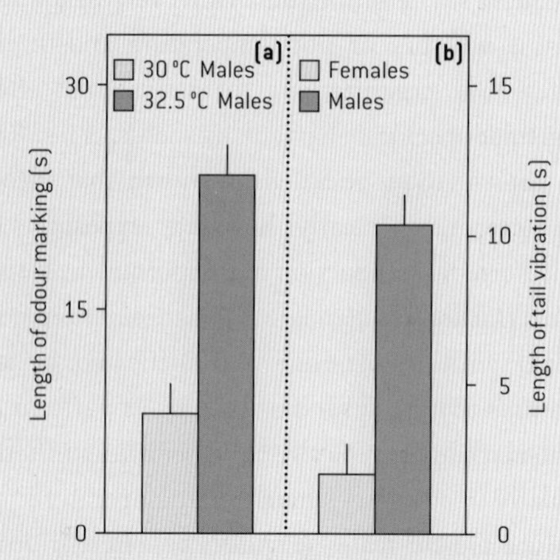


Figure 6.7 **Determination of sex and sexual behaviours in the leopard gecko (*Eublepharis macularius*)**

**a.** Effect of embryonic developmental temperature on the odour marking behaviour in castrated leopard geckos supplemented with testosterone. An incubation temperature of 30 °C produces mostly females (but also some males), whereas a temperature of 32.5 °C produces mostly males. Data for females that never odour mark are not shown. Incubation temperature has a significant effect on marking behaviour (ANOVA, $F_{1,458} = 12.1, P = 0.0005$). *Post hoc* comparisons have shown that males from the two incubation temperatures respond differently to hormonal treatment (Dunn-Sidák method $\alpha' = 0.005$). This result reveals that incubation temperature influences adult behaviour.

**b.** Effect of gonadal sex (before gonadectomy) and of testosterone treatment of adults on tail vibration behaviour, taken as a measure of male-typical display behaviour. There is a significant effect of gonadal sex on the duration of tail vibrations ($F_{1,458} = 139.1, P < 0.0001$), which were activated by the hormonal treatment ($F_{3,458} = 18.7, P < 0.0001$). Contrary to odour marking behaviour, embryonic temperature has no effect on display behaviour ($P > 0.05$). *Post hoc* comparisons show that gonadal males exhibit more display behaviours than gonadal females in response to testosterone treatment. Therefore, gonadal sex itself influences adult behaviour. The data in the two figures are means plus the standard deviation.

After Rhen and Crews (1999).

carry-over from the ancestral species, a relict that no longer has a role in *Cnemidophorus* life history. Rather, the behaviour serves an important function, because females that exhibit receptive behaviour and are mounted release more eggs during ovulation than do females that do not engage in pseudosexual behaviour (Crew *et al.* 1986). Note that it is progesterone, a sex steroid hormone generally associated with female reproductive endocrinology, and not one of the androgens, which activates male-like behaviour in these species. This reflects the plasticity of pathways by which hormones can affect behaviour, despite the conservative nature of the hormones themselves across taxa (Crews 1997).

It would be reasonable to think that endocrine organizational effects may be unnecessary in species that develop into only a single sex, such as parthenogenic *Cnemidophorus* lizards. However, one must keep in mind that these parthenogenic species evolved from sexually reproducing species, and elements of some ontogenetic processes of the ancestral species that were hormonally mediated may have been incorporated into developmental pathways of the parthenogenic ones. Indeed, this seems to be the case in at least one of the parthenogenic *Cnemidophorus* species. *C. uniparens*, a unisexual species, has the same pattern of hormone secretion as one of its sexual ancestral species, suggesting that the evolution of parthenogenesis involved changes in the response to endocrine secretions rather than changes in the pattern of hormone secretions themselves (Moore and Crews 1986). Furthermore, treatment of developing *C. uniparens* embryos with an aromatase inhibitor (which prevents the conversion of testosterone to oestrogen) results in the production of male offspring in this normally all-female species (Wibbels and Crews 1994), demonstrating that genes involved in the production of males are still present in *C. uniparens*.

### 6.2.3 Phenotypic plasticity within a sex, or how does the environment influence phenotype?

A variant of the organizational/activational hypothesis has been used to explain phenotypic plasticity within a single sex. Numerous species have discrete intra-sexual phenotypes that differ in morphology, physiology, and behaviour. This phenotypic plasticity, most commonly seen in males, typically is associated with the adoption of alternative reproductive strategies.

*Lizards*

In lizards, males of some species may exhibit territorial behaviour whereas others may adopt a more covert strategy, sneakily courting females as they approach territorial males (Moore *et al.* 1998). Moore (1991) suggested that differences in exposure to hormones may be important in the development of within-sex phenotypic differentiation. For example, tree lizards (*Urosaurus ornatus*) exhibit morphological polymorphism in the coloration of their dewlap (a flap of skin that hangs down from the throat) (Hews *et al.* 1997). Males possessing orange dewlaps with a blue central spot are territorial, while those with orange dewlaps but no central spot are non-territorial (Hews *et al.* 1997). There is an endocrine basis for these different behavioural and morphological phenotypes, for orange-blue territorial males are exposed to relatively high levels of testosterone and progesterone as juveniles, while orange non-territorial males experience, at the same age, relatively low levels of the two hormones (Hews *et al.* 1994, Moore *et al.* 1998). Notably, the adrenal glands appear to be the source of much of this progesterone (Jennings *et al.* 2004), raising the possibility of a relationship between neonatal exposure to stressors and phenotypic development in this species.

The ability to alter dewlap coloration in response to hormonal manipulations before 60 days of age indicates the existence of a critical period within which organizational effects must occur (Hews and Moore 1996). Interestingly, the non-territorial, orange dewlap males can be subdivided in adulthood into two categories: those that are sedentary and those that are nomadic. Orange dewlap males exposed to stressful conditions, in which their plasma testosterone levels are low while their corticosterone levels are high, become nomadic. Conversely, non-territorial males in non-stressful environments have elevated testosterone and low corticosterone levels and become sedentary (Moore *et al.* 1998). Orange males can move back and forth from nomadic to sedentary states, depending on environmental conditions and their hormonal milieu (Knapp *et al.* 2003), but never change their territorial status (Moore *et al.* 1998). Thus, the activating effects of these hormones are reversible, unlike their organizational effects.

Another species, the side-blotched lizard (*Uta stansburiana*), has three male morphs, each with a different behavioural and hormonal profile. Orange-throated males have high testosterone levels and are 'ultradominant,' in the sense that they have large territories that they defend very aggressively (Sinervo *et al.* 2000). Blue-throated males have smaller territories that they defend less aggressively, and yellow-throated males do not defend territories at all and are more secretive than the other two morphs. Blue and yellow morphs have similar testosterone levels. The role hormones play in organizing the development of these different phenotypes is unknown, but testosterone appears to be important in activating behaviours associated with the different morphs. For example, if testosterone levels in blue- and yellow-throated males are artificially elevated to the level of orange-throated males, then the treated males defend territories of similar size as the orange-throated males (DeNardo and Sinervo 1994). Increased testosterone also enhances endurance levels, which may enhance their ability to defend territories successfully

(Sinervo *et al.* 2000). Interestingly, yellow-throated males can transform into quasi blue-throated territorial males late in the breeding season if a territory holder should die. This is accompanied by an increase in plasma testosterone and a partial change in throat colour (Sinervo *et al.* 2000), again suggesting an activational effect of testosterone on behaviour and physiology. This also shows that social interactions can strongly influence hormone secretion: males with yellow throats, although capable of secreting testosterone when the social context permits, do not do so as long as the competition resulting from the presence of territorial males is strong. The reader can refer to Chapters 5 and 10 for more details about the various strategies in that lizard.

*Fish*

A similar phenomenon occurs in teleost fish. For example, male plainfin midshipman fish (*Porichtlys notatus*) have two reproductive morphs: type I and type II. Type I males are larger than type II individuals and, also unlike their type II counterparts, they build nests, court females with a distinctive 'hum' vocalization, and guard eggs (Brantley and Bass 1994; McKibben and Bass 1998). The smaller type II males resemble females in their morphology (Bass 1995). This female mimicry apparently allows type II males to enter the territories of type I males and to release their milt there. Thus, the two morphs have very different reproductive strategies. Furthermore, males that adopt the type II developmental trajectory become mature sooner than type I males (Bass *et al.* 1996), allowing them to begin breeding earlier.

As we saw with tree lizards, male plainfin midshipman fish are characterized by differences in endocrine profiles that may explain the different patterns of growth and behaviour (Brantley *et al.* 1993, Bass 1996). In addition, the two morphs also differ in the number and/or size of neurons in brain areas associated with the hormones gonadotropin releasing hormone and arginine vasotocin (Foran and Bass 1999). Although causal relationships are still to be determined, this species may represent another example of early organizational effects of hormones on development of discrete phenotypes within a single sex. Other plastic reproductive tactics in fish, such as sex change or reproductive suppression, may also have endocrine bases, and are reviewed by Foran and Bass (1999). The existence of differing tactics is common in fish, but such intra-specific variation in reproductive strategies is relatively common in most animal groups. Although these may not always entail clear morphological differences, it is probable that even subtle variations involve different developmental pathways, as in the examples discussed above.

## 6.3 Environmental effects on phenotypic development

Individuals pass through different developmental stages in their lives, e.g., from embryo to immature young to sexually mature adult. Although the genome directs this general progression, environmental factors play a major role in shaping individual phenotypic features. That is, individuals with similar genotypes are capable of a certain amount of phenotypic plasticity. We have seen in Chapter 2 how the simple presence or absence of predators in the environment can lead to two appreciably different adult forms in daphnids, clonal cladocerans. Individuals do not develop a prominent rostrum unless they are in the presence of predators or are exposed to indicators of predators. A prominent rostrum strongly reduces the probability of being depredated, suggesting that this phenotypic plasticity can be adaptive. More generally, the source of such plasticity involves environmental stimuli, including those of maternal origin, which modify phenotypic development, most often through hormonal signals.

The preceding sections describe a few examples of phenotypic plasticity. Another example of this type of plasticity is provided by the western spade-foot toad (*Scaphiopus hammondii*), which breeds in ephemeral ponds in arid deserts of North America, where tadpoles must metamorphose into adults before their ponds dry out. In a fascinating set of studies that combine behavioural and physiological approaches, Robert Denver (1997, 1998, 1999) has shown that tadpoles accelerate their metamorphosis as pond water levels decline, and that this acceleration is driven by hormonal changes (Figure 6.8). Thyroid hormones underlie the overall process of metamorphosis, although increased corticosterone secretion from the inter-renal glands synergizes with thyroid hormones to speed-up the process. In amphibians, CRH controls the secretion of the thyroid hormones and corticosterone, by controlling the release of ACTH and TSH by the anterior pituitary gland. The effect of CRH on the secretion of thyroid hormones and corticosterone in different environmental conditions may depend on changes in the number or type of CRH receptors in the anterior pituitary.

Recently, reduced food availability has been shown to stimulate increased corticosterone secretion in spade-foot toad tadpoles and to accelerate metamorphosis (Boorse and Denver 2003). This may represent a physiological link connecting resource availability with flexibility in the timing of the onset of metamorphosis (Crespi and Denver 2005).

Rapid metamorphosis occurs at a cost because the resulting adults are smaller than those produced with a slower rate of metamorphosis. Large frogs may have greater reproductive success than small frogs (Berven 1981), but the advantage of this **trade-off** between the two aspects of fitness, i.e., survival and reproduction, is that small individuals at least have at a chance to reproduce, a chance that would have been lost had their pond dried out before they transformed into adults.

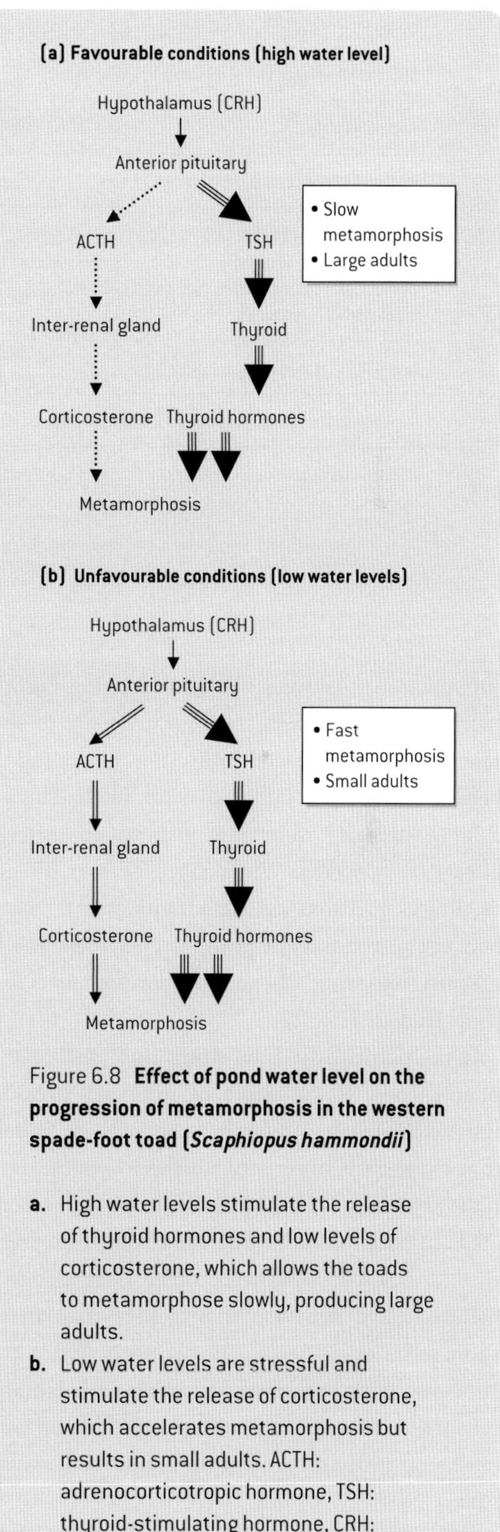


Figure 6.8 **Effect of pond water level on the progression of metamorphosis in the western spade-foot toad (*Scaphiopus hammondii*)**

**a.** High water levels stimulate the release of thyroid hormones and low levels of corticosterone, which allows the toads to metamorphose slowly, producing large adults.

**b.** Low water levels are stressful and stimulate the release of corticosterone, which accelerates metamorphosis but results in small adults. ACTH: adrenocorticotropic hormone, TSH: thyroid-stimulating hormone, CRH: corticotropin-releasing hormone, Inter-renal gland: amphibian homologue of the adrenal gland.

### 6.3.1 Maternal effects: another means of transmitting information on the state of the environment

The possibility that maternal factors influence offspring early in their development is considerable and occurs in many taxa. This is obviously the case in placental mammals, where the exchange of metabolites across the placenta results in intimate and prolonged

| Short-term effects (from minutes to a few hours) | Chronic effects (from days to weeks) |
|---|---|
| Reduces libido | Suppresses reproductive activities |
| Reduces appetite | Increases/decreases food consumption |
| Increases glucogenesis | Depletes muscle protein |
| Stimulates immune system; prepares body for 'fight-or-flight' response | Reduces resistance to illness and infections |
| Improves memory consolidation | Impairs memory consolidation |
| Increases retention of sodium and blood pressure | Stimulates hypertension |
| Reduces territoriality and increased foraging behaviour | Reduces reproductive success |

Table 6.2 **Immediate and chronic effects of glucocorticoids on physiology and behaviour**

communication between mother and young. Furthermore, mammalian young continue to depend on their mothers for nurture and protection, extending the period of maternal influence after birth. There is a large literature addressing mammalian maternal effects on offspring. While much of it is directed at clinical questions associated with stress, disease, and aging in adult offspring, a great deal can be gleaned regarding phenotypic development from these studies.

One particularly interesting and well-studied phenomenon is the effect of maternal stress on the development of offspring phenotypes. Animals that experience stress activate their hypothalamic-pituitary-adrenal (HPA) axis and release glucocorticoids in an effort to restore homeostasis (Sapolsky *et al.* 2000). Glucocorticoids orchestrate a suite of metabolic and behavioural responses that, in the short term, act to restore homeostasis (Table 6.2). However, with prolonged activation or heightened sensitivity of the HPA, glucocorticoids have negative effects on many systems of an organism (Sapolsky 1992).

The effects of maternal stress on offspring may involve (1) prenatal effects, in which mothers are subjected to stressors whose effects are transmitted to the developing young, or (2) perinatal effects, in which young are temporarily separated from their mothers. Many studies have shown deleterious effects of prenatal maternal stress on offspring. For example, female offspring of rodent mothers subjected to repeated handling have an increased stress response compared with controls (McCormick *et al.* 1995). That is, when daughters of stressed mothers are themselves restrained, they secrete more corticosterone during the restraint than do females in a control group, indicating heightened sensitivity of the HPA axis. Interestingly, there may also be a sex difference in the response because the response of sons of stressed mothers is similar to that of control males, although other studies have shown that prenatal stress affects both sexes (see, for example, Lordi *et al.* 1997). HPA axis activation in offspring of stressed mothers has been seen in other mammals; for example, in guinea pigs (see Sachser 1998), pigs (see Haussmann *et al.* 2000), and primates (see Schneider *et al.* 1999), including humans (see O'Connor *et al.* 2005), highlighting the importance of the gestational period for phenotypic development of offspring.

Perinatal maternal effects also have been studied in detail, especially in rodents, and involve the separation of newborns from their mothers. Brief (less than 15 minutes) handling of pups reduces the magnitude of hormonal and behavioural responses to stress in adulthood, whereas prolonged manipulation (several hours) enhances these responses (Francis and Meaney 1999). That is, neonates that are briefly handled secrete lower amounts of corticosterone into the blood when stressed as adults than do young that are handled for several hours or than do control pups. This results from permanent differences in the development of brain structures, glucocorticoid receptor density, and modification of the glucocorticoid receptor gene through differences in DNA methylation patterns in newborns from the different groups (Meaney *et al.* 1991; Meaney and Szyf 2005). The precise timing of the manipulation during the neonatal period is important, for the response to separation during the first few days of life is different from that produced by the same manipulation performed a week later (van Oers *et al.* 1998).

It must be noted that the key factor in the development of these different offspring phenotypes is the behavioural response of mothers to handled pups. Mothers of briefly handled pups spend significantly more time licking and grooming them than do mothers of unhandled pups, although the total time attending litters is the same for mothers in both groups (Liu *et al.* 1997a, b). It is this maternal licking/grooming behaviour that induces the developmental change in the young. There is considerable natural variation in the amount of licking/grooming done by mothers, and unhandled offspring of mothers that nonetheless provide high levels of licking/grooming to their young respond to restraint stress in a manner similar to that of handled offspring (Liu *et al.* 1997a, b).

The results of cross-fostering experiments support the importance of perinatal maternal influences on phenotypic development of young. Mouse pups from strains bred for high HPA reactivity exhibit a reduced stress response if reared by control mothers (Anisman *et al.* 1998). In the converse experiment, newborns show increased HPA reactivity when cross-fostered to females with a strong HPA-responsiveness. It is critical to note that the latter females manifest reduced licking/grooming behaviour compared with the control females. It is also noteworthy that not all of the phenotypic variation in offspring is explained by maternal behaviour. In other words, offspring from high HPA-responsive females still showed an enhanced response when compared with the young of control females. Thus, there is a genetic component and an environmental (maternal) component to the variation observed in these behaviours (see McCormick *et al.* 2003).

#### 6.3.1.1 *Maternal effects among various taxa*

Recent studies have shown that such maternal effects are not limited solely to mammals. For example, Schwabl (1993) demonstrated that female birds differentially deposit steroid hormones into their eggs, which affects the physiology and behaviour of the resulting young. In some species, young from eggs with high testosterone levels in the yolk grow faster and, once fledged, they achieve higher social rank than young with lower levels of yolk testosterone (Schwabl 1993; von Engelhardt 2006). The amount of testosterone added to yolk varies with breeding conditions (Schwabl 1997), which enables maternal effects to reflect the state of the local environment. Different yolk layers also can have different concentrations of some hormones (Lipar *et al.* 1999), suggesting that exposure of the embryo to specific combinations of hormones is not constant during development. That is, different embryological stages may be exposed to different hormonal cocktails. Furthermore, maternal effects may span generations in birds, as in mammals, as has been demonstrated in the zebra finch. Williams (1999) found that daughters of oestrogen-treated adult zebra finch females, breeding for the first time, produced larger eggs than daughters of untreated females.

In the common lizard (*Lacerta vivipara*), body condition and the stress level of adult females affect the dispersal behaviour of their young (Léna *et al.* 1998; de Fraipont *et al.* 2000). Another example is provided by the daphnia of Figure 2.6 in which the quality of the environment encountered by an individual (the presence or absence of predators) influences the phenotype of its offspring. In this same example, the authors also report grandmother effects, by which environmental effects experienced by a female influences the phenotype of her grandchildren. Finally, these maternal effects can involve substances other than hormones. For example, Gasparini and his collaborators (2001) have demonstrated that female kittiwakes transfer into their eggs the antibodies against parasites transmitted by ticks (*Ixodes uriae*) as a function of the local density of parasites. This confers passive immunity to the young in areas of high tick infestation.

#### 6.3.1.2 *Maternal effects and adaptation*

From an evolutionary point of view, the value of such phenotypic plasticity is that it allows individuals to integrate features of their environment into their specific developmental trajectory. This results in adult phenotypes that are able to survive and reproduce in that environment. In a sense, it allows organisms to fine-tune their physiological and behavioural response mechanisms. For example, in certain environments a strong, rapid response to stress may be useful, whereas in other environments it may be better to have an attenuated stress response, especially given the deleterious effects that prolonged glucocorticoid exposure can have on the brain (McEwen 1999). Mothers who have survived in their environment and are in sufficiently good body condition to have produced offspring are in a position to transmit information about that environment to their young at a time when the latter can integrate that information into their own phenotype during their development. This, in turn, would affect offspring behaviour, including their parental behaviour when they become adults, which would affect phenotypic development of their own offspring, etc. Thus, maternal effects can be incorporated into the phenotypes of succeeding generations (see, for example, Wang and vom Saal 2000). Moreover, we have seen in Chapter 2 how this can pose a problem for the measure of **heritability** (Box 2.3).

Recent intriguing studies indicate that some female birds increase the amount of testosterone they deposit into egg yolk if paired with attractive mates (Gil *et al.* 1999, 2004; von Engelhardt 2006). Females paired with attractive males produce more male offspring and add significantly more testosterone to their eggs, and young with the additional testosterone grow faster than young with less testosterone. Interestingly, male attractiveness is not restricted to physical features; females paired with males that sing attractive songs also produce eggs with more testosterone (Gil *et al.* 2004). These results add the possibility that paternal effects may also be incorporated into phenotypic development of the young. This example is related to the problem of investment of resources into the production of one sex or the other, which will be developed in Chapter 13.

The question of whether or not these different phenotypes achieve different fitness levels is poorly documented. Thus, despite numerous studies of maternal effects in relation to phenotypic plasticity, few studies have explored the adaptive significance of these effects. Most studies on this subject have been conducted in highly controlled clinical settings, where maternal effects are applied experimentally and do not result from environmental variability. Indeed, variation from perceived optima, such as reduced exploratory behaviour, fear of novel food types, or enhanced HPA reactivity, typically is interpreted as a pathological condition. However, as we saw in the case of metamorphosis in desert spadefoot toads, optimal behaviour is strongly context dependent. That is, tadpoles that accelerate their rate of metamorphosis in the face of declining pool water depth may be small as adults, but at least they reach adulthood. Tadpoles that select the 'optimal' developmental path, that of maximizing larval growth before undergoing metamorphosis, are more likely to die before reaching adulthood because they do not complete metamorphosis before their pools disappear. Thus, from an evolutionary perspective, responses such as fear of novel food items or enhanced HPA reactivity may be adaptive rather than pathological, for they may prepare an individual to survive in its natural environment, and the integration of maternal effects early in life may help in this preparation.

The fitness of different phenotypes can vary greatly, and this variation is due, at least in part, to environmental conditions. Additional empirical data are needed to elucidate the extent to which maternal effects are adaptive. An implication of all these examples is that animals constantly extract information about environmental conditions that influences the developmental pathways taken by various phenotypes. The question of information is developed in Chapter 4.

## 6.4 Important transitions in life history strategies

An organism's phenotypic development is marked by several transition points, at which they change morphologically, physiologically, and behaviourally. Many, if not all, of these transitions are regulated by hormonal changes. We already have discussed one of these transition points, that from an immature juvenile to a sexually mature adult. We have seen how hormones experienced early in life act to shape subsequent sex-specific behaviours that are triggered by increased secretion of hormones at puberty. The metamorphosis of desert spade-foot toads from tadpoles into adults is another example of a hormonally induced transition. Other examples include avian fledging behaviour, whereby a young bird increases its overall locomotor activity in conjunction with the development of independence by or natal dispersal of juveniles, in which young animals move from their natal area to an area where they will settle and breed.

Some of these transitional phenomena occur only once in an individual's life: amphibian metamorphosis and natal dispersal, for example. However, some transitions are recurrent, and mark the predictable passage from one life stage of the annual cycle to another (Jacobs and Wingfield 2000). For example, seasonally breeding animals come into reproductive condition annually, and migratory species change behaviourally and physiologically in conjunction with movements to and from the breeding grounds, etc. We will explore some of these transitions in more detail and will examine ways in which the development of these behaviours is affected by environmental factors.

### 6.4.1 The first transition: birth

In humans and in mammals generally, the first transition is perhaps the most important in an organism's life history. It is the transition by which an individual moves from the protective environment of the womb into the 'real world.' The endocrine changes associated with parturition are well-known, and involve dramatic drops in plasma levels of progesterone and oestradiol, along with an increase in secretion of oxytocin that stimulates the contraction of the uterus and propels the fetus through the birth canal (reviewed in Nelson 2000).

The timing of parturition is crucial, for it should not occur before physiological systems are sufficiently developed for survival of the fetus outside the womb. The triggering mechanism for the onset of parturition revolves around on the levels of CRH in the blood (McLean *et al.* 1995, Wadhwa *et al.* 1998). CRH is secreted by the placenta, as well as by the hypothalamus of the mother, and maternal plasma CRH levels increase at parturition. Glucocorticoids, such as cortisol (the principal glucocorticoid in humans), are released in response to elevated CRH and normally have a negative feedback effect on hypothalamic CRH release. However, cortisol stimulates CRH release by the placenta. Furthermore, recall that steroid hormones are transported in the blood by binding proteins, but that it is the unbound fraction of the hormone that is biologically active. The major binding protein for cortisol is called corticosteroid-binding globulin (CBG). The levels of CBG decline as parturition approaches, which further increases the amount of biologically active cortisol in the blood. The end result is an increase in the amount of active CRH produced by the placenta, which helps to trigger parturition.

This cascade of effects explains why maternal stress, which also results in enhanced cortisol production, can trigger the premature onset of parturition (Majzoub *et al.* 1999). Interestingly, in a pregnancy in which the female is chronically stressed, advancement in parturition date may benefit both mother and offspring, despite the numerous developmental difficulties associated with premature birth (Knackstedt *et al.* 2005). From an evolutionary perspective, mothers and fetuses may have conflicting interests when it comes to resource allocation: increased energy allocation to a fetus may put at risk the mother's future reproductive success, and increased allocation to the mother may affect the fitness of the offspring (Stearns 1992). Speeding up the developmental process to the point where the offspring no longer relies directly upon the mother for resources may benefit both mother and child. Limiting investment in that particular offspring may improve the mother's lifetime reproductive success, and early parturition, even with its inherent risks, is infinitely preferable to a miscarriage (see Pike (2005) for a detailed discussion of the evolutionary aspects of maternal stress and preterm delivery).

### 6.4.2 Fledging

In altricial birds, which are dependent on their parents for food and protection until they are sufficiently developed to care for themselves, the transition from sedentary nestling to relatively mobile fledgling can be abrupt, because nest departure typically is an all-or-nothing event and young generally do not return to the nest once they have left. Little is known about the endocrine events associated with this transition. Heath (1997) found that baseline corticosterone levels increase in young American kestrels (*Falco sparverius*) as fledging date approaches. Sims and Holberton (2000) found age-related increases in the corticosterone stress response in young Northern mockingbirds (*Mimus polyglottos*), although it is unknown if these changes are related to fledging. The hormonal basis of fledging behaviour clearly needs to be explored in more detail.

### 6.4.3 Natal dispersal: a condition-dependent process

Almost all young animals leave their place of origin to search for a suitable location to breed (Stenseth and Lidicker 1992; Zera and Denno 1997). This process is called **natal dispersal** (Clobert *et al.* 2001). The evolutionary significance of this behaviour is discussed in Chapter 10. There is great variety among and within species in the patterns of natal dispersal behaviour, and we will see in Chapter 10 that there are a variety of ultimate and proximate reasons why not all individuals in a given species disperse at the same time or travel the same distance.

As with many other behaviours, there are hormonal correlates to natal dispersal behaviour. Because mammals generally show male-biased dispersal (Greenwood 1980, Pusey 1987), much of the work on the proximate mechanisms of mammalian dispersal has focused on changes in androgen secretion. For example, female grey-sided voles (*Clethrionomys rufocanus*) from litters with a high proportion of males may be exposed prenatally to elevated testosterone levels (vom Saal 1984). More females disperse from such litters (Ims 1989, 1990) and they initiate dispersal behaviour sooner than do females from control litters (Andreassen and Ims 1990). However, neither exposure to male-biased litters nor direct neonatal application of testosterone induces male-like natal dispersal behaviour in female voles of the genus *Microtus* (Bondrup-Nielsen 1992; Lambin 1994; Nichols and Bondrup-Nielsen 1995).

#### 6.4.3.1 *The importance of body condition*

In Belding's ground squirrels (*Spermophilus beldingi*), androgens may have organizational but not activational effects on natal dispersal behaviour (Holekamp *et al.* 1984, Holekamp and Sherman 1989). Young female ground squirrels treated with testosterone exhibit a male-like pattern of natal dispersal. However, castration of juveniles of either sex before dispersal has little effect on natal dispersal patterns, suggesting that testosterone does not produce activational effects. Instead, body condition or, more specifically, percent body fat may play the role of an 'ontogenetic switch' in stimulating natal dispersal behaviour (Nunes and Holekamp 1996; Nunes *et al.* 1998). These authors suggest that the threshold percentage of body fat needed to initiate natal dispersal changes with the approach of post-dispersal hibernation. Juveniles active early in the year would disperse with a low percentage of body fat because there is ample time to put on additional fat reserves to support hibernation. However, late in the active season dispersal may be delayed or inhibited altogether because more and more of the fat stores are targeted to survive hibernation. Thus, phenotypic plasticity in natal dispersal behaviour represents a trade-off between energetic demands needed to support natal dispersal and those needed to support hibernation.

The mechanism whereby achievement of a particular percentage body fat triggers natal dispersal behaviour is not known, but recent studies suggest that another hormonal messenger may be involved. Leptin, a protein hormone discovered in 1994, is released primarily by adipose cells (reviewed in Ahima and Flier 2000). This hormone primarily acts on energy balance, inhibiting food intake and increasing energy expenditure (Friedman and Halaas 1998; Ahima and Flier 2000) and locomotor activity (Pelleymounter *et al.* 1995). Increased leptin secretion has been linked to the onset of puberty in mammals (Ahima *et al.* 1997), and leptin levels correlate with levels of body fat reserves (Considine *et al.* 1996). Thus, it may serve as a signal that the body has achieved sufficient energy stores to support reproduction. It is not unreasonable to suggest that leptin may serve a similar role in natal dispersal. That is, it may signal the presence of adequate fat reserves, thus stimulating dispersal behaviour in male ground squirrels whose brains were organized earlier by androgens to be receptive to the activational effects of leptin.

#### 6.4.3.2 *Interactions among various factors*

Note that the model of Holekamp and her colleagues (1984, 1989) presented above for ground squirrels is based on the interplay among endocrine signals, body condition, and ecological factors. An individual's percentage of body fat is assessed, through leptin or some other means, in comparison with the base level for that stage of the annual cycle. A level of body fat that triggers natal dispersal early in the active season is insufficient to trigger this behaviour later in the season. Thus, phenotypic variation in body composition as well as variation in the seasonal onset of activity would affect the ultimate expression of dispersal behaviour.

Other taxa may use different combinations of hormones and/or environmental cues to trigger the transitional behaviour of natal dispersal. For example, Belthoff and Dufty (1998) suggest that changes in corticosterone secretion, in conjunction with achievement of a sufficient body mass, may stimulate natal dispersal behaviour in western screech-owls (*Otus kennicottii*; Figure 6.9). Support for such a model comes from correlational data showing that there is an endogenous increase in plasma corticosterone in captive western screech-owl fledglings at the time that their free-living siblings are dispersing (Belthoff and Dufty 1995; Dufty and Belthoff 2001). Furthermore, Silverin (1997) showed that juvenile willow tits (*Parus montanus*) disperse if implanted

Figure 6.9 **The western screech-owl**

Photo provided by Jim Belthoff.

with corticosterone during winter flock formation, the time when natal dispersal normally occurs. However, similar implants given at the same time to adults, or to juveniles after winter flocks have stabilized, have no effect on movements, again highlighting the importance of environmental factors (in this case, the annual cycle) on the development of behaviour.

As a final example of how hormonal changes interact with maternal and other environmental factors to develop natal dispersal behaviour, consider the work of Jean Clobert and his co-workers on the common lizard (*L. vivipara*). Although an egg-laying species, females retain eggs internally for most of the 2.5 month incubation, and young hatch within an hour of oviposition. There is no parental care in this species, and natal dispersal occurs within 10 days of hatching. While opportunities for postnatal maternal effects are limited in these lizards, prenatal maternal characteristics influence offspring attributes, including dispersal. For example, young lizards from well-nourished mothers disperse at a higher rate than those from poorly nourished females (Massot and Clobert 1995). Similarly, Ronce and her collaborators (1998) analysed theoretically how the state of the female can influence the dispersal phenotype of her offspring: they expected that daughters of old females would disperse less than daughters of young females. This is due to the fact that young mothers would probably still be alive when their offspring become mature, increasing the risk of competition between related individuals (see Chapter 10). So young females, but not young males, should disperse farther when produced by young mothers, a prediction that is supported by data from the common lizard (see Figure 10.7). In this species, prenatal stress, as simulated by the application of corticosterone to the skin of pregnant females (Meylan *et al.* 2003), interacts with maternal condition to affect natal dispersal of offspring in ways that appear to reduce kin competition (de Fraipont *et al.* 2000; Meylan *et al.* 2003) and even increases survival, at least in males (Meylan and Clobert 2003; Cote *et al.* 2006). Similarly, maternal parasite load influences the life history traits of offspring (Sorci *et al.* 1994; Sorci and Clobert 1995), and environmental stressors, such as high levels of agonistic interactions, increase the intensity of such parasite infections and elevate corticosterone levels in adults (Oppliger *et al.* 1998). Thus, parasite load, through its effects on maternal condition and/or on corticosterone levels, may have an effect on natal dispersal in common lizards.

### 6.4.4 Migration

**Migration**, or the seasonal movement of animals between their breeding and wintering grounds, is an example of a recurring life history transition. It is an activity that takes an individual from one ecological and behavioural environment to another. This behaviour is most easily seen in birds, where spectacular movements in spring and autumn have been noted (Able 1999), although it is also a feature in the life history of many other animals (Dingle 1996).

#### 6.4.4.1 *A genetic component*

Migratory behaviour, like other events of the avian annual cycle, can have a strong genetic component (Pulido *et al.* 2001). For example, blackcaps (*Sylvia atricapilla*) from different populations show different orientation patterns if held in captivity during the migratory period, and they display these patterns for different lengths of time that correspond to the various distances travelled by individuals of these diverse populations. The heritability of these behaviours is high, and hybrids between parents of different populations adopt intermediate characteristics between those of their parents in terms of duration and orientation (Berthold 1990). The precise onset and duration of migration can be modified by environmental cues, such as photoperiod, climate, light intensity, and food availability, but these do not alter the overall pattern of migration (Gwinner 1996). Seasonal changes in diet are also regulated by a circannual programme (Bairlein 1990), as is selection of the direction of migration (Gwinner and Wiltschko 1980).

#### 6.4.4.2 *A cascade of major changes*

Avian migration necessitates that birds make some major behavioural and physiological adjustments. Energy demands are high at a time when great distances may be travelled, often without stopping to feed or rest en route. Birds preparing for migration can lower their body temperature to reduce basal metabolism, which facilitates the accumulation of additional fat deposits that are used later to fuel migratory flight (Butler and Woakes 2001). Digestive organs are reduced in size just before and during migration, especially in individuals that embark on long migratory flights without refuelling (Piersma 1998). Indeed, changes in morphology and in the efficiency of digestive organs influence the patterns of migratory activity that different species can produce (McWilliams and Karasov 2001). Additional phenotypic flexibility is exhibited by flight muscles, which increase in mass before migration (Piersma *et al.* 1999) in the absence of any obvious increase in muscular activity (Dietz *et al.* 1999), the additional muscle protein being then used to fuel flight (Schwilch *et al.* 2002).

*Autumn migration and spring migration*

What do we know about hormones and the development of the migratory phenotype? At present, the answer is: not much (Holberton and Dufty 2005). Changes in daylength are known to induce changes in hormone secretion (Farner and Follett 1979; Nicholls *et al.* 1988), and to a certain extent we can track endocrine changes that occur during migration. Furthermore, seasonal migration of birds occurs simultaneously with significant changes in life history strategy. On the one hand, autumn migrants in the Northern Hemisphere are post-reproductive and are moving away from areas that, while typically providing abundant food, will soon become inhospitable. They are moving into wintering areas that differ from breeding sites, both ecologically and socially. On the other hand, spring migrants are moving toward the breeding grounds that may still have poor food supplies and inclement weather, but where the birds must quickly find a mate and begin reproductive activities. Given the different ecological and physiological features experienced by migrants at different times of the year, it should not be surprising that spring and autumn migrants have different endocrine profiles. That is, migratory phenotypes differ between spring and autumn and do not simply represent the same phenomenon, pointed in two opposite directions (O'Reilly and Wingfield 1995; Holberton and Dufty 2005).

*Preparatory hyperphagia*

Although the endocrine changes associated with development of a migratory phenotype are yet to be fully elucidated, some information is available. Most avian species engage in a period of premigratory hyperphagia, consuming a large quantity of food that is stored as the fat that fuels the journey. The regulation of food storage and utilization is a complex process (Blem 1990; Ramenofsky 1990), and no overall endocrine pattern for migration has yet to emerge. Hyperphagia and lipogenesis have been linked to the secretion of prolactin and corticosterone, during both the non-migratory (Buntin 1989; Berdanier 1989) and migratory (Boswell *et al.* 1995; Holberton 1999; Landys *et al.* 2004) periods. These two hormones may synergize with each other, although differences in photoperiod may alter the synergistic relationship in the two seasons (Meier and Farner 1964; Meier and Martin 1971; but see Boswell *et al.* 1995).

*Gonadal hormones, etc.*

Gonadal hormones also are important in fat deposition in vernal (spring) migrants, whose gonads become active as they return to the breeding grounds (Wingfield *et al.* 1990b; Deviche 1995). For example, ovariectomy has been shown to interfere with vernal fattening but not with autumnal fattening (Schwabl *et al.* 1988). In autumn migrants, whose gonadal hormone levels are very low, glucagon and insulin are involved in regulating fat deposition (Totzke *et al.* 1997; Hintz 2000). These hormones, which are intimately linked to the general regulation of food processing (Hadley 1996), may be involved in these activities in vernal migrants, too. Basal corticosterone levels are elevated in migrating birds (Schwabl *et al.* 1991; Holberton *et al.* 1996; Piersma *et al.* 2000), and corticosterone probably is involved in regulating the metabolism of lipid and protein energy reserves for fuel (Jenni *et al.* 2000). As discussed in more detail below, migrating birds suppress their glucocortical response to stress (Holberton *et al.* 1996), probably to avoid catabolizing muscle proteins that are used for flight. However, the nature of the adrenocortical response depends on the bird's body condition when it is stressed (Jenni *et al.* 2000; Long and Holberton 2004).

Finally, other hormones and neuropeptides, theoretically less closely linked to migratory behaviour, merit additional attention in this field. For example, thyroid hormones have been shown to be important in the development of migratory behaviour in some species (Nair *et al.* 1994), and may be responsible for selectively increasing the aerobic capacity of flight muscles (Bishop *et al.* 1995). In addition, sensitivity to neuropeptide Y, a powerful appetite stimulant, increases during the period of pre-migratory fattening in the European house sparrow (*Passer domesticus*; Richardson *et al.* 1995).

Much of the available information linking hormones to migration addresses the physiological and morphological adjustments that occur, rather than the behavioural changes themselves. Fundamental questions about what endocrine changes influence the onset, direction, and duration of migration, remain unknown. It is highly likely that such hormonal regulation exists, but it awaits documentation.

## 6.5 Adult phenotypic plasticity

Conventional thinking suggests that once an individual has reached sexual maturity, then its behavioural phenotype is fixed and not susceptible to change. That is, once the transition to adulthood has been made, no further important changes occur. We have seen already that this is not the case; for example, adult male lizards are known to change from one morphological and behavioural phenotype to another (Moore *et al* 1998; Sinervo 2000). We are now going to discuss additional examples of adult behavioural plasticity and their hormonal correlates.

### 6.5.1 Birdsong

Many animals breed at specific times of the year, resulting in the seasonal expression of certain behaviours. For example, singing behaviour in male passerine birds increases in the spring as males aggressively attempt to establish territories and to attract females (Kroodsma and Byers 1991). The vernal increase in daylength stimulates increased singing through activation of the HPG axis and increased secretion of testosterone and other androgens (Farner and Wingfield 1980). Testosterone stimulates singing, courtship, and aggression in male birds, whereas castration greatly reduces these behaviours (Arnold 1975).

The brain areas that control song learning and production have been mapped in birds (Reiner *et al.* 2004), and these involve neural connections among several brain nuclei. In species where females do not sing or sing very little, the song-control nuclei exhibit sexual dimorphism, with greater development in males than in females (Nottebohm and Arnold 1976). In contrast, in species where females sing regularly (such as dueting with males), their central song control systems show considerable development (MacDougall-Shackleton and Ball 1999).

These brain regions show seasonal variation in size, growing larger in spring and shrinking at the end of the breeding season (Brenowitz 2004). Testosterone appears to be responsible for these seasonal changes (Bottjer *et al.* 1986), although it may first be converted to oestradiol at the target tissues (Schlinger 1997). Interestingly, factors independent of testosterone also influence changes in the size of brain areas that control avian song. For example, environmental factors, such as the photoperiodic condition of the bird, play a role in regulating the extent of neural growth in response to testosterone (Bernard *et al.* 1997; Bernard and Ball 1997).

Photosensitive (i.e., responding to long daylengths with increased volume of song control areas) European starlings (*Sturnus vulgaris*) and photorefractory (i.e., not responding to long daylengths with increased volume of song control areas) starlings were implanted with exogenous testosterone and the size of one of the song-control nuclei subsequently was measured. The photosensitive group showed greater neural growth than did the photorefractory group, indicating that testosterone-independent effects associated with photoperiod also play a role in determining volume of song control nuclei (Figure 6.10). In addition, Tramontin *et al.* (1999) demonstrated that a social cue, exposure to a female, increases the size of some of the song control nuclei and the singing rate, even though testosterone levels were similar in experimental and control groups. Furthermore, Hamilton *et al.* (1998) found that courtship success in male brown-headed cowbirds is correlated with two brain areas: one song control area and one visual area. Male cowbirds respond vocally to subtle behaviours exhibited by females (West and King 1988), and the effects of these visual cues are integrated into the overall mechanism of song development and production. Finally, singing itself may lead to changes in the volume of song control nuclei (Sartor and Ball 2005; but see Adkins-Regan 2005).

Our understanding of how hormones interact with non-endocrine signals to regulate seasonal and individual variation in brain nuclei in birds and other taxa is far from complete and this field offers many exciting research opportunities. For example, conspecific birds living in different habitats apparently rely on different proximate environmental cues to stimulate gonadal recrudescence and development of brain song control areas (Caro *et al.* 2005; Perfito

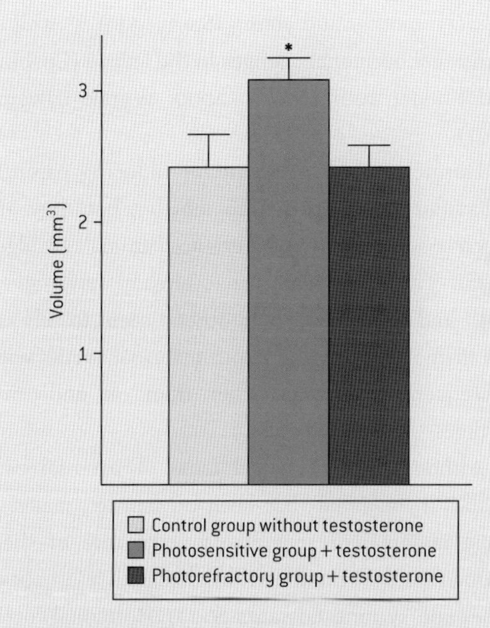


Figure 6.10 **Existence of testosterone-independent factors in the development of song control areas of the brain in the European starling (*Sturnus vulgaris*)**

Mean (+ standard deviation) volume of the 'high vocal centre' in the brain of male European starlings belonging to three different treatment groups containing photosensitive, photorefractory, and control individuals. The birds were rendered photorefractory by raising them on long days (16 hours of light and 8 hours of darkness) for 9–16 weeks. Photorefractory birds regressed their gonads and remained in that state until they were returned to long days. The birds in the photosensitive group were exposed to seven weeks of short days (8 hours of light and 16 hours of darkness), which makes them photosensitive; that is, their gonads will enlarge if they are again exposed to long days. The results show that the photoperiodic state itself can affect responses of song control nuclei to testosterone stimulation, because photosensitive birds respond more strongly to testosterone implants than do individuals in either the photorefractory or control groups. Mean volumes of nuclei of different groups were compared by a one-factor ANOVA ($F_{2,15} = 5.13, P < 0.02$) and by Fisher post-hoc tests. The asterisk * indicates means that there are significantly different among groups at $P < 0.05$.

After Bernard and Ball (1997).

*et al.* 2005). While ambient temperature may play a role (Perfito *et al.* 2005), the mechanisms are poorly understood.

### 6.5.2 Parental care

Vertebrates have evolved a wide range of parenting behaviours. The care of offspring obviously is a very important process for species whose young depend on their parents for protection and/or food, yet it is an activity in which most young parents participate with little or no real experience. Nonetheless, these inexperienced parents are remarkably adept at caring for offspring, exhibiting such complex behaviours as nest-building, maternal aggression against intruders, and nursing behaviour. As one might expect, parental behaviour has strong hormonal underpinnings, although, as we have seen throughout this chapter, non-endocrine variables also affect its expression.

#### 6.5.2.1 *Mammals*

In mammals, parental care involves remarkable physiological changes, but also the activation of behaviours, primarily in females, that are specific to the care of offspring. Female mammals produce milk from the mammary glands in response to the combined effects of two protein hormones, prolactin and oxytocin. The plasma concentration of these two hormones is elevated as parturition approaches (Rosenblatt *et al.* 1979; Fuchs and Dawood 1980), and they also are involved in the induction of maternal behaviour. Prolactin, which is produced by the anterior pituitary, induces the rapid onset of maternal behaviour if given to nulliparous female rats (see Bridges and Mann 1994). However, for this effect to occur, the animals must be 'primed' by previous exposure to oestradiol and progesterone, mimicking the endocrine changes that occur naturally during pregnancy. That is, oestradiol and progesterone themselves do not stimulate maternal behaviour, but if the animals are not exposed to these hormones then prolactin will not do so, either. Oestradiol and progesterone have a permissive effect in this case. Lactogen, a hormone similar to prolactin that is produced by the placenta, also induces parental behaviour in females (Bridges *et al.* 1997). Insofar as the placenta develops from fetal cells (as well as from maternal cells), this suggests that the fetus itself, acting through placental secretions, may help to regulate the onset of its own parental care (Bridges *et al.* 1997). Thus, this production of lactogen can be viewed as a 'physiological weapon' used by progeny in the parent-offspring conflict that occurs during gestation in mammals: It is in the interest of the offspring to divert as many maternal resources as possible for its own use, even if this reduces the prospects for survival and reproduction for the mother. Oxytocin, a hormone released by the posterior pituitary, also plays a role in the induction of parental behaviour in females; again, in association with priming by steroid hormones (Keverne and Kendrick 1994). Once parental behaviour has begun, its maintenance involves tactile interactions between mother and young (Stern 1996), as well as additional hormonal changes in the female (Cushing and Kramer 2005; Nelson 2000; Wynne-Edwards 2001).

Hormones are important for inducing parental behaviour in inexperienced females, but they are much less important in experienced animals (Bridges 1996). That is, females that already have given birth and experienced parenthood seem to rely on the memory of those previous maternal events with subsequent offspring, rather than on endocrine changes.

Although the historical focus of parental care studies in mammals has been on females, males of some mammalian species also take care of their young (see, for example, Gubernick and Teferi 2000; Jones and Wynne-Edwards 2000). Hormonal correlates of paternal behaviour are under investigation in several mammalian species, including humans (Storey *et al.* 2000). Results so far suggest that some of the same endocrine mechanisms occur in males and females that exhibit parental care and that testosterone, which interferes with paternal care in other taxa (see below), declines at parturition in male mammals that exhibit paternal care (Wynne-

Edwards and Reburn 2000; Young *et al.* 2001). Although not all of the data are consistent, part of the difficulty may arise from the fact that endocrine measurements typically are made on blood collected peripherally, and peripheral hormone levels may not accurately reflect central (i.e., brain) processes (Schum and Wynne-Edwards 2005).

#### 6.5.2.2 *Birds*

Unlike the situation in mammals, parental care in birds frequently involves both sexes. However, many examples can be found where only females take care of the young and, even in species where both sexes contribute to parental care, females generally provide most of the care. In some role-reversed species only paternal care is provided, and in obligate brood parasites, which lay all their eggs in nests of other birds, the genetic parents provide no parental care at all.

##### a *Species where females invest the most*

Avian parental care has both similarities and differences with that of mammals. In altricial species adults typically build nests, incubate the eggs, brood the chicks, and feed and protect the young. In birds, males are able to participate in these activities more fully than in mammals because of obvious physiological differences between the two taxa: gestation and milk production are the exclusive purveyance of the female in almost all mammals (see Francis *et al.* (1994) for an exception).

As we saw with mammals, prolactin plays a major role in parental behaviour in birds (reviewed in Goldsmith 1983; Ball 1991). Bird species where females provide the majority of parental care have elevated plasma prolactin levels at the end of clutch completion (Goldsmith 1982). Prolactin patterns are less closely linked with male parental care (Silverin and Goldsmith 1983). Seiler *et al.* (1992) found that prolactin increases during incubation in males and females of an aseasonally breeding, biparental finch, indicating that non-photoperiodic environmental cues, such as those provided by the eggs themselves, can stimulate the increase in prolactin. Cues from eggs and/or young also affect the temporal progression of prolactin secretion. Silverin and Goldsmith (1990) experimentally increased (or decreased) the length of the early nestling period of pied flycatchers (*Ficedula hypoleuca*) by exchanging nestling among nests, so that females were in contact with very young nestlings during a period of time that was longer (or shorter) than normal. Females naturally brood young nestlings during the first few days of life to keep them warm, and the manipulations resulted in the predicted prolongation (or shortening) of the period of elevated prolactin secretion in the females.

Although environmental cues affect individual differences in prolactin secretion and parental care in some species, other species are less sensitive to these changes. For example, both members of Adelie penguin (*Pygoscelis adeliae*) pairs incubate eggs and brood and feed young, and prolactin levels are elevated in both sexes during the incubation and chick-rearing periods (Vleck *et al.* 1999). Manipulation of incubation periods in these penguins has little effect on prolactin levels, suggesting that sustained high levels of prolactin may support parental behaviour of adults, who must be away from the nest for days to forage for their young (Vleck *et al.* 2000). On the other hand, as we have seen with mammals, experience modifies the effects of prolactin on parental behaviour. In ring doves (*Streptopelia risoria*), not only do experienced, non-breeding female doves exhibit more parental care than do inexperienced females when presented with chicks, but again, prolactin injections increase the level of parental care in non-breeding, experienced females more than that of inexperienced females (Wang and Buntin 1999).

##### b *Species with sex-role reversal*

Avian species with sex-role reversal, where males assume most responsibility for parental care, offer an opportunity to examine whether males also have patterns of prolactin secretion that can be modified as in other species (see Oring *et al.* 1986, 1988). Indeed, role-reversed males tend to have higher prolactin levels than females, especially during incubation. In addition, certain brain areas in these males may become more sensitive to prolactin during incubation (Buntin *et al.* 1998), which could also help to explain differences in parental behaviour. However, not all species with unusual breeding strategies have correspondingly unusual patterns of prolactin secretion. As brood parasites, brown-headed cowbirds (*Molothrus ater*) perform no parental care, so one would not expect to see an increase in prolactin in the course of the breeding season. However, both sexes show such increases in prolactin secretion (Dufty *et al.* 1987). Prolactin has functions in addition to parental care, such as inhibiting the secretion of reproductive hormones (Buntin *et al.* 1999) and stimulating the development of photorefractoriness (Sharp and Blache 2003), so a seasonal change in prolactin secretion, independent of the level of parental care, should not be surprising.

### C *Evolutionary explanations and proximate mechanisms*

In Chapter 12 we will examine evolutionary explanations for why males of so many avian species choose to engage in parental care rather than attempting to increase their fitness by mating with additional females. One of these explanations is that raising a clutch of small, helpless, young is a demanding job, and unaided females may raise fewer young or produce poorly developed young compared with a pair of adults (see Meek and Robertson 1994). The proximate mechanism involves testosterone, where elevated levels in males facilitate singing and courtship behaviour, and reduced levels occur when males care for young. For example, male song sparrows (*Melospiza melodia*) have elevated testosterone levels during territory establishment and courtship, and testosterone declines thereafter and is low when the males help to care for the young (Wingfield 1984a). However, males supplemented with endogenous testosterone during the chick-rearing period attempt to attract a second female to their territory, and this activity reduces the rate at which they feed young (Wingfield 1984b). Even if successful at attracting a second mate, overall male reproductive success declines because female song sparrows have difficulty raising young by themselves. Conversely, a reduction in reproductive success in testosterone-implanted dark-eyed junco (*Junco hyemalis*) males is compensated for by an increase in extra-pair fertilizations by these males (Raouf *et al.* 1997). However, there may be other disadvantages of maintaining elevated testosterone that have not been explored fully (Ketterson and Nolan 1999). In species with short reproductive seasons, such as Arctic-breeding Lapland longspurs (*Calcarius lapponicus*), male parental care is initially reduced by exogenous testosterone, but the level of paternal care subsequently recovers (Hunt *et al.* 1999). This is, perhaps, a result of the limited advantages afforded by continued sexual activity late in the short, Arctic breeding season.

Male pied flycatchers (*F. hypoleuca*) exhibit variability in their pattern of testosterone secretion that correlates with differences in parental investment. Many male pied flycatchers have elevated testosterone levels during territory establishment and mate attraction, but have reduced testosterone when caring for nestlings (Silverin and Wingfield 1982). However, after producing a clutch of eggs with one female, some males then establish another territory elsewhere and mate with a second female. Testosterone levels in these polygynous males do not decline the way they do in the monogamous conspecifics. Instead, testosterone levels remain high until the second female has also produced eggs, at which point they do decline, and the male returns to the first female to help her care for the young (Silverin and Wingfield 1982). Other patterns of testosterone secretion have been described in males, yet there appears to be a consistent relationship among testosterone secretion, the amount of parental care provided by males, and mating systems (Wingfield *et al.* 1990a). For instance, males of brood parasitic or highly polygynous species provide little or no parental care, and these males maintain high testosterone levels for much longer than do males of species that care for their offspring (Dufty and Wingfield 1986a; Beletsky *et al.* 1995).

#### 6.5.2.3 *Fish*

In teleost fish, paternal care is the most widespread type of parental care of young. We will see in Chapter 12 the evolutionary explanation for such a pattern, but here we discuss the mechanisms involved in the behaviour. As with other taxa, male androgen levels decline when they engage in parental care. For example, black-chin tilapia (*Sarotherodon melanotheron*) males brood their fertilized eggs in their mouth for over two weeks, during which time androgen levels decline significantly. If the eggs are removed shortly after mouth-brooding begins, then the androgen decline is forestalled, indicating that some signal from the eggs triggers the reduction in hormone secretion (Specker and Kishida 2000). Some male plainfin midshipman fish (*P. notatus*) build nests and guard nests in which females deposit their eggs (DeMartini 1988; Bass 1996). Males accumulate eggs sequentially from several females, which leads to a conflict for the males. That is, males continue to defend a territory and to court additional females, even as the eggs from earlier females develop and require a certain level of paternal care. As we have seen, the former behaviour is facilitated by androgens, while the latter is hindered by androgens. Plainfin midshipman fish strike a hormonal compromise that does not contradict either of these behaviours (Knapp *et al.* 1999). Males guarding only eggs in their nest possess androgen levels that are as high as those found in males with an empty nest, suggesting that these males continue to seek additional females at the same time that they exhibit paternal guarding behaviour. However, once males have embryos in their nest, androgen levels decline, which may reflect a behavioural shift away from courtship behaviour and toward parental care (Figure 6.11; Knapp *et al.* 1999).

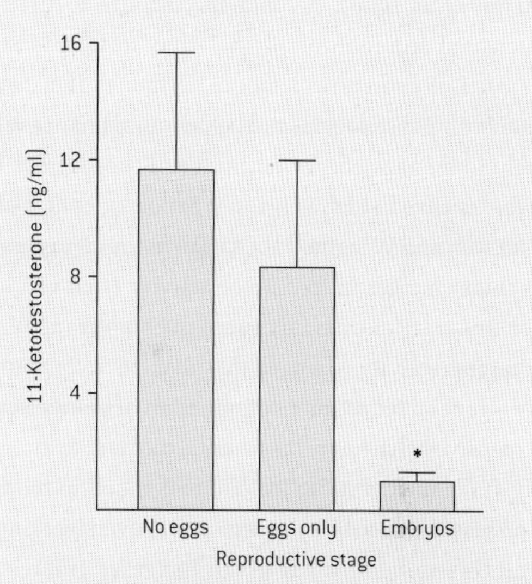


Figure 6.11 **Testosterone and behavioural conflict in male plainfin midshipman fish (*Porichthys notatus*)**

Mean (+ standard deviation) level of 11-ketotestosterone, the principal androgen in this species, during three stages of parental care. Males whose nests contain eggs continue to secrete androgens at the same level as males having empty nests. This suggests that the former continue to court females. However, androgen levels decline when embryos appear, suggesting that the males reduce courtship with the onset of parental care. Significant differences in androgen levels were determined by Kruskal–Wallis tests ($H = 12.07$, df $= 2$, $P = 0.002$). Means in each category then were evaluated by Dunn multiple comparison tests. The symbol * indicates that the mean is significantly different from that of the other groups (embryo versus eggs: $Q = 2.52$, $P < 0.05$; embryo versus no eggs: $Q = 3.40$, $P < 0.002$).

After Knapp *et al.* (1999).

### 6.5.3 Adrenocortical response

Stressors, or aversive stimuli that reduce (or threaten to reduce) fitness, can disrupt homeostasis in animals and thus affect their behaviour, their physiology, and, ultimately, their fitness (see, for example, Sapolsky 1982; Hofer and East 1998; Korte *et al.* 2005). Stressors include such things as increased population density, adverse weather conditions, reduced availability of food, low temperatures, etc. (Christian *et al.* 1965; Wingfield 1984; Bronson 1989;

Akana *et al.* 1999). A common response to stress is increased secretion of adrenal glucocorticoids, such as cortisol and corticosterone, hormones that facilitate both physiological and behavioural changes that help to manage energy output until the stressor has been resolved (McEwen 1998; Sapolsky *et al.* 2000; McEwen and Wingfield 2003). Behavioural responses include changes in locomotor activity (that can lead to temporary movements out of the home range) and foraging activity, changes that may be accompanied by a general reduction or elimination of energetically expensive behaviours like parental care and territoriality (Wingfield *et al.* 1998). There is much individual and seasonal variability in response to aversive stimuli, and this variability has led to increased interest in the interactions between stress-related hormones and behaviour.

Adults may respond to potential stressors with different patterns of glucocorticoid secretion, depending, for example, on the stage of the reproductive cycle. In seasonally breeding birds, individuals living in the Arctic experience short breeding seasons and are frequently exposed to inclement weather, such as late-spring snowstorms that occur during nesting. Such species have been selected to suppress their adrenocortical response during the breeding season (Wingfield *et al.* 1995). That is, a stressor that induces a rapid and robust release of corticosterone in the non-breeding season may have little or no effect on corticosterone release during the breeding season (Wingfield *et al.* 1992). This reduces the behavioural changes induced by corticosterone (such as reductions in territoriality or parental care) that interrupt breeding activity, which would jeopardize the reproductive effort in a habitat where there may be little or no opportunity to nest again in the same season (see Figure 6.4 for an example).

There is geographic variability in strength of the adrenocortical response to perturbations (Silverin *et al.* 1997). At lower latitudes, where longer breeding seasons afford increased re-nesting opportunities and where selection to maintain a breeding effort in the face of adverse conditions is weak, birds exhibit a stronger adrenocortical response than those at higher latitudes. Recent work has demonstrated that latitudinal differences in adrenocortical responsiveness can be mediated through differences in steroid binding globulins and receptor densities rather than through differences in hormone secretion. Breuner *et al.* (2003) showed that levels of corticosteroid-binding globulin, the protein that transports glucocorticoids in the blood, vary latitudinally among populations of white-crowned sparrows (*Zonotrichia leucophrys*). Similarly, the number of glucocorticoid receptors on target tissues also varied among populations. Thus, individuals in these populations could vary in their responsiveness to stressors, even though their baseline and stress-induced plasma glucocorticoid levels were similar.

There also is phenotypic variation in adrenocortical activity within the same season at the same location. In species breeding in harsh habitats where only one sex provides parental care, members of that sex modulate their adrenal response to stressors while members of the other sex do not, presumably to avoid corticosterone-induced reduction in parental behaviour in the sex providing parental care (Wingfield *et al.* 1992).

Changes in the pattern of adrenal responsiveness also occur in birds during autumn migration. Corticosterone appears to play a role in regulating energy use to enable birds to fuel their long-distance movements (Jenni *et al.* 2000; Holberton and Dufty 2005). Autumn migrants often have higher baseline corticosterone levels during migration than during the non-migratory period, and corticosterone levels in these migrants may not increase further when the birds are subjected to handling stress (Holberton *et al.* 1996; Jenni *et al.* 2000; Long and Holberton 2004; but see Romero *et al.* 1997). This increase is related to body condition: birds with large fat stores have lower baseline corticosterone levels than those with small fat reserves (Jenni *et al.* 2000; Piersma *et al.* 2000; Long and Holberton 2004). Such elevated baseline levels in migrants may facilitate activation of behavioural and metabolic responses that help the birds to meet the costs of migration (Holberton 1999; Holberton *et al.* 1999; Piersma *et al.* 2000; Landys-Ciannelli *et al.* 2002). Dampening the adrenocortical stress response probably prevents the catabolic effects of sustained elevated corticosterone on skeletal muscle protein (Holberton *et al.* 1999; Jenni *et al.* 2000) at a time when such protein is needed for flight (Schwilch *et al.* 2002). These changes in endocrine responsiveness may also be characterized by changes in the types or numbers of hormone receptors at target tissues, both of which vary seasonally (Breuner and Orchinik 2000), although these relationships are poorly understood.

Although working with juveniles rather than adults, Heath and Dufty (1998) found that body condition also affects the adrenal stress response, with animals in poor condition maintaining elevated corticosterone levels for a longer period of time than animals in good condition (Figure 6.12). Similarly, adult behaviour is also affected by the interplay between adrenal secretions and body condition. For example, the American redstart (*Setophaga ruticilla*), a species that spends the winter in habitats that contain high densities of either males or females, shows no differences in adrenal responsiveness upon their arrival on the wintering grounds in autumn (Marra and Holberton 1998). In contrast, in spring, individuals that overwintered in habitats with a high density of males were in good body condition and showed a strong adrenal response to handling compared with individuals that spent the winter in habitats containing a high density of females, which exhibited reduced adrenoresponsiveness and were in poorer body condition. The latter individuals maintained an elevated baseline level of corticosterone that changed little in response to handling. These differences were due to variability in habitat quality and not sex-specific factors. The disparity in habitat-linked body condition resulted in differences in spring migratory behaviour and in arrival dates on the breeding grounds (Marra *et al.* 1998), which could have an influence on reproductive success and, more generally, on phenotypic quality. Studies such as these that combine endocrine measurements with other ecological measures, may prove useful in addressing issues of importance to conservation biology (Walker *et al.* 2005).

Furthermore, king (*Aptenodytes patagonica*) and emperor (*A. forsteri*) penguin males fast for weeks at a time while incubating their egg (Cherel *et al.* 1988). During this phase, lipid stores become depleted and protein begins to be catabolized. This metabolic shift is accompanied by an increase in corticosterone secretion which, in turn, is part of an internal signal that stimulates the resumption of feeding behaviour. This re-feeding signal prevents the animal from

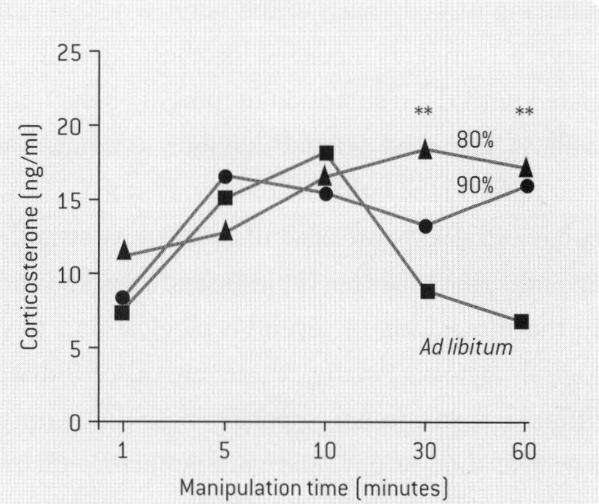


Figure 6.12 **Body condition and adrenal responses to a stressor**

Body condition affects the adrenal response to a stressor in American kestrels (*Falco sparverius*) juveniles in captivity. Individuals belonged to one of three treatment groups: individuals in the first group were fed *ad libitum*; those in the second were fed at 90% and those of the third group at 80% of their estimated needs. A standardized handling-stress protocol was applied for one hour. Blood samples were taken at regular intervals. Birds fed *ad libitum* adapted to the manipulation after 30 minutes, while underfed birds did not adapt to handling and still had elevated corticosterone levels at the end of the handling period. The patterns of corticosterone secretion differed significantly (two-factor ANOVA – feeding regimen and time – with repeated measures: $F_{8,98} = 3.03$, $P = 0.0045$, after logarithmic transformation of the data). The asterisks ** indicate that the means differ significantly at the level of $P < 0.001$.

After Heath and Dufty (1998).

reducing its body mass to the point where it cannot survive. It triggers increased locomotor activity associated with the resumption of foraging behaviour, as well as increased vocal behaviour associated with nest relief by the female (Robin *et al.* 1998). Thus, the behavioural response of adults to aversive stimuli has an endocrine basis that is plastic and varies according to the integration of several ecological and physiological factors. Finally, there are also individual differences in adrenal responsiveness that vary with age, sex, time of day, and genetic factors (Schwabl 1995; Romero and Remage-Healey 2000; Sims and Holberton 2000).

Variation in adrenal responsiveness is by no means restricted to birds (Moore *et al.* 1991; Dunlap and Wingfield 1995; Wendelaar Bonga 1997). For example, olive ridley sea turtles (*Lepidochelys olivacea*) frequently oviposit in group nests. Nesting females exhibit a low sensitivity to the presence of conspecifics, and this change in their typical behaviour is mediated by a temporary reduction in adrenal responsiveness (Valverde *et al.* 1999). Males and non-breeding females do not show this reduced adrenal responsiveness, suggesting that it is part of the mechanism underlying group-nesting behaviour. In non-human primates, social rank affects both the baseline level of corticosterone and the strength of the stress response. Subordinate males have chronically elevated levels of corticosterone compared with dominant individuals, and the adrenocortical response to stressors is also different (Abbott *et al.* 2003). Chronic corticosterone secretion, such as that often exhibited by the subordinates, has serious negative effects on an organism's well-being (Sapolsky 1992, 1996). However, there are distinct differences in adrenocortical responsiveness among males, and these are related to the social 'trajectory' of individual males (Virgin and Sapolsky 1997). That is, males moving upward in the social hierarchy and assuming more high-ranking positions have dominant-like adrenocortical responses, while those mired in low status positions maintain subordinate-like responses. Again, the interplay among hormones, behaviour, and environmental (social) conditions is striking.

## 6.6 Conclusions and future directions

### 6.6.1 Two major conclusions

#### 6.6.1.1 *The state of the phenotype is highly condition-dependent*

A recurring result in the study of phenotypic development (which we have considered here in its broadest sense, by including the adult phase) is that the state of the phenotype is highly condition-dependent: for a given genotype, the observed phenotype can vary greatly depending on conditions encountered by the individual. This means that the majority of morphological and behavioural traits are very probably strongly condition-dependent and thus reflect the current state of the individual. This characteristic has important consequences for evolution. Indeed, most models of communication, habitat choice, or mate choice show an evolutionary advantage for signals and strategies that take into account the phenotypic condition and the genotype of the individual (e.g., for sexual selection see Andersson (1994); one will see many examples throughout this book). If we generalize this conclusion that condition-dependent strategies have higher fitness, then we can expect condition-dependence to be common in nature, both in plants and animals. We saw in Chapter 4 how condition-dependence is involved in information gathering and use in decision-making and how this has profound implications in the role of information in evolution.

#### 6.6.1.2 *There is great structural stability but great functional plasticity among species*

Another important conclusion emerges: physiological mechanisms among vertebrate taxa show an impressive structural and functional consistency for many hormones. The most obvious example is that of sex steroid hormones, which are found to be unchanged as much in terms of structural as in terms of function in many vertebrates. Others, like protein hormones, modify their amino-acid sequences, but these alterations are much less important than are changes in their function. For example, the protein hormone prolactin has a role in stimulating, among other things, growth and secretion of seminal vesicles in fish, pre-breeding searching behaviour in amphibians, inhibition of gonadal growth in reptiles, development of the brood patch in birds, and production of milk in mammals! In the same way, insulin, another protein hormone, is found in many species, from bacteria to mammals (see Norman and Litwack 1987). Its function varies greatly, but its structure is largely unchanged. It seems that evolution modifies the consequences more than the mechanisms. Evolution is constrained by the availability of raw materials. That is, it is more likely that pre-existing structures, like hormones, are used to produce a new behaviour or a new physiological response than that an entirely new mechanism will be created to accomplish this new function. In this sense, comparative behavioural endocrinology can add much to our understanding of evolutionary processes and the taxonomic relationship among species and taxa.

### 6.6.2 What is the future for evolutionary physiology?

This chapter is limited to the study of endocrine processes of development in vertebrates. But it is highly likely that such complex and varied processes also occur in invertebrates. Support for this contention is the regular discovery of new hormones in invertebrates and new functions for hormones already known to exist. There is no *a priori* reason to believe that phenotypic developmental processes are simpler in invertebrates. The impression of a greater simplicity in these organisms is, in fact, based on our greater ignorance of the endocrinology of invertebrates.

More generally, it is clear that evolutionary physiology is currently in its infancy. There are only a few laboratories in the world that span the border between a pure physiology approach and a pure evolutionary approach. With the growth of the scientific community in this field, one can expect that our knowledge of the mechanisms underlying behaviour will greatly increase in the years to come. All in all, investigations of the hormonal bases of phenotypic development probably will become both narrower and broader, the apparent contradiction in that statement notwithstanding. Narrower, in that as techniques become more and more refined, we shall be able to focus on the effects on behaviour of events that occur at the molecular level. For example, we can now look at differences in numbers of receptor sub-types in animals that express one behaviour or another (Stamatakis *et al.* 2006). Furthermore, variation in receptor density also can be a powerful mechanism in regulating hormonal and behavioural responses (Shaw and Kennedy 2002), and warrants additional investigation. In addition, the ability to manipulate genetic information to produce chimeric animals that exhibit very specific behaviours, either by inhibiting or by adding behaviours to the natural patterns of the species, is a research tool that has great potential in behavioural studies (Balaban 2005). Furthermore, we are coming to appreciate more fully how neuronal stimulation can have subtle effects on hormone synthesis (Balthazart and Ball 1998). Indeed, the endocrine, nervous, and immune systems, once considered as separate entities, are now known to affect one another in intricate and significant ways (Ader 2000; Straub *et al.* 2001). Additional knowledge of their interplay will be necessary to advance our understanding of behavioural processes.

The field also will become progressively broader as the endocrine bases of additional behaviours are examined. For example, female songbirds often prefer to mate with males that have large song repertoires (Searcy 1984), and large repertoires may be characteristic of large males that have a high-quality phenotype (Doutrelant *et al.* 2000). Brain areas associated with song production undergo development early in a bird's life and may be sensitive to nutrition they are provided while in the nest (Nowicki *et al.* 2002). It thus appears that females may choose a male based on the size of his song repertoire, because it reflects body condition of the male during development. Given the association between hormones and song in birds and between pre- and perinatal events and phenotypic development under hormonal control, it is reasonable to ask what role(s) endocrine secretions might play in these phenomena? Another example concerns the plasticity of the hippocampus, an area of the brain associated with spatial memory. The size of the hippocampus increases over time in London taxi drivers, who memorize many street locations and of alternative routes to various destinations (Maguire *et al.* 2000). Clearly, environmental factors must be involved in this development (in this case, the continued navigation of the taxi drivers through the streets of London), but what endocrine factors are involved in such plasticity of the adult human brain? Furthermore, recent work demonstrates that environmental factors experienced by mothers can affect hippocampal development and spatial learning and memory in their offspring (Son *et al.* 2006).

Finally, although we have been discussing hormonal and, to a lesser extent, environmental factors that shape the behavioural development of phenotypes, we want to emphasize, as we did at the beginning of this chapter, the importance of taking into account the role of genes in the development of behaviour. Although the genotype determines the kinds of proteins an individual is capable of producing, the relationship between genes and behaviour is bi-directional. That is, not only do genes affect behaviour, but behaviour can affect gene expression. For example, genes that are normally inactive in 12-day-old rat pups can be turned on if the pups are deprived of maternal contact (Smith *et al.* 1997). Similarly, Honkaniemi *et al.* (1994) found that rats in an environment rich with social contact showed enhanced activity of certain genes compared with control rats. In birds, males attending to conspecific vocalizations during the song learning period, or those engaged in vocal motor development, increase their level of expression of a gene called *zenk*, which is associated with memory consolidation (Mello *et al.* 1992; Jin and Clayton 1997). Song learning and production both have strong hormonal involvement (Marler *et al.* 1988), although the precise relationships among the *zenk* gene, hormones, and vocal behaviour are still under investigation. One intriguing study showed that singing stimulates transcription of the *zenk* gene in zebra finches, but that translation of the mRNA into protein requires social interactions, indicating that sensory and motor processes must interact to produce *zenk* expression (Whitney and Johnson 2005). Such recently described interactions linking genes and song learning have already led to new theories regarding the maintenance of vocal learning behaviour in general (Lachlan and Slater 1999).

More and more investigators are seeking to understand the hormonal mechanisms underlying behaviour, and the kinds of investigations that can be pursued are limited only by our imaginations and by our knowledge of behaviour, once the necessary techniques become available. Knapp (2003, p. 664) put it well, noting that approaching questions of phenotypic plasticity

> '. . . from behavioral ecological and molecular mechanistic perspectives simultaneously will advance our understanding greatly, as an understanding of the ultimate and proximate mechanisms underlying such behavioral variation cannot arise from addressing each level in isolation.'

With a fundamental understanding of the behaviour in question, and with the careful design of controlled experiments, exploration of the endocrine changes associated with development of the phenotype can be both rewarding in and of itself and can clarify our understanding of the evolution of behaviour. We hope that this chapter will have convinced the reader that the close and inescapable links between physiology and behaviour bind the physiological dimension tightly to the study of the evolution of behaviour.

## » Further reading

> *The interested reader can find additional information in the following variety of articles and works:*

> *The basics of behavioural endocrinology can be found in Nelson's (2000) book.*

**Nelson, R.** 2000. *An Introduction to Behavioral Endocrinology*. Sinauer Associates, Sunderland, Massachusetts.

> *For additional information on how the organization of brain structures is affected by steroid hormones and their receptors, see Kawata (1995).*

**Kawata, M.** 1995. Roles of steroid hormones and their receptors in structural organization in the nervous system. *Neuroscience Research* 24: 1–46.

> *Emlen and Nijhout (2000) provide an excellent review of phenotypic plasticity in a taxon (insects) little discussed in this chapter.*

**Emlen, D.J. & Nijhout, H.F.** 2000. The development and evolution of exaggerated morphologies in insects. *Annual Review of Entomology* 45: 661–708.

> *The rich variety of sexual plasticity in teleost fish is discussed by Bass and Groberb (2001).*

**Bass, A.H. & Groberb, M.S.** 2001. Social and neural modulation of sexual plasticity in Teleost fish. *Brain, Behavior and Evolution* 57: 293–300.

> *Finally, Agrawal (2001) examines the role played by inter-specific interactions on phenotypic plasticity and the evolution of new species.*

**Agrawal, A.** 2001. Phenotypic plasticity in the interactions and evolution of species. *Science* 294: 321–326.

## » Questions

1. In Figure 6.3 do you think that the control group is absolutely necessary? Explain your answer.

2. In Section 6.1.6 we interpret the difference in behavioural reaction of two sparrow species to stress in terms of differences in their ecology. However, the two species certainly differ in many other aspects that could equally well explain the observed difference. We thus propose three questions to help readers think about this general problem. (1) Propose some alternative parameters that may explain the observed difference. (2) Try to find in this chapter other situations that could also lead to the conclusion that differences between species' reactions to stress can be explained by differences in ecology. (3) To what extent does this accumulation of facts all pointing in the same direction lead us to be more confident in our ecological interpretation of variation in species reaction to stress?

3. This chapter mainly deals with the relationship between hormones, the development of the phenotype and behaviour. Can you propose some topics that could also have been treated under a chapter on the ontogeny of behaviour?

three

Part Three

# Exploiting the Environment

Development and growth are obviously deeply impacted by an individual's capacity to find itself at the right place at the right moment for the right function in relation to environmental changes. There are two major issues along this line: foraging for food and finding a place to breed. Part three is about these questions.

Historically, foraging issues have constituted the core of behavioural ecology at the time of its emergence. It still is a major domain of this discipline. In contrast, and despite the fact that the fundamental model of ideal free distribution was developed in the context of breeding site choice, the study of breeding habitat selection appears to be only emerging from a long phase of studying animal distribution in order to infer the choice mechanisms that were assumed to produce such patterns. The future of this domain of behavioural ecology is likely to lie partly in the development of interdisciplinary interactions with environment-centred disciplines that study the dynamics of environmental systems in space and time.

Chapters 7 and 8 present the whole spectrum of issues concerning foraging, separated into two chapters to highlight the distinctive methods used to model social and non-social foraging.

Chapter 9 deals with issues of breeding habitat choices. The spatial and temporal scales of breeding habitat choices may differ so profoundly from those of foraging that the information that is likely to be used and the strategies that are likely to be selected may differ sharply from those involved in foraging. Furthermore, at least in animals, the evolution of breeding habitat choice strategies is directly linked to that of dispersal and gene flow among populations.

Chapter 10 thus deals with dispersal and its consequences in terms of population structuring, eventually setting the stage for speciation.

7

# Solitary Foraging Strategies

Luc-Alain Giraldeau

## 7.1 Introduction

An Eastern chipmunk (*Tamias striatus*), a small North American diurnal rodent, searches in the leaf litter for maple samaras, acorns, and beechnuts – packing them into its extendable cheek pouches. At some point, it stops searching, returns to its burrow and caches its load. A few seconds later it reappears at the surface and starts once again searching for seeds. This foraging behaviour raises several questions. For example, does the chipmunk collect any seed it finds, or is it selective? Does it choose where to search for seeds or does it search at random? Does it return to its burrow when its cheek pouches are completely full, or when the foraging location is depleted? All of these questions concern foraging behaviour, the theme of this chapter. We will deal primarily with the foraging of solitary individuals like the chipmunk, and leave social foraging for the next chapter. This distinction is necessary because the modelling techniques used for the two situations are relatively different. Solitary foraging relies primarily on **simple optimality** models while, as we will see in the next chapter, social foraging requires the use of **game theory** and the concept of **evolutionarily stable strategies** (see Chapter 3).

## 7.2 What are resources?

Survival and reproduction require the exploitation of a variety of elements, some of which, such as water, food, time and space, may be in short supply. When the exploitation of these elements leads to their exhaustion, we designate them as **resources**, distinguishing them from elements that, despite their usefulness, are inexhaustible, like the wind, air, or temperature. The notion of resources encompasses a wide variety of elements that contribute directly to the **fitness** of an individual. Begon *et al.* (1990) recognize three broad categories of resources for living beings: the elements that compose their bodies, the energy they require for their activity, and the spaces that are necessary for the completion of their life cycles. Behavioural ecologists recognize a fourth: sexual partners and their gametes. So for males, a female's unfertilized eggs are a resource whose exploitation (i.e. their fertilization) reduces their availability for other males. Similarly, for monogamous passerine females, the territories defended in the spring by males of several species are resources that decline in availability each time a female pairs with one of the available males.

By **foraging**, we mean all the activities related to the search for and exploitation of resources. For instance, the first optimal patch exploitation model was applied to the analysis of optimal copulation times in male dung flies (*Scatophaga stercoraria*; Parker 1978). Nonetheless, in this chapter and the next the word 'foraging' will be used uniquely to describe the exploitation of food resources. It is important to remember, however, that the models we will present can be applied with relatively few modifications to the exploitation of other resource types.

## 7.3 The modelling approach

Confronted with foraging behaviour, behavioural ecologists seek to understand the selective forces that have given rise to the precise forms they have today. For example, why does the chipmunk accept some types of beechnuts and not others? To answer such a question, given that food resources contribute directly to the fitness of an organism, we will assume that the foraging decisions that in the past have not contributed maximally to the **phenotypic fitness** of the animal were selected out and so are now absent from current populations. The behaviours that can be observed, then, are those that contributed maximally to the fitness of the animal's ancestors. In other words, we envision current foraging strategies as **adaptations** (see Chapter 1).

Because of this fundamental **assumption** the behavioural ecologist's task is to discover how the current behaviour can maximize the fitness of an animal displaying it. To do so, he or she will explore the relationship between different foraging strategies and their fitness consequences. Economic theories are quite useful in this type of analysis. These theories assume rational consumers that choose between available options in order to maximize **utility**. In economic theory, this notion of utility poses an important problem in that it can vary according to the history or conditions of individuals. In the case of evolution, this is not a problem because the utility of a behaviour can be defined as its effect on fitness. It becomes possible then to analyse the effect of a behaviour on the fitness of an animal using mathematical models borrowed from the economic sciences (see Chapter 2).

The traditional economic analysis of foraging behaviour partitions the foraging cycle into a series of **decisions**. The individual chooses where to search for **prey**. When it detects them, it chooses whether or not to attack them. If it captures one, it decides how much time to spend exploiting it before looking for another. Two of these decisions, the choice of prey to attack and the optimal exploitation time, have been the target of most theoretical as well as empirical advances (Stephens and Krebs 1986).

## 7.4 The prey model

The origin of the prey model (sometimes called the optimal diet or diet breadth model) lies in an ecological question about competitive exclusion. Two species of predators that exploit the same prey are said to be in competition, and generally one species excludes the other from the environment. Several factors can contribute to determining which of the species is more susceptible to being excluded. MacArthur and Pianka (1966) proposed that species that are more flexible in their foraging habits will be more likely to persist in a variable habitat. Therefore, species for whom foraging is particularly stereotyped or specialized (e.g. the koala that consumes only eucalyptus leaves, or the panda that eats only bamboo shoots) will be greatly affected by the reduction or disappearance of their prey, while others such as omnivores (e.g. seagulls or bears) can easily adjust to reduced availability of one prey type by exploiting another. MacArthur and Pianka were the first to model the question of prey selection to understand the ecological conditions under which specialist or generalist foragers would do best and hence exclude the other. By doing this they constructed a model that paved the way for an economic analysis of behaviour based on the **principle of optimality**. All optimality models have since been characterized by a specific logical structure which is depicted in Box 7.1. Let us first explore the current version of MacArthur and Pianka's prey model.

The objective of the prey model is to establish the prey selection strategy that maximizes a hypothetical **currency of fitness** and so ultimately fitness itself. While foraging, most predators encounter a sequence of prey of unequal value. For example, imagine a pair of insectivorous birds such as black-capped chickadees (*Poecile atricapillus*) that feed their brood with two species of caterpillars. One caterpillar species is rare but individuals are quite large, and so provide more energy than individuals of the other smaller but more common species. The prey selection model therefore calculates which of taking both types of prey as encountered or accepting only the larger one maximizes some hypothetical currency of fitness. We discuss in the next section why it is never optimal to accept only the smaller species.

### 7.4.1 Analysis of optimal prey choice

Our hypothesis for the prey model is that the currency of fitness is the long-term rate of energy intake. In this context, the ecological variables that are likely to affect the currency include the energy content of the prey as well as the time required to find, capture, and consume them.

The first element of our verbal analysis is the prey types' **profitability**. Profitability is given by the ratio of the prey's energetic content ($E$ in joules) and its **handling time** ($h$ in seconds) including the time required to capture and consume it. An important assumption that simplifies the analysis quite a bit is that while it is handling, a predator cannot search and hence detect, chase, or capture another prey item. The profitability of a prey item therefore represents the rate of energy acquisition realized while consuming the item.

The abundance of each prey species influences the length of the interval between successive prey encounters and so the second element of the analysis is the prey **encounter rate** ($\lambda$ expressed in prey items per unit time). An important assumption concerning prey encounters is that they always occur sequentially and are never simultaneous. Naturally, the prey problem is of interest when the most profitable prey item happens to be the less abundant prey of all such that the question for the animal concerns whether it should also accept some of the more abundant but less profitable prey types. Because it is the least abundant, search intervals between successive encounters with the large prey will be longer than intervals between successive encounters with smaller more abundant caterpillars.

If **natural selection** favoured birds that adopted selection policies that maximized fitness, then our job now is to determine which choice policies maximized long term energy intake?

## Box 7.1 **The structure of optimality models**

An optimality model consists of a decision, a currency, and a set of constraint assumptions. The decision explicitly specifies the choice to be analysed, the currency of fitness expresses the hypothetical consequences of a given course of action on the animal's fitness, and the constraint assumptions provide a framework within which the model applies.

The decision:

In the prey selection model is: **attack an encountered prey item, or ignore it and search for another**. The model analyses the value of each alternative in terms of a currency of fitness.

The currency:

It is the model's hypothesis. It represents the means through which the consequences of foraging decisions affect the animal's fitness. We assume that when confronted with a choice, animals opt for the alternative that maximizes fitness and hence that corresponds to the maximum value of the currency of fitness. In foraging models the net rate of energy intake is a commonly used currency of fitness as behaviours that increase the energy intake rate are hypothesized to also increase fitness (see Figure 7.1).

Constraint assumptions:

The constraint assumptions are a series of specific circumstances under which the model applies. There are typically two types of constraining assumption: those that are related to the characteristics of the forager, such as its ecology, anatomy and cognitive abilities; and those that result from the mathematical formalizations used to analyse the problem.

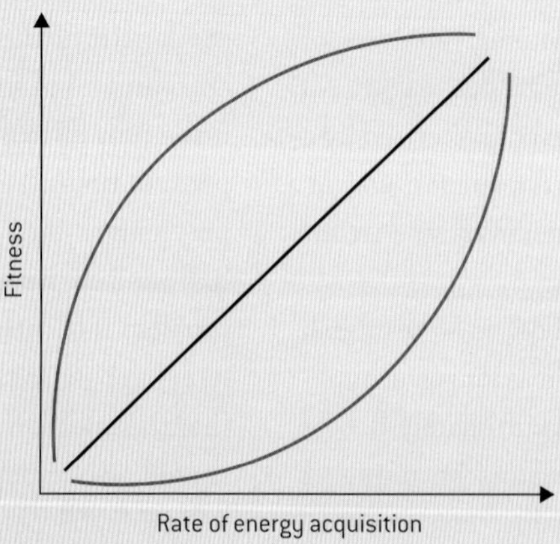


Figure 7.1 **Three hypothetical relationships of the currency of fitness (here the net rate of energy acquisition) and the animal's fitness**

The shape assumed in this relationship, linear, concave or convex, is part of the constraining assumptions of the model. The first models that we will consider assume a linear relationship. The risk models that we will see later assume a more complex, curved relationship between fitness and rate of energy acquisition.

*A priori* there are only three choice possibilities and only two of these are of any interest:

1. attack only the most profitable prey items;
2. attack all prey encountered;
3. attack only the least profitable prey items.

The first option (attack only the most profitable prey) maximizes the profitability of prey and so the rate of energy acquisition during consumption. Its disadvantage is that it imposes long search intervals between encounters with acceptable prey items. The second option (attack all prey encountered) has the advantage of reducing the search intervals between acceptable items but provides a lower average intake rate during consumption because it includes some high and many low profitability prey items. The third option (attack only the least profitable prey items) is irrational in the sense that it can never be advantageous to ignore a more profitable prey when it is encountered. We will not consider it any further. For the moment we can also ignore for the same reason situations where a given prey type is attacked sometimes but not always (called **partial preference**) because in the model's logic if it is profitable to attack that prey type, then it should always be attacked, whereas if it is not profitable to attack it, then it should always be ignored leading to the **all-or-nothing rule** characteristic of predictions of the prey model.

It becomes clear that the animal's optimal choice policy will represent a **trade-off** between maximizing the intake rate while eating and minimizing the search interval between prey. Although it is easy to understand such a verbal model, it also has the weaknesses inherent to all verbal models: its logic is implicit and difficult both to analyse and to criticize, and it makes only qualitative predictions that make it difficult to reject alternative hypotheses. If we wish to test the model more rigorously it is necessary to formalize it mathematically. This makes the logic explicit and thus open to scrutiny and criticism. Additionally, it generates precise quantitative predictions that can be more easily contrasted with those of alternative hypotheses. But, before starting this more formal analysis, it is useful to identify some of the constraint assumptions that characterize the classical analysis of prey choice as presented in Box 7.2.

We assume a predator in a world with two types of prey: prey types 1 and 2 and assume that prey type

Box 7.2 **The most commonly invoked constraint assumptions for the prey model**

**Related to the animal**

1. Searching for and handling prey are mutually exclusive activities.
2. Prey are encountered sequentially, never simultaneously.
3. The animal has all necessary information regarding the available prey, its encounter rate with each prey type, and their respective profitability.

**Related to the mathematical formalizations**

4. The mean rate of encounter with prey is constant and encounters occur at random.
5. The energy content and handling time is constant for all prey of a given type.
6. The encounter of a prey item not followed by an attack incurs no cost.
7. Each prey type is recognized instantaneously and with 100% accuracy.
8. Prey types are discontinuous categories and each prey of a given type is identical to all others of that type.

1 is more profitable than prey type 2. The rate of energy intake that characterizes a predator that systematically attacks any of the two prey types it encounters depends on the energetic content of those prey, their handling time, and the search time required to locate them. For two prey types energy acquisition $E$ during a search time of $T_s$ is given by the expression:

$$E = T_s\,(\lambda_1 E_1 + \lambda_2 E_2)$$

where $\lambda_i$ ($i = 1, 2$) is the encounter frequency with prey type 1 and 2, respectively and $E_i$ ($i = 1, 2$) is the amount of energy extracted from each. The total time $T$ that the predator spends foraging is the sum of the search intervals $T_s$ and the handling times:

$$T = T_s + T_s\,(\lambda_1 h_1 + \lambda_2 h_2)$$

The total handling time depends on the handling time $h_i$ ($i = 1, 2$) required for the consumption of each prey type and the total number of each prey type that is exploited ($T_s(\lambda_i)$). This number is a function of time spent searching and the encounter rate that characterizes each prey type. Naturally, the more abundant a prey type is, the more often it is encountered and if consumed, increases the total time spent handling this type of prey. The long-term rate of energy intake is therefore the ratio of these two equations:

$$\frac{E}{T} = \frac{T_s\,(\lambda_1 E_1 + \lambda_2 E_2)}{T_s + T_s\,(\lambda_1 h_1 + \lambda_2 h_2)}$$

This expression can be simplified,

$$\frac{E}{T} = \frac{\lambda_1 E_1 + \lambda_2 E_2}{1 + \lambda_1 h_1 + \lambda_2 h_2}$$

expressing therefore in our currency ($E/T$) the output obtained by the generalist strategy that consists of attacking all encountered prey. The only rational alternative to this rule is never to attack the less profitable prey. This more specialized strategy could be more profitable if the following condition is satisfied:

$$\frac{\lambda_1 E_1}{1 + \lambda_1 h_1} > \frac{\lambda_1 E_1 + \lambda_2 E_2}{1 + \lambda_1 h_1 + \lambda_2 h_2}$$

The first term of the inequality represents the rate attained when only the most profitable prey are attacked. The second term corresponds to the rate attained for a systematic attack of both prey types. This inequality can be simplified to obtain:

$$\frac{1}{\lambda_1} < \frac{E_1}{E_2}(h_1 - h_2)$$

When the inequality holds, the time required to encounter the more profitable prey is short and so the optimal strategy is to attack only the most profitable prey and always ignore prey type 2. When the equality does not hold, the interval between successive encounters with profitable prey is longer and so the optimal strategy is to generalize, that is, to attack both prey types whenever encountered. The algebraic simplification leads therefore to a rather counter-intuitive prediction. The diversity of prey attacked by a predator does not depend on the relative abundance of all the prey types. Rather the inclusion of a less profitable item depends **uniquely on the absolute abundance of the more profitable prey types**. Proof of this is given by the disappearance of $\lambda_2$ during the simplification above. Therefore, if the abundance of the most profitable prey is sufficient to warrant specialization on the more profitable prey, no matter how many carloads of less profitable prey are added to the habitat this choice strategy remains optimal.

### 7.4.2 Tests of the model: two classic examples

Before tackling the tests of the models, it is worth deciding beforehand how negative results are usually interpreted. For example, if after an experiment we find that the model's predictions were wrong, what should we conclude? It would be unwise to conclude that natural selection did not contribute to shaping the animal's foraging behaviour or that the animal is not optimal. Our economic approach assumes *a priori* that the animal *is* optimal because its behaviour was shaped over generations by natural selection. So tests of optimality models do not really address this assumption. What they test, instead, is the hypothesis concerning the currency of fitness. One possible conclusion, therefore, is that the currency is inappropriate. However, given that the model's predictions hold only in a world where all its constraint assumptions are true, it is more common to call into question first whether some of the constraint assumptions of the model were violated by the test situation. This is the approach we will adopt.

Two early experiments have become classic textbook examples of tests of the prey model. These examples are not recent, but they have the advantage of being simple and of demonstrating the experimental approach as it is applied to the prey model.

*Crabs and mussels*

Elner and Hughes (1978) were among the first to test the prey model. They performed an experiment to test whether crabs (*Carcinus maenas*) presented with mussels (*Mytilus edulis*) of varying sizes would preferentially consume only the most profitable ones. They noted the crabs' choice, taking care to replace each consumed mussel with another of the same size (so the encounter rate remains constant). To eat a mussel, the crab must first break its shell. The amount of flesh available and the handling time necessary to do this and eat the flesh depends on the size of the mussel and also on the size of the crab. However, due to nonlinear effects, profitability, the quantity of flesh extracted per unit time spent handling the mussel ($E/T_h$), is maximized for mussels of intermediate size. Elner and Hughes' observations demonstrated that crabs did indeed prefer mussels of intermediate size, that is to say that they made up a greater proportion of the crabs' diet than expected by their representation in the environment (Figure a).

In addition, over a period of three days, they recorded the choices made by crabs encountering habitats with different mussel densities, while maintaining the size ratio by replacing eaten mussels. Each habitat contained three categories (I, II, and III) of mussels according to their profitability that were available in given proportions. In the low-density environment, the crabs behaved as predicted by the model, they were unselective and attacked mussels of the three categories in proportion to their abundance in the environment (Figure 7.2b). In the intermediate environment, they also behaved as predicted by the model and ceased to attack the least profitable mussels (III). In the richest environment the crabs did not behave exactly as predicted by the prey model. While the largest mussels were clearly overrepresented in the crabs' diet, they nonetheless continued to eat the intermediate-sized mussels even though the model predicts these should always be ignored (Figure 7.2b).

Given these results, should we reject the prey model? No, because an examination of the conditions under which Elner and Hughes observed the choices of crabs shows that they were unlikely to meet some of the prey model's constraint assumptions. For example, the conditions of the test allow the simultaneous encounter of several prey, which is counter to the assumption of sequential encounters of prey (see Boxes 7.2 and 7.3). Additionally, it is difficult to measure the exact encounter rate with different-sized mussels. For example, was an uneaten mussel actually

Box 7.3 **Prey model's main predictions**

1. The inclusion of a prey in the diet does not depend on its own abundance but on that of the prey that immediately precedes it in the profitability hierarchy.
2. An increase in the absolute abundance of all prey in a habitat leads to a narrowing of the optimal diet because the most profitable prey will be more numerous.
3. Under the assumptions of the model there cannot be partial preferences for prey. A prey type is either always included or always excluded from the diet; this is the all-or-nothing rule.

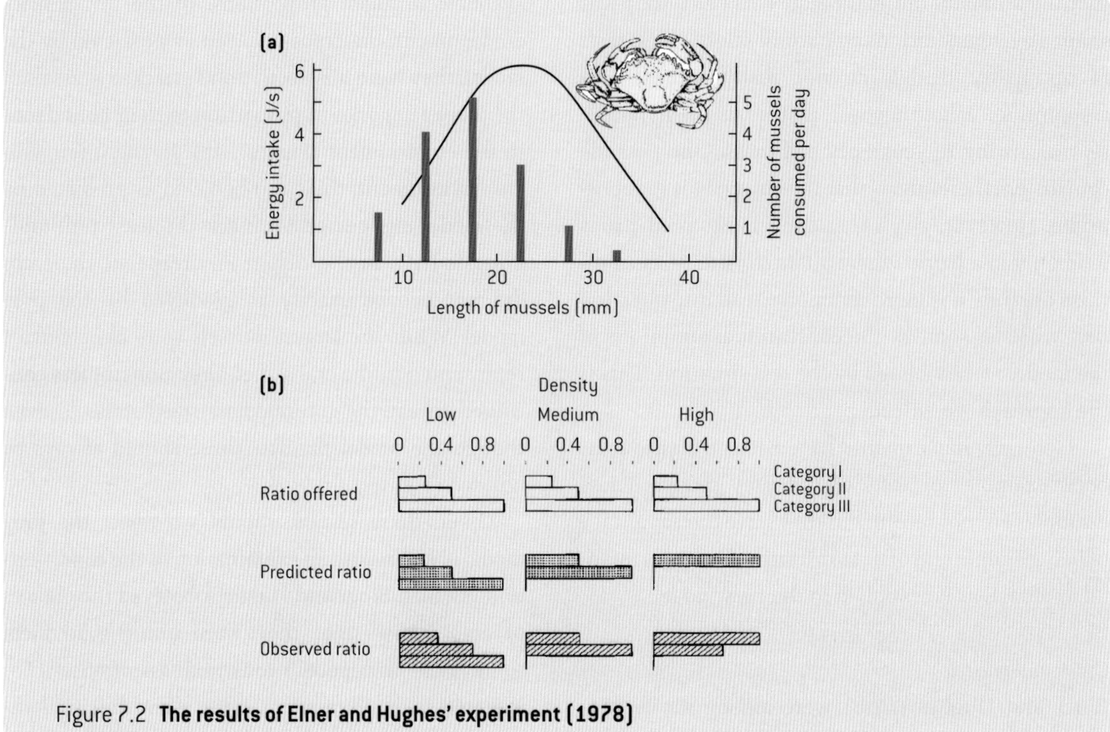

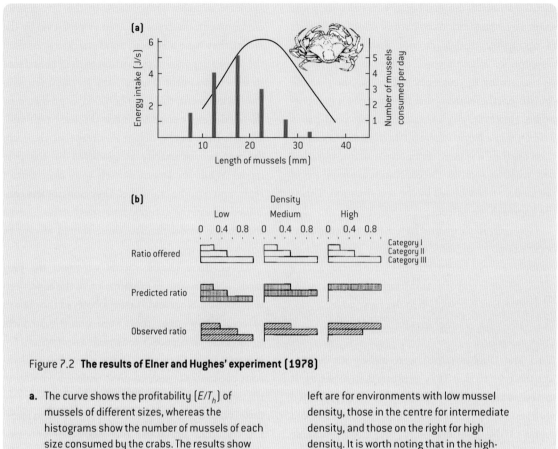

Figure 7.2 **The results of Elner and Hughes' experiment (1978)**

**a.** The curve shows the profitability ($E/T_h$) of mussels of different sizes, whereas the histograms show the number of mussels of each size consumed by the crabs. The results show that the crabs *Carcinus maenas* prefer to eat the most profitably sized mussels; that is, those that maximize the quantity of energy ($E$) per unit time spent in manipulation ($T_h$).

**b.** Histograms showing the proportions of mussels of high (category I, top bars), intermediate (category II, middle bars) and low (category III, bottom bars) profitability presented to the crabs, as well as the predicted and observed frequency that each size class was observed in the diet of the crabs. The histograms on the left are for environments with low mussel density, those in the centre for intermediate density, and those on the right for high density. It is worth noting that in the high-density environment, the model predicts a specialization on the most profitable mussels, but the crabs continue to include mussels of intermediate profitability. Elner and Hughes (1978) attributed this partial preference to a strategy whereby the crabs accept a prey of inferior profitability when they encounter a prey of this type for the second time in a row. According to the authors, it is a strategy that is adapted to a possible exhaustion of the supply of the most profitable prey type.

encountered by a crab that simply decided not to eat it? Elner and Hughes' test confirms that prey profitability is an important factor in the preferences of the animal. However, a test able to control precisely the encounter rate with each type of prey is needed to evaluate the ability of the model to predict prey selection strategies. Krebs *et al.* (1977) designed an experiment to do exactly this.

*Great tits and mealworms*

The strongest prediction of the prey selection model is that the abundance of the least profitable prey type has no effect on its inclusion in the diet (Box 7.3). To test this prediction, it is essential to be able to manipulate the predator's rate of encounter with its prey. This necessity poses a technical problem, particularly in the field, in that it is always difficult to determine with certainty the precise moment at which a predator encounters a prey.

John Krebs of Oxford University came up with a clever experimental device capable of manipulating the encounter rate of a predator with its prey. Inspired by the luggage conveyor-belt system in airports, he designed an apparatus that consisted of a cage and a conveyor belt (Figure 7.3). The predator (a great tit, *Parus major*) must stand at a window overlooking the conveyor belt when it wants to

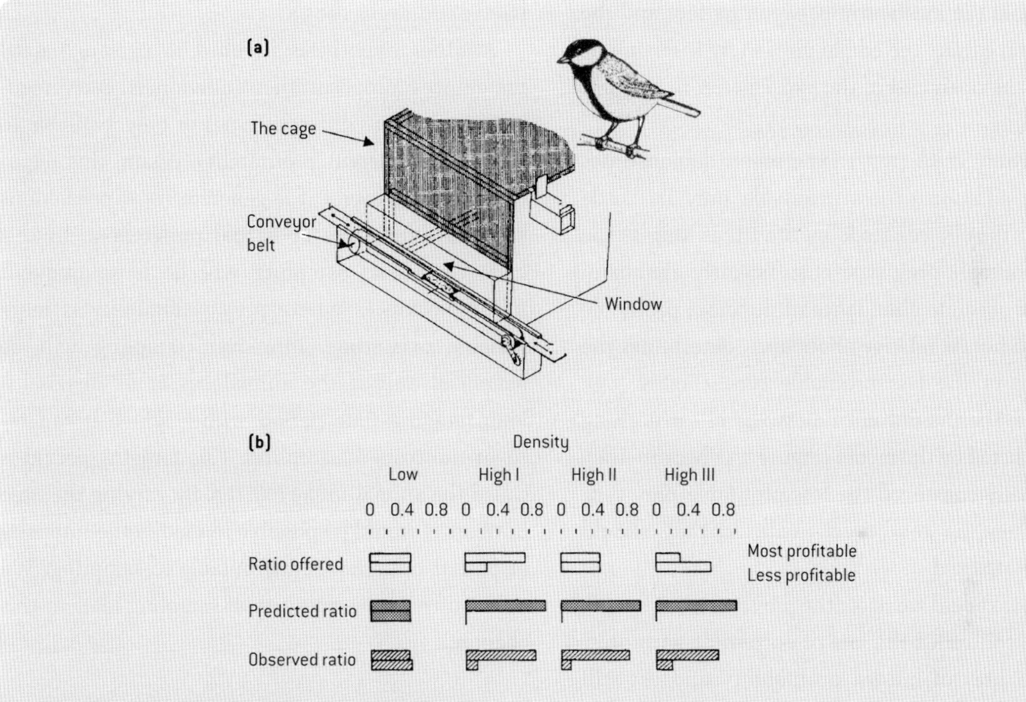


Figure 7.3 **The Krebs *et al.* experiment (1977)**

**a.** The top diagram (modified from Krebs and Davies 1987) shows a sketch of the experimental apparatus. The great tits (*Parus major*) are placed one at a time in a cage where they can, by standing at a window overlooking a conveyer belt, watch the passage of prey. An assistant sitting at the end of the conveyor-belt places the two prey types, four or eight segment pieces of mealworm, according to a predetermined order and time interval. The prey that are not attacked fall off the end of the conveyor belt. After attacking a prey, the bird must return to a perch at the back of the cage to eat it, by holding it under its foot. This apparatus meets two important constraint assumptions. During prey manipulation at the perch, the tit cannot search at the window, and the size of the window over the conveyor belt restricts the tit to seeing a single prey item at a time.

**b.** The lower panel illustrates the results of the experiment (taken from Krebs and Davies 1978). In the low-density environment (the histograms on the left), the tits show low selectivity, attacking the two prey types proportionally to their availability in the environment. The model predicts the exclusive selection of the most profitable prey in the three high-density environments. However, as is the case for the crabs, the birds continue to attack the less profitable prey, although at a low frequency. This inclusion of the less profitable prey indicates that they do not follow the all-or-nothing rule. What the results reject very clearly, however, is the alternative prediction that the birds should accept the prey according to their relative availability. When prey density is elevated, the tits continue to favour the most profitable prey, even if the less profitable prey are either just as abundant (High II) or even much more abundant (High III) than the most profitable prey.

eat. By controlling the speed of the conveyor belt and the spacing of prey items on it, the experimenter can manipulate the encounter rate between the predator and two types of prey of different profitability (four- or eight-segment pieces of mealworm). To consume the prey, the tit must then take it to a perch, hold the mealworm between its foot and the perch, and extract the edible portion by inserting its beak in the open end of the exoskeleton to pull out its contents.

The Krebs *et al.* (1977) experiment demonstrates that the addition of less profitable prey has little effect on the choices made by the tit, as long as the more profitable prey type is sufficiently abundant. Only the absolute abundance of the more profitable type, and not its relative abundance, determines the inclusion of the less profitable type (Figure 7.3). However, the tits did not follow the all-or-nothing rule predicted by the model (Figure 7.3) because they occasionally consumed the less profitable prey type even when the conditions predicted they should never be attacked.

So now should we reject the prey model? After all, we have controlled the rate of encounter with prey, provided prey of known profitability and yet still the model fails to predict all the details of great tit behaviour. Speculation abounds about the origin of the partial preferences observed in tits. For example, the currency may be wrong because energy is not the only useful element in food. However, this interpretation is not likely to apply in either study because both prey types had the same composition. Most research has focused on the constraint assumptions. For example, it is possible that animals cannot recognize prey instantaneously and perfectly such that partial preferences resulted from identification errors violating assumption 7 of Box 7.2. The error hypothesis is unsatisfactory from a purely epistemological point of view in that the error can always be invoked *a posteriori* as a fudge factor capable of explaining away any deviation from the predictions of the model. A more productive approach would be to formulate hypotheses regarding the causes of these errors. If partial preferences really do result from identification problems, there must be conditions that reduce the frequency of these errors, such as by accentuating the visual differences between prey types. Similarly, it is also possible to increase the resemblance of prey types and to predict that there is likely to be a critical resemblance after which it is no longer profitable to distinguish among them (Getty and Krebs 1985).

Another hypothesis invoked to explain partial preferences proposes that individuals must regularly sample the different types of prey available to assess their current profitability. Again, this hypothesis is only useful if it generates new predictions about the intensity of partial preferences. In this case, we expect sampling will be more frequent in circumstances where prey profitability is more variable over time. Moreover, sampling will be beneficial only if the animal has at its disposal a long foraging time period within which it can recover the costs of sampling. This foraging period is called the animal's **time horizon**. By varying the time horizon, it would be possible to modify the intensity of partial preferences attributable to sampling. We will return to the question of sampling later in this chapter.

## 7.5 The patch model

Generally, prey are distributed heterogeneously in the environment such that they are found aggregated in patches that are separated by zones that are more or less empty (Figure 7.4). When exploitation leads to the exhaustion of a patch, there arrives a time when it becomes profitable to abandon it in order to search for another. This is the case for instance when European starlings (*Sturnus vulgaris*) exploit invertebrates in a lawn; they must choose the moment when it would be preferable to leave the current lawn in search of a new one rather than continue searching on one that may be exhausted or nearly so. In all foraging theory, it is without a doubt this decision that has been the subject of the most models and experimental tests. The model presented here is its first version, also known as the **marginal value theorem** (Charnov 1976).

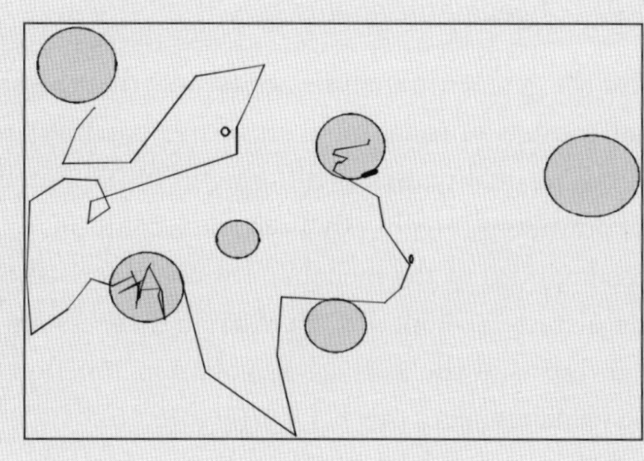

Figure 7.4 **Diagram of a hypothetical patchy habitat**

The prey are found only in patches (circles) and the spaces between the patches are empty. The line indicates the trajectory of a predator that is searching for patches at random. The diameter of the patches indicates their richness. This diagram represents the kind of simplified world that is modelled by the patch model.

### 7.5.1 The model

Like all optimality models, the patch model has three parts. The decision is to continue exploiting a patch or to abandon it to search for another. The currency is the same as for the prey model, the long-term net rate of energy intake. Finally, the constraint assumptions which once again fall into two groups.

*Constraints linked to the animal*

The model assumes that the predator has all the information necessary for a rational decision, it can recognize a patch instantaneously, it knows the average travel time between patches in the habitat, as well as the expected quality of those patches. The prey are distributed randomly within a patch and the predator cannot do better than to search for them at random.

*Constraints from the mathematical formalizations*

The model stipulates that the prey are concentrated in patches and that the rate of encounter between prey and predator in a patch declines as prey density declines. Hence, the longer a predator spends exploiting prey in a patch, the lower the density of prey in the patch becomes (Figure 7.5). The cumulative gains expressed as a function of the time spent exploiting the patch is given by a decelerated exponential function known as the **exploitation function**.

The model assumes that the time spent moving from one patch to another, the **travel time**, is exclusively a function of the distance between patches.

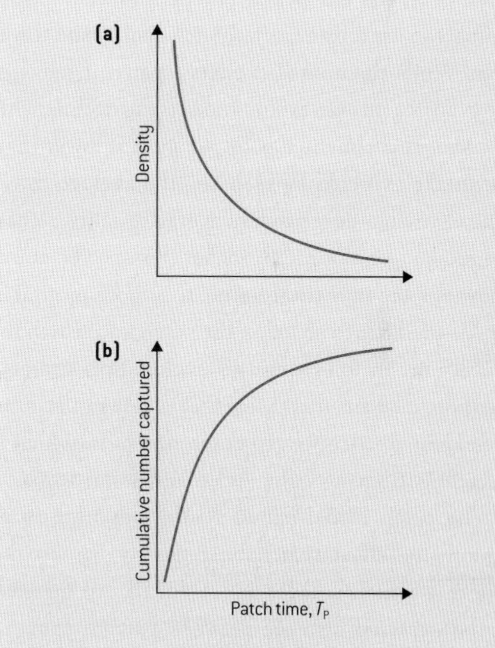


Figure 7.5 **The main constraint assumptions of the patch model**

**a.** The density of prey in a patch decreases exponentially as the predator exploits them because the prey are distributed haphazardly and the predator searches randomly.

**b.** The diagram shows the consequence of the above constraint: the cumulative number of prey captured by the predator increases rapidly at first and more slowly as the density of prey decreases. It is a decelerated exploitation function.

The predator's travel speed is therefore fixed and is part of the constraining assumptions and not the decision.

### 7.5.2 Analysis of the patch model

#### 7.5.2.1 *A verbal analysis*

The best patch residence time will represent a compromise between the immediate energetic gain resulting from the exploitation of a patch that is depleting and the expected gain by moving to the next patch, including the travel time necessary to reach it. In the absence of precise information, the animal can use only expected patch quality and travel time. While the animal exploits a patch it removes prey causing its own exploitation rate to decline. This rate should decline to a point where it is lower than the intake expected by spending the average travel time to reach a new patch of average quality. When the intake rate at a patch reaches this specific point, known as the **marginal value**, it is more profitable for the predator to abandon the patch and search for a new one. In following this exploitation strategy, the individual ensures a yield that is never inferior to the expected value for the environment in which it is feeding (Charnov 1976; Parker and Stuart 1976).

The patch model allows us the opportunity to explore the effect that habitat quality has on the extent to which patches are exploited. An increase in the expected distance between patches leads to an increase in travel time, reducing the intake rate anticipated by leaving a patch to find a new one. Consequently it lowers the marginal value of the habitat and delays the moment during patch exploitation when it becomes profitable to abandon the exploitation of a particular patch. The consequence is that patches will be more thoroughly exploited in habitats with longer travel times.

The effect of patch richness in a given habitat is not as easy to predict with such a verbal model because it depends on the precise shape of the exploitation curves. A quantitative approach using a geometric analysis will allow us to tackle this question.

#### 7.5.2.2 *A geometric analysis*

For the geometric analysis, we assume that the habitat consists of equal numbers of five patch types of differing quality; that is differing in initial numbers of prey items (Figure 7.6). Each patch conforms to the assumptions of the model and is characterized by a decelerated exploitation curve. From these curves, we can estimate the average exploitation function for the patch types of this habitat (Figure 7.6).

The patch model predicts that for each average travel time ($T_T$) characteristic of a habitat, there corresponds a patch time that maximizes the net rate of energy acquisition ($E/[T_P + T_T]$) for the average patch of that habitat. Figure 7.7 illustrates how to estimate this optimal patch time using the tangent method. This method is used in Figure 7.8 for the situation where expected travel times vary from one habitat to another. The geometric analysis confirms the verbal reasoning above: an increase in travel time leads to an increase in optimal patch time and hence an increase in the degree of patch exhaustion. The geometric analysis allows us to go even farther and predict that this increase in patch time is not linear but rather given by a decelerated curve (Figure 7.8).

The effect of the mean richness of patches on optimal patch time depends on the precise shape

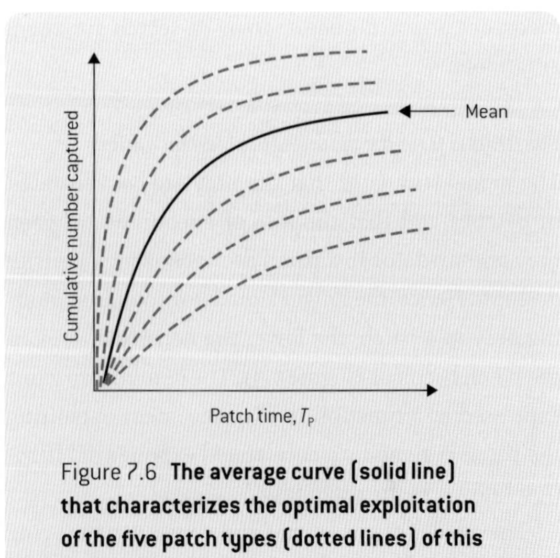


Figure 7.6 **The average curve (solid line) that characterizes the optimal exploitation of the five patch types (dotted lines) of this habitat**

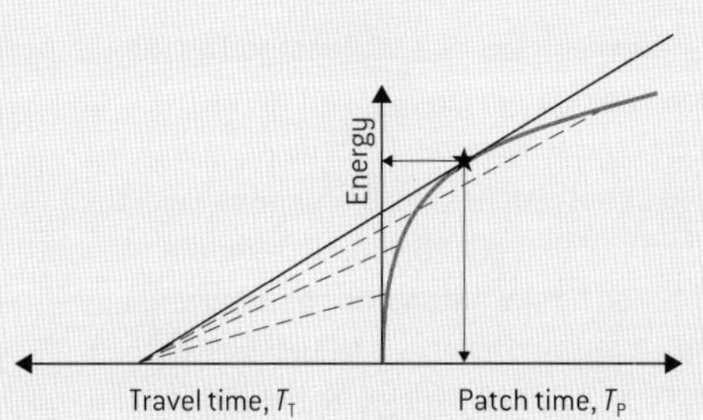


Figure 7.7 **Graphical illustration of the tangent method**

The *x*-axis to the right of the *y*-axis represents patch exploitation time, and to the left of the *y*-axis represents travel time increasing towards the left. The expected travel time ($T_T$) in this habitat is found at the intersection of the four lines. The curve in the right half of the diagram represents the expected value of the exploitation curve for this habitat. The decision that is being modelled is the optimal patch exploitation time of the patch for this habitat, i.e. the patch time that maximizes the slope of a line connecting the expected travel time and a point on the exploitation curve. The three dotted lines connect the travel time to three possible patch times but where the yield is not optimal. The line that is tangent to the exploitation curve is the optimal solution because its slope (energy per second) is the greatest and therefore corresponds to the maximal intake rate that can be obtained for the patches in this habitat. This rate characterizes the habitat and is called the 'marginal value'. The model predicts that an animal should abandon any patch in this habitat when its instantaneous exploitation rate drops to the marginal value of the habitat. If all patches in the habitat are identical, they will all be exploited for the same optimal amount of time $T_{P*}$.

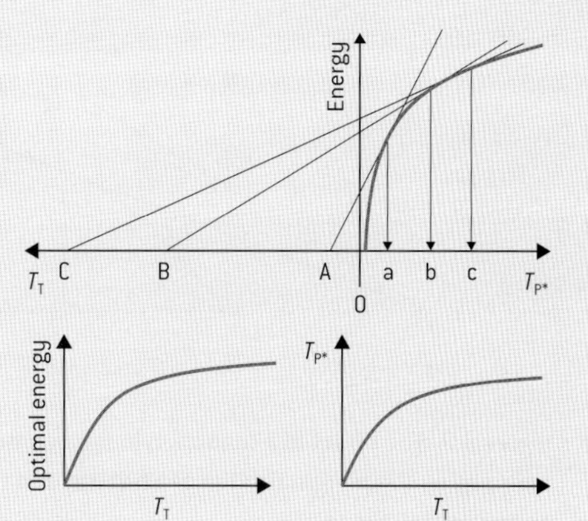


Figure 7.8 **The consequences of increased expected travel time on a habitat's marginal value**

At each expected travel time (A, B, and C) there is a corresponding and distinct tangent point that determines the optimal patch time for each patch (a, b, and c) and therefore also the different marginal values. As long as the expected travel time increases, the marginal value (here the slope of the strait lines) for that habitat decreases. The bottom graphics show that for habitats composed of identical patches, the quantity of energy extracted and the optimal patch time of a patch ($T_{P*}$) increases nonlinearly with the expected distance and hence travel times between patches.

of the exploitation functions. Therefore, a reduction in patch quality in a habitat can result either in an increase or a decrease in optimal patch time (Figure 7.9). To test the effect of patch richness, it is necessary, therefore, to measure precisely the exploitation functions of the tested animals. It is also necessary to measure these functions to test the predictions drawn from variations in travel time quantitatively.

### 7.5.3 Tests of the patch model

There are several tests of the patch model, and we present an example illustrating **central place foraging** (Orians and Pearson 1979), a variant of the patch model that applies to situations where prey are transported to a central location (such as a nest or burrow) rather than being consumed where they are captured. At the central location, the prey can

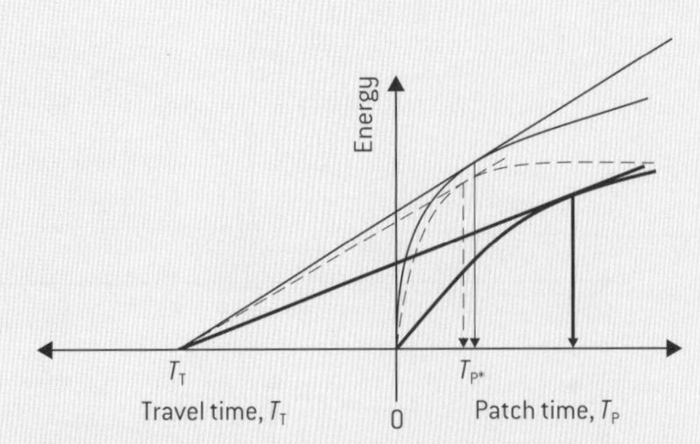


Figure 7.9 **The effect of a reduction in mean patch richness on the marginal value of a habitat and the optimal patch times**

The reference richness is represented by the highest exploitation curve (solid lines). In the first case, a reduction in modal richness of patches (dotted lines) reduces the marginal value of the habitat and consequently predicts shorter optimal patch times for the patches of the impoverished habitat. In the other case, a reduction in patch richness also reduces the marginal value of the habitat, but this time the shape of exploitation curve (bold lines) is such that the patches of the impoverished habitat are exploited for longer than those of the reference habitat.

either be eaten, fed to young, or stored for later use. Central place foraging is relevant to the transport of single prey items, such as an ant that carries a seed to its nest as well as multiple prey loads, such as a starling carrying several insect larvae to its brood (Kacelnik 1984). Here we will examine the case of multiple-prey loads.

For multiple-prey loads the exploitation function is called the loading function. The deceleration of the loading function could be caused either by the filling of the prey carrying apparatus, a bird's bill, a chipmunk's cheek pouches, and/or, as before, the depletion of prey in the patch. Whatever the cause of the deceleration, the consequences are the same. The travel time is now the time required for a return trip between the patch and the central location including any mandatory time spent unloading in the central place. When the richness of a patch is such that the animal makes several return trips to the same patch, the set of return trips corresponds to the set of patches of a habitat in the conventional patch exploitation model. In this case, comparing two habitats with different expected travel times is similar to comparing the return trips of an animal exploiting two patches placed at different distances from the central location and each requiring several return trips. The model predicts an increase in loading time, and therefore also of the size of the loads transported, with an increasing distance between the patch and the central location. This prediction has been tested several times, but we will give just one example here.

The eastern chipmunk, as was noted above, is a small, terrestrial, diurnal rodent of eastern North America (Figure 7.10a). A solitary forager, it spends a good part of its time in autumn filling its extendable cheek pouches with seeds that it transports to its burrow for storage. It will eat the seeds later in the winter during brief periods of activity interspersed among bouts of torpor that last four or five days, during which body temperature drops to nearly zero degrees Celsius.

By placing trays of sunflower seed at different distances from a chipmunk's burrow, it is possible to test the predictions of the central place foraging model quantitatively (Kramer and Nowell 1980; Giraldeau and Kramer 1982; Giraldeau *et al.* 1994; Lair *et al.* 1994). An observer can easily watch the chipmunk in the process of gathering its load and note the instant at which each seed is taken. From these observations, it is possible to reconstruct the loading functions for those seeds and verify that they are curvilinear and decelerated as the model assumes (Figure 7.10b). From these functions, the model predicts that the size of loads transported to the burrow will increase with the distance to the patch. This is exactly the result observed (Figure 7.10c). However, it is worth noting that despite the qualitative resemblance with the model (in the sense that the tendencies observed correspond well to the

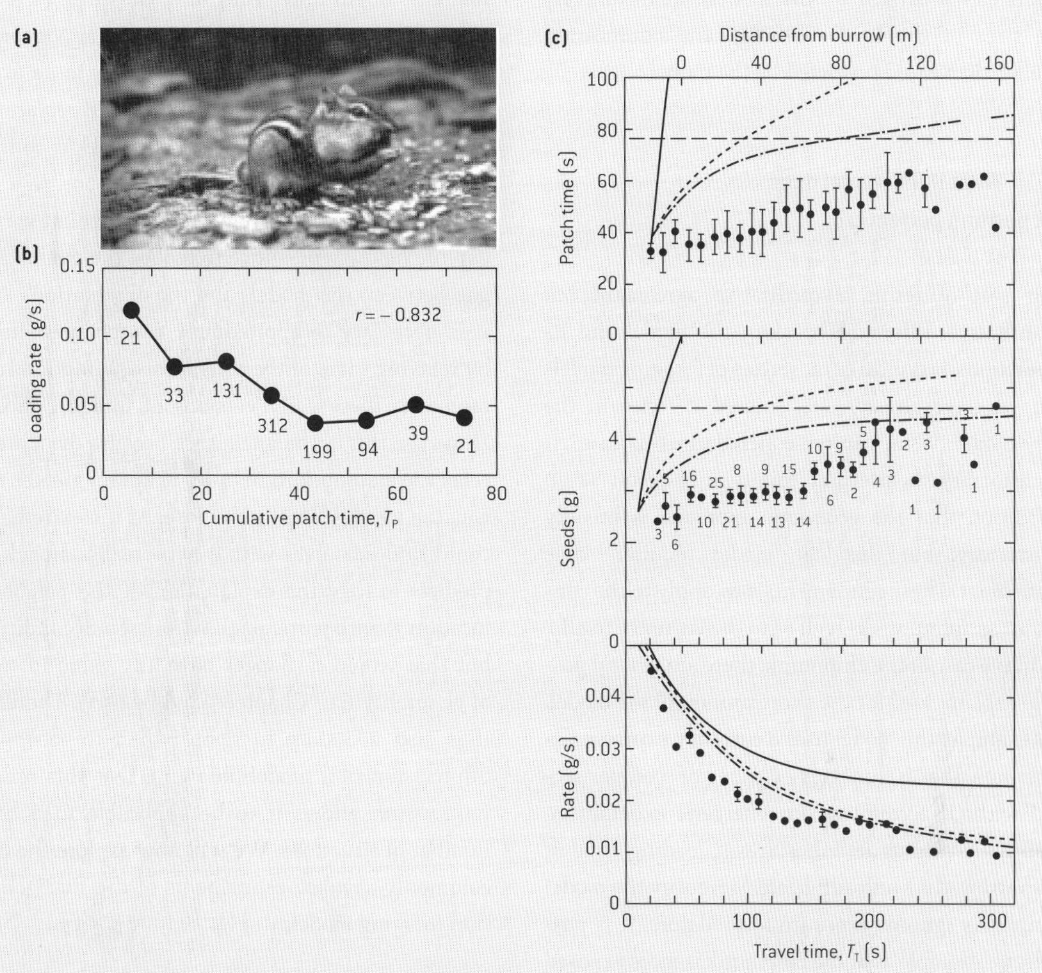


Figure 7.10 **Results from Giraldeau and Kramer (1982)**

- **a.** Photo of an Eastern chipmunk (*Tamias striatus*) filling its cheek pouches with sunflower seeds (Photograph courtesy of Dominique Proulx).
- **b.** The rate of sunflower seed loading measured in Eastern chipmunks interrupted at different moments during seed loading. The rate decreases as a function of the cumulative time spent loading in the patch ($T_P$), which shows that the constraint assumption of a decelerated exploitation curve holds in this situation.
- **c.** The results of the test of the foraging model from a central location with Eastern chipmunks. The top graph shows the patch times observed at trays of sunflower seeds placed at different distances from the burrow and therefore requiring different travel times. The middle graph shows the mass of seed loads collected at different travel times. The bottom graph represents the foraging rate (in grams per second) according to the travel time ($T_T$). The curve on each graph shows the predictions made from an estimate of the loading curve. It is worth noting that the patch times and the load sizes are all less than the model's predicted values.

After Giraldeau and Kramer (1982).

tendencies predicted by the model), quantitatively the sizes of the loads observed are significantly smaller than those predicted by the model.

### 7.5.4 What if the model does not perform perfectly?

This sort of result, a qualitative agreement but quantitative difference between the model and its predictions, is typical of most tests of foraging models and we have already encountered it twice in the prey model. What can we conclude in this case? It is customary to interpret this type of result as an indication that the economic approach is broadly satisfactory, but that the model requires some refinement. These refinements may apply to the constraint assumptions as well as to the currency itself.

In the case of the chipmunk, there are several possible explanations for the shortcomings of the model. For example, the model uses a currency expressed as gross rate, that is, in terms of grams of seed per second without accounting for energetic expenditure during the various activities: travel, loading, time in the burrow, time unloading, etc. Moreover, the model completely ignores exposure to predators. It is possible, for example, that the chipmunk is more exposed to predators during seed loading than during travel. This would mean that time spent in the patch is more costly than travel time, and this would be compatible with a reduction in time spent in loading seeds. It is also possible that the chipmunk is affected by the presence of competitors such that it prematurely leaves the patch when a competitor is present, probably to defend its seed caches in the burrow against possible intrusion (Ydenberg *et al.* 1986; Giraldeau *et al.* 1994). It is evident that no simple economic model can encompass all the biological complexity of the animals on which they are tested. Considering the simplicity of the models that we have presented so far, it is not very surprising to see that their quantitative predictions do not correspond exactly to the results from experiments involving animals.

It is worth noting, however, that despite their simplicity, these models have had an undeniable heuristic importance. They have caused us to reconsider the earlier view that animals are devoid of decisional processes, that they eat prey more or less at random during their encounters, and replace this view with one where it is permissible to consider animals as capable of subtle distinctions. We will see, however, whether the refinements that we will explore below will suffice to reduce the quantitative gap between the models and the observations. Now it is necessary to approximate reality more closely by relaxing some of the constraint assumptions contained in these early versions of foraging models. However, this is a balancing act, because the more we relax the assumptions in order to obtain greater realism, the more generality we lose. At the extreme, we could find ourselves with a hyperrealist model that predicts exactly the exploitation of the patches in location X with particular environmental conditions, but that would no longer have any value in terms of generality (see Chapter 3 and Box 3.1). Therefore, it is also necessary to know when to accept the imprecision of a model, knowing that this relative imprecision allows it to be applicable to a larger number of situations. We will now explore the most commonly invoked refinements of the first generation foraging models.

### 7.5.5 A refinement about information

The models presented up to now take for granted that individuals confronted with a choice make their decision based on several explicit parameters. For example, take the case of the chipmunk gathering sunflower seeds. The patch model assumes that the chipmunk's decision takes into account the expected travel time to the next patch. How does the chipmunk know this travel time? The model also assumes that the chipmunk knows the expected quality of patches in this habitat. But how did it get this information? Finally, the model also supposes that the chipmunk lacks certain information. For example, it assumes it is incapable of evaluating the quality of a patch during exploitation and that is why it cannot do better than to base its decision on the mean expected patch value and travel times in the habitat it is exploiting.

It is evident that animals will not always be in possession of all the information presumed by the models. In this case, they should either make more approximate decisions and therefore make errors, or as we will see below invest effort in acquiring the missing information by sampling the environment at a cost of lowering its foraging efficiency.

#### 7.5.5.1 *How do we know that an animal is sampling?*

The rules that govern the behaviour of an animal that must involve some effort in sampling are quite different from those of an animal that already has all the available information. A foraging efficiency that is below an optimal efficiency may therefore be the sign of a sampling behaviour, but it is not sufficient because several other factors, like errors for example, could also result in a reduction in foraging efficiency. To recognize sampling activity, we must be able to predict its expected characteristics; these can be generated *a priori* from an economic analysis. This is the approach used in the first test of a sampling model proposed by Krebs *et al.* (1977).

Krebs and colleagues offered to several great tits the choice between two patches with different but fixed foraging rates. Naturally, the tits did not know which of the two patches would provide the higher rate, and they would inevitably have to invest in sampling to find out. But what form would it take? Based on calculations that we will not go into here, Krebs and his colleagues predicted the form of an optimal sampling strategy in an experimental apparatus known as a **two-armed bandit**.

If we define a sampling behaviour as all alternations between patches, then we expect that the rate of alternation should be higher at the start of a session and become less frequent as information regarding the quality of the patches is obtained, leading to a phase of optimal exploitation of the patch judged to be more profitable. The length of the sampling period should depend partly on the total time available to the animal for foraging in the apparatus (i.e. its time horizon), and partly on the magnitude of the difference in yield between the patches. The effect of the time horizon is explained by the fact that sampling is only profitable if its initial cost can be recuperated during exploitation. In a long time horizon, the tit will have a longer patch time in which to absorb the cost of its sampling. By contrast, a short time horizon leaves little time for recovering the costs of sampling. An increase in the similarity of yields requires more samples and therefore more time to differentiate between patches.

In a two-armed bandit, the tits sample as predicted by the model. Alternation is more frequent at the start of a trial, it decreases with a reduction in the time horizon, and it increases with the similarity of the rates of yield (Krebs *et al.* 1977). Therefore, when they are placed in an unfamiliar situation, the birds are capable of sampling, and they follow certain well-established rules in doing so. In consequence, foraging models should take into account the necessity for an animal, in certain situations, to invest part of its time in sampling before entering the exploitation period. The shortcomings of the all-or-nothing rule, which we examined with the tests of the prey model, could well be due to such a sampling process.

#### 7.5.5.2 *The addition of stochasticity*

There are additional problems with the 'Charnovian' representation of the **marginal value** theorem: it rests on determinist expectations calculated from continuous exploitation curves. It is probably more realistic to expect that the patches encountered are of variable quality and that the prey consumed are more truly represented by step functions than by a continuous curve. For example, a patch may be empty, such as in a lawn where all the earthworms are so deeply buried that they cannot be reached by a starling's beak. In the conventional model, the prediction is that the animal leaves a patch when its acquisition rate falls below a critical level. However, if the starling does not encounter any worms, it is difficult to use this rule. Steve Lima (1984), in the USA, was the first to examine this exploitation problem of an environment containing empty patches and step exploitation functions.

Lima explored this problem in the field by observing the foraging of downy woodpeckers (*Picoides pubescens*). He drilled 24 holes in each of 60 logs (patches) that he suspended from trees in a forest. He placed a prey item in some of these holes, and then covered all of the holes with masking tape. The woodpeckers had to pierce the masking tape with their beaks to find out if there was any prey inside. He constructed three types of habitat that he presented successively to the woodpeckers, all of which contained two types of patches, totally empty or food-containing patches: poor and rich patches respectively. In all cases, the poor patches contained 24 empty holes, and it was the distribution of prey in the rich patches that differentiated the habitats. In the first habitat, all 24 holes of each rich patch contained a prey item. In the second, half of the holes of the rich patches contained a prey (12/24), and in the third, a quarter of the holes contained a prey (6/24).

In the first habitat, it is easy to distinguish a rich patch from a poor patch; it is sufficient to probe a single hole. An empty hole identifies a poor patch without error. In the second and third habitats, the task is more difficult because both types of patches have empty holes. Even though a hole containing a prey indicates with certainty that a patch is rich, a single empty hole does not allow the distinction of the quality of a patch. In fact, according to Lima's calculations, a woodpecker could maximize its rate of energy acquisition by leaving a patch after encountering three empty holes consecutively in the 12/24 habitat, and 6 in the 6/24 habitat. These numbers of holes maximize the probability of correctly recognizing the quality of a patch while economizing the number of empty holes explored.

The observations confirm the hypothesis that woodpeckers sample to distinguish the quality of patches they exploit (Figure 7.11). The birds had several days to experience the different patch types, and while presenting the three habitat types sequentially, Lima noted the number of empty holes a bird sampled before abandoning a log. In the habitat with patch types 24/24 versus 0/24, the woodpeckers sampled an average of 1.7 empty holes before abandoning a log, which is a bit more than the number predicted by the model (1.0). In the habitat with 12/24 versus 0/24 patch types, the number of empty holes tolerated before departure was between 4 and 5, whereas the model predicts 3.0. Finally, in the habitat with 6/24 versus 0/24 patch types, 6.3 empty holes were tolerated before departure, quite close to the 6.0 predicted by the model (Figure 7.11). These results demonstrate that an economic approach can also be used in a case where the environment is more uncertain than that of the first models established by Charnov. In the case of the downy woodpecker, the exploitation of a patch that is completely empty, a situation excluded by Charnov's classical approach, is predicted with relative success by an economic approach that is similar in the sense that it calculates the strategy that maximizes the energetic yield. The downy woodpeckers behave as though they are sampling the patches in order to determine with acceptable certainty whether they are empty or not.

The behaviour of the woodpeckers is quite extraordinary in the sense that it shows that these birds are capable of rapidly evaluating and memorizing the quality of the patches that they exploit. Moreover, each time that the conditions changed, they adapted nearly immediately. The phenomenon is even more remarkable in that it assumes that the same event can have, in different habitats, different informational content. In essence, the experience 'empty hole' is combined with prior information, 'in this habitat, there are two types of patches, partly full or completely empty', to allow the animal to decide if the patch that is being exploited is rich or poor. In more technical language, we say that the animal combined **current information** (this hole is empty) with **prior information** (the patches are either completely empty or partially full) in order to obtain an estimation of the quality of the patch, i.e. the animal may be capable of **Bayesian** estimation (the combination of two types of information, prior and current, to obtain an estimation of the value of the habitat as it exploits the environment; Giraldeau 1997). In other words, animals seem to be constantly sampling their environment and thus acquiring information about it. We have dealt with the question of information in Chapter 4.

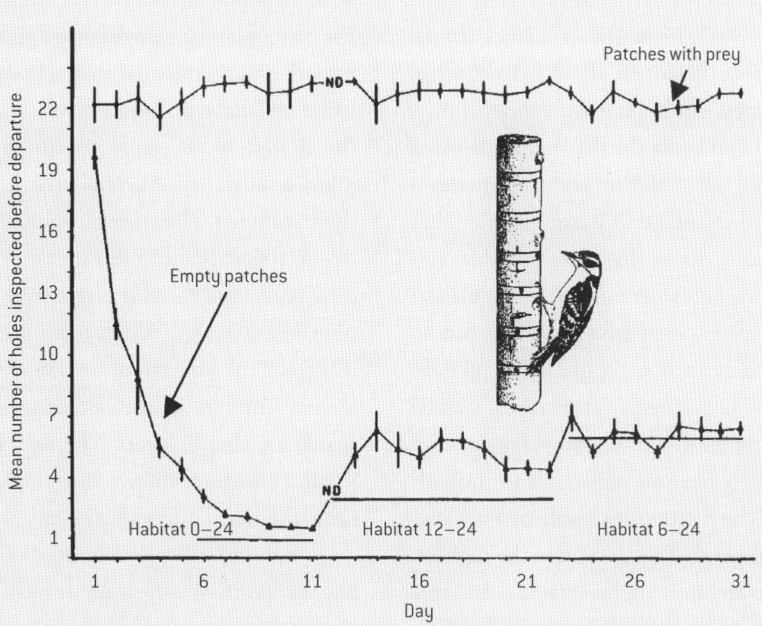


Figure 7.11 **The number of holes probed by downy woodpeckers (*Picoides pubescens*) during the three phases of Lima's study (1984)**

From Lima (1984), Downy woodpecker foraging behaviour: efficient sampling in simple stochastic environments, *Ecology*, 65: 166–174. © Ecological Society of America.

The woodpeckers probe more empty holes before abandoning a log in which they found a prey (top line) than a log where they did not (bottom line). The number of holes probed without success before abandoning a log depends on the experience of the animal. From day 1 to 31 the woodpeckers were exposed to three types of habitats. All the habitats had logs that were completely empty, but the logs that contained prey could have 0, 12, or 18 empty holes. The horizontal straight lines indicate the exposure time in days in each habitat type, as well as the optimal number of holes to probe without success before abandoning a log. We see that the woodpeckers quickly adjust their tolerance to failure when the number of empty holes in a log containing prey changes. At the beginning of the experiment, the woodpeckers tolerated long sequences of failure before abandoning. After 6 days, they had learned that a log with some empty holes was surely completely empty and they abandoned it after only a few fruitless probes. Towards the end of the experiment, the woodpeckers seemed to have learned that the logs with prey also had several empty holes. They therefore developed a tolerance for longer sequences of fruitless probing, as predicted by the model.

### 7.5.6 Refining the currency of fitness: the effect of risk

Until now, we have assumed that the fitness of an animal increases as a function of its average rate of foraging: the higher this rate is, the more phenotypic fitness increases. In the early 1980's, the American Thomas Caraco thought that this view may be incomplete. One problem is that the view assumes a linear relationship between fitness and an animal's energy intake rate. Another is that variation in the intake rate associated with a given foraging choice is assumed implicitly to have no effect on fitness. It is possible that fitness increases nonlinearly with energy intake rate (in contrast to what is represented in Figure 7.1). For example, an animal that is on the verge of starving will likely receive very little increase in fitness with increases in energy intake rate that remain insufficient to allow it to survive. In contrast, its fitness could increase very abruptly, passing from zero (probable death) to a high value (probable survival) when the intake rate crosses a critical requirement threshold. It is also possible that an additional increase above this threshold does not always result in an equivalent increase in fitness, since the animal, having assured its survival, can do no better, particularly in non-reproductive periods. Given such curvilinear currency functions, Caraco proposed that variance (or uncertainty) in energy intake rate will likely have an impact on an individual's fitness. Take the fictitious example of a bat that must consume $R$ insects before the end of the night in order to survive to the next night. Imagine that this animal has the choice between two locations that offer the same expected foraging rate of $R/2$. The first, near a stream, invariably provides $R/2$ insects. The bat that forages there will certainly die because it will not find the $R$ insects necessary to survive until the next night. The alternative, a small wood, provides it with an equal probability of either $R$ or 0 insects. Even though the mean expected rate ($R/2$) is the same in both patches, the bat that chooses to forage in the wood will have one out of two chances of obtaining the $R$ insects necessary to survive until the next night, a much better probability than that offered by the stream (0). The animal should therefore be sensitive to this variability in rates and, in this particular case, prefer the variable alternative to the one that is not (Figure 7.12). Evidently, this is an extreme case to illustrate the impact of resource variability on an animal's fitness. In the longer term, the bat should search for a better location because even in the wood, it runs a serious chance of dying. But in the short term, the wood provides the best future prospect.

Variability of provisioning rate can therefore have a positive effect on fitness, but this is not always the case. Imagine the same scenario with the bat, but this time the animal only needs $R/2$ insects to survive. The invariable site near the stream now ensures survival, but the wood, despite having the same average provisioning rate, entails a risk of death of 50%. The variability introduced is now a danger

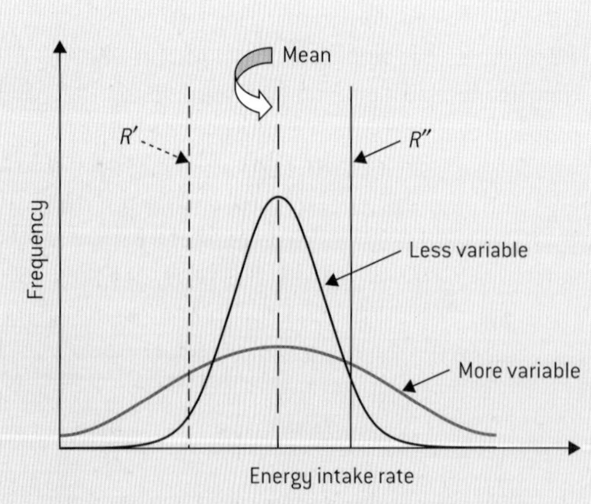


Figure 7.12 **The frequency distributions of rates of energy acquisition offered by two alternatives**

The two alternatives supply the same expected energy intake rates. However, one distribution has rates that are more variable than the other, although the areas under the curves are equal. $R$ represents the minimum rate of energy acquisition necessary to insure survival, and it can be either low $R'$ or high $R'$. For a given value of $R$, the animal must choose the distribution that presents the smallest area in the region that threatens the animal's survival. In the case of $R'$, where the threshold is inferior to the expected value, the less variable alternative presents the smallest area below the threshold. The animal that opts for the less variable alternative is risk-averse and therefore maximizes its probability of survival. In the case of $R''$ where the threshold is greater than the expected value, the inverse is true. In this case it is the more variable alternative that presents the smallest area below the threshold. The animal that chooses in this case the more variable alternative is risk-prone and maximizes its chances of survival. These two cases illustrate the **energy budget rule**.

and therefore an unfavourable element for the fitness of the animal (Figure 7.12). Caraco remarked that variability can affect an animal's phenotypic fitness differently depending on the ratio of the required and the current energy intake rates. This type of modelling approach is known as **risk-sensitive foraging theory** where the term 'risk' is borrowed from microeconomic theory where it is used to mean uncertainty or variability. It would be wrong in this context to use risk to mean danger.

> The essence of the predictions of risk-sensitive foraging theory can be summarized by the energy budget rule that states that: when the energy intake rate $R$ required to avoid some fitness cost exceeds the expected current rate, it is advantageous to prefer the more variable alternative; the animal is **risk-prone**. On the other hand, when the required rate $R$ is equal to or less than the expected current rate, it is advantageous to prefer the less variable alternative; the animal is **risk-averse**.

### 7.5.7 A study of risk-sensitivity in the shrew

An example of a test of this prediction of risk-sensitive foraging in the common shrew (*Sorex araneus*) comes from the University of Nottingham in the United Kingdom. Chris Barnard and C.A.J. Brown (1985) presented seven captive shrews with a choice between two sources of mealworm pieces, both providing the same overall foraging rate but one with a variable and the other a fixed rate. First, they calibrated the number of mealworm pieces required by each individual to maintain their body weight, and therefore an equilibrium energy budget ($R$). They exposed each individual to conditions where the patches provided expected energy acquisition rates that were sometimes less than and sometimes more than $R$. The relationship between the proportions of choices directed towards the more variable alternative depended on the average rate offered by the two sources (Figure 7.13). The energy budget rule explained the behaviour of the shrews: they showed risk-aversion when their experienced rate exceeded $R$, but were much more inclined towards risk when the experienced rate was less than $R$ (Figure 7.13).

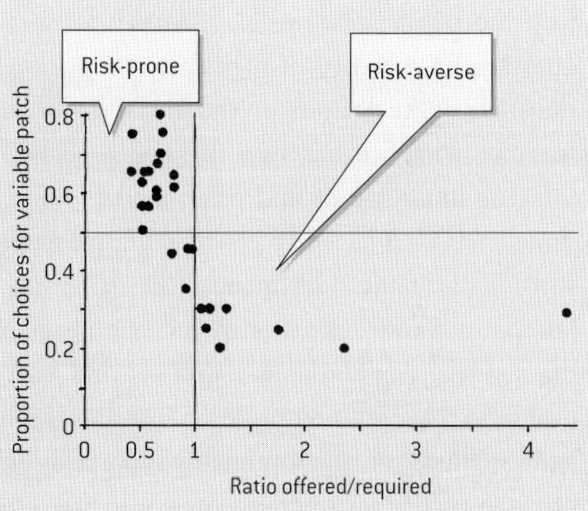


Figure 7.13 **The results of Barnard and Brown's study (1985)**

The proportion of visits to the variable patch observed for seven common shrews (*Sorex araneus*) confronted with a choice between two patches offering the same average energy intake rate as a function of the ratio between the energy intake rate and the rate required to maintain a positive energy balance. The vertical line separates the graph into two zones: to the left the rate obtained in the patches is inferior to the rate required to maintain a positive energy balance while to the right the inverse is true. The horizontal line corresponds to indifference: the points above this line indicate a preference for the variable alternative, and those below a preference for the invariable alternative. The model of risk sensitivity predicts that the shrews should be risk-prone when they are in the left hand zone, but be risk-averse in the right-hand zone. The points in the left zone are higher than those in the right, which indicates that the shrews preferred the more variable alternative when their intake rate was inferior to the required rate. The results agree with the energy budget rule.

Modified from Barnard and Brown (1985).

The shrews demonstrated that they are risk-sensitive. This ability was also shown in black-capped chickadees (*P. atricapillus*), in yellow-eyed juncos (*Junco phaenotus*) and in bumblebees (*Bombus* ssp.) (see Real and Caraco 1986).

### 7.5.8 From solitary foraging . . .

The first generation of foraging models was deterministic and assumed that the world was continuous rather than discrete (Sections 7.4–7.5.4). Each foraging strategy corresponded to a single fixed value of currency of fitness. Lima added uncertainty where some patches had to be sampled to be sure they were empty (Section 7.5.5). Then Caraco showed that animals were not insensitive to the stochasticity in rewards associated with any choice that affect their fitness (Section 7.5.6). To increase the realism of foraging models, the deterministic fitness functions of first generation models must be replaced by appropriate stochastic reward distributions. However, the formal presentation of such stochastic models is beyond the scope of this book (see Giraldeau and Caraco 2000). Suffice to say that overall, stochastic models offer the advantage of integrating the effects of uncertainty on foraging decisions and because ultimately the currency of fitness is given by a probability of survival, it becomes possible to mix in the effect of other factors such as the danger of predation associated with a given foraging alternative.

### 7.5.9 . . . to social foraging

We have until now confined ourselves to situations of foraging solitarily and so have not dealt with competitive foraging interactions. The presence of competitors, however, changes many things. Consider, for example, the prey model where it was possible to specify for a solitary forager its encounter rate with prey, the profitability of the prey types, and from these calculate the attack strategy that maximizes the currency of fitness. However, if the predator has a competitor, then the encounter rate with prey may well depend on the competitor's attack strategy. If one predator specializes on attacking only the most profitable prey, for example, the rate of encounter with these prey will likely be reduced for all other competitors. The strategy of one of the competitors thereby influences the yield of the strategies available to the others. These situations are characterized by the **frequency-dependence** of payoffs and cannot be analysed by the optimality methods that we have used so far; they require a different economic approach, called **game theory**. We will come to these problems in the next chapter.

## 7.6 Conclusion

By exploring the logic of optimality models, we learned that each model is made of three parts: decision, currency, and constraint assumptions. More importantly, we learned that the objective of the optimality approach is not so much to prove that an animal's behaviour is optimal but rather to test a hypothesis about the survival value of foraging behaviour. So, when constraint assumptions are correct, the failure of a model's predictions calls into question its currency of fitness, an indication that the effect of the decision on the fitness of the animal was not what we had hypothesized it to be. Two foraging models, prey and patch models, were presented and each have had good qualitative empirical success but much less quantitative support. This last point forces scientists to re-examine many of the models' constraint assumptions and currency of fitness. This revision process illustrates the heuristic richness of the optimality approach which has allowed behavioural ecologists to move away from the view of stereotyped animals deprived of much cognitive capacity towards one where they show plasticity and the capacity of integrating sometimes rather subtle information to make adaptive decisions.

## » Further reading

> *The following books and articles provide a general review on the topic of solitary foraging:*

**Begon, M., Harper, J.L. & Townsend, C.R.** 1990. *Ecology: Individuals, Populations and Communities*, 2nd edn. Blackwell Scientific Publications, Boston, MA.

**Charnov, E.** 1976. Optimal foraging, the marginal value theorem. *Theoretical Population Biology*, 9: 129–136.

**MacArthur, R.H. & Pianka, E.R.** 1966. On optimal use of a patch environment. *American Naturalist* 100: 603–609.

**Stephens, D.W., Brown, J.S. & Ydenberg, R.C.** 2007. *Foraging Behavior and Ecology*. The University of Chicago Press, Chicago.

**Stephens, D.W. & Krebs, J.R.** 1986. *Foraging Theory*. Princeton University Press.

## » Questions

1. In the prey model, try to explain why the abundance of a given prey item has nothing to do with its inclusion in the optimal diet. One way to do this would be to redo by yourself the algebra presented in the prey model section.

2. Can you imagine what would happen to the rate of exploitation of a patch if the prey, instead of being distributed randomly, were spread in a regular and predictable fashion, allowing the predator to search for them systematically such that exploitation functions would become linear up to a horizontal plateau? Hint: you will need to use the tangent method to solve this one.

3. You have no doubt noticed that the currency used in the models presented was essentially the gross energy intake rate. Try to think out and draw what the exploitation functions presented in the patch models would really look like if the currency of fitness was the net energy intake rate taking into account the energetic cost of foraging.

8

# 8

# Social Foraging

Luc-Alain Giraldeau

## 8.1 Introduction

Nine starlings are foraging on a lawn. You watch as they probe their beaks into the humid earth, here and there, sometimes pulling out a worm or an insect larva, which they immediately swallow. You notice that they often raise their heads and seem to be paying attention to the others. Every now and then one of the starlings rushes over to a companion that has just unearthed something to eat, and probes the surrounding area, or even tries to snatch away the food. Only moments ago, all of the birds were on the alert to some invisible danger and were about to take flight, then they continued searching for food. After a few moments, during which time none of the birds seems to find anything to eat, they all fly over to a different lawn, landing just a few metres away from four other foraging starlings.

This scene is not particularly unique. We have all seen it dozens of times in different species. However, it effectively illustrates the decisions that characterize social foraging. First, we might ask why there are nine starlings on the lawn rather than four, or fifteen. We might wonder if the starlings that raised their heads were actually paying attention to their companions, and if so, why? Were they using their companions as lookouts so as to be forewarned of a potential danger? Were they assessing the success of their companions so as to evaluate the quality of the resource patch more efficiently? When they decided to leave the patch in search of another, were they taking the others into consideration? All of these questions are related to a burgeoning field of study in behavioural ecology: social **foraging**. Contrary to solitary foraging, described in the preceding chapter, this field is relatively new and consequently its works are fewer and more recent. The subject of social foraging developed in the present chapter is closely related to that of the evolution of group-living, which is described in Chapter 14, and to the question of information, which is developed in Chapter 4.

*A distinctive approach: evolutionary game theory*

In the preceding chapter, we were able to calculate the payoff of a foraging strategy without taking into account the other members of the population. For example, based on the **profitability** of an assemblage of prey items and the encounter rate with each of them, we were able to compare the payoff of specialist and generalist strategies, and then decide which was most profitable. Social foraging theory, on the other hand, does not allow for this type of calculation because the presence of competitors and the strategies that they employ modify the payoff of any given strategy. For example, to calculate the payoff of a strategy that involves exploiting the discoveries of other group members, we must first know how many individuals in the group use this same strategy. If few individuals choose to exploit the discoveries of others, then its payoff will be high; whereas, if all members of the group choose this strategy, then its payoff will be very low.

This interdependence of the strategies' payoffs is precisely the type of situation that calls for the use of evolutionary **game theory**. In classic game theory (Davis 1970), economists are able to specify the criteria with which the strategies are analysed. For instance, the winning strategy could be that which maximizes the losses of the adversary, or that which maximizes the gains of both players, etc. In evolutionary game theory, on the other hand, we do not have the luxury of choosing the criteria for selecting a strategy (Maynard Smith 1982; Sigmund 1993). According to **natural selection**, the winning strategy and thus that which is expected, is the strategy that, if adopted by an entire population, cannot be invaded by any plausible alternative. That is, any invader would have lower **fitness** than individuals playing the resident strategy. This strategy is stable from an evolutionary stand point; in other words, it is an **evolutionarily stable strategy** (ESS), as termed by Maynard Smith (1982; see Chapter 3).

The passage from optimization to evolutionary game theory is not achieved via a simple process of modification. We must view the consequences of evolution in a new light. The link between optimization and foraging is replaced by that between game theory and ESS. We must keep in mind that, contrary to an optimal solution that exists because it maximizes payoffs while taking into account precise constraints, an ESS exists because no alternative strategy can do better. This change in the principle used to seek out the solution favoured by selection introduces an important nuance to the concept of optimization, resulting in multiple potential consequences. For instance, natural selection often leads to less advantageous states for the population, simply because these states are the most stable from an evolutionary stand point when compared to more advantageous, but unstable, states. We will reiterate this point many times throughout this and other chapters.

In what follows, we discuss the consequences of foraging in a group. As in the preceding chapter, we divide the behaviour into a series of decisions which we then analyse to understand the governing ecological factors. The foraging cycle is thus divided into a logical hierarchy of decisions: first, deciding whether or not to join a group, second, deciding whether to search for food or to exploit the discoveries of others, third, choosing prey, and finally, deciding how long to exploit a patch of prey.

## 8.2 Joining a group: deciding where and with whom to eat

At the beginning of the chapter, we described a scene in which nine starlings were searching for food on a lawn. Why nine? Were they foraging together by chance, because there were too few lawns to go around, or because they benefit from being together? In solitary foraging, deciding where to eat amounts to simply choosing a site, or a patch. However, in social foraging, deciding where to eat often also entails choosing with whom to eat, at least for a moment. This decision will depend upon the effect of the presence of others on each individual's fitness: this presence may be costly or beneficial.

### 8.2.1 The costs of group-living

The presence of others may be costly, and thereby have a negative effect on a patch's payoff. For example, individuals that are already present may act as competitors by reducing the resources' availability. We can categorize these competitive effects into two classes: **scramble** and **interference competition** (Smith and Smith 1998, p. 178; Begon *et al.* 1990, p. 198).

#### 8.2.1.1 *Scramble competition*

Competitors can reduce the payoff of a habitat by reducing resource availability through **exploitation**, or in other words, by scrambling for a share of the resource: what is known as scramble competition. This type of competition would apply to the starlings described in the beginning of this chapter, if the prey consumption of individual starlings reduced the amount available to the others. The negative effect of this form of competition on the animals' intake rate is **long-lived**. In other words, the competitors' departure does not enable the intake rate of the remaining individuals to re-establish the level attained before competition. The effect persists simply because the density of the available prey does not increase when the competitors depart.

#### 8.2.1.2 *Interference competition*

Interactions between competitors can also reduce the intake rate of individuals: what is known as contest or interference competition. This type of competition would apply, for example, if the presence of socially dominant individuals drove subordinates to be more on the alert so as to avoid attack; or if the competitors' presence scared the prey away more quickly, reducing the intake rate of their companions. Indeed, interference competition may arise by several different mechanisms, but this form of competition usually has a **shorter-lived effect** than does scramble competition. For instance, the departure of the aggressive individuals in a group might reduce the need for vigilance, enabling the remaining individuals to resume intake rates at levels attained before competition.

Competition, whether it occurs through exploitation or interference, reduces the value of a resource patch. If two patches are of equal **intrinsic habitat quality**, then we expect individuals to prefer the one with fewer competitors. Therefore, competition has a dispersive effect on individuals; in this type of situation, individuals are choosing where to forage in a **dispersion economy**.

### 8.2.2 The benefits of group-living

Alternatively, per capita growth rate of a population may increase with increasing density, at least initially. This outcome is often referred to as the **Allee effect**, in honour of the American Warder Clyde Allee who proposed this effect as an explanation for animal aggregations. A variety of hypotheses address how increasing population density can be beneficial and these can be divided into two classes: those advantages related to a reduction in the threat[1] of predation, and those related to an increase in the efficiency of resource exploitation. Here we will only briefly describe these two classes, given that they are more fully explained and illustrated in Chapter 14.

#### 8.2.2.1 *Reduction of predation threat*

The presence of others can counteract predators in several ways. Primarily, companions serve as alternative targets for the predators, and can thereby reduce the probability of being attacked: referred to as the **dilution** effect (Hamilton 1971). During an attack, companions may also serve as a shield when the group is made up of individuals that avoid attack by hiding behind others: referred to as a **selfish herd** (Hamilton 1971). Furthermore, a group of prey might present a difficult choice between alternative moving

1 We use the term 'threat' here to avoid any confusion with the word 'risk', which has a specific definition in relation to foraging (see Chapter 5).

targets when a predator approaches, causing a **confusion effect** and thereby reducing the efficiency of attack (Bertram 1978).

The presence of others also increases the probability that at least one of the group members detects the approaching predator and broadcasts an early warning, allowing for greater chance of escape (Pulliam 1973). Groups may also engage in mobbing or united defence against a predator's attack (Bertram 1978).

#### 8.2.2.2 *Advantages related to resource exploitation*

In the preceding section, we discussed how gregariousness can help prey escape predators. Now let us consider the predators by discussing how gregariousness can aid them in capturing prey more efficiently.

The presence of others enables predators to engage in a united attack. This type of attack can allow access to prey that are more difficult or dangerous to catch compared with prey that are manageable through a solitary attack. For example, this advantage may explain the disproportionately large size of prey taken down by lions (*Panthera leo*), spotted hyenas (*Crocuta crocuta*), wolves (*Canis lupus*; Pulliam and Caraco 1985), and social spiders (Vollrath 1982), compared with prey captured by phylogenetically related, similar sized, solitary predators.

When animals search for food within a group, each individual has access to the discoveries of others. This **information sharing** results in each individual achieving a much higher encounter rate with food patches in comparison to that which is attainable when searching alone. An increase in the encounter rate with food patches can have two consequences on foraging efficiency: a reduction in the risk of not finding food within a specified time and a potential increase in individual feeding rate.

The first consequence of a higher encounter rate with food patches resulting from information sharing is an increased probability per unit time for each group member of exploiting a patch. In effect, information sharing reduces the chances for each group member of coming up empty handed. The group therefore provides a less risky option, which is advantageous when conditions favour **risk aversion** (see Chapter 7). The second consequence is an increased intake rate; however, this increase is not guaranteed. For example, searching as a trio may yield the discovery of three times as many patches, but competition results in each of these patches being shared three ways. Therefore, the trio must discover and consume three times as many patches as a solitary individual just to break even. At best, in the absence of competition, an individual's long-term intake rate is simply unaffected by an increase in its encounter rate with patches. However, if a single patch contains enough resources to satisfy the needs of all of the group members, then an increase in the speed at which this patch is discovered results in an increase in the individual's long-term intake rate. Therefore, for an information-sharing group to increase the intake rate of its members, the patches must either be extremely rich or extremely ephemeral so that the portion obtained from each patch remains independent of (or only weakly dependent on) the number of individuals in the group. In effect, information sharing is more likely to reduce foraging risks than to increase intake rate.

Whether the benefit of group membership is due to avoiding predators or to increasing resource exploitation efficiency, the presence of others appears to be attractive. If two patches offer identical resource quality, we expect that individuals will prefer the one with the largest number of conspecifics because being with others is beneficial. Because the benefits promote aggregation, we say the animals are choosing patches in an **aggregation economy**.

### 8.2.3 Choosing where to eat in a dispersion economy

In a dispersion economy, there is no advantage of being with others and so individuals strive to spread out as much as possible among the available patches to minimize competition. In such an economy, an aggregation within certain patches is attributable to the availability of too few patches, rather than to an advantage of group-living (this type of reasoning will be further discussed in Chapter 14). Americans Stephen Fretwell and Henry Lucas (1970) explored how passerine birds would distribute over habitats of differing suitability in cases of dispersion economies and came up with **ideal free distribution** (IFD) theory (Fretwell 1972). Although the theory was developed to explain the distribution of animals within habitats, behavioural ecologists have also used it to explain the distribution of a population of consumers among a series of resource patches. Accordingly, we shall discuss the ideal free distribution model in the context of the exploitation of patches of prey. We will provide an example in the context of the exploitation of breeding patches in Chapter 10.

#### 8.2.3.1 *Ideal free distribution*

##### a *Assumptions*

The model assumes that all of the available patches can be characterized by an intrinsic quality that corresponds to a patch's exploitation rate when the density of consumers approaches zero (i.e. in the absence of competition). The model also assumes that all of the animals will proceed without hindrance to the patch of highest value (i.e. they are free), and that they know the value of each patch (they are ideal), hence the name ideal free distribution. Furthermore, given that the model is set in a dispersion economy, it assumes that adding an individual to a patch reduces the patch's value, through scramble or interference competition, or both. Finally, it assumes that individuals are competitively equal, that is they all possess the same ability to exploit the patches.

##### b *Predictions*

Consider a population of individuals that meet the **assumptions** listed above and that must distribute themselves among two patches of equal intrinsic quality. The IFD theory predicts that they will distribute themselves equally between the two alternatives (Figure 8.1). Any imbalance in the distribution of individuals can only be transitory given that a member of the overexploited patch, given it is ideal and free, will migrate to the alternative patch. This advantage attained via migration exists as long as one of the patches remains overexploited in relation to the resources it provides. Once the two patches contain the number of individuals proportional to the resources that they offer (in this case, an equal number given that the resources are equal), migration is no longer advantageous and therefore the individuals are in a **Nash equilibrium**: in other words, the situation is such that any unilateral modification in the choice of an individual can only be disadvantageous for that individual. Hence the IFD is an ESS (see Chapter 3).

The same principle applies to two patches of unequal intrinsic quality (Figure 8.2). The first individuals to colonize the patches will prefer the one with the greater intrinsic quality. They will choose to exploit this patch until its value decreases (because of competition between foragers) below that of the alternative. Some individuals will then switch to this second, now more profitable patch, reducing its value for those individuals that have not yet made their choice. The ideal free individuals will partition themselves among the two patches, always migrating to the most profitable alternative at the time of the decision. Using this decision rule, once the whole population has distributed itself among the two patches, it will reach a Nash equilibrium, the distribution that is proportional to resource availability. When the ratio of individuals at the patches corresponds to the ratio of the resources available in those patches, the population has attained an IFD through the **habitat matching effect** (Figure 8.2). Refer to Box 8.1 for a more formal presentation of this argument. Despite the difference in the intrinsic values of the patches, once the IFD is attained **all of the individuals in the population will have the same feeding rate**.

##### c *Tests of the ideal free distribution*

*Sticklebacks*

A German scientist, Manfred Milinski (1979), was the first to publish a test of the IFD model, using

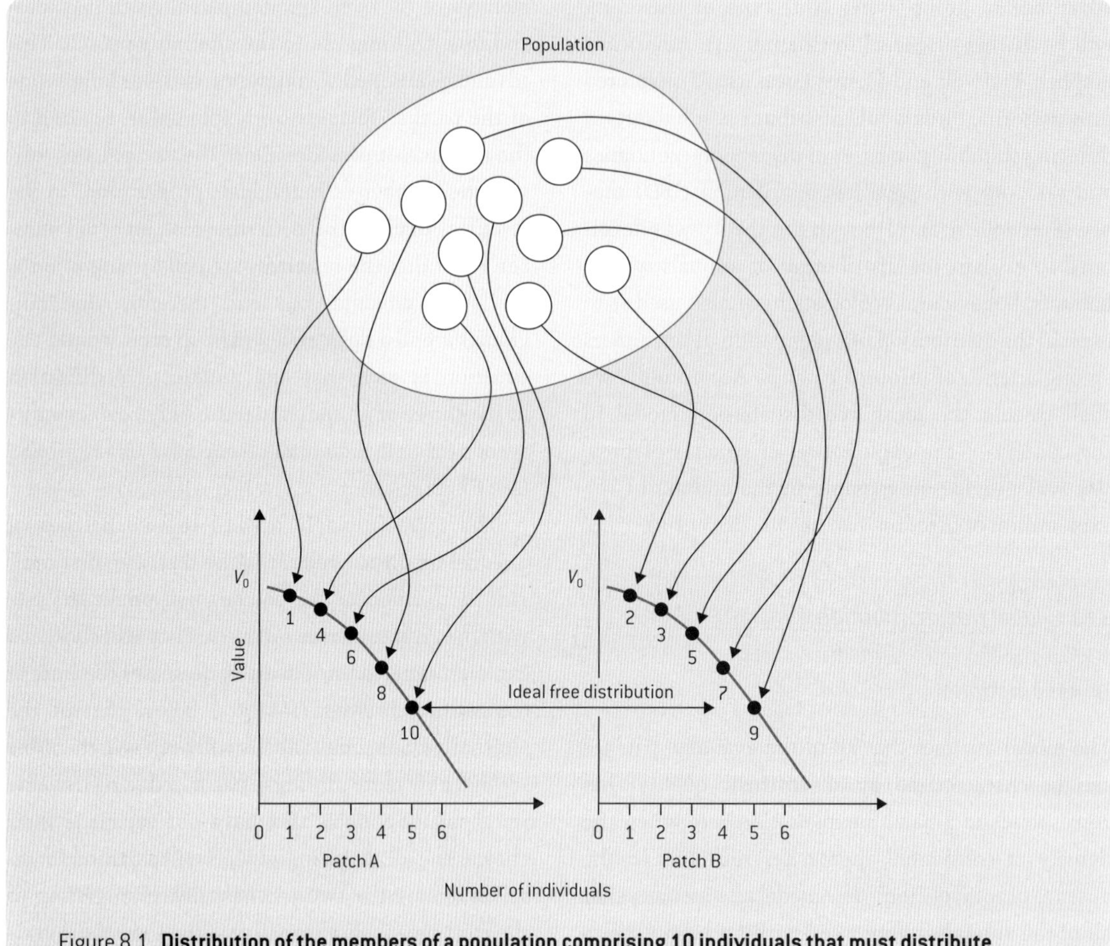


Figure 8.1 **Distribution of the members of a population comprising 10 individuals that must distribute themselves among two patches of equal value**

The curved arrows indicate the decision of each individual and the numbers indicate the sequence of each decision. On each occasion, the individual chooses the patch that offers the highest profitability at that moment. The distribution that is attained once the population reaches equilibrium is an ideal free distribution. The number of individuals found at each patch is such that it is disadvantageous for any individual to migrate to the alternative patch. The population has reached a Nash equilibrium, in which all individuals obtain equal benefits.

small groups of stickleback fish (*Gasterosteus aculeatus*). His assistants, hidden behind blinds, dropped daphnia (*Daphnia magna*) onto the water's surface at both ends of a 43.5 cm × 20 cm × 23 cm aquarium while videotaping the position of the six sticklebacks. In the first experiment, the two patches each offered daphnia arrival rates of 0.5 per second and 0.1 per second; then in the second experiment the arrival rates were changed to 0.5 per second and 0.25 per second. The results (Figure 8.3) show that the sticklebacks are able to distribute themselves between the two patches according to the predictions of the IFD. They also show that this distribution is not attained instantaneously. On the contrary, the sticklebacks seem to sample the two patches several times before choosing which patch to exploit. Furthermore, even after they choose a patch, they continue to sample the alternative periodically. Taking into consideration Milinski's results, among others, it becomes apparent that there is a subtle difference between the IFD

Box 8.1 **The ideal free distribution**

Suppose there are $Z$ patches in an environment, and the **intrinsic value** of each patch can be expressed as the rate of prey arrival $k_i$ ($i = 1, 2, \ldots, Z$) which is patch specific and constant for each patch. The expected feeding rate at patch $i$ ($W_i$) increases with the rate of prey arrival at patch $k_i$ and decreases with the number of competitors at the patch $G_i$:

$$W_i = \frac{k_i}{G_i}$$

IFD theory predicts that all of the consumers have the same feeding rate at equilibrium. Therefore:

$$W_i(G_i) = C$$

for all individuals, where $C$ is a constant. At equilibrium the feeding rates will be:

$$\frac{k_i}{G_i} = C$$

and, at the Nash equilibrium, the ratio of the number of competitors at patches $i$ and $j$ should be:

$$\frac{G_i}{G_j} = \frac{k_i}{k_j}$$

We can predict the proportion of the population at each patch:

$$\frac{G_i}{\sum_{i=1}^{Z} G_i} = \frac{k_i}{\sum_{i=1}^{Z} k_i}$$

such that:

$$G_i = k_i \left\{ \frac{\sum_{i=1}^{Z} G_i}{\sum_{i=1}^{Z} k_i} \right\}$$

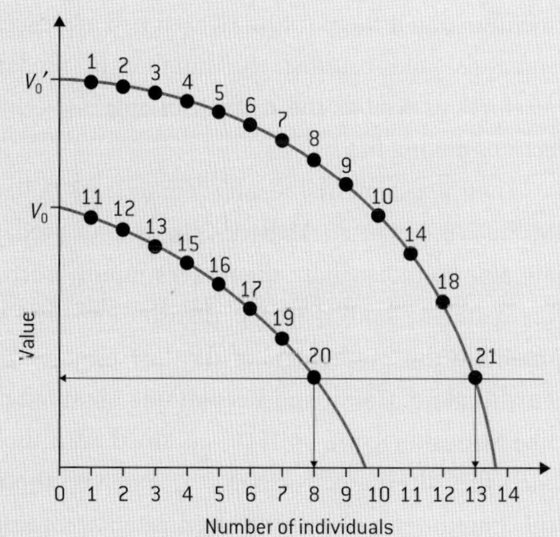


Figure 8.2 **Distribution sequence of the members of a population comprising 21 individuals that must distribute themselves among two patches of unequal value**

The first 10 individuals choose the richest patch. Then the remaining individuals distribute themselves among both patches. Once the entire population is distributed, the population has attained an IFD with 8 individuals at the poor patch and 13 individuals at the rich patch. Despite the imbalance in the number of competitors at each patch, once a Nash equilibrium is attained, all of the members of the population obtain the same benefits and there is no advantage to be gained from migrating to the alternative patch.

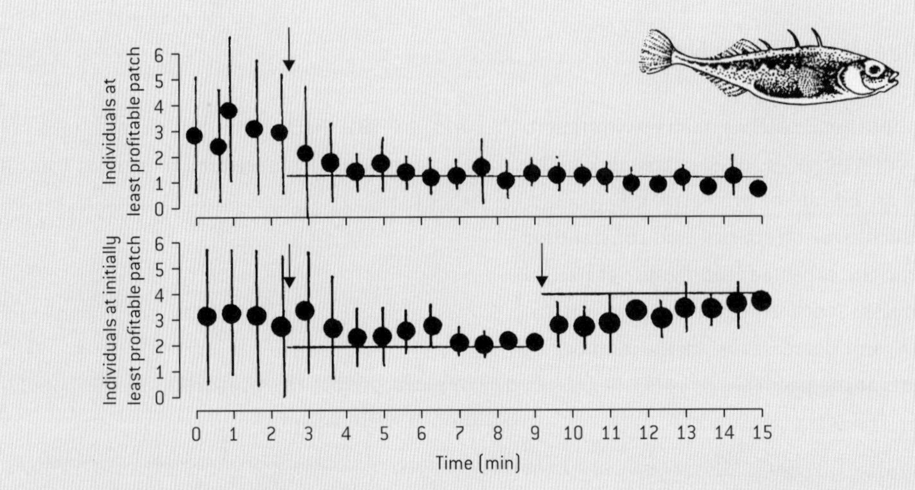


Figure 8.3 **Milinski's (1979) experiments using groups of six stickleback fish (*Gasterosteus aculeatus*)**

Each point represents the mean of eight trials, the vertical lines give the standard deviations of the means, and the horizontal lines among the points indicate the number of individuals predicted by the IFD. In the top graph, the patches offer a 5:1 profitability ratio and the results show the number of individuals observed at the poorest patch. The arrows indicate the onset of the dissemination of food at the patches. In the bottom graph, the patches offer a 2:1 profitability ratio. Here a patch starts out as the poorest, but after five minutes, indicated by the second arrow, its becomes the most profitable.

Modified from Milinski (1979).

predictions and the observed distributions. For example, there are slightly too many fish at the least profitable patch and a shortage of fish at the richest patch. This difference occurs in almost all tests of the IFD, whether using insects, fish, or birds (see Kennedy and Gray 1993; Tregenza 1995).

*Cichlids*

In a repeat of Milinski's experiment, this time using cichlid fish (*Aequidens curviceps*), Canadians Jean-Guy Godin and Miles Keenleyside (1984) explored the consequences of a violation of the assumption of equal competitive ability on the IFD. They placed six cichlid fish in a rectangular basin and created two patches of prey via an influx of *Herotilapia multispinosa* larvae (another cichlid fish) at each end of the basin. The basin was filmed with a video camera located over the top which allowed them to determine the position of the predators. The intrinsic value of the patches was modified by changing the influx rate of larva at the two patches. Godin and Keenleyside provided three different profitability ratios: 1:1, 2:1 and 5:1. The richest patch always yielded one larvae every 6 seconds. The poor patches yielded one larvae every 6, 12, and 30 seconds in accordance with the treatment. During the trials, the experimenters noted the position of each fish at 15 second intervals, and counted the number of predation attempts as well as the number of aggressive interactions among fish.

Their experimental results demonstrate that, like Milinski's (1979) sticklebacks, the cichlid predators are able to redistribute themselves rapidly between the patches, and that the distribution seems to resemble the predictions of the IFD (Figure 8.4a). Furthermore, Godin and Keenleyside observed that, like Milinski (1979), an IFD is attained even though each fish alternates between the patches, allotting the amount of time spent at each patch according to the patch's prey arrival rate (Figure 8.4b).

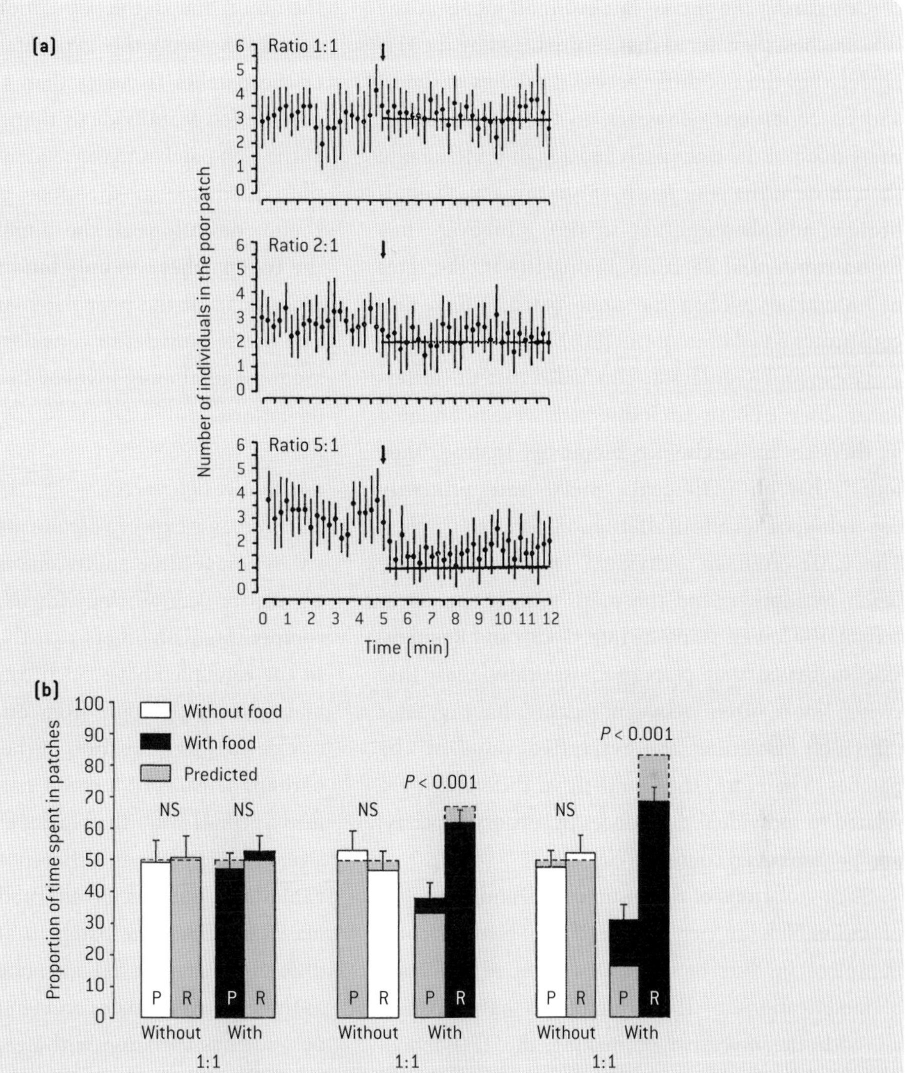


Figure 8.4 **Godin and Keenleyside's (1984) experiment**

The experiment was performed using groups of six cichlid fish *Aequidens curviceps* foraging among two patches which contained fish larva as the food source.

**a.** Mean number (and 95% confidence interval) of individuals observed at 15 second intervals within the less profitable patch for three different profitability ratios between the two patches (1:1, 2:1, 5:1).

**b.** Proportion of the total time spent at the poor patch (P) and at the rich patch (R) for each profitability ratio. For each of these profitability ratios, the time spent at the patches is divided into the time 'without' food (before the addition of prey) and 'with' food.

Modified from Godin and Keenleyside (1984).

Comparing the intake rates of individuals at the two patches, they noted that, as predicted by the IFD model, they did not differ between patches in the 1:1 and 2:1 conditions. However, in the 5:1 condition, individuals at the rich patch had a higher intake rate than those at the poor patch; this is not a Nash equilibrium, and therefore not an IFD. Moreover, they found substantial variance among the intake rates of individuals within the same patch, which also constitutes a violation of the IFD.

Godin and Keenleyside concluded that the differences observed between individuals were not related to their social dominance hierarchy. Instead, they found that an individual's intake rate primarily depends upon its rate of alternation between patches. The more often an individual switches from one patch to another, and the more time it spends in transit, the lower its intake rate. Godin and Keenleyside suggested that perhaps those individuals that switch most often between patches have greater difficulty discerning the respective value of the patches. Therefore, the competitive differences are related to individual differences in sensory capacity, and perhaps even cognitive capacity.

Such violations of the assumptions and shortages of individuals at rich patches have become commonplace in tests of the IFD. Hence, a multitude of modifications of the assumptions have ensued, all with the objective of making the IFD model more realistic and true to observation. The primary modifications are discussed in the next section.

### d Modifications of the assumptions

All modifications of the assumptions of the IFD have the same objective: to explain why there is a shortage of individuals at the richest patches and an excess at the poorest. First and foremost, it is important to understand why a departure from the initial assumptions always has the same consequence on the fit between prediction and observation. Imagine an extreme case, in which the individuals do not have the capacity to choose patches according to their intrinsic value. These individuals would have no preference for one patch over the other, and so the population would divide itself equally between the two patches. Given that one of the patches is richer than the other, we would observe fewer individuals than predicted by the IFD at the rich patch and consequently, more individuals than predicted at the poor patch. Therefore, all chance deviations in the number of individuals at the two patches can only tend towards an excess of individuals at the poor patch and a shortage at the rich patch. Now let us consider the hypotheses that are most commonly invoked to explain the observed deviations.

*Competitive differences*

In reality, we expect a certain amount of heterogeneity in competitive ability among the members of a population: certain individuals are probably better competitors than others, as is the case for the cichlids in the example above. Such competitive differences may remain constant from one patch to another, or else vary as a function of the patch. For example, perhaps females of a given fish species are able to feed 1.7 times faster than males owing to a larger body size. This 1.7 times difference may occur at all exploitable patches, or alternatively, the difference may vary from one patch to another. Competitive differences may be discontinuous and discrete, such as those between males and females, or else continuous, such as variation with age, experience, or size. Regardless of whether or not the competitive differences are constant, and whether they are based on discrete classes or on continuous variables, they have profound consequences on the predictions of the IFD. Box 8.2 provides a simplified example of constant competitive differences arising from discrete phenotypic differences.

*Incomplete information*

The IFD assumes that individuals have perfect knowledge about the value of the available patches and that they always proceed to the most advantageous patch. However, the model does not address how the individuals acquire knowledge about the value of the alternatives. Ever since Milinski's (1979) experiment, it has become evident that individuals

## Box 8.2 **Ideal free distribution with unequal competitors**

Imagine a population of 12 fish in which 6 are small and 6 are large. Suppose that the large fish are able to ingest prey exactly twice as fast as the small fish. Hence, the large individuals possess exactly twice the competitive weight of the small individuals. Now suppose that there are two patches (1 and 2) of unequal intrinsic value $k_1 = 2k_2$ with constant prey arrival rates. When competitors are unequal, the IFD model assumes that it is the competitive capacities of the individuals, rather than the individuals themselves, that are distributed in proportion to the patches' values. Once the total competitive capacity is distributed between the patches in correspondence with their intrinsic values, no individual can switch patches without decreasing its intake rate, and therefore a Nash equilibrium has been attained. More specifically for the example in question, because the intrinsic value of patch 1 ($k_1$) is exactly double that of patch 2 ($k_2$) the model assumes that at equilibrium the total competitive capacity of the individuals at patch 1 will be twice that at patch 2. This condition would hold, for example, if we found 6 large fish at patch 1 ($6 \times 2 = 12$) and 6 small fish at patch 2 ($6 \times 1 = 6$) (see Figure 8.5). However, this is not the only way to succeed in matching the competitive capacity at each patch with its intrinsic value. For instance, this condition would also hold if we found 5 large fish and 2 small fish at patch 1 ($(5 \times 2) + (2 \times 1) = 12$), and 1 large fish and 4 small fish at patch 2 ($(1 \times 2) + (4 \times 1) = 6$). There are also several other assortments of fish that meet the assumption of correspondence between the patches' competitive capacity and intrinsic value (cf. Figure 8.5). However, even though several possible combinations are Nash solutions, they are not all equivalent, given that the difference in the average intake rates between the patches and between the individuals depends upon the combination. When the total population is small, all of the combinations are not equally probable.

The most likely combination is that in which the individuals of each competitive type distribute themselves according to the IFD model (the combination represented in Figure 8.5c). In this case, there are twice as many small fish at patch 1 as at patch 2, and twice as many large fish at patch 1 as at patch 2. If we simply consider the number of individuals at each patch, we count 8 at patch 1 and 4 at patch 2, which is precisely the prediction of the original IFD model, despite a serious violation of an assumption, specifically we have unequal competitors. However, note that for all of the other combinations that are also Nash equilibriums, there is an excess of individuals at the poor patch and a shortage at the rich patch. **In all cases, the intake rates remain unequal between individuals**[2].

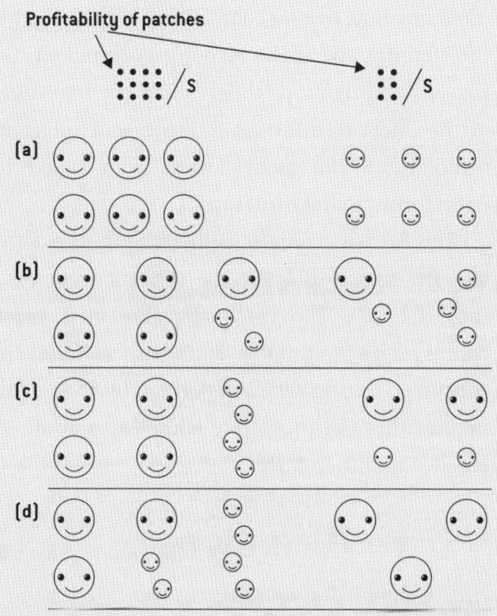


Figure 8.5 **Schematic representation of the IFD with unequal competitors (portrayed here by the size of the circles)**

Inspired by Sutherland and Parker (1985).

2 Readers who wish to know more about consequences of asymmetric competitors should consult Sutherland and Parker (1985).

must invest time and effort into sampling the alternatives. This sampling activity implies that an individual will be found at the poor patch more often than predicted by the IFD, simply because an estimation of its value is required. As a corollary, the individual will be absent more often from the rich patch so as to sample the poor patch. Therefore, sampling will inevitably result in an under-representation of individuals at the rich patch.

Sometimes individuals only have access to partial, biased, or even false information, despite engaging in sampling activity. In this type of situation, the outcome may depend on the nature of the information. If false information over-rates the good habitat then perhaps too many individuals will aggregate in the richer.

*Sensory limitations*

The original theory assumes that individuals are able to detect any qualitative difference between the available patches. In reality, however, there are almost certainly limitations to the discriminative capacities of the individuals of a population. In the most extreme case, a population without any ability to discriminate will distribute itself completely at random, resulting in lesser use of the best habitats. The better the discriminative capacity of the individuals, the smaller will be the divergence from the predictions of the IFD.

#### 8.2.3.2 *Variable interference*

In most early tests of the IFD, patches offered constant prey arrival rates. Hence the patches' resources are renewed in a continuous manner: referred to as a **continuous input model**. For example, Milinski (1979) added daphnia to the ends of the aquarium at different rates. Here, the model assumes that the resources are consumed as soon as they appear in the patch. There is no longer-term accumulation or decline in food availability. When an individual consumes a prey item, that item is thereby made unavailable to the competitors. The more competitors are present in a patch the fewer the prey available per competitor, the longer the waiting interval in prey consumption, resulting in a fairly intense form of **scramble competition**.

In many situations, the assumption of continuous renewal of resources does not apply. In fact, it is more likely that the patches already contain the majority of the exploitable resources, meaning that competitors simply harvest food. For example, a herd of herbivores that have a choice between two pastures of different quality are not faced with a continuous renewal of resources, but a standing crop. One advantage of using a standing crop approach is that it allows for potential variation in the intensity of competition among individuals at a patch and so one can explore the consequences of a broader range of competitive intensities on consumer distribution. Under the assumption of continuous input, a patch that supplies $k$ prey items per unit of time will offer $k/2$ when two equal competitors share the patch and $k/3$ when there are three competitors, etc. In a standing crop situation, on the other hand, a lone individual that is exploiting the patch can potentially ingest $k$ units, and still ingest the same amount when joined by one, or two competitors, if they exert little interference and if the resource is abundant. Given that a standing crop model allows the investigation of the effect of different intensities of interference competition, the models have been called **ideal interference distribution** (IID) models.

The British ecologist William Sutherland (1983) was the first to propose the conversion of the IFD into the IID. The intrinsic value of a patch is expressed by the number of prey items $K$ that can be consumed during time interval $T$. The ESS condition implies that the input rate is the same at all of the patches, and therefore:

$$\frac{K}{T} = c = \kappa_i (G_i)^{-m} \text{ for any } i$$

where $\kappa_i$ is the prey density at patch $i$, $m$ is a parameter that quantifies the intensity of the effect of each competitor on the intake rate of a given individual, hence an index of interference between the individuals, and $G_i$ is the number of individuals at patch $i$. It follows that the number of competitors in habitat $i$ is:

$$G_i = \left[\frac{K_i}{c}\right]^{1/m}$$

And if we know the total number of competitors in the population and the total number of prey items available at all of the patches, then we can calculate the proportion ($g_i$) of consumers expected at patch $i$:

$$g_i = (\gamma \alpha_i)^{1/m};\ 0 < g_i < 1$$

where $\alpha_i$ is the proportion of all of the prey in the habitat contained within patch $i$, and $\gamma$ is a constant of standardization calculated as:

$$\gamma = \frac{\sum_{i=1}^{Z} g_i}{\sum_{i=1}^{Z} \alpha_i^{1/m}}$$

for all of the $Z$ patches in the habitat. Note that there is a specific value of $\gamma$ for each assortment of prey within the $Z$ patches, which implies that there is no general quantitative prediction about the distribution of consumers. Each distribution of prey in the habitat gives rise to a particular prediction about the distribution of consumers. It goes without saying that, even though this approach seems more realistic, it is certainly less convenient. This does not mean that the approach is totally without value, given that we can use it to understand, at least qualitatively, the effect that the intensity of interference has on the distribution of consumers in a population.

It is important to note that the parameter $m$ always equals 1 in an IFD with continuous prey input. Such a strong level of interference arises given that the addition of each competitor results in the others losing a portion of the resource. Now let us verbally consider the consequences of a level of interference so low that it is almost nonexistent ($m \rightarrow 0$), for a population in a habitat with two patches (1 and 2) of unequal intrinsic value. Since the addition of competitors does not affect the individuals' intake rate, the distribution will be completely independent of consumer density, and based solely on the intrinsic value of the patches. The entire population should choose the patch with the highest intrinsic value. Now let us imagine the alternative situation: the competitors have a very strong negative effect on the others' intake rate, such that interference is very strong ($m \rightarrow \infty$). Consequently, the intrinsic value of the patch has a negligible effect on the intake rate of the individuals, and the distribution will be purely determined by the number of competitors at the patch. Hence, the individuals should distribute themselves such that they avoid each other as much as possible, resulting in a uniform distribution, independent of the difference between the patches' intrinsic values. Of course, these two extreme cases are unrealistic. Nevertheless, they illustrate the effect of the intensity of interference on the distribution of consumers. The lower the value of $m$, the greater the number of competitors at the rich patch compared with expectations of the IFD. The higher the value of $m$, the greater the number of competitors at the poor patch compared with expectations of the IFD. Therefore, it is important that $m$ be measurable. Given that most studies so far find an excess of individuals at poor patches, it is conceivable that the most common values of $m$ are greater than 1. However, we will see in Chapter 14 that the case of **coloniality** is an exception. In colonial species, good patches are often overexploited (leading to true **aggregation**), indicating that interference at the colony site may be low ($m < 1$).

According to Sutherland (1983), we can estimate $m$ by measuring the intake rate of a patch's consumers at different consumer densities where $m$ corresponds to the slope of a linear regression between the log of intake rate and the log of consumer density.

*An example with zebra fish*

Once again using fish as the experimental organism, this time zebra fish (*Brachydanio rerio*) one of the first tests of the concept of variation in intensity of interference in the context of the IFD was conducted by Canadians Darren Gillis and Donald Kramer (1987). In this experiment, the fish had access to three patches of prey, nauplii larvae of *Artemia salina*, which offered a ratio of arrival rates of 1:2:4. In their novel experimental design, the fish were able to occupy the space between the patches that was void of food (Figure 8.6a). The experimenters examined the distributions of fish at four different competitor densities, 30, 60, 120, and 240 individuals, under the assumption that the higher the number of competitors, the higher the level of interference will be and the more the distribution of predators will deviate from that which is predicted by the IFD. This is precisely what they observed (Figure 8.6b). The higher competitor density, the more the observed distribution deviates from the predictions of the IFD. However, they noticed that the rate of aggressive chases between individuals decreased with an increase in fish density. Therefore, the effects of interference were not attributable to aggression between competitors. Indeed, the precise mechanism responsible for the interference remains to be discovered.

*Preliminary conclusions*

So far we have seen that in a dispersion economy, the expected distribution of individuals among an assemblage of patches of given intrinsic values depends on competitor density and on the patches' quality. Initially, the IFD model predicted that the proportion of consumers at a patch would be commensurate with the proportion of resources offered at that patch. We have learned that most tests of the model find an excess of individuals at poor patches and a shortage at rich patches. These deviations can be explained by violations of the initial assumptions, as well as by a strong level of competitor interference. Now let us turn to the distribution of competitors among patches in an aggregation economy.

### 8.2.4 When the presence of others is advantageous

To illustrate an aggregation economy, here we consider killer whales (*Orcinus orca*) hunting a variety of prey in the Pacific Ocean along the North American coastline (Baird and Dill 1996). Two types of killer whale inhabit this part of the Pacific: residents and

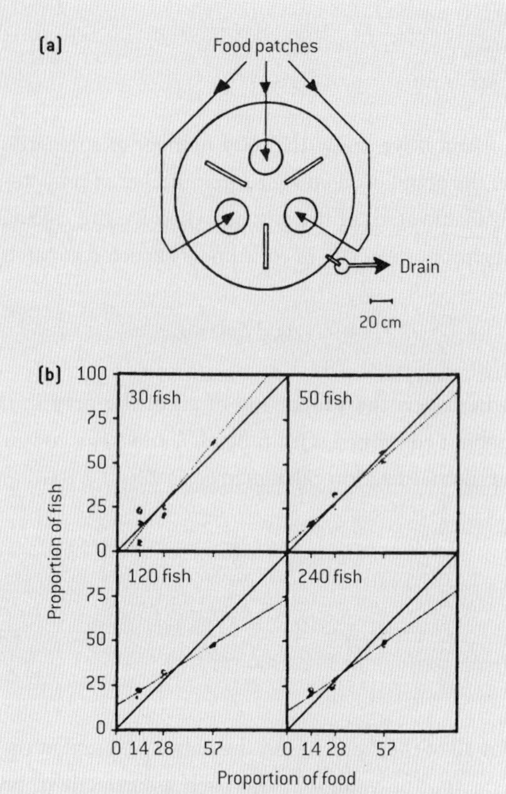


Figure 8.6 **Gillis and Kramer's (1987) experimental test using zebra fish (*Brachydanio rerio*)**

**a.** Top view of the experimental apparatus, which consists of a main basin containing three patches, each partially isolated by small partitions of simulated vegetation. The prey items arrive at the centre of each patch via tubes and the water flows towards a drain.

**b.** Proportion of fish observed versus proportion of food offered at the three patches for four population densities (30, 60, 120, and 240 fish). Each point represents the mean of six replications. The solid line represents the prediction of the IFD model and the dotted line represents the linear regression calculated using the observations.

Modified from Gillis and Kramer (1987).

transients. Resident-type whales form matrilineal pods (see Chapter 12) whose size changes as a function of births and deaths. When these small-sized whales hunt, they form temporary groups of one to fifteen individuals that sometimes originate from more than one pod. Are these killer whales in an aggregation economy? Canadians Baird and Dill (1996) say yes, arguing that grouping allows for increased hunting success (Figure 8.7). However, it is important to keep in mind that the more hunters present, the smaller the share consumed by each. Given that the portion of prey available continually decreases with increasing group size, hunting efficiency reaches an upper limit at an intermediate group size; adding hunters cannot increase the payoff indefinitely. Therefore, the net payoff increases up to an intermediate group size and then decreases (Figure 8.7).

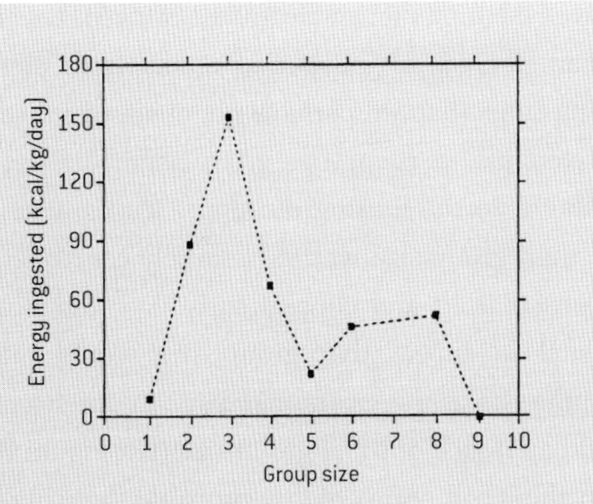


Figure 8.7 **Group size and feeding success of killer whales (*Orcinus orca*)**

From Baird, R.N. and Dill, L.M., Ecological and social determinants of group size in transient killer whales, *Behavioural Ecology* (1996), volume 7: 408–410, by permission of Oxford University Press.

Quantity of energy taken in daily as a function of hunting group size. Intake efficiency is maximized when the group contains three individuals.

Taken from Baird and Dill (1996).

Baird and Dill (1996) tested the hypothesis that killer whales hunt in groups to maximize hunting success. They calculated the productivity of different sized hunting groups for three sizes of prey (Figure 8.7). However, testing this hypothesis requires deciding in advance which group size will arise from a combination of each whale's attempt to maximize its own hunting success by joining a hunting group.

#### 8.2.4.1 *Expected group size: optimal or stable group size?*

The type of graph relating individual success to group size, shown in Figure 8.7, has long led to the prediction that animals should form groups of optimal size; that is, the size that maximizes the payoff for all group members. However, if one applies to aggregation economies the same set of assumptions of ideal free individuals as was presented in dispersion economies we realize this may not be the case. Fretwell and Lucas (1970) had already considered this problem in what they called Allee-type distributions but the consequences for aggregation economies were never generally appreciated. Application of ideal free assumptions and the Nash equilibrium to aggregation economies leads to the prediction that **groups of optimal size are simply not stable solutions** to group formation games (Caraco and Pulliam 1984; Clark and Mangel 1984; Sibly 1983). Instead, the expected more stable group size is often greater than the optimal size. Here is why this is so.

Let us imagine the ideal benefit curve taken from the killer whale example (Figure 8.8). Two solitary whales hunting a seal meet. Both will benefit from joining forces in the hunt (Figure 8.8). Then a third whale meets the pair and is faced with the choice of hunting alone or joining the group. For the moment, imagine that the pair cannot prevent the third whale from joining. In this case, the solitary individual will benefit from joining the group (Figure 8.8). A fourth whale appears, and will also benefit from joining the group, because, even though its membership will decrease the uptake of the first three, its own uptake will be better than that which is attainable by hunting

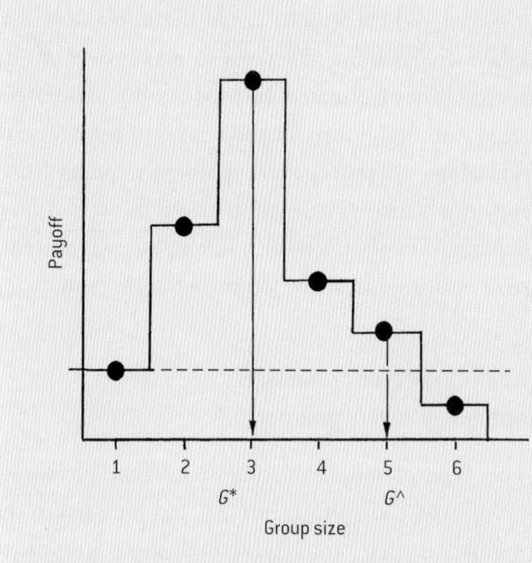


Figure 8.8 **An example of the instability of optimal group size (*G**) in favour of stable group size (*G^*)**

The payoff of a solitary individual is improved by joining another individual, and therefore both will be tempted to join forces. A third individual will also benefit from joining the first two to form a group of three instead of remaining alone. Even though the trio enjoys maximum payoff, another solitary individual will benefit from joining the group making it grow to four. Solitary individuals will be attracted to the group so long as their payoff is superior to that which is attainable when hunting alone (the dashed horizontal line). In this graph, this occurs at a group size of five individuals, at which point a sixth member would do worse than a solitary individual. Solitary individuals cannot benefit from joining the group and the members of the group cannot benefit from leaving to hunt alone. Therefore, a group size of five is a Nash equilibrium and hence the stable group size (*G^*).

alone. Let us continue the process so long as the solitary individual will benefit from joining the group. A Nash equilibrium is reached at group size *G^*, that is, the group size at which a solitary individual does not benefit from joining the group and a member of the group does not benefit from leaving to hunt alone (Figure 8.8). This large group is the expected solution and leads to the paradox of group size: **the selective advantage of group membership results in group sizes for which group membership is no longer advantageous**. This paradox is a perfect example of a point we discussed above: **natural selection does not always lead to maximization of benefits**.

#### 8.2.4.2 *Down with the paradox of group size*

*From killer whales . . .*

When we apply the reasoning described above to Baird and Dill's (1996) data, we expect that the whales will associate in groups of sizes between eight and nine individuals (Figure 8.7). However, instead Baird and Dill observe whales in groups that are closer to the optimal group size. Therefore, either we revise our argument or we conclude that killer whale hunting groups do not form because whales are attempting to maximize hunting success. Once again, it is worth going over the assumptions that allowed us to predict stable groups of size *G^*.

The **paradox of group size** stems from several assumptions that are perhaps not always applicable. For example, perhaps the group members oppose resistance to the joining of others. If this were the case then group size would be largely controlled by the group itself and consequently it would not be unusual to expect that group size would reflect the interests of the group members; in other words, would approach optimal group size (*G**; Figure 8.8). Even though it seems highly probable that there be a certain resistance on the part of the group to intruders, it is not always clear how this resistance could organize itself. Indeed, which group member would be willing to pay the costs associated with defence against intruders?

*. . . To lions . . .*

This issue of the coordination of defence against intruders is elegantly illustrated with female lions. Robert Heinsohn and Craig Packer (1995) played a recording of the roaring of unfamiliar females to groups of lionesses holding a territory. The lionesses reacted by approaching the speakers aggressively.

During the experiments, Heinsohn and Packer (1995) noticed that certain individuals tended to lag behind the others whereas others systematically took on the role of leader. Clearly not all of the lionesses are willing to pay the cost of defence, and such dissension within a group might explain why observed group sizes tend to inflate beyond the optimal size.

Communal territorial defence illustrates a problem that is central to many studies of social behaviour: the exploitation of the investment of companions. In the case of the lionesses, for instance, certain members of the group are exploiting the defensive behaviour of others. These individuals still gain access to a defended territory but without paying the full cost required to maintain exclusive access. This problem of exploitation among groups is recurrent and will reappear whenever members of a group exploit a resource.

*. . . Back to killer whales*

Returning to the killer whale example, the stable group size predicted by applying ideal free assumptions remains larger than the group size reported by Baird and Dill (1996). Therefore the ideal free assumptions are probably not met. Because the actual groups are smaller than the ones expected on the basis of ideal free whales, it is perhaps the case that the hunting group size is largely controlled by individuals within the group rather than by intruders that wish to join. This conclusion suggests that it would be interesting to explore further the social mechanisms responsible for the formation of killer whale hunting groups, because group controlled entry implies the coordination of efforts on the part of the group members. For example, it would be interesting to see if, like the lionesses, certain individuals are exploiting the efforts of others. Alternatively, it may be that individuals in a social species like killer whales may coordinate group leaving decisions, invalidating the assumption of the theory that foragers act singly. Lastly, it may be that group size is also affected by issues unrelated to foraging. We will return to this issue in Chapter 14.

## 8.3 Searching for patches within groups

In the previous sections, finding patches of food was not a problem. We simply assumed that individuals knew the precise location of patches as well as their respective intrinsic values, as in the IFD model for example. Now we shall consider the case in which the patches are hidden, and therefore the animals must search to find them. When the members of a group must search for food, the discoveries of some are often exploited by many. In other words, when an animal finds food, its finding often informs others of the food's location, either voluntarily or inadvertently, thereby attracting competitors. What follows is either scramble competition, in which each individual hurries to gain its share of food, or interference competition, in which certain individuals use their competitive advantage to exclude others aggressively from the discovery. This exploitative phenomenon begs the question: do all of the group members participate in the search for food or alternatively, do certain individuals specialize in exploiting the discoveries of others? Here we address this problem associated with animals' having to search for food patches within a group.

### 8.3.1 The basic model: information sharing

When the discoverer of a patch is immediately joined by all other group members, we reason that there must be sharing of information within the group. In the simplest case, all the individuals search for food while at the same time surveying the other group members, which they join whenever they notice that they have found food. In this case, the only possible ESS requires that all group members join immediately upon detecting the discovery of a companion. This is the only possible ESS because there is nothing to gain from not joining. Imagine, for example, an individual within this type of group that decides to join only some of the uncovered patches, say, 50% of the discoveries. This individual will only have access to half of the food discovered by the other group members, whereas all of the others will have access to twice as many discoveries, in addition to monopolizing a part of that individual's personal discoveries. Therefore, there is no advantage to be gained from not always joining: joining is a Nash equilibrium and an ESS. This means that if one is observing a group of size $G$ and noting the frequency of information sharing, the expected frequency is $(G - 1)/G$, the entire group minus the discoverer. The larger the group's size, the more frequently food sharing will occur. We will come back to the role of information sharing when we address the subject of aggregation in Chapter 14.

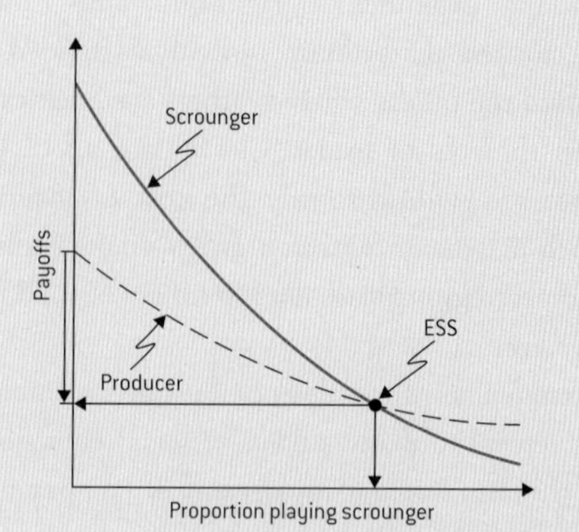


Figure 8.9 **The payoffs of the producer and scrounger strategies as a function of the proportion of scroungers**

Scroungers do better than producers when the proportion of scroungers is low. Scroungers do worse than producers when scroungers are common. Therefore, the payoff curves intersect at an intermediate frequency of scroungers. At this frequency, the payoffs of the two strategies are equal, and the population has reached an evolutionarily stable state composed of a mixture of the two strategies, since neither is an ESS on its own. The vertical arrow pointing downwards on the $y$-axis indicates the reduction in the payoff of a population as it shifts from a composition of all producers to the evolutionarily stable state. Evolution leads to the ESS even if the payoff at this point is less than that of a population composed entirely of producers.

## 8.3.2 The producer–scrounger game

Now let us imagine the same scenario of searching for food within a group, but this time let us say it is impossible for an individual to search for food and survey the others concurrently. Such an incompatibility can arise from constraints that are environmental (e.g. searching in tall vegetation), sensory (e.g. eyes cannot focus on close and far objects at the same time) or cognitive (e.g. it is difficult to concentrate on two tasks at once). Whatever the cause of the incompatibility, each individual is faced with a choice between either searching for food or else looking for a companion that has already found food. In such a scenario the solution is not as simple as in the information-sharing game; there may be something to gain now from not joining, you could search and find food. This type of situation raises a question: is it more profitable to invest in finding resources oneself, or is it better to rely on the investment of the other group members? The answer to this question depends on the proportion of group members that are already searching for food. If this proportion is low, then it is more profitable to search for food oneself. On the other hand, if the proportion of individuals that are searching is high, then it is more profitable to rely on others to find food (Figure 8.9). This frequency-dependency of the rewards leads us to an evolutionary game: the **producer–scrounger** game, first proposed by the British researchers Chris Barnard and Richard Sibly (1981).

In the producer–scrounger game two strategies play against each other: the **producer** strategy, which consists of searching for one's own food and never partaking in the discoveries of others, and the **scrounger** strategy, which consists of seeking out and consuming only food that was discovered by individuals using the producer strategy. The payoff of these strategies is negatively frequency-dependent. The producer strategy does better than the scrounger strategy when the scrounger strategy is common, and the opposite is true when the scrounger strategy is rare (Figure 8.9).

The payoff curves of the two strategies (Figure 8.9) adequately illustrate the game's solution. On the one hand, the scrounger strategy can never qualify as an ESS because all of the individuals would be waiting in vain for the discovery of a resource patch. They would all starve! On the other hand, the producer strategy cannot qualify as an ESS either. A population composed entirely of producers would be vulnerable to invasion by a few rare mutant scrounger strategists. When scroungers are rare they do better than the producer strategy and so we expect the strategy to increase in the population. Two mechanisms can be involved in this increase. The scrounger strategy may be **heritable** and increase over generations as a result of natural selection favouring individuals with the highest fitness. Alternatively, it could be the result of behaviourally plastic individuals choosing to increase their investment in the use of whichever alternative pays more than its alternative. Whatever the mechanism involved, given the frequency-dependence illustrated in Figure 8.9, there is a frequency of scrounger strategists that results in a stable equilibrium at which the two strategies receive identical payoffs. If it is achieved by the operation of natural selection then evolution becomes fixed at this frequency, and the population attains a mixed ESS in which the two strategies coexist. If it is achieved by behavioural plasticity, then the populations reaches a stable equilibrium frequency which has also been called a **developmentally stable strategy** or DSS.

The stable mixture of strategies can take on several different forms. For instance, the group might consist of individuals that are entirely dedicated to either producer or scrounger, a situation that is highly unlikely to occur in cases of behavioural plasticity. In that case, natural selection would act on the number of individuals playing each strategy and an equilibrium would be reached after several generations. The group might also consist of individuals randomly alternating between episodes of producing and scrounging at the ESS frequency (like Godin and Keenleyside's (1984) cichlids described above as an example of the IFD). Such a scenario would be possible in a case of behavioural plasticity. In this case, if we took a snapshot of the group, we would see each individual playing either producer or scrounger at that moment. However, if we accumulated several of these snapshots over time, we would notice that the individuals are actually sometimes playing producer and sometimes scrounger. There also exists a third means of attaining the ESS frequency of producer–scrounger. The group might consist of a mix of the two previously mentioned solutions: a combination of specialists, which always play either one strategy or the other, and flexible individuals, which switch between strategies according to local conditions.

#### 8.3.2.1 *Two important consequences of game theoretical analysis and the evolutionarily stable strategy*

In a simple optimality analysis of the type we discussed in the previous chapter, we would justify the existence of scroungers in groups by looking to see whether individuals using this strategy do better than those that do not. However, here this type of reasoning does not hold because once an equilibrium is reached, scroungers do no better than producers. Therefore, the scrounger strategy does not exist because it provides an advantage. Scroungers are present simply because a population without them is evolutionarily unstable in a majority of ecological conditions.

The producer–scrounger game also reveals a second characteristic of the ESS. In the previous chapter, individuals maximized their fitness by choosing optimal solutions. Hence a population comprising individuals all adopting the optimal strategy would benefit from a higher mean fitness than a population of individuals adopting non-optimal strategies. Here, this is no longer the case. In Figure 8.9 we notice that the individuals of a population without scroungers would have a higher mean fitness than the members of a population at the ESS. Hence, natural selection leads to a situation in which all members of the population suffer a substantial decrease in fitness as a result of the presence of individuals using the scrounger strategy. The existence of scroungers within the group is a historically inevitable consequence of natural selection given that selection acts on individuals rather than on the group itself, and therefore does not generate solutions that maximize payoffs to all, but solutions that are evolutionarily (or developmentally) stable.

#### 8.3.2.2 *How many scroungers? Testing the model*

The above analysis leads us to predict that scroungers will be present in many groups, however to test the game we need a quantitative prediction, and therefore a model. Several different versions of models of the producer–scrounger game exist (Giraldeau and Caraco 2000), but for the moment we will consider the simplest version (see Box 8.3). This version predicts that the frequency of scroungers in a group depends on group size $G$ and on the **finder's share** $a/F$, where $a$ is the number of prey consumed by the producer before the arrival of the scroungers and $F$ is the total number of prey at the patch. The prediction that the frequency of scroungers depends on the finder's share has been tested twice using nutmeg mannikins (*Lonchura punctulata*), small estrildid finches from southeast Asia (Giraldeau and Livoreil 1998; Coolen *et al.* 2001).

*Nutmeg manikins adapt their strategy according to the finder's share*

Nutmeg mannikins are very social birds and live in groups. They rarely engage in aggressive interactions when foraging, peacefully sharing patches of seed. These birds are commonly kept as pets and so are

**Box 8.3 A deterministic model of the producer–scrounger game**

A model is constructed according to a plausible scenario which determines its initial assumptions. Here, we assume that a group of $G$ individuals are searching for patches each containing $F$ indivisible prey items. Among these individuals, $pG$ adopt the producer strategy and $qG = (G - pG)$ adopt the scrounger strategy. A producer discovers a patch at frequency $\lambda$. The $pG$ producers uncover patches at a total frequency of $pG\lambda$. When a producer discovers a patch, it consumes $a$ prey items before the arrival of the scroungers: this is the **finder's advantage**. The producer remains at the patch after the scroungers arrive and then consumes its portion of the $A = (F - a)$ prey items remaining to be shared between the producer and the $qG$ scroungers.

We can estimate the payoff of the producer ($W_p$) and scrounger ($W_s$) strategies while taking into account the alternative's frequency in the population. We can calculate the equilibrium frequency of producers $\hat{p}$ algebraically given the prediction of equal payoffs at equilibrium.

$$W_p = W_s$$

$$\lambda\left(a + \frac{A}{qG}\right) = pG\lambda\left(\frac{A}{1+qG}\right)$$

$$\hat{p} = \frac{a}{F} + \frac{1}{G}$$

According to this analysis the proportion of producers in the group increases with the **finder's share** ($a/F$), the part of each patch that is exclusively consumed by the finder and decreases with group size.

easily kept in captivity. To test the effect of the finder's share on the frequency of scroungers in a group, we must be able to manipulate this variable. The finder's share depends on patch richness (the number of seeds per patch) and on the number of patches available. Actually, numerous small patches containing few seeds each offer a higher finder's share than do richer patches that are fewer in number. Accordingly, Barbara Livoreil, a French scientist, varied the size and number of available seed patches to test the model using three flocks of nutmeg mannikins foraging for a total of 200 seeds distributed in three levels of patch clumping: high (10 patches), medium (20 patches), and low (40 patches) (Giraldeau and Livoreil 1998). Each flock of birds fed five times a day for six consecutive days at a given level of clumping before going to the next level. Each flock experienced a different sequence of the three levels of clumping. At each of the five daily feeding events Livoreil noted the behaviour of one member of the flock, counting the number of times it fed from a patch that it discovered itself *vs* a patch discovered by another member of the flock. She found that the proportion of patches exploited by producers vs. scroungers changed as a function of the level of clumping (Figure 8.10). That is, the birds invested proportionately more into the scrounger strategy when the patches were richer and less numerous and thus offered a lower finder's share.

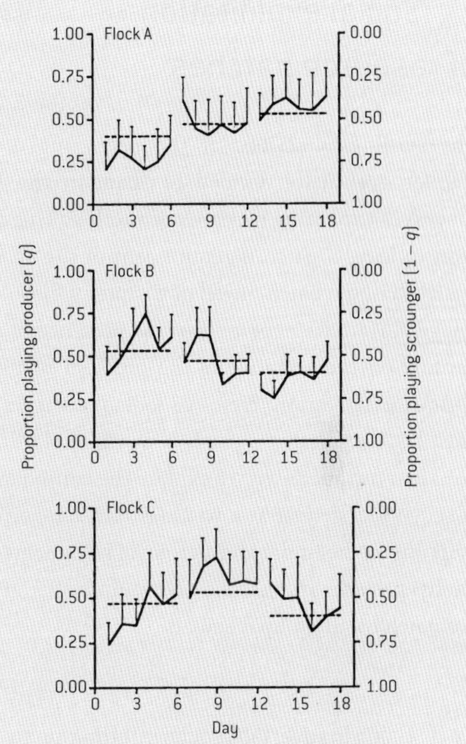


Figure 8.10 **Effect of the finder's share on the proportion of patches exploited by producers**

Three flocks of five nutmeg mannikins (*Lonchura punctulata*) were used. Each flock experienced the same three levels of patch clumping generating three different finder's shares; the sequence of the levels was distinct for each flock. The points represent the mean (and standard deviation) of the five individuals of the flock tested that day. The dashed lines represent the predictions based on the finder's share.

Modified from Giraldeau and Caraco (2000).

It is important to contrast the prediction of the producer–scrounger game with that of the information sharing model. Actually, the latter always predicts the same frequency of scroungers for a given group size, $(G - 1)/G$, or in this particular case, $(5 - 1)/5 = 0.80$. The birds' behaviour seems to fit the prediction of the producer–scrounger game better than that of the information sharing game. This means that for nutmeg mannikins, searching for one's own food (playing producer) and detecting opportunities to partake in the discoveries of others (playing scrounger) are probably incompatible activities (Coolen *et al.* 2001).

## 8.4 Social exploitation of resource patches

The starlings described at the beginning of this chapter eventually decided to abandon the lawn they were exploiting in search of another. This **decision** to leave a patch is reminiscent of the decision analysed in the patch model of Chapter 7 for solitary foragers. Here we explore how the presence of competitors on the lawn alters the problem related to optimal exploitation. First, we look at foragers that must decide whether a patch is worth exploiting or not, then further down we explore the consequences of sequential arrival at a patch as two categories of individuals, the finder and a group of scroungers, and finally provide an analysis of simultaneous arrival at a patch by all group members.

### 8.4.1 Using inadvertent social information to assess whether a patch: is worth exploiting

One of the first major differences of being with others concerns the type of information that becomes available for making patch decisions. When foraging alone, animals only have access to information acquired through their own interactions with the environment, namely personal information. However, when foraging in a group, they also have access to information acquired by watching other group members interacting with the environment, a form of information we call inadvertent social information (ISI) in this book (but see Chapters 4 for a historical perspective about the terminology of the various forms of biological **information**). The availability of ISI makes for more reliable estimation of the quality of a given patch. For example, say one of the starlings on the lawn is having a string of bad luck as it probes the lawn and so concludes that the lawn is of lower value than it really is. However, observing a companion that may be having better luck and using that information could allow it to refine its estimation of the lawn's value and perhaps gain a more realistic estimate of the lawn's real value. If it did this, the starling would be using inadvertent social information (here extracted from the success of a companion) added to its personal information (success/failure by its own trial and error) to obtain an estimation of a patch's value.

*Starlings can use inadvertent social information*

A Canadian scientist, Jennifer Templeton, experimentally demonstrated the use of ISI[3] by starlings in the course of her doctoral research (Templeton and Giraldeau 1996). Inspired by the problem Lima presented to woodpeckers (see Chapter 7) she offered captive starlings two potential patches that looked exactly the same and were made up of 30 wells covered by piece of opaque rubber. Each rubber cover had a slit through which the starling could probe with its bill to determine whether it contained a mealworm or not. For an experimental test the starlings were always exposed to a pair of empty patches but their training lead them to expect that one of them contained food in some fraction of the 30 wells. As was true in Lima's experiment (Chapter 7) probing an empty hole did not say whether the starling was at the empty or full patch. The starling needs to cumulate a sequence of negative probes before deciding it is at the empty patch to switch to the other. During these tests, Templeton counted the number of wells that the starling probed before switching to the other patch. She reasoned that if the starlings could use ISI to estimate patch quality they would probe fewer holes before switching when sampling the patch with a companion. Furthermore, she thought that if the **foraging success and not just presence of a companion** accelerates the starling's departure decision then the amount of information provided by the companion would have an effect on patch departure. She predicted that patch departure would occur after fewer probes when sampling with a

3 Note that in their paper Templeton and Giraldeau (1996) used the expression 'public information' *sensu* Valone (1989) to qualify this form of information. In this book we call it inadvertent social information or ISI. Please refer to Figure 4.3 and the attached text for more details about the conventions used in this book concerning information.

high-information partner than with a low-information partner.

To test these predictions Templeton offered her experimental starlings two types of companions. High-information companions were previously trained to probe all of the wells in a patch and so during experiments sampled a large number of wells. Low-information companions, however, were trained to probe only a few wells. Templeton found that the number of wells probed by a subject in such an experiment was affected by the presence and the quality of the companion; as expected if the starlings were capable of using the foraging performance of conspecifics as a source of inadvertent social information, they probed most holes when sampling alone, slightly fewer holes when sampling with the low-information partner and yet fewer holes still when sampling with the high-information partner (Figure 8.11). Similar results were later obtained when contrasting the presence versus absence of ISI (see Figures 4.4 and 4.5).

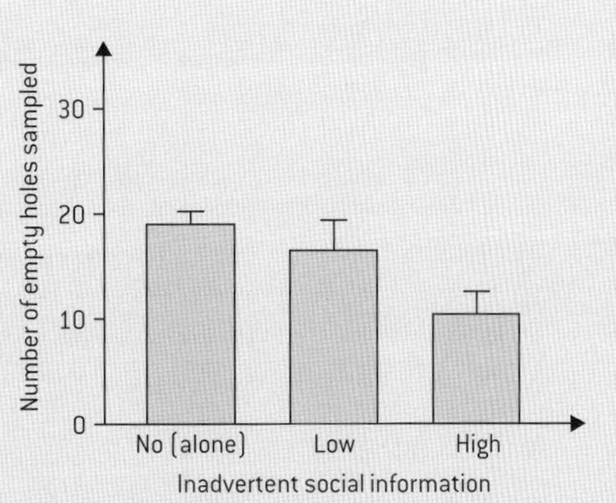


Figure 8.11 **Influence of inadvertent social information on the number of holes of an empty patch a starling (*Sturnus vulgaris*) samples before departure when confronted with having to evaluate the quality of patches**

Mean (+standard error) number of empty holes sampled by a starling before it abandons the patch, when tested alone or with a companion providing either a 'low' or 'high' amount of ISI. A lone bird probes the highest number of empty holes before departure. A bird tested with a companion, probes fewer wells when the companion provides a high amount of ISI compared with a low amount.

Modified from Templeton and Giraldeau (1996).

### 8.4.2 Sequential arrival of competitors

When an individual that is part of a group discovers a patch it is often joined by other group members. These new arrivals probably cause some interference at the patch and so reduce the forager's exploitation rate. Guy Beauchamp, a Canadian scientist, was the first to propose a model that examines the effect of the arrival of scroungers on the producer's patch exploitation time (Beauchamp and Giraldeau 1997). His graphical model inspired from the marginal value theorem (see Chapter 7) allowed the group arrival of scroungers after a brief period of exclusive patch exploitation by the discoverer (Figure 8.12). His optimality analysis predicts that the intensity of interference and the time required to search and discover a new patch should interact in determining a producer's patch time. If interference is intense or if there are a large number of patches in the surrounding area, then the discoverer should leave the patch as soon as the scroungers arrive. Alternatively, if interference is weak or if there are few patches in the surrounding area, then the discoverer should remain at the patch after the scroungers arrive. Beauchamp tested his model using flocks of nutmeg mannikins (*Lonchura punctulata*) and found that discoverers behaved precisely as his model predicts.

### 8.4.3 Simultaneous arrival of competitors

Since at least 1978, several models have examined the optimal exploitation of patches when competitors arrive simultaneously. The British behavioural ecologist Geoff Parker (1978) was the first to propose a model of the social exploitation of resource patches, using the case of the insemination of female yellow

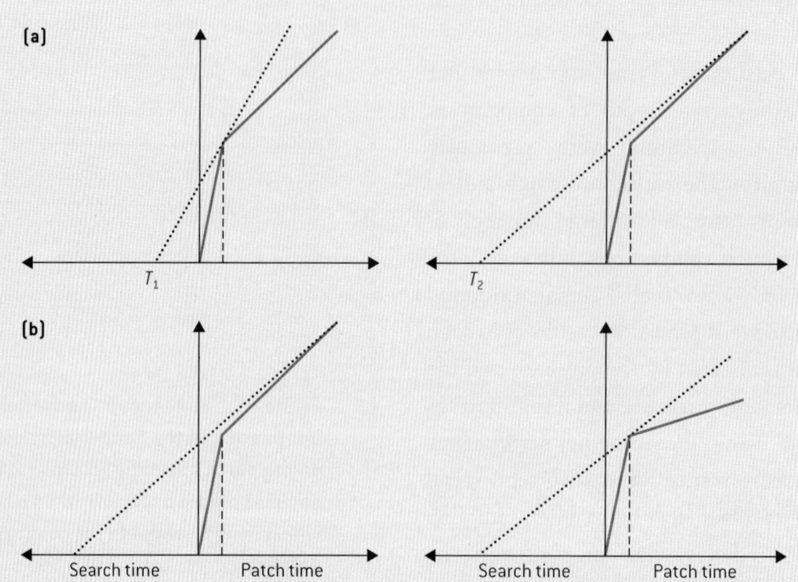


Figure 8.12 **Effect of the arrival of competitors on the time spent exploiting a patch by its discoverer**

Graphical model of the optimal time spent at a patch when a group of scroungers join the discoverer after a brief period of delay. The vertical dashed line indicates the moment of the competitors' arrival. The model uses the tangent method described in Figure 7.7. The abscissas are divided by the ordinate into two sections: time spent at the patch, increasing towards the right, and time spent searching for a patch, increasing towards the left. The ordinate represents the amount of food exploited at the patch.

**a.** Effect of an increase in the time spent searching for patches on the optimal departure time. When this time is of short duration ($T_1$) the discoverer of the patch should leave as soon as the competitors arrive. When this time is of long duration ($T_2$) the discoverer should remain at the patch after the competitors arrive.

**b.** Effect of the intensity of interference experienced by the discoverer owing to the arrival of competitors. When interference is weak (left panel), the discoverer should remain at the patch after the competitors' arrival. Interference decreases the slope of the line representing the profit gained per unit time after the competitors' arrival. The discoverer should abandon the patch upon the competitors' arrival when these new individuals exert a strong level of interference (right panel).

Adapted from Beauchamp and Giraldeau (1997).

dung flies (*Scatophaga stercoraria*). His model is based on a situation in which patches deplete over time, and thus is also well suited to the exploitation of food patches (Giraldeau and Caraco 2000). Since then, Parker has presented a series of **social patch exploitation models** for a series of different foraging conditions (see Giraldeau and Caraco 2000). Two scientists from the Netherlands, Marianne Sjerps and Patsy Haccou (1994), transformed the problem into a war of attrition game, a type of waiting game between competitors. They reasoned that competitors that leave too soon have the advantage of exploiting the next patch while it is either full of food or exploited by only a few competitors. However, leaving too soon also leaves many good resources behind for others to exploit.

Analyses of the game proposed by Sjerps and Haccou is rather technical and beyond the level of an introductory text. However, it predicts that the ESS solution to this game is to pick a residence time from a stochastic distribution of residence times whose mean depends on the extent of interference among foragers. It is important that each forager be unpredictable in its exploitation time or else competitors could easily come up with a better alternative. These social patch exploitation games have yet to receive the kind of empirical scrutiny that was given to solitary patch models.

## 8.5 Choosing prey in a competitive situation

A solitary animal exploiting a patch is free to choose prey according to the payoff provided by its strategy. However, when two individuals simultaneously exploit a patch, the strategy chosen by one of the individuals has a direct impact on the payoff of the other individual's strategies. For example, the payoff obtained by specialising on the most profitable prey, in a patch containing two types of prey of unequal profitabilities, depends on whether the competitor also chooses the specialist strategy or chooses a generalist strategy.

Models of the simultaneous exploitation of a patch by more than one individual are few and far between since the problem calls for complex mathematics. Which strategy an individual chooses is affected by both patch depletion rate, an effect that is ignored in the classic prey models presented in Chapter 7, and by the competitors' choice of strategy. So far, only Heller (1980) and Mitchell (1990) have modelled the effect of adding competitors on the choice of prey in depleting patches. The details of these models once again surpass the level of an introductory text. However, the message is that the two models both predict the same effect of adding competitors, an effect that is initially counterintuitive but can nevertheless be understood.

A lone animal is free to select only the best prey at a patch until the density of this prey type declines so much that generalizing on both prey types is more profitable. The critical prey density at which point the animal extends its choice to include both prey types depends on the same principles of the prey model in Chapter 7. When competitors are added to the model, Mitchell (1990) reveals that the critical density marking passage to the generalist strategy is delayed. The explanation is simple: when the animal is alone, it has the luxury of generalizing at a higher density of profitable prey because in the end all of the prey remain available for its exclusive use. This is no longer the case when the animal is faced with competitors. An individual that extends its choice to include both types of prey will necessarily spend a certain amount of time exploiting the less profitable prey, time its competitor might be using to deplete the supply of the more profitable prey. The presence of competitors therefore results in an extension of the time during which individuals ought to specialize on the more profitable prey type. This strategy is stable, even though the two competitors are less efficient, because all alternative strategies are vulnerable to invasion. Once again, we are faced with a situation in which evolution by natural selection does not lead to the best strategy for all, but instead to the most stable strategy which is least vulnerable to replacement by an alternative. When a patch contains two types of prey with distinct profitabilities, we predict that social exploitation compared with solitary exploitation will lead to overuse of the more profitable prey type.

## 8.6 Conclusion

Chapter 7 marked our first attempt to apply evolutionary reasoning to the study of animal behaviour. We began with a very simplistic vision of food searching behaviour that has been profoundly changed by much of the information presented in the two chapters about foraging. We found that there is a remarkable qualitative correspondence between many of the details of foraging behaviour and predictions based on optimality analyses that assume that behaviour is shaped by a process of selection favouring those individuals best able to transmit their genes. Of course, in most cases quantitative discrepancies between behaviour and predictions led to more research into the assumptions of the models. This research often led us to realize that the animals are capable of rather impressive behavioural plasticity which implies no less impressive cognitive capacities. Naturally, using an optimization approach does not assume that animals perform the complex calculations described by the models. However, the animal's behavioural system has clearly been designed to compute automatically near-optimal solutions to complicated problems, much like the nervous systems of two tennis players that automatically solve with little error the complex algebra required to predict ball trajectories. Throughout this book we will have many more opportunities to illustrate the incredible predictive power of the evolutionary approach behind optimization and game theory. Such predictive power is undoubtedly the main argument in favour of the adaptationist view of behaviour.

## » Further reading


> *About the producer–scrounger game:*

**Barnard, C.J. & R.M. Sibly.** 1981. Producers and scroungers: a general model and its application to captive flocks of house sparrows. *Animal Behaviour* 29: 543–555.

> *About the ideal free distribution:*

**Fretwell, S.D.** 1972. *Populations in a Seasonal Environment.* Princeton University Press, Princeton, NJ.

**Fretwell, S.D. & H.L. Lucas.** 1970. On territorial behaviour and other factors influencing habitat distribution in birds. *Acta Biotheoretica* 19: 16–36.

> *For a general perspective on foraging theory:*

**Caraco, T. & Pulliam, R.** 1984. Sociality and survivorship in animals exposed to predation. In *A New Ecology: Novel Approaches to Interactive Systems* (P.W. Price, C.N. Slobodchikoff & W.S. Gaud eds), pp. 179–309. Wiley Interscience, New York.

**Clark, C.C. & Mangel, M.** 1984. Foraging and flocking strategies: information in an uncertain environment. *American Naturalist* 123: 626–641.

**Giraldeau, L.-A. & Caraco, Th.** 2000. *Social Foraging Theory.* Princeton University Press, Princeton.

**Maynard Smith, J.** 1982. *Evolution and the Theory of Games.* Cambridge University Press, Cambridge.

**Hamilton, W.D.** 1971. Geometry for the selfish herd. *Journal of Theoretical Biology* 31: 295–311.

> *For a discussion about optimal group size:*

**Pulliam, R.H. & Caraco, Th.** 1985. Living in groups: is there an optimal group size? In *Behavioural Ecology: An Evolutionary Approach*, 2nd edn (J.R. Krebs & N.B. Davies eds), pp. 122–147. Sinauer Associates, Sunderland, MA.

**Sibly, R.M.** 1983. Optimal group size is unstable. *Animal Behaviour* 31: 947–948.

For a textbook example of the use of social information in a foraging context (see also Chapter 4):

**Templeton, J.J. & Giraldeau, L.-A.** 1996. Vicarious sampling: the use of personal and public information by starlings foraging in a simple patchy environment. *Behavioral Ecology and Sociobiology* 38: 105–113.

## » Questions

1. Prey are distributed in an ideal free manner in an environment in which some resource patches are richer than others and in which predators are also distributed in an ideal free manner according to their prey's density. In such a habitat, which of the prey in the rich or in the poor resource patch experience the highest per capita predation pressure?

2. In an aggregation economy, what would happen to the expected group size if group size were completely under the control of intruders? In your opinion, which of the two games, producer–scrounger or information sharing, results in the highest foraging payoff for all members of the group? Can you explain your answer?

3. Use reasoning to predict whether the exploitation of a patch by simultaneously arriving individuals would be more or less intense than the exploitation of the same patch by a solitary individual.

9

# 9

# Choosing Where to Breed: Breeding Habitat Choice

Thierry Boulinier, Mylène Mariette, Blandine Doligez, and Étienne Danchin

## 9.1 Introduction

A fundamental **assumption** of the approaches developed in the two previous chapters is that the environments in which individuals live are heterogeneous at various spatial and temporal scales. In this context, individuals capable of occupying the most favourable areas for activities linked to **fitness** (i.e. survival and reproduction) have an obvious selective advantage. Spatial and temporal environmental heterogeneity thus leads to selective pressures favouring the evolution of strategies of habitat choice. This reasoning holds whatever the type of behaviour being considered (**foraging** for food, searching for a mate, finding shelter from predators, locating a spot to breed, etc.). The two previous chapters focused on choice of foraging locations. Here we analyse the choice of a breeding location.

### 9.1.1 What differs between breeding and foraging habitat choice?

There may be several fundamental differences between choosing a breeding and a foraging site. It can, however, be noted here that in some species, for instance some insect parasitoids, searching for food and reproductive sites (i.e. hosts) must often need to be done concurrently and both will be traded-off against one another. In such species mechanisms involved in both foraging and **habitat** selection may therefore interact. In many, if not most other species, however, selecting a breeding and a foraging location will be distinct non-interacting processes that occur at different times during the life cycle of individuals.

*Potential differences in time and spatial scales*
Often habitat choice is distinct from the choice of foraging location but nonetheless constrains foraging decisions at least during the whole breeding period. This is particularly the case in species that are

spatially constrained during reproduction such as many marine invertebrates with sessile adults, and any territorially breeding animals such as some fish, reptiles, birds, and many mammals. In such species, breeding habitat choice has consequences over a period that is much longer relative to an individual's lifespan than those of foraging decisions. Estimating the quality of a foraging patch may involve only the time required to search for and find a food item in that patch, an investment in the order of a few seconds to minutes, depending on the biology of the species. In breeding habitat choice, however, the equivalent rule for assessing patch quality would require much more time. It would include an attempt to breed in the patch, the equivalent of food searching, which may represent a significant portion of the lifespan, especially in species where the number of breeding attempts is limited (e.g. many vertebrates). In the case of seasonally breeding species, for instance, breeding habitat choice implies a time scale that is measured in years rather than minutes. In marine invertebrates with a sessile adult stage, this type of trial and error tactic is impossible as individuals can only choose a site once. In other words, foraging decisions are more dynamic than breeding habitat decisions. In many species, once the breeding place has been selected, individuals are stuck in that place for the whole breeding season, perhaps even their entire life.

These differences in time scale between foraging and breeding **decisions** imply that the type of trial and error tactics that are so useful when looking for foraging spots, are unlikely to be of much use in the context of breeding decisions such as choosing the best breeding habitat. Finally, foraging for food and selecting a breeding site often imply different habitat requirements. These two categories of decision may thus follow different decision rules.

*Potential differences in the relative impact on fitness*

Foraging and breeding decisions also differ in their link to fitness. Foraging decisions will usually affect survival components of fitness while breeding habitat choice will be mostly linked with reproductive components. Because breeding habitat choice involves time and energy **trade-offs** with other activities, it can be more tightly linked to the evolution of life history strategies than foraging decisions. Both types of decision will affect the dynamics of the distribution of individuals in the environment, but often at different time scales, and breeding habitat choice will be directly linked to the long-term persistence of local populations through reproduction, and to the exchanges of individuals (and thus genes) between populations (dispersal).

*Selecting a foraging or a breeding site are choice processes*

Despite these differences, foraging and breeding habitat choices are behavioural processes of choice and thus share several properties with other choice processes, such as the choice of a mating partner. In all these cases, information about potential opportunities may be critical and the choice of each individual may have important consequences for the rest of the population. For instance, we will discuss in detail in Chapter 14 recent hypotheses about the evolution of **coloniality** that involve consequences of individual choices at the scale of the population (see also Sections 9.4.2 and 9.6).

Because **information** is a major common thread for all choice processes, it is possible that comparable, or even common, information-gathering behaviours have evolved in all of them (Chapter 4).

### 9.1.2 Breeding habitat choice, a decision tightly linked to fitness

The choice of a breeding site can have a major impact on fitness. For instance, seals that settle on a beach that is flooded during high spring tide fail in their reproduction. But breeding habitat choice can also affect survival to adulthood, or even adult survival. For instance, in sessile marine invertebrates, a larva that would settle and metamorphose in the wrong place would not survive to reach adulthood. An error of a few meters can be fatal in such a case. Mussels (*Mytilus edulis*) for instance, will only develop if they are attached at the right depth relative to the tide. If placed too high or too low relative to low tide, they do not survive and thus do not breed. More generally, the choice of a breeding habitat or site will be all the more critical when breeders' movements are strongly constrained in space, e.g. restricted around a particular location, during reproduction. Such constraints can be due to the necessity of attending non-mobile young (e.g. eggs in birds, nestlings in altricial birds, new born young in many mammals) or the impossibility of changing location (e.g. sessile marine invertebrates, plants), and may vary temporally at different scales within and between breeding attempts. In these cases, the habitat used for breeding also determines the conditions to which breeders will be exposed during this period of life, which can sometimes represent most of an individual's lifetime. In such highly constrained situations breeding habitat choice will not only affect reproductive success, but also adult survival, and the choice of a good breeding site is thus critical for the fitness of individuals: once settled, adults cannot adjust their choice easily.

Selective pressures on breeding habitat choice can therefore be strong, and differ from those acting on foraging decisions. We can thus expect living organisms to have evolved breeding habitat choice behaviours allowing them to choose optimal habitats at a given time. In this context, the use of any kind of information by individuals allowing them to improve their choice should be favoured by selection. Therefore, we should not be surprised by the existence of sophisticated behaviours of breeding habitat choice in the animal kingdom as well as in plants.

### 9.1.3 Some definitions

Before going further, we need to define more clearly the meaning of several concepts we have already used several times.

#### 9.1.3.1 *Environment, habitat, and habitat patches*

By habitat, we mean the type of environment whose characteristics are suitable enough for the activities of the species of interest. For a given species, a habitat can thus encompass several meanings. It can thus be applied to various activities of individuals (Figure 9.1). At a relatively large spatial scale, the habitat can be defined by the **general type of environment used by a species**, which can be, for instance, rivers or lakes for fresh water fish, or woods for forest birds. At a finer scale, the habitat is defined by the portion of the **environment devoted to a particular activity**. For instance, individuals of a species can breed in areas of dense vegetation, but forage in more open areas where predation risk is higher. Usually, the term habitat does not encompass conspecifics. When we want also to encompass the social components of an individual's requirements, we use the term environment that comprises the habitat and conspecifics.

A habitat patch can be defined as a continuous and homogeneous portion of a habitat. For instance, for a forest living species, the habitat is the forest and a habitat patch is a delimited portion of a forest. Patches may differ by some intrinsic characteristics, and individuals can settle on patches of different quality for a given activity. Habitat quality (or intrinsic habitat quality) of a habitat patch is defined by its consequences in terms of individuals' fitness at low density: a habitat patch in which individuals achieve high fitness is defined as being of high quality (relative to other patches). For instance, individuals may breed in the same habitat, but on patches differing in breeding sites or food availability, density of competitors, predation risk, etc. A habitat of a given quality may be more or less crowded. People use the term **habitat suitability** to describe habitat quality after accounting for its density of occupation by competitors. Within the framework of breeding habitat choice, the habitat suitability is often evaluated in terms of reproductive success and more rarely in terms of adult and juvenile survival rate.

#### 9.1.3.2 *What do we mean by 'choice'?*

Because of environmental heterogeneity, individuals often face alternatives with different fitness outcomes. For instance, a female may have the opportunity of mating with males of either type *x* or *y*, or of ignoring both males and continuing to forage for food. If male *x* is of higher quality than male y, the female may gain higher fitness benefits by mating with him; if both

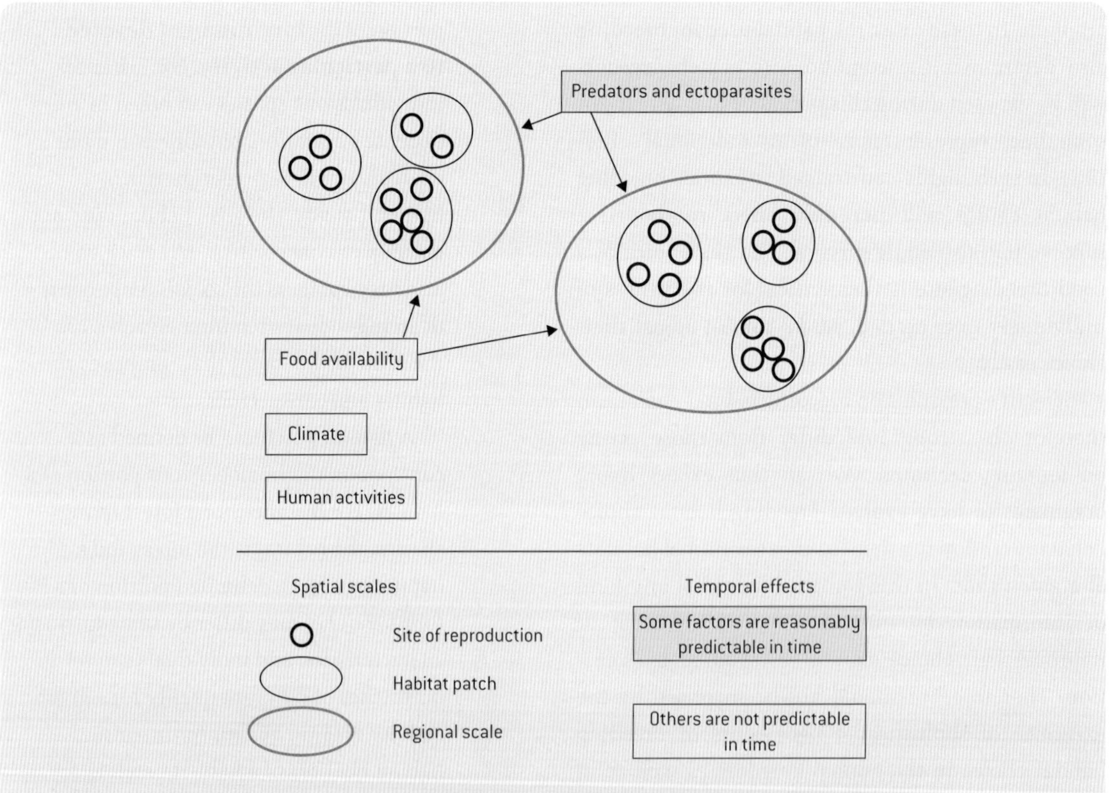


Figure 9.1 **Spatial scales of environmental variability potentially affecting breeding performance**

The ellipses represent the spatial structure of the environment at various scales, from the breeding site to the habitat patch and to the habitat at a higher (regional) scale. Several factors may act on spatial variability including food availability, nest predation, infestation by ectoparasites, etc. Depending on circumstances, these factors may be more or less spatially and temporally autocorrelated and hence predictable. Furthermore, the temporal autocorrelation may differ among factors and according to spatial scale, illustrating how spatial and temporal scales of environmental heterogeneity may interact in a complex way.

Adapted from Boulinier and Lemel (1996).

males are of poor quality or if the female has not secured enough reserves to engage in reproduction, she may gain higher fitness by continuing to forage for food. When confronted with such multiple alternative situations, animals eventually select one of them: it has exhibited **choice**. For instance, in the example above, our focal female may end up copulating with male *y* and not with male *x*. This detectable fact is referred to as a choice. The word 'choice' in this context does not imply a conscious mechanism. It encompasses decisions rules, including the processes by which the individual potentially acquires information about existing opportunities and decides to use the one that, given the situation, is expected to maximize its fitness.

From a practical point of view, choices can be studied at the scale of a series of independent individuals confronted by the same type of choice. Owing to such replication, we can estimate the observed probability of choosing one option to that predicted by chance. If the occurrence of one choice is significantly more common than expected by chance, we can state that the individuals chose or preferred one option (i.e. an operational definition). In the example of the female confronted by two male types, *x* and *y*, if we observe that when given the choice between *x* and *y* males, females copulate with male of type *y* in 75% of the cases and that this distribution differs significantly from the expected 50% by chance, then we can say that females of that population choose to copulate with male *y* rather than male *x*. Such a conclusion can then be tested by experimentally modifying the characteristics of males. For instance, in the above example, experimenters might modify the **phenotype** of *x* males in order to make them look like *y* males and vice versa. If females then start to copulate with *x* males looking like *y* males and ignore *y* males looking like *x* males, we can ascertain that females use the characteristic we **manipulated** to choose mates.

*Two meanings of the word 'selection'*

It is important to note that the word 'selection' has two very different meanings. In the expressions **natural selection** or **sexual selection**, it denotes a mechanism that does not involve individual behaviour. As soon as individuals of one phenotype have more offspring than other phenotypes, and as long as this difference is to some degree **heritable**, frequencies of genes will change in the population, possibly leading to a population comprising descendants of the most successful phenotype only.

In the classic expression 'habitat selection' that is often used to qualify 'habitat choice', the term 'selection' describes an actual process of individual choice, with living organisms gathering information about the environment in one way or another and deciding accordingly. In order to avoid confusion we use the term 'choice' to describe this type of individual behavioural selection in this chapter. Thus the expression 'habitat choice' used here is equivalent to the expression 'habitat selection' that is common in the literature.

*Do plants choose?*

When living organisms arrive and settle in habitat patches of different qualities natural selection favours those individuals that occupy the best patches by chance, despite the fact that they did not perform any actual choice. For instance, the spatial distribution of the offspring of a plant that disseminates its seeds at random would of course not appear to be random, because only seeds falling on good quality patches may develop. This apparently non-random distribution would nevertheless only involve natural selection by the environment: seeds that would have fallen in favourable sites would have germinated and grown into adult plants, while others would not. In such a case, even if the spatial distribution pattern of offspring obtained could have been identical if seeds had 'chosen' favourable sites, we do not define this as a choice since the pattern is only the result of natural selection by the environment. Conversely, if the mother plant produces variable proportions of dispersing and non-dispersing seeds depending on the local environmental conditions, we could define this strategy as a form of choice. In such a case however, the 'choice' would be made by the mother plant using some information about the environment, not by the seeds. More generally, our definition of choice implies that plants do perform choices. For instance, when plants grow toward a light source (i.e. phototropism), we can consider that they made a choice, as some sort of information was gathered and used to orientate growth.

#### 9.1.3.3 *Personal information or inadvertent social information*

##### a *Choosing implies using some information*

In common language, the term 'choice' encompasses the use of some kind of information about alternatives. The minimal possible information is knowledge of the existence of alternatives. Like humans, animals only choose among alternatives about which they have the minimum information that they exist. Thus, as we have defined it above, the concept of choice inherently incorporates the notion of information gathering and use.

To make an optimal choice between different potential breeding sites or patches, individuals are likely to benefit from having *a priori* information. This raises the question of what type of information about environmental variability should be used by individuals in choosing a breeding site. This is not a simple question in the context of breeding habitat choice, because the breeding **performance** of individuals often depends on many factors. A possible strategy could be to evaluate all characteristics affecting breeding success in each site independently of each other. This could clearly become prohibitive in terms of time, energy and cognitive mechanisms when many factors influence reproduction. Furthermore, some factors may act at the beginning of the season, while others are critical at the end. We may thus speculate that individual strategies for choosing a breeding site or patch using integrative cues that reveal the effect of various factors on expected fitness may be especially favoured when many factors may affect reproduction.

##### b *Various forms of information*

Animals can use many types of environmental information. It can be extracted from the physical and biological characteristics of the habitat. It may also be extracted from conspecifics. The relative role of the various sources of information will greatly vary depending on species biology and the question under study. For instance, we have seen in Chapters 7 and 8 that in the context of foraging, different strategies involve gathering information via a direct sampling of the environment by individuals. This type of information has been called **personal information** (often misleadingly called **private information**; see Figure 4.3 for more details).

Individuals may also gather information about current environmental conditions from conspecifics. Such **social information** may be provided either intentionally through **signals**, or inadvertently. In the former case we talk of true communication (see Chapter 16). In the latter case we talk of **inadvertent social information** (ISI) that arises from the monitoring of the behaviour and performance of other individuals. Signals often reveal individual quality reliably (see Chapter 16) and are thus used in choices of partners. Signals, such as alarm signals may also inform on environmental conditions. When individuals use ISI, they benefit from environment sampling performed by other individuals. Thus ISI involves **social cues**. A common thread of signals and social cues (i.e. ISI) is that they all involve condition-dependent traits as the source of information that reveals individual performance in the environment. Altogether they thus provide **performance information**. Chapter 4, and particularly Figure 4.3, provides a general definition and 'taxonomy' of biological information. In this chapter, we chose to focus on social information because several recent studies in birds, mammals, and fish have demonstrated the importance of this type of information in the context of breeding habitat choice.

We mentioned above that, although natural selection may favour trial and error strategies in foraging, these strategies are unlikely to be favoured for the choice of a breeding site in many situations.

Trial-and-error tactics boil down to using personal information as the sole source of information, which implies settling at random and using the information gained from the breeding attempt to decide about future habitat choice (e.g. leave after a failure or stay after a success). This might be very costly because the total number of breeding attempts is limited. Therefore, we might expect individuals to use ISI more commonly in the context of breeding habitat choice than in the foraging context, despite the lack of empirical support for the moment.

This chapter deals with breeding habitat choice, often called breeding habitat selection. Section 9.2 discusses the sources of temporal and spatial variability of the environment affecting the expected fitness of individuals, which creates the selective pressures for habitat choice. This will lead us to define the conditions in which animals choose their habitat and specify how we can demonstrate the use of a given type of cue in breeding habitat selection in Section 9.3. Section 9.4 and 9.5 review the sources of information that could be used by individuals to choose a habitat. Section 9.6 discusses some of the consequences of breeding habitat choice mechanisms on the dynamics of the distribution of individuals in the environment.

## 9.2 Patterns of habitat variability in space and time in relation to breeding habitat choice

### 9.2.1 Spatial and temporal heterogeneity of factors affecting reproduction

Spatial heterogeneity and temporal predictability of habitat quality are required conditions for breeding habitat choice to evolve. In a homogeneous and equally exploited environment, there is no need for individuals to choose because the expected fitness will be the same wherever they settle (Orians and Witenberger 1991). Moreover, if the environment is not predictable at the relevant time scale, then it would also be pointless to choose a breeding location based on some characteristics as these characteristics could have randomly changed in the time between information gathering and decision.

#### 9.2.1.1 *A matter of scale*

The detection of spatial heterogeneity and temporal predictability are tightly linked with the scale considered. Indeed, the environment may be homogeneous at a given spatial scale (e.g., within patches), but heterogeneous at another scale (e.g., among patches). The same is true for temporal predictability, which can occur at a scale varying from minutes up to years. Furthermore, the scale at which an individual of a given species may perceive spatial heterogeneity depends on its movement ability in the environment. Individuals can detect spatial heterogeneity, and may thus select their breeding site, only if their movement ability allows them to visit sites of varying qualities. Individual movement ability thus constrains habitat choice by defining the upper scale at which breeding habitat choice actually occurs.

Breeding habitat choice likely involves a cascade of nested scales: for instance, individuals can first choose among habitat types, then within a given habitat type, a general area in which a habitat patch may be chosen that comprises several potential breeding sites (Orians and Winttenberger 1991). One thus needs to identify the pertinent spatial and temporal scales at which habitat choice needs to be investigated. These scales are constrained by habitat heterogeneity itself, but also by the ability of individuals to gather information on this heterogeneity, as will be developed in Sections 9.4 and 9.5.

#### 9.2.1.2 *Factors affecting habitat quality*

Environmental factors relevant to habitat choice are those that affect fitness and that vary in time and space at the scales individuals can explore. Environmental heterogeneity at various spatial and temporal scales comprises **abiotic, biotic, and social environmental characteristics** that may affect fitness differentially.

### a *Variations in abiotic habitat*

The physical characteristics of the environment that are relevant for breeding habitat choice can vary from climatic conditions (rain, wind, and temperature regimes), soil type (for instance for species that dig burrows), stability of the substratum (when on a slope or in a tree), salinity (for marine species), etc. All these factors may affect fitness in very different ways depending on the species and not all of them need to be considered in each case. For instance, temperature may have relatively little influence on the reproduction of an endotherm (within a reasonable range of variation), but can greatly affect reproductive output of ectotherms, such as marine turtles (see Chapter 6).

The importance of a factor for habitat choice can vary with its **spatial and temporal variability**. For instance, when a marmot chooses a breeding site, it is unlikely that the level of precipitation will affect its choice between two sites 0.3 km from each other, simply because this factor is probably homogeneous at that scale. Nevertheless, in a mountainous environment, it could snow on the first site but rain on the other, if one is located a little lower in the valley. Thus, if the nature of precipitation (rain or snow) affects fitness, precipitations could then affect the choice of a hibernation site by marmots, because precipitations would be spatially heterogeneous at the scale considered. In this example the same burrows are used in winter and during breeding, so that burrow characteristics likely affect all components of fitness, and thus are relevant to breeding habitat choice.

### b *Sources of variation in biotic habitat*

*Quantity and quality of resources*

Some biotic resources, such as food and nest building materials, are critically needed for the survival and/or reproduction of individuals of many species. They can thus directly define the quality of a breeding site. Such resources may be required in sufficient quantity, but also quality. For instance, specifically required nutrients may only be available in specific food items. This can complicate the appraisal of the quality of a breeding site both for animals and researchers.

*The need to account for all ecological requirements*

In many species, the breeding niche is multidimensional, and individuals not only need food but also critical resources such as mates, nesting, oviposition, displaying, or settling sites. Each type may be a prominent component of breeding territory quality. Competition for each of these resources may also affect the density of breeders as predicted by the **ideal free distribution** (IFD) model (see Chapter 8). A suitable patch must contain all the resources needed to breed. Thus, accounting for only part of the fundamental ecological requirements for breeding may be misleading. For instance, a forest may provide large amounts of food, but may not be used for breeding because of the lack of other important requirements, such as tree holes for hole-nesting forest species.

*Predation or parasitism risks*

Two environmental biotic factors that are often spatially heterogeneous are predators and parasites. Predator territoriality makes predation risk spatially heterogeneous. Nevertheless, if predators are numerous and/or diverse, or if their territories are larger than the area explored by prey during breeding habitat choice, it can be difficult for prey to settle in a predator-free zone, and predator pressure may appear relatively homogeneous (Clark and Shutler 1999). This provides another example of the importance of the scale of environmental variation relative to individuals' movement capacities. Lack of territoriality in predators can make predation risk spatially homogeneous and non-predictable in time. In birds, nest predation has long been shown to be a major factor of breeding habitat quality that can strongly affect **life history strategies** (Martin 1992). The actual presence of a predator is however not always synonymous of increased local risk of predation. For instance, some predators may breed on territories of species which they are known to prey upon, but may predate individuals only in areas further away, possibly benefiting from the local presence of heterospecifics, in terms of detection of danger. Individuals may also adopt breeding strategies limiting the risk of nest predation. For instance, azure-winged magpies (*Cyanopica cyana*) may gain benefit from settling close to sparrowhawks (*Accipiter nisus*) and breeding synchronously with them, because sparrowhawks provide protection by defending their own nests from potential magpie predators (Ueta 2001).

Spatial heterogeneity in the distribution of parasites is often linked with their mode of transmission between hosts and with the hosts' distribution (Combes 2001). Parasitism has also been identified as a major factor of breeding habitat quality, potentially strongly affecting life history strategies. For instance, in colonial birds, such as the cliff swallow (*Hirundo pyrrhonota*) and the black-legged kittiwake (*Rissa tridactyla*), negative effects on breeding success of high infestation by ectoparasites have been reported (see Figure 14.2), and parasite infestation can vary at the scale of a few meters. The local level of infestation being temporally autocorrelated between successive years, this factor may directly or indirectly be taken into account by birds when selecting a breeding site (Brown and Brown 1996; Danchin *et al.* 1998).

*The importance of subtle differences in foraging strategies, diet, and seasonality*

Furthermore, different phenotypes within the same population may adopt different foraging strategies during breeding showing that different solutions may exist to fulfil foraging requirements. In the herring gulls (*Larus argentatus*) breeding in eastern Canada, for instance, some individuals forage from the sea whereas others from the same breeding colony primarily forage in rubbish dumps (Pierotti and Annett 1991). As these two types of food resource differ in nutritive value, predictability and availability, this shows that various strategies may be adopted to solve the problem of accessing to sufficient quantities and qualities of food.

The spatial distribution of suitable resources may also depend on the specificity of the diet of the species considered, with higher constraints for specialist species. For instance, the wildebeest (*Connochaetes taurinus*) forages on a wide range of herbaceous plants and can exploit vegetation over a wide area unlike the dik-dik (*Rhynchotragus kirkii*), that feeds only on shoots of some specific plants, and thus can only exploit a patchy portion of the same habitat. In this case, the definition of the habitat itself may in fact be considered to differ between the two species. Finally, the predictability of food availability may vary in time. In seasonally breeding species, it may be easier to predict environmental conditions from one year to the next than from the beginning to the end of the breeding season.

### C *Social characteristics*

*Density of conspecifics: conspecifics as competitors*

Conspecifics are often considered as competitors because they require the same resources and their presence may reduce the fitness of a given individual. **Competition** among conspecifics can be direct, such as when they compete for a limiting resource, or indirect, such as through the attraction of common predators. Because conspecifics also have to select and secure a breeding site, they will affect the relative quality of potentially available breeding sites depending on their distribution among breeding patches. We are thus dealing here with a dynamic system where the decisions of conspecifics will affect fitness gains expected by individuals choosing a particular site (i.e. a frequency-dependent process).

Because of competition, all individuals may not be able to settle in the highest quality, preferred patch. This may be either because individuals are prevented from breeding there by dominant competitors despite attempting to secure a breeding site, or because individuals somehow evaluated the intensity of competition beforehand and chose not to settle on these best patches. The result is that individuals may not be aggregated on the best patches of habitat, their distribution depending on the local availability of resources and density of conspecifics. Competition is in fact likely to be the main process leading living organisms' distribution towards an IFD (see Chapters 8 and 10 for details on this concept). In some cases, competition can be negligible, such as among organisms that filter water like mussels, for which food availability depends upon the concentration of plankton in the water but little on the local density of conspecifics.

For a given species, intra-specific competition levels can vary in time and space, notably when it depends on environmental conditions such as variations in resource availability. Competition can also exist with individuals of other species exploiting the same resource (inter-specific competition; Petit and Petit 1996).

*Conspecifics as sources of information*

The presence of conspecifics can also have positive effects on fitness. For instance, in some species individuals interact with each other to detect and capture prey, to defend themselves against predators, to build nests, etc., so that when conspecific density decreases below a certain value, individuals' fitness decreases: a process known as the **Allee effect**. Allee effects are not limited to cooperative or social species. For instance, even in solitary species, a minimum density is required for sexual partners to meet. Finally, the presence of conspecifics could be beneficial not only after the settlement of individuals, but also by allowing individuals to choose optimal breeding sites. Indeed, density can be a simple and integrative cue used by **prospecting** individuals to distinguish among sites that are more or less suitable, the presence and breeding success of conspecifics indicating site suitability. Thus density may provide ISI, a point we return to later in this chapter and that are discussed in Chapters 4 and 14.

In addition to a mere quantitative Allee effect, the quality of potential partners may vary with density and thus affect breeding habitat quality. Conspecifics can differ in their ability to detect and/or deter predators, to capture preys, to built nests, etc. Several studies have thus shown that patch quality encompasses not only the presence of conspecifics, but also their phenotypic quality, and that these two components are taken into account by individuals in breeding habitat choice (see below).

*Kinship*

Competition among unrelated individuals can be costly, but competition among related individuals can be even more damaging as it also involves the cost of competition among individuals carrying identical genes present in related individuals. In other words, competition among non-kin only affects the direct component of **inclusive fitness**, whereas competition among kin affects both the direct and indirect component of inclusive fitness. Therefore, species in which kin are spatially aggregated are expected to take into account the kin structure of local populations in their settlement decisions. Kin structure therefore is an essential part of the social component of environmental quality (see Chapter 10).

Mechanisms involved in avoiding kin competition are diverse. They can for instance imply kin recognition, which may involve complex cognitive processes. Alternatively, they can be indirect and much simpler. In many species, because of the spatial viscosity of movements of individuals in the habitat, the probability of interacting with kin decreases dramatically with distance between birth and settlement sites. We will come back on these important considerations in Chapter 10, and later in Chapter 15 about their impact on the evolution of cooperation.

### d *Interactions between factors (spatial locations, temporal constraints . . .)*

We have seen that the quality of a site depends on many factors. Nevertheless, it is likely that spatial and temporal variations of important factors occur at different scales and have different patterns. These differences generate **trade-offs** between factors (see Chapter 5), because the values of the different factors that maximize individual fitness may not occur in the same locations at the same time. Furthermore, some factors of habitat quality are likely to be negatively correlated. For instance, the stability of the substrate for building a nest is often negatively correlated with the risk of exposure to predators: e.g. birds breeding in vertical cliffs or on high tree branches limit nest exposure to ground predators, but may have higher risk of experiencing the fall of their nest.

Patterns of temporal variation of factors affecting fitness can also differ. For instance, the infestation level of a site by ectoparasites may follow a multi-year dynamic, while the presence of an active predator may only be predictable over a shorter time scale. Depending on the relative importance of these two factors on fitness, individuals may achieve a higher success either by leaving a breeding patch once a predator starts being active there, or by staying while parasites are not so abundant.

Interactions between factors can be detected by comparing situations where several of them act together with situations where one of them is absent. Returning to the example of birds nesting in inaccessible sites, such a choice is likely to be advantageous on continents where predators are numerous. But on isolated islands where ground predators are absent, individuals may gain higher fitness by breeding on flat ground and avoiding cliffs. The trade-offs experienced by individuals may themselves differ in time and space. The spatio-temporal variability of trade-offs should be tracked by individuals to adjust their breeding site choice.

### 9.2.2 Constraints on habitat choice: from the general model to the choice process

We have seen how the capacity of acquiring information and choosing a breeding site accordingly may increase fitness in a spatially heterogeneous and temporally predictable environment. As we will see in the next chapter, the process of choosing a site includes two important steps. First, deciding whether to leave the current patch must occur at least once in the lifetime of most organisms, i.e. when they decide to disperse or not from the natal site (natal site fidelity versus **natal dispersal**; see Chapter 10). In mobile **iteroparous** species (i.e. species that can reproduce several times), individuals can change breeding sites between breeding events (**breeding dispersal**). Second, if individuals decide to leave, then they must also choose where to settle. We will see that these two decisions may be based on different criteria of habitat quality, and may be either independent or linked: individuals may decide to leave before having decided where to settle next; alternatively they may decide to leave because they have already chosen their next breeding patch.

Breeding habitat choices are thus relatively complex processes constrained by many parameters (Figure 9.2). Furthermore, as we have seen, these constraints can be linked to: (1) the characteristics of the species, such as cognitive capacities, ability to move in the environment, etc.; (2) the life history strategies of the species and the trade-offs involved; (3) the characteristics of the individuals; and (4) particularities of the environment.

#### 9.2.2.1 *Constraints linked to the species' characteristics*

As mentioned previously, habitat choice is only possible at the spatio-temporal scales at which individuals can explore their environment. This constraint is important in the context of fragmented habitats: good quality habitat out of reach of individuals because of isolation may remain empty.

To be effective, habitat choice may require complex cognitive capacities (Klopfer and Ganzhorn 1985). For instance, in time limited marine species, individual larvae may need to track the time already spent searching for a breeding site, because they can fix on a substratum only during a given time window after which they die if not fixed (Ward 1987). More generally, memory, and in particular spatial memory, seems to be important in breeding habitat choice. Individuals need to be able to remember the sites already visited, which of them are occupied or free, which patches are of high quality, etc. In the specific case of migratory species, **philopatry** (i.e., individuals breeding on the site or patch where they were born or previously bred) implies the memorization of birth or previous breeding site location. All these abilities constitute factors for which a given individual can show limits. For instance, spatial memory in vertebrates has been found to be associated with the

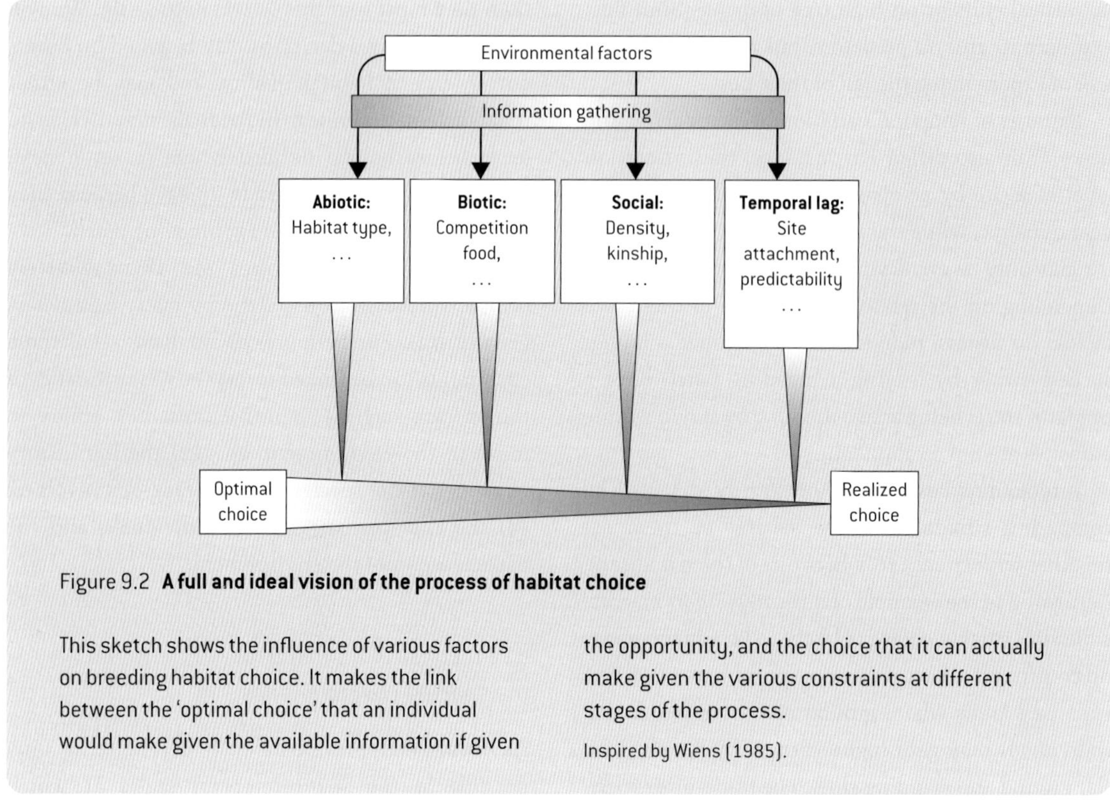


Figure 9.2 **A full and ideal vision of the process of habitat choice**

This sketch shows the influence of various factors on breeding habitat choice. It makes the link between the 'optimal choice' that an individual would make given the available information if given the opportunity, and the choice that it can actually make given the various constraints at different stages of the process.

Inspired by Wiens (1985).

size of the hippocampus, a region of the telencephalus. Bird species that cache food show a higher relative size of the hippocampus than species that do not cache. Because brain size does not necessarily increase in such species, the development of the hippocampus occurs at the expense of other brain structures.

Thus, as for any choice process, individuals have to acquire and process information about opportunities in order to make optimal choices. Limitation in cognitive abilities may constrain habitat choice. However, despite their apparent complexity, breeding habitat choices do not necessarily involve complex cognitive mechanisms. For instance, marine invertebrates without complex neural structures and with sessile adults are capable of sensing subtle differences in environmental chemicals that reveal habitat quality. In such species, habitat choice probably only rests on molecular recognition which triggers changes in metabolism leading larvae to fix on the substratum (Meadows and Campbell 1972). Few people would claim that such mechanisms belong to what is usually called cognition.

#### 9.2.2.2 *Constraints due to species' life histories*

Breeding habitat choice is a key component of life history strategies not only because of its direct effects on fitness, but also because individuals may have to trade-off the investment in various activities with investment in habitat choice (see Chapter 5). We have shown the importance of movement capacity in relation to the scale of breeding habitat choice. However, the **mobility** of individuals may be crucial in a different way for species where a change of stage during the life cycle is accompanied by a change in mobility. This is especially the case of species with reduced range during the breeding season. For marine invertebrates with sessile adults, the mobility before settlement is even more critical because adults do not have the possibility to change after a bad habitat choice. A wrong choice of fixation site in this case can basically reduce fitness to zero.

The ability to choose a breeding site is also limited by the costs of prospecting for potential breeding sites. Such costs can be in terms of time and energy.

Because of time constraints in prospecting, individuals may end up settling in sub-optimal habitat patches because they have not been able to fully explore the environment and identify higher quality patches. Prospecting costs and benefits may vary with species' life expectancy: long-lived species may be able to spend more time gathering information during the immature phase, before settling. When prospecting occurs at the end of the breeding season, individuals have to trade investment in current reproduction against choosing a nesting site for the following year. Costs can also result from a higher predation pressure on prospecting individuals that may move through riskier habitats, though evidence of this increased risk is scarce.

Finally, the choice of a site or territory can be constrained by the functions associated with the use of this site or territory. In some species individuals defend a multi-purpose breeding territory that should thus contain the necessary food resources, while in others, individuals occupy a territory providing no other resource than a breeding site. As we have seen, the different factors affecting habitat quality may rarely be optimal simultaneously. Thus, the higher the number of functions of a site, the stronger the trade-offs among factors of habitat quality, and the less likely that the best sites will be optimal for all functions. Natural selection will then favour individuals adopting strategies allowing them to optimize overall fitness rather than specific fitness components linked to some characteristics of the site. This problem is important when searching for correlations between fitness components and specific breeding site characteristics. Finding no correlation may not mean that the study factor is unimportant, but rather that there are more important ones.

#### 9.2.2.3 *Constraints linked to characteristics of individuals: phenotype–environment interactions*

Habitat choice models often assume that a habitat patch or a site has an absolute or **intrinsic habitat quality**, and thus that all individuals of a species rank habitat patches or sites according to that quality in the same way. However, an increasing number of studies suggest that a given habitat may not necessarily have the same value for all individuals (review in Stamps 2001). Such inter-individual differences can sometimes reduce competition because all individuals will not attempt to settle on the same sites.

In particular, the '**habitat training**' hypothesis suggests that prior experience of an individual with a given habitat type increases its fitness in that habitat type by modifying its behaviour and/or physiology. As a consequence, the individual should settle on the same type of habitat if it chooses to disperse (Stamps 2001). Habitat training may for instance occur via acquired resistance to parasites present in the individual's natal habitat. When exposed to the same parasites in a similar type of habitat, this individual may achieve a higher fitness than individuals not previously exposed to these parasites. The presence of specific parasites in a patch could thus lead to different breeding habitat choices by individuals depending on their past history. Although it is clear that choosing a patch with parasites will always be the 'best-of-a-bad-job' strategy (Boulinier and Lemel 1996), the impact on fitness will depend on the individual's past experience with these parasites. Similarly, individuals may use some characteristics of their natal site to identify patches of high quality for them ('**habitat cueing**' hypothesis). The phenotypic plasticity underlying habitat training or cueing processes could allow a better use of given types of habitat, but at the same time, may reduce the potential use of other, yet favourable, habitat types. Such a specialization can constrain habitat use, especially if the natal habitat type is rare in the environment.

Individuals can also differ in their propensity to fight predators, their capacity to defend a territory, their current breeding investment, etc. so that the main criteria for choosing a habitat patch may differ among individuals. For instance, the most competitive individuals may try to settle on the most attractive areas, while less competitive ones may settle on sites of lower quality, for which competition is lower. Moreover, individuals may give priority to one habitat characteristic or another depending on their own phenotype, in particular in the case of multi-purpose territories, and may consequently

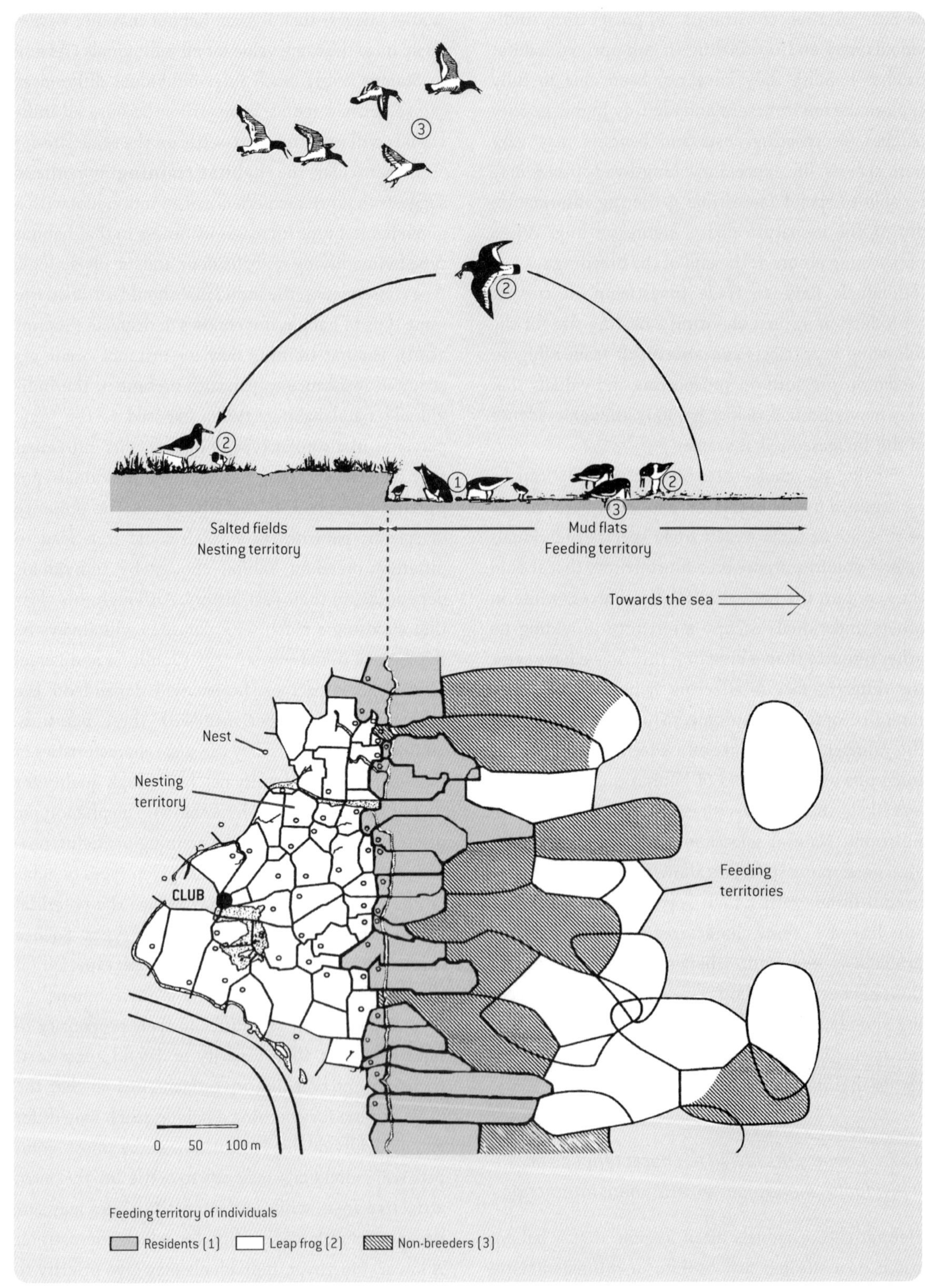
③
②
②
①
②
③
Salted fields
Nesting territory
Mud flats
Feeding territory
Towards the sea
Nest
Nesting territory
CLUB
Feeding territories
0 50 100 m
Feeding territory of individuals
Residents (1)
Leap frog (2)
Non-breeders (3)

rank sites differently. Thus, inter-individual differences may lead to diverging habitat preferences.

*Two breeding strategies in oystercatchers*

A nice example is the study of the oystercatcher (*Haematopus ostralegus*) which showed that individuals follow two different strategies: some aim at settling on the best breeding territories (those next to feeding territories), whereas others, less competitive, choose to breed on territories of lower quality (those further away from the foraging area). The first type of individual may have to wait for several years before securing one of the high-quality territories, because of high intra-specific competition, whereas the second can start breeding as soon as they reach sexual maturity, but they achieve a lower breeding success for each breeding attempt (Ens *et al.* 1995; Heg 1999; Figure 9.3).

*Local adaptation, genotype–environment interactions, and mate choice*

Individuals may make different use of the same information in breeding habitat choice for many other reasons than the ones described above. For instance, they may have different gene combinations that give them different resistance levels to parasites. More generally, genotype–environment interactions (see Chapter 2) may lead to different optimal choices. We have seen that a way to resolve risk of kin competition may be to disperse over a significant distance. However, dispersing over too long a distance may lead to the breaking of favourable gene complexes, thus generating a cost of outbreeding.

Finally, males and females are obviously subjected to different constraints (see Chapters 11 and 12) and may thus be prone to making different habitat decisions. Sexual selection in general is likely to influence breeding habitat choice. For instance, males with high male–male competitive abilities will be more likely to settle in highly competitive patches than less competitive males in polygynous species, whereas the difference in females may be much lower.

## 9.3 Showing that choice occurs

It is important here to distinguish **processes of choice** from **patterns of space use**. The processes of habitat choice are the mechanisms by which individuals select their habitat. The patterns of space use

Figure 9.3 (*opposite*) **Two strategies for becoming a breeder in the oystercatcher (*Haematopus ostralegus*)**

In the study population at Schiermonnikoog, the Netherlands, two types of breeding territory exist, which differ greatly in quality. The territories situated immediately along the coastal mud flats are of high quality because breeders can defend a breeding territory (a portion of salty fields along the coastal line) and a feeding territory (a portion of the mud flat) that are contiguous. This proximity diminishes the cost of movements between territories and the costs of defence of the territories, and it offers better protection to the brood against predation. Breeding territories further away from the mud flats are of lower quality because breeders have to defend a feeding territory further away (grey areas on the mud flats) and thus pay extra costs. Pairs that settle along the coastline therefore produce more offspring each year, and the two types of territory are used by two types of individual depending on their competitive ability. However, the overall lifetime reproductive success does not differ between individuals using the two types of territory. This is because access to low-quality territories is much easier (lower competition), and thus individuals can start breeding on those territories much earlier than on the high-quality sites, where individuals have to queue for several years for a vacancy before settling. Individuals may thus choose between the two different settling strategies as a function of their own competitive ability.

Modified from Ens *et al.* 1995.

are the distributions of individuals in the environment, which result from every individual making decisions (Wiens 1985; Figure 9.2). Patterns are often used to infer processes.

### 9.3.1 Patterns of distribution of individuals: the ideal free distribution and its limits

Determining whether individuals are actually choosing where to settle is not so easy (van Teeffelen and Ovaskainen 2007). Much of the work on habitat choice has not been conducted on behavioural processes of choice, but rather on the distribution of individuals in the environment, or sometimes on the spatial variation in fitness. These descriptive elements are used in a second step to infer potential choices by individuals. We have already detailed several examples in this book showing the risks of using patterns to infer processes, despite the popularity of this approach. This has especially been illustrated by the concept of IFD (already explained in detail in Chapter 8, predictions and their tests are provided in Chapter 10), proposed by Fretwell and Lucas (1970). The results of studies using the IFD concept, while obtained mostly in the context of foraging patch choice, have nevertheless been used to discuss the choice of any habitat type (see for instance Bernstein *et al.* 1991).

#### 9.3.1.1 *A theoretical concept*

As we saw in Chapter 8, IFD is the distribution of individuals among habitat patches that is expected under the **assumptions** that individuals (1) distribute themselves so as to maximize their fitness, (2) are free to move among habitat patches, i.e. without any cost or constraint, and (3) have a perfect (ideal) and instantaneous knowledge of the relative quality of habitat patches and local density dependence functions (see Box 8.1). At equilibrium, (1) mean individual fitness is the same on all patches and (2) individuals cannot increase their fitness by changing patch. The simplest IFD model assumes a negative density-dependence function (see Chapter 8), i.e. the fitness of individuals decreases with increasing local density of conspecifics. A model including an Allee effect (i.e. a non-monotonous density dependence function, with fitness first increasing with local density up to a threshold beyond which fitness decreases with increasing density) had already been proposed by Fretwell and Lucas (1970). All IFD models, however, predict that the gain of each individual is equal to the ratio of resource availability divided by the number of competitors on each patch if all individuals have the same competitiveness in accessing resources (see Chapter 8).

#### 9.3.1.2 *The limits and strengths of the ideal free distribution concept*

Although the IFD concept has proved very useful in ecological studies (see Chapters 8, 10, and 14), the two main assumptions, i.e. 'free' and 'ideal' individuals, will rarely be met in natural situations. The first assumption of the IFD states that individuals make ideal choices because they are omniscient, i.e. they have full and up-to-date knowledge of the relative quality of all available patches. Individuals will, however, rarely achieve such knowledge, in particular because of rapid environmental changes. For instance, the sudden arrival of a predator on a patch may dramatically increase local predation risk, and thus decrease patch quality. Tracking such sudden changes in factors affecting patch quality will be very difficult for individuals without a high and constant effort devoted to gathering information. The choice of a patch will thus rarely be 'ideal'. Moreover, determining the shape of density dependence functions within habitat patches may be difficult (Wiens 1985), in particular because of the multiplicity of factors affecting habitat quality and because of phenotype-environment interactions (see Section 9.2).

The distribution of individuals should also rarely be 'free', contrary to the second main assumption of the IFD, because movements are always constrained, e.g. by environmental features or conspecifics, and costly, e.g. in terms of time, energy, and social interactions. For instance, in the context of breeding habitat choice, the acquisition of a new breeding site by dispersing individuals after changing patch can represent a significant cost (Switzer 1993; see Chapter 10). More complex IFD models integrate differences among individuals in competitiveness, some individuals being able to monopolize access to a resource or territory. In this case, all individuals do not achieve the same fitness, with more competitive individuals being able to monopolize a larger fraction of resources and achieving higher fitness (see Chapter 8). Inter-individual differences in competitive ability impose constraints on individual movement between patches, implying that the second assumption of 'free' individuals is not met anymore. In this case, therefore, the distribution of individuals in the environment follows an **ideal despotic (or dominant) distribution** (Milinski and Parker 1991).

The limited ability of IFD models to predict the distribution of individuals among patches in the environment is notably illustrated by the frequent observation of high-quality patches left unoccupied or very sparsely occupied. We will come back to the question of the limits of IFD models in Chapters 10 and 14. The important message here is that the IFD represents a useful theoretical framework because it is a null model that describes the spatial distribution of individuals in the environment that maximizes fitness at the population scale given the distribution of various types of habitat. The IFD thus constitutes an 'ideal' distribution to which the distributions generated via different habitat choice processes incorporating constraints on information accessible to individuals and their movements can usefully be compared (Bernstein *et al.* 1988; Doligez *et al.* 2003).

#### 9.3.1.3 *What can really be inferred from observed patterns?*

Besides the work directly based on the IFD concept, several empirical studies analysed habitat choice processes by comparing the patterns of site use in different types of habitat. Occupied sites are indeed expected to be of higher quality for various factors than unoccupied sites or sites picked up at random if habitat choice occurs. Site characteristics have therefore been compared between either (1) occupied sites and all unoccupied sites (independently of their suitability), or (2) occupied sites and all suitable sites for the species considered (Johnson 1980). The second type of comparison is more appropriate because it focuses only on sites that can actually be used for breeding. Studies based on such comparisons may however be impaired by problems of data independence and the fact that patterns may result from many processes other than actual choice. We have for instance discussed this problem for plants in Section 9.1.3.2. Moreover, it can often be difficult to determine whether a given site is actually suitable as well as accessible.

Inferring habitat choice solely from patterns of habitat use by individuals can be misleading because relatively high densities of individuals may be observed on low-quality habitat patches. Such mismatches may be due to constraints for individuals in obtaining reliable information about the quality of potential habitat patches or in selecting patches in general, or to a restricted use of high-quality habitat patches for a given activity (Van Horne 1983). For instance, individuals could spend only a few hours each day foraging in the best quality habitat patches, and spend the rest of the day in patches with low food availability but providing resources for other activities (e.g. refuges from predators or arenas for mate choice).

The use of sub-optimal patches by individuals has been especially investigated within the framework of **source–sink metapopulations model** (Pulliam 1988). In this framework, habitat patches are assumed to vary dramatically in quality: source populations are found on high-quality patches, where reproduction exceeds mortality, whereas sink populations, characterized by a demographic deficit (i.e. mortality exceeds reproduction), are found on low-quality patches. Sink populations are maintained in the long-term only through immigration of individuals from source habitats. In a source–sink system, a large fraction of a population may thus be found in low-quality habitat patches if the difference in quality between sources and sinks is high, thus allowing small source populations to provide enough immigrants to maintain large sink populations. This example stresses that the local density of individuals is not always a good indicator of its quality (Van Horne 1983). Another neglected factor is the pattern of temporal variability in habitat quality (Arthur *et al.* 1996): in the previous example, source populations can for instance become sinks over time and vice versa (Pulliam 2000). Unless individuals can rapidly track variation in habitat quality, a delay will be observed between the change in habitat quality and the resulting change in the local density of individuals, during which density will not reflect habitat quality (see Chapter 14). The issue of mismatches between individual choices and the resulting local density and actual patch quality is also largely discussed in the context of 'ecological traps' (see below).

This is why, instead of using the presence or local density of individuals as an indicator of patch quality, some authors have proposed that animals use the reproductive success of others as a source of information about patch quality (Valone 1989; Templeton and Giraldeau 1996; Boulinier and Danchin 1997; Danchin *et al.* 2001; Doligez *et al.* 2002, 2003). It is indeed the success of individuals in a given activity that reveals the actual value of a habitat patch for this activity. If individuals produce twice as many offspring per breeding attempt in patch A compared to patch B, one can say that the quality of patch A is twice better than that of patch B. As the breeding performance of others provides information on habitat quality inadvertently, it belongs to ISI (Danchin *et al.* 2004).

It is thus important to remember that the same patterns can result from very different processes. The temporal dimension of the spatial distribution of individuals should notably be considered carefully, in terms of both the time scale of observations underlying patterns, and temporal variability (instability or predictability) of the environment. Current work developed in the context of landscape ecology focuses more and more on identifying individual **strategies** that may have evolved in temporally variable environments and under different constraints (Lima and Zollner 1996).

### 9.3.2 Studying how individuals sample the environment: prospecting behaviour

Another aspect of the study of habitat choice is the analysis of how individuals sample the environment: individuals must gather information to make a choice, because individuals can only choose between alternatives they are aware of. Such information gathering can be done through prospecting behaviour, involving visits to potential breeding patches by an individual that does not currently breed there. Individuals present in a breeding site but not currently breeding there are thus often called **prospectors** (although they may not really be gathering information).

Prospecting has been described in many species of different taxa, and such behaviour potentially allows individuals to gather information on the relative quality of different breeding habitat patches (Reed *et al.* 1999; Danchin *et al.* 2001). For instance, prospectors are recorded sequentially visiting several current breeding patches (e.g. larvae of marine invertebrate species); they are described to land on unoccupied nests, such as in the black-legged kittiwake (*Rissa tridactyla*; Cadiou *et al.* 1994), or visit occupied nest boxes such as in collared flycatchers (*Ficedula albicollis*; Doligez *et al.* 2004a).

Prospecting is best known in bird species because individual behaviour is often more conspicuous and easily observed compared to other taxa (e.g. nocturnal or cryptic mammal species, small-sized insect species, etc.). Even in birds, however, observations and data on prospecting are fragmentary (Danchin *et al.* 1991; Reed and Oring 1992; Boulinier *et al.* 1996; see Reed *et al.* 1999 for a synthesis), despite its major impact on breeding habitat choice. For instance, few studies clearly identified the status of many prospecting individuals. These studies show that prospectors usually belong to three categories of individual (Cadiou *et al.* 1994):

1. young immature individuals before recruitment into the breeding population;
2. non-breeding adults during the current season;
3. adults that have started to breed during the current season, but have failed their

reproduction early and afterwards visit other sites where conspecifics are breeding.

These three categories of individual are those most likely to be looking for a breeding site for the following year (Monnat *et al.* 1990). However, the links between prospecting, the type of information gathered by prospectors, and subsequent breeding habitat choice is still poorly investigated. The study of prospecting is still in its infancy, but it has recently received increased interest because it is a critical component of choice processes such as breeding habitat choice. Constraints acting on prospecting can for instance determine which types of information will be available to individuals, and thus which strategies of breeding habitat choice are to be expected in nature (Pärt and Doligez 2003). Prospecting may also shape the evolution of life-history traits, such as age at first breeding, when individuals have to prospect before settling (Boulinier and Danchin 1997).

### 9.3.3 Studying choice processes

A direct investigation of habitat choice is often more appropriate than approaches based on inferences from observed patterns. A direct approach aims at identifying the cues used by individuals to choose a breeding site and determining the extent to which habitat choice strategies affect the fitness of individuals (Jones 2001). Such an approach thus allows linking **proximate** and **ultimate** factors involved in habitat choice. Proximate factors are the elements of the environment that are directly (or in a mechanistic way) taken into account by choosing individuals. The abundance of a resource (food, nest sites, mates), for instance estimated through the encounter rate, or cues produced by conspecifics may be such proximate factors. Ultimate factors are the evolutionary causes of individual choices, i.e. those linked to the relative fitness of individuals adopting different habitat choice strategies. For instance, individuals may be selected to choose a patch or territory that provides access to a specific type of food needed for survival, or in which the local breeding performance of their conspecifics, and thus their own expected breeding success, is high. In these examples, proximate and ultimate factors are considered simultaneously: individuals choose a patch because food is available or conspecifics are successful (proximate factors) and, this way, individuals achieve higher fitness (ultimate factors).

The study of breeding habitat choice involves two steps focusing (1) on **information gathering** behaviour, and (2) on the decision rules followed by individuals given the information gathered. As in most cases, the study of the choice processes can use theoretical, observational and experimental approaches that provide complementary information. Here we briefly detail the interest of each approach.

*From theoretical approaches . . .*

Theoretical approaches have attempted to identify the rules and mechanisms involved in the evaluation of site quality and the decision to leave or stay on a patch (Giraldeau 1997). Direct theoretical approaches to breeding habitat choice remain scarce; most theoretical studies on the use of information for habitat choice processes have indeed been done in the context of optimal foraging, but, as we have seen above, their conclusions are likely to be irrelevant in the context of breeding habitat choice (Valone and Templeton 2002). In particular, animals are more likely to use social information in breeding habitat choice than in foraging (see Section 9.1.1).

In the context of breeding habitat choice, theoretical work has helped in explaining the higher probability for failed individuals to change breeding sites, frequently observed in empirical studies. A first major step has been to identify the levels of temporal predictability of habitat quality as a prominent component of breeding habitat choice (Switzer 1993, 1997). Then, other models were proposed that used an optimality approach (Boulinier and Danchin 1997; Schjørring 2002). However, such optimality models were of limited relevance because they ignored the impact of the choices of other individuals in the population. By ignoring the social component of the environment such models estimated fitness only on the basis of the intrinsic quality of the habitat as if that quality was not affected by the choices of other individuals. Thus, **evolutionarily stable strategy** (ESS) approaches were developed (Doligez *et al.* 2003), which by accounting for competition between strategies, showed the importance of temporal autocorrelation of intrinsic habitat quality in determining which habitat choice strategy is likely to be selected. Such models make explicit predictions that can be tested empirically (see Section 9.4.2).

*. . . to observations . . .*

Theoretical work can be used to predict which habitat choice strategies may be selected in different environmental conditions given the species life history, under specific assumptions. These assumptions and *a priori* predictions can then be tested using observational data from natural populations. The direct observation of the behaviour of individually marked prospectors can for instance be used to test whether prospecting potential breeding sites actually leads to local recruitment of those individuals according to a choice process based on the use of information about relative site quality (Reed *et al.* 1999). More specifically, one can test predictions concerning time spent prospecting or number of habitat patches prospected. Patterns of distribution of individuals in the environment that should result from different habitat choice strategies can also be predicted, and tested against observational data. For instance, many studies have long tried to link habitat quality to observed distribution of animals in the environment. This is the case of most of the studies that tried to analyse whether animals are distributed in ideal free way (see examples in Section 9.5 and Chapters 8, 10, and 14).

*. . . and experiments*

To identify with certainty the cues used by individuals for habitat choice, and also show causal relationships between variations in these cues and individuals' choices, **experimental designs** are needed. Experimental **manipulations** of potentially identified cues permit ascertaining whether they are actually used by individuals in the breeding habitat choice. Such experiments can be conducted in the laboratory (e.g., Templeton and Giraldeau 1996; Schuck-Paim and Alonso 2001) or in the field (e.g., Boulinier *et al.* 2002; Doligez *et al.* 2002). Many types of cue have been manipulated, such as direct factors of habitat quality or more integrative cues reflecting habitat quality. For example, an experimental increase of the perception of nest predation risk induced an increase in the subsequent probability of shifting nest site and in breeding dispersal distance in male Tengmalm's owls *Aegolius funereus* (Hakkarainen *et al.* 2001). Similarly, an experimental reduction of individual mating probability or breeding success induced an increased probability of leaving one's current site in both invertebrates (eastern amberwings, *Perithemis tenera*; Switzer 1997) and vertebrates (American robins, *Turdus migratorius*, and brown thrashers, *Toxostoma rufum*; Haas 1998). These experiments showed that individuals directly use the manipulated cue to make breeding habitat decisions. We will come back later to the results of experiments testing for the role of local success of conspecifics in breeding habitat selection (see Section 9.5.3). Our message here is that **only experimental manipulations of the suspected cues** used by animals in their decision can allow one to demonstrate that this cue is indeed used in the choice. Correlative approaches are useful in detecting which cues are likely to be used, but they remain subject to uncertainty because animals may actually use other habitat characteristics that are correlated to the suspected cue.

## 9.4 What type of information should be used to select a breeding habitat?

Before discussing the reliability and validity of cues for selecting a breeding habitat, we need to discuss briefly the types of information that organisms may use to evaluate potential breeding patches quality. Chapter 4 provides a taxonomy of biological information (Figure 4.3), here we briefly present concepts that are relevant to the question of breeding habitat choice.

### 9.4.1 The various types of information

There are many potential cues available to individuals. Animals may use non-social or social cues.

#### 9.4.1.1 *Non-social cues*

Animals may evaluate all the potential resources and constraints affecting breeding success. Some are of prime importance such as food availability or protection from predators. Habitat quality assessment could involve the direct assessment of these factors. However, such factors may be numerous and some may be difficult to assess, as for instance predation risk. Furthermore, the information about some factors may not be available at the time when information gathering is possible.

The alternative is to use cues revealing the effect of important factors indirectly. For instance, it may be much easier to use indices of the presence of predators (odours, presence of a burrow or a nest...). Nevertheless, integrating all of these parameters and evaluating their importance to individuals' fitness is likely to be difficult in many cases.

Individuals may also use their own performance (personal information). In Chapters 7 and 8, we have seen several examples of such information use. Nevertheless, we have seen that in the context of breeding habitat choice the use of personal information is relatively unlikely (Section 9.1.1).

#### 9.4.1.2 *Social cues*

Another strategy consists in using conspecifics as a source of information, that is social information.

*Social attraction*

The presence of conspecifics on a patch reveals conditions that are good enough for a local population to persist. The presence of others thus reveals habitat suitability. Nevertheless, as wc have seen earlier the mere presence of conspecifics may be misleading (see for instance the source–sink discussion). It has also been suggested that the presence of members of other species sharing the same ecological needs may also lead to inter-specific **social attraction**.

As the presence of conspecifics is often confounded with their local breeding success, the impact of social attraction may have been overestimated. This is suggested by theoretical work that showed that conspecific attraction is possibly not the best strategy in some situations (Doligez *et al.* 2003).

*Inadvertent social information*

The activity of conspecifics and their success in that activity probably better reveal habitat suitability. For instance, a ground squirrel observing a conspecific going back and forth between a foraging patch and its nest would easily learn about local conditions. Similarly, in the case of breeding habitat choice, conspecific breeding success could be used as a cue of habitat suitability.

In fact, when choosing a breeding site an organism is facing the following evolutionary question: 'Where can I achieve the highest reproductive success?'. An answer is 'Where conspecifics are the most successful'. The success of conspecifics is a form of ISI (Danchin *et al.* 2004; Figure 4.3). As conspecifics share similar needs, the fact that they are successful shows that their patch is likely to have the required qualities for breeding. Prospectors thus use their conspecifics as environment samplers.

Conspecific breeding success integrates in a single parameter the effect of any component of environmental quality including social interactions such as competition and the quality of potential mates. ISI use is also more precise than personal information as it can be based on a much larger sample than personal information. However, ISI is based on the sampling by others individuals which may differ from the observer in their requirements and capacities. Such phenotype–environment interactions may limit the value of ISI in some circumstances. ISI is based on the sampling of more individuals, but that information may be less valuable because of phenotype–environment interactions.

*Mix of information*

In reality, it is probable that animals combine several sources of information. The likely importance of the information used will depend also on its accessibility. It is, for instance, relatively difficult to estimate the breeding success of species nesting in burrows and that come back to feed their young only at night.

*Any choice strategy involves learning about the environment*

In any case, the behaviour by which an individual samples its environment and thus acquires information is a learning process (see Figure 4.1). Prospecting allows animals to learn about their environment. We will see later that this comparison makes sense; theoretical models about learning and about habitat choice share comparable characteristics, notably regarding the importance of environmental predictability.

### 9.4.2 Comparing strategies based on different sources of information

Many theoretical approaches have modelled the use of different types of information in the context of optimal foraging or mate choice, but until recently few studies had investigated strategies of information use in the context of breeding habitat choice. This discrepancy may seem surprising with regard to the numerous models addressing the evolution of dispersal and the implications of dispersal on (meta)population dynamics (Clobert *et al.* 2001), but is possibly the result of the long predominance of ultimate models of the evolution of dispersal, with emphasis on intra-specific competition and kin interactions, ignoring proximate mechanisms involved in breeding habitat choice (see Chapter 10). Furthermore, until recently, the paucity of good empirical data due to the large spatial and temporal scales of breeding compared to foraging habitat or mate choice did not encourage the development of theoretical approaches since predictions could only be tested in very limited situations.

It is sometimes possible to transpose conclusions from models developed in the framework of optimal foraging to breeding habitat choice. Nevertheless, specific modelling is required to account for both the spatio-temporal scales involved in breeding habitat choice and the interactions with individuals' life history traits. For instance, in iteroparous species, it can be relevant to separate the habitat choice strategies used to select the first breeding site from subsequent breeding sites (Switzer 1993; Boulinier and Danchin 1997). Such modelling has allowed consideration of the evolution of habitat choice strategies within the classic framework of life history evolution (Stearns 1992). The different characteristics of breeding habitat choice, such as time spent prospecting or number of habitat patches visited, can indeed be seen as life history traits, in the same way as age at first reproduction, clutch size, longevity, etc. In the remainder of this section, we describe recent models proposed to investigate the relative success of a strategy based on local reproductive success of conspecifics compared to strategies based on other types of information, as an illustration of the way models can help investigating processes of breeding habitat choice. Many other types of model have recently been built in the breeding habitat choice framework (e.g. the evolution of philopatry in cooperatively breeding species (see Kokko and Lundberg 2001; Kokko and Ekman 2002)).

#### a *An optimization model . . .*

With optimization models, Boulinier and Danchin (1997) investigated the trade-off between (1) early reproduction without patch choice (the **Random** strategy, i.e. settlement at random) and (2) delayed reproduction with patch choice based on ISI revealing patch quality (the **Prospecting** strategy). The latter strategy involves skipping breeding to take time to prospect a varying number of patches (1–10) to gather ISI at the end of a breeding season when ISI is readily available. Thus, this strategy imposes a cost as individuals postpone breeding for at least one year, and so on until they find a suitable patch as revealed by conspecific breeding success. In these models, ISI was quantified as the proportion of successful pairs, on each visited patch (Boulinier and Danchin 1997). The model compared the mean lifetime reproductive success of two breeding habitat choice strategies in an environment with varying proportion of high-quality patches. Patch quality varied in a more or less predictable way.

Results showed that individuals have a higher fitness in the Prospecting than the Random strategy when high-quality patches are rare and the temporal predictability of patch quality is high. A similar selective advantage of information gathering despite the costs involved had been obtained in the context of the evolution of learning (Stephens 1989). A very similar model specifically addressed the evolution of age at first reproduction in relation to prospecting and environmental spatio-temporal variability (Schjørring 2002).

These different models however, did not explicitly consider social interactions. We have seen in Chapters 3, 7, and 8 that, when the benefit associated with a behavioural strategy depends on the strategies adopted by the other members of the population (i.e. a frequency-dependent process), very different results may be obtained depending on whether or not interactions between individuals are considered. In socially based habitat choice, the value of social information probably directly depends on the strategies used by conspecifics. A model ignoring such frequency-dependence could thus be unable to predict the functioning of natural populations.

### b ... *followed by game theory models*

Game theoretic models, therefore, were proposed (Doligez *et al.* 2003), that pit five breeding habitat choice strategies against each other in a simple two-patch system in which patch quality varies with some predictability (Figure 9.4). The five strategies used were: (1) **Random** settlement; (2) **Philopatry**; (3) **Presence** (habitat choice based on the presence of conspecifics; (4) ISI obtained from the proportion of successful pairs (this does not account for density-dependence, **Success 1**); and (5) ISI obtained from the mean number of offspring produced by a breeding pair on each patch (thus accounting for negative density-dependence, **Success 2**).

Conclusions were that the evolutionary stability of a strategy depends on environmental predictability as well as the costs induced by density-dependence. The effect of environmental predictability on evolutionary stability was through its impact on the capacity of the various strategies to track temporal changes of patch quality. Costs linked to density were related to the relative tendency for a strategy to spatially aggregate individuals on one of the two patches. The two successful strategies performed best when the environment is neither constant nor totally unpredictable, i.e. when breeding patch quality is variable but temporally autocorrelated (Figure 9.4). In other words, the model predicts that strategies based on ISI use are likely to be selected except in exceptional situations at the extremes of the gradient of environmental predictability. In a constant environment, the 'philopatry' strategy is selected. This is a classic result of models on the evolution of dispersal (Chapter 10). In a randomly varying environment, Random is the best strategy because it does not aggregate individuals and thus pays the lowest density-dependence cost (Figure 9.4).

Furthermore, the Presence strategy managed to coexist with the most efficient strategy under the simulation conditions, for instance with Success 1 and 2 in intermediate values of autocorrelation (Figure 9.4). However, the frequency of individuals of the Presence strategy remains low (Doligez *et al.* 2003). In fact, individuals of the Presence strategy parasitize the information produced by individuals of the other strategy. Therefore, the fitness of individuals of the Presence strategy is frequency-dependent because the quality of the information found in the relative presence of individuals on each patch diminishes when the frequency of the Presence strategy increases, as fewer and fewer individuals rely on information directly reflecting patch quality, such as breeding success. This result suggests that a strategy

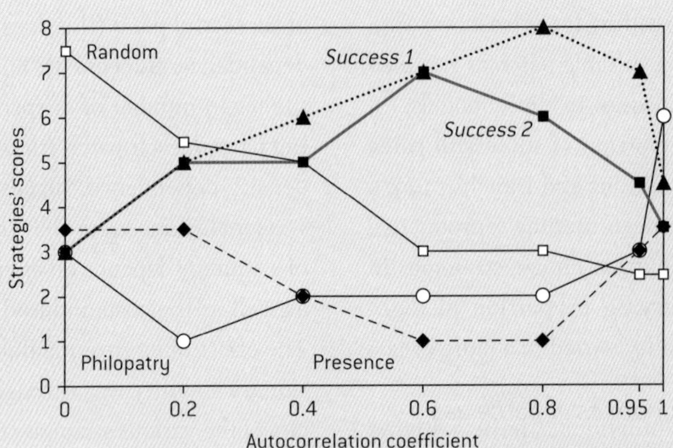


Figure 9.4 **Evolutionarily stable strategies of breeding habitat choice and temporal environmental autocorrelation**

The figure shows the performance of five strategies of habitat choice in invading other strategies and resisting invasion as a function of environment predictability (level of temporal autocorrelation of breeding patch quality from one year to the next) when strategies are confronted two by two, in a two-patch environment. When the autocorrelation coefficient is zero, patch quality varies randomly from one year to the next. When the autocorrelation is 1, the environment is constant in time. When the autocorrelation is between 0 and 1, the environment varies in a more or less predictable way depending on the value of the coefficient. The *y*-axis shows a score quantifying the proportion of individuals adopting each strategy at the end of the confrontations, summed over all confrontations involving this strategy, computed from 100 simulations under the same conditions for each confrontation. The score can vary from 0 to 8 because each strategy is involved in 8 confrontations in total (4 as a resident strategy and 4 as a mutant). Simulations used a matrix population model, two patches of varying qualities and a negative local density-dependence function. The results presented correspond to a short-lived species (e.g., small passerine bird), but the results are qualitatively similar when considering a long-lived species.

The five habitat choice strategies confronted are:

**Random**: individuals do not choose a patch (i.e. same probability to settle on each of the two patches);
**Philopatry**: individuals always breed on their natal patch;
**Presence**: individuals choose their patch as a function of relative densities of conspecifics on the two patches (i.e. the probability of settling on a patch is proportional to the relative density of individuals in the patch the previous year);
**Success 1**: individuals choose their patch as a function of relative breeding performance on the two patches, estimated as the proportion of pairs that fail (i.e. the probability of settling on a patch is proportional to the relative failure rate of individuals in the patch the previous year);
**Success 2**: individuals choose their patch as a function of relative breeding performance on the two patches, this time estimated as the mean number of offspring produced per breeding attempt (i.e. the probability of settling on a patch is proportional to the relative mean number of offspring per breeding attempt in the patch the previous year), which in the model is a function of the local densities of breeders.

From Doligez *et al.* 2003.

using solely the relative numbers of individuals present in a patch (usually called social attraction) is unlikely to be frequent in a population. Its existence is conditioned to other strategies that are more efficient in tracking environmental variation. This casts new light on the importance of social attraction in natural populations, a strategy frequently reported in empirical studies (e.g. Stamps 1988, 1991; see Doligez *et al.* (2003) for a full discussion).

### c *Implications for the dynamics of patch use*

We have already considered above how individual breeding habitat choice is linked to spatial **aggregation** processes. These links between individual choices and population patterns can be generalized. In particular, such theoretical approaches have provided elements to solve an interesting paradox: in some species, breeders aggregate on a fraction of suitable patches, and on these patches, only a fraction of the individuals manage to secure a breeding site, the rest of the individuals remaining non-breeders because of the lack of available sites on these patches (Forbes and Kaiser 1994). Some individuals are thus prevented from breeding because they do not find a site while breeding sites are available on other patches. The presence of individuals prevented from breeding because of patch saturation has been revealed by the immediate replacement of experimentally removed breeders (Manuwald 1974). To explain this paradox, Forbes and Kaiser (1994) proposed that individuals use their conspecifics as a source of information about the quality of breeding patches. Therefore, individuals are not willing to colonize empty patches, but aggregate on current breeding patches where density can reach saturation, forcing individuals to queue for a breeding site. Different levels of competition to access territories of different qualities can lead to variation in habitat choice, as shown in the case of the oystercatcher (Figure 9.3). It is thus interesting to understand how floating and queuing strategies may have evolved under such constraints (Forbes and Kaiser 1994; Kokko and Sutherland 1998). Individual strategies based on conspecific cues have also been suggested to be at the origin of the evolution of coloniality (see Section 9.6 and Chapter 14).

### d *Conclusion*

Theoretical approaches comparing the success of strategies using different information types, and the conditions under which they are likely to be selected, are useful to predict the strategies expected to evolve in natural conditions. Clearly, among important factors are not only the absolute quality of a habitat at a given point in time, but also its temporal predictability, affecting the reliability of information gathered for future decision-making, especially if the time interval between information gathering and use is long (e.g. in the previous year). Many natural or managed environments have intermediate patterns of predictability, which could lead to selecting strategies based on inadvertent social information under the assumptions of the models described above. This observation underlines the importance of exploring the temporal dynamics of habitat quality to assess information reliability, while also accounting for interactions with conspecifics through density dependence (including competition for breeding sites) and frequency dependence.

## 9.5 What information is actually used?

Theoretical studies predict which sources of information should be used in a given environment, and which breeding habitat choice strategy is optimal given the spatial and temporal variability of the environment. Nevertheless, as we have seen in Section 9.2.2, information gathering can be constrained by factors ignored by models. Tests of predictions derived from theoretical work tailored to specific situations are thus critical. Observational and experimental approaches are needed for this purpose.

### 9.5.1 Constraints on the characteristics of potential information cues

All potential sources of information cannot be used by any species or any individual in a population. As we have seen, to be informative a cue should allow individuals (1) to predict the quality of a breeding site (i.e. predict the fitness expected locally), and (2) to compare among potential sites.

The reliability of a cue will strongly depend on the temporal variability of patch quality. In the case of breeding site choice, the delay between information gathering and settlement decision may be fairly long, and may depend on the cue used. A cue of site quality may be available at a given time, but used only later, e.g. in the next breeding season for seasonally breeding species. The environment must thus be stable enough so that relative local patch quality does not change too much during the time lag between prospecting and settlement. The environment should thus be predictable enough (see Section 8.4.2). Moreover, variations in the cue used should reflect environmental variation immediately, so that there is no discrepancy between actual site quality and that assessed via the cue. Finally, to be of any use, the cue needs to be available to prospectors over a sufficient period of time (Boulinier *et al.* 1996).

Furthermore, the costs of information gathering must be considered. Such costs depend on the spatio-temporal variability of the environment and species life history (see Section 9.4). Depending on (1) the characteristics of the species breeding biology, such as the length of the breeding period, the synchronization of breeding events among and within patches, the mobility of breeders, etc., (2) the inter-seasonal variability of the environment, and (3) the cue(s) used, it may be more rewarding for individuals to gather information during a specific time window. For instance, the local reproductive success of conspecifics (i.e. ISI) is readily available towards the end of breeding in species with high reproductive synchrony, but all along the breeding season in asynchronous species. Other cues may be available just before settlement. In any case, prospecting is expected to be optimized because the time spent prospecting is traded off against other activities. For instance, if patch or site quality is predictable from one year to the next, then failed breeders which do not need to care for offspring and that cannot engage in a new breeding attempt may be able to spend more time prospecting after failing than individuals still caring for young.

### 9.5.2 Can breeding habitat choice be inferred from conspecific distribution and performance?

Many studies have shown correlations between various environmental factors and the presence or breeding performance of individuals. For instance, Petit and Petit (1996) reported that the patches with the highest food availability were those occupied first and with the highest densities of breeders of their study species, the prothonotary warbler (*Protonotaria citrea*). This distribution was in agreement with predictions from the Ideal Free Distribution. However, this study ignored potential information gathering processes. Such a pattern may indeed result from individuals assessing food abundance directly or environmental parameters correlated with food availability. More generally, most 'habitat choice' studies do not consider the decision process followed by individuals, and focus on whether individuals occupy the most favourable patches.

Other studies mention the need for individuals to gather information in order to choose where to settle, without going into details about potential processes. For instance, Orians and Wittenberger (1991) suggested that females of the yellow-headed blackbird (*Xanthocephalus xanthocephalus*) choose their nesting marsh as a function of prey densities, and choose a territory within the nesting marsh according to the vegetation providing protection against nest predation. These hypotheses rely on correlations between environmental factors and individual densities at different spatial scales. Despite the fact that authors mentioned that some habitat factors, such as prey density, may be difficult to evaluate, they implicitly assumed that the preference for certain patches directly resulted from the environmental characteristics to which densities were correlated. Nevertheless, correlations do not imply causality (see Chapter 3). In many cases, the choice may rely on other, more indirect, cues, themselves correlated to environmental factors. The study by Orians and Wittenberger (1991) nicely emphasizes the necessity of considering different spatial scales, but neglects the question of the actual cues used by animals to select their breeding habitat.

### 9.5.3 Inadvertent social information, as a source of information for breeding habitat choice

We have already mentioned various elements suggesting that the breeding performance of conspecifics may constitute a source of ISI in breeding habitat choice (Valone and Templeton 2002). Such a cue may be of high information value because it encompasses the fitness component that individuals are likely to maximize in breeding habitat choice. For this reason, we now present some evidence that individuals actually use such information.

#### 9.5.3.1 *Timing of prospecting and the value of information*

In many bird species, individuals are usually observed prospecting towards the end of the breeding season, when conspecifics are still engaged in breeding activities (Boulinier *et al.* 1996; Doligez *et al.* 2004a; review in Reed *et al.* 1999). At that time, the proportion of breeding territories or nests containing offspring, or the mean number or quality of offspring, or social correlates such as parental activity around the breeding site, can reveal the patch-specific relative breeding success, which informs about current patch quality. In the kittiwake (*Rissa tridactyla*) for instance, the peak of prospecting occurs when the local breeding success can be estimated the most reliably in a single prospecting visit (Boulinier *et al.* 1996; Figure 9.5). Moreover, prospecting behaviour at the end of the breeding season can be directly related to the future breeding success of individuals (Schjørring *et al.* (1999): great cormorants (*Phalacrocorax carbo*) breeding for the first time but having been recorded prospecting in the patch the

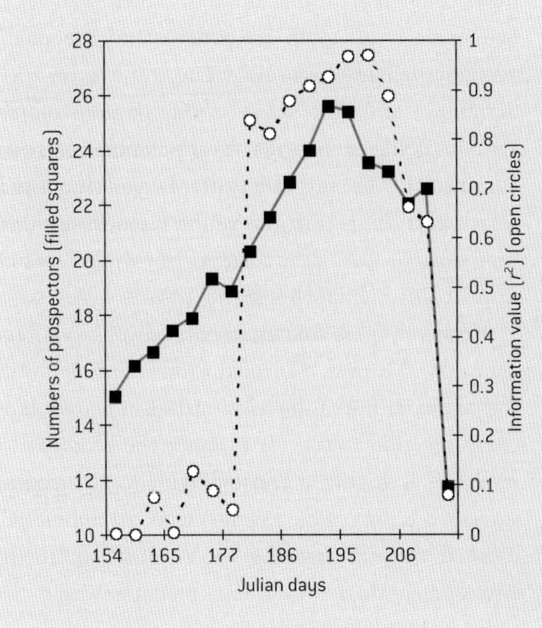


Figure 9.5 **Estimating breeding patch quality through ISI and timing of prospecting.**

In the kittiwake (*Rissa tridactyla*), most prospecting at breeding patches (filled squares) occur at the time when the information on local conspecific breeding success that can be gathered reliably in a single visit (open circles). The reliability of the information on the relative quality of breeding patches was measured here by the correlation coefficient between the proportion of successful nests at a given date and the actual proportion of successful nests in the year considered. This result is expected under the hypothesis that individuals use conspecific reproductive success as inadvertent social information for breeding habitat choice in the subsequent year.

Modified from Boulinier *et al.* (1996).

year before, achieve a higher breeding success than individuals not previously recorded prospecting (Schjørring *et al.* 1999).

Such a temporal pattern of prospecting suggest that individuals prospect to acquire ISI for deciding where to settle in the following year (Boulinier *et al.* 1996; Boulinier and Danchin 1997; Reed *et al.* 1999; Doligez *et al.* 2004a).

#### 9.5.3.2 *Assumptions and predictions are supported by correlations . . .*

The strategy of using the breeding success of conspecifics as a source of inadvertent social information for breeding habitat selection has been termed **habitat copying** (Wagner *et al.* 2000; Wagner and Danchin 2003). Correlative analyses on individual movement data first tested several predictions derived from habitat copying. In the kittiwake (*Rissa tridactyla*), several assumptions and predictions of habitat copying were tested using observational data from a long-term survey of a natural breeding population in Brittany, France (Danchin *et al.* 1998). Two key assumptions of habitat copying were met in this population: the mean breeding performance (as measured by the average number of fledglings per nesting pair) varied among breeding patches (i.e. portions of breeding cliffs), and was predictable from one year to the next (positive temporal autocorrelation of local breeding success).

Furthermore, in many species, individuals having experienced a breeding failure in a given year (thus with personal information indicating low site quality) are more prone to leave their breeding site in the following year (Switzer 1993, 1997). This process is often considered as an absolute rule followed by individuals. However, if individuals use the breeding performance of their neighbours as complementary information on breeding patch quality, a failed breeder can be expected to remain faithful to its breeding site when surrounded by highly successful neighbours (Boulinier and Danchin 1997). This is what was found in the kittiwake (Figure 9.6): individual site fidelity from one year to the next was related to both personal and inadvertent social information in an interacting way. More specifically, individuals that failed in their breeding attempt at an early stage (e.g. at or before the egg stage) remained faithful to their breeding patch if their neighbours had a high average breeding success. Conversely, individuals that failed at the same early stage but were surrounded by neighbours that also failed had a lower probability of being faithful to their breeding patch.

This study on kittiwakes also tested a demographic prediction of habitat copying. As expected, the local breeding **population growth rate** was higher in years after a high local reproductive success relative to years of low local success. In kittiwakes young first breed when four years old on average. Thus, local changes in numbers of breeders between subsequent years cannot result from the local recruitment of young hatched in the preceding year (Danchin *et al.* 1998). Natal philopatry would predict a positive relationship between local breeding success in a given year and local rate of increase on the same patch four years later. Such a relationship was not observed, which shows that local growth rate was mainly due to emigration and immigration decisions (Danchin *et al.* 1998).

A series of similar studies have tested such assumptions and predictions of habitat copying in other colonial (Erwin *et al.* 1998; Brown *et al.* 2000; Schjørring *et al.* 1999; Frederiksen and Bregnballe 2001; Suryan and Irons 2001; Oro and Ruxton 2001; Serrano *et al.* 2001; Safran 2004) or non-colonial species (Doligez *et al.* 1999, 2004b; Blums *et al.* 2002; Ward 2005; Nocera *et al.* 2006). Many of these studies support the predictions of the use of conspecific reproductive success, or some social correlates, as ISI for breeding habitat choice.

However, these studies are based on observational data and it is thus difficult to infer with certainty the causal nature of the relationships reported. If an environmental factor, for instance the activity of a predator, is highly correlated with local breeding success, it is impossible to know whether individuals respond to local breeding success itself, or to the environmental factors (here, predators) that are correlated with breeding success.

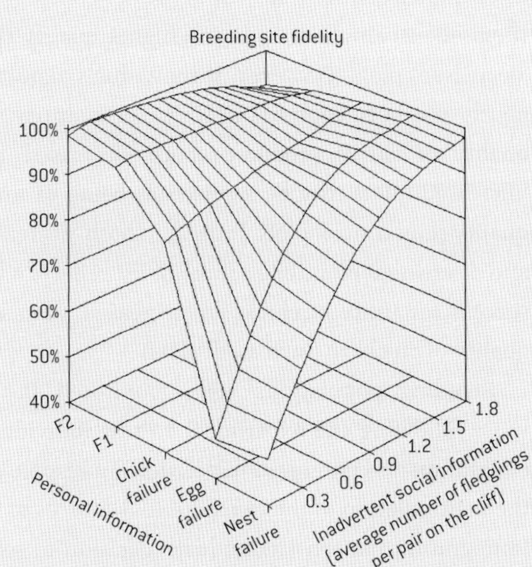


Figure 9.6 **Breeding site fidelity as a function of personal and inadvertent social information**

In the kittiwake (*Rissa tridactyla*), breeding patch fidelity is a function of both individual performance (personal information) and mean number of chicks fledged per breeding pair in the cliff (public information) in the previous year. Predicted values based on a multiple logistic regression model fitted to the actual data.

For a given level of individual breeding performance (for example failure at the egg stage), the probability for individuals of being faithful to their breeding patch increases with increasing performance of the neighbours, except for successful individuals (which have very high fidelity to their previous breeding cliff). This result based on correlative data, suggests that individuals use ISI in their decision regarding breeding site fidelity.

Levels of individual breeding performance:

F2: individuals having produced two or three fledglings (i.e. highly successful individuals);
F1: individuals having produced 1 fledgling;
Chick Failure: individuals having failed in their reproduction at the chick stage (no hatched chick fledged);
Egg Failure: individuals having failed in their reproduction at the egg stage (no egg hatched);
Nest Failure: individuals having failed at the nest stage (nest built, but no eggs laid).

The interaction between personal and inadvertent social information effects was significant ($P=$ 0.0056), indicating that the effect of one variable on site faithfulness depended on the other variable (this is why the plane is curved): individuals were more likely to use ISI when they failed at an early breeding stage.

Adapted from Danchin *et al.* (1998).

#### 9.5.3.3 . . . *and by experiments*

Only manipulations of the cues suspected to be used by individuals can confirm the causal relationships between ISI and subsequent breeding habitat choices (Valone and Templeton 2002). Such an experiment was conducted on a kittiwake population breeding on an island in Northern-Norway. Nine pairs of breeding study plots were used to investigate individual responses to the manipulation of local breeding success on future emigration and immigration. On one of the plots within each pair, all nests were experimentally failed by removing all eggs, and on the second plot of each pair, breeding pairs nests were kept experimentally successful either by leaving nests unmanipulated if already successful or by adding eggs to nests that had been victims of predation. As detailed in the previous section (Figure 9.6), correlative results on this species suggest that only failed breeders are sensitive to ISI. In successful plots, therefore, a few nests were experimentally failed at the egg stage. The goal was to compare prospecting and future emigration of early failed breeders surrounded either by successful or failed neighbours. Failed breeders on both types of plot were individually marked before treatment. Massive breeding failure events such as the one mimicked by removing eggs from an entire plot are not uncommon in nature where predators such as ravens or gulls can catch every nestling in some patches. Individual behaviour was subsequently monitored during the breeding season and in the next year in order to contrast prospecting and emigration of early failed breeders on the two types of plot. The results showed both a higher prospecting activity in the experimental year and a higher breeding patch fidelity in the subsequent year on successful plots compared to those put into massive failure, thus confirming one prediction of habitat copying in this species (Boulinier *et al.* 2002).

A comparable experiment was performed in a fragmented population of collared flycatchers (*Ficedula albicollis*) a small hole-nesting passerine breeding in woodlots on the island of Gotland, Sweden (Doligez *et al.* 2002). The experiment involved 12 plots and 4 treatments: in 'reduced' (R) plots (three replicates), all chicks from 30% of the nests were removed, mimicking nest predation, and were transferred to 60% of the nests of the 'increased' (I) plots (three replicates). Two types of control plot were used: in 'manipulated control' (MC) plots (two replicates), all chicks of 20% of nests were transferred to 40% of other nests within the same plots, thus leaving the mean number of chicks per breeding attempt unchanged at the plot scale but not at the nest scale; and in 'non-manipulated control' (CC) plots (four replicates), none of the nests were manipulated. This experiment was performed identically for three consecutive years. The manipulation successfully changed the mean number of chicks produced per breeding attempt in the expected way (Figure 9.7a). In particular, the mean number of chicks produced did not differ between the two types of control plot. However, the treatment also affected the mean body condition of chicks at fledging, since adding chicks in a nest imposes a higher feeding effort for parents: chicks from nests with increased brood size in 'increased' and 'manipulated control' plots were on average in worse condition than those from non-manipulated control and 'reduced' plots (Figure 9.7b). Thus, the manipulation resulted in conflicting inadvertent social information about the quantity and quality of chicks produced in the experimental plots. In natural situations, these two parameters are positively correlated; here, in breaking that natural correlation, the manipulations allowed the authors to make separate predictions depending on which component of ISI individuals may use as a cue. If individuals use only the mean number of chicks locally produced to assess habitat quality, 'increased' plots should be assessed as of higher quality than both types of control plot, themselves being assessed as of higher quality than 'decreased' plots. However, if individuals simultaneously use both sources of ISI (mean quantity and quality of chicks produced) both 'decreased' and 'increased' plots may be assessed as being of lower quality than control plots. Indeed, in both 'decreased' and 'increased' plots, one source of ISI reflects a low success (quantity of chicks in 'decreased' plots and quality of chicks in 'increased' plots).

Breeding populations were monitored in the years following the experiment to test individuals' emigration and immigration decisions in response to the manipulation of the two potential sources of ISI. Immigration pattern (individuals' decision to settle on a plot) matched the pattern expected if birds only use the mean number of chicks per breeding attempt (or some social correlate such as parental activity or feeding rate) as ISI on patch quality (Figure 9.7c). Emigration pattern, (individuals' decision to leave a plot) matched the pattern expected if birds use information on both the mean quantity and quality of chicks locally produced as ISI on patch quality (Figure 9.7d).

The difference in patterns of emigration and immigration suggests different ability of individuals to gather information on their current versus other breeding plots, probably revealing differential costs of prospecting close versus far from the current breeding site. To make a decision about leaving their current breeding plot, individuals only need information on this plot. It may be relatively easy for breeders to accumulate valuable information on their current patch, including both the mean number and condition of chicks locally produced. Conversely, to choose a new breeding patch, individuals need to gather and compare information from different potential plots, which may be costly (see above). The possibilities to gather detailed ISI may thus be limited and only one source of information, here the mean number of chicks produced per breeding attempt, seems to be gathered and used.

In any case, these two experiments in species with contrasted life histories clearly show that individuals can use the breeding performance of their

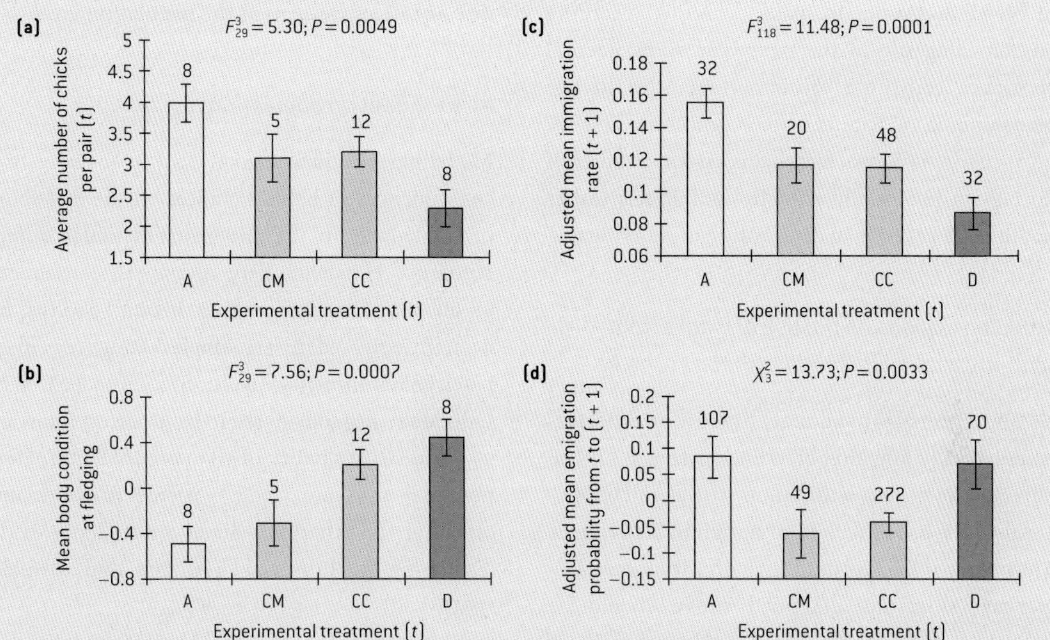


Figure 9.7 **Experimental manipulation of inadvertent social information and breeding habitat choice in the collared flycatcher (*Ficedula albicollis*)**

R: plots where the mean number of nestlings per breeding attempt was reduced; I: plots where the mean number of nestlings was increased; MC: control plots where nestlings were swapped between nests within the plot (thus without affecting the mean number of nestlings); CC: control plots where no manipulation was performed on nestlings.

**a.** The mean number (±standard error) of nestlings fledged per breeding attempt was significantly differed between treatments in the expected way, thus showing that the experimental manipulation was significant ($P = 0.0049$).

**b.** The mean body condition (±standard error) of nestlings fledged differed significantly between treatments ($P = 0.0007$). Manipulating the number of nestlings thus affected two components of ISI, nestling quantity and quality, in an opposite way.

**c.** The plot immigration rate (±standard error) observed the year following the manipulation differed between treatments ($P = 0.0001$). The *y*-axis represents immigration probability corrected for the other significant effects (sex, age of individuals, etc.) This pattern of immigration fits predictions of habitat copying if individuals use the mean number of nestlings per pair for selecting their breeding patch in the subsequent year.

**d.** The plot emigration rate (±standard error) in the year following the manipulation also differed between treatments ($P = 0.0033$). The *y*-axis represents the emigration probability corrected for the other significant effects (sex of individuals, etc.). This pattern of emigration fits predictions of habitat copying if individuals use both mean quality and number of nestlings fledged per pair to assess habitat quality.

From Doligez *et al.* (2002).

conspecifics as a source of ISI gathered at the end of the breeding season, to make decisions regarding their breeding site in the next year. Experimental approaches constitute an important step in the understanding of habitat choice processes. Indeed, they focus on the actual fact that is used as a source of information potentially used by individuals rather than the sole patterns of distribution of individuals.

#### 9.5.3.4 *Conclusions on the use of inadvertent social information for breeding habitat choice*

As we have seen, ISI integrates all environmental factors potentially affecting breeding success, a critical component of fitness, at different spatial and temporal scales. The use of such information by individuals is thus likely to be favoured in many situations. The experimental results described above should be reproduced in other taxa to assess the generality of habitat copying (i.e. breeding habitat choice based on ISI). Theoretical approaches stressed the importance of the spatial and temporal scales of environmental variability. Furthermore, as discussed above, it is likely to depend on life history constraints, potential heterogeneity among individuals and type of density dependence. Ultimately, a full understanding of the role of ISI in breeding habitat choice may require a comparative approach to determine more precisely the conditions that favour the use of ISI (or some social correlates) in breeding decisions. This field has recently received increased interest and will likely continue developing in the future.

## 9.6 Consequences of breeding habitat choice: dynamics of the distribution of individuals

### 9.6.1 Individual strategies generating different distributions

Habitat choice directly affects the distribution of individuals in the environment and their use of more or less suitable habitat patches through dispersal (see Chapter 10). Such individual decisions will thus directly affect patterns at the population level.

#### 9.6.1.1 *Effect on population regulation*

Modelling provides a good way to investigate the consequences of habitat choices on the distribution of individuals in the environment under different scenarios. However, very often, the consequences of an individual's breeding habitat choices have been investigated in very simple settings (e.g. in two-patch environments, as in Section 9.4.2), where individuals are considered to have perfect knowledge of the relative quality of the potential sites. Hence, potential sites are often expected to be occupied one after the other as a direct consequence of their decreasing quality rank (Pulliam and Danielson 1991).

Simple models are nevertheless useful in studying population consequences of individual strategies of breeding habitat choice. Rodenhouse *et al.* (1997) have shown how simple habitat choice decision rules could participate in **site-dependent regulation of populations**. They considered a system comprising a set of patches varying gradually in quality. When the population increases due to high individual fitness on good quality patches, an increasing proportion of individuals start settling on patches of lower quality. Thus the average fitness at the scale of the population decreases, which reduces the overall population growth rate. This may lead the population to start decreasing. During such declines, only the highest quality patches tend to remain occupied, which, by raising the average reproductive success of individuals, may lead the overall population growth rate to increase again. This regulatory effect can function even in the absence of negative density dependence at the level of individuals: the population growth rate varies without any variation in the fitness of individuals settled on the best sites, but just because of the variation in the mean quality of occupied sites. Via this simple mechanism, habitat choice not only conditions the distribution of individuals in the environment, but also participates in population regulation.

#### 9.6.1.2 *Habitat choice and local population viability*

As we have seen, individual choices are constrained by the accessibility of reliable information on the quality of potential breeding patches. Strategies of habitat choice based on different sources of information may generate different distributions of individuals among patches and different temporal dynamics. For instance, under social attraction animals cue on the relative densities of conspecifics in the various habitat patches. That density is itself the result of previous-year relative densities, etc. Hence, once a patch has started to get more individuals than others, its density can only continue increasing in a kind of autocatalytic process. The result is that, in a population in which all individuals only use the presence of conspecifics to select their breeding habitat, at some point all the population will end up in a single patch, independently of the current relative qualities of the patches (Doligez *et al.* 2003). Social attraction thus leads to **informational cascades** (Giraldeau *et al.* 2002) in which all individuals only cue on social information and no one is actually sampling the habitat anymore. In contrast, when all individuals use the **breeding performance** of others to assess habitat quality, they in fact cue on the recent habitat sampling by conspecifics. Thus, habitat copying does not lead to informational cascades, and leads to much lower levels of true aggregation (Doligez *et al.* 2003, see Chapter 14 for a definition).

In a metapopulation (i.e., a set of populations connected by dispersal), the aggregation of individuals on some of the patches leaving others empty may increase the overall extinction probability of the metapopulation by increasing the probability of simultaneous extinction of all sub-populations (Ray *et al.* 1991). In addition, if individual fitness negatively depends on local density, aggregated distributions further increase extinction probabilities. These mechanisms may nevertheless be limited by natural selection favouring (1) habitat choice strategies allowing individuals to settle on the best patches, thus leading the population to be distributed in an ideal free way, which minimizes extinction probabilities, and (2) mixed strategies of habitat choice (using a combination of cues) or strategies depending on circumstances.

#### 9.6.1.3 *Habitat choice and the evolution of coloniality*

We have seen that inferring mechanisms from the observed patterns of animal distribution is a real challenge (van Teeffelen and Ovaskainen 2007). In this chapter we have taken the opposite approach by focusing on the mechanisms of habitat choice that generate animal distribution. Such an approach considers the proximate and ultimate factors affecting individual choices, which in turn has potential consequences for the dynamics of animal distribution among habitat patches. Chapter 14 will discuss how the use of social information in breeding habitat choice may have led to forms of group living, such as coloniality, i.e. breeding within aggregated nesting territories that contain no resource other than nesting sites. Individuals basing breeding site choice on the mere presence or breeding performance of conspecifics tend to settle on already occupied patches, which should intuitively lead to the spatial aggregation of breeding sites (see Figure 14.9). Furthermore, breeding close to conspecifics may favour the gathering of social information on habitat quality, thus amplifying the tendency to aggregate.

The general message is that the link between habitat choice and the spatial distribution of individuals clearly shows that the study of habitat choice strategies is important in understanding the evolution of group living. It further underlines the necessity to always consider population level consequences of individual strategies (Sutherland 1996).

### 9.6.2 Habitat choice and conservation biology

Conservation biology makes use of scientific knowledge and methods to analyse questions related to the conservation of biodiversity, such as the maintenance of endangered populations. The dynamics of small populations is strongly affected by the probability of local extinction. Habitat choice behaviours are thus likely to play a critical role in the dynamics and prospects of small populations. The role of behavioural ecology in conservation biology is specifically addressed in Chapter 18, but we stress here a few points directly related to breeding habitat choice.

#### 9.6.2.1 *Small and fragmented populations*

Threatened populations are usually small and subdivided within fragmented habitats. Thus, habitat choice behaviours are critical for conservation issues because they affect both the exploratory and prospecting movements of individuals among suitable patches, which then affect the distribution of individuals in the environment. Individuals may for instance be exposed to greater mortality risk when prospecting than once settled, and such additional mortality increases with distance between patches (i.e. the degree of fragmentation). Moreover, the distribution and movements of individuals among patches directly affect the dynamics and viability of the metapopulation, which often encompasses small sub-populations (see Section 9.6.1).

The study of breeding habitat choice is thus critical for the monitoring and management of threatened, re-introduced or re-enforced populations. Knowledge of the factors affecting habitat choices has direct implications in such situations, and may greatly influence the design and monitoring of protected areas as well as the assessment of subdivided populations' viability (Smith and Peacock 1992; Reed and Dobson 1993). As mentioned previously, individuals may end up settling on low-quality habitat because of constraints on mobility and information gathering. Furthermore, individuals may settle on sites of decaying quality because of human activity, if they get lured for instance by high local densities of conspecifics, usually indicating good quality sites in a natural situation (see the discussion on 'ecological traps' in the next section). Such issues can be critical for the conservation of threatened populations.

#### 9.6.2.2 *Environments under human influence*

Naturally selected habitat choice strategies use environmental components as cues of habitat suitability because such components reveal habitat quality in natural situations. However, human activities can alter the structure and quality of habitats. In particular, they often break the natural correlations among habitat components. Thus, naturally selected habitat choice strategies may become maladaptive in environments modified by human activity: individuals may be lured to unsuitable patches because cues that reveal habitat quality in natural setting do not anymore in environments altered by humans.

Such situations can arise when some habitat characteristics affecting fitness deteriorated but do not affect the cues used by individuals to assess site quality: such a mismatch between the value of the cues used for habitat choice and the actual habitat quality defines an 'ecological trap' (Battin 2004), which could eventually lead to population extinctions (Delibes *et al.* 2001; Kokko and Sutherland 2001). In a population of the northern wheatear (*Oenanthe oenanthe*) breeding in a farmland landscape in Sweden, individuals have been shown to prefer certain territories, but this preference was not for predictors of reproductive success. Thus breeding success achieved on attractive and selected territories was not higher than on less attractive territories, i.e. breeding habitat choice was non-ideal (Arlt and Pärt 2007). The study of habitat choice in relation to human-related changes in the environment is thus critical for the management of small and subdivided populations. Because such populations have often greatly decreased in the recent past, a large proportion of potentially suitable patches may be unoccupied, but may nevertheless need to be preserved to allow individuals or even populations to move. Such situations require managing habitat in terms of **meta-reserves:** that is, groups of reserves that aim at protecting a certain type of habitat, independently from their current occupation by the species of interest.

#### 9.6.2.3 *Reintroduced populations*

Finally, a thorough understanding of habitat choice behaviours is useful for increasing the efficiency of population re-introduction or re-enforcement. The rearing conditions of individuals may affect their tendency to choose specific types of habitat, and habit of using a site can contribute to the early settlement of individuals after release. Conversely, a large mismatch between rearing and release habitats may result in individuals being unable to make optimal habitat choices. Social interactions have also been recognized as critical in several situations. In particular, attraction to active breeding conspecifics can be used to help fix a group of individuals on a patch, thereby compensating for adverse Allee effects. The success of the re-introduction of Griffon vultures (*Gyps fulvus*) in the Cévennes area of France (see Figure 18.6), has for instance been linked to the early constitution of a 'nucleus' of individuals that lead to the fixation of individuals released later on (Sarrazin *et al.* 1996). In some cases, the use of visual and/or sound decoys can attract individuals to sites identified as suitable by managers. Decoys have for instance been used to create new seabird colonies, mimicking successful (incubating) conspecifics (see Veen 1977; Kress 1998); decoys of predators can also be used to deter focal individuals from settling in areas identified as low quality by managers. In other words, understanding the cues used by individuals for selecting a breeding habitat patch allows manipulation of these cues to alter individuals' choices.

## 9.7 Conclusion

We have seen that breeding habitat choice is a critical process from both an evolutionary and ecological point of view. It has numerous direct and indirect effects on fitness (e.g. through securing a high-quality breeding site), but also implications in the evolution of other behaviours (e.g. prospecting) and life history traits (e.g. age at first reproduction, investment in reproduction). The choice of a breeding habitat patch by individuals affects the distribution, social structure and regulation of populations, and may thus lead to the evolution of social traits (e.g. coloniality, cooperative breeding). One of the goals of this chapter was to stress the importance of focusing on the source of information used by individuals to choose among the potential alternatives of varying quality rather than on the distribution of individuals among the potential patches.

In the next chapter, dedicated to the evolution of dispersal, we will still deal with direct consequences of breeding habitat choice. Habitat choice processes may indeed induce movements of individuals between breeding habitat patches, potentially leading individuals to settle and breed in other places than their natal or previous breeding sites. Such movements can lead to gene flow, which can have a major role in the evolution of populations at various spatial and temporal scales, and may influence speciation.

## » Further reading

> *For theoretical considerations on breeding habitat choice:*

**Bernstein, C., Krebs, J.R. & Kacelnik, A.** 1991. Distribution of birds amongst habitat: theory and relevance to conservation. In *Bird Population Studies* (Perrins, C.M., Lebreton, J.-D. and Hirons, G.J.M., eds), pp. 317–345. Oxford University Press, Oxford.

**Doligez, B., Cadet, C., Danchin, E. & Boulinier, T.** 2003. When to use public information for breeding habitat selection? The role of environmental predictability and density dependence. *Animal Behaviour* 66: 973–988.

**Fretwell, S.D. & Lucas, H.L. Jr.** 1970. On territorial behaviour and other factors influencing habitat distribution in birds. *Acta Biotheoretica* 19: 16–36.

**Stephens, D.W.** 1989. Variance and the value of information. *American Naturalist* 134: 128–140.

**Switzer, P.V.** 1997. Past reproductive success affects future habitat selection. *Behavioral Ecology and Sociobiology* 40: 307–312.

> *For general reviews on information and animal choice:*

**Clobert, J., Danchin, E., Dhondt, A. & Nichols, J.D.** 2001. *Dispersal.* Oxford University Press, Oxford.

**Valone, T.J. & Templeton, J.J.** 2002. Public information for the assessment of quality: a widespread phenomenon. *Philosophical Transaction of the Royal Society of London Series B* 357: 1549–1557.

> *For a review on prospecting behaviour:*

**Reed, J.M., Boulinier. T., Danchin, E. & Oring, L.** 1999. Informed dispersal: prospecting by birds for breeding sites. *Current Ornithology* 15: 189–259.

> *For an experimental test based on the manipulation of ISI in the context of breeding habitat choice:*

**Doligez, B., Danchin, E. & Clobert, J.** 2002. Public information and breeding habitat selection in a wild bird population. *Science* 297: 1168–1170.

> *For a review on ecological traps:*

**Kokko, H. & Sutherland, W.J.** 2001. Ecological traps in changing environments: ecological and evolutionary consequences of a behaviourally mediated Allee effect. *Evolutionary Ecology Research* 3: 537–551.

## » Questions

1. Design experiments to test the existence of one mechanism of breeding habitat choice. Give particular care to sampling and replication issues.

2. To what extent does the study of animal distribution inform about actual choice mechanisms? More generally, to what extent do patterns allow one to study the mechanisms at the origin of that pattern?

3. What is the interest of correlational approaches in the study of breeding habitat choice? This question is linked to the previous one.

4. Does a population distribution that is highly skewed for a given type of habitat always demonstrate the existence of an actual preference for that type of habitat?

# 10

# Evolution of Dispersal

Jean Clobert, Michèle de Fraipont, and Étienne Danchin

## 10.1 Introduction

Every species, at some point in its **life cycle**, goes through a phase of movement. The phase can be short, as for numerous plants or sessile organisms, or long as for most herbivores and nomadic species. Movement is essentially linked to the presence of some local problem because at any given moment, individuals must answer the following evolutionary question: **'Do I have all that I need here and now?'** Asking the question this way, shows that two types of answer are required: one that deals with space and the other with time.

### 10.1.1 Movement in time as an alternative to movement in space

From a temporal point of view, a temporary shortage of **resources** at a given location can be overcome in two ways: either by leaving in search of greener pastures or by adapting one's body to this shortage. Torpor or hibernation as well as dormancy of larval stages or spores are all tactics that help an individual cope with a temporary shortage of resources, here mostly food. Other available tactics are more behavioural and aim to mitigate the effects of the shortage. For example, in the Eurasian oystercatcher (*Haematopus ostralegus*; see Figure 9.3), some individuals do not hesitate to delay their own reproduction in order to obtain a higher-quality territory. Two researchers from the Netherlands, Bruno Ens and Dik Heg (Ens *et al.* 1995; Heg 1999) found that in oystercatchers the territories located on the edge of the mudflat are of higher quality because they offer a nesting as well as a feeding zone (see Figure 9.3). **Competition** for these good territories is strong and individuals virtually 'line-up' for them, some juveniles, for instance, waiting several years before getting a good one. Other individuals simply choose to reproduce at a younger age on poorer territories simply to avoid the long wait for good territories. Thus, there coexists within the same population two different temporal tactics to gain reproduction and the effectiveness of both are more or less equal (Ens *et al.* 1995; Heg 1999).

### 10.1.2 Movement in time or in space

Although temporal movement can always be seen as an alternative to spatial movement in response to a local problem, we will not discuss it any further in this chapter for several reasons. Firstly, many of these temporal movements presuppose a morphological or physiological adaptation that is not within the scope of this book. Secondly, species that have developed temporal movement tactics are also capable of movement in space. This suggests that a spatially localized constraint cannot always be resolved by spatial movement. The evolution of a given response to a local shortage is conditioned by the cost of construction and the maintenance of any of the morphological, physiological and/or behavioural systems necessary for these types of response (see cost of plasticity in Chapter 5). In the present chapter, we discuss temporal movement only in cases where it is necessary for the understanding of spatial movement.

### 10.1.3 What do we mean by dispersal?

An individual may try to escape local conditions for several reasons. It is cold and no refuge can be found, it is hungry and food is absent, there are too many predators and not enough places to hide, there are too many conspecifics and an insufficient number of reproduction sites, etc. In this chapter, we will cover only movement that concerns reproduction because some types of movement, such as those strictly linked to feeding, have already been treated in Chapters 7 and 8. Moreover, the movements linked to reproduction are special in that they are the only ones that result in gene flow between groups or populations. This does not mean that the other types of movement are not correlated with reproductive movement. For example, nomadic species, like many insects, reproduce while searching for food. However, the choice of a suitable partner always imposes a movement that cannot be optimized only by foraging or by avoiding predators. This is often shown by the presence of sexual pheromones that allow an individual to detect, sometimes at long distances, the presence and location of compatible sexual partners.

> We call breeding dispersal the movement between two reproduction sites and natal dispersal the movement between the site of birth and the site of first reproduction. The distinction between these two types of dispersal is important because their evolution seems to be driven by different selective pressures.

From a behavioural point of view, dispersal events involve several distinct steps triggered by different behavioural decisions (Figure 10.1). First, movement is initiated by a **decision to leave** which is followed by a movement phase that ends by a **decision to settle** in a new reproduction site. During the move, the individual makes a great many decisions and in the end we say that the organism 'has dispersed'.

Several nuances, however, must be added to this scheme. For example, in the case of plants, the natal dispersing phase is achieved by the seed or the fruit (pollen dispersal being like breeding dispersal for animals). Some plants create seeds that are adapted to dispersal and others seeds that are not adapted to dispersal. It is thus the plant (i.e. the parent), not the offspring (i.e. the seed) that takes the dispersal decision. A similar argument applies to the settlement decision when dispersal is mostly passive, like for example, when the seeds are transported by wind or by animals.

In this chapter we first present several evolutionary causes of dispersal (Section 10.2), describe some mechanisms of dispersal (Section 10.3), analyse some of its consequences in terms of fitness and population dynamics (Section 10.4) and conclude by considering the multiple types of dispersal and the potential for future development of dispersal

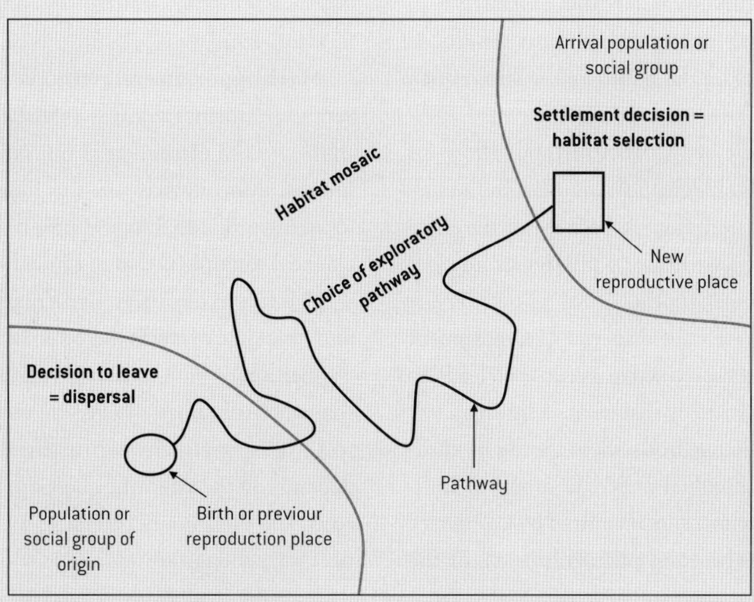


Figure 10.1 **The principal steps of dispersal**

Most often in animals, dispersal event starts with the decision to leave either the site of birth (natal dispersal) or the current reproduction site (breeding dispersal) and ends with a decision to settle at a new site: the settlement decision. Between these two decisions, there is the movement phase during which the individual must make successive decisions. In general, when studying the decision to leave, we talk about dispersal; when we are interested in the settlement decision, we talk about habitat selection (see Chapter 9).

research. One of our goals is to convince readers that dispersal is deeply rooted in behaviour (e.g. breeding habitat choice, see Chapter 9) and as such belongs to behavioural ecology.

## 10.2 The ultimate causes of dispersal

Population geneticists were the first to realize the importance of dispersal in explaining population genetic structure, and more generally evolution. The usual term used by geneticists is **migration**, which is unfortunate because the same term is used by ecologists to describe seasonal or repetitive movements between two or more **habitats** that for the most part do not involve reproduction, and thus no gene flow. The focus on genes rather than individuals kept geneticists from paying any attention to the cause of the movement, and so dispersal, the behaviour behind the geneticists' migration, was seen more as a correction factor or a source of variation rather than a phenomenon worth studying in its own right (Wright 1932). The importance of the numerous environmental constraints for the evolution of dispersal was recognized only relatively late and the same is true for the population dynamical consequences of dispersal. It is particularly striking to see the extent to which the literature on dispersal and on **habitat selection** have ignored and continue to ignore each other (Figure 10.2). This may be due to the difficulty of acquiring data on dispersal and on the classic, though unverified assumption that the number of immigrants to and emigrants from a population must be equal. The development of genetic and telemetric techniques as well as capture–mark–recapture methods (see Box 10.1), in conjunction

Box 10.1 **Methods used to measure dispersal and the information they provide**

This short overview is far from exhaustive, in particular with regard to technical aspects. It focuses mainly on the nature of the information gathered.

The **genetic** or **indirect** methods allow for a partial or total identification of an individual or a group of individuals, because of the large number of genetic markers (allozymes, nuclear DNA and mitochondria). They allow for the measuring of 'effective dispersal' (i.e. movement followed by a reproductive event that leads to surviving individuals that create a lineage).

**Telemetric methods** allow for the localization of an individual and/or a group of individuals, owing to a signal emitted by an embark device (Argos beacons, microchip, etc.), or the deformation of an emitted signal (radar, echolocation, etc).

**Marking** or **direct methods** allow for observation during physical or visual re-capture, thanks to the placement of internal or external identification devices on the individual, thus allowing individual follow-ups in the field. The classical example is that of bird rings.

The recent advances, particularly in genetic and telemetric methods, often allow for the acquisition of data as precise, and often richer, than those obtained by classical marking (Table 10.1). The final selection of the best method depends on the scientific question, the financial costs, and the amount of data, as well as the time necessary to monitor individuals in nature.

| Type of method | Individual Identification | Spatial monitoring | Temporal monitoring | Other advantages | Other limitations |
|---|---|---|---|---|---|
| | | | **Monitoring using genetic markers** | | |
| **Gene genealogy** | No | Limited | No | Access to very old events | Not appropriate for population dynamical studies |
| **Allelic frequencies and linkage disequilibrium** | No | Limited | Limited | Access to recent historical events (colonization, extinction, etc.) | Confusion between selection and dispersal |
| **Genotyping** | Yes | Yes | Yes | Allows measuring relatedness between individuals | Very costly |
| | | | **Monitoring individuals in the field** | | |
| **Radar, echolocation** | No | Difficult | Yes | Possible in hard to reach environments | Need a recognition of shapes |
| **Radio or satellite transmitters** | Yes | Yes | Limited | Other parameters, such as physiological parameters | Can affect the development or the behaviour of the individual |
| **Numbered or coloured tags** | Yes | Limited | Yes | Easy identification for detailed behaviour | May sometimes affect the behaviour of the tagged individual |

Table 10.1 **Advantages and disadvantages of different methods for studying dispersal in the field**

with the choice of either low mobility or **colonial** species have allowed dispersal studies to flourish. The mid-1960s and the beginning of the 1970s saw the first theoretical studies of dispersal, which were followed towards the end of the 1970s and the beginning of the 1980s with the first experimental work.

In the first theoretical studies, three mains causes were hypothesized for the evolution of dispersal, the quality of the: 1) non-social (**physical, predators, parasites, etc**); 2) **social;** and 3) **genetic environments** (Box 10.2). We treat each in order.

### 10.2.1 The role of the non-social environment's quality

The recognition that space and its heterogeneous nature are important drivers of dispersal only occurred during the late 1960s (Levins and MacArthur 1966; Van Valen 1971). The first models took into account the possibility that small local populations could go extinct for different reasons linked to demographic or environmental **stochasticity** (i.e. the effect of random processes). These models highlighted the importance of dispersal rate, through colonization, and extinction rate for the long-term persistence of populations and the evolution of dispersal. They provided the foundation of two important concepts: **metapopulation** and **risk spreading**, the fact that spreading propagules into several populations reduces the risk of extinction of the whole lineage.

> A metapopulation is a group of populations (i.e. units of reproduction composed of a group of individuals that are more likely to reproduce with one-another than with other members of different reproductive units) loosely connected to one another by dispersal, subjected to recurrent extinction and that could be colonized by propagules coming from other populations but that belong to that metapopulation. In other words, a metapopulation is an ensemble of populations connected by dispersal.

#### Box 10.2 The main evolutionary causes of dispersal

Three principal groups of causes were suggested in the literature to explain the evolution of dispersal.

**Variation in the non-social environment**: the quality of the habitat can vary in space and time. Breeding sites, food, parasites, predators, etc. are all components that can lead to a decision to leave and/or settle.

**Variation in the social environment.** Can be of two types:

- variation linked to competition between individuals (inter-age-group, intra-sex, and inter-sex);
- variation due to competition among genetically related individuals, between kin or between parents and offspring.

**Variation in the genetic environment.** These can have two different sources with opposite effects:

- on the one hand, the avoidance of **inbreeding** prompts dispersal to avoid mating with genetically related partners.
- on the other hand, the presence of co-adapted genes will prompt the avoidance of dispersal over great distances because this may break the associations between co-adapted genes.

**Multiple causes.** Most often, a combination of these diverse constraints are involved in the evolution of dispersal behaviour.

If a species that lives within a metapopulation is unable to disperse, it will undergo extinction at some point because every population of the metapopulation will become extinct with certainty at some point in time. Thus, individuals that disperse offspring to many populations within the metapopulation spread risks as this will in fact ensure the survival of at least one while minimizing the probability that they will all go extinct with the extinction of a single population. The positive effect of dispersal increases with the increasing independence of each population's extinction probabilities. In other words, dispersal allows a lineage to **spread the risks of extinction** by sending offspring into several patches of the environment. This is an important concept in conservation biology (see Chapter 18).

#### 10.2.1.1 *Two important, sometimes contradictory, types of historical approach*

From the above models, several different situations were explored while taking into account the age structure, the local dynamics, the structure of and distance to other populations, the quality of the sites, etc. In fact, two important types of model, different in their goals, were developed in parallel.

The first type focuses on the **demographic** consequences (type of population dynamics on a local and global scale, persistence of metapopulations, etc.) of a variation in dispersal or extinction rates, and site quality, etc. It is the demographic approach to dispersal.

The second type focuses on the **evolution** of dispersal rates in a heterogeneous habitat and on the consequences of the evolution of extinction probabilities as a function of site quality, etc. This is the evolutionary approach to dispersal.

We can contrast these two types of model with the behaviour-centred models discussed in Chapter 9. Often the demographic and evolutionary approaches produce contradictory predictions. The best known example is dispersal in source–sink systems. A source–sink system consists of two types of population: source populations produce individuals in excess relative to the number of individuals that need to be replaced (they have a positive population growth rate, see Chapter 5); whereas, sink populations have a negative demographic balance (population growth rate less than 1). In such situations, demographic models predict a viable system if the surplus individuals of the sources that disperse towards the sinks provide enough individuals to compensate for the low production in the sink populations while the reverse movement is precluded. However, evolutionary models predict that dispersal could not evolve in that particular case: on an evolutionary level, for such a system to work, some descendants of the lineages that left the source to join sink populations must return to source populations.

This can be easily understood if we compare the fitnesses of philopatric ($W_p$) and dispersing individuals ($W_d$). For the sake of simplicity, imagine a two-patch system, one being a source (population 1), the other a sink (population 2). It follows that:

$$W_p = f_1 - m_1$$

and

$$W_d = (1 - d)(f_1 - m_1) + d(f_2 - m_2)$$

where $f_1$ and $f_2$ are the numbers of new reproducers that are produced by an individual in populations 1 and 2, $m_1$ and $m_2$ the numbers of reproducers that died in populations 1 and 2, and $d$ is the dispersal rate. By definition, $f_2 - m_2$ is negative (this is the sink population) and $f_1 - m_1$ is positive (the source population). We see that $W_p$ is superior to $W_d$ for every value of $d > 0$. Indeed,

$$W_d = (f_1 - m_1) - d(f_1 - m_1) + d(f_2 - m_2)$$

where the two last terms are negative.

This led many authors to claim that, in spatially variable environments, dispersal is counter-selected and should not be observed. This can be easily understood because dispersal, if it is fixed, will usually result in a greater flow of individuals toward the bad than the good habitats, the latter having a superior occupation rate. Some authors (see Anderson (1989) for a review) have even developed a theory in which philopatry is the only winning strategy, dispersal representing the only available alternative for individuals of lower quality who use it as a 'best of a bad job' alternative.

However, temporal variation in **habitat quality**, whose most extreme form can be reflected in the local extinction rate, would select for a greater dispersal rate. The degradation of the habitat of origin being certain on a long temporal scale, this would inevitably cause extinction or a decrease in the fitness of a strictly philopatric **genotype**. The key observation is that in a spatially structured environment that varies over time there is always a given patch of habitat, not always the same, that is favourable and that could be reached by dispersal. In other words, the source-sink approach ignores the fact that the location of sources and sinks often changes over time within the metapopulation. The result is that the evolution of dispersal rates in temporally variable environments depends largely on the type of habitat temporal variability. In particular, if all the available habitats suffer from the same temporal variation (spatial autocorrelation of 100%), there is a constant difference between habitats: in other words, the good habitats remain totally **predictable** in space. On the other hand, recent studies show that in habitats with too large a temporal variation (in particular, a large extinction rate), dispersal can be counter-selected.

#### 10.2.1.2 *Models that ignore behaviour*

Several models fail to recognize dispersal as a behaviour pattern (mainly by neglecting its cognitive and plastic aspects) and take for granted its innate, genetically fixed nature that manifests itself independently of the conditions that animals encounter over the time scale in which these models are supposed to occur.

However, there are numerous reasons to challenge this implicit independence of dispersal from ambient conditions. Indeed, the three steps of dispersal (Figure 10.1) rely heavily on behaviour: the decision to leave, the choice of pathway and the length of movement, the choice of the final location of reproduction, etc. Each of these steps requires **information** to evaluate alternatives. We have seen many examples of these conditional choice processes in Chapters 4, 7, 8, and 9. Moreover, the existence of relatively simple behavioural responses such as negative or positive **taxes** (Box 10.3) indicate that most organisms are able to perceive intrinsic differences in habitat quality and can take them into account in their movements. Last, but not least, the quality of a habitat is dynamic because it depends largely on the number of individuals that use it. That means that even if the intrinsic quality of the various patches of the habitat is constant, the simple movement of individuals among patches creates a real variation of quality in time by the sole effect of density dependence. Note that the **ideal free distribution** (IFD) model explained in Chapter 8 rests on this key **assumption**.

It is thus unrealistic today to think of dispersal has a genetically fixed trait that is invariably expressed independently from the animal's ambient conditions. We saw in Chapter 9 how environmental conditions strongly influence the habitat choice strategy that is selected for during evolution. In other words, **dispersal needs to be studied as behaviour**.

### 10.2.2 The role of the social environment

#### 10.2.2.1 *Ideal free distribution and intra-specific competition*

In the early 1970s, Fretwell and Lucas (1970) proposed the ideal free distribution (IFD) model to explain the distribution of individuals among patches of variable quality (see Chapter 8, Section 8.2.3a, for details). Under certain hypotheses (like an individual's perfect knowledge of the quality of all the habitat patches), this model shows that the optimal distribution is the one that allows equal fitness among all individuals in all patches. Under the assumption that individuals tend to optimize their fitness (Chapter 3), we expect them to follow the IFD. This model is the basis of numerous studies in behavioural ecology in particular in foraging (see Chapters 7 and 8) and habitat selection (Chapter 9).

##### a *A forgotten model in the study of dispersal*

Surprisingly, the notion of IFD has been used only relatively recently for the study of the evolution

Box 10.3 **Taxes, as simple behaviour patterns of habitat choice**

Taxes are innate attraction or repulsion responses to coarse elements of the habitat. A taxis is either positive (towards) or negative (away from) a given environmental element (light, gravity, odour, sound, etc.). For example, positive phototaxis is the tendency to move towards light, and negative phototaxis the tendency to move away from light towards shade. A response to gravity is referred to as geotaxis, whereas a response to sound is a phonotaxis. A more unusual example of a taxis concerns animals that inhabit large plains and that show avoidance of closed habitats such as forests. When they move, individuals of such a species systematically avoid entering closed habitats, even when doing so would save them time and energy. For these species, the forest edges are true reflective borders. If we ignored this type of taxis, we could wrongly conclude that the animals are engaged in a random walk dispersal path. However, a more precise knowledge of taxis allows for a significant reduction of path randomness. The constraint imposed by the structure of the landscape can significantly modify the dispersal distances (see Figure 10.2), and thus by modifying the cost of the path, influence the dispersal decision. It is particularly important to study these mechanisms if we want to predict the increase in a species' range, in particular in conservation biology.

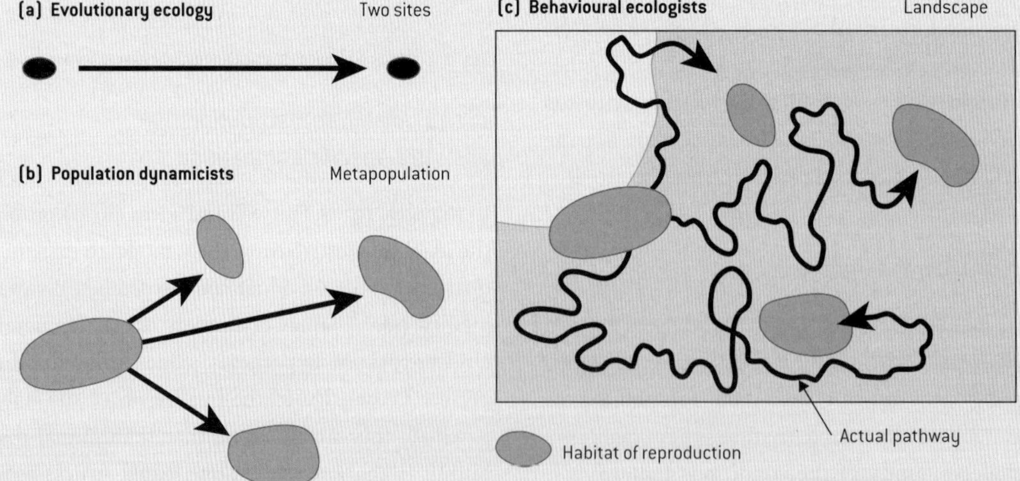


Figure 10.2 **The three visions that researchers have had concerning dispersal in relation to their specialization and era**

**a.** evolutionary ecologists are mainly interested in the fact that individuals switch populations. That is what will influence the genetic structure, and thus the evolution of a population.
**b.** population dynamicists introduced space: they conceive these movements in a habitat matrix, i.e. in a metapopulation, without considering individual motivations.
**c.** behaviourists are interested in the decision making that occurs throughout the process and in interaction with the mosaic of habitats that shape the environment.

of dispersal rate. This is unexpected because (1) movement is the core process by which individuals can distribute in an ideal free manner and (2) Fretwell and Lucas' original model was about the distribution of breeders among potential breeding habitats. One possible reason for this delay may be that paradoxically, it was the research on foraging that first used the IFD. Another reason may be that foraging involves movements with low amplitudes in time and space whereas dispersal involves decisions between successive reproductions over distances and times that can be much larger. It was thus easier to test the predictions of the IFD in a foraging than a dispersal context. Consequently it was only at the beginning of the nineties that the idea that dispersal could homogenize fitness among heterogeneous patches was used again in the context of the evolution of dispersal rates. Models by Mark A. McPeek and Robert D. Holt (McPeek and Holt 1992), generalized by Jean-Yves Lemel and collaborators (Lemel *et al.* 1997), were the first efforts to reintroduce this concept in the study of dispersal. They showed that for a large range of types of environmental variation, a genotype with a crude capacity to take into account the conditions met will invade and resist any combination of genotypes coding for a fixed rate of dispersal (and thus condition-independent). Although the evolution of dispersal rates only equalizes fitness among patches in specific cases – most often when dispersal is condition-dependent – these studies highlighted important phenomena acting on dispersal, like plasticity (the dispersal of an individual depending on local conditions), the variability of local conditions (in particular density dependence) and the cues used as sources of information (see Chapter 4). In other words, from this moment on, dispersal was once again considered as a pattern of behaviour.

### b *Can we find populations that are ideally distributed?*

One recurrent question in the literature consists of looking for natural populations distributed (at least partly) in an IFD way. In the foraging context, we saw several examples in Chapter 8. We saw that the predictions are often globally true but that recurrent discrepancies still exist in the details. These discrepancies helped improve the model. In the context of breeding habitats, the search for an IFD is much more complicated and examples are sparse. In fact, the time frame involved is in the order of a year at minimum, as opposed to the minutes needed in the case of foraging. Lastly, manipulating the intrinsic quality of habitat patches is much more difficult to achieve in a breeding context.

*A story of fast and slow clerks*

IFD can be understood intuitively from a simple example. Imagine a ticket office in a train station. Suppose there are counter clerks working at two different speeds: some are fast and serve twice as many customers in a minute than the slow ones. In that system, we can consider that the cost in time and energy of moving from a waiting line to another is almost non-existent, because the distance is short between parallel lines. This corresponds to the assumed freedom of movement of the IFD.

In such a situation, the **utility** is time and customers wish to minimize waiting time. So the inverse of waiting time is the **currency of fitness** (Chapter 2) and if we assume that the customers know each clerk's speed (the 'ideal' assumption of the IFD) then the waiting line in front of the fast clerk will be twice as long as the one in front of the slow clerk. Once this equilibrium line-up is attained, we can make some predictions about the state of the system.

1. The mean customer waiting time will be the same in each queue. This is the fundamental prediction of the IFD.
2. If, having reached the equilibrium some individuals nonetheless change lines (perhaps because they erroneously think the other is faster) the distribution will remain ideal so long as every line switch leads to another in the opposite direction. This means that at equilibrium, for a given line, the number of individuals coming in that line equals the number leaving it.

3. This also means that for a given pair of lines, the number of individuals going from line *i* to line *j* must be equal to the number of individuals going in the opposite direction.
4. As a corollary to this last prediction, the rate of renewal of individuals in waiting lines (i.e. the number of new individuals in the line divided by the total number of individuals in that line) must decline with an increase in the length of the waiting line.

When applying IFD to animals it is clear that there is little chance that the animals would know about the relative quality of every patch of the environment. This would correspond to a situation where the customers do not know the difference between the speeds of the two clerks. What would happen then? In our system of queues and customers, we know by personal experience that the waiting lines will eventually have different lengths, the longest being in front of the fast queues. In the case of animal populations numerous authors expect that the IFD is unlikely to be reached. We will see that this is not always so.

*Breeding flycatchers that are distributed in an ideal free way?*

A way to test whether a metapopulation is distributed in a ideal free way is to test the four predictions mentioned in the preceding paragraph on a natural population. This was done by the British scientist Patrick Doncaster and his collaborators (Doncaster *et al.* 1997), from a data set obtained in Sweden on the island of Gotland on a population of collared flycatchers (*Ficedula albicollis*). The metapopulation consisted of 11 forests in which these little passerines breed in nest-boxes provided by the Swedish researchers. Doncaster *et al.* (1997) used the data on movements between those different patches of forest for six consecutive years. Their results fulfilled the predictions mentioned above (Figure 10.3). First, the average fecundity in the 11 woodlands varied from one year to the other in a manner that showed an interaction between the effects of the forest and of the year. This indicates that in each year, some forests would show a better average performance than others but the best forests were not always the same. Despite that result, the average individual fecundity, over the six years of the study did not differ among the forests, as expected under the IFD (note that this prediction is not illustrated in Figure 10.3). Second, the numbers of individuals migrating to and away from a given forest were significantly correlated (Figure 10.3a). Third, the number of individuals dispersing in either direction between two forests was significantly correlated (Figure 10.3.b). Fourth, the emigration rate for a given forest was negatively correlated to the size of the forest (Figure 10.3.c). The authors concluded that breeding flycatchers on Gotland were distributed in an ideal free way. A similar finding was found recently in pike (*Esox lucius*; Haugen *et al.* 2006).

*On which scale should we test the ideal free distribution?*

The flycatcher example above presented statistically significant results because the authors used the data collected over a certain period of time. The fecundity results they report after six years, for example, provides a good example of the problem this can raise. Had the authors obtained data for a single year they could have found that fitness differed significantly among the different forests and this would have led to a rejection of IFD. In fact, given the assumptions of IFD, there is little chance that populations will be exactly at the IFD at all times. It may then be surprising to observe that the distribution fulfils the predictions of IFD when tested over a relatively short period of time. In the example of the ticket counter clerk used earlier, if we looked over a single period of one minute for instance, there is a good chance that we would find significant differences in the waiting times between lanes. However, over a longer period such as half an hour the fast clerks will have delivered roughly twice as many tickets as the slow ones. The conclusion is that we should not expect the populations to be at the IFD at all times but rather over a certain period of time. In fact, the cases in which significant deviations from IFD are expected are numerous (Holt and Barfield 2001; Leturque and Rousset 2002; and see Chapter 14).

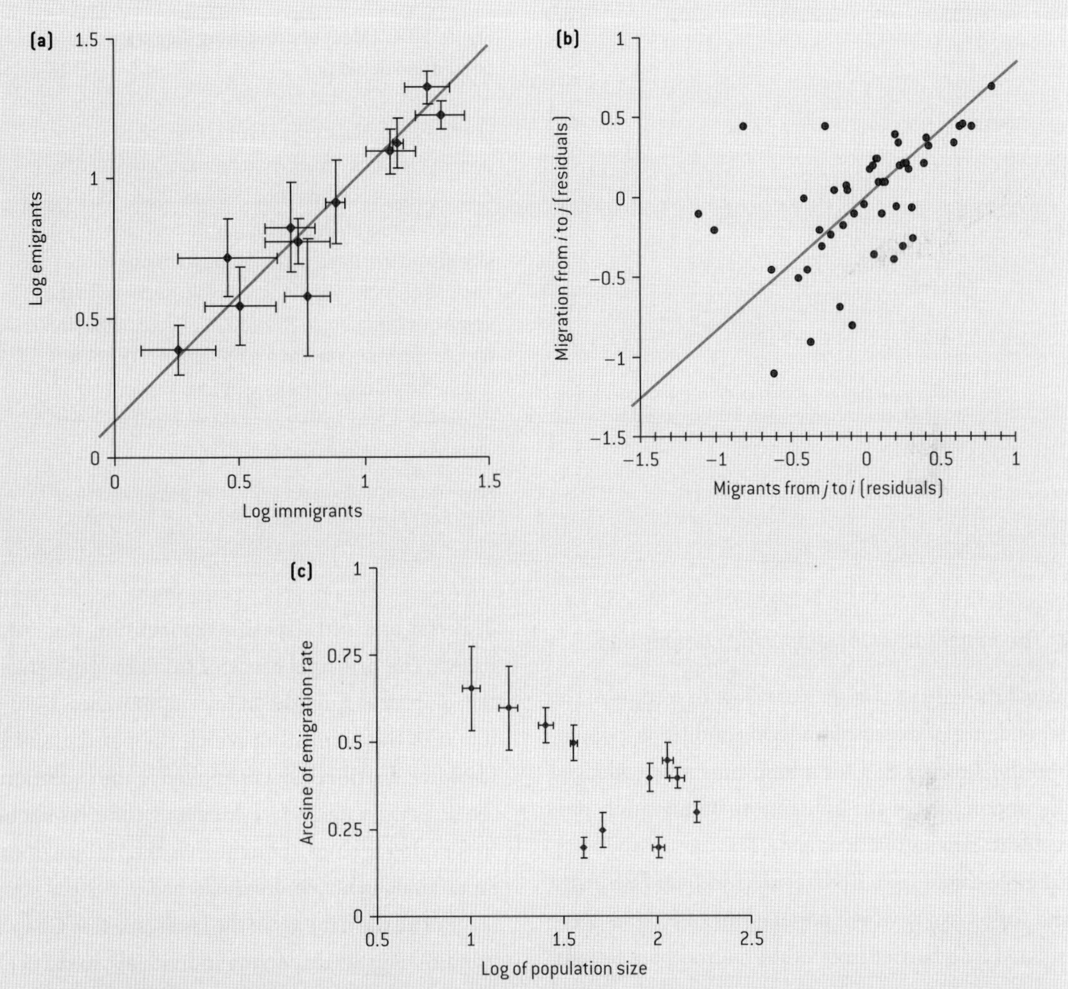


Figure 10.3 **Ideal free distribution in breeding collared flycatchers**

**a.** The number of individuals immigrating into a given woodland was correlated with the number of emigrants leaving the same woodland ($P < 0.0001$). This is the expected result of the second prediction of IFD.

**b.** For a pair of forest *i*, *j* the number of migrants from *i* to *j* is correlated to the numbers going from *j* to *i* ($P < 0.0001$). Note that the simple effect of distance between forests could explain this result, without using assumptions of the IFD. Authors thus corrected for the distance between forests by using the residuals of the regression of numbers of individuals to the distance between patches. This fulfils the third prediction of IFD.

**c.** The emigration rate (number of individuals leaving the patch divided by the total number of individual in the patch) decreased with the size of the population ($P < 0.004$). This fulfils the fourth prediction of IFD. The *x*-axis represents log transformed data and the Y axis arcsinus transformed numbers of migrants that were used in the analyses in order to fulfil the assumption of normality of the variable.

Derived from Doncaster *et al.* 1997.

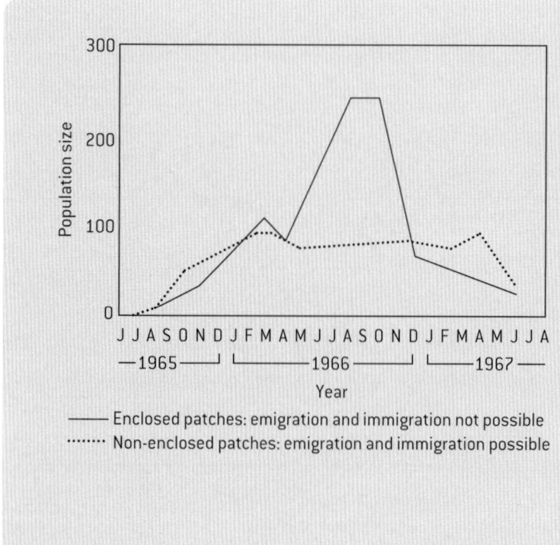


Figure 10.4 **Effect of preventing dispersal on population dynamics**

Changes in populations of voles (*Microtus pennsylvanicus*) in two patches: one enclosed to prevent dispersal, the other not enclosed thus allowing emigration and immigration. There was a significant increase of rodent populations in the enclosed patch, which then rapidly declined to the verge of extinction. In the non-enclosed patch, the population increased at a reasonable rate, then reached a plateau for a long period. Preventing dispersal seems to increase the numerical instability of populations.

Simplified from Krebs *et al.* (1969). See also Boonstra and Krebs (1977).

### c *The major role of intra-specific competition*

If evolutionary ecologists were slow to recognize the dependence of dispersal on local conditions, population dynamicists had for a long time acknowledged the importance of the emigration/immigration processes in population dynamics. The most convincing example comes from Krebs *et al.* (1969) and Boonstra and Krebs (1977) where populations of rodents were isolated by a barrier from other natural populations. In a first step, population sizes grew and reached sizes not observed in natural adjacent populations (Figure 10.4). Later, the enclosed populations collapsed and went almost extinct. Although this phenomenon-was not found in every species and although the mechanisms involved are still ill understood, the **experiments** demonstrated the prominent role of dispersal in population regulation. It is thus not surprising that, because of these pioneering studies, numerous authors have tried to demonstrate a direct link between dispersal and density of conspecifics for organisms belonging to various taxa (Box 10.4).

### d *Why should we expect to observe metapopulations at ideal free distribution?*

It was not until the end of 1990 that the theoretical confirmation of the link between the evolution of dispersal and intra-specific competition was established. The principal cause of this time lag between the empirical and theoretical approaches is to be found in the moment at which the paradigm of the ideal free distribution was applied to the evolution of the dispersal rate. In fact, for a long time, researchers in foraging ecology thought that the IFD could only be attained when individuals had a perfect knowledge of all the habitats in the landscape (see Chapters 7 and 8). For the dispersing individuals, in particular those that leave their natal site for the first time, this assumption is unlikely to hold. Without having any information on the degree of intra-specific competition occurring elsewhere, the general wisdom was that the IFD was impossible to achieve. However, these conclusions did not take into account several important facts.

For example, if the distribution was not ideal and free, then some patches may be overexploited and others underexploited. This means that any individual in an overexploited patch capable of detecting that it is in a low-quality (overexploited) patch and is able to move to a better (underexploited) patch will be favoured by **natural selection**. Every such movement will lead the distribution closer to the IFD. Thus, we should expect to see distributions that tend towards the IFD as long as the environment has not been perturbed too recently and/or the

### Box 10.4 **Dispersal and intra-specific competition**

In many insects that show a dimorphism for the presence of wings (macropterous), the percentage of winged-insects in the population is linked to the density of conspecifics. Some examples exist in grasshoppers, crickets, and aphids.

An elegant experiment was conducted by Herzig (1995), which experimentally created high and low density larval populations in the beetle *Trirhabda virgata*. Such density manipulations strongly affected the percentage of defoliated plants, showing that intra-specific competition varied across treatments. The results showed that long-distance movements and movements involving flying increased in high-density population.

Moreover, the aphid *Aphis fabae* in its apterous form produces winged descendants when placed in overpopulated conditions (estimated among other things by the number of contacts with others). However, the winged individuals disperse only if they meet high-density conditions. This shows that the conditions work at two different moments: during development that drives individuals towards different morphotypes; then on the winged adult, by initiating dispersal or not, depending on whether the density met by winged individuals is high or low (Roff 2001).

However, the response to density seems to depend on sex, the most philopatric sex being generally the most sensitive to density (example in tits, reviewed by Lambin *et al.* (2001)). In other species, however, the response to density seems reversed (example in voles, Lambin (1994)) without any clear reason. Such differences may arise because the measure of dispersal varied among studies: some studies used effective dispersal (i.e. accompanied by reproduction), whereas others only used dispersal attempts without taking into account their actual efficiency. In high density environments, dispersal attempts are more frequent but the ones that succeed are those made over short distances. However, there are alternative explanations to such behavioural differences. For example, the landscape structure (presence and knowledge of other populations), the structure of kinship, the meaning of density for the species (if the density is considered as reflective of habitat quality) are other potential explanations for the differences in the results of such studies (Lambin *et al.* 2001; Clobert *et al.* 2004).

environment is not too heterogeneous and as long as dispersal is not only to find a good habitat.

On the other hand, density dependence maintains populations within a range of sizes where **demographic stochasticity** still holds, in other words where some temporal and spatial variation can be created by random deaths and births (Figure 10.5). Travis *et al.* (1999) and Cadet *et al.* (2003) showed theoretically that intra-specific competition could in itself cause the evolution of non-zero rates of dispersal under the sole action of demographic stochasticity. In fact, a patch has only a small chance of having the same number of survivors every year, which creates temporal heterogeneity. In addition, two close patches have very small chances of having the same number of survivors even though they have the same **carrying capacity**, which *ipso facto* brings spatial heterogeneity (Figure 10.5).

As we have seen before, it is the combination of two sources of heterogeneity that selects for the evolution of dispersal. Even if the causal link between density and dispersal has been known empirically for several years, it was still important to demonstrate it theoretically, because this changes the classic view that intra-specifically caused dispersal leads to competitive exclusion, implying that dispersers are of

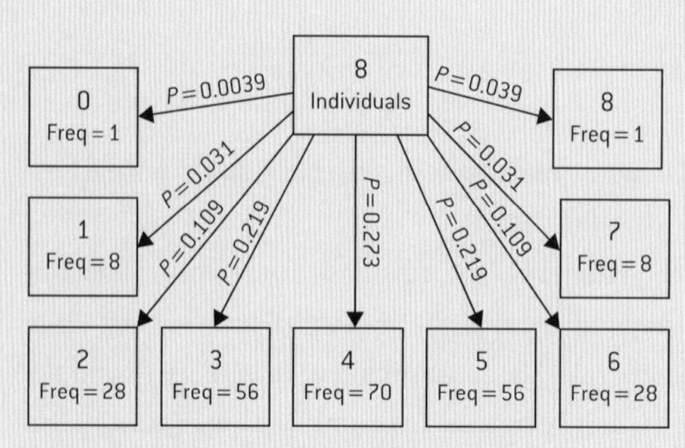


Figure 10.5 **Demographic stochasticity**

Consider a population consisting of eight patches containing eight individuals each. Each has a chance $p$ of surviving to the next reproductive season, with $p = 0.5$. The probability that $x$ individuals among $y$ stay alive in any patch equals

$$[(y!/x!)(y-x)!]p^{x}(1-p)^{y-x}.$$

Here the probabilities of having three, four, or five survivors in a given year are similar, bringing *de facto* temporal variation in the numbers of individuals, and thus variation in the number of free patches. The chance that two close patches will have the same number of individual is only 0.135 in this example. Density dependence accentuates this effect, because it results in having to keep the size of the population to a maximum of eight individuals and from then on, to keep the same probabilities in time to have three, four, or five, etc. live individuals. The probabilities of observing zero or eight, one or seven, two or six, and three or five individuals are identical. Statistically, the probability of having four individuals surviving is 70 times higher than that of having none survive.

lower quality. In this context, dispersal is a strategy of response to competition, and it becomes plausible that dispersers are individuals of at least equal quality to that of the philopatric ones.

Finally, the density of conspecifics carries information that can be used differently depending on the individual and on the situation. In particular, the presence of individuals in a habitat can attest not only to the possibility of competitive interaction, but also the quality of the habitat. For instance, in the common lizards (*Lacerta vivipara*) juveniles of the year tend to disperse in the presence of adult females, whereas they tend to stay in the presence of adult males (Figure 10.6). The logic behind this difference in response of juveniles is linked to the fact that laying females must replenish their energy reserves rapidly before hibernation and because their larger size provides them with a competitive advantage in collecting resources. Because of this, laying females will be fiercer competitors against juveniles than adult males which have longer time to replenish their reserves and are not very aggressive despite their high social dominance at that time of year. Population density also reveals habitat quality such that there is often a negative relationship between density and dispersal. In another species of lizard (*Anolis*) settlement site choice is made both on the basis of the presence and the density of conspecifics (Stamps 1987) confirming the double information (about competition and habitat quality) carried by density. This can explain, in part, why in numerous species of rodents sometimes we find positive and sometimes negative relationships between dispersal and density (Gundersen *et al.* 2002).

#### 10.2.2.2 *Searching for a partner, intra- and extra-sexual competition, and inbreeding depression*

##### a *Sexual conflict can generate dispersal biases according to sex*

The characteristics of attending conspecifics can further constitute a source of information for the individual that must decide whether to stay or disperse. Thus, in sexually reproducing species, males and females are logically attracted to high densities of the other sex and repulsed from high densities of their own sex. In such species, we can predict that dispersal motivated by the quest for a mate might depend on the population **sex ratio**. However, males and females have divergent interests. Females are limited in their reproductive abilities by the number

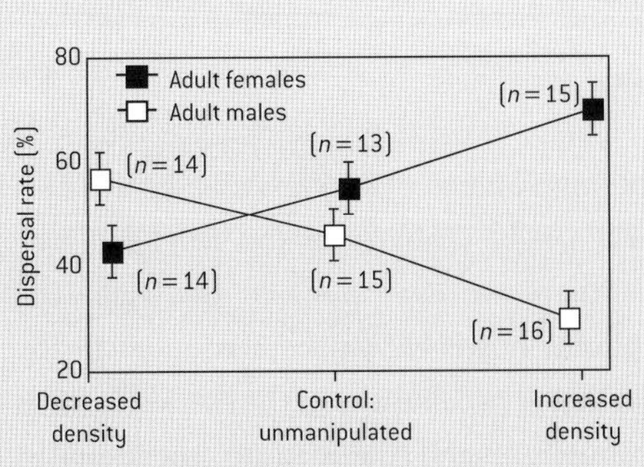


Figure 10.6 **Adult density and natal dispersal in the common lizard (*Lacerta vivipra*)**
From Léna, J.-P. *et al.*, The relative influence of density and kinship on dispersal in the common lizard, *Behavioural Ecology*, volume 9: 500–507, by permission of Oxford University Press.

In this experiment, juveniles faced either adult females or males and their tendency to leave when adults were present was measured over a five day period. Dispersal of juvenile common lizards was positively influenced by the density of female adults (filled squares), and negatively by the density of adult males (open squares; significant interactions).

At the population level, a negative relationship between density and dispersal was also found (Le Galliard *et al.* 2005).

of zygotes they can produce (Chapter 11) and so the best they can do to increase their fitness is to maximize the quality of their partner(s) in terms of genes, or parental care when necessary, and/or social position. Males for their part, are limited by the number of females they can inseminate except when they provide parental care. We will see in Chapters 11 and 12 how such conflicts of interest push males and females toward morphological, physiological and behavioural specializations, which very often result in a partial segregation of their ecological niches, or the development of **mating systems** that vary according to the environment. It is partly on such ideas that the current theory of the evolution of sex-biased dispersal was elaborated.

Sex-biased dispersal is common in mammals and birds, and poorly documented in other taxa, such as reptiles and amphibians and more generally in invertebrates (except ants, see Box 10.5).

In mammals there are more species in which males disperse than females, whereas the reverse is true in birds. Birds and mammals also differ in their mating systems and in the way they monopolize resources. Male mammals are typically polygynous in that they monopolize females, but are not territorial in that they do not monopolize resources. The resource-holding role is allotted to females. In birds, males are typically monogamous and responsible for monopolizing resources, whereas females only evaluate the resources monopolized by the males (Box 10.6).

Such differences in mating systems and resource defence between males and females led Greenwood (1980) and Greenwood and Harvey (1982) to propose that the sex that monopolizes food resources would

Box 10.5 **Biases in dispersal rates between sexes in ants**

In the ant *Diacamma cyaneiventre* (Doums *et al.* 2002), the future queens do not have wings and are condemned to stay in the nest where they hatched. To be fertilized, they must wait for a winged male from another colony to find and enter their colony. The flow of genes between colonies across significant distances is achieved only by males, whereas the swarming of colonies only occurs by budding (division of colonies by digging new galleries and establishing a queen in this new network of galleries). These two processes operate at two clearly different spatial scales. For other species, like fire ants (*Solenopsis invicta*), wings are produced in both sexes, which greatly facilitates dispersal, and which may be at the root of the invasive nature of such species (Ross 2001).

Box 10.6 **Possible explanations for sex-biased dispersal**

In two reviews of natal and breeding dispersal in mammals and birds, Paul J. Greenwood (Greenwood 1980; Greenwood and Harvey 1982) showed that in birds females disperse more than males for most species and families, whereas the opposite is true in mammals (Table 10.2 and 10.3).

Mating systems of mammals and birds also differ strongly (Table 10.2 and 10.3) but naturally there are exceptions to this rule. The exceptions, however, can be accounted for by the biological peculiarities of the species. In birds for example, several Anatidae are exceptional in that pairs form on the wintering quarters, and ▶

| | Birds | | | | | | Mammals | | | | | |
|---|---|---|---|---|---|---|---|---|---|---|---|---|
| | Natal dispersal | | | Breeding dispersal | | | Natal dispersal | | | Breeding dispersal | | |
| | Male | Female | Both | Male | Female | Both | Male | Female | Both | Male | Female | Both |
| **Species** | 3 | 21 | 6 | 3 | 25 | 1 | 45 | 5 | 15 | 21 | 2 | 2 |
| **Family** | 1 | 11 | 5 | 1 | 14 | 1 | 23 | 4 | 7 | 6 | 2 | 2 |

Table 10.2 **Sex-biased dispersal in birds and mammals**

| Birds | Mammals |
|---|---|
| Defence of resources by males | Defence of females by males |
| High male investment in resources in the absence or presence of the other partner | Low male investment in resources particularly in the absence of the partner |
| Low investment of females in resources | High investment of females in resources |
| Competition between males for resources | Competition between males for females |
| Mainly social monogamy | Mainly polygyny |
| Male philopatry | Female philopatry |
| Larger natal and breeding dispersal by females explained<br>**1)** Inbreeding avoidance<br>**2)** Increased reproductive success by choosing males defending high-quality resources | Larger natal and breeding dispersal by males explained<br>**1)** Inbreeding avoidance<br>**2)** Increased reproductive success by defending a greater number of females |
| Evolution toward a patriarchal system | Evolution toward a matriarchal system |

Table 10.3 **The relation between mating system and dispersal**

Summary of the differences in mating systems of mammals and birds, in relation to resource and sexual partner defence, as well as dispersal in both sexes.

From Greenwood (1980).

males follow females that return to their natal site. In these species, territory choice occurs after mate choice such that males defend only their sexual resource (access to females) and this is consistent with the hypothesis that predicts higher dispersal rates among males, the sex that defends reproductive resources. In mammals, territorial species provide examples of exceptions to the general pattern. In pica (*Ochotona princeps*), an American lagomorph inhabiting mountain ranges, males and females defend a territory but competition for territories is more intense among males. Young males tend to remain close to their natal site presumably taking advantage of their knowledge of the area in order to monopolize its resources more effectively once they can obtain a territory when one becomes vacant (Peacock and Ray 2001). Females pair up only with territorial males and must search for them over an area much greater than any one male's territory leading to higher female natal dispersal rates.

be philopatric, whereas the sex that monopolizes reproductive resources (i.e. members of the other sex) would disperse. The idea behind this hypothesis is that familiarity with a given area is beneficial to food resource monopolization such that philopatry is more profitable and hence selected in the territorial sex. The sex that does not monopolize food resources can thus move without suffering an additional cost and hence search for appropriate partners. However, what would more certainly create a dispersal bias between the sexes (after all, the sex that is free to move could also choose to stay in place) is the fact that, by moving the dispersing sex can avoid inbreeding depression. It is thus the combination of the monopolization of resources and the avoidance of inbreeding that best explains sex-bias dispersal. Certain exceptions (Box 10.6) to this general framework, as in ducks in birds, and picas, badgers, and some shrews in mammals, confirm this hypothesis. However, recent theoretical and experimental evidence has somehow challenged these explanations for the observed sex-biased dispersal (Perrin and Goudet 2001; Le Galliard *et al.* 2005).

### b *Inbreeding avoidance, a cause often proposed but rarely demonstrated*

Although the avoidance of inbreeding can, at least theoretically, account for the evolution of dispersal (Bengtsson 1978; Motro 1991; Gandon 1999), experimental or empirical evidence of this is limited and questionable. Most studies that have looked into the issue have only assessed inbreeding depression and, from this apparent cost, concluded on the importance of inbreeding for dispersal. For example, a recent fragmentation of butterfly habitats in Finland (Hanski and Thomas 1994) allowed for comparisons among the butterflies living in populations of differing degrees of isolation, such that populations living in small fragments suffered numerous malformations that reduced individual fitness (Saccheri *et al.* 1998). Similarly, in daphnia (*Daphnia*, Haag *et al.* 2002) populations resulting from the colonization of a small pond by a single clone in the Spring suffer from inbreeding depression during the bout of sexual reproduction that takes place at the beginning of the following winter. Other studies correlated the degree of genetic proximity to the local dispersal rate. For example, in the song sparrow (*Melospiza melodia*), the dispersal apparently did not result from inbreeding avoidance, and other factors like territory defence (Arcese 1989a, b) were most probably involved. This lack of inbreeding avoidance might result in a significant number of individuals that show a high rate of inbreeding with a high loss of fitness (Keller *et al.* 1994).

However, most studies are correlational and evidence remains debatable. There are only a few experiments on the role of inbreeding depression conducted in settings that would allow more robust interpretations of the results in terms of inbreeding.

One of the most convincing is an experimental study by Jerry Wolff (1992). He found that, in white-footed mice (*Peromyscus leucopus*), each parent expels the pups of the opposite sex from its home range, with greater intensity the more numerous they are. This result is all the more surprising in that intra-sexual competition would make the opposite prediction, that individuals would concentrate on expelling pups of their own sex. Hence the result was interpreted as demonstrating the existence of a parental strategy whose objective is avoiding breeding with one of your own offspring. Nonetheless important information is still lacking in that study. First, the likelihood that parents and offspring mate together is unknown. Second, the pattern of eviction may be due to reduction of intra-sex offspring competition within the more numerous sex by adults of the sex not directly concerned with this competition. It is also possible that individuals of the evicting sex are indirectly affected by the competition within the sex they evict such that by reducing this competition they gain benefits such as for example increased investment in parental care by their mate.

Theoretical studies concluded that inbreeding avoidance alone predicts contrasted rates of dispersal between individuals of the sexes, with one sex having predicted dispersal rates close to 0% (Perrin and Mazalov 2000; Perrin and Goudet 2001). Only the action of other causes can lead to non-zero dispersal rates for both sexes. Thus inbreeding avoidance as an evolutionary explanation of dispersal may only apply to cases where dispersal rates differ markedly between the sexes. It is the case in several social species like many queenless ants (Peeters and Ito 2001), and several primates (Pusey 1987). However, because habitat affects both sex ratio and dispersal, it is possible to be misled into falsely accepting the existence of significant between sex dispersal differences when animals are studied in only one habitat. Those dispersal differences would not necessarily have evolved to reduce inbreeding (Julliard 2000) but rather because of differential dispersal costs for each sex that can vary across habitats. Generally speaking, the sex differences in dispersal observed in the wild are more moderate than those predicted by inbreeding avoidance alone. Other factors, therefore, likely play a paramount role.

In the next section we review a list of the potential problems group members may encounter when there is some non-zero probability of genetic proximity.

### 10.2.3 Interactions between kinship and the evolution of individual recognition

Several antagonistic forces work for and against movement by members of a group in which some are genetically close (Table 10.4). Inbreeding avoidance constitutes a centrifugal force, but could be achieved by mechanisms other than movement. For example, recognition of genetic kin would allow individuals to avoid inbreeding while still living together.

| Benefits of philopatry | Costs of philopatry |
|---|---|
| + Knowledge of the habitat | + Reduced choice of habitat |
| + Possibility of obtaining indirect fitness | + Possibility of parental or familial manipulation:<br>– Reproductive suppression<br>– Forced cooperation |
| + Reduction of competition owing to familiarity | + Risk of only reaching a low dominance rank |
| + Partner of close and known genetic quality | + Reduced choice of partner:<br>– Inbreeding depression<br>– Partner too similar |

Table 10.4 **Cost and benefits of life in a kin group**

#### 10.2.3.1 *Helping should favour the evolution of kin recognition*

Kin recognition was studied extensively in the evolutionary framework of **eusociality** (see Chapter 15). In fact, societies where some members are specialized in feeding tasks, defence, or maintenance at the cost of their own reproduction can evolve only under special conditions. This is the case in particular when the individuals that do not reproduce help (**altruism**) close kin (**kin selection**, see Chapter 2). In such situations, it is really important to detect **cheaters** and to possess an effective mechanism of kin recognition in order to direct altruism preferentially towards kin. As we saw in Chapters 2 and 5, Richard Dawkins (1976) named such possible recognition systems 'green beards'. Historically, the kin selection hypothesis was the first genetically based solution to the evolution of altruism and so many following theories simply assumed genetic proximity among social mates (we will see in Chapter 15 that this assumption is unnecessary), and the existence of kin recognition mechanisms. As a result many studies searched for kin recognition mechanisms in species with division of labour and/or cooperative behaviour and documented their existence in species such as the isopod *Hemilepistus reaumuri*, a sub-social arthropod (Linsenmair 1987). Similar results were obtained in non-cooperative species such as green iguanas (*Iguana iguana*; Werner *et al.* 1987), and numerous anurans (see Blaustein and Waldman 1992; synthesis in Fletcher and Michener 1987 and Hepper 1991).

#### 10.2.3.2 *Other factors promoting kin recognition*

Mechanisms of kin recognition were also documented in species that do not have helpers or division of labour and so it must serve in contexts other than **cooperation**. Theoretically it can serve to recognize genetically compatible mates (but few examples support this possibility). The case of collective defence mechanisms of some tadpoles (Box 10.7) suggests that kin recognition may also play a role in the control of spatial distribution of close kin.

Indeed, even in a homogeneous habitat, there is a selective force, namely the competition between kin or genetic relatives that generates the evolution of a non-zero dispersal rate (Hamilton and May 1977). In that model, kin competition often causes more than half of the pups to disperse even in the presence of a substantial cost of dispersal. More recent models (Perrin and Goudet 2001; Gandon and Michalakis 2001) have further demonstrated the potential role of kin competition in the evolution of dispersal in several different situations. However, despite the predicted importance of this factor, only recently has empirical evidence been found to support this hypothesis. The first evidence, surprisingly, came from species that withhold genetic relatives (kin). In some rodents, for example, pups settle in their

> **Box 10.7 Kin groups of tadpoles**
>
> The tadpoles of many anuran species form compact groups in the presence of predators. These groups reduce by **dilution** each individual's vulnerability to predation. The individuals forming such groups are often close kin. Because individuals coming from different clutches (thus not genetically close) share the same pond, aggregates of individuals from the same clutch are open to exploitation by cheaters from other clutches. These cheaters would benefit from group protection without having to pay the costs of grouping, such as kin competition. Kin recognition is one of the mechanisms that allows clutches to avoid being 'parasitized' and thus invaded by selfish genotypes. Examples of this type have been reported in several species.

parents' territory where a tolerance linked to familiarity or the active recognition of kinship allows individuals to waste less time in agonistic behaviour, and/or to develop a certain level of mutual aid (Lambin and Yoccoz 1998). In some bird species, young stay in the parental territory and help rear their parents' new offspring (scrub jay, *Aphelocoma coerulescens* (Woolfenden and Fitzpatrick 1990); Seychelles brush warbler, *Acrocephalus sechellensis* (Komdeur 1996)). In the latter case, the lack of available territory constrains a certain number of the young to stay on their parent's territory, and helping allows individuals to acquire certain skills indirectly (in helping to rear its half-sibs), to gain some experience for future reproduction, and/or inherit the parental territory at the death of the parent of the same sex (see Section 13.5.4 for additional information). These last examples thus lead to the hypothesis of an antagonism between the evolution of dispersal and that of cooperation. As we will see in Chapter 15, the cooperation/dispersal relationship is more complex than previously thought.

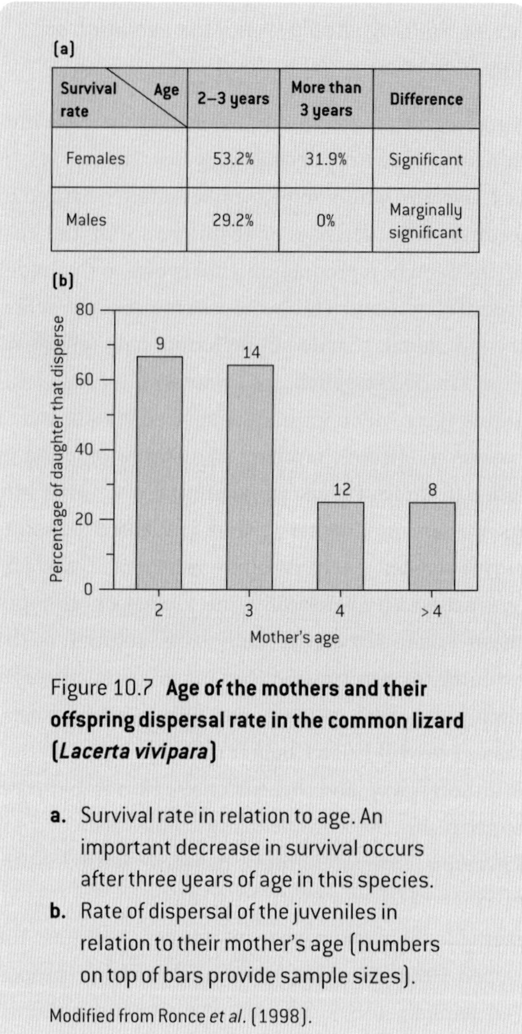

(a)

| Survival rate \ Age | 2–3 years | More than 3 years | Difference |
|---|---|---|---|
| Females | 53.2% | 31.9% | Significant |
| Males | 29.2% | 0% | Marginally significant |



Figure 10.7 **Age of the mothers and their offspring dispersal rate in the common lizard (*Lacerta vivipara*)**

**a.** Survival rate in relation to age. An important decrease in survival occurs after three years of age in this species.
**b.** Rate of dispersal of the juveniles in relation to their mother's age (numbers on top of bars provide sample sizes).

Modified from Ronce *et al.* (1998).

### 10.2.3.3 *Natal dispersal, age of the parents, and kin influence*

The probability of inheriting the parental territory depends of course, on the life expectancy of the parents. If they are young, there is little chance the young will inherit the territory in the short-term. If the parents are old and senescent, however, its odds are better. Because of this age effect, older parents should produce more philopatric young than younger parents. Using this prediction, Ophélie Ronce and her collaborators (Ronce *et al.* 2001) have examined the relationship between the onset of senescence (decrease in parental survival after a given age; Figure 10.7a) and pup dispersal rate (Figure 10.7b). Even though this result was not found in males, complementary experiments later showed the causality of this relationship and the role of kin recognition in dispersal (de Fraipont *et al.* 2000; Le Galliard *et al.* 2003).

The forces promoting helping between kin are potentially antagonistic to those promoting kin competition. Hence, we should expect a different response of dispersal rate, leading to an opposite spatial distribution of kinship, when one of these forces dominates. We can thus hypothesize that evolution would only lead to one form or another of response in different species acccording to the equilibrium between these antagonistic forces. A recent example shows, however, that both forms of responses can coexist in the same species and in the same population (Box 10.8). These recent discoveries have since led to the hypothesis that, like any other **life history trait** (i.e. size, sex, age at sexual maturity, etc.), altruism, kin recognition, and dispersal are three traits that can be the object of parental investment. Furthermore, this investment can result in different **trade-offs** in relation to the internal and external conditions of the parents.

**Box 10.8 Helping and competition among kin: both mechanisms can exist side by side in the same species when related to dispersal**

*Uta stansburiana* is a little North American iguanid lizard that lives in rock clusters dispersed in open spaces. The colour of the males' throat can be blue, orange, or yellow (see Chapters 5 and 6). Each colour is associated with a given reproductive strategy (see Chapter 5 for more details): blue males are monogamous, tolerant, and defend a small territory; orange males are polygamous, aggressive, and defend a large territory; yellow males are 'sneakers' and do not defend a territory. Generally, orange aggressive males dominate blue monogamous males and mate with their females, resulting in blue males having multi-fathered broods. Blue monogamous males dominate yellow sneaker males and are able to keep their females away from them; however, yellow males also behave and look like females and because of that have some success in circumventing the vigilance of aggressive orange males and copulate with some of their females. This situation, where each strategy both dominates and is dominated by another, and where the success of one strategy depends on the frequency of the other strategies in the population, is analogous to the famous rock–paper–scissors game. Recall that in this childhood game paper dominates rock but is dominated by scissors, whereas rock dominates scissors. It is easy to show that the optimal tactic is to play equiprobably and unpredictably rock, scissors, or paper strategies.

In this iguanid lizard, dispersal of the young is influenced by paternal genotype. Kin competition should be more intense among orange than among blue males because of the orange males' greater aggressiveness and polygynous tendencies. Not surprisingly, therefore, the reproductive success of orange males decreases when the number of kin in the neighbourhood increases. Blue males' reproductive success, on the other hand, increases as a function of the number of genetic kin in the neighbourhood (Figure 10.8). Thus, helping between blue throat kin seems to exist. It not too surprising, therefore, that the young of orange males tend to disperse further from each other than expected by chance whereas the young of blue males attract one another more than expected by chance.

In this lizard, throat colour is used as a recognition cue (analogous to a green beard effect, see Chapter 2) and the rhythm of the head nodding allows for genotypic recognition. We thus see that in this species, habitat selection is based in part on the presence of kin, and, depending on strategies, dispersal leads individuals to choose or to avoid a habitat where the presence of kin is detected. We can see that the fitness consequences of making mistakes when choosing a habitat can be quite important. The presence of complex mechanisms and of polymorphism in the response to kinship in what is otherwise a relatively non-social species suggests that social behaviour may evolve more easily than was once thought. For more information on this system see Sinervo and Lively (1996); Sinervo *et al.* (2001, 2006); Zamudio and Sinervo (2000); Sinervo and Clobert (2003).

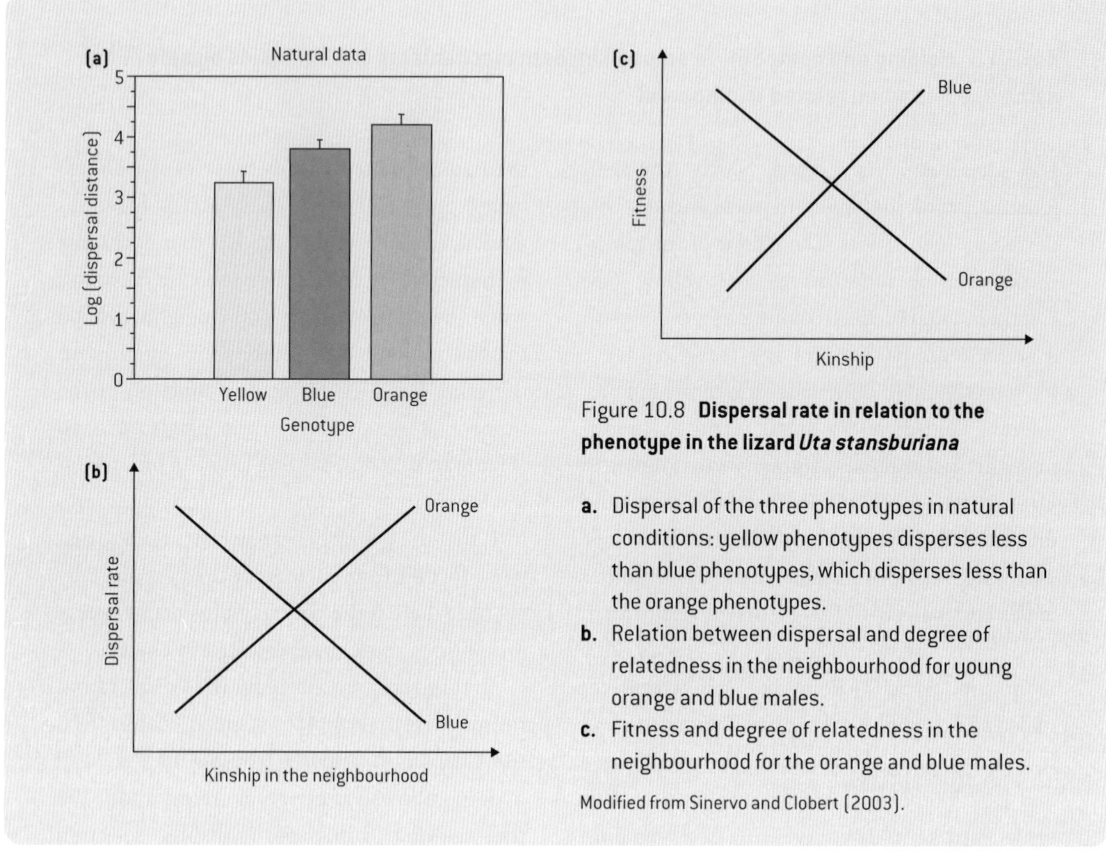


Figure 10.8 **Dispersal rate in relation to the phenotype in the lizard *Uta stansburiana***

**a.** Dispersal of the three phenotypes in natural conditions: yellow phenotypes disperses less than blue phenotypes, which disperses less than the orange phenotypes.
**b.** Relation between dispersal and degree of relatedness in the neighbourhood for young orange and blue males.
**c.** Fitness and degree of relatedness in the neighbourhood for the orange and blue males.

Modified from Sinervo and Clobert (2003).

### 10.2.4 Dispersal: a multi-purpose behaviour under multiple selective pressures

As we have seen in the preceding sections, there are numerous theoretical reasons for the evolution of dispersal, each supported by many empirical examples. Moreover, although they are still scarce, the examples of co-occurrence of different proximate causes of dispersal in the same species are now starting to accumulate. For example, one species of aphid increases the production of winged morphs in its lineage when the rate of predation or intra-specific competition increases or when the quantity of food available decreases. Similarly, in common lizards, dispersal increases with both the number of conspecifics and the number of kin.

Is this to say that a dispersal pattern in a given species had multiple evolutionary causes from its origin or instead that it first was elicited by a single selective pressure and over time was pre-empted to respond to additional selective pressures. This question might be premature, however, but the study of the evolution of dispersal under the influence of multiple selective pressures can help us progress. For instance, the Swiss scientists Nicolas Perrin and Jérôme Goudet (Perrin and Goudet 2001) have built a model that included inbreeding avoidance, kin competition, as well as local competition for resources or sexual mates. They concluded that inbreeding depression could not be the only force behind the evolution of dispersal when both sexes disperse. In the same way, kin competition could not be responsible for a disequilibrium between male and female dispersal rate. Only asymmetries in the selective forces working on males and females (those that concern habitat, reproduction, partner quality, or competition) can induce a sex-biased dispersal, not inbreeding avoidance per se.

In the same way, Sylvain Gandon and Yoannis Michalakis (2001) studied the role of inbreeding avoidance, kin competition, and environmental stochasticity on the evolution of dispersal. They concluded that the selective pressure resulting from temporal variability is more influential than social forces in determining dispersal rates, and focused on the numerous interactions between these selective pressures that often lead to difficult and counterintuitive predictions. Indeed, the changes induced by a single dispersal event jointly affect the level of local competition among kin, among individuals of the same sex as well as the likelihood of an inbred mating, making it difficult to test empirically their separate effects, their interaction and the extent of their respective influences.

Most of these theoretical results were obtained under several simplifying assumptions the list of which would be too long to discuss here. However, these assumptions make the quality of the predictions arguable. Notwithstanding, the predictions have the merit of attracting attention to the different social causes for the evolution of dispersal, the role of resource temporal variability, as well as the important potential interactions that may exist among all of these causes. The few experimental studies on this subject published so far seem to be in accordance with these predictions. For instance, in the springtail (*Onychiurus armatus*) a small insect, the dispersal rate obtained for a given density largely depends on habitat quality (Bengtsson *et al.* 1994; synthesis in Ims and Hjermann 2001). In the common lizard (*Lacerta vivipara*) temperature and humidity in the environment of pregnant females have interactive effects on dispersal of new-borns (Massot *et al.* 2002).

These first research efforts thus seem to orient us towards an integrated multi-dimensional vision of the evolution of dispersal rather than towards an accumulation of independent mechanisms. All these results suggest that it would be inadequate to imagine dispersal reflecting mechanisms that have evolved many times independently for each selective pressure promoting dispersal. However, because of the important diversity of factors that cause dispersal as well as the large number of the habitat's biotic and abiotic parameters that must be assessed to make a unique dispersal behavioural decision, it is normal to expect the existence of a complex integrative process that likely involves most of the organizational systems of the individual in order to allow the multiple forms of information to be acquired and processed.

Just like the integration of available habitat information is desirable for successful dispersal decision, we expect that dispersing individuals will bare morphological, physiological, or behavioural competencies that will help them face the dangers inherent to crossing hostile habitats and adapting to a new one, and/or to facilitate their integration into a new group of individuals.

### 10.2.5 Life history traits and dispersal behaviour

Although they are not often measured, the risks encountered during dispersal and those linked to the establishment in a new habitat, constitute costs of dispersal. The encounters with predators, a lack of familiarity with new habitats or conspecifics, the unsuitable habitats crossed, etc. are all problems that dispersers must face to have a non-zero fitness. For example, in the spider *Latrodectus revivensis* living in the Negev desert, only 40% of individuals survive the crossing between two bushes, the only places where their webs can be cast. The few theoretical studies that have compared the fitness of philopatric and dispersing individuals (see Murren *et al.* 2001 and Box 10.9) concluded that fitness should be equal for both types of behaviour. In other words, if there is a dispersal cost, it should be compensated by the existence of adequate strategies or by initial advantages that allow them to compensate for dispersal costs. The only case when the equality of fitness is not expected is when dispersal results from kin competition.

The existence of individuals, in a family or in a population, that have specialized morphological structures for dispersal strongly supports the existence of adapted dispersal strategies. In the thistle *Crepis sancta* (Imbert 1999), seeds in the distal part of the capitulum are equipped with winglets that insure long-range wind dispersal whereas central seeds of the same capitulum have no winglets. In the aphid *Acyrthosiphon pisum* (Weisser 2001), the proportion of winged imagos in a lineage is a function of several environmental factors. Although it is not clearly demonstrated that such morphological specialization effectively leads individuals who possess them to disperse more (a difference between having the

## Box 10.9 **Phenotypic difference between dispersing and resident individuals**

In many species, individuals that disperse have morphological, physiological, or behavioural traits that differ from those of philopatric individuals (see Murren *et al.* 2001). The classical examples are those of winged individuals in insects and winged seeds in plants. However, the phenotypic differences between dispersing and philopatric individuals are often more quantitatively subtle than that. In the lizard *Uta stansburiana*, juvenile orange males that disperse are, on average, bigger than philopatric juveniles, but the relation is reversed in juveniles produced by yellow males (Figure 10.9a). This is related to the fact that, on average, orange individuals disperse, a lot more than yellow ones (Figure 10.9a). To demonstrate the causal character of this relationship, an experiment was undertaken to modify the size of juveniles at birth. Follicular ablation reduced the size of the clutch thereby increasing the investment put by manipulated females into each remaining egg leading to gigantic juveniles. If the link between juvenile size and dispersal is causal, we should find that the giant juveniles of orange males disperse more, whereas the giant juveniles of yellow males disperse less. This is exactly what was found (Figure 10.9b). For more details, see Sinervo *et al.* (2006).

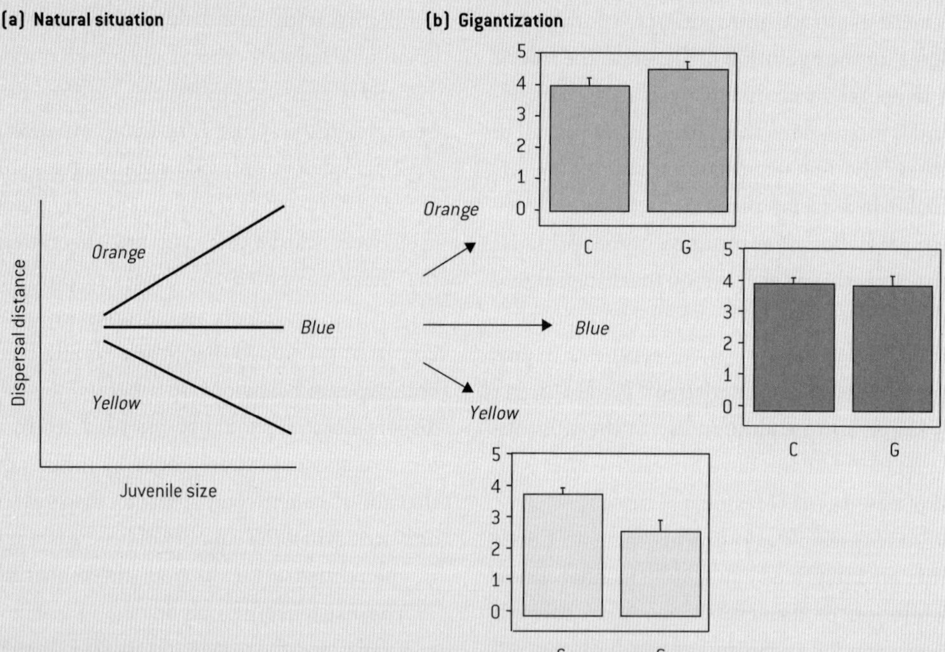


Figure 10.9 **Dispersal in relation to the body size in the lizard *Uta stansburiana***

C: Control; G: Gigantization by follicular ablation.

**a.** Relation between the phenotype of the dispersing juvenile (body size) and dispersing distance.

**b.** Results of the experiment of gigantization. The *y*-axis represents dispersal distance, the bars the mean (+ standard error) of treatment groups. Female follicular ablations decreased clutch size which in turn produced gigantic juveniles, because the female invested more in the remaining eggs. This confirmed the causal relationship between juvenile size and dispersal in young of orange and yellow males. This also shows that the phenotype of dispersing and philopatric individuals can differ significantly.

Modified from Sinervo *et al.* (2006).

abilities and using them), it is most probable that it significantly influences dispersal. Indeed, in the mole rat, it was demonstrated that dispersing individuals are bigger than philopatric individuals (see O'Riain and Braude 2001).

Morphological differentiations can be thought of as an extreme form of specialization for dispersal, which suggest that more subtle types of differentiation can also develop. Indeed, in the domestic cat, males with orange fur (a genetic trait) are more aggressive and disperse in greater proportions than cats of other colour types. In the great tit (*Parus major*), individuals that leave their family relatively early show higher exploratory capacities and are dominant in two-by-two interactions (but not in larger groups). In the common lizard, philopatric individuals are attracted to their mother's odour and, in stressful situations, this familiar odour reduces their level of reaction to stress. However, in that species, this behavioural difference is linked to a morphological difference only in the context of kin competition. A similar situation can be found in *Uta stansburiana* (Box 10.9). Orange-throated males that disperse are bigger than philopatric ones. Furthermore, orange-throated males are the most intolerant to one another. On the contrary, in yellow-throated males, which are the most tolerant **phenotype** relative to others, the philopatric individuals are the biggest (Figure 10.9).

Such examples suggest the existence of a relationship between different morphological and behavioural traits that would minimize the cost of the chosen strategy (dispersal or philopatry). Thus, we could sketch the following scenario: dispersing individuals would have stronger reactivity and lower neophobia and so would be better able to take advantage of the physical rather than the social characteristics of the habitat. Philopatric individuals, on the other hand, would be shyer in unfamiliar environments but would be better able to take advantage of their social rather than their physical environment. The morphological differentiation between individuals following these two strategies would evolve in relation to the advantage conferred by body size in either situation. The existence of links between dispersal and other life history traits has been demonstrated by many theoretical and empirical studies (Ronce and Olivieri 2004). We can therefore expect to find similar associations at the level of morphological and/or behavioural traits (also called syndromes). Such associations have been found in birds (see Dingemanse *et al.* 2004a, b; Dingemanse and Réale 2005; Both *et al.* 2005) and reptiles (see de Fraipont *et al.* 2000; Cote and Clobert 2007), but their generality has still to be demonstrated more explicitly and their characteristics studied in more detail. It is expected, however, that their nature will depend on the costs and benefits associated with philopatry and dispersal and we can predict that such behavioural syndromes will likely differ depending on the causes of dispersal.

More generally, the direct consequence of the maintenance of such within-species variation of behavioural strategies may create the conditions for the recurrent evolution of altruism and sociality, or instead, a reinforcement of individualism and territoriality. This evolutionary potential will of course depend on the determinism (genetic or phenotypic plasticity) of this variation.

We have now seen that dispersal can have several **ultimate** causes. However, dispersal implies important behaviour and metabolic changes before, during and after the movement itself. Moving animals for instance need to store energy just before departing. Then they need to orientate, as well as to redirect their energy on movement rather than on any other activity. We have seen in Chapters 5 and 6 that **hormones** are likely to be involved in such changes in activities. Thus, having studied the multiple origins of dispersal, we will now develop some of the mechanisms that are involved in the behaviour of dispersal.

## 10.3 Mechanisms of dispersal

The study of the mechanisms of orientation during migration and dispersal has fascinated researchers for a long time. It was demonstrated that the initiators of movement (often linked to photoperiod), their amplitude, and their direction are largely genetically conditioned, as is the case in the blackcap (*Sylvia atricapilla*) or the monarch (*Danaus plexippus*) butterfly (Berthold and Pulido 1994). More rarely, as in some geese, these parameters are learned by imitation, often from the parents. Orientation can involve numerous mechanisms. For long displacements, the use of terrestrial magnetic fields, as well as stellar maps and the position of the sun have been demonstrated. For smaller-scale displacements, animals use geographic cues like conspicuous landmarks, the configuration of landscapes, and also olfactory cues, as in the case of newts that recognize their reproduction pond by its peculiar odours. In many situations, taxes play an important role by deterring individuals from entering hostile habitats (Box 10.3).

### 10.3.1 The importance of conditions

Most certainly, all of the mechanisms involved in migration also occur during the displacement between the natal and future reproductive sites or between two successive reproductive sites (i.e. during dispersal). However, as those mechanisms do not constitute the crux of dispersal we will not discuss them any further. In fact, during dispersal, and unlike the case of migration as defined above, there is high individual variability within populations or families in the conditions that influence departure, direction and duration and the criteria used to choose a new habitat. Consequently, the mechanisms by which these three groups of decisions are made are likely to be more complicated than those that prevail in migration.

#### 10.3.1.1 *Empirical argument*

If migration frequently appears to be under direct genetic control, it is much less often the case with dispersal. In fact, only a few species show a strict genetic determinism for dispersal capacity. One example is the cricket *Gryllus firmus* (Roff and Fairbairn 2001) that presents two morphotypes: one apterous, the other with wings, each of which is associated with a given genotype. Macaques (*Rhesus* sp.) provide another example in which some males disperse before their first sexual experience, whereas others disperse long after (Trefilov *et al.* 2000). This difference seems to be associated with the presence of an allele on the coding gene for serotonin, a hormone linked to the regulation of the internal clock.

However, even if there are examples of genetic influence on the morphological or physiological capacities that are required during dispersal (see Roff and Fairbairn (2001) for a synthesis), in most cases the response largely depends on the environment, even when the response is morphological or physiological. This is particularly the case with numerous insects. For example, in the coleopterous insect *Tetraopes tetraophtalamus*, dispersal of males increases with their proportion in the population. The most convincing example comes from drosophila. In that species, numerous experiments have shown that traits linked to dispersal, such as flight duration, power of flight, locomotor behaviour, and the activity or type of habitat chosen are all highly dependent on genes (synthesis by Roff and Fairbairn (2001)). However, when one experimentally partitions the dispersal rate's different sources of variation (genetic and environmental), the results most frequently show a high contribution of the environment encountered by individuals whether larvae or adults, and an apparently weak genetic contribution (Box 10.10). A series of such experiments provide inconsistent results concerning the role of genes, ranging from a significant effect on walking capacities to a non-significant effect on flight capacities. Two similar factorial experiments, one in rodents (see Ims 1990), the other in reptiles (see Massot *et al.* 2002), also found a dominant role for the environment, and a potentially weak role for genetics.

Is this to say that the genetic determinism of dispersal is weak? No, it is more likely that genes are involved in the determination of response thresholds to environmental variations from which individuals decide to disperse. Furthermore, the genetic determinism may be hidden by an unrecognized strategy polymorphism. In the lizard *U. stansburiana* (Box 10.9) for instance, the colour of the males' throat is

**Box 10.10 Genetic or environmental determinism of dispersal capacity in *Drosophila***

The question of the genetic determinism of locomotor activity in *Drosophila* is controversial (van Dijken and Scharloo 1980), in part because some pre-imaginal effects can influence the measure of **heritability** of these traits. In a multifactorial experiment, Lefranc (2001) selected 30 iso-female lines (genetically different) and had them lay in two types of environment: one axenic, the other containing a blend of ethanol and acetic acid in a proportion similar to that found in decaying fruit (the favourite food of *Drosophila melanogaster*). Adults that emerged from those larvae were given the choice of laying their eggs on one of the types of environment used to raise the larvae. One of the environments was set near the point of release of the adults, the other at a certain distance. All of the possible combinations of the three factors (rearing and laying environments, close versus far from the point of release) were tested for 30 iso-female lines. The genetic contribution was then measured statistically as the 'iso-female effect' (i.e. the fact that all the offspring from the same iso-female were more similar than offspring from different iso-females) on the choice of laying-site by the offspring once adults. The pre-natal effect was estimated statistically by the 'rearing environment effect' (i.e. the effect of the environment from which adults emerged). The choice of laying site was estimated statistically by the effect of the position in relation to the point of release. Results show that the habitat choice is mainly (70%) determined by the position relative to the point of release, and to a lesser degree by interactions between the larvae's rearing environment and the type of environment, as well as by the distance of this environment relative to the point of release. As no significant difference among iso-female lines was found, these results do not support the existence of any direct genetic effects.

genetically determined, as are their reproductive strategies. These different reproductive strategies impose different patterns of dispersal. Sinervo *et al.* (2006) showed that whatever the maternal environment, the genotype of the male significantly influences offspring dispersal. In that species, it is the existence of a difference in male colour that allowed the measurement of the genetic determinism of dispersal: this could not have been determined if males of different genetically based strategies could not be distinguished.

#### 10.3.1.2 *Theoretical reasons: the importance of environmental predictability*

There are theoretical reasons to expect dispersal to be influenced more by environmental conditions (i.e. plastic) than by migration. Migration, more than dispersal, expresses itself in highly variable environments but at a temporal scale that makes such a variability highly predictable (e.g. the succession of seasons). In such a case, theory predicts that some fixed, unconditional strategies are likely to be selected for (see Figure 4.1). Dispersal, even when it tends towards being nomadic, is also expressed in variable environments but at temporal and spatial scales that lead to much lower environmental predictability (from one year to the next), because of a lower autocorrelation of the variations of the many environmental factors that govern it. If predictability at the concerned scales still remains significant, mixed strategies (coexistence of many strategies or genetic polymorphism) or conditional strategies (phenotypic plasticity) are thus preferable (see Figure 4.1). Evidence suggests that conditional strategies of dispersal prevail.

#### 10.3.1.3 *The role of hormones*

How are environmental variations perceived and how do they alter dispersal rates? Though the answers to this question were already discussed in Chapters 6 and 9, we can provide some relevant components of an answer here. In the aphid, the production of wings in the lineage depends on both predation rate and conspecific density. Predation risk is perceived through the number or odour of carcasses encountered that allow individuals to adjust their offspring dispersal rate. The density of conspecifics is assessed through the number of antennal contacts. The female spruce budworm (*Choristoneura fumiferana*: Lepidoptera, Tortricidae) produces more dispersers when she mates with a low-quality male (McNeil *et al.* 1995).

In most cases, though the proximate factors are identified, the proximate mechanisms (i.e. the physiological or behavioural cascades) remain largely unknown. In vertebrates, however, natal dispersal often occurs in combination with hormonal modifications, essentially steroids hormones like testosterone and corticosterone. We have seen numerous examples in Chapter 6 (Section 6.4.3) so we now only provide a few complementary examples. For instance, grey-sided vole (*Clethrionomys rufocanus*) females born in male-biased litters (i.e. litters with more males than females) have higher testosterone levels and disperse more than females raised in female biased litters (Ims 1989). In the western screech owl (*Otus kennicottii*), corticosterone rates increase just before natal dispersal, and in the willow tit (*Parus montanus*), young individuals with experimentally elevated corticosterone levels leave their winter group more often than individuals that received a placebo (Belthoff and Dufty 1998; Silverin 1997). It is possible that corticosterone in birds and testosterone in mammals are directly involved in the initiation of dispersal movement through their determining role in sexual differentiation (see Chapter 6). Indeed, the absence of territorial defence, the increase in activity and reserve mobilization are all functions that are affected by variations in steroid hormones levels in the blood, which in turn may also be implicated in dispersal (Dufty *et al.* 2002).

Although corticosterone is better known for its involvement in situations of stress (activational role), it also plays an important role, in combination with testosterone, in the construction of the phenotype (organizational role, see Chapter 6). In the rat, subjecting embryos to an increased maternal plasmatic level of corticosterone results in offspring that show a greater reaction to stress, that tend to flee unfamiliar situations and that have a reduced capacity for exploration. In mice, this same experiment affects brain development during embryogenesis, in particular the hypothalamic region, an area of the brain linked to emotions.

We saw earlier that natal dispersal could be affected by parental experience. In the aphid, predation risk, conspecific density, and food availability experienced by the parents all determine the number of winged descendants that are produced. In the common lizard, the place of birth of the mother, the food received, temperature, and humidity conditions during pregnancy all affect offspring natal dispersal (Figure 10.10). Natal dispersal is thus largely influenced by events that occur both before and during development (in aphids, even grandmother effects have been demonstrated). From a theoretical point of view this is not surprising considering that dispersal involves great risks and requires the use of as much of an animal's capacities of integrating environmental information in order to optimize dispersal decisions. Because steroid hormones are known to affect development of phenotypes (see Chapters 5 and 6, and Figure 10.11), especially those linked to exploration and resistance to stress, it is tempting to suppose that they are also implicated in the pre-natal determination of natal dispersal. To test this hypothesis, viviparous female lizards were given a daily dose of corticosterone during the second half of their pregnancy. The dose was calculated to increase their levels of circulating hormones to levels classically observed in stressful situations. Juveniles born from corticosterone-treated mothers showed an increased attraction towards the odour of their mother, an increased tendency to flee unknown situations and an increased degree of philopatry, compared with juveniles from placebo females (Figure 10.11). However, the effect of corticosterone treatment also

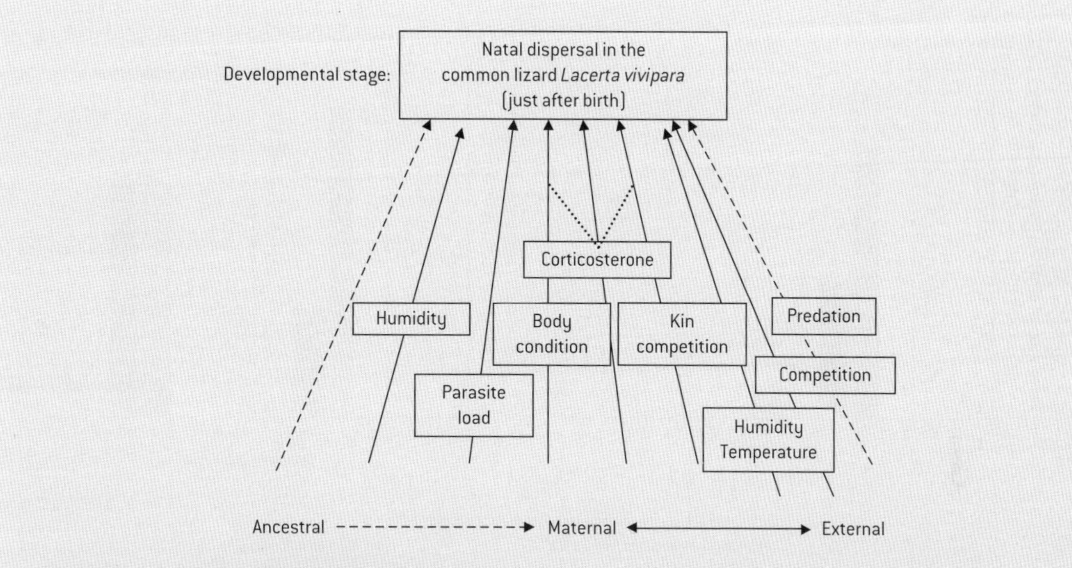


Figure 10.10 **Pre-natal (maternal) and post-natal factors that influence the dispersal phenotype of the young common lizard (*Lacerta vivipara*)**

This example illustrates the complexity of the agents acting on dispersal. It was shown in this species that several parameters, including abiotic factors such as temperature, biotic factors such as parasitism and social factors such as density or kinship, modulate natal dispersal. These factors can act during the pre-natal (humidity, parasitism, kinship) or post-natal (density, temperature) stages. Some like humidity act on dispersal both during the pre- and post-natal periods but do so in different directions. As we can see here, the dispersal rate is thus the result of the complex interactions of multiple factors at different developmental stages of the behavioural phenotype.

Modified from Dufty *et al.* (2002).

depended largely on the mother's characteristic size and weight, which suggests that other factors, likely other hormones, are also important. The evolutionary interpretation of these results is as follows: a chronic elevation of corticosterone is associated with a pathological state similar to a parasite infection, lack of food, etc., and would thus affect the health status of the mother; a mother in poor condition has a reduced chance of surviving until the next reproductive episode, which reduces the probability of competition between mother and juveniles in the future. Thus, a chronic elevation of corticosterone would inform juveniles of reduced kin competition and would thus reduce their tendency to disperse. Later experiments supported that interpretation.

The study of the role of hormones in the organization of a phenotype (Figure 10.12, and see Chapters 5 and 6) is still developing, but dispersal, like all other behaviours, will not escape a detailed study of its onset during ontogeny, and more generally of the influence of the genotype–environment interactions in its determinism. In particular, this will involve a thorough study of the influence of temporal and spatial autocorrelation patterns in abiotic and social environments. In fact, dispersal being subject to numerous influences and entailing significant risks, one expects natural selection to bring species to develop a maximum number of information acquisition mechanisms, either maternally or directly. However, to demonstrate the adaptive character of those mechanisms and the responses they bring, it is necessary to know the fitness consequences, and in particular, the feedback induced by dispersal at the level of the population dynamics.

Beyond the multiple causes of dispersal and beyond the many mechanisms involved in the movement of

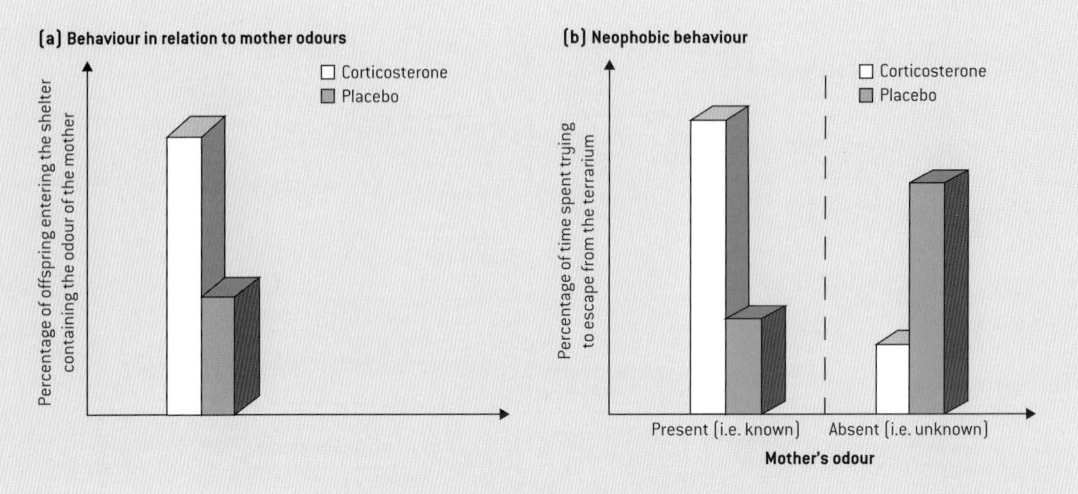


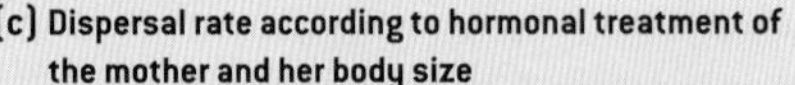


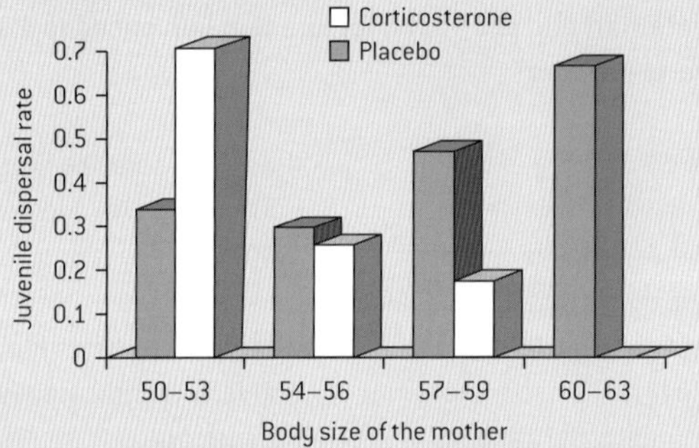


Figure 10.11 **Corticosterone and natal dispersal in the common lizard (*Lacerta vivipara*)**

An experiment in the common lizard evidenced the role of corticosterone in the mother on the dispersal phenotype of its offspring.

**a.** Pregnant females treated with corticosterone have juveniles that are more attracted to their mothers' odour after birth, and

**b.** that are more neophobic (i.e. that avoid unknown situations), than juveniles coming from non-corticosterone-treated females.

**c.** Juveniles coming from corticosterone treated females are also more philopatric than those coming from placebo females. This is especially true for large (old) females, whereas the opposite is seen in small (young) females. The interaction between the effect of mother's size and the treatment is significant ($P < 0.0001$), indicating that the effect of the treatment varies significantly depending on mother size.

The interpretation of this result is that corticosterone is associated with the health state of mothers only when they are large – internal environment – whereas with young mothers, corticosterone is associated with the degree of stress produced by the environment in which she lives – external environment of the mother. In the first case, young stay because the mother has little chances of surviving until the next year, thus decreasing the probability of kin competition. In the second case, young leave because the environment in which the mother is living is of a bad quality, without it really affecting the health status of the mother.

Extracted from Fraipont *et al.* (2000), Meylan *et al.* (2002) and Meylan *et al.* (2004).

The reader will see the similarity between the interpretations used to explain this result and the one in Figure 10.7. Compare also these results to the low and fast explorers described in the great tit by Dingemanse *et al.* (2004a, b).

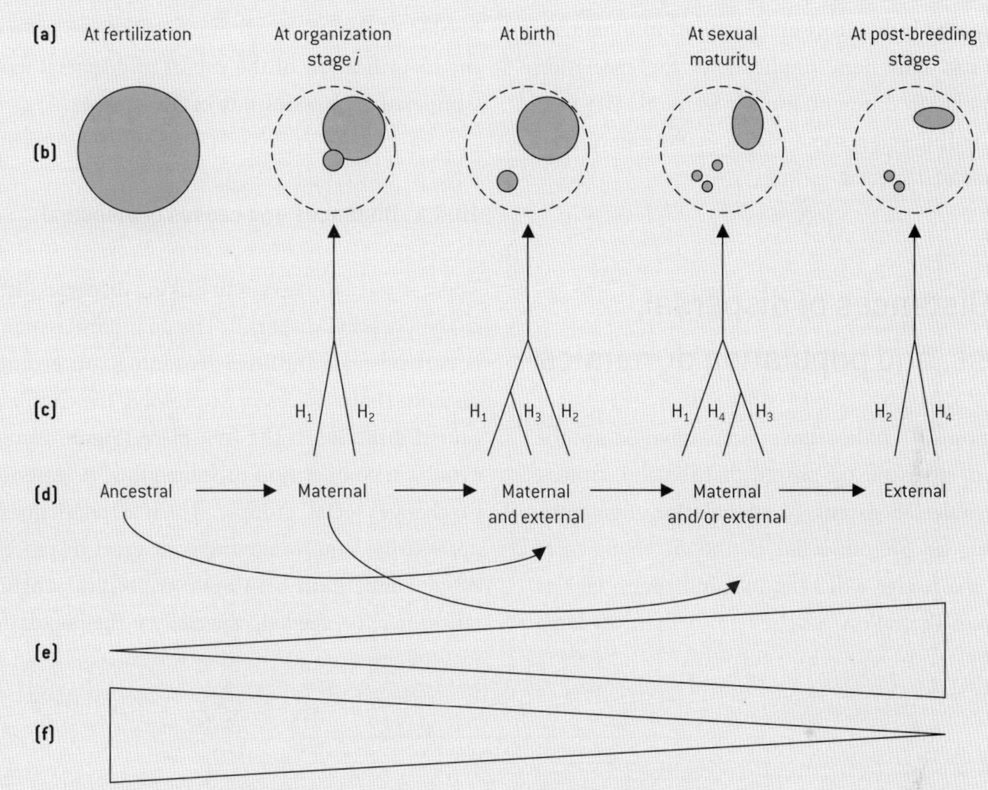


Figure 10.12 **A general mechanism of development of the behavioural phenotype**

During the whole duration of development both past and current information can influence the choice of the phenotype's developmental pathway.

**a.** The important steps of development.
**b.** Phenotypes that are still possible (in dark) for the genotype *j* at each developmental stage. We see that the range of possibilities narrows during development and can become discrete (i.e. separated in the space of initially possible phenotypes).
**c.** Integration of environmental information by hormones. The internal and external environment of the mother or of the egg is expressed by hormonal variations (Hi) that directly affect the development of the embryo. These hormonal responses and their effects on the development of the phenotype can be of adaptive origin.
**d.** Origin of the information taken into account at every step of development: Only the ancestral information transmitted to the juvenile at fertilization is of a genetic nature. All the other information is of an environmental nature.
**e.** Informative content of the information source. This will increase during development as recent information updates the information on the current state of the environment. The integration in time of all the information acquired during development also allows for more precise information.
**f.** Usefulness of information. This decreases in time because, as the end of life approaches, the information acquired will be less and less useful, or more exactly, maintaining information acquisition mechanisms becomes more costly in relation to their potential future benefits.

For more details, see Dufty *et al.* (2002).

dispersers, dispersal has prominent implications for the fitness of individuals and population dynamics. It also has important implications for speciation and population differentiation in general. The next section will develop some of the evolutionary consequences of dispersal.

## 10.4 Distances of dispersal, fitness, and population dynamics

Few empirical studies have been devoted to the impact of dispersal on spatial distribution, fitness, and to a lesser extent on population dynamics. However, we present what is known about each, hoping the reader will keep in mind that a lack of techniques that allow researchers to follow a large number of individuals, the size of the study system needed to document all movement types, as well as the identification of the causes of dispersal have all hampered empirical study of these issues.

### 10.4.1 Distances and causes of dispersal

There is virtually no study linking dispersal distance to the causes of dispersal. The dominant assumption is derived from intuitive reasoning concerning the distance that is necessary to resolve the conflict that caused dispersal in the first place (Figure 10.13). To avoid kin competition or inbreeding, it is necessary to change social groups, which *a priori* does not necessarily require moving over great distances. To avoid competition between conspecifics, longer distances may be needed, especially if the population is

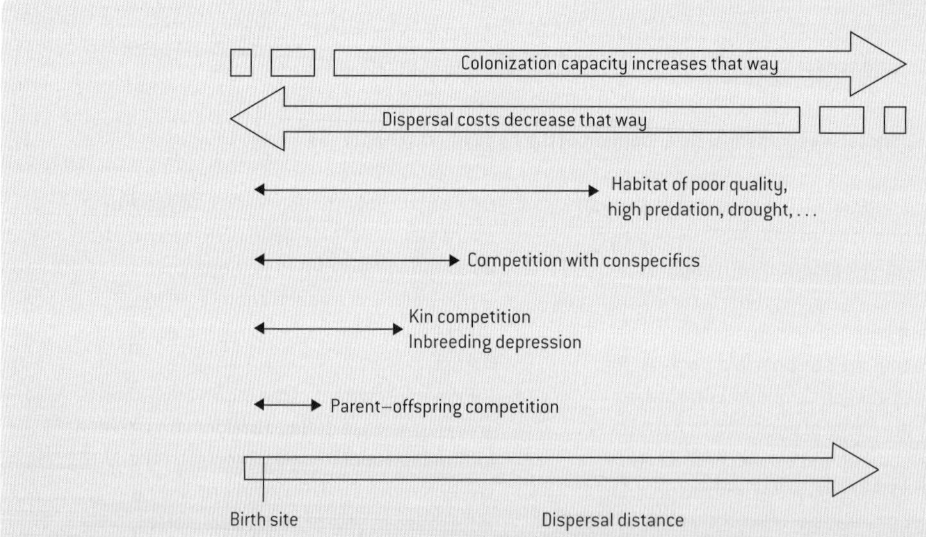


Figure 10.13 **Current assumptions on the relationship between dispersal cause and distance**

Parent–offspring competition should generate smaller distances of dispersal because it would suffice to leave the territory of the parents to resolve this problem. Kin competition and inbreeding depression should generate slightly greater distances of dispersal, because it is necessary to leave the neighbourhood to avoid such risks. Competition between conspecifics should generate even larger distances of dispersal because, in this case, it is necessary to leave the original population in order to escape this type of competition. Finally, to find a better habitat, it seems necessary in most cases to cover even larger distances to leave the zone of unfavourable environmental conditions; when such conditions are of a climatic nature, the distances involved can then become very large.

Modified from Ronce *et al.* (2001).

made up of many groups, or if the unit of kinship is of a smaller size than the social unit. As we can see, this will probably depend highly on the species and its social organization. Finally, the distances required to change habitat types is, *a priori*, much larger than the distances required to change units of kinship or social units, because the heterogeneity of the habitat is generally expressed on a larger scale than the other two.

The simple logic of this view makes it attractive. However, two facts may challenge it: the phenotype of the dispersers can change in relation to the cause of dispersal, and habitat selection can differ in relation to the cause of dispersal. For example, we know that the phenotype of dispersers may differ from the phenotypes of philopatric individuals. Differences may be in terms of morphology and physiology (size, condition, energetic reserves, resistance to stress), or behaviour (exploration ability, competitive capacities). In at least one study, it was shown that this was only true for some causes of dispersal, in particular those that bring about smaller distances of dispersal. Although no measure of distances of dispersal was made, it is likely that the phenotypes more adapted to dispersive movements are in fact adapted only to accomplish relatively short distances. In the same way, although empirical studies are lacking, if different causes of dispersal produce different behavioural phenotypes in terms of competitive or exploration capacities, one can imagine that this may influence arrival habitat selection (**habitat matching** and **silver spoon effect**; Stamps 2006). In other words, if there is a correlation between the cause of departure and arrival habitat choice (see Chapter 9), the patterns of dispersal produced by different causes of dispersal could very well stray radically from the patterns predicted on the sole basis of the scale of spatial heterogeneity in the concerned cause, an approach we used to build Figure 10.13.

### 10.4.2 Fitness of dispersers and philopatrics

With the exception of kin competition, most models comparing the fitness of philopatric and dispersing individuals according to different causes of dispersal suggest equality of fitness even in the presence of a temporary cost to dispersal (for example, only those induced by the movement). Dispersal cost can be permanent (i.e. have a lifetime effect on the individual) only if there is (at least partial) compensation of this cost (Lemmel *et al.* 1997; Murren *et al.* 2001). For example, a decreased reproduction induced by an unfamiliar habitat must be compensated for one way or another by increased survival. A certain number of empirical studies have thus compared the fecundity and survival of philopatric and dispersing individuals once they are established in their new population. Generally speaking, such comparisons do not take into account the cost incurred during dispersal itself. Most of these studies report fitness differences between philopatric and dispersing individuals, but most often these results are based on only one life history trait, fecundity, or survival rather than complete measures of fitness. Either the philopatric or dispersing individuals showed the highest fitness, the direction of the difference varying according to the categories of individuals considered within the population. Some of these studies have shown compensation between traits, the survival of dispersers once established in their new population being usually higher, but their reproduction worse than that of philopatric individuals.

However, these comparisons between philopatric and dispersing individuals suffer from numerous methodological problems. One of the most important problems is that the estimation of survival rates is often taken to be the rate of observing individuals returning to a population, which assumes that individuals that leave it have died. If the rate of dispersal is not the same for both strategies, then the comparison of the return rates from local data results from both true survival and differences in emigration rates simultaneously. Interpreting it as survival only is invalid. A few rare experiments have attempted to use discrete patches of habitat that individuals could reach either by taking the corridors between patches or by crossing a matrix of hostile habitats (in which the individual cannot establish). These habitat patches could be either occupied or empty. Not enough **replicates** of these experiments have been conducted to infer general results (Gundersen *et al.* 2002; Lecomte *et al.* 2004). During colonization of empty sites, it seems that dispersers have better growth and better chances to reproduce (rodents and reptiles). When the habitats of origin and arrival are occupied, few or no differences between strategies are found (except in predation during the transition phase in rodents, but this result may have been caused by the experimental design). When dispersers are prevented from dispersing, results are ambiguous as two experiments, one with rodents and one with reptiles, provided opposing results (see references in Clobert *et al.* 2004).

This situation of opposing results may be not surprising because the designs of these experiments provide limited opportunities to assess the differences between the two strategies. To compare the fitness of dispersing and philopatric individuals experimentally, one must take into account both habitat differences (dispersing and philopatric individuals end up in different habitats). It is also important to take into account the consequences that dispersing individuals will have in the patch of origin. This means that the fitness of philopatric individual may depend strongly on their interactions with dispersers. To do all this, we need to equalize in real time the components of habitat (for instance density) that are affected by the movements of dispersers. One way to proceed would be to prevent dispersing individuals from dispersing by putting them back in their original population while forcing philopatric individuals to disperse by transplanting them in a new population. By constructing all the possible combinations between philopatric and dispersing individuals forced or prevented (Figure 10.14) for different causes of dispersal, we should be able to progress in the comparison of the fitness of the two strategies.

### 10.4.3 Dispersal, fitness, and population dynamics

As we have seen in this chapter, dispersal plays an important role in population dynamics. Dispersal plays a central role in the processes of population regulation, colonization, or isolation, as well as in the

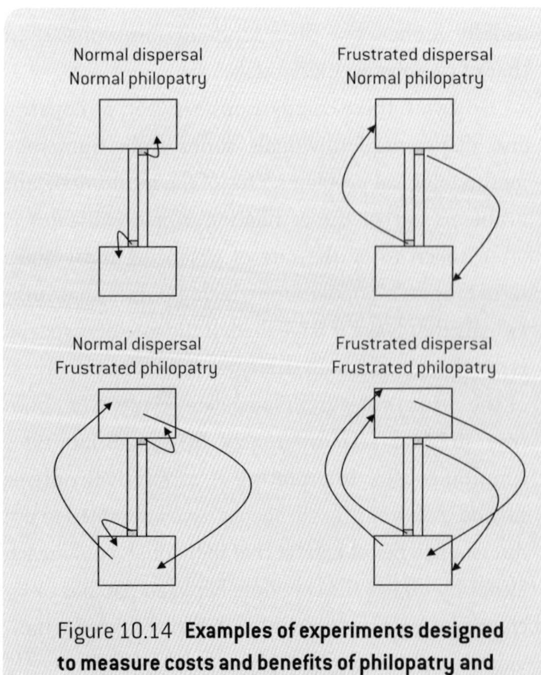


Figure 10.14 **Examples of experiments designed to measure costs and benefits of philopatry and dispersal at an individual level**

This can be achieved by using two population systems connected by uni-directionnal corridors of dispersal ending on a trap (small squares at the end of corridors). This allows for the identification of those individuals that travel the whole corridor (from population 1 towards population 2, or from population 2 towards population 1). Corridors should have lengths at least equal to the minimum distance travelled by dispersers. By capturing the individuals in the populations of origin and at the end of the corridors, we can identify the philopatric or dispersing status of the individuals. By preventing dispersers from reaching the population towards which they were going or by forcing philopatric individuals to leave their population (and all the possible combinations), we can measure fitness loss (or gain) of both strategies by measuring the survival and the reproduction of both categories of individuals. See Lecomte and Clobert (1996) for an example.

Modified from Clobert *et al.* (2004).

processes of exchange of genetic, social, or cultural information. Moreover, as we have seen earlier, dispersal can destabilize a population, as it can standardize connected populations, thus eroding biodiversity. Hence, its role in increasing or diminishing the entropy of a system will essentially depend on the cause or causes that promoted its evolution and thus on the direct or indirect selective pressures (i.e. on the related traits) that influenced it. An example concerning the joint evolution of cooperation and dispersal is developed in Chapter 15. In this context, it should be noted that the evolution of dispersal does not always lead to the maximization of population size. For instance, kin competition can sometimes put a population in danger of extinction because competition is a function of the quality of its members not their quantity. In the same way, colonization success can be superior to the success expected under pure demographic stochasticity if colonizers are endowed with attributes that reduce the risks incurred during colonization. More generally, to describe today a disperser as a particle taken randomly from a group of particles with the same characteristics, moving randomly in space is a misleading image, no-matter what scale is adopted. Several recent books have focused on the non-trivial role of behaviour in conservation biology (see Chapter 18). The multiplicity of forms of dispersal, as well as of the nature of dispersers and their fitness once set in a new population or habitat, should encourage us (1) to abandon this reductionist view of the effects of dispersal on population persistence and evolution and (2) to measure more clearly its importance by taking into account the real and well-documented characteristics of living creatures.

## 10.5 Conclusion

Dispersal is probably the most important but least understood behaviour of all the life history traits of a species. The links with behaviour such as mate or breeding habitat choices are fundamental. Dispersal behaviour is highly involved in determining the persistence of species. Although extensive theoretical literature is devoted to dispersal, it is not always immediately applicable because most of these theoretical studies are far too reductionist (which boils down to ignoring dispersal's behavioural nature), and because quality empirical data (in particular experimental data) are rare, which prevents any firm attempt at generalization. The reasons for this lack of data are purely practical: the study of dispersal involves significant spatial and temporal scales.

Today, dispersal is increasingly perceived as an ***omnibus* behaviour**, that is as a solution to multiple problems: kin competition, conspecific competition, mate choice, and breeding habitat selection. **We can thus consider dispersal to be a response to any factor that creates spatial or temporal heterogeneity whether of biotic or abiotic origin**.

*Two conceptual models for the evolution of dispersal*

Even though dispersal is multi-causal in many species, it is not always clear whether we should consider dispersal as a unique behaviour with multiple causes or as a family of behaviours that strongly resemble one another, with each having its own different evolutionary causality. Thus, we could consider that dispersal first evolved to avoid kin competition, because this problem is faced by every living creature; or as an answer to environmental variability because this is also intrinsic to the systems that living creatures exploit. To the contrary, we could conceive the origin of dispersal as polyphyletic, i.e. one (or several) type(s) of dispersal exists for each cause, combining different elements of the behavioural repertoire of the species.

*Dispersal, a family of behaviours?*

Dispersal behaviour is essentially composite: it associates different, more elementary behaviours, such as kin recognition, reaction to stress, exploration, mate, and habitat choice, etc., which may or may not constitute a behavioural syndrome. In this sense, dispersal could be defined as a '**super**' or **meta-behaviour**. The study of dispersal brings us back to the existence of trade-offs between different components of behavioural fitness (see also Chapter 5) at the phenotypic (set up during the individual's ontogeny) and/or genotypic levels (genetic covariance).

In most cases, dispersal appears to be **highly plastic**, in other words highly dependent on the environment. Even when morphological specialization confers a larger power of dispersal to a fraction of the population or of the offspring, the appearance of these specializations is also very frequently partly determined by the environment. This does not mean that there is no genetic control, but that this control is more likely to be found in the different morphological, physiological and/or behavioural elements that constitute the 'meta-behaviour' of dispersal and in the threshold values of responses to environmental factors. **Dispersal thus appears to be a behavioural pattern that prominently depends on the information that is available in the environment.**

For natal dispersal at least, the information to be acquired is often transmitted by parents by way of learning, imitation, or prenatal effect (see Chapters 12 and 20). This leads us to assume that parents can easily manipulate their offspring and that parent-offspring conflicts are usually won by parents. The evolution of a plastic rate of dispersal is conditioned by the existence of autocorrelation in the environment (see Figure 4.1). Such patterns of autocorrelation are likely to guide the evolution of a given sensitivity to the environment at a given stage in the development of the dispersal phenotype.

*Breeding dispersal: a behavioural pattern that may be less complex*

Breeding dispersal seems to respond to fewer environmental factors: mate choice and habitat selection are probably the two most important. However, the individual's history of movement interacts at least partially with these decision-making processes. The literature on the evolution of life history strategies has been interested in the distribution of reproduction throughout an individual's lifespan (see Chapter 5). We could also generalize this question by using a broader definition of **reproductive investment** that would include the temporal and **spatial** distribution of reproductive effort all along the lifetime. This definition would provide the framework for the comparison of the different strategies of dispersal.

*Inter-specific competition, predation, parasitism, and dispersal*

We have spoken little about the influence of interactions with other species on the evolution of dispersal, in particular according to the type of interaction: competition, predation, or parasitism. However, dispersal can allow animals to escape competitors, predators, or parasites. The last two types of species are expected to react to the movements of their prey or host. Thus, the rates of dispersal of species sharing functional links may co-evolve. We can imagine that predators and parasites, depending on their strategy of finding a prey or a host (active searching versus sit and wait) will influence their prey's or host's dispersal, and vice versa. If dispersal in one species can be influenced by those of other species, we can also predict that inter-specific competition, anti-predator, and anti-parasite behaviour is also linked to dispersal behaviour thus leading to trade-offs among these different series of behaviours.

The reader will perhaps have understood that this chapter was conceived with the deliberate *a priori* assumption that kin competition probably constitutes the most structuring factor of dispersal. The reader is invited to challenge this assumption. By choosing this starting position, we hoped that a critical view of the evolution of dispersal behaviour, of the method of study, and of the conclusions based on still largely fragmented data and experiments, would generate the desire to question and go beyond the knowledge that we currently have in this domain.

The study of dispersal is thus a field of research in behaviour that is still **largely in progress**, but that is essential if we are to understand and predict the reaction of species to the major changes that affect and/or will affect our planet, such as the fragmentation of habitats and climate change.

## » Further reading

> *To learn more about the evolution of dispersal, it is best to read the available multi-author research books that provide a general review on the causes, mechanisms, and consequences of dispersal:*

**Stenseth, N.C. & Lidicker, W.Z. Jr** (eds) 1992. *Animal Dispersal: Small Mammals as a Model.* Chapman and Hall, London.

**Dingle, H.** 1996. *Migration: The Biology of Life on the Move.* Oxford University Press, Oxford.

**Clobert, J., Danchin, E., Dhondt, A.A. & Nichols, J.D.** 2001. *Dispersal.* Oxford University Press, New York.

**Bullock, J.M., Kenward, R.E. & Hails, R.S.** 2002. *Dispersal Ecology.* The British Ecological Society, Blackwell, Oxford.

> *As well as single reviews such as:*

**Clobert, J., Ims, R. & F. Rousset** 2004. Causes, mechanisms and consequences of dispersal. In *Metapopulation Biology* (Hanski and Gaggiotti eds), pp. 307–355. Academic Press.

**Bowler, D.E. & T.G. Benton** 2005. Causes and consequences of animal dispersal strategies: relating individual behaviour to spatial dynamics. *Biological Reviews* 80: 205–225.

> *Specific reference for insects:*

**Woiwood, I.P., Reynolds, D.R. & Thomas, C.D.** (eds) 2001. *Insect Movement: Mechanism and Consequences.* CAB Publication, Wallingford, UK.

**Zera, A.J. & Denno, R.F.** (1997). Physiology and ecology of dispersal polymorphism in insects. *Annual Review of Entomology* 42: 207–231.

## » Questions

1. In this chapter and the preceding one, we have considered that dispersal can be conceived of as the simple product of the process of choosing a habitat of reproduction. Is this true in every case and every organism?

2. We have concluded that dispersal behaviour can be regarded as a 'meta-behaviour' in the sense that this same behaviour (here, the movement) can have multiple causes and mechanisms in permanent interaction. Is this a situation unique to dispersal or are there other traits classically called behaviours that are also multi-faceted?

four

Part Four

# Sex and Behaviour

Individuals, once adult and having acquired the necessary foraging skills and settled for reproduction, must reproduce. In sexually reproducing species, this entails another important choice process, that of selecting a reproductive mate. Indeed, mates constitute one of the necessary reproductive commodities that may strongly vary in quality. It should be noticed that most often the choice of a breeding site and that of a reproductive mate are concomitant processes. However, despite the fact they interact, these choices are of very different nature, mainly because the resources selected in habitat selection are passive whereas mates can themselves make a choice. In mate choice, both partners are selected to optimize their own fitness. In many species, such choices become a major agent of selection on individuals of the other sex. The goal of this fourth part is to develop basic and current issues of sexual selection.

Historically, sexual selection emerged as part of the core of behavioural ecology in the 1980s. In the past decade, the seeking of a better understanding of mate quality has led the field of sexual selection to develop interdisciplinary interactions with several disciplines such as physiology, endocrinology, immunology, and medicine.

Chapter 11 develops the fundamental evolutionary consequences of sexual reproduction. It is the field of behavioural ecology that has developed the most in the past two decades.

Chapter 12 presents the major types and ecological determinants of mating systems, as well as the general principles that allow understanding the logic of the various mating systems observed in nature.

Chapter 13 deals with the questions of sexually dimorphic parental investment in offspring of either sex. It tackles questions of why parents may have fitness interests in investing more in the production of offspring of one sex or the other according to their own condition and environmental situations.

11

# 11

# Sexual Selection: Another Evolutionary Process

Étienne Danchin and Frank Cézilly

## 11.1 Introduction

Among the problematic species of the classification established by Linné in 1758 are two ducks. One, *Anas platyrhynchos*, has a mottled brown plumage and an iridescent-blue patch on the wings. The other is light grey with a chestnut breast, has an iridescent-dark-green neck and head as well as an iridescent-blue wing patch. Linné named it *Anas boschas*. It was only later that they were identified as females and males of the same species, the mallard duck. Many such cases exist where males and females differ so much in appearance that even eminent taxonomists once classified them as different species. There arises the question of the evolutionary origin of such morphological differences between males and females of the same species. In particular, the very existence of secondary sexual traits (i.e. traits not directly involved in the production and fertilization of gametes, such as gonads and copulatory organs) is an evolutionary wonder: if favourable to individuals of a given gender, then why are they absent in the other? *A priori*, it is reasonable to assume that both males and females of a given species share similar environmental constraints. For instance, if antlers allow male deer to fight predators, then why are females deprived of these appendages if they incur the same risk of predation?

Darwin addresses this problem explicitly in *The Origin of Species*, as early as 1859. He acknowledges that the existence, in many extant animals, of odd forms and of extravagant characters such as particularly bright colours, extremely long feathers, colourful cockscombs or other secondary sexual traits, is a challenge to his theory of evolution by **natural selection** (Pomiankowski 1988). Indeed these traits, generally present in males, do not seem to contribute to the survival of those who express them. On the contrary, many of these traits, like the peacock's tail, seem instead to hinder males, making them more conspicuous and more vulnerable to predators. Likewise, these traits do not seem to serve offspring survival as do some other gender-restricted traits such as human breasts, avian brood patch, or mammalian placenta. In his 1871 book *The Descent of Man and Selection in Relation To Sex* and in its second edition in 1874, Darwin explained the evolution of secondary sexual traits by the theory of **sexual selection**. Constituting, to some extent, the counterpart of **natural selection**, this original theory proposes that secondary sexual traits evolved through the advantage conferred to males when competing against other males to fertilize females. According to their nature, these traits are of central importance in determining the winner of contests against other males, or in attracting and encouraging females to mate.

Right from the start, mechanisms of natural and sexual selection have been closely tied. However, although most of the major principles of sexual selection had been stated as early as the end of the 1870s, little research was conducted on the subject until the 1960s. It is thus not surprising that, whereas the general public has, for the most part, already heard about natural selection, few, even among non-behavioural biologists, can say that sexual selection evokes a familiar concept. This chapter aims to explain the guiding principles of contemporary research in sexual selection, presenting the various theoretical alternatives and related empirical work, as well as illustrating its various implications for population biology and evolution. By way of introduction, a brief historical perspective will show how the current state of this area of behavioural ecology is rooted in the history of its development. We will see how sexual selection is a fundamental mechanism of evolution. This chapter is unusually long, but its length is due mostly to the diversity of issues that must be mastered to grasp the importance of sexual selection.

## 11.2 From Darwin to modern times: history of sexual selection studies

### 11.2.1 Disagreement between Darwin and Wallace

In the 19th century, the debate about the importance of sexual selection mainly took place between Charles Darwin and Alfred Russel Wallace. Although the two protagonists were initially in full agreement, their discussion then polarized, from the 1870s on, around two apparently irreconcilable alternative conceptions (Pomiankowski 1988). Indeed, Darwin thought that males' extravagant **ornaments** had evolved by sexual selection simply owing to the fact that females systematically preferred to mate with the most attractive and ornamented males. Today, this conception is considered unsatisfactory for two main reasons. First, the father of natural selection invoked the existence of a true aesthetic **choice** in females, without questioning the origin of this arbitrary preference. Darwin possibly ascribed, at least to higher vertebrates, aesthetic feelings similar to those of human beings. Secondly, in his enthusiasm, Darwin omitted to provide an explanation for the broad prevalence of bright colours and extravagant traits in males.

Wallace, for his part, granted a predominant role to the process of natural selection. For him, females choose some males over others only because the former present traits favoured by natural selection. For Wallace (1891), sexual selection was only a secondary process whose effect was, at the most, to reinforce that of natural selection. To explain how natural selection could have promoted female preferences, Wallace considered that they served mainly in species recognition. According to him these preferences evolved to enhance pairing efficiency through synchronizing the release of gametes, and to enable the choosing of males of highest quality and/or having the best **resources**. In fact, it is partly the prevalence of Wallace's view that caused sexual selection to be regarded only as one minor process for nearly a century.

### 11.2.2 The contribution of Fisher

The debate remained dormant until the publication of two influential articles by Ronald A. Fisher in 1915 and 1930. Fisher immediately identified the two problems left unsolved by Darwin. Preferences, like any other trait, are shaped by the selective advantages they provide. In particular, Fisher noticed that such preferences could evolve because they affect the reproductive success of male offspring. The advantage is thus expressed in the next generation. Fisher then proposed a process composed of two stages.

Assume that a genetic variation arises in one male trait, say tail length, and those males with a tail longer than average have a slight survival advantage. Assume, also, that females choose among males and that females' tendency to pair with males of various tail length has a genetic basis. Then females preferring males with a long tail tend to have male offspring with higher survival rates. Therefore the alleles coding for a long tail in males will invade the population. The same goes for the alleles that make females more sensitive to male tail length. At this stage, the male trait, which was initially only supported by natural selection, also becomes favoured by female choice. The weight of this second component increases with the frequency of female preference in the population. Moreover, the relative advantage of the female preference itself increases with the increased advantage of the male ornament in terms of pairing success, i.e. with the strength of the female preference. Thus both female preference and preferred male trait will increase in frequency because of these positive feedback effects. The process can also be formulated by saying that female preference results in non-random pairing, which leads to a positive genetic correlation between exaggerated female preferences and exaggerated male traits (see Section 11.5.2a for a more detailed and formalized explanation. See also further considerations of this issue in Chapters 4 and 20). At this point a second phase begins, where the male trait is further exaggerated without being necessarily supported by natural selection. This **runaway process** can continue even after the male trait itself becomes counter selected because of its negative consequences on survival (e.g. if moderately long tails were beneficial but very long tails were not). At this stage, female choice alone keeps the system going. This runaway process can operate until the disadvantages in terms of male viability exceed the benefits linked to female preference. It is considered today that the Fisherian runaway process can affect not only the evolution of sexual signals, but also that of mating behaviours in males, of genital organs, and of other contact organs used during copulation (Eberhard 1993).

Initially the Fisherian runaway process went mostly unnoticed (Pomiankowski 1988), despite its advantages of (1) reinstating sexual selection's place in the debate on the evolution of species, and (2) reconciling Wallace and Darwin's views by involving natural and sexual selection equally. Fisher's writings had the potential to allow an earlier revival of sexual selection as a major topic of research in behavioural ecology. Yet, it took more than 30 years before sexual selection regained its legitimate place. In this respect, it is noteworthy that even the famous evolutionist Julian Huxley, among the rare researchers of the time having published on sexual selection, mentions the runaway process in a first article published in 1938 (Huxley 1938a) but neglects to quote Fisher's writings in his review on sexual selection published in the same year (Huxley 1938b), or in his contribution to the 'modern synthesis', or 'neodarwinism' (Huxley 1942).

### 11.2.3 The contribution of Lande

The Fisherian runaway process was gradually rediscovered during the 1960's, 1970's and 1980's (Pomiankowski 1988; Andersson 1994). This involved mainly theoretical attempts to check quantitatively the reality of the changes in genetic frequency suggested by Fisher. Although O'Donald (1962, 1967) was the first to adopt this theoretical perspective, the definitive demonstration of the Fisherian runaway process is generally ascribed to Lande (1981). O'Donald (1980) had, however, shown that an allele coding for a preference can invade a population at the same time as an allele coding for a preferred trait improving survival thus leading to genetic **epistasis** (see Chapter 5). Moreover, when the preference has become sufficiently common in the population, the trait can continue to increase, even if its exaggeration reduces survival. In other words, the alleles producing an even larger trait continue to be favoured, even after the trait becomes disadvantageous in terms of natural selection, as was suggested by Ronald Fisher. In addition, Lande (1981) showed that male traits and female preferences can co-evolve under some conditions through a runaway process and that the final state of the system can depend partly on stochastic events occurring at the beginning of the process. This is particularly true in small populations exposed to **genetic drift**, and can explain why some very closely related species differ mainly (if not only) in male secondary sexual traits (see Section 11.9). As a result, the question of the emergence of male traits and female preferences is no longer an evolutionary problem.

### 11.2.4 The contribution of Zahavi: the handicap principle

The main alternative mechanism to the Fisherian runaway process proposes that female preference evolved because it allows females to mate with males of high **fitness**. According to this concept, the preference for handicapping ornaments evolved because these traits indicate good health and vigour. The debate on this alternative theory generated several complementary hypotheses. Andersson (1994) suggested referring to these mechanisms as 'indicator mechanisms'. All these mechanisms rest on the assumed existence of genes conferring greater viability (**good genes theory**), indicated by traits that disadvantage males somehow and thus that can be developed only by males of high quality.

In two papers (1975, 1977) provocative for the time, Amotz Zahavi particularly defended these ideas when presenting the **handicap principle**. This principle proposes that females prefer males with costly heritable ornaments, in terms of survival, because such handicaps reveal, without possible cheating, the male's capacity to survive despite the handicap. So females choosing the ornamented males are in fact favoured because their offspring tend to be of higher than average viability. However, although these ideas are commonly attributed to Amotz Zahavi, the first explicit steps appear in the writings of various earlier authors. In particular, Ronald Fisher (1915) discussed, very early on, the possibility that ornaments may indicate the quality (i.e. the fitness) of males. This other aspect of Ronald Fisher's contribution is ignored even more than his contribution to the runaway process. The ideas of Fisher were then taken up and developed by George Williams (1966).

Zahavi's papers raised lively criticism because the first genetic models of the handicap principle concluded that it was unlikely to work. However, later models combining the **heritability** of differences in viability with a mating advantage suggested that this mechanism could, in fact, contribute to the evolution of male ornaments. Overall, the theoretical problems raised by the formalization of the handicap principle proved to be very difficult to solve. More than 15 years passed before the question of whether such a mechanism could work was fully answered in the affirmative (Pomiankowski 1988; Grafen 1990a, b; Maynard Smith 1991). In parallel, the accumulation over time of various empirical data in support of the handicap principle gave rise to a convincing synthesis in the mid 1990s (Johnstone 1995).

### 11.2.5 The revival: current prevalence of sexual selection in behavioural ecology

Although sexual selection had long been regarded as a minor process, it became one of the main study subjects of behavioural ecology in the 1980s and still is (Figure 11.1). Over the same period the relative share of research relating to optimal **foraging** (Chapter 7)

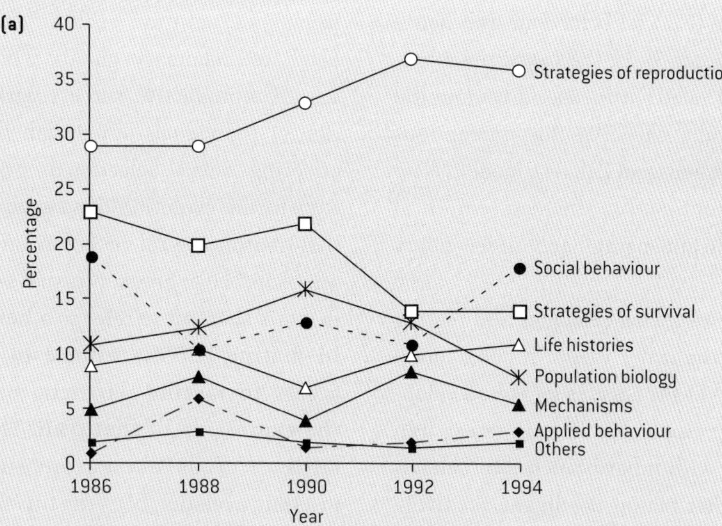


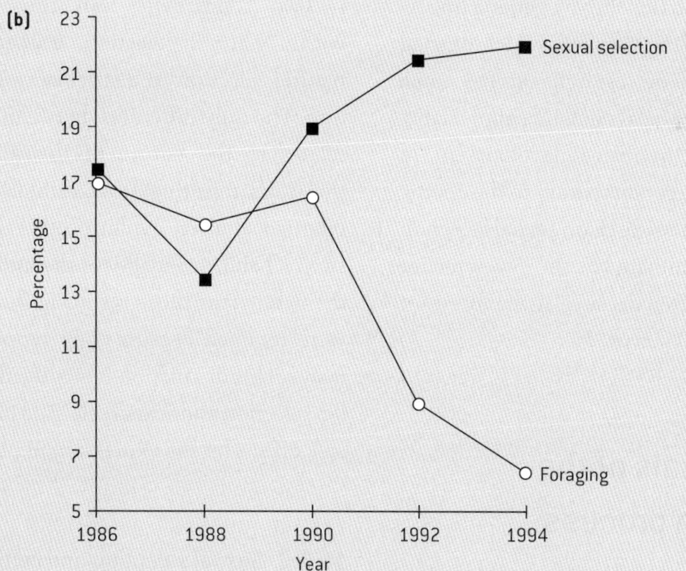


Figure 11.1 **The increasing importance of sexual selection during the 1980s**

Change in the percentage of communications relating to various topics over the first five meetings of the International Society for Behavioral Ecology (ISBE).

**a.** The increase in communications relating to mating strategies is significant ($R = 0.92$; $P = 0.027$). The reduction in communications on survival strategies is also significant ($R = 0.90$; $P = 0.04$).

**b.** These two tendencies are especially due to the increase in presentations relating to sexual selection (mating strategies) and to the reduction in those on optimal foraging.

After Gross 1994.

declined sharply (Gross 1994) while foraging interest has shifted more to social behaviour and hence social foraging (Chapter 8). In the 1990s the attention given to sexual selection increased beyond the problem of empirically checking the assumptions and predictions of Fisher's and Zahavi's models. New models arose and continue to stimulate empirical research on the origin and maintenance of secondary sexual traits.

Today, the study of sexual selection and mating strategies is among the major topics in behavioural ecology. Perhaps one of the reasons for this development resides in the considerable delay in research in this field compared with other branches of behavioural ecology. Another reason lies in the relatively recent appearance of methods allowing for the study of both social and genetic (kinship by genetic fingerprinting) aspects of mating systems, which opened a completely new field of research. At the same time, concerns raised by social demands appreciably changed. Whereas the intensive development of optimization models was contemporary to the energy crisis of the 1970s, that of research on sexual selection occurred in a context dominated by the development of new sexually transmitted diseases at the forefront of which figured the AIDS virus.

## 11.3 Fundamentals of the sexual selection process

### 11.3.1 Relation between sexual selection and natural selection

In a species with sexual reproduction, the fitness of individuals depends on their ability to survive and their fecundity, but also on their ability to access mating partners. The sexual selection process is rooted in the latter aspect. Regarded today, by some, as a component of natural selection and by others as a process at the same hierarchical level, it 'depends on the advantage which certain individuals have over other individuals of the same sex and species, in exclusive relation to reproduction' (Darwin 1871, Part 1, p. 256). This definition discards from the onset all primary sexual traits such as genitalia, avian brood patch or mammary glands whose function is to ensure reproductive success without the intervention of individuals other than the parent and the offspring. Sexual selection is primarily invoked to explain the evolution of secondary sexual traits, be they morphological or behavioural. As for natural selection, the expression sexual selection of a trait is in fact a shortcut to refer to a **heritable difference in mating success caused by competition for access to mating partners and in relation to the expression of that trait**. The three conditions necessary to any selection process are thus: variation in a trait, a relation between this trait and fitness, and heritability of this trait.

Two components can thus be distinguished within natural selection: utilitarian selection (or natural selection in a strict sense) and sexual selection. The only difference between the two processes relates to the nature of the sorting carried out: the traits resulting from utilitarian selection were retained during evolution because they support survival and/or fertility of the individuals who carry them, whereas those resulting from sexual selection were retained because they support their access to mates. This distinction is significant because utilitarian selection and sexual selection often have antagonistic effects on the expression of phenotypic traits.

### 11.3.2 Sexual selection and competition

Given that the origin of the sexual selection process lies in competition for mates, we should now clarify what we mean by competition and the conditions that generate this competition. Sexual selection can only exist in species with **sexual reproduction**, i.e. those where reproduction requires the combination of genetic material from two parents to obtain an offspring. The question of the origin of sexual reproduction pertains to another field than that of sexual selection. The large literature on this topic (e.g. Hurst and Peck 1996, and references therein) is beyond the scope of this book.

#### 11.3.2.1 *Anisogamy and its consequences*

*Males and females*

What criterion allows us, whatever the species studied, plant or animal, to distinguish males from females? The only universal criterion is that males have small, generally mobile, gametes called sperm, whereas females have large, generally immobile, gametes called ova. The fact that gametes that fuse to produce a new individual differ in size is called **anisogamy**.

*The origin of anisogamy*

The origin of anisogamy is relatively easier to explain than that of sexual reproduction. Although isogamy is a rare phenomenon today (even in unicellular organisms), sexual reproduction probably first appeared under the form of isogamous reproduction, i.e. between cells similar in every aspect. Anisogamous reproduction would have appeared only secondarily. Most models of the evolution of anisogamy assume the existence of two selection pressures: one for increasing the size of zygotes (insofar as that increases their survival), and another to increase the number of gametes (insofar as that increases parents' fitness). However, resources being limited, these two selection pressures oppose one another. A **trade-off** (see Chapter 5) lies in the evolution of the two different sexes: one producing few large zygotes, the other producing many small zygotes. Parker *et al.* (1972) suggested that once sexuality appeared, anisogamy was highly probable through disruptive selection (see Box 2.3). If gamete size varies (highly probable by mere chance), and if embryo survival increases with size, then selection should favour gametes that merge with large gametes. The shape of the relation between gamete size and survival strongly influences the system expected. For various shapes, the evolutionary stable strategy is anisogamy, with large and small gametes (i.e. females and males) in the population (Maynard Smith 1982). Tests of these predictions have been proposed, but there are many exceptions and many aspects remain to be tested, in particular the relation between gamete size and survival (Andersson 1994). Another explanation would be that the existence of small gametes prevents the transmission by the latter of cellular organelles and/or of cytoplasmic parasites to the offspring. Indeed, significant **conflicts** could emerge between the organelles of the two parents which would lower offspring fitness (see Andersson (1994) for a synthesis).

*A cascade of consequences*

Once anisogamy exists and whatever its origin and the conditions of its maintenance, sexual conflicts inevitably appear both between individuals of the same sex and between individuals of different sexes. Anisogamy triggers a cascade of consequences that lead to sexual conflicts. In this section, we will discuss the case of an **anisogamous species that does not provide any care to its offspring** once gametes are produced (**parental care** will be considered in Section 11.3.2.2).

In such a species, females produce gametes rich in energy, the ova whereas males produce small inexpensive gametes, **sperm**. Consequently, for a given quantity of resource, males can produce many more gametes than females. Therefore, at any given time, reproduction at the level of the population is limited by the availability of female gametes, whereas many male gametes will always be wasted in the sense that they will not lead to fertilization. In this situation, female gametes are a **rare resource**, and thus limiting for males. If female fitness is mainly determined by access to energetic resources in order to produce ova, male fitness depends mainly on access to fertilizable females.

Thus, there is competition among males to access females. The direct outcome of this competition is that male reproductive success becomes more variable than that of females, because some males manage to fertilize more than one female. Variations in terms of fitness will thus be greater among males than among females. However, as seen in Chapter 2, it is variation that constitutes the raw material for selection. Selection among males will thus mainly depend on the consequences of this variation on their ability to monopolize fertile females. For their part, females have an important selective advantage if they can distinguish and choose to mate with males of higher quality. We thus expect females to be **choosy** and more discriminative than males in their mate choice. This greater discrimination does not necessarily imply that females have more elaborate cognitive abilities, but simply that they are more attracted to certain male traits, or express more interest for males presenting traits revealing quality, so that they have a higher probability of mating with some males than with others. In contrast, considering the low availability of fertile females and the strong competition to obtain a mating partner, males should generally not be very selective but rather try to mate with any fertile female encountered. Female choice can then generate an important selection pressure for the evolution of male traits whereas, due to anisogamy, the opposite is generally much rarer.

The nature and consequences of this competition for mates are the core of sexual selection theory.

#### 11.3.2.2 *Generalizing to investment*

What matters in anisogamy is the differential **parental investment** between males and females **in each offspring**. By definition, females of anisogamous species invest more per gamete than do males. It is this difference in investment that, by a cascade of consequences, leads females to be the limiting sex, to be demanding when it comes to mate choice, and thus to exert a selection pressure on members of the opposite sex.

However, we will see in the following chapter that in many species investment in offspring is not limited to the mere production of gametes. Many species also provide very expensive care to their offspring. For such species, costs related to the care of offspring can be much more important than those related to the mere production of gametes. In some species, however, males invest more in the care of offspring than females. This then creates conditions for:

1. Either balancing the investment by the two sexes, thereby leading to mutual sexual selection.
2. Or, when males invest much more than females, an inversion of the usual effects of sexual selection whereby males become the limiting and demanding sex, leading to a sex role reversal (Clutton-Brock and Vincent 1991). In such a case, males exert, through their mate choice, a strong selection pressure on females. The exaggerated secondary sexual traits are then seen in females and not in males (Clutton-Brock and Vincent 1991).

> We can thus generalize sexual selection by saying that it is the sex that invests the most in each offspring that constitutes the limiting sex, and thus that exerts a selection pressure on members of the other sex through its choosiness during mate choice.

#### 11.3.2.3 *Types of sexual selection*

##### a *Intra- and inter-sexual competition*

Sexual selection takes on two forms: **intra-sexual selection** that involves interactions between individuals of the same sex for access to members of the opposite sex, and **inter-sexual selection**, in which interactions occur between males and females.

Here, the term competition has the same meaning as in ecology: competition exists because the use of a resource (gametes) by an individual makes that resource unavailable to all other individuals. Rivals need not meet for competition to occur (see scramble and interference competition in Chapter 8). A female agreeing to mate with a male and have her eggs fertilized makes these eggs unavailable to other males. The same distinction exists in ecology. In **interference** (or contest) **competition**, each candidate seeks to access a limited resource by excluding the other candidates physically. In **scramble competition**, the candidates need not meet as long as they all exploit the same resource, and their exploitation reduces the quantity of resources available to the other competitors.

The traits selected in intra-sexual selection (Table 11.1) are often referred to as **armaments** because they often though not always serve either as weapons or shields or as signals during male contests.

| Mechanism | Field | Traits favoured in the competing sex |
|---|---|---|
| **Competition by exploitation** More or less equivalent to inter-sexual competition | Physical abilities | Early reproductive age and efficient localization of mates<br>Well developed locomotion and sensory organs<br>Endurance: ability to remain reproductively active during a large part of the season |
| | Mate choice | Morphological traits or behaviours that attract and stimulate partners<br>Nuptial gifts, monopolization of territories, nest sites or any other high quality resource necessary for reproduction<br>Alternative mating strategies such as coercive mating |
| **Sperm competition** | | **Mate guarding**, mate confinement, frequent copulations, production of copulatory plugs or kamikaze sperm, or any other means of preventing rivals from mating with the partner<br>Ability to exceed rivals' sperm, e.g. the production of very abundant sperm, female stimulation so that she rejects competitors' sperm, etc. |
| **Competition by interference** More or less equivalent to intra-sexual competition | | Traits that increase the chances of winning a fight (body size, force, armament, agility, and impressive threat signals).<br>Alternative tactics for lesser competitors to avoid fighting with higher competitors |

Table 11.1 **The various mechanisms of competition for mates, and examples of traits likely to be selected in the competing sex**

Modified after Andersson (1994).

They are also referred to as **badges** revealing the status of the individuals bearing them. Intra-sexual selection can also involve direct fights, which can be very violent, as in deer, or highly ritualized, as in some insects such as *Cyrtodiopis dalmanni* (see Figure 16.8), a southeast Asian fly whose eyes are located at the end of long stalks. When males fight for access to a female, they face 'eye to eye', and the individual with the largest horizontal extension of the eye stalks wins the contest. Contests occur only among males of similar eye-stalk spread magnitude. Intra-sexual selection can also take more subtle forms in individuals physically unable to win such contests and who resort to sneaky mating tactics to reach the other sex. This is particularly the case in many fishes where young adults will fertilize some of the female's eggs even in the presence of older and thus physically dominant males, growth being continuous in fish.

The traits selected in inter-sexual selection are often referred to as ornaments because they are used to attract females. However, inter-sexual selection can take many forms (Table 11.1). It can imply elaborated courtships such as visual or vocal displays, or aggressive courtship pursuits. It can also imply subtle behaviour as in the dunnock (*Prunella modularis*) where the male pecks at the female's cloaca before mating, causing her to eject the sperm received in previous matings (this type of behaviour relates to **sperm competition**, see below).

### b *A third type of sexual selection: sperm competition*

There is a third form of sexual competition (Table 11.1). Indeed, after mating (or matings), the numerous sperm will compete among themselves, inside the female genital tract, for fertilization. This is **sperm**

**competition** (Table 11.1). Moreover, females have a whole range of possibilities enabling them to choose among received sperm, whether received from only one male or from several males. This is **cryptic female choice**. Typically, these interactions do not belong to either intra-sexual or inter-sexual competition. Although it is related to these two processes, sperm competition tends today to be clearly separated from the two other processes of sexual competition because the nature of the mechanisms is basically different. Besides, historically, sperm competition has not been identified and studied as such until relatively recently: the first book devoted to this subject was published in 1984 (Smith 1984), followed by the book from Birkhead and Møller (1992) on sperm competition in birds, and more recently by a more general review on the subject by the same authors (Birkhead and Møller 1998). We will return to this topic in Section 11.6.

The remainder of this chapter will revolve around these three types of competition (summarized in Table 11.1), because these various processes probably act sequentially during evolution and also over a reproductive season. We will start with intra-sexual competition, then turn to inter-sexual competition and finally develop sperm competition. However, before that, two important questions must be addressed: first, that of the relation between intra- and inter-sexual selection; and secondly, that of the direction of sexual selection.

#### 11.3.2.4 *Armament or ornament?*

##### a *The dual utility of sexual traits: a meta-analysis*

Although the dichotomy between inter- and intra-sexual selections is relatively useful, it is not always easy to know if a given trait evolved because of the advantages it provided in one or the other form of competition. These two functions very often seem closely related. In a meta-analysis on studies on the intra- or inter-sexual function of secondary sexual traits, Anders Berglund *et al.* (1996) showed that 37 out of 48 visual, acoustic, chemical, or electric signals, thus more than three quarters (77%), had both functions. Such a link between these two functions cannot therefore be casual. Even when considering only the studies supported by strong statistics, the link between these two functions remained significant. This result is all the more significant as most studies focused on only one function, the link emerging only when synthesizing the results of various studies. It is thus likely that the proportion of signals fulfilling both functions of armament and ornament is underestimated. Indeed, far from having generated a spurious link, this sampling bias rather impaired the ability to detect it. The result is thus said to be **conservative** because presumed biases can only have decreased the link highlighted.

Moreover, of the 11 studies documenting only one of the two functions, nine reported an armament function. However, here, more so than in the above paragraph, sampling biases are likely to create statistical effects without biological meaning. Failing to find a given function in a signal may be because this function has not been investigated, or because it was more difficult to show than the other (Berglund *et al.* 1996).

##### b *A scenario: first armament, then ornament*

Following their meta-analysis, Berglund *et al.* (1996) suggested that secondary sexual traits usually do not evolve through female choice, but rather originate in male–male competition. The underlying idea is that signals used in contests are honest because they are expensive to produce for males in poor condition and because they are constantly probed during male contests. Females would benefit from using the **information** conveyed by these signals to select high quality males. Evidence for this scenario is detailed below.

1. The scenario assumes that females benefit from mating with the best males. There is a consensus on the fact that, all else being equal, females should choose a mate of high status. The benefits of such a strategy for females can be **direct**: males' success in contests has been shown to indicate their success during life in various

| Type of support | Species (taxon) | Latin name | Reference |
|---|---|---|---|
| Females prefer winning males | Pronghorn | *Antolocapra americana* | Byers *et al.* 1994 |
| | Guianan cock-of-the-rock | *Rupicola rupicola* | Trail 1985 |
| | Domestic chicken | *Gallus domesticus* | Graves *et al.* 1985 |
| | Mosquito fish | *Gambusia holbrooki* | Bisazza and Marin 1991 |
| | Siamese fighting fish | *Betta splendens* | Doutrelant and McGregor 2000 |
| | Fly | *Physiophora demandata* | Alcock and Pyle 1979 |
| Females elicit male–male contests | Pronghorn | *Antolocapra americana* | Byers *et al.* 1994 |
| | Sea elephant | *Mirounga angustirostris* | Cox and LeBoeuf 1977 |
| | Domestic chicken | *Gallus domesticus* | Thornhill 1988 |
| | Sailfin molly | *Poecilia latipinna* | Farr and Travis 1986 |
| | Fish | *Padogobius martensi* | Bisazza *et al.* 1989a |
| | Mosquito fish | *Gambusia holbrooki* | Bisazza *et al.* 1989b |
| | Fly | *Scatophaga stercoraria* | Borgia 1981 |
| | Spider | *Linyphia litigiosa* | Watson 1990 |

Table 11.2 **Arguments in favour of the main assumption of the scenario 'first armament then ornament' proposed by Berglund *et al.* (1996): females prefer males winning male–male contests, and cause such contests**

Extracted and completed after Berglund *et al.* (1996).

activities such as foraging, resistance to parasites, and the ability to avoid predators and injuries (Borgia 1979). Benefits can also be **indirect** if the trait is heritable. If so, females choosing to mate with the males of greatest competitive ability will produce sons inheriting their father's competitive abilities (Alexander 1975). The females of many species prefer winners of contests, or even cause such contests (Table 11.2; and see Chapter 20).

2. As Berglund *et al.*'s (1996) meta-analysis suggests, secondary sexual traits seem to have frequently evolved initially as an indicator of social status: the ornament and armament functions are very often associated. Moreover, the fact that signals having apparently only one function are generally of the armament type suggests that the origin of secondary sexual traits occurs more often in an aggressive context than in that of female choice.
3. Other arguments suggest that traits having a dual function do not lose their utility because of exhaustion of the variance in male quality (Berglund *et al.* 1996). First of all, there seems to be substantial variance in armaments. In *Drosophila* males, success in territorial contests shows considerable genetic variation (Hoffmann 1988). In honeybees, worker dominance is strongly heritable (Moritz and Hillesheim 1985). In sticklebacks, aggressiveness and dominance are heritable in natural populations (Bakker 1986), and in cockroaches (*Nauphoeta cinerea*), social dominance shows moderate to high levels of additive genetic variance (Moore 1990). In addition, the mechanisms proposed to maintain variation in inter-sexually selected traits (i.e. of ornament type), such as mutation or spatial and temporal environmental heterogeneity (Hamilton and Zuk 1982), can also contribute to the maintenance of variation in armaments.

Moreover, other models show that males' genetic variance can be maintained through selection generated by female choice (Andersson 1994). However, no model seems to have investigated the maintenance of genetic variation in traits selected concomitantly by intra- and inter-sexual competition.

4. It seems that traits used in contests are probably more honest than traits used only in mate choice. Indeed, males should be better than females at detecting cheats insofar as they can immediately test their rivals' quality during contests. Moreover, falsifying a signal is likely more costly in a male–male contest, where it could result in injuries, than in courtship where the risk of being uncovered before mating is to lose the mating opportunity. This argument is supported by an **experiment** in brown-headed cowbirds (*Molothrus ater*) where males experimentally induced to emit loud songs attracted more females but underwent more attacks by other males, sometimes leading to the death of the 'cheater' (West and King 1980; West *et al.* 1981).
5. Lastly, the stability of an honest status signal seems to be maintained even if the status badge acquires a functional role outside contests. Johnstone and Norris (1993) analysed the maintenance of an honest signal of aggressiveness using an **evolutionarily stable strategy** (ESS) approach. Insofar as the aggressive individual pays a cost according to the context, and insofar as this cost differs among individuals, selection can concurrently maintain both variance in the signal trait and its honesty. They also showed that this result is unchanged when badge size conveys a benefit outside contests, that is when the badge is also used in female attraction.

All these points support the evolutionary hypothesis suggested by Berglund *et al.* (1996): intra-sexual selection should generally be ancestral; inter-sexual selection would be only secondary, but would continue to promote the development of the secondary sexual trait. Insofar as males already have honest signals revealing their quality in male–male contests, females exploiting this information rather than using any other arbitrary trait would be favoured and the armament in question would then continue to evolve as an ornament. This potential transition from an armament to an ornament function of secondary sexual characters parallels that of the transformation of **cue** into **signals** that we develop in Chapters 4, and 16.

It should be noted that the strong association between armaments and ornaments could also be interpreted in the reverse direction: males could parasitize the information from rivals' quality signals that have evolved through female choice. However, the numerous points developed above appear to support Berglund *et al.*'s (1996) interpretation. Also, their hypothesis could be tested through a comparative approach, for instance using Mark Pagel's general method of comparative analysis for discrete variables (1994, 1997; see Chapter 3). In such an analysis, the transition towards an armament function is expected to prevail over the transition towards an ornament function of secondary sexual traits.

#### 11.3.2.5 *The direction of inter-sexual selection*

Through the process of inter-sexual selection, one sex will generally exert a strong selective pressure on the other. Inter-sexual selection is said to be exerted on the chosen sex. However, we saw that the direction of inter-sexual selection can vary among species, mainly because of parental care. Most often, females invest more per offspring and inter-sexual selection is then directed towards males; but in some cases, males provide more parental care and invest more per offspring. The question of the direction of inter-sexual selection is thus the subject of many debates still in progress. In a first step, these debates allowed the clear identification of the factors influencing the direction of inter-sexual selection. It is only very recently that a genuine theoretical framework allowing formal analysis of the respective weight of these various factors has been proposed.

### a *Factors influencing the direction of inter-sexual selection*

*Operational sex ratio*

The fact that the **sex ratio** is generally balanced in populations, with one male for every female, does not mean that there are always as many females and males available for mating. The difference in investment in reproduction implies that most of time, the members of the sex investing most will not be available for reproduction whereas members of the other sex will almost always be available. This difference translates into the operational sex ratio (OSR), i.e. **the ratio of males and females that are sexually receptive (and thus operational in terms of reproduction) at a given time**. In most species, it is strongly biased towards males (OSR greater than 1 as the sex ratio is usually computed as the number of males per females). Of course, if there is a role inversion, it will be biased towards females (OSR less than 1). Operational sex ratio has thus been proposed as one of the key factors in inter-sexual selection (Emlen and Oring 1977) as it integrates the effects of both the global population sex ratio and that of differential investment by the sexes.

*Potential reproductive rate and time of non-receptivity*

The operational sex ratio is difficult to estimate in the field. Indirect estimates have thus been proposed. For instance, the ratio of maximum potential reproductive rate (PRR, measured as the maximum number of independent offspring that parents can produce per time unit) of males to females has been proposed as a more easily assessable parameter (Clutton-Brock and Vincent 1991). This rate is positively related to the difference in number of offspring between successful males and females (Shuster and Wade 2003). The period during which individuals are not receptive to a new reproduction due to their current reproductive load (which is called **time out**) constitutes one of the major determinants of both the OSR and the PRR (Clutton-Brock and Parker 1992). For the remainder of the time (called **time in**) individuals are receptive to mating opportunities. OSR, PRR, and time of non-receptivity are not *a priori* independent from one another. For instance, the period of non-receptivity in females is positively related to the OSR.

*Roles of life history traits*

Several studies have insisted on the fact that other factors influence the direction of inter-sexual selection. For instance, **differences in mortality** between the sexes will play an important part in sex roles because they directly influence which is the limiting sex (Clutton-Brock and Parker 1992). Consider a population in which females invest more per offspring, but in which males are particularly vulnerable to predation (for instance, during the mating season), leading to a low survival of males compared with females. If this differential survival is quantitatively important, it may be that, although investing little in reproduction, males become the limiting sex, thus forcing females to accept the first willing male encountered, allowing surviving males to become choosy. These differences in mortality between individuals of the two sexes are generally due to differential costs of reproduction, themselves related to differences in male and female roles.

In addition, the relative advantage of obtaining a high quality partner, which itself depends on the amount of variation in quality among potential mates (Clutton-Brock and Parker 1992), strongly influences the costs and benefits (and thus the evolution) of choosiness in a sex. In a population where the variance in quality is high in females and low in males, natural selection should, all else being equal, favour the evolution of active choice in males because discriminating between the various females may provide large benefits.

Lastly, discrimination and competition are not necessarily two exclusive strategies in individuals of a given sex (Kokko and Monaghan 2001). Some combinations of parameters can render the most competitive sex choosier.

### b *Towards a synthetic theory of the direction of inter-sexual selection*

The debate on the direction of inter-sexual selection illustrates the limitation of verbal argumentation.

Several parameters likely to influence the direction of sexual selection have been identified. Their relative importance varies from author to author, each acknowledging *a priori* a major role to one or another factor. In particular, most authors attribute a central role to the OSR and the PRR. In addition, the fact that most suggested factors are strongly interlinked, so that any variation in one involves a variation in several others, makes any verbal reasoning particularly tenuous. An inextricable situation quickly emerges, where the only way out is through mathematical modelling. Clutton-Brock and Parker (1992) proposed a first theoretical approach that stressed the importance of the OSR, as well as the importance of other factors in determining the direction of inter-sexual selection. In particular, their model underlined the importance of differential mortality between sexes.

More recently, however, Kokko and Monaghan (2001) proposed a global theory of the roles of each sex. This approach includes, in a single theoretical framework, the combined effects of parental investment, mortality, sex ratio at maturity and variation of potential mate quality. Kokko and Monaghan (2001) use **lifetime reproductive success** as a measure of fitness. In their model, reproduction occurs continuously and males and females can be in either of two states: receptive or non-receptive.

The significance of their formalism lies in the prior identification of the fundamental components binding the various factors commonly acknowledged to act on the direction of inter-sexual selection. This allowed them to reduce the number of parameters in the model and, above all, to express each factor in question (OSR, PRR, etc.) as a combination of these fundamental parameters. They thus did not have *a priori* assumptions on the relation between OSR and the direction of inter-sexual selection.

Kokko and Monaghan's (2001) conclusions question several classically admitted ideas. According to them, the most influential parameter on the direction of sexual selection would be the differential costs of reproduction for males and females. This model does not deny, however, that the OSR and the PRR are good predictors of sexual competition. However, this is because most of the parameters involved co-vary. Indeed, all else being equal, a difference in PRR biases the OSR, and these two parameters 'draw' the mating system in the same direction.

However, Kokko and Monaghan's (2001) approach does not allow the various parameters to evolve: parental care and specific mortalities of the sexes were kept fixed. Only a game-theoretic approach applied to mutual mate choice where mortalities depend on individual parental strategies, while taking into account the possibility of pairing for each sex, would allow one to study the evolution of the direction of inter-sexual selection in a general context.

### 11.3.3 Quantitative measures of sexual selection

For various reasons, researchers may be interested in assessing which traits contribute to individuals' mating success, as well as quantifying the intensity of sexual selection exerted on each one of these traits, and comparing these intensities among natural populations. There is no specific method for measuring selection in studies of sexual selection. There is however a whole set of methods available to quantify natural selection in populations (see Manly 1985; Endler 1986; Brodie *et al.* 1995), methods whose detailed account would exceed the scope of this book. Among the commonly used methods, it is noteworthy to quote those developed by Lande (1979), Lande and Arnold (1983), and Arnold and Wade (1984a, b). In particular, these methods allow one to: determine whether a relationship exists between fitness (measured within the sexual selection framework as the number of mates obtained) and the degree of expression of one or more traits; determine the form of this relationship; and predict the modifications generated by selection in the distribution of traits from one generation to the next. These methods are by essence regression methods.

The univariate **linear selection differential** describes the linear component of selection. It corresponds to the mean slope of the regression of fitness on the value of the trait considered. This selection differential is mathematically equivalent to the co-variance between fitness and the studied trait (Endler 1986) and measures the force of displacement of the mean value of the trait considered in one generation under the effect of directional selection. The **nonlinear selection differential** allows one to describe the possible curve of the relationship between fitness and the trait. It expresses the change in the variance of the trait induced either by the action of stabilizing selection, or by that of diversifying selection. Dividing the selection differential by the standard deviation of the distribution of the trait before selection provides a standardized estimate of the **intensity of selection**, which can be compared between populations. However, univariate selection methods do not enable distinction between direct and indirect effects of selection. This distinction is possible only by using multiple regression methods that allow calculation of selection gradients, whether linear or quadratic (Arnold and Lande 1983).

## 11.4 Intra-sexual selection

### 11.4.1 Evolution of size dimorphism

#### 11.4.1.1 *Some theoretical considerations*

Physical contests involving individuals of the same sex are frequent in nature and favour the evolution of offensive or defensive structures, or of signals allowing individuals to threaten rivals. The viability cost (related to an energetic investment or to an increased predation risk) associated with the development of such structures is not a problem if it is compensated for by a substantial advantage in access to mates. However, a male's competitive ability under competition by interference does not depend so much on his absolute size (or on the absolute size of his armaments) but rather on his size (or that of his armaments) relative to other males in the population. This can lead to an evolutionary escalation in the development of male body size, or in that of some armaments, and partly explain why, in some species, males are far larger and heavier than females and have disproportionate armaments. If armaments have some degree of heritability, directional selection will, over generations, favour the largest males, or those having the most fearsome armaments, and, at the same time, decrease the mean viability of males in the population. *A priori* such a process can lead to different situations (Harvey and Bradbury 1991). The evolutionary arms race engaged in by males might perhaps continue until males become so rare that the population might finally be doomed to extinction. Sexual selection was thus invoked to account for the tendency of increasing body size seen within several fossil lines of mammals, perhaps at the origin of their extinction (see Ghiselin 1974; Maynard Smith and Brown 1986). A second possibility is the emergence within populations of cyclic variations in male body size. The largest males would initially invade the population until adult males become so rare that smaller males would become favoured and then be able to invade, thus restarting the process. The third solution consists of a distribution of male size corresponding to a balance between the costs and benefits associated with the exaggeration of size or armaments.

Various types of model were built to study the conditions of evolution of secondary sexual traits through arms races. They can involve evolutionary game theory for either haploid (Parker 1983a), or diploid systems (Maynard Smith and Brown 1986). These models concur in their general conclusions that at stability, male traits show polymorphism, the mean value of which differs from the utilitarian optimum. This stability has all the more chance of being reached when (1) survival costs associated with male trait exaggeration increase exponentially for each increment in trait value, (2) fighting costs among similarly sized males are important, and (3) there is an environmentally based variation in trait development. When these conditions are not met, extinction events or cyclic variations in male size can occur. However, these last predictions hold only for the haploid model. These theoretical results thus support Darwin's assertion that intra-sexual competition can lead to the stable exaggeration of some male traits despite a reduced mean viability for males.

Parker (1983a) specified which environmental factors were likely to alter males' investment in size or armament development. From an evolutionary point of view, the stable level of investment in size or armaments should increase with the relative number of females requiring defence compared with the number of males defending females. It should however decrease when phenotypic variance of environmental origin ($V_E$) is important. Lastly, it should be higher when the distribution of trait values is biased towards low values. In vertebrates, size and weight, along with the development of armaments, frequently depend on the male's age. As old individuals are rarer than young ones, the trait distribution is generally biased towards low values. In insects, however, adult size generally follows a normal distribution. Therefore, according to Parker, the mean investment in size and armaments should be greater in vertebrates than in insects.

The evolution of size dimorphism, under intrasexual selection, however, does not necessarily imply an arms race and under some conditions can even lead to mean adult size that is lower in males than in females. This can happen when there is strong exploitation (i.e. scramble) competition for the chance to inseminate females. In such a case, it can be advantageous for a male to mature early to meet and inseminate females before the larger males that invested a longer time in growth and maturation. In the long term, this advantage can induce protandry, which occurs in some invertebrates where males have a shorter growth time, emerge before and are thus smaller than females (see Wicklund and Fagerström 1977, Singer 1982). A smaller size can also be advantageous when competition for access to females occurs in a three-dimensional space (water, air) where displacement abilities, manoeuvrability, and agility play a crucial role (Ghiselin 1974). Lastly, if the evolution of large size, being more effective in physical contests, is favoured by numerous male–male interactions occurring at high density, the ability to detect and locate receptive females quickly can prove to be of paramount advantage at low density. This should favour the amplification of males' sensory and locomotory abilities. The advantage of larger size is then better explained by scramble competition rather than by interference competition.

#### 11.4.1.2 *Empirical studies*

##### a *Advantage of a larger size in male vertebrates*

The importance of competition among males in the evolution of size dimorphism is well illustrated in some species of pinnipeds. In the mating season, elephant seal (*Mirounga angustirostris*) males fight for control of female groups in aggressive contests. Females gather on a few beaches that are fairly devoid of predators in order to mate and give birth, and remain there during the three-month mating season. The species is polygynous, and a single male can monopolize the access to several tens of females. Contests between males involve bites and violent head and chest collisions, and thus only the largest males succeed in taking control of a group of females. The smallest males and young males are regularly chased away by the dominant males, and have few chances to mate (LeBoeuf 1974; Deutsch *et al.* 1990). Large size provides a second advantage as it increases the endurance of those males who are thus able to maintain their control throughout the whole mating season despite severe energy losses during contests. The capacity to store energy increases faster with body size than do metabolic costs (Calder 1984). Fewer than one third of the males present on the same beaches as females manage to mate during a mating season. Male–male competition is so intense that males frequently die before having had the opportunity to mate. Conversely, those managing to access females generally experience disproportionate mating success. LeBoeuf and Reiter (1988) reported that eight males alone had fertilized 348 females! Such a situation generates a strong selective pressure in favour of large male size. Large size will not be reached before a certain age, and variance in mating success is much higher for males than for females (Figure 11.2). In this species, size dimorphism between sexes is particularly marked, an adult male being on average three times heavier than an adult female. However, the same correlation between male mating success and body size is found in several other vertebrates where sexual size dimorphism is less marked (see Clutton-Brock *et al.* 1982; Poole 1989; Madsen *et al.* 1993; Fisher and Lara 1999).

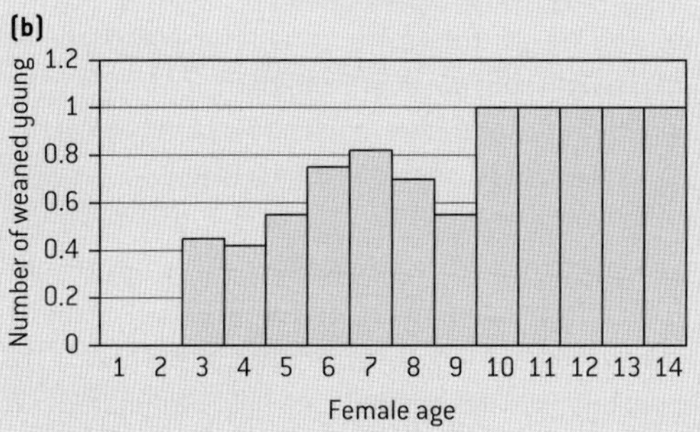


Figure 11.2 **Mean reproductive success in elephant seals (*Mirounga angustirostris*)**

Male (a) and female (b) reproductive success in relation to age. Note the difference in the $y$-axes in the two graphs. Reproductive success for a given male corresponds to the estimated number of weaned offspring whose paternity could be assigned to him. This same value corresponds for each female to the number of weaned offspring she produced. In males, success is concentrated in the older age classes and in the largest individuals, whereas in females, the effect of age is far less pronounced.

After Le Boeuf and Reiter 1988.

### b *Body size or armament size? The example of earwigs*

In practice, it is not always easy to determine whether the observed advantage of larger size is directly related to body size itself or to that of a particular trait. Positive directional selection on body size will generally cause a correlated increase in the size of other morphological traits. So the relation between body size and that of armaments is often allometric (Gould 1974; Harvey and Pagel 1991; see Chapters 3 and 5). This is the case for the European earwig (*Forficula auricularia*). In this arthropod, adult size is determined by the size at emergence. The species is sexually dimorphic, with males having forceps of different form and of larger size than females (Radesäter and Halldorsdottir 1993; Tomkins and Simmons 1996). Moreover, there is male dimorphism, with small males having short forceps and large males having long forceps. Serious fights occur between males for access to females in the various earwig species, the outcome of which seems to be influenced by both body size and length of the forceps (Moore and Wilson 1993; Radesäter and Halldorsdottir 1993; Briceno and Eberhard 1995). The relative role of body and size of the forceps in determining males' competitive ability is not easy to assess because the two traits are strongly correlated (Eberhard and Gutierrez 1991; Radesäter and Halldorsdottir 1993). Pär Forslund (2000) tackled this problem by in a series of laboratory experiments, placing a female and two males together and recording which male won the contest, i.e. copulated longest with the female (Figure 11.3). In the 'forceps' experiment, males were matched for weight (which is tightly correlated with body size), but differed in forceps length. In the 'weight' experiment, males had similar forceps length but differed in weight. Lastly, in the 'size' experiment, small males (low weight and short forceps) competed with large males (high weight and long forceps). Males with long forceps in the 'forceps' experiment were not more efficient at mating than males with short forceps. In the 'weight' experiment, the heaviest males were significantly more efficient. In the 'size' experiment, for which a male was both heavier and equipped with longer forceps than his rival, the probability of winning was significantly and positively related to body weight but not to forceps size. The conclusion is that it is the relative weight, rather than the relative size of armaments, that is crucial for the outcome of the contest.

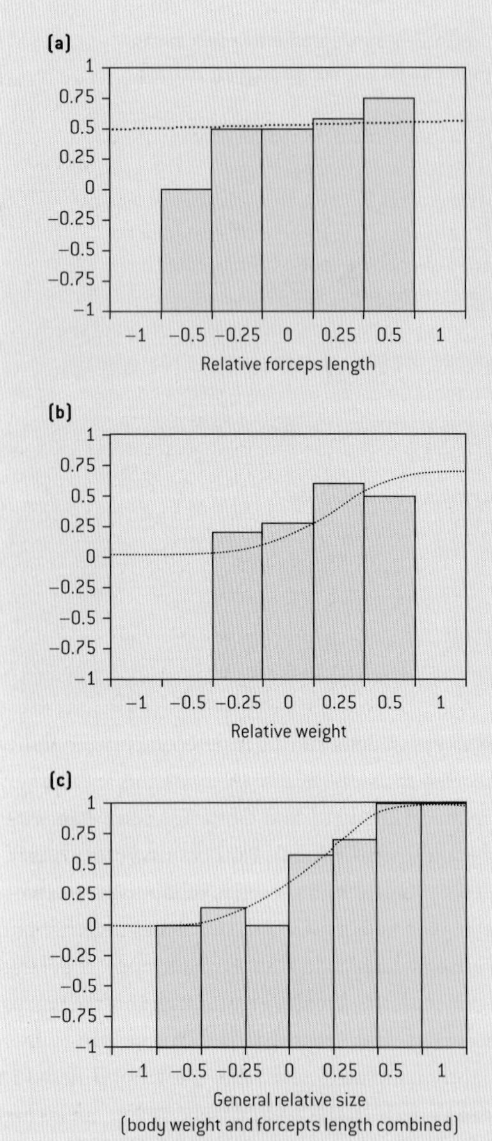


Figure 11.3 **Body size and armament size and outcome of contests in the earwig *Forficula auricularia***

Results of the experiments of competition for access to a female in male earwigs. The three graphs represent the probability of winning in relation to:

**a.** The 'forceps' Treatment: variation in forceps length, but not in body weight.
**b.** The 'weight' treatment: variation in body size (weight) but not in forceps size.
**c.** The 'size' treatment: one male is both heavier and with longer forceps than the other.

For each graph, histograms show real observations (percentages of wins per increment of 0.25 of the relative forceps length) and dotted lines show predictions of the winning probability estimated from logistic regression.

Modified from Forslund (2000).

### C *Advantage to small males: the case of the spider* Misumena vatia

In various spider species there is a strong sexual size dimorphism, some females in genera *Misumena* and *Misumenoides* being up to twice the size of males and nearly 100 times heavier. Legrand and Morse (2000) studied the factors involved in the maintenance of this dimorphism in *Misumena vatia*. Adult males, being much smaller than females, are equipped with relatively longer legs and can move faster. This difference ensues from protandry, the nymphal development being faster and involving fewer stages in males. This protandry is advantageous because of the very low population size and density. Although larger males are favoured in physical contests, such encounters are very rare. Moreover, females are seldom mobile, very dispersed and do not advertise their presence. In this context, the relatively longer legs of males, conferring greater speed, constitute a crucial advantage in locating females. In fact, Legrand and Morse (2000) showed that in *M. Vatia*, males can cover up to 13.5 metres in 30 minutes, a distance that would take several days for females.

### d *Evolution of size dimorphism in the absence of intra-sexual selection*

Although male–male competition is regularly reported to account for size differences between males and females in various species, sexual size dimorphism does not necessarily involve intra-sexual selection. Sexual dimorphism increases as a sex becomes larger or the other smaller, thus selection resulting in decreased female size contributes to increased sexual size dimorphism. Karubian and Swaddle (2001) highlighted this in a comparative analysis of a clade of birds. Using reconstruction of ancestral character states (see Chapter 3), they showed that size dimorphism could actually be attributed to selection for a reduction in female size.

The origin of sexual size dimorphism should also be studied in terms of the mechanisms at play. Size dimorphism results from a combination of sex-specific growth patterns and selection acting on individuals during their growth. Yet most studies focus on sexual dimorphism in adults. However, a study analysed the ontogenesis of size dimorphism in a North American passerine, the house finch (*Carpodacus mexicanus*; Badyaev *et al.* 2001). In this species, growth rates differ between males and females depending on the trait considered. Most female traits show faster growth, but males' growth occurs over a longer period of time. In addition, sexual dimorphism of various traits emerges at different stages of ontogenesis. Natural selection acting on morphological traits after growth was also important and eventually able to cancel or reverse the dimorphisms resulting from growth asymmetries between the sexes. However, adult size dimorphism was, to a very large extent, due to selection acting on juvenile stages. This study is interesting because it suggests that differences observed between distinct populations in the degree of dimorphism measured in adults could be partly produced by contrasting environmental conditions occurring during the growth of organisms rather than during adulthood.

These two studies therefore call for caution in the interpretation of factors responsible for sexual size dimorphism. The selective forces likely to generate such a dimorphism are certainly multiple and may vary between species. However important, intra-sexual selection does not act independently of other selective forces and developmental constraints.

## 11.4.2 Evolution and consequences of precopulatory guarding

In many animal species, males remain as long as possible within close proximity of their female while she is fertile. This behaviour is called mate guarding. Depending on the taxon, such behaviour can take multiple forms. In monogamous birds, males try to remain near their mate, whatever her activity, until the clutch is completed. Accordingly, some males manage to spend almost 100% of their time beside their mate. By doing this, they can prevent the female from mating with any other male and thereby avoid raising chicks that are not genetically theirs. In some insects such as damselflies, the male that copulated last with a female has an advantage in terms of fertilized eggs (last male precedence). To secure paternity, males use appendages at the extreme tip of their abdomen to clasp the female they have just fertilized securely behind her head, and hold her until she lays all of her eggs. In other invertebrates, males engage in precopulatory guarding by clinging to a female's back. We will consider this behaviour briefly now and in more detail in Chapter 12.

**Precopulatory guarding**, or **amplexus**, consists of a male remaining in the vicinity of or clinging temporarily to a female until she becomes receptive. After mating, the male leaves the female. Reported in various vertebrate and invertebrate species, this competitive strategy seems to match certain conditions (Parker 1974; Grafen and Ridley 1983; Ridley 1983). It is typical of species in which (1) female receptive periods are cyclic and limited in time, (2) female receptive periods are not synchronous among females, and (3) males are available to reproduce almost continuously. When there are equal numbers of sexually mature males and females in the population, an asymmetry between sexes ensues, resulting in a bias of the operational sex ratio in favour of males, and thus strong intra-sexual competition. If males can assess how far a female is from being receptive, they can monopolize the female closest to this state through precopulatory guarding, and thus optimize their search for mates (see Figures 2.1 and 2.2).

In various invertebrate species, precopulatory guarding is often associated with a strong positive correlation between the size of males and females in amplexus (see Figure 2.1). Even if various factors can *a priori* explain this correlation (see Chapter 2), it seems to result mainly from the competition between males for access to the most fertile females. Indeed, in several species where precopulatory guarding is observed, female fecundity tends to increase exponentially with size. Thus, males mating with the largest females have higher fitness. Two types of competition can disrupt males. For instance, in amphipods, small males in amplexus can be displaced by larger males (Ward 1983): this is interference competition. However, in the same species, scramble competition can also occur. Indeed, some studies showed that the costs of being in amplexus for amphipods increases with female size (Robinson and Doyle 1985; Plaistow *et al.* 2003). Elwood and Dick (1990) assumed that this cost increases with guarding duration so that only large males may have sufficient energy resources to engage in amplexus with large females far from a moult (i.e. receptivity). Thus, large males are expected to monopolize large females and leave small males with no other choice than mating with small females. This type of competition does not involve direct conflict between males but can in the long term produce the same pairing patterns. Which form of male–male competition prevails in populations likely depends on the operational sex ratio and density.

## 11.5 Inter-sexual selection

We have seen in Section 11.3.2.5 that the direction of inter-sexual selection is controlled by several factors, such that the sex likely to drive the evolution of traits in the opposite sex varies according to the species' life history traits and ecological conditions. Nevertheless, on average, female constitutes the choosy sex. This predominance will be seen in the following examples, although it should be kept in mind that inter-sexual selection can also drive the evolution of female traits as a result of male preference.

It should also be remembered that inter-sexual selection does not necessarily imply an active choice on behalf of either sex. In some cases, the mating season is so short that it is impossible to devote time to assessing various potential mates. In such cases, an optimal strategy for females will be to mate with the first male encountered. This does not prevent inter-sexual selection from operating. Indeed, male traits enhancing female encounters will be favoured, whether these traits allow males to locate females better (sensors) or females to detect males more quickly (bright colours of males, calls, etc.). The **passive attraction** of females can then produce the same effects as an active choice on their part, namely that females will tend to pair more often with some male phenotypes than with others (Parker 1983b; Arak 1988). In practice, it is not always easy to distinguish between active choice and passive attraction. Nevertheless, in many cases, it is possible to demonstrate some form of active choice on behalf of females. As in the case of prey choice described in Chapter 7, we will consider that if a choice occurs, some value should be maximized through this process. Ultimately, the maximized value must be the fitness of the individual making the choice. The maximization of fitness in the context of mate choice can take varied and subtle forms, and involve direct or indirect benefits.

### 11.5.1 Gaining direct benefits

Models addressing the question of the evolution of mate choice strategies generally assume that secondary sexual traits and preferences are heritable. Such an approach amounts to assuming the existence of **indirect benefits**, i.e. that will be expressed in future generations. However, we will see in this section that empirical support for the existence of such indirect benefits remains scarce. One of the reasons for the lack of documented cases could be because of the relative difficulty of showing the existence of indirect benefits.

Nevertheless, the emphasis put on the necessary heritability of secondary sexual traits and preferences in the various models tends to override the importance of **direct** benefits individuals can obtain (i.e. at the time of the current mating event) through their choice of a mate (Andersson 1994). Yet, the direct consequences of individuals' choice on the production of offspring can affect fecundity, quality of parental care, access to resources (food, territory), or protection with respect to various forms of aggression (predation, sexual harassment).

#### 11.5.1.1 *Male insemination abilities and female fecundity*

If the main function of mating is fertilization, females may benefit from ensuring that their mates have enough sperm to fertilize all of their eggs. In various species, males cannot produce sperm in unlimited quantity and may face sperm shortage (Dewsbury 1982; Birkhead and Møller 1992). It has been suggested that the risk of incomplete fertilization could affect female mating decisions, but empirical support for this hypothesis remains scant (Andersson 1994).

Some experiments, however, demonstrated a link between female choice and increased fertility and fecundity. In a species of Coleoptera, *Stator limbatus*, females prefer to mate with the largest males, a preference that results in increased fecundity (Savalli and Fox 1998). In the southern green stink bug (*Nezara viridula*), females having the opportunity to choose their mates produced more fertile eggs over their life than those that were allocated a mating partner randomly (McLain 1998). Similar results were obtained in the Japanese quail (*Coturnix coturnix japonica*; see Chapter 20). Moreover, the uncertainty of male insemination abilities (or compatibility) should lead females to mate with several males to avoid infertile matings. Female mating with multiple males (i.e. sexual **polyandry**) occurs in many species, and several hypotheses have been suggested to explain this phenomenon (Andersson 1994; Arnquist and Nilsson 2000). Recently, Baker *et al.* (2001) concluded that the main function of multiple matings in *Cyrtodiopsis dalmanni* females, a stalked-eyed fly species, was to provide them with sufficient quantities of sperm to fertilize their eggs. This may be due to the reduced size of male ejaculates compared with female sperm storage capacity. This asymmetry does not seem to be an isolated phenomenon (Eberhard 1996) in insects, where it has also been shown that a single copulation is often insufficient to maximize female fecundity (Arnquist and Nilsson 2000).

In contrast, many studies showed the existence of a preference in males for the most fertile females. Whenever females differ substantially in fecundity and mating opportunities are limited (e.g. because of low male insemination ability, low female density, long pairing time, or increased predation risk during mating), males should also be choosy when mating (Trivers 1972). This is particularly true in arthropods, fish, and amphibians, where female fecundity tends to increase exponentially with their size, so that a small difference in size translates into a marked difference in fecundity. In these species, males have a clear preference for the largest females (see Gwynne 1981; McLain and Boromisa 1987; Coté and Hunte 1989; Olsson 1993).

#### 11.5.1.2 *Protection and safety*

The choice of a mate with particular characteristics can reduce predation risk. In many bird species, this risk is increased when visual attention must be divided between the execution of a particular task (pecking ground, grooming, courting) and monitoring the **environment**. Decreasing predation risk by investing more time in vigilance implies reduced efficiency in the current task (see Chapter 14). This trade-off can be partly solved within pairs when each sex benefits from the monitoring efforts of its partner. To associate with a particularly vigilant male may thus allow a female to devote more time to foraging and resource acquisition, which can affect fecundity directly by for instance producing larger clutches. In such species, females should be particularly attracted to the most vigilant males. This prediction was tested in the grey partridge (*Perdix perdix*), a monogamous galliform with low sexual dimorphism. A chestnut horseshoe marking on males' chests constitutes the only obvious sexually dimorphic trait. However, female preference is little affected by this trait (Beani and Dessi-Fulgheri 1995) and females appear more sensitive to males' level of vigilance. In this species vigilance consists of a stereotypic alarm position, raised head, stretched neck, and wings tucked against the body. It is more commonly used by males, which can allocate up to 65% of their time in vigilance (Dahlgren 1990). In experiments conducted in controlled conditions, Dahlgren (1990) showed that grey partridge females strongly prefer the most vigilant males.

A male having particular qualities does not provide protection solely against predators, but also in interactions with congeners. During the mating season, females of some species can be subjected to sexual harassment by males seeking to force copulations or request them with disproportionate eagerness. Such harassment may continue after a stable pair is formed, leading to violent interruptions by rival males of within-pair copulations, enhancing risks of female injury. Choosing a socially dominant or large male (Trail 1985) less exposed to attacks from other males (Borgia 1981) protects females from such harassment risks.

#### 11.5.1.3 *Access to resources*

##### a *Nutritive contributions through nuptial gifts*

Female choice can be driven by the acquisition of food or nutritive substances produced by males. In various species, males provide females with **nuptial gifts** in the form of nutritive substances before, during, or after mating (Thornhill and Alcock 1983). This food contribution can take various forms and is found in a wide range of species. In birds and insects and at least one species of spider, males bring captured prey or other kinds of food items as nuptial gifts. In some insects, males present parts of their anatomy, whereas in other species, cannibalistic females devour their mate during copulation. This phenomenon, famous in European mantids, is also found in various spiders, scorpions, and copepods. Nuptial gifts can also encompass male secretions, on occasions related to sperm transfer (Andersson 1994; Vahed 1998).

*Two alternative hypotheses*

Various adaptive functions have been suggested for nuptial gifts. Males could increase their offspring's fitness, which would constitute a kind of parental investment on their behalf. Alternatively, males may use nuptial gifts to stimulate the female and prolong the mating, perhaps allowing the transfer of greater quantities of sperm, an advantage in a sperm competition context. In this case, the nuptial gift represents an **additional mating effort** granted by the male. In a literature review, Vahed (1998) found evidence in insects in support for the latter hypothesis, whereas empirical evidence for the former is scant.

*Empirical tests of the significance of nuptial gifts*

An experimental study recently examined these two hypotheses in *Pisaura mirabilis*, the only species of spider where males are known to provide nuptial gifts (Stålhandske 2001). In this species, nuptial gifts consist of a prey item rolled up in silk. The male presents his gift during courtship. The acceptance of the prey by the female indicates her consent to mate. Males introduce their copulatory organ and sperm is transferred while the female eats the prey. The interruption of the mating remains under the control of the female who can depart with the gift. Stålhandske (2001) observed the mating success of males with and without nuptial gifts and experimentally manipulated the size of the gifts. The percentage of successful mating attempts increased from 40% without to 90% with a gift.

However, the nuptial gift did not affect the number or size of offspring, which rejects the hypothesis of parental investment. Gift size influenced mating duration and the number of eggs fertilized by the male (Figure 11.4), which supports the hypothesis of male copulatory effort. The absence of effect upon female fecundity raises the question of what benefit she obtains from the nuptial gift. Actually, females of

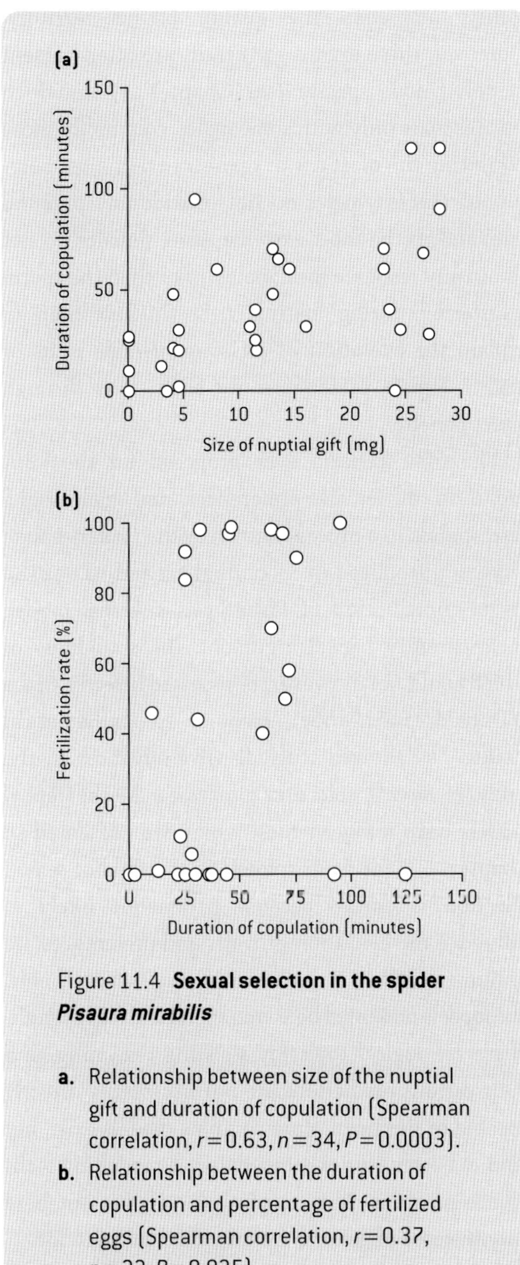


Figure 11.4 **Sexual selection in the spider *Pisaura mirabilis***

**a.** Relationship between size of the nuptial gift and duration of copulation (Spearman correlation, $r = 0.63$, $n = 34$, $P = 0.0003$).

**b.** Relationship between the duration of copulation and percentage of fertilized eggs (Spearman correlation, $r = 0.37$, $n = 33$, $P = 0.035$).

Modified from Stålhandske (2001).

this species are polyandrous and engage in multiple matings. The benefit of a nuptial gift may be negligible, but the benefit accumulated from several gifts from successive males could be significant for female fecundity or survival. By controlling mating duration according to the size of the nuptial gifts, females exert a strong selective pressure on males.

Nuptial gifts also exist in several species of birds, such as the common tern (*Sterna hirundo*), in which males feed their partner before egg laying. In this species, the rate of courtship feeding is a good predictor of the future paternal provisioning rate to offspring (Nisbet 1973; Wiggins and Morris 1986). The best-fed females tend to lay earlier in the season and produce larger clutches. However, it has not been directly shown that female terns select males on the basis of the quality of their nuptial feeding. In another marine bird, the red-billed gull (*Larus novaehollandie scopulinus*), nuptial feeding during courtship increases the probability of copulation (Tasker and Mills 1981).

*Other nutritive contributions*

In various species of insects, males' seminal liquid contains nutritive substances, which constitute a food supplement for females before egg-laying (Markow 1988). In Orthoptera, for instance, females obtain from males sperm and nutritive substances in the form of a spermatophore applied to their genital opening (Gwynne 1984; Butlin *et al.* 1987; Wedell 1994). Those females mated to males producing spermatophores of better quality or in greater number have an increased fecundity.

The most spectacular form of food contribution occurs when males are devoured by mating females. As this contribution can increase female fecundity, it was suggested that such 'suicide' could be adaptive for males that cannot provide parental care and are unlikely to find another mate (Buskirk *et al.* 1984). In the case of European mantids, Roeder (1935) suggests that male decapitation by the female is necessary for successful insemination. This suggestion was later refuted (Liske and Davis 1987) as males of such species approach cannibal females cautiously and try to flee as soon as copulation is completed (Elgar 1992), which contradicts Roeder's hypothesis. Moreover, in some spider species, females can devour males before they have a chance to initiate copulation. This behaviour can benefit females having a high chance of mating with another male (Elgar and Nash 1988).

#### b *Territory quality*

In territorial species, females can discriminate between males on the basis of the size or quality of the territory they defend. In some cases, territory characteristics can be more influential in female choice than male characteristics (Alatalo *et al.* 1986). Territory quality does not depend necessarily on the food resources it contains. For example, in the fish *Pseudolabrus celidotus*, females prefer males whose territories are located in deep water where eggs are protected against predators (Jones 1981). In the amphibian *Rana catesbeiana*, females prefer males whose territories have an optimal water temperature for embryo development and low predation risk by leeches (Howard 1978).

#### 11.5.1.4 *Parental care*

In many species with biparental care (see Chapter 12), females widowed or abandoned by their partner often endure decreased reproductive success (Sasvári 1986), which suggests that the male's contribution is particularly crucial for breeding success. Consequently, females are likely to prefer males providing the best parental care, thereby favouring the evolution of male traits advertising their paternal capacities (Hoelzer 1989). As a matter of fact, various studies in several bird species showed that some male traits, such as the intensity of plumage colour, are good indicators of their paternal performance and that these same traits are preferred by females (see Norris 1990; Hill 1991; Palokangas *et al.* 1994; Wiehn 1997; Linville *et al.* 1998; Keyser and Hill 2000).

### 11.5.2 Gaining indirect benefits

Direct benefits can explain female preferences for many male traits. However, it is more difficult to interpret the evolution of male secondary sexual traits exclusively in these terms. Secondary sexual traits may be particularly extravagant, and strongly influence female choice without necessarily indicating paternal care capacity, nor a male's competitive ability, nor even the quality of his territory. Moreover, these extravagant traits are often present in polygynous species (see Chapter 12) in which females mate only with some males, commonly those exhibiting the most extravagant traits. In most polygynous species, males do not provide any parental care, and in the most extreme cases, females do not obtain any resource from males other than the sperm transferred during mating. It is difficult then to explain the evolution of extravagant male traits or that of female preferences for such traits through direct benefits.

The first scenario ever proposed for the joint evolution of female preferences and male traits corresponds to a process known as the Fisher–Lande process, introduced in Sections 11.2.2 and 10.1.3. The initial contribution of Fisher was a verbal model, an apparently logical argument, the soundness of which could not, however, be assessed directly. Since O'Donald (1962, 1967), recourse to mathematical models has allowed truly decisive advances in the study of sexual selection processes. Mathematical models have many virtues, which we presented in Chapter 3. Their development in the field of sexual selection is crucial because parameters likely to influence the evolution of traits and preferences are so numerous that intuitive reasoning is not reliable. The models presented here are general and fairly simple. These are logical tools for assisting reasoning, and their various predictions are not necessarily directly testable in the field. More complex models exist, but tend to become quickly out-dated as modelling techniques progress. This is why we will present here only the core of genetic models of sexual selection.

#### 11.5.2.1 *The Fisher–Lande process*

The models built by Lande (1981) and Kirkpatrick (1982) have several common characteristics. They consider two traits: a secondary sexual trait expressed only in males and a trait representing a preference for this male trait, expressed only in females. The genes influencing these traits are assumed to be located on autosomes; their transmission is thus not related to sex. Males are assumed to mate whenever the opportunity arises, not to provide any parental care and not to practice any discrimination among females. It is assumed that there is no direct selection on female traits and that all females have the same fecundity and viability. In these two models, the evolution of female preference is correlated to selection acting on the male trait. The two models differ, however, in the genetic determinism considered. Lande (1981) considers a polygenic model (i.e. the traits is determined by several genes), whereas Kirkpatrick (1982) developed a haploid model with two loci. We will develop the latter model first, because of its simplicity and its didactic nature, even though it is certainly not the most realistic.

### a *Two-locus model*

In Kirkpatrick's (1982) model, variation of a male trait is determined by a locus T with two alleles. The hypothesis of haploid determinism considerably simplifies formalizations but does not affect the conclusions of the model. The same predictions are obtained with a diploid model (Kirkpatrick 1982; Gomulkiewicz and Hastings 1990).

Males having allele $T_1$ do not express the trait whereas those having allele $T_2$ do, which reduces their survival to $1 - s$ compared with 1 for males $T_1$. Female preference depends on a locus P with two alleles. Females carrying the $P_1$ allele mate randomly whereas those carrying the $P_2$ allele express a preference for males $T_2$. The intensity of this preference amounts to $a_2$, which means that when given a choice between males $T_1$ and $T_2$, $P_2$ females will mate with a male $T_2$ $a_2$ times more often than with a male $T_1$. Let $t_1$ and $t_2$ represent the frequencies of the two male types in the population. Males $T_2$ obtain a proportion $a_2t_2/(t_1 + a_2t_2)$ of the matings from females $P_2$. It is noteworthy that in that model no direct selection occurs on the preference alleles: they do not have any consequence for female survival or fecundity.

The key point (which is found in other formalizations of the runaway process) is that a linkage disequilibrium emerges between the allele coding for the female preference and that coding for the male trait. At first, the two alleles are not associated with each other and are distributed among males and females independently. However, although the locus coding for female preference is expressed only in female behaviour, it is nevertheless also present in males who can thus have one or other allele. Likewise, the locus coding for the male trait is present in females, and each one of them inevitably has one of the two alleles (Kirkpatrick's model is a haploid model).

(Note that the idea that a gene is not expressed in an individual is not a departure from general genetic theory. In humans, only men are bald despite the fact that this trait is present, but not expressed, in women. Genes can thus be present in an organism without being expressed.)

Let us return to the runaway process. As the preference and the trait increase in frequency, the two respective alleles will be more and more linked within individuals for a simple reason. As soon as the process starts, there is a strong probability for any male carrying the trait that his father has the same trait and that his mother has the allele for the preference, which he may have also inherited. Likewise, any female carrying the preference allele will probably have a mother of similar genotype that has mated with a male expressing the trait. After some time, the allele coding for the preference and that coding for the trait are found together within individuals more often than if they were distributed randomly among population members. This is referred to as **linkage disequilibrium**. It does not involve a physical linkage of the genes on the chromosomes (see Chapter 5 for general considerations on this notion of linkage disequilibrium).

In Kirkpatrick's (1982) model, the frequency change of the allele coding for the trait between two generations is given by the formula:

$$\Delta t_2 = \tfrac{1}{2}\beta^T,$$

where $\beta^T$ represents a measurement of the direction and intensity of selection exerted on locus $T$ (formally corresponding to a differential of selection, defined as the difference between the frequency of allele $T_2$ after selection, $t_2'$, and that before selection, $t_2$; $\beta^T = t_2' - t_2$). The intensity and direction of selection are the result of the combined action of natural selection, which penalizes the expression of the trait, and of sexual selection, which favours it: $\beta^T = \beta^T_{NS} + \beta^T_{SS}$.

The frequency change of the allele coding for the preference between two generations is given by the formula:

$$\Delta p_2 = \tfrac{1}{2}[D_{TP}/t_1t_2]\beta^T,$$

where $D_{TP}$ represents the linkage disequilibrium between loci $T$ and $P$, with $D_{TP}$ = frequency $(T_2P_2) - t_2p_2$. The linkage disequilibrium is always positive because females $P_2$ tend to mate with males $T_2$, making genotype $T_2P_2$ more frequent in the population than expected under random mating. Therefore if $\beta^T > 0$, both the female preference and the male trait increase in frequency in the population, whereas if $\beta^T < 0$, the female preference and the male trait decrease in frequency in the population. An equilibrium is reached when a balance between natural and sexual selection on the male trait is reached, i.e. for $\beta^T = 0$. This condition specifies an **equilibrium line** (Figure 11.5) along which, for an intensity $p_2$ of the preference, there is a corresponding frequency $t_2$ for the male trait, such that

$$p_2 = [s/(1-s)(a_2-1)]\ (1+Vt_2)$$
$$\text{with } V = a_2(1-s) - 1.$$

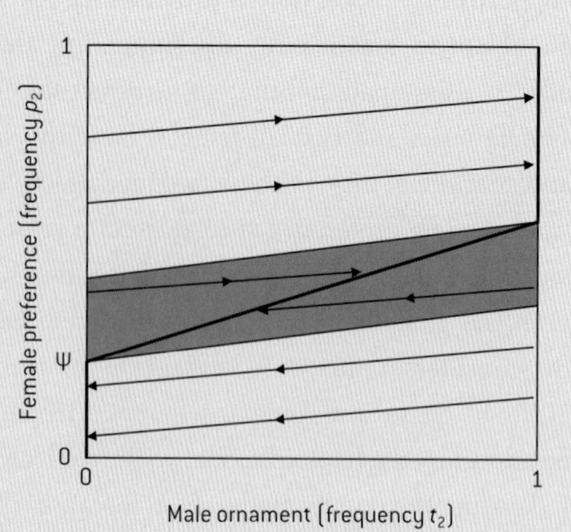


Figure 11.5 **Graphic representation of Kirkpatrick's (1982) model with two loci**

The $x$-axis represents the values taken by the frequency of allele $T_2$ coding for the expression of the male trait. The $y$-axis represents the values taken by the frequency of allele $P_2$ coding for the female preference towards the male trait. In the grey area, any deviation from the equilibrium line returns to another point on the same line. Outside this area, any deviation leads to the fixation or loss of the trait. $\psi$ represents the threshold frequency of female preference necessary for the process to start.

Redrawn from Pomiankowski (1988).

As soon as the preference is strong enough to oppose utilitarian selection, i.e. when $s < (1 - 1/a_2)$, the male trait frequency differs from zero. If allele frequencies move away from this line, they tend to return towards an equilibrium value located on this same line, without necessarily returning to their original positions (this is illustrated by the direction of the arrows in Figure 11.5). If this displacement exceeds a certain threshold, frequencies can then reach one of the two limits where the trait $T_2$ is fixed ($t_2 = 1$) or lost ($t_2 = 0$).

## b *Polygenic models*

Models using a reduced number of loci are far from realistic because secondary sexual traits and preferences are most probably influenced by several loci. Lande's (1981) model considers that both male traits and **female preferences** are influenced by an infinite number of genes, each one having a moderate effect, so that the heredity of the trait is polygenic. Contrary to Kirpatrick's model, variation of the trait is continuous (the study of the evolution of traits with continuous variation, implying cumulated effects of several genes, is known as quantitative genetics). In Lande's model, males can express the trait to a greater or lesser extent (trait size, for instance, varies among males) and females can prefer various degrees of development of the trait (different females can prefer different values of the trait). At the beginning, male trait and female preference are assumed to follow two normal distributions, with means $\bar{z}$

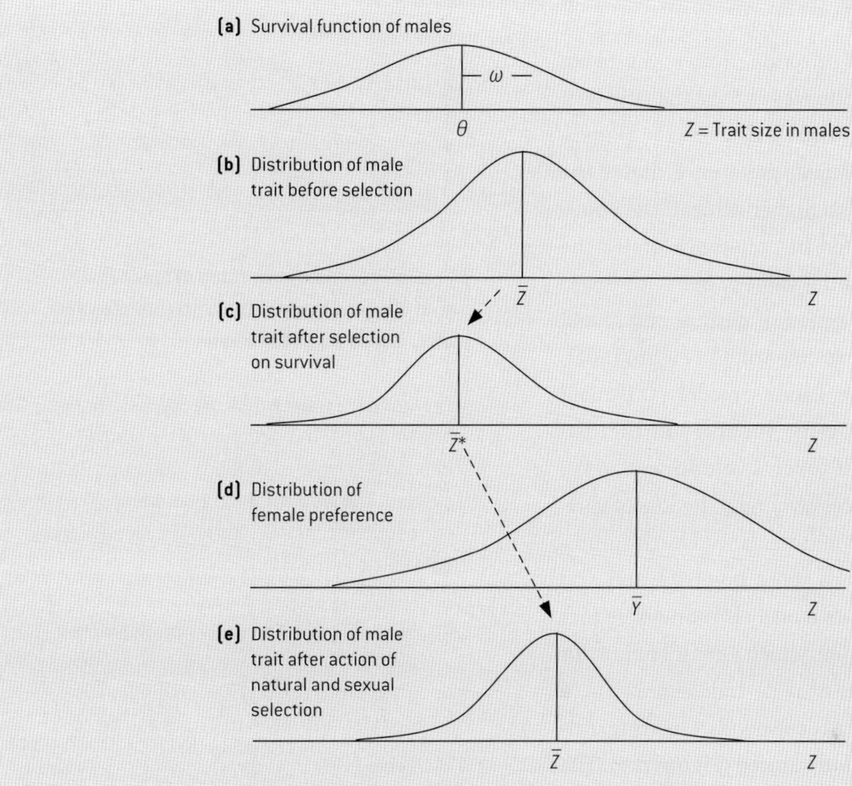


Figure 11.6 **Graphic representation of Lande's (1981) polygenic model**

Male trait size (Z) is maintained at an equilibrium value by the opposing forces of natural and sexual selection.

- **a.** Survival distribution of males as a function of their trait size.
- **b.** Distribution of male trait size in the population before the action of selection.
- **c.** Distribution of this same trait after action of selection through survival.
- **d.** Distribution of preferences within the female population.
- **e.** Distribution of male trait size after the action of natural selection and sexual selection.

After Lande (1981).

and $\bar{y}$ and of additive genetic variances $G$ and $H$, respectively.

The male trait is considered to be under weak stabilizing selection. Let $\theta$ be the utilitarian optimum value for the male trait (i.e. the optimal trait value with respect to the sole action of natural selection). Mortality risk increases as one moves away from this value. The distribution of survival values can thus be represented as a function of male trait value through a normal (also called Gaussian) distribution of mean $\theta$ and variance $\omega$ (Figure 11.6a). The smaller $\omega$ is, the more natural selection imposes constraints on the trait. Female preference can also be represented by a normal distribution centred on the mean $\bar{y}$ and of variance $v$ (Figure 11.6d). A low value of $v$ means that female preference tends to be very pronounced. The process of sexual selection can start when $\theta$ and $\bar{y}$ differ.

As in Figure 11.6 let $\bar{y} > \theta$ (female preference tends to favour increases in trait value; the reverse is also possible and trait size would then be reduced by sexual selection). At equilibrium, the mean genotypic value of a male trait, after the combined action of natural and sexual selection, $\bar{z}$, is higher than $\theta$ (Figure 11.6b). Initially, natural selection 'draws' the mean trait value towards the value $\theta$. Let $\bar{z}^*$ be the new trait value after the action of natural selection (Figure 11.16c). So long as $\bar{y} > \bar{z}^*$, female preference 'draws' the mean trait value towards a higher value. If the population is at equilibrium, female preference 'draws' the trait value back to its initial value $\bar{z}$, but with lower variance; the combined action of natural and sexual selection thus discards extreme values (Figure 11.6e). A crucial assumption of Lande's model (O'Donald 1983), however, is that genic variances of the two traits remains constant over evolutionary time, which amounts to a process whereby the loss of variance due to selection is counterbalanced by the mutations affecting the various genes involved, and also by recombination. Lande's model also concludes that there is an equilibrium line, which joins all points for which, for a given level of male trait expression, a level of female preference corresponds, ensuring utilitarian selection is counterbalanced (Figure 11.7). The stability of the equilibrium line depends on its slope value, $(v^2/w^2) + 1$.

If the process has not reached an equilibrium point, mean male trait size, $\bar{z}$, and mean female preference value, $\bar{y}$, change between generations. Over time a genetic coupling between the male trait and the female preference develops. This coupling emerges also because females with the most extreme preferences (in one direction or the other) reproduce preferentially with males having developed the corresponding trait value and results in linkage disequilibrium between the various genes involved in female preferences and male traits. The pairing of individuals according to their degree of preference and their degree of trait expression thus results in a genetic correlation between the degree of preference in sisters and the degree of trait development in brothers produced by the same pair. This correlation corresponds to additive genetic co-variance, $B$ (Figure 11.7). Lande (1981) showed that as long as the inequality $B/G < (v^2/w^2) + 1$ holds, the system reaches equilibrium (Figure 11.7a). Otherwise (Figure 11.7b), the system follows a runaway process in accordance with Fisher's (1930) predictions. This tends to occur more when females have a very pronounced and stereotyped preference (low value of $v^2$) and when the force of utilitarian selection is weak (high value of $\omega^2$). The evolution rates of the trait and the preference then follow a geometric progression (Lande 1981).

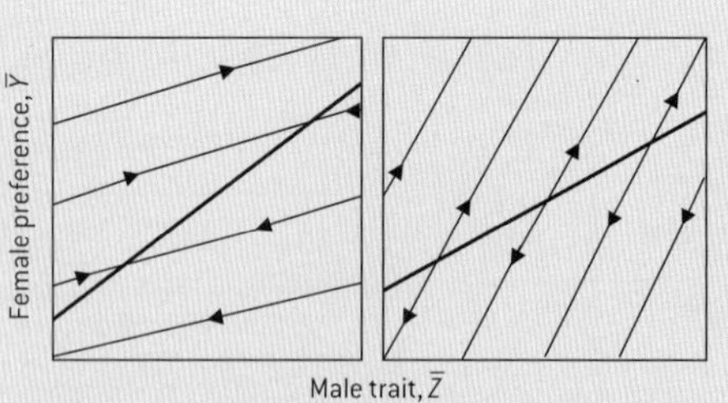


Figure 11.7 **Joint evolution of mean male trait size and mean female preference value, after Lande's (1981) model**

Evolutionary trajectories are represented by arrowed lines. The direction of the arrows indicates in which direction evolution is exerted. The slope of the trajectories corresponds to the $B/G$ ratio. The slope of the bold line is $(v^2/w^2) + 1$.

**a.** When $B/G < (v^2/w^2) + 1$, equilibrium is stable.
**b.** When $B/G > (v^2/w^2) + 1$, equilibrium is unstable and leads to a runaway process.

After Lande (1981).

#### 11.5.2.2 *Where does the runaway process start and stop?*

Kirkpatrick's (1982) model succeeds in predicting a runaway process using simple assumptions and genetic determinism. However, it also stipulates that as long as the female preference remains rare (thus for low values of $p_2$), the equilibrium frequency of the trait remains zero. No runaway process occurs, and the male trait does not spread. Likewise, Lande's (1981) model assumes that a threshold value is required for the female preference before the process can start. The necessity of a high threshold value poses the problem of how such a strong preference can emerge in the first place (Pomiankowski 1988).

### a *Utilitarian selection*

Fisher (1930, 1958) proposed a solution to this problem: he assumed that initially, the male trait is in fact favoured by natural selection. Initially the female preference is thus explained simply by the fact that her male offspring will benefit from the trait favoured by natural selection. Once such a preference is established in females, more attractive variants that may even suffer from a reduction in viability may then evolve. Some empirical arguments support this assumption.

*An empirical test: barn swallows again*

Tail streamers, the secondary sexual trait of barn swallows (*Hirundo rustica*), were the subject of so many publications in the field of sexual selection that they became a classic example. In both males and females of this species, the outermost tail feathers are much longer than the other feathers (see Figures 3.1 and 3.3), and these tail streamers are also notably longer in males than in females. Experiments reported in Chapter 3 drew the conclusion that this sexual difference results from female preference for long streamer males. Many arguments of this type led several authors to suggest that this trait has in fact evolved mainly through sexual selection.

However, Rowe *et al.* (2001) have manipulated the length of male and female tail streamers and studied the effect on manoeuvrability during flight. After manipulation, birds were released in a flight maze in the second part of which flight was hindered by strings suspended from the roof. This experiment confirmed that streamers unaltered in size constitute a handicap for flight and that a 12 millimetre reduction in streamer length improves flight performances. Unaltered streamers thus reduce swallows' flight abilities. However, this experiment showed a similar effect in both sexes, which would indicate a mutual sexual selection in males and females, a common phenomenon in monogamous species like swallows. In addition, Rowe and her collaborators noted that when streamers were shortened beyond 12 millimetres, birds took longer to fly through the maze (Figure 11.8) indicating a loss of manoeuvrability. So flight performance is better with streamers than when streamers have been reduced to the same length as other feathers. This shows that the presence of streamers of a given size enhances flight performance in barn swallows.

This result was confirmed in another experiment: Rowe *et al.* (2001) added streamers to sand martins (*Riparia riparia*), a closely related, streamer-less, species. They found out that the addition of streamers improved flight performance. Rowe *et al.* (2001) suggest that, contrary to what was formerly thought, streamers originally evolved through natural selection, and have then been elongated beyond the optimum length by sexual selection. In this second phase, female choice would have been the only factor favouring streamer lengthening. The evolution of the very long streamer in male barn swallows would thus be the result of a typical runaway process.

### b *Two other possible mechanisms*

Alternatively, other genetic mechanisms can increase female preference frequency without the requirement for an initial selective advantage. The first such mechanism is genetic drift: within small populations, mere chance can lead to a high frequency of the preference allele, which can possibly suffice to initiate the process. A second mechanism is based on the **pleiotropic** effects of genes. Here the preference allele codes for other phenotypic traits in addition to preference. If these traits increase female fitness, they will allow an increase in preference frequency, which can also, under some conditions, trigger the runaway process. We will see in Section 11.5.4 a behavioural mechanism by which such a preference can be acquired in the population because it confers a selective advantage under another function such as, for instance, foraging.

### c *When does the runaway process stop?*

Although various scenarios have been proposed to start the runaway process, the models of Lande (1981) and Kirkpatrick (1982) do not stipulate under

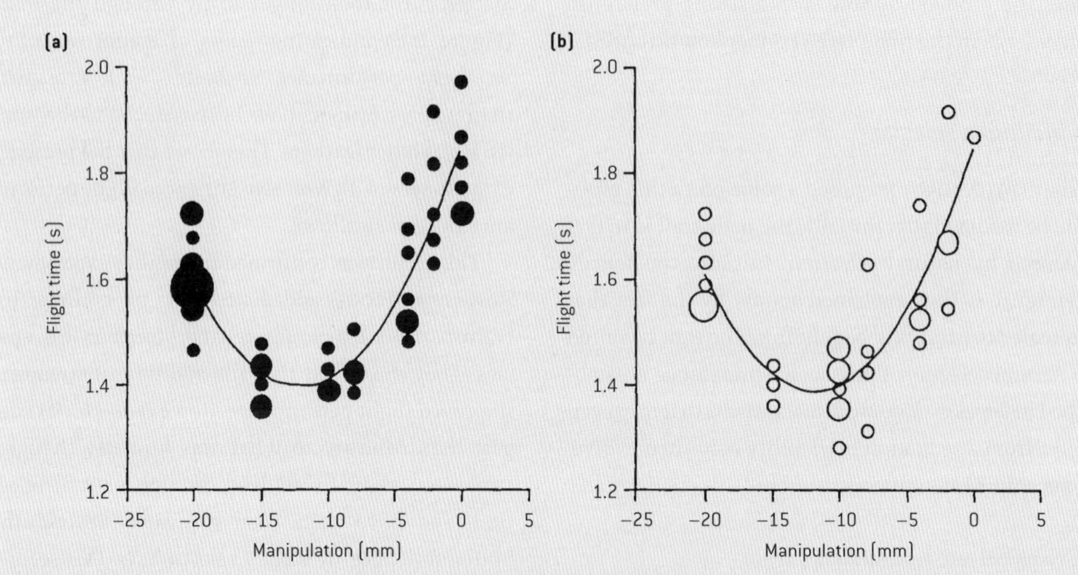


Figure 11.8 **Manoeuvrability and tail streamer length in barn swallows**

**a.** Effect on flight manoeuvrability of manipulating tail length in male barn swallows (*Hirundo rustica*) (measured here as the time to fly through a maze).

**b.** Effect of the same manipulation in barn swallow females. After manipulation, individuals were released at the entrance to a maze made up of strings stretched across a tunnel that was left open at the opposite end. The *y*-axis represents the time taken by the individuals to fly through the maze. An increase in the time taken to fly though the maze indicates a loss of manoeuvrability. Results did not show a difference between males and females: in both cases, the time to exit the tunnel was shortest when streamer length was reduced by about 12 millimetres. This shows that in both sexes, the natural length of streamers is greater than the optimum for flight manoeuvrability. This excess length can thus be explained by mutual sexual selection. However, when streamers were shortened beyond this optimum, the time taken to fly through the maze increased again, suggesting that streamers originally appeared because they improved flight.

After Rowe *et al.* 2001.

which conditions this process can be completed or stop. The extravagance of the male trait may reach such values that the intensity of the utilitarian counterselection would then increase abruptly (Fisher 1930, 1958). Alternatively, the viability costs related to male trait exaggeration eventually become so high that females with the most extreme preferences fail to find a suitable mate (Fisher 1958).

These suggestions remain very speculative. The runaway process however may be slowed down by psychophysical constraints. The runaway process assumes that females are perfectly able to detect differences in size or in colour intensity. However, sensory systems may restrain the sexual selection process (see Chapter 16). For instance, we know from psychophysics that the relation between the increase in stimulation intensity and the increase in sensory intensity increases logarithmically. Consider a simple example. Imagine that you have to compare two weights, one of 1 kilogram and the other of 1.1 kilograms. You will probably be able to determine which is heaviest easily. However, if the weights have much larger masses, say 10 and 10.1 kilograms, now even though the difference between the two is the same 100 grams as in the previous example you will likely have a much harder time telling which of the two is heaviest. This is because now the 100 gram difference only represents a 1% difference between the two weights compared with 10% in the prevous case.

Cohen (1984) suggested that this simple perceptive constraint may suffice to slow down the runaway process. Indeed, as the male trait increases, the same increase in male traits represents a diminishing proportion of the actual male trait and females are expected to become less efficient at discriminating males. Consider, for instance, a bird species in which females select males on the level of elongation of their caudal feathers. At the beginning, male tail measures 5 centimetres. If to be perceived a difference in length must reach at least 10%, females can discriminate males with tails differing by 0.5 centimetres. Such an elongation reduces male viability by a value *s*. Then, imagine that male tails elongate up to 30 centimetres through the runaway process. At this stage, a 0.5 centimeter increase will pass unperceived, and females can now only detect elongations amounting to 3 centimetres. If each increment is equally costly, the elongation cost will then be 6*s*. This means that over time the relative benefit of an increasingly greater extravagance decreases; the benefit in terms of access to females does not change, whereas the production cost of a larger ornament increases. This phenomenon can thus set a strict limit upon trait development. Unfortunately, no study has yet analysed, in detail, female perception abilities or variation in development of male secondary sexual traits in natural populations.

Lastly, up to now, we have assumed that choosing between various potential mates imposes no cost on females. This assumption is likely to be unrealistic. In nature, the search for a mate can expose females to greater predation risks, or energy and time losses. For females, preferring some male traits can, for example, increase search times and costs. However, it is also possible that the evolution of extravagant male traits actually facilitates their location by females and thereby reduces the costs of searching for a mate (Kirkpatrick 1987). The consequences of these costs for the dynamics of the runaway process have been considered by several authors (Kirkpatrick 1987; Pomiankowski *et al.* 1991), but remain unresolved.

#### 11.5.2.3 *Testing the Fisher–Lande process*

The genetic models mentioned earlier do not prove that all extravagant sexual traits found in nature are the result of a Fisherian runaway process. They only show that, **under some conditions**, such an evolution is theoretically possible. Contrary to the foraging models described in Chapter 7, these models are extremely difficult to test empirically because they generally incorporate many *a priori* plausible assumptions, but whose relevance in nature is not always easy to check. Moreover, they are testable only between generations, hence complicating the task. In addition, some assumptions or predictions of Fisherian models are also compatible with other hypotheses for the evolution of secondary sexual traits (Andersson 1994).

##### a *Heritabilities of traits and preferences*

There are, however, two points on which it is possible to assess the relevance of the runaway model. Whatever the model considered, the genetic variability of the trait and of the preference must be maintained for the process to continue. It was also seen that models predict a genetic co-variance between the trait and the preference and that the two are supposed to co-evolve. Various studies have shown that both the variation observed in secondary sexual traits and in preferences have a genetic basis (Bakker 1990; Moore 1990; Ritchie 1992). Genetic analyses tend to establish that secondary sexual traits are influenced by several loci (Andersson 1994), and often have substantial levels of heritability (Andersson 1994; Pomiankowski and Møller 1995). This maintenance of a strong heritability puzzled several authors (Taylor and Williams 1982; Kirkpatrick and Ryan 1991) because directional sexual selection should in the long run erode the additive genetic variance of the selected traits and preferences. Some recent work, however, proposed mechanisms by which additive genetic variance on a secondary sexual trait, subjected to strong selection, can be maintained (Rowe and Houle 1996; Moore and Moore 1999). One of them resides in the fact that a significant part of female preference may be under social influences during development. We have seen for instance in Chapter 4 how females may copy the sexual preferences of other females in the population. Such a mechanism may participate in maintaining significant levels of additive genetic variance (see Chapter 20 for a general discussion), both in female preference and in male trait.

### b *Existence of genetic correlations*

But what about the genetic correlations between male traits and female preferences? This is an important prediction of the Fisher–Lande process. At this level, various analyses are possible.

*Experimental selection*

An initial approach consists of subjecting males to a specific selection based on the size of their trait while keeping females unselected (randomly chosen). Male offspring resulting from the first generation are measured anew and the experiment is repeated for several generations. After several generations, females are tested to determine whether their preference differs from that of the population of origin, that of another line selected in the opposite direction, or that of a control line. Such experiments revealed the existence of genetic correlations in various species (see Houde 1994; Wilkinson and Reillo 1994; see Bakker and Pomiankowski (1995) for a result synthesis). Unfortunately, the procedure is not perfect because experimenters do not really control the pairings of the individuals nor female reproductive investment (Gray and Cade 1999a), thus weakening this kind of experiment. Furthermore, we will see in Chapter 20 that another reason why these results may be weakened is because none of these experiments controlled for the impact of social influences.

*Trait and preference distribution among siblings . . .*

An alternative approach aims at analysing the genetic correlation between the degree of female preference and the development of the male trait among siblings. This procedure was used by Bakker (1993) in three-spined sticklebacks (*Gasterosteus aculeatus*). Male sticklebacks develop a red belly of varying brilliance during the mating season. Females prefer brighter males. At the same time, the visual system's sensitivity to the red colour increases in females but not in males (Cronly-Dillon and Sharma 1968), suggesting that female preference evolved in direct relation with the male trait. Bakker (1993) crossed males of differing belly colouration with females from the same natural population and obtained, for six crossings, a positive correlation between the mean preference of sisters and the mean trait value of brothers. A similar result was obtained in field crickets, *Gryllus integer* (Gray and Cade 1999b). We will see in Chapter 20 how such results may partly be explained by uncontrolled social influences leading to cultural transmission in female preferences. Lastly, Blows (1999) hybridized two species of *Drosophila* flies to break down the natural systems of mate recognition, and then studied the joint evolution of recognition systems. As expected, a genetic correlation between the male and female components of mating systems quickly evolved. Nevertheless, this study relates to recognition systems and not specifically to the evolution of an exaggerated male trait counterselected by utilitarian selection.

*. . . or among populations*

Another approach consists of studying the covariation between a male secondary sexual trait and female preference for this trait in several populations. The advantage, relative to artificial selection, is that this allows the examination of the co-evolution of the trait and female preference over longer time spans and under natural conditions. The finding of a correlation across populations (Houde and Endler 1990) does not constitute an absolute proof of the existence of a genetic correlation (Houde 1994). However, obtaining negative results indicates that a genetic correlation does not exist. Morris *et al.* (1996) studied the degree of variation in female preference in relation to the variation in male traits in the fish *Xiphophorus pygmaeus*. Adult males of this species are particularly small compared with other species of the same genus. Females are, however, attracted to larger males, to the point of preferring males of a congeneric species in experimental conditions. However, a rare phenotype of large males seems to have recently appeared and spread in some populations of X. *pygmaeus*. This unique opportunity allowed Morris *et al.* (1996) to test the joint evolution of female preference and male trait. In a five-year study, the authors found a surprising negative correlation within populations between female preference and the frequency of the male trait. Where the 'large male' phenotype became frequent, female preference for large males tended to decrease or disappear.

Considering the central role played by the prediction of a genetic co-variance between male secondary sexual traits and female preferences, it is surprising that so few studies have directly addressed the question. As a result and despite the enthusiasm triggered by the models developed in the 1980s, empirical evidence in favour of the runaway process remains limited. We will come back to this important question in Chapter 20.

### 11.5.3 The handicap principle

The handicap principle was formulated by Zahavi (1975) who proposes a particular interpretation of female choice. According to Zahavi, selection tends to favour female choice when it relates to traits that 'handicap' the male, i.e. that *a priori* substantially decrease male survival rate. Here, females directly assess male viability through the degree of development of the handicap because only high-quality males could afford the cost of the full blown handicap. No individual of poor quality could afford to develop a secondary sexual trait that would handicap his survival too much, therefore increasing the signal's enforced **honesty** (see chapter 16). Since Zahavi, the idea spread that the logic behind female choice lies in obtaining the best possible genetic contribution for her offspring. The relevance of this hypothesis, known as the 'good genes' hypothesis, however, continues to divide researchers.

The handicap principle stimulated many theoretical models. As a result, some authors (Eshel 1978; Andersson 1982; Pomiankowski 1987a) concluded that the reasoning is logically valid and can explain the evolution of costly preferences (Andersson 1986; Pomiankowski 1987b; Heywood 1989; Grafen 1990b; Iwasa *et al.* 1991), whereas several others claimed that the general principle of the handicap does not work or has only moderate effects (Maynard Smith 1976; Bell 1978; Kirkpatrick 1986; Tomlinson 1988). Some, initially sceptical, authors then acknowledged the potential validity of this principle. The debate on the validity of the handicap principle is tricky, if not confused, and again it is difficult to determine whether the predictions emerging from various models must be regarded as general or remain narrowly dependent on some assumptions. To clarify the situation, it is useful to specify exactly what is meant by handicap as the concept varies with authors and models.

#### 11.5.3.1 *Types of handicap*

There are in the literature three major variants of the handicap principle. The simplest variant is known as the **fixed handicap** (Maynard Smith 1985), also called 'epistatic handicap'. In this model, all males having a certain allele *h* express the handicap, but males of low viability are penalized by it to a greater extent than males of high viability. Once selection has operated, there is a greater proportion of high viability males among those having the handicap than among those that do not have the handicap. Females choosing a male having the handicap are thus more likely to obtain 'good genes' for their offspring. There is thus no correlation between the development of the handicap and male quality. The expression of the handicap depends only on the presence of the allele *h* (hence the name of 'epistatic handicap'). The name 'Zahavi's handicap' given by Maynard Smith (1985) to this variant oversimplifies the original idea of Zahavi and is thus not very appropriate (Iwasa *et al.* 1991, Collins 1993).

Indeed, in his original proposal, Zahavi (1975, 1977) considered that males of higher quality (i.e. of high viability) invested more in the development of the handicap, than did males of lower quality. This version has been labelled the 'conditional handicap' (Iwasa *et al.* 1991; Collins 1993). It assumes that the degree of handicap expression increases with the condition of the individual, which is correlated with his viability, itself dependent on his genotype (Kodric-Brown and Brown 1984; Andersson 1986; Zeh and Zeh 1988). Thus, males of poorer condition do not develop the handicap even if they have the corresponding allele.

The third variant is known under the name of **revealing handicap** (Hamilton and Zuk 1982; Hasson 1991). All males develop the trait initially, whatever their genetic quality. However, over their lifetime the ornament functioning as a handicap advertises the quality or condition of males. For instance, sick or weakened males will have more difficulty in maintaining their plumage. As this propensity to get sick depends on their genetic quality, females choosing males with intact ornaments obtain an indirect benefit, in the form of good genes, for their offspring.

Various formalizations were developed to assess the soundness of these three types of handicap. It appears that the conditional and the revealing handicap are theoretically valid and can thus lead to the exaggeration of traits and preferences (Maynard Smith 1985; Pomiankowski 1987a; Hasson 1989; Iwasa *et al.* 1991). The two models differ only slightly in their assumptions and predictions, such that the utility of maintaining a distinction between these two versions of the handicap principle may seem debatable (Collins 1993). However, it should be stressed that the term 'handicap' is used in a very broad sense to define both very labile structures, such as behaviours, and complex morpho-anatomical structures whose developmental biology remains little documented. The concepts of conditional or revealing handicaps, though close, can be useful in distinguishing between traits whose development obeys various types of constraint. This line of research has yet to be explored. As for the fixed handicap, it was first rejected because the models developed initially (Maynard Smith 1985; Pomiankowski 1987a; Iwasa *et al.* 1991) concluded that female preference could not be exacerbated over time in the absence of a direct relation between the degree of trait expression and male viability. However, more recently, Siller (1998) added some credibility to the fixed handicap using various game theory and quantitative genetics models. It is thus difficult to conclude definitively whether this type of handicap is valid, especially as the debate centres around the problem of the origin and maintenance of additive genetic variance on the trait and the preference under natural conditions, a problem that does not seem close to being solved (see Section 11.5.2.3).

#### 11.5.3.2 *Do male traits function as handicaps?*

The central assumption of handicap models is that an ornament is costly (to be produced or maintained) and that this cost is more affordable for an individual of high viability than an individual of low viability (see Figure 16.8). What does happen in nature? Are male secondary sexual traits costly and does this cost depend on the condition of the individuals? Various types of cost associated with the exaggeration of male traits have been measured. They can be put into two main categories.

##### a *Predation-related costs*

Bright colours, loud vocalizations, and odorant signals allow males to be located more easily by females. However, they also expose males to predators able to exploit these cues to locate their prey. Various studies on a wide range of sexually dimorphic species highlighted differential predation between sexes, generally in disfavour of males (see Andersson 1994). These studies are difficult to analyse, however, because it is not always easy to determine which trait or combination of traits makes males more vulnerable to predation. The weight of secondary sexual traits can easily be overestimated because males and females can also differ in their diet or in their foraging mode, different traits that may involve more or less important predation risks. However, some studies established a direct link of causality between searching for a mate and an increased predation risk. For instance, the little blue heron (*Florida caerulea*) can locate males of short-tailed crickets (*Anurogryllus celerinictus*) through their stridulation trills (Bell 1979). An exemplary study was performed by John

Endler (1980, 1983, 1987) with the guppy (*Poecilia reticulata*). Males of this species show a strong polymorphism. The number of colour spots on their body greatly varies among individuals and populations. Endler sought to understand to what extent this variation could result from antagonistic selective pressures. On the one hand, females prefer the most colourful males (Houde 1997), and on the other hand colour spots expose males to predation by other fishes (Endler 1978). By combining studies on natural populations and on experimental populations, subjected to a greater or lesser predation pressure over several generations, Endler (1980, 1983) showed that the number of colour spots decreased in the populations most exposed to predation, suggesting that there is a cost to the expression of sexual signals in terms of predation.

#### b *Physiological costs*

The development of extravagant sexual traits or particularly intense courtship behaviour can also involve physiological costs. Indeed, the energy available to organisms is limited, just as are several nutrients involved in both the development of some traits and essential physiological functions. Measuring such costs generally requires sophisticated techniques. For instance, in male sage grouse (*Centrocercus urophasianus*), the use of the 'doubly labelled water' technique highlighted an energetic cost of courtship (Vehrencamp *et al.* 1989). This technique allows the assessment of the daily $CO_2$ production of free-living wild animals over several days, a measurement then used to determine their energy expenditure. The daily energy expense of vigorously courting males was twice that of non-courting males and four times higher than the species basal metabolic rate. Energy losses during courtship were 13.5 to 17.5 times the basal metabolic rates depending on individuals. Other studies measured the energy cost of courtship behaviour and showed a substantial cost of vocalizations in males of some anouran species (Taigen and Wells 1985).

If secondary sexual traits do seem to involve a cost for the individual expressing them, the relation between such costs and individual quality is less easy to analyse. However, if male investment in secondary sexual traits does not vary with their ability to afford such costs, then male survival ability should generally be negatively related to the degree of expression of their secondary sexual traits. Recently, Jennions *et al.* (2001) performed a meta-analysis using 122 samples taken from 69 studies conducted on 40 different species of birds, arachnids, insects, and fish. It appeared that irrespective of the level of analysis considered (sample, study, species), there was a significant and positive correlation. Males developing the largest ornaments or exhibiting the most vigorous courtship displays were generally also those having the best survival ability and the greatest longevity. This suggests that the degree of development of sexual traits depends on male condition. However, the biological mechanisms underlying this phenomenon remain unknown. In the future, the development of studies accounting for the development of secondary sexual traits should bring crucial elements to our understanding of the relation between individual quality and handicaps.

### 11.5.3.3 *The paradox of 'good genes'*

The handicap hypothesis is extremely tempting because it allows us to explain, *a priori*, the existence of a choice by females even when they do not seem to obtain any direct benefit from males. This situation is typical in species forming **leks**.

Leks are gatherings of males in areas that do not contain resources of any kind (see Chapter 12). Each male defends a small display territory, known as a court, and engages in a courtship display combining visual, acoustic, or olfactory signals, depending on species. Females visit leks and generally visit several males before mating with one of them. Females seem not to choose their mate randomly and most copulations are usually performed by a minority of males present on the lek. Females then leave the lek and proceed to the raising of offspring without the contribution of males (Bradbury and Gibson 1983; Davies 1991; Hoglund and Alatalo 1995; see Chapter 12).

A question thus arises: what can females obtain through such a choice if the males' sole contribution to reproduction is insemination? The handicap principle considers that female benefits are in terms of finding the best genes for their offspring. Note that this argument is valid only if male fitness is heritable. However, the theoretical calculations performed by Fisher (1930) showed that the heritability of traits contributing strongly to fitness is inevitably weak. It is easy to understand why. Over generations, as female choice for traits, indicating a better viability, spreads, males having the alleles determining the best quality should be increasingly numerous. In the long term, these alleles should become fixed in the population. Consequently, all males are equivalent and no selection pressure maintains female preference anymore. If making a choice proves to be costly, for instance in energy or time spent examining males, the preference will soon be counterselected and disappear. Hence the **paradox of good genes** (often called the **lek paradox**; Kirkpatrick and Ryan 1991; Andersson 1994).

To solve the paradox of good genes we must determine how the heritability of fitness can be maintained in natural populations. In fact, the analysis of empirical data tends to establish that various species often show high heritabilities both in male secondary sexual traits (Pomiankowski and Møller 1995) and female preferences (Bakker and Pomiankowski 1995). A first possibility supposes the accumulation, over time, of deleterious mutations, which would tend to restore additive genetic variance to the traits indicating male quality. The actual importance of this phenomenon remains unclear (Andersson 1994; Kirkpatrick 1996). An alternative possibility supposes that selection pressures can vary greatly over space and time and that this variation is sufficient to restore additive genetic variance to the traits subjected to sexual selection (Andersson 1994). A final and often ignored alternative is that females do not seek good genes in an absolute meaning, but rather relative to their own genes. This hypothesis that can be called the compatibility hypothesis however would not explain why in a classic lek only a few males obtain all the matings.

#### 11.5.3.4 *The Hamilton–Zuk hypothesis*

In 1982, Bill Hamilton and Marlene Zuk proposed a new mechanism to solve the lek paradox. Their hypothesis (Hamilton and Zuk 1982) is based on the existence of co-evolutionary cycles between parasites (in the broad sense) and their host species. Parasites, viruses, and other pathogens must regularly face host immune defences. Favourable mutations occur regularly, which allow parasites to circumvent or resist the defenses developed by the hosts. For their part, hosts are subjected to a strong selective pressure to defend themselves against pathogens. What follows is a true evolutionary arms race capable, according to Hamilton and Zuk (1982), of maintaining some additive genetic variance of the traits subjected to sexual selection. Because the degree of development and the maintenance of male extravagant traits will generally depend on the condition of individuals, these traits signal the ability to resist the pathogenic agents present in the environment. As parasites and viruses mutate regularly, alleles conferring a better resistance are likely to vary in time, which maintains some heritability of male genetic quality. This hypothesis thus states that female preference for extravagant male traits evolved because it allows females to identify males in good health so as to transmit resistance alleles to their offspring.

The Hamilton–Zuk hypothesis was very favourably received and stimulated many studies. These studies can be divided into two main groups. The first approach used comparative studies of the degree of extravagance of male sexual traits in relation to the risk of parasitic infestation undergone by the species (Read 1987, 1991; Read and Harvey 1989; Clayton 1991, Johnson 1991; Weatherhead *et al.* 1991; Pruett-Jones *et al.* 1991). These studies produced ambiguous, sometimes contradictory, results. The second group of studies consists of case studies. Here too results are ambivalent (see Andersson (1994) for a synthesis). Some studies found a negative correlation between the degree of development of a secondary sexual trait and male parasitic load, whereas others showed an absence of relation. Moreover, some studies highlighting a female preference for the least parasitized males are also compatible with the more parsimonious hypothesis that females are actively avoiding parasitic infection (see Loehle 1997).

#### 11.5.3.5 *The immunocompetence handicap hypothesis*

The relative failure in the validation of the Hamilton–Zuk hypothesis did not lead scientists to forsake this hypothesis, but rather to reformulate it. Living organisms are seldom exposed to a single type of parasite or pathogen, but rather a suite of pathogens, successively or simultaneously. In this fight, the individuals' immune system plays a prominent role. A better way to assess the relevance of the Hamilton–Zuk hypothesis may be, therefore, to study the relation between the development of male secondary sexual traits and their immune system's ability to fight infections. The existence of a relation between the **immunocompetence** of individuals and the extravagance of their secondary sexual traits was thus suggested by Folstad and Karter (1992). According to the immunocompetence handicap hypothesis, testosterone, a **hormone** involved in the development of male secondary sexual traits, usually involves a reduction in the efficiency of the immune system. Consequently, only individuals equipped with a particularly powerful immune system would be able to pay the costs associated with the development of secondary sexual traits, whose expression depends on testosterone levels (see Chapter 6).

This hypothesis has received increasing attention over the past decade and some empirical support (Zuk *et al.* 1995; Saino and Møller 1996; Saino *et al.* 1997a, b, 1999; Zuk and Johnsen 1998; Gonzalez *et al.* 1999; Verhulst *et al.* 1999; Owens and Wilson 1999; Duffy *et al.* 2000; Faivre *et al.* 2003; for a review see Møller *et al.* 2000). However, results are once again difficult to interpret as a whole. Some studies, such as that of Zuk *et al.* (1995) on red junglefowl, *Gallus gallus*, found a negative correlation between the size of a secondary sexual trait and a measure of immunocompetence. In contrast, other studies such as that performed by Gonzalez *et al.* (1999) in house sparrows observed the opposite relationship.

One difficulty in comparing the various results is that the measure of immunocompetence itself is debatable, particularly in vertebrates (Siva-Jothy 1995; Norris and Evans 2000). Vertebrates have both humoral and cell-mediated immunity. The former is responsible for the detection and elimination of specific pathogenic agents by antibodies. The second is non-specific and involves T-lymphocytes. Norris and Evans (2000) insisted on the need to assess both components of immunity simultaneously.

Recently, Faivre *et al.* (2003) performed such an investigation in blackbirds (*Turdus merula*). This species exhibits strong sexual dimorphism. Females are brownish and have a light coloured bill. In contrast, males have entirely black plumage and a yellow–orange beak, the saturation of which varies between individuals. Males with the brightest bills mate first and do so with the females that are in the best condition (Faivre *et al.* 2001), suggesting that females prefer males with colourful beaks. Faivre *et al.* (2003) results were conclusive. Orange-billed males showed a weaker humoral response than yellow-billed males, whereas the reverse was observed for the cellular response. The relation between immunocompetence and the degree of expression of a secondary sexual trait seems thus to depend of the component of immunity (see also Zuk and Johnsen 1998). Moreover, the negative influence of testosterone on immunocompetence claimed by Folstad and Karter (1992) is not always verified (Hasselquist *et al.* 1999; however, see Duffy *et al.* 2000). These ambiguous results thus call for caution. It is likely that the degree of complexity involved in vertebrate immunity has been underestimated in behavioural ecology. In the future, more sophisticated techniques should allow better tests of the immunocompetence handicap hypothesis.

### 11.5.4 The sensory exploitation hypothesis

Both the fisherian process and the handicap principle assume that the male trait chronologically precedes the appearance of the female preference. These two processes, however, cannot be easily invoked to account for the existence of heterospecific preferences. Several studies showed that, in various species, individuals express clear preferences for traits that are not naturally expressed in their own species (Ryan and Wagner 1987; Jones and Hunter 1998). These observations led to the development of the sensory exploitation hypothesis that involves intersexual selection.

The **sensory exploitation** hypothesis was introduced during the 1990s (Basolo 1990; Ryan 1990; Ryan and Rand 1990; Ryan *et al.* 1990; Endler and Basolo 1998). It considers that the evolution of male secondary sexual traits is influenced by pre-existing sensory biases in females. For instance, females of a species may have sensory equipment that renders them particularly sensitive to a given colour. A mutation appearing in males leading them to develop a spot of that same colour would facilitate their detection by females or increase their attractiveness. In a way, males would 'exploit' a sensory bias already present in females. In this scenario, female preference precedes the appearance of the male trait, contrary to the runaway or good genes models. The sensory exploitation hypothesis also differs from the other models of sexual selection by considering that traits and preferences are not necessarily coupled (Shaw 1995).

#### 11.5.4.1 *Four criteria to detect a case of sensory exploitation*

Basolo (1990, 1995a, b) proposed four criteria that allow us to test whether a male secondary sexual trait evolved through sensory exploitation.

1. The species has both the trait and the preference, and the trait is used in mate choice.
2. The trait was absent (or in a precursor state) in the ancestral species.
3. The preference for the trait is ancestral.
4. There is a bias in the sensory system or the brain, which allows the precise prediction of the direction of female preference.

The testing of points 2 and 3 requires a phylogenetic approach (see Chapter 3).

#### 11.5.4.2 *Examples of sensory exploitation*

The first study on sensory exploitation mainly involved two taxa: the frogs of the species complex *Physalaemus pustulosus* (Ryan 1997) and swordtails, fish of the genus *Xiphophorus* (Basolo 1990, 1995a, b). In the first case, a trait common to all males of the various frog species of the *Physalaemus* group is a vocal signal described as a 'whine'. In some species, this whine-like call has been enriched by chucks, which seem to stimulate females further. In the species *P. coloradorum*, males do not emit chucks, but when chucks are added artificially to their whine call, females prefer the modified signal over the normal call of males of their species. The most parsimonious reconstruction of this trait on the phylogeny of this group suggests that the preference already existed in the common ancestor of the two species (Ryan 1997).

Swordtails have an elongation of the caudal fin that resembles a sword (hence their name). Females of *X. helleri* prefer males whose caudal fin is lengthened. Other species in the same genus do not exhibit the sword. When the caudal fins of males of the platy species *X. maculatus* and *X. variatus* are artificially lengthened, they are preferred by females of their species over normal males (Basolo 1990, 1995a). Here again, the phylogenetic relationships support the hypothesis of trait evolution through sensory exploitation.

#### 11.5.4.3 *The origin of the sensory bias: an open question*

An important question is that of the origin of the female sensory bias. Few studies have been conducted on this subject. A recent study (Rodd *et al.* 2002) suggests that ecological constraints can shape female preferences and thereby influence the evolution of male traits. Female guppies prefer males with larger, more intense orange spots. In a series of experiments and field studies, Rodd *et al.* (2002) suggested that female preference results from a sensory bias for the orange colour that appeared in the context of food detection. Indeed, Rodd *et al.* (2002) observed that both male and female guppies in natural populations are attracted to and voraciously consume orange-coloured food items, in particular some fruit rich in proteins, sugar, and carotenoids. Two major results of their study support the sensory bias mechanism. First, field and laboratory experiments outside a sexual context showed that male and female guppies are more attracted to orange than identical non-orange objects. This attraction to orange thus appears innate. Second, across population variation in attraction towards orange objects of males and females explained 94% of the variation in female preference when tested with males of varying colour as produced by diets incorporating more or less carotenoids.

A similar result has been obtained in the three-spined stickleback (*Gasterosteus aculeatus*; Smith *et al.* 2004). In most populations of the three-spined stickleback, females show a preference for for males that develop red nuptial colourations on their throat and jaws. However, not all sticklebacks species dipslay nuptial colorations, and the most likely state for male nuptial colouration in the stickleback lineage appears to be black. In laboratory trials, male and female three-spined sticklebacks responded most strongly to red objects outside a mating context. Interestingly the same preference for red objects was observed in male and female nine-spined sticklebacks, a species belonging to a clade that does not exhibit red coloration. Overall, the results suggest that a red feeding bias in the sticklebacks preceded the evolution of the red nuptial coloration. Again, the preference for red food items shown by three-spined and nine-spined sticklebacks seems to be related to a requirement in their diet for carotenoid-rich food items. Given the ubiquity of carotenoid-based secondary sexual traits in the animal kingdom, similar tests could be run in birds or other fish to assess the generality of the phenomenon.

#### 11.5.4.4 *A place for sensory exploitation?*

The importance of sensory exploitation in the evolution of secondary sexual traits is still debated. Some authors consider that female preference for heterospecific traits does not necessarily involve a particular sensory bias, but relates to a general property of learning mechanisms that results in a preference for novelty (Enquist and Arak 1993; Weary *et al.* 1993), or greater apparent size (Rosenthal and Evans 1998). In addition, testing sensory bias hypotheses depends on the reliability of phylogenies. It is possible to consider that, in some cases, the male trait did not evolve after female preference but rather before and was subsequently lost on several occasions. This scenario is not the most parsimonious in several taxa and is thus dismissed. Nevertheless, a male trait may repeatedly disappear through various selective pressures, for instance related to predation, whereas female sensory biases may not regress at the same rate. The importance of the loss of male traits subjected to sexual selection was long ignored (Wiens 2001). Yet, an attentive evaluation of available data reveals that in some clades, the rate of extravagant sexual trait losses can be five times higher than that of gains (Burns 1998). These results suggest that sexual selection can frequently be countered by other selective forces or that its intensity is not necessarily stable over time. The consequences of these observations for our understanding of the evolution of secondary sexual traits remain unexplored.

### 11.5.5 Conclusion: is inter-sexual selection a pluralist process?

Since Darwin and Fisher, the attention paid to inter-sexual selection has greatly increased. Hypotheses were first viewed as contradictory but tend today to be viewed as compatible and it seems that a synthesis could emerge soon. Inter-sexual selection should no longer be seen as a unique process, and it may become commonly accepted that various processes may have acted in concert in the evolution of traits and preferences. Some recent developments testify to this new attitude.

#### 11.5.5.1 *Fisherian process versus good genes hypothesis: towards a reconciliation?*

The abundant literature on the evolution of secondary sexual traits may suggest that the Fisherian runaway process and the good genes hypotheses are irreconcilable alternatives. A positive correlation between male attractiveness and its offspring's fitness is generally interpreted as supporting the good genes hypothesis (Norris 1993; Petrie 1994; Møller and Alatalo 1999; Jennions *et al.* 2001). The opposite relation, plus the existence of some heritability of male attractiveness, is instead presented as support for the Fisherian runaway process (Etges 1996; Wedell and Tregenza 1999; Brooks 2000). However, a female preference for extravagant handicapping male traits can conceivably indirectly benefit the female through the fitness of her offspring. Kokko *et al.* (2002) assessed the theoretical soundness of this argument and suggested that the increase in mating success or in offspring survival are both valid indirect benefits, the relative importance of which varies with the costs of female choice. The indirect benefits of female preference for some male traits thus constitute a continuum. A consequence of this model is that it no longer appears valid to consider only one component of fitness at a time when seeking to determine whether female preference can evolve through indirect benefits. Some questions imply the necessity of accounting for both components simultaneously. Moreover, a negative relation between survival and attractiveness does not dismiss the good genes hypothesis. Kokko and Monaghan (2001; see also Mead and Arnold 2004) conclude that it is no longer useful to keep distinguishing between the Fisherian and good genes hypotheses. Only the future will tell whether this position becomes prevalent.

#### 11.5.5.2 *Can the relative importance of direct and indirect benefits be distinguished?*

Direct benefits are often regarded as most important and most obvious in mate selection. In fact, contrary to indirect benefits, direct benefits are immediate and do not depend on a mechanism of maintenance of genetic variability. Explaining individual choices through direct benefits can thus seem trivial, and this undoubtedly explains why there are few formalizations of the evolution of preferences through direct benefits (however see Grafen 1990a; Price *et al.* 1993; Kirkpatrick 1996).

Is it possible, however, to compare the relative importance of direct and indirect benefits? According to Kirkpatrick and Barton's (1997) model, indirect benefits tend to produce relatively weak selective forces on preferences. This is because the effect is likely to depend on a long causal pathway, from preference to ornament to overall fitness. Any weak link in the pathway would necessarily make the whole pathway weak. In addition, environmental influences on mate choice are likely to be important in the wild. It has been argued that, together with behavioural flexibility (Hunt *et al.* 2005), environmental influences may considerably reduce the potential for indirect selection on mate choice. The best way to evaluate the exact strength of indirect selection is thus to explore the genetics underlying the co-evolution of mate choice and ornaments in natural populations. Qvarnström *et al.* (2006) have recently undertaken such an analysis in the collared flycatcher (*Ficedula albicollis*). They relied on quantitative genetics to analyse data from population monitored for 24 years and comprising 8,500 marked individuals. Significant genetic additive variance was found for male ornament (forehead patch size), female mate choice for the ornament, male fitness, and female fitness. However, after taking into account the genetic correlations between these components, the strength of indirect sexual selection on female mate choice was found to be negligible. Although this study suggests that genes coding for preference for an ornament may evolve independently of genes coding for the ornament, the generality of the result remains to be assessed in other natural populations.

The issue is by no means settled as a recent literature review suggests that the impact of direct benefits on fitness is not necessarily higher, and may even be lower than that of indirect benefits (Møller and Jennions 2001). However, there is no incompatibility between direct and indirect benefits, and both may be acting simultaneously. Many studies established that females do not necessarily base mate choice on a single trait, and choose on the basis of various traits or cues (Iwasa and Pomiankowski 1994; Johnstone

1996). Such cues may indicate direct or indirect benefits. Candolin and Reynolds (2001) studied female mate choice in the European bitterling, *Rhodeus sericeus*, a cyprinid freshwater fish. In this species, females deposit their eggs inside the cavity of a mussel where development occurs. Females can lay several clutches and several females can spawn in the same mussel. Egg survival varies with mussel species and with the number of eggs already present. Males court females by exhibiting their red colouration to induce egg laying in the mussel that they vigorously defend. A females' initial decision to approach a male is related to his behaviour and red colouration but the final decision to spawn is related to the quality of the spawning site. Male colouration likely reflects his genetic quality or body condition, but highly coloured males sometimes induce females to spawn in mussels of lesser quality. Females, however, are particularly demanding about the quality of spawning sites, which they carefully inspect before laying. The female choice in this species thus depends on various criteria used sequentially, but a greater importance seems to be given to cues related to direct benefits in terms of egg survival. Finally, a cue can inform females equally about the potential direct and indirect benefits. For instance, carotenoid-based colourations can signal both male foraging efficiency and its immune system quality (see Section 11.5.3.3).

#### 11.5.5.3 *Good genes or genes that fit together?*

Until recently, only the intrinsic genetic quality of males was considered when discussing the importance of indirect benefits for the evolution of female mate choice. However, genetic compatibility between mates might be just as important. In sexually reproducing organisms, offspring fitness will largely depend on the quality of the match between female and male genotypes, as particular combinations of maternal and parental alleles may be most efficient in masking deleterious alleles. Indeed, much evidence exists to support a link between overall heterozygosity and fitness (Amos *et al.* 2001; Hansson and Westerberg 2002), suggesting that females may prefer to mate with males that are optimally genetically dissimilar to themselves (Mays and Hill 2004). A first line of evidence in favour of female choice based on genetic compatibilty comes from inbreeding avoidance. In various species, females refuse to engage in mating with genetically similar mates, thus avoiding the cost of inbreeding depression (Simmons 1991; O'Riain *et al.* 2000; Stow and Sunnucks 2004). Still, active inbreeding avoidance through mate choice is not observed in all species (see Cohen and Dearborn 2004; Lampert *et al.* 2006; Viken *et al.* 2006), even in the case of strong inbreeding depression in the population (Hansson *et al.* 2007). In addition, inbreeding depression is difficult to measure in nature, and may have to exceed a substantial threshold before any avoidance behaviour can evolve (Kokko and Ots 2006). Another line of evidence for female choice based on genetic dissimilarity comes, however, from patterns of extra-pair paternity. In several monogamous species (see chapter 12), paired females engage in copulations with extra-pair males (Birkhead and Møller 1992). Because the interaction between females and extra-pair partners is most often limited to copulation, females are supposed to obtain only indirect benefits through extra-pair copulations. Although some evidence has been provided for female choice of extra-pair partners based on directional preference for more ornamented males (supporting the 'good genes' hypothesis; Kempenaers *et al.* 1992; Møller 1994), an increasing number of studies point to a role of genetic dissimilarity (Blomqvist *et al.* 2002; Foerster *et al.* 2003; Freeman-Gallant *et al.* 2003). For instance, Tarvin *et al.* (2005) found that levels of extra-pair paternity in broods of the splendid fairy wren (*Malurus splendens*), increased with genetic similarity between social mates, suggesting that females benefit from extra-pair mating because it leads to lower levels of inbreeding and increased heterozygosity in their offspring.

Evidence thus suggests that both mating with males with good genes and mating with males that are optimally genetically dissimilar to females can provide indirect benefits to females. However, for most females the most ornamented male may not be the most genetically compatible, whereas the most compatible male may not be the one who carries the best genes. This apparent conflict between female preference based on the degree of development of male ornaments and female preference based on genetic dissimilarity is the subject of increasing attention (Colegrave *et al.* 2002; Mays & Hill 2004; Neff and Pitcher 2005; Roberts *et al.* 2006). One possbility is that females may use different criteria for different sort of males. Alternatively, female choice may consist of a trade-off between acquiring good genes for their offspring and maximizing offspring heterozygosity. Variation in both abolute and relative male quality in the environment may then play an important role in shaping mating patterns (see Roberts and Gosling 2003). Experimental work, manipulating opportunities to mate with high-quality males versus compatible males is of crucial importance to assess to what extent females can prioritize between absolute and relative quality of potential mates.

## 11.6 Sperm competition and cryptic female choice

Between mating and the production of viable offspring, a whole series of processes occur upon which selection can act in many ways. First of all, sperm compete among themselves irrespective of their origin. In addition, the conditions met inside the female, or later during parental care, can continue to influence the outcome of reproduction.

Two main processes are classically defined: sperm competition (expression coined by Parker 1970), which implies competition between the sperm of two males, and cryptic female choice (expression coined by Thornhill (1983, 1984)) which implies direct or indirect actions on the part of females. The sole topic of sperm competition would justify a complete book. We will thus present only some general aspects, without seeking exhaustiveness. The arguments in favour of cryptic female choice are much fewer. We will limit ourselves to enumerating the various mechanisms that can lead to such an active or passive female choice.

### 11.6.1 Sperm competition

#### 11.6.1.1 *Definition*

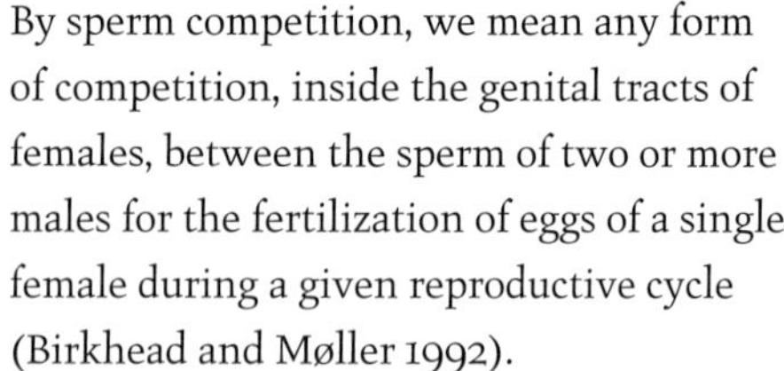

By sperm competition, we mean any form of competition, inside the genital tracts of females, between the sperm of two or more males for the fertilization of eggs of a single female during a given reproductive cycle (Birkhead and Møller 1992).

Sperm competition is a common process in animals (Smith 1984b) occurring when a given female copulates with more than one male during a single reproductive event. In the narrow sense, sperm competition involves only the physiological processes occurring inside the female's genital organs after multiple matings. In the broad sense, sperm competition involves a large range of morphological, behavioural, and physiological attributes including sperm size, number, and structure, morphology of male and female reproductive apparatus, sperm storage processes and structures, courtship, and copulation behaviour. We will use the expression here only to refer to the processes occurring during or after mating, excluding processes related to courtship that are covered elsewhere in this chapter, as well as in Chapter 12.

Marler (1956) provided the first description of a typical process of sperm competition in chaffinches (*Fringilla coelebs*). At that time, only the physiological aspects of such behaviour were studied, whereas today the stress is put on its adaptive function.

#### 11.6.1.2 *Which type of competition?*

Part of the debate on sperm competition relates to the type of competition involved. Is it competition by interference or scramble competition? Within the sperm competition framework, scramble competition would imply that sperm do not interact directly to reach the resource (eggs of the female). The winner is simply the one arriving first. The process can then be compared to a lottery or a race. In both cases, all else being equal, a male will have more chances of fertilizing eggs the more he contributes sperm to the female's tract, which is equivalent to getting more lottery tickets or having more runners racing in his team. Resources being limited for males, the only way to increase the number of sperm is to decrease their size. However, there are strong variations in sperm size both between and within species, and within individuals, thus raising the question of why such a polymorphism exists in the first place.

One possible answer lies in the hypothesis that interference sperm competition exists. This type of competition requires that the sperm from one male's ejaculation interfere with the sperm from other males' ejaculations, preventing any of them from reaching the unfertilized egg(s) first. For this type of intra-ejaculatory cooperation to emerge, a certain number of conditions must be met: (1) the egg resources must be defendable, (2) the sperm of different males must interact and be able to recognize each other as competitors, and (3) competition must result in death of the opponents. When these conditions are met it is possible for sperm within one ejaculate to altruistically die in combating sperm from another ejaculate in order to allow sperm from its own ejaculate to reach the egg. The condition that requires death of the opponent amounts to saying that this behaviour must be regulated at the level of the diploid organism not the haploid sperm. The competition will be by scramble if none of these conditions are met.

Most authors consider that sperm competition is of the scramble type rather than by interference. Scramble competition, however, cannot explain the recurring polymorphism of sperm observed in many species. In fact, Baker and Bellis (1988, 1995) argue that this sperm polymorphism may well be the outcome of interference sperm competition with some types of sperm acting as 'kamikaze' sperm that locate and destroy the sperm of other males. However, the argument, developed mainly for humans by Baker and Bellis, rests mostly on a series of observations that can be interpreted without reference to interference competition (Gomendio *et al.* 1998). For instance, many sperm are malformed and their malformations may hinder their movements. Baker and Bellis proposed that such sperm are kamikaze sperm. However, because of their malformation, they tend to stick together, forming a kind of copulatory plug that blocks the female genital tract, and thus seriously slows down the sperm of (later copulating) competitor males. The evolution of this type of sperm is still interference competition but it avoids the need to invoke altruistic 'seek and destroy' behaviour required for kamikaze-type sperm. Thus the debate remains open.

#### 11.6.1.3 *Some examples of male adaptations*

Males have everything to gain from adaptations that increase the competitive success of their sperm, which increases the number of eggs they can fertilize. They can also benefit from adaptations decreasing the rate of female remating, even if these benefits are obtained at the expense of female fitness. Many such conflicts involve specific morphological structures and molecules transmitted to females with the seminal fluids.

##### a *Copulatory organs shaped by sperm competition*

Caleopterygid damselflies constitute an excellent illustration of sperm competition. Females mate regularly with several males successively, and males have developed a whole suite of morphological and behavioural adaptations increasing their chances of fertilizing the eggs of females with which they copulate. In damselfly females, sperm is stored in two specialized organs, the spherical bursa copulatrix and the tubular spermathecae. Copulation typically takes place in two steps. In several species, the male starts by removing sperm from previous males, before transferring its own sperm. To that end, the male's penis carries stiff hairs and other structures that remove sperm from previous males (Waage 1979; Hayashi and Tsuchiya 2005). This is not, however, the case of all species.

Recently, Alex Córdoba-Aguilar (1999, 2005) suggested that in the damselfly *Calopteryx haemorrhoidalis*, males exploit female sensory mechanisms normally associated with fertilization to make females eject most already stored sperm. At egg laying, the egg stimulates two lateral sclerotized vaginal plates, each bearing campaniform sensilla. The passage of the egg thus triggers a series of reflexes leading to sperm ejection into the oviduct, thereby ensuring egg fertilization. Sperm is ejected into the genital tract in the absence of egg is lost for further fertilization. For mechanical reasons, in *C. haemorrhoidalis*, the removal of sperm stored in the spermathecae is unlikely to result from the penetration of the aedeagus, the male intromitent organ. Córdoba-Aguilar (1999) proposes that during copulation, the aedeagus stimulates the vaginal plates through frictional movements mimicking the stimulation resulting from an egg during laying, and thereby triggers the ejection of sperm from previous matings. In this species, copulation typically involves up to 80 rhythmical abdominal flexions, each one stimulating a reflex of sperm ejection in the female. Córdoba-Aguilar (1999, 2005) provides a series of arguments in favour of this mechanism. Indeed, during copulation, the quantity of sperm stored in the spermathecae and the bursa copulatrix decreases until around the fiftieth abdominal flexion. At this time, only very few sperm from former matings remain. Then, the quantity increases again as sperm resulting from the current mating are stored. In addition, a negative relation was observed between the quantity of sperm remaining in the spermathecae and penis width (Figure 11.9), suggesting that penis size is correlated to sperm ejection efficacy, as a wider penis is likely to stimulate the vaginal plates more than a small one. Córdoba-Aguilar (1999) further showed that there is a positive relationship between asymmetries in sensilla numbers on the left and right vaginal plates and asymmetries in sperm ejection: when there are more sensilla on the left, the left spermathecae contains less sperm after the fiftieth abdominal flexion, and vice versa.

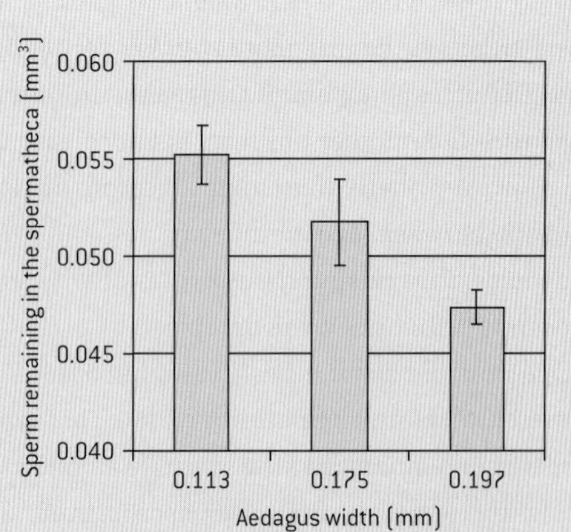


Figure 11.9 **Width of the aedeagus borne by the penis and sperm ejection in the odonate *Calopteryx maculata***

Copulations were chosen by the observer. Copulations were interrupted just after the fiftieth abdominal flexion, i.e. when the spermathecae is almost emptied from previous males' sperms. At that time, the wider the penis, the less sperm remained in the female's spermathecae. This supports the hypothesis that the aedeagus stimulates females to eject sperm already present in her spermathecae, by exploiting the females' reflex associated with egg fertilization.

After Córdoba-Aguilar (1999).

Note that, although the interpretation proposed by Córdoba-Aguilar (1999) may be correct, this result can also be interpreted differently by saying that females use their sensory abilities to select the best males (those having developed the widest penises and the most efficient *aedeagus* to stimulate them), thereby favouring their sperm (Pitnick and Brown 2000). If this male ability is heritable, then females favouring these males will have a better fitness through their sons. These two interpretations lead almost to the same predictions, despite the fact that the evolutionary pathways involved might strongly differ.

### b *Sperm competition and postmating female mortality*

Female behaviour can be strongly modified by matings. In particular, in some species, females with an initial high mating rate become less attractive and/or less receptive to males though time. In addition, such females have a higher laying rate but die earlier than females with a lower mating rate (Wolfner 1997). A series of experiments in fruit flies (*Drosophila melanogaster*) has shown that this increased mortality is directly due to mating and not to the consequences of mating (Chapman *et al.* 1993, 1995). The experimental protocol consisted in keeping female clutch size, female exposure to males outside mating, and mating rate constant while varying the exposure of females to seminal fluids. Chapman *et al.* (1995) found that seminal fluid proteins, produced by male accessory glands (not sperm themselves) were responsible for increased female mortality and suggested that the primary function of accessory gland toxic proteins is to outcompete the sperm of other mates (**sperm offence ability hypothesis**; Harshman and Prout 1994). However, some empirical evidence also supports a role of the accessory gland in enhancing sperm resistance to displacement by sperm of subsequent males (Clark *et al.* 1995; Civetta and Clark 2000). In addition, accessory gland proteins trigger a series of postmating responses in the female, including increased egg laying rate and lower remating propensity, but, as a pleiotropic effect, they also increase female death rate (Eberhard 1996; Wolfner 1997; Civetta and Clark 2000). The cost in terms of survival for the female thus appears to be a side effect of competition between males.

### 11.6.2 Possibilities of female cryptic choice

After one or more matings, females have many opportunities to choose the sperm that will fertilize their egg(s) (Wedekind 1994; Birkhead 1998). These processes are called cryptic because they take place inside the female's body and cannot be observed directly. Eberhard (1996) identified at least 20 different ways by which females can make such choices.

The existence of such cryptic female choice is shown by the fact that inter-specific matings usually lead to a reduction in female fecundity. We will return to this point in Section 11.9. However, similar mechanisms occur between the sperm of conspecifics, and this is the subject of this section.

There are two categories of mechanism of cryptic female choice: those occurring between mating and fertilization, and those acting after fertilization (Birkhead 1998). Cryptic female choice has been much less documented than sperm competition because (1) it was often regarded as less powerful than sperm competition, and (2) its study requires an understanding of the mechanisms associated with insemination, sperm storage, and fertilization, areas that are in general poorly known by behavioural ecologists. Consequently, although female cryptic choices may play an important role in the competition between sperm of various, as well as a single male, they are still often ignored in the context of sexual selection. Moreover, these mechanisms may also be involved in the choice of an offspring's sex, a problem whose evolutionary importance will be seen in Chapter 13. It is thus important at least to list the various means by which females can cryptically choose the genotype of their offspring.

Cryptic female choice has evolutionary implications only if it provides a fitness benefit to the females. One of the particularly well studied aspects relates to the major histocompatibility complex (MHC) in mammals (Wedekind 1994). These are groups of genes having a strong polymorphism both within populations and individuals. Among other things, they act on immune defences and thus on resistance to parasites. Females of viviparous species have the most possibilities of cryptic choice. We thus use the case of mammals to illustrate the various mechanisms of female cryptic choice. In ovoviviparous animals, some of these mechanisms cannot, by definition, exist.

#### 11.6.2.1 *Pre-fertilization cryptic choice*

##### a *Ejecting sperm, just after mating*

A female can first of all choose by ejecting all or part of the sperm from a given male immediately after mating and hence before fertilization. When this happens, the male's sperm is no longer involved in sperm competition. The existence of this behaviour offers all the possibilities of choice if females systematically discriminate against the sperm of some male types.

This behaviour has been described so far only in a few species in which females mate with more than one male: domestic chickens (*Gallus domesticus*; Pizzari and Birkhead 2000), dunnocks (*Prunella modularis*; Davies 1983), and some mammals and insects (review in Eberhard 1996). However, recently the behaviour has also been described in the kittiwake (*Rissa tridactyla*; Wagner *et al.* 2004), a species that is **genetically monogamous** (see Chapter 12). The last results suggest that in this species sperm ejection is related more to sperm viability than to issues of sperm competition. Because it is hard to observe, this behaviour is probably more widespread than it would appear on the basis of existing empirical arguments.

#### b *Sperm choice in the female genital tract*

A female having received sperm can also choose among the sperm inside her genital tract. The female reproductive tract constitutes a very hostile environment for sperm. Physiochemical and immunological factors in the vagina and the cervix of mammals influence sperm transport and survival. In many mammals, most sperm of an ejaculate do not even pass the cervix. In mammals at least, to reach the uterus, sperm must survive strong selection by physical and chemical barriers, phagocytosis by leucocytes, and a strong concentration of spermicidal antibodies that surround most sperm. Consequently maternal secretions favouring some haplotypes and produced by various means may be effective from the vagina to the oviduct. This supposes that sperm signal their haplotype on their surface.

Various mechanisms allowing such signalling have been highlighted in mammals, one of which involves the MHC genotype, which is detectable on the sperm membrane. A very strong influence of female genotype on sperm transport has been shown in two mouse strains (review in Wedekind 1994), suggesting that females can favour some sperm genotypes according to their own MHC genotype. Given the positive impact of MHC heterozygosity on parasite resistance, high offspring MHC heterozygosity is expected in nature.

#### c *Choice by the egg of a given sperm*

Once a sperm has reached an ovum other choices occur (review in Wedekind 1994). Ovocytes are normally surrounded by an envelope called the pellucid zone in mammals or the vitelline envelope in amphibians, reptiles, and many invertebrates. These two types of envelope seem to contain specific receivers that bind to sperm thus providing the basis for intense selection to occur.

The pellucid zone of mammals is important in the initial phases of fertilization by preventing polyspermy, and, in some species, by capacitating sperm, thus enabling the sperm to fertilize the egg (Wedekind 1994). Moreover, the sperm must bind to the pellucid zone and penetrate it before merging with the ovocyte membrane. Several antigens influence the interaction between the sperm and the egg. Although these mechanisms are still not well understood, there is clearly a possibility for female choice at this level.

Gametes of many unicellular organisms (algae, yeasts, protozoa) seem to choose their mate on the basis of pheromones. In more complex organisms, however, ovocytes may even be able to choose a given haplotype. Some colonial ascidians (*Botryllus* sp.) are prone to natural tissue transplantations. The ability of *Botryllus* sp. to fuse together is controlled by a polymorphic locus analogous to the vertebrate MHC (references in Wedekind 1994). It has been shown that this locus does not control only allo-recognition but also the fusion of gametes: the eggs of *Botryllus* resist fertilization by sperm of the same colony for longer than by sperm having foreign alleles at this locus. This suggests that ovocytes can choose sperm that are heterozygous at this very locus.

### 11.6.2.2 *Choice at the time of fertilization: end of meiosis influenced by the sperm haplotype*

Ovocytes of most vertebrates and invertebrates have not completed their meiosis at the time of fertilization (review in Wedekind 1994). In mammals, the second division of maturation is completed only when the sperm create a path through the pellucid zone and penetrate the vitelline membrane of the ovocyte. It is only then that the first or second polar body is ejected into the vitelline membrane space and the other preserved. The meaning of this pause in meiosis is unclear, but one supposition is that it affords females the possibility to continue choosing among the various options that arise after a mating. The choice by the ovocyte of the haplotype that will indeed become the zygote and of those that will be lost at the time of meiosis completion could very well be influenced by the haplotype of the sperm that succeeded in entering the ovocyte. This choice could have an impact on the fitness of the future individual, in particular through its heterozygosity.

#### 11.6.2.3 *Post-fertilization cryptic choice*

##### a *Cleavage of the zygote and its implantation in humans*

Females can benefit from selecting among embryos after fertilization (review in Wedekind 1994). However, this is expected to occur soon after fertilization, before nourishing the embryo, to minimize costs. Zygote formation also marks the beginning of a conflict between the zygote and its mother, and the zygote can develop mechanisms of self-protection. The oviduct epithelium is in close contact with the embryo, and seems to produce secretions that support the metabolism of the young embryo, whose descent is strongly under the control of the oviduct. Premature arrival of the embryo in the uterus often leads to its degeneration. After the blastocyte enters the uterus, it must implant tissues in the uterus to establish the necessary contacts with its mother during gestation. Many embryos are lost at these various stages. It is not clear whether one of the reasons for these losses resides in a choice by the mother of the genotype of her offspring, but the possibility exists. Moreover, embryos remain wrapped in the pellucid zone until entering the uterus, which may protect it against its mother. Indeed, even though the MHC genotype of the sperm can be identified on the membrane of the embryo as early as the stage of eight cells, none of the concerned molecules are detectable on the pellucid zone.

##### b *Embryonic growth, spontaneous abortion, and resorption*

Maternal selection can still occur later during embryonic growth, with the effect of choosing offspring having particular genotypes at loci that, for instance, confer advantages in terms of parasite resistance, as is the case for genes coding for the MHC. In humans, from 10% to 25% of identified pregnancies end in miscarriage. Some of these miscarriages seem to be due to immunological factors. Couples suffering from repeated miscarriages share a greater proportion of fragments of MHC on average than control couples (review in Wedekind 1994). Moreover, the babies born to couples having similar MHC are lighter at birth. However, it should not be forgotten that MHC heterozygosity is probably correlated with heterozygosity at other loci, which may provide a general advantage to offspring other than in the sole context of resistance against pathogens.

##### c *Selective infanticide*

Mothers can continue to select their offspring during the whole rearing phase. Bad parental care or cannibalism is well documented in rodents for instance. Maternal cannibalism can have selective value by killing the least vigorous offspring, the female prevents additional investment in an offspring that would not be viable anyway. More indirectly, females can also allocate their resources selectively towards some of their offspring. When the environment is unfavourable, mothers often kill and eat the youngest newborn of a litter, thereby minimizing the investment and allowing the most vigorous offspring to obtain more resources (see examples in Section 19.3.2.3).

It thus appears that mothers have countless possibilities of keeping on choosing among sperm, embryos and then offspring, well after mating. This is one of the forms that conflicts between males and females can take. Such cryptic choice allows females to react in real time to the constant variations of environmental pressures. After all, one of the major benefits of sexual reproduction is to allow organisms to react quickly to the never-ending environmental variation by recombining the various alleles present in the population. We will see in Section 11.9 that these post-mating choice processes, including sperm competition, play a greater role than previously thought in speciation.

### 11.6.3 Link between sperm competition and cryptic female choice

It is clear that during sperm transfer within the female, sperm competition is strongly influenced by the conditions met in the genital tract. Females can influence these conditions, actively or passively, thereby exerting a cryptic choice. Female behaviour and physiology are shaped by selection to reduce the fitness costs resulting from male adaptations for sperm competition, as well as to benefit from having their eggs fertilized by a given male. These two mechanisms, sperm competition and pre-fertilization cryptic female choice, are thus strongly linked, and the mere observation of the final result (for instance, proportion of offspring sired by a given male) does not really allow one to infer which mechanisms produced this pattern. It is thus difficult to distinguish between mechanisms that relate to sperm competition per se from cryptic choice by the female (Birkhead 1998, 2000; Kempenaers *et al.* 2000; Pitnick and Brown 2000). Indeed, these two processes always occur simultaneously, with the conditions inside the female tract constituting the environmental context in which sperm competition occurs (Eberhard 2000). Only fine protocols, carefully thought out to fit specific criteria will make it possible to study the impact of cryptic female choice. Therefore the evidence for the existence of cryptic female choice remains slight (Birkhead 1998, 2000; Kempenaers *et al.* 2000; Pitnick and Brown 2000).

There is also debate as to whether cryptic female choice concerns sexual or natural selection (Birkhead 1998; Eberhard 2000). Indeed, the few results showing the existence of cryptic choice did not show that the choice favoured the sperm of attractive males, but that the selection criterion favoured particular compatible genotypes that prevented genetic proximity or other types of genetic incompatibility. The question thus remains open.

## 11.7 Sexual conflict: causes and consequences

The strong asymmetry between the evolutionary interests of males and females regarding the mode and frequency of mating (Trivers 1972) inevitably results in conflicts between mates. The importance of inter-sexual conflict for the evolution of male and female traits has been evoked several times from a theoretical point of view (Parker 1979, 1983b), but it is only recently that empirical studies have demonstrated it, as in the case of toxic sperm (Chapman *et al.* 1995). A reduction in the quantity of toxic substances in the seminal liquid, or in their harmfulness, would be advantageous to females, but would undermine male competitive ability. Multiple matings can thus be extremely dangerous for females, whereas males seek to mate as much as possible. The same asymmetry exists for the other elements of mating behaviour that imply a cost for one or both sexes. The receptivity threshold of females determines, for instance, the courtship effort necessary for males. A lower receptivity threshold in females would directly benefit males by enabling them to save time and energy in courtship. The optimal value of reproductive traits thus differs between males and females and this difference may be greater in highly promiscuous species (Figure 11.10).

### 11.7.1 Empirical arguments

Direct evidence for inter-sexual conflict has recently been obtained in dipterans (Rice 1996; Holland and Rice 1999; Hosken *et al.* 2001; Pitnick *et al.* 2001a, b).

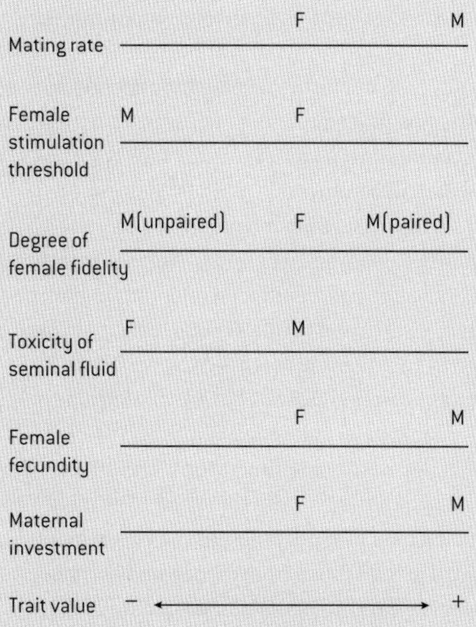

Mating rate — F — M

Female stimulation threshold — M — F

Degree of female fidelity — M(unpaired) — F — M(paired)

Toxicity of seminal fluid — F — M

Female fecundity — F — M

Maternal investment — F — M

Trait value − ⟵⟶ +

Figure 11.10 **Conflicts between males and females in non-monogamous species**

The optimal value of a trait is not necessarily the same for males and females. For each case, what matters is the relative position of the optimum for males and females. The horizontal distance quantifies the intensity of the conflict of interest: on the left the trait is poorly developed, on the right it is strongly developed. For instance, in the case of female stimulation threshold (i.e. the level of solicitation required before they become receptive), males would benefit from a low threshold, whereas females would benefit from being selective, and thus having a higher threshold. Conversely, females would benefit from low or even no sperm toxicity, whereas males benefit from a relatively high toxicity resulting in destruction of rival sperm. The same kind of diagram can be drawn for monogamous species.

After Holland and Rice (1998).

These experiments used the same methodology. In usually polygynous species, lines of males and females have been maintained for several generations in a strictly monogamous system (each female and each male mating only once), with random pairing of the individuals. In such an artificially imposed mating system, male and female evolutionary interests are exactly the same, and there is no more conflict over who fertilize the eggs. Thus, any trait present in a sex that decreases the reproductive success of the opposite sex decreases by as much as the success of the sex having the trait. Strict monogamy must thus lead to the reduction of inter-sexual conflict over fertilization. Holland and Rice (1999) observed that after 47 generations of forced monogamy, fruit fly males became less harmful for females and females less resistant to toxic sperm substances. Females maintained in a polygamous mating system (control line) laid more eggs when mated with males of the monogamous lines than when mated with males of the control lines. In addition, the survival rate of females from the monogamous lines strongly decreased when mated with males of the polygamous line. Lastly, in males that were held with females of their own line, those from the monogamous line courted less frequently than those from the control lines. In fruit flies still, Pitnick *et al.* (2001b) observed that, within monogamous lines, male body and testes size as well as sperm production strongly decreased. Despite this, females paired with males from monogamous lines had more offspring of greater viability than females paired with males from control lines. These results indicate that sexual selection favours on the one hand the production of a greater quantity of sperm in males and on the other hand favours male traits that may generate a direct cost on female fecundity.

Inter-sexual conflicts are not limited to fruit flies. Hosken *et al.* (2001) highlighted the importance of such conflicts in relation to sperm competition in the coprophagous fly *Scatophaga stercoraria*. These authors experimentally studied the effects of sperm competition by selecting polyandrous (each female being mated with several males) or monogamous lines over ten generations. Males from polyandrous lines evolved larger testes, whereas females evolved larger accessory sex glands that produce spermicidal secretions. Larger glands could thus increase female ability to influence paternity. In fact, the second male's success was reduced when mating with females from the polyandrous line (Figure 11.11).

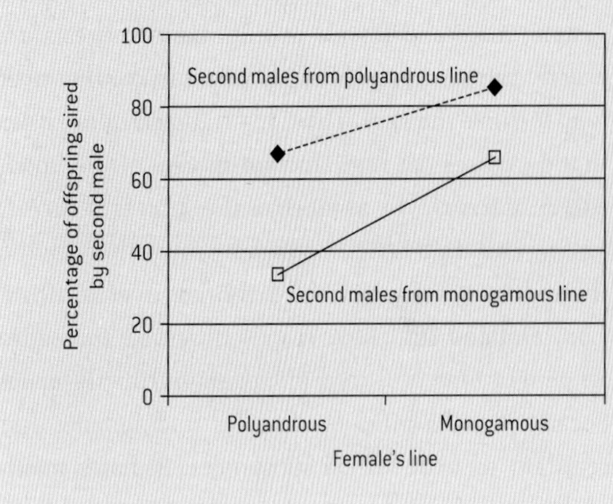


Figure 11.11 **Sexual conflict in the coprophagous fly *Scatophaga stercoraria***

After selection over ten generations, with females that mated with several males (polyandrous line) and females that mated with only one male (monogamous line), the ability of males to fertilize eggs and that of females to choose among males' sperm were tested by looking at the percentage of offspring sired by the male copulating second. In all cases, the first copulating male was from the monogamous line. All the possible cases were tested: females from monogamous lines mated with a second male from monogamous or polyandrous lines, and vice versa. The results are expressed as the percentage of offspring (mean ± standard error) sired by the second male that was from either polyandrous (dotted line and filled diamonds) or monogamous lines (solid line and open squares). For statistical analyses, the percentages were transformed. Both the effects of male and female lines were significant. The male effect indicates that polyandrous males are more competitive than those from monogamous lines; the female effect indicates that polyandrous females exert a stronger cryptic choice than females from monogamous lines.

After Hosken *et al.* 2001.

On the other hand, males from polyandrous lines achieve higher paternity when competing with males from monogamous lines.

### 11.7.2 Inter-sexual conflict and chase-away sexual selection

Holland and Rice (1998) proposed a new model to account for the evolution of extravagant secondary sexual traits that involves inter-sexual conflicts, sensory exploitation, and female resistance. The model considers that the evolution of sexual conflicts brings a relative balance within which courtship level and the development of male secondary sexual traits are to some extent 'adjusted' to the level of resistance of females. However, the possibility for males of exploiting a sensory bias in females can disrupt this balance. After the appearance of a new male trait matching a female sensory bias, females become more vulnerable to males' courtship and will copulate at a rate that is suboptimal for them (too often, or for too long). This new situation generates a strong selection pressure resulting in females evolving a resistance to (in the form of a lower attraction), rather than a preference for the trait. In turn, a higher receptivity threshold in females induces a selection pressure favouring a greater amplification of the male trait. This process can be maintained over many iterations. Males and females are then engaged in **chase-away sexual selection** (Figure 11.12).

Empirical arguments showing the existence of sexual conflicts, of sensory biases in females, and of a resistance of females to certain male traits have accumulated over the years (Andersson 1994; Eberhard 1996; Holland and Rice 1998). However, it is still difficult to assess the relevance of the chase-away sexual selection hypothesis. Pitnick *et al.* (2001b) tested the hypothesis in *Drosophila melanogaster*. The selection of monogamous and polyandrous lines over 84 generations allowed comparison of the levels of divergence between lines in the time interval separating two consecutive matings in females, and in the relative effects of seminal fluids and male courtship on female receptivity. No evidence of an antagonistic co-evolution between courtship signals and female

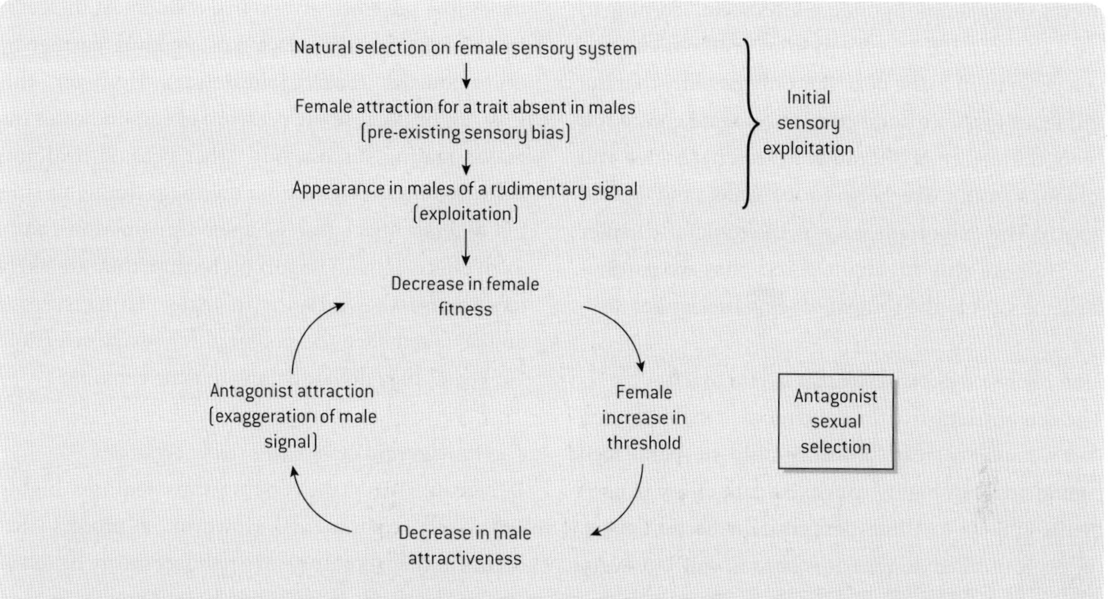


Figure 11.12 **Evolution of extravagant male traits through the chase-away sexual selection model**

Natural selection acts on the sensory system of females to make them more sensitive to certain features of their environment (for instance, in a foraging context). This creates a sensory bias in females for a trait independently from its presence in males. Hence, if this trait appears in males, such male mutants will be preferred by females. This is the sensory exploitation mechanism (see Chapter 16). However, chances are faint that this trait is related to male genetic quality. Thus, by choosing males having the trait in question females do not choose the best males, and their fitness decreases. This selects for females that ignore this trait (these females are said to be resistant to the trait). This decreases males' attractiveness, but female choice will continue to exaggerate the trait in question. Once such a process of sexually antagonistic selection is initiated, it has no reason to stop as long as the conditions do not change significantly, for instance when the trait becomes so exaggerated that males are counter-selected.

After Holland and Rice 1998.

receptivity threshold could be found. In contrast, results supported the hypothesis of a reduction in harmfulness of the seminal fluid from males of monogamous lines. Further studies are required to determine whether chase-away sexual selection can account for the co-evolution of sexual signals and preferences.

## 11.8 Sociocultural influences on the process of sexual selection

So far, mate choice was considered in this chapter as an individual process, independent of the social context. In particular, variations in female preferences were considered as resulting solely from allelic variations at specific loci coding for sexual preferences. However, many species are social, so that individuals can often observe and even copy the choices made by conspecifics. We have seen in Chapter 4 that such social effects can be particularly influential in the development of female sexual preferences, thus challenging the view that sexual preferences are solely transmitted by genes. Such social influences are called **mate choice copying**, or **eavesdropping** (see Section 4.3.3).

Conforming to decisions made by other individuals can be adaptive if assessing the quality of potential mates involves a cost and/or if some females are less efficient than others in this task. Moreover, the overlap of generations often allows social learning to play a role in the transmission of preferences from one generation to the next. It is only recently that the influence of the cultural context on mate choice has been the object of thorough research (Avital and Jablonka 2000; Freeberg 2000; Galef and White 2000). This important issue is the topic of Chapter 20. Here we only highlight points that are particularly relevant to the context of sexual selection.

*Mate copying: a phenomenon that is hard to test . . .*

The demonstration of the occurrence of mate copying (see Chapter 4) can probably only be performed convincingly in highly controlled laboratory experiments similar to those described in Section 4.3.3. In less controlled designs, it is hard to rule out every alternative explanation. For instance, a certain degree of mutual attraction between females can exist by which the presence of a female on the territory of a male can suffice to attract other females, thus increasing that male's mating success. Such a phenomenon has been observed in fallow deer (*Dama dama*; Clutton-Brock and McComb 1993).

Alternatively, in mate copying experiments, females may not be attracted to the males that were with a female because they remember that they were with a female but rather because males that were with a female do not behave in the same way. This was controlled for in several experiments. For instance, White and Galef (1999) replicated the classic mate copying experiment with a slightly modified design in which the focal female could see the interacting male but not the model female. In such circumstances, the demonstration had no effect on focal female preference. White and Galef (1999) thus showed that it is the fact of observing the model female interacting with the target male that modified the preference of the focal female. The mere sight of the courting male did not alter the preference of focal females.

Furthermore, in natural situations, the first female may leave some attractive cues in the male's territory. For instance, in fish with external fertilization in which parental care is exclusively provided by males, the presence of eggs in a male's nest increases its chances of being selected as a mating partner by other females (Pruett-Jones 1992). However, this may not be because the presence of eggs indicates the male's mating success, but rather because it diminishes predation risks for her own eggs through a simple **dilution** effect (see Chapter 14; Jamieson 1995). Such effects remain important because they are likely to increase the variance in male mating success, and thus enhance the potential for sexual selection. However, they do not belong to mate copying.

*. . . However field experiments can be performed*

Demonstrating mate copying thus requires highly controlled experiments. However, a process that occurs in the laboratory may not necessarily be relevant in the field. In fact, most early studies undertaken in the field that claimed to have shown the existence of mate copying (Sikkel 1989; Gibson *et al.* 1991; Hoglund and Alatalo 1995) could not really rule out alternative hypotheses. However, they led to a series of laboratory experiments that proved more conclusive. More recently, convincing field experimental demonstration of mate copying was provided by Witte and Ryan (2002) in the sailfin molly (*Poecilia latipinna*), by transposed laboratory designs into field conditions. They concluded that both males and females mate copy in the field, suggesting that, at least in that species, mate copying is not a laboratory artefact. Similarly, convincing field evidence for eavesdropping was proposed by Mennill *et al.* (2002) in the black-capped chickadee (*Poecile atricapilla*; Figures 4.14 and 4.15). These experiments suggest that mate copying and eavesdropping are not simply artefacts of the laboratory conditions.

*Mate copying and eavesdropping are probably taxonomically widespread processes . . .*

Today, experimental evidence of mate copying and eavesdropping (see Chapter 4) have been found mainly in fish (see Dugatkin 1992, 1996a, 1998; Dugatkin and Godin 1992, 1993; Doutrelant and McGregor 2000; Doutrelant *et al.* 2001; Witte and Noltemeier 2002; Brown and Laland 2003; Godin *et al.* 2005) as well as in polygynous (Galef and White 1998, 2000; White and Galef 1999) and monogamous birds (Swaddle *et al.* 2005), despite the fact that some authors failed to find mate copying in some species (Brooks 1998). Evidence also exists in crustaceans (see Shuster and Wade 1991).

*... that have been claimed to demonstrate cultural transmission in animals*

Several authors reporting on mate copying claimed that this was evidence for **culture** in animals (Dugatkin 1996b). Some theoretical models suggested that this could be the case (Kirkpatrick and Dugatkin 1994; Laland 1994). However, in Chapter 20 we provide four criteria that need to be fulfilled simultaneously to be able to talk of a cultural trait in an evolutionary context. So far, none of the published evidence has fulfilled these four criteria simultaneously, so that for the moment we feel it is not really possible to claim that mate copying convincingly demonstrates the existence of animal culture.

## 11.9 Sexual selection and speciation

An ornament need not reflect better quality to increase in frequency in a population. It can increase in frequency because it enhances species recognition. Since 1889, Wallace spoke about traits that allowed the sexes to recognize their conspecifics and how this avoids the cost of non-fertile hybrids. He also suggested that such a need for species recognition could explain the incredible diversity of form and colour found in birds and insects. Likewise, Fisher (1930) noticed that the worst blunder in sexual preference that we can conceive consists of mating with an individual of another species. Indeed, such behaviour would lead to a drastic reduction of fitness, particularly for females, which may explain why often it is males that bear the species recognition signal.

It is now widely recognized that the development of elaborate sexual signals contributes to the genetic isolation of natural populations. Mechanisms of mate recognition can influence sexual selection, which can in turn affect the evolution of such mechanisms. In this sense, the very concept of species is closely related to that of reproduction (Gouyon *et al.* 1997). This is, however, a matter of debate within the theories of speciation (Andersson 1994). For instance, divergent sexual selection between populations can increase the differences in traits used for mate recognition, thereby reducing hybridization (Doutrelant *et al.* 2000). In contrast, the Fisherian processes and those that rely on revealing male quality can start on traits and preferences that were initially favoured because they reduced risks of hybridization. Such traits do not differ fundamentally from traits involved in mate choice. The traits used in species recognition are a subset of those used in the choice of an adequate mate, a crucial aspect of which is species identity. There is thus a structural link between sexual selection and speciation processes. Historically though (Section 11.2.1), one of the reasons for the delay in the rise of interest for sexual selection lies in the erroneous idea that mate choice and secondary sexual traits in general mainly play a role in species isolation. It is now clear that such an explanation is not sufficient. However, it took several decades before a revival of interest in sexual selection uncovered important problems, specific to sexual selection, that had hitherto been hidden behind the question of species isolation (Andersson 1994).

### 11.9.1 Mechanisms relating sexual selection to speciation

There are two main mechanisms of divergence of mate and trait preferences during speciation: **allopatric divergence** and **reinforcement**. These two groups of mechanisms are mutually exclusive and are often said to act successively.

#### 11.9.1.1 *Allopatric divergence*

The hypothesis of allopatric divergence suggests that isolation traits diverge by chance within geographically separated populations through the accumulation of mutations, as well as the effect of different selective pressures and/or genetic drift. A side effect is that mate recognition can diverge so much that the two forms no longer breed when, for any reason, they later come back into contact. Some arguments support this mechanism. For instance, in Coleoptera of the genus *Epicauta*, twin species of allopatric origin show species-specific courtship behaviours. When put into contact, males do not court the females of the twin species (Pinto 1980).

#### 11.9.1.2 *Reinforcement*

According to this mechanism, traits involved in isolation continue to evolve in the zones of secondary contact between two forms having started to diverge allopatrically. The individuals that most resemble the other form, or those that do not discriminate among the two forms, are likely to mate with the other form. Such individuals are counter-selected if such hybrid matings produce less-viable offspring. What follows is reinforcement or **reproductive character displacement**, thereby enhancing differences between the two forms in the contact zones. In the process of reinforcement suggested by Dobzhansky (1940), there is always gene flow between populations because isolation is not complete when the diverging populations come back into contact. Divergence is reinforced by selection against hybrids. The mechanism of reproductive character displacement takes place later when isolation becomes total, i.e. between distinct species. Displacement reduces the risks of wasting gametes in unfertile matings and reduces the risks of interference between species by making the signals more differentiated and thus more effective.

### 11.9.2 Can sexual selection favour speciation?

#### 11.9.2.1 *Mate recognition: a source of pre-reproductive isolation . . .*

Beyond the debate on sympatric or allopatric speciation and on the reinforcement or displacement of characters, the question of the respective roles of natural and sexual selection in speciation remains open. The role of ecological differences was underlined by the founders of the synthetic theory of evolution (as Dobzhansky 1940, or Mayr 1963), but other authors insisted on the fact that sexual selection can promote character divergence between species. In crickets and fruit flies of the Hawaiian archipelago, for instance, many twin species living either in sympatry or very close to one another seem to have identical ecologies (Otte 1989). Actually, these species differ mainly in their behaviour, and in particular in their sexual behaviour. For example, two cricket species of the genus *Anaxipha* have nearly identical songs, but one sings at night hidden under the bark of a tree, whereas the other sings during the day in the ferns located between these same trees. In other cases, only song characteristics allow for species differentiation.

Sexual selection can lead to rapid changes in sexual signals (Andersson 1994). As a consequence they can diverge quickly among isolated populations. Accordingly, in Lande's (1981) model on the Fisherian runaway process, small differences in the initial conditions between populations can lead them to very different equilibria, in particular if the process becomes unstable (Section 11.5.2.2). Many processes can promote differences in initial conditions of populations that split off: founder effects, genetic drift or simply differences in local conditions. Such differences can thus generate great differences in secondary sexual traits among populations, even when isolated from one another for a relatively short time. The Fisherian runaway process can thus explain why close species can differ mainly in male secondary sexual traits, in a way that does not appear especially adaptive in a context of natural selection. Examples can be found in ducks, birds of paradise, or the colours of some lizards, the calls used in mate attraction in amphibians, birds and insects, or courtship behaviour that often differ greatly between close species.

Lande (1982) used an extension of his female choice model to study the potential link between sexual selection and speciation along a geographical cline. His approach added spatial structure and **dispersal** (Chapter 10) to his previous model. He assumed that mating preferences evolved jointly with the evolution of male geographical variants. The conclusion was that the runaway process, owing to sexual selection, can amplify geographical variation and increase the difference in male traits among populations, hence leading to behavioural isolation. However, he did not find any cline reversion at the level of the ecological frontier zone, suggesting that sexual character displacement does not occur under the conditions of his model. This suggests that the Fisherian runaway process can more easily lead to reproductive isolation than to selection against hybrids.

#### 11.9.2.2 . . . *or an adaptation as a mechanism of pre-reproductive isolation?*

So far, we have presented the view that, during speciation, mate recognition tends to evolve first under the effect of drift or divergent adaptations, secondarily leading populations to reproductive isolation. Speciation could then continue through reinforcement and reproductive character displacement. Recently, however, Podos (2001), studying Darwin's finches, suggested a very different route towards speciation: in some cases, the adaptation itself would produce the divergence of sexual signals as a by-product, thereby leading to a pre-reproductive isolation. Darwin's finches, which live in the Galapagos archipelago, have largely been studied after Darwin was inspired by them when building his theory of evolution by natural selection. These closely related species differ mainly in their beak size and shape. Many studies, starting with those of Darwin, interpreted these differences as adaptations to different food: species with large beaks eat large, hard-to-break seeds, whereas at the other end of the spectrum, species with slender beaks eat insects.

However, many studies also showed that the size and structure of the various organs of the vocal tract constrain the type of sound that can be emitted. In particular, beak size strongly influences the pitch of vocal signals. First, Podos (2001) showed that this is true among individuals of the same species: intraspecific variation in beak size is correlated with song pitch in the expected way. A comparative approach further showed that this is also true across species. This led him to propose a very different mechanism of speciation: species first diverge by adaptation to different food sources; this secondarily forces them to emit such different sounds that individuals of the various forms no longer recognize each other as potential mates. This hypothesis is supported by the fact that many of these species can interbreed in the laboratory and that hybrids are viable and fertile. Their strong isolation in nature thus results mainly from pre-reproductive isolation processes.

As Ryan (2001) emphasizes, this is a new and unique view that requires further support than mere comparative (hence mainly correlative) observations. This hypothesis also underlines the fact that individuals must be viewed as a coherent whole, any change to a phenotypic trait potentially influencing other traits, the whole potentially leading to speciation.

### 11.9.3 Some case studies

It has often been stressed that sympatric divergence has only seldom been shown, despite much research. However, several studies have produced positive support (synthesis in Gerhardt (1994)). The debate relates to whether the mechanisms of species recognition that diverge in sympatry do so primarily by character displacement. Theoretical arguments suggest that the mechanisms of reinforcement or displacement are unlikely to be common (Andersson 1994); but other authors have come to different conclusions on the basis of empirical evidence, for example using extensive comparative studies of fruit flies and Hawaiian crickets (see Coyne and Orr 1989; Otte 1989) and New World frogs (Gerhardt 1994).

#### 11.9.3.1 *Reinforcement process*

The processes of reinforcement can intervene when populations having diverged allopatrically come again into contact for some reason, and hybridizations are still possible. A nice example is given by the evolution of fruit flies.

*Speciation in Hawaiian species of* Drosophila

Sexual selection seems to have played a crucial role in the explosive speciation of Hawaiian fruit flies (genus *Drosophila*). This probably occurred because of the numerous opportunities for geographical isolation in this archipelago, with large altitudinal variations and the existence of dispersal barriers such as the ocean or large lava flows. It is currently estimated that the archipelago harbours from 800 to 900 species of *Drosophila*, which represents a quarter of the total number of species of this group in the world. Another reason for this high species richness may be the absence of many taxa in the archipelago, which may have left several empty niches for new species. As in the case of several other groups, the most remarkable differences between these *Drosophila* generally involve mating systems and secondary sexual traits, suggesting an important role of sexual selection in speciation (Andersson 1994). One of the well-documented Hawaiian speciations occurred between *Drosophila silvestris* and *D. heteroneura*, two species almost always observed in sympatry and that share almost the same ecological niche. They can be differentiated through the use of a series of traits defining a syndrome particular to each species. Two studies (Ahearn and Templeton 1989; Carson *et al.* 1989) independently suggested that these species diverged by allopatric sexual selection after a founder event. According to these authors, when these two forms came back into contact for an unknown reason, sexual selection kept them distinct despite the existence of hybrids only observed in populations that underwent a strong disturbance either after the arrival of a lava flow, or after human activities.

A detailed comparative study using 119 pairs of closely related *Drosophila* species allowed analysis of the evolution of reproductive isolation within this group (Coyne and Orr 1989). Data on the geographical distribution, mate discrimination, viability, and strength of the sterility of hybrids for each pair of species were obtained from the literature. The genetic distance between two species had been estimated by proteic electrophoresis. Assuming the existence of a constant 'molecular clock' (which amounts to assuming that the evolution rate of proteins is constant between the various lines), the genetic distance between species represents a measure of the time elapsed since these species diverged. This assumption allowed comparison of the rates of evolution of pre-zygotic isolation (i.e. processes of mate discrimination at mating) and post-zygotic isolation (i.e. sterility and viability of hybrids), between sympatric and allopatric species.

Coyne and Orr (1989) make several predictions. First, the rates of evolution of pre- and post-zygotic processes of isolation should not differ between pairs of allopatric species: when species are not in contact, the two forms of isolation are equally likely to occur by chance as by-products of genetic divergence between species. Second, post-zygotic processes of isolation that lead to incompatibilities between forms, and thus to the waste of gametes, occur by chance as a consequence of the genetic divergence of isolated populations. There is thus no reason why such incompatibilities should be selected for. As a result, evolution rates of post-zygotic processes of isolation should not differ between allopatric and sympatric situations. Third, in allopatry, there is no reason, as such, for the processes of pre-zygotic isolation to be favoured because individuals of the two forms are never in contact; whereas in situations of sympatry, the processes of pre-zygotic isolation by reinforcement are strongly favoured because of the loss of fertility or viability of hybrids. The processes of pre-zygotic isolation should thus evolve more rapidly in situations of sympatry.

The comparative analysis supported all three predictions (Coyne and Orr 1989). The rates of evolution of the pre- and post-zygotic processes did not differ in allopatric pairs of species (Figure 11.13), and the processes of post-zygotic isolation did not evolve at different rates between pairs of sympatric and allopatric species. Lastly, pre-zygotic discrimination developed more rapidly in species living in sympatry than in isolated species: it is already maximal in most of the species having diverged recently (as shown by their low genetic distance; Figure 11.14 for genetic

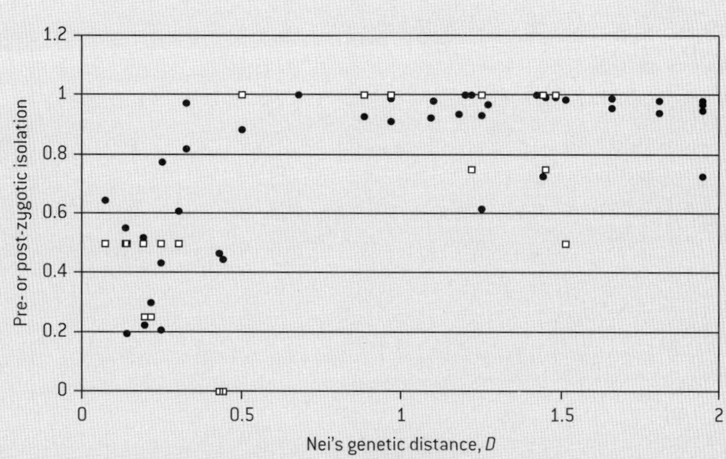


Figure 11.13 **Rate of pre- (filled circles) and post-zygotic (open squares) isolation as a function of Nei's genetic distance (*D*) in species in situations of allopatry**

The figure presents raw data, but the significance of the relationships has been tested by taking phylogeny into account using the contrasts method (Chapter 3). The two forms of isolation significantly increased with genetic distance. However, the force of isolation did not differ significantly between pre- and post-zygotic isolation (Wilcoxon test on cases where genetic distance $D$ is lower than 0.5: $Z = 1.19$, $N = 11$, $P > 0.20$).

After Coyne and Orr (1989).

distances lower than 0.5). This difference remained significant when the molecular clock was not assumed to be constant: more pairs of sympatric species than pairs of allopatric species showed greater pre-zygotic than post-zygotic isolation. Selection must thus favour mate recognition in sympatry, leading to faster species recognition than in allopatry.

Coyne and Orr (1989) also used their data to refute some possible alternative explanations. The main one was that only species having already developed fine recognition mechanisms are likely to remain in sympatry: otherwise, species may either hybridize, hence preventing us from distinguishing them as different species, or compete until the extinction of one of the other or both species. For this alternative to be supported, the situation observed in the various species should have involved a certain number of fusions and/or extinctions of species after a contact between species that have not developed mate discrimination. This is in fact contradicted by Coyne and Orr's (1989) results. Indeed, such a fusion and/or extinction process would concern all insufficiently differentiated pairs of species when they come into contact once again. As pre- and post-zygotic processes evolve at the same rate, both processes of pre- and post-zygotic isolation should be stronger in situations of sympatry. Yet, this is not what Coyne and Orr (1989) found. In addition, if there were processes of extinction/fusion, the degree of isolation observed in sympatry should only constitute a sub-part of the whole range of degrees of isolation. As Figure 11.14 shows, this is not the case: situations of weak isolation or of small genetic distances in situations of sympatry are rather common.

Hence, mate preferences or preferred traits in many fruit fly species seem to have been partly shaped by counter-selection of inter-specific matings. The mechanism responsible for this situation could be either reinforcement, or of reproductive character displacement for species that can no longer hybridize when they meet again. These two processes imply sexual selection.

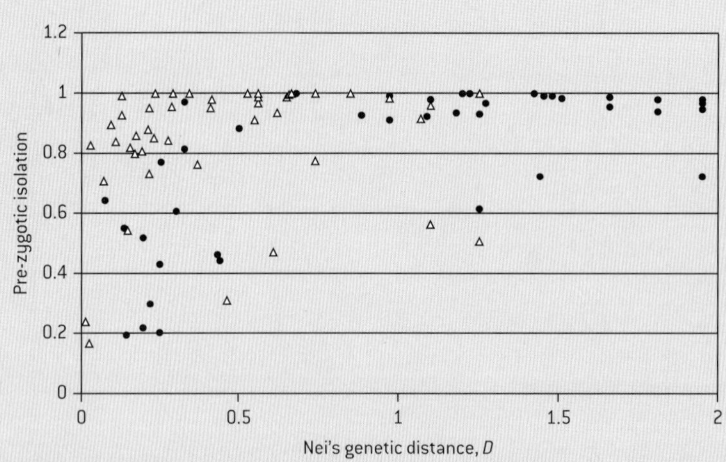


Figure 11.14 **Level of pre-zygotic isolation between the situations of allopatry (filled circles) and sympatry (empty triangles) as a function of genetic distance**

This figure shows raw data, i.e. without taking phylogenies into account, but the significance of relations was tested by taking phylogeny into account using the contrasts method (Chapter 3). The processes of mate discrimination and of post-zygotic isolation evolve at the same pace in situation of allopatry (filled circles). Indeed, because they are not subjected to selection, genetic divergence between species represents a measure of their post-zygotic isolation. For small genetic distances then, the level of pre-zygotic isolation is more or less proportional to genetic divergence in pairs of allopatric species. In contrast, pre-zygotic isolation very rapidly reaches its maximal level in pairs of species living in sympatry (empty triangles): even for genetic divergences lower than 0.5, sympatric species have a high level of mate discrimination (Mann–Whitney *U* test: $Z = 2.89, N_1 = 17, N_2 = 7, P < 0.01$). It follows that in species living in sympatry genetic divergence does not have any effect on pre-zygotic isolation.

After Coyne and Orr (1989).

#### 11.9.3.2 *Displacement of reproductive characters*

*The speciation of New World frogs*

An example of displacement of reproductive characters comes from frogs of the genus *Hyla* documented by H. Carl Gerhardt (1994). Having observed that the relatively low number of cases where this mechanism had been shown may be because most studies focused on male signals, Gerhardt (1994) focused on female selectivity. In the eastern United States, there are two frog forms so close to one another that they are almost indistinguishable. However, *H. chrysoscelis* is diploid whereas *H. versicolor* is tetraploid. The distributions of the two species overlap in the western part of southeast USA, whereas only *H. chrysoscelis* occurs in all southeastern USA (Figure 11.15). Laboratory crosses revealed strong mortality during early larval stages and the rare triploid hybrids that reached sexual maturity are sterile. Transmission through hybrids is thus no longer possible. It is not surprising then that hybrid adults are rather rare in nature, although hybrid matings are not rare in some parts of Missouri. Such mating mistakes occur despite the tendency of males of each species to select singing sites at different heights in sympatric populations. There must thus be a strong selective pressure favouring behaviours minimizing mating mistakes, particularly on females because most of them reproduce only once a year.

In both species, only males sing and females initiate sexual contacts; males do not defend any oviposition site and provide only sperm. Once a male and female meet, they form an amplexus, with the

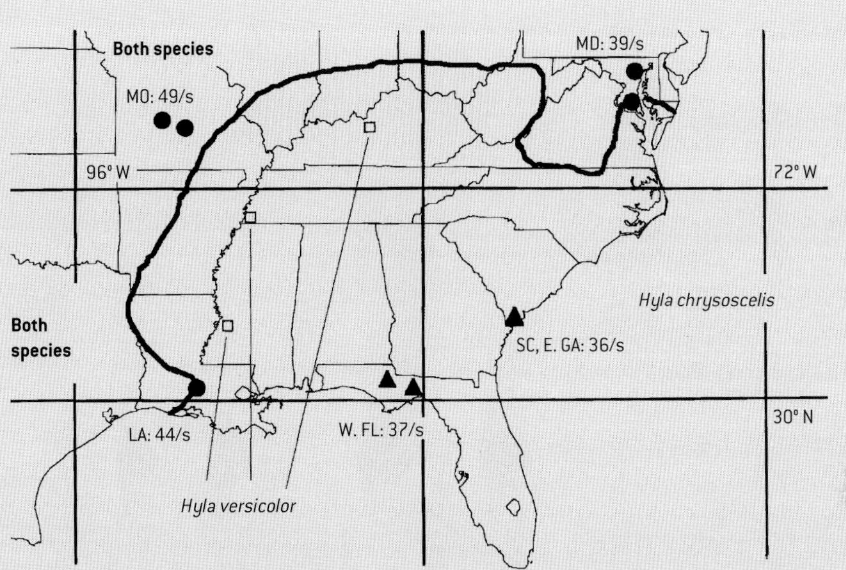


Figure 11.15 **Distribution of the two frog species studied by Gerhardt (1994) and localization of the study populations**

The thick line shows the limit of *Hyla versicolor* distribution (occurring northwest of the line, except for some isolated localities in the Mississippi Basin, represented by open squares). Only *Hyla chrysoscelis* occurs in practically the whole zone depicted. Points indicate the localization of sampled populations where the two species are sympatric. The triangles indicate the populations of *H. chrysoscelis* studied in situations of allopatry. The numbers associated with these localities indicate the average male pulsation rate of *H. chrysoscelis* populations.

Modified after Gerhardt 1994.

male climbing on the female's back. Male songs are composed of repeated pulsations that differ in their fine structure and in the rate at which they are emitted (Figure 11.16). In a first stage, Gerhardt showed that females of both species strongly prefer synthetic songs whose fine-scale temporal properties were typical of conspecific calls, over songs having the properties of heterospecific calls (Gerhardt 1982). He also showed that females of *H. versicolor* from Missouri prefer long over short songs (Klump and Gerhardt 1987), although the distributions of song durations overlap much in the two species.

Next, Gerhardt (1994) showed that for each species there is considerable geographical variation in pulsation rates, but that despite this variation the species can still be distinguished on this sole criterion. The average song durations are similar both within and between species, and there is no geographical variation in the pulse shape, which is thus species-specific. In addition, when pulse rate is constant and typical of their species (short basic pulses, Figure 11.16b), females of *H. chrysoscelis* prefer long over short songs. Furthermore, over 70% of females coming from an area where only *H. chrysoscelis* occurs chose the synthetic songs having the typical pulse rate of *H. chrysoscelis* males, even when they had the choice with a song of the same duration but presenting a pulse rate slower by 31% and four times louder. In contrast, under the same conditions, 90% of *H. chrysoscelis* females coming from Missouri, where both species are sympatric, preferred the stimulus mimicking that of their conspecifics. Thus, females of *Hyla chrysoscelis* are more selective when coming from populations where *H. versicolor* is present, than

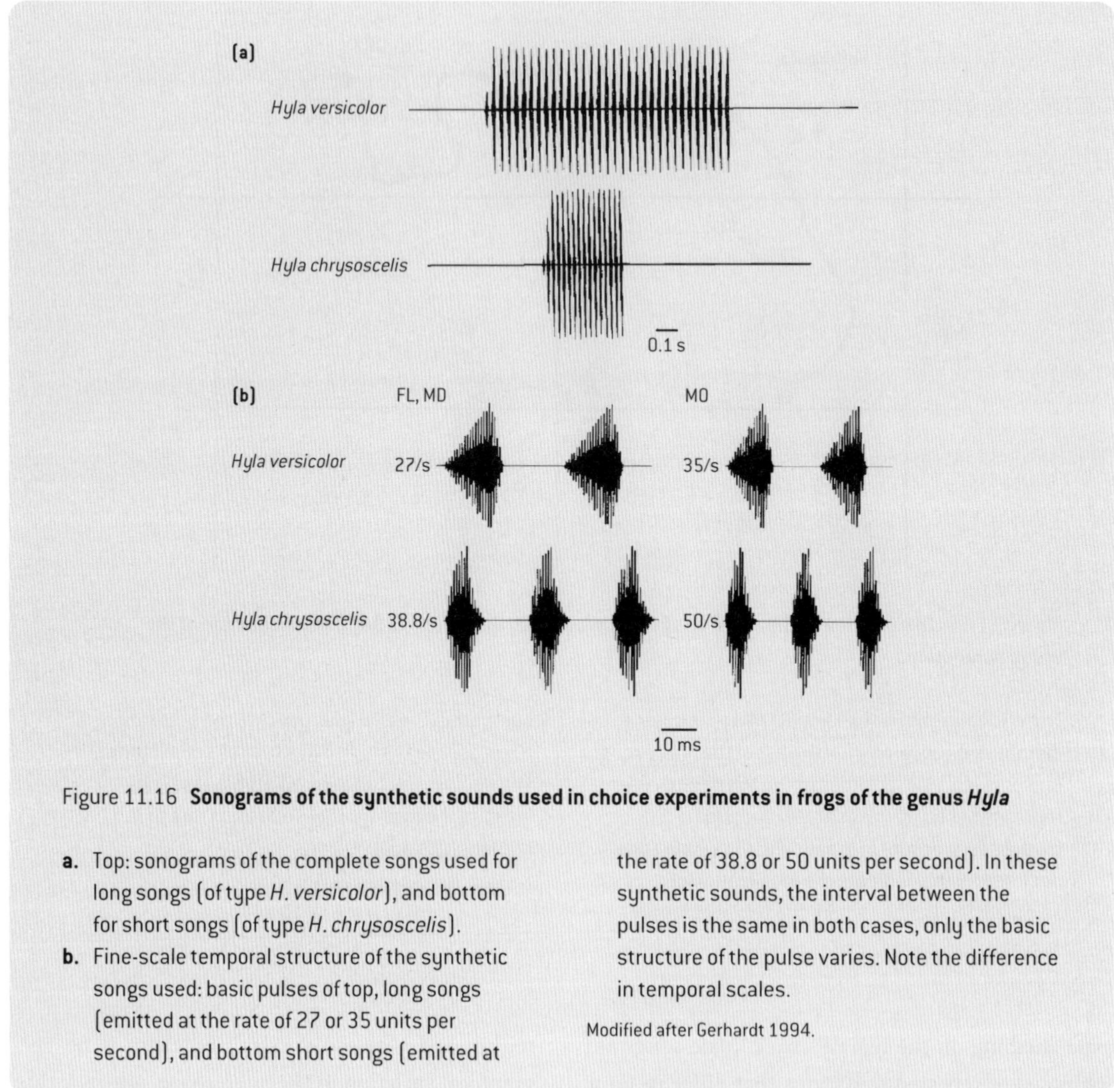


Figure 11.16 **Sonograms of the synthetic sounds used in choice experiments in frogs of the genus *Hyla***

**a.** Top: sonograms of the complete songs used for long songs (of type *H. versicolor*), and bottom for short songs (of type *H. chrysoscelis*).

**b.** Fine-scale temporal structure of the synthetic songs used: basic pulses of top, long songs (emitted at the rate of 27 or 35 units per second), and bottom short songs (emitted at the rate of 38.8 or 50 units per second). In these synthetic sounds, the interval between the pulses is the same in both cases, only the basic structure of the pulse varies. Note the difference in temporal scales.

Modified after Gerhardt 1994.

from populations where the latter species is absent (Figure 11.17). Consequently, females risking infertile crossings acquired greater discriminatory abilities than females of the same species not subjected to such a risk. A displacement of reproductive characters thus seems to have occurred.

### 11.9.4 Post-mating sexual competition and speciation

All the mechanisms of speciation mediated by sexual selection relate to processes occurring before mating, mainly through mate choice. A quite different route to speciation has been suggested, however. In general it is differences in male and female evolutionary interests (see Section 11.7), that can lead to diversification between isolated populations, thus setting the stage for speciation. Göran Arnqvist and his collaborators (2000) tested, in insects, the idea that post-mating sexual conflicts could lead to speciation. Such conflicts, implying sperm competition, could lead to perpetual antagonistic co-evolution between males and females and thus produce fast evolutionary divergences in reproductive traits. Post-mating conflicts are almost ubiquitous and relate to male–male competition for the fertilization of the ovocytes of the females with which they have both mated.

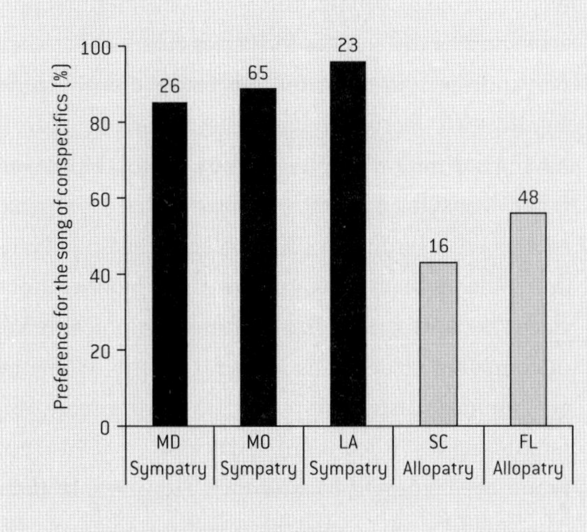


Figure 11.17 **Mating preference of *Hyla chrysoscelis* females suggests reproductive character displacement**

Females coming from populations where the other species is present (black bars) show a greater selectivity than females coming from populations where the other species is not present (shaded bars). The stimulus used was a short song exhibiting a pulse rate typical of the *Hyla chrysoscelis* males of the zone of origin of the female tested. Numbers tested are given above bars. MD, Maryland; MO, Missouri; LA, Louisiana; SC, South Carolina; FL, Florida.

Modified after Gerhardt 1994.

Whenever females' interests are compromised by males, their reproductive strategy will evolve to reduce these costs. This will create perpetual (or episodes of) post-mating sexual selection (by sperm competition and/or cryptic female choice) which in turn produce novel adaptations in males by biasing their post-mating fertilization success, thus favouring males most able to manipulate female reproduction to their own benefit. Such sexually antagonistic adaptations will generate rapid co-evolution between male and female reproductive physiology and morphology, eventually resulting in reproductive isolation between allopatric populations. Indeed, such co-evolutionary sequences are strongly context-dependent so that the same adaptations will not necessarily occur in two distinct populations.

#### 11.9.4.1 *Species richness in polyandrous versus monoandrous insects*

Arnqvist *et al.* (2000) assessed the importance of such post-mating sexual conflicts in speciation by comparing extant species richness in pairs of related clades of insects differing in the opportunity for post-mating sexual conflicts. To test whether the intensity of post-mating sexual conflict co-varies with speciation rates, they analysed a series of paired phylogenetic contrasts. In each contrast, they compared the number of described extant species in a clade where females typically mate with many different males (sexual polyandry) with the number in a closely related clade where females typically mate with only one male (monoandry). In polyandrous species, the ejaculates of several males compete over fertilization. As a result, male traits that aid in such competition are favoured, even if they convey costs to females. In polyandrous species, there is thus ample opportunity for post-mating sexual conflict and therefore for antagonistic co-evolution between the sexes. In monoandrous species, male ejaculates do not compete and male reproductive success increases with any post-mating increase of his mate's fitness. In the absence of parental care, male and female interests are then identical after mating in monoandrous species. Post-mating sexual conflicts, as well as any resulting post-mating sexual selection, are thus absent or low in monoandrous species. Because both clades in a given contrast share a common ancestor, the relative number of extant species in these clades depends only on what occurred after these clades diverged and reflects differences in speciation rate in these two clades. Arnqvist *et al.* (2000) thus expected polyandrous clades to have more species than monoandrous clades.

As comparative analyses can have several methodological pitfalls (see Chapter 3), Arnqvist *et al.*

(2000) took several precautions when collecting and analysing data. They obtained 25 phylogenetic contrasts, representing five different orders of insects, all of which were independent, because no clade was represented in more than one contrast. On average, polyandrous clades contained 3.98 times more species than monoandrous clades. The null hypothesis of equality in the number of species in the two types of clade was thus strongly rejected. Among the 25 contrasts, only nine involved true sister groups, i.e. clearly sharing a common ancestor, and 16 involved more distantly related groups. Results did not differ significantly between these two types of contrast. In particular, an analysis including only sister group contrasts, led to the same results. Furthermore, there were no significant differences between the results obtained in the five orders.

Several variables known as having an important impact on species richness could have had a confounding effect on these results. The three main ones are trophic ecology of the species, range of geographical distribution, and latitude. Some of the contrasts of Arnqvist *et al.* (2000) involved clades with similar trophic ecology, others, clades with different trophic ecology. The differences in species richness between polyandrous and monoandrous clades did not differ significantly between contrasts involving species having a different trophic ecology or not. Moreover, when considering only contrasts that compared species with similar trophic ecologies, again polyandrous clades had more species than monoandrous clades. Regarding the size of the geographical range of compared clades, it is known that the wider the geographical distribution of a taxon, the more different species it contains. Arnqvist *et al.* (2000) found this expected relationship. However, in their data, polyandrous clades occupied on average 3.88 (standard deviation 1.88) biogeographic regions, against 3.84 (standard deviation 2.01) for monoandrous clades, a non-significant difference. Moreover, Arnqvist *et al.* (2000) found that, when controlling for the size of the geographical distribution, polyandrous clades still showed a higher species richness than monoandrous clades (Figure 11.18). For latitude, some comparative studies documented a stronger speciation rate toward the equator, which could

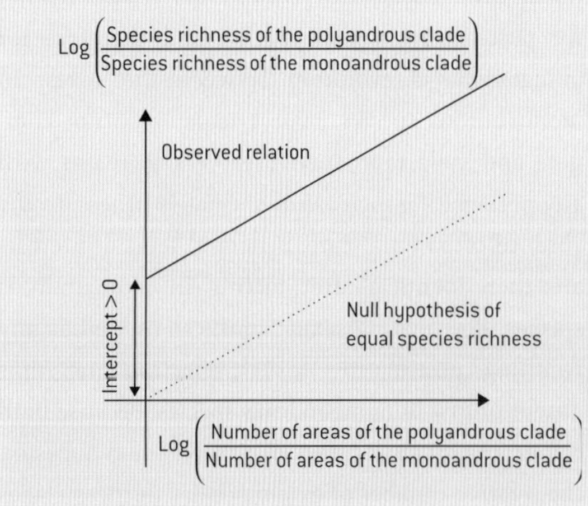


Figure 11.18 **Specific richness and geographical distribution**

Logarithms of the ratio of species richness of polyandrous versus monoandrous clades as a function of the logarithm of the ratio of their respective geographical distributions. This analysis corrects for a possible effect of geographical distribution on the specific richness of clades. According to the null hypothesis of equal number of species per clade in polyandrous and monoandrous clades, polyandrous and monoandrous clades should contain similar numbers of species when occupying the same number of geographical areas. Hence the intercept should not differ significantly from zero (thus corresponding to a ratio of species richness of 1). However, Arnqvist *et al.* (2000) found that the intercept was significantly higher than zero, whether they used their whole data ($N = 25, P = 0.009$), or only contrasts where both clades have similar trophic ecology ($N = 20, P = 0.001$). This shows that the ratio of species richness in polyandrous to monoandrous species is significantly higher than predicted if the number of species in the two types of clade were the same once corrected for geographical distribution.

contribute to the latitudinal gradient of biodiversity. When controlling for latitude in their analyses, Arnqvist *et al.* (2000) found that polyandrous clades contained relatively more species when inhabiting more equatorial areas. They calculated a latitude index for each clade and noted that the average of this index did not differ between polyandrous and monoandrous clades. They applied the same method as for geographical distribution detailed in Figure 11.18 and again rejected the null hypothesis of equal number of species when the clades were equally equatorial.

#### 11.9.4.2 *Sperm competition can lead to speciation*

All of these results support the idea of an effect of sperm competition on insect speciation rate: clades having strong sperm competition (polyandrous) show species diversity approximately four times higher than clades with weak sperm or no competition (monoandrous). This conclusion was robust with respect to the three main potentially confounding variables known to influence species diversity. Moreover, accounting for these potential confounding variables actually reinforced the relationship between species diversity and sperm competition. Another important element that must be considered is species extinction rate. Sexual conflicts are known to increase the risk of species extinction (see Chapter 18). Therefore, species extinction rates should be stronger in polyandrous than in monoandrous clades. Arnqvit *et al.*'s (2000) results are thus conservative because such differences in extinction rates can only have reduced species diversity in polyandrous clades. Likewise, the processes of reinforcement when diverging populations meet cannot explain the results. Indeed, in polyandrous clades, fitness costs of mating mistakes affect only some offspring, whereas in monoandrous clades costs of mating mistakes affect the whole progeny. The process of reinforcement is thus stronger in monoandrous than in polyandrous clades, leading to faster isolation and thus speciation in monoandrous clades, which is the opposite of the observed trend.

This result thus stresses the potential importance of inter-sexual conflict and sperm competition in speciation, a process largely neglected so far. More generally, this result shows the importance of all processes of sexual selection in evolution.

## 11.10 Conclusion

Here we consider that the general process of natural selection can take two major forms: utilitarian and sexual selection. Each of these processes act mainly on one of the two major fitness components: sexual selection mainly generates variation in reproductive success (access to mates in particular), utilitarian selection mainly translates to variation in survival rate. We hope to have convinced the reader that sexual selection is one of the fundamental mechanisms of evolution, as is the process of natural selection, with which it must be put on equal footing. Given its equal footing with utilitarian natural selection and that the rest of this book is almost entirely based on utilitarian natural selection, it may not be so surprising that the single chapter devoted to sexual selection is unusually long. Indeed, sexual selection contributes to and can explain many evolutionary processes, ranging from exaggerated traits to speciation. Although long neglected, sexual selection has been the focus of many developments since the 1980s when this question started to become a major theme of behavioural ecology. Today, mate selection clearly appears as one of the major mechanisms in evolution.

## » Further reading

> *For a general review on sexual selection, Andersson (1994) constitutes a very general and complete synthesis. It does not present, however, the latest developments in the field of sexual conflicts.*

**Andersson, M.B.** 1994. *Sexual Selection. Monographs in Behavior and Ecology.* Princeton University Press, Princeton.

> *Several synthesis articles have been published in recent years in the journal* Trends in Ecology and Evolution. *In addition, excellent reviews of the historical aspects of sexual selection can be found in:*

**Pomiankowski, A.** 1988. The evolution of female mate preferences for male genetic quality. *Oxford Surveys in Evolutionary Biology*, no. 5, pp. 136–184. Oxford University Press, Oxford.

**Otte, D.** 1989. Speciation in Hawaiian crickets. In *Speciation and its Consequences* (Otte, D. & Endler, J.A. eds), pp. 482–526. Sinauer, Sunderland, Massachusetts.

> *For cryptic female choice, the current state of the question can be grasped by reading the five articles on this question that appeared in the journal* Evolution:

**Birkhead, T.R.** 1998. Cryptic female choice: criteria for establishing female sperm choice. *Evolution* 52: 1212–1218.

**Birkhead, T.R.** 2000. Defining and demonstrating female sperm choice. *Evolution* 54: 1057–1060.

**Pitnick, S. & Brown, D.** 2000. Criteria for demonstrating female sperm choice. *Evolution* 54: 1052–1056.

**Kempenaers, B., Foerster, K., Questiau, B., Robertson, B.C. & Vermeirssen, E.L.M.** 2000. Distinguishing between female sperm choice *versus* male sperm competition: a comment on Birkhead'. *Evolution* 54: 1050–1052.

**Eberhard, W.G.** 2000. Criteria for demonstrating postcopulatory female choice. *Evolution* 54: 1047–1050.

## » Questions

1. What do we mean by 'choice'? What definition would you give of the mate (or any other resource) choice process? What criteria allow us to show the existence of a choice? This question is to be analysed in relation to Chapters 7–9.

2. How can we test the existence of sensory biases and their importance in the evolution of secondary sexual traits?

3. Is sexual selection inoperative in species not exhibiting sexual dimorphism? (See Pomiankowski 1988, Section 5.)

12

12

# Mating Systems and Parental Care

Frank Cézilly and Étienne Danchin

## 12.1 Introduction

Before the advent of behavioural ecology, the social organization of breeding was treated descriptively in the main (Tinbergen 1953; Bourlière 1967). Differences between taxa were set out in detail but there was no general principle with which to reveal any overall logic. Accordingly the influence of ecological factors on male–female interaction long remained virtually unknown and parental behaviour was generally thought of as a harmonious interaction between breeding individuals and their offspring.

A radical conceptual advance came about in the 1970s with the publication of work by Trivers (1972) and the seminal paper by Emlen and Oring (1977) introducing the idea of **mating systems**. This expression encompasses the way individuals of a species or population gain access to mates, the number of mates they interact with in a breeding season, the duration of social bonds between mates, and the relative involvement of each sex in providing **parental care** (Davies 1991; Reynolds 1996). Over the past 30 years, empirical and theoretical research in this field has revealed the complexity of different mating systems and progressively refined the early concepts put forward by Emlen and Oring (1977).

There is a close connection between mating systems and **sexual selection**. Mating systems are determined by the ability of one sex (usually males) to monopolize mates of the opposite sex either by direct association or by controlling access to essential resources. This asymmetry between the sexes arises initially from the asymmetrical investment of the sexes in producing each gamete, a topic covered in the previous chapter. In this scheme of things, which emphasizes the close relation between mating systems and sexual selection, the sex with the lowest potential breeding rate (in most cases females: cf. previous chapter) is a scarce resource for the other. It is predicted therefore that females occupy space depending on the availability of food resources, on predation risk and on the distribution of suitable breeding sites whereas the distribution of males supposedly adjusts to that of females (Figure 12.1). In other words, **the distribution of females is governed by the distribution of resources and the distribution of males by that of females** (Emlen and Oring 1977).

This initial scheme is oversimplified, of course. Just as sexual selection becomes more complicated when it is acknowledged that male investment may not be limited to producing gametes, things become complicated in understanding mating systems when one considers that in some species males may contribute to parental care. Parental care **benefits** offspring but necessarily **costs** time and energy for the caregiver. The care provided by males may therefore represent a significant resource for females. Within breeding pairs, each sex will benefit from the parental care efforts of the opposite sex. This induces strong selection pressure for each sex to adjust its own investment in parental care in relation to its mate's investment. Maximization of each parent's **fitness** does not necessarily result, therefore, in equal sharing of parental care (Trivers 1972). In addition, the level of parental care provided does not necessarily correspond to the level maximizing the survival rate and health of their offspring.

This chapter is an introduction to mating systems. Situations found in nature are so diverse that it would be pointless trying to be exhaustive here. Our exposition is confined, then, to the selective constraints favouring one or the other type of association between males and females and determining the way parental care is shared between the sexes. Similarly, empirical studies serving as illustrations shall cover only certain zoological groups that are especially representative of one or other mating system.

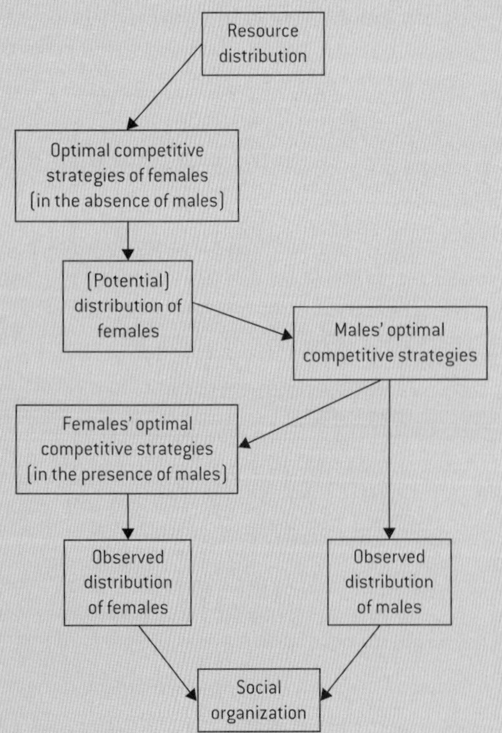


Figure 12.1 **Relationship between the distributions of females and males**

Being limited above all by access to resources, females are assumed to occupy space depending on resource distribution, predation risk and the cost of sociality. The distribution of males is guided by that of females. Males may compete for access to females either directly by confrontation or indirectly by controlling access to resources sought after by females.

## 12.2 Some general principles

If people not involved in studying evolution are asked whether in their view of the relationship between a male and female in a pair is one of **cooperation** or a situation of **conflict**, the vast majority will answer cooperation. However, if the same people are asked to choose a term that best describes the relationship between individuals of the same species in general, a large number will choose conflict or **competition**. The number will be even larger if the question is framed in a context of exploitation of limited resources.

And yet, why should we view these two situations as sufficiently different as to lead to such different answers? In both cases, two individuals interact in a context of exploitation of limited resources and have an interest in maximizing their own benefits, even at the other protagonist's expense. Of course, in the case of two mates, in evolutionary terms, they share a common interest in the success of their descendants, which *a priori* contain half of the genes of each of parent (we shall see that this, *a priori*, may be wrong). The success of descendants is for each of the genetic parents the only guarantee of their own evolutionary success, the only way for them to increase their phenotypic fitness. However, if either mate can, for one reason or another, exploit the other's investment in the offspring to keep the maximum of its own resources for a possible further chance at breeding, then, it will clearly be favoured by **natural selection**.

In fact, if one is trying to understand the forces of evolution having led to the appearance of one or other mating system, to start from the principle that the relationship between males and females is more akin to collaboration than to a conflict of interest where both parties seek to maximize their own fitness would be to deny the very principles of natural selection. Naturally enough, things may seem like harmonious cooperation, but that does not preclude conflicts of interest from governing the system. We shall see in this chapter how important it is to bear this in mind when studying mating systems. We shall also see what large discrepancies there may be between the way things look and what they are really like beneath the surface.

One of the fundamental principles of our analysis remains that of the **economic approach** whereby we try to assess the various costs and benefits associated with any particular strategy, depending on the conditions encountered. A second fundamental principle of our analysis is that the interests of the two mates are often very different. In the previous chapter we saw that **anisogamy** or more generally asymmetric investment by males and females in their offspring may have cascading evolutionary effects. Mating systems are a part of this cascading effects. We must always endeavour to analyse the situation from two points of view: that of the females and that of the males, keeping in mind that these points of view may be diametrically opposed.

Lastly, one must keep in mind that the mating systems on the whole are highly condition-dependent. The optimal mating system may not be the same for each sex and can vary with the state of the **environment** and the strategies adopted by other members of the population. Mating systems therefore constitute an eminently dynamic property. We should not be surprised to observe variations among species, among populations and even within a single population. It is this dynamic process we attempt to analyse in this chapter. We shall begin by describing the major types of mating system. Then we shall show in what way they are ambiguous. Lastly, we shall analyse the role of parental care in determining mating systems.

## 12.3 The main types of mating system

We start here with an elementary classification of mating systems that ranks the various ways males and females of a population can organize themselves during the breeding season among four major categories. This general classification is essentially based on the number of mates that individuals of each sex experience on average during a single breeding season. Refined classifications of mating systems may involve a larger number of parameters that affect sex difference in the opportunity for parental care (Davies 1991; Shuster and Wade 2003).

### 12.3.1 Sexual promiscuity

Sexual **promiscuity** is in some sense an unrestricted mating system. In any one breeding season, both males and females mate with several partners and there is no rule as to which sex provides parental care. This system can be observed in gastropods (Baur 1998), one species of megapode bird (Jones *et al.* 1995) and, in some guises, in several primate species (Dixson 1998).

*Gastropods . . .*

In most opisthobranchs (marine gastropod molluscs) and in various species of lunged fresh-water or land gastropods, individuals copulate with several partners and eggs are generally fertilized by sperm from several individuals (see Hadfield and Switzer-Dunlap 1984; Rollinson *et al.* 1989; Baur 1994). Intertidal or land gastropod species, however, are subjected to different constraints from those encountered by marine or freshwater species. They are slow moving and they risk desiccation while looking for mates. Accordingly, in these species, copulation is less frequent and only occurs when environmental conditions are favourable. The risk of not finding a mate promotes sexual promiscuity and mating between individuals seems to depend on chance encounters (Baur 1998).

*. . . birds . . .*

Megapodes are a group of bird species endemic to Australia and New Guinea (Jones *et al.* 1995). They are thought to be the most primitive of galliforms and have particularly long and powerful toes. Another characteristic of the group is the very large size of the eggs. Depending on species, eggs are laid in pits or in a mound of earth, plant litter, twigs and branches that may be several metres high. Incubation is a very slow process performed by heat from the ground or heat given off by the respiration of the many micro-organisms living inside the mound of decomposing organic matter (del Hoyo *et al.* 1994). Upon hatching the young can fend for themselves and live alone. There is no parental care apart from building and watching over the mounds, both of which tasks the males carry out. The case of the Australian brush turkey (*Alectura lathami*) is unique. Females mate with several males and lay their eggs successively in different mounds. Males inseminate several females so the chicks from eggs incubated in the mound of the same male are therefore by different females.

*. . . and monkeys*

In various primate species, sexual relations are organized in mixed groups comprising several sexually active males and females. No long-term relation binds individuals of opposite sexes except for short-lived associations between a male and a female during the female's period of oestrus, without implying sexual exclusivity. And yet the system is not strictly promiscuous. Copulation between males and females is not random but is influenced by various factors such as the age of individuals, their social rank, their ties of kinship or the existence of individual preferences. Such a mating system has been observed in several species of macaques (see Lindburg 1983; Van Noordwijk 1985; Ménard *et al.* 1992), in baboons (*Papio ursinus*; Seyfarth 1978) and in chimpanzees (*Pan troglodytes*; Tutin 1979; Goodall 1986). In all, Dixson (1998) numbered 21 primate species with 'pseudo-promiscuous' systems.

### 12.3.2 Polygyny

In **polygyny** in the strict sense, one male breeds with several females whereas each female breeds with just a single male. In the course of a single breeding season, a male may associate with several females either at the same time (simultaneous polygyny) or in succession (sequential polygyny). In polygynic systems, parental care is usually provided by the females. The main difference among polygynic systems lies in whether male–female interaction is lasting or transient.

#### 12.3.2.1 *Polygyny based on monopolization of resources (resource-defence polygyny)*

The distribution of sexually receptive females in space and time directly determines the males' capacity to engage in a system of simultaneous polygyny. More often than not resources such as food, water, or breeding sites are not evenly distributed in space but are concentrated in certain points. A male can then take control of enough space to attract and maintain several females. We then speak of a **mating system based on resource monopolization**. The more heterogeneous the environment in terms of resource distribution, the more likely it is that polygyny will occur.

##### a *Female sub-tenancy of a male's territory*

Polygynic systems based on resource monopolization are observed in a wide variety of species, both in vertebrates (see Cronin and Sherman 1977; Schoener and Schoener 1982; Dixson 1998) and invertebrates (see Shelly *et al.* 1987; McVey 1988). In some species several females coexist on the territory of a single male but each of them holds and defends her own plot of the male's territory against other females. Males confront each other to control the adjoining territories of several females. Such a polygynic system involving female territoriality has been termed 'sublease territory' system (Gould and Gould 1989).

*From cleaner fish . . .*

This polygynic system based on resource monopolization has been observed for a species of cleaner fish (see Chapter 17), the striped cleaner wrasse (*Labroides dimidiatus*; Robsertson and Hoffman 1977). Females of this species occupy a particular site that they defend against other females. The males, which are larger than the females, compete for control of the territories of several females, the largest of them managing to monopolize up to six females.

*. . . to tigers*

Female tigers live year round on a territory of about 20 square kilometres whereas males' larger territories cover 70–100 square kilometres (Smith *et al.* 1987; Sunquist and Sunquist 1988). Outside the breeding periods, these big cats live solitary existences. Controlling a large territory implies substantial energy costs for males, who must also regularly cope with attempted intrusions by other males. Given the spatial scale of these territories, encounters between males and females are infrequent. The oestrus female signals her condition with roars the male can hear a long distance away. Male–female interaction is brief, being limited to copulation. After a pregnancy of about three and a half months, the female gives birth to a litter of two to three cubs. Once independent, the young females attempt to divide up or take over the mother's territory or settle nearby. The young males scatter over great distances until they are able to confront and oust an older male from its territory. The cost of daily defence of a large territory and the effects of age lead to regular renewal of the males controlling territories occupied by females.

##### b *Does polygyny have any costs?*

When males provide no aid for females, polygyny may imply no cost for females and the mating system which is established is the direct consequence of the capacity of some males to monopolize a greater or lesser share of resources. However, under some circumstances, it may be disadvantageous for females to have to share the resources controlled by a single male. Polygyny is then imposed on females by a small proportion of males having taken over all the suitable breeding sites. Females then have no other choice other than either to exploit communally the resources held by a single male or to forego breeding.

*The importance of a heterogeneous environment*

The situation is different if most males are able to defend a territory but the quality of territories varies markedly from one male to another. Females then have the choice of setting up alone on a low-quality territory with an unpaired male or of joining an already paired male with a higher-quality territory. Depending on the variation among males in territory quality, the female may find it advantageous to settle as a second female on a territory that is particularly amenable for breeding despite the costs of polygyny. It is sufficient for the difference in quality between territories to exceed a threshold value for the cost to be offset and polygyny to be economically favoured. This argument can be displayed graphically as a **polygyny threshold model** (Figure 12.2).

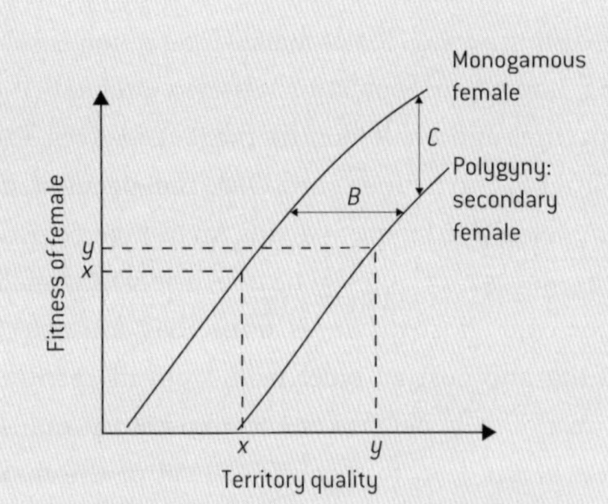


Figure 12.2 **The polygyny threshold model**

The two curves represent the variation in breeding success of females as a function of the quality of the male's territory, depending on whether they live alone on the territory (monogamous females) or share the territory with another female (polygyny: secondary females). Success increases with the quality of the territory. The model assumes a fixed cost *C* of sharing the territory with another female such that the success curve for polygynic females lies below that of monogamous females. When the difference in quality between the two territories exceeds a certain value (the polygyny threshold, arrow B) it is in the female's interest to settle on the very high quality territory of an already paired male rather than being monogamous on a lower quality territory.

After Verner and Wilson (1966); Orians (1969).

*Allowing for alternative hypotheses*

Although the polygyny threshold model has attracted much interest (Andersson 1994), experimental tests have been concerned with a few bird species (see Pleszczynska 1978; Ewald and Rohwer 1982; Searcy and Yasukawa 1989; Slagsvold *et al.* 1992; however, see Borgerhoff Mulder (1990) and Marlowe (2000) for an application of the model to humans). Moreover, it is difficult to test the model rigorously because many parameters must be measured concurrently and the gains for females cannot be readily evaluated (Andersson 1994). For example, being the first or second female to settle in a male's territory may have consequences on the females' survival or their capacity to breed in the next season, two parameters that are not always easy to measure in the field. Resident females may also actively oppose an additional female moving in if their own breeding success is affected. Lastly, understanding how females settle on the territories of different males presupposes knowledge of the information they have about the quality of the territories, the quality of the males and the presence of other females already settled on the territory. Although the polygyny threshold model is an important step towards understanding the factors that favour polygyny, there are several alternative hypotheses (Weatherhead and Robertson 1979; Alatalo *et al.* 1981; Lightbody and Weatherhead 1988; Stenmark *et al.* 1988) that should also be considered to explain how territorial polygyny arises and is maintained in a given species (Table 12.1).

#### 12.3.2.2 *Polygyny based on monopolization of females: harems*

Harems are another form of territorial polygyny in which males monopolize the females themselves. This strategy is found especially in equids (horses and zebras; Rubenstein 1986), lions (Bertram 1975), some species of deer (Clutton-Brock *et al.* 1982) and pinnipeds (seals). A group of females resides on the territory of a single male that actively excludes any other males. In some cases, the territory defended by the male corresponds to the females' home range. Over time, different males successively take control

| A. No cost of polygyny for females |
|---|
| Females occupy space independently and at random. No particular benefit from polygyny. Neutral model, polygyny is produced by chance. |
| Females benefit by clustering (collective defence of nests, synchronization of egg laying). Polygyny is maintained because of these fitness benefits. |
| B. Costs for females induced by polygyny |
| No possibility of choice for females: unpaired males are rare or difficult to locate or identify. Polygyny is imposed upon females by conditions. |
| Costs offset by direct benefits (quality of territory, male's contribution to parental care) or indirect benefits (genetic quality of male). Corresponds to the polygyny threshold model. |

Table 12.1 **Different modes of female settlement leading to territorial polygyny in birds**

of a harem each ousted in turn by what is often a younger, more vigorous rival.

### a *Infanticidal lions*

Infanticide is typical of African lions (*Panthera leo*; Bertram 1975; Packer *et al.* 1988). Prides of lions are made up of several adult females, their pre-adult daughters, cubs, and one to six males. There is generally close kinship between females, either because they have grown up in the same group which has persisted over several generations, or because they and their sisters and cousins left a group that became too large and settled in a new territory. The breeding activity of females lasts on average a dozen years with, when cubs survive until they are weaned, a litter every two years. However, juvenile mortality is high and 80% of cubs generally die before reaching the age of one year. The cubs' deaths result in the females returning into **oestrus**. Females reside for a long time in the same territory and have several breeding opportunities during this period.

Things are very different for males. Driven out of their natal group at puberty, brothers and cousins band together to hunt in areas of low food quality where no pride is settled. During this period several males die. Those that survive to the age of four years may then try to take control of a group of females either by confronting the male(s) controlling a harem alone or by forming coalitions with other young lions. Controlling a harem alone allows a male to inseminate a larger number of females but renders him particularly vulnerable to attacks by coalitions of lions. Moreover, kinship between brothers and cousins within coalitions tends, in evolutionary terms, to reduce competition among males of a same group for access to oestrus females. Fights between males within the same group are even detrimental in the long term for males particularly if the males are related because they involve a risk of injury and would reduce the ability of the males in the group to ward off attacks by other coalitions. The period of control exercised by a coalition over a harem is brief, of the order of thirty months. That and the high death rate of cubs explain the behaviour of males when they take control of a harem. In most cases the new masters of the harem kill the cubs, especially the males, and drive out any juvenile males still present. The lionesses of the pride hardly oppose this murderous behaviour (see, however, Grinnell and McComb 1996) and gestating females abort more often than not after a change of dominion over the group (otherwise a young lion born soon after the take-over of the harem would be killed anyway by the males). The benefit is obvious for the males: the rapid return of females to oestrus gives them a chance to leave descendants during their short period of breeding activity.

#### b *Polygyny based on monopolization of resources or females: the importance of a heterogeneous environment*

One of the most extreme cases is the polygynic system of elephant seals (*Mirounga angustirostris*; Le Boeuf 1974). Females copulate and breed on land, on beaches that are largely safe from predators, where they congregate sometimes in large numbers. A single large male can take control of this resource by defending the site against other males at the start of the breeding season before the females settle. Several factors make it easier for a single male to monopolize a large number (up to a hundred) of females. First, the females do not feed ashore but at sea and need little room to give birth and rear their young. Secondly, the breeding season is short, leading many females to look for a suitable breeding site at the same time. The situation is very different for seals that breed on the ice where suitable sites seem abundant and females tend to be more scattered. Males then control only a limited number of females, or sometimes only one (Le Boeuf 1978).

A polygynic system similar to harems is observed also in some bird species such as tinamous and some pheasants (see Bennett and Owens 2002). A male defends a group of females against other males and mates with each of them. After mating, the females leave the harem to lay their eggs and rear their young alone. Unlike in mammals, the harem in birds is a transient structure.

### 12.3.2.3 *Leks*

Polygynic systems do not necessarily imply any direct or indirect constraint on females. In various insect (Shelly and Whittier 1997) and vertebrate species (see Höglund and Alatalo 1995), males congregate in breeding arenas that are visited by females solely for the benefit of copulating with males. Such a peculiar mating system is called a **lek**. It was initially suggested that leks evolved in species where males were unable to monopolize either females or the resources they needed (Emlen and Oring 1977). Classically, a lek exists when:

1. males do not provide any parental care, their contribution being strictly limited to inseminating females;
2. males congregate in 'arenas' where virtually all mating occurs;
3. males defend tiny display territories with no vital resources for females or they are unable to control access of females to any resources in their territory;
4. females have a free choice of mate when they visit arenas;
5. most copulations involve a small proportion of males displaying on the lek (Bradbury 1985; Shelly and Whittier 1997).

#### a *Two major types of lek*

Leks have been particularly well studied in birds where they have evolved independently on several occasions (Ligon 1999) even though it is a very uncommon avian mating system (less than 1% of species; Höglund and Alatalo 1995; Jiguet *et al.* 2000). Two major types of lek can be distinguished: **concentrated** or classical and **exploded** leks. Classical leks are gatherings of males in small areas implying a high density and short inter-male distances. Such assemblies do not go unnoticed, particularly as the same sites are generally re-used from year to year. Such leks are observed in some tetraonids, some shorebirds, and birds of paradise. By contrast, exploded leks are not easy to spot. They involve some degree of crowding of male display grounds, but on a much larger spatial scale. Individuals remain large distances apart and usually interact through their calls alone. Unlike in classical leks, females may exploit the resources on the male's territory and even nest there. However, because males' territories are adjacent, females may explore several of them in a short period of time. In birds, this type of lek site is traditional, for example, for the little bustard (*Tetrax tetrax*; Jiguet *et al.* 2000). The evolutionary transition from one type of lek to another is still uncertain, but it is likely that exploded leks generally preceded rather than followed classical leks (Théry 1992).

### b How do leks arise?

Understanding how leks have evolved involves evaluating the advantages and drawbacks for each sex separately. *A priori* the system seems costly to males as it implies fierce competition for access to mates. And indeed, the success of males at a lek generally varies greatly, with a significant number of males not having a single opportunity to mate during a season (Beehler 1983; Wiley 1991). The success of males is said to be highly skewed in favour of certain males who have access to most copulations.

Some fitness benefit must therefore at least partly offset this major drawback, or perhaps the aggregate behaviour of males is merely the outcome of a constraint exerted by the females. The males of lekking species often perform vigorous displays and are generally very **ornamented**. Both these characteristics may make them more vulnerable to predators. The safety provided by aggregating (see Chapter 14) might then explain the evolution of classical leks. And it is true that cases of predation observed at leks remain the exception (Ligon 1999). However, it remains difficult to verify the causal role played by protection from predators in the evolution of leks. Another advantage for males in grouping to display may derive from the attraction the gathering of males exerts over females. It has been confirmed in various species that the larger leks draw the largest numbers of females and consequently the average number of copulations per male is higher (Höglund and Alatalo 1995). This interpretation is only partly satisfactory as it leaves unanswered the question of the actual origin of females' preference for gatherings of males. Two major hypotheses have been put forward to account for this. Females might benefit from leks by copying the choice of their conspecifics (see Chapter 4). Such behaviour might reduce the time spent choosing a mate and so be of particular benefit to inexperienced young females. However, evidence of **mate choice copying** in leks remains contradictory (Ligon 1999). Conversely, it is generally accepted that females benefit at leks from the opportunity afforded to her to make rapid comparisons among several males. This advantage is arguably decisive for species where a single copulation is enough to fertilize the entire clutch of eggs, as seems to happen in tetraonids forming leks or in species like birds of paradise that lay just a single egg.

### c Lek formation models

Several models have been developed to account for the evolution of leks.

*Females prefer males displaying in groups . . .*

The **female preference model** (Bradbury 1981) considers that females' mating strategies (being more attracted by a group of males than by an isolated male) force males to congregate. By their preference, females are thought to constrain males to cluster because solitary males would have no chance of breeding. This model may explain leks in species where males of different quality are difficult to tell apart. This model is plausible, but to explain why leks are generally located in precise spots and why that location usually remains stable over time requires reference to the existence of traditions or of places that are more suitable as display grounds.

*Hotspots . . .*

The **hotspot model** (Bradbury and Gibson 1983) attempts to answer this question by suggesting that leks form at the instigation of males in locations that are particularly frequented by females: 'hotspots'. These exceptional sites are said to be, for example, at the crossing points of the overlapping home ranges of several females. This proposal has its limits too. In practice it is difficult to differentiate between cause and effect. Is the site exceptional for males because females ordinarily frequent it or is the more marked presence of females at the site the direct result of the congregation of males? The critical point remains the great spatio-temporal stability of leks. The environment being generally somewhat unstable, females tend to modify the boundaries of their home ranges more or less continuously, which is at odds with the observed stability of leks. Again the existence of traditions can be evoked: by returning to places traditionally used for display, both males and females arguably increase their chances of finding potential mates.

*... or hotshots ...?*

Beehler and Foster (1998) in turn proposed a third model known as the **hotshot model**. According to this model, the marked asymmetry observed in mating success of males reflects far-reaching differences in the attraction some males exert over females. This model may be relevant to species in which variation in male quality is easy to detect. Males of the most attractive type are called 'hotshots' and are supposed to be relatively rare. It is in the interests of less attractive males (known as secondary males) to remain next to hotshots in order to intercept any females visiting the lek. In this model it is the females' preference for hotshot males that constrains other males to congregate around them. A major argument for the hotshot model comes from evidence of the deterioration of the lek after experimental eviction of the dominant males (Robel and Ballard 1974).

An alternative version of the hotshot model has been advanced by Kokko and Lindström (1996) who propose the involvement of kinship selection in the evolution of leks. Unattractive males would gain an indirect benefit by grouping with other more attractive related individuals whose breeding success they would help to enhance. Some work has indeed revealed a degree of kinship structure within leks in several bird species (see Höglund *et al.* 1999; Petrie *et al.* 1999; Shorey *et al.* 2000). Even so, the existence of such structuring is not an irrefutable argument for the involvement of kin selection in the evolution of leks and other alternative explanations may also account for the phenomenon (Sæther 2002). For example, the concentration of related males within the same lek might simply reflect a heterogeneous distribution of female preference through the various leks of a given species (Sæther 2002). The involvement of kin selection in the evolution of leks certainly deserves greater attention in future to determine whether it is a general phenomenon or is limited to a few species and to judge what direct or indirect benefits males derive from clustering.

*... or then again black holes?*

Ungulates mammals usually form harems. However, in several species living in herds whose composition is unstable, oestrus females frequently leave their usual group to join males defending aggregated mating territories. It has been suggested that these movements result from an excessive level of harassment by harem-controlling male or after the intrusion of young males (Clutton-Brock *et al.* 1992). Stillman *et al.* (1993) showed that when females join the territory of a neighbouring male, it becomes particularly advantageous for males to aggregate their territories. Thus, in such species, harassed oestrus females deserting harems are captured by the attraction of clusters of males that are called **black holes**.

*Models that are compatible*

Each of the different models proposed to explain the evolution of leks receives some empirical (Ligon 1999; Jiguet *et al.* 2000; Jones and Quinnell 2002) and theoretical support but none of them can be considered a universal explanation. The relevance of each model varies with the species considered. Until now, special attention has been given to birds and ungulates whereas the formation of leks in insects has received less attention (see though Jones and Quinnell 2002). Although no study so far has been able to test the **predictions** of the various models simultaneously, it is likely that for many species several factors concur in establishing and maintaining leks. Lastly, the evolution of leks being above all a matter of spatial distribution, the significance of the mobility of males and females in the various scenarios of lek evolution should be addressed more closely in the future (Jones and Quinnell 2002).

### 12.3.3 Polyandry

The term **polyandry** is used in a general sense to designate the association of one female with several males in the course of a breeding season. In most cases, the males provide parental care. Polyandry is an infrequent mating system in the animal kingdom. It concerns a limited number of species of fish and birds, and a few primate species.

#### 12.3.3.1 *Two types of polyandry*

Two variants of polyandry are distinguished (Ligon 1999). In **classical polyandry** there is a role reversal between the sexes. During a breeding cycle one female mates sequentially or simultaneously with two or more males whereas a male forms a bond with just one female. Each male has its own nest where it incubates the eggs and cares for the brood with the female having little if any involvement in parental care. This peculiar system is observed in some insects (see Smith 1980; Choe and Crespi 1997), a few fish (see Berglund *et al.* 1989; Jones *et al.* 2001), and some birds (Ligon 1999; Bennett and Owens 2002). In the case of **cooperative polyandry** two or more males are associated with one and the same female during a breeding attempt or throughout a breeding season. This system is observed in some birds (see Ligon 1999; Bennett and Owens 2002) and few primates (Dixson 1998).

#### 12.3.3.2 *Females that play the role of males*

**Classical polyandry** is doubtless the most puzzling mating system in the animal kingdom (Clutton-Brock 1991; Ligon 1999). One of its essential characteristics is that females confront each other for access to males. The fierce competition among females is probably at the root of the reversal in sexual dimorphism: females of polyandrous species are generally larger and more brightly coloured than males (Jehl and Murray 1986). Such sex-role reversal is typical of several pipefish species, where females have a higher reproductive potential than males and increase their fitness through producing eggs for several males (see Wilson *et al.* 2003). Classical polyandry has also been described in 11 bird species, but is particularly well represented in the order Charadriiformes and especially among small waders. Despite several comparative studies (Erckmann 1983; Szekely and Reynolds 1995; Ligon 1999; Bennett and Owens 2002; Andersson 2005; see also Section 12.4) it has proved very difficult to identify the historical or ecological factors that might have favoured the evolution and persistence of classical polyandry, which continues to intrigue biologists.

*Females jacana that fight for male territories . . .*

An excellent example of simultaneous classical polyandry is found in jacanas (Jenni and Collier 1972; Butchart *et al.* 1999; Emlen and Wrege 2004a, b). Jacanas are tropical birds that nest in marshlands. The jacana's mating system corresponds to sublease territory where female territories are superimposed upon the mosaic of male territories. Males defend small territories against other males. They build a floating nest and they alone incubate the eggs. Females court males and confront each other to control larger zones covering the territories of several males that each incubates separate clutches of eggs. In the bronze-winged jacana (*Metopidius indicus*) female territories can encompass from one to four male territories, depending on female competitive ability, and males in larger groups defend smaller territories (Butchart *et al.* 1999). Therefore, both male and female territory size contribute to the degree of polyandry in this species. Sexual dimorphism favours larger female jacanas: they are about 50–60% heavier and have more developed spurs on their wings than males (Davison 1985; Butchart 2000; Emlen and Wrege 2004a). In this sex-role-reversed species, females typically contribute very little to parental care.

*. . . Females that are infanticidal*

In the wattled jacana (*Jacana jacana*), for instance, males were the sole caretakers of chicks for 97% of broods ($n = 252$; Emlen and Wrege 2004b). Females seemed to engage in parental duties only when the male was unavailable to provide care himself. This may happen when females lay new clutches for males that are still tending dependent young or when males are killed by a predator, leaving the brood undefended. A particularly interesting aspect of the mating system of wattled jacanas is the existence of infanticides performed by females. Steve

Emlen and co-workers experimentally removed two polyandrously breeding females and then observed the behaviour of the incoming females that competed to establish themselves on the removed females' territories (Emlen *et al.* 1989). After the removal of each resident female, female neighbours quickly attempted to extend their territories to encompass the vacant area and the resident males they contained. Replacement females actively searched for and attacked the chicks, causing the death of five out of nine chicks, and evicting two others. Males tried to oppose females' aggression unsuccessfully. Three of three incoming females adopted infanticidal behaviour and all four broods present in the vacant territories were attacked. Although the sample size is limited (for obvious ethical reasons, see Emlen (1993)), the similarity between the behaviour of female jacanas and that of male lions is quite striking and provides further support for the adaptive value of infanticide in nature.

*Females that prevent males from accessing other females*
In other bird species such as the dotterel (*Charadrius morinellus*) and the red-necked phalarope (*Phalaropus lobatus*) sequential classical polyandry is observed (Erckmann 1983; Schamel *et al.* 2004). A single female pairs successively with different males, whereas each male generally mates with just one female. During the association, the female prevents her mate from having access to other females. This type of polyandry is less stringent though, than simultaneous polyandry. Although incubation and parental care are provided mostly by males, females, depending on species, may also participate. Moreover, the polyandrous system is not uniformly distributed within a given species but varies in frequency among populations.

*Males that are pregnant*
In pipefish (Syngnathidae), females transfer eggs to the male brood pouch, where he then fertilizes them. Embryos remain in the brood pouch where they are nourished and oxygenated for several weeks before birth (Berglund *et al.* 1986). This mode of fertilization ensures 100% paternity confidence to males (Jones *et al.* 1999). Pregnancy is, however, costly to males as they feed less and grow more slowly than females during this period. Berglund *et al.* (1989) showed that the male pipefish (*Syngnathus typhle*) has a limited brooding capacity, of about half the size of the clutch produced by a large female. Consequently, a female can raise her fitness by dividing her clutch among several males. Molecular studies have shown that in the Gulf pipefish, *S. Scovelli*, up to four males can simultaneously bear the clutches of the same female (Jones *et al.* 2001). In species where males are large enough to accommodate a complete clutch, females are still able to produce another clutch before the embryos from her previous clutch have completed their development. Thus, female reproductive potential is always higher than that of males. Typically, male pipefish are more **choosy** than females, whereas females compete more intensively than males for access to mates. In accordance with sex-role reversal, females tend to allocate more time to mating activities, whereas males tend to allocate more time to feeding, thus increasing their parental ability (Berglund *et al.* 2006).

*Unusual situations that have evolutionary explanations*
Various factors may contribute to explanations of classic polyandry in natural populations. Evolutionary interpretations of polyandry in shorebirds first focused on **energy limitations of females** and the **frequency of clutch losses**.

According to the 'replacement clutch hypothesis' (Erckmann 1983; Oring 1986), high rates of clutch loss and subsequent re-nesting, favoured female emancipation from incubation and brood care to restore energy reserves for additional clutch production. The importance of clutch loss might be relevant to the evolution of polyandry in some shorebirds. Indeed, predation on nests and young can be important, and might have had a direct influence on sex-role reversal in jacanas. In southern India, for instance, over 90% of the clutches of the bronze-winged jacana can be depredated before hatching (Butchart 2000). Only females able to lay multiple clutches at short intervals might then be able to compensate for egg loss due to heavy predation and regular flooding of nests. This may have selected for larger females, while maintaining clutch size constant. Larger females eventually became dominant over males, resulting ultimately in a polyandrous mating system.

Energy limitation on females, however, may not be a general explanation for the evolution of polyandry in shorebirds, as evidenced by a study of red-necked phalaropes (see Schamel *et al.* 2004). In an individually marked population where the rate of clutch loss was experimentally manipulated, females that produced a second clutch were found to take the same amount of time to complete their clutches compared with females producing their first clutch at the same time period. Moreover, egg size was larger in second clutches compared with first ones, making energy limitation unlikely. Detailed analysis of mating patterns showed that the proportion of females that became polyandrous was actually limited by males choosing to re-nest with their original females following breeding failure. Males may choose to remain with the same female in order to decrease the probability of caring for eggs potentially fertilized by a female's previous mate. Female propensity to become polyandrous might therefore be limited by male mate choice.

The evolution of polyandry in pipefish is probably linked to a major change from male care for eggs in the nest, as observed for instance in sticklebacks, to pregnancy where males carry eggs on or in their body (Wilson *et al.* 2003). At an intermediate stage, eggs might have been glued to the male's body rather than to the nest (Berglund and Rosenquist 2003). It has been proposed (Andersson 2005) that this change in type of male care may have evolved in relation to a change in feeding behaviour. Compared with sticklebacks, to which they are phylogenetically closely related, pipefish are ambush predators that rely more on manoeuvring and crypsis than on swiftness in order to capture their prey. Reduced swiftness may have rendered male pipefish more vulnerable to predators and more at risk when defending a nest and its eggs, hence the habit of carrying the eggs on or in their body, even at the expense of reduced fecundity.

#### 12.3.3.3 *Males that cooperate in raising the young of a single female*

**Cooperative polyandry** differs radically from classical polyandry in the stable character of the association between one female and at least two males and in the sharing of parental care among all the protagonists of the male–female association. This mating system is rare and has evolved independently of classical polyandry (Ligon 1999). There are few reliable data on the sharing of paternity among the various males in a polyandrous association. Generally all the males mate with the female, even if there is a social hierarchy among males. What might induce a dominant male to share a female with other males? The solution to this evolutionary puzzle lies in the ecological conditions that cooperative polyandrous species face (Ligon 1999). In most known cases, an alliance among males seems more effective than an isolated male in defending an important resource. When the resource is very valuable, the benefits of an alliance among males are supposedly greater than the costs implied by shared paternity.

The best studied case of cooperative polyandry in birds is that of the Galapagos hawk (*Buteo galapagoensis*; Faaborg *et al.* 1995; DeLay *et al.* 1996). In this species the average number of (unrelated) males associated with one female varies between two and three but may be as many as eight! The groups are territorial and it seems that the largest groups are better able to defend or even extend their territory. No aggressive interaction has been observed among males to copulate with the female and molecular markers reveal that paternity is indeed shared among the males of the group. Typically the productivity of groups is low with an average of one or two young that fledge each year. The number of young to be raised is therefore always lower than the number of adult males in the group, meaning that more often than not males rear and take care of young they have not sired. The longevity of adults seems to allow the association to stabilize through reciprocal cooperation.

In primates polyandry has been observed in some species of marmosets and tamarins (Dixson 1998) although with an optional character. Terborgh and Goldizen (1985) suggested that polyandry is adaptive in the saddle-back tamarin (*Saguinus fusicollis*) because of constraints inherent to raising twins which, when weaned, represent 50% of the female's body weight. The addition of a second male to a breeding pair can therefore reduce the cost of parental care borne by each individual. The costs and benefits of polyandrous association for males are not precisely known. The occurrence of multiple paternity within polyandrous associations of marmosets and tamarins has neither been proven nor disproven and kinship structures among individuals remain poorly known. However, it seems that sexual promiscuity of females tends to extend beyond conception, which favours the persistence of bonds between the female and males and means males cannot ensure their possible paternity. In golden lion tamarins (*Leontopithecus rosalia*) groups contain two adult males that are unrelated to the breeding female (Baker *et al.* 2002). Although both males copulate with her, access to the female during oestrus appears to be monopolized by the dominant male (Baker *et al.* 1993). Cost of polyandry to subordinate males might be reduced in this species through association between related males (i.e. brothers or father and son), although 25% of male duos involve two unrelated individuals (Baker *et al.* 2002). However, association between two unrelated males tend to persist for less time and to be less cooperative than those between related males. The extent and adaptive value of polyandry among marmosets and tamarins still need to be clarified.

### 12.3.4 Monogamy

**Monogamy** has always been the subject of special attention. Two reasons may explain this interest (Wickler and Seibt 1983; Reichard 2003). First, the debate about monogamy as the ideal or even 'natural' family structure for humans, even if the discussion remains largely open (Cartwright 2000; Cézilly 2006), bestows a certain value on monogamous species as 'biological models' for the study of emotional relations and social bonds within human couples (Carter *et al.* 1999). Second, the asymmetrical costs and benefits of sexual promiscuity for males and females (see Chapter 11) make monogamy a paradoxical mating system which *a priori* would seem to be counter-selected in males (this argument also applies to polyandry).

From the point of view of social organization, monogamy corresponds to a mating system where one male form a bond with one female. This bond may last for all or part of a breeding season or extend over several successive breeding seasons, or even be maintained for an entire lifetime. In most cases both sexes provide parental care. In genetic terms, however, monogamy implies an exclusive sexual relation between the two mates or at least that the offspring raised by the pair is indeed the outcome of copulation between the two mates exclusively. In practice, these two facets of monogamy are only rarely presented at the same time. Accordingly today we distinguish social from genetic monogamy (Wickler and Seibt 1983; Gowaty 1996). Until the late 1980s, for want of confirmation, it was implicitly considered that if two mates formed a social association to raise young then that automatically implied genetic monogamy; that is, that all the young were the genetic offspring of both members of the pair. With the advent of genetic fingerprinting techniques in the late 1980s it became obvious that the two aspects of mating systems were not necessarily connected. In the case of monogamy, the social and genetic systems are often completely dissociated from each other to the extent that genetic monogamy (also called strict monogamy) is now considered an exception.

Social monogamy is infrequent and unevenly distributed in the animal kingdom. It concerns a limited number of species although they are spread over a wide range of zoological groups. In invertebrates it is encountered, for example, in some opisthobranch molluscs (Rudman 1981), various coleopterans

(Klemperer 1983; Trumbo 1992), and some crustaceans (Wickler and Seibt 1983; Matthews 2002). In vertebrates it occurs in fish (Fricke 1975; Barlow 2000), amphibians (Gillette *et al.* 2000), reptiles (Bull 2000), and mammals (Kleinman 1977; Dixson 1998; Runcie 2000) but is only really dominant in birds (Ligon 1999; Bennett and Owens 2002). However, the association between one male and one female takes on very different aspects depending on the species. In invertebrates in particular the association is usually short-lived and limited to one breeding episode. Long-term monogamy, implying the repeated association of the same partners during successive breeding episodes, is rarer and is only commonly observed in birds, some rodents and a few primates (however, see Bull (2000) for an example in a lizard species).

#### 12.3.4.1 *Monogamy within the breeding season*

The fact that monogamy is not a ubiquitous mating system suggests that it depends on specific environmental conditions. Different hypotheses have been advanced to explain the appearance and persistence of monogamy in various zoological groups, with birds remaining the most studied group.

##### a *Ecological constraints on males*

One hypothesis is that monogamy is in some sense a 'default' mating system for males (Emlen and Oring 1977; Davies 1991). This hypothesis applies when the capacity of a male to breed concurrently with more than one female depends directly on the resources he can control. Where resources are evenly spread over space, it would be impossible for males to control a sufficient quantity to ensure breeding with more than one female. In a sense, monogamy would be the only possible system until the polygyny threshold is attained (see Section 12.3.2.1b). This argument seems to hold for some species. In the burying beetle (*Nicrophorus defodiens*) the mating system varies within a given population, with some males being monogamous and others polygynous. Breeding in this species depends on a critical resource, the carcass of a small vertebrate. Small carcasses are hardly enough to ensure the development of a single maximum sized litter (Trumbo 1992); they are usually exploited by a single pair. The mating system is more variable on larger carcasses. Trumbo and Eggert (1994) have shown that the mating system is largely controlled by males. The males are able to attract females to a carcass they control by giving off a sexual pheromone. Only 6% of males on a small carcass (10–15 grams) with a single female and 13% of males on a large carcass (45–60 grams) with four females emitted the pheromone. Conversely, more than 60% of males on a large carcass with a single female attempted to attract additional females (Figure 12.3). This experiment shows that males estimate both the value of the resource and the number of females before giving off their pheromone. The establishment of a monogamous or polygynic system in N.

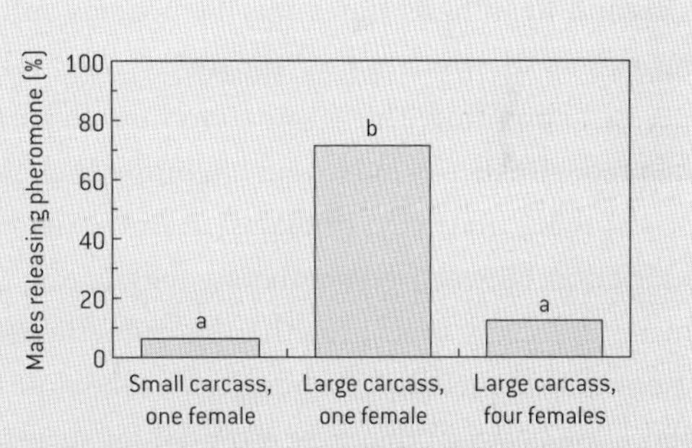


Figure 12.3 **Percentage of male beetles *Nicrophorus defodiens* observed to emit the pheromone for sexual attraction by experimental treatment**

The letters above the histograms show statistical differences between treatments (bars with the same letters are not statistically different). On small carcasses, males do not emit the sexual pheromone, whereas on large carcasses with few occupants (with just one female present) they seek to attract other females. However, if the large carcasses already have many occupants (four females present), then the males do not try to attract other females.

After Trumbo and Eggert (1994).

*defodiens* seems therefore to depend directly on males' immediate breeding interests (Trumbo and Eggert 1994).

Even so, it does seem that in several species monogamy is independent of the polygyny threshold. For example, Veiga (1992) tested experimentally whether male house sparrows (*Passer domesticus*) were forced into monogamy because they were unable to defend more than one nesting site successfully. The experiment involved placing new nest boxes next to those used the previous year and observing the consequences on mating systems. Although the males defended more nest boxes when these were placed next to each other, defending several boxes had no effect on the males' mating success. This outcome suggests that monogamy of male sparrows is not related to resource defence.

### b *Importance of paternal care*

A second hypothesis states that monogamy reflects the extent to which **parental care** is essential for successful breeding. This is said to account in part for the predominance of monogamy in birds (Lack 1968; Orians 1969) and some primates. In birds, the close association between homeothermy, egg-laying, and the need for very rapid growth to shorten the period of exposure to predation is thought to make close collaboration between the sexes essential for successful breeding (Oring 1982), and this constraint may have been crucial in birds' evolutionary history (Ligon 1999). However, today, homeothermy, oviparity, and predation do not preclude polygyny and polyandry in some species. In primates, monogamy largely concerns small species where the weight of the litter is high compared with the mother's body weight (Dunbar 1988), which supports the view that the need for paternal care is an important component of monogamy.

The essential character of parental care in monogamous birds has been appraised in various experimental studies. The protocol usually involves removing the male at various stages of breeding and observing the repercussions of his absence on the brood's success. A review of all such studies (Bart and Tornes 1989) concluded that, overall, the presence of males is indeed essential to fully successful breeding. However, the importance of the male's presence varies with the stage of breeding and with species. In some passerines in temperate regions, it even seems that the presence of males during the feeding period is not essential (Bart and Tornes 1989). Anyway, although the importance of paternal care is generally clear in birds, it cannot be considered a universal explanation for the evolution of social monogamy (Van Schaik and Dunbar 1990; Dunbar 1995) because monogamy does not necessarily imply the existence of paternal care. A study by Komers (1996) on the monogamous antelope Kirk's dikdik (*Madoqua kirki*) has shown that although males do not provide parental care they do not attempt to become polygynic either, even if the opportunity arises.

### c *Monogamy and territoriality*

Social monogamy may even exist without any form of parental care as in the Chaetodontidae (butterfly fish), a family of marine fish that form stable long-term bonds where neither the male nor female care for their offspring (Wickler and Seibt 1983). It seems that the benefit of monogamy lies in better territorial defence. Several socially monogamous species are territorial and it may be that, in some cases, monogamy represents a cooperative strategy that allows them to defend their common territory more effectively (see Mathews (2002) for an example in a monogamous crustacean).

### d *Synchronized breeding*

The polygenic potential of males may also be limited by the extent to which females synchronize their availability. Reproductive synchrony may result from an unstable environment, particularly in some opportunistic species that rely on the sudden appearance of favourable environmental conditions to breed. While the synchronization hypothesis cannot account easily for the origin of monogamy, it may nonetheless contribute to its persistence in some wild populations.

### e Female vulnerability to predation and infanticide

It has been suggested that the risk of predation and/or infanticide could be behind the evolution and persistence of monogamy (Clutton-Brock 1989; Van Schaik and Dunbar 1990). In some antelopes, males do not provide parental care but do seem particularly vigilant and better able than females to detect predators (Dunbar and Dunbar 1980). The importance of predation risk for other monogamous species, particularly primates, seems much less obvious though (Van Schaik and Dunbar 1990).

In primate species whose lifestyle does not favour association among females, Van Schaik and Dunbar (1990) suggested that monogamy evolved as a result of the risk of infanticide. As seen in lions in Section 12.3.2.2, in some primates, lactation induces a period of amenorrhoea during which females are not sexually receptive. Under such circumstances, the establishment of a lasting bond with just one mate would be favoured as it would ensure the males' protection of the females and their offspring against the assaults of other males (see Section 12.3.2.2). This hypothesis is, however, far from being applicable to all monogamous primates. Although cases of infanticide have been observed in different primate species, particularly with changes of dominant male in polygynic species, they are poorly documented and their importance for the evolution of primate mating systems seems debatable (Dixson 1998).

### f The active role of females in maintaining monogamy

The previous hypotheses tend to consider the balance between monogamy and polygyny essentially from the males' point of view. This point of view overlooks the importance of females' behaviour in maintaining monogamy. It has been suggested that in some species monogamy can be imposed on males through the synchronization of females' fertile periods (Knowlton 1979), but this hypothesis has not yet received strong empirical support. The tendency of males to engage in polygyny may also be countered by females' behaviour, particularly their aggressiveness towards their rivals. In various monogamous bird species, already mated females may prove extremely aggressive towards other females looking to enter the pair's territory (Arcese 1989; Dale *et al.* 1992, see Cézilly *et al.* 2000b for a review). Aggressiveness between females may have various functions that are difficult to tease apart. In particular, it may serve to ward off the risk of intra-specific nest parasitism. In several bird species, some females may try to lay all or part of their clutch in another female's nest, particularly if their own nest has been destroyed or fallen victim to a predator before all the eggs have been laid (Yom-Tov 2001; and see Chapter 17 on nest parasitism). Whatever the evolutionary origin of female aggressiveness, it seems to contribute to limiting the opportunities for polygyny in several bird species. Female aggressiveness may also play a decisive role in establishing a monogamous system in other zoological groups, particularly among certain primates (French and Inglette 1989; Dunbar 1995; Dixson 1998).

*Not necessarily mutually exclusive hypotheses*

As we have just seen, none of the hypotheses currently advanced can by themselves explain all cases of monogamy throughout the animal kingdom. Then again, several of these hypotheses are not exclusive such that in many cases monogamy may have arisen out of the combined action of several of these hypotheses simultaneously.

### 12.3.5 Lasting pair bonds between breeding seasons

Social monogamy comes into its own with long-lived **iteroparous** species in which individuals' reproductive lives extend over several breeding episodes or seasons. This raises the problem of choosing a mate at each breeding attempt. Studies conducted within populations of marked individuals of birds (Black 1996), primates (Dixson 1998), lizards (Bull 2000), and fish (Vincent and Sadler 1995; Matsumoto and Yanagisawa 2001) show that in some species there is genuine social mate fidelity. However, this social fidelity varies greatly from one species to another and even between populations of the same species, or among individuals within a population, which requires seeking out the causes of such variation. Once again, on the subject of monogamy most studies of social fidelity concern birds.

Various hypotheses have been put forward to explain the maintenance or dissolution of bonds between mates in monogamous birds.

### a *Site or mate fidelity?*

A simple hypothesis connects mate fidelity to breeding site fidelity and considers the former a simple by-product of the latter. Males and females, because they return seasonally to the same nesting site, end up finding the same mate without there being any adaptive advantage to such temporally persistent bonding. Fidelity to the breeding site is thought to be conditioned by reproductive success at that place.

Conversely, it may be argued that site fidelity in fact evolved to make it easier to find the same mate at each breeding episode when mates are not associated outside of the breeding period. Cézilly *et al.* (2000a) evaluated the likelihood of these statements in a comparative study of the relationship between site fidelity and mate fidelity in bird species of the order Ciconiiformes (seabirds and waders). The results showed that, in these species, mate and site fidelity are significantly correlated independently of phylogeny. Reconstruction of the joint evolutionary scenario of the two variables (see Chapter 3) is depicted in Figure 12.4. The most likely ancestral state corresponds to the absence of both site and mate fidelity. Site fidelity seems to have arisen before mate fidelity, which would be consistent with available knowledge of fossil species and their environments (Cézilly *et al.* 2000a). However, it seems that more recently site and mate fidelity evolved independently. Over the course of evolution, the advantages associated with maintaining the pair bonds (through joint experience of mates, see Black 1996) probably favoured the appearance of other mechanisms allowing mates to recognize one another and to maintain their bonds from one reproductive episode to another.

### b *A matter of compatibility*

Historically, the first truly adaptationist interpretation of divorce in birds was made by the British researcher John C. Coulson (1966) who argued that the persistence of bonds between mates is related to their degree of compatibility, independently of the intrinsic quality of the individuals. In support of this hypothesis, Coulson (1966) noted that in the black-legged kittiwake (*Rissa tridactyla*) mate changes or 'divorces' occurred essentially among young individuals after a reproductive failure. For him, mating of individuals had to proceed by trial and error leading eventually to a compatible pair forming a lasting bond. In this scheme, divorces are in the interest of both mates, with neither gaining anything by remaining associated with an incompatible individual. However, the notion of compatibility, as introduced by Coulson (1966) in terms of the mates' ability to coordinate their parental activities, remains difficult to measure. Moreover, the higher divorce rate observed in younger age groups may simply reflect the inferior quality of young individuals, related, say, to inexperience.

### c *Divorce prompted by the search for a better option*

The compatibility hypothesis was called into question by the 'better option' hypothesis (Ens *et al.* 1993), which considers that divorce is generally a 'unilateral' action undertaken by one of the two mates to improve its breeding status. In monogamous species, the mating process necessarily occurs within a limited time during which it is scarcely possible to appraise the quality of all potential mates in a population (Real 1990; Sullivan 1994). However, quality (health, vigour, resource potential, ability to provide parental care) varies among individuals. If the mating process implies a mutual choice by males and females, one expects to observe positive assortative mating for quality within pairs. In the absence of any time constraint for mate selection, the best males and best females should be the first to mate together, leaving no other choice for the next best individuals

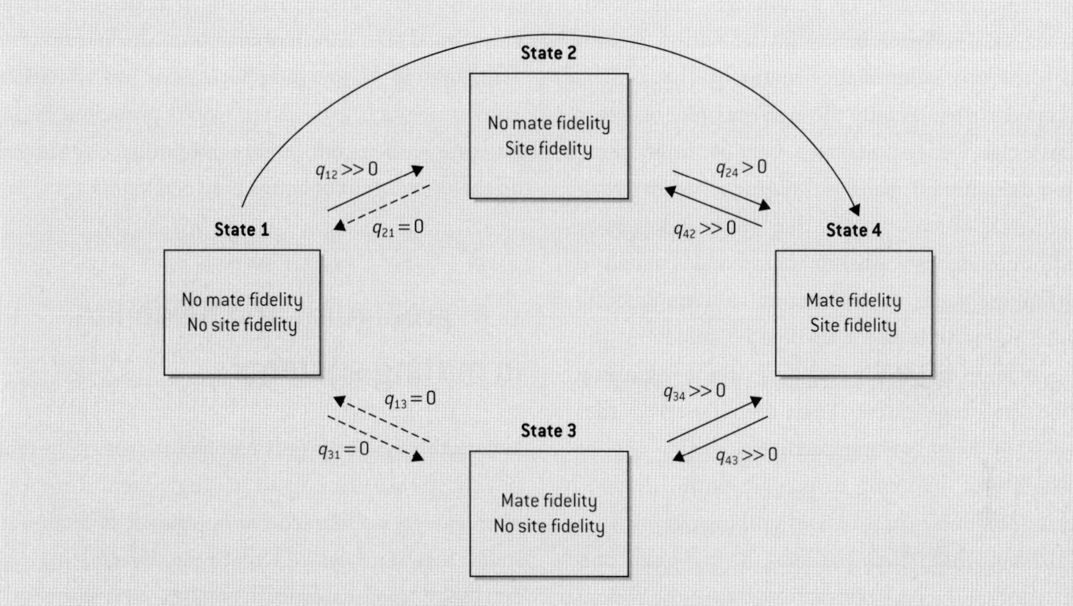


Figure 12.4 **Scenario of evolution of site and mate fidelity in Ciconiiformes**

The ancestral state is one of no fidelity to either site or mate, in the left-hand box. The arrows show the transitions between different possible states of the two variables (here mate fidelity and breeding-site fidelity). The dashed arrows are not significantly different from zero, indicating the corresponding transition did not take place significantly along the phylogenetic tree. The > 0 arrows indicate transitions that are significant, whereas the >> 0 arrows indicate transitions that are highly significant. The commonest evolutionary pathway between the ancestral state with no fidelity and the derived state with both mate fidelity and breeding-site fidelity is via an intermediate state with site but no mate fidelity (arrow above the flow diagram). This indicates that mate fidelity developed after site fidelity, probably as a consequence of breeding-site fidelity.

Modified from Cézilly *et al.* (2000a).

but to pair up, and so on. In practice, however, having limited time to choose a mate inevitably leads some individuals to pair with mates of lower quality (Johnston and Ryder 1987). Divorces are supposedly instigated by those same individuals when an opportunity arises to pair up with a better quality mate. Such opportunities might arise, for instance, after the death of the mate of a good quality individual.

*Two testable predictions*

Unlike the compatibility hypothesis, the best option hypothesis predicts that after a divorce one individual alone improves its reproductive status (the individual that prompted the separation), with the other individual being in some sense the victim of the divorce. It also predicts that the divorce rate should be higher in populations or species with higher death rates. In practice, it is often tricky to say which mate prompts the divorce (see Cézilly *et al.* (2000b) for a review of this topic).

However, such predictions have been tested by correlational, experimental, and comparative tests. In blue tits (*Parus caeruleus*; Dhondt and Adriaensen 1994) and willow tits (*P. montanus*; Orell *et al.* 1994), females but not males improved their breeding success after a mate change. In the great skua (*Catharacta skua*), Catry *et al.* (1997) observed that most divorces seemed to be initiated by females that mated with a new male soon after separation, whereas their previous mates needed more time to attract a new female. However, the most convincing study remains that of Otter and Ratcliffe (1996) of the American black-capped chickadee (*P. atricapillus*). This monogamous species forms bands in the winter within which there is a dominance hierarchy that can be evaluated experimentally through the order of access to artificial feeders. Females benefit from being mated to dominant males by gaining a higher priority of access to food. Hence, they are in better body condition and survive better than females mated to low-ranking males. Ken Otter and Laurene Ratcliffe created opportunities for divorce by removing the mates of high- or low-ranking males at the beginning of the breeding season. They then observed that females of neighbouring territories had a higher probability of divorcing to pair up with a higher-ranking 'widower' than a lower ranking one.

Although these studies support the first prediction inferred from the best-option hypothesis, the second prediction concerning an association between divorce and death rates could not be validated. A comparative study (Ens *et al.* 1996) accounting for the effect of phylogeny covering 76 species of monogamous birds failed to find a link between the divorce and death rates. Among very long-lived species, mate fidelity from year to year may vary from close to 0% in the Greater flamingo (*Phoenicopterus ruber roseus*; Cézilly and Johnson 1995) to close to 100% in some albatross species (Warham 1990). However, another comparative study (Dubois *et al.* 1998) finds that in Ciconiiformes the degree of **coloniality** influences the duration of pair bonds: divorce rates are higher in species living in larger or very dense colonies, a result consistent with the best-option hypothesis assuming that there are more mate switching opportunities in dense or large colonies.

*The effect of reproductive success on the divorce rate*

The immediate determinants of divorce are still poorly known. Both the incompatibility and the best-option hypotheses suggest that the probability of divorce should rise after an unsuccessful breeding attempt. A quantitative literature review (Dubois and Cézilly 2002) finds that, overall, a failed breeding attempt significantly increases the probability of divorce. However, most studies are correlational and few have controlled for the age of mates. Experimental tests in species with contrasting demographic strategies (brood size, length of breeding season) allowing **manipulation** of breeding success of individuals of known age and experience would provide a better estimate of the real influence of reproductive success on the dissolution of bonds between mates.

## 12.4 Ambiguity and flexibility in mating systems

Monogamy is not the only ambiguous mating system. In fact, the partitioning of mating systems into four main categories soon comes up against its limits. In some cases, it proves difficult to classify a mating system into a single category depending on whether one takes the point of view of the males or of the females (Ligon 1999; Shuster and Wade 2003). Thus in some species, the females are free to move around alone or in groups and to cross territories defended by different males in succession. Depending on their physical strength and stamina, males are able to defend portions of space that may attract varying numbers of females. When a female enters oestrus on a male's territory, the male is at leisure to inseminate her. The set of more or less contiguous zones defended by males forms a **male matrix system** (Gould and Gould 1989). This matrix is generally unstable because territorial boundaries change regularly as a result of confrontations between neighbouring males in more or less ritualized form depending on species. Such an arrangement is found in species as different as insects (see Fincke *et al.* 1997; Greenfield 1997), birds (Ligon 1999), and mammals (Owen-Smith 1977).

### 12.4.1 Differences from male and female points of view

In the damselfly (*Calopteryx splendens xanthosthoma*), females lay their eggs in water and are on the lookout for stretches of streams where the current is strong enough to ensure good oxygenation of their eggs (Siva-Jothy *et al.* 1995). Males confront each other to obtain control of the zones most favourable for egg development and establish their territory there. Females patrol the male matrix and then settle on a particular zone where they mate with the resident male. Copulation occurs in the immediate vicinity of the female's oviposition site (Gibbons and Pain 1992). In the rhea (*Rhea Americana*) males establish their territories at the beginning of the breeding season; females move in groups over the males' territories. They lay their eggs in a male's nest to whom they leave all the work of incubation and chick rearing, and then, if environmental conditions are favourable, they may leave the first male to lay in another male's nest. In both instances, there is a mixed mating system, combining resource-monopolization polygyny (from the males' point of view) and sequential polyandry (from the females' point of view; Oring 1986; Ligon 1999). Such mixed mating systems can be further classified as **polygynandry** (when both sexes have several mates, but males are more variable than females) or **polyandrogyny** (when both sexes have several mates, but females are more variable than males; Shuster and Wade 2003). However, distinguishing between true polygyny, true polyandry, polygynandry, and polyandrogyny might not always be easy in practice.

### 12.4.2 Mating systems are dynamic

The division of mating systems into four separate categories also has the major drawback of obscuring their often dynamic character. In many species, the mating system is not at all fixed and may be flexible depending on circumstances. It may vary between populations and within a population of the same species (Zabel and Taggart 1989; Davies 1992; Roberts *et al.* 1998; Thirgood *et al.* 1999; Jiguet *et al.* 2000).

#### a Variations between populations

In many deer species, males form leks when the population density is high but defend harems or resource-rich territories when the density is low (Clutton-Brock *et al.* 1988). Competition among males is probably less intense with low population densities, so reducing the cost associated with resource defence. However, increased density does not necessarily have the same effect in other organisms subjected to different ecological conditions. Thus, at low population densities, the males of some dragonfly species patrol large areas in search of females, whereas at higher densities, they defend small territories (Sherman 1983).

#### b Variations within a population

The best example of intrapopulation variability in mating system is given by a small passerine bird, the dunnock (*Prunella modularis*). Dunnocks show a flexible mating system that reflects the various outcomes of sexual conflict (Davies 1992). Females benefit from copulating with two males, because each one will then participate to brood care. Conversely, males benefit most from polygyny through fertilizing several broods. The resulting mating system is a sort of compromise between the competitive abilities of each sex and the mating options available in the environment. Monogamous males are those that are successful at driving off second males. If they have failed to expel their rival, dominant ($\alpha$) males will make the best of a bad job by attempting to prevent the second ($\beta$) male from copulating with the female. Females, however, regularly attempt to escape the $\alpha$ male's **mate guarding**, and actively solicit copulations from the $\beta$ male. Females will also try to chase off any second female that would attempt to settle on the territory. Sometimes neither sex is successful at obtaining its preferred mating option, and the outcome is cooperative polygynandry, in which two males are simultaneously mated to two females. Social monogamy, polygyny, polyandry, and cooperative polygynandry can all coexist in a single population of dunnocks (Davies 1992).

#### c Categories that poorly describe system diversity

Even within a 'sub-system', the sheer diversity of situations often eludes any attempt at too fine a categorization. This is the case for leks, particularly in insects (see Shelly and Whittier 1997). Strict application of the criteria for a classic lek set out above would exclude many insect species that nonetheless meet several (though not all) of the criteria (Shelly and Whittier 1997). Faced with such a difficulty, Bradbury (1985) recommends considering instead that, aside from absence of paternal care, which is an absolute and necessary criterion, the other criteria that define a classic lek form instead a multidimensional space within which a diverse set of species may fit depending on their ecological conditions and taxonomic position.

#### d *Mating systems are shaped by environmental conditions*

In many cases, social organization during the breeding period is the outcome of an interaction between ecological conditions, population dynamics and certain demographic parameters. The study by Carlson and Isbell (2001) of the patas monkey (*Erythrocebus patas*) illustrates this point. This species is particularly sexually dimorphic because the body weight and canine length of males averages 1.8 times that of females. Males provide no parental care, which in primates is typical of a harem-type mating system. Field studies in Africa have shown that during the breeding season there may be groups of females associated with one or several males. Carlson and Isbell (2001) studied a troop of patas monkeys in Kenya over four consecutive years. During this time, the troop's social organization proved to be flexible: several males resided in the troop and copulated with females in one breeding season, whereas in two other seasons, various males monopolized access to females in turn.

Paradoxically, the simultaneous presence of several males was observed at a time when the troop counted just 10 receptive females, whereas in the other years there were 14 or 15 females ready to mate. The authors related their observations to other studies of different species (genus *Cercopithecus*) where the females showed some flexibility in their mating system, to produce a general causal model clearly bringing out the interaction between population dynamics (female group size), ecological factors (mainly food resource availability) determining both the density of males around the troop and the capacity of females to produce offspring, and a particular demographic parameter, the length of time between litters. In species where this interval is short (less than 12 months) there is generally a fairly large proportion of sexually active females in each breeding season, which increases the likelihood of additional males arriving in the troop. In this complex system, the variability of ecological conditions over time and **demographic stochasticity** contribute directly to the flexibility of the mating system.

### 12.4.3 The gap between apparent and real mating systems

As we saw earlier for monogamy, there may be a considerable gap between the social and the genetic mating systems. In evolutionary terms, what matters ultimately for gene transmission is indeed the genetic system more than the social system. The example of the **extra-pair paternity** (EPP) rate in socially monogamous birds clearly shows the scale of this gap: a review finds that in socially monogamous birds, from 0 to 55% of chicks are 'illegitimate' and the proportion of broods containing at least one 'illegitimate' chick may reach 87% (Griffith *et al.* 2002).

The gap between these two facets of mating systems is so large that is often leads to rather surprising situations. For instance, the razorbill (*Alca torda*) described in detail in Section 14.3.4 was considered typically monogamous with high among-season mate fidelity. However, Richard H. Wagner (1998), a behavioural ecologist from the United States and now senior scientist at the Konrad Lorenz Institute for Ethology in Austria, reported that in fact two different systems worked in parallel: a classic and quite visible social monogamy in which mates share the reproductive load; and a genetic system strangely similar to a classic lek hidden behind the social system; hence the name 'hidden-lek' hypothesis

(Wagner 1998). The existence of such a double system can be understood as the outcome of conflicting interests between males and females (see Chapter 14).

Such a phenomenon may call for a reversed attitude towards monogamy: rather than trying to assess the benefits of mixed breeding strategies, it might be more rewarding to identify the ecological factors and breeding strategies of the two sexes that can lead to the existence of populations or species with strict monogamy without **extra-pair copulations** (EPCs). In particular, one might wonder how males of genetically monogamous populations manage to control their paternity fully. Although true monogamy clearly remains an exception, understanding its evolution and maintenance may greatly improve the understanding of sexual conflict as a whole.

## 12.5 Parental behaviour and mating systems

Involvement of each sex in parental care is a key feature of mating systems. The extent of care provided to offspring varies very widely from one species to another, in part jointly with the species' biodemographic strategies (**semelparity** versus **iteroparity**), their modes of reproduction (viviparity versus oviparity), and the number and size of eggs laid or of young born (Clutton-Brock 1991; Rosenblatt and Snowdon 1996). Within a species, among-individual variation is often related to the parents' age and experience (Pugesek and Diem 1990; Clark *et al.* 2002) as well as the offsprings' age (Maynard Smith 1977; Cézilly *et al.* 1994).

Compared with earlier approaches, behavioural ecology studies parental care in terms of balancing the costs and benefits of both parents and their progeny. Depending on mating systems, costs and benefits may vary widely among protagonists (male, female, and offspring). Identifying the various costs and benefits is therefore a first step towards an overall understanding of the division of parental care and its repercussions for mating systems.

### 12.5.1 Costs and benefits of parental care

The terminology used in the literature may sometimes be confusing and make it difficult to appraise the costs and benefits associated with parental behaviour. To clarify matters, it can by useful to distinguish between three components: parental care, **parental expense**, and **parental investment** (Clutton-Brock 1991).

#### 12.5.1.1 *Parental care*

> Parental care covers any instance of parental behaviour that increases the fitness of the offspring. It includes the preparation of nests and burrows, production of eggs provisioned with reserve substances, care for eggs and young both within and outside the parent's body, feeding of the young before and after birth, and care that may be provided after the young become able to feed themselves.

It should be noted that some parental care, such as the provisioning of food, is necessarily shared among the young of a single litter but may be appropriated by only some of them. The efficacy of care declines when there is a large number of young. Such care is termed **depreciative care**. It is contrasted with **non-depreciative care**, such as guarding against predators, the effectiveness of which is largely independent of the number of young (Lazarus and Inglis 1986; Clutton-Brock 1991). The relevance of this distinction has been tested in a species of duck, the goldeneye (*Bucephala clangula*), in which the chicks are nidifugous, leaving the nest just 48 hours after hatching. Thanks to their rapid development, they quickly find their food on their own. Parental care, provided by the female alone, consists essentially of watching over the young and defending the family feeding territory. The size of the territory tends to increase with the number of chicks in the brood. Ruusila and Pöysä (1998) reported that females caring for a brood spent more time watching over the environment and defending their territory against

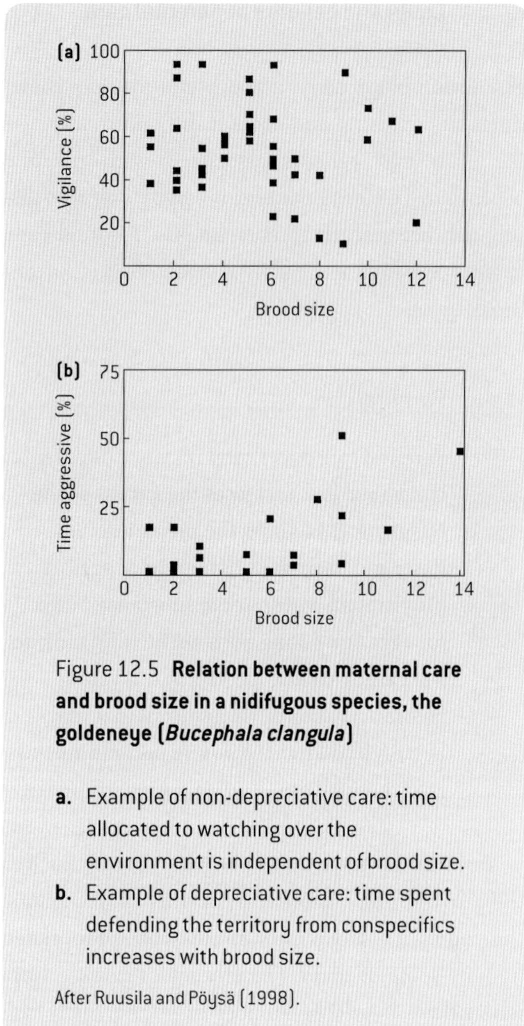


Figure 12.5 **Relation between maternal care and brood size in a nidifugous species, the goldeneye (*Bucephala clangula*)**

**a.** Example of non-depreciative care: time allocated to watching over the environment is independent of brood size.
**b.** Example of depreciative care: time spent defending the territory from conspecifics increases with brood size.

After Ruusila and Pöysä (1998).

conspecifics than females with no broods. The level of supervision was independent of brood size (Figure 12.5a), while time spent defending the territory against conspecifics increased with brood size (Figure 12.5b). The non-depreciative character of supervision was confirmed by the fact that a reduction in brood size (after the death of one or more chicks) did not entail any reduction in the level of supervision in females.

#### 12.5.1.2 *Parental expense*

The notion of parental expense reflects the share of parental resources (in time or energy) invested in parental care for one or more young. Relative parental expense corresponds to the fraction of parental resources allocated to parental care.

Parental expense is not necessarily constant over the breeding cycle. In mammals and birds, care provided for young after birth is generally much greater than the costs associated with egg production or gestation (Drent and Daan 1980). Parental expense is not necessarily related to parental input received by each of the young. When environmental conditions are particularly unfavourable, parental expense may be high, as a result of the difficulty of finding resources, even if each of the young receives a limited or even insufficient amount of resources. For example, Martin and Wright (1993) experimentally controlled brood size in several pairs of swifts (*Apus apus*). The body weight of the parents declined with the increase in size of the brood and with parental effort. However, whereas the quantity of food brought back to the nest did indeed increase with brood size, the food ration of each chick, and their average weight, varied inversely.

#### 12.5.1.3 *Parental investment*

Parental investment is defined by the repercussions of parental care on the parents' fitness. Their expense may have various consequences for their survival and reproduction in the short-, medium- or long term. For example, providing more food for the young may imply more time during which the parents are exposed to predators. In some species, postponing the date of weaning may jeopardize the females' chances of mating again quickly. **Parental investment is therefore defined as any parental expenditure that is beneficial for the offspring at the expense of the parent's chances of breeding in the future** (Trivers 1972).

In practice, measuring parental investment may be subtle because it presupposes correctly measuring all the consequences of parental expenses for organisms. In the zebra finch (*Taeniopygia punctata*) for example, the supply of proteins to eggs may induce a reduction in muscle mass (Veasey *et al.* 2000). This melting away of the muscles can increase the females' predation risk in the wild. Although crucial, it is not always easy to estimate the costs associated with parental care (cf. Clutton-Brock 1991). In particular, correlational studies that relate the number of young produced one year with the likelihood of the parents surviving and breeding in a subsequent season cannot readily control all of the factors that can affect these variables. Experimental approaches are more demonstrative. For instance, Strohm and Marliani (2002) estimated the costs associated with parental behaviour in the digger wasp (*Philanthus triangulum*) by experimentally manipulating the individuals' **phenotype**. This solitary hymenopteran feeds exclusively on honey bees (*Apis mellifera*), which it attacks and paralyses when they collect nectar from flowers. It deposits in a storage chamber of the underground nest one to five paralysed honey bees and lays an egg in one of them, and proceeds in this way for every egg. Each larva then develops while feeding on the live but paralysed bee. The cost of transporting prey to the nest is particularly high; the wasp, which weighs about 100 milligrams, must carry prey that weigh 80–160 milligrams. Strohm and Marliani experimentally increased or reduced a wasp's hunting cost both in the laboratory and in the field and found that an increase in hunting costs reduced the number of bees carried to the nest the next day whereas the opposite was observed when hunting cost was reduced. It seems therefore that the number of bees allotted to one larva has a direct effect on ability to provide for other larvae subsequently. Hunting cost is therefore a parental investment in the sense of Trivers (1972).

## 12.5.2 Parental behaviour and life history strategies

### 12.5.2.1 *Long- or short-lived species*

If investment by parents implies a reduction in their residual reproductive value, it must be expected that individuals adjust their parental behaviour so as to maximize their fitness (Williams 1996). Long-lived species in particular, whose reproductive careers extend over several seasons, are supposed to vary their investment depending on costs perceived and to avoid compromising their chances of future reproduction (Linden and Møller 1989). In such species, the level of investment should be inversely proportional to the parents' risk of dying during the period of raising the young. Most seabirds are long-lived and generally few adults die outside the breeding season. However, depending on breeding colonies or nest position within colonies, the risk of predation by various predators (large gulls, birds of prey) may vary greatly. Several studies (Harris 1980; Watanuki 1986; Harfenist and Ydenberg 1995) of different species of seabirds have shown that breeding pairs reduce their parental effort when confronted with a high risk of predation. Similarly, when energy costs associated with parental care are artificially increased, long-lived seabirds tend to reduce their level of care for the young and to maintain their own body condition (Sæther *et al.* 1993; Mauck and Grubb 1995).

### 12.5.2.2 *The role of parents' age*

The costs associated with producing and raising young must also vary with parents' age (Trivers 1972; Beauchamp and Kacelnick 1990). In fairly long-lived iteroparous species, natural selection should in theory favour more substantial investment by older individuals, because they have fewer chances of being able to engage in a new breeding attempt in the future compared with younger individuals that *a priori* have multiple opportunities to reproduce in the future (Drent and Daan 1980; Clutton-Brock 1991). Moreover, for a given reproductive **performance**, the costs associated with reproduction are often higher for young individuals and novices than for older ones. A long-term study of the Greater flamingo (*Phoenicopterus ruber roseus*) illustrates this point. In this species, one egg is laid regardless of the age or experience of the breeding pair. The cost of the first reproduction in terms of reduction of the probability of survival until the next breeding season is greater in young than in older females (Tavecchia *et al.*

2001). That this difference is not observed in males, plus other observations of the same species (Cézilly 1993), strongly suggest that the cost of first breeding is essentially related to egg production. Lastly, in species that continue to grow after reaching sexual maturity, too great a reproductive effort at too early an age may alter growth and jeopardize future reproductive potential.

*A recurrent methodological problem . . .*

Several correlational studies made from observations in the wild (Clutton-Brock 1984; Pugesek 1995) confirm that parental investment is higher in older individuals than in younger ones. However, once again, correlational studies may be biased by several factors. In particular, the progressive and non-random appearance or disappearance of categories of phenotypes within the fraction of breeding individuals of a population may, by a simple selection process, give the impression that reproductive investment varies with age whereas this is not at all the case (Forslund and Pärt 1995).

The effect of such processes of within generation selection over time, although ubiquitous in any life history approach is often overlooked. The impact of individual covariation is so strong that it may even lead to a reversal of the apparent trend in the relation between age and a given life history trait depending on whether it is analysed at the population scale (without allowing for any selection process) or individual scale (allowing for individual variations). A fine example of this type concerns variation in kittiwake (*Rissa tridactyla*) adult survival in Brittany, western France (Cam *et al.* 2002). In this study, the age effect on adult survival was first analysed without allowing for possible intrinsic variations among individuals as to their survival capacity. This first analysis revealed a significant and constant increase in survival rate with age. However, when allowance was made for possible natural selection of individuals related to their intrinsic differences in survival capacity (individuals with intrinsic low survival expectation die sooner), the opposite trend was observed: adult survival steadily decreased with age. At the population scale the fact that the proportion of individuals with high intrinsic survival ability (i.e. good quality) increases regularly over time quite simply because individuals with lower survival potential die sooner is responsible for the apparent increased survival with age at.

At the scale of parental investment, the same outcome will be observed if best-quality individuals first breed at an advanced age (progressive appearance of the best phenotypes) or if the lower-quality individuals do not survive as long (progressive disappearance of inferior phenotypes). In addition, in long-lived monogamous species with a low divorce rate, the apparent improvement in reproductive performance and parental investment with age may then result from improved coordination of the partners as pairs gain more experience (Cézilly and Nager 1996; Green 2002; van de Pol *et al.* 2006).

*. . . but few experimental studies*

Paradoxically, there are very few experimental studies comparing the parental behaviour of young and old individuals. The laboratory study by Clark *et al.* (2002) on Mongolian gerbils (*Meriones unguiculatus*) provides a rare example. It compared the parental behaviour of unmated females randomly assigned to four groups of 12 individuals differing only in the age (35, 70, 90, or 120 days) at which they were placed individually in a cage in the company of a sexually active male. The age of the females had a significant positive influence on: (1) the likelihood of their returning young to the nest that had been moved for the experiment; (2) the time spent in contact with and caring for the young; and (3) the growth rate of the young. Older females were also less likely than younger females to gestate, and if they reproduced a second time they experienced a longer gap between the two litters and produced less numerous second litters (Table 12.2). Moreover, between the birth of the young and their weaning, the older females lost more weight than the younger ones (Clark *et al.* 2002). These results are compatible with Trivers' (1972) hypothesis that older mothers invest more in their reproduction because of their lower residual reproductive value than younger females. In addition, this study provides many prospects for further

| Age of females (in days) on mating | 35 | 70 | 90 | 120 |
|---|---|---|---|---|
| Age (in days) of females on giving birth | 69.7 ± 1.5 | 106.1 ± 2.1 | 135.8 ± 3.2 | 178.9 ± 4.2 |
| Weight loss (%) between 1 and 30 days after giving birth | 2.9 ± 1.8 | −2.2 ± 1.1 | −3.3 ± 1.2 | −3.1 ± 1.7 |
| Latency time (in days) before giving birth again | 27.3 ± 0.3 | 28.1 ± 0.3 | 31.0 ± 1.9 | 31.9 ± 1.9 |
| Size of second litter at birth | 7.9 ± 0.4 | 5.4 ± 0.5 | 6.4 ± 0.6 | 5.2 ± 0.6 |

Table 12.2 **Reproductive performances of female Mongolian gerbils according to their age**

After Clark *et al.* (2002).

development. It would be particularly interesting to compare different rodent species to evaluate whether the difference in investment between younger and older individuals is directly related to their difference in residual reproductive value.

### 12.5.3 Sharing of parental care between sexes

Although parental care is often essential for successful breeding, its distribution between males and females within pairs varies largely from one species to another. In particular, the constraints on the division of parental care between the two sexes differ between ectothermal and homeothermal species (Clutton-Brock 1991). In terrestrial invertebrates and reptiles, maternal care is predominant. Males and females are involved in single-parent care in roughly the same frequencies in amphibians. Paternal care is particularly frequent in fish only.

#### 12.5.3.1 *The importance of the mode of insemination*

Generally, both within and between various zoological groups, single-parent care is provided by males in ectothermal species with external fertilization whereas females tend to provide parental care in ectothermal species with internal fertilization, even if several exceptions to this rule can be found (cf. Clutton-Brock (1991) for fuller treatment of this issue).

Various interpretations have been proposed for the preponderance of paternal care in externally fertilizing species. Trivers (1972) suggests that a male's paternity is better ensured with external fertilization because a female cannot keep the sperm of several males within her, which explains why these males are more inclined to care for the young. Trivers' hypothesis, however, has not been confirmed empirically (Baylis 1981; Beck 1998). In fact, it seems that in external fertilizing species single-parent paternal care is related more to particular circumstances such as sequentially or simultaneously reproducing females in a single male's territory, low female density or short breeding season that limit males' breeding opportunities than to certainty of paternity (Clutton-Brock 1991).

In internally fertilizing species, however, it seems that the predominance of maternal care can be explained by evolutionary constraints. Once internal fertilization has evolved, the way would be open for the evolution in turn of egg retention by females. The costs of egg retention are presumably low for females, whereas the benefits for the young are substantial (Gross and Sargent 1985). Lastly, bi-parental care is rare in most ectotherms probably because it would be scarcely more effective than single-parent care. However, when it is essential to guard and feed the offspring, parental care is depreciative, and there is thus strong intra-specific competition for access to essential resources for the proper development of the young and bi-parental care is then more common. Thus in invertebrates, bi-parental care is observed in 50% of land arthropod orders where the young depend on food regurgitated by the parents against only 14% in orders where care is limited to looking after eggs (Clutton-Brock 1991).

#### 12.5.3.2 *The role of homeothermy*

Homeotherms are confronted with a double problem: they must both feed their young and provide the thermal environment that is essential to their survival. However, parental strategies are highly contrasted between birds and mammals. In birds, bi-parental care, which is ordinarily associated with social monogamy, dominates. Conversely in mammals, males contribute to parental care in fewer than 5% of species, and single-parent male care is non-existent (Clutton-Brock 1991). Here again, bi-parental care is associated with social monogamy (Kleinman 1977; Runcie 2000). The scarcity of paternal care in mammals might be related to the mode of development of the young inside the mother and the production of milk by the mother to feed the young. Under these conditions, the contribution of males to raising the young would be marginal (Orians 1969) outside of particular ecological conditions, for example in the event of strong predation pressure (Kleinman and Malcolm 1981; Clutton-Brock 1989). In birds, because the acquisition of flight likely implied rapid development and growth, incubation and care for the young generally seems to require cooperation between both parents.

#### 12.5.3.3 *Bi-parental care*

Within mating systems where there is bi-parental care, an individual's optimal investment depends not just on the consequences of its own parental effort but also on its mate's effort (Williams 1966; Trivers 1972). A conflict between the sexes arises then whenever the optimal level of investment differs between the sexes. The parents' investment should then reflect an equilibrium between the interests of each sex (Westneat and Sargent 1996).

Various models have been developed to predict how the investment of one parent should vary with its mate's investment when there is bi-parental care (Chase 1980; Winkler 1987; Lazarus 1989). In some situations, it is predicted that the investments of the two mates should be negatively correlated, and that the shortcomings of one should be compensated for by the other. However, empirical data do not always comply with these predictions. In birds, studies that remove one of the two mates (Whillans and Falls 1990; Dunn and Robertson 1992; Markman *et al.* 1996) or constrain it experimentally to reduce its provision of care (Wright and Cuthill 1990; Markman *et al.* 1995) show that parents generally only partly make up for their mates' shortcomings. Depending on the species, compensation may be total (Wolf *et al.* 1990; Saino and Møller 1995) or nil (Lozano and Lemon 1996; Schwagmeyer *et al.* 2002; Mazuc *et al.* 2003).

Møller (2000) studied the origin of these differences among species. His comparative study suggests that the differences observed among species in the level of compensation when a mate is removed are due in part to the size of the mate's contribution. Where the contribution is modest, an isolated parent would be able to make good the mate's absence, or compensation may not even be required to ensure the survival and proper development of the chicks. This may also explain why the level of compensation may also vary with ecological conditions within a single species (Dunn and Robertson 1992).

##### a *Sex specialization in parental care*

Other factors can also influence a parent's ability to compensate for the absence of its mate or the reduction of its parental effort. Compensation may be more difficult for example if each sex specializes in certain aspects of parental care. This is particularly true of some insects where the parental tasks of each parent are rather different. Hunt and Simmons

(2002) studied the distribution of care in a species of bi-parental coleopterans, dung beetles (*Onthophagus taurus*). In this species, both mates extract small portions of excrement from mammal dung, which they deposit in tunnels made in the ground beneath the dung. Each egg is laid in a cell, which is furnished with a ball of excrement and then sealed. The quality of the excrement provided is a crucial factor in the reproductive success and fitness of the progeny. The balls of dung are placed in the cells mostly by the female (probably in direct relation with oviposition behaviour), whereas the male stores fragments of dung in the tunnels. Hunt and Simmons (2002) observed that in the absence of a male, a female beetle provides less total parental care than a pair and provides its larvae with less dung. The females were thus unable to compensate fully for the male's absence. Moreover, contrary to some theoretical predictions, there was a positive relationship between male and female investment within pairs.

### b *Importance of the timing of provision of parental care*

Hunt and Simmons (2002) also suggest that the existence of negative or positive correlations between each sex's investment depends on the type of parental care provided. In dung beetles, the investment is provided before egg laying, whereas in birds a large proportion of bi-parental care is provided after the young have hatched. Solicitation by the young might then allow the parents to evaluate their true needs and so adjust their effort depending on their mate's effort. Such adjustment would be impossible in species where parents do not interact directly with their young. This hypothesis deserves further examination.

### c *Differential allocation depending on the mate's attractiveness*

According to the **differential allocation hypothesis** (Burley 1988; Sheldon 2000), variation in maternal effort may be related to an equilibrium between the attractiveness of the male mate and the amount of care it can provide for the young. A review of the literature (Møller and Thornhill 1998) reveals that in birds, male parental care varies largely with the attractiveness of males for females. At one extreme, the females modify their parental effort as a function of their mate's attractiveness, females paired with the most attractive males being those that invest most. The costs of increased parental effort would then be offset by the advantage of producing sons as attractive as their father. At another extreme, the attraction of males is said to be directly related to their ability to provide parental care.

In support of this hypothesis, Møller and Thornhill (1998) found that, at the inter-specific level, the correlation coefficient between paternal care and the degree of expression of secondary sexual characteristics of males is negatively correlated with the frequency of EPP. In other words, where females look for indirect benefits (cf. Chapter 11), the secondary sexual characters do not indicate paternal fitness of males but rather their genetic quality. In these species, the parental effort of females should be more variable depending on their mate's quality. This prediction has been confirmed in the mallard duck (*Anas platyrhynchos*) where Cunningham and Russel (2000) observed that hens laid larger eggs when mated with the more attractive drakes.

However, a stringent test of the differential allocation hypothesis implies experimental manipulation of the degree of attractiveness of males. Mazuc *et al.* (2003) conducted such an experiment by manipulating the expression of the secondary sexual character in male house sparrows. The manipulation involved treating males with testosterone implants, a **hormone** that is responsible for development of the black bib on the males' breasts. A control group received empty implants. The study showed no effect of the increased attractiveness of males on the different components of female investment. Although the actual importance of the differential allocation phenomenon still needs to be clarified, its multiple potential implications for the evolution of parental behaviour call for extra attention (Sheldon 2000).

## 12.6 Sperm competition and mating systems

Apart from the opposition between external and internal fertilization, the direct role of **sperm competition** (cf. Chapter 11) in the evolution of mating systems has only recently been taken into account. Two aspects are particularly important. First, fundamental differences in the mechanisms of sperm competition could be responsible for the predominance of one or other type of mating system, at least in homeotherms (see Gomendio and Roldan 1993). Secondly, the consequences of sperm competition could directly influence the males' parental effort. Indeed the existence of bi-parental care is not dissociated from the search for extra-pair copulations. Many examples in monogamous birds (see Birkhead and Møller 1992) and primates (see Reichard 1995) support the importance of sperm competition in shaping mating systems.

### 12.6.1 Forms of sperm competition and mating systems: fundamental differences between birds and mammals

Gomendio and Roldan (1993) have emphasized the importance of differences in sperm competition between birds and mammals to explain the marked contrast between their respective mating systems.

In birds, females lay their eggs in a chronological sequence in a few days and are able to store sperm in their genital organs for several weeks. Sperm competition occurs in such a way that it confers a preponderant advantage on the last male (last male precedence) to copulate with the female in terms of probability of fertilizing the next egg produced. Indeed, the frequency of copulation peaks just before the onset of laying. To ensure their paternity, males must then remain beside the female and prevent her from copulating with other males. Such mate guarding could explain the evolution of stable social bonds between males and females. The advantage conferred on the last male with which the female copulates is thought to underlie guarding behaviour and also the males' investment in producing multiple ejaculates. To obtain paternity of a rival's brood, a single copulation at the optimal time may suffice. Similarly, females may exercise quite wide control over paternity of their brood through one-off, stealthy copulation. Both aspects are thought to have largely contributed to the high incidence of extra-pair copulation in socially monogamous birds.

This situation differs radically from that found in mammals (Figure 12.6). Females remain fertile for very short periods because all the eggs are produced simultaneously and remain viable for a brief period of about 24 hours. Moreover, mammals do not have special sperm storage organs. A sperm's lifespan is short and the sequence order of copulation has little effect on the probability of fertilizing the ova, or, in species where ovulation is induced by copulation, it actually favours the first male (Gomendio *et al.* 1998). The probability of fertilizing ova depends usually on the number of sperm per ejaculate, their mobility and the time when copulation occurs in the female's cycle. The best strategy for males is therefore to copulate at the time of the fertile period, but without there being any need thereafter to prevent the female from copulating with other males. As the female carries the young and lactates, the male then has few opportunities to increase the fitness of its progeny by remaining with the female and more to gain by seeking to mate with other females.

### 12.6.2 Extra-pair paternity and paternal behaviour

*A theoretical framework . . .*

Sperm competition has substantial repercussions on the evolution of parental care. Clearly enough, natural selection should favour males providing no parental care to the young they have only a slight chance of having sired. Several researchers have thus hypothesized a straightforward correlation between the probability of male paternity and the level of care they provide for offspring. Westneat and Sherman (1993) proposed a conceptual framework combining

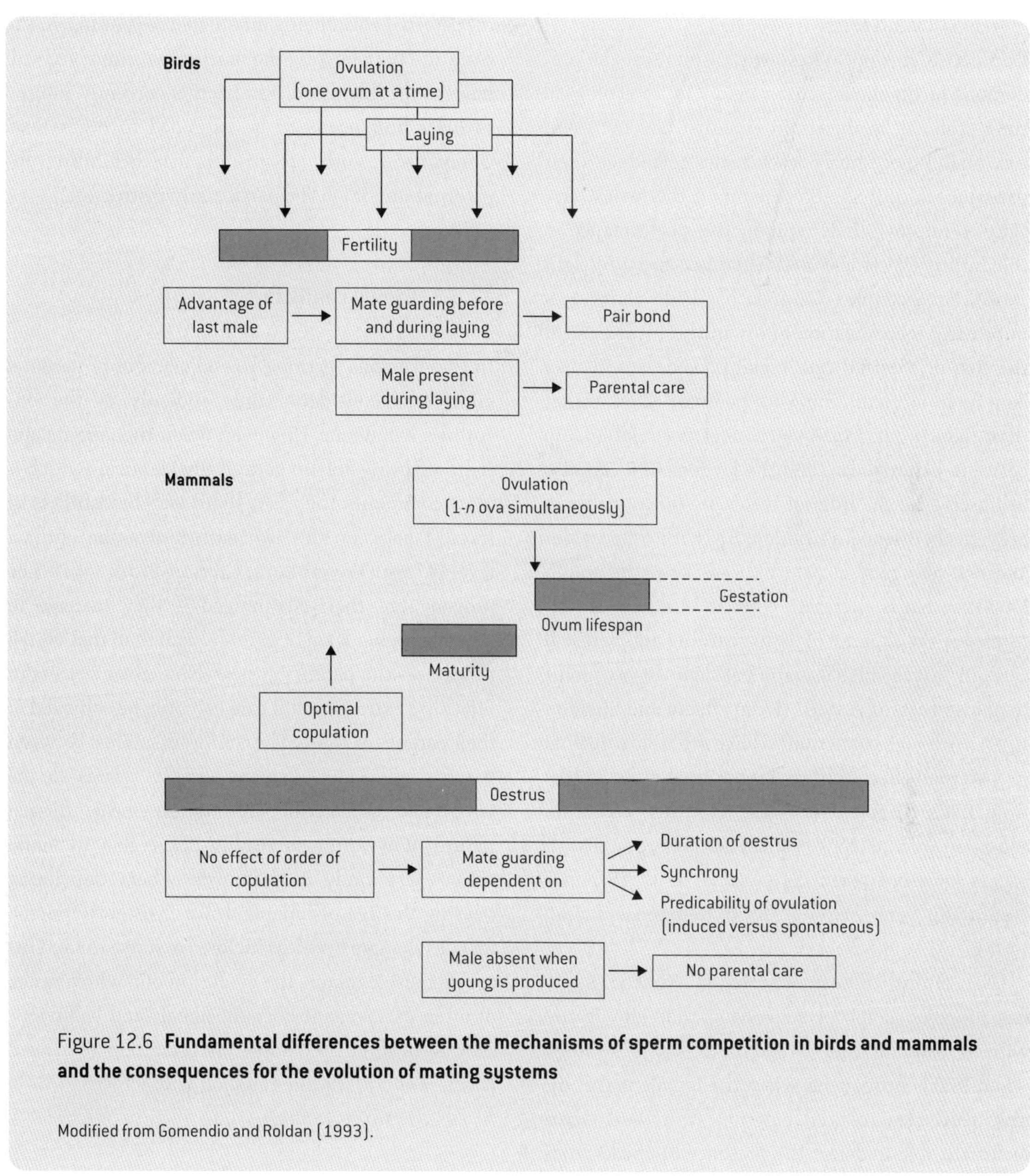


Figure 12.6 **Fundamental differences between the mechanisms of sperm competition in birds and mammals and the consequences for the evolution of mating systems**

Modified from Gomendio and Roldan (1993).

parental effort by males, their copulation effort (that is, their effort to copulate with as many females as possible) and their maintenance effort (that is, the effort males make to ensure their survival until the next breeding season). These three efforts are interdependent: resource allocation (time or energy) in any one effort is to the detriment of the other two. An optimal **trade-off** fixing allocation to each effort is supposed to evolve by natural selection. In evolutionary terms, the existence of EPP reduces the **profitability** of parental effort and should lead to an increased copulation effort and/or maintenance effort. At the same time, an increased copulation effort may increase the chances of paternity (within the brood raised by the male or through EPC) but necessarily reduces parental effort.

*... that is difficult to test*

In practice, it is not necessarily easy to test the predictions of the model in the field. On an ecological time scale, it is uncertain whether an increased risk of EPP necessarily leads males to reduce their parental effort: it should depend on the males' capacity to improve their reproductive performance in the future (Westneat and Sherman 1993; Wright 1998). For example, if a male has little chances of improving its copulation effort in the future and if the cost of paternal care remains moderate, it may benefit from maintaining its parental effort rather than reducing it. Moreover, a negative relationship between paternal care and the level of EPP can also be observed if the inferior males are both unable to prevent their females from mating with other males and not very good at paternal care, or if the search for EPC (that is, increased copulation effort) is performed at the expense of mate guarding and paternal care. In fact, available empirical data do not point to any consistant pattern (Kempenaers and Sheldon 1997). Although some studies have reported a decline in paternal effort within broods including chicks from EPCs (Dixon *et al.* 1994; Weatherhead *et al.* 1995; Lifjeld *et al.* 1998; Chuang-Dobbs *et al.* 2001), others have not reached the same result (Wagner *et al.* 1996; Yezerinac *et al.* 1996; Kempenaers *et al.* 1998).

The Finnish researcher Hanna Kokko (1999) has proposed a model suggesting that monogamy with bi-parental care would be a stable strategy when EPP is infrequent, when the female makes up only partly for reduced paternal effort and when males are able to detect the presence of chicks from EPCs. This last point certainly deserves further research. Although the opinion is that males are unable to distinguish their own young in a brood from those sired by another male (Kempanaers and Sheldon 1996), the information could be obtained indirectly, for example through the male's effort to ensure paternity, its perception of its mate's involvement in EPC, or through the density of breeding pairs and the synchronization of reproduction. Even so, the use of such information by males remains speculative. Although formally there are arguments for paternal effort being modulated depending on the certainty of paternity, the mechanisms involved still have to be elucidated. Research is currently limited to birds and needs to be extended to other taxa. Comparison with species of fish exhibiting bi-paternal care might be particularly instructive.

### 12.6.3 Family conflicts

In the previous sections, the adjustment of parental effort was considered independently of the offspring's behaviour. However, when the parents and their offspring are not genetically identical (which is most commonly the case) there may be conflicts of interest between the two parties over the optimal level of care (Trivers 1974; Clutton-Brock 1991). The progeny are then advantaged if they manage to obtain a higher level of parental care than that which maximizes the parents' fitness. Now, their success in hijacking extra parental care can only be achieved at their parents' expense. The conflict of interest between parents and their offspring usually relates to the length and intensity of parental care. The conflictual nature of weaning is readily observable in many species, especially in mammals, where the young nearing the age of independence insistently pursue females that are ever less inclined to nurse them. This conflict also explains the evolution of a whole range of often extravagant begging signals and behaviour that go beyond the simple manifestation of a need (Lyon *et al.* 1994; Kilner and Johnstone 1997).

### 12.6.4 Trivers' model . . .

Trivers (1974) was the first to propose a simple model to account for the genetic origin of parent–offspring conflict. He envisaged a situation where females produce a single offspring at each breeding episode that they raise alone and in which the cost of her parental care is measured in terms of reduced future reproductive potential. Initially, the offspring benefits from the care provided by its mother while the mother obtains an indirect benefit through the increased fitness of her offspring. The mother can therefore continue investing in the young she is raising, but at some point, it may be more advantageous for her to stop and produce a new offspring. The mother is equally genetically related to all her offspring (50%) including the one she is currently raising and the ones she will eventually produce. For her, both the benefits and costs of parental care are measured in the same way, total number of offspring produced. The interest of the offspring, however, is not completely consistent with its mother's interest because the benefits it obtains from his mother's care are measured as an advantage to itself (genetic relatedness of 100%) but the costs are in terms of the brothers and sisters its mother would fail to produce. An offspring shares on average half of its genes with a full sibling (one with the same father and mother) and a quarter with half-siblings (different fathers). For the offspring, therefore, the costs and benefits from extending maternal care do not have the same weight as for the mother. For the offspring, the costs are half (or a quarter) of those the mother must bear and so the point at which costs and benefits of maternal care balance for female correspond to a point where the offspring can still afford to extort extra care from the mother.

Let us consider the costs and benefits related to the maintenance of maternal care during the offspring's development. The female should end her investment when the ratio of costs to benefits is greater than 1. The conflict over the age at which the young should be weaned will continue, though, until the cost of care provided to the offspring is twice as high as the benefit the female derives when the next offspring is a full sibling. The benefit would have to be four times higher when the next offspring is a half-sibling (calculations follow Hamilton's rule (1964a, b)). At this stage the female's interest and that of the young converge; the offspring will gain more if the mother invests in producing a sibling than if she continues to care for it. Weaning is then in both parties' interest. Parents should therefore try to wean the young once the costs of care exceeds the benefits, while the offspring should accept to be weaned only when the benefit for the parent is twice (or four times) greater than the cost for the offspring. The conflict will vary in duration depending on how the cost-benefit ratio varies with the offspring's age (Figure 12.7).

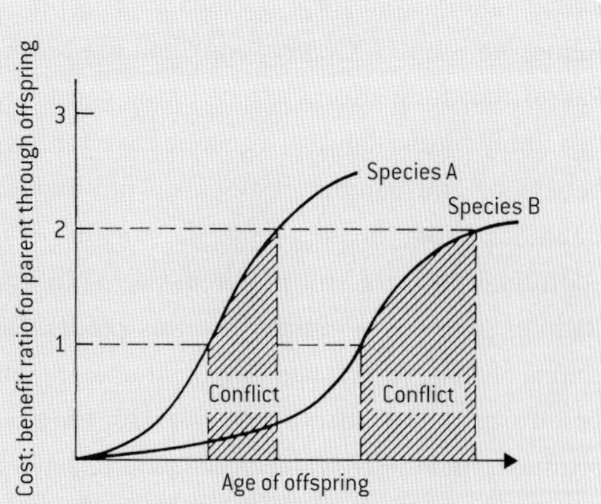


Figure 12.7 **Graph of the parent–offspring conflict model**

The cost–benefit ratio of parental care a parent obtains through its offspring rises with the age of the young (the older the offspring, the greater the care required and the less it contributes to the survival of the offspring). Differences between species may arise depending on how quickly the young develops, the left-hand growth curve relating to a more rapidly developing species. Selection favours the parent that stops providing care when the ratio is 1, that is, when costs exceed benefits. However, selection favours an offspring that stops soliciting care when the ratio reaches or exceeds 2, because the costs it imposes on its parent diminish as a function of the kinship coefficient it has with its mother's other offspring. **Parent–offspring conflict** occurs between the cost–benefit ratios for parents of 1 and 2. The faster the young develops, the sooner the conflict zone is reached.

Redrawn from Trivers (1974).

### 12.6.5 . . . and its generalizations

Trivers' (1974) original model and the various developments thereof (Alexander 1974; Parker and Macnair 1978; Parker 1985; Lazarus and Inglis 1986;

Parker and Mock 1987; Tokuda and Seno 1994; Johnstone 1996; cf. Godfray 1995 and see Mock and Parker 1997 for a critical review of the various models) imply the existence of a genetic basis for begging behaviour and parents' responses. This hypothesis now has quite broad empirical support even if additional data are still required (see Kölliker and Richner 2001 for a summary of this point). In the course of time, several additional parameters such as brood size, the number of care givers, or the age and residual reproductive value of the parents have been taken into account. Depending on the hypotheses about parent–offspring behavioural interactions and the genetic bases of solicitation by the young and parents' responses, the models predict different types of **evolutionarily stable strategy**.

### 12.6.6 Parent–offspring conflicts in harriers

In certain circumstances, the parents can retain control of weaning as in Montagu's harriers (*Circus pygargus*; Arroyo *et al.* 2002). For this species, the critical period comes after the young's first flight, when they are not effective enough to fend for themselves. Arroyo *et al.* (2002) observed that the end of the period of dependence of the young was preceded by a gradual reduction in the provisioning of prey by adults, while the capture success of the young is still moderate. Moreover, aggression directed by the young at adults was frequent and increased when the parents were reducing their feeding effort, suggesting that the reduced feeding of the young by the adults was not due to disinterest in the prey by the young but was indeed the outcome of a change in adult behaviour.

### 12.6.7 Conflicts that go as far as siblicide

It is possible too, in theory that the young manage to induce a greater parental effort, with begging behaviour of the young then being interpreted as attempts by young to 'manipulate' parents (Stamps *et al.* 1978; Parker and Macnair 1979). Alternatively, equilibrium may be achieved within the conflict, with the young then being supposed to signal their true needs **honestly** through their begging (Kilner and Johnstone 1997). The situation becomes highly complicated when the interaction no longer concerns one parent and one offspring but two parents and several young. Another form of family conflict may arise. It opposes the young and the parents, and the young among themselves, over the brood or litter size (Mock and Parker 1997). Although increased brood size may be to the parents' advantage, it tends to increase the level of competition among the young. In its most acute form, such competition may extend to siblicide, which is often termed **brood size reduction**: in several bird species where the eldest attacks and kills and may even devour its siblings. Such siblicidal behaviour has been particularly well studied in birds, especially in some birds of prey and in gannets and ardeids (herons, egrets; cf. Mock and Parker 1997, and Drummond 2001 and Drummond *et al.* 2003 for a review of the matter). Formally, it seems that competition within the broods may have a direct influence on the level of solicitation of the young (Royle *et al.* 2002) and on the evolutionary stability of parent–offspring conflict (Rodriguez-Gironés 1999). Notably, competition among chicks may influence begging intensity as satiety level does (Royle *et al.* 2002).

### 12.6.8 Do young emit honest begging signals to their parents?

In this context how honest are the young's **signals** (cf. Chapter 16)? Several workers have concluded that nestlings' solicitation reflected their food needs fairly accurately (Kilner 1995; Cotton *et al.* 1996; Price *et al.* 1996; Iacovides and Evans 1998; Lotem 1998) and that parents distributed their effort as a function of the intensity of the solicitation (Kilner 1995; Leonard and Horn 1996; Price 1998). However, other studies (Redondo and Castro 1992; Clark and Lee 1998) came up with different results. These works mainly concerned passerine species where broods number several nestlings. It is then difficult to know whether the begging is determined exclusively by the chicks' needs or is influenced by the level of competition within the brood. A study (Quillfeldt 2002) has managed to avoid this trap by considering parent–offspring interactions in a species whose clutch size is a single egg. In Wilson's storm petrels (*Oceanites oceanicus*) the chicks seem to modify their calls in their parents' presence depending on their body condition and the parents respond to more intense solicitation by increasing the quantity of food regurgitated for the chick. However, a previous study (Granadeiro *et al.* 2000) of another seabird species also raising a single chick did not report the same finding. Kilner (1995) showed that signals other than the nestlings' calls may be involved. When begging, young passerines also stretch their necks and gape. In many species, the inside of the bill is brightly coloured and Kilner showed that in canaries (*Serinus canaria*) the inside of the beak grows redder in food-deprived birds. Not surprisingly the parents preferentially fed the young, whose gaping bills showed redder insides, suggesting that they respond to a signal that honestly reflects their chicks' needs.

However, data for the barn swallow (*Hirundo rustica*) suggest an alternative hypothesis (Saino *et al.* 2000). The colour of the inside of the beak of young swallows is due to the presence of carotenoids and allegedly reflects their level of immunocompetence (cf. Chapter 11). The parents are said to feed the healthiest chicks first because they have the highest residual reproductive value. The colour of the inside of the beak is supposedly always an honest signal informing the parents of the chicks' health but not necessarily of its nutritional condition. Furthermore, the informative character of colouring has recently been called into question, at least for cavity-nesting bird species. The bright colour of the inside of the chicks' beaks is said to have evolved in these species because it makes the chicks easier to see in a light-poor environment. Heeb *et al.* (2003) manipulated the colouring of the inside of the beak of great tit chicks (*Parus major*) at two lighting levels. The inside of the beak was painted either red or yellow. In bright light, the colour of the inside of the beak had no effect on the chicks' weight gain. However, in dim light, the chicks with the insides of their beaks painted yellow gained more weight than those with the insides of their beaks painted red. Similar results were later obtained in the starling (*Sturnus vulgaris*) where chick's mouth and body ultraviolet reflectance seems to influence parental investment as UV contrast may help the parent in detecting hungry offspring (Jourdie *et al.* 2004). These results invite some caution in the matter of the honest character of coloured signals involved in parent–offspring interaction.

## 12.7 Conclusion

After Chapter 11 on sexual selection, this chapter illustrates the ubiquity of conflicts of interest between mates involved in any social interactions. It would be difficult to understand the evolution of mating systems without considering that the apparent harmony one so often sees in such relationships is only the outcome of a compromise (trade-off) between mates whose interests may diverge at any time. Likewise, it would be difficult to understand the behaviour of brood size reduction so often observed in some bird species if one started from the principle that the young of a single brood have similar evolutionary interests. Whatever the interactions, the very presence of aggressive behaviour reveals the existence of conflicts between those involved. In parent–offspring conflicts in conjunction with conflicts between the young within a litter, we are still a long way from understanding the intricacies of the processes in play, to such an extent that the honesty of begging behaviour of young towards their parents remains unclear.

## » Further reading


> *For a review on the importance of sexual selection, social behaviour, and sperm competition in the evolution of mating systems:*

**Alexander, R.D.** 1974. The evolution of social behavior. *Annual Review of Ecology and Systematics* 5: 325–383.

**Birkhead, T.R. & Møller, A.P.** 1992. *Sperm Competition in Birds. Evolutionary Causes and Consequences.* Academic Press, London.

**Dubois, F. & Cézilly, F.** 2002. Breeding success and mate retention in birds: a meta-analysis. *Behavioral Ecology and Sociobiology* 52: 357–364.

**Emlen, S.T. & Oring, L.W.** 1977. Ecology, sexual selection and the evolution of animal mating systems. *Science* 197: 215–223.

**Griffith, B., Owens, I.P.F. & Thuman, Ka.** 2002. Extra-pair paternity in birds: a review of interspecific variation and adaptive function. *Molecular Ecology* 11: 2195–2212.

**Mock, D.W. & Parker, Ga.** 1997. *The Evolution of Sibling Rivalry.* Oxford University Press, Oxford.

**Weatherhead, P.J. & Robertson, R.J.** 1979. Offspring quality and the polygyny threshold: 'the sexy son hypothesis'. *American Naturalist* 113: 201–208.

> *For a review on the evolution of parental care and parent–offspring conflicts:*

**Clutton-Brock, T.H.** 1991. *The Evolution of Parental Care.* Princeton University Press, Princeton.

**Kokko, H.** 1999. Cuckoldry and the stability of biparental care. *Ecology Letters* 2: 247–255.

**Maynard Smith, J.** 1977. Parental investment: a prospective analysis. *Animal Behaviour* 25: 1–9.

**Sheldon, B.C.** 2000. Differential allocation: tests, mechanisms and implications. *Trends in Ecology and Evolution* 15: 397–402.

**Trivers, R.L.** 1974. Parent–offspring conflict. *American Zoologist* 11: 249–264.

**Wagner, R.H.** 1997. Hidden leks: Sexual selection and the clustering of avian territories. In: *Extra-Pair Mating Tactics in Birds* (Parker, P.G. & Burley, N. eds), pp. 123–145. Ornithological Monographs, American Ornithologists' Union, Washington, DC.

> *For reviews on specific mating systems:*

**Bradbury, J.W. & Gibson, R.M.** 1983. Leks and mate choice. In *Mate Choice* (P. Bateson ed.), pp. 109–138. Cambridge University Press, Cambridge.

**Höglund, J. & Alatalo, R.V.** 1995. *Leks.* Princeton University Press, Princeton.

## » Questions

1. What are the merits and the major drawbacks of dividing mating systems into four main categories? On what bases might an alternative classification be established?

2. Within which variants of polygynic mating systems is sexual selection the most intense? Why?

3. On what conditions might social monogamy be optional or compulsory?

4. What must be checked to establish whether solicitation by young is honest signalling?

# 13

# 13

# Sex Allocation

Michel Chapuisat

## 13.1 Introduction

When came the flood, Noah took into the ark seven couples of every clean beast, and one couple of every beast that was not clean, two and two, the male and the female (Genesis 7). Similarly, the number of males and females is often, but not always, balanced in nature. In this chapter, we will examine how **natural selection** influences the relative number of each sex.

Proximately, one might think that even **sex ratios** simply result from the sex determination mechanism. If the presence or absence of a particular chromosome determines the sex of the individual, the fair segregation of sex chromosomes during meiosis will lead to one individual out of two developing into a female. However, sex determination mechanisms are very diverse (Bull 1983). Many species with non-chromosomal sex determination have balanced sex ratios, whereas species with chromosomal sex determination sometimes show highly biased sex ratios. In short, proximate constraints do not suffice to explain variation in sex ratio.

Sex ratio might also be naively seen as an **optimum**. For example, if the father and mother rear offspring together, identical numbers of males and females might appear to be optimal both at the population and individual levels. However, obligate **biparental care** and monogamy are too rare (see Chapter 12) to explain the widespread occurrence of even sex ratios in the animal kingdom.

In most cases, the sex ratio is not optimal for the population, nor for any particular individual or gene. If a male can inseminate 20 females, why aren't there 20 females for one male? Why waste resources in the production of males that will compete together for access to females, rather than maximizing the number of females and thus the **population growth rate**? We have seen that traits do not evolve for the good of the population or species (Chapters 1 and 2). **Natural selection** simply favours the fittest individuals, those that will contribute most to the genetic composition of future generations. The sex ratio is a remarkable example of the process of natural selection. Twenty females for a male might maximize population growth; however, we will see that such a sex ratio is evolutionarily unstable, simply because male-producing individuals are fitter than female-producing individuals in such conditions. Ultimately, the sex ratio does not settle at an optimum, but at an equilibrium corresponding to an **evolutionarily stable** solution (see Chapter 3) that is often an even sex ratio.

## 13.2 Fisher's sex ratio principle: an equal allocation to each sex

### 13.2.1 Sex ratio: the basic logic

**Sex allocation** is a measure of how parental resources are partitioned between male and female offspring, whereas sex ratio simply refers to the number of males and females. In his book *The Genetical Theory of Natural Selection*, Ronald Aylmer Fisher (Figure 13.1) expressed the basic principle describing how natural selection shapes both sex ratio and sex allocation. In a concise and somewhat cryptic argument, Fisher stated '. . . the sex ratio will so adjust itself, under the influence of natural selection, that the total parental expenditure incurred in respect of children of each sex, shall be equal . . .' (Fisher 1930). Fisher's principle marks the start of the modern study of sex allocation and **parental investment**. However, Charles Darwin had already provided the main elements of Fisher's principle in the first edition of his book on *The Descent of Man, and Selection in Relation to Sex* (1871), and Carl Düsing had expressed it mathematically in 1884 (Edwards 1998, 2000).

We will now develop Fisher's condensed idea. Let us first consider a simple case that satisfies the following five **assumptions**:

1. Male and female offspring request the same amount of parental investment.
2. The relative **fitness** of male offspring, compared with female offspring, does not vary among families.
3. The population is large and **panmictic** (i.e. matings are equally likely among any possible pair of adults of different sexes). Mating occurs at random, **competition** for mates is global, and there is no **inbreeding** or population genetic differentiation.

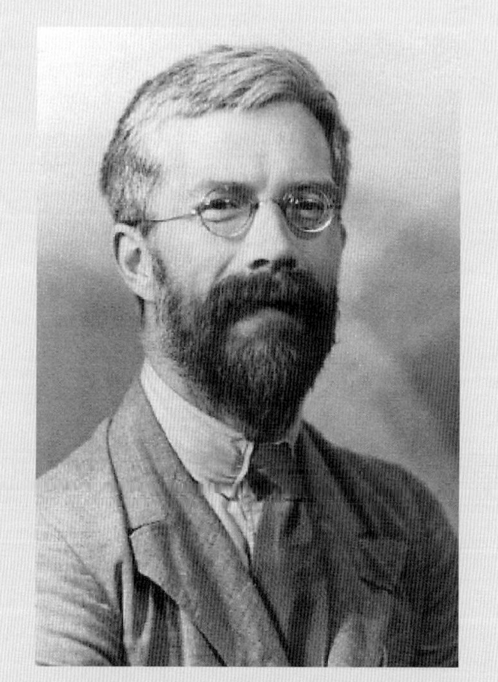

Figure 13.1 **R.A. Fisher at the time of the 1932 International Congress of Genetics**

Picture kindly provided by James F. Crow

4. The parents have full control over the sex ratio of their offspring. They are equally related to their sons and daughters.
5. Autosomal genes in parents control the sex ratio, and there is enough genetic variability to ensure that all strategies occur in the population.

This simple case is pictured in Figure 13.2. Later on, we will examine what happens when some of the above assumptions are relaxed (Table 13.1). We will see that the theory becomes more complex, but also generates interesting and testable predictions. When sticking to the simplest case, Fisher's principle can be presented as follows:

1. Suppose that males are less numerous than females (Figure 13.2a).
2. Each individual has a father and a mother. Therefore, each sex contributes to half of the genes in the next generation. A newborn male offspring will on average have a higher fitness than a newborn female, just because it is less frequent. A male offspring will thus transmit more copies of his genes to the next generation.
3. Parents genetically disposed to produce males will have more grandchildren.
4. Genes favouring male production increase in frequency in the population.
5. Male births become commoner. The sex ratio becomes less male biased, and the advantage associated with male production fades away.
6. A symmetrical process starts if females become less frequent than males.
7. The sex ratio is at equilibrium when there is exactly the same number of males and females in the population.

This is the only evolutionarily stable sex ratio in such conditions because any departure from an even sex ratio is immediately corrected so that the sex ratio is pulled back to one male for one female. When the number of males and females is equal, both sexes are equally efficient at transmitting genes to the next generation (Figure 13.2b). As long as the population sex ratio sticks to 1:1, natural selection has no effect on the tendency to produce one sex or the other and parents become indifferent to the sex of their offspring. However, as soon as the population sex ratio deviates from 1:1, frequency-dependent selection favours the parents that produce the rare sex, which pushes the sex ratio back towards the robust equilibrium of 1:1. In Box 13.1, a simple numerical example illustrates Fisher's sex ratio principle.

### 13.2.2 Equal investment, different numbers

In his verbal argument, Fisher made clear that it is the total amount of **parental expenditure** (i.e. what we called parental investment in Chapter 12) in each sex that has to be equal at the population level, rather than the number of males and females. When different amounts of resources are requested to produce a male or a female offspring, numbers should balance the per capita investment, so that the total investment in each sex is equal (Figure 13.3).

Consider, for instance, that a female offspring requests more resources than a male offspring, maybe because she needs more energy reserves. If the population contains the same number of males and females, male-producing parents are favoured by natural selection. Indeed, the fitness of males and females is on average equal, whereas male-producing parents produce more offspring, as males are cheaper to produce. The genetically determined propensity to produce males will spread in the population, and males increase in frequency to the point where the higher cost of production of females is exactly balanced by their higher fitness due to rarity. The equilibrium sex ratio is reached when the total amount of resources invested in each sex is equal (Figure 13.3).

### 13.2.3 Fisher's sex ratio principle: algebraic formulation

Fisher's verbal argument can be put into simple algebra. Assume that the production cost of a female is $c$ times the production cost of a male. At the population level, the proportion of resources invested in

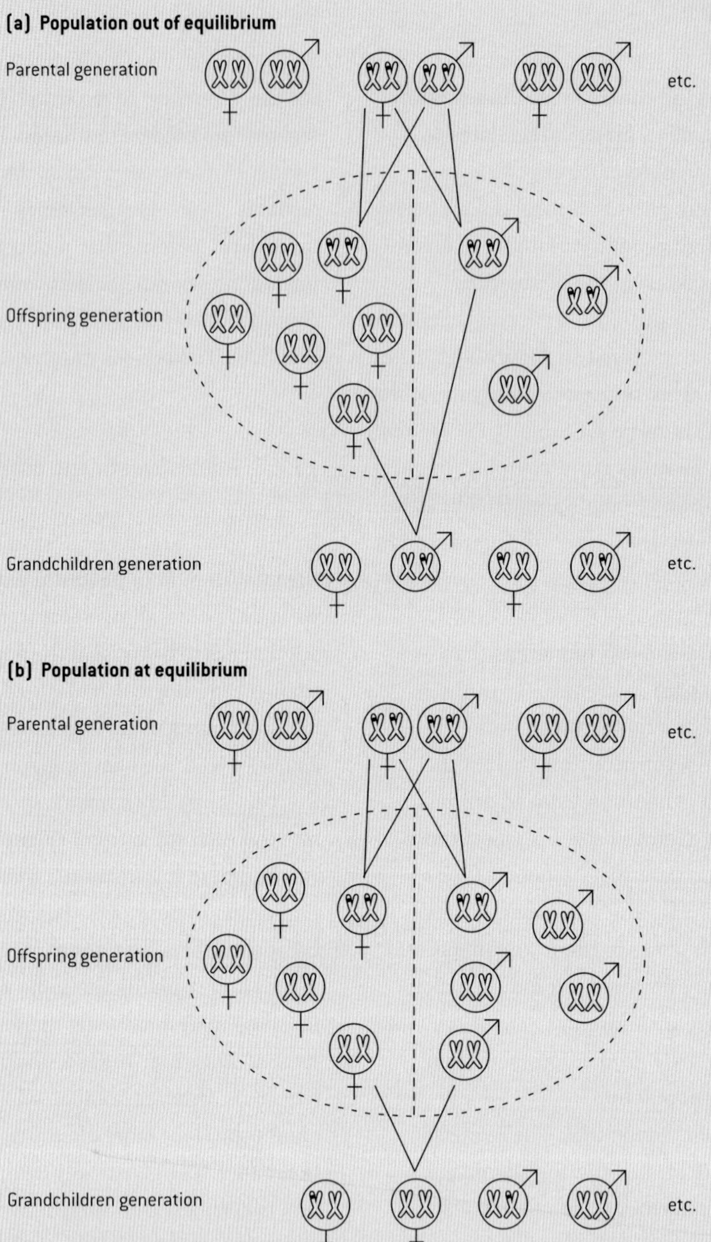


Figure 13.2 **Sex ratio in a simple case (see text)**

**a.** In this example, the population of the offspring generation contains two times more females than males. Because of this asymmetry in numbers, male offspring will on average have twice the fitness of females. Male offspring are therefore twice as valuable as female offspring for transmitting copies of their parents' genes. Hence, parents producing relatively more males will transmit more copies of their genes to grandchildren.

**b.** In this population, the number of males is equal to the number of females. The sex ratio is at equilibrium because male and female offspring are equally good at transmitting copies of their parents' genes. Black dots on the chromosomes indicate autosomal genes present in the focal parents, and transmitted to the third generation with a probability that depends on the sex ratio in the offspring of the focal couple, relative to the sex ratio in the population (see equation 1). Samples of individuals are indicated at each generation.

## Box 13.1 **A numerical example of Fisher's sex ratio principle**

Let us start with a population in which there are 20 females for one male. The fitness of a male is on average 20 times higher than that of a female, because each offspring has a father and a mother. The fitness of a mutant producing only males would thus be 10.5 times greater than the fitness of the average individual in the population that produces 1 male for 20 females (the population sex ratio).

Why 10.5? In that population, on average a male will fertilize 20 females. Thus males' fitness will be on average 20 times higher than that of females. Setting the female fitness to 1, then male fitness will be 20. The male-producing mutant will have on average N male offspring with a fitness of 20, whereas individuals with a sex ratio strategy of 20 females for a male will have $N \times 1/21$ male offspring with a fitness of 20, plus $N \times 20/21$ female offspring with a fitness of 1. Hence, the fitness of the male-producing mutant, relative to the one of average individuals in the population, is $(N \times 20)/(N \times 40/21)$, which simplifies to 10.5.

With such a higher fitness, the genes from the male-producing mutant will spread in the population, and male production will increase. As males become more frequent, their fitness decreases. When the population sex ratio reaches 1:1, the fitness of males and females is equal, and the fitness of the male-producing mutant is equal to the fitness of all other individuals, whatever their sex ratio.

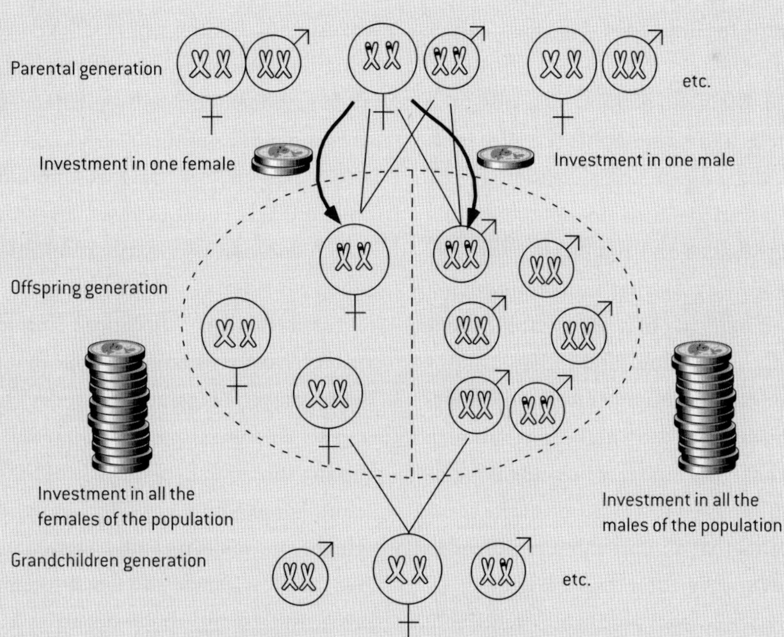


Figure 13.3 **Sex allocation when the production cost of each sex differs**

In this example, producing a female requires twice the investment needed to produce a male, as illustrated by the number of coins required for each. The figure depicts a situation where the equilibrium is exactly one female for two males. The equilibrium corresponds nonetheless to equal investment in each sex at the population level. At this point, there are two males for one female such that females transmit twice as many genes as males do to the next generation, a difference that exactly compensates for their doubled production cost.

females is $F$, whereas the proportion of resources invested in males is $(1 - F)$. Hence, the numerical proportion of each sex is $F/c$ for females and $(1 - F)$ for males. In a diploid species, half of the autosomal genes are inherited from the mother, and half from the father. On average, the fitness of each sex is inversely proportional to its frequency in the population. It is $c/F$ for females and $1/(1 - F)$ for males.

We can calculate the fitness $Wi$ of a parental **phenotype** $i$ that allocates a proportion $fi$ of its resources to daughters and $(1 - fi)$ to sons. The phenotype $i$ produces a numerical proportion of $fi/c$ daughters that have a fitness of $c/F$ and $(1 - fi)$ sons that have a fitness of $1/(1 - F)$, which sums up to:

$$Wi = \frac{fi}{F} + \frac{(1 - fi)}{(1 - F)} \tag{13.1}$$

This formula is usually referred to as the Shaw and Mohler equation (Shaw and Mohler 1953), even if Düsing had already formulated it 69 years earlier (Edwards 2000).

The Shaw and Mohler equation describes the fitness of a parent as a function of his sex allocation strategy ($fi$) and the average sex allocation in the population ($F$). It can be used to determine the allocation strategy that is the **evolutionarily stable strategy** (ESS) (Figure 13.4): that is, a strategy that if adopted by all individuals in a population cannot be invaded by any competing alternative strategy (Chapter 3). When the allocation strategy of the parents matches the mean allocation in the population, $fi = F$ and $Wi = 2$, whatever $F$ and $fi$. When the population sex allocation is biased towards one of the sexes ($F \neq 0.5$), the parents that allocate more resources to the rare sex have a fitness greater than 2 (Figure 13.4). For example, if $F = 0.6$, parents producing exclusively sons ($fi = 0$) will have a fitness $Wi$ of 2.5, as opposed to 2 for the parents with the allocation strategy matching the population's allocation. These male-producing parents do better than all other phenotypes that produce some females (Figure 13.4). The phenotypes producing the rare sex spread in the population, bringing the population allocation back to 1:1, which is the only evolutionarily stable solution. When the same amount of resources is invested into males and females in the population, $F = (1 - F) = 0.5$, and $Wi = 2$ whatever the allocation strategy $fi$ adopted by the parents (Equation 1 and Figure 13.4). This is the equilibrium sex allocation: all strategies are equally fit, as shown by the horizontal line when $F = 0.5$ (Figure 13.4). Interestingly, the equilibrium sex allocation is a characteristic of the population that can result from any combination of individual strategies, whereas the single evolutionary stable strategy is an equal investment in each sex ($fi = 0.5$).

### 13.2.4 Significance of Fisher's sex ratio principle

*Important predictions*

Fisher's sex ratio principle generates important and sometimes counter-intuitive predictions. First, sex allocation should not be affected by sex-biased mortality after the end of the period of parental investment. For the parents, the cost associated with a higher probability that offspring of one sex will die after emancipation will be balanced by the higher fitness of adults of that sex, because higher mortality also makes the sex rarer as an adult. Fisher's principle predicts equal parental investment in each sex, which does not necessarily result in equal allocation when considering the adult stage.

Similarly, sex allocation should be generally insensitive to variation in the **mating system** (e.g. monogamy, polygamy, polyandry, or promiscuity), and to all the factors that increase the variance in mating success independently of the parental investment. For example, consider a polygamous species in which one male typically monopolizes a **harem** of five females. If male and female offspring request the same investment, the sex ratio at birth and thus at emancipation, should be 1:1. Males will only have a one out of five chances of obtaining a harem and reproducing as adults, but the few successful ones will have a mating success that is five times higher.

Fisher's principle also predicts that there should be a **trade-off** between the cost and the number of offspring. If one sex requests more parental resources, it should be less frequent in exactly the

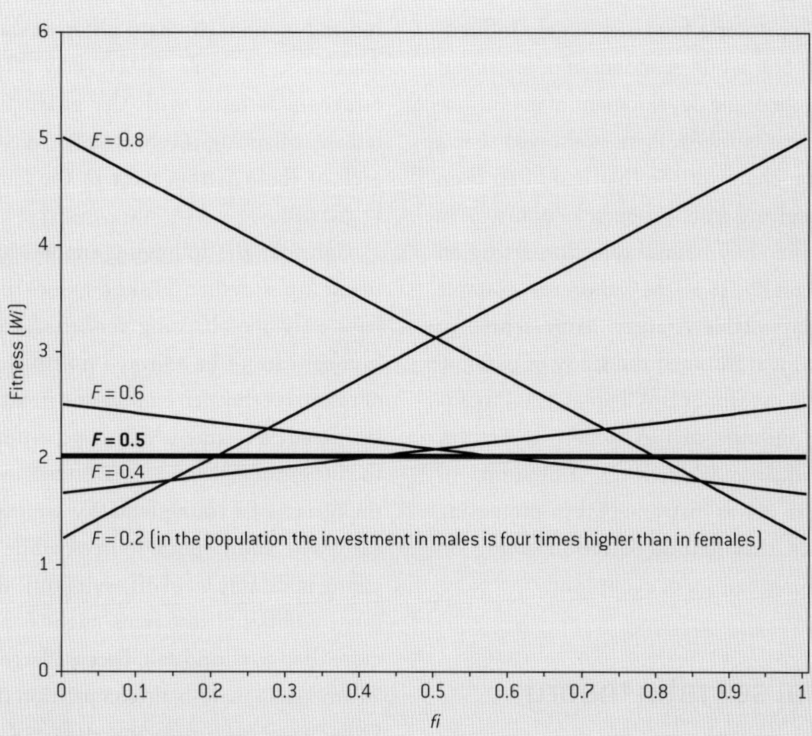


Figure 13.4 **Fitness of the phenotype i that allocates a fraction *fi* of its resources to daughters, as a function of the population sex allocation to females (*F*)**

Equation (1) can be rearranged to obtain *Wi* as a function of *fi* and *F*:

$$Wi = \frac{1}{1-F} + fi\frac{1-2F}{F(1-F)}$$

When the population sex allocation is biased towards one sex ($F \neq 0.5$), the strategies allocating more resources to the rare sex are favoured. The figure shows that an even sex allocation at the population level ($F = 0.5$) is an evolutionarily stable equilibrium, not an individual optimum. Indeed, some individuals have a higher fitness when the population sex allocation is not at the equilibrium.

After Crozier and Pamilo (1996, p. 32).

same proportion. However, only parental investment counts in this trade-off. For example, no sex ratio bias is expected if one sex becomes larger by acquiring resources independently of the parents, after emancipation.

*Empirical support*

Overall, sex allocation data from many species support Fisher's sex ratio principle. In most animal species, the same parental investment is needed to produce a son or a daughter. This investment is often limited to an ovum and sperm. As predicted by Fisher's principle, sex ratio at birth is close to 1:1 in many insects, reptiles, amphibians, birds and mammals (Trivers 1985). This general trend is found across species with highly diverse mating systems. Species with sex-biased mortality after the end of the period of parental investment, such as many mammals, generally do have even sex ratios at birth, whereas the sex ratios in adults are often highly biased. Finally, species such as solitary wasps show a trade-off between the number and the cost of production of offspring, resulting in an equal allocation to each sex (Trivers 1985).

*Historical importance*

Fisher's sex ratio principle has been hugely influential. This simple but subtle economic-like argument played a major role in the development of evolutionary biology (Edwards 1998). First, it showed that a major population characteristic that had often been presented as resulting from group selection was in fact fully determined by natural selection acting on individuals. Secondly, it is the canonical example of an evolutionary stable strategy, as presented in Chapter 3. Thirdly, it initiated the study of parental and reproductive investments (Chapter 12). Finally, by influencing authors such as Robert L. Trivers, George C. Williams, William D. Hamilton, or Richard Dawkins, it was instrumental in promoting the modern idea that natural selection acting on genes is a major mechanism of evolution.

## 13.3 General sex ratio theory

### 13.3.1 The limits of Fisher's sex ratio principle

Fisher's principle is built on the assumption that fitness returns are linear: if a parent increases its investment in one sex, the number of copies of its genes that will be transmitted by this sex will increase in proportion (Frank 1990). Fitness returns are a central concept in optimization models applied to behavioural ecology. With respect to sex allocation, fitness return is the increase in the genetic contribution to future generations that results from a given investment in new males and females.

Linear fitness returns are likely when the parental resources are divided between many offspring that will disperse and mate randomly in a large population. In such a case, the number of offspring can be precisely adjusted to the level of parental resources, and the competition is global. For example, a parent will have doubly high fitness if by doubling its investment in males it produces twice the number of sons that will mate at random in a large population.

However, fitness returns are not always linear. Moreover, the relationship between parental investment and parental fitness can differ between male and female offspring. In other words, the fitness return functions can have different shapes for sons or daughters. Multiple factors may affect the fitness returns associated with the production of male or female offspring (Table 13.1). We will examine each of these factors later in the chapter (Sections 13.5–13.8).

Consider the following simple example of non-linear fitness returns. Take a species in which females have only one offspring per brood, so that offspring number cannot be adjusted precisely to the amount of resources that can be invested in brood rearing. An increased investment can have different effects on the fitness of male and female offspring. For example, the fitness of a young female wasp may increase linearly with additional resources, while the fitness of a young male may level off asymptotically. To the contrary, additional resources might benefit a stag calf more than to a hind calf, because large males have an advantage in male-male competition (Section 13.5.2). Fisher's sex ratio principle does not apply to these examples, because the fitness return functions are not linear and differ between the sexes. **More generally, Fisher's sex ratio principle does not consider all factors and constraints that affect the fitness return through male and female offspring.**

### 13.3.2 Equal fitness return

#### 13.3.2.1 *A generalization of Fisher's sex ratio principle*

The criterion of **equal marginal value** is a very general principle that takes into account all of the factors affecting the fitness return through male and female offspring. The basic idea is to compare how many genes of the party that controls sex allocation will be transmitted through either new males or new females for a similar investment. Following Fisher's logic, it is possible to predict that **the population sex allocation is evolutionarily stable when a marginal investment of the limiting resource in either male or female offspring yields an identical inclusive fitness return for the party controlling sex allocation** (Figure 13.5).

| Factor | Main predictions |
|---|---|
| Parental condition, local ecological factors | Families specialize in the production of the sex with greater fitness return according to parental and local conditions (Section 13.5) |
| Relatedness asymmetry | At the population level, sex allocation is biased towards the sex that is more related to the party controlling sex allocation. In social Hymenoptera, bias towards females if workers control sex allocation. When relatedness asymmetry varies among families, bias towards females in families with relatively high relatedness asymmetry and towards males in families with relatively low relatedness asymmetry (Section 13.6) |
| Competition among related males (Local mate competition) | Bias towards females (Section 13.7) |
| Competition among related females (Local resource competition) | Bias towards males (Section 13.7) |
| Cooperation among related males (Local mate enhancement) | Bias towards males (Section 13.7) |
| Cooperation among related females (Local resource enhancement) | Bias towards females (Section 13.7) |
| Non-mendelian inheritance of the genes controlling sex allocation | Bias towards the sex transmitting the genes controlling sex allocation (Section 13.8) |
| Parasites and selfish genetic elements | Bias according to the prevalence, efficiency and transmission of parasites (Section 13.8) |

Table 13.1 **Predicted effects of several factors that affect the fitness returns associated with male and female production**

These factors may result in biased sex allocation at the population level. If they vary among families, they may also lead to sex ratio specialization among families (Section 13.4 and Figure 13.6).

The sex allocation in the population is at equilibrium because **the marginal value of males and females is equal**. The marginal value of a male or a female is simply the fitness return coming from a very small additional investment in that sex. The criterion of equal marginal value was expressed mathematically by Charnov (1979) and presented in various ways by numerous authors such as Trivers (1985), Frank (1990, 1998), or Bourke and Franks (1995). The mathematical formulation is close to the Shaw and Mohler equation. However, the criterion of equal marginal value is more general and can be applied to fitness return functions of all shapes, including nonlinear ones (see Frank 1998 for example). Mathematically, the marginal value is simply the derivative of the function at a given point describing how the fitness increases with respect to investment at that point.

Sex allocation is evolutionarily stable when the controlling party gains the same fitness returns when investing a small additional amount of resources in either male or female offspring. In other words, when sex allocation is at equilibrium in the population, the fitness gained by investing some resources in one sex is exactly balanced by the fitness lost by not investing these resources in the other sex. Frequency-dependence, which is the heart of Fisher's principle, is maintained. However, the stable allocation at the population level may deviate from 1:1 if the fitness return functions take different shapes for males and females (Frank 1990).

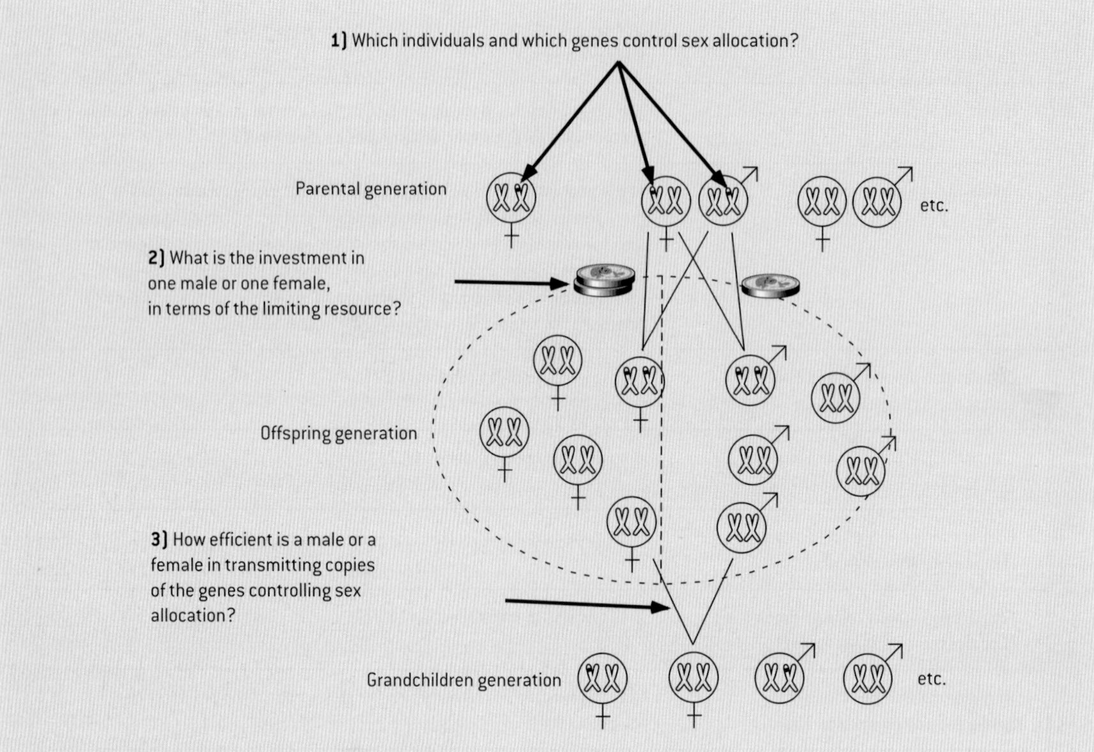


Figure 13.5 **Sex allocation and fitness return**

The population's sex allocation evolves in response to the fitness gained when investing resources in males or females. The basic question is to examine how many of the genes controlling sex allocation will be transmitted to future generations by either males or females for a given investment of the limiting resource. For the party that controls sex allocation (which can be genes or individuals, question 1 on the figure), the fitness return will depend on the investment that is necessary to produce either one male or one female (question 2), and on how good this new male or new female will be at transmitting copies of the genes present in the controlling party to future generations (question 3). Many factors can affect the fitness returns for a given investment in either males or females (Table 13.1). Black dots indicate some potential genes controlling sex allocation that are present in the focal individuals of the first generation and are transmitted to subsequent generations.

#### 13.3.2.2 *The advantages of this generalization*

The criterion of equal marginal value has two main advantages (Figure 13.5). First, it can be applied to cases where sex allocation is not controlled by autosomal genes in the parents (thus relaxing the assumptions 4 and 5 mentioned in Section 13.2.1 above). The criterion examines the fitness of the party controlling sex allocation, be it the mother, the father, other members of the social group (Section 13.6), or genetic elements with special modes of inheritance (Section 13.8).

Second, this more general model takes into account all factors and peculiarities that affect the fitness return through males and females. It can therefore also be applied to cases where the above-mentioned assumptions 1, 2, and 3 are not satisfied (Section 13.2.1). The model allows for nonlinear and sex-specific fitness return functions. Fitness return can in turn depend on multiple genetic, social, and ecological factors, as well as on **life history**, mating system, and interactions among relatives (Table 13.1). In short, this model is much more general and opens a new avenue into the study of sex allocation.

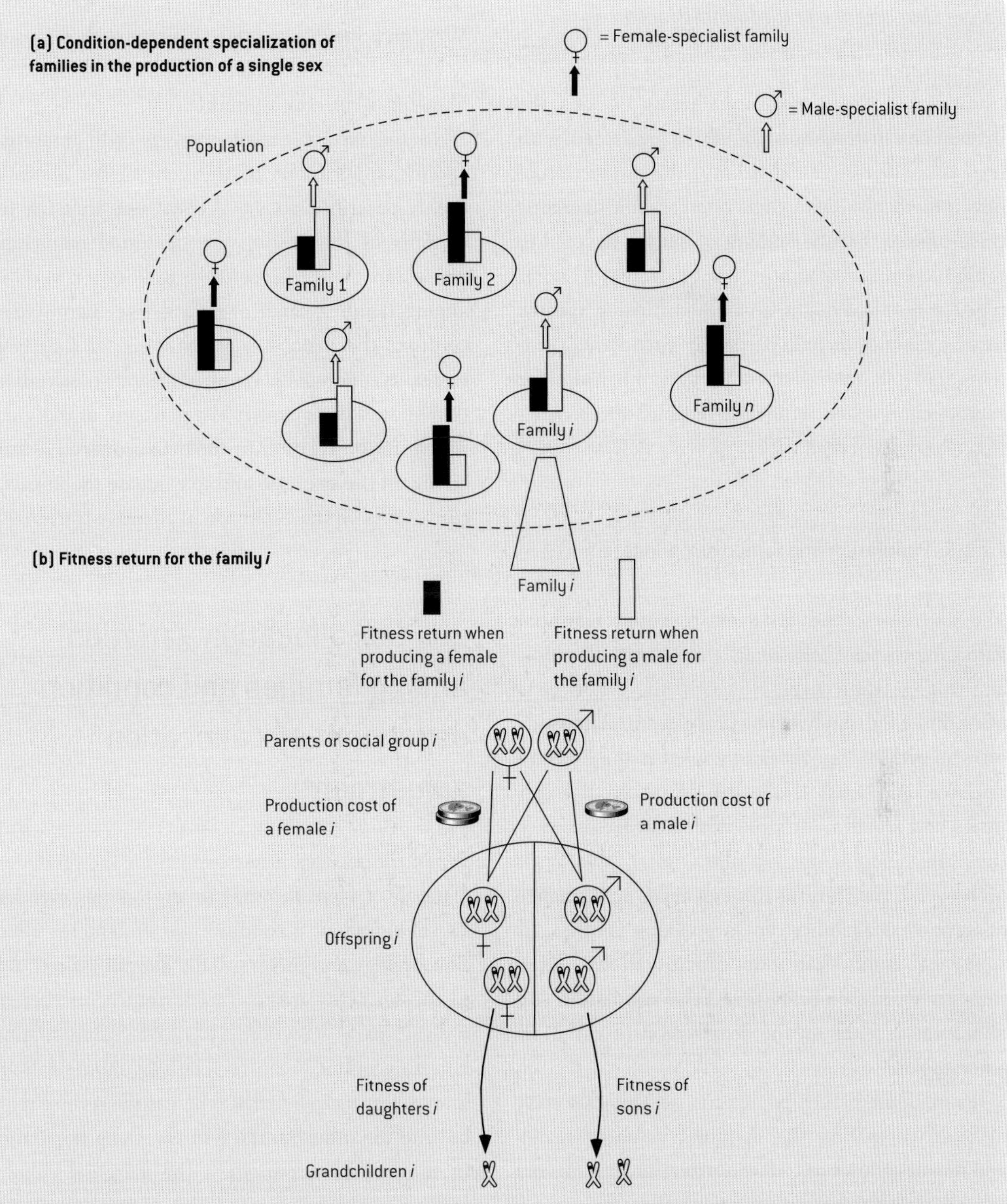


Figure 13.6 **Condition-dependent sex allocation according to variation in the fitness returns through males and females**

**a.** Sex allocation is expected to vary among families if the fitness returns for producing a male, relative to the one for producing a female, varies among families. Each family should specialize in producing the sex with highest relative fitness returns, compared with the population mean.

**b.** The fitness returns depends on the production cost and fitness of sons and daughters, as well as on the relatedness between the controlling party and the offspring. If these factors vary among families, it can lead to a condition-dependent specialization of families in one sex or the other.

#### 13.3.2.3 *Importance of the limiting factor*

Several types of resource are invested in offspring. In all sex allocation models, including Fisher's, the investment in the sexes should be measured with respect to the resource that limits the production of new individuals. Researchers often assume energetic trade-offs, so that the amount of energy devoted to reproduction is the limiting factor. However, other types of resource might become limiting in specific cases. Examples include particular nutriments (proteins, minerals, trace-elements), egg number, sites for reproduction, space in the nest, or time for reproduction (Rosenheim *et al.* 1996).

### 13.3.3 Population and family sex allocation

The above-described criterion of equal marginal value defines the evolutionarily stable sex allocation at the population level. However, the criterion is often difficult to apply in practical cases, as it depends on who controls sex allocation, what is the limiting resource, and what is the fitness return through males and females. Multiple ecological factors may thus affect population sex allocation (Table 13.1). Moreover, fitness returns through males and females, or resource availability can vary among families. In such cases, families can adopt conditional sex allocation strategies (see below), and the population sex allocation depends on the distribution of variation among families.

Even if predicting population sex allocation is difficult, comparing the marginal value of males and females allows us to make strong predictions on sex allocation variation among families or groups (Figure 13.6). If the relative marginal value of each sex varies among families, **each family should specialize in the sex providing the highest fitness return with respect to family condition and population sex allocation.** The logic of this approach is further explained in Figure 13.6, Section 13.3, and Section 13.4. Predictions on sex allocation variation among families are more robust and more straightforward to test than predictions on population sex allocation (Chapuisat and Keller 1999; West and Sheldon 2002). This approach represents one of the most powerful and fruitful current developments of sex allocation theory.

In the rest of the chapter, we will investigate specific applications of the general sex allocation theory described above. We will see that the predictions of the theory can be tested empirically, particularly with respect to sex allocation variation among families. Such empirical tests permit us to confront the logic and assumptions of the theory to reality. Moreover, empirical data on sex allocation allow us to quantify the precision of individual adaptation in relation to the evolutionary pressures acting on the sex ratio and to evaluate the impact of constraints on the evolution of adaptive traits.

## 13.4 Sex allocation variation among families and condition-dependent sex allocation adjustment

When the relative fitness return for the production of a male or a female varies among families, each family should adopt a condition-dependent sex allocation strategy and preferentially allocate its resources to the sex with greater fitness return for the controlling party (Section 13.3). For some families, investing in males can be relatively more profitable than investing in females, whereas the reverse might be true for other families in the same population. In economical terminology, the marginal value of males is greater than the value of females in some families, whereas the marginal value of females is greater than that of males in others. Hence, the first type of family should specialize in the production of males, and the second type in the production of females (Figure 13.6).

Trivers and Willard (1973) were the first to propose that parents should conditionally adjust the sex allocation of their progeny in response to their resource level. If an increase in parental investment benefits one of the sexes more, then resource-rich parents should specialize in this sex. For example, if a few of the largest males monopolize most of the reproduction, the mothers in exceptionally good condition and able to produce very large offspring should preferentially produce sons. The case of the Kakapo developed in Chapter 18 (Section 18.2.3) provides an excellent example of this situation. Trivers and Willard's model was originally proposed for polygamous mammals with a single young per litter. The model assumes that fitness return functions take different shapes for males and females, resources vary among females, and offspring number is strongly constrained.

More generally, many genetic, ecological, and social factors can generate between-family variation in the relative fitness return through each sex (Table 13.1 and Figure 13.6). For example, families may vary in one or more of the following characteristics: the quality of the mother or father, access to resources, costs and benefits of interactions among relatives, or relatedness towards each sex. Such variations among families may lead to conditional specialization of families in producing the sex with the greater fitness returns. Precise predictions depend on the particularities of the system under investigation, and can be affected by the interactions among multiple factors (e.g. resource distribution, litter size, mating system, and the impact of current reproductive investment on future reproductive prospects). Conditional allocation might also depend on the cost of manipulation and constraints of the system. Despite these complications, the study of sex allocation variation among families has provided remarkable tests of sex allocation theory.

## 13.5 Parental condition and local ecological factors

In this section, we will examine four situations in which the relative fitness return through each sex varies among families. Specifically, sex allocation theory predicts a conditional sex allocation specialization of families (Section 13.3.3) according to mate attractiveness (13.5.1), maternal condition (13.5.2), quantity of resources (13.5.3), or need for **helpers** in cooperatively breeding species where offspring of one sex help their parents (Section 13.5.4).

### 13.5.1 Mate attractiveness in birds

In birds, females should adjust the sex ratio of their clutch according to the attractiveness of their mates. They should produce more sons when mated to an attractive male, because males show greater variation in reproductive success than females do (Chapters 11 and 12). Hence, if attractiveness is **heritable**, sons will benefit more than daughters from having an attractive father. Indeed, sons from attractive fathers are likely to obtain many more mating opportunities than sons from unattractive ones, whereas the fitness of daughters will be less affected by the attractiveness of their fathers. This logic is a kin to the one of the sexy son hypothesis.

The relationship between family sex allocation and mate attractiveness has been tested in at least 21 studies involving 12 bird species. Male attractiveness was estimated in very diverse ways, such as testosterone level, size of song repertoire, song characteristics, shape, colour, or reflectance under ultra-violet light of various plumage, and even the colour of human-added leg rings! In 14 out of 21 studies, the sex ratio was significantly male-biased in clutches from attractive fathers. For example, the proportion of sons was correlated with the size of the father's frontal patch in collared flycatchers (*Ficedula albicollis*) (Figure 13.7; Ellegren *et al.* 1996). Female peafowl (*Pavo cristatus*) produced significantly fewer males when mated to males whose attractiveness had been experimentally reduced by removing eyespot feathers from the tail (Pike and Petrie 2005). A meta-analysis (an analysis of all the published results that can be found) revealed that the proportion of males in a clutch indeed increases significantly with mate attractiveness across 11 studies (West and Sheldon 2002; Figure 13.15). However, male attractiveness explained only 4% of the variance in sex ratio among families (West and Sheldon 2002). Moreover, the

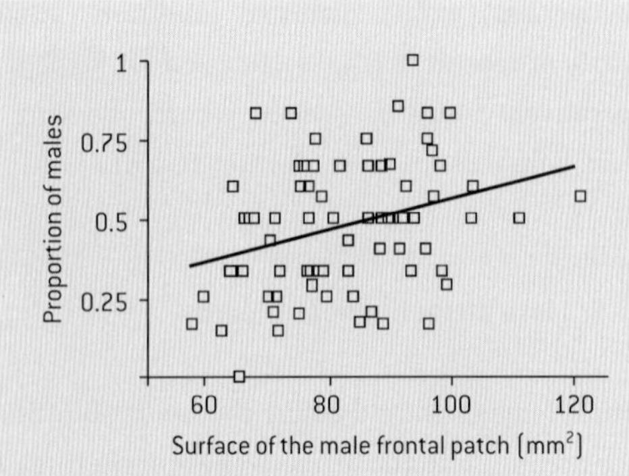


Figure 13.7 **Sex ratio according to mate attractiveness in the collared flycatcher (*Ficedula albicollis*)**

The white frontal patch of adult males is a secondary sexual character. Males with a larger frontal patch have a higher mating success and a higher probability of being polygamous. The proportion of sons in the clutch is positively correlated with the size of the paternal frontal patch ($R^2 = 0.08$, $P < 0.01$, $N = 79$ clutches).

After Ellegren *et al.* 1996.

results sometimes vary among different populations of the same species (Svensson and Nilsson 1996; Sheldon *et al.* 1999; Leech *et al.* 2001; Rosivall *et al.* 2004; Dreiss *et al.* 2006), or over years in the same population (Griffith *et al.* 2003).

### 13.5.2 Maternal condition in ungulates and primates

Maternal condition can also affect sex allocation, but precise predictions depend on the mating system and life history. In some cases, an increase in maternal investment may profit male more than female offspring, particularly when there is strong **sexual selection** on male size (Trivers and Willard 1973; Chapter 11). If litter size is strongly constrained, mothers in good condition should specialize in male production, and those in bad conditions in females. This prediction has been tested in many ungulate species and corresponds to the case of the kakapo developed in Chapter 18. The results show a large variability both within and among species, suggesting that many **environmental** and life history factors may affect the relationship (Hewison and Gaillard 1999). However, a meta-analysis based on data from 37 studies of 18 ungulate species shows a weak but significant positive correlation between maternal condition and male-biased sex ratio (Sheldon and West 2004). The correlation is stronger when maternal condition is measured before conception or estimated from behavioural dominance data. The relationship also depends on **life history** characteristics: it is stronger in species with long gestation periods and large sexual dimorphism.

A long-term field study illustrates the complexity of the relationship between maternal condition and offspring sex ratio. On the island of Rum in Scotland, a population of red deer (*Cervus elaphus*) has been monitored since 1971. Initially, mothers with high social ranks produced more males (see Clutton-Brock *et al.* 1984). In agreement with Trivers and Willard's hypothesis, males born from dominant mothers had a much higher reproductive success than males born from subordinate mothers, whereas female reproductive success was not correlated with the social rank of their mothers. However, the relationship between maternal social rank and sex ratio disappeared over time, possibly because of correlated variation in demographic and climatic factors (Figure 13.8; Kruuk *et al.* 1999).

In some species, the life history and social systems may be such that additional resources profit more to female than male offspring. Mothers in good condition should thus produce more females. For example, in many primates female offspring inherit the territory or social rank of their mothers, whereas male offspring disperse. The mothers with good territories or high social rank should thus preferentially produce females. There have been early claims that the proportion of females increase with the social rank of their mothers in some species of baboons and macaques (see Clutton-Brock 1991). However, recent detailed analyses did not find this pattern in baboons (see Packer *et al.* 2000; Silk *et al.* 2005), and

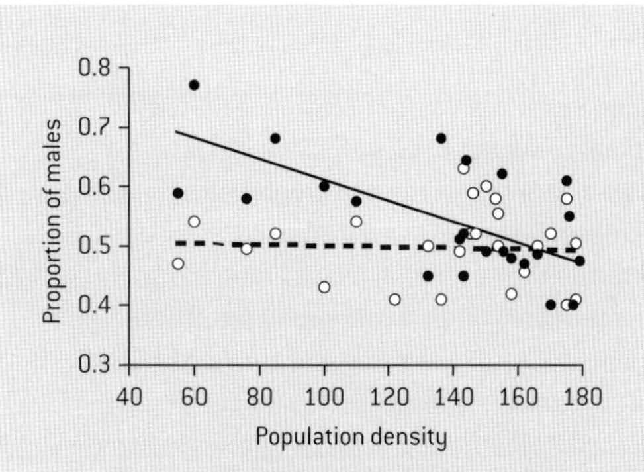


Figure 13.8 **Sex ratio according to maternal social rank in the red deer (*Cervus elaphus*)**

Dominant females produced more males when population density was low, but not when density was high (filled circles and continuous line; each dot is the mean of a year). In contrast, subordinate females had less variable sex ratios (empty circles and dashed line). Kruuk and colleagues proposed that the increase in deer density and winter rainfall resulted in an increased mortality of male embryos.

After Kruuk *et al.* 1999.

a meta-analysis of 35 data sets from 15 species indicates that there is no simple and general relationship between maternal dominance rank and offspring sex ratio in primates (see Brown and Silk 2002).

Overall, it seems that Trivers and Willard's (1973) hypothesis only applies to particular groups with specific life history characteristics. In ungulates, maternal condition explains only 0.2–2% of the variation in offspring sex ratio, and correlated factors may also play some role in generating the observed pattern (Sheldon and West 2004). Clearly, many environmental and social factors affect the relationship between maternal condition and offspring sex ratio.

Interestingly, an understanding of sex ratio biasing according to maternal condition is highly relevant for the conservation of some endangered species, because supplemental feeding can result in a biased sex ratio that increases extinction risks (Tella 2001). In Chapter 18, we will see how sex allocation theory helps guide the conservation programme of a critically endangered New Zealand parrot.

### 13.5.3 Host size in parasitoid wasps

When supplementary resources profit one of the sexes more, mothers should conditionally adjust the sex of their offspring to the available resources. The best experimental tests of this prediction were performed with parasitoid wasps, solitary insects that lay their eggs in the body of other arthropods. In many species, the female lays a single egg per host. The larva feeds on the host body, which is often killed or paralysed. Offspring laid in small hosts will have few resources and develop into small adults. In contrast, offspring laid in large hosts will develop into large adults.

In wasps, a larger body size profits females more than males. For example, in the braconid wasp *Heterospilis prosopoidis*, a standardized increase in host size results in a 20 times higher fecundity for females but only a three times higher number of matings for males (Charnov *et al.* 1981). Female parasitoid wasps should thus adjust the sex ratio according to host size. Specifically, they should lay male eggs in relatively small hosts and female eggs in relatively large hosts. Such condition-dependent sex ratio adjustment might be facilitated by the haplo-diploid sex-determination mechanism of Hymenoptera. Haploid males develop from unfertilized eggs, whereas diploid females develop from fertilized eggs. By controlling fertilization, the mother can determine the sex of her offspring (Section 13.9).

Overall, many studies have found that parasitoid wasps adjust offspring sex ratio in response to host size. A meta-analysis based on 65 studies using 56 species revealed that wasps lay significantly more female eggs in larger hosts (right part of Figure 13.15; West and Sheldon 2002). Host size explained 19% of the variance in offspring sex ratio in wasp species that kill or paralyse the host, and 5% in species that keep the host alive. In the latter group of species, the host continues to feed and grow, so that it might

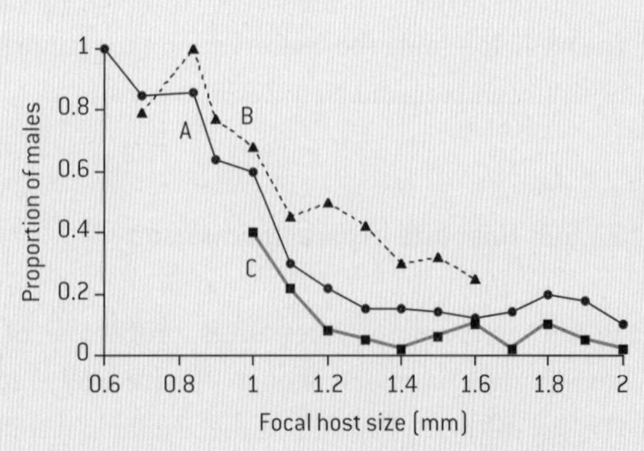


Figure 13.9 **Sex ratio adjustment in response to host size in the parasitoid wasp *Lariophagus distinguendus***

The sex ratio (proportion of males emerging) is plotted against the size of the weevil larva host, which was estimated from the width of the tunnel in grains of wheat. A new host is offered every 2.5 hours. Each circle of the A curve corresponds to the sex ratio laid by female wasps that were presented sequentially with 20 focal hosts of a single size. As expected, the proportion of males decreases when host size increases. The B and C curves demonstrate that the sex ratio is also adjusted in response to the size of the other available hosts. Each triangle of the B curve is the sex ratio when the focal host is presented in an alternating sequence with a 0.4 mm larger host. Each square of the C curve corresponds to the sex ratio when the same sized focal host is presented in alternating sequence with a smaller 0.4 mm host. The wasps lay more male eggs when the focal host of a given size is relatively small compared with the other hosts encountered (B curve), and more females when the focal host is relatively large (C curve).

After Charnov *et al.* 1981.

be more difficult to predict precisely the amount of resources that the offspring will obtain (Section 13.10).

Experimental work has revealed that wasps adjust offspring sex ratio precisely in response to environmental variation in host size. The pteromalid parasitoid wasp *Lariophagus distinguendus* lays one egg per weevil larva. In an elegant experiment, Charnov *et al.* (1981) manipulated the size of the host offered to the wasps. As predicted, the wasps laid more males in small hosts and more females in large hosts. The condition-dependent adjustment of sex ratio was very strong, with more than 80% of males in hosts smaller than 0.8 mm, and more than 80% of females in hosts larger than 1.2 mm (Figure 13.9). Moreover, by manipulating host size distribution, Charnov and his colleagues showed that the wasps adjusted sex ratio in response to the host size relative to other hosts available. They presented the wasps with a focal host of a given size and an alternative host that was either large or small. When the focal host alternated with a larger host, wasps laid more males in the focal host. In contrast, when the focal host alternated with smaller host, wasps laid more females in the focal host (Figure 13.9). Adjustment of sex ratio in response to environmental variability is remarkably flexible and in agreement with expectations. For example, when only 1.4 mm hosts were presented sequentially, the wasps laid 15% of male eggs. When 1.4 mm hosts alternated with 1.8 mm hosts, the wasps laid 30% male eggs. In contrast, the sex ratio dropped to 2% when the focal 1.4 mm hosts alternated with 1.0 mm hosts (Figure 13.9).

All in all, these many impressive studies consistently show that many animal groups have the capacity to adjust the sex ratio of their offspring according to resource levels. They thus show the predictive power of the general theory of sex allocation that we developed in Section 13.3.

### 13.5.4 Need for helpers in cooperatively breeding birds and mammals

In some species of birds and mammals, offspring of one sex help their parents to rear the next brood. Parents should thus preferentially produce one sex or the other depending on their need for additional helpers. If helpers are few, parents should produce more of the helping sex. In contrast, if there are many helpers, additional 'helpers' might actually be costly for the parents, because they will use up resources or disturb brood rearing. In such cases, parents should produce more of the non-helping sex that disperse away from the territory.

Adjustment of sex ratio in response to the need for helpers is quite variable across cooperatively breeding birds and mammals, but is more pronounced when helpers provide larger benefits (Griffin *et al.* 2005). Across four species of birds, sex ratio is biased towards the helping sex when helpers are rare, with the presence or absence of helpers explaining 16% of the variance in sex ratio (Figure 13.15; West and Sheldon 2002).

One of the most remarkable examples of conditional sex ratio adjustment was found in the Seychelles warbler (*Acrocephallus sechellensis*; Komdeur *et al.* 1997, Komdeur 1998). Seychelles warblers lay a single egg per year. Male offspring usually disperse, whereas females stay in the parental territory and help the parents rear the next brood. Breeding pairs remain on the same territory for several years. On high-quality territory, one or two helpers increase the fitness of parents. In contrast, helpers on low-quality territory or helpers in excess of two on high-quality territories actually decrease parental fitness. Female Seychelles warblers are able to adjust precisely the sex of their single egg in response to local conditions (Figure 13.10) and produce more of the helping female sex when helpers are needed. When no helper is present, pairs on low-quality territories produce 80% males, whereas pairs on high-quality territories produce almost exclusively females. When they have two helpers, pairs on high-quality territories switch to producing 85% males. Helper need was also experimentally manipulated. Male-producing pairs on low-quality territories switched to producing females after being experimentally transferred to high-quality territories. Male-producing pairs with two helpers started to produce females after experimental removal of helpers. This quick and precise adjustment of sex ratio was not due to differential mortality. Females are thus able to control the sex of their eggs before egg laying, which is particularly surprising given the chromosomal sex determination of birds (Section 13.9). In Seychelles warblers, cooperation and competition among females vary

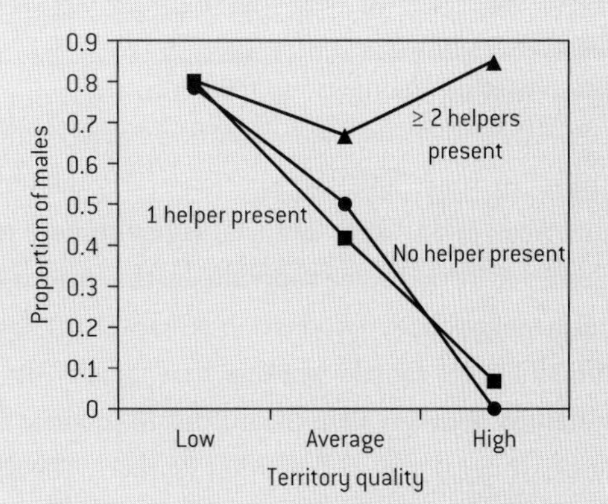


Figure 13.10 **Sex ratio adjustment in response to the need for helpers in the Seychelles warbler (*Acrocephallus sechellensis*)**

Females are the helping sex. Helpers on low-quality territory decrease parental fitness, whereas up to two helpers on high-quality territory increase parental fitness. Pairs with no helper produce mostly males on low-quality territory and females on high-quality territories (circles). The same pattern is found when there is one helper (squares). When there are two or more helpers, pairs produce males whatever the quality of the territory (triangles).

After Komdeur *et al.* (1997).

between families owing to local differences in two ecological variables, number of helpers, and territory quality. In section 13.7 we will further examine how **cooperation** and **competition** among relatives affect sex allocation.

## 13.6 Social control and relatedness asymmetry

### 13.6.1 Kin selection and variation in relatedness

Fitness returns depend on the efficiency of new males and females at transmitting genes from the party controlling sex allocation (Figure 13.15). So far, we have considered that the mother controlled the sex allocation. We will now see that in social species, other members of the social group may influence sex allocation to match their own genetic interest, generating a **conflict** among group members.

Social conflicts over sex allocation have been well studied in some **eusocial** species of Hymenoptera (ants, bees, and wasps). Eusocial Hymenoptera form colonies in which some individuals, the queens and males, monopolize reproduction (Chapter 15). Other individuals are female workers that usually do not reproduce. Workers help the queens to produce new queens and males as well as other workers. The **altruistic** behaviour of workers, helping reproductives produce offspring rather than reproducing themselves, is best explained by **kin selection** (Chapter 2). Colonies are typically closed family units, and workers are genetically related to the brood they rear. Hence, by helping their mother or other related individuals to produce individuals to whom they are genetically related, workers indirectly transmit copies of their own genes to the next generation.

Kin selection not only promotes cooperation among related individuals, it also generates potential conflicts when colony members are genetically diverse (Keller and Chapuisat 1999). When the group is not composed of identical clones, each individual may try to maximize the transmission of copies of its own genes to the detriment of other colony members. This simple logic explains much of the competitive interactions and conflicts over reproduction occurring in social groups.

Conflict over sex allocation is a particular type of kin conflict. Hymenoptera have a male-haploid, female-diploid sex determination system: fertilized eggs develop into females, whereas unfertilized eggs develop into males. Hence, females are diploid, but males only harbour maternal chromosomes (Figure 13.11). In colonies with one queen that has mated with a single male, workers are three times more closely related to their sisters ($r = 0.75$) than they are to their brothers ($r = 0.25$). This relatedness asymmetry is traditionally expressed as the relatedness between workers and new queens divided by the relatedness between workers and new males. Because of such a relatedness asymmetry, workers are selected to allocate more resources to sisters (new fertile queens), the sex that carries more copies of their own genes (Trivers and Hare 1976). In contrast, queens are always symmetrically related to daughters and sons ($r = 0.5$), and should invest equally in each sex. This variation in the relatedness towards males and females generates a potential conflict between workers and queens, because workers are selected to bias sex allocation towards females whereas queens are selected to invest equally in each sex.

When the queen has mated with multiple males, when there are multiple related queens, or when workers lay unfertilized eggs that develop into males, the relatedness asymmetry decreases. Thus, when the queen has mated on multiple occasions if workers control sex allocation, the skew towards females should diminish.

Kin selection yields the interesting prediction that sex allocation should depend on the extent of relatedness asymmetry when workers control sex allocation. Incorporating kin selection into sex allocation theory generates new quantitative predictions at both the population and colony levels. By studying sex allocation empirically, researchers can test whether there is an actual conflict between queens and workers, and determine which party controls sex allocation. This is a powerful approach to study kin conflict and conflict resolution, and more generally to test both kin selection and sex allocation theories.

### 13.6.2 Sex allocation variation among populations

Trivers and Hare (1976) were the first to combine Fisher's sex ratio principle with kin selection theory in order to study the queen–worker conflict over population sex allocation in social Hymenoptera. If workers fully control sex allocation, female bias should be 3:1 in populations where all colonies are headed by a single queen that has mated with one male. This is because workers are always three times more related to sisters than to brothers, and relatedness asymmetry is 3:1 in all colonies (Figure 13.11). Female bias is expected to decrease concomitantly with a decrease in relatedness asymmetry (Boomsma,

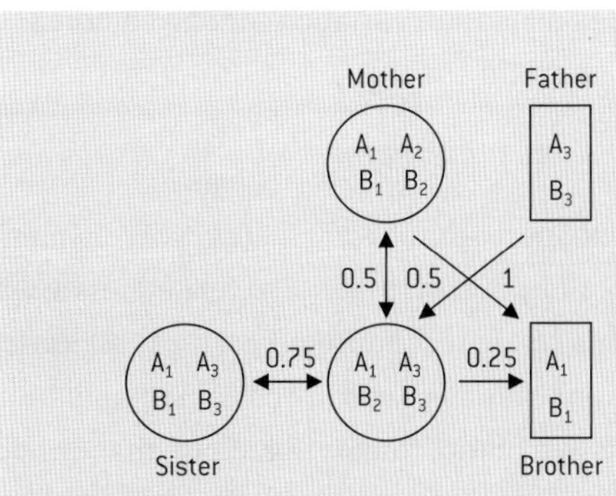


Figure 13.11 **Haplo-diploidy and relatedness asymmetry in Hymenoptera**

In Hymenoptera, the haplo-diploid sex determination system generates relatedness asymmetries. Females (the circles in this figure) develop from fertilized eggs, and males (the rectangles) from unfertilized eggs. Letters represent loci (particular segments of DNA) and numbers identify alleles (genetic variants) at these loci. At each locus, a diploid individual has a gene from maternal origin (e.g. $A_1$), and a gene from paternal origin (e.g. $A_3$). Haploid males transmit all their genes to daughters, without meiotic reduction or recombination. However, males never have sons. They also have no father, and only receive one set of chromosomes from their mother. As a result, females are three times more related to sisters ($r=0.75$, because three quarters of their genes are identical by descent) than to brothers ($r=0.25$, because only one quarter of the sister's genes are identical by descent with copies in the brother). In contrast, the queen is equally related to her daughters and sons ($r=0.5$, because daughters and sons receive exactly half of the genes from the mother). See Figure 15.6 for the calculation of coefficient of relatedness in polygynous societies.

1989). For example, if every colony in the population has one queen that has mated with two males sharing reproduction equally, relatedness asymmetry is 2:1 in all colonies. Indeed, such colonies contain an equal mix of full sisters ($r = 0.75$) and half sisters ($r = 0.25$), and workers are on average two times more related to sisters ($r = 0.50$) than to brothers ($r = 0.25$). Under worker control, the bias towards reproductive females should now be 2:1 at the population level. In contrast, if queens control sex allocation, the allocation to each sex should be equal whatever the social and genetic structures of the colonies (Figure 13.11).

#### 13.6.2.1 *A cross-species comparative analysis*

Trivers and Hare (1976) compared the population sex allocation among ant species that differ in social structure and thus presumably in their degree of relatedness asymmetry. Subsequently, this analysis has been refined and extended by many authors. The most striking result of this approach is that population sex allocation is globally female-biased (1.7:1) in 40 species of ants with one queen per colony, and slightly male-biased (1:1.25) in 25 species of ants with multiple queens per colony (Bourke and Franks, 1995).

Female-biased sex allocation in populations of ants with a single queen per colony (and thus high relatedness asymmetry) strongly suggests that workers manipulate sex allocation in favour of their sisters to the detriment of their brothers. Lower female bias in populations with multiple queens per colony bolsters the argument, because the presence of several related queens decreases the average relatedness asymmetry, and nestmate queens are generally related in ants. The lower than 3:1 sex allocation in ant species with predominantly single-queen colonies is consistent with workers being in full control, as long as a significant part of the colonies have reduced relatedness asymmetries due to multiple mating by queens, occasional presence of multiple related queens, or worker reproduction. Alternatively, queens and workers may share partial control over sex allocation, with the population sex allocation settling between the queens' and workers' equilibriums.

#### 13.6.2.2 *Some problems with this cross-species comparison*

The contrast across species suggests that workers frequently control population sex allocation, at least partly. This result has been extremely important historically, triggering further theoretical and empirical research in the field (Bourke and Franks 1995; Crozier and Pamilo 1996). However, several correlated factors may affect this pattern of population sex allocation variation across species (Crozier and Pamilo 1996; Chapuisat and Keller 1999).

A first problem is that cross-species variation in relatedness asymmetry is typically associated with major changes in the breeding system, life history, and mode of colony reproduction that are also expected to influence sex allocation. For example, in ant species with multiple queens per colony, young queens often return to their parental colony after mating, and new colonies frequently arise in the proximity of old ones. Such limited **dispersal** of females may result in **local resource competition**, which in turn promotes male-biased sex allocation (Section 13.7). Moreover, in ant species with multiple queens per colony, workers and queens often found new colonies together, by colony budding. In this case, the production of workers represents an investment into females, and sex allocation estimated from new queens and males may underestimate the actual female bias (Pamilo 1991).

A second problem arises from the difficulty of estimating the relative cost of female and male production (Boomsma 1989). Costs are usually measured in terms of energetic investment, estimated from dry mass. However, queens are often larger than males in ants, and they contain more lipids and fewer sugars. Therefore the metabolism and maintenance costs per unit of mass are lower for females. Because queens and males are generally more dimorphic in ant species with a single queen per colony, the bias towards females might have been overestimated in these species. Another potential problem is that the factor limiting the production of new queens and males may not always be energy, so that it becomes difficult to assess the relative cost of females and males across species. Finally, each species may not represent an independent data point, because of the phylogenetic relationships among species (Chapter 3).

### 13.6.3 Sex allocation variation among colonies: the 'split sex ratio' theory

#### 13.6.3.1 *Precise predictions*

A powerful approach to test whether workers manipulate sex allocation in response to relatedness asymmetry consists in examining how sex allocation varies among colonies within the same population. The 'split sex ratio' theory predicts that, under worker control, colonies with relatively high or low relatedness asymmetry should specialize in producing females or males, respectively (Boomsma and Grafen 1990, 1991). Precise quantitative predictions depend on the frequency distribution of colonies in classes with different levels of relatedness asymmetry (Figure 13.12). Usually, colonies belonging to one class of relatedness asymmetry should produce only one sex, whereas colonies belonging to the other class should produce mostly the other sex (Figure 13.14).

Let us examine the logic of the split sex ratio theory using an example. Consider a population in which most colonies are headed by one single-mated queen. Under worker control, the population sex allocation equilibrium is 3:1, because workers are three times more related to sisters than to brothers (Figure 13.11). When three times more resources are allocated to females at the population level, the three times lower mating success of females per unit of investment exactly balances their three times higher relatedness. Now imagine that the population contains some rare colonies headed by one double-mated queen, in which the relatedness asymmetry is 2:1. Workers in these colonies have a higher fitness return when producing males. Because the population sex allocation is 3:1, the mating success of males per unit of investment is three times that of females, which overcompensates their two times lower relatedness. Hence, workers from colonies headed by one double-mated queen should invest only in the rearing of males. In contrast, workers from colonies headed by one single-mated queen maximize their inclusive fitness by investing preferentially into females in such a way that the population sex allocation stays close to 3:1 (Figure 13.12).

The split sex ratio theory applies the general principle of sex allocation theory to the special case

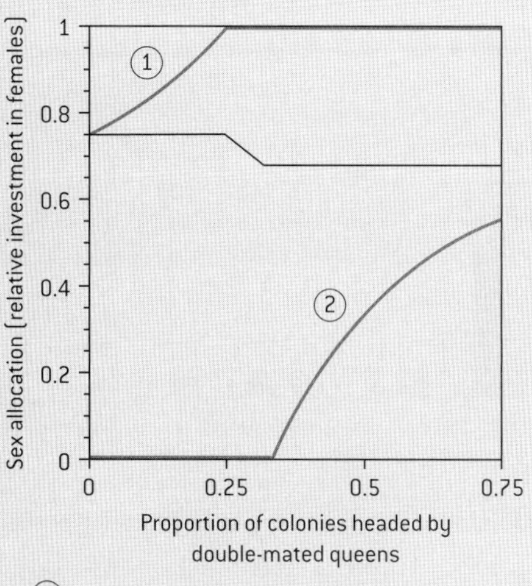


Figure 13.12 **Split sex ratio theory in social Hymenoptera**

Split sex ratio theory (Boomsma and Grafen 1990, 1991) predicts how sex allocation should vary among colonies that differ in relatedness asymmetry within a population. Assumptions are that workers fully control colony sex allocation and that mating is random at the population level. Workers manipulate sex allocation in response to both their relatedness towards new queens and males (the relatedness asymmetry in their colony, Figure 13.11) and the sex allocation at the population level. In this example, all colonies have a single queen. In part of the colonies, the queen has mated with a single male, and the relatedness asymmetry is 3:1. In the remaining colonies, the queen has mated with two males that share reproduction equally, and the relatedness asymmetry is 2:1. The bold lines numbered 1 and 2 show the proportion of females that should be produced by colonies headed by one single-mated or double-mated queen, respectively, according to the proportion of colonies with a double-mated queen (horizontal axis). Colonies with relatively high relatedness asymmetry specialize in females, and colonies with relatively low relatedness asymmetry specialize in males. Depending on the frequency of colonies with relatively low relatedness asymmetry (which is given by the proportion of double-mated queen on the horizontal axis), all colonies belonging to one relatedness asymmetry class should specialize in one sex, whereas colonies belonging to the other class should produce both sexes in such a proportion that the population sex allocation matches the relatedness asymmetry of their class, which thus becomes the 'balancing class'. The fine line in the middle is the population sex allocation, which is 0.75 (three females for one male) when colonies with one single-mated queen are the balancing class, and 0.67 (two females for a male) when colonies with one double-mated queen are the balancing class.

After Boomsma (1996).

of social Hymenoptera populations in which relatedness asymmetry varies among colonies. The theory simply predicts that workers should favour the sex with higher fitness return (Section 13.3.3). Fitness return varies among colonies because of variation in the kin structure, which affects the relative relatedness of workers to females and males. Hence, split sex ratio theory is a special case of condition-dependent variation in sex allocation among families (Section 13.4).

Split sex ratio theory makes quantitative predictions that can be tested within populations where colonies differ by a single factor altering their genetic structure, but are similar in other characteristics. Such intra-specific tests are more powerful than cross-species comparisons, because they do not rely on precise estimates of the cost of producing males or females, and avoid much of the confounding variation in breeding system (Section 13.6.2). Moreover, experimental manipulations can be performed. Hence, studies of sex allocation variation among colonies have emerged as powerful tests of kin selection and sex allocation theory.

#### 13.6.3.2 *Empirical tests*

Overall, a meta-analysis review showed that split sex ratio correlated with variation in relatedness asymmetry in 19 out of 25 study cases in ants, bees and

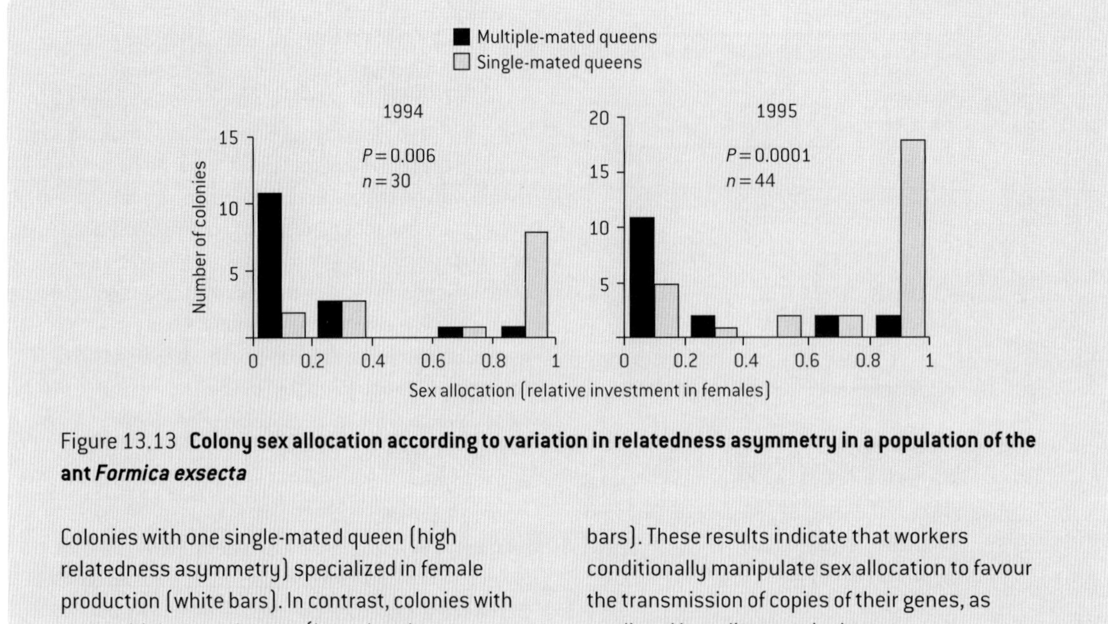


Figure 13.13 **Colony sex allocation according to variation in relatedness asymmetry in a population of the ant *Formica exsecta***

Colonies with one single-mated queen (high relatedness asymmetry) specialized in female production (white bars). In contrast, colonies with one multiple-mated queen (low relatedness asymmetry) specialized in male production (black bars). These results indicate that workers conditionally manipulate sex allocation to favour the transmission of copies of their genes, as predicted by split sex ratio theory.

Redrawn from Sundström *et al.* (1996).

wasps (Bourke 2005). In all these cases, colonies with relatively high relatedness asymmetry produced more females than colonies with relatively low relatedness asymmetry, as predicted by split sex ratio theory (Boomsma and Grafen 1990, 1991). These empirical data indicate that in many cases workers conditionally manipulate colony sex allocation according to their selfish genetic interest.

Sex ratio specialization occurred when relatedness asymmetry varied among colonies for several reasons: (1) variation in the number of matings by queens in ants with a single queen per colony (Sundström 1994; Sundström *et al.* 1996; Figure 13.13); (2) variation in the number of queens per colony in ants and wasps (Queller *et al.* 1993; Chan and Bourke 1994; Deslippe and Savolainen 1995; Evans 1995; Hastings *et al.* 1998; Henshaw *et al.* 2000; Walin and Seppä 2001; Hammond *et al.* 2002); and (3) replacement of the mother queen by one of her daughters in halictid bees (Boomsma 1991; Mueller 1991; Packer and Owen 1994). In the eusocial halictid bee *Augochlorella striata*, relatedness asymmetry has been experimentally manipulated by removing the foundress queen, which results in the queen being replaced by one of her daughters. Such a queen turnover decreases the relatedness asymmetry from 3:1 to 1:1 (the workers being equally related to nieces and nephews). As predicted by the split sex ratio theory, sex allocation was significantly more female-biased in control colonies than in colonies where the foundress had been replaced (Mueller 1991).

*Split sex ratios often reveal worker power . . .*

The frequent sex ratio specialization of colonies according to variation in relatedness asymmetry demonstrates that workers often control sex allocation. More generally, these data not only illustrate the major role played by kin selection and relatedness in hymenopteran societies, they also demonstrate the predictive power of modern, gene-centred evolutionary theory. By biasing sex allocation according to relatedness levels, workers increase the transmission of their own genes. This conditional response depends on the relatedness asymmetry in their colony and on sex allocation in the other colonies of the population, which is a complex problem to solve (Figure 13.12). Natural selection over millions of years must have fine-tuned the sex allocation biasing behaviour of workers. However, the exact way by which workers react to the relative relatedness asymmetry in their colony remains enigmatic. Workers might use relatively simple cues and decision rules, such as response thresholds to queen number or level of intracolony genetic diversity as sources of **information**.

*. . . in spite of ongoing conflicts*

Despite the striking sex allocation patterns described above, worker control is far from universal. In some cases, queens strongly influence colony sex allocation by laying a large proportion of haploid eggs or worker-destined diploid eggs (Section 13.9; Helms *et al.* 2000; Passera *et al.* 2001; de Menten *et al.* 2005; Rosset and Chapuisat 2006). Altogether, these data provide evidence for an ongoing conflict between queens and workers.

To conclude, workers often influence sex allocation in social Hymenoptera, but they do not have full control. Sex allocation represents an area of ongoing conflict between queens and workers, with each party trying to manipulate sex allocation (Section 13.9). The outcome of the conflict depends on the power of each party, and colony sex allocation can thus vary between the queen and worker equilibriums.

## 13.7 Competition and cooperation among relatives

Interactions among same-sex relatives may affect the fitness return through male and female offspring. Four types of interaction can occur depending on the sex involved and whether the interaction is competitive or cooperative (Table 13.2). In such cases, the prediction is that sex allocation should be biased towards the sex that provides higher fitness return by avoiding competition or cooperating with relatives of the same sex.

### 13.7.1 Competition among related males: local mate competition

W.D. Hamilton pioneered this field when he proposed that sex allocation biases should be expected when individuals do not mate randomly in the population. If related males compete locally for mating, sex allocation should become female-biased (Hamilton 1967). Consider the extreme case in which mating occurs exclusively among siblings, for example because offspring mate before dispersal (Figure 13.14). All females therefore mate with brothers and males compete exclusively with brothers, never with unrelated males. For the mother, this competition among related males decreases her fitness returns through

| Sex | Type of interaction among relatives | Effect on sex allocation |
|---|---|---|
| **Males** | Competition for mating (Local mate competition) | Bias towards females |
| **Males** | Cooperation increasing mating (Local mate enhancement) | Bias towards males |
| **Females** | Competition for resources (Local resource competition) | Bias towards males |
| **Females** | Cooperation increasing resources (Local resource enhancement) | Bias towards females |

Table 13.2 **The four types of interaction among same-sex relatives**

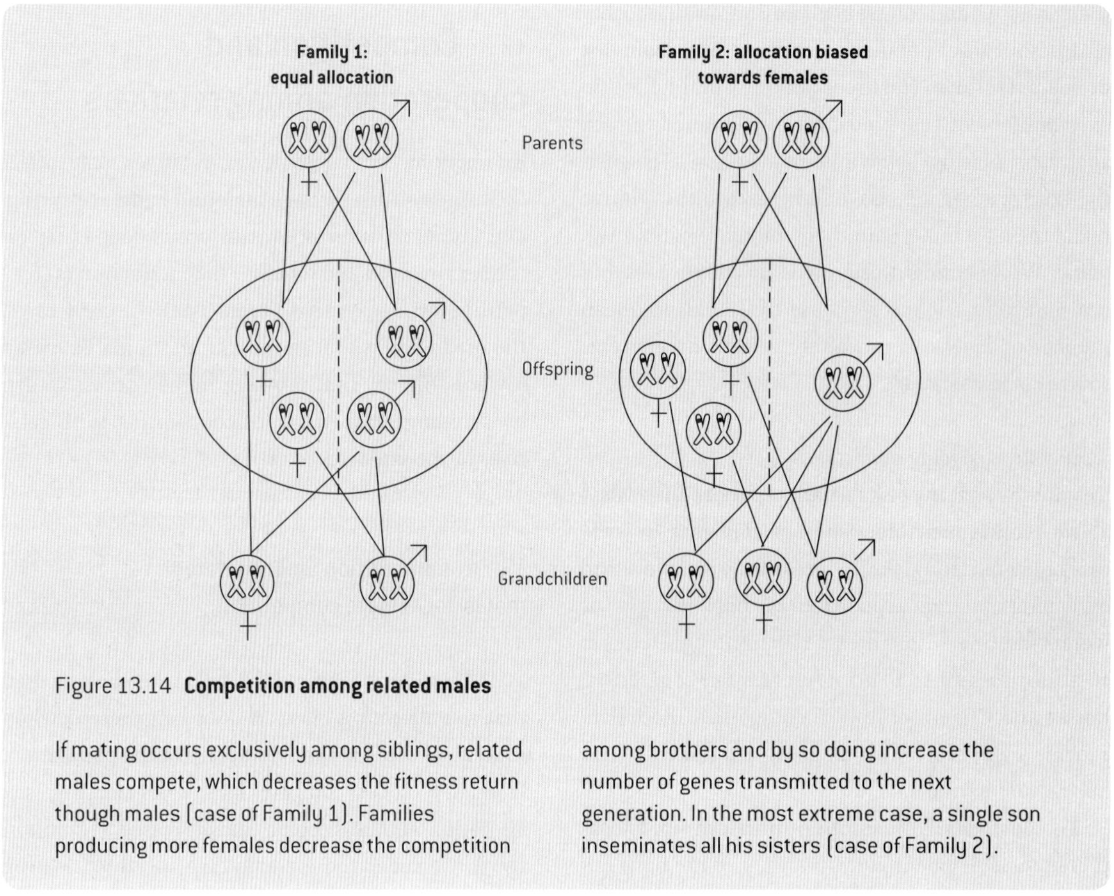


Figure 13.14 **Competition among related males**

If mating occurs exclusively among siblings, related males compete, which decreases the fitness return though males (case of Family 1). Families producing more females decrease the competition among brothers and by so doing increase the number of genes transmitted to the next generation. In the most extreme case, a single son inseminates all his sisters (case of Family 2).

sons, because copies of the mother's genes in one son compete with copies of the same genes in other sons (Figure 13.14, Family 1). If one male suffices to inseminate all females, the mother would maximize her fitness by producing a single son and investing all other resources in daughters (Figure 13.14, Family 2).

A spectacular example of **local mate competition** occurs in the mite *Acarophenax tribolii*. Offspring develop within the body of their mother, and feed by eating her from the inside. Brothers and sisters mate while they are still within the mother. Usually, there is only one male, who mates with his numerous sisters and dies before being born (Hamilton 1967).

The effect of competition among related males has been particularly well studied in fig and parasitoid wasps. Fig wasps pollinate fig flowers and lay eggs that will develop within the fruit. The young then mate within the fig, before the females disperse. When the fig contains the offspring of a single wasp, there is a strong competition among related males, and the sex ratio is highly female-biased (about nine females for a male). In contrast, when the fig contains the offspring of multiple wasps, the competition among related males decreases and sex allocation becomes more balanced (Herre 1985, 1987). Similarly, females of the parasitoid wasp *Nasonia vitripennis* lay a less female-biased sex ratio in response to both the presence of other females on the patch and the presence of eggs laid by other females (Shuker and West 2004).

### 13.7.2 Cooperation among related males: local mate enhancement

Genetically related males can also cooperate to obtain more matings or more resources. In such cases, a mother's fitness returns through sons increases when multiple sons can be produced simultaneously.

For example, male lions that form coalitions are more successful at displacing older males from a pride of females. Hence, the individual fitness of males increases sharply when the number of males in the same cohort increases, whereas no such effect is detected for females. As predicted by theory, more males are produced in large cohorts than in small ones (Packer and Pusey 1987).

### 13.7.3 Competition among related females: local resource competition

The same types of process described for males also apply to females. Genetically related females may compete for resources (Clark 1978), particularly when females do not disperse and hence are philopatric. Competition among related females decreases a mother's fitness returns through her daughters, which selects for male-biased sex allocation. In primates, the sex ratio at birth is less female-biased in species where females are philopatric (Johnson 1988). In ants with multiple queens per colony, young queens often stay in their mother's nest, and sex allocation is generally male biased (Bourke and Franks 1995). In honeybees and army ants, colony fission occurs. The worker force is split into two parts, one for the mother and one for a single daughter queen. New queens thus strongly compete with their sisters, whereas males disperse and mate at random. In these species, as expected, colonies produce a few new queens and thousands of males (Crozier and Pamilo 1996; Bourke and Franks 1995).

### 13.7.4 Cooperation among related females: local resource enhancement

Related females may also increase each other's fitness, for example by cooperating to enhance their local resources. Local resource enhancement seems to occur in the allodapine bee *Exoneura bicolour*. The female fitness is larger when a few related females associate to rear their offspring together, and small broods are more female-biased than large ones (Schwarz 1988). In cooperatively breeding birds with female helpers, such as the Seychelles warbler, sex allocation is also conditionally biased towards females in families that need helpers (Section 13.5.4).

In summary, competitive and cooperative interactions among same-sex relatives can result in biased sex allocation at both the family and population levels. A meta-analysis based on 87 studies of 64 species shows that such interactions among relatives have a very strong impact on sex allocation, with mothers consistently biasing sex allocation as predicted by theory (West *et al.* 2005). Overall, local mate competition and local resource competition explain as much as 33% of the variance in sex allocation, whereas **local resource enhancement** has less impact, explaining about 8% of the variance.

## 13.8 Non-Mendelian inheritance of the genes controlling sex allocation

Sex allocation strongly depends on the inheritance of the sex allocation control genes. Standard models consider that autosomal genes in the parents control sex allocation. These genes segregate in a Mendelian fashion and are transmitted in the same way by males and females. However, if the genes controlling sex allocation are passed on with different probabilities to each sex, they will be selected to bias sex allocation towards the sex that transmits them more efficiently. We have seen such a case in Section 13.6. In social Hymenoptera, genes present in workers are better transmitted by sisters than by brothers, because of the non-Mendelian haplo-diploid mode of inheritance (Figure 13.11). If the workers' genes can control the sex allocation of their mothers, they should bias sex allocation towards the production of sisters (Section 13.6).

A more extreme bias is expected if the genes controlling sex allocation are transmitted by a single sex. Sex chromosomes, some kinds of supernumerary chromosomes and cytoplasmic elements are all characterized by such single-sex transmission. If these genetic elements can gain control over sex allocation, all resources should be invested in the sex that transmits them. Interestingly, an intra-genomic conflict over sex allocation may then occur between genetic elements that have different modes of inheritance.

Sex chromosomes, supernumerary chromosomes, and cytoplasmic elements are indeed known to affect sex allocation in particular cases. Their precise impact depends on their mode of inheritance, means to manipulate sex allocation, and population dynamics.

*Sex chromosomes*

If genes on sex chromosomes can influence sex allocation, they should promote extreme biases (Hamilton 1967). As an example, consider the familiar XY-male XX-female sex-determination system found in humans and most other mammals. The Y chromosome is only transmitted through males. If a gene on the Y chromosome was able to influence the sex ratio, it should bias sex allocation towards males (Hamilton 1967). A mutant Y chromosome with full control over sex ratio should favour its own transmission over that of the X chromosome, thus generating a strong meiotic drive (distortion of meiosis). The mutant Y will quickly spread in the population, which will become increasingly male-biased and smaller, until extinction (Hamilton 1967). The mutant thus wipes himself out. The mutant is also in conflict with genes on other chromosomes, which are under selection to suppress this type of meiotic drive. Therefore, meiotic drive by sex chromosomes is bound to be transient and difficult to observe. Nevertheless, examples of sex chromosomes distorting meiosis and selfishly biasing sex allocation have been found in flies, butterflies, and rodents (Werren and Beukeboom 1998; Jaenike 2001).

*Supernumerary chromosomes*

Some insects harbour a supernumerary chromosome that affects sex allocation. *Psr* (= paternal sex ratio) is a small chromosome found in some individuals of the parasitoid wasps *Nasonia vitripennis* and *Trichogramma kaykai* (Nur *et al.* 1988; Werren and Stouthamer, 2003). These wasps have the usual haplo-diploid system of all Hymenoptera. The *Psr* chromosome, however, is only found in some males and never in females. This is surprising, because males normally do not have sons and never transmit genetic material to other males (Figure 13.11). *Psr* is transmitted through sperm, but it causes the loss of all other paternal chromosomes early in the development (Nur *et al.* 1988). All the paternal genome is eliminated, except *Psr* which is now associated with the maternal chromosomes. The diploid fertilized egg (that would normally have developed into a female) thus becomes a haploid male carrying *Psr.* In short, *Psr* converts females into males and thus biases sex allocation towards males. *Psr* is an ultra-selfish genetic element: at each generation, it promotes its own transmission by destroying the entire genome with which it is associated.

*Cytoplasmic elements*

Cytoplasmic elements such as mitochondria or intracellular parasites are only transmitted by mothers. This is because mothers contribute cytoplasm to their ova, whereas sperm cells only transmit nuclear material, to the egg, no cytoplasm. Cytoplasmic elements in a male are in a dead end because they will not be transmitted further. Cytoplasmic elements, therefore, are under strong selection to bias sex allocation towards females. *Wolbachia* bacteria are such intracellular symbionts present in many arthropod species (Werren 1997). *Wolbachia* manipulate the reproduction of their host in order to produce more infected females (see Chapter 17). They can induce male killing, feminization of genetic males, female parthenogenesis, and reproductive incompatibility that is such that non-infected females mated with infected males produce few or no females (Werren 1997).

All the examples of biased sex ratio in this section show that strong conflicts may exist within genomes, that is within individuals. They also show that the ultimate unit of selection is the gene rather than the individual. All those examples would be impossible to understand, and would probably be used to invalidate the theory if we had not uncovered the genetic mechanisms that are responsible for such biases. In fact, those apparent exceptions support the theory rather than invalidate it.

## 13.9 Mechanisms to manipulate sex allocation

The degree of adaptive sex allocation adjustment may depend on mechanisms that permit and constrain sex ratio biasing. When a conflict over sex allocation occurs among genetic elements or individuals, the outcome of the conflict also depends on the means by which each party can manipulate sex allocation. It is thus of interest to examine the **proximate** mechanisms of manipulation in diverse sex-determination systems.

### 13.9.1 Chromosomal sex determination

In species with chromosomal sex determination, such as birds or mammals, the Mendelian segregation of sex chromosomes during meiosis normally results in a 1:1 sex ratio at conception. However, sex ratio can still be biased by distorting the segregation of sex chromosomes during meiosis or favouring certain gametes before or during fertilization (Chapter 11, Section 11.6). In birds, mechanisms of sex ratio biasing are poorly known, but the precision and timing of sex ratio adjustment during egg laying suggests that precise, pre-ovulation control mechanisms do exist in some species (see Komdeur *et al.* 1997; Badyaev *et al.* 2002). Females are the heterogametic sex, and they may influence the segregation of sex chromosomes during the first meiotic division by manipulating the amount of steroid **hormones** they deposit in their eggs (Petrie *et al.* 2001; Pike and Petrie 2005). In mammals, the sex ratio at conception partly depends on copulation frequency and time lag between insemination and ovulation (Krackow 1995). This is because sperms bearing the X or Y chromosome differ in motility and survival. Interestingly, hormonal concentration in females, and in particular the surge of lutein hormone during ovulation, can influence these differences.

### 13.9.2 Environmental sex determination

Sex is determined by environmental factors in many reptiles (Chapter 6), some fish, and several invertebrates. For example, offspring sex depends on nest temperature in many species of turtles and crocodiles. But parents can still influence the sex ratio by choosing the nest site and controlling incubation conditions. Of course, environmental variation strongly influences the sex ratio, and a large part of the variance in sex allocation is not explained by classical models (Bull and Charnov 1988; Freedberg and Wade 2001). However, it is striking that even in such species the population sex ratio remains close to being balanced.

### 13.9.3 Haplo-diploid sex determination

Haplo-diploid species comprise all Hymenoptera and many other invertebrates. In such species, females can determine the sex of their offspring by controlling fertilization. Queens of social Hymenoptera mate at the beginning of their adult life and store sperm in a specialized organ, the spermatheca. When the queen needs to fertilize an egg, she releases a few sperms before she lays the egg. Honeybee queens seem to be in full control of the fertilization process, laying haploid eggs in drone cells and diploid eggs in worker cells (Ratnieks and Keller 1998). Female parasitoid wasps are also able to adjust offspring sex to host size (Section 13.5.3).

### 13.9.4 Post-fertilization manipulation

Sex allocation can still be manipulated after fertilization. First, mothers may selectively abort one sex, and both parents may commit sex-specific infanticide. Second, resources can be differentially allocated to males and females. For example, one sex may receive more food than the other.

In many mammal species, abortion is more frequent for male offspring, particularly under stressful conditions (Clutton-Brock 1991). Females of coypu (*Myocastor coypu*) seem to exert some control over sex allocation by selectively aborting entire litters. Abortions of entire litters are particularly frequent when the mother has large energetic reserves but carries a small, female-biased litter. After abortion, the female coypu produces a larger or male-biased litter, as males profit more from extra resources than females do (Gosling 1986).

Parental infanticide is rare and seems largely unbiased with respect to sex in most vertebrates, with the sad exception of some human populations. Sex-biased mortality is often observed in sexually dimorphic species, but this mortality might be largely independent from parental behaviour (Clutton-Brock 1991). In the same vein, resources obtained by male and female offspring often differ, but this does not constitute direct evidence for active parental discrimination (Clutton-Brock and Iason 1986; Clutton-Brock 1991).

### 13.9.5 The case of social Hymenoptera

In social Hymenoptera, the study of manipulation mechanisms revealed that there is often an actual conflict between queens and workers (Section 13.6; Chapuisat and Keller 1999; Ratnieks *et al.* 2006). Queens and workers use diverse mechanisms to manipulate sex allocation before or after fertilization. Queens control the sex ratio in their eggs, but workers take care of the brood. Hence, a powerful way to study the queen–worker conflict is to compare the sex ratio at the egg and pupa stages.

In some species queens have a strong impact on colony sex allocation. In the fire ant *Solenopsis invicta*, colonies headed by a single queen specialized in producing either males or females, and the experimental exchange of queens reversed the sex allocation biases of adopting colonies (Passera *et al.* 2001). In *Formica selysi*, queens in female-specialist colonies laid a high proportion of diploid eggs, whereas queens in male-specialist colonies laid almost exclusively haploid eggs, but the change in sex ratio between the egg and pupa stages also suggested that workers eliminated some male brood (Rosset and Chapuisat 2006) thus showing the existence of conflicts. In two ant species of the genus *Pheidole*, queens also influenced sex allocation by laying a high proportion of haploid or worker-destined eggs in part of the colonies (Helms 1999; de Menten *et al.* 2005). Overall, these data indicate that queens can force workers to raise males by limiting the number of female eggs in some circumstances.

However, in most species workers manage selfishly to bias sex allocation according to variation in relatedness asymmetry (Sections 13.6.2 and 13.6.3). In the ant *Formica exsecta*, queens lay a similar proportion of haploid eggs in all colonies, but workers selectively destroy male offspring in colonies with high relatedness asymmetry (Sundström *et al.* 1996; Chapuisat *et al.* 1997). *Leptothorax acervorum* ant workers do not eliminate males, but conditionally influence the proportion of females developing into queens or workers (Hammond *et al.* 2002). Finally, the wasps *Polistes dominulus* have found an original way to bias sex allocation towards females. Workers stuff males headfirst into empty nest cells when foragers are returning to the nest, so that the resources are preferentially distributed to females and larvae (Starks and Poe 1997).

The emerging picture is that the outcome of the conflict is variable, with a prominent influence of workers. The study of sex allocation manipulation may help to understand how conflicts are resolved, and more generally how conflicts affect the evolution of sex allocation. This area of research is still in its infancy. The outcome of the conflict depends on the power of each party, and in particular on access to information and means of manipulation (Ratnieks *et al.* 2006). The expression of the conflict is also constrained by the risks of errors and by the costs of manipulation and counter-manipulation (Keller and Chapuisat 1999).

## 13.10 Constraints and precision of adaptation

Constraints are important in evolution, but their role is often difficult to assess. Sex allocation offers an excellent opportunity to evaluate whether constraints set by the mechanism of sex determination limit the amount of sex allocation adjustment and hence the precision of adaptation (West and Sheldon 2002; West *et al.* 2005).

### 13.10.1 Weak constraints

Chromosomal sex determination has long been considered to limit the magnitude of sex allocation biasing because the Mendelian segregation of chromosomes during meiosis might constrain the sex ratio at fertilization to be 1:1 (Bull and Charnov 1988). In the absence of genetic variation for the sex ratio, the trait simply cannot evolve. There are indeed signs of constraints. Animal breeders have long tried to modify the sex ratio in cattle and poultry, with very little success. Moreover, the sex ratio at birth is generally much less variable in species with chromosomal sex determination than in species with environmental or haplo-diploid sex determination (Clutton-Brock and Iason 1986; Bull and Charnov 1988).

Studies do, however, show that chromosomal sex determination does not prevent condition-dependent sex allocation biasing (Sections 13.5.1 and 13.5.4). In a meta-analysis, West and Sheldon (2002) compared the magnitude of adaptive sex ratio adjustment in species with chromosomal and haplo-diploid sex determination (Figure 13.15). Overall, they found that the degree of conditional sex ratio biasing was similar in birds and parasitoid wasps when clear *a priori* predictions about the direction of the adjustment could be made. Specifically, birds with helpers at the nest (Section 13.5.4) and parasitoid wasps that kill or paralyse the host (Section 13.5.3) showed large and similar degrees of adjustment. Birds reacting to male attractiveness (Section 13.5.1) and parasitoid wasps that do not kill or paralyse the host (Section 13.5.3) also showed significant adjustment in the predicted direction, but the magnitude of this adjustment was reduced (Figure 13.15). Another meta-analysis revealed that the chromosomal sex determination

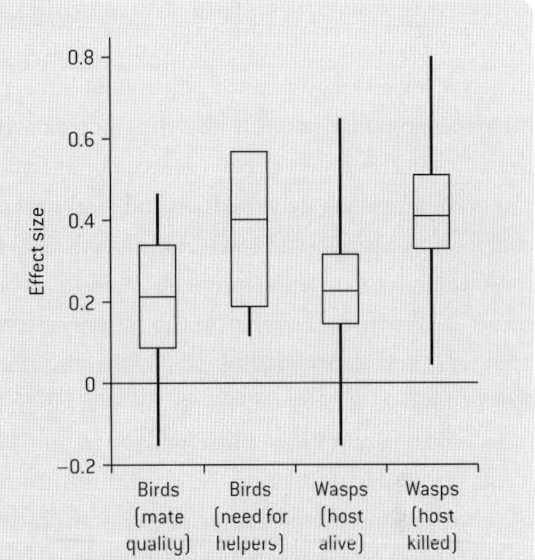


Figure 13.15 **Magnitude of sex ratio adjustment in birds (chromosomal sex determination) and parasitoid wasps (haplo-diploid sex determination)**

Mean (horizontal lines), 95% confidence interval (boxes) and range (vertical lines) of the effect size when sex allocation varies with respect to mate attractiveness in birds, need for helpers in birds, host size in parasitoid wasps in which the host can still grow, and host size in parasitoid wasps that kill or paralyse their hosts. The effect size is the correlation between the sex ratio in offspring and the environmental factor. A positive effect size indicates a sex ratio adjustment in the direction predicted by the theory. All groups show adaptive sex ratio biasing. The magnitude of the adjustment varies significantly among groups and seems to be more affected by the type of selective pressure and environmental predictability than by the mechanism of sex determination.

After West and Sheldon 2002.

system does not prevent facultative offspring sex ratio biasing in vertebrates and invertebrates, and that the lower magnitude of adjustment in vertebrates is confounded with the different selective pressures in this group (local resource enhancement rather than local mate competition and local resource competition, Section 13.7; West *et al.* 2005). Overall, the mechanisms of sex determination seem to be a rather weak constraint, whereas the type of selective pressures and **predictability** of the environment appear to have a large impact on the degree of sex ratio adjustment (West and Sheldon 2002; West *et al.* 2005; Figure 13.15). This counter-intuitive result underlines the predictive power of evolutionary thinking. When theory predicts that selective pressures are strong, even constraints that seem to be rather compelling such as the chromosomal sex determinism in relation to sex ratio are in fact circumvented in one way or another.

The evolution of condition-dependent sex allocation adjustment depends on multiple constraints and selective pressures that vary from case to case (Bull and Charnov 1988; West *et al.* 2005). Hence, precise and powerful predictions crucially depend on a good knowledge of the sex determination mechanism, genetic elements that may influence sex allocation, means of manipulation by each party, and ecological pressures shaping sex allocation.

### 13.10.2 Challenges

We hope that the reader is now in a position to appreciate the success that sex allocation theory has had in predicting major patterns of sex allocation across a large range of taxa. There seems to be no limit to what organisms can sense in their environment and react to in terms of sex allocation. Parasitoid wasps appear capable of counting the frequency of females they encounter before laying and adjust the sex ratio of their offspring accordingly. Female birds react to the brightness of their mates' plumage and overcome the constraint of chromosomal sex determination to bias the sex ratio in their clutch. We will see in Chapter 19 that even humans may perhaps be influenced by such processes.

These striking cases of adaptive sex allocation and the success of the general theory should, however, not conceal the fact that a great part of the variation in sex allocation remains unexplained. In many of the results detailed in this chapter, the predicted factors had significant effects but only explained a small proportion of the observed variance in sex allocation. Moreover, the causes of sex allocation biases at the population level are often still poorly understood, and sometimes sex allocation seems to vary stochastically, possibly because of environmental fluctuations (Liautard *et al.* 2003). This suggests that future studies may need to integrate the potential effect of patterns of environmental variation in the theory. In fact, we have seen in Chapter 4 (Figure 4.1) that accounting for patterns of environmental variation is a crucial issue in most decision-making. This is mainly because the value of the information that animals may use to make adaptive decision is tightly linked to the patterns of environmental variation. For instance, if the environment varies randomly, the current state of the environment does not inform about its future state, and this lack of information may prevent accurate sex allocation decisions.

A large part of the variation among families also remains unexplained, and might simply not be adaptive. For example, mate attractiveness and need of helpers explained only 4% and 16% of the variation among families in sex allocation, respectively (Sections 13.5.1 and 13.5.4; West and Sheldon 2002). In mammals, and particularly in primates, variation in sex allocation among families are complex and do not appear to be caused by any single general factor leading to facultative sex allocation adjustment (Clutton-Brock and Iason 1986; Packer *et al.* 2000; Brown and Silk 2002). Beyond the obvious success of sex allocation theory, the field still offers a wide range of open questions that need to be addressed before we fully understand all the details of how and why sex allocation varies among breeding episodes, families, populations, and species.

## 13.11 Conclusion

The study of sex allocation is a fruitful area of behavioural ecology and evolutionary biology. The strength of the approach comes from combining impeccable logic with powerful empirical tests. The basic principle of sex allocation theory is that natural selection acts on the number of male and female offspring in such a way that sex allocation reaches an equilibrium at which the fitness return for a marginal investment in a male or a female is equal. This stable equilibrium often corresponds to an equal investment in each sex, and in the absence of sexual dimorphism, to a similar number of males and females. However, there are interesting exceptions. When ecological, genetic, social, or behavioural factors affect the fitness returns a parent obtains through sons and daughters, the theory predicts allocation biases at the population level or among families within populations. Such factors include variation in resource availability, parental quality, competitive and cooperative interactions among relatives, relatedness towards male and female offspring, or even individuals or genetic elements controlling sex allocation.

Over the years, sex allocation theory was expanded to become a rich edifice based on the most basic principle of modern evolutionary theory, the natural selection of genes transmitted from one generation to the next. Hundreds of empirical studies, experimental or correlational, in the lab or in the field, have tested some aspects of the theory in very diverse contexts and organisms, from hermaphroditic plants to human beings. Overall, the empirical results have provided strong support to the predictions of sex allocation theory. The basic logic has been largely supported, and the effect of many potential factors has been assessed. In particular, some cases of facultative, condition-dependent sex allocation constitute beautiful examples of the value and predictive power of modern evolutionary theory. However, a large part of the variation in sex allocation remains unexplained. The current challenge is to understand how multiple factors, selective pressures, and constraints interact in shaping sex allocation. By considering environmental, genetic, behavioural, and social factors simultaneously, as well as by better understanding the relevant selective pressures, constraints, and mechanisms involved, theoreticians and empiricists should be able to explain a larger part of the variation in sex allocation, as well as the lack of adaptive adjustment in specific cases.

## Further reading


> *Summary on sex allocation:*

Seger, J. 2000. Natural selection: sex ratio. *Encyclopedia of Life Sciences*, Nature Publishing Group, London, http://els.wileycom.

> *General presentations of sex allocation theory:*

Charnov E.L. 1982. *The Theory of Sex Allocation*. Princeton University Press.
Bull J.J. & Charnov E.L. 1988. How fundamental are Fisherian sex ratios? *Oxford Surveys in Evolutionary Biology* 5: 96–135.
Bulmer, M. 1994. *Theoretical Evolutionary Ecology*. Sinauer, Sunderland, MA.
Frank, S.A. 1998. *Foundations of Social Evolution*. Princeton University Press.

> *Pioneer synthesis of theory and empirical data:*

Trivers, R.L. 1985. *Social Evolution*. Benjamin/Cummings Publishing Company, Menlo Park, CA.

> *Theory on conditional sex allocation in birds and mammals:*

Frank, S.A. 1990. Sex allocation theory for birds and mammals. *Annual Review of Ecology and Systematics* 21: 13–55.

> *Synthesis of empirical data on sex allocation manipulation, mostly in vertebrates:*

Clutton-Brock, T.H. 1991. *The Evolution of Parental Care*. Princeton University Press.

> *Synthesis of theory and empirical data on sex allocation in social Hymenoptera:*

Crozier, R.H. & Pamilo, P. 1996. *Evolution of Social Insect Colonies: Sex Allocation and Kin Selection*. Oxford University Press.

*Synthesis of theory and empirical data on sex allocation in ants:*

Bourke, A.F.G. & Franks, N.R. 1995. *Social Evolution in Ants*. Princeton University Press.

## » Questions

1. Does the stable sex ratio correspond to an individual's optimum value?

2. In a large random mating population with a sex ratio of 1:1, does the fitness of a parent depend on the sex ratio among his offspring?

3. What would happen if mitochondria could control the sex ratio?

4. What is the expected sex ratio in a hypothetical social animal living in isolated colonies with no migration, endogamy and colony fission?

5. Male elephant seals weigh up to 4000 kg, in contrast to females that on average weigh 500 kg. A dominant male may have a harem of up to 100 females, but only 2–3% of the males obtain harems. At the time of weaning, the weight difference between male and female offspring is negligible. What is the expected sex ratio at birth, assuming maternal control?

6. You are a freshly born female *Polistes* wasp. You now have the choice to stay at home and help your mother or leave the nest to breed independently. If you help your mother, she will be able to rear three more offspring, and you can decide whether these offspring are going to be males of females by selective egg eating. If you leave your mother to breed independently, you will be able to rear two offspring, and you can also determine their gender by controlling the fertilization of your eggs. The population sex ratio is biased towards females, with two females for one male.
- What are you going to do if your mother had mated with a single male?
- And if she had mated with two males?
- Will there be a conflict with your mother? If yes, how could she influence you?
- Would you need to be mathematically literate to take decisions?

7. Nine million men died during the First World War. What would be the expected qualitative variation in birth sex ratio after the war, according to Fisher's sex ratio theory, and independently of the proximate factors influencing the sex ratio? What would happen in the case of a perpetual war?

# Part Five

# Social Interactions Among Individuals

In most species, every individual is bound to interact with others, whether they be conspecifics or not. Such interactions are of a social nature. This part concerns questions of the role of social interactions in evolution. The term 'social' is used here in its broad sense to encompass any situation in which individuals interact regularly with the same kind of partner.

The question of social interactions is central to behavioural ecology, in part because in such interactions the optimal strategy often depends on that of social partners, not only the one an individual is interacting with, but also that of the bulk of the population. This generates dynamics that are particularly involved as any individual adopting a strategy changes the game for all other individuals with which it is likely to interact. We have seen several times before in this book that the solutions to such problems are called evolutionary stable strategies (ESS): that is, strategies that can resist the invasion by (and invade) any other strategy. ESS can be determined by game theoretical approaches. The question of the social dimension of the phenotype is at the origin of interdisciplinary interactions with other disciplines such as anthropology, psychology, and human social sciences in general.

Chapter 14 is about the question of the evolution of group living through the parasocial pathway, which is as a consequence of individuals choosing to live together at some stage in their life.

Chapter 15 deals with the general dilemma of cooperation. How is it that individuals selected to behave selfishly can nonetheless cooperate at some stage or in some circumstances? Cooperation has been raised as an obvious case against evolution by natural selection. We will see that cooperation no longer raises intractable issues to evolutionary biologists.

Chapter 16 places a particular emphasis on the environmentally induced physical constraints that influence the evolution of communication using different sensory channels.

Chapter 17 develops behavioural questions in the context of long-lasting inter-specific interactions such as parasitism and mutualism. The former type of interaction is a relatively well-studied form of long-lasting interaction, but its behavioural aspects are sometimes neglected. The latter type of interaction appears to be particularly unstable and thus poses interesting evolutionary questions.

# 14

# 14

# Animal Aggregations: Hypotheses and Controversies

Étienne Danchin, Luc-Alain Giraldeau, and Richard H. Wagner

## 14.1 Introduction

Upon discovering a breeding colony of seals on a rocky beach (Figure 14.1) or a colony of terns where nests are just a few centimetres apart even though there is plenty of adjacent free space, you might ask, 'Why are they grouping together like this to reproduce?' In this chapter, we will examine the question of why many animals live in groups. This question has long been a topic of controversy. A first observation is that various forms of group life exist in all animal taxa, from invertebrates (Meadows and Campbell 1972) to mammal societies, social insects, fishes that form schools, certain species of dinosaurs that probably bred in colonies (Horner 1982; Morastalla and Powell 1994), and **colonial** birds. It thus seems that there exists a general tendency towards **aggregation** and group living in the animal world.

This chapter presents the various hypotheses that have been put forward to explain the evolution of animal aggregations. Because it is closely linked to Chapters 7 and 8 on solitary and group **foraging**, we recommend reading those chapters first.

There are two main evolutionary pathways towards social life: the **parasocial** path, which is a consequence of individual decisions to live together, and the **quasi-social** path, in which parents keep their young with them to form groups. Both paths lead to the evolution of societies. This chapter only concerns the animals that have become social through the parasocial path. We are only interested here in the evolutionary processes involved in the **formation** of groups, or aggregates of individuals in space, which constitutes a kind of primary form of social life. Once such groups exist, more elaborate forms of social interactions can develop in the course of evolution. This development of social life itself, as well as evolution towards social life through the quasi-social path, will be the subject of the next chapter. It will include, amongst others, the special case of social insects because the mechanisms involved in this group are of a very specific nature. The general message of the present chapter is that it is not necessary to bring up the processes of cooperation to explain the initiation of group formation, while cooperation constitutes the fundamental mechanism at the origin of the evolution of insect and human societies.

*A priori*, living in dense groups should incur numerous **fitness** costs. These costs are mostly related to higher risks of pathogen transmission or **conflicts** with other individuals of the group. Considering the existence of such costs, the main approach that has been used to explain the evolution of animal aggregation has consisted of finding **fitness benefits** that may balance the **costs** (Wittenberger and Hunt 1985). In other words, the main approach assumes that aggregation has evolved because it serves a function. We will thus call it the **functional approach**. However, more recently, hypotheses of an entirely different nature have been proposed. This second framework views groups as the by-product of decision-making by many individuals. We will thus call it the **by-product approach.**

This chapter comprises two main sections: one that reviews the hypotheses of the evolution of animal aggregation within the functional approach, and another that discusses hypotheses that have been proposed under the by-product framework in the past decade. We will see that the latter approach has the potential to transform our perception of the evolution of aggregation as a particular form of social living.

(a)

(b)

(c)

Figure 14.1 **Examples of aggregated species**

a. A colony of Atlantic puffins (*Fratercula arctica*), Grimsey, north of Iceland.
b. A colony of black-legged kittiwakes (*Rissa ridactyla*) in Brittany, France.
c. California sea lions (*Zalophus californianus*) breeding in a colony on the peer of Monterey harbour (California).

Photographs by Étienne Danchin.

## 14.2 The traditional functional approach

The existence of animal aggregation poses an evolutionary problem because living in a group seems to impose fitness costs to the individuals that use this strategy. The costs identified in the literature are part of diverse categories, with higher risks of pathogen and parasite transmission (Brown and Bomberger-Brown 1986, 1996), cuckoldry (Møller and Birkhead 1993; Westneat and Sherman 1997), conspecific **competition** for food, breeding sites, and mates (density dependence; Møller 1987), and higher risks of cannibalism and infanticide (Wittenberger and Hunt 1985; Møller 1987). As shown by studies of the cliff swallow, the existence of these costs is well documented (Figure 14.2).

(a)

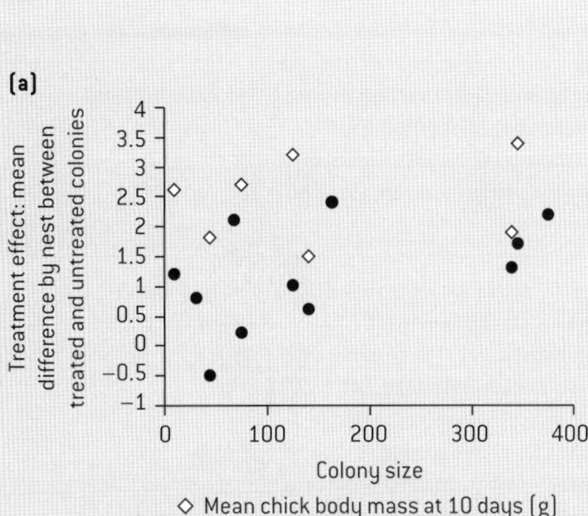


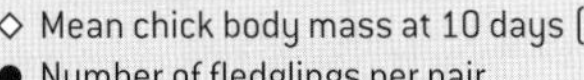


(b)

Figure 14.2 **Aggregation and parasite transmission**

In a classic experiment, Charles Brown and Mary Bomberger Brown showed the reality of fitness costs of aggregation associated with parasites in the cliff swallow (*Hirundo pyrrhonota*). Starting from the observation that (1) the number of parasites per chick and per nest increases with colony size, and that (2) body weight at fledging as well as chick survival in the nest decreases when colony size increases, these two US researchers sprayed some randomly chosen colonies with pesticides to remove parasites, and compared them to similarly sized untreated colonies. Their goal was to show that parasites are, at least partly, responsible for the lower reproductive success of pairs breeding in big colonies. Effects were impressively strong:

**a.** Both the number of chicks at fledging and their body mass at 10 days of age were higher in treated than untreated colonies of similar size (most values in the graph are positive). It should be noted that the apparent increase of the treatment effect is non-significant in both the number of fledglings ($P=0.59$, $n=12$), and for chick body mass at 10 days ($P=0.61$, $n=8$). This may, however, be due to the relatively small sample sizes (modified from Brown and Bomberger Brown 1996).

**b.** Comparison of same age chicks from an untreated (left) and treated (right) colony. Both colonies comprised about 350 breeding pairs, and both chicks were 10 days old.

Photograph kindly provided by Charles Brown.

Obviously, if only costs existed, this aggregative behaviour could not have been selected over the course of evolution. The fact that we observe many species living in groups leads to the **assumption** that individuals that use this strategy must gain some fitness benefits, and that these benefits must at least balance the costs associated to this strategy. It is around this fundamental **economic** statement that the functional approach was developed to explain aggregation and this approach has essentially been the only one considered until the early 1990s.

The hypotheses about the benefits of aggregation can be divided in three broad categories: the benefits related to the structure of the habitat itself, to safety from predators, and to foraging.

### 14.2.1 Spatial aspects

One form of group life that has led to many studies is colonial breeding, which can be found in practically all vertebrate groups (Figure 14.1) and, albeit in slightly different forms, in many invertebrates (Meadows and Campbell 1972).

> A species is considered colonial when reproduction occurs in territories densely grouped in space, and when these territories do not contain resources other than the breeding sites themselves. Breeders must leave the breeding site regularly to search for food outside their territory.

| Species | Argument | Reference |
|---|---|---|
| Bank swallow (*Riparia riparia*) | This species reproduces in burrows dug by breeders in sandy cliffs on river shores or gravel pits. This gives each pair the opportunity to choose the location of their nest within the available space. However, several favourable nesting sites within a very large study area are not used at all by this species. | Stutchbury 1988 |
| Bank swallow (*Riparia riparia*) | Four observations: (1) fewer than 1% of the pairs nest more than 100 metres away from other pairs; (2) in many colonies, nests are aggregated in only one part of the available space; (3) each year, swallows only use a small part of the favourable sand banks; (4) within colonies that use all the available space, neighbouring nests are very synchronized, meaning that birds that have settled simultaneously have aggregated themselves while other favourable areas (evidenced by the fact that they were used that same year by late pairs) remained vacant. | Hoogland and Sherman 1976 |
| Colonial birds | In most species, there remains available space that is not used. This is also the case for marine birds that do not use certain islands or certain suitable sections. | Wittenberger and Hunt 1985 |
| Barn swallow (*Hirundo rustica*) | In years of high population, the proportion of swallows that nest in large colonies is higher than in years of low population. Furthermore, the distribution of distances between nests is significantly shorter than if nests were distributed randomly in the available space. | Møller 1987 |
| Barn swallow (*Hirundo rustica*) | A precise definition of a favourable nesting habitat leads to the conclusion that such habitats are sufficiently abundant not to be a limiting factor. | Shields *et al.* 1988 |
| Marine birds: European shag, black-legged kittiwake, common guillemot, razorbill, northern fulmar | A detailed description of a portion of the coastline near Aberdeen (Scotland) shows that for each of these species, there are numerous favourable zones that go unused. Such unused areas are located within the colonies themselves, in the surroundings, or further away from the colonies. | Olsthoorn *et al.* 1990 |
| Cliff swallow (*Hirundo pyrrhonota*) | The use of the abundant artificial structures was not followed by a decrease in the density of nests in colonies, or by an increase of the number of colonies. Many favourable areas are unoccupied. | Brown and Bomberger Brown 1996 |
| Harbour seal (*Phoca vitulina*) | A discriminant analysis based on a detailed description of the sites used or left vacant by this species shows that they only use 64% of the favourable sites. Additionally, no physical difference could be detected between used and unused sites. | Krieber and Barrette 1984 |

Table 14.1 **Various arguments against the role of spatial constraints in the evolution of colonial breeding in birds and mammals**

One of the oldest hypotheses is that coloniality in marine mammals, marine reptiles, and marine birds in particular results from the low availability of favourable breeding sites (oceanic islands) compared with the vast surfaces of feeding sites available across the oceans (Wittenberger and Hunt 1985, Cairns 1992, Post 1994). For example, in southern elephant seals (*Mirounga leonine*), the availability of long stretches of coastline seems to induce a lower density in breeding females, which in turn influences numerous behavioural patterns and many components of fitness (Baldi *et al.* 1996). However, even if food is not the limiting factor, this cannot explain why breeding territories are often unnecessarily aggregated despite the existence of many sites adequate for reproduction that remain unoccupied nearby. In many species, the potential breeding sites do not seem to constitute the limiting factor to explain the aggregation of individuals (Table 14.1).

Another hypothesis is that in heterogeneous and unpredictable **environments**, breeders should concentrate at the barycentre of food resources, the spot that minimizes the distances they have to travel to access food. This is the **central place foraging** hypothesis (Horn 1968). However, authors such as Brown *et al.* (1992) were led to seriously reconsider this geometrical model of coloniality based on their long-term studies of cliff swallows (*Hirundo pyrrhonota*; Figure 14.3). Furthermore, the prerequisites of this model are probably rarely fulfilled in natural populations (particularly for marine vertebrates such as reptiles, mammals, and birds), so this hypothesis is unlikely to explain the evolution of aggregation in general (Brown *et al.* 1992).

Figure 14.3 **Cliff swallows (*Hirundo pyrrhonota*) on their nest**

This species nests within colonies that may comprise up to three thousand pairs. Nests are made of mud and saliva and are often attached to each other. The natural site for colonies is on cliffs, but today, they are often built under bridges and artificial human structures. Thus, favourable nesting sites seem to be unlimited and every year only a small proportion of the potential sites are used while pairs may aggregate under a single bridge by the thousand.

Photo kindly provided by Charles Brown.

## 14.2.2 Aggregation and predation

The aggregating role of predation is one of the oldest hypotheses for the evolution of group living (Darling 1938). Today, predation is still thought to be one of the major evolutionary forces that led to the evolution of group living (Bertram 1978; Endler 1995), particularly of coloniality (Darling 1938; Veen 1977; Brown and Bomberger-Brown 1987, 1996; Rodgers 1987). There are several ways in which aggregation could provide protection against predators: vigilance, dilution, confusion, and collective defence.

### 14.2.2.1 *The vigilance effect*

#### a *Group size and detection of predators*

For many predators, hunting success depends on their ability to catch prey by surprise, and they must be able to approach prey without being detected. This is the case of the northern goshawk (*Accipiter gentilis*) hunting pigeons. Their attack success diminishes when the size of the flock of common wood-pigeons increases, and this effect seems mainly due to earlier detection by larger flocks of pigeons (Figure 14.4).

Prey can protect themselves from predators by vigilance behaviour. However, the time spent in vigilance can represent a costly proportion of time that is taken away from other activities such as foraging, courting, or care of young. Group living in common wood-pigeons allows them an earlier detection of predators by a simple effect of increasing numbers (Figure 14.4a, b). Similarly, in barn swallows, the time needed to detect a stuffed little owl (*Athene noctua*) presented to a colony decreases rapidly when colony size increases (Figure 14.4c).

Thanks to the vigilance of other group members, each individual can spend less time in vigilance to detect approaching predators, and allow more time

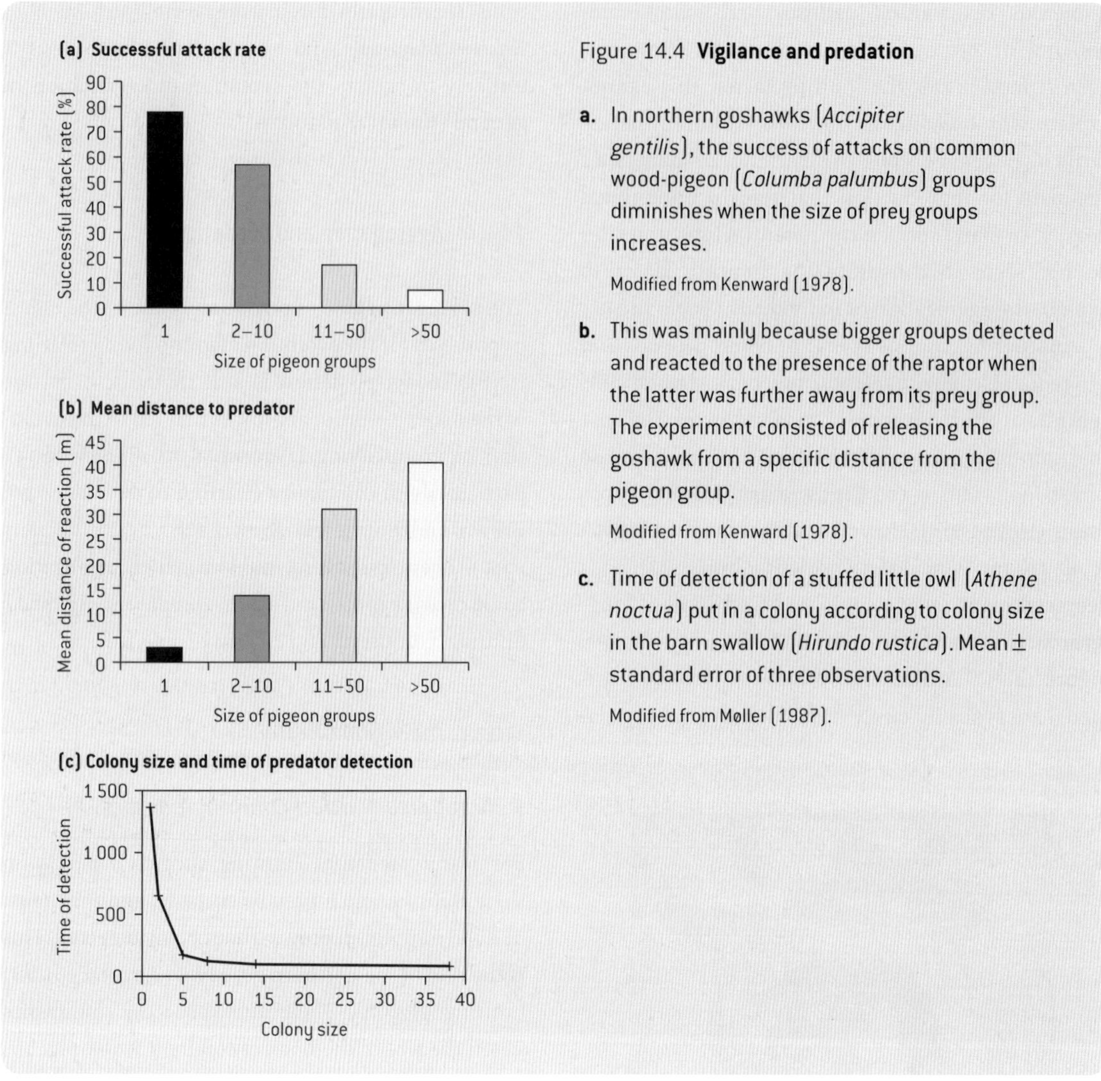


Figure 14.4 **Vigilance and predation**

**a.** In northern goshawks (*Accipiter gentilis*), the success of attacks on common wood-pigeon (*Columba palumbus*) groups diminishes when the size of prey groups increases.

Modified from Kenward (1978).

**b.** This was mainly because bigger groups detected and reacted to the presence of the raptor when the latter was further away from its prey group. The experiment consisted of releasing the goshawk from a specific distance from the pigeon group.

Modified from Kenward (1978).

**c.** Time of detection of a stuffed little owl (*Athene noctua*) put in a colony according to colony size in the barn swallow (*Hirundo rustica*). Mean ± standard error of three observations.

Modified from Møller (1987).

for other activities. This is an example of how social life may change the **trade-off** for each group member (see Chapter 5). For instance, in the harbour seal (*Phoca vitulina*), once the effect of an individual's position in the group is removed (individuals in the centre of the group are less vigilant than those at the periphery), each individual's time spent in vigilance decreases as the size of the colony increases (Figure 14.5).

## b *Chaotic dynamics of shifts between vigilance and feeding . . .*

On the other hand, the dynamics of the shifts between feeding phases (pecking with head lowered) and vigilance phases (head held upright) can represent an adaptation in itself. For example, an animal exhibiting a high degree of **predictability** in its frequency of shifts between these two phases could enable a potential predator to approach, whereas an unpredictable pattern of shifts would make a predator's approach much more difficult (Ferrière *et al.* 1996). When examining such shifting predictability over an increasing period of time, a clear dichotomy appears between some species, such as the purple sandpiper (*Calidris maritima*; Figure 14.6a–c) or the Eurasian collared-dove (*Streptopelia decaocta*; Figure 14.6d, e) and the red-billed chough (*Pyrrhocorax pyrrhocorax*; Figure 14.6f, g). The profiles of predictability of sandpipers and doves are chaotic (high degrees of predictability over the short term,

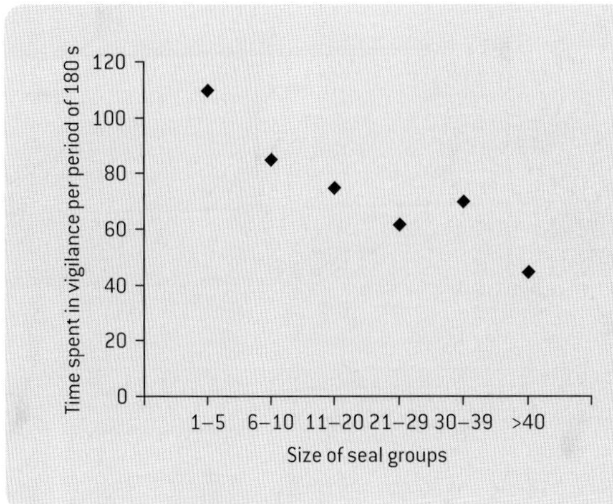


**Figure 14.5 Vigilance and group size**

Time spent in vigilance in harbour seals (*Phoca vitulina*) according to group size. The time spent in vigilance was estimated over a period of 180 seconds.

Modified from Terhune and Brillant (1996).

which decrease exponentially over time), whereas the choughs shift between pecking and vigilance periodically (flat predictability profile).

In that study, Ferrière *et al.* (1996) showed with mathematical simulations that the coordination, even a loose one, between group members based on a single pecking period, can dramatically decrease the predictability of the behaviour of an individual, while increasing the rate of vigilance of the group as a whole. These characteristics generate a chaotic dynamic. Thus, the characteristics of a chaotic dynamic can provide a real selective advantage for prey when the risk of predation is high (which does not seem to be the case for the red-billed chough): a high degree of predictability over a short period allows the different individuals in the group to time their own behaviour to that of the other group members and optimize the level of vigilance of the whole group. A low predictability over a longer period of time prevents predators from learning to predict a succession of non-vigilance phases over a sufficient period to allow it to approach its prey without being detected.

### c ... *do not show the existence of dynamic coordination of group member*

However, the fact that these simulations create a pattern of vigilance that resembles the pattern observed in nature doesn't prove that animals actively coordinate their time spent in vigilance. It merely shows that the observed dynamics of vigilance could result in a certain degree of coordination between individuals. To test this, Steve Lima (1995) added previously starved individuals to a flock of dark-eyed juncos (*Junco hyemalis*) that he observed in the field, and quantified the vigilance of all group members. He observed that the starved individuals almost never contributed to vigilance and spent most of their time feeding. However, the other group members did not adjust their degree of vigilance to account for the presence of these non-vigilant birds. This suggests that, at least in dark-eyed juncos, group members do not pay attention to the vigilance of others, and only react to their presence or number in the group (Lima 1995).

### d *The complexity of the group vigilance game*

The evolutionary stability of the vigilance behaviour is far from simple. In fact, within a group that reaches a collective degree of vigilance of 100% (there is always at least one vigilant individual at all times), any given individual could **cheat** by spending all its time feeding and relying only on the vigilance of others. This strategy is equivalent to that of the **scrounger** seen in Chapter 8, and the solution to this problem is covered in Box 8.3.

It is also possible that cheaters would not be able to invade a group of smarter individuals that only watch for predators when their fellow group members also spend time watching for predators (Pulliam *et al.* 1982). In other cases, vigilance can have direct benefits, which is the case in Thompson's gazelle (*Gazella thomsoni*), where individuals that happen to be in vigilance at the time of a predator's approach

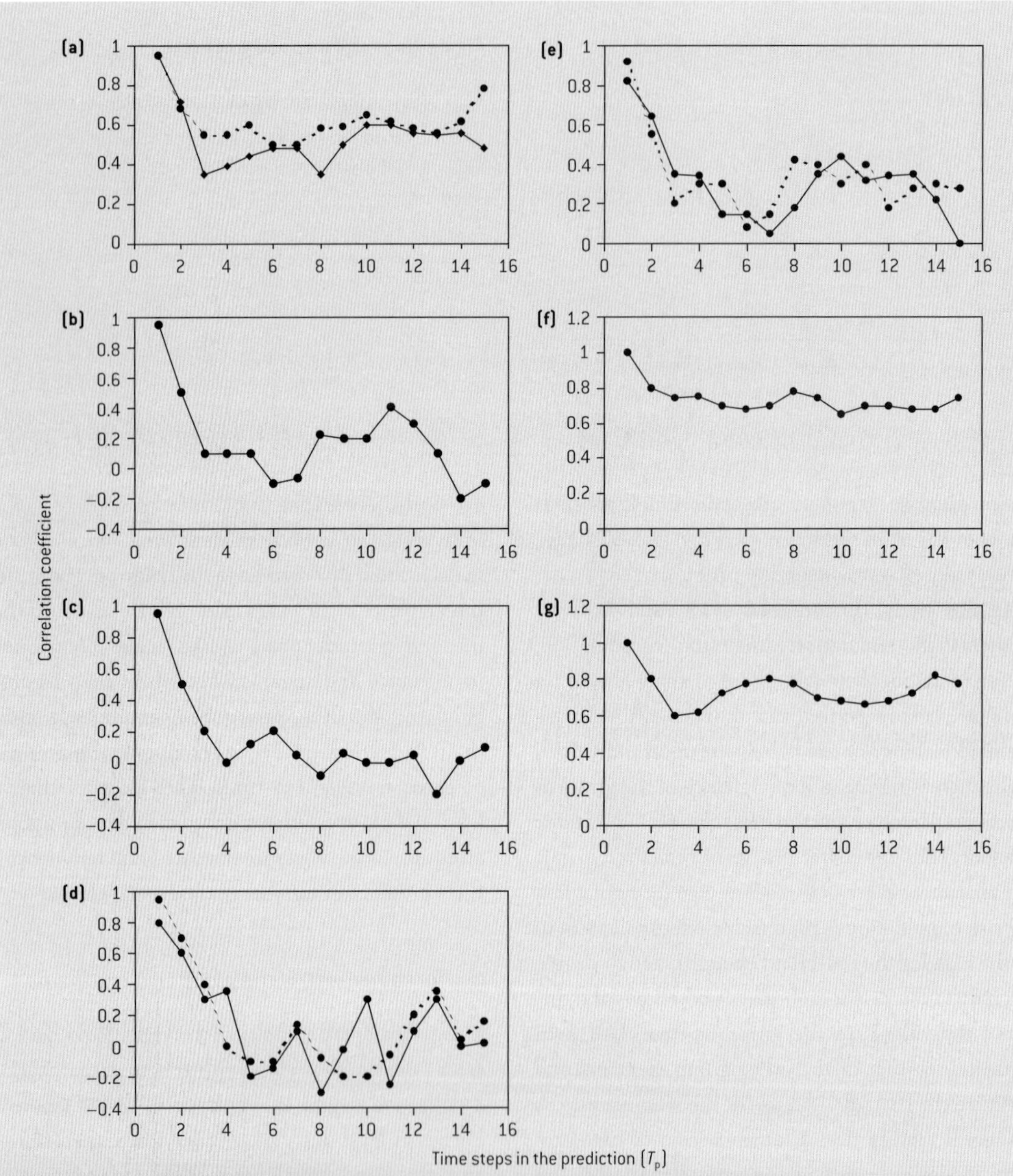


Figure 14.6 **Predictability profile of the vigilance behaviour**

Observations of a purple sandpiper (*Calidris maritime*) (**a**–**c**), two Eurasian collared-doves (*Streptopelia decaocta*) (**d**, **e**) and two red-billed choughs (*Pyrrhocorax pyrrhocorax*) (**f**, **g**). The dynamic of the behaviour is characterized by the chain of successive lengths of pecking and vigilance bouts. The predictability of the behaviour is measured by a coefficient $\rho$ ($y$-axis) as a function of the number of time steps $Tp$ on which the predictions are made ($x$-axis). The solid and dotted lines correspond to different prediction methods. For the sandpiper, the lengths of pecking and vigilance were analysed together (**a**) and then separately for periods of vigilance (**b**) and pecking (**c**). A high predictability on the short term that declines exponentially as predictions are made further away in time reveal a chaotic dynamic. A flat profile that shows a constant and strong quality of predictions is indicative of a periodic rhythm (see main text).

Redrawn from Ferrière *et al.* (1996).

have a greater chance of escape (Fitzgibbon 1989). This direct benefit of vigilance can also protect a group from being invaded by cheaters.

In all models of vigilance we have seen thus far, we suppose that individuals flee instantly when they detect a predator. This is how, for example, we interpret the faster flight of a large flock of pigeons attacked by a goshawk, which we described previously (Kenward 1978). However, Canadians Ronald Ydenberg and Lawrence Dill (1986) applied the cost–benefit approach to the choice of fleeing, and proposed that it may not be beneficial to flee immediately when a predator is detected. In fact, the ideal moment to flee can be strongly influenced by the state of the animal. For example, it can be profitable for a very hungry individual to delay flight so it can eat for a longer period of time. Thus, if solitary pigeons tend to be individuals that are hungrier, or in poorer condition, it is possible that they detect the predator as quickly as individuals in a group, but flee later. Several studies have now shown that detection and flight are two distinct processes, and that the moment of flight is a function of several other variables aside from the timing of predator detection (Ydenberg and Dill 1986; Bonenfant and Kramer 1996).

#### 14.2.2.2 *Dilution effects*

##### a *The classical example of water striders*

As most predators are unable to capture more than one prey at a time, because of a simple dilution effect individual prey can have an advantage by staying in groups that lower each individual's chance of being captured. In a group of $n$ prey, one individual's chance of being caught at each attack is simply $1/n$. This is what we observe in water striders, insects that are predated by fish: the probability for one insect to be captured decreases when group size increases (Figure 14.7).

##### b *Dilution and breeding synchrony*

Owing to the effect of dilution, we expect that group size will vary positively with predation intensity. This is the case for guppies (*Poecilia reticulata*), where individuals that live in rivers with few predators form smaller groups than those that live in rivers where predators are at high density (Seghers 1974). Dilution effects also explain the tendency of several species of birds and fishes to form very dense flocks or schools when a predator approaches.

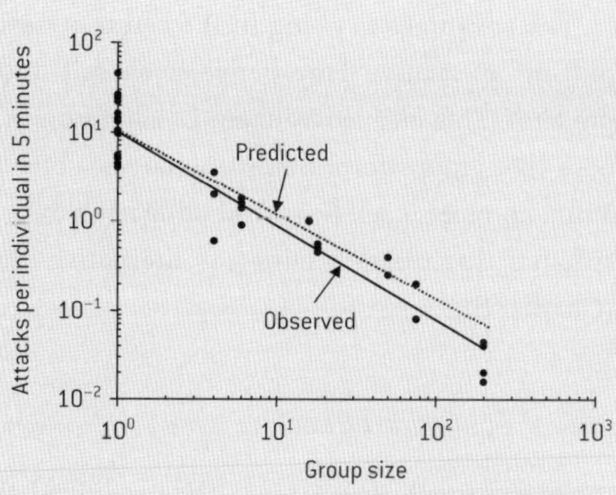


Figure 14.7 **Group living and the dilution effect**

It is difficult to clearly show the effect of dilution because, first, predator attacks are relatively rare, which makes observations and statistical analyses difficult, and second, the benefits of dilution are often confounded by other advantages such as those related to vigilance and predator confusion.

Foster and Treherne (1981) have very clearly shown this effect in a water strider species, the ocean skater (*Halobates robustus*) that is predated by fish (*Sardinops sagax*). Indeed, in this system, because of optical phenomena associated with the water surface, striders can't see the predators approaching from below, which means that vigilance cannot be involved. Additionally, the attacks are frequent and it is thus fairly easy to obtain reliable data. The attack rate per group did not vary with prey group size, so the attack rate per individual only varied through a dilution effect. Consequently, the observations (solid line) corresponded to the rates predicted by the effect of dilution alone (dotted line with a slope of −1 in logarithmic coordinates). The slope of the curve of observations does not significantly differ from the predicted slope ($-1.118 \pm 0.123$, 95% confidence interval).

Redrawn from Foster and Treherne (1981).

In certain species, the effect of dilution is heightened by breeding synchrony in space and time (Darling 1938), and could explain the remarkably long life cycle of some species of cicadas (13 and 17 years). These insects remain in the ground at the nymph stage and millions emerge simultaneously as adults after 13 or 17 years, depending on the species or the location (Dybas and Loyd 1974). These massive and synchronous emergences can literally flood the environment with potential prey, which greatly reduces the chance of a given individual being captured. There has been much speculation over the significance of the 13- or 17-year periods (Simon 1979). A long period of dormancy between periods of simultaneous emergence leads to long periods during which cicadas are not present in the environment, which forces specialist predators and parasites to either exploit other prey or also enter a state of dormancy. These very long cycles would thus be the result of a true arms race between predators and prey, in which cicadas and their predators have gradually lengthened their life cycles until cicadas have finally won. The reason why the life cycles last 13 and 17 years could be that, because they are prime numbers, it would be impossible for predators or parasites to synchronize with the cicadas by having life cycles that are sub-factors of the life cycles of their prey. If the prey had life cycles of 15 years, predators with life cycles of 3 or 5 years could exploit these prey every five or three generations.

These are all speculations that would be difficult to test, but nonetheless some cases of finely tuned breeding synchrony are found in nature. Such synchrony could be the result of social stimulation resulting from interactions between group members that lead to the synchronization of breeding pairs (Darling 1938; Gochfeld 1980). In this context, Coulson (1986) suggested that the synchronization of breeding pairs could be the result of faster development rates inside than outside the groups, which would allow individuals to reproduce simultaneously at the most favourable moment. However, although such processes of synchronization (i.e. the 'Fraser Darling effect') could explain how existing groups can maintain themselves, they are unlikely to be the evolutionary force that caused the formation of reproductive groups in the first place (Hoogland and Sherman 1976).

#### C *Dilution and position within a group*

The probability of a prey to be captured by a predator must vary depending on its position within the group, with those in the middle of the group being less vulnerable. Thus, we expect that the individuals on the periphery are responsible for a greater proportion of the group's vigilance since they are more exposed to predators than individuals in the centre. This is the case for the harbour seal (*Phoca vitulina*) where peripheral individuals are vigilant an average of 38.5 ± 17.4 seconds per period of 3 minutes, versus only 17.2 ± 9.1 seconds for central individuals (significant difference, $t = 5.99$, $df = 41$, $P < 0.0001$; Terhune and Brillant 1996). Thus, we expect individuals exposed to imminent predation risk to try to move towards the centre of the group. This preference for the centre can thus explain the high increase of density in starling flocks or fish shoals when a predator approaches. When each individual attempts to get to the centre, it results in a strong contraction of the space occupied by a group and therefore an increase in density.

The advantage of being in the centre consists of putting alternative prey between oneself and the predator. This idea led William D. Hamilton to propose that groups form 'selfish herds' in which each animal protects itself by surrounding itself with conspecifics that create a buffer of alternative victims (Chapter 8).

### 14.2.2.3 *Confusion effect*

During an attack, a predator can be strongly hindered by the presence of numerous prey. A good analogy would be to imagine throwing several balls simultaneously at one person. Their chance of catching even one is much lower than if you had thrown them only one. It seems that predators suffer from the same confusion effect when they attempt to capture one prey in a group, with each prey simultaneously fleeing in an unpredictable direction (Neil and Cullen 1974). This lowers the capture rate of predators and can explain why they preferentially attack prey on the periphery of a group, where this effect is reduced, and focus their attack on individuals that they managed to separate from the group.

An effect of such confusion was observed by Foster and Treherne (1981) in their dilution **experiment** with marine water striders, and is detailed in Figure 14.7. In fact, they found that the mean success of a fish attacking a lone water strider is 20.5 ± 3.3% (mean ± two standard deviations) whereas the success rate drops to 6.6 ± 3.8% for attacks directed to groups of 15–17. This difference in the success rate was probably due to the continuous movements that water striders make to maintain their position in the group on the surface of the water and due to continuous contacts among individuals. However, the confusion effect in this example was less important than the dilution effect because individuals in groups of 16 were 16 times less likely to be captured than a solitary individual by the dilution effect, compared with 3 times less by the confusion effect (20.5/6.6).

#### 14.2.2.4 *Group defence*

Prey can also attempt to defend themselves actively by harassing approaching predators. It is not uncommon to see a large raptor, for example the common buzzard (*Buteo buteo*), aggressively attacked by the much smaller European kestrel (*Falco tinnunculus*), which itself can be harassed by house sparrows (*Passer domesticus*). When animals live in groups, they all have an interest in making the predator leave the area and often participate collectively in these harassment activities, an activity known as mobbing. The predator then finds itself confronted with several individuals that, even though much smaller, make it difficult to hunt successfully. This collective defence behaviour has been described in vertebrates and invertebrates, and through the participation of several individuals, the group can manage to chase away predators much larger than each group member (Hoogland and Sherman 1976). Anyone who has ever entered a tern colony, for example Arctic terns (*Sterna paradisea*), has experienced the effectiveness of collective defence which consists of multiple terns striking the intruder on the head with their sharp bills, defecating on him or her, all the while emitting loud, shrill vocalizations.

### 14.2.3 Group life and the search for food

The third type of benefit associated with group living concerns the ability of animals to obtain **social information** (see Chapter 4 for a definition) about the environment when joining a group.

#### 14.2.3.1 *The information centre hypothesis*

The first hypothesis put forward in this context is the ‘information centre hypothesis’ (ICH), which became perhaps the most heavily debated idea about group living. Ward and Zahavi (1973) suggested that one of the main reasons colonies and bird roosts are formed is because these communal areas allow individuals to obtain information about their environment from other birds. The reasoning behind this hypothesis mainly concerned the search for food; an individual that had failed to find food could return to the colony and wait for conspecifics to return with food. Unsuccessful individuals had only to follow successful individuals on their next foraging trip to find the current source of food.

Despite a long search for empirical support, every time evidence for the ICH was reported, it was criticized because the results obtained could be explained by alternative mechanisms. Other authors have tried to clarify the basic foundations of this mechanism. We first have to suppose that it is possible to recognize which individuals have succeeded in finding food. We have to suppose that these populations live in an environment where food is abundant and where competition during foraging isn't important. These food sources must appear randomly in the landscape, otherwise individuals could develop different strategies to find food by themselves. Finally, we have to suppose that these food sources last long enough to allow individuals that discover them the chance to return to the food sources at least once. However, food sources must be exhausted quickly otherwise all individuals would learn the food locations and information gathering at the colony would become useless. It is important to note that these conditions are the same as those proposed for information transfers in social foraging. On the other hand, these conditions are met in several species that reproduce in colonies. This is the case for marine mammals, like cetaceans and pinnipeds, as well as many seabirds that feed on fish shoals (Figure 14.1). The shoals can be very difficult to locate, but once found, they remain in approximately the same area for a certain time, and their typically large sizes make competition costs during foraging relatively low. These same conditions are fulfilled in colonial birds such as swifts and swallows that feed on large swarms of aerial insects.

*A controversial hypothesis*

The information centre hypothesis has been the subject of many debates, especially during the 1990s when many authors openly argued against it. In Ward and Zahavi's (1973) paper, the initial formation of a meeting place is not clearly explained. Rather, the authors seem to imply that meeting places already exist. Thus, this mechanism does not appear to lead to the initial formation of groups, but would intervene later, after groups already exist. Douglas W. Mock of the University of Oklahoma and his collaborators (1988) have since insisted that most of the predictions of the ICH are shared by other more parsimonious mechanisms. In other words, for these authors, this hypothesis is not falsifiable (i.e. it is impossible to prove it wrong) and thus is of no interest. Amotz Zahavi himself, and more explicitly Heinz Richner from the University of Bern and Phillip Heeb (1995, 1996), have insisted on the fact that to work, such a mechanism would require strict reciprocity, so that an individual that helps another will in turn receive help from that same individual in the future. They implied that individuals need to be capable of recognizing each other so that individuals that never return help can be excluded from further interactions (Richner and Heeb 1995; Danchin and Richner 2001). Such conditions have little chances of being met if the composition of the group changes rapidly, which is the case in night roosts of birds.

*Another problem of cheaters*

Zoltán Barta and Luc-Alain Giraldeau (2001) have used a theoretical approach to address this problem. They imagined a population in which all individuals depart in the morning to search for food. Next, they suppose the appearance of a mutant that never leaves in the morning and merely waits in the communal place until the return of successful individuals, and then follows them out to the current food source. This individual would not need to spend any energy in locating food in the morning. Additionally, it could spend this free time and energy in other activities, like courting mates etc. Such an individual would then have higher fitness than its conspecifics and this should, at least initially, increase the frequency of mutant 'cheaters' in the population. This 'cheater' strategy would then spread in the population and the benefits gained by the finders and cheaters would correspond exactly to the curves described for the producer–scrounger game seen in Chapter 8. The number of cheaters (scroungers) in the population would increase until the benefits of each strategy became equal. This point of equilibrium would occur when the proportion of cheaters is very high because the finder's share, the fraction of the food patch belonging exclusively to the finder, would be low.

If Barta and Giraldeau are right, it becomes difficult to reject the ICH solely on the basis that few cases of information exchange at the colony site have been reported. At best, their model predicts that these exchanges would be rare. They suggest another way to proceed, based on a distinction between flights of finders (producers) and flights of recruited individuals (cheaters). The first flights of the day to depart from the colony probably represent searching flights of finders characterized by frequent changes of direction. When individuals that have found food return to the colony after that first flight, they must return to that food patch directly. The cheaters recruited in that second flight must also head directly towards the food patch. In a system of information exchange, Barta and Giraldeau (2001) thus expect that the number of direct flights would always be greater than the number of searching flights. If we were to observe the opposite, which would be a higher number of searching flights than direct flights, we could reject the information centre hypothesis. So far, this important prediction has never been tested in the field.

Interestingly, Barta and Giraldeau (2001)'s approach also assumes that the communal places already exist at the beginning. Their general message is thus that when some meeting point already exists, social games like the one proposed by the ICH may sometimes emerge. This continues to leave unresolved the question of the origin of the aggregation.

#### 14.2.3.2 *An alternative hypothesis: 'the recruitment centre hypothesis'*

Several alternative hypotheses have been proposed to correct the drawbacks of the ICH. One of these was offered by Heinz Richner and Philipp Heeb in 1995 and 1996. According to them, benefits derived from feeding in groups, rather than benefits resulting from a transfer of information at the communal site, are more likely to favour collective foraging in communal areas. We have seen in Chapter 8 how **individuals can benefit from hunting in groups, either because being in a group increases each individual's capture rate, or because it allows them to defend themselves against predation** (see Section 14.2.2). These are what we call **Allee effects** (Section 8.2), occurring here again in the context of foraging. Allee effects could surpass the costs incurred by providing information on the location of food patches. This would explain why the individual that found food returns to the communal area: because it would likely to be able to recruit conspecifics to hunt with, which would increase the rate of food intake and/or its protection against predators. This question is central when studying non-breeding communal aggregations because in such situations individuals have no reason to return to the communal roost. The conditions to develop this mechanism are in fact very similar to those of the ICH and of those favouring group foraging (see Chapter 8): relatively abundant **resources** that are spatially and temporally unpredictable. In such conditions, the individuals that are searching for food are often very far from each other, implying that the return to and the recruitment at the communal site can become a more efficient strategy than one that would consist of waiting at the feeding site. In this context, the communal area would play the role of a recruitment centre leading Richner and Heeb to call their idea 'the recruitment centre hypothesis' (RCH).

##### a *The existence of recruitment calls*

According to this mechanism, the individuals that have found food can benefit from returning to the communal site to recruit conspecifics. Thus, the RCH predicts that when individuals that have found food return to the colony, they will communicate their discovery to conspecifics to attract them. Examples of this kind of recruitment behaviour have been observed in the colonies or roosts of several species, and had not been explained by any other hypotheses. This recruitment behaviour represents a strong argument in favour of the recruitment centre hypothesis. For example, Stoddart (1988) described a good example of recruitment behaviour in cliff swallows (*Hirundo pyrrhonota*). He repeatedly observed the return of an adult flying straight towards the colony, which would start making a sound resembling an alarm call when it got within about 10 metres of the colony. Such individuals did not land, so obviously they did not return to feed their young, and would immediately return back to the direction they came from, while repeating the same conspicuous alarm call two or three times. All the adults in the colony would then leave their nests and head in the same direction as the individual that had made the call. Since it is possible in this species to follow individuals over a great distance and to observe by their behaviour whether they are feeding or not, Stoddart was able to find out that these swallows would rapidly start feeding together, probably because they had found a large swarm of insects. About three minutes later, the adults would return and feed their chicks, and then depart immediately towards the insect swarm. The parents would make several trips until the resource appeared to run out. It is clear that a mechanism such as the information centre cannot explain such a phenomenon.

### b *Why would successful individuals return to the communal site?*

In the case of breeding colonies, explaining why successful individuals return to the breeding colony is straightforward because they have to come back to the nest. However, in the case of roosting animals, this return needs to be explained because there is no specific resource at the roost. The ICH fails to solve this problem, whereas the RCH provides such an explanation: individuals return to the roost to recruit foraging mates. The RCH also explains why certain individuals come back to recruit at the communal area whereas others do not, as well as why returning individuals signal their discovery with varying intensities. Because the optimal size of a group would vary according to conditions (such as the quantity of available resources), we expect that there would be differences in the **motivation** to recruit by various individuals. The intensity of the recruitment signal can vary as a function of the potential net benefit for the finder, which is directly related to the size of the group. An example of this kind of variation is given by Australian scientist Mark A. Elgar for the house sparrow (*Passer domesticus*). In fact, a solitary sparrow that finds a food patch usually 'chirrups' once before it starts to feed, which recruits other sparrows to the patch. Using very simple experiments, Elgar (1986) showed that the rate at which sparrows chirrup varies depending on circumstances: when the food source is divisible (like bread crumbs), the sparrow that finds it chirrups; when the food source is not divisible (the same amount of bread, but in one piece), the finder doesn't chirrup. Additionally, Elgar (1986) reports that the latency from food discovery to arrival of new sparrows is inversely proportional to the chirrup rate. By using playbacks of sparrow chirrups, it is possible to attract sparrows to an empty feeder (Table 14.2). Therefore, this sound does appear to have a recruitment function.

Elgar (1986) suggested a simple interpretation of this behaviour. Because sparrows suffer predation risks when feeding, an individual that finds a divisible food patch can benefit from attracting conspecifics (Section 14.2.2). Conversely, when the food patch is not shareable, attracting conspecifics may simply increase direct competition with conspecifics perhaps gaining no food at all.

An individual that has not found food, can compare the variable intensities of the recruitment signals of others that have found food. This situation is reminiscent of the language of bees in which the intensity of the recruitment signal is modulated according to the potential global quality of the food source.

| | Sparrow chirrups | Human whistles | No sound |
|---|---|---|---|
| Number of bouts during which sparrows came to the feeder | 14 | 5 | 6 |
| Number of bouts during which no sparrows came to the feeder | 6 | 15 | 14 |

Table 14.2 **A recruitment call: the chirrup of the house sparrow**

Attraction of house sparrows to a bird feeder that has been empty for at least two days and placed close to a speaker. Three treatments are applied during the experiment (columns): (1) playing back sparrow's chirrup calls; (2) playing back human whistles; and (3) no sound. Each bout lasted five minutes during which the treatment was applied.

The predictions of the recruitment centre hypothesis can be tested in various ways (see Table 14.3). By providing animals with ephemeral food sources that simulate a natural situation we expect: (1) that the net benefit of a forager on the feeding site will be lower at the time of discovery than when it returns with conspecifics; and (2) that birds will stop recruiting other individuals at the colony when no further benefits can be gained from increasing the group size. The recruitment signals must be intense in the beginning, and nonexistent in the subsequent returns to the colony (Table 14.3).

In summary, although similar to the information centre hypothesis, the RCH adds a fundamental assumption that could explain the benefits obtained by leaders and followers. Leaders can benefit by an increased feeding rate and protection from predators and followers can benefit from receiving information on food sources. The benefits of leaders and followers are immediate, and reciprocity is unnecessary. Richner and Heeb (1996) and Danchin and Richner (2001) have suggested that this creates favourable conditions for the evolutionary stability of this mechanism.

#### 14.2.3.3 *An ongoing debate*

The debate about the ICH continues. In 1996, Marzluff, Heinrich, and Marzluff published experimental results obtained for raven (*Corvus corax*) roosts. **Their study was very impressive because it was one of the only real experimental approaches applied to this hypothesis**. They found compelling evidence that some information transfer was involved in foraging ravens. This article led to a forum in the same journal (*Animal Behaviour*) a few years later (Danchin and Richner 2001; Marzluff and Heinrich 2001; Mock 2001; Richner and Danchin 2001). The forum emphasized the importance of Marzluff *et al.*'s (1996) results, but insisted, again, that despite the authors' interpretation, the results were not decisive evidence because identical results could be expected from the recruitment centre hypothesis. To distinguish hypotheses, **it is necessary to find situations where alternative mechanisms lead to contrasting predictions** (see Table 14.3). In their paper, Barta and Giraldeau (2001) predict measurements that are more likely to allow the rejection of the information centre hypothesis without any ambiguity.

##### a *The importance of subtle nuances in proposed mechanisms*

Even though the differences among the mechanisms implied by each of the three hypotheses implying information transfers seem tenuous, it remains that this type of subtle differences are often crucial for the evolutionary stability of a given behaviour (Richner and Danchin 2001). It is important to note that Marzluff and collaborators (1996) described various behaviours that closely resemble recruitment signals, and these signals suggest a mechanism like the recruitment centre rather than an information centre. This is not so unexpected because, as stated above, the RCH is better adapted to the case of avian roosts than to colonial breeding. Furthermore, Wright *et al.* (2003) have clearly stated that in ravens, as in vultures, the advantage of cooperative recruitment accrues at the feeding site on the carcass. Indeed, 'local territorial pairs can successfully defend any carcass from one or two juveniles, but give way only once six or more juveniles gather together (Heinrich 1990, Marzluff and Heinrich 1991, Heinrich *et al.* 1993)' (Wright *et al.* 2003). This clearly suggests that by recruiting foraging mates individuals are likely to exploit food patches better, which suggests that raven roosts likely function more as recruitment than information centres.

However, Marzluff and Heinrich (2001) continue to believe that raven roosts correspond to an information centre mechanism, as do other authors. For instance, Wright *et al.* (2003) published another very interesting study of raven roosts in which they placed carcasses baited with colour coded plastic beads in the environment and studied the distribution of the plastic beads in the pellets regurgitated by the birds at their roosts. This method demonstrated that the distribution of baited pellets at the roost consistently reflected the geographical location of bait sites. Furthermore, they found that aggregations of

| | Information centre | Recruitment centre | Producer–scrounger game |
|---|---|---|---|
| *Reciprocity* | Necessary | Not necessary | Not necessary |
| *Type of benefit* | Reciprocal and of the same nature | Mutual, but not necessarily of the same nature | None. Parasitism of information is the result of its evolutionary stability and implies a cost |
| *Nature of benefits* | Food | Food and possibly protection from predators | None at equilibrium |
| *Variation in net benefits between individuals* | Possible variation | No variation, equivalent benefits are maintained by density dependence | None at equilibrium |
| *Phenotypic types present at communal area* | All individuals are equally efficient in locating food which allows reciprocity | There may be differences in the capacity to locate food, but it is not necessary | Individuals can alternate between 'finder' (producer) and 'follower' (scrounger) strategies |
| *Variation in the feeding success of the finder when it returns to the feeding site while followed* | Decreases compared with the first visit | Increases compared with the first visit | Decreases compared with the first visit |
| *And/or predation risk during the return to the feeding site* | No predictions made by this hypothesis | Decreases compared with the first visit | No predictions |
| *Stability of the feeding group and of the community* | Required because to function, reciprocity implies that individuals recognize each other | Not necessary. In other words, individuals do not need to recognize each other | Not necessary |
| *Function of the behaviour seen at the communal area* | – Indicator of the position of the communal area (Ward and Zahavi 1973)<br>– Indicator of the 'mood' of the community (Ward and Zahavi 1973), meaning the amount of information on feeding sites available in the community<br>– Evaluation of the competition once on the feeding site (Zahavi 1986) | Recruitment of conspecifics to feed together and thus benefit together from a better feeding rate and an increased protection against predators | No function, simply an inevitable ESS |

Table 14.3 **Synopsis of the assumptions and predictions that allow us to discriminate three of the principal hypotheses of the evolution of aggregation in relation to foraging: the information centre hypothesis, the recruitment centre hypothesis (modified from Danchin and Richner 2001), and the producer–scrounger game hypothesis (Barta and Giraldeau 2001)**

beads at the roost grew daily with an increasing radius around the first pellets in a way that mirrors the linear increase in the size of the groups flying between roost and carcass each morning. Furthermore, rates of increase were greater for carcasses closer to the roost. Wright *et al.* (2003) concluded that their results support the ICH, despite the fact that such results would also be predicted by the RCH. The debate is far from over.

#### b *The necessity to state clearly all the assumptions and predictions of a hypothesis*

This example illustrates the need to state clearly the assumptions and predictions of proposed mechanisms. As we have seen in Chapter 3, it is one of the fundamental merits of the hypothetico-deductive approach to formalize the assumptions of each mechanism. As soon as several parameters are involved, we cannot limit ourselves simply to verbal reasoning. Only a formal, explicit mathematical approach can clarify the question. Historically, the debate on the role of information in the emergence of aggregations has greatly suffered from a lack of formalism (Barta and Giraldeau 2001; Richner and Danchin 2001). Considering this, Barta and Giraldeau's model (2001) marks a serious move forward in understanding the role of information gathering in the formation and maintenance of animal aggregations (see also Section 14.3.5.2).

### 14.2.4 Some comments about the functional approach to animal aggregation

#### 14.2.4.1 *Predation and the evolution of breeding aggregations*

It is clear that if some of the above hypotheses for the evolution of group living can apply to some kinds of aggregations, this is not the case for others. As underlined by Varela *et al.* (2007), an important distinction is whether group members can instantaneously adjust their positions relative to other group members. Clearly, members of foraging flocks or night roosts can react rapidly to the occurrence of a predator. Such groups are aggregations of mobile individuals. As we have seen, one common response is to rush towards the centre of the aggregation. However, the situation differs sharply in the breeding colonies of vertebrates (birds, reptiles) and in the aggregations of marine invertebrates with sessile adults in which the individuals exposed to predation are fixed on the substratum. In such species, eggs or offspring, or invertebrates fixed to the substrate, cannot relocate in reaction to a predator. Predators can thus more easily prey on the immobile individuals. **Colonies are thus aggregations of breeding sites containing immobile eggs or offspring**. As colonial breeding is highly conspicuous, colonies may just as well attract more predators just as they also reduce the danger they pose.

A recent comparative study of the Ciconiiformes, a large avian order comprising over 1000 species, showed that exposure to nest predators probably led to the breaking up of colonies over evolutionary time (Varela *et al.* 2007). In particular, colonial species nesting in situations exposed to predation usually either evolved back to solitary nesting or evolved to enclosed or inaccessible nest sites that provide protection against predators. In this bird taxon, the colonial/exposed state was relatively unstable along the phylogeny. This study thus supports the hypothesis that coloniality increases predation risk and contradicts the classical idea that predation has favoured the evolution of coloniality.

It is interesting that most of the examples in Section 14.2.2 of group living and predation involve groups of mobile, not immobile, individuals. This was clearly the case for the examples provided to illustrate vigilance, dilution, and confusion effects. These considerations question the role of predation in explaining the evolution of colonial breeding.

#### 14.2.4.2 *Food finding and the evolution of animal aggregation*

As we have seen, the second major group of hypotheses used to explain the evolution of animal aggregation involves enhanced food finding for group living individuals. One of the major arguments in favour of a role of enhanced food finding in groups resides in the fact that most seabirds are colonial. The usual interpretation of this correlation is that birds feeding in the marine environment are secondarily constrained to become colonial in order to exploit this particularly unpredictable environment efficiently through the benefits of mechanisms such as information transfer.

This interpretation implies an evolutionary scenario in which species first become marine, and then, secondarily, become colonial. However, when testing predictions of this hypothesis with directional comparative methods (see Chapter 3), Rolland *et al.* (1998) found strong evidence for the opposite scenario, in which extant seabirds first became colonial and only secondarily moved to the marine environment (see Figure 3.12 and Section 3.5.2.3b for more details). Although this result does not rule out the possibility that enhanced food finding in groups played a role in the evolution of coloniality in the first step, it contradicts the main empirical arguments that coloniality evolved in relation to food finding behaviour.

## 14.3 The new by-product approach

This section presents another way of approaching the evolution of animal aggregation. In fact, it consists of a generalization of a relatively recent approach to all forms of animal aggregation. This approach was developed to explain the evolution of a particular form of aggregation: breeding coloniality (Danchin and Wagner 1997). It is called the **by-product approach** to differentiate it from the functional approach. **The term by-product means the direct and inadvertent consequence of another process**. This approach thus views coloniality as the direct and inevitable consequence of fitness enhancing processes of choice.

### 14.3.1 The basic statement

By the end of the 1980s, the debate over the evolution of coloniality had reached a dead end. Researchers had investigated the importance of the various benefits covered in the previous section. One researcher working on a certain species could conclude that a specific benefit, such as reduced predation, accrued in colonies, whereas another author working on a different population of the same species could find that predation was not important, but that enhanced foraging apparently occurred. Such variability led certain authors to suggest that coloniality is not a unique phenomenon but more like a family of phenomena and that there are no general explanations of it. Wittenberger and Hunt (1985), reviewing the various hypotheses, concluded that despite decades of research, clear generalizations had failed to emerge. However, they proposed to continue measuring costs and benefits under different circumstances in order to estimate their balance, with the hope of one day understanding the evolution of coloniality overall.

A fundamental reason for this dead end is that it is extremely difficult to obtain a full balance of the costs and benefits of coloniality. First, the intensity of potential costs and benefits is likely to vary, both in space and time. Because the assessment of the economic balance of aggregation would take several years and would require several samples, it is impractical if not impossible to establish such a balance for a given population in time. Furthermore, such an economical balance probably varies among individuals. For example, the costs of competition related to increased density in certain habitats probably differ among individuals of varying competitive abilities. We have also seen that predation risks vary according to the position within the group (see Section 14.2.2b). Moreover, parasite costs probably vary greatly according to each individual's resistance, genetic or otherwise, to the pathogen, and this would of course vary depending on the pathogen. It would then be necessary to derive a balance of the costs and benefits of coloniality for each category of individual rather than an average for the population. The difficulty of that task may explain why most researchers on the issue at the end of the 1980s were stuck in a dead end. In retrospect, it seems clear that new approaches were necessary to answer this question.

## 14.3.2 The emergence of new approaches

### 14.3.2.1 *Defining aggregation*

Several ideas have led to the emergence of a new approach to studying the evolution of aggregation, and more specifically coloniality. First, it was necessary to define clearly what we mean by aggregation. It is surprising that the debate on the evolution of coloniality has not led to a clear definition of this concept. To do this, we must use the ideal free distribution (IFD) as a reference (see Chapter 8). This idea arose almost simultaneously in the UK with Richard Silby (1983), in Canada with Colin Clark and Marc Mangel (1984), and in the USA with Thomas Caraco and Ronald Pulliam (1984). They all thought of applying the assumptions of the IFD to the formation of aggregations. Later, Donald Kramer (1985), and William Shields and collaborators (1988), introduced the assumptions of the IFD in the context of coloniality, with their hypothesis of aggregation in traditional zones. A paper by Charles R. Brown, Bridget J. Stutchbury, and Peter D. Walsh (1990) then detailed the link between the IFD and coloniality. They stressed the importance of the shape of the relationship between density and fitness. Subsequently, Danchin and Wagner (1997) proposed a definition involving two types of aggregation (Figure 14.8).

Aggregation is a pattern, a distribution of individuals in space. We have seen in Chapters 8, 9, and 10 how the IFD is a distribution that, although purely theoretical, can provide a reference for the description of any distribution of individuals that gain no benefit in being together. The IFD predicts that all individuals in the population have access to the same resources. Thus, this distribution leads to an even sharing of all the resources in the environment among all the individuals of the population. It is the distribution that would be reached by theoretical

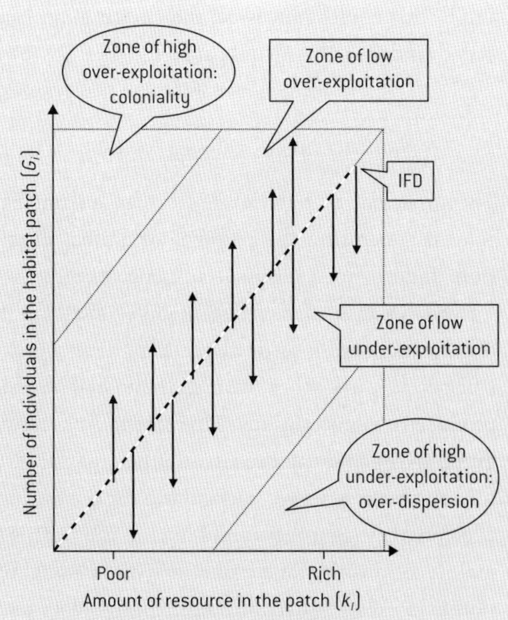


Figure 14.8 **A definition of aggregation (this figure uses the terms described in Box 8.1)**

Aggregation is a pattern that we can define if we can measure both the number of individuals $G_i$ in each patch and the amount of resources $k_i$ it contains. In an ideal free distribution, the data points will all be aligned (dotted line) in this graph relating population size and patch carrying capacity. The slope of this straight curve varies as a function of the global size of the population. The bigger the global population, the steeper the slope. Patches located below this line correspond to patches where individuals have on average more resources than they would if the population was at the IFD (under-exploited patches); those located above the dotted line correspond to patches where individuals have on average fewer resources than they would if the population was at the IFD (over-exploited patches). The further away vertically from the dotted line, the stronger the aggregation. At the upper left-hand corner, we find the extreme case of coloniality. However, when the habitat is strongly heterogeneous (i.e. when it is made up of patches of very different quality), at the IFD, an important part of the population may be found in the few richer patches. This aggregation is only apparent and results entirely from habitat heterogeneity. It can be called **habitat-mediated aggregation** (Danchin and Wagner 1997). Any deviation from the IFD implies **real aggregation**. Aggregation is thus a skew in animal distributions compared with IFD.

organisms that would possess three characteristics: (1) a perfect knowledge of the environment (hence the word 'ideal'); (2) they can freely move among the different patches of the environment without any costs or constraints (hence the word 'free'); and (3) they all have the same competitive abilities and gain no benefits in being together (i.e. a **dispersion economy** see Chapter 8). In this case, as we have seen in Box 8.1, at equilibrium, the number of individuals in each habitat patch matches the quantity of resources in this patch. If we draw the curve of the number of individuals in each patch as a function of the quantity of resources in each patch, we obtain a straight line (the dotted line in Figure 14.8).

Let us imagine a habitat that contains many poor patches and a few rich patches, and that the population is spread out in that environment according to the IFD. Even though the population is distributed in an ideal free way, an observer in the field could get the impression that the distribution is aggregated because a large proportion of the population is located in the few rich patches. However, this distribution would only reflect the variation in patch quality. The amount of resources per individual would not vary among the various patches. Danchin and Wagner (1997) have dubbed this habitat-mediated aggregation. In this case, aggregation is only apparent because all individuals from all patches have the same quantity of resources. Conversely, any deviation from the IFD implies that all the individuals from different patches do not have access to the same quantity of resources. The area above the line in Figure 14.8 corresponds to patches where individuals have access to fewer resources than in an IFD (these patches are over-exploited); in the area below the line individuals have access to more resources, and the corresponding patches are thus under-exploited considering the total size of the population. Danchin and Wagner (1997) have called this real aggregation.

With aggregation defined in this way, it appears that **aggregation is predominately a pattern**, a skewed distribution of individuals in space relative to an IFD (Wagner and Danchin 2003). We can then ask what causes such skews in animal distributions. The question is important because when a distribution shows real aggregation, the partitioning of resources is unequal among the individuals of different patches. Any individual that moves from an over-exploited patch to an under-exploited patch would benefit in terms of fitness. We thus expect **natural selection** to favour individuals able to do so, which, as we have seen in Chapter 8, should tend to push the distribution towards an IFD and to the disappearance of aggregation (see also specific examples in Chapter 10). Obviously, in certain species like colonial birds or various other animal groups, the distribution does not tend towards the IFD. Therefore, there must be other reasons why the distributions of these species remain so highly aggregated.

The question of the evolution of animal aggregation can then be reformulated. It requires the understanding not of costs and benefits of group living, but rather that of the processes that generate skewed patterns of distributions of individuals in space. What are the individual behaviours that lead to the observed skewed distributions? These behaviours imply two choice processes: the choice of a place to breed, and that of a sexual partner.

#### 14.3.2.2 *Habitat choice and aggregation*

The processes of habitat choice play an important role in generating the patterns of animal distribution. Indeed, in a group living species, groups are formed because each group member after the first one has **chosen** to settle closer than necessary to its conspecifics. In other words, individual decisions produce the aggregations. This strong link between group living and the processes of habitat choice has too long been ignored, particularly for the evolution of coloniality (Boulinier and Danchin 1997). This was mainly because no one had clearly defined aggregation as a skewed distribution of individuals within the available space (Wagner and Danchin 2003).

In Chapter 9, we covered the question of breeding habitat choice. We have seen that habitat choice is expected in heterogeneous environments. Another important point is that animals can only choose among alternatives on which they have some information, even if that knowledge is incomplete. This implies that any individual that can obtain information about patch quality, and that can compare the various patches and choose the one that is most beneficial in terms of fitness, will be favoured by natural selection. Heterogeneous environments generate strong selection for individuals capable of choosing the best habitats. In other words, selection favours efficient **habitat selection**. The information used to make this choice can be very different in nature. We have dealt with such questions in Chapter 9 and the reader will find in Chapter 4 a general taxonomy of biological information.

#### 14.3.2.3 *Mate choice and aggregation*

Choosing a sexual partner is another choice that affects the final distribution of breeders. Indeed, potential partners are resources of varying quality that are necessary for reproduction. An individual that would choose its breeding habitat solely on environmental quality would risk finding itself in a suitable habitat but possibly without a suitable mate. Although performance information (see Figure 4.3), which is extracted from the performance of others, integrates a lot of information on the global quality of the environment (Chapter 9), it may contain little information on the quality of potential sexual partners (except if high-quality individuals settle in high-quality habitats). Therefore, habitat choice must interact with mate choice. Mate choice is part of sexual selection, which was detailed in Chapter 11.

In this context, it seems obvious that sexual selection must play an important role the evolution of group living. Still, until the beginning of the 1990s, sexual selection was rarely mentioned in the abundant literature about, for example, the evolution of coloniality. In Section 14.3.4, we will analyse the reasons for this missing piece, but one reason is again that coloniality, and group living in general, has rarely been perceived as a skewed distribution. Because of this, researchers have not focused on the mechanisms that produce these skews, that is the processes of choice, one of them being mate choice. Mate choice involves information exchanges between potential mates. It implies sender–receiver signalling, that is **true communication**, which has complex implications, as we will see in Chapter 16.

### 14.3.3 Habitat copying

We have seen in Chapter 8 that even in situations that were very simplified for experimental goals, the obtained distributions rarely matched the ideal free distribution exactly. In these very simple and controlled experiments, there is always a slight excess of individuals in poor patches, which implies a slight under exploitation of high-quality patches. In other words, there is often real aggregation (Figure 14.8) at any given time. (However, in Chapter 10, in the context of dispersal, we have seen that when measured over certain periods of time, animal distributions could on average become surprisingly close to the IFD.) This occurs because in nature, the conditions of the IFD are never completely fulfilled: (1) there often are differences in the competitive abilities of individuals; (2) the movements between patches are costly in energy and time; and (3) individuals never have complete knowledge of the current quality of the different patches available, either because this requires a lot of time and energy to acquire, or because they lack the cognitive abilities to measure differences that can be very subtle (a complete review of these issues and their consequences for IFD can be found in Giraldeau and Caraco (2000)).

#### 14.3.3.1 *The importance of information*

The question of a lack of knowledge is probably the hardest one to deal with. It can seem negligible in highly controlled situations comprising only two patches that do not vary in quality, as in the experiments we described in Chapter 8. However, when the environment comprises many patches that vary in quality in space and time, the lack of information has crucial consequences for animal aggregation.

In the context of the by-product approach of coloniality, the major question is to know whether the evolutionarily stable strategy (ESS) for habitat choice generates a certain amount of aggregation as a by-product. This was analysed with a spatially explicit model designed to determine what breeding habitat choice strategy is evolutionarily stable (Doligez *et al.* 2003).

In their simulation study (see Chapter 9), these authors were able to study the aggregation effects generated by various strategies according to environmental predictability (Figure 14.9). Their model explicitly integrated the existence of two patches that varied independently from one another. When analysing the distribution obtained at the end of simulations, they found that the five basic strategies they modelled can produce a certain degree of real aggregation. However, levels of aggregation differed according to the strategy and environmental predictability (Figure 14.9). In particular, a strategy they called **Presence**, because it uses the ratio of the numbers of breeders in the two patches as information about patch quality, leads to substantially higher levels of aggregation (Figure 14.9). As the 'Presence' strategy (which corresponds to **social attraction**) was able to maintain itself at low frequency when confronted by **habitat copying** strategies (definition in Chapter 9, Figure 9.4, and Section 14.3.3.3 below), this might generate some aggregation at the population level. The two experimental tests of this hypothesis (Doligez *et al.* 2002; Parejo *et al.* 2007) showed that **prospectors** were influenced in their habitat

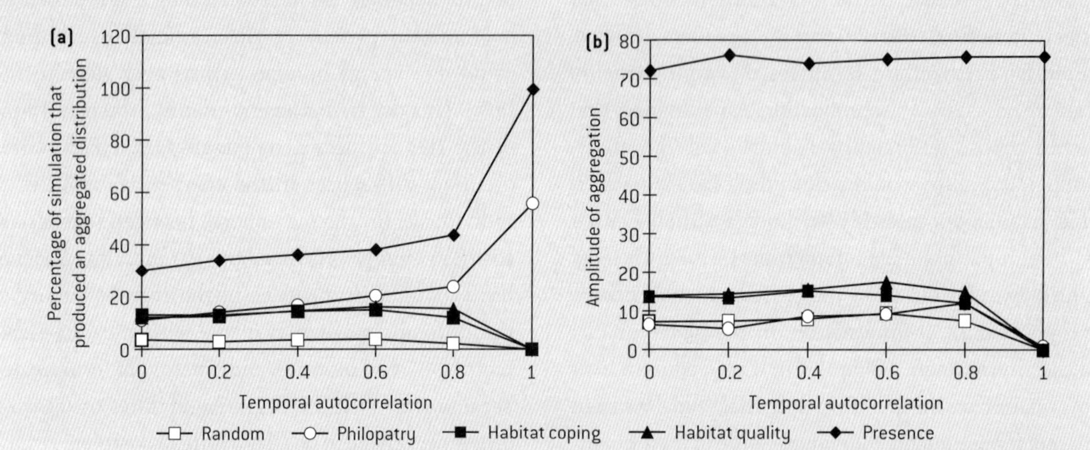


Figure 14.9 **Strategies of habitat choice and aggregation**

Comparison of the effect in terms of real aggregation of five major strategies of habitat choice. For each strategy, only individuals that fail in their last breeding attempt change patches according to rules that depend on each strategy. In the **Random** strategy, individuals that fail have an equal probability of entering one of two patches of the simulated environment. In the **Philopatry** strategy, individuals return to the patch where they were born or where they had previously reproduced. In the strategy **Presence**, they spread out according to their relative density during the previous year (this is called 'social attraction'). In the strategy **Habitat quality**, they distribute as a function of the intrinsic quality of the patches in the previous year (translated in the model in terms of the probability of total reproductive failure). Finally, in the **Breeding success** strategy, individuals settle according to the mean reproductive success of the previous year. The last two strategies correspond to two forms of 'habitat copying'. The individuals in the last three strategies use information about the environment collected on a given year for settlement decisions in the next year.

**a.** Percentage of times when a real aggregated distribution occurred.

**b.** Mean magnitude of the aggregation, i.e. mean number of additional individuals on the good patch (given that the total population size varied around 200 breeding pairs) compared with the IFD.

Modified and completed from Doligez *et al.* (2003).

choices by manipulations of local reproductive success in the previous year. Those experiments, however, were of a non-colonial species, and were not designed to address whether habitat copying could generate nesting aggregations. However, in the correlative study of kittiwakes (Danchin *et al.* 1998b) that we saw in Chapter 9, it was suggested that the increases and decreases in between-year recruitment could lead to colony extinction and formation. Subsequent correlative analyses of cliff swallows by Brown *et al.* (2000) showed similar results to kittiwakes, which also immigrated and emigrated to colonies according to their local reproductive success in the previous year. These similar findings in such ecologically and taxonomically contrasting species may be considered as evidence for the generality of the habitat copying mechanism.

#### 14.3.3.2 *The processes of habitat choice can still produce colonies*

It appears that all these strategies lead to a certain degree of skewing in animal distribution, meaning that some aggregation occurs in practically any circumstances. For the most part, this aggregation is due to the quality of the information collected: individuals choose their breeding area according to the information collected the previous year concerning environmental quality. However, the environment may well have changed since the previous year. The probability and importance of such changes depend on the environmental predictability, i.e. its temporal autocorrelation. Because of this one-year lag, individuals may have a hard time tracking the variations of the environment in real time.

In the case of the 'Presence' strategy, this lag is even more important because individuals use the relative population sizes in the patches of the environment as a source of information. However, the relative number of breeders on the patches resulted from the relative size the year before, and so on. Thus, the value of this information varies according to the relative history of the patches. In short, as soon as one patch accumulates more breeders than the other, it recruits more individuals of the 'Presence' strategy, and thus it can only grow larger, independently of the changes in patch quality. In the end, all individuals end up in one of the patches, which is not necessarily the currently best one. In other words, social attraction is prone to informational cascades (Giraldeau *et al.* 2002). This is why the strategy 'Presence' is the one that leads to the highest rate and levels of aggregation (Figure 14.9). The conclusion is that social attraction is likely to generate high levels of aggregation.

One of the important conclusions of Chapter 9 was that the degree of aggregation generated by the different strategies is mostly responsible for the evolutionary stability of the various strategies. Roughly, it can be said that the less aggregation is produced by a strategy, the more this strategy is stable from an evolutionary standpoint (Doligez *et al.* 2003). However, aggregation is not the only factor that influences the evolutionary stability of these strategies (see Chapter 9). Another important conclusion of these models is that the degree of aggregation produced by these strategies strongly depends on environmental conditions. Aside from the strategy of social attraction that most frequently (Figure 14.9a) and most strongly (Figure 14.9b) produces aggregation, the other strategies seem mostly to produce aggregation in the environments that do not show a very strong autocorrelation.

We can then ask whether animals could instead rely on information collected just before settlement, during the current year, to choose their breeding site. However, most studies that have searched for these strategies have concluded that the characteristics of the environment at the beginning of the breeding season rarely allow individuals to make reliable predictions of the future quality of the environment for the rest of the season. This seems to be because environmental predictability is in fact higher across years than across months. Because of this, it seems that strategies that use information from the previous year probably lead to better choices. Still, very little is known about the predictability of natural environments.

Thus, even relatively sophisticated strategies of breeding habitat choice that imply a whole series of prospecting and information gathering behaviour do not allow reaching a perfect IFD. This generally means that we should expect to find a certain degree of aggregation in most natural populations. It remains to be seen whether this degree of aggregation is sufficient to generate the high levels of breeding aggregation observed in colonies or social groups. Current models do not answer this question. Models conceived specifically to test this question are necessary to confirm that simple habitat choice processes can generate sufficient aggregation to lead to the formation of dense social groups.

#### 14.3.3.3 *The 'habitat copying' hypothesis*

In Chapter 8, where we went over the choice of foraging habitat, we saw how individual selection can lead to much greater group sizes than the one that maximizes fitness (Figure 8.8). According to Figure 8.2, it clearly appears that one of the fundamental reasons for this lag is the existence of environmental heterogeneity: the more the habitats differ in quality (i.e. the more the fitness curves in Figure 8.2 are at different levels), the more we expect the observed group size to be higher than the optimal size in the best habitats. That very simple example shows how general and natural the tendency of individuals to aggregate spatially is, as soon as the environment shows some heterogeneity. However, in these examples only habitat-mediated aggregation is generated.

In Chapter 9 and in the previous sections, we have seen that strategies of breeding habitat choice that use **inadvertent social information** (ISI) have very good chances of being used in conditions of intermediate predictability, when the environment is neither completely unpredictable nor completely predictable. This is probably the most common case in nature. Because such strategies produce a certain degree of aggregation (Figure 14.9), this led to the habitat copying hypothesis. According to this idea, coloniality is simply a consequence, or a by-product, of a habitat choice process, particularly a habitat choice based on ISI provided by the reproductive performance of conspecifics (see Boulinier and Danchin 1997; Danchin *et al.* 1998a, b; Danchin and Wagner 1997; Wagner *et al.* 2000). Furthermore, Figure 14.9 shows that most of the main strategies of breeding habitat choice that exist in nature generate a certain degree of aggregation of individuals in space. This suggests that this natural tendency to aggregate may set the stage for the evolution of various types of group living. Indeed, once clusters of individuals are formed, these individuals must interact. Any individual that can profit from this unavoidably aggregated situation will then be favoured by selection.

### 14.3.4 Hidden leks

#### 14.3.4.1 *Long-neglected sexual selection*

The reader may have noticed that sexual selection has been lacking from this chapter so far. This absence reflects the literature on the evolution of group living in general, and of coloniality in particular. Until the end of the 1980s, surprisingly, it was rarely considered that sexual selection could play a role in the aggregation of breeding territories. An early connection between coloniality and sexual selection was made by Collias and Collias (1969), who suggested in village weaverbirds (*Ploceus cuculattus*) breeding in Senegal that larger colonies appeared to provide more opportunities for mate choice. Other early suggestions that a fitness advantage of coloniality may stem from the ease of finding mates were formulated by Hamilton (1971) and Alexander (1974).

Much later in 1988, Dirk Draulans, a young Belgian researcher working at the Edward Grey Institute at Oxford, further developed this question. After commenting on the strange absence of sexual selection in the debate on the evolution of coloniality, he used a comparative approach on herons to examine the factors that correlate to coloniality in this taxon (Draulans 1988). The study analysed the relationship between colony size and variables linked to behaviour: four variables quantifying the visual aspect of the morphology of these species, and three variables quantifying flight performance, as well as visual and auditory behaviours not occurring in flight. The results showed a positive link between colony size and the visual aspect of colour signals, diurnal activity, and the intensity of visual signals at the scale of genus, within families. These results were supported in a more specific analysis on the genera *Egretta* and *Ardea*, leading Draulans to propose that 'coloniality improves the attraction between partners and the probability of encounters with potential partners, which increases the opportunities of reproductive partner choice'.

Most colonial birds are **socially monogamous** (Chapter 12) and **parental care** usually requires effort by a male and a female. Accordingly, the hypothetical connection between coloniality and mate choice had been expressed in terms of more easily finding breeding partners in a group. Before the late 1980s and the widespread use of genetic techniques such as DNA fingerprinting to determine parentage, socially monogamous (i.e. bi-parental) species were typically regarded also to be genetically monogamous (Box 14.1). However, as we saw in Chapter 12, it is now well known that the prevalence of **extra-pair copulations** (EPCs) has resulted in many socially monogamous species being genetically **promiscuous**. The common existence of **extra-pair paternity** in the broods of socially monogamous species dramatically increases variation in male mating success and thus the intensity of sexual selection. Most studies show marked skews in extra-pair mating or fertilization success, making many socially monogamous species similar to promiscuous species when viewing mating success independently from parental care.

*Can females force males to aggregate?*

The discovery that socially monogamous species often pursue copulations just like promiscuous species helped lead to the formulation of the **hidden lek hypothesis** (Wagner 1993, 1998), which is based on the distinction between the social and genetic mating systems of bi-parental species. The hypothesis predicts that the pursuit of EPCs by monogamous

Box 14.1 **Coloniality and sexual selection in monogamous species**

The debates on the evolution of coloniality have been dominated by researchers working on birds. Coloniality is in fact most common in this class. It just so happens that most colonial birds are monogamous. This is particularly true in marine birds, the group for which coloniality was the most extensively studied. As we have seen in Chapter 12, until the development of molecular methods to study paternity, mainly in the late 1980s, it was generally thought that sexual selection only played a minor role in monogamous species. In fact, the low **apparent** variation in the reproductive success of males seemed to imply a low potential for sexual selection. Because of this, there was no reason to think that sexual selection could explain the evolution of monogamous species, and thus of colonial species.

However, different studies, like those of the Danish researcher Anders P. Møller on barn swallows (*Hirundo rustica*), which were published from 1986 and which we have mentioned several times before, have led to the idea that in monogamous species, sexual selection can be sufficient to explain the evolution of exaggerated secondary sexual traits. Additionally, DNA fingerprinting has shown that in socially monogamous species, social mate paternity was far from certain (see Chapters 11 and 12). Within a population, paternity can vary among nests from 100% to 0%. The idea that there is a low variation in the reproductive success of monogamous males is thus incorrect. This has led to the distinction of the social and genetic **mating systems** (see Chapter 12). When we make this distinction, monogamous species are often similar to promiscuous species such as polygynous and lekking species.

females produces aggregations of male-defended nesting territories just as the pursuit of copulations by females in promiscuous species causes males to aggregate into leks. Thus the same mechanisms that form leks in promiscuous species may contribute to colony formation in socially monogamous species. For example, the hotshot model (Chapter 12; Beehler and Foster 1988) predicts that the most dominant and attractive male, the 'hotshot', attracts not only females, but also less attractive secondary males that attempt to access females near the hotshot, thus forming a lek. In a lek, where males provide no parental care, most females can copulate with the hotshot, whereas in a bi-parental species, only one female can form a pair bond with the top local male. Many females, however, can obtain EPCs from him, and females may pursue this strategy by pairing with less attractive males that nest near the top male. Such a female strategy may induce outlying males to switch to territories near top males. The logic is that by breeding near more attractive males, secondary males increase their chances of obtaining mates. Although the cost may be the sharing of paternity with the more attractive neighbour, most secondary males obtain at least partial paternity by breeding close to it. Another variation of the hotshot model has been proposed where females do not necessarily pursue extra-pair fertilizations but can benefit from breeding near a more attractive male in order to court him as a mate for a subsequent breeding attempt (Wagner 1999). This female strategy may also have the potential to compel secondary males to aggregate around higher-quality males

#### 14.3.4.2 *Socially monogamous birds copulating in leks*

The hidden lek hypothesis was developed from the mating system of razorbills (*Alca torda*), a monogamous, colonial seabird. The species had not previously been known to engage in EPC, and their relatively sparse breeding densities appeared to limit the possibility of such social interactions occurring frequently. In the study colony on Skomer Island, Wales, most pairs nested underneath boulders that afforded protection from predators, but limited social interaction with neighbours. However, razorbills attended crowded ledges outside the boulder colonies where they aggregated at densities 50 times greater than in the colony (Wagner 1992a; Figure 14.10). These ledges served as mating arenas in which paired males and females engaged in EPCs when their mates were absent, or copulated with their mates when the pair was present. Nearly all females were subjected to EPC attempts in the arenas, and 50% of females accepted one to seven EPCs before egg-laying. Females typically appeared to resist extra-pair mountings; however, their visits to an arena away from their nesting areas revealed that females often resisted copulations as a ploy to evaluate males for EPCs (Wagner 1991).

The discovery of this 'secondary mating system' elucidated several features of razorbill behaviour that are applicable to many other monogamous species. The most important finding is that the mating arenas shared all of the features of leks (Figure 14.11). Females visited for copulations with males who did

Figure 14.10 **Two mating arenas of razorbills**

The two boulders represent distinct arenas. Individuals on each boulder belong to two different parts of the colony. Individuals thus find themselves in display arenas with their immediate nesting neighbours.

Photograph by R.H. Wagner.

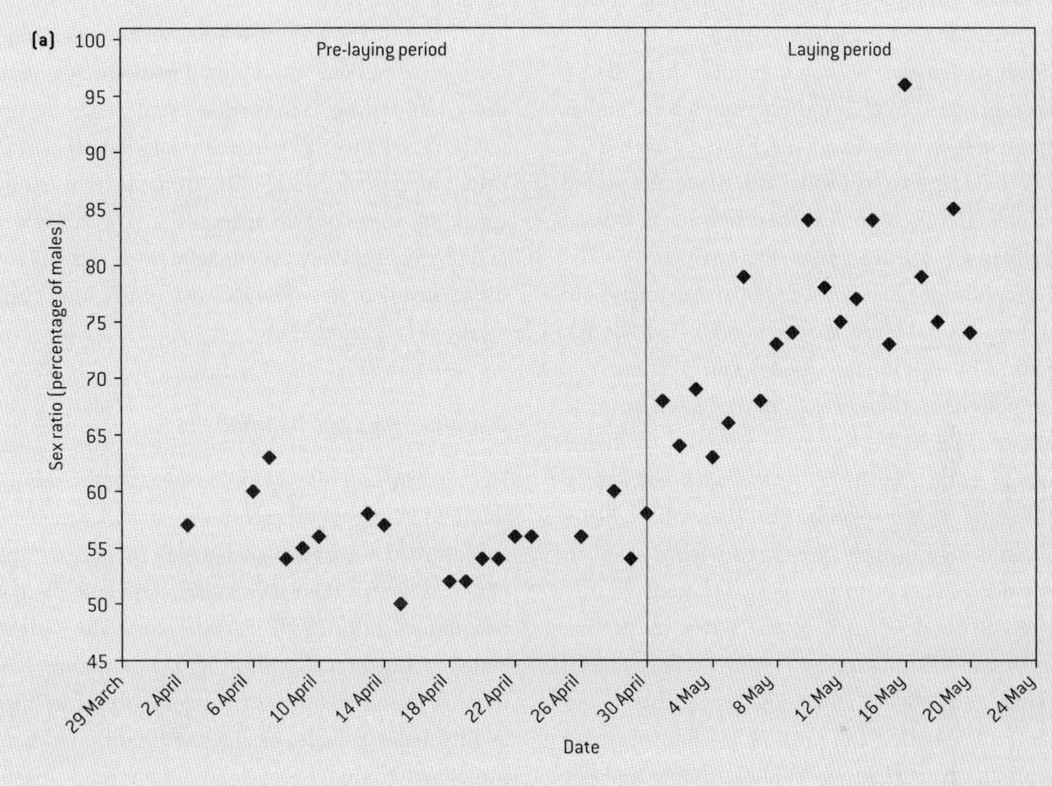


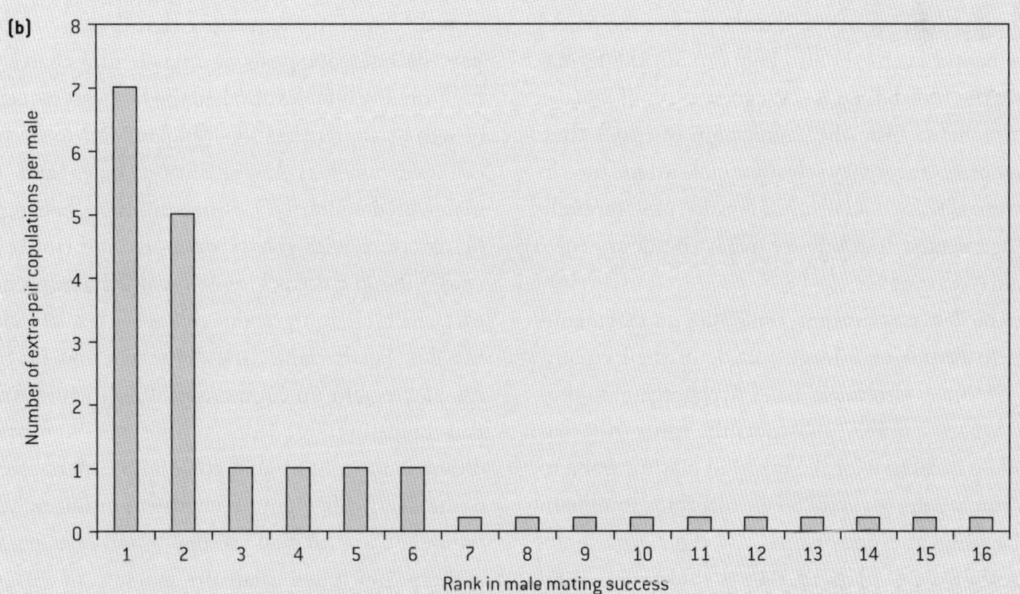


Figure 14.11 **Breeding arenas of razorbills resemble leks**

**a.** Variation in the sex ratio over the season in the breeding arena. Globally, the sex ratio increases over time ($r_s = 0.84$; $n = 39$; $P < 0.0001$). However, there was no significant change in the period preceding egg laying. In fact the sex ratio increases significantly ($r_s = 0.69$; $n = 20$; $P < 0.003$) after egg laying.

**b.** Individual variation in the total mating success of EPCs of the various males that frequent an arena.

Redrawn from Wagner (1992b).

not provide parental care for their offspring (males only provided care for their own offspring and never to those of females with whom they had EPCs). Males aggregated to obtain these copulations, and as is typical of leks, males fought aggressively over EPCs, timing their fights to coincide with the fertile period of females, and also as in leks there was a male-biased **sex ratio** in the mating arenas (Figure 14.11a).

The conditions in the razorbill mating arenas bore a striking resemblance to those predicted for lek formation by the hotshot model (see Chapter 12). As in many leks, there was a marked skew in male dominance, with the top male winning 93 fights compared with a mean of 13, and as predicted by Beehler and Foster (1988), the skew in fighting ability was significantly correlated with a skew in male mating success. In one year, for example, the top fighter obtained 44% of the EPCs, much like males in some lekking species (see Pruett-Jones 1988; Hoglund and Alatalo 1995; Figure 14.11b). These findings suggest that the hotshot mechanisms may have produced aggregations of monogamous males.

The mechanism of the female preference model (see Chapter 12) might also help explain the formation of the razorbill arena (Wagner 1993). If females attain benefits from the aggregation of males then female preferences may select for male attendance in the arenas. This could result if females preferentially visit aggregated over solitary males (Bradbury 1981; Alatalo *et al.* 1992).

If the lek mechanisms can cause monogamous males to cluster on a ledge outside of their colony, then they may also cause males to aggregate in locations where colonies subsequently form. All that would be necessary for the razorbill mating arena to form into a colony would be for razorbills to remain on the ledge and once paired, lay their eggs there. If razorbills were to do so, they would be breeding densely on an open ledge, much like many other colonial species. By visiting the mating arenas, where there were no resources other than males, females clearly demonstrated receptivity to EPCs. Although razorbills are the only monogamous species known to mate in arenas, the behaviours they perform in the arenas are typical of those performed by other species of monogamous birds within their nesting colonies. The razorbill arena, therefore, serves as a model for the social behaviours of colonial species, minus the confounding factors associated with nesting-territory defence. The hidden lek hypothesis has been supported by results from several studies including those of two species that are ecologically and phylogenetically distant from razorbills and each other: bearded tits (*Panurus biarmicus*) and purple martins (*Progne subis*).

### a *Hidden leks and bearded tits*

Herbert Hoi and Maria Hoi-Leitner (1997) examined the hidden lek hypothesis in the socially monogamous bearded tit, in which females incite extra-pair males, as well as their own mates, to pursue them for copulations (Hoi 1997). Females copulate with the male that outcompetes the other males, thereby maximizing the probability of being fertilized by high-quality males. Females in superior condition initiated prolonged flight chases more often than females in poor condition, suggesting that they were able to pay the energetic costs of inter-sexual chases. The high-quality females also had the highest frequencies of extra-pair fertilizations and tended to breed in colonies, whereas low-quality females and their mates bred solitarily. Confirming their behavioural observations with genetic data, Hoi and Hoi-Leitner (1997) found that 43% of colonial females produced extra-pair offspring compared with 0% of solitary females. Furthermore, the authors found no benefits of breeding in higher density: foraging success and predation frequencies were not significantly different for solitary and colonial breeders. Because of the overt behaviour of females in initiating sexual flight chases, as well as the other evidence, the authors concluded that the pursuit of EPCs by females may have led to the aggregation of nests of the colonial breeders.

### b *Hidden leks and purple martins*

Purple martins have also been used to examine the hidden lek hypothesis in a study built upon the work of Eugene S. Morton and colleagues (1990) at the US Smithsonian Institution. They proposed that EPC may be a benefit to males that could contribute to colony formation. This idea stemmed from their discovery of a marked relationship between male age class and paternity. While yearlings lost an astonishing 61% of their paternity through extra-pair fertilizations, older males achieved nearly complete paternity, losing only 4%. This remarkable finding represents the widest paternity difference in age class reported in any avian species and is associated with a conspicuous feature of male purple martins, that of delayed plumage maturation. Males of the adult age class (two years and older) attain the definitive glossy purple plumage that gives the species its name. In contrast, one-year-old males, although undergoing a complete moult at their wintering grounds (Stutchbury 1991), moult back into a drab brown and white juvenile-like plumage. Despite their juvenile appearance, one-year-old males are sexually mature breeders who contribute parental care to the same degree as older males (Wagner *et al.* 1996a). Yearling males are thus referred to as 'subadults' in reference to their plumage rather than their reproductive status.

Morton and colleagues (1990) observed that adult males could increase their fitness by recruiting subadult males to the colony and fertilizing their mates, and suggested that the benefits adult males accrued from EPCs selected for coloniality. This idea was based on another striking feature of purple martin life history, that of a marked age class difference in spring migration schedules, with adults arriving and nesting a full month earlier than subadults (Morton and Derrickson 1990). When their mates complete egg-laying, adult males perform a loud pre-dawn song which appears to attract migrating subadult males and females to the colony. At this time, adult males are emancipated from nest-building and mate-guarding and can pursue the mates of the subadult males for EPCs, which were described as being forced (Morton 1987).

Thus, the key assumptions of the hidden lek hypothesis are met by purple martins, with one crucial exception: EPCs appeared to be forced. For the hidden lek hypothesis to explain colony formation, it is necessary that females willingly accept EPCs, despite appearing to resist male attempts to inseminate them. If EPCs are forced, then some adult males would clearly benefit by obtaining EPCs while subadult males and their mates would suffer costs. The mates of subadult males should therefore avoid nesting near adult males to avoid EPCs and subadult males should avoid nesting near adult males to prevent losing paternity to them. Thus, if EPCs are forced, it is difficult to envision a causal link between EPC and colony formation.

Alternatively, it is possible that females prefer to pair with adult males, and if none are available then they pursue a mixed mating strategy of pairing with a subadult male and accepting EPCs from adult males. If so, then unpaired females could be drawn to adult males in the colony and subadult males could be drawn to the unpaired females, producing nesting aggregations. Although sharing paternity with adult males would be disadvantageous to subadult males, they could achieve greater fitness by breeding in a colony near adult males who might fertilize their mates than by forgoing breeding altogether.

*Evidence that female purple martins control extra-pair fertilizations*

The question of whether lek mechanisms operate in the purple martin colony, then, depends upon whether EPCs are forced or subtly accepted by receptive females. Because of the potential of the purple martin mating system to provide tests of the lek hypothesis, Wagner collaborated with Eugene Morton and Malcolm D. Schug to determine whether males or females control extra-pair fertilizations. DNA fingerprinting confirmed the marked relationship between paternity and male age class reported by Morton *et al.* (1990). Whereas adult males lost only 4% ($n$ = 85 offspring) of their paternity, subadult males lost 43% ($n$ = 53; Wagner *et al.* 1996). As predicted, all assignable extra-pair fertilizations were obtained by adult males. What was striking was the distribution of extra-pair fertilization success, with one adult male obtaining 88% the first year and another obtaining 83% the second year, matching the skew in mating success of many lekking species (see

Pruett-Jones 1988; Hoglund and Alatalo 1995) and the skew in EPC success by males in the razorbill mating arena (Wagner 1992b).

The huge age class difference in paternity could be caused either by adult males forcibly inseminating the mates of subadult males or by those females allowing EPCs. EPC attempts typically occur on the ground when females alight to collect nesting material (Morton 1987). EPC attempts appear aggressive and females resist, but their resistance occurs in a wide range of forms. Some fly away before males can approach near enough to attempt mountings, whereas others continue foraging for material until the male is able to mount, at which point females emit an alarm call and struggle to escape (Morton 1987). This range in responses could be caused by variation in female experience in avoiding EPCs or female receptivity to EPCs. In the latter case, receptive females may subtly allow mountings while also resisting as a ploy to test males (Westneat *et al.* 1990), as female razorbills have illustrated in the mating arena (Wagner 1991). Because such behaviours are extremely difficult to interpret (McKinney *et al.* 1984; McKinney and Evarts 1997; Gowaty and Buschhaus 1998), the authors searched for objective methods to determine whether EPCs were actually forced. The male control hypothesis predicts that the probability of a female being subjected to forced EPCs is determined by the number of males available to pursue her (Morton *et al.* 1990). Since the operational sex ratio (i.e. the number of sexually active males per fertile female on a given day; Emlen and Oring 1977) increases with date, with more males becoming available to chase later laying females (Morton *et al.* 1990), the male control hypothesis predicts a negative correlation between laying date and paternity, regardless of male age. Alternatively, the age-related pattern of extra-pair paternity (EPP) could be explained by females controlling fertilization and accepting EPCs when paired to subadult males and refusing EPCs when paired to adult males. The female control hypothesis predicts that male age alone determines paternity.

Contrary to the male control hypothesis, there was no relationship between laying date and paternity. It is also possible, however, that the much higher paternity of adult males was caused by their guarding more intensely than subadult males. However, adult males guarded significantly *less* intensely, escorting their mates during 53% of female departures, versus 72% by subadult males.

The female control hypothesis predicts that a female's receptivity to EPCs depends upon the age of her mate and not her own age. The male control hypothesis predicts that inexperienced females are less able to escape forced EPCs than adult females, and therefore many subadult males lose paternity because they are more often paired to subadult females (Morton and Derrickson 1990). Female age, however, was not a factor because adult and subadult females were equally likely to have EPP when paired to subadult males, and equally unlikely when paired to adult males. The authors therefore concluded that EPCs were rarely or never forced, and that females paired with subadult males pursued a mixed mating strategy whereas females paired with adult males avoided EPCs. These results in the context of the species' breeding system suggests a scenario in which the pursuit of EPCs may lead to colony formation.

### C *Aggregations as a by-product of mate choice*

In this model, the aggregation of secondary males is a by-product of mate choice by females, a behaviour that incurs costs for females themselves. Females do not seek aggregation but attractive males, which produces a chain of events that induces costs for all actors (Wagner *et al.* 2000; Wagner and Danchin 2003). However, aggregation still happens. In such a model, it makes no sense to ask what the benefits of aggregation are. And if we were able to measure everything, we would find that there is none: nobody really benefits from aggregation in and of itself, but this does not prevent aggregation from happening because it is the result of a process (mate choice) whose stability is only reached by the formation of an aggregation. A similar consequence occurs in the producer–scrounger game seen in Chapter 8.

### d *The recurrent lag between the optimal and evolutionarily stable solutions*

We can compare this with the problem of the difference between the optimal strategy that ignores the conflicts of interest between the different possible phenotypes and the ESS which explicitly takes into account these conflicts (see Chapter 3). Because of these conflicts, the solution that is selected is often very different and leads to an average fitness that is lower than the fitness associated with the optimal strategy. In the case of the producer–scrounger game, for example (see Chapter 8), at the evolutionarily stable frequency, because of the presence of the scrounger strategy, all individuals end up with a fitness that is lower than they would obtain in an optimal situation without scroungers. Still, the scrounger strategy exists, and is inevitable because it is part of the ESS. The general message is that it is not because a situation is optimal that it is retained in the course of evolution. Often, the observed strategy will have a lower performance at the scale of the group than the optimal strategy. This is a recurring theme in behavioural ecology that is not generally appreciated.

#### 14.3.4.3 *From leks to colonies: the 'hidden lek' hypothesis*

*From leks . . .*

In a chapter from a book published in 1998, Wagner presented a figure to illustrate his hypothesis (Figure 14.12). Let us start from the existence of leks, which are densely aggregated male display territories. There are several models to explain how sexual selection can lead males to aggregate their display territories and attract females (Chapter 12). First, there is the **hotshot model** (see also Section 12.3.2.3). There also exists the **hotspot model**, which predicts that in species for which female territories overlap, males should tend to regroup to display in the places where territories of several females overlap. There is also the **female preference** model, according to which it is easier for females to compare and evaluate different males when males are together. In all these models, it is female behaviour that forces males to aggregate.

All these different models of aggregation involve sexual selection as the driving force for aggregation. Each of these models is probably applicable to different situations, and they are not mutually

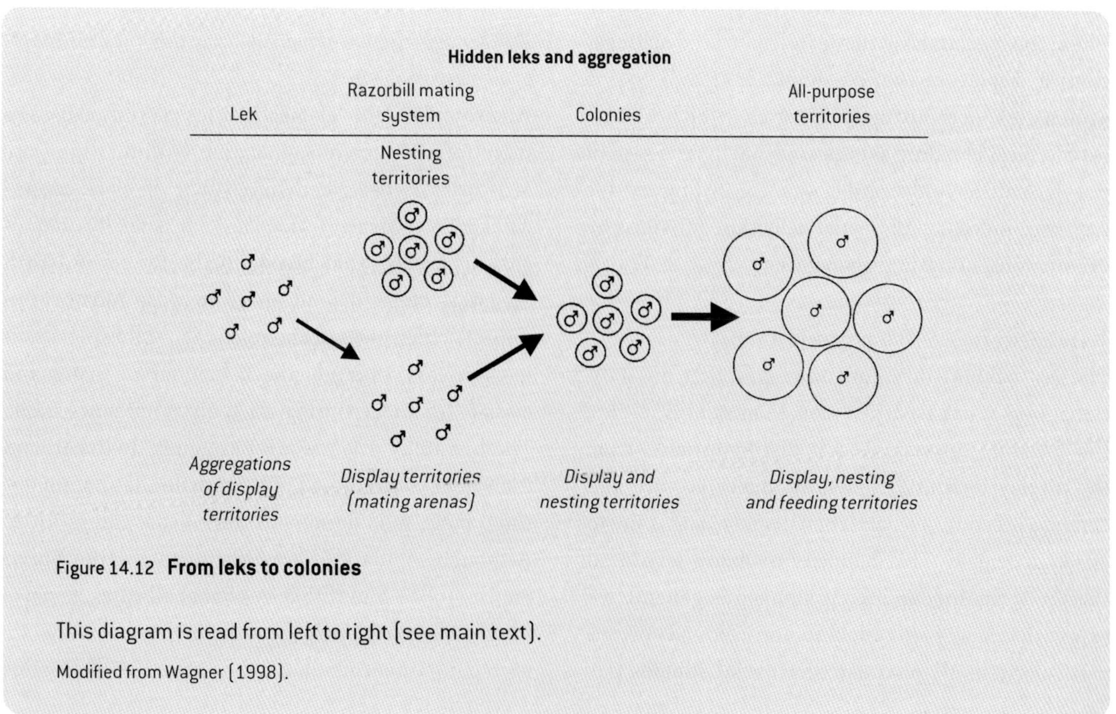


Figure 14.12 **From leks to colonies**

This diagram is read from left to right (see main text).

Modified from Wagner (1998).

exclusive. All the lek models and the hidden lek hypothesis assume the importance of sexual selection as a driving force for the aggregation of male display territories (Figure 14.12).

*... to razorbills ...*

The next step (second column from the left in Figure 14.12) is the one we described with razorbills, where the two systems occur simultaneously but in different places: one is social monogamy occurring at the breeding colony, and the other is genetic promiscuity and takes place at the mating arenas that function like leks. Such a mechanism was discovered in the razorbill because in this species, the social and genetic systems take place at different locations. The display territories are found on boulders around the breeding colony, whereas the breeding territories comprise the colony itself, in rocky hills, where razorbills lay and incubate their eggs and rear the chick.

*... to colonies ...*

Let us imagine the next step with a species where display and breeding occur at the same place (third column from the left in Figure 14.12). It is in fact the case of most bird species, including colonial species. Then, the mechanisms that can lead to the aggregation of display territories should also lead to the aggregation of breeding territories, which should form typical breeding colonies, as may be the case in purple martins and bearded tits. According to this scenario, processes of sexual selection provide the driving force of nesting aggregation.

*... to aggregations of all-purpose territories*

The last column in Figure 14.12 proposes an additional step. As we have just seen, in most bird species, the breeding territory is not distinguishable from the display territory. Generally, males acquire and defend a breeding site, and then display on this site to attract a female. This happens in many territorial species. If the hidden lek mechanism is general, we expect that it will also apply to non-colonial species, particularly in **all-purpose territorial species** (i.e. with territories that fulfil all the necessary functions for breeding: feeding sites, nesting sites, singing spots, etc.). Breeding territories for these species should also be more aggregated than is necessary. For instance, the hidden lek hypothesis predicts the existence of some level of aggregation of breeding territories in forest species while other areas of equivalent quality remain unoccupied.

The clustering of all-purpose territories in an apparently homogenous habitat has been described in several bird species. Scott Tarof and colleagues (2005) of Queen's University, Canada, explicitly examined this in least flycatchers, *Empidonax minimus*. They reported evidence that the highly aggregated all-purpose territories of that species conformed to certain predictions of the hidden lek hypothesis such as that (1) males in clusters had higher and faster success in attracting mates than those that bred solitarily, and (2) early settling males in clusters were in better body condition and ended up in the centre of clusters, suggesting that later males were drawn to breed near them. Partial DNA sampling did not, however, support the prediction that central males achieved a skew in extra-pair fertilization success.

*Further aggregation of individuals within the constraint of all-purpose territories*

An example of the lek mechanisms possibly causing a more refined type of aggregation within all-purpose territories is described in a study of black-capped chickadee (*Poecile atricapillus*), a socially monogamous species that breeds in the forests of North America. Males defend an all-purpose territory. In winter, chickadees form relatively stable flocks. It is then easy to study the behaviour of males and classify them according to their dominance rank. The hierarchy of dominance is generally linear, and each flock is composed of an $\alpha$ male that dominates the $\beta$ male, who dominates the $\gamma$ male, and so forth. Although copulations are difficult to observe, Susan Smith (1988) found that in almost all cases, females sought EPCs from neighbouring males that were of a higher dominance rank than their mates. Subsequent

DNA studies revealed a substantial frequency of extrapair paternity in this species. Finally, it is the female that chooses the nesting site within her mate's territory. The Canadian researcher Scott M. Ramsay and his collaborators have analysed factors influencing female choice of a nest site (Ramsay *et al.* 1999). They proposed four alternative hypotheses and tested them on a chickadee population at the Queen's University Biological Station at Lake Opinicon, Ontario.

1. First, females can choose the location where food is the most abundant. They tested this hypothesis by sampling food in the area surrounding the nests and in randomly chosen zones in the rest of the male's territory. No significant differences were found, and so the hypothesis was rejected.
2. Then, females can choose locations where the vegetation has certain characteristics that make it favourable for breeding. Using the same method, they found that the vegetation around the nest did not possess different characteristics than other parts of the male's territory. This hypothesis was thus rejected.
3. A third hypothesis was that females choose the nest site using their past experiences or the history of this particular territory. For example, females could choose the same site as the previous year or avoid it. Here again, the tests showed that there was no association between the sites chosen in two consecutive years, which led the authors to reject this hypothesis as well.
4. The fourth hypothesis was that females build the nest near the border of the male's territory to be close to an attractive neighbouring male, as was predicted by the hidden lek hypothesis. Under this hypothesis, we expect females that are paired with a low-ranking male to attempt to build their nest close to the border of a higher-ranking neighbouring male, whereas this tendency would not be as strong for females paired with a high-ranking male. More generally, the hidden lek hypothesis predicts that all females, regardless of the mate's rank, should still tend to build their nest closer than is necessary to the limit of their males' territory. The results obtained for this hypothesis are intriguing (Figure 14.13a). In the first year, 1996, the two predictions were not rejected: the females that were surrounded by neighbouring males of lower rank than their mate tended to build their nests farther away from the border of the territory than females that had a least one neighbouring male of higher rank than their own mate. However, the next year, there was no significant difference in the distance to the border between these two categories of females, which would lead to the rejection of the first prediction. However, during that particular year, the distances between the nests and the territory borders were on average very short (Figure 14.13a), so the second prediction cannot be rejected. As a result, although we are dealing with a strictly territorial species, the distances between pairs are in fact much shorter than the mean size of the territories would lead us to calculate. In certain zones, we can even see real aggregations of nests (circles on Figure 14.13b).

The conclusion of this study is that the four hypotheses that were tested, only the hidden lek hypothesis was not rejected by the data obtained in 1996, and is only partly rejected by the data of 1997. These results are striking because one of the factors that seems to explain the presence of these small aggregations, at least in 1996, is linked to differences in the phenotypic quality of males, which suggests that sexual selection plays a role. These results support the hidden lek hypothesis (1993, 1998; see Figure 14.12) about the importance of sexual selection as a driving force of aggregation in species breeding within all-purpose territories. This study also implies that even within the constraints of defending large all-purpose territories, aggregations can be achieved at a finer scale.

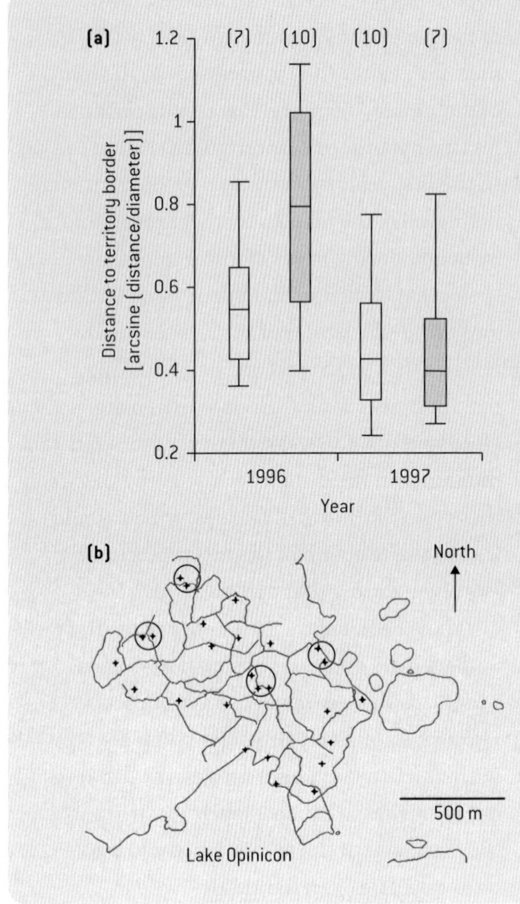


Figure 14.13 **Factors explaining the position of the nest within a male's territory in black-capped chickadees**

**a.** Mean distance from the border of the territory for females that have at last one neighbour that is of higher rank than their mate (open boxes), and females that are mated to a male of higher rank than their neighbours (grey boxes). Results were obtained in two consecutive years. In 1996, the difference is significant: females nested closer to the territorial border when their closest neighbour was a male of higher rank than their mate; $P = 0.04$. In 1997, all females nested closer to the territory border than in 1996, independently of the social rank of their mate. The numbers in brackets represent sample sizes.

**b.** Map of territories and nest locations in 1997. The circles surround small aggregates of a few nests very close together.

Modified from Ramsay *et al.* (1999).

## 14.3.5 The importance of information

It appears that, like in the functional approach, concepts of information (see Chapter 4) are central to the habitat copying and hidden lek hypotheses. According to these two hypotheses, we cannot understand the evolution of group living without integrating the mechanisms of habitat and mate choice. These two mechanisms can generate aggregation as a by-product (see Section 14.3.3). In fact, any choice process by definition involves the use of information on the different alternatives. The nature and value of that information probaly greatly determines the aggregative consequences of these individual choice processes. These two characteristics, nature and quality, essentially depend on the species biology.

### 14.3.5.1 *Information sharing*

Information is shared among members of an aggregation when group members can use the measures of the environment made by other members of the aggregation as information for their own decisions. Information sharing qualitatively differs from the sharing of physical resources in the sense that when an individual shares information, there is no reduction of its own information, as would be the case for a physical resource (see definition in Chapter 7). In other words, information sharing does not directly reduce the amount of information available. Thus, information sharing usually does not diminish the value of information. On the contrary, when an individual shares a physical resource, it loses exactly the quantity of resources that it surrenders.

### a *Excludability: drawing a limit between private and public information*

An important notion is that of the **excludability** of information, a concept that we did not directly discuss in Chapter 4, and which was introduced by Lachmann *et al.* (2000).

> Information is excludable if the individual that possesses it can hide it from conspecifics. Thus excludable information corresponds to the part of information that remains **private**. Information is non-excludable if the individual that possesses it is unable to hide it from others. Then the information is public (see Figure 4.3 for definitions).

Information excludability depends on the biology of the species. In certain bird species, such as terns (*Sterna* sp.) and Atlantic puffins (*Fratercula arctica*), breeders return to the nest with fish hanging from their bill (Figure 14.14). In such species, information on foraging success is non-excludable, i.e. it is **public** (or social). In other bird species, the food that is brought back to the nest is hidden. Food may even be partly digested and metabolized before bringing it back to the young, like in petrels and albatrosses. These species hunt at sea, very far from their nests, and are therefore forced to digest and transform the food they catch into oil, a highly condensed form of energy. This allows them, in a single trip, to bring back a large quantity of calories and nutrients to the chick. In such species, the information on food is more excludable (i.e. private) because individuals can hide information about whether they found food. In such cases, conspecifics can have a difficult time acquiring information on the food sources. Clearly, it appears that only the part of **personal information** that is non-excludable can become ISI and be shared. However, the limit between private and social information is not always as easy to draw. For instance, it may well be that in albatrosses a successful forager has a heavier flight than an unsuccessful forager, a subtle difference that may be detected by conspecifics. It is thus often easier to rank various pieces of information in terms of their relative excludability rather than to decide in absolute terms whether such piece of information is excludable or not.

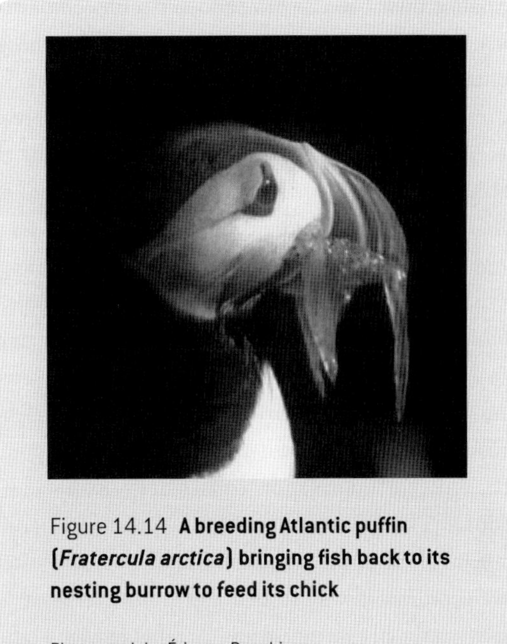

Figure 14.14 **A breeding Atlantic puffin (*Fratercula arctica*) bringing fish back to its nesting burrow to feed its chick**

Photograph by Étienne Danchin.

#### 14.3.5.2 *Is there a link between information sharing and aggregation?*

Most of the mechanisms developed in Section 14.2 imply information sharing between individuals (for example, vigilance, group defence and all the hypotheses linked to resource exploitation). Additionally, information is crucial for habitat and mate choice. The debate on the relationship between information and group living has been addressed by Michael Lachmann, Guy Sella, and Eva Jablonka (2000), who used a formal approach to answer the general question of the role of information in aggregation.

These authors raise the fact that many organisms' aggregations involve information sharing. For instance, myxobacteria have adopted multicellularity as a survival strategy. Before the aggregation phase, solitary bacteria assess their local environment. Afterwards, some interaction occurs that involves specific signals concerning hundreds of millions of individuals. This allows a more reliable estimate of whether a lack of nutrients is local or occurs at a larger scale. If the conclusion of these interactions is that resources are disappearing, cells build multicellular organisms of fructification (Shapiro and Dworkin 1997; see Chapter 15). This process resembles a step towards pluricellularity. Similar mechanisms exist in ants where colonies can respond efficiently to information collected by many individuals (Theraulaz *et al.* 1998). The great number of ants involved in the process allows the colony to act in a coordinated way so that the group can react to environmental contingencies in a reliable way. There are also many examples in marine invertebrates with a sessile adult phase, where the information provided by the simple distribution of conspecifics has long been shown to constitute the principal cause of the important aggregation that characterizes this group (Meadows and Campbell 1972).

Lachmann and his collaborators (2000) developed three models to examine the effect of group size (N) on the fitness of group members.

### a *Assumptions of the model*

Lachmann *et al.*'s (2000) models make four major assumptions.

1. Individuals live in a two state environment, $E_1$ and $E_2$. The dynamic of shifts between these two states is described by a Markovian process of probability of change equivalent to $n$ units of time.
2. Individuals' perceptions of their environment are prone to errors. The measurement error probability is $e$ for each measurement. Thus, the probability of a correct estimation of the environmental state is $1 - e$. This value is assumed to be less than 0.5. An individual determines its own phenotype on the basis of $M$ independent measures of the environment. $M$ thus quantifies memory size.
3. At each time step, individuals can be in one of two phenotypic states, $F_1$ and $F_2$, which are respectively adapted to environments $E_1$ and $E_2$.
4. A generation lasts for $T_g$ units of time. An individual's relative fitness is proportional to the fraction of time during which it is in the correct state relative to the environment. Time scales are such that $T_g >> T_e >> 1$, where $T_e = 1/n$ is the mean duration of an environmental state. Then, environmental change occurs several times during an individual's lifespan. It can thus infer this state and adapt accordingly. Thus an individual's capacity of being in the right state at the right moment provides a proxy of fitness.

When estimations are not too costly and independent from one another, individuals can make enough measures to eliminate error effects. They are thus always in the correct phenotypic state. As this particular case is not likely in nature, Lachmann *et al.* (2000) examined situations where the number of measures made by individuals is limited. They built two complementary versions of their model. In the first one, individuals could make only one measure each time step at no cost. In the second model, individuals could make many measures each time step, but each with a fitness cost $q$. They then sought the optimal strategy in each condition: the strategy that leads to the highest fitness, independently of the strategy of other individuals.

### b *Individuals can only make one free measure each time unit*

When individuals only take into account their personal information (i.e. group size of $N = 1$), the important variable is memory size, $M$. When only the last estimation can be used, i.e. $M = 1$, the optimal strategy is to choose the phenotype that is adapted to the last estimation of the environmental state. Then mean fitness is $1 - e$. When memory $M$ increases, individuals estimate the current state of the environment better and fitness increases (Figure 14.15, bottom

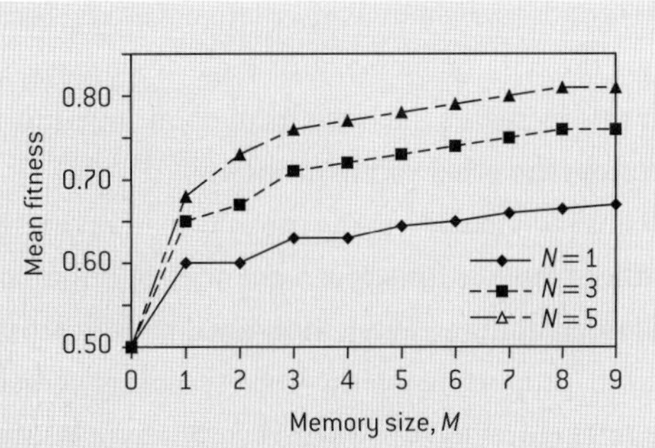


Figure 14.15 **One of the advantages of information sharing for individuals in a group**

In this figure, the individuals can only make one measure per time unit. The fitness of the optimal strategy is related to the capacity of the memory ($M$, $x$-axis) and group size ($N$, the three curves). The optimal fitness increases with memory capacity and with group size because each individual has access to more information. In this figure, the error is $e = 0.4$, and the rate of environmental variations is $U = 0.05$.

Redrawn from Lachmann *et al.* (2000).

curve, $N = 1$). When $M$ becomes very high, fitness levels off because the oldest measures become less and less useful to determine the current state of the environment because it may have changed since then.

Now, let us consider an additional assumption:

**5.** In an aggregation of $N$ individuals, each individual measures the environment once each time unit, and these measures, as well as all the measures made by the other individuals of the group, are available to all (information is non-excludable).

In such conditions, the mean fitness of the optimal strategy increases with group and memory size (curves for $N > 1$ in Figure 14.15). This is because when the size of an aggregation increases, each individual of the group has more information about the recent state of the environment, through its own measures (personal information) and those made by the other group members (inadvertent social information). In these conditions, fitness is lower in solitary than in group living individuals, which should favour the evolution of aggregation.

However, we may ponder the biological pertinence of such a model. For instance, what would happen if, as is likely in nature, memory is costly? If this were the case, fitness would not increase indefinitely with memory size, but would reach some maximum at a given memory size. However, the existence of a memory cost also increases the benefits of group living because information sharing allows individuals to learn the outcomes of many measures of the environment each unit of time (but see Giraldeau *et al.* 2002). Aggregation thus allows an individual to obtain more information without paying any costs. Another important problem is that of the risk of error in the transmission of information between individuals of the group. As errors can happen in the estimation of the environmental state (parameter $e$), errors can happen in the transmission of the results of these estimations between individuals ($e_s$). Lachmann *et al.* (2000) show that information sharing becomes non-advantageous if $e_s \geq e$. In fact, there is no reason for there to be a physical link between the code and the message itself. For example, the alarm call made when a predator approaches does not have to become more difficult to interpret when the predator becomes more difficult to see. Thus, $e$ and $e_s$ can vary independently from one another, and for information sharing to be useful, $e$ (the error in the measure) must be higher than $e_s$ (the error in the transmission of information). This is what postulates the second version of the model of Lachmann *et al.* (2000).

### C Can information be shared if it is costly to obtain?

To account for the cost $q$ of each measurement, the model must integrate the individual's rate of measurement each unit of time. If all members of a group of $N$ individuals behave in the same manner and share the outcomes of all their $n$ independent measurements of the environment each unit of time, each group member pays a cost of $qn$, but benefits from the information of $Nn$ measurements each unit of time.

The optimal value of $n$ in these conditions can be determined as follows. Let us call $g(x)$ the function expressing the fitness gain in which $x$ is the number of measurements an individual can access each unit of time, either personally or socially. Let us assume that this function increases monotonically with $x$. The fitness of a solitary individual is of $w(x) = g(x) - qx$. Finding the value of $n$ that maximizes this function implies finding the value that makes the derivative of fitness relative to $x$ equal zero. This derivative $w'(x) = g'(x) - q$ is equal to zero when $g'(x) = q$. For an individual in a group of $N$ members, the fitness of each individual is $w(x) = g(Nx) - qx$. This function is maximized when $g'(x) = q/N$. We can illustrate the main results with the case where memory size is infinite and has no cost (Figure 14.16). This corresponds to the most conservative scenario because when the cost of memory increases, we have seen that this increases the benefits of obtaining information from the other group members.

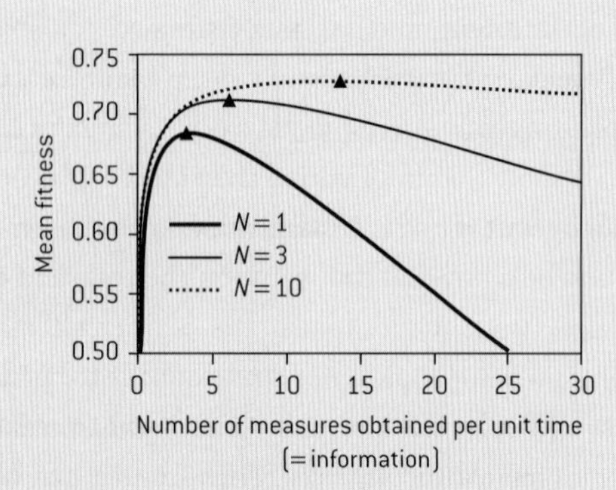


Figure 14.16 **Mean fitness in an aggregation as a function of the information received each time unit and group size when information acquisition is costly**

In these conditions, there are two advantages to group living: the individuals within groups have more information in total, and this information is less costly to obtain. In the case shown here, the memory size is infinite, the cost $q$ per measure is 0.01, the error of the measure $e$ is 0.4, and the environment varies at a rate of $U = 0.05$. The curves show that for various group sizes, fitness is maximized for a certain value of the information. The available information increases with group size (filled triangles).

Redrawn from Lachmann *et al.* (2000).

Thus individuals leaving the information accessible to others pay less for a given quantity of information because in doing so every individual pays the costs of a lower proportion of the corresponding measurements (Figure 14.16). Thus, by living in a group, individuals can access more information while only paying a small fraction of the information acquisition costs. Individuals living in groups have two complementary benefits: they access more information, and pay less for this information (Figure 14.16). As group size increases, the cost paid by each individual decreases, which should favour aggregation. Naturally, this model assumes that individuals can gather social information and personal information simultaneously. Otherwise, the number of group members producing personal information at any given time diminishes every time one individual gathers social information. When the gatherings of social and personal information are incompatible activities, the situation turns into a producer–scrounger game. In this case, aggregation may not provide any net benefit in terms of information (Giraldeau *et al.* 2002). It is thus important to study such incompatibilities.

### d *Is information sharing an evolutionarily stable strategy?*

We have seen several times that the fact that a behavioural strategy is advantageous to an individual does not guarantee its evolutionary stability. The main problem is the possible existence of a selfish phenotype that would use (or parasitize) the information inadvertently produced by the findings of other group members without ever performing sampling itself. Such a phenotype would have the higher fitness because it would not pay any information gathering cost while having access to the information gathered by all the other group members. In fact, the stability of the information sharing behaviour depends mainly on the **excludability** of the information. If we assume that information on the environment is non-excludable, then information sharing within groups is stable from an evolutionary viewpoint: individuals within a group have a higher fitness than solitary individuals (Lachmann *et al.* 2000). Then the game becomes similar to the producer-scrounger game (Chapter 8). Conversely, there are situations where information is excludable and sharing is costly, either because of the competition it can cause on resource exploitation or because of the production of costly signals. In such cases, it is necessary for other mechanisms to be involved, such as **kin selection** or reciprocity so that cheaters cannot invade information sharing groups.

It is notable that Lachmann *et al.* (2000) do not mention the ICH in their paper. This was probably purposefully to avoid entering into endless polemics on that matter and to place their model at a more general level. However, in fact, their model also offers an answer to the ongoing debate: it seems that when, because of the biology of species, information on the foraging success is non-excludable, i.e. biological constraints prevent individuals from hiding information broadcast by their behaviour, the mechanism of the information centre can be efficient and favour the formation of aggregations. However, the application of this reasoning to the case of the information centre hypothesis calls for some more thinking about what we mean by excludable or **non-excludable** information.

### 14.3.6 A synthesis: aggregation as a by-product of commodity selection

#### 14.3.6.1 *The commodity selection framework*

We now describe a synthesis of the recent approaches to the evolution of aggregation. Both the 'habitat copying' and 'hidden lek' hypotheses share the assumption that the individual choices of the commodities necessary to reproduce can generate aggregations of breeding territories as a mere by-product. This is why Danchin and Wagner (1997) proposed to unify these two hypotheses into a single framework that they called 'commodity selection'. The two most fundamental commodities required by sexually reproducing organisms are habitat and mates. Variation in reproductive success among breeding patches provides an **integrative cue** that summarizes information about patch quality. Prospectors (and researchers!) need not perform complicated analyses to balance the costs and benefits of various patches in terms of predation, food, nest site availability, parasites, etc.: all that is required for adaptive habitat choice is to compare the average quantity, and perhaps quality, of offspring previously produced among patches. The integrative cue of mean conspecific reproductive success therefore appears to comprise the long sought common currency of costs and benefits. As we have proposed, aggregations such as leks and colonies may also be generated as by-products of mate choice. This effect might be enhanced where high-quality prospective mates coincide with high-quality breeding patches.

The commodity selection approach explicitly integrates the role of social information among group members as the primary mechanism that generates skewed distribution of animals (see Wagner and Danchin 2003; Figure 14.17). The habitat copying and hidden lek hypotheses concern the effect on aggregation of the use of the information gathered from conspecifics. Both hypotheses concern the nature of the information involved: social information on the state of the environment in the case of habitat copying, and on potential partners in the case of hidden leks (Figure 14.17). Both hypotheses involve social information among group members. The 'information sharing' model studies the evolutionary stability of the sharing of information among group members (Figure 14.17). In fact, Lachmann *et al.* (2000) show that the commodity selection framework is evolutionarily stable under a wide range of natural conditions. Their results suggest, however, that when information is excludable, aggregation through information sharing on the current state of the environment cannot occur without the involvement of other processes.

It happens that in a general way the information involved in both hypotheses used in commodity selection are non-excludable by nature. In the case of the inadvertent social information available from the

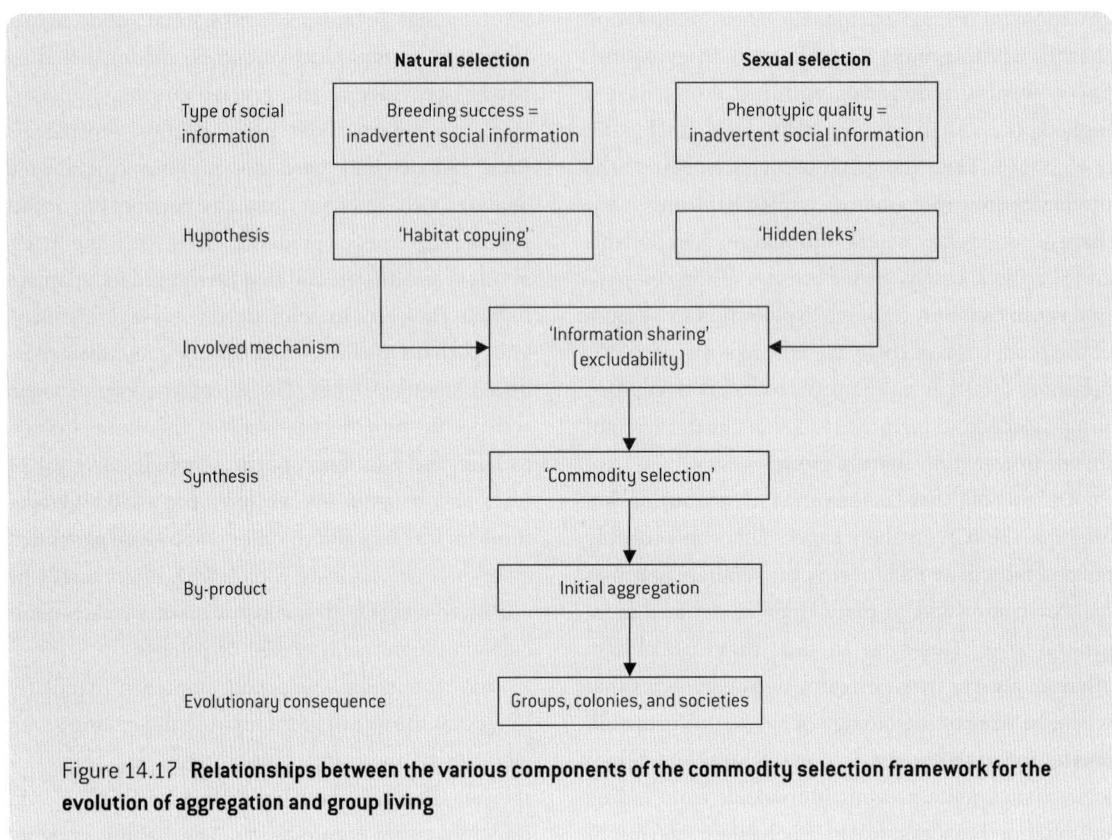


Figure 14.17 **Relationships between the various components of the commodity selection framework for the evolution of aggregation and group living**

reproductive performance of conspecifics, it would be very costly in most cases, if not practically impossible, for individuals to hide or modify their reproductive success. Individuals have been long selected to maximize their fitness, of which reproduction is an important component. Individuals could not afford to lower their reproductive success only to prevent providing correct information about the current state of the environment. Individuals are thus too constrained to cheat by modifying their reproductive success in order to deceive others about the current state of the environment. Most often, that information is public, because it is available to all. Likewise, for the hidden lek hypothesis, the individual signals of quality involved in mate choice are favoured by sexual selection to be visible. These also belong to the public domain. Thus, we can conclude that it is the use of social information that is in essence public that favours the aggregation of individuals and, from there, leads to group living.

#### 14.3.6.2 *Are aggregations and group living still an evolutionary enigma?*

The commodity selection framework considerably modifies the approach to study the evolution of aggregation through the **parasocial** path (i.e. through the consequence of individual decisions to live together). This way of thinking has numerous consequences for our perception of the evolution of group living.

*If aggregation is the natural state . . .*

The evolution of diverse forms of social life resulting from the parasocial path appears as an inevitable consequence of the natural tendency of populations to form skewed distributions, because of the evolutionary advantages provided by the processes of choice of the commodities necessary to animals. This readjusts the way we should study the evolution of aggregation and group living. In fact, according to the by-product approach, group living represents an inevitable, hence natural, state of organisms, simply because the selected processes of habitat choice and mate choice (two major fitness-affecting decisions) create aggregation, or deviation from the IFD as a by-product. One of the main reasons behind this discrepancy from the IFD resides in the fact that the assumptions of ideal and free are rarely fulfilled in nature.

*. . . why then are not all animals aggregated?*

If an aggregated distribution represents the natural state of populations, we are then entitled to ask ourselves why animals are not always aggregated. This means that we must find out why, when the general tendency to aggregate is probably universal, there exist so many species that do not seem to present an aggregated distribution. **In other words, rather than asking why so many species are highly aggregated, we should ask why so many species do not seem to aggregate**.

A possible answer to this question is that, in fact, it is probably mainly a question of spatial scale. The case of the black-capped chickadee presented in Section 14.3.4f is a particularly striking example. This species is known to be territorial, i.e. supposed not to show any spatial aggregation, yet we have seen that the inter-individual distances are in fact much shorter than what we expect from looking at the average size of their territories. Thus, at a very small scale, there is a certain degree of aggregation of breeding pairs. On the other extreme, as proposed in Figure 14.12, if we tried to analyse the distribution of reproductive territories within the span of available habitats, we might notice that chickadee territories are unnecessarily grouped in only one part of the habitat, and that a large portion of the habitat remains unoccupied. Results obtained in many species (for instance in the least flycatcher of Section 14.3.4e) seem to indicate that this might be the case. Thus, aggregation is probably more widespread than we think. We will come back to the question of whether group living still remains an evolutionary enigma in the next chapter, in a context of the evolution of cooperation that also encompasses the evolution of social life through the quasi-social path.

*Commodity selection uses economic principles*

The commodity selection approach does not contradict the functional approach that aims to weigh the costs and benefits of a given strategy. Rather, it uses the general principle of the economic approach to behaviour. The only difference is where we should look at the costs and benefits. Commodity selection suggests that we should not look for the costs and benefits of aggregation, but rather those of the choices of commodities that are at the root of aggregation. The idea was implied by Caraco and Pulliam (1984), Sibly (1983), and Clark and Mangel (1984) in the study of optimal group size. There is no real biological reason to focus on the costs and benefits of aggregations in themselves because commodity selection suggests that, at least in the beginning, aggregation was not subjected to real selection. Initially, selection would only have operated on the fitness enhancing mechanisms of choice of commodities necessary for fitness.

## 14.4 Conclusion

The general message of this chapter is that searching for benefits of coloniality has yielded a large body of interesting data, but is not necessarily explanatory. If colonies are by-products then the costs and benefits that are measured may be valuable for understanding the stability of aggregations once they have formed but not for explaining their origin. Although it is possible that identifiable benefits have selected for colony formation, decades of applying this approach have not led to general explanations or to the discovery of a common currency to allow inter-specific comparisons of cost:benefit ratios. In contrast, recent developments in this field, (as we have stressed in Chapter 4 for behaviour in general) call for an information-driven approach to animal aggregations as the most promising way forward. Adopting this approach entails focusing on the information that is used by individual animals to make decisions, rather than on hypothetical group properties such as group size. In particular, performance information appears to be the closest we have achieved to a common currency because the outcomes of decisions of other individuals provide a simplified integrative cue to the quality of reproductive commodities. In conclusion, the question of animal aggregations may best be advanced by combining the application of new ideas about public information with the prediction that aggregations are by-products of the decision-making of multiple individuals at the same time in the same place. This general approach can underlie further phylogenetic analyses and experiments which, among other goals, may achieve new insights into the factors that produce aggregations.

## » Further reading

> *For a general review on the functional approach:*

**Wittenberger, J.F. & Hunt, G.L.** 1985. The adaptative significance of coloniality in birds. In *Avian Biology*, vol. 8 (Farner, D.S., King, J.R. & Parkes, K.C. eds), pp. 1–28. Academic Press, New York.

> *For a deeper study of the information centre hypothesis:*

**Ward, P. & Zahavi, A.** 1973. The importance of certain assemblages of birds as 'information centres' for food finding. *Ibis* 115: 517–534.

**Clark, C.W. & Mangel, M.** 1984. Foraging and flocking strategies: information in an uncertain environment. *American Naturalist* 123: 626–641.

**Mock, D.W., Lamey, T.C. & Thompson, D.B.A.** 1988. Faslifiability and the information centre hypothesis. *Ornis Scandinavica* 19: 231–248.

**Richner, H. & Heeb, P.** 1996. Communal life: honest signaling and the recruitment center hypothesis. *Behavioral Ecology* 7: 115–118.

**Lachmann, M., Stella, G. & Jablonka, E.** 2000. On advantages of information sharing. *Proceedings of the Royal Society of London Series B* 267: 1287–1293.

**Barta, Z. & Giraldeau, L.-A.** 2001. Breeding colonies as information centers: a re-appraisal of information-based hypotheses using the producer-scrounger game. *Behavioural Ecology* 12: 121–127.

> *To understand the emergence of the habitat copying hypothesis:*

**Shields, W.M., Crook, J.R., Hebblethwaite, M.L. & Wiles-Ehmann, S.S.** 1988. Ideal free coloniality in the swallows. In *The Ecology of Social Behaviour* (Slobodchikoff, C.N. ed.), pp. 189–228. Academic Press, San Diego, CA.

**Brown, C.R., Stuchbury, B.J. & Walsh, P.D.** 1990. Choice of colony size in birds. *Trends in Ecology and Evolution* 5: 398–403.

**Boulinier, T. & Danchin, E.** 1997. The use of conspecific reproductive success for breeding patch selection in territorial migratory species. *Evolutionary Ecology* 11: 505–517.

**Danchin, E., Boulinier, T. & Massot, M.** 1998. Conspecific reproductive success and breeding habitat selection: implications for the study of coloniality. *Ecology* 79: 2415–2428.

**Giraldeau, L.-A., Valone, T.J. & Templeton, J.J.** 2002. Potential disadvantages of using socially acquired information. *Philosophical Transactions of the Royal Society of London Series B* 357: 1559–1566.
**Doligez, B., Cadet, C., Danchin, E. & Boulinier, T.** 2003. When to use public information for breeding habitat selection? The role of environmental predictability and density dependence. *Animal Behaviour* 66: 973–988.

> *The hidden lek hypothesis and implications of sexual selection:*
**Wagner, R.H.** 1998. Hidden leks: sexual selection and the clumping of avian territories. In *Extra-Pair Mating Tactics in Birds* (Parker, P.G. & Burley, N. eds), pp. 123–145. Ornithological Monographs, American Ornithologists' Union, Washington, DC.
**Fletcher, R.J. & Miller, C.W.** 2006. On the evolution of hidden leks and the implications for reproductive and habitat selection behaviours. *Animal Behaviour* 71: 1247–1251.

> *For the emergence of the commodity selection approach:*
**Caraco, T. & Pulliam, R.H.** 1984. Sociality and survivorship in animals exposed to predation. In *A New Ecology: Novel Approached to Interactive Systems* (Price, P.W., Sloboschikoff, C.N. & Gaud, W.S. eds), pp. 279–309. Wiley Interscience, New York.
**Danchin, E. & Wagner, R.H.** 1997. The evolution of coloniality: the emergence of new perspectives. *Trends in Ecology and Evolution* 12: 342–347.

## » Questions

1. Test the impact of group living on the risks of pathogen transmission: investigate among your classmates the frequency of sore throats and colds in general within two groups, those who use public transport and those who do not.

2. What are the conditions that make information excludable or not? What is the link between the notion of performance or private information on one hand, and the notion of excludable information on the other? Refer to Figure 4.3 for definitions of these terms.

3. A little field experiment: observe in foraging birds the shifts between vigilance and feeding phases. Measure the length of each of these phases. Think about the different analyses that are possible for this type of temporal series.

4. Design a comparative study to test some of the predictions of the commodity selection approach. What would be the variables you would quantify? How would you code them? Do not hesitate to refer to Chapter 3 about the comparative approach.

# 15

# 15

# The Adaptive Evolution of Social Traits

Jean-François Le Galliard and Régis Ferrière

## 15.1 Introduction

Slime moulds (*Dictyostelium discoideum*) are usually solitary unicellular micro-organisms that inhabit the litter of temperate forests. However, when critical food resources become rare, the cells start swarming collectively, ultimately aggregating into a colony through a complex process that requires involved communication. The colony of cells then grows quickly and engages in a unique developmental differentiation. Myriads of cells turn into a stalk supporting a 'fruiting body' that protects numerous spores whose survival and **dispersal** will ensure the future of the colony (Figure 15.1). Stalk cells are doomed to a premature death: they apparently sacrifice themselves for the benefit of the selfish spores. Why do stalk cells suppress their own reproduction for the good of spore cells?

In the same forest, insects forage on a germinating plant. The plant reacts by producing costly chemicals that inform the neighbouring plants that herbivores are present. These signals induce defence reactions that protect the neighbouring plants against the hungry insects. Functionally similar alarm calls have been described in birds and mammals. Sometimes, alarm calls imperil the signaller by increasing its detectability by the predator. What could explain the evolution of such costly alarm calls that are only profitable to conspecifics?

High up in the canopy, two conspecific birds carefully groom each other. During each grooming bout, one individual removes the ectoparasites of its partner. The two individuals invest in a subtle form of mutual **cooperation** that implies a series of reciprocal behaviours. Reciprocal cooperation has been described in birds, mammals, and in some fish species, and it is also a fundamental property of human interactions. What guarantees the resistance of **reciprocal** cooperation against a strategy that would exploit the kindness of its partners without returning the favour?

All these behaviours involve cooperation among individuals that apparently indulge in **altruism**. The origin and evolution of diverse forms of cooperation is a fundamental issue of behavioural ecology that will be studied in this chapter.

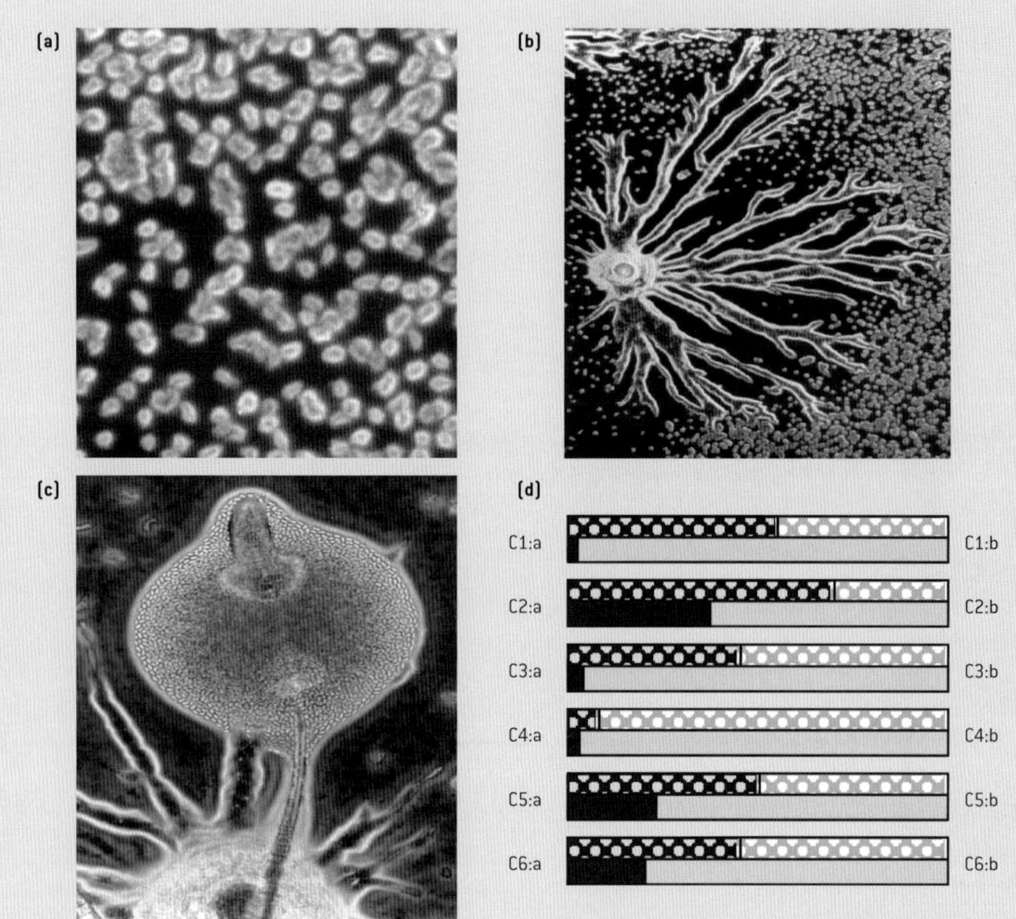


Figure 15.1 **The social behaviour of the amoeba (slime moulds, *Dictyostelium discoideum*) varies with environmental conditions**

**a.** Solitary stage. Solitary cells are dispersed when food resources are abundant.

**b.** Colonial stage. A colony results from the **aggregation** of solitary cells when critical food resources are lacking.

**c.** The 'fruiting body' stage. The fruiting body stems from the developmental differentiation of the colony into a stalk (a long and stretched structure of dried cells) and sporangia of resistant and dispersive spores.

Photographs courtesy of Thomas Tully.

**d.** Proportion of cells located in the stalk (lower plain line) and in the sporangia (upper dotted line) in experimental mixtures of two slime mould clones (on the left, clone illustrated with black; on the right, clone illustrated with grey). In some experimental mixtures, cells are not equally represented between the stalk (somatic cells) and the sporangia (germ cells): six selfish clones are overrepresented in the germ line from the sporangia.

Adapted from Strassmann *et al.* (2000).

## 15.2 Standing problems

In this section, we define the concepts used throughout this chapter: altruism, cooperation, and **sociality**. These concepts, some of which have already been touched upon in Chapter 2, will lead us to define the standing problems considered in the forthcoming sections: the genetic basis of altruistic behaviours, the identification of selection pressures acting on altruistic traits, the adaptive dynamics of altruistic traits and their coevolution with other traits, related to either behaviour or life history.

### 15.2.1 Altruism, cooperation, sociality

An individual behaviour is altruistic and an activity is cooperative when in a single species population, all else being equal:

1. For a solitary individual, the expression of the behaviour carries a distinct negative effect on the reproductive success of the bearer of the altruistic trait (the direct **cost**).
2. For a group of interacting individuals, the expression of the behaviour carries a net positive effect on the reproductive success of the recipients of the altruistic act (the indirect **benefit**).

This means that cooperation implies a collective activity that benefits all partners (Connor 1995). A functional classification of different forms of altruistic behaviour is provided in Table 15.1. Cooperation is called symmetric when indirect benefits are shared equally among group members. Cooperation is asymmetric in a honey bee (*Apis melifera*) colony, because workers are responsible for tasks such as defence or feeding while reproduction is monopolized by the queen. Mutual grooming in the impala (*Aepyceros melampus*), an African ungulate, is symmetric because it takes the form of mutual and reciprocal grooming trials.

A population of social animals is structured into **social groups** of individuals, and different types of **social systems** are associated with different forms of cooperative behaviours (Crespi 2001). For example, in birds, social systems have been classified according to how individuals share breeding sites and **parental care** (Cockburn 1998). The solitary breeding mode implies the most basic form of reproductive cooperation: parents usually cooperate on the same nest for the sake of their progeny. Sometimes, several birds outside the breeding pair can also cooperate for tasks such as territory defence and even parental care. This cooperative breeding mode has been observed in about 3% of bird species and takes three basic forms (Brown 1987). In **plural breeding** species, several breeding pairs share a territory where they cooperate for **foraging** and defence against competitors and predators. Several breeding pairs may also occasionally share the same nesting site or the same nest, such as in the North American acorn woodpecker (*Melanerpes formicivorus*). In **polygynandrous** breeding species, several females lay their eggs in a communal nest that they abandon to the male. It is the male's responsibility to incubate the eggs and care for the hatchlings. Lastly, in **cooperatively breeding** species *sensu stricto*, one breeding pair is assisted by auxiliaries (called **helpers**) that forego their own reproduction. This diversity of social systems in birds and other taxa can be described by a hierarchical classification including five basic forms of social system (Table 15.2; Crespi and Choe 1997; Crespi and Yanega 1995; see also Box 15.1).

### 15.2.2 Selection pressures

An altruistic behaviour carries a direct cost to the bearer and benefits the recipients. Section 15.4 will detail the critical empirical support for these costs and benefits, an evaluation that is a prerequisite to any proper evaluation of selection pressures acting on altruistic traits. However, we must keep in mind that it is variation in the type and degree of altruistic behaviour that provides the raw material for its evolution by **natural selection**, and such quantitative variation in altruistic investment seems to be common (Roberts and Sherratt 1998). For example, there is strong variation in the effort that impalas devote to mutual grooming activities (Hart and Hart 1992). **Cheaters** in this context are individuals that invest

| Indirect benefit | Direct cost | Examples |
|---|---|---|
| | **Grooming behaviour** | |
| Lower parasite infestation | Higher exposure to parasites | Mutual grooming in the impala |
| | Lower vigilance towards predators | Collective grooming in bees |
| | Energy and time expenditure | |
| | **Sentinel behaviour** | |
| Lower predation risk | Higher susceptibility towards predators | Marmot sentinels |
| | Lower feeding rate | Alarm calls in birds |
| | Energetic and time expenditure | Alarm signals in plants and insects |
| | **Defence behaviour** | |
| Lower predation risk | Higher exposure to predators | Ant soldiers |
| | Lower feeding rate | Collective defence in colonial birds |
| | Energy and time expenditure | |
| | **Collective foraging** | |
| Higher group feeding rate | Lower individual feeding rate | Collective foraging in social mammals |
| | | Collective feeding in micro-organisms |
| | **Food sharing** | |
| Higher feeding rate | Lower individual feeding rate | Blood sharing in bats |
| | | Food exchanges in ants |
| | **Parental care** | |
| Higher current reproductive success | Lower future reproductive success | Sterile workers in insects societies |
| | | Helpers in vertebrate families |

Table 15.1 **Examples of hypothetical altruistic behaviours**

| Society | Parental cares | Shared breeding sites | Cooperation | Specialized casts | Example |
|---|---|---|---|---|---|
| Solitary breeding | Yes | No | No | No | Blue tit |
| Colonial breeding | Yes | Yes | No | No | Colonial sea birds |
| Communal breeding | Yes | Yes/No | Yes | No | Queen wasp foundress |
| Cooperative breeding | Yes | Yes | Yes | No | Extended families in lions |
| Eusociality | Yes | Yes | Yes | Yes | Ants, bees, termites |

In cooperatively breeding and eusocial species, cooperation is usually asymmetric.

Table 15.2 **A classification of social systems**

Box 15.1 **A hierarchical classification of social structures**

A hierarchical classification of social structures was proposed by Crespi and Choe (1997) and Crespi and Yanega (1995) who distinguished five basic forms of sociality (see Table 15.2):

**Solitary breeding**. Solitary life is characterized by no cooperation, or in the case of monogamy cooperation between parents within a shared nesting site defended against neighbouring breeding pairs.

**Colonial breeding**. Typically, coloniality describes a situation where several breeding pairs share the same breeding site without active cooperation (see Chapter 14).

**Communal breeding**. In communal breeding, colony members cooperate for reproduction and participate equally in cooperation. Communal breeding is observed in some wasp or ant species when young queens cooperate to establish a new colony. For example, unrelated queens can cooperate to dig their nest and provision the first generations of workers, although only one dominant queen will eventually lead the colony.

**Cooperative breeding**. Cooperative breeding implies task sharing between individuals specialized in breeding (typically a breeding pair) and helpers (typically offspring of the breeding pair). Specialization is based on reversible behavioural differences between breeders and helpers.

**Eusociality**. Eusociality is the most 'complex' form of sociality (Crespi and Yanega 1995; Wcislo 1997; Wilson 1971). It involves specialization between a reproductive and one or several sterile casts. Sterility is based on irreversible developmental changes and further specialization is observed in the sterile cast. For example, some individuals may be responsible for foraging and parental care (workers) while others are responsible for defence or attack (soldiers). Eusociality has been observed in numerous Hymenoptera (wasps, bees, ants; Wilson 1971) and Isoptera (termites; Shellman-Reeve 1997; Thorne 1997). Eusociality has also evolved in the beetle (*Austroplatypus incompertus*), in several gall thrips (Crespi 1992), or in some aphids (see Benton and Foster 1992), and has been recently discovered in *Synalpheus* shrimps (Duffy 1996) and in two mammalian species of mole-rats (Jarvis *et al.* 1994).

less time in the grooming of their partners while purely selfish individuals benefit from the grooming behaviour of their partners without ever returning the favour. The heritability of this behavioural variation is discussed in Section 15.3.

### 15.2.3 Origin and persistence of altruistic traits

The individual disadvantage of altruistic behaviours raises two major evolutionary problems:

1. How can a new altruistic strategy produced by mutation gain a foothold in an ancestral selfish population?
2. How can an altruistic strategy resist the threat of new selfish strategies produced by mutation?

The **prisoner's dilemma** is a classical example in evolutionary game theory that yields an appropriate framework to understand both evolutionary problems. The prisoner's dilemma is a game between two individuals whose payoffs determine the gains or losses of one individual depending on its own and its partner's strategy. In an evolutionary context, a round of the game corresponds to a behavioural interaction between two individuals, the strategy describes the behaviour (assumed heritable) of one individual, and the payoffs of the game translate into fitness gains or losses for each individual. The

## Box 15.2 **Non-iterated game between cooperation and defection**

Here, we consider a one-round game between two partners playing two strategies, altruism C (cooperation) and selfishness D (defection). The payoffs of the game depend on the strategy of the actor and its partner and are given by the payoff matrix:

| Payoff to the actor of the interaction between | Partner C | Partner D |
|---|---|---|
| Actor C | $R$ | $S$ |
| Actor D | $T$ | $P$ |

where $R$ is the reward for mutual cooperation, $S$ is the sucker's payoff, $T$ is the temptation to defect, and $P$ is the punishment of mutual selfishness (notations follow the conventions established by Axelrod and Hamilton (1981)). The biological formulation of the game implies that $R > S$ and $T > P$ (the player gets the rewards of a cooperator), and $T > S$ (the defector derives the benefits of cooperation while the cooperator only pays the cost), and for referencing purposes we chose $P = 0$. We assume that altruism results in an indirect benefit (called $a$) that affects the fitness of the recipient and in the direct cost $c$ that affects the fitness of the altruistic actor.

**Frequency dynamics and evolutionary stability.** We assume a large and well-mixed population of cooperators and defectors. The dynamics describing the frequency of C at time t ($P_{c,t}$) in the population is called the **replicators**' dynamics and reads:

$$p_{c,t+1} = p_{c,t}\frac{Rp_{c,t} + S(1 - p_{c,t})}{p_{c,t}(Rp_{c,t} + S(1 - p_{c,t})) + (1 - p_{c,t})(Tp_{c,t} + P(1 - p_{c,t}))}.$$

The evolutionary stability describes the resistance of one strategy when common against the other strategy when rare and can be studied with the replicators' dynamics. If $R > T$ or if $R = T$ and $P < S$, then C is evolutionarily stable. If $P > S$ or $P = S$ and $R < T$, then D is evolutionarily stable.

**Investment in altruism affects payoffs additively.** Assuming that the benefit of altruism $a$ additively affects the payoffs of the game, we get $R = a - c$, $S = -c$, $T = a$ and $P = 0$. Thus, a population of cooperators can be invaded by a population of defectors ($T > R$) while a population of defectors is immune to invasion by cooperation ($P > S$): defection D is the only **evolutionarily stable strategy**. In the case where $R > P$, the game is a prisoner's dilemma (see Section 15.2.3) and cooperation can evolve if the game is iterated (see Section 15.6.1).

**Investment in altruism affects payoffs non-additively.** Assuming that the benefit of altruism $a$ non-additively affects the payoffs of the game, $f(a) > a$ is the function describing the benefits of mutual cooperation. Now, we get $R = f(a) - c$, $S = -c$, $T = a$ and $P = 0$ and we can distinguish between two cases. If the non-additive effects of mutual cooperation are weak ($f(a) - c < a$), we get to the same situation as before and the game is a prisoner's dilemma when $0 < f(a) - c < a$. However, the game behaves differently when the non-additive effects of mutual cooperation are strong ($f(a) - c > a$): C is now also evolutionarily stable ($R > T$). In this case, an analysis of the replicators' dynamics shows that altruism could take off in a selfish population if the initial frequency of the cooperators is higher than some threshold.

original version of the prisoner's dilemma involves two prisoners charged for a minor theft without any direct evidence. Each prisoner is interrogated separately by the police. If both prisoners refuse to incriminate their partner they are released after only a short period of detention. If both prisoners choose to incriminate their partner, then both go to jail for some medium sentence. However, the outcome gets interesting when the prisoners use different strategies. If one prisoner remains silent while his partner incriminates him, the one that incriminates the other is released immediately while the incriminated partner gets the longest jail sentence of all (see Box 15.2). Here, denial and squealing symbolize the notions of altruism and selfishness!

More generally, call $R$ the payoff (reproductive success) of both individuals when both act selfishly. Selfishness does not cost the perpetrator nor does it benefit the partner. However, an interaction between two altruistic individuals would cause each a direct fitness cost $-c$ and an indirect fitness benefit $+b$, hence a reproductive success $R = b - c$ for each individual. Furthermore, when one selfish and one altruistic individual interact, the reproductive success of the selfish and altruist are respectively $R + b$ (the selfish individual gets the benefit without paying the cost) and $R - c$ (the altruist pays the cost without getting the benefit). A selfish strategy is thus inevitably favoured. Or, more precisely, if a mutant altruistic strategy appears in a large, stationary ($R = 1$) and well-mixed (i.e., interactions occur between randomly chosen individuals) population dominated by a selfish strategy, the reproductive success of the mutant strategy will be only $1 - c$ and the mutant cannot spread and eventually disappears. On the other hand, if a mutant selfish strategy appears in a large, stationary ($R + b - c = 1$) and well-mixed population dominated by an altruistic strategy, the reproductive success of the mutant strategy will be as high as $1 + c$ and the mutant strategy will invade and eventually replace the resident population. Therefore, the origin and persistence of altruism proves to be a fundamental enigma whose key solutions are explored in Sections 15.5–15.7. Section 15.8 then calls into question the idea that the evolution of altruism is irreversible and proceeds only in a direction of ever increasing social complexity.

## 15.3 Genetics and plasticity of altruism

The evolution of behavioural traits through natural selection as we saw in Chapter 2 requires inter-individual variation, heritable behavioural variation and a consistent relationship between behaviour and fitness (Cockburn 1991; Endler 1986). Before analysing the fitness costs and benefits of altruistic traits (Section 15.4), we examine here the genetic determinism and **phenotypic plasticity** of these traits.

### 15.3.1 Genetic determinism

Although most evolutionary models assume that a simple genetic architecture underlies natural variation in altruistic traits, the genetic determinism of altruistic behaviour has been poorly investigated, especially in vertebrates. A few illustrative examples are given below.

*Cooperation in viruses*

Phages are RNA viruses that infect bacteria and produce simple catalytic substances to replicate within their host. Turner and Chao (1999) have compared the metabolic properties of the phage $\phi 6$ and a mutant strain $\phi H2$. When these two strains co-infect the same bacteria, the catalytic substances produced by one strain benefit the replication, and hence the fitness, of the other strain. However, the mutant strain $\phi H2$ behaves selfishly by producing less of these costly catalytic substances needed for mutual replication. An experimental assessment of the payoffs of the behavioural interactions between these two strains indicates that these two viruses play a prisoner's dilemma! Using the notations of this game as defined in Box 15.2, the replication rate of the selfish strain $\phi H2$ in a bacteria dominated by the phage $\phi 6$ is almost twice the replication rate of the phage $\phi6$ alone ($T$ = 1.99 versus $R$ = 1). In an infected cell dominated by the selfish strain, the replication rate of $\phi$H2 reduces to $P$ = 0.83, which is still higher than the replication rate of $\phi6$ ($S$ = 0.65). Thus, these RNA viruses exhibit genetically determined altruistic and selfish strategies that are engaged in a prisoner's dilemma.

*Cheating in social amoeba*

As we saw in the introduction to this chapter, the slime mould (*D. discoideum*) shows striking social behaviour. The life cycle of the species alternates between solitary and colonial stages where thousands of cells cooperate to form a 'fruiting body'. The colony differentiates into a stalk and a fruiting body producing resistant spores (Figure 15.1), a developmental change akin to the differentiation between the soma and germ lines of multicellular organisms. On average, 20% of the stem cells differentiate into stalk cells that die and thus forego their own reproduction for the benefits of spore cells. Yet, this apparent cooperation is a battleground between naïve altruistic and perfidious selfish strains. According to Strassmann *et al.* (2000), who constructed several artificial colonies involving two clonal lines taken from the wild, half of the clonal constructions revealed an altruistic clone that is over-represented in the stalk, and a selfish clone, that is under-represented in the stalk (Figure 15.1d). Similar selfish clones can be obtained through directional mutation in the laboratory, which indicates that the social behaviour of amoebas was determined, at least partly, by a few genes controlling cellular mobility (Ennis *et al.* 2000).

*Social insects*

Various studies have successfully documented genetic variation in cooperative behaviours that underlie complex social systems in insects, most notably in bees and ants (see Keller and Ross 1998; Moritz *et al.* 1996; Olroyd *et al.* 1994). For example, controlled breeding experiments in the honey bee (*A. mellifera*) indicate that the grooming and nest cleaning behaviour is controlled by two di-allelic genes (Rothenbuhler 1964). More recently, Ross and Keller (1998) also reported on a simple genetic system controlling facultative polygyny and social structure in the imported red fire ant (*Solenopsis invicta*).

### 15.3.2 Plasticity

While studies on genetic polymorphism remain exceedingly rare, social and phenotypic plasticity of altruistic behaviour has been observed in numerous species. In insects, worker status can be controlled by pheromones produced by the queen (Keller and Nonacs 1993), by food provisioning during early developmental stages (Wilson 1971), by the age of the individual (Stern and Foster 1997), or by the social **environment** (Abbot *et al.* 2001). This is so, for example, in the social aphid (*Pemphigus obesinymphae*) a species whose colonies feed on sap. A typical colony results from the asexual reproduction (parthenogenesis) of a clonal line. Young individuals can reproduce, whereas older individuals undergo a moult and develop into sterile soldiers defending the colony against predators and parasites. Winged individuals are produced by sexual reproduction and they disperse to establish new colonies. A colony is usually started by a single winged individual, enabling peaceful cooperation among genetically similar individuals (Hamilton 1972). However, winged individuals sometimes attempt to immigrate into a foreign colony. If successful, the immigrant clone adopts a selfish behaviour by slowing down its development and thus over-producing breeding individuals (Abbot *et al.* 2001). This adaptive ontogenic flexibility is an example of the plasticity of social behaviours.

Another case study of plasticity is provided by a long-term monitoring of family dynamics in the rare and endemic Seychelles warbler (*Acrocephallus sechellensis*; Komdeur 1992). In this species, the reproduction mode—either solitary involving one breeding pair or cooperative breeding involving one breeding pair and some helpers—is sensitive to habitat quality and saturation. Social groups of this species are called 'extended families' because they are formed mostly by local recruitment of offspring who stay on the

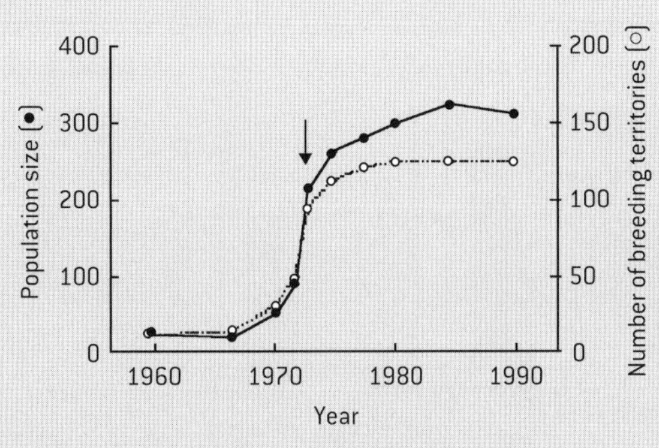


Figure 15.2 **Changes in population size and number of breeding territories after a restoration programme conducted on an endangered population of the endemic Seychelles warbler (*Acrocephallus sechellensis*)**

After the number of breeding territories reached a plateau because of habitat saturation around the start of the 1970s (arrow), the population size continued to increase owing to the emergence of extended families where a mean number of about 0.5 to 0.8 individuals cooperate with the breeding pair for territory defence, foraging, and parental care.

Adapted from Kondeur (1992).

parental territory beyond sexual maturity. Helpers assist the breeding pair by defending their territory, building their nest, incubating their eggs or feeding their chicks. Since the early 1960s, a small population of Seychelles warblers on Cousin Island has been subjected to a conservation programme. The initial population consisted of 26 solitarily breeding individuals on independent territories, but the population has rapidly grown reaching 300 individuals by the late 1980s. As early as 1973, extended families started to form on high-quality territories, and then over the whole island around 1982 (Figure 15.2). The emergence of extended families happened in parallel to the saturation of the habitat. Experimental transfers of helpers into empty neighbouring islands confirmed that solitary reproduction was inhibited by strong competition for space. Thus, habitat quality and saturation interact to determine social tactics, and cooperative breeding is a plastic response to high habitat quality and saturation making independent reproduction a less valuable option (Komdeur 1992).

## 15.4 Costs and benefits of altruism

Altruistic interactions involve costs and benefits that depend on the behaviour of the altruist, the response of the recipient, and the context of the social interaction. The costs and benefits of altruistic behaviours have been well characterized in cooperatively breeding bird species (see Cockburn 1998; Heinsohn and Legge 1999). In this group, non-breeding assistant birds (helpers) are distinct individuals whose helping behaviour can be quantified. Short-term physiological consequences of helping can be measured, as well as more long-term fitness consequences on future survival or reproduction.

### 15.4.1 Direct costs

Quite surprisingly, the costs of altruism have rarely been quantified, even in very well-studied species (Clutton-Brock *et al.* 1998). In cooperatively breeding birds, field measurements of these costs have relied on correlative approaches. In the social white-winged chough (*Corcorax melanorhamphos*) helpers are inexperienced individuals that assist the breeding pair (Heinsohn and Cockburn 1994). During the breeding season, one-year-old helpers lost body mass proportionally to their investment in helping, while the body mass of breeding individuals remained stable (Figure 15.3a). This correlation suggests physiological costs of helping in young individuals, but more long-term effects are also possible. For example, helping effort is negatively correlated with future survival in the stripe-backed wren (*Campylorhynchus nuchalis*; Rabenold *et al.* 1990; Figure 15.3b).

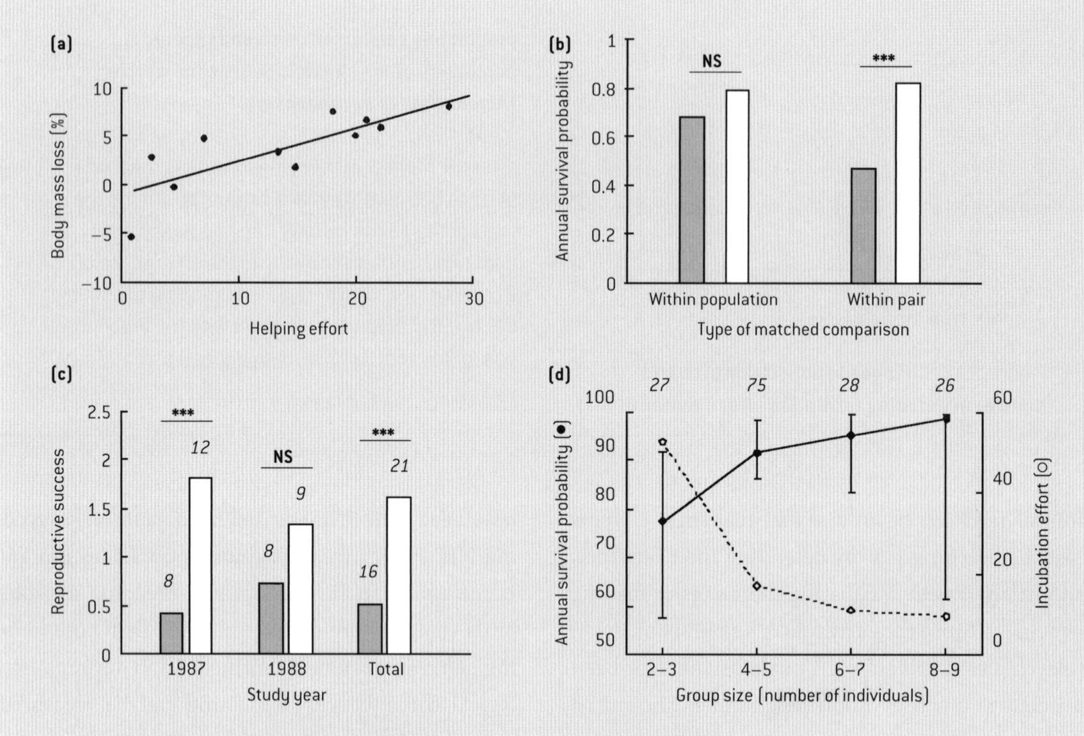


Figure 15.3 **Costs and benefits of helping in cooperatively breeding birds**

**a.** Relationship between helping effort (percentage of time spent incubating eggs of the breeding pair) and body mass loss in helpers during the breeding season. Study species: white-winged choughs.

Modified from Heinsohn and Cockburn (1994).

**b.** Annual survival probability of helping birds that invest more (filled bars) or less (empty bars) than average in helping. Helpers were compared within the same population including all social groups or within social groups including both categories of helpers. Study species: stripe-backed wren.

Modified from Rabenold (1990).

**c.** Reproductive success (number of chicks 60 days old) in control unmanipulated families (empty bars) and in families where helpers had been experimentally removed (filled bars) during two different study years. Study species: Florida scrub jay (*Aphelocoma c. coerulescens*).

Modified from Mumme (1992).

**d.** Annual survival probability (continuous line) and individual investment in cooperative effort (percentage of time spent incubating eggs, dotted line) according to group size. Study species: groove-billed ani (*Crotophaga sulcirostris*).

For all figures, statistics read as: *** $p < 0.001$, ** $p < 0.01$, * $p < 0.05$, NS (non-significant).

Adapted from data in Vehrencamp *et al.* (1988).

### 15.4.2 Direct benefits

Direct costs of investment in altruism can be compensated later in life by direct benefits if, for instance, partners reciprocate their help. Even in cases without reciprocation, for example in cooperatively breeding birds, potential direct benefits can involve the inheritance of the parental territory (Stacey and Ligon 1991), the inheritance of the breeding status (Rabenold 1990, Sherley 1990), the learning of breeding skills (Heinsohn 1991; Komdeur 1996), the formation of long-term social bonds such as kin coalitions (Zahavi 1990), or even increased social **prestige** (Box 15.3).

### Box 15.3 Social prestige in Arabian babblers

In a controversial essay, the Israeli scientist Amotz Zahavi proposed that the **handicap principle** can explain the evolution of altruism (Zahavi 1995). According to the handicap principle (see Chapter 11), some behaviours are **signals** that honestly reflect the genetic quality of the signaller. **Honesty** makes sense if signals are costly and hence impose a handicap on the signaller such that the signal becomes disproportionately costly for low-quality individuals (see Chapters 11 and 16).

The handicap principle could in theory also apply to altruistic behaviour: altruism is by definition costly and so it is possible that its costs are disproportionately larger for low-quality individuals. Thus, Zahavi argued that altruistic behaviour can be a signal that the animal can afford the cost of the behaviour. Hence, altruism serves as an advertisement for 'social prestige' (Zahavi 1990) which indicates value as a partner for future cooperation (**indirect reciprocity**, see Section 15.6.2) as well as its quality as a future mate (Nowak and Sigmund 1998, Zahavi and Zahavi 1997).

Observations of social behaviour in Arabian babblers (*Turdoides squamiceps*), a bird species studied by Zahavi since the 1970s, seem to support the 'social prestige' theory. Adults compete for altruistic donations, such as chick or adult feeding, and altruistic behaviours are distributed according to a social hierarchy. Dominant individuals defend privileged access to altruistic tasks and refuse the donations of subordinates (Carlisle and Zahavi 1986; Zahavi and Zahavi 1997). The social prestige theory seems to provide a simple and parsimonious explanation of these behavioural interactions, but there are some potential caveats:

- Wright (1997) argued that behavioural data equally well support a model of **parental investment** (where investment in cooperation should optimize the reproductive success of individuals) than a model of social prestige (where investment in cooperation should optimize the social prestige of individuals).
- The potential for **kin selection** cannot be excluded since behavioural interactions often occur among kin birds inside extended families (Wright 1999).
- Very few studies have been able to document the type of competition for altruism observed in Arabian babblers, thus questioning the generality of the social prestige theory (Boland *et al.* 1997; Reyer 1984; Wright 1999).

### 15.4.3 Indirect costs

An altruistic individual that improves the survival or reproduction of his neighbours may exacerbate local competition for critical resources like food or mates (Griffin and West 2002). Theoretical studies suggest that indirect costs caused by such increased competition with social partners may negate the benefits of altruism (Taylor 1992a). In fig wasps, males emerging from the same fig (including brothers) compete overtly for mates and competition among kin male wasps seems to be strong enough to preclude the evolutionary repression of aggressive fights among relatives (West *et al.* 2001). An indirect cost due to competition among kin is also probably present in cooperatively breeding mammals and birds who compete for higher social ranks and for breeding opportunities within the same family (Clutton-Brock 2002), or between colonies of social insects produced by budding from the same colony (Thorne 1997).

### 15.4.4 Indirect benefits

Indirect benefits of helping have been assessed by experimentally removing or adding helpers in social groups of cooperatively breeding birds. Mumme (1992) removed helpers in extended families of the Florida scrub jay (*Aphelocoma c. coerulescens*) and compared the **performances** of families without helpers with those of unmanipulated families. Helping was found to improve chick survival in this species (Figure 15.3c): helper jays provision chicks with food and protect the nest against predators. A more subtle form of indirect benefits is that helping can decrease investment in reproduction by the breeding pairs, and thus improve the future survival and reproduction of the breeding couple. This type of indirect benefit has been documented in the groove-billed ani (*Crotophaga sulcirostris*), a species where several females can lay their eggs in the same nest (Vehrencamp *et al.* 1988). In large social groups, females are seen to reduce their incubation effort and enjoy higher future survival (Figure 15.3d).

## 15.5 Evolution of unconditional altruism

Our analysis of the prisoner dilemma in Section 15.2 hypothesized a large and homogeneous population. In this case, the probability of interacting with a strategy is given by the frequency of that strategy over the whole population. In a very large population, cooperation between two rare altruistic mutants is thus extremely unlikely. The **assumption** of a homogeneous population is, however, not very realistic. Several factors could explain spontaneous heterogeneous spatial structuring of individuals: individual mobility is often limited and social groups are usually small and so social interactions occur preferentially among neighbouring individuals or, more generally, along social networks connecting a finite set of individuals. Under these conditions, encounter rates of individuals belonging to a minority can reach high values.

### 15.5.1 Kin selection and Hamilton's rule

#### a *Rediscovering Hamilton's rule*

Consider a very large stationary population initially dominated by a selfish strategy and denote by $\rho$, the mean encounter rate between altruistic individuals drawn from an initially rare population. The growth rate of the rare mutant population is thus $1 - c + \rho \times b$, while growth rate equals 1 for the population of the dominant selfish strategy if we neglect initially rare interactions with the mutant strategy. In other words, the population of the rare mutant can grow and invade the selfish population if $1 - c + \rho \times b > 1$, or, stated differently, if the mean encounter rate between altruists is higher than the cost/benefit ratio: $\rho > c/b$.

Hamilton used a somewhat different theoretical approach to reach a similar conclusion (see Chapter 2). Hamilton (1964a, b) derived equations describing the frequency dynamics of an allele in a kin-structured population of a diploid organism. He found that the quantity maximized by natural selection was a metric that measures the fitness of relatives of individuals bearing the allele (including the individual itself) due to the total effects of that allele on genes identical by descent in relatives. He called that metric the **inclusive fitness** and showed that the direction of evolution was given by the sign of the inclusive fitness effect (see Chapter 2). For a kin-structured population of altruistic individuals where the relatedness $r$ measures average co-ancestry among social partners, the inclusive fitness effects defined by Hamilton can be written as $r \times b - c$, where $r \times b$ measures the indirect benefits derived from helping relatives. Thus, conditions for the growth in frequency of an altruistic allele are subsumed in Hamilton's famous rule: $r > c/b$. The selection mechanism revealed by Hamilton somehow sorts kin groups and has thus been called kin selection. It is worth noticing that a kin selection approach yields similar results to a direct fitness approach (Taylor and Frank 1996): a fitness measurement bookkeeping of what individuals give (and to whom) is equivalent to a fitness measurement bookkeeping of what individuals receive (and from whom). Furthermore, Day and Taylor (1998) have shown that relatedness was equivalent to the probability of encounter between two mutants in a haploid genetic system.

### b *Structure of kinship*

Hamilton's rule was a major breakthrough and has deeply influenced our understanding of social behaviour (Hamilton 1995). One merit of this rule is that it provides a simple and intuitive evolutionary explanation of altruistic traits: genes that cause a fitness cost to their altruistic bearers can be rewarded if they contribute to enhance the replication of genes related by descent in recipients (Dawkins 1976). The classic expression of Hamilton's rule, however, applies only when a set of strict assumptions, such as a very large population or a constant relatedness $r$ are met (Hamilton 1964a, b). Imagine a more realistic alternative situation where the population is spatially structured such that a small group of altruists gets established somewhere within the larger selfish population (Ferrière and Michod 1995, 1996). The expansion of this small group depends on the growth of a core dominated by altruists and the diffusion of a front. Here, the relatedness is highly structured in space: its value is close to 1 in the core of the mutant group and is probably variable along the front. Furthermore, the mixed population of altruists and selfish individuals along the front makes it more difficult to calculate the fitness of both strategies. Nowak and May (1992) used computer simulations to show that this small group of altruists can invade a grid inhabited by selfish individuals. In their simulations, spatial structure enables long-term coexistence between selfish and altruist. This stresses the fact that population viscosity through limited mobility (Hamilton 1964a, b) can favour the emergence of altruism through kin selection.

## 15.5.2 Ecological context

Kin selection comes up against a paradox that was recently pointed out by several authors (Queller 1994; Taylor 1992a, b; Wilson *et al.* 1992). Indeed, the same factors that enhance cooperation among relatives may also increase local competition for space and food between relatives, a form of competition called **kin competition** (see Section 15.4.4). Kin competition can potentially negate the benefits of kin cooperation and thus selects again altruism (West *et al.* 2002). However, it remains possible for altruism to evolve if the competitive neighbourhood does not overlap entirely with the cooperative neighbourhood (Queller 1994). This is the case, for example, when individuals compete over large spatial scales (Queller 1992), when dispersal occurs after the cooperative stage and before the competitive stage (West *et al.* 2002), or when free space opens for the reproduction of altruistic groups (Le Galliard *et al.* 2003; Mitteldorf and Wilson 2000). In these cases, relatedness within the cooperation neighbourhood may remain higher than relatedness within the competition neighbourhood. The evolution of the altruistic strategy is then driven by a modified Hamilton's rule involving two relatedness metrics, one measuring co-ancestry within the competition neighbourhood and another measuring co-ancestry within the cooperative neighbourhood (Frank 1998, Queller 1994).

One important ecological trait that influences both the degree of kin competition and kin cooperation is individual mobility. Now, mobility is sensitive to the same selection pressures that influence altruism (Hamilton and May 1977) and changes in altruism can affect the very selection pressures that affect mobility (Le Galliard *et al.* 2003). Evolutionary interactions between altruism and mobility are thus keys to a better understanding of the ecological context of the evolution of altruistic traits (Perrin and Lehmann 2001). General rules have emerged from kin selection models that investigate the joint evolution of altruism and mobility (Le Galliard *et al.* 2005):

1. The cost of mobility is crucial to explain the emergence of altruism and the persistence of high levels of altruism requires strong costs of mobility.
2. Starting from a selfish and mobile ancestor, the evolution can proceed through a first stage where

low mobility evolves in a selfish population before altruism secondarily takes over.

The African mole-rats, or Bathyergidae, are excellent model systems for analysing the influence of mobility on the evolution of altruism. Mole-rats are fossorial mammals living inside subterranean cavities. The Bathyergidae family encompasses a total of 18 species including 4 solitarily breeding, 12 cooperatively breeding, and 2 eusocial species. The eusocial organization of African mole-rats is unique in mammals and eusociality has evolved independently in the naked mole-rat (*Heterocephalus glaber*; Jarvis 1981) and in the Damaraland mole-rat (*Cryptomys damarensis*; Jarvis *et al.* 1994). In African mole-rats, cooperative breeding and eusociality have evolved in the most arid habitats where dispersal is limited (Faulkes and Bennett 2001; Jarvis *et al.* 1998; Jarvis, O'Riain *et al.* 1994). Aridity is thought to increase the cost of **dispersal**, the costs of independent breeding and the benefits of group living (Jarvis *et al.* 1994). A similar co-variation between habitat harshness, dispersal and sociality has been observed within the same species. In the cooperatively breeding common mole-rat (*Cryptomys hottentotus hottentotus*; Spinks *et al.* 2000), dispersal is lower, plural reproduction is rare, and breeding couples are more stable in arid than in mesic habitats (Figure 15.4). Thus, the phylogenetic distribution of sociality in African mole-rats supports a scenario where strong costs of dispersal induce philopatry and reproductive altruism (Jarvis *et al.* 1994).

### 15.5.3 Genetic context

The influence of kinship structure on costs and benefits of altruism is highly dependent upon the genetic and **mating system**. A classical example of a genetic system promoting kin selection is the haplo-diploid karyotype of Hymenoptera (wasps, bees and ants), a taxonomic group containing the most highly social species and characterized by extensive reproductive altruism among females. In 'simple' societies, the colony is made up of few breeding females, most often sisters, that share equally breeding in the same nest and collectively provision their offspring (Peeters 1997). At the other end of the social spectrum, large societies of some ants and bees are made up of one dominant breeding female (the queen) and hundreds of female workers that may have irreversibly lost their capacity to breed. The commonality and complexity of social systems in Hymenoptera has been classically explained by their peculiar karyotype (Hamilton 1964a, b, 1972). Hymenopteran males are haploid and come from the development of unfertilized eggs, while females are diploid and come from the development of fertilized eggs. In a colony, sex ratio can be controlled in two different ways. The queen controls fertilization of

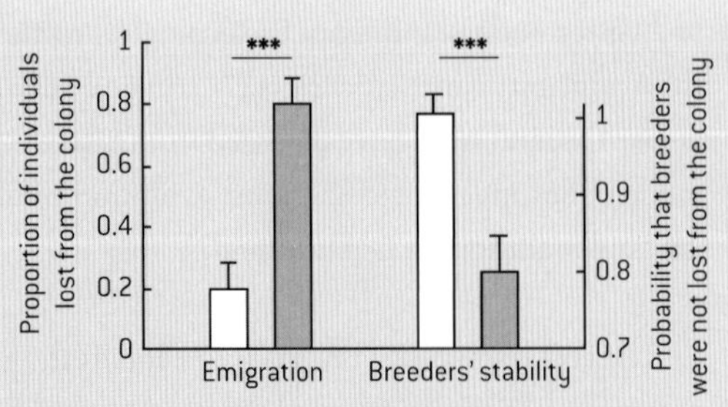


Figure 15.4 **Individual emigration and breeders' stability in cooperatively breeding colonies of the Common mole-rat from optimal mesic habitats (empty bars) and suboptimal non arid habitats (filled bars)**

Emigration was measured as the proportion of individuals lost from the colony between successive capture periods. The stability of the breeding individuals (usually a breeding pair) was measured by the probability that breeders were not lost from the colony between successive capture periods.

After Spinks *et al.* (2000).

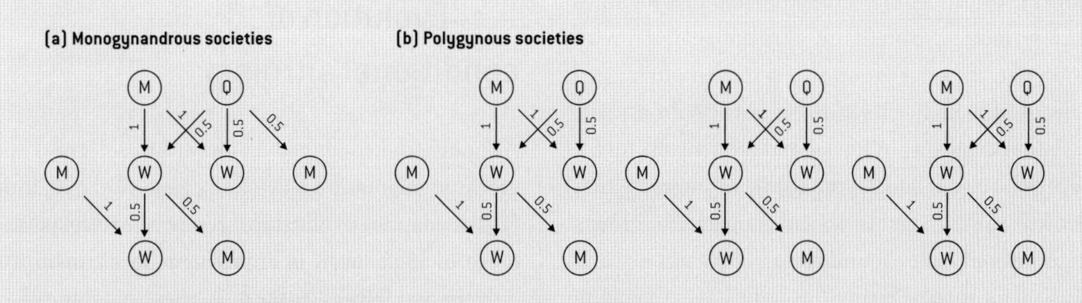


Figure 15.5 **Relatedness coefficients in hymenopteran societies**

**a.** Co-ancestry in a monogynandrous colony.
**b.** Co-ancestry in a polygynous colony with three breeding queens. Values along each arrow indicate genetic identity by descent. Letters indicate queens (Q), workers (W), and males (M).

her ova by the sperm that is stored in a special organ, the abdominal spermatheca, thus producing male or female eggs as required. Female workers can lay their own unfertilized eggs to produce male offspring, or selectively provision and kill the eggs laid by the queen to control the sex ratio of the queen's offspring.

We wish to calculate relatedness coefficients in a primitive society where all females are able to breed and produce only females. In a monogynandrous colony the coefficient of relatedness between sisters ($r$ = 0.75) is higher than between a female and her daughters ($r$ = 0.5, Figure 15.5a). According to Hamilton's rule, higher relatedness among sisters than between mother and daughters should make it easier to evolve reproductive altruism among sisters (Hamilton 1964a, b). Thus, the haplo-diploid karyotype of Hymenoptera has probably been a strong contributor to the evolution of sociality.

Relatedness between sisters can be much lower in polygynous colonies where several females share breeding (up to a hundred queens in some ant species). For example, if three sister queens share breeding equally within the same colony, the mean relatedness coefficient between sisters plummets to $r$ = 0.375 (Pamilo 1991), which is now lower than the relatedness between a female and her daughters (Figure 15.5b). Similarly, relatedness between sisters collapses when several males sire offspring in the same colony (up to 17 different males in the honey bee). For example, if the queen is mated with three different males that contribute equally to paternity, the relatedness coefficient between sisters goes down to $r$ = 0.42, which should thus promote selfish reproduction in females. In short, the haplo-diploid genetic system is not sufficient to explain the maintenance of reproductive altruism within multiple-queens or multiple-male colonies. Keller (1995) proposed that cooperation could persist if the benefits of altruism were stronger or if female workers had irreversibly lost their ability to breed independently. A better understanding of the evolution of sociality in Hymenoptera has now been reached by considering interactions between multiple ecological and genetic factors, the haplo-diploid genetic system being one among others (see Choe and Crespi (1997) for further discussion).

### 15.5.4 Group augmentation effect

Group augmentation is an alternative to kin selection that can explain the evolution and persistence of altruism among unrelated individuals (Bernasconi and Strassmann 1999; Clutton-Brock 2002; Emlen 1997; Jarvis *et al.* 1994). Clutton-Brock (2002) suggested that group augmentation operates in most cooperatively breeding vertebrates and invertebrates because indirect benefits of altruism increase with group size. A larger group size is indeed often associated with a higher foraging success (Wilson 1971), predation avoidance (Queller and Strassmann 1998), dispersal (Ligon and Ligon 1978), or reproduction (Brown 1987). This is so in cooperative groups of unrelated ant foundresses (Bernasconi and Strassmann 1999). During the early stages of colony establishment, queens must rely on their own body reserves to lay their eggs and provision the first generations of workers. Coalitions of unrelated queens fare better than solitary queens: they lay eggs and reach large colony sizes earlier (Bernasconi and Strassmann 1999).

Group augmentation may also increase the direct benefits of altruism, for example because helpers inherit a larger and more productive group. Another mechanism that could promote the group augmentation effect is competition among social groups, which seems to be common in cooperatively breeding birds (Cockburn 1998) and also in social insects (Wilson 1971). Recent modelling work suggests that group augmentation favours the evolution of higher investment in altruism, provided that the benefits generated by the group augmentation effects are shared equally among group members (see Kokko *et al.* 2001; Roberts 1998). Group augmentation may have more limited value in explaining the persistence of altruism in species where individuals adjust their investment in altruism selfishly according to group size: investing less *per capita* at larger group size to diminish the costs of altruism (Kitchen and Packer 1999). Furthermore, the indirect benefits generated by group augmentation alone are not sufficient to explain the evolution of altruism (Box 15.2). Firstly, indirect benefits must be a non-additive function of group size, such that the collective rewards of altruism are more than the sum of individual investments in altruism (see Kitchen and Packer (1999) for a critical discussion of this). Secondly, even when the first condition is met, the group augmentation effect alone will not be sufficient for altruism to evolve in a population dominated by selfish individuals.

## 15.6 Evolution of conditional altruism

Although the combined action of kin selection and group augmentation may offer a general explanation for the adaptive evolution of cooperation, the persistence of high levels of investment in altruism can only be explained in limited cases, notably when individual mobility is very restricted (Le Galliard *et al.* 2003). Furthermore, even in this case, a mobile selfish strategy would avoid deleterious interactions within an individual's own group and could threaten the persistence of a population of sessile, altruistic individuals (Dugatkin and Wilson 1991; Enquist and Leimar 1993; Houston 1993). In the face of such a dilemma, behavioural plasticity – cooperating only under specific conditions – can guarantee a potential resistance of cooperation towards selfishness. Here, we investigate individual mechanisms and evolutionary consequences of behavioural plasticity depending on whether altruism is sensitive to the state of the donor or to the state of the recipient.

### 15.6.1 Sensitivity to the donor

When behavioural interactions are iterated, an individual can change its behaviour according to previous interactions and thus flexibly adjusts its degree of investment in altruism. In the late 1970s, Robert Axelrod led the first worldwide computer tournament of an iterated prisoner's dilemma (see Box 15.2) that opposed numerous conditional strategies, some very complex using information derived from numerous previous iterations to predict the future behaviour of the individual (Axelrod and Hamilton 1981). A surprising result of this tournament was that a quasi-systematic winner was the simple strategy Tit-For-Tat (TFT), which cooperates during its first move and then imitates the behaviour of its previous opponent. TFT is swift to retaliation and is thus not a naive altruistic strategy that helps unconditionally; yet, a pure population of TFT would behave in appearance like a population of cooperators.

A simple calculation demonstrates that a TFT population can resist invasion by selfish mutants and cooperation evolves by **reciprocity** if the turnover of partners in the game is low, i.e. most iterations of the game occur between the same partners (see Axelrod and Hamilton 1981). Several factors may oppose this reciprocity, such as differential mortality or mobility between altruistic and selfish partners. For example, TFT is unable to invade a highly mobile selfish population that would be homogenized after each round of the game. The invasion and persistence of TFT is nonetheless possible at intermediate levels of mobility. Intermediate mobility allows TFT to spread from a source population and maintains its ability to retaliate against mobile selfish partners (Ferrière and Michod 1995, 1996). Behavioural errors in the tournament may also undermine reciprocal cooperation: a single misunderstanding in a repeated interaction between two TFT players can indeed induce a burst of selfish behaviour. With regard to this problem, Nowak and Sigmund (1993) discovered a new strategy (called Win–Stay–Lose–Shift or WSLS), which is more robust than TFT against errors. WSLS copies the partner's last move or adopts an opposite strategy according to whether its payoff is positive or zero at most, respectively. Thus, a selfish behaviour played by mistake between two WSLS partners cause one round of mutual retaliation followed by a bilateral revival of mutual cooperation. Although WSLS is very resistant against behavioural errors, it is not much capable of invading an ancestral selfish population. Its competitive superiority is only obvious when more ruthless strategies, like the concession free TFT, have paved the way and wiped out unconditional defectors (Nowak and Sigmund 1993).

Cognitive constraints on reciprocal altruism may limit the generality of reciprocal altruism (Stephens *et al.* 2002). TFT is dependent upon the state of the agent at the close of each interaction: if the payoff is positive, TFT behaves like an altruist; if not, TFT retaliates. Thus, in a united interaction between two TFT partners, no particular cognition is needed. However, in a more realistic tournament where pairs of players can shift, TFT demands individual memory. It can be as complex as the capacity to remember individual partners encountered during previous interactions (Brown *et al.* 1982; Ferrière and Michod 1996), or the more simple aptitude of 'keeping an eye on the neighbours': an individual would then be forgotten as soon as it left its circle of acquaintance (Hutson and Vickers 1995). Lindgren and Nordahl (1994) reviewed several computer simulations investigating the effects of memory capacities on the evolution of reciprocal altruism.

Another limit stems from the fact that, despite a plethora of evolutionary games, empirical examples of reciprocal altruism are highly controversial (see, for instance, Stephens *et al.* 2002 and Kefi *et al.* 2007). Reciprocal altruism seems restricted to few goods and services markets like grooming or feeding activities. Data showing that individuals have the capacity to retaliate are few (Clutton-Brock 2002), and the same empirical examples of reciprocal altruism go round in the specialized literature (Dugatkin 1997). It remains also unclear whether animals do not often 'prefer' the short-term advantage of selfishness to the long-term benefits of reciprocal cooperation. Reciprocity is nonetheless a fundamental characteristic of human societies, and evolutionary game theory has contributed to a much better understanding of cooperation in our social and economic markets (Sigmund and Nowak 1999). For example, experimental prisoner's dilemma tournaments conducted with human subjects confirm the preferential use of a WSLS-like strategy (Wedekind and Milinski 1996). In agreement with theory, the performance of this strategy is influenced by the memory capacity of players (see Milinski and Wedekind 1998).

### 15.6.2 Sensitivity to the recipient

#### 15.6.2.1 *Cooperation*

Cooperation may also be dependent upon the state of the partner. Eshel and Cavalli-Sforza (1982) already envisioned this behavioural plasticity in the form of a preferential cooperation according to the degree of altruism of the partner. Whether individuals may signal directly their degree of investment in altruism is still controversial, although recent quests for 'green beard genes'– termed by Dawkins (1976) as we saw in Chapters 2 and 5 to describe a gene that would cause individuals to behave more or less altruistically and at the same time promote discrimination among altruistic partners – have been successful (Keller and Ross 1998). Direct genetic signals of the 'green beard' type seem rare, however, and two common alternatives have been found:

- Discrimination based on kinship (see Box 15.4);
- Discrimination based on reputation, or 'image scoring'.

#### 15.6.2.2 *Cooperation among relatives*

The evolution of cooperation may be facilitated if benefits of altruism are distributed preferentially to relatives (Agrawal 2001, Hamilton 1964a, b). Such kin discrimination can allow the evolution of altruism in situations where competition among kin would negate the benefits of cooperation (Perrin and Lehmann 2001). This is so because kin recognition allows restriction of social interactions among the most related individuals within the competition neighbourhood (see Section 15.5.2). Differential treatment of relatives, or **nepotism**, is therefore

### Box 15.4 **Kin discrimination**

**Functions**. Kin recognition serves four major purposes:

- avoidance of diseases transmitted during physical contacts with relatives;
- avoidance of competition with relatives;
- preferential cooperation with relatives or **nepotism**;
- and **inbreeding** avoidance.

**Mechanisms**. Three main kin recognition mechanisms have been recognized and only one can be considered as kin recognition *sensu stricto* (Hepper 1991). The true kin recognition mechanism involves **recognition alleles** and a direct assessment of kinship, i.e. genetic resemblance by descent (Grafen 1990). Indirect kin recognition mechanisms are based on **phenotype matching** or **spatial distance**. In the first case, genetic distance can be assessed with phenotypic proxies, such that a large phenotypic distance is a cue for low kinship (Lacy and Sherman 1983). When this cue is learned during early development, the mechanism is called **associative learning** and kin discrimination relies on familiarity. The latter case corresponds to a minimalist process where spatial distance between conspecifics serves as a cue for kinship.

**Evolution of kin recognition**. Kin recognition is widespread in both solitary and colonial species, suggesting that the evolution of kin recognition has often preceded that of altruism (Hepper 1991; Waldman 1988). For example, tadpoles use kin recognition to avoid cannibalizing relatives (Pfennig *et al.* 1993). It is therefore unlikely that the ability to discriminate among kin has been a strong constraint for the evolution of altruism. However, the evolution of sociality may feedback on kin recognition mechanisms. Complex families that occur in cooperatively breeding vertebrates may favour extended contacts across generations and thus select for elaborate kin recognition mechanisms (Emlen 1997; Emlen and Wrege 1994). Comparative studies across species should be conducted to understand better the joint evolution of kin recognition and social systems (Perrin and Lehmann 2001). A major difficulty with this approach is that kin recognition mechanisms are highly flexible and it may thus be difficult to describe the average behaviour of a species (Waldman 1988).

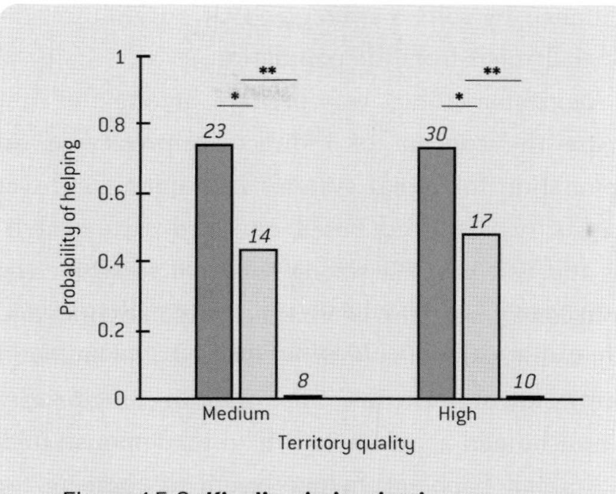


Figure 15.6 **Kin discrimination in cooperatively breeding Seychelles warblers**

Data shown are the probability of helping full-sibs (filled black bars), half-sibs (shaded bars), or less related chicks (empty bars) observed in medium-quality and high-quality territories.

fundamental in explaining the evolution of altruism by kin selection.

Nepotism has been described in cooperatively breeding birds (see Clarke 1984; Curry 1988; Emlen and Wrege 1988; Marzluff and Balda 1990; Mumme 1992) and mammals (see Holmes 1986; Owens and Owens 1984; Sherman 1981). For example, in the Seychelles warbler, investment in helping rises with relatedness between helpers and chicks (Figure 15.6). Helpers direct their help preferentially to the breeding pair that reared them as a chick rather than to other related individuals that did not care for them as a chick, suggesting that kin discrimination in this case is based on associative learning (Box 15.4). Hatchwell *et al* (2001) also found out that helper birds in cooperatively breeding long-tailed tits preferentially join the social groups of their parents, irrespective of true kinship. In the eusocial naked mole-rats, kin discrimination is also based on **familiarity** rather than true kinship (Clarke and Faulkes 1999; O'Riain and Jarvis 1997). Thus, help cannot be distributed differentially according to paternal lineages, although several males are seen to breed within the same colony (Lacey and Sherman 1991; Reeve and Sherman 1991). Familiarity-based discrimination seems the norm in social vertebrates (see Komdeur and Hatchwell 1999); although it should be more prone to errors than direct recognition (Box 15.4). However, direct kin recognition may occur in small social groups with complex genealogies and strong benefits of cooperation (see Blaustein *et al.* 1991; Emlen and Wrege 1994; Petrie *et al.* 1999).

Nepotism is also common in eusocial insects. Social ants, bees, and aphids usually discriminate colony members from foreigners (see Gamboa *et al.* 1986; Getz 1991; Pfennig *et al.* 1983). Colony recognition relies on a learned chemical recognition cue, most often a cuticular product passed on during physical contacts between colony members (Jaisson 1991). More elaborate mechanisms of direct recognition based on genetic signals would allow workers to discriminate among paternal or maternal lineages in polyandrous or polygynous colonies respectively (Keller 1995). Earlier studies in honey bees suggested that paternal lineages could be treated differently by workers (Frumhoff and Schneider 1987; Page *et al.* 1989), but these results have been criticized because these experiments involved a poor diversity of paternal lineages (Carlin and Frumhoff 1990). In realistic situations involving numerous paternal lineages and more natural conditions, workers do not seem to discriminate among kin within the colony (Keller 1997).

This brief overview of kin discrimination mechanisms suggests that kin recognition relies mostly on learned colonial or familial cues, rather than on true genetic signals, which led some to question whether animals do really recognize kin (Grafen 1990). To our knowledge, only two studies demonstrated unequivocally direct recognition mechanisms. One involved fusion mechanisms in colonies of a marine invertebrate (Grosberg and Hart 2000; Grosberg and Quinn 1986), and the other studied queen recruitment in a polygynous ant (Keller and Ross 1998). Limited abilities to discriminate among relatives may be explained by several factors:

1. Published results might be biased because it is more difficult to provide evidence for direct

recognition based on genetic cues than familiarity-based discrimination (Grafen 1990).
2. In genetically diverse societies, a direct recognition mechanism may cause rejection errors and thus be selected against (Getz 1991).
3. Dominant individuals could control kin discrimination mechanisms and impose social norms that limit social conflicts within their colony. Such conflicts could indeed compromise the reproduction of dominants (Keller 1997).

#### 15.6.2.3 *Image scoring: when social reputation becomes important*

Nowak and Sigmund (1998) imagined an image scoring mechanism to explain the evolution of reciprocal altruism when it is highly unlikely that beneficiaries of an altruistic act could return the favour to the benefactor. In this case, indirect reciprocity could still evolve: an altruistic individual could be watched and scored by the rest of the population, forge a good reputation, and then be helped on the basis of his good 'image score'. Cooperation would emerge provided individuals can discriminate partners by helping players with a good image score and refraining from helping those with a bad reputation.

The game nonetheless suggests a new dilemma. Assume that a cooperator refrains from cooperating with a defector whose image score is poor. The cooperator then compromises his own image score and runs the risk of being rejected by his former cooperative partners. Despite this dilemma, the mathematical analysis of the game demonstrates that indirect reciprocity can evolve and supplant ambient selfishness (Nowak and Sigmund 1998). The reputation of the individual is a score that is updated after each move and individuals discriminate partners according to a threshold score that can evolve. An unconditional altruist uses the highest possible threshold; an unconditional egoist the lowest, and a conditional cooperator an intermediate value. The evolutionary game dynamics show alternating waves of unconditional altruism, unconditional selfishness and conditional altruism. The persistence of a discriminating strategy based on reputation is then facilitated by selfish strategies that purge unconditional altruists from the population.

Alexander (1986) was perhaps among the first to use the concept of indirect reciprocity as an explanation for moral systems in humans. Indirect reciprocity is indeed based on third-party watching and can prove a strong selective force for the evolution of complex communication networks and judgement systems in humans (Nowak and Sigmund 2005). However, the relevance of indirect reciprocity in non-human societies is still to be demonstrated. In Arabian babblers, helpers seem to compete for a good reputation (see Box 15.3). One reason for the rarity of image scoring tactics may be that even the simplest image-scoring model requires strong cognitive abilities, because individuals need the capacity to observe and memorize the reputation of their partners. Factors that diminish the perfection of this information (e.g. limited group size, imperfect neighbourhood watching, or mistakes) would threaten indirect reciprocity. Yet, cooperation could be restored if these factors cause new communicational tools to evolve (Ferrière 1998; Riolo *et al.* 2001).

## 15.7 Regulation of social conflicts

Buss (1987), Maynard Smith and Szathmary (1995), and Michod (1999) each argued that the hierarchical organization of life (genes, chromosomes, cells, multicellular organisms, and societies) resulted from major evolutionary transitions driven by cooperation: cooperation among genes on the same chromosome, cooperation among cells within a multicellular organism, and cooperation among individuals within a cooperative group. Every individual entity from one hierarchical level (e.g. genes on chromosomes) behaves altruistically: it pays a direct cost but withdraws benefits from the harmonious working of the superior level it belongs to (e.g. the chromosome). In this context, the metaphor of the 'tragedy of the commons' is useful (Hardin 1968): benefits are maximized if everybody invests in the common good, but cooperation can be undermined by cheaters and then no one will benefit from the common good.

The tragedy of the commons is associated with an essential conflict among hierarchical levels of life: selection should favour selfish individual behaviour, although the viability of the whole group requires cooperation. Such conflicts could be regulated by:

1. **Volunteering**, in the form of a cost to integration (see Section 17.7.1), may be imposed upon all individuals, which reduces the benefits of selfishness (Hauert *et al.* 2002).
2. A form of **specialization**, where reproduction is monopolized by few individuals, the probability that selfish individuals take off and may stimulate a stronger competition among groups that tends to favour more cooperative units (Michod 1999).
3. A group may impose some sort of **social control** on the selfish tendency of others (Ratnieks 1988) via coercive actions like group eviction (Johnstone and Cant 1999) or positive actions like bribes (Reeve and Keller 1997).

To understand conditions under which these mechanisms can evolve, one needs to invoke the effects of selection both within and between social groups.

### 15.7.1 Volunteering promotes cooperation

'Public goods games' are simple models of the tragedy of the commons. Here, individuals do not interact directly but can invest in a common pot. The capital of the common pot can rise and be distributed equally and independently from individual investment among contributors. Like in the tragedy of the commons, the optimal gain is obtained when all contributors cooperate and mean gains dwindle when selfishness gets common (Michor and Nowak 2002). Hauert *et al.* (2002) suggested that a form of volunteering could, however, explain the persistence of a mutual cooperation. Volunteering takes here the form of a costly 'ticket' (participation in the common pot would cost the player) that individuals can decline (a strategy called 'loners' in this game). Selfish individuals dominate altruists, but loners outcompete their selfish partners and pave the way for the re-emergence of cooperation. Of course, altruism can only persist in a mixture with selfish individuals and gains are less than optimal; yet, the volunteering mechanism prevents the invasion of selfishness.

### 15.7.2 Specialization

In numerous social species, reproduction is shared unequally among individuals. The degree to which reproduction can be monopolized, also called 'reproductive skew', varies a lot between species and also within species depending on environmental conditions (Emlen 1982; Keller and Reeve 1994; Reeve *et al.* 1998; Sherman *et al.* 1995; Vehrencamp 1983; Figure 15.7). High reproductive skews can be observed in species where a minority monopolizes reproduction (Figure 15.7). Such reproductive skew may regulate social conflicts.

#### 15.7.2.1 *Germ/soma differentiation*

Eusocial species are extreme examples of task specialization where reproduction can be monopolized by a single individual. The organization of multicellular organisms is very similar: sexual reproduction is monopolized by a lineage of cells called the germ line and the efficient reproduction of germ cells relies on the cooperative activities of the somatic line. A tension between self-reproduction and cooperation underlines this germ/soma differentiation, like (once again) a tragedy of the commons. Here, the benefits of mutual cooperation could take various forms. For example, Volvocales are green algae that exhibit both solitary and colonial stages where cooperative cells forego their own reproduction and develop into mobile cells: cooperation then brings the benefit of a more mobile colony (Michod 1999).

In a multicellular organism, cheaters would be cells from the somatic line that would show less restraint in their asexual reproduction and participate less in cooperative tasks. Cheaters have been described in colonies of Volvocales, and Bell (1985) for instance found that cooperative colonies replicate faster than non-cooperative ones, thus confirming the basic assumption of the tragedy of the commons

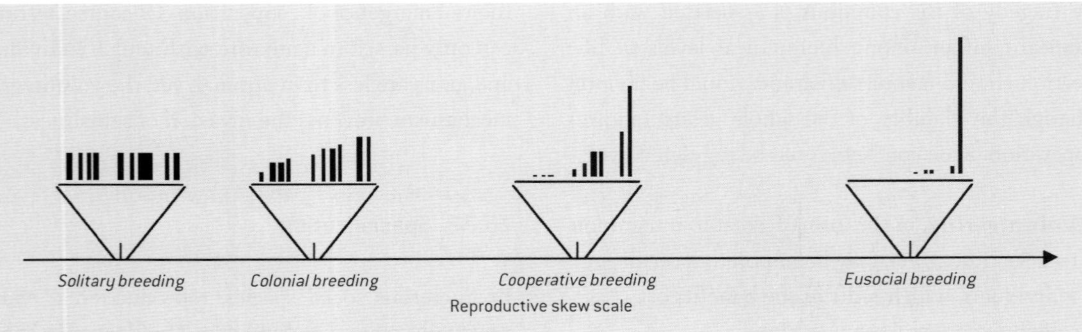


Figure 15.7 **Reproductive skew in animal societies**

The reproductive skew measures how reproduction is shared among adult group members and can be used to scale species along a sociality continuum (Sherman *et al.* 1995). Some species have a low reproductive skew so that most members obtain direct reproduction success, e.g. solitarily breeding species (e.g., a great tit, *Parus major*) or colonial breeding (e.g., a paper wasp, *Polistes* sp.). In others, skew is high and only a few individuals monopolize breeding, e.g. cooperatively breeding species (e.g., Seychelles warbler *Acrocephalus sechellensis*) and eusocial species (e.g., the naked mole-rat, *Heterocephalus glaber*). In practice, the reproductive skew index measures the asymmetry of the distribution of reproduction success, here represented by the breeding success (black bars) of 10 representative individuals scaled along a social ranking gradient from the left to the right. In solitarily breeding or colonial species, the distribution is uniform and the skew is low; in cooperatively breeding or eusocial species, the distribution can be extremely asymmetric and the skew is high.

(Hardin 1968). What could fix this system in a cooperative state and avoid evolutionary cycles predicted by the 'public goods games' (Hauert *et al.* 2002) mentioned before? Multi-level selection is probably essential to understand the persistence of cooperation (Wilson 1997). The common goods in this case takes the form of a higher-order unit (organism) whose replication is ensured by the cooperation of lower-order units (cells). The differentiation between germ and somatic lines confers a strict heritability to the higher-order unit, because each organism can only be founded by a single germ cell. Selfishness can result either from a germinal mutation, the mutant germ cell then establishes a non-cooperative organism, or from a somatic mutation, the mutant somatic cell then develops its group inside the organism. The germ/soma differentiation allows regulating of selfishness:

1. by limiting the risks of germinal mutations and reducing the growth of non-cooperative colonies;
2. and by eliminating the risks that somatic mutations propagate: somatic mutant cells are doomed to disappear after the death of the higher-order organism they develop in.

#### 15.7.2.2 *Repression of social parasites*

In social species, specialization is much more complex than a simple differentiation between germ and soma. In eusocial animals, worker sterility is usually accompanied by numerous cast specializations (e.g. soldiers, foragers, cleaners) that make the colony more efficient (Wilson 1971). A model describing the evolution of metabolic networks allows understanding of why such complex and specialized societies can be more immune to selfish parasites (Czaran and Szathmary 2000; Szathmary and Demeter 1987). In the metabolic network, each class of enzymes contributes to the synthesis of a small molecule (monomer) towards a common metabolic pot. In return, enzymes benefit from complex products synthesized by the metabolic network with the common metabolic pot, but only if all necessary monomers have been sent to the common pot. This cooperative specialization enables the metabolic

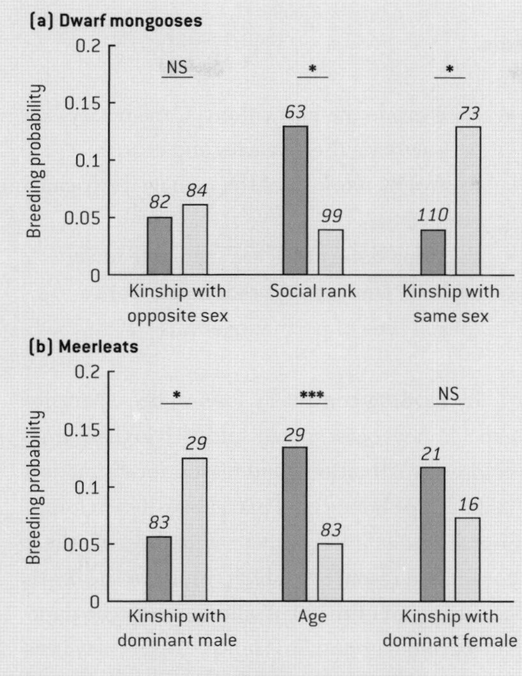


Figure 15.8 **Reproductive skew in social mongooses**

Data shown are the breeding probability of subordinates, with high values indicating shared breeding and low reproductive skew (see Figure 15.7).

**a.** Breeding probability in subordinate dwarf mongooses (*Helogale parvula*) according to kinship with opposite-sex dominant individual (dark bar: dominant is a close kin; grey bar: dominant not a close kin), social ranking of the subordinate (dark bars: second and third rank; grey bars: lower ranks), and kinship with the same-sex dominant individual (dark bars: dominant is a close kin; grey bars: dominant not a close kin).

Data kindly communicated by B. Keane and published in Keane *et al.* (1994).

**b.** Breeding probability in female subordinate meerkats (*Suricata suricatta*) according to kinship with the dominant male (dark bar: dominant is a close kin; grey bar: dominant not a close kin), age of the subordinate (dark bars: same generation as the dominant; grey bars: subordinate younger than the dominant), and kinship with the dominant female (dark bars: dominant is a close kin; grey bars: dominant not a close kin). Dark bars: high values; grey bars: low values.

After Clutton-Brock *et al.* (2001).
*, $P > 0.05$; ***, $P < 0.001$.

network to resist invasion by parasitic enzymes (Czaran and Szathmary 2000). A frequency-dependent selection is responsible for this: the parasite cannot grow when it is abundant because its metabolic neighbourhood is incomplete. Furthermore, the persistence of a cooperative metabolic network is highly dependent upon population mixing, and thus high diffusion of enzymes. This counter-intuitive effect (see Section 15.5.2) is because the benefits of cooperation are obtained only if metabolic networks are complete, which is more likely when enzymatic products move a lot. However, the metabolic network cannot completely wipe out parasites. A possible alternative is that parasites are domesticated in new metabolic networks, for example because of the parasites contributing to novel monomers. Domestication of social parasites may be a singular, yet widespread, solution to reduce conflicts within groups and could lead to more efficient 'cooperation markets' (Ferrière *et al.* 2002; Noe and Hammerstein 1995).

### 15.7.3 Social control

Task sharing is often controlled by competition among dominants and subordinates. In humans, social and moral norms can be enforced by religious or administrative authorities. In non-human social species, this control can be exerted in three different ways:

1. A control benefiting subordinates (see Ratnieks and Visscher 1989; Ratnieks 1988). Dominants destroy eggs produced within their cast and a

Box 15.5 **Immigrant bees evade control mechanisms**

In eusocial Hymenoptera, workers are usually unable to breed in the presence of the queen. Anarchist workers can produce unfertilized male offspring that other workers usually kill, and their behaviour is thus unlikely to threaten the colony (Barron *et al.* 2001; Ratnieks 1988). In 1990, bee-keepers from South Africa transplanted colonies of the wild Cape honey bee (*Apis mellifera capensis*) towards the northern parts of the country. After this transfer, the Cape honey bee started to live as a parasite on local colonies of the managed African honey bee *A. m. scutellata*, which led to a major collapse of the honey production in this area. Martin *et al.* (2002) showed that Cape honey bees were able to evade social regulation by the managed African honey bee:

- immigrant Cape honey bees invade foreign colonies without being expelled by guard bees (Downs and Ratnieks 2000);
- once inside the host colony, Cape honey bees activate their ovaries and escape control by workers of the managed African honey bee;
- the Cape honey bees are then able to lay diploid (female) eggs that result from a special type of parthenogenesis.

By evading control by foreign bees, parasite cape honey bees spread and drive their host colony to extinction. One reason for extinction is that Cape honey bees participate less in foraging activities and compete with host workers for limited resources, which results in the death of the queen of the host colony when the parasite becomes abundant. Hence, Cape honey bees develop like a 'social cancer'. Yet, a fundamental difference between this cancer and that of multicellular organism is that 'tumorous' workers originate from outside the host colony.

failure of this mechanism can cause a form of 'cancer' that has now been described in honey bees (Martin *et al.* 2002; see Box 15.5).

2. A remarkable joint control between dominants and subordinates, as in 'queenless' ants (Monnin and Peeters 1999).
3. A control benefiting dominants through repression (Bennett and Faulkes 2000; Heinze *et al.* 1994; Röseler 1991) and concessions (Reeve *et al.* 1998) that refrain reproduction in subordinates.

#### 15.7.3.1 *Cooperative control in 'queenless' ants*

Ponerine ants have lost their queen cast during evolution: all females are workers that maintain the ability to breed, and breeding workers are called **gamergates**. In the species *Dinoponera quadriceps*, colony size averages 80 adult workers and only one dominant female gamergate is able to breed. However, two to four high-ranking workers can be pretenders to the breeding status: these pretenders usually work less and eventually replace the gamergate after she dies. Only one male fertilizes the gamergate, and workers are all daughters or sisters of the gamergate (Monnin and Peeters 1999).

A high-ranking female worker can increase her inclusive fitness (see Section 15.5.1) by replacing the gamergate. When a high ranking female undertakes to challenge the gamergate, the gamergate skims the candidate with her sting and marks her opponent chemically. The end result is immobilization of the candidate female by low-ranking workers, an immobilization that can last for several days and after which the candidate has lost her high-ranking position (Monnin and Peeters 1999; Monnin *et al.* 2002). Cooperation between the gamergate and low-ranking females makes sense here because (1) both are more related to the offspring of the gamergate than they would be to the offspring of the high-ranking female, and (2) the costs for the colony associated with the eviction of the gamergate can be substantial (Monnin and Peeters 1999). This behaviour is an example of conflict regulation by punishment (Clutton-Brock and Parker 1995).

#### 15.7.3.2 *Dominant control*

In cooperatively breeding species, monopolization of breeding by dominant individuals can be incomplete and two processes may favour a more symmetric distribution of reproduction: dominants may have limited control on subordinates (Reeve *et al.* 1998), or dominants may benefit from conceding reproduction opportunities (Clutton-Brock 1998). Concessions are called incentives for philopatry when they act to reduce departures of subordinates, and peaceful incentives when they avoid harmful conflicts (Reeve and Ratnieks 1993). Concessions are modulated by three main factors:

1. **Competitive ability of subordinates**. When subordinates compete on equal grounds with dominants, dominants may make peaceful concessions.
2. **Ecological constraints on independent breeding**. The option to disperse and breed solitarily would pay more when ecological constraints are weak, which should promote incentives for philopatry (Vehrencamp 1983).
3. **Relatedness between dominants and subordinates**. A lower relatedness would select for more concessions, because the subordinate gets less indirect genetic benefits from cooperating with dominants (Keller and Reeve 1994) and the risk of inbred mating is lower (Emlen 1996).

Contrasted mechanisms of social control can exist even in phylogenetically close species.

#### 15.7.3.3 *Concessions in dwarf mongooses*

Dwarf mongooses (*Helogale parvula*) are cooperatively breeding mammals living in small colonies of 3 to 18 individuals in savannas and open forest from Central Africa. Small colonies comprise one breeding pair and their offspring, whereas larger colonies form after the fusion of smaller related groups. Breeding in subordinates can be suppressed by the dominant breeding pair (Creel *et al.* 1992) and patterns of breeding in subordinates are well predicted by a model of optimal concessions (Vehrencamp 1983). Subordinates mate together irrespective of inbreeding (Figure 15.8a) and inbred matings are not deleterious (Keane *et al.* 1990). However, older subordinates have a preferential access to breeding because they are competitively superior; this might be caused by peaceful incentives from dominant. Also, subordinates gain preferential access to breeding when they are less related to the dominant (Figure 15.8a; Keane *et al.* 1994). Thus, this species is characterized by a dominant breeding pair that concedes breeding opportunities to subordinates when this reduces potential conflicts within the group.

#### 15.7.3.4 *Limited control in meerkats*

Meerkats (*Suricata suricatta*) are also social mongooses. Their small colonies (2–30 individuals) live in semi-arid habitats in southern Africa and comprise one breeding pair, several subordinates, and year-born offspring. Group members cooperate for vigilance and for feeding offspring, but reproduction is monopolized by the dominant breeding pair (Clutton-Brock *et al.* 1998, Clutton-Brock *et al.* 1999). The long-term monitoring of a single population from the Kalahari area has allowed a detailed description of conditions under which subordinates may breed (Clutton-Brock *et al.* 2001). The end result is unequivocal evidence that subordinate reproduction is limited by inbreeding risks in this species and limited control by the dominant female (Figure. 15.8b).

## 15.8 Evolutionary loss of sociality

The advance of molecular phylogenies has greatly contributed to our understanding of sociality (Choe and Crespi 1997) and recent phylogenetic data have unambiguously revealed that complex social structures

| Study group | Number of evolutionary losses | Reference |
|---|---|---|
| Bees (Halictine, Allodapine, and Auglocorhine) | 5–6 | Wcislo and Danforth 1997 |
| Aphids | 1–2 | Stern and Foster 1997 |
| Thrips | 1–2 | Crespi 1996 |
| Shrimps (*Synalpheus*) | 1 | Duffy *et al.* 2000 |

Table 15.3 **Examples of evolutionary loss of eusociality**

may be lost, suggesting that evolution towards eusociality is not irreversible (Wilson 1975). Here, we discuss the factors that might cause evolutionary reversals of sociality.

### 15.8.1 Phylogenetic evidence

The hierarchical classification of social structures (Table 15.2) has been criticized since it implies a step-by-step increase in complexity of social structures and suggests that evolution proceeds towards the evolutionary summit of eusociality (Wilson 1975). Evolutionary reversals of eusociality are indeed still to be discovered in highly social taxa like Isoptera and some Hymenoptera (Formicidae, Apini, or Bombini) and in the mammalian group of mole-rats (Bathyergidae), suggesting that eusociality is a robust endpoint in these groups. Yet, evolutionary losses of eusociality have now been observed in several other taxa (see Crespi 1996; Table 15.3 updated after Wcislo and Danforth (1997)).

When comparing evolutionary transitions in and out of eusociality, it appears that evolutionary losses have been observed in the history of social life, although relatively rarely (8–11 compared with the 25–27 hypothesized transitions towards eusociality; see Duffy *et al.* (2000)). Similarly, other aspects of sociality such as cooperative breeding, collective foraging behaviour or colonial life may have been lost repeatedly (Rolland *et al.* 1998; Wcislo and Danforth 1997; Varela *et al.* 2007). For example, a reconstruction of the phylogenetic distribution of cooperative breeding in birds by Arnold and Owens (1999) suggests two independent evolutionary losses of cooperation in lyrebirds and bowerbirds, while several losses may have occurred in the Corvoidea superfamily (Edwards and Naeem 1993). Detailed phylogenies within restricted groups of birds gave similar results (Edwards and Naeem 1993). Future comparative analyses should focus on trying to identify the selection pressures that caused these evolutionary losses of social traits (Rolland *et al.* 1998; Varela *et al.* 2007). Mathematical modelling gives useful tips for future empirical research on this topic and identifies three evolutionary mechanisms that may cause evolutionary reversals in sociality (Le Galliard *et al.* 2003, 2005):

1. Environmental changes may cause a reversion in selection pressures.
2. Evolution of social traits may feedback on **ecological dynamics** and cause a shift towards an ecological equilibrium where sociality is lost.
3. Social traits may co-evolve with other life history traits that cause evolutionary reversals of sociality.

### 15.8.2 Environmental changes

Environmental changes that diminish the direct benefits of altruism (e.g. changes in habitat quality, predation risk) or the indirect benefits of altruism (e.g. changes in habitat structure) could cause secondary evolution of selfishness. This might explain the loss of sociality observed along climatic gradients in bees (see Danforth and Eickwort 1997; Eickwort *et al.* 1996). Such environmental changes caused adaptive reversals in the social structure of the Myxobacteria (*Myxococcus xanthus*) a biological model amenable to artificial evolution experiments in the laboratory. The experiment contrasted the social habits of bacteria maintained in a rich, liquid, well-mixed culture medium (unstructured habitat) with those of bacteria maintained in a poor, solid, spatially structured culture medium (structured habitat; Velicer *et al.* 1998). After 10,000 generations of evolution in the laboratory, the strains maintained in the unstructured habitat had lost their social behaviour (collective foraging and production of fruiting bodies). Asocial **phenotypes** struck back and strived in a rich and homogeneous habitat that tended to favour selfishness. Detailed analyses suggested that these changes were caused by changes in a genetic system controlling social mobility (Velicer *et al.* 2000).

### 15.8.3 Ecological feedbacks

Bistability (i.e. existence of two stable ecological equilibriums) is a common feature of social species because of an **Allee effect** induced by individual investment in altruism (Courchamp *et al.* 1999). Starting from a small initial group, the growth rate of a population of altruist individuals is negative: low group size implies low performances and a lack of mating opportunities (see Section 15.3.4). However, if the group starts with a sufficiently large size, the population may reach a viable equilibrium. In highly mobile species, Le Galliard *et al.* (2003) have shown that this ecological bistability can be lost under a threshold level of investment into altruism where the population is doomed to extinction whatever the initial population size. Furthermore, selection pressures favour the evolution towards lower altruistic investment in that case and thus cause an 'evolutionary suicide'! An initially viable and altruistic population would then be driven by adaptation towards extinction, involving the loss of sociality. To understand the generality of these dynamics, the consequences of ecological feedbacks should be explored more systematically.

## 15.9 Conclusion

After decades of involved debates over the evolutionary origin of altruistic behaviour (see Chapter 1), the mere identification of the selection pressures acting on altruistic behaviour remains problematic (Clutton-Brock 2002). The development of numerous analytical models has shed light on the fine balance of costs and benefits that may be involved in the evolution of altruism. Nowadays, it is admitted that both direct benefits and indirect costs of altruism have been neglected and that the whole series of costs and benefits depends on a genetic and ecological context that can hardly be ignored (Clutton-Brock 2002; West *et al.* 2002). Long-term studies of extended families in cooperatively breeding birds exemplify these modern economic studies of altruistic behaviour (Cockburn 1998), that are now considered in the more general framework of the comparative analysis of behavioural and life history syndromes (Arnold and Owens 1998, 1999).

The apparent weakness of this adaptionist programme as originally conceived by sociobiology (Wilson 1975) has long been due to our ignorance of the genetic bases of altruistic traits as well as to a poor understanding of their plasticity. Recent advances in molecular genetics of social insects or micro-organisms and experimental studies on behavioural flexibility have allowed us to overcome these serious limitations (Crespi 2001; Keller and Chapuisat 1999).

The contribution of studies first designed to understand the evolution of social behaviours has indeed been prolific. The behavioural ecology of social traits has developed largely beyond the scope of animal ecology and metamorphosed into a much broader research programme that now provides functional solutions for the origin and evolution of hierarchical levels of life themselves: from the origin of genetic information up to the evolution of languages in human societies (Maynard Smith and Szathmary 1995). Cooperation thus appears as a major evolutionary force that allows the emergence of integrated units capable of self replication and where conflicts between lower level units can be repressed (Frank

2003). Studies on altruistic behaviour have pointed out that it is fundamental to recognize the action of natural selection throughout hierarchical levels of life in order to understand better how competition can be regulated. This notion of regulation is probably not the most appropriate concept to describe the precarious maintenance of cooperation in front of never ending attacks by selfish elements that can only be restrained within the limits allowed by recognition, repression and domestication mechanisms.

Whereas Darwin laid down the principle of speciation by natural selection (Darwin 1859), Buss (1987), Maynard Smith and Szathmary (1995), and Michod (1999) have all turned the evolution of cooperation into a central question for the study of the increasing complexity of life. In fact, these two issues may not be too distant if one follows Margulis and Dorion's (2002) interpretation of speciation as a result of cooperation among **genomes**. This remains an open question.

## » Further reading


> *Recommended books:*

**Cockburn, A.** 1991. *An Introduction to Evolutionary Ecology*. Blackwell Science, Oxford.

**Gouyon, P.-H., Henry, J.-P. & Arnould, J.** 1997. *Les Avatars du Gènes*. Belin, Paris.

**Maynard Smith, J. & Szathmary, E.** 1995. *The Major Transitions in Evolution.*

**Michod, R.E.** 1999. *Darwinian dynamics – Evolutionary Transitions in Fitness and Individuality.* Princeton University Press, Princeton, NJ.

**Wilson, E.O.** 1975. *Sociobiology: The New Synthesis*. Harvard University Press, Cambridge, Massachusetts.

**Zahavi, A. & Zahavi, A.** 1997. *The Handicap Principle: A Missing Piece of Darwin's Puzzle.* Oxford University Press, Oxford.

> *Key references on game theory and the evolution of cooperation:*

**Axelrod, R. & Hamilton, W.D.** 1981. The evolution of cooperation. *Science* 211: 1390–1396.

**Ferrière, R.** 1998. Help and you shall be helped. *Nature* 393: 517–518.

**Milinski, M., Semmann, D. & Krambeck, H.-J.** 2002. Reputation helps solve the 'tragedy of the commons'. *Nature* 415: 424–426.

**Nowak, M. A. & Sigmund, K.** 1998. Evolution of indirect reciprocity by image scoring. *Nature* 393: 573–577.

**Wedekind, C. & Milinski, M.** 2000. Cooperation through image scoring in humans. *Science* 288: 850–852.

> *Key references on kin selection and kin recognition:*

**Charnov, E.L. & Krebs, J.R.** 1975. The evolution of alarm calls: altruism or manipulation? *American Naturalist* 109: 107–112.

**Griffin, A.S. & West, S.A.** 2002. Kin selection: fact and fiction. *Trends in Ecology and Evolution* 17: 15–21.

**Keller, L.** 1997. Indiscriminate altruism: unduly nice parents and siblings. *Trends in Ecology and Evolution* 12: 99–103.

**Komdeur, J. & Hatchwell, B.J.** 1999. Kin recognition: function and mechanism in avian societies. *Trends in Ecology and Evolution* 14, 237–241.

**Sherman, P.W., Reeve, H.K. & Pfennig, D.W.** 1997. Recognition systems. In: *Behavioural Ecology: An Evolutionary Approach* (Krebs, J.R. & Davies, N.B. eds), pp. 69–96. Blackwell Science, Oxford.

> *Key references on cooperative breeding:*

**Clutton-Brock, T.** 2002. Breeding together: kin selection and mutualism in cooperative vertebrates. *Science* 296: 69–72.

**Cockburn, A.** 1998. Evolution of helping behaviors in cooperatively breeding birds. *Annual Review of Ecology and Systematics.* 29: 141–177.

**Emlen, S.T.** 1982. The evolution of helping. I. An ecological constraints model. *American Naturalist* 119: 29–39.

> *Key references on population dynamical consequences of cooperation:*

**Courchamp, F., Clutton-Brock, T. & Grenfell, B.** 1999a. Inverse density dependence and the Allee effect. *Trends in Ecology and Evolution* 14: 405–410.

## » Questions

1. There is a long-standing controversy over the relative importance of kin selection versus reciprocity to explain the evolution of cooperation in animal societies. Highlight the critical differences between these two explanations of the evolution of social behaviour and suggest relevant theoretical or empirical studies that would help to disentangle these two explanations.

2. Use the replicators' dynamics described in Box 15.2 to reconstruct the frequency dynamics of cooperators and defectors for the games described in this box. Once you have analysed these frequency dynamics for well-mixed populations, conduct a thought experiment about the frequency dynamics for spatially structured populations and make predictions about the evolution of cooperation.

3. In Section 15.4, we showed how field studies can be used to measure the costs and benefits of helping in cooperatively breeding birds and suggested that such studies are critical if we are to distinguish altruism from other seemingly cooperative behaviours. Quite recently, Clutton-Brock (2002) reviewed the direct and indirect benefits of helping behaviour in a broad range of cooperatively breeding vertebrates. Summarize the main conclusions of that study.

4. Spiteful behaviour is social behaviour that imposes costs to both the actors and the recipients. Use Hamilton's rule defined in Section 15.5.1 to derive conditions for the evolution of spite and discuss these conditions.

5. What types of information would be needed to measure relatedness in the wild and test Hamilton's rule for the evolution of altruistic behaviours?

# 16

# 16

# Communication, Sensory Ecology, and Signal Evolution

Marc Théry and Philipp Heeb

## 16.1 Introduction

Animal **communication** has always been a central theme for ethologists as well as for psychologists, behavioural ecologists, and neurophysiologists. People have long recognized that **information** is being transmitted in the song of birds, the roaring of deer, and the mobbing of foxes by crows. They have also observed that information was being used by sexual partners, competitors, and predators. We will see that studying the way in which animals communicate and the nature of the **signals** they produce, and coupling all this with studies of their physiology and sensory ecology will provide many valuable insights into evolutionary processes (Maynard Smith and Harper 2003). Examples are provided in the next paragraph and throughout this chapter. There are at least three important reasons for behavioural ecologists to study and understand the evolution of communication.

First, the evolution of sociality has relied in one way or another on communication among individuals living in the same group. The formation and functioning of animal societies depend on the exchange of information required to maintain group cohesion, interactions with other species and constant gathering of information. We have already seen in Chapter 4 that information can be extracted from very diverse detectable facts. Furthermore, it can reveal numerous motivational states and be used by receivers as diverse as conspecifics, predators, or prey. Information content is often a trade-off between the forces of natural and sexual selection and its value must be measured in terms of its fitness costs and benefits (Maynard Smith and Harper 2003). Finally, the communication strategies of other group members can also influence the value of theses strategies and this means that a game theoretical approach will have to be used (Johnstone 1998).

Secondly, the nature and extent of communication among animals can be used to answer evolutionary questions. For example, because of phylogenetic inertia it becomes possible to use the comparative analysis of signal transmission and receiving systems to determine the phylogenetic relationships between species. Signals also result from multiple adaptations to the environment: predators and the physical properties of the environment exert important constraints on the nature and shape of signals. As a result, the study of communication signals constitutes an important field of study on the action of both natural and sexual selection on individual behavioural phenotypes.

Thirdly, communication is one of the first components allowing the emergence of cultural transmission, leading to the transmission of information across generations in parallel with genetic transmission (see Chapter 20). It is thus necessary to take into account the process of communication so as to understand the relative contribution of cultural transmission in evolutionary processes.

Finally, by their function of long-distance communication, signals are very useful for researchers and managers of natural resources when censusing species, evaluating the state of the environment by the presence of indicator species, or in fighting against parasitic pests.

## 16.2 Concepts in the study of communication

### 16.2.1 Different ways to understand communication

Through time and according to different authors, the scientific definition of communication has varied greatly (Figure 16.1).

> Communication is considered as any interaction taking place between an individual acting as a **sender** who delivers some information to another individual that acts as a **receiver** who uses this information to make a decision. This is **communication in the broadest sense**. We have seen in Chapter 4 that information that does not benefit the sender can be generated inadvertently by individuals and be used either by conspecifics, predators or parasites. Detectable facts from which bystanders can extract information but that do not benefit the sender are called **cues**; when they benefit the sender they are called **signals** (see Chapter 4 for more details on the concept of information).

> A stricter definition adds two conditions to this definition. The first is **intentionality**, meaning that the sender benefits, in terms of fitness, by the signal emission without implying that it is conscious of its decision. The second condition concerns the benefits to the receiver: communication has to be **honest** so that the receiver can also benefit from the information being transmitted in order to make the best decision. When these two conditions are fulfilled we can speak of true communication (Marler 1977).

Often during an exchange of information between individuals the interests of the two parties differ, leading to conflicts that generate different forms of communication. For example, when information sharing entails a cost to the sender, a form of **disinformation** from the sender towards the receiver may take place. In turn, the receiver can adjust its response to the signal produced by a sender in relation to the benefits generated by the use of that information. There is **eavesdropping**, when an individual that is not involved in an action or an interaction (often called the bystander) extracts and uses the information carried by cues or signals produced. When eavesdropping provides an advantage to the bystander at a cost to the actors of the action or

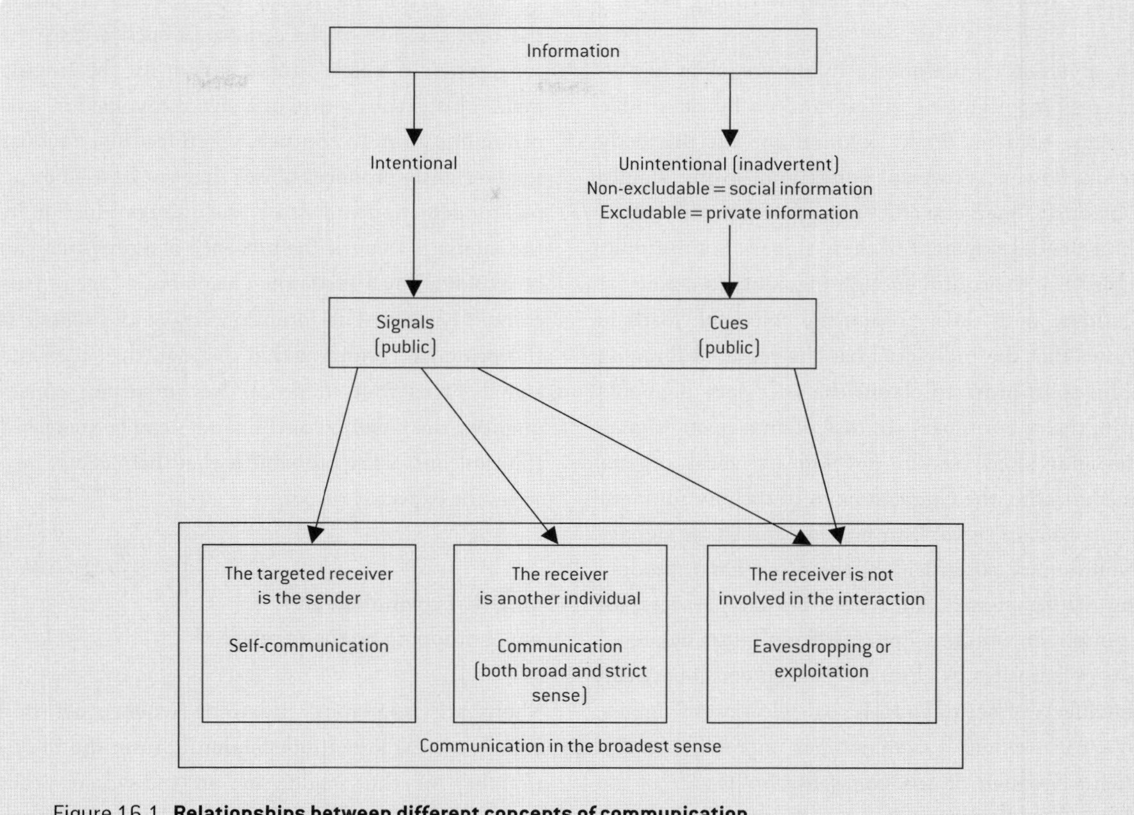


Figure 16.1 **Relationships between different concepts of communication**

Certain terms used in this figure are defined in Chapter 4. For a detailed definition of the various forms of biological information, see Figure 4.3.

interaction, we speak of **exploitation**. In general, and contrary to exploitation, eavesdropping does not imply direct costs for the sender.

The study of communication has long examined intra-specific relationships (Munn 1986). Recently, however, researchers have recognized that alarm calls, for example, can be used by other species to avoid predators (Cheney and Seyfarth 1985). In both situations, disinformation has the potential to evolve (Munn 1986; Møller 1988). We also distinguish two categories of signals (Green and Marler 1979). **States** are signals permanently perceptible such as colourful plumages and persistent body odours. **Events** are produced over a short time span. Examples of events are songs, alarm calls, electrical discharges or the short and sharp release of odour. Finally, the notion of **auto-communication** covers the mechanisms of echolocation and electrolocation and can be interpreted as a refined form of cue detection. Nevertheless, signals of auto-communication fit within the general laws of communication and their evolution is also addressed in this chapter.

In summary, communication in the broad sense includes the intentional or inadvertent production of information that is used by individuals of the same or different species, as well as the gathering of information on the properties of the environment (Bradbury and Vehrencamp 1998). The term **information transfer** found in the literature is even more general than communication in the broad sense since it describes all situations where information passes between individuals without considering any assumption on the benefits obtained by the participants of the interaction.

### 16.2.2 Natural and sexual selection of signals

To advertise its presence a living organism has an interest in producing strong and easily detectable signals, such as bright colourations on the body or on flowers, powerful songs and strong smells. Obviously, energetic and environmental constraints limit the development of these signals. For example, a bird has to acquire enough pigments to colour its feathers, and thus presenting colourful feathers shows that the individual has the required foraging abilities in order to obtain the necessary pigments from the environment. In case of an acoustic signal, the individual has to develop its vocal organs. Furthermore, the characteristics of the environment can constrain signal transmission and shape signals' evolutionary potential. On the other hand, producing strong signals, to attract sexual partners for example, exposes the sender to exploitation by predators or parasites. A close association exists between the effects of natural and sexual selection of signals, to a point where it is sometimes impossible to distinguish which of the two selection modes is the most important. Signals are considered as being under the action of **natural selection** when they provide an advantage in terms of survival for the sender. In contrast, **sexual selection** for signals is considered for signals providing an advantage in terms of access to reproductive opportunities including mating displays, mate choice, copulations, and fertilizations (Chapter 11). This distinction seems clear enough, but we will see that sexual selection of signals is often constrained by natural selection. Below we expand and illustrate the principal concepts presented above and summarized in Figure 16.1.

### 16.2.3 Intentionality: benefits for the sender

The first criterion generally proposed to characterize true communication is the existence of some fitness benefit for the sender: signal production is not incidental; it is intentional because the response of the receiver affects the sender's **fitness** positively. For example, the song of a passerine in spring has the effect of attracting sexual mates, thus favouring the reproduction of the sender. In such a case, the song provides a clear fitness benefit for the singing male. This is thus a case of a signal involved in true communication. In contrast, when feeding, a mouse inadvertently produces noises that can be used by an owl for detection and attack: the noise production by the mouse is a cue of the presence of a prey that can be exploited by a predator. This is not a case of true communication. The intentionality of the sender is the principal condition that distinguishes cues and signals. This first condition has important consequences: the benefits for the sender can be increased if it can control the probability that the receiver provides the expected response.

### 16.2.4 Exploitation and eavesdropping by the receiver

There are numerous situations where cues and signals are exploited independently from the intentionality of the sender by an individual other than the receiver initially targeted by the signal. When the sender suffers no cost from this we are dealing with eavesdropping (also called indiscretion; Doutrelant and McGregor 2000; Doutrelant *et al.* 2001; McGregor and Peake 2000) and when sender suffers a cost it is exploitation (Figure 16.1). Certain examples, in the context of sexual selection, of predator–prey or host–parasite interactions have demonstrated the occurrence of indiscretion or exploitation by the receiver (see for instance Chapter 17).

#### 16.2.4.1 *Predators exploiting prey and parasites exploiting hosts*

The exploitation of cues of many kinds by predators is particularly frequent. For example, the common kestrel (*Falco tinnunculus*) can detect urine and faeces left by voles (*Microtus agrestis*; Figure 16.2) during their territorial markings. The odour-based marks left by the voles reflect in the ultraviolet (UV) wavelengths UV that can be perceived by the falcons. Accordingly, predators concentrate their hunting

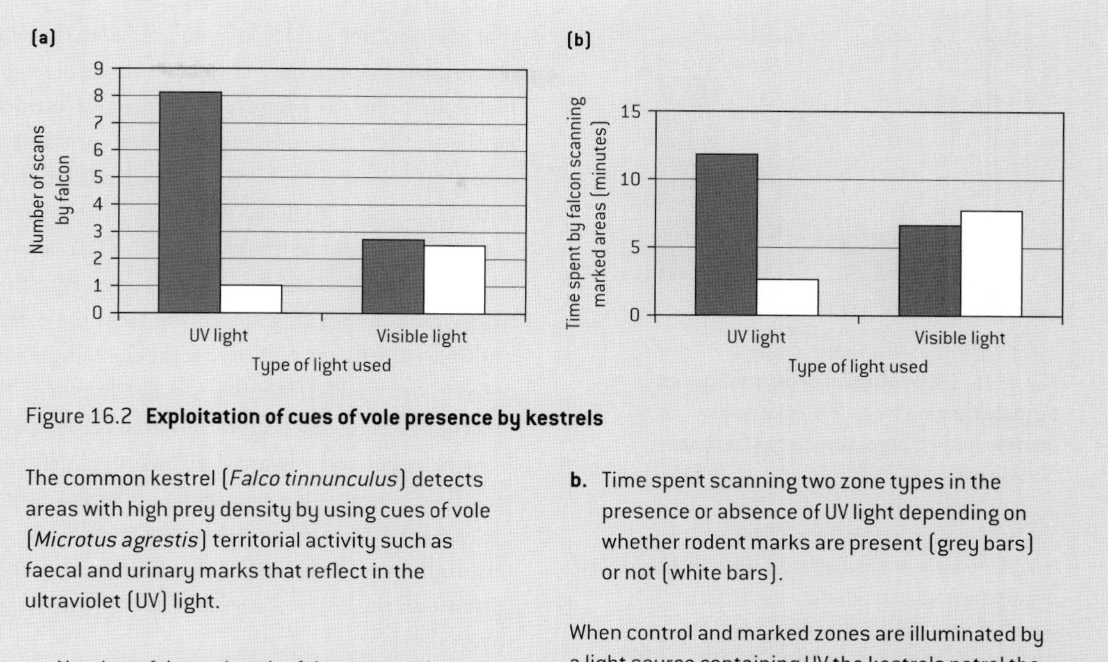

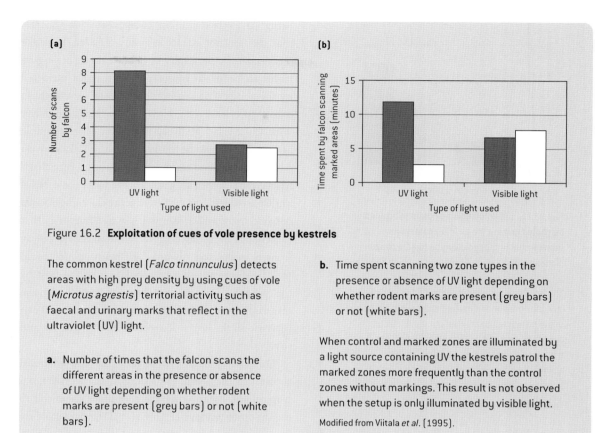

Figure 16.2 **Exploitation of cues of vole presence by kestrels**

The common kestrel (*Falco tinnunculus*) detects areas with high prey density by using cues of vole (*Microtus agrestis*) territorial activity such as faecal and urinary marks that reflect in the ultraviolet (UV) light.

**a.** Number of times that the falcon scans the different areas in the presence or absence of UV light depending on whether rodent marks are present (grey bars) or not (white bars).

**b.** Time spent scanning two zone types in the presence or absence of UV light depending on whether rodent marks are present (grey bars) or not (white bars).

When control and marked zones are illuminated by a light source containing UV the kestrels patrol the marked zones more frequently than the control zones without markings. This result is not observed when the setup is only illuminated by visible light.

Modified from Viitala *et al.* (1995).

effort on high UV reflecting zones as it reveals high prey density. Predators can also use certain cues from the visual appearance of prey in the form of **search images** that allow them to be more efficient in their search. The search image is developed by trial and error and leads to a change in the animal's perceptual system, making it capable of detecting hidden prey more effectively (Shettleworth 1998).

There are numerous cases of signal exploitation by predators, the most famous being the predation of male tungara frogs, *Physalaemus pustulosus* by predatory bats *Trachops cirrhosus*. These bats locate male frogs based on their calls, in particular they use one sound that Michael Ryan and his collaborators call the 'Chuck call' (Ryan *et al.* (1982); Figure 16.3). In response to this signal exploitation by a predator, male frogs reduce the emission frequencies of the 'chuck', in particular if they are isolated from other males. We may ask, why do male use such a dangerous call? Shouldn't we expect natural selection to eliminate the production of a call that increases predation risk? Part of the solution to this paradox resides in the fact that females are also preferentially attracted by the calls of males containing a 'chuck' at the end of a 'whine'. Males that do not use the 'chuck' at the end of the 'whine' remain safe but do not attract females, initially the main function of call production by males. To solve this problem, males sing with a low frequency of 'chuck' when they are alone and increase their 'chuck' frequency when calling in groups where their predation risk is diluted (see Chapter 14).

Predators and parasitoids often exploit acoustic signals, but exploitation can also be based on visual and olfactive signals. For example, the parasitoid flies *Euphasiopteryx ochracea* lay eggs on the crickets, *Gryllus integer*, that they localize by their songs (Cade 1975; Figure 16.4). Silent males have lower risks of

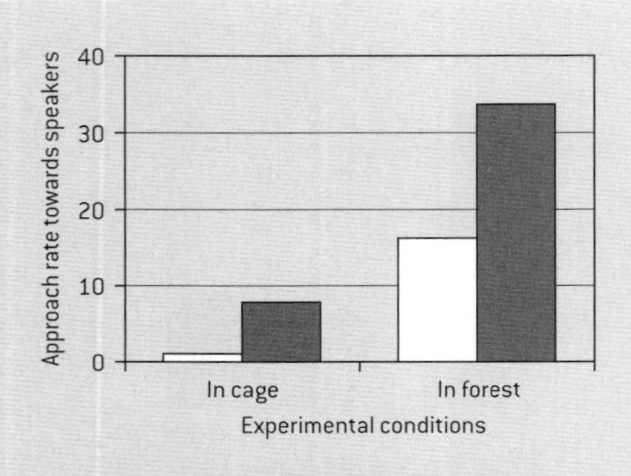


Figure 16.3 **Exploitation of calls produced by male tungara frogs (*Physalaemus pustulosus*) by predatory bats (*Trachops cirrhosus*)**

The predatory bats detect male frogs by a 'chuck' component in their calls. Female frogs are also attracted by the 'chuck'. Bats, in captivity or in the wild, approach more frequently loudspeakers diffusing the complex calls 'whine-chuck' (grey bars) than speakers with only the 'whine' (white bars).

Based on Ryan *et al.* (1982).

being parasitized and to compensate for the fact that they do not attract as many females they take position at the periphery of the territories of singing males in order to intercept the females attracted by the songs. The acoustic organ used by the fly to detect the cricket song shows a remarkable similarity with the cricket auditory organ. Another example of exploitation can be found in the flies *Colcondamyia auditrix*, which find cicadas by their songs. Parasitized male cicada cannot sing and this reduces the cases of multiple parasitisms while reducing the reproductive behaviours of males. We will see in Chapter 17 that this change in resource allocation from the host away from reproduction towards self-maintenance could well be the result of a manipulation of the host by the parasite. The parasite benefits directly by the interruption of the hosts' reproduction because it can then use all the host resources for its own growth and survival.

Other flies detect their hosts by the chemical substances they emit. For example, the workers of a tropical ant, *Paraponera clavata*, produce chemicals to recruit other workers (Figure 16.5). The parasitoid fly

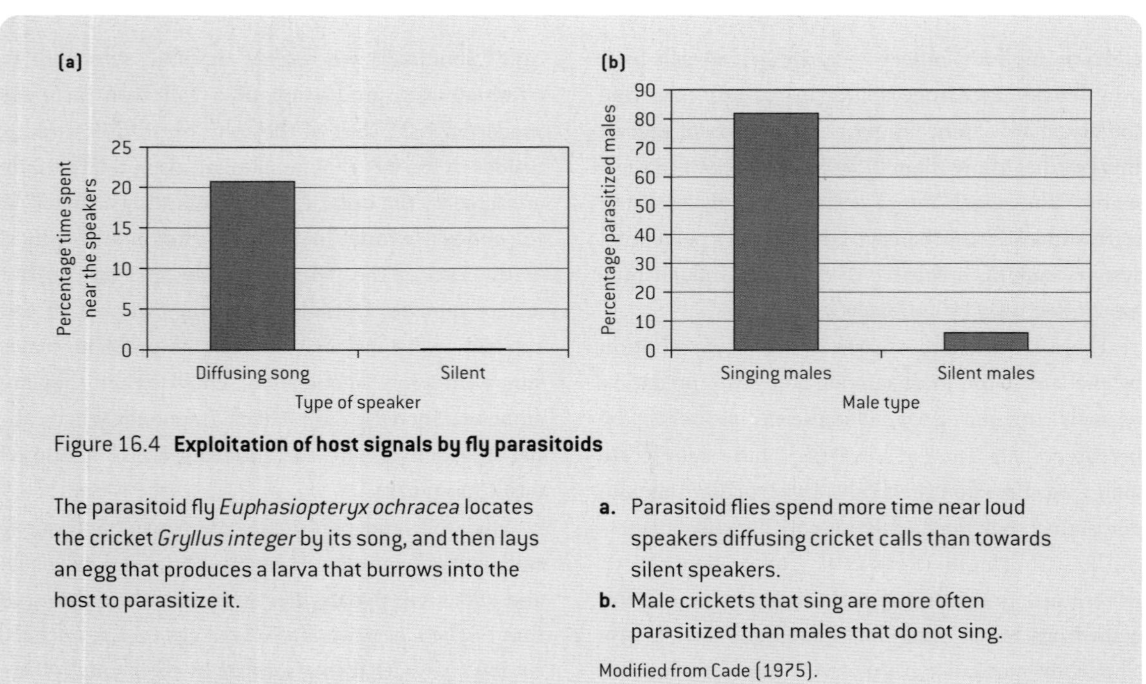


Figure 16.4 **Exploitation of host signals by fly parasitoids**

The parasitoid fly *Euphasiopteryx ochracea* locates the cricket *Gryllus integer* by its song, and then lays an egg that produces a larva that burrows into the host to parasitize it.

**a.** Parasitoid flies spend more time near loud speakers diffusing cricket calls than towards silent speakers.
**b.** Male crickets that sing are more often parasitized than males that do not sing.

Modified from Cade (1975).

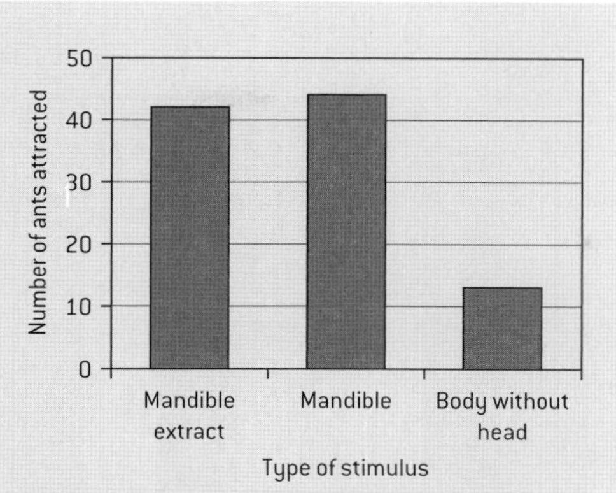


Figure 16.5 **Exploitation of host signals by parasitoids**

The fly *Apocephalus paraponerae* detects ants *Paraponera clavata* by a ketone and alcohol odour, two molecules produced by their mandibular glands. The purified products are nearly as attractive as pieces of ant mandibles, whereas their body is much less attractive. Ant pheromones are exploited by the flies in order to find their hosts on which they lay their eggs. Based on Feener *et al.* (1996).

*Apocephalus paraponerae* exploits the olfactory signal produced by these recruitment pheromones in ants.

The exploitation of cues and signals is also observed between members of the same species. We have seen various examples in Chapters 4, 8, 9, and 14. For instance, we can consider the functioning of an information centre as an exploitation process of non-excludable cues produced by the finder of a temporary resource by the follower who then exploits the information produced by the finder (Chapter 14). Indeed, by following the finder, the follower can have a negative impact on the food provisioning of the finder.

#### 16.2.4.2 *Nosy conspecifics*

Cases of eavesdropping have only been described as taking place during interactions within the same species, for instance in the common nightingale (*Luscinia megarhynchos*; Naguib and Todt 1997), the Siamese fighting fish (*Betta splendens*; Oliveira *et al.* 1998; Doutrelant and McGregor 2000; Doutrelant *et al.* 2001), and the great tit (*Parus major*; Otter *et al.* 1999). An individual can thus obtain important information simply by observing the interactions of other individuals (see Chapter 4). For example, in the fighting fish, females searching for a sexual partner observe agonistic interactions among males and then show a preference for the male winning these interactions (Figure 16.6; see other examples in Figures 4.14–4.16). After observing males during fights, females display their solicitation colours to winners for longer than they do to losers.

Females can extract useful information from male signals that are designed for intra-sexual communication. Thus females can obtain information on the quality of potential sexual partners by observing their success during agonistic interactions. In Chapter 4 we have seen that this is a case of **performance information** (Figure 4.3).

On the other hand, Doutrelant *et al.* (2001) highlighted the fact that the presence of an **audience** of conspecifics has a strong effect on the evolution of behaviour and signals. Individuals in the audience can have interests that are different from those of the intended signal receiver (the competing males in the experience shown in Figure 16.6), and the sender may also attempt to transmit information towards these other potential receivers as well. For example, during the aggressive interactions of two male fighting fish, the presence of a male audience does not change anything about the fight (Figure 16.7). However, a female audience changes the males' behaviour; they interact less aggressively, using more visible signals that provide more information about themselves to the female.

More generally, because most signals used by males are flashy or can be transmitted over long distances, they can be perceived by others such as competitors (other males) or potential sexual partners. Such audience effects and eavesdropping are probably more frequent than has been previously thought. Doutrelant *et al.* (2001) further proposed that the generalist aspect of male signals (they can be used both in male–male and male–female communication),

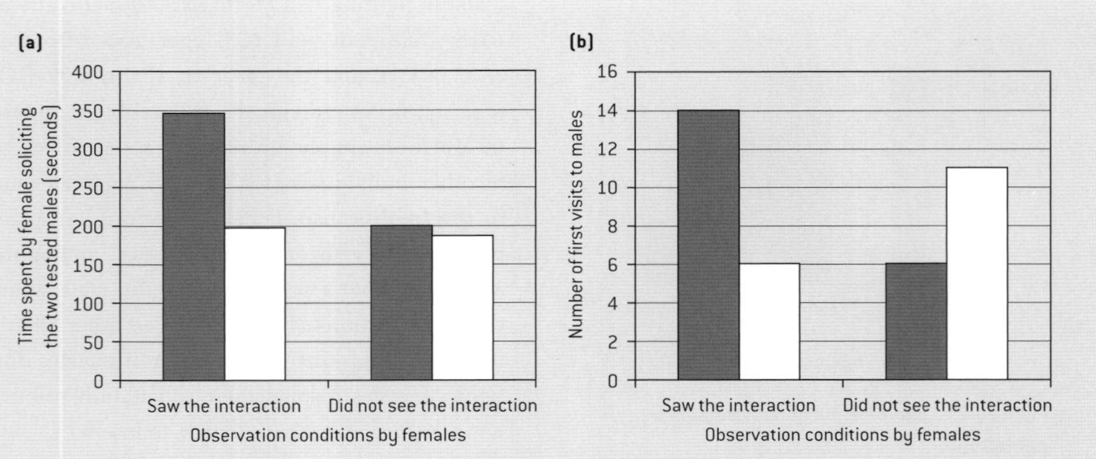


Figure 16.6 **Female eavesdropping in Siamese fighting fish (*Betta splendens*)**

Females of fighting fish preferentially direct their solicitations towards the dominant male when they had observed fights between males. The experiment was as follows: two males were placed in two transparent compartments of a tank in such a way that they could see each other and interact aggressively. Their respective behaviours allowed determination of the winner of the interaction. A female was placed in a third compartment where she could watch the males through a transparent partition. After the interaction, an opaque partition was inserted between the males who could no longer see each other and hence stopped fighting. The behaviour of the females towards the two males was then recorded. To confirm that it was the direct observation of the interaction that influenced the female' behaviour (and not a difference in male behaviour following the interaction) a control group was used in which females responded to males that had interacted while she had been prevented from seeing the interaction.

**a.** Time spent by the female presenting solicitation colours towards the winning (grey bars) or losing (white bars) male. Females did more solicitations towards the winners if they could observe the interaction between males (Wilcoxon paired test, $P = 0.009$), whereas this effect was absent when they did not observe the interaction (P = 0.67). This result is thus not due to the fact that winning and losing males behave differently.

**b.** Furthermore, after the fight, the tested female made more first visits to the winning male (grey bars) than towards the male having lost the interaction (white bars). Again, this result is obtained only if the female has observed the interaction among males.

Based on Doutrelant and McGregor (2000).

can be traced to the audience effect because most sexual signals can be received and used by individuals of both sexes. According to these authors, in order to understand the evolution of signalling it is important to conceive communication not only as a dual interaction but rather as interactions within a **communication network** where several individuals with different interests are involved. A challenge for future studies in communication will be to consider its social dimension.

### 16.2.5 Sources of information, decision-making, and behavioural responses

In its strictest sense, communication implies an information transfer by means of a signal between the sender and the receiver for the benefit of the two parties (see Marler 1977; Bradbury and Vehrencamp 1988). The resulting benefits can be considered as the driving function for the exchange of information. The receiver uses the information contained in the

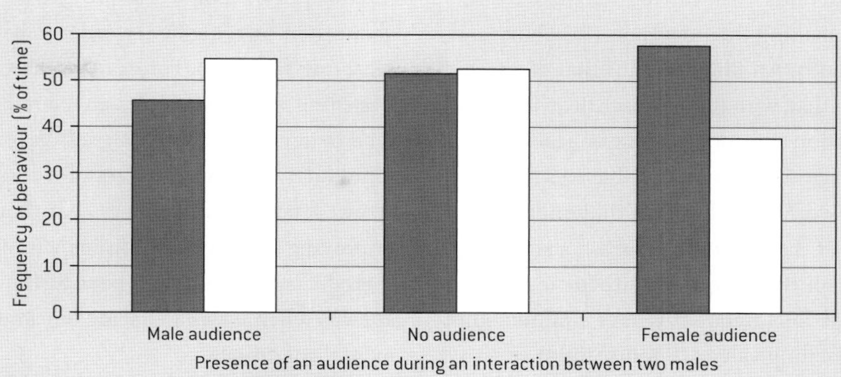


Figure 16.7 **The audience effect: a female audience modifies male–male communication**

The presence of a male audience does not modify male–male communication in the fighting fish *Betta splendens* (left part of the figure, the difference is non-significant). In contrast, in presence of a female audience (right part of the figure), males produce more ritualized signals, for example shaking their tails ($P = 0.003$) and opening of their opercules ($P = 0.03$) in the direction of the other male and the female (grey bars). They also produce fewer aggressive signals towards the other male (white bars) like biting ($P = 0.016$) and spending time close to the glass separation ($P = 0.028$). See Chapter 4 for complementary information.

Adapted from Doutrelant *et al.* (2001).

signal in order to make a behavioural decision. Depending on the context, this signal informs the receiver on different aspects of the relationship between individuals: is the opponent going to attack? Is the female going to allow copulation? When several possible discrete responses (that are discontinuous and exclusive) to each question exist, these are called **alternative conditions** (for example, in an aggressive context the responses can be either: attack highly probable, 50% chances of attack, attack highly improbable). To transfer more accurate information, the signaller modulates his signal according to a code that associates the signal to the condition: in wolves upright ears and apparent teeth signal aggressiveness while flattened ears indicate submission. Before deciding, the receiver interprets the signal or makes a **direct** assessment of the current condition (Bradbury and Vehrencamp 1998). It can also use other sources of information.

For example, let us consider a species in which a breeding couple defends a breeding site. Let us also suppose that when an individual without a territory encounters a territory holder, it can attack it and usurp the territory if it displays a larger body size. The intruding individual could then directly evaluate the size of the territorial individual. Nevertheless, a direct evaluation of condition (in our case, body size) is often imprecise, impossible or risky. The intruder can then use one of four sources of secondary information: (1) a knowledge of its probability of dominating the interaction (it has discovered that generally it is of a large size and can thus attack); (2) an evaluation based on **indirect** cues (for example, turbulences generated by the individual cleaning its nest can be a cue of its size): (3) the exploitation of **amplifiers** (a trait that facilitates or renders more precise the direct evaluation of a cue, for example the contrasting lines that limit the body of fish allow an accurate estimation of their size (Zahavi 1987)); finally (4) the reception of communication signals emitted by the resident.

The actions of the sender can function as signals, but it is often difficult to determine whether an exchange of information involves a signal or not. For example, actions can be **tactical**, that is they change the surrounding conditions and thus the properties of alternative actions that can be adopted by the receiver before making their decision (Bradbury and Vehrencamp 1998). An illustration is that of a territorial individual that, with the approach of an intruder, changes its position and adopts a new one facilitating attack: it performs an action that influences the decision of the receiver while changing the conditions of the perception of the action. In contrast, a signalling action informs on the condition without modifying it. In practice, actions frequently have simultaneous tactical and signalling values (Enquist 1985).

These distinctions, which can seem subtle, are important for expanding the concept of communication to behaviours that are not signals. We have described these nuances for the sender of the information, but they also apply to the receiver. We have also dealt with these issues in Chapter 4. In a general way, the notion of communication should not be restricted just to the process of emission/reception/action generated by a signal. It is necessary to include the use of cues and amplifiers in the context of strategic decisions. These notions are essential to better understand the evolution of signals.

### 16.2.6 Communication and honesty

The strictest definition of communication supposes that not only the sender benefits, but also the receiver benefits from the information. This implies that the sender provides it with accurate and honest information allowing it to improve its decision-making. However, there are numerous situations where the interests of senders and receivers diverge to such a point where the sender can benefit by providing false information (review by Bradbury and Vehrencamp 1998). As seen in Chapter 11, everything else being equal, selection is supposed to favour males that multiply matings, irrespective of female quality. In contrast, females should prefer to mate with males fulfilling certain quality criteria expressed by signals such as bright colours, exaggerated morphological traits or the vigour of their mating displays. A conflict of interest can thus exist between males (senders) and females (receivers) such that male senders can be tempted to provide false information about their own quality in order to be chosen by the receiving female(s). For example, in the stickleback, *Gasterosus aculeatus*, males in poor body condition tend to develop colourful signals beyond their actual reproductive capacities (Candolin 2000). Such males attract females that lay eggs that they can eat, thus improving their body condition before starting a new reproductive cycle. This type of dishonest signal only appears when males are facing unfavourable environmental conditions.

The risk of facing **lies** is inherent to all communication (Johnstone 1998). When the sender lies to the receiver, it is called **cheating** or **deceit** (review by Bradbury and Vehrencamp 1998). Lies can take many forms. A first type concerns signals that are important to the receiver but that can only take on one or a few intensities (as opposed say to a continuously graded signal). In this case the sender can produce the signal in order to deceive receivers. Alarm signals, for example, are like this; they are discrete and vary little in intensity. A bird that produces a false alarm call is lying: it says a predator is detected when it is not. The liar may benefit, for instance by causing others to leave a site and abandon their food leaving it all to the benefit of the liar. A second type of cheating involves withholding information: the sender hides the truth to the receiver. For example an isolated animal that finds a new source of food may decide not to inform others of its discovery (see Section 14.2.3.4). A third type of cheating is possible with continuously graded signals. In this case it becomes possible to lie either by exaggerating or minimizing the information. For example the growl of a dog represents a graded signal that can be weak or strong. Dogs can lie, therefore, either by using a growl intensity that indicates a motivation to fight that exceeds its real motivation to fight. That kind of cheating is called **exaggeration** or **bluff**. Bluffing can also occur when the expression of the signal is weaker than the existing intention, for example when a dog growls weakly despite its strong intention of attacking. We have seen above that amplifiers facilitate the evaluation of the sender by the receiver. The cheating version of an amplifier is called an

**attenuator**, which renders more difficult signal evaluation, as for example the disruptive lines or colours that make more difficult body size estimation in birds more difficult (Hasson 1994). Inversely, cases when the receiver lies to the sender are called exploitation by the receiver. In such cases, the information provided by the sender is honest, but the receiver uses it to make a decision that benefits himself at a cost to the sender.

#### 16.2.6.1 *The handicap principle*

The question of signal honesty remained a fundamental problem in the study of animal communication for a long time (Maynard Smith and Harper 2003). Lies are cheap and so in that context it was hard to understand how a signal could ever evolve when senders could lie. When liars are common, the receivers should simply ignore the signal because it does not really contain valuable information (as in the case of false alarm calls; see Figure 4.1). In other words, taking into account a signal would bring no benefit to the receiver. It was expected that signals that could be cheated would simply disappear allowing only the evolution of signals that could not be cheated. It was the Israeli biologist Amotz Zahavi (1975, 1977) who first proposed a mechanism for the maintenance of signal honesty: any signal that is costly to produce will be in essence honest because the costs associated with its production will prevent cheating; that is, prevent its use by individuals who cannot pay the cost of using it (Figure 16.8). According to Zahavi, an individual must balance the fitness benefits obtained by producing a signal against the costs involved in its production. A key aspect of Zahavi's idea is that the rate of increase of the costs of signalling is steeper for poor-condition than good-condition individuals (Figure 16.8a). It follows that, for the same fitness benefits, individuals in good conditions can produce an optimal signal intensity that is greater than that of poor-condition individuals. This is the **handicap principle** where the optimal signal produced by an individual results from the fitness benefits obtained by producing the signal minus the costs of producing it (Johnstone 1997, 1998). A variation of this hypothesis proposes that the costs are the same for all individuals but the benefits obtained can be greater for certain individuals in poor-condition (hungry nestlings, for example). Under this variant, individuals with greater needs will signal at greater intensity because they obtain greater benefits (Figure 16.8b). As simple as it may sound, the implications of the handicap principle are extremely important (see Chapter 11), and it led to numerous and strong negative reactions about whether such signals could ever evolve until mathematical modelling by Pomiankovski (1987) and Grafen (1990a, b) showed they could (Johnstone 1997, 1998; Maynard Smith and Harper 2003; see Chapter 11).

#### 16.2.6.2 *Examples of handicaps . . .*

Numerous signals are honest indicators of the quality of the sender. For example, a close relationship between body size, muscular mass, and physical power has been found in different species (see Le Bœuf 1974; Whitham 1979; Dodson 1997). The larger size of certain physical attributes are often more costly to produce and carry. They are difficult to imitate by small or weak individuals and can thus inform honestly on the competitive ability of an individual bearing them. Classic examples of such signals are the antlers in deer (Clutton-Brock *et al.* 1979) or the distance between the eyes of certain male flies from Malaysia and New Guinea (Wilkinson and Reillo 1994; Wilkinson and Dodson 1997; Figure 16.9).

In the same manner, the bellowing of red deer or the croaking of toads are honest signals of male body condition. Only male deer in excellent condition can roar during long time periods practically without taking any food (Clutton-Brock and Albon 1979). Only the biggest toads can croak at low frequencies, thus low frequency croaks are honest signals of body size allowing rivals to assess their chances of winning a fight (Figure 16.10).

#### 16.2.6.3 *. . . But dishonest signals exist*

Nevertheless, examples of dishonest signals do exist, in particular when given between different species.

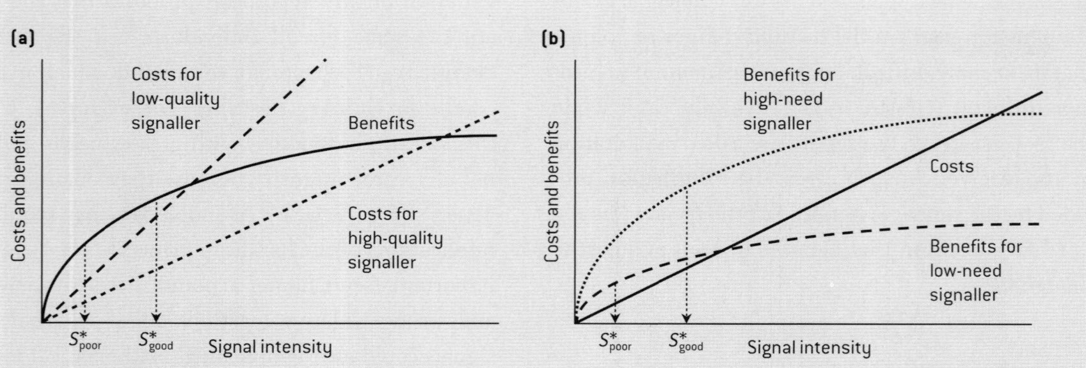


Figure 16.8 **The handicap principle: costs and benefits of signal optimization and maintenance of honesty**

Two graphical models to understand how different optimal signal intensities (e.g. maximizing net benefits) may be achieved by signallers of different phenotypic condition. Variation in phenotypic condition can affect the slope of the relationship between either the costs (**a**) or the benefits (**b**) or both (not represented here) of signalling with different intensity.

**a.** The rate of cost increases faster with signal intensity in poor-condition individuals (top dashed line) than in good-condition individuals (bottom dashed line). The continuous line shows the increase in the benefits with signal intensity. If all senders gain the same fitness benefits, individuals in poor condition pay higher costs for a given intensity of signalling and the optimal signal intensity for a poor-condition individual, $S^*_{poor}$, is thus lower than for a good-condition individual, $S^*_{good}$.

**b.** In this case, it is the benefits that a sender may get from signalling that vary in relation to the condition of the individuals. Poor-condition individuals gain fewer benefits from a given signal intensity and their optimal signal intensity, $S^*_{poor}$, is thus lower than for good-condition individuals, $S^*_{good}$. In this scenario, the rate of cost increases in relation to signal intensity is the same for all senders. These two conditions can be combined.

Modified from Johnstone (1997).

A well-known example is the case of the fireflies studied by Lloyd (1965, 1975). Females of predatory fireflies of the genus *Photuris* respond to the signals of male fireflies of the genus *Photinus* and attract them in order to eat them. These luminous signals have specific codes and certain *Photuris* females can imitate the response of three *Photinus* species. There may also be dishonest signals among members of the same species, for example, great tits may produce false alarms with the effect of chasing their conspecifics away from discovered food sources (Møller 1988). These cheating signals can only be maintained in populations if their frequency is low because at high frequencies the costs of the cheating signal would be too high and potential receivers would be selected to ignore them. This type of frequency dependent problem is similar to the producer-scrounger game described in Chapter 8. This is the 'never cry wolf' principle.

### 16.2.7 Broadening the idea of communication

As we have seen above, the idea of communication can be expanded to include environmental cues and auto-communication (see Figure 16.1). If we ignore intentionality by the sender, then the environment in its broadest sense provides numerous cues concerning current environmental conditions. These environmental cues are constantly used by organisms to synchronize their biological rhythms, control their reproductive investment or select their habitat.

Figure 16.9 **Examples of handicaps: size indicators and male quality**

The spacing between the eyes placed at the tip of stalks varies among males and is a honest indicator of the relative size and male competitive ability in male stalk-eyed flies, *Cyrtodiopis whitei*. Males fight with their legs and not with their heads or their eyes.

Photograph generously provided by Mark Moffett.

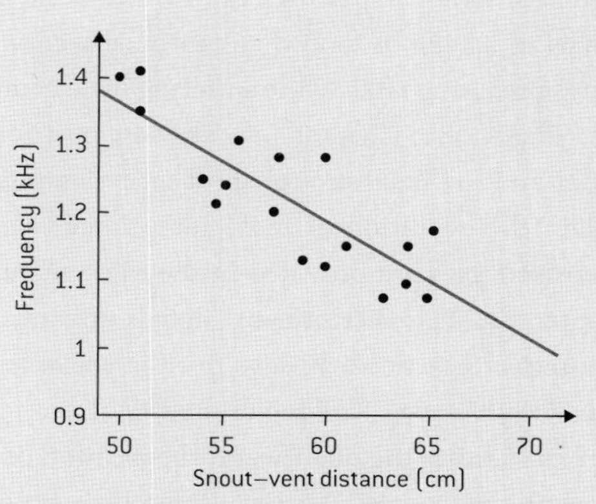


Figure 16.10 **The croaks of toads are honest signals**

Only large toads, *Bufo bufo*, can croak at low frequencies. The low frequency croaks are thus honest indicators of body size: they are costly to produce in terms of energy and time spent producing the croak. Males cannot cheat by producing croaks at lower frequencies than their size allows because the link between body size and croak frequency is determined by the physics of sound production.

Based on Davies and Halliday (1978).

There are two types of information obtained from the environment according to whether information is extracted from signals and/or cues produced by other species of the same or different trophic level. Among species at the same trophic level, signals and cues are generally of a cooperative type as they benefit both senders and receivers. For example, this is the case of *Cercopitecus* monkeys living in multispecies groups and responding to alarm calls of different species in the group (Gautier and Gautier 1977). The individual that produces the alarm pays the potential associated cost whereas those who take it into account obtain a benefit by protecting themselves against the detected predator. If we observed only one interaction, we could conclude that we are in the presence of a case of information parasitism. However, in such a group of primates, cooperation arises through reciprocity because the roles in detecting predators and giving alarm calls alternate regularly (see Chapter 15). Signals and cues between species at different trophic levels tend to inform more on the presence of prey or on the risks associated with the presence of predators. We will now examine these two types of signal.

#### 16.2.7.1 *Communication among species of the same trophic level*

Within species, certain signals reveal the presence of food resources. For example, domestic cocks advertise the presence of food by a high-pitched call which indicates the attractiveness of the food and attracts the hens (Marler *et al.* 1986a, b). This signal allows the rooster to attract sexual partners and increase his mating opportunities. In contrast this signal is rarely produced in the presence of another rooster thus avoiding male–male competition for fertilization. Another example is provided in Chapter 14, Section 14.2.3.4 in the house sparrow, *Passer domesticus*, where individuals finding a new source of abundant food that cannot be carried away to a safe place produce a 'chirrup' call that attracts other sparrows.

In such situations sparrows can attract other individuals to a type of food that can be shared which dilutes predation risks while suffering little competition (Elgar 1986; Chapter 14). The low cost for the sender and the insurance of short term benefits can explain the evolution of these recruitment signals. Other examples are provided in the same section of Chapter 14. The recruitment signals towards the resources are particularly developed in social insects (von Frisch 1967; Hölldobler and Wilson 1990), and can indicate the quality and the location of food or nesting resources. They are sometimes observed between different species, as in the example of honey guides in Africa (*Indicator indicator* and *I. variegatus*) which attract people and other honey-eating mammals (like the honey badger *Mellivora capensis*) by their calls and particular behaviours towards hymenopteran nests that they cannot open (see Chapter 17). Depending on the example considered, cooperation arises through different mechanisms from the fact that the two partners benefit from the interaction. In the example of the honey guides and badgers, the indicator benefits from the badger's action as it allows it to exploit an otherwise unavailable resource, while the badger obtains information on the location of a valuable food source.

A second category of environmental signals between species within trophic groups concerns alarm signals that indicate the origin or identity of a predator. For example, domestic roosters produce different alarm calls when facing aerial or terrestrial predators (Evans *et al.* 1993). The vervet monkey *Cercopithecus aethiops* produces four different alarm calls depending on whether they have detected a bird of prey, a snake, a leopard or another predatory mammal (Cheney and Seyfarth 1990). In all cases, the costs of producing the signal have to be compensated by direct or indirect benefits. Alarm signals can have multiple usages depending on the context in which they are produced. They can serve to help a conspecific escape the claws of a predator, coordinate groups during flight or swimming so as to remain grouped and reduce the risks of predation, to protect its partner or offspring and finally to maintain an optimal group size (Bradbury and Vehrencamp 1998).

In intra-specific interactions we have seen in Chapters 8 and 9 how performance information (i.e. information arising from the performance of other individuals) can constitute an important and revealing cue of environmental conditions. Nevertheless, it is possible that individuals use the reproductive or feeding success from individuals of another species sharing the same ecological resources in order to assess the quality of diverse environments.

#### 16.2.7.2 *Environmental information exchanged among trophic levels*

##### a *Flowers attract pollinators . . .*

The signals produced among trophic groups lead to the attraction or the repulsion of the receiver. The attractive signals are for example those produced by brightly coloured flowers or fruits that select or attract pollinators or seed dispersers. Such colourful and olfactory signals are generally honest since they are correlated with the presence of a food source like the nectar in flowers or ripe fruit pulp. However, sometimes they can also be cheats and imitate the colour and form of edible structures. For example, the seeds of *Ormosia coccinea* (a tropical plant) present a red design over a black background that imitates well the external pulp of other species but are dry and toxic if an animal destroys their envelope (Figure 16.11).

##### b *. . . spiders that use camouflage . . .*

The possibility that all signals produced by a sender can be exploited creates a conflict between the necessity of communicating and becoming detectable and the necessity of remaining undetected by potential communication exploiters. This conflict creates suitable conditions for the evolution of mimicry. A particular example of camouflage is aggressive **mimicry**, which is used by predators that imitate an attractive substrate in order to fool their prey into a fatal approach. An interesting example is given by female crab-spiders (*Misumena vatia*, *Thomisus onustus*), which adopt the colour of flowers on which they settle and

(a)

(b)

Figure 16.11 **Plant signals attractive for pollinators or dispersers**

Many flowers and fruits have bright colours that attract their diurnal pollinators and dispersers.

**a.** A hummingbird in French Guiana (the Fork-tailed woodnymph, *Thalurania furcata*) visits an orange flower of *Pitcairnia geyskessi* to collect nectar. When doing so it pollinates the flowers of this species.

Photo Marc Théry.

**b.** Eleven seeds of *Ormosia coccinea* that can cheat dispersing birds by mimicking the presence of external pulp (by the rounded form which contrasts with the black background of the seed) but not providing any food resource to the individual that ingests and disperses it.

Photo Pierre Charles-Dominique.

Figure 16.12 **Aggressive mimicry in the crab-spider *Misumena vatia***

This spider presents strong colour mimicry with the flower where she sits for hunting. Here the spider is eating a bee she has just captured on a daisy where she mimics the colour of the petals. See Figure 16.17 for more details.

Photo graciously provided by Roger Le Guen.

feed from the pollinating insects attracted by the colour and smell of the flowers. Their colours make them difficult to detect by birds that eat them and by pollinating insects on which they themselves feed. This outcome is particularly striking given that birds and insects have very different colour vision systems (Figure 16.12). We will come back to this example later on (see Figure 16.17).

### C *. . . repulsive signals . . .*

In contrast, repulsive signals are generally aimed at potential predators. A potential prey can modify its behaviour and **signal to a predator that it has been detected** thus indicating to the predator that it has lost the advantage of surprise. The classical examples concern stotting, the jumps and sounds made by numerous African ungulates, the tail movements of certain lizards, the foot stomping of kangaroo rats and the alarm calls given by birds when they have detected a predator (review in Bradbury and Vehrencamp (1998) and Ruxton *et al.* (2004)). By following the logic of the handicap principle, the ability to perform difficult and energetically costly movements like the stotting by gazelles could honestly indicate that the potential prey is in such good body condition that the predator is dissuaded form attacking it (Caro 1994, 2005). Furthermore, numerous species produce acoustic signals that recruit other individuals (from the same or other species) around a predator where they will observe and mob it with the result of the predator abandoning its hiding position, stalking or chase (Curio 1978).

### d ... *bright colours that identify a non-edible prey*

Another example of signals between two species at a different trophic level consists of the bright colours shown by insects that contrast with the background where they live. These highly visible signals associated with a real toxicity or unpalatability of the prey are often termed **warning** or **aposematic** signals (review in Ruxton *et al.* 2004). The evolution of these signals can take place when predators learn to associate the bright signals with the unpalatability of the prey. Within a community of species living in the same environment, cases of mimicry among unrelated species that share the same unpalatability have been observed. This convergent evolution of bright colourations is called **Müllerian** mimicry. The evolution of this type of mimicry relies on the fact that the costs of individual production of aposematic signals are probably low and that the benefits obtained in terms of reduction of predation risk increases with the frequency of the aposematic signals within the prey community. Interestingly, this honest system is exploited by palatable species mimicking the warning signals of unpalatable species and thus reducing their predation risks. This situation has been described as **Batesian** mimicry. It is worth pointing out that the frequency-dependent producer –scrounger problem described in Chapter 8 is again relevant to this situation: if palatable mimics (scroungers) become too frequent the value of the aposematic signal (producers) will decrease to a point where predators may well choose to stop taking it into account.

#### 16.2.7.3 *Auto-communication*

Auto-communication involves the broadcasting and reception of signals by the same individual. Signals can be acoustic (echolocation) or electric (electrolocation). Auto-communication is considered by some researchers like sensory physiologists, as a refined form of a direct estimation or assessment of environmental cues. This form of communication responds to the same laws of production, transmission and reception as conventional signals (review in Bradbury and Vehrencamp 1998). Nevertheless, because these signals provide information to the sender itself there cannot be a conflict of interest between sender and receiver and the information will be honest in all cases of auto-communication. Auto-communication is used as a substitute for vision when light conditions are weak; it is thus found in nocturnal or subterranean animals as well as in some living in deep and turbid aquatic environments. Some simple forms of auto-communication are used by cave birds like certain martins and the oilbird *Steatornis caripensis* (Figure 16.13) and several nocturnal mammals like shrews, rats, and tenrecs.

More complicated echolocation systems are used by bats and certain cetaceans (review in Bradbury and Vehrencamp 1998). Echolocation allows the detection and localization (distance and angle) of distant objects, and the estimation of their nature (shape, texture, composition), and relative speed and trajectory. The megachiropterans of Africa and Asia produce ultrasounds by clicking their tongue on a side of the mouth whereas microchiropterans produce sounds in their larynx.

Electrolocation is based on the production and reception of moderate voltage (review in Bullock and Heiligenberg 1986). Fast electric bursts are produced by an electric organ which also analyses the distortion of the electric fields caused by obstacles in a similar way as echolocation. This type of

Figure 16.13 **Echolocation in the oilbird (*Steatornis caripensis*)**

A female on her nest in a cave in Trinidad. This species uses echolocation for orientation in the dark. The wavelengths used, greater that those produced by bats and whales probably do not allow them to detect small objects.

Photo graciously provided by Louise Emmons.

auto-communication is found in numerous marine rays, naked-back knifefish (family Gymnotidae) in South America and the elephant fish in Africa (family Mormyridae).

## 16.3 Physical and physiological constraints on the evolution of signals

### 16.3.1 Sensory ecology: the physical properties of sound, light, chemical, and electrical signalling

There are four modes of communication corresponding to the use of sound, light, chemical, or electric signals. Observing the distribution of the different communication modes among animal taxa in relation to their habitat suggests that environmental constraints likely exert strong selection on the signals that can be used. For example, electrical communication is only used in aquatic environments because they are the only ones able to carry the weak electric fields produced by animals. In the same way, animals in total darkness do not generally communicate visually, and if they do, they must generate their own light, like, for example, fireflies.

Whenever analysing the structure and type of signals produced it is worth remembering that the physical properties of the habitats used for communication constitute a rigid framework constraining the four types of communication signals (review in Bradbury and Vehrencamp 1998). Signal propagation differs drastically among the four main types of signals. Light and sound propagate like wave pulses moving in a straight line from the emission source. Chemical signals also move away from their source but in a more irregular manner given that molecules can move back and forth from the emission source. The speed of propagation also differs greatly: it can be fast for light and sound but much slower for odours. A sound transmits mechanical perturbations and no molecule is transmitted, whereas odours are transmitted directly by molecular propagation. The variation and temporal coding of sound and light signals are largely conserved during their transmission, whereas this is not the case for odours because molecules are not displaced synchronously and in a linear fashion.

Besides strong environmental constraints, there are also numerous phylogenetic and morphological constraints that do not allow the evolution of certain communication modes. We are now going to examine some of these constraints, in particular those acting on sound and visual signals since they are better understood than the constraints on chemical or electric signals. This description will allow us to better understand the adaptive value of the various signals used. In certain cases, a good understanding of the physical properties of signals can allow us to understand the exact nature of the information involved in communication.

### 16.3.2 Production, transmission, and reception of signals

#### 16.3.2.1 *Sound signals*

A detailed presentation of the production, propagation, and reception of sound can be found in Bradbury and Vehrencamp (1998).

### a The production of sound signals

The emission of sounds requires the production of vibrations within the medium of propagation. An object vibrating in a single direction is called a **dipole**, whereas a **tetrapole** is an object vibrating along two perpendicular axes, both produce directional sounds.

Sound pressure depends on the amplitude with which the vibrator moves a given volume of the propagation medium. In this way, small animals can only produce sounds of low intensity. Furthermore, their small size forces them to produce short wavelengths and thus sounds at high frequency (high-pitch sounds). The frequency of the sounds produced by a body element depends uniquely on the natural vibration frequency of this element. With the same physico-chemical properties, the bigger an element the slower its vibration frequency and the lower the sound.

Muscles cannot contract more than 1000 times per second and the production of high frequency sound requires the action of frequency amplifiers, for example the stridulating organs of insects. In a grasshopper for example, the internal face of the rear legs is called the 'plectrum' and is made of a sharp ridge or blade that is rubbed against the 'file' (a row of small cuticular teeth) on the abdomen. The grooves allow a multiplication of sound frequency produced during rubbing. Thanks to such frequency multipliers, each muscular contraction produces numerous vibrations reaching frequencies of 90,000 vibrations per second. Arthropod stridulating is facilitated by their external skeleton, where the stridulatory organ is fixed and numerous articulations can be moved in pairs in opposite directions. Thus, arthropods can use practically all body parts to stridulate. Lobsters rub their antenna against their head, some coleopteran and shrimps rub their heads against their body, other coleopteran rub together their thoracic segments, ants rub their abdominal segments whereas some butterflies rub their wings on their thorax and others their legs against their wings.

Each structure involved in stridulating vibrates at its own frequency (called natural mode) based on its mechanical properties and not at the initial excitation frequency. Numerous animals use these properties: for example, the cicadas that contract and release a tympanic membrane with its own natural frequency.

Aquatic animals have fewer problems to link their sound signals to the surrounding environment. They use stridulatory organs and produce snaps with their teeth or other hard parts of their bodies. Resonance is harder to obtain in water due to the speed of sound transmission in this media. In certain fish, this problem is solved by using the swim bladder (which has the main function of controlling swimming depth) as a resonance organ or even as an organ for sound production. Because vibrations are produced directly by muscular contractions the maximal fundamental frequencies are relatively limited but can be enriched by higher harmonic frequencies. It seems that several whales and porpoises use the equivalent of a horn to produce sounds. This 'horn' has transmission properties different from those of the body and the surrounding water (which share similar properties). The corresponding morphological structure is an acoustical lens constituted by a sac filled by an oily matter (the spermaceti) that acts as a horn by focalizing the sounds. Vertebrates with aerial respiration use muscles to expel air and produce vibrations. The sounds of amphibians and most mammals are produced by the larynx acting as a door controlling the air flux. Mammals use the vibrations of their vocal chords acting as membranes in the glottis whereas the vocal chords of amphibians are separated from the glottis which controls the air flux after the sound production. In contrast to mammals that expulse air and the sound by the mouth and the nose, amphibians create a closed air circuit that allows amplifying the sound in resonance sacs. The larynx is also used as a resonance case in howler monkeys, gibbons, orang-utans and certain *Hypsignathus* bats; the lips forming a 'horn' allow linkage of the sound to the transmitting media. Birds, in contrast use a structure called the syrinx which consists in a narrowing of the airways situated before the bronchus. It vibrates with air passage and the tension of the membranes modifies simultaneously the frequency and amplitude of the vibrations. Both mammals and amphibians produce sounds by expelling air from the lungs whereas birds produce sounds by breathing in. The junction of the two bronchi is modified before arriving in the trachea thus giving the possibility to the birds to produce two sounds with different harmonics simultaneously. In numerous bird species, the opposite side of the vibrating membrane has a protuberance on its surface, called the labium which controls of the opening of the conduct and thus sound amplitude.

## b *Sound propagation*

The speed of sound propagation varies depending on the environment. Sound travels at 344 m/s in air, approximately 4.4 times faster in water and between 13 and 17 times faster in solids. The absorption of the sound energy reduces sound intensity with distance. This absorption is hundreds of times more important in air than in water. It is also 1050 times greater in salt than in freshwater and generally it is high in solids. In the air, the losses are more important when the temperature increases and humidity diminishes. In general, high frequencies are more easily attenuated by absorption than low frequencies. However, despite the low frequency of their sounds (less than 1–2 kHz) most insects can only communicate to distances of 1 to 2 metres (Ichikawa 1976).

*Distortions vary according to frequency*

Sound frequencies can also be distorted by reflection effects caused by environmental objects that are smaller than the wavelength of the signal. This phenomenon is called **distortion**. The greatest sources of distortion in nature are the vegetation (distortion is greater in a green leafed forest compared with a pine forest (Marten and Marler 1977)), as well as differences in temperature between air layers and vortexes caused by the wind and generating hot and cool air bubbles. The only way to reduce such distortions is to use low frequency sounds, less sensitive to distortion. Sounds can also be reflected by surfaces like the ground, the bottom and the surface of the oceans. The propagation of sounds is a process too complex to be discussed here and we will only describe the effects of variations in temperature in aquatic and aerial environments. However, it is worth remembering that these physical processes play an important role in the natural selection of sound signals.

*Waves trapped in sound channels can travel long distances*

A general principle in acoustics is that a sound wave is always sidetracked from the medium that transmits the sounds best towards the medium transmitting them less efficiently. Differences in the speed of sound transmission can be caused by temperature and density gradients: the higher the temperature or the pressure the faster the sounds are transmitted. For example, on a sunny day a bird that sings close to the ground finds itself in a layer of hot air, its song will be deviated upwards away from the ground. Close to the bird there will thus be a low lying sound shadow. In contrast, a bird singing on a clear night in an open area (or on the ground in a forest during daylight) will find itself in a area with cold air under a sound shadow that will reflect its song towards the ground allowing him to be heard further away at ground level. The cold air layer situated between the ground and the hot air layer above plays the role of a sound channel where sounds tend to be trapped and where they can be transmitted over long distances.

In the oceans we find the same effects of temperature inversions on sound propagations. For example, in winter the superior layers of a medium depth ocean are colder, the intermediate layers are warmer and the lower layers are cold. For the animals using acoustic signals close to the surface there is a sound shadow at a certain depth and these animals can use the sound channel close to the surface to transmit long-distance sounds. In summer, the best transmission channel is found at mid-depth in the oceans. At the surface the hot water allows high sound speed. With increasing depth sound speed slows down (lower temperature but moderate pressure) whereas at great depth sound speed increases again (low temperature but high pressure). There are thus sound shadows close to the surface and at great depth. The sounds produced at mid depth travel more slowly but are reflected by the two shadows. Sounds can thus

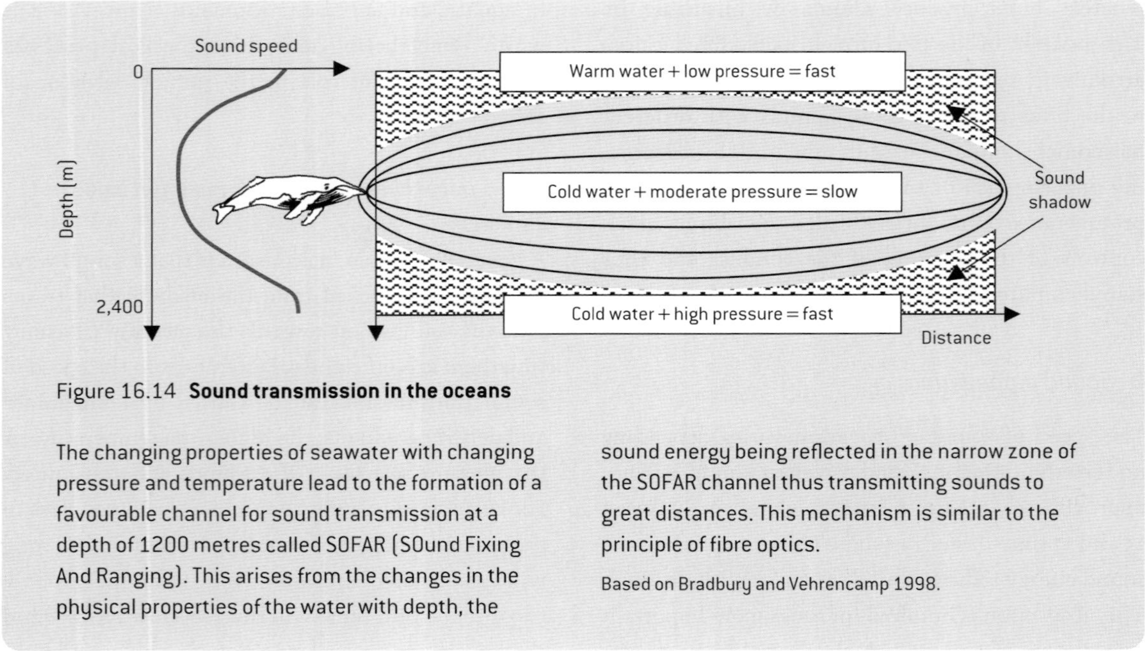


Figure 16.14 **Sound transmission in the oceans**

The changing properties of seawater with changing pressure and temperature lead to the formation of a favourable channel for sound transmission at a depth of 1200 metres called SOFAR (SOund Fixing And Ranging). This arises from the changes in the physical properties of the water with depth, the sound energy being reflected in the narrow zone of the SOFAR channel thus transmitting sounds to great distances. This mechanism is similar to the principle of fibre optics.

Based on Bradbury and Vehrencamp 1998.

travel hundreds or thousands of kilometres and still be detectable, as has been observed for the songs and signals of whales (Payne and Webb 1971; Winn and Winn 1978; Thompson *et al.* 1979). The sound channel is called the SOFAR (SOund Fixing And Ranging) and is found at a depth of 1200 metres at mid-latitudes but closer to the surface at the poles (Figure 16.14).

*The role of noise*

Noise also alters the propagation of sounds by adding new frequencies and/or reinforcing some of them. The principal source of noise in the air is the wind as well as the turbulences on the vegetation, soil, and the receiver's body. Noise comprises mostly low frequencies (less than 2 kHz), though insects may be responsible for high frequency noise (greater than 4 kHz). A relatively calm window can be found between 1 and 4 kHz in various terrestrial environments. This may explain why numerous bird and mammal vocalizations are produced in this range of frequencies (Klump and Shalter 1984; Ryan and Brenowitz 1985). A similar window exists in the oceans. Fish, however, produce sounds between 200 and 900 Hz which corresponds to a zone rich in noises. Potentially this may be explained by the fact that with their swim bladder fish cannot produce sounds with frequencies higher than 1 kHz. An alternative explanation could be that most fish communication takes place at short distances, so that the presence of environmental noise does not interfere with their communication.

*The reverberation on vegetation varies according to frequency*

Sound reverberation can also represent a constraint for the animals, in particular in forest habitats (Richards and Wiley 1980). Low-frequency sounds (less than 1 kHz) are reflected by the forest canopy and the ground, which produces intense echoes that superimpose on the sounds that travel more directly to the receiver. High frequencies are reflected by the vegetation and form numerous, though less intense, echoes. The frequencies found between 1 and 3 kHz undergo less reverberation in forests. Forest-dwelling birds produce efficient signals in this range of frequencies that, as seen above, is also the one with the lowest noise.

Such physical considerations predict that birds living in densely vegetated or open milieu should produce different acoustic signals in order to be better adapted to the acoustic properties of their habitat. A test of this hypothesis was performed by comparing the propagation of alarm calls made by a woodland-dwelling species (the blackbird, *Turdus merula*) with those of a species living in open areas (the European starling, *Sturnus vulgaris*; Mathevon *et al.* 1997). As expected, the reduction of the high frequencies after long distance propagation tends to concentrate the signals in a range between 1.5–4 kHz independently of the initial spectrum, and starlings' alarm calls have a broad spectrum (0.8–7 kHz) and are thus greatly modified, whereas those of the blackbird are narrower (2–5.5 kHz), and suffers only little distortion. From a spectral stance, blackbird calls are well adapted to propagating in dense vegetation whereas those of starlings are not. Although the comparison is based on only two species, the repetition of this type of observations (Wiley 1991; Boncoraglio and Saino 2007) tends to validate the interpretation. It is thus possible to correlate habitat acoustic constraints with differences in the spectral structure of the songs produced in them. Similar relationships have also been found between song characteristics and songpost height in the common wren, *Troglodytes troglodytes* (Mathevon *et al.* 1996).

#### c *Sound reception*

Signal receivers can use the degradation of acoustic signals with distance as information about the distance that separates it from the sender. For example, in a North American passerine bird, the eastern towhee (*Pipilo erythrophtalmus*), the level of sound pressure from a song is correlated with different acoustic parameters, like the duration or the frequency of calls, and can be used to estimate the localization of the source (Nelson 2000). Male towhees have a precise perception of the distance of a speaker broadcasting songs with the same amplitude and structure as natural songs. The distance perceived can be modified experimentally by reducing or amplifying playbacks of natural songs leading to expected changes in the flying behaviours of territorial males. By manipulating song parameters, experiments show that birds are able to evaluate the distance from the sender by using the constraints imposed by the environment on sound movement. A major problem for the receiver is to be able to discriminate a given signal in a noisy environment. This problem is even more important when the background noise is made by numerous conspecific individuals producing similar calls, as is found in seabird colonies like the huge king penguin (*Aptenodytes patagonicus*) colonies. Young penguins that have to be fed by their parents can recognize their parents' calls even if the latter are much weaker than the background noise made by other adults (Aubin and Jouventin 1998). Researchers have found that penguins use a system with two voices, each of the branches of the syrinx producing an independent sound. The interaction of these low and fundamental frequencies generates a pulsation that participates in individual recognition (Bremond *et al.* 1990). Playback experiments and modification of the pulses by Aubin *et al.* (2000) showed that adult and young penguins recognize each other use these two voice signals. These low frequency signals suffer little distortion while travelling in dense penguin colonies.

### 16.3.2.2 *Light signals*

#### a *Production of light signals*

The evolution of light signals is strongly constrained by the range of frequencies detectable by biological systems. Electromagnetic wavelengths range from radio waves that have a wavelength between 1 metre and 1000 kilometres to cosmic rays that can have wavelengths inferior to the size of atoms. Between these two extremes we find microwaves, infrared, the 'visible' (to humans) and ultraviolet spectra, X- and gamma rays. Radio wavelengths are not detectable by animals because they have such low frequencies and energies that they can pass through or around biological and non-metallic objects without being absorbed. Microwaves accelerate the rotation of atoms and are dissipated in the form of heat in the environment. This is practical to warm a meal but inadequate for vision because these radiations are strongly absorbed by atmospheric humidity. Infrareds are also perceived in the form of an increase in temperature and are easily absorbed by biological tissues. In animals, infrared detectors have thus low sensitivity and a viper that has infrared detectors on its head can hardly detect a mouse at more than 30 centimetres. Visible and ultraviolet radiations increase the energy of the electrons that is then dissipated as heat. The main problem is that UV rays have high energy, and so they produce a strong dissipation of heat that can damage biological molecules, notably DNA. This problem is even greater for X-rays and cosmic rays where their high energy can ionize atoms and destroy chemical links and molecules.

Visual systems are generally constrained by the range of radiation available on land and in the oceans. Long and short wavelengths are not reflected by objects but pass directly through them, and they cannot be used for vision. Given all these physical properties, only radiations in the 'visible' spectrum (between 400 and 700 nanometres) and in the near ultraviolet (between 300 and 400 nanometres) are compatible with the development of visual systems. Moreover, it is in these domains uniquely that solar radiations are available after their passage through the atmosphere: most solar radiation lies between 300 and 1000 nanometres with more than 80% within the visible spectra. There is thus congruence between the available radiation and the range of variation that are used by organisms to communicate visually.

Particles that are suspended in the atmosphere diffuse the polarized light found at a 90° angle from the sun. At dawn this band is at the zenith following a north–south orientation. The position of this band of polarized light varies in the sky as a function of the hour of the day; numerous invertebrates, for example flying insects, use it for orientation. Cephalopods, which can be highly coloured, do not seem to see colours but are probably sensitive to polarized light. Each colour perceived by the retina corresponds to a particular vibration of polarized light.

*Three physical mechanisms at the origin of colours*
Animals and plants use three different mechanisms to reflect and absorb wavelengths selectively from light and produce colours: pigmentation, interference, and diffusion.

1. **Pigments** are formed by molecules that absorb certain wavelengths while transmitting others. For example, a pigment absorbing violet, blue, green, and yellow wavelengths only transmits and reflects light in the red wavelengths, the colour we see. There are different types of pigment, of which carotenoids are only found in food. The molecule β-carotene transmits the green and the red which when combined appear as a yellow or orange colour. Bigger carotenoid molecules absorb blue and green and appear either as orange or red. Carotenoids bound to proteins absorb the green and transmit violet and red and appear purple. Pigments can also be constituted of pterins (white, yellow, red), of quinones (yellow, red, orange), verdins (blue-green), porphyrins (for example red haemoglobin with a central iron ion), green chlorophyll with a magnesium ion, and red, violet, or green pigments with a copper ion. Finally, the term 'melanin' groups numerous dark pigments used by plants and animals. Guanine reflects all wavelengths and the accumulation of its crystals produces the silvery aspect of fish scales. Blue pigments are relatively rare in animals (a carotenoid in some crustaceans like lobsters and haemocyanin in molluscs, crustaceans and arachnids) and blue colours are mostly obtained by diffusion. Specialized cells allow the reflection of wavelengths that have not been absorbed by pigments. The **chromatophores** produce certain pigments that are stored in granules called **chromatosomes**. Intense colours can be produced with a layer of specialized chromatosomes, the iridosomes, situated just under pigmented chromatosomes. The **iridosomes** play the role of mirrors and reflect the light transmitted by the pigments. Finally, **leucophores** are placed deeper than the

chromatophores and contain small cells in a pear shape that reflect and disperse light giving a white colouration.

2. **Interference of thin layers** is the second mechanism for the production of colours. A thin layer of transparent matter with a high index of refraction like wax or keratin covers feathers, scales or the external skeleton of animals. Part of the light is then reflected by this matter whereas the light passing through is in turn reflected by the lower layers that send the light back through the thin layer towards the exterior. As a function of the thickness of the thin layer and, depending on the index of refraction and view angle, the primary and secondary reflections will be in phase for certain light colours. When these thin layers are stacked intense colours can be produced.
3. **Scattering of light waves** by particles is the third mechanism for the production of colours. The theory of **Rayleigh scattering** shows that particles with a diameter smaller than 300 nanometres scatter short wavelengths more than long wavelengths. This is the mechanism that explains why the sky is blue. Certain plant and animal colours are produced by this principle with the existence at their surface of a transparent matter containing a matrix of small dense particles or air bubbles much smaller than 300 nanometres. Violets, blues, and greens can be scattered while, longer wavelengths, are absorbed by an underlying layer of melanin for example. This type of scattering associated with interference in thin layers generates blue and ultraviolet colours in birds. A carotenoid layer can also be superimposed thus absorbing the violet and the blue while scattering a green colour. In contrast, the **Mie scattering** is when particles much larger than light wavelengths scatter all colours of the ambient light and produce a white colouration. This case is observed in birds where the plumage is formed by large particles like air bubbles, fat molecules, proteins, keratin, or crystals diffusing white light. A detailed presentation of the mechanisms involved in colour production can be found in Bradbury and Vehrencamp (1998), Hill and McGraw (2006), and Berthier (2007).

*Animals that change their colouration*

Once the colour is produced, most animals cannot modify it quickly and often have to wait until a new moult to produce new colourations. Nevertheless, when colours are restricted to patches and/or parts that are retractable or can be hidden, the presentation of colours can be controlled to produce or hide a signal. This is, for example, the case of erectile crests in birds, butterfly wings that can be folded to hide bright colours, the throats of lizards that can be extended for displays, etc. In contrast, certain amphibians, reptiles (like chameleons), numerous fish, crustaceans, and cephalopods as well as insects can rapidly change colours by the movement of pigment granules (often melanin) in chromatophores. These colour changes are used for camouflage, threats, or displays. In invertebrates, the chromatophores can contain various pigments whereas in vertebrates they only contain one. Cephalopods and fish are the animals that are known to change colour the quickest.

*Luminous organisms*

Many organisms living at night or in dark environments produce their own light in photophores by using ATP as a source of energy and the enzyme luciferase to oxidize the luciferine that produces light photons at a wavelength of 562 nm. This efficient but relatively expensive communication system is used by most deep-sea marine life. The second system of light production is by using bacteria to produce **bioluminescence**. Bioluminescence can be used to attract sexual partners like in fireflies, or as a lure to attract prey as in anglerfish and, finally, by sea fish that illuminate their underside in a backlit environment making them less detectable by predators from below. This countershading camouflage is used at depth between 350 and 800 metres with weak light or alternatively at the surface during nighttimes.

### b *Light transmission*

Colours can be modified, as we have shown for sounds, by the absorption, reflection, and the differential filtration of certain wavelengths. These phenomena are particularly active in marine and forest environments. Water absorbs preferentially ultraviolet and infrared radiations. The deeper we go down in the water column the less there are ultraviolet, red, and orange, leaving only blue light, giving its characteristic colour to the sea. Organic particles or phytoplankton can also dye water green or yellow whereas tannins can tint water red (or in infrared, seen as black by humans since they are insensitive to infrareds). Imagine a bright red carotenoid coloured spot, just under the surface: it will appear red, at 5–10 metres deep it will appear orange, and as we go deeper it will appear darker and darker. From 10 metres and below the spot will appear black.

Similarly, in forested habitats, light conditions alter the conspicuousness of plant and animal signals (Endler 1993; Théry 2001; Gomez and Théry 2004). In tropical rainforests, the mean light intensity reaching the ground is only between 0.1% and 2% of the light in the open, highest values being found in sun patches that constitute most of the light in tropical forest understory (Bazzaz and Pickett 1980). Owing to the reflection and the selective transmission by leaves, this reduction in available light is paralleled by important changes in the shape (i.e. colour) of the ambient light spectra (Endler 1993; Théry 2001; Figure 16.15). Ultraviolet and visible light are both nearly totally absorbed with the exception of the light between 520 and 620 nanometres (Endler 1993), while a more important fraction of infrared radiation is transmitted by leaves (Gates 1965). The shading modifies the spectral distribution of the light through selective filtration of blue and red light by chlorophyll. Despite the large spatial and temporal variability of light conditions in forested environments, forest geometry, meteorological conditions, and the inclination of the sun determine six types of ambient light spectrum (Figure 16.15):

1. Large gaps and open areas present white light spectra.
2. Small gaps with an angle smaller than 2 degrees (1 metre diameter with a 30 metre high canopy) are enriched in yellow orange.
3. The forest shade is enriched in green or yellow green.
4. The shade is enriched in blue when the canopy is discontinuous and a greater proportion of the sky is visible from the ground forming 'the woodland shade'.
5. At dawn and dusk, when the sun is less than 10° above the horizon, light is enriched in purple. The deficit in middle wavelengths is caused by the absorption of the solar radiation by atmospheric ozone.
6. In cloudy weather, the spectra of small gaps or of the shades converge towards the white colour of large openings.

The woodland shade is particularly rare in tropical forests but frequent in temperate areas or in dry forests with thin canopies.

We will see below that the existence of ambient light colouration has a strong influence on the evolution of coloured signals in different habitats. A colour highly visible under direct sunlight (with white light), can become much less conspicuous when placed within a forest. It follows that certain sexual signals can be very inconspicuous in their common undergrowth habitat but become brightly coloured as soon as they are exposed to the sunlight (or inversely).

Ambient light is also polarized. If animals can perceive it they can use it for their orientation, predation and communication. This is for example the case of some insects, birds, crustaceans, cephalopods. and fish in the ocean. Shaded areas within forests and open areas are dominated by sky polarization whereas small gaps are non-polarized. Cloud cover has little effect on light polarization in small gaps and shaded areas but depolarizes the luminous field in large openings. Like in the aerial environment, the angle of polarization is directly associated to the position of the sun above the oceans. Insects detect

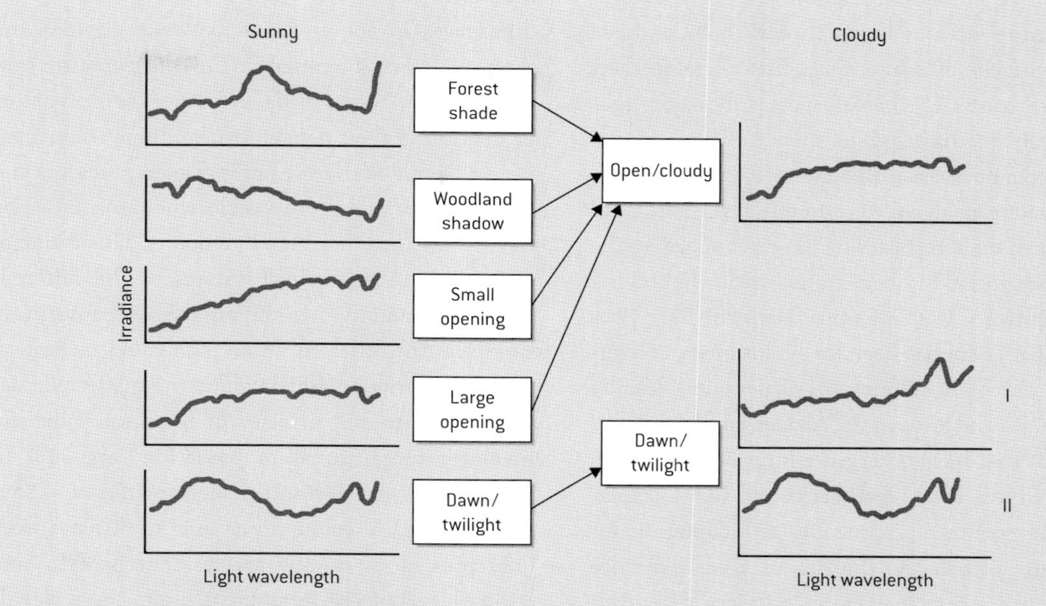


Figure 16.15 **Habitat light colouration in forests**

The direct light from the sun is white, and its spectrum in the visible light is relatively flat (like spectra from large openings for example). The colour of the white light in the open is modified by the filtration and reflection of the vegetation. To the left, in a sunny situation the forest shade is green (the spectra has a peak in the green wavelengths, that is in the middle of the spectra); the woodland shade is blue (spectra with an energy peak in the short wavelengths); the small gaps are yellow/orange (spectra with more energy in the long wavelengths); at dawn and dusk the spectra are purple (a mixture of blue and red, that is a peak in the blue and in the red extremes). To the right, in a cloudy situation the colour of dawn or sunset is clearly red due to the presence of clouds (situation 1) and then purple (situation 2). During cloudy days, light in the shade and in small gaps converge towards the white colour of the open areas.

Modified from Endler (1993).

polarized light in the UV (rich in direct sunlight), whereas aquatic animals detect polarized light in green wavelengths (common in water).

### C *On the use of light signals in communication*

The efficiency of the transmission of a light signal can be increased by four types of contrast against the visual background: **contrast of brightness**, **colour**, **pattern**, and **movement** (Bradbury and Vehrencamp 1998). Changes in the intensity of these contrasts can lead either to an increase in the visibility and detectability or to a reduction in contrast that increases mimicry. In each case, the brightness, the colour of the ambient light, and the light reflected by the visual background modify the conspicuousness of coloured signals (Hailman 1977; Lythgoe 1979; Endler 1986, 1990, 1993; Endler and Théry 1996; Fleishman *et al.* 1997). These visual contrasts can affect the communication process between animals (predation and camouflage, species recognition, age and sex determination, sexual selection), and between plants and animals (herbivory, pollination, seed dispersal, etc.).

Contrasts in brightness are particularly efficient when the visual background is extremely dark or bright and when it is influenced by the direction of the light source. For example, the Himalayan warblers (small songbirds) of the genus *Phylloscopus* exploit different habitats that differ in their brightness. Warbler species living in dark habitats have a large number of white spots that increase their brightness contrast, whereas species living in light habitats do not have white spots and use instead acoustic communication (Marchetti 1993). In animals illuminated from above, for example at the surface of oceans or in the forest canopy the camouflage on the visual background can be obtained with a lightening of the ventral side (Cott 1940; Hailman 1977; Rees *et al.* 1998). There has been an evolutionary convergence in the body colouration of penguins (family Sphenicidae) and birds in the Alcidae family (puffins, auks, etc.) that have diverged a long time ago: both groups of birds have black backs and white bellies. This colour pattern is probably an adaptation for the capture of fish and the risk of predation since both require a certain level of mimicry to remain unnoticed (Cairns 1986).

Colour contrasts can be obtained by differences in hue or chroma (saturation of a hue) between the visual signal and the background (Endler 1990). It follows that the most visibly coloured signals are those rich in the wavelengths of ambient light but poor in background wavelengths. If the ambient light is very coloured as it is frequent in forests, visible signals have to involve colours identical to that of ambient light and be surrounded by complementary colours so as to maximize colour contrast. In a general way, the contrast in brightness is mostly used for the detection at long distances and it is replaced by colour contrast at short distance.

Several studies have examined the use of environmental light by animals. One of the most detailed is probably the study by John Endler (1987, 1991) who showed that guppies (*Poecilia reticulata*), small fishes from tropical forest streams, display essentially at dawn and at dusk, when predation risk is lowest. The guppies present maximum contrast of brightness and colour in the purple light of dawn and dusk and minimum contrasts in forest light conditions particularly in forest shade. These characteristics increase male conspicuousness during displays while decreasing their conspicuousness towards predators at other times of day. This colour variation in function of different light environments is also increased by the differences in colour vision between guppies and their predators. If we analyse the conspicuousness of guppy signals to the visual system of their predators in relation to their dangerousness, these colours are nearly cryptic to the eye of their main predators and become more and more visible when encountering the less and less dangerous predators. This is due to differences in the visual systems of the different predators in relation to the number and absorption spectra of the different visual pigments (see Section 16.1). Furthermore, when comparing guppy populations, the size and vivacity of nuptial colours are inversely proportional to predation intensity. In *Anolis* lizards in Puerto Rico, Fleishman *et al.* (1993) showed that UV colourations used in displays were only present in canopy species that benefit from strong levels of UV in ambient light. These signals contrast with the green vegetation that reflects little in the UV. This study, and the ones by Marchetti (1993) and Gomez and Théry (2007) on birds, show that ambient light characteristics can influence the evolution of visual signals in association with habitat selection.

The influence of light on the displays of lek forming species has been studied in manakins and the cock-of-the-rock, *Rupicola rupicola* (Théry 1987, 1990a, b; Théry and Vehrencamp 1995; Endler and Théry 1996; Théry and Endler 2001). These birds maximize their conspicuousness to conspecifics and minimize their conspicuousness towards predators by choosing light habitats that generate differences in visibility. Each species optimizes its conspicuousness by different combinations of display movements, colours, and light habitats. In five species of manakins in French Guyana, the distribution of colour patches on the superior or inferior parts of the body is correlated with the display height and the light incidence in relation to forest geometry. In the white-throated manakin, *Corapipo gutturalis*, the light characteristics of display sites and the duration of male presence determine the attractiveness of displays towards females (Figure 16.16). As for the guppies, the colour contrasts and/or brightness are maximal during the hours and places of displays and

Figure 16.16 **The display of the white-throated manakin, *Corapipo gutturalis***

In the white-throated manakin each male defends a mossy log where it displays and attempts to attract females. A male (at left) is performing a display behaviour called 'wing-shiver display' towards the female. After a fast and spectacular flight display, the male walks backwards towards the female while opening its wings and laying its throat on the moss covering the trunk, thus exposing the white spots on throat and wings. By performing so, the spots on the plumage exposed to sunlight create a sharp visual contrast (Endler and Théry 1996). Females accept males by touching the male's wing tip with her bill. The behavioural characteristic of males and the light conditions at display sites are important for female attraction.

From Théry (1990a), Théry and Vehrencamp (1995). Photograph provided by Marc Théry.

these colourful signals are used in the choice of sexual partners.

Another example of the use of light environments is given by the colour variation in the web of the golden silk spider, *Nephila clavipens*, a widely distributed species in warm parts of the new world (Craig *et al.* 1996). Yellow is a very attractive colour for herbivorous and pollinating insects and this spider attracts and intercepts more insect prey by building its webs with yellow pigments (these appear golden under strong light). The colour of the silk is adapted to different light colours and intensities and seems to have the same colouration in different light environments. Spiders appear to be modulating the production of certain pigments adjusting them in function of the ambient light colouration. In contrast to guppies and manakins that choose a light environment to expose their specific colours, this spider adapts its colour signal to a large range of light environments. In this sense, it is potentially less sensitive to changes in light conditions and this could explain its broad distribution in forested environments, forest edges and non-forested areas.

The example of this spider is interesting since it shows that certain flexible behavioural traits are important in term of adaptation to differences in the light environment. We see here how living organisms specialized for a specific light environment can be more susceptible to light modifications with all the potential consequences in terms of survival and conservation. Forest-dwelling animals can disappear if their natural light conditions are disturbed, for example by transforming a primary forest into a secondary and exploited forest, where certain types of light condition do not exist or are spatially distributed in a very different way. In contrast to specialist species that are restricted to very specific conditions, generalist species have a greater plasticity, as for example the golden silk spider, and are probably more adaptable to changing light conditions.

### d *Visual systems and contrast perception*

For a long time, researchers have described and evaluated colourful signals in nature through the perceptual filter of our own vision. The colour plates of numerous nature guides reproduce the way humans see them. In some circumstances this approach is suitable. However, there can be problems as soon as we have to deal with species having different vision spectra than ours. Progressively, the development and combinations of quantitative methods allowing the measurement of the actual sensitivity of photoreceptors, plus the ambient and reflected light, as well as the use of discrimination models have allowed accounting for the visual systems of the study species (as well as that of their predators and prey).

These new techniques allowed the recent discovery that the blue tit (*Parus caeruleus*), a bird species well known for its lack of sexual dimorphism was in fact strongly dimorphic when UV was taken into account. Most birds can detect UV light; in blue tits the colour of the blue feathers on the head cap is richer in UV in males than in females (Andersson *et al.* 1998; Hunt *et al.* 1998). That species must thus appear to be dimorphic to all individuals seeing in the UV. Subsequent experiments showed that females probably base their choice of sexual partner and their strategy of reproduction on the intensity of this UV colouration (Hunt *et al.* 1999). For instance, in certain birds like blue tits, females modified the sex ratio of their young in relation to the visual attractiveness of their mate (see Chapter 13, Section 13.5.1).

We had to wait until 1970 for a detailed description of the different bird photoreceptors and visual cones. Studies (review in Kelber *et al.* 2003) have shown that, besides having receptors similar to humans (that is, that are sensitive to blue, green, and red), most of the 30 bird species with known visual systems have cones sensitive to UV. These birds have thus a greater sensitivity and a different colour perception than humans. Measurements and analyses of animal colours (or of colour signals produced by plants that attract or repel animals) have to account for the visual systems involved in the interactions. A recent study illustrates this problem, the case of female crab-spiders that we have seen already in Figure 16.12 and considered further in Box 16.1.

Box 16.1 **The crab-spider *Thomisus onustus* on flowers is cryptic to her predators and prey.**

Females of the crab-spiders *Thomisus onustus* and *Misumena vatia* use a sit-and-wait strategy when hunting on flowers visited by insects. It is striking to observe that individuals of the same species can use flowers of different colours but are generally cryptic on the particular flower on which they stay (Figure 16.12). This mimicry had been interpreted as providing a double advantage since the spider could not be seen either by its avian predators or its insect prey. However, nothing had yet demonstrated that what appears cryptic to human eyes was also true for two groups of species with very different visual systems.

To test this possibility, Marc Théry and Jérôme Casas (2002) measured the reflectance spectra of flowers and spiders *Thomisus onustus* on two plants: the common ragwort *Senecio jacobea* (with yellow petals to the human eye) and spearmint *Mentha spicata* (with pink petals). By using the detailed knowledge currently available on the visual systems of insectivorous birds (four photopigments ranging from UV to red) and hymenopteran insects (three photopigments from the UV to green) it was possible to analyse the colour contrasts and brightness of the spider on its flower. Colour vision spaces were constructed separately for predators and prey based on the spectral absorption of the pigments of their visual systems. It was then possible to project the measured spectra for the spider and the corresponding flower in these two visual systems. The colour contrasts were then measured as the Euclidian distance between the colour of the flower and the spider in both visual spaces. The comparison of the minimal distances, calculated as necessary for the birds and insects to discriminate, showed that colour mimicry functions with the same efficiency both in the visual systems of predators and prey: spiders are cryptic both for the predatory birds or their insect prey (Figure 16.17). At a greater distance, prey and predators can be attracted by differences in brightness between spiders and the flowers, but at short distance the colour contrast does not allow them to distinguish the spiders from the flowers.

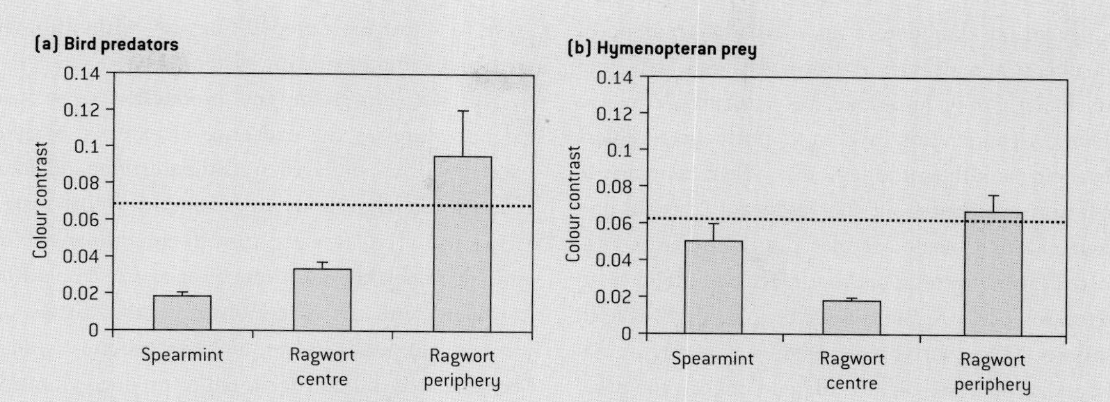


Figure 16.17 **Colour contrast of the crab-spider *Thomisus onustus* on flowers**

Colour contrast between the crab-spider *Thomisus onustus* and spearmint flowers *Mentha spicata* and the ragwort *Senecio jacobea* measured in the visual systems of **a**. predatory birds and **b**. bee preys. Figures represent the Euclidian distance between the spider and occupied flowers spectra; the distance is measured in the visual space of **a**. birds and **b**. bees. The horizontal dotted lines indicate the minimal thresholds required by predators or prey to be able to detect a colour difference: below this line, colours are perceived as identical, above, the colours are perceived as different. Two contrasts are calculated: between the spider and the central zone and between the spider and the petals at the periphery. The spider is only mimetic on the central zone where it usually stands; it will appear with greater contrast when it is placed at the periphery of the petals of ragwort.

Based on Théry and Casas 2002.

#### 16.3.2.3 *Chemical signals*

From the first unicellular micro-organisms, the detection of food has been the first function of chemical receptors. Ever since, animal chemoreceptors have evolved and are involved in a great diversity of types of chemical communication. The molecules acting inside an organism and allowing the communication between its different organs are called **hormones** (see Chapter 6). The molecules allowing communication among conspecifics are called **pheromones**. Those that are detected between species are **allelochemical** substances separated between **allomones** or allohormones that benefit the sender and **kairomones** that benefit the receiver. Pheromones and allelochemical substances can be detected by olfaction (detection by smell in air and water of chemical substances from a distant source) and by taste (which requires a direct contact of the receiver with the chemical compound). The pheromones can be produced by excreting glands or by organs involved in digestion or reproduction. **Endocrine** glands discharge hormones in the blood flow, thus helping regulate the metabolism of the organisms. **Exocrine** glands, on the other hand excrete pheromones or allomones.

The methods for chemical marking are very diverse. Liquid secretions can be emitted as a squirt directed towards a specific target, for example the formic acid squirt of an ant in response to the presence of a predator, the scent marking with urine by wolves or the reciprocal marking with urine in rabbits. A substrate like a tree or one's own fur can also be coated with products from exocrine glands. Certain species can produce an air or water flow that favours the movement of chemical compounds produced by the glands. Finally, morphological structures in the shape of brushes favour the diffusion of pheromones. The diffusion of chemical compounds follows the characteristics of fluid dynamics.

Information on source location is obtained by organs of reception that detect an intensity gradient of the chemical molecules. Arthropods smell, taste, and touch with their antennae; they can also perceive chemical compounds by contact with the receivers on their mandibles and legs. The olfactive organs of fish are separated from the respiratory system and their organs of taste are placed around the mouth. In terrestrial vertebrates the olfactory system is an extension of the respiratory system that feeds it with airflow. Many vertebrates possess a vomero-nasal organ, an olfactive organ situated between the nasal cavity and the palate that allows the tasting of odours in a liquid phase.

### 16.3.3 Constraints for signal broadcasting and receiving

#### 16.3.3.1 *Physical and phylogenetic constraints*

We have seen in Chapter 3 how phylogenetic inertia can pose an important methodological problem when comparing species. In the case of the evolution of communication the same problem arises: the evolutionary history of the species constrains the possibilities of evolution of the signal producing organs. In other words, an ancestral species that communicated principally by sound cannot easily evolve into a species that communicates by chemical or visual signals if these senses organs were absent or unused by the ancestral species. This restriction adds to the physical constraints imposed by the sensory channels of communication. In order to understand how signals have been selected during evolution it is thus necessary to identify the phylogenetic constraints that limit and constrain the evolution of adaptive traits. This approach requires the use of tools and concepts of the comparative method (Harvey and Pagel 1991; Martins 1996; see Chapter 2).

Generally, physical and phylogenetic constraints appear to be particularly strong in the case of acoustic signalling, relatively weaker for colour signals and weakest for chemical and electric signals. Acoustic communication seems much more restrictive for the production than for the reception of its signals. This is supported by the fact that many animals use audition to detect their predator or prey without having evolved a sound emitting system.

As we have seen, the production of acoustic signals requires the existence of external skeleton and mobile appendages to stridulate, tympanic membranes like the cicada or a complex vocal system. Thus, only crustaceans, insects, amphibians, birds, and mammals communicate by acoustic signals. Most reptiles (with the exception of geckos, turtles, and crocodiles), salamanders, and fish are mostly silent because it is difficult for them to produce vibrations and link them efficiently to the milieu in which they live. Low frequencies and high intensity signals are limited by body size and the energetic costs of producing them. On the other hand, visual communication is dependent primarily on eye performances, which are constrained by the zoological group as well as eye size that limits its resolution power. Eye sensitivity is also adapted to the intensity and colour of ambient light, which are obvious constraints for the evolution of visual signals. Large size and the existence of body structures as well as neuromuscular predispositions allowing eye movements favour the development of visual signals. Some pigments, for example carotenoids, cannot be synthesized by animals and need to be obtained from food. The development of pigment colours can thus be constrained by pigment provisioning through foraging ability and diet composition. Hence, the production of visual signals carries costs in terms of time and energy and increased susceptibility to predators and may interfere with the original function of the structures used in the signal.

Chemical and electric communication modes appear as being the less constraining for both senders and receivers, but their efficiency strongly depends on the environmental conditions that limit signal transmission. The transmission distance of odour molecules in air and the persistence or odour marks depend on their molecular weight. The power and potential for detection of electric signals are essentially limited by body size and this means of communication appears strictly limited to aquatic environments.

#### 16.3.3.2 *A general mechanism for the evolution of signals*

Ethologists think generally that signals have evolved from pre-existing behavioural, physiological, or morphological traits in the sender that have been modified through a ritualization process. During a first phase the pre-existing traits can play the role of cues. When the presence of these traits provides useful information for the receivers that can detect it, the receivers sensitive to this information will have an advantage in comparison with non-sensitive receivers. It is expected that during evolutionary time receivers should evolve sensitivity towards this trait and use it as a cue to make behavioural decisions (Johnstone 1998). At the origin, this situation is similar to cue exploitation (or to the use of signals in another context form which it appeared). Many signals evolving from already existing traits have been described (Bradbury and Vehrencamp 1998). For example, visual signals have evolved from three origins:

1. Intention movements are preparatory and incomplete feeding, escape, attack or self-maintenance behaviours. They inform on what the sender is going to do next.
2. Motivational conflicts appear when two opposed motivational systems like aggression and fear are strongly stimulated and give rise to ambivalent behaviours often leading to ritualization.
3. Stressful conditions can trigger physiological responses in the autonomous nervous system leading to changes in behaviour. Reflex responses can initially provide reliable information to the receiver of the state of fright or excitation of the sender. When these responses become more ritualized they can be uncoupled from their source and used to signal different information. Such ritualized behaviours can act as substitutive activities, for example simulating sleep as conflict appeasement in the Pied avocet, *Recurvirostra avosetta* (Tinbergen 1951), or activities involving a re-direction of behaviours towards an inadequate object if the main object cannot be attained.

In the first step, revealing cues of the state of the sender can be exploited by receivers that obtain a benefit. This first phase creates favourable conditions for a second phase in which cues can be ritualized according to the fitness benefits that the sender may obtain. At that stage, the former cue has become a real signal involved in true communication. We will see in Section 16.4 that theoretical approaches suggest that such a process takes place in two phases.

In the same way, acoustic signals and visual and tactile displays can have their origin in different behaviours because they imply exaggerated movements that can produce a percussion sound or vibration generating information that can be used from a distance. Powerful sounds originally used to chase predators, or normal locomotion movements as well as movements associated with feeding or breathing, can be re-directed in the context of communication.

Numerous chemical substances can have acquired their role as olfactive signals by recycling products of metabolism. Specialized secretory structures and chemical secretions or specific behaviours to disperse odours are good indicators for the potential existence of a chemical signal.

Characteristics of the receiver can also play an important role as a background for signal evolution. For example, animals can be sensitive to certain stimuli selected for prey or predator detection and senders can then exploit that natural sensitivity of the receiver. Two proximate mechanisms have been proposed for this scenario.

##### a *The sensory drive model*

The **sensory drive** model (Endler 1992; Endler and Basolo 1998; Ryan 1998) is where the physical and social environments favour the transmission of specific types of signal (for example long wavelengths in forested habitats) and impose perception biases on the receiver. Physical and social constraints shape the physiological and behavioural characteristics of the receivers acting as a filter on the sender, thus selecting for certain signal characteristics. Clearly, a visual signal has no chance of evolving in opaque environments, whereas in the same medium a sound signal will be favoured. In the same way, a green signal has no chance of being conspicuous in a green environment because it will not be distinguishable from the background. If these constraints are strong enough and differ between taxa, signals can then diverge rapidly along a phylogeny.

### b *The sensory exploitation model*

The **sensory exploitation** model (Ryan 1990; see Chapter 11) proposes that characteristics of the receivers that have evolved in a context other than in communication can be exploited by senders in a signalling context. In this second model, the sensory biases of the receiver are pre-existent because they provide a benefit in another biological function. For example, a nocturnal hunting organism will evolve a sensitive hearing giving it a predisposition to detect subtle sounds produced by conspecifics during daytime. In a similar way, individuals from a species feeding on yellow prey or fruits will be more sensitive to this colour and males displaying a yellow spot may have a better chance of being chosen as mates by females. For example, female guppies show a preference for males with larger, more chromatic orange spots. Rodd and collaborators (2002) proposed that the origin of the preference in female guppies could have arisen in the context of food detection. In their experiments, guppies showed a strong preference for orange objects outside a mating context. They propose that the origin for the preference for orange males could be due to a pleiotropic effect of a sensory bias for the colour orange and their results provide support for the hypothesis of a **sensory bias** for the evolution of mating preferences. Note, however, that these results could also be interpreted in the opposite direction: the preference for orange objects outside of a mating context may result from a long history of preference for orange males in that species because the orange colour of males is costly to develop and thus honestly signals male quality. The senders will exploit this bias for their own benefit by mimicking the characteristics of the trait to which the receivers are sensitive. In this model, the sender exploits a specific sensitivity of potential receivers. Sensory exploitation can generate new signals if the sender and the receiver both benefit from the response of the receiver to the signal as could be the case in mate choice situations (Johnstone 1998). An important lesson is that new signals can evolve without having a historical link with the old context, for example a sensory bias in food searching can lead to the evolution of a new trait to be a signal in the context of mate choice. In this scenario, signals can take totally arbitrary forms because the link between the signal and the expressed condition arise afterwards. This scenario also provides an explanation for the emergence of preferences for new traits in a population.

#### 16.3.3.3 *The example of light signals: the evolution of colour vision*

We will illustrate the issue of phylogenetic constraints with the example of vision. The phylogenetic history of colour vision in vertebrates can be described on the bases of the gene sequences coding for cone and rod pigments as well as using recent evidence concerning retina anatomy and behavioural responses to colours (Bradbury and Vehrencamp 1998). Cones sensitive to colours seem to have appeared first: the oldest living vertebrate, the lamprey, has an eye comprising only two cones leading to a dichromatic vision (Yokoyama and Yokoyama 1996). The rods could have then developed from cones during fish evolution and could have provided a better visual sensitivity to deep-dwelling species. Most fish species have one type of rod and two types of cone, whereas some fish have lost all their cones to the benefit of rods. A few freshwater species have acquired a third and, sometimes, a fourth cone type providing better colour vision. To face diurnal and nocturnal constraints, amphibians have conserved a retina with rods and a small number of cones. In contrast, numerous birds and reptiles are diurnal and appear to rely less on the high sensitivity of rods that are most useful under weak light conditions. With a reduced number of rods and an increased number of cones, diurnal reptiles and birds have a strong sensitivity to different colours and levels of saturation.

Indeed, various lizards have completely lost their rods. The geckos, a large group of nocturnal lizards, were submitted to strong obscurity constraints without having rod precursors. They have in fact evolved, from cones sensitive to blue, a new system having properties similar to those of rods. Snakes, originating from lizards, could also have developed a similar system.

Early mammals were probably endowed with a limited dichromatic vision because they arose from ancient reptiles that probably still had rods and probably had not lost their cones. They were small nocturnal animals dominated by reptiles and dinosaurs with a visual system containing rods (review by Martin and Ross 2005). With the disappearance of dinosaurs, mammal radiation was possible in newly available diurnal ecological niches and, accordingly, the proportion of rods decreased. Nevertheless, even in animals like squirrels with a large proportion of cones, the number of cone types remained stable. The original nocturnal habits of mammals seem to have restrained the developmental possibilities of colour vision in mammals. Only recently, the evolution in primates of a trichromatic system with the duplication of the opsin gene, sensitive to middle range (green) wavelengths and its sensitivity slightly shifted towards yellow, allowed an increment of colour vision in these mammals (see Martin and Ross 2005). In summary, phylogenetic constraints can have strong effects on the evolution of specific sensory systems. However, if benefits are important both for senders and receivers, constraints can be sidestepped towards a new solution in communication and signalling.

## 16.4 Conclusion

We have shown in this chapter that the evolution of signals depends on different types of constraint (whether phylogenetic or environmental) and was the result of a trade-off between natural (physical characteristic of the environment, predation) and sexual selection (intra-sexual competition, mate choice, social structure). The production and detection of signals are associated with their costs and benefits and it is to be expected that natural and sexual selection favour certain signals over others because they maximize the fitness benefits of participants. The optimization of signalling systems raises a central question in evolutionary biology: if two parties cooperate (see Chapter 15), a necessary requirement for a real communication, what types of signal do they need to adopt jointly in order to maximize fitness?

### a *Communication as a form of cooperation*

An important point to answer this question is to remember that all true communication involves a minimum of cooperation between at least two individuals: the intentionality implies the existence of a benefit for the sender and the honesty implies a benefit for the receiver. True communication is thus an interaction with mutual benefits. However, this situation is open to cheating, bluffing, and other trickery where certain individuals exploit the cooperation of others. This is a recurrent question in the evolution of social behaviour. We have seen in various examples of cheating how the two parties, sender and receiver, can exploit communication at the expense of the other. We have also seen in this chapter (and in Chapter 4) how individuals not directly involved in communication can nonetheless extract benefits from other individuals' communication and by doing so can either impose costs for the actors leading to exploitation, or not impose costs in which case we are in the presence of eavesdropping or indiscretion. We have seen various examples where the presence of a predator that can exploit a communication act as a constraint on the evolution and maintenance of signals with, in the extreme case, a situation where communication should be absent altogether.

Such a situation where unintended receivers impose a cost to the sender can be highly unstable, cheaters being favoured by selection because they can increase their fitness at the expense of other individuals (Johnstone 1998). The evolution of communication thus raises a series of interesting evolutionary questions:

1. Under which set of circumstances can the mutual benefits be maintained without invasion from cheaters? Answers to this question can be found in Chapters 8, 13, and 15, the last being specifically dedicated to the evolution of cooperation and providing various examples of similar situations. For example, models of direct and indirect reciprocity explained in Chapter 15 can be applied in the context of communication.
2. Under which set of conditions should the receiver take into account the information carried by a signal if the sender may send false information? We have seen in Section 16.2.6 that one of the possible answers to this question could be based on the handicap principle proposed by Zahavi (1975): if the intensity of a signal is associated with increasing costs for the sender then information can be honest because the differential costs of the signal does not allow the sender to cheat by producing a signal beyond its optimum. This is in fact a common output of models involving information gathering processes. As long as cheaters are not too frequent, cooperation (and thus communication) can still be maintained at the same time as a low frequency of cheaters.
3. How did precise encoding of information appear? A certain precision in the coding of the signal is required for the sender to obtain the participation of the receiver. This question is tightly linked to the next.
4. How can the receiver decode this information? The signal has to be detectable with enough sensitivity by the receiver, but this sensitivity is always limited by a maximal value (corresponding to the ideal receiver). Even if the signal is the most precise possible, receivers make mistakes as an inevitable component of their decision process. The optimum value between the precision of the sender and the sensitivity of the receiver is generally situated at an intermediate level of costs for the two participants while allowing an efficient communication.

### b *General considerations about decision-making*

We have seen in Figure 4.1 how general are the above considerations about communication for decision-making. Several models in various contexts showed that excessive precision in information (either by the accurate encoding–decoding of the information, or because the information predicts the future perfectly) probably make information use unnecessary, because selection then favours lineages that incorporate that information into their genes (or transmit it culturally, see Chapter 20). In other words, a certain level of error (or uncertainty) is necessary for the maintenance of decision-making processes. As communication is a form of decision-making, the same holds for communication.

In a recent model, Jack Bradbury and Sandra Vehrencamp (2000) stressed the necessity of the existence of a decision-making process by the receiver before the existence of the signal. They show that it is unlikely that the receivers have made random choices in the phase preceding the true signal (see Section 16.3.3.2). That preceding evolutionary stage probably involved the use of **inadvertent social information**, i.e. cues. Most of the receivers would then use mechanisms such as an estimation of the probability of benefits (probability to improve its condition and/or fitness) and estimate the value of cues to make a decision (see Section 16.2.5). A new signal will only evolve if it provides enough information so that the receiver is able to adopt a more adaptive response. On the other hand, the sender expecting a different answer than the one provided by the receiver can still benefit from the interaction by using imperfect signals with some level of cheating. Economic models of animal communication provide widespread support for the theory of **sensory drive** (see Section 16.3.3) and the handicap principle (see Section 16.2.6.1 and Chapter 11). All these theoretical considerations strongly support the scenario for the evolution of signals developed in Section 16.3.3.2.

## » Further reading

> *Apart from Chapters 4, 8, 9, and 14, complementary information on simple and dynamic optimization applied to communication can be found in:*

**Mangel, M. & Clark, C.W.** 1988. *Dynamic Modelling in Behavioral Ecology*. Princeton University Press, Princeton.

> *A book deals specifically with the evolution of animal signals:*

**Maynard Smith J. & Harper, D.** 2003. *Animal Signals.* Oxford University Press.

> *For an early review of the possibility of cheating in animal communication:*

**Krebs, J.R. & Davies, N.B.** 1978. *Behavioural Ecology: An Evolutionary Approach.* Blackwell Scientific Publications, Oxford.

> *A chapter by Johnstone provides a review on the use of game theory in the study of communication:*

**Johnstone, R.A.** 1998. Game theory and communication. In: *Game Theory and Animal Behavior* (Dugatkin, L.A. & Reeve, H.K. eds), pp. 94–117.

> *An approach using game theory and an essential work on communication is this book:*

**Bradbury, J.W. & Vehrencamp, S.L.** 1998. *Principles of Animal Communication.* Sinauer Associates, Sunderland, Massachusetts.

> *A synthesis on the economical approaches in communication in:*

**Bradbury, J.W. & Vehrencamp, S.L.** 1998. *Principles of Animal Communication.* Sinauer Associates, Sunderland, Massachusetts.

**Bradbury, J.W. & Vehrencamp, S.L.** 2000. Economic models of animal communication. *Animal Behaviour* 59: 259–268.

> *For those who would like to read more about birdsong:*

**Catchpole, C.K. & Slater, P.J.B.** 1995. *Bird Song: Biological Themes and Variations.* Cambridge University Press.

**Kroodsma, D.E. & Miller, E.H.** 1996. *Ecology and Evolution of Acoustic Communication in Birds.* Comstock Publishing Associates, Cornell University Press, Ithaca, New York.

> *The two volumes edited by Hill and McGraw (2006) explain in detail the mechanisms of colour production, function, evolution and analysis in birds.*

**Hill, G.E. & McGraw, K.J.** 2006. *Bird Coloration*, vol. 1. *Mechanisms and Measurements*. Harvard University Press.

**Hill, G.E. & McGraw, K.J.** 2006. *Bird Coloration*, vol. 2. *Function and Evolution*. Harvard University Press.

## » Questions

1. Analyse the relationship existing between the different concepts presented in this chapter (indiscretion, signals, cues) and the various types of information and the notion of excludability introduced in Chapters 4 and 14.

2. Describe the different steps of ritualization and explain how this type of communication can be free, or not, of cheating. We suggest reading Chapter 4 before answering that question.

3. Based on the information included in this chapter, what characteristics do you expect from a mobbing call (a sound signal) made by a bird to attract the largest number of individuals and thus attack a predator? In what way would this signal differ if produced in open or closed habitats?

4. Speculate on the environmental conditions that favour or do not favour olfactive communication. The same exercise can be done with visual communication.

# 17

# 17

# Inter-Specific Parasitism and Mutualism

Gabriele Sorci and Frank Cézilly

## 17.1 Introduction

Interactions among individuals belonging to different species are diverse, varied, and often complex. They play a fundamental role in structuring communities and help determine their richness and temporal dynamics (Begon *et al.* 1996). For instance, inter-specific interactions are particularly important in regulating energy flows within trophic networks, as is the case for instance of predation, an especially acute form of inter-specific interaction that has positive consequences for the predator and negative ones for the prey. Likewise, the sharing of space among species within ecosystems often implies competition by interference (Keddy 1989) through aggressive or avoidance behaviour, which may induce true inter-specific territoriality (Murray 1971; Wilson 1971). The concept of inter-specific interaction does not necessarily imply, however, any direct interaction or relationship between the protagonists. Straightforward competition between two separate species for the exploitation of the same resource may be enough to entail reduced availability of the resource for both species without any actual confrontation.

### 17.1.1 Long-lasting interactions

An essential characteristic for classifying inter-specific interactions is their persistence or duration in time. Some are ephemeral or even instantaneous by their very nature. A predatory interaction is brief, lasting the time required for a predator to detect, pursue, capture and ingest its prey (such forms of interaction are addressed in Chapter 7). But other forms of interaction are longer-lasting and extend over time, bringing together the same players either permanently or at regular intervals. Host–parasite systems are an excellent illustration of this. Most of the parasite's life cycle generally occurs inside the host that provides full board and lodging. The French parasitologist Claude Combes (1995) proposed the term **long-lasting interactions** for such lasting interactions between species. It should be specified that the duration of a long-lasting interaction is to be read at two levels. First, in terms of individuals, where the interaction is usually ended by the death of one of the interactants. Secondly, in terms of species history, where continued interaction leads to specialization over the course of evolution, in which at least one of the two partners has become entirely dependent on the other, thus ruling out any possibility of independent existence, as in the case of some flagellates that can only live in the gut of a termite. Some of the physiological functions of one partner species may have regressed to a point where they are accomplished entirely by the other partner species (Combes 1995): several species of termites are, for instance, completely dependent on their symbiotic flagellates for the degradation of ligocellulose and are unable to survive after their elimination (Cleveland 1926).

### 17.1.2 Long-lasting interactions, extended phenotype, manipulation, co-evolution

From an evolutionary standpoint, the relationship between two species can involve an interaction where the genotype of one influences the phenotype of the other. For instance, in a host–parasite interaction, the parasite's genotype can extend its expression into the host's phenotype, through various physiological changes. This is an example of an **extended phenotype**, a concept first introduced by Richard Dawkins (1982). The host's phenotype is generally changed to the parasite's advantage but to the host's detriment. When this is the case we speak of **manipulation** of the host by the parasite. The shared influences of the two genotypes, however, are not necessarily antagonistic and interaction between genotypes may sometimes verge on **cooperation** (see Chapter 15 for examples of such situations). Whether the interaction between the two species is antagonistic or synergetic, its persistence over generations allows the two genomes to co-evolve in time. In an antagonistic interaction, such **co-evolution** may take the form of a start-stop **arms race**, marked by periods of relative stability interspersed by events where each protagonist vies to manipulate the other or to resist manipulation.

### 17.1.3 The main types of long-lasting interaction

Intimate interactions can be arranged schematically into two main categories: **parasitism** and **mutualism** (Boucher *et al.* 1982). In the first category, one partner, the parasite, develops and reproduces at the expense of the other partner, the host. Their interaction is beneficial for the parasite and harmful to the host. In mutualism, both partners benefit from the interaction *a priori*.

However, the combination of costs and benefits of each partner varies widely from one long-lasting interaction to another (Smith 1992; Bronstein 1994; Combes 1995). Long-lasting interactions can be categorized by the straightforward cost–benefit evaluation for each partner (Table 17.1). However, the drawback of such a static typology is that it conceals the essentially dynamic character of interactions between species. Long-lasting interactions can vary in intensity in space and time and mutualistic interactions can veer towards parasitism (and vice versa) if the cost–benefit balance of either of the partners shifts on the scale of ecological or evolutionary time.

Depending on the organisms involved, long-lasting interactions take on different forms. Its physiological and genetic aspects are covered by several reviews (Boucher *et al.* 1982; Cushman and Beattie 1991; Bronstein 1994; Combes 1995; Begon *et al.* 1996). In this chapter we look exclusively at the role of the behaviour of the protagonists in such interactions and endeavour to illustrate the very wide range of organisms and the great subtlety of behaviours involved.

| | | Second protagonist's balance | | |
|---|---|---|---|---|
| | | *Negative* | *Zero* | *Positive* |
| **First protagonist's balance** | *Negative* | Unstable | Amensalism | Parasitism |
| | *Zero* | | Neutralism | Commensalism |
| | *Positive* | | | Mutualism |

Table 17.1 **Categorization of long-lasting interactions between individuals of different species by the cost–benefit balance of each protagonist**

Situations where both partners pay a cost seem unstable *a priori* because the interaction is then directly counter-selected and the two species cannot interact closely on a sustainable basis.

## 17.2 The role of behaviour in mutualistic interactions

Mutualism corresponds *a priori* to an interaction between species from which each partner derives some net benefit. It is widespread in nature and covers interactions between two animal species as well as those between plants and animals such as in pollination. Some authors use the term **symbiosis** instead of mutualism whereas others reserve symbiosis for highly specialized obligate mutualistic interactions between partners of which neither can survive without the other. We shall refer to the term mutualism here to refer to an outcome of an interaction, regardless of its evolutionary history and the size of the benefit derived by the protagonists. Such an interaction consists in mutual exploitation between two species and implies some mutual benefit. In this sense, it cannot be put down either as cooperation or an act of altruism as these concepts are meaningful only in the case of intra-specific interactions.

### 17.2.1 Mutualism, not so simple a relation

While some mutualistic interactions do genuinely appear to provide reciprocal benefits to both partners, it now seems that this state of affairs is probably an exception rather than the rule. The fact is that the boundary between mutualism and parasitism is often less clear-cut than might be thought at first sight and, in many instances, the old idea of stable mutualism, based on mutual and equivalent benefits for both parties, does not resist close scrutiny. The stability and equilibrium of mutualistic interactions depend, in fact, on many factors that may alter the associated costs and benefits (Bronstein 1994). For example, the abundance and identity of the other species with which mutualist species interact may exert substantial influence on their interaction. Depending on the ecological context, the 'service' rendered by a mutualist species may be of variable importance for the partner species. To illustrate this conditional aspect of mutualistic interaction, we develop two examples involving vertebrate species. The first interaction is facultative while the second may be obligatory.

### 17.2.2 Honeyguides, honey badgers, and humans

Honeyguides form a group of bird species that is unusual in more ways than one. Most of them live in the African savannah (Bakyono 1988). Several species can feed on and digest wax which they gather mainly from bees' nests, which is quite a rare feat in nature. They also eat adult bees and larvae. They have even been reported, on occasion, to eat the wax of candles on church altars (Friedmann and Kern 1956; Diamond and Place 1988)! But their interactions with humans are not limited to plundering places of worship. Their very name evokes an odd form of behaviour forming the basis of a rather exceptional mutualistic interaction. Some species of honeyguide are famous for leading honey-eating mammals such as the honey badger or ratel (*Mellivora capensis*) and humans to wild bee nests they are unable to open up for themselves. Stories of honeyguides directing gatherers to bee swarms were long thought to be legend or at least had remained anecdotal. That is until a three-year study in northern Kenya (Isack and Reyer 1988) provided hard evidence of a genuine mutualism between humans and honeyguides.

#### 17.2.2.1 *Of birds and humans*

The region of Northern Kenya in which the study was conducted is inhabited by a nomadic people, the Borans, who seek out swarms of wild bees to collect their honey. The Borans are often helped in their quest by the greater honeyguide (*Indicator indicator*), a bird about 20 centimetres long with brown plumage except for its tail, which looks to have white stripes in flight. The honeyguides' help is largely beneficial to the Borans as on average it reduces the time required to seek out a swarm from 8.9 hours without a guide to just 3.2 hours with a guide. The honeyguides too benefit from the interaction. Smoking of the wild hives by the Borans reduces the risk of being stung by bees and the use of tools to open up the nests makes more food available for the honeyguides. Isack and Reyer (1988) thus estimated that 96% of nests were accessible to honeyguides only through their cooperation with humans.

It is difficult to establish exactly when the mutualistic interaction between humans and honeyguides started, but rock paintings in the Sahara or southern Africa attest that humans collected honey there at least 20,000 years ago. During this period, the evolution of mutualistic interactions has gone along with stereotyped recruitment behaviour of honey badgers or humans by the honeyguides. After locating a swarm, the birds set about attracting the attention of an individual of either of the partner species. They then fly from perch to perch fanning out their tails and giving a characteristic call made up of double notes. When the partner shows an interest by approaching the bird, it flies a few metres on, and repeats its ploy until the swarm has been reached (Bakyono 1988; Isack and Reyer 1988). People too have developed a communication system to attract honeyguides. The Borans have made various instruments for the purpose (pierced shells, hollowed out pieces of wood) with which they can give out a shrill whistle that can be heard within a radius of about 1 kilometre. The birds seem to interpret the whistling as a call signal as it doubles the encounter rate between Borans and honeyguides (Isack and Reyer 1988).

#### 17.2.2.2 *Conditions for sustaining the interaction*

Mutualism between honeyguides and their partners relies, then, on reciprocal and honest communication (see Chapter 16). Examination of honeyguides' behaviour (Isack and Reyer 1988; see also Bakyono 1988) has revealed that, as reported by the Borans, the style of flight and the direction the bird takes are reliable information for localizing bee swarms. There seems to be little scope for cheating because the honey-eating mammals are mainly interested in the honey whereas the honeyguides have more of a liking for the wax and insect larvae. However, the mutualistic interaction must be regularly reinforced. In parts of Africa where honey collected from wild hives has been replaced by apiculture products or by other sugar-based substitutes, honeyguides seem to have stopped directing their recruitment behaviour at people (Bakyono 1988; Isack and Richner 1988).

Various factors may account for this swift disappearance. First, relatively recent interactions in evolutionary terms might be more labile. Secondly, it has been suggested (Bronstein 1994) that facultative interactions are less stable. Lastly, the tripartite character of the interaction (honeyguide, honey badger, human) might also make it less stable than obligate ones (Bronstein 1994). In this respect, it would be interesting to know whether in areas where people have stopped collecting honey from wild hives, honeyguides have redirected their recruitment efforts to honey badgers or more generally whether the frequency of recruitment directed to humans is negatively correlated with the local density of honey badgers. The absence of guiding behaviour in some honeyguide species might then be a recent loss, related to a change in the costs and benefits associated with the interaction. In the absence of any precise phylogeny of the group and of reliable data on the occurrence and frequency of the guiding behaviour in the different species, the question remains an open one.

### 17.2.3 Cleaning symbiosis

#### 17.2.3.1 *Oxpeckers: cleaners or vampires?*

The honeyguide is not the only bird species to include wax in its diet. Oxpeckers (*Buphagus*) also feed on wax but get it by 'cleaning' the ears of various species of cattle and antelopes. This aspect of their diet is little known and they are generally described as insectivores, feeding mainly on ticks and other ectoparasites they forage for on the hides of various wild ungulates. The sight of antelopes placidly carrying on their backs a group of oxpeckers busy ridding them of their ectoparasites has often been presented as a perfect example of mutualistic interaction (Dickman 1992; Connor 1995).

However, the prosaic image of the bird which, in foraging for its food, cleans the vertebrate and relieves it from the bites, stings and other bloody meals of a vast array of parasitic arthropods might actually be a long way from the truth. The fact is that oxpeckers do not confine themselves to taking wax or insects from ungulates (Weeks 1999). They also feed on blood from their wounds and delay their scarring. The painful or at least unpleasant character of these blood feasts for the host is evidenced by the reactions of the mammals, which writhe and sometimes manage to shake off the blood-eating birds. Weeks (1999, 2000) studied the feeding behaviour of red-billed oxpeckers, *B. erythrorhynchus*, on domestic cattle on a Zimbabwe farm. Ingestion of ticks made up only a small proportion of the feeding activity, which was dominated by exploitation of cattle wounds. In one experiment, Weeks (2000) prevented the oxpeckers from gaining access to part of the herd, while the other part was exposed to the birds. The number of ticks on each animal was counted at the beginning and end of the four-week experiment. No significant difference in the variation in the number of ticks was found between the start and end of the experiment whether the cattle were exposed to oxpeckers or not (Figure 17.1a). However, scarring of wounds was

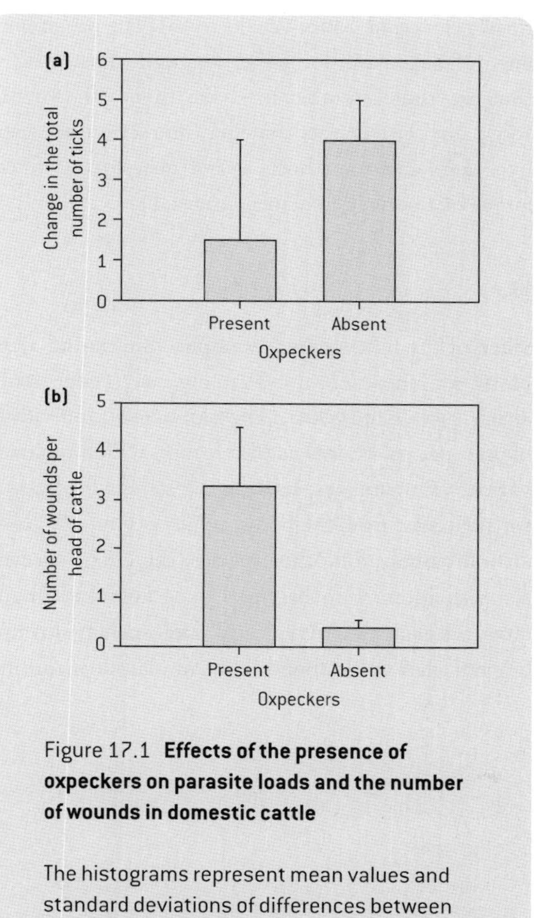


Figure 17.1 **Effects of the presence of oxpeckers on parasite loads and the number of wounds in domestic cattle**

The histograms represent mean values and standard deviations of differences between the end and the beginning of the experiment for: **a.** number of ticks ($P > 0.60$) and **b.** number of wounds per animal. The difference was significant for the number of wounds only ($P = 0.003$). See text for details.

Modified from Weeks (2000).

significantly longer in the group exposed to oxpeckers (Figure 17.1b).

It is hard to know to what extent these results may be generalized to interactions between oxpeckers and wild ungulate species, because domestic cattle did not co-evolve with oxpeckers. However, they do indicate that the costs and benefits of interaction may vary with the host species. We can know, then, generally whether oxpeckers behave more often as vampires or as honest cleaners. It seems they exclusively exploit wounds on hippopotami (Olivier and Laurie 1974), but their relation with impalas or rhinoceroses might be less of a disadvantage to their hosts (Hart *et al.* 1990; Weeks 1999). The net preference of oxpeckers for large, less agile host species that are thus less able to shake them off (Koenig 1997) further suggests that the interaction between oxpeckers and their hosts is not necessarily an example of a perfectly balanced mutualism.

#### 17.2.3.2 *Cleaner fish, client fish*

Some of the best-studied examples of cleaning symbiosis are those observed in the sea (Feder 1966; Arnal 2000; Côté 2000). They involve an individual of one species, known as the cleaner, which removes various ectoparasites, bacteria, or shreds of dead or rotting tissue from the body, mouth cavity, or gills of an individual of another species, the client species. As with interaction between oxpeckers and ungulates, the cleaners are typically small such as shrimps or small fish while their clients are larger including some species of ray and shark (Côté 2000; Sazima and Moura 2000).

The distribution of cleaning behaviour across the different taxa indicates that it evolved independently several times among crustaceans and fish (Côté 2000). Whereas cleaning behaviour has been described in more than 100 fish species, it remains a sporadic activity for most of them and only 18% can be called obligate cleaners in that their food comes mostly from cleaning (Côté 2000). Two families of fish are particularly relevant here, the Gobiidae and the Labridae. It is among the species belonging to these two families and their clients that quite original morphological and behavioural characteristics have been observed. First, there is the similar colour of different species of cleaners in different parts of the world (Côté 2000). Although cleaners vary greatly in shape from one species to another, there is a convergence in colour, particularly the occurrence of lateral stripes along the length of the body. In cleaner gobies of the genus *Elactinus*, a survey (Côté 2000) indicates that the lateral stripes are broader compared with body size in species heavily involved in cleaning interactions than in other species (Figure 17.2). However, no particular colour pattern is found in facultative cleaner species (Côté 2000).

##### a *Clients and cleaners*

Interactions involving an obligate cleaner species typically take place in a fixed territory known as a 'cleaning station' (Potts 1973; Arnal and Côté 1998).

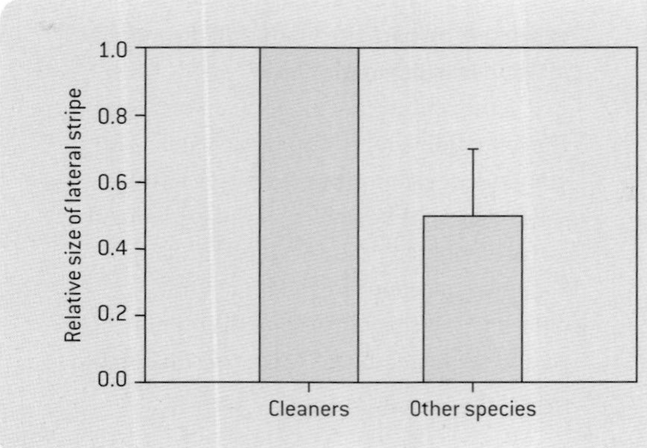


Figure 17.2 **Colour convergence in various cleaner gobies**

Relative size of the lateral stripe in goby species of the genus *Elactinus* depending on whether they engage in cleaning ($n = 6$) or not ($n = 7$). The difference is significant ($P = 0.04$).

After Côté (2000).

The interaction begins when a client fish freely enters the station or is actively recruited by the cleaner fish, which may swim in a zigzag pattern to make an approach. The client usually responds by adopting a stereotyped 'pose' (Losey 1979; Arnal 2000). It holds its body vertically, head down or head up, holds out its fins and opens its gills. In some client species, poses may also involve sudden colour changes (Côté 2000). Such poses generally trigger the inspection and cleaning activity (Côté *et al.* 1998). During the preliminary inspection, the cleaner strikes the client's body several times with its fins. It then sets about picking off various items from the surface of the client's body or interrupts the interaction and returns to the centre of the cleaning station. The cleaning interaction may also be terminated by the client, which often performs a series of somersaults before leaving the cleaning station (Eibl-Eibesfeldt 1955; Randall 1958).

### b *What does the cleaner gain?*

Determining what proximal mechanisms underlie the interaction between cleaners and clients and evaluating the resulting adaptive benefits for each of the parties is a major challenge in the study of cleaning symbioses (Côté 2000; Bshary and Würth 2001; Grutter 2001). For cleaners, it is largely accepted that they do benefit from the interaction by collecting an important food source from the client's body. Examination of the stomach contents of various cleaner species has revealed that, for most of them, ectoparasites were the main food source collected from clients. Several cleaner species, however, are not dependent exclusively on their clients and are able to capture invertebrates directly on the substrate. Within a given species of cleaner fish there may even be a geographical variation in trophic dependence for its client species (Côté 2000). It has also been suggested that cleaners might derive an additional benefit from mutualistic interactions in terms of protection from predators, which are kept at bay by the clients, but the limited empirical data available fail to support this hypothesis (Arnal and Côté 1998).

### c *What does the cleaner lose?*

The net benefit of a mutualistic interaction may be reduced by various costs. Several observations suggest that the cleaner's risk of becoming a client's prey is not entirely negligible. Several species concentrate their inspection and cleaning effort on parts of the client's body that least expose them to the risk of predation, such as the tail and fins, avoiding the danger zones of the head and mouth (Francini-Filho *et al.* 2000) even though that is precisely where ectoparasites are generally more abundant (Rohde 1980). Moreover, various cases of predation of client fish species on cleaner fish species have been reported (cf. Côté 2000 for a review of the data) even if they seem infrequent. Under some conditions, there would even appear to be a risk for cleaners of becoming contaminated by parasites in turn through contact with clients (Hobson 1971), although this drawback does not seem to concern all species (Bron and Treasurer 1992).

### d *What does the client gain?*

At first sight, the benefits for clients seem obvious enough. By removing ectoparasites, the cleaners must reduce the various costs related to the pathogenic effects that parasites induce. However, experiments where cleaner fish are removed locally to detect (by comparison with control areas) any effect of their presence on the parasite load of client species have not always yielded consistent results (Grutter 1996; Côté 2000). This is partly due to methodological problems. It seems that some cleaners concentrate their predation effort on larger parasites, which causes a concomitant increase in the number of small parasites. The total density of ectoparasites is therefore not directly influenced but the total biomass is reduced, which may be a benefit for the clients. Other studies (Grutter 1999; Grutter and Hendrikz 1999) suggest that, for some species, the time of day when the parasite load is measured may greatly influence the outcome. Some cleaner species are exclusively diurnal whereas ectoparasites colonize hosts day and night. Grutter (1999) observed that clients of the species *Hemigymnus melapterus* placed in cages on reefs with cleaner wrasse of the species *Labroides dimidiatus* harboured fewer ectoparasitic isopod crustaceans than individuals placed in reefs with no cleaners, but only if the parasite load was examined at the end of the day.

Visiting cleaning stations may also involve real or potential costs for clients. First, several client species are territorial. To get to the cleaning stations, the clients must therefore temporarily abandon their territory and run the risk of being supplanted by an intruder. Arnal and Côté (1998) have shown that clients of the species *Stegastes doropunicans* with a cleaning station in their territory get cleaned almost twice as often as individuals with no station (Figure 17.3), suggesting that having to leave their territory to get cleaned restricts their use of cleaner fish.

Secondly, travelling to cleaning stations may force clients to cross areas that are particularly exposed to predation or to go through the territory of other individuals where they may be attacked (Arnal and Côté 1998). Thirdly, the posture adopted by clients during a cleaning interaction may make them vulnerable to predation by other client species visiting the same station.

### e *Not so clean cleaners*

The most direct costs may result from the actual cleaning interaction. Most cleaner species do not confine themselves to picking off ectoparasites: they may also ingest scales and above all mucus (Gorlick 1980; Grutter 1997; Arnal 2000). Mucus is an important source of glycoproteins for cleaners. Moreover, whereas the biomass of ectoparasites available may vary greatly from one client to another, the presence of mucus is guaranteed. Loss of mucus may be important (Arnal 2000) and represents a cost for clients. Mucus is, in a way, the immune system of fish skin. Any reduction in it may lead to increased infestation by pathogenic agents and parasites. The mucus removed is replaced but the new mucus tends to contain fewer defensive substances than the old (Svensden and Bogwald 1997). Arnal and Morand (2000) showed that, in the Mediterranean cleaner wrasse *Symphodus melanocercus*, the intensity of cleaning was directly related to the clients' parasite load, but also to the quality of their mucus.

### f *Another hypothesis: the quest for tactile stimulation*

Confronted with the uncertainty as to the net benefit of actual cleaning for the client, some authors (Losey and Margules 1974; Losey 1987) have suggested that the primary motivation of clients is not to get rid of their ectoparasites but to obtain tactile stimulation. On this assumption, the removal of parasites by cleaners is not an indispensable component of the interaction. Cleaner fish are said to take advantage rather of the need for tactile stimulation felt by clients to exploit access to a food resource. And, indeed, tactile sensations resulting from cleaners' activity do seem to have a soothing effect on the clients (see Potts 1973; Lemaire and Maigret 1987). It has been suggested (Côté 2000) that such contacts could be useful to cleaners in localizing any prey on the surface of the client's body. However, a recent

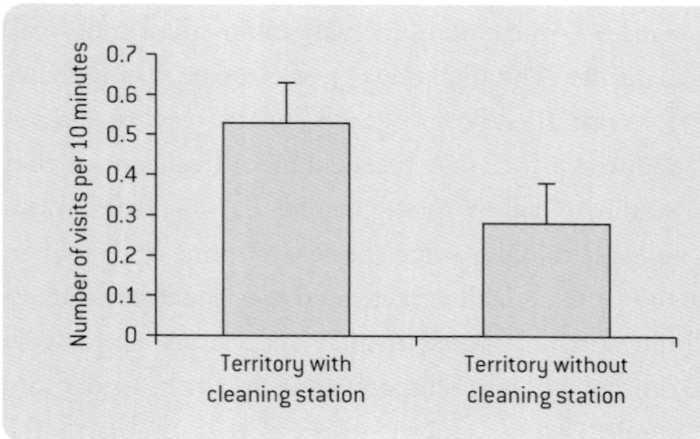


Figure 17.3 **Territoriality and frequency of visits to a cleaning station**

Frequency of visits to cleaning stations by clients depending on whether their territory contains a cleaning station or not. The difference is significant; one-tailed test: $P < 0.10$ ($n_1 = n_2 = 6$).

From Arnal and Côté (1998).

study (Bshary and Würth 2001) indicates that during tactile stimulations, the relative positions of client and cleaner are stable, with the cleaner's head typically pointing away from the client's body, which seems incompatible with prey capture.

However, apart from the fact that the actual physiological effect of tactile stimulation remains to be demonstrated, evidence against the hypothesis of tactile stimulation has recently emerged. Grutter (2001) studied the behaviour of client fish faced through a glass in an aquarium either with a cleaner species or with a non-cleaner species as a control. The experimenter manipulated the parasite infestation of the clients to check the effect this had on the client's tendency to seek out contact with cleaners. Infected client fish spent more time near cleaner fish than non-infected individuals did. Moreover, infected client fish spent more time close to cleaners than close to control fish, whereas no preference was observable in non-infected individuals. Grutter (2001) concluded that parasite infestation rather than the search for tactile stimulation lay behind the client's behaviour. This does not mean, though, that tactile stimulation has no role in the cleaning interaction. From field observations, Bshary and Wurth (2001) showed in *L. dimidiatus* that tactile stimulation of cleaners extended the posture time of clients particularly by soothing clients after reaction to a bite inflicted by the cleaner. The same authors also observed that tactile stimulations were used more often by cleaners in interactions with predatory client species than during interactions with non-predatory species.

## g *And the bottom line of the interaction?*

Detailed cost–benefit analysis of the client–cleaner interaction reveals that the mutualistic interaction of cleaning involves an easily upset and delicate balance with different costs and benefits. The stability of the relation once again seems dependent on the more or less ambiguous role played by each protagonist. If the cleaner, tied to its cleaning station, tends to exploit its clients' mucus rather than to relieve clients of their parasites, there is a real risk that the clients will desert the station. Conversely, if the client fish turn up for cleaning even if they have few parasites, they run the risk of the cleaners sanctioning an unprofitable inspection by taking a large helping of mucus. Even if various factors suggest that cleaning symbioses imply some degree of honesty between protagonists (Arnal 2000), particularly because clients visit stations more readily when they are heavily infected, the benefits derived by each party may vary between species and for a single species from one geographical area to another. Moreover, such honesty is sometimes corrupted by mimetic species (Côté 2000) such as the blenny *Aspidontus taeniatus*, which imitates the wrasse *L. dimidiatus*, a common and widespread cleaner species (Figure 17.4). The mimetic species deceives clients by preferring to feed on scales and tissue rather than removing ectoparasites. The effect of mimetic species on the stability of the interaction between cleaner fish and clients remains to be appraised, however. The

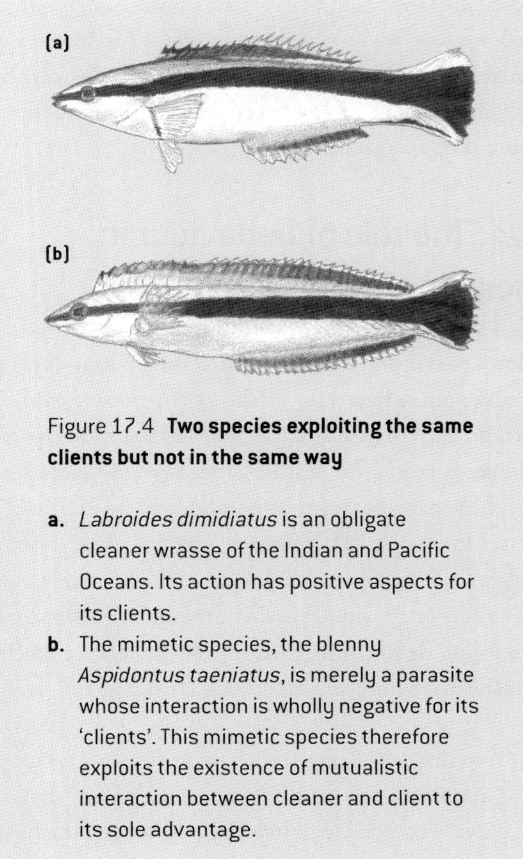


Figure 17.4 **Two species exploiting the same clients but not in the same way**

- **a.** *Labroides dimidiatus* is an obligate cleaner wrasse of the Indian and Pacific Oceans. Its action has positive aspects for its clients.
- **b.** The mimetic species, the blenny *Aspidontus taeniatus*, is merely a parasite whose interaction is wholly negative for its 'clients'. This mimetic species therefore exploits the existence of mutualistic interaction between cleaner and client to its sole advantage.

complexity of such interaction very likely requires the use of a quantitative modelling approach for understanding all the ins and outs.

#### h *Where next?*

The question of how cleaning symbioses evolve and are sustained in the marine world is therefore far from being resolved. Future progress may come from comparative analyses. Some fish species behave like true parasites, feeding on scales and mucus they pick off from other species (Hoese 1966; Major 1973). The hypothesis of cleaning symbioses evolving from non-specialized ectoparasitic forms (Gorlick *et al.* 1978) is worth appraising from phylogenetic data. This same approach would be useful for a better comparison of the characteristics of interactions between facultative and obligate cleaner species. Unfortunately, phylogenetic relationships among cleaner species and client species are not well established. Moreover, because of the complexity of the interaction, theoretical approaches would be very useful for specifying on what conditions mutualistic interactions can be sustained. Chapter 15 provides examples of the input from such theoretical approaches.

## 17.3 The role of behaviour in host–parasite relations

Host–parasite relations are a form of long-lasting interaction where one of the two partners gains a benefit at the expense of the other. By definition, parasitism is costly for the host because the resources used by parasites cannot be allocated to the host's vital functions. Therefore, parasites exert strong selection pressures on their hosts, which, on the scale of evolutionary time, has favoured the appearance of the hosts' defence mechanisms. These defence mechanisms have multiple functions. They may involve:

1. Avoiding encounters with parasites.
2. Containing the proliferation of pathogens that have succeeded in contacting or entering the host.
3. Limiting the adverse effects on host's fitness.

Behaviour may have a significant role in each of these stages. Avoiding encounters with parasites may be the result of a series of behavioural choices such as the decision to settle in one habitat rather than another, to copulate with one mate rather than another or to choose a given prey. Even though containing the parasite population (point 2 above) is often achieved by the immune system, some behaviours may affect the efficiency of the immune function. The ingestion of certain foodstuffs rich in carotenoids may, for example, stimulate immune responses (Olson and Owens 1998). Other examples show the existence of self-medication in animals such as birds and mammals. Some avian species such as the European starling, *Sturnus vulgaris*, or the blue tit, *Parus caeruleus*, choose plants with anti-microbial and anti-parasitic properties to line their nests (Clark 1990; Lafuma *et al.* 2001; Petit *et al.* 2002), a behaviour that ultimately reduces the risk of ectoparasites or other pathogens developing in the nest (Clark and Mason 1985).

In the remainder of the chapter we look at the central role played by behaviour in host–parasite interactions. As a consequence of the arms race that opposes hosts and parasites, some parasites have acquired the ability to 'manipulate' their hosts' behaviour and physiology, going as far as hastening their death. The adaptive character of such manipulation is discussed in the next section. We then address a complex phenomenon where the entire host–parasite interaction, whether the strategy of parasitism or the host's defence mechanisms, is based on behaviour: **brood parasitism**. Lastly, the consequences for sociality and group life, in terms of risk of parasitism, are covered in the final part of the chapter.

### 17.3.1 Parasitic manipulation

The concept of the extended phenotype (Dawkins 1982) includes the capacity of an organism's genes to create phenotypes that induce phenotypic modifications in another organism (see Chapter 2 on the relation between genes and their vehicles, that is, the avatars formed by individuals whose behaviour ensures their continued existence on the evolutionary scale). The modifications induced in one organism by the genes of another are adaptive if they help to increase the selective value of these extended phenotype genes via the increase in the vehicle's phenotypic fitness. For the time being, the genes of parasites conferring on them the capacity to 'manipulate' their host's phenotype have not been clearly identified. However, there is an abundant literature describing changes in the phenotype of hosts infected by some parasitic species and quantifying the consequences of such alterations in terms of survival and development of the parasite (Combes 2001; Moore 2002). These modifications are often gripping and unexpected.

One particularly fascinating case of parasite manipulation has recently been observed in palm tree plantations in Cost Rica (Eberhard 2000). It involves a parasitoid wasp, *Hymenoepimecis* sp. and an orb-weaving spider, *Plesiometa argyra*. The interaction begins when a wasp lays its egg (oviposits) into the spider's abdomen. During the first two weeks after laying, the larva develops by sucking the haemolymph of its host, without causing any noticeable change in the behaviour of the spider, which continues to weave its geometrical web. However, on the eve of the spider's death, its spinning behaviour is suddenly changed. It produces a new structure, unrelated to what the spider usually wove before. Instead of the classic web usually produced (Figure 17.5a), the spider constructs a sort of cocoon fastened to the vegetation by two to eight threads (Figure 17.5b). Once the cocoon has been spun, the parasitoid undergoes a moult before killing and devouring the spider. It then rolls itself in the cocoon, which it suspends vertically from the network of threads. According to Eberhard (2000), the change in host behaviour clearly benefits the parasitoid as the modified, stronger web provides better protection against heavy rain, a cause of high mortality in *Hymenoepimecis* wasps.

Spectacular as it may be, this example of manipulation is not unique. Other parasite species are able to manipulate their host's phenotype to their advantage in surprising ways. Moreover, a single parasite is often able to cause several changes to its host's phenotype. We detail here two particular aspects of manipulations induced by parasites. The first concerns the effect of parasites on their host's reproduction. The second deals with modifications induced by parasites whose development cycle implies a stage of trophic transmission from an up- to a down-line host.

(a)

(b)

Figure 17.5 **Example of the impact of a parasite on its host's behaviour**

**a.** Web normally constructed by the spider *Plesiometa argyra*.
**b.** Outcome of manipulation by the parasitoid wasp *Hymenoepimecis* sp. The modified 'web' being stronger, it allegedly provides protection against hard rain which is fatal to the parasitoid insect.

After Eberhard (2000).

#### 17.3.1.1 *Manipulation of host reproduction*

##### a *Horizontal transmission of parasites*

Natural selection favours different sorts of interaction between parasites and the reproduction of their hosts depending on the way parasites are transmitted in space and time. **Horizontal transmission** is the commonest form of transmission and the only known one in metazoan parasites. We speak of horizontal transmission of parasites when parasites are transmitted from a host either during contact between hosts (as with ectoparasites and sexually transmitted parasites) or by sexual or asexual production of free propagules, which will in turn infect new host individuals.

Parasites provide for their own subsistence at their hosts' expense from the resources they manage to divert. Generally, a host, at the adult stage, allocates its resources, like any organism, to two essential functions: its maintenance and its reproduction. While the host's maintenance is paramount for the parasite's survival and growth, whatever energy the host puts into its own reproduction is no longer available for the parasite. Moreover, living organisms have to face the **trade-off** between survival and reproduction (Roff 1992; see Chapter 5). Any extra investment in either of these components of phenotypic fitness is made at the expense of the other. Symmetrically, reducing the energy allocation to one component may make it possible to increase the investment in the other component. *A priori*, it is in the parasite's interest that their hosts should not favour their reproduction at the expense of their survival. Accordingly, many parasites have a negative effect on their host's reproduction. Parasitic infection often involves castration of their hosts, which may be total or partial depending on the host–parasite system considered (Poulin 1998; Hurd 2001; Bollache *et al.* 2002). In terms of mechanisms, castration may be brought about by mechanical destruction of gonads or may involve physiological disruption of the host's gametogenesis.

The term 'behavioural castration' is used when the parasite's presence merely reduces its host's ability to gain access to sexual partners, without affecting its physiological ability to reproduce. This may be because of a reduction in the sexual attractiveness of the infected hosts, or in their capacity to compete with rivals, or in their degree of reactivity to a mate (Hurd 2001).

For example, males of various amphipod species infected by acanthocephalans have markedly less mating success than their non-infected conspecifics in the wild (Figure 17.6; Zohar and Holmes 1998). This deficit is apparently related to the reduced competitive ability of the infected males and their lower reactivity when in the presence of a sexually receptive female (Bollache *et al.* 2001). As the investment of male amphipods in reproduction comes at a physiological price (Robinson and Doyle 1985, Plaistow *et al.* 2003), their behavioural castration

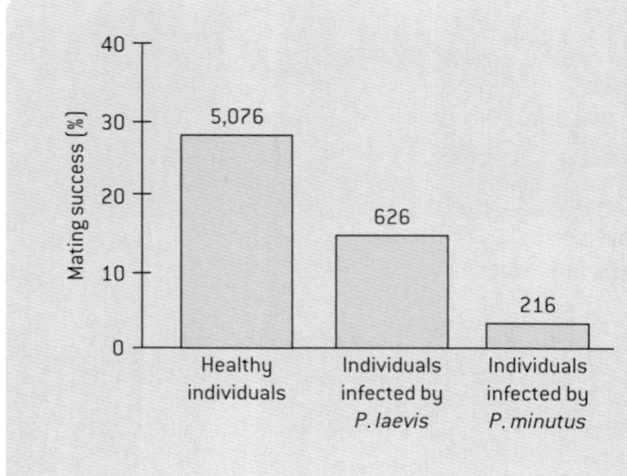


Figure 17.6 **Parasite infection and reduction in mating success**

Mating success of males in the amphipod crustacean *Gammarus pulex* infected by two species of acanthocephalan parasite, *Pomphorhynchus laevis* and *Polymorphus minutus*. The bars show the percentage of individuals mated by category. The numbers above the bars indicate sample sizes. The difference between the three situations is significant ($P = 0.001$).

Redrawn from Bollache *et al.* (2001).

could therefore benefit the parasites by preventing partial exhaustion of the host's resources.

### b *Vertical transmission of parasites*

We say that a parasite is **vertically transmitted** when it is transmitted from parents to offspring. Various intracytoplasmic symbionts have a vertical mode of transmission (Dunn *et al.* 1995; Rigaud 1997). They are transmitted from generation to generation via the cytoplasm of maternal origin contained by each egg. Females whose cytoplasm is infected produce infected eggs.

Vertically transmitted parasites are an even more demonstrative example of manipulation of host reproduction. For example, intracytoplasmic symbionts may increase their transmission from one generation to the next, even in the absence of any other advantageous effect and even if the infection implies a cost for the host, by biasing the sex ratio of host populations towards a surplus of females (Werren and O'Neill 1997). Such situations are not rare.

### c *All too feminine woodlice*

This capacity has been demonstrated, for example, in several intracellular parasitic micro-organisms of crustaceans (Rigaud 1997) and particularly by the symbiotic bacteria *Wolbachia* in various species of terrestrial isopods (see Bouchon *et al.* 1998; Rigaud *et al.* 1999). In terrestrial isopods (woodlice), the presence of the feminizing bacteria transforms males into functional neo-females, capable of mating with unaffected males and producing offspring. *A priori*, the feminizing bacteria *Wolbachia* should rapidly invade woodlouse populations whose sex ratio it is able to manipulate. However, it seems that feminization is not perfect. In the species *Armadillidium vulgare*, males interact more and make more attempts to mate with genetic females than with neo-females and the latter copulate less and receive less sperm than uninfected, genetic females (Moreau *et al.* 2001).

The feminization of males is not the only mean by which *Wolbachia* bacteria can bias their hosts' sex ratio to favour their own transmission. In various species of butterfly of the genus *Acracea*, another *Wolbachia* bacteria, also inherited from the mother but different from that in isopods, produces the same effect by being lethal for male eggs in the course of their development (Jiggins *et al.* 1998). Selection favours propagation of the intracytoplasmic parasite because the death of males benefits their sisters, either by reducing competition among siblings or by providing a source of nutrition, as the females devour their dead brothers (Hurst and Majerus 1993). Surprisingly, the distortion of the sex ratio caused by *Wolbachia*, which may be very significant in some butterfly populations where more than 90% of females are infected, seems to be the cause of a change in the host species' breeding system. In nature, a large proportion of females remain unmated, suggesting their reproductive success is limited by their difficulty in meeting up with males (Jiggins *et al.* 2000). In some places these females form swarms rather like reversed **leks** (cf. Chapters 11 and 12) and exhibit behaviour that seems to be designed to induce males to copulate. Such leks are not observed in populations where the butterfly sex ratio is only weakly biased in favour of females (Jiggins *et al.* 2000).

#### 17.3.1.2 *Host phenotype manipulation and trophic transmission of parasites with complex cycles*

Cases of parasite manipulation are particularly well documented in various parasite species having heteroxene cycles, that is, involving more than one host (Combes 2001). In such cycles it is common for the parasite to undergo various stages of development, some of which imply an asexual phase of reproduction. The final sexual reproductive stage, however, can only be achieved in the last host, known as the final host. In most species, transmission from an upstream intermediate host to a final host downstream is by trophic means: the intermediate host is a prey of the final host. In this context, any alteration of the intermediate host's phenotype making it more vulnerable to predation by the final host benefits the parasite. Since the 1970s, many examples of phenotype alteration of infected hosts have been interpreted as manipulation by the parasite as they facilitate its trophic transmission to the final host. The phenomenon has been studied in different groups of parasites, but is particularly marked in interactions linking trematodes and acanthocephalans to their invertebrate intermediate hosts (Moore 2002).

*Gammarus manipulated by its parasites*

Cézilly *et al.* (2000; see also Bethel and Holmes 1973, 1977) provided evidence for a connection between the type of phenotypic change observed in parasitized hosts and the feeding behaviour of the final hosts. The amphipod *Gammarus pulex* is regularly infected by two acanthocephalan species, *Pomphorhynchus laevis* and *Polymorphus minutus*. The two parasites differ markedly in the type of final host they must reach to complete their life cycle. *P. laevis* can only mature and reproduce in fish species, whereas only bird species constitute an appropriate final host for *P. minutus*. Cézilly *et al.* (2000) showed that the two parasites bring about contrasted alterations in the behaviour of their common intermediate host. Uninfected gammarids are strongly photophobic (they tend to avoid lighted areas) and show positive geotaxis (i.e. they tend to stay close to the bottom of the water column). By contrast, infected individuals of *P. laevis* are attracted to light and infected *P. minutus* individuals show negative geotaxis and tend to swim near the surface, where they become particularly vulnerable to predation by aquatic birds.

A recurring limit in studies of parasite manipulation is that they rely mostly on experiments with infected hosts collected from the wild. Strictly, this requires considering an alternative to the manipulation hypothesis: the modified behaviour could, in fact, not be the consequence but rather the cause of the parasite infestation. For example, the least photophobic amphipods might also be those most exposed to parasites. This is an unlikely hypothesis, though. For one thing, at least with acanthocephalans, the intermediate host is infected by eating the eggs of the parasite which are released into the water with the final host's faeces. Being more or less photophobic does not seem to predispose amphipods to ingest parasites. Moreover, phenotypic alterations only become apparent when the parasite has reached the stage of development at which it becomes infectious for the final host (Maynard *et al.* 1998).

#### 17.3.1.3 *Is manipulation adaptive?*

Showing that changes in phenotype are the consequence of infection does not in itself demonstrate that they are the outcome of parasite manipulation, nor that they are adaptive for the parasite. Some alterations in phenotype may simply be the by-product of a physiological defensive reaction by the host to the parasite infection. For instance, fever, the elevation of body temperature, may be seen as an adaptive reaction to infection because elevated temperatures can harm perhaps even eliminate parasites (Moore 2002). However, fever is also sometimes quite debilitating and it is possible that in the end it exposes the sick individual to greater threats from predators, which perhaps are the parasite's final host. Another hypothesis is that the infected hosts commit a sort of 'adaptive suicide' to slow down the parasites' demographic onslaught (Smith Trail 1980). However, the evolution of such behaviour can be envisaged only through **kin selection**: that is, if this behaviour favours kin. Now, in most interactions between helminthic parasites and their arthropod intermediate hosts, the mode of dispersal of the parasite and the time it takes to develop make the adaptive suicide hypothesis unlikely (Moore 1984, 2002).

### a *Criteria for speaking of manipulation of the host by the parasite*

Not all changes in the infected host's phenotype can be regarded as extended phenotypes. The Canadian scientist Robert Poulin (1995) has listed several criteria by which 'true' cases of parasitic manipulation can be distinguished from by-products of infection:

1. Phenotypic alteration in the infected intermediate host must be of a complex character.
2. Phenotypic alteration in the infected intermediate host must favour encounter with an appropriate final host.

3. Similar changes in phenotype must have evolved independently in different host and parasite lineages (convergent evolution).
4. Parasite's fitness must increase as a direct consequence or the changes brought about in the host.

### b *Relative value and application of these criteria*

The first two criteria are designed to distinguish changes in the host's phenotype from simple pathological effects. A general weakening of the host, for example, may be the direct consequence of the parasite exploiting the host's resources. This loss of stamina may explain both the host's reduced capacity to react to predators and the observed lower mating propensity of infected males. In the two cases, the alteration of host's behaviour is not an adaptation for the parasite's benefit and should be regarded as the simple expression of its direct pathological effect. Some changes, however, cannot readily be put down as simple by-products of infection. For example, in various animal species, healthy individuals ordinarily tend to avoid encounters with their predators. Chemical signals betraying the presence of predators have a strong repulsive effect and are enough to cause them to flee or to seek shelter. In parasitized intermediate hosts, this aversion to predator odours is often altered. The absence of response might be analysed as a pathological disorder of the host's nervous system. However, in some cases, the deterioration is not just an absence of response but implies an actual **reversal** of the response. What was a repulsive predator's odour becomes an attractive one.

*Prey attracted by their predator . . .*

Such a phenomenon has been reported in rodents carrying toxoplasmosis. *Toxoplasma gondii* is an intracellular protozoan with a complex cycle, able to infect all mammals (Webster 2001). However, cats are the only known final hosts appropriate for the parasite: *T. gondii* eggs are found in cat faeces only, and never in those of other mammals infected by the parasite. When the eggs are ingested by a mammal, such as a rodent, the parasite encysts in the tissues, mainly in the brain. A cat may therefore be contaminated either directly by ingesting the parasite eggs or indirectly by ingesting the cysts of a previously contaminated prey. The brown rat, *Rattus norvegicus*, is a particularly important intermediate host for *T. gondii*, with an average prevalence of 35%. Rats usually avoid places where they perceive evidence that cats are present. The manipulation by *T. gondii* transforms the innate aversion of rats for the smell of cats into a (probably fatal) attraction (Figure 17.7). This shows that the change in behaviour induced by the parasite is both subtle and specific. Only the reaction to the smell of cats seems to be altered, while all other behaviours of the infected rats and their general state of health remain unchanged. A similar example has been reported in *Gammarus pulex* parasitized by *Pomphorhynhus terreticollis* (Perrot-Minnot *et al.* 2007; see also Hechtel *et al.* 1993): infected gammarids were significantly attracted towards the scent of a fish that was the parasite's final host whereas healthy gammarids were repulsed by the same odour.

*. . . prey that forget to hide*

In the preceding example, the fact that two similar manipulation phenomena evolved independently of each other in host–parasite interactions involving phylogenetically very remote organisms but sharing the same requirement for trophic transmission helps to establish their adaptive character, in accordance with the third criterion laid down by Poulin (1995).

The similarity of the changes produced in the same type of host by different parasites exploiting the same type of final host is equally demonstrative. For instance, the bird trematode *Microphallus papillorobusts* alters the geotaxis of its intermediate host, the amphipod *Gammarus insensibilis* (Helluy 1983, 1984). This is exactly the same type of behavioural change that the one reported above in the amphipod *G. pulex* parasitized by the acanthocephalan *Polymorphus minutus* whose final host is also a bird (Cézilly *et al.* 2000). The adaptive character of the manipulation may be judged even better by comparing the changes induced by a given parasite within a group of related host-species with the phylogeny of the group. This approach indicates whether such a

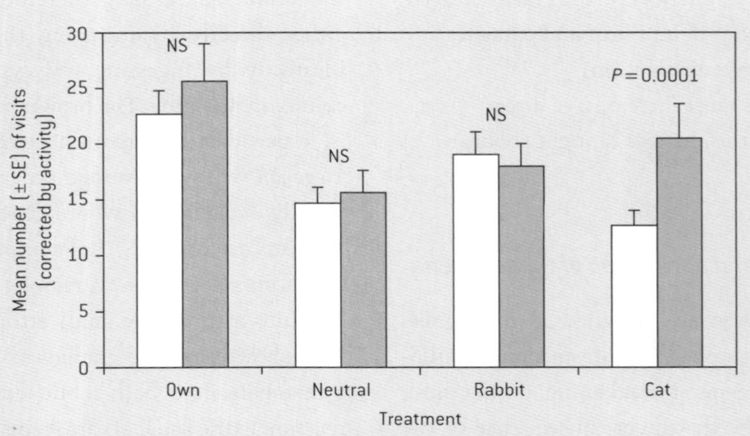


Figure 17.7 **Parasite infection and behaviour towards a predator**

Behaviour of healthy rats (white bars) and rats infected by *T. gondii* (dark bars) towards the smell of cats. Four different odours were presented simultaneously, each at four different corners of a 2 m × 2 m enclosure.

Own: straw marked by the rat's own odour; neutral: fresh straw soaked with water; rabbit: straw marked by rabbit urine; cat: straw marked by cat urine.

The histograms show the number of visits by healthy or infected rats to the four corners marked by one of the four odours. Infected rats differ from healthy rats only in their attraction to the cat odour.

Redrawn from Berdoy *et al.* (2000).

manipulation of behaviour appeared only once or several times independently in different taxa (Figure 17.8). In the absence of phylogenetic inertia, it is then possible to conclude that the manipulation is adaptive. This historical perspective is often missing from studies of manipulation because of the difficulty of experimentally controlling complex cycles of infestation and, in part, because of the lack of reliable phylogenetic information. This gap has been partly filled by Moore and Gotelli's study (1996) of the susceptibility of 29 species of woodlouse to manipulation by the acanthocephalan *Moniliformis moniliformis*. Species susceptibility to the manipulation varied with the subfamilies considered but did not concord with phylogeny (Figure 17.8), confirming the adaptive character of the manipulation for the parasite (Moore and Gotelli 1996).

*. . . Does that facilitate transmission of the parasite?*

The final criterion for establishing the adaptive character of manipulation consists in checking that the infected hosts are indeed more susceptible than healthy specimens to predation by the final hosts. In a recent review, Moore (2002) counted 25 studies having concluded that there was increased predation of infected hosts by the parasite's final host. Overall, Poulin's (1995) fourth criterion seems to be confirmed. However, closer scrutiny of published studies urges caution. First, some studies (Urdal *et al.* 1995; Webster *et al.* 2000) have failed to prove differential predation on infected intermediate hosts although their behaviour had been deeply affected by the infection. Secondly, most studies generally presented equal numbers of healthy and infected individuals to predators, whereas in nature the prevalence of manipulating parasites is often low, of the order of 5–15%. Oddly enough, the significance of the relative density of infected hosts in the phenomenon of increased vulnerability to the final host has not yet been considered. Lastly, the experiments all consisted of testing the vulnerability of intermediate hosts and healthy individuals for the parasite's

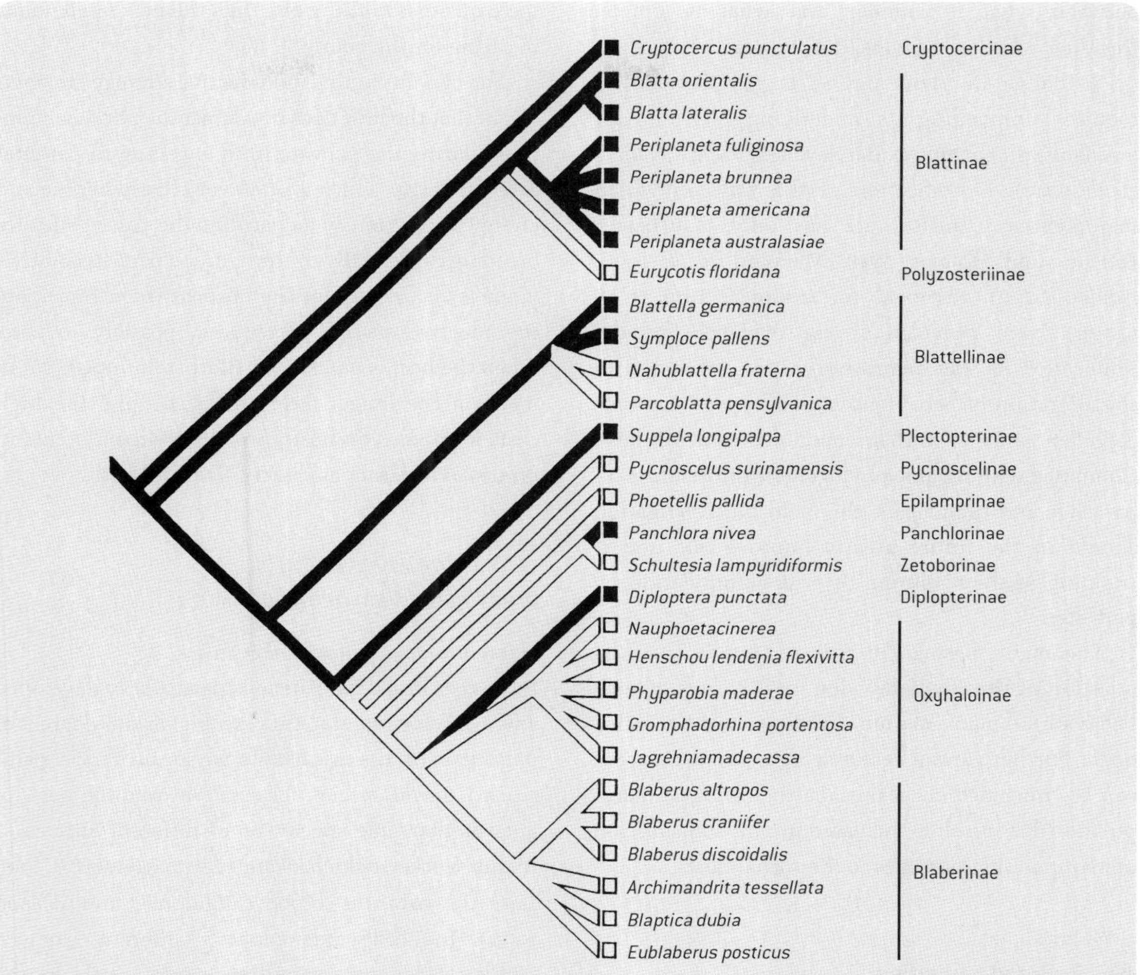


Figure 17.8 **Phylogeny of woodlouse species, showing susceptibility and resistance to manipulation by the acanthocephalan parasite *Moniliformis moniliformis***

Black branches are branches of the phylogeny for which the most parsimonious reconstruction implies that the species were susceptible to pathogen manipulation. White branches are the parts reconstructed as being resistant to parasite manipulation. The ancestral state in this group is therefore the non-resistant state. It appears that acquisition of resistance to manipulation probably occurred three times in different branches of the phylogeny. In addition, resistance was lost at least twice among the resistant taxa.

After Moore and Gotelli (1996).

appropriate final hosts. However, phenotypic changes induced by the infection might equally well make infected hosts more vulnerable to other predators within which the parasite would be unable to complete its cycle (Mouritsen and Poulin 2003). The benefit of manipulation for the parasite would then be considerably reduced.

### C *What conclusions can be drawn?*

In some respects, parasite manipulation seems, then, to extend beyond the simple pathological consequences ordinarily related with infestation. It should not systematically be accepted as adaptive, however, because it is so seldom easy in biology to distinguish between what is complex and what is simple. Deciding whether a reversal of the reaction to light or a change in pigmentation is complex ideally requires unravelling its underlying physiological mechanisms (Thompson and Kavaliers 1994; Kavaliers *et al.* 1999). Identification of the ways in which manipulative parasites act has only just begun (Helluy and Holmes 1990; Maynard *et al.* 1996; Adamo 2002). Recently, for instance, Tain *et al.* (2006) have provided strong evidence for the implication of the neurotransmitter serotonin in the alteration of behaviour caused by two acanthocephalan parasites in their common crustacean host. Combined with reliable phylogenetic information on parasites and their hosts, this field of investigation should in the future greatly improve our understanding of the evolution of host manipulation by parasites.

The main message of this section is that the question of the adaptive value for parasites of such behavioural manipulation in their hosts remains open. Further careful research is required before it can be concluded decisively that behavioural manipulation does result in increased trophic transmission of manipulating parasites to their final hosts.

### 17.3.2 Brood parasitism

Brood parasitism has been and still is one of the big puzzles in evolutionary biology. It can be defined as a reproductive strategy where an individual (the parasite) exploits the parental care provided by another unrelated individual (the host) to complete its reproduction. Such brood parasitism may involve either individuals of two separate species or individuals of the same species. In the latter case we speak of intra-specific brood parasitism; in the former case we speak of inter-specific brood parasitism or just of brood parasitism. Obviously enough, brood parasitism can arise and evolve only in species that provide parental care, which restricts the taxonomic groups that may harbour brood parasites. Brood parasitism is therefore parasitism in which the **resource** exploited by the parasite is nothing else than the host's behaviour in caring for the young.

The benefits of this reproductive strategy are obvious in that the host does not detect the phenomenon, thus sparing the parasite from engaging in parental care behaviour which is often very costly in time and energy and seriously jeopardizes the parent's future breeding capacity. By contrast, brood parasitism behaviour is very costly for the hosts as their immediate reproductive success is reduced, possibly to zero, when the host is parasitized (Rothstein 1990), not to mention the longer-term effects. Despite the high costs it imposes on hosts, brood parasitism occurs in groups as different as insects, fish, and birds.

#### 17.3.2.1 *The wide variety of brood parasitism behaviour*

*Slaver ants or parasites of other ants . . .*

In insects, brood parasitism is particularly developed among Hymenoptera. Ants exhibit various degrees of parasitism (in the case of ants, we speak of social parasitism). Slaver species, for example, raid the nests of ants of the same species or of different species to recruit workers which, changed into true slaves, then raise the parasites' brood (Hölldobler and Wilson 1990). Inquiline ants adopt a different strategy. Unlike slaver ants, inquiline species settle in the hosts' nest and parasitize it by diverting the care provided by the host ants which then rear just their parasites' breeding individuals (Hölldobler and Wilson 1990). In both cases, though, the parasitized resource is indeed care for progeny.

*. . . Gentian-eating and then ant-parasitizing butterflies . . .*

Other insect species exhibit highly elaborate parasitic behaviour. Some butterflies of the genus *Maculinea*, for example, are tremendous parasites of ants of the genus *Myrmica*. The female lays eggs on gentian buds (Thomas *et al.* 1989). The caterpillars feed on the buds for a few days and then drop to the ground. At this stage, the caterpillar's fate depends exclusively on it encountering workers of the host ant species. By imitating the chemical and acoustic characteristics of the host ants' larvae, the caterpillar triggers recruitment behaviour in the workers, which, so deceived, carry it back to their colony as if it were a stray larva of their own (Akino *et al.* 1999). Inside the colony, the caterpillar is fed by the workers at a sustained rate. Competition between the ant larvae and the caterpillar leads to a marked reduction in the growth of the colony (Thomas and Elmes 1998). In some cases, the caterpillar may even feed directly on the ant larvae (Thomas and Elmes 1998).

*. . . Fish that imitate and parasitize other fish . . .*

Surprising as it may seem, there is a catfish species, *Synodontis multipunctatus*, which has a somewhat similar reproductive strategy to that of *Maculinea* butterflies (Sato 1986). *Synodontis* parasitizes mouth-brooding cichlid fish in Lake Tanganyika in East Africa. The female cichlids spawn some 50 eggs in all, in sequences of two or three. The eggs are released into the water at the same time as the male's sperm. The female then takes the spawn into her mouth where fertilization takes place. The parasite's strategy consists in imitating the host's behaviour to perfection. The female *Synodontis* approaches the host female and, with impressive timing, lays her eggs at the same time as the host cichlid. Likewise, the male *Synodontis* produces his milt simultaneously so that when the female host cichlid picks up her and her mate's spawn in her mouth, she also picks up some of the parasite's spawn. As with the host, the *Synodontis* eggs are fertilized in the female host cichlid's mouth. However, the parasite has a considerable advantage over its host: the parasite's eggs hatch before the host's. Like the *Maculinea* caterpillar that devours the ant larvae in its host colony, the *Synodontis* fry devours the host's eggs and fry in its adoptive mother's mouth. Needless to say, the reproductive success of parasitized cichlids is markedly reduced compared with non-parasitized individuals.

*. . . Fish that parasitize mussels . . .*

The brood parasite behaviour of fish is not limited to host species belonging to the same class. There is, for instance, a close association between bitterling species of the genus *Rhodeus* (cf. Section 11.4.5.2) and several mussel species of the genera *Unio* and *Anodonta* (Smith *et al.* 2000; Mills and Reynolds 2002). Male bitterling defend a territory containing one or more mussels and court female bitterling. The females insert their long ovipositors into the excurrent siphon of a mussel in the male's territory and lay their eggs in the host's gills. After the eggs have been introduced into the mussel's gills, the male releases milt which is sucked in with the host's respiratory current. Fertilization therefore takes place inside the host mussel. Incubation lasts for two to four weeks inside the mussel, before the fry leave their host through the mussel's excurrent siphon.

Even if some authors had initially ranked this association in the symbiosis category (Reynolds *et al.* 1997) recent results suggest it is rather a host–parasite relation. For one thing, the parasitized mussels suffer from reduced breathing efficiency, which seems to vary with the number of eggs released into their gills; another point is that there are defence mechanisms that tend to expel the *Rhodeus* eggs laid in their gills. However, this defence mechanism is not found in all of the mussel species *Rhodeus* exploits (Mills and Reynolds 2002). It is important to realize that this expulsion behaviour may not have evolved in response to parasitism but to ensure other functions such as the expulsion of foreign matter accidentally entering the gills or the release of the larval stages (glochidia) of the mussel itself.

*. . . and birds that parasitize other birds*

While brood parasitism occurs in insects and fish, it is in birds that it has been best studied. About 1% of all living bird species are inter-specific brood parasites.

Intra-specific parasitism behaviour is also relatively common in birds, but the number of intra-specific parasite species is not yet fully known (Johnsgard 1997; Davies 2000). Unlike inter-specific parasites, which do not build nests and so depend entirely on another species to complete their reproduction, intra-specific parasites use what is termed a mixed strategy: they take care of some of the eggs laid and they lay a few additional eggs in the nests of other females of the same species. In some cases, such as the cliff swallow (*Hirundo pyrrhonota*) the parents even begin by incubating all their eggs and near the end of incubation may take an egg in their bill and deposit it in a nearby nest that is lagging slightly behind theirs (Brown and Brown 1988). So when the young parasite hatches, it has a head start in the competition for food with its adoptive siblings, giving it a good chance of flying the nest in good shape. Seen through an evolutionary eye, the meaning of such behaviour is obvious. As this chapter is about long-lasting interactions between species, we shall not go into the case of intra-specific brood parasitism, even if the phenomenon is relatively common in birds and if there may be an evolutionary connection between intra- and inter-specific brood parasitism.

Both intra- and inter-specific parasitism arose several times independently in the course of bird evolution. For example, inter-specific parasitism is found in five families (Anatidae, Indicatoridae, Cuculidae, Icteridae, and Estrilidadae) but molecular phylogenies suggest that in some of these taxa the evolution of brood parasitism occurred twice independently (Aragon *et al.* 1999). These independent evolutionary events clearly raise the issue of why and how brood parasitism evolved. In the next two sections we discuss the hypotheses proposed to explain the evolution of brood parasitism in birds and we shall see how, subsequent to its appearance, hosts and parasites engaged in co-evolutionary cycles.

#### 17.3.2.2 *The origin of brood parasitism*

How can we explain, by reference to the mechanisms of biological evolution, the appearance of the series of fine and often subtle behaviours that make obligate brood parasitism behaviour possible? This question has haunted many an evolutionary biologist since Darwin. The favourite model for studies of brood parasitism is without doubt the cuckoo (*Cuculus canorus*).

##### a *A cuckoo's tale*

The female cuckoo arrives at her European breeding ground in late April after spending the winter months in sub-Saharan Africa. She settles in a breeding territory where she seeks out nests of the host species. The cuckoo is very choosy as to the host species. In fact, within *C. canorus* there are races known as gentes (singular gens), each of which exploits a single host species. When the female cuckoo finds a host nest at a suitable stage (one where the clutch is not yet complete) she lays one of her own eggs. The egg is laid while the host is absent and it takes just a few seconds, reducing the chances of the host female noticing the act of parasitism (Davies 2000). During this time, the female swallows one of the host's eggs in the nest (Figure 17.9a).The parasite's egg is usually a very good mimic of the host's eggs and only an expert eye can differentiate it from the host's eggs (Figure 17.9b; Brooke and Davies 1988). After a few days' incubation, the parasite egg hatches first and the chick sets about ejecting the other eggs from the nest (Davies 2000). This ejection behaviour is made more effective by a morphological structure that is unique to cuckoo chicks and consists of a cavity above the rump allowing them to 'spoon up' the eggs (Figure 17.9c). The cuckoo chick alone remains and monopolizes all the care provided by the adoptive parents. Here again the parasite exhibits an amazing capacity for manipulation, because its calling reproduces that of a whole brood of the host chicks, urging the adoptive parents to bring enough food back to the nest (Figure 17.9d; Davies *et al.* 1998; Kilner *et al.* 1999). After several weeks of intensive work, the hosts will have reared just one parasite chick and consequently their fitness gain from this breeding attempt will be zero (Figure 17.9e).

##### b *The major hypotheses*

What evolutionary scenario can be envisaged to explain the appearance of all these adaptations favouring the parasite's success? In particular, how could such a system have arisen in the course of

Figure 17.9 **The strategy of the cuckoo *Cuculus canorus***

**a.** Female cuckoo laying in the nest of a reed warbler (*Acrocephalus scirpaceus*).
**b.** Parasitized brood (the cuckoo's egg is slightly larger than the host's eggs).
**c.** Newly hatched cuckoo ejecting the host's eggs.
**d. and e.** Reed warbler feeding a young cuckoo.

Photographs from Davies (2000).

evolution; implying as it does a series of different and highly specific behaviours, each seeming to have a substantial impact on the cuckoo's fitness? Two types of hypothesis can be distinguished: those invoking an accidental origin, and those presupposing that some selective process underlies the appearance of parasitism.

### c *An accidental evolution*

Hamilton and Orians (1965) proposed that the evolution of brood parasitism is simply the consequence of nest loss at the moment females lay their eggs. This idea relies on the consideration that if a female loses her nest during the laying stage, when she is physiologically compelled to lay the eggs already formed in her oviduct, the behaviour of laying in another female's nest could be favoured. Studies to test the model's predictions have failed to provide support for them (Rothstein 1993). However, there is a whole debate about the relevance of experiments undertaken at any particular time (now) with the purpose of inferring macro-evolutionary processes that took place a long time ago when the feature was evolving (Yezerinac and Dufour 1994; Rothstein 1994).

### d *Evolution resulting from a selection process*

More recently, other models have formalized the selection pressures that might be involved in the evolution of brood parasitism. Several life history traits have thus been identified as key factors potentially involved in the evolution of brood parasitism, notably clutch size (Lyon 1998; Robert and Sorci 2001), the incubation period, and the size difference between parasites and hosts (Slagsvold 1998), as well as the degree of relatedness between species (Andersson 2001). As for the Hamilton and Orians' (1965) model, the predictions are hard to test as they aim to reconstruct evolutionary scenarios. The accumulation of data on the ecology and life history traits of parasites and phylogentically related species, together with the growing availability of phylogenetic relationships, establishing phylogenetic relatedness within various groups of birds, should allow us to tackle these questions using a **comparative approach** for analysing processes that have arisen on a macro-evolutionary time scale.

#### 17.3.2.3 *Host–parasite co-evolution*

Whatever the selective process that led to the emergence of brood parasitism, it has to be observed that the appearance of this form of reproduction has been associated with an impressive number of phenotypic changes in both the host and parasite, favouring resistance in the host and success in the parasite. The array of phenotypic modifications associated with host-brood parasite interactions is considered one of the finest examples of co-evolution (Rothstein 1990) for two main reasons: first, because of the specific nature of the interaction (in most cases a single host species is exploited by a single parasite species, although there are notable exceptions); secondly, because of the specific nature of the adaptations involved.

Probably the best-known example of specific adaptation within host-brood parasite interaction is the evolution of discrimination and mimetism of eggs. When the parasite has managed to cross the first line of defence, which involves preventing laying the parasite egg, the host has three options:

1. To continue incubating the parasitized clutch with the cost that engenders.
2. To abandon the clutch (involving variable costs depending on the ecological characteristics of the species under consideration).
3. To recognize the parasitized egg and throw the intruder out.

This final strategy seems to be the least costly *a priori* and indeed many host species are found to exercise great discrimination towards eggs placed in their nests experimentally (Davies 2000). Such recognition, together with casting out of the egg, imposes very strong selection pressures on the parasite, who in turn has only two possible counter strategies:

1. To switch to a new host species (a strategy that may not prove very rewarding).
2. To adopt a mimetic **strategy** consisting in laying eggs that resemble those of the host as closely as possible (Brooke and Davies 1988).

The benefits of the second option are obvious enough: reducing (or eliminating) the chances of the host recognizing the parasite egg and/or markedly increasing the rate of error of a host who undertakes regardless to throw out eggs whose phenotype is slightly different from the average for the clutch (Marchetti 1992). Therefore we observe, on the evolutionary scale, a real arms race where the hosts are selected for their increasing discrimination and the parasites for their ever more effective mimetism. This arms race may lead to true co-evolutionary cycles, where the host or the parasite have in turn the upper hand for one or other phase of the cycle (Robert *et al.* 1999).

##### a *Evolutionary lag . . .*

The hypothesis of co-evolutionary cycles also provides an explanation for what is at first sight a surprising observation. Given the costs imposed by brood parasites and the existence of defence mechanisms in hosts, it should be expected that alleles conferring such resistance would be rapidly fixed so that no individual in the host population might be successfully exploited by a parasite (Rothstein 1975; Kelly 1987; Takasu 1998). The persistence of susceptibility despite the obvious selection advantages of resistance might therefore reflect the absence of the necessary genetic variability for the trait to evolve or in other words a phase in the cycle where the parasite has the advantage over the host. This hypothesis, known as **evolutionary lag**, emphasizes the time lag between the appearance of parasitism strategies and defence strategies.

### b ... *or evolutionary equilibrium?*

The evolutionary lag hypothesis is not the only one, though, to explain the co-existence of resistant and susceptible phenotypes. For some authors, such co-existence reflects rather a situation of equilibrium between the benefits of resistance and the costs associated with it (Marchetti 1992; Lotem and Nakamura 1998). This is the evolutionary equilibrium hypothesis. Discriminating the parasite egg involves risks for the host, risks that may take the form of recognition errors (ejection of its own eggs) or manipulation errors (damaging its own eggs when the parasite egg is ejected; Davies *et al.* 1996). In this case, while selection pressures exerted by brood parasitism are not strong enough (low probability of parasitism, slight decline in reproductive success, etc.) the best strategy for the host would be simply not to take the risk of damaging its own eggs (Davies *et al.* 1996).

### c *Weaverbirds and cuckoos*

Several empirical studies have provided results in agreement with the hypothesis of evolutionary equilibrium (Rohwer and Spaw 1988; Lotem *et al.* 1992; Marchetti 1992; Davies *et al.* 1996; Brooker and Brooker 1996; Brooke *et al.* 1998). In particular, a colonization event caused by man in historical times has made it possible to test the hypothesis of evolutionary equilibrium virtually experimentally (Cruz and Wiley 1989; Robert and Sorci 1999; Lahti 2006). The village weaverbird (*Ploceus cucullatus*) is a passerine widespread in sub-Saharan Africa where it co-exists with a cuckoo species, the Diederick cuckoo (*Chrysococcyx caprius*), which parasitizes its nests. In agreement with the prediction of co-evolutionary models, village weavers living sympatrically with the Diederick cuckoo have great powers of discrimination and eject from their nests any eggs differing from the phenotype of their own eggs (Victoria 1972). In the 18th century, village weaverbirds from West Africa were introduced by humans to Hispaniola, in the West Indies (Moreau de Saint-Méry 1797). On Hispaniola, the weaverbirds found environmental conditions similar to those of their place of origin and successfully colonized the entire island. Unlike West African populations, which must cope with parasitism from the Diederick cuckoo, the weaverbirds of Hispaniola enjoyed, after their colonization, a brood parasite-free environment since Hispaniola was not home to any brood parasite species.

In the absence of selection pressures from cuckoos, have the weaverbirds of Hispaniola maintained their ability to spot and eject foreign eggs from their nests? The evolutionary equilibrium model, one of the main assumptions of which is the existence of a cost of defence in the absence of parasitism, predicts the gradual reduction of defence. Cruz and Wiley (1989) demonstrated that in 1982, some 150 years after their arrival on Hispaniola, the weaverbirds had indeed lost their ability to discriminate foreign eggs experimentally placed in their nests.

During the 1970s a new factor appeared on Hispaniola. The shiny cowbird, *Molothrus bonariensis*, progressively invaded the entire arc of the West Indies from the Atlantic coast of South America. The cowbird was observed for the first time on Hispaniola in 1972 (Post and Wiley 1977). Being a highly generalist parasite it soon began to exploit the village weaver bird as a host (the proportion of parasitized nests being 1.3% in the period 1974–1977 and 15.7% in 1982 (Cruz and Wiley 1989)). After the cowbird's natural colonization, the cost–benefit ratio of the defence mechanisms was altered again. This colonization event created the conditions for a near-natural experiment. Questions about the effect of restoration of selection pressure on the expression of discrimination of parasite eggs could then be addressed. Another study provided an answer to this question. Robert and Sorci (1999) showed that in 1998, just 16 years after Cruz and Wiley's experiments, the weaverbirds were able to recognize the foreign eggs with the same intensity and precision as the African populations living sympatrically with the Diederick cuckoo. Whereas these results were interpreted as evidence that village weaverbirds had recovered their ability to detect and reject foreign eggs, a recent study has suggested that differences in rejection rate are probably not due to variable selection pressures acting on rejection ability, but rather on within clutch variability in egg colour and spots (Lahti 2005, 2006). Lahti (2005) investigated the within- and between-clutch variation of African populations of weaverbirds as well as of two introduced, insular, populations (Hispaniola and Mauritius). He found that within clutch egg appearance was more variable in the two introduced populations compared with native African populations, whereas between clutch egg appearance was less variable in the introduced populations. This finding is consistent with the idea that brood parasites exert strong selection pressures on hosts as to lay eggs with very little variance in colour and spots which makes the detection of alien eggs easier and less risky. Taking into account this novel finding, Lahti (2006) subsequently showed that the differences in rejection rate reported by Cruz and Wiley (1989) and by Robert and Sorci (1999) were probably due to the relative difference in colour and spot patterns between the eggs in the clutch and the eggs added by the experimenters. These results, therefore, suggest that the intrinsic capacity of village weaverbirds has not changed between native, parasitized, and introduced, non-parasitized, populations. Instead, relaxed selection due to the absence of parasitism seems to have affected the variability of egg appearance.

### d *Why not discriminate among chicks?*

The need to take account of the risks of parasitism resides, as we have already seen, in the risk of errors related to mimetism between the host's eggs and the parasite's eggs. But what about discrimination of the parasite chick? While the cuckoo's eggs seem to be subjected to strong selection pressure to resemble those of the host, the chicks have all the characteristics (size, shape, colour) making them obviously different from the host's chicks (Figure 17.9d). It might be expected, then, that hosts capable of recognizing and discriminating eggs on the basis of subtle differences in colour and shape might also be able to recognize a chick weighing sometimes four times as much as its adoptive parents (Figure 17.9d).

Paradoxically, and as surprising as it may seem, discrimination of parasitic nestlings is extremely rare (Langmore *et al.* 2003; Grim 2007). How can this paradox be explained?

The solution might be once again in the cost/benefit ratio of ejection at the chick stage. Using an elegant theoretical model, Lotem (1993) demonstrated that if discrimination is based on imprinting and learning, then ejection at the chick stage might prove maladaptive for the host. The idea Lotem develops is as follows. Imagine the host needs to learn the phenotype characteristics of its own eggs in order to recognize and identify any parasite egg. Learning can then only occur during the first reproductive event in the host's life. Two possibilities arise (Figure 17.10a): (1) either the clutch is not parasitized and the host learns the phenotype of its own eggs correctly; or (2) the clutch is parasitized and the host integrates the phenotype of the parasitized egg in the range of possible variation of its own eggs.

What are the consequences of the two events for the host's fitness? If it learns the phenotype of its own eggs correctly, it will be able to eject the parasite egg during future breeding events and its reproductive success in the case of parasitism will always be equal to the reproductive success in the absence of parasites, less the possible cost of manipulation of the parasite egg (Figure 17.10b). However, if the host is parasitized during its first reproduction, it will learn to recognize the parasite's and its own eggs as its own, the breeding success of parasitized broods will invariably be zero (only the parasite chick is raised), while non-parasitized broods will continue to produce several chicks that is independent of learning to recognize the parasite egg.

Let us imagine now that the same learning phenomenon is required for recognizing the parasite chick. Once again, there are two possible scenarios:

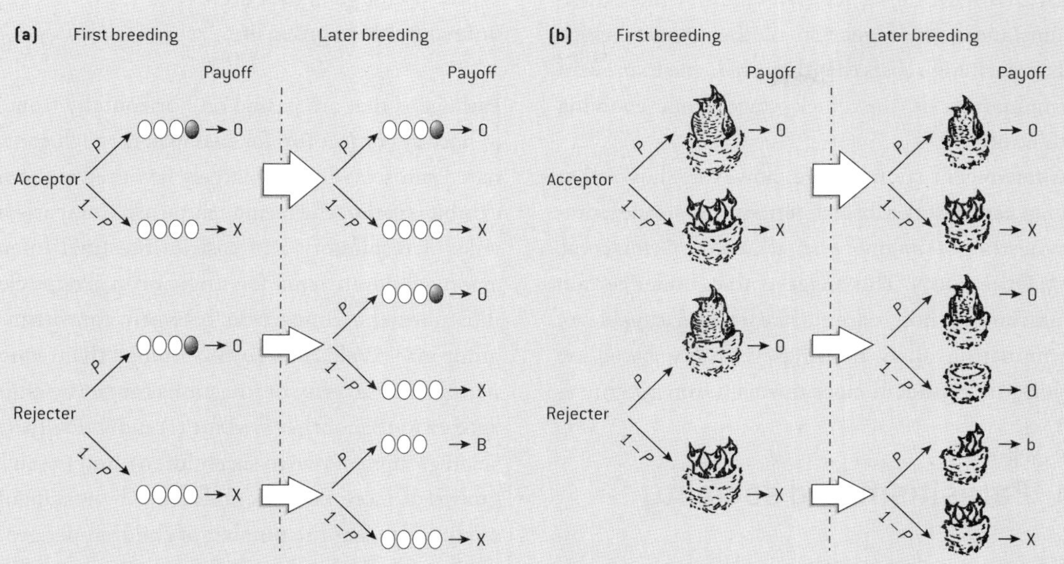


Figure 17.10 **Why spot subtle differences in eggs and not chicks that are so obviously different?**

Reproductive success of a parasitized versus non-parasitized host depending on whether ejection takes place at the egg stage or at the chick stage.

**a.** Fitness if the parasite's egg is ejected.
**b.** Fitness if the parasite chick is ejected. *P* is the probability of a nest being parasitized. **X** is the mean reproductive success. **B** represents the benefit of rejecting the parasite eggs, that is the mean reproductive success of ejectors; because of costs related to ejection and because of the egg removed from the host clutch by the cuckoo when laying its own egg, *B* is less than $X-1$.

This model involves two important assumptions: (1) first-time breeders learn to recognize the eggs and young of their own species by an imprinting process; (2) upon hatching the parasite chick eliminates all the other young of the host species. Accordingly, when first-time breeders are parasitized, they impregnate both their eggs and that of the parasite but only the parasite chick (their own chicks have been eliminated very quickly by the parasite). Discriminatory behaviour for chicks would therefore reduce the fitness of individuals whose first brood was parasitized throughout their life time (they would always reject their own chicks if they were not parasitized subsequently and would otherwise accept those of the parasite). However, egg discrimination does not reduce the host's fitness when subsequently it is not parasitized. It should be noted, however, that the assumptions of the model may be questioned as they are themselves subject to selection pressure. One might ask, for example, why a species would not have a genetically fixed mechanism for recognizing its own chicks. In addition, many brood parasites do not eliminate all the host's eggs. So this model applies only to specific situations, but under those circumstances it provides a convincing explanation for the absence of discrimination at the chick stage.

After Lotem (1993).

(1) when the first brood is not parasitized, the host learns to recognize its own chick and is able to eject the parasite chick in the future; (2) conversely, when the first brood is parasitized, given that the cuckoo chick hatches first and ejects the host's eggs from the nest before they can hatch, the host learns only to recognize the cuckoo chick as its own. In this case, the aggregate reproductive success over the host's lifetime will be zero, because when parasitized, it will recognize the parasite chick as it own and when not

parasitized it will eject its own chicks. Under these circumstances, impregnation of the parasite chick leads to such high costs that they are liable to prevent discrimination at the chick stage from evolving (Lotem 1993).

A few recent studies have, however, shown that parents can desert a brood containing a lone parasitic nestling (Langmore *et al.* 2003; Grim *et al.* 2003; Grim 2007). This suggests that, under certain circumstances, host parents are indeed capable of discriminating alien nestlings, possibly based on learning-independent mechanisms (Grim 2007).

## 17.4 Parasitism and sociality

Defence mechanisms against parasites cannot evolve without costs for the hosts, if only because such resistance necessarily entails allocating resources to the defence activity. In this context, natural selection can be expected to have favoured strategies conferring maximum protection at least cost. Behavioural defence strategies fit into this framework perfectly because they aim to minimize the up-line cost of parasitism by reducing exposure to pathogens. The behaviours most commonly involved in avoiding pathogens are habitat choice, prey selection, and social behaviour (Moore 2002). We shall address this issue very succinctly here. Readers can find complementary information in Chapter 15.

Whether individuals in a population are evenly spread across space or form clusters has far-reaching repercussions on the rate of transmission of parasites (probability of transmission from one host to another and/or probability of the parasite in its infectious stage encountering a host). Whether an organism lives a solitary life or a group life (formation of socially stable or unstable groups) may then very largely affect its risk of contracting pathogens. Although it seems obvious that the evolution of group life and sociality is influenced by a wealth of factors from resource distribution to reproductive systems (see Chapters 14 and 15), some authors have proposed parasitism as a force that can both constrain and promote group life.

### 17.4.1 Pathogens that are unfavourable to group life . . .

Pathogens that are passed on horizontally from host to host by contact (or, for example, through contaminated aerosols) are very largely favoured when hosts exhibit spatial clustering patterns. Comparison of risks of respiratory and gastric/intestinal infection among children cared for singly or in groups clearly illustrates the connection between parasitism and group size. Wald *et al.* (1988) studied the frequency, nature, and severity of infections caught by children aged 12 to 18 months cared for (1) individually, (2) in small groups (of two to six children), and (3) in large groups of more than six children. The results of the study show that the number of children per group is positively correlated with the number of respiratory infections developed, their duration, and their severity (Wald *et al.* 1988). When faced with pathogens that are passed on in this way, sociality clearly has a cost. Pathogens should therefore constrain species to reduce their social interactions.

### 17.4.2 . . . and pathogens favouring group life

What happens when the parasites are not intimately associated with a host but on the contrary are able to move from one host to another? Hamilton (1971) and others argued that, when confronted with mobile parasites, hosts forming large groups have an advantage because the probability of being attacked by the parasite is diluted within the group (see Section 14.2.2b). This argument is similar to the one given to explain the advantage of the group faced with predation: admitting that each individual in the group has the same probability of being attacked by a predator, this probability is lower in a larger group.

In terms of the probability of being parasitized, some parasites, such as certain blood-sucking Diptera, behave like predators. If there is a single mosquito at a given location, the probability of an individual being bitten is ten times lower if in a group of 10 people (obviously this is true if the mosquito bites once only). A literature review has confirmed the general character of the results presented above. When parasites are mobile and actively seek out their hosts, their intensity generally falls with the size of the host group, whereas with contagious parasites, their intensity is positively correlated with group size (Côté and Poulin 1995).

The reduced risk of contagion may also have direct consequences for social behaviour. Mutual grooming is one example of social interaction that very probably evolved under the influence of parasitism (Moore 2002). It has even been proposed that the xenophobic behaviour observed in some primate species with regard to outsiders trying to mix with a new social group could have the function of maintaining a form of quarantine which allegedly limits the group members' exposure to new parasites (Freeland 1976, 1977; Loehle 1995).

## 17.5 Conclusion

In this chapter we have developed some of the behavioural aspects associated with several long-lasting interactions, most notably host–parasite relationships. We have seen in particular how brood parasitism is a genuine form of parasitism of the host's behaviour in that the resource parasitized is not food but rather an individual's investment in care for young. More generally, all of the examples developed in this chapter are clear illustrations of the complexity of many long-lasting interactions, true theatres of interactions between the biological and ecological characteristics of living organisms that have to co-evolve in a changing environment.

Through their population dynamics, parasites are one of the important factors behind the dynamics of spatio-temporal heterogeneity of the environment. If a place is parasitized today, there is a strong possibility it will be for some time to come. In the context of the evolution of group life, which we rapidly addressed at the end of the chapter, working with the framework of commodity selection developed in Chapter 14, it appears that the important role of parasites in engendering environmental heterogeneity makes behaviours for selecting breeding locations all the more necessary. We saw in Chapter 14 how the same behaviour could itself generate clustering in space. We come to a paradox in that parasitism, one of the costs of group life classically raised in the literature, may, in fact, have indirectly contributed to engendering group life, for which it may subsequently be a cost.

We also saw at the end of this chapter an example involving human behaviour in the risk of pathogen transmission. It is clear that in view of the challenges related to understanding host-pathogen relations, particularly in terms of public health, and of the obvious role of behaviour in this context, more attention should be paid in future to long-lasting interactions in behavioural ecology.

## » Further reading

**Bollache, L., Gambade, G. & Cézilly, F.** 2001. The effects of two acanthocephalan parasites, *Pomphorhynchus laevis* and *Polymorphus minutus*, on pairing success in male *Gammarus pulex* (Crustacea: Amphipoda). *Behavioural Ecology and Sociobiology* 49: 296–303.

**Bouchon, D. Rigaud, T. & Juchault, P.** 1998. Evidence for widespread *Wolbachia* infection in isopod crustaceans: molecular identification and host feminization. *Proceedings of the Royal Society of London Series B* 265: 1081–1090.

**Brown, C.R. & Brown, M.B.** 1988. A new form of reproductive parasitism in cliff swallows. *Nature* 331: 66–68.

**Cézilly, F., Grégoire, A. & Bertin, A.** 2000. Conflict between co–occurring manipulative parasites? An experimental study of the joint influence of two acanthocephalan parasites on the behaviour of *Gammarus pulex*. *Parasitology* 120: 625–630.

**Combes, C.** 2001. *Parasitism. The Ecology and Evolution of Intimate Interactions*. Chicago University Press, Chicago.

**Côté, I.M.** 2000. Evolution and ecology of cleaning symbioses in the sea. *Oceanography and Marine Biology: An Annual Review* 38: 311–355.

**Davies, N.B.** 2000. *Cuckoos, Cowbirds and Other Cheats*. T. & A.D. Poyser, London.

**Dawkins, R.** 1982. *The Extended Phenotype*. Oxford University Press, Oxford.

**Grim T.** 2007. Experimental evidence for chick discrimination without recognition in a brood parasite host. *Proceedings of the Royal Society of London Series B* 274: 373–381.

**Langmore, N.E., Hunt S. & Kilner R.M.** 2003. Escalation of a co-evolutionary arms race through host rejection of brood parasitic young. *Nature* 422: 157–160.

**Lotem, A.** 1993. Learning to recognize nestlings is maladaptive for cuckoo *Cuculus canorus* hosts. *Nature* 362: 743–745.

**Lotem, A. & Nakamura, H.** 1998. Evolutionary equilibria in avian brood parasitism. In: *Parasitic Birds and their Hosts* (Rothstein, S.I. & Robinson, S.K. eds.), pp. 223–235. Oxford University Press, Oxford.

**Poulin, R.** 1998. *Evolutionary Ecology of Parasites. From Individuals to Communities*. Chapman and Hall, London.

**Rigaud, T., Moreau, J. & Juchault, P.** 1999. *Wolbachia* infection in the terrestrial isopod *Oniscus asellus*: sex ratio distortion and effect on fecundity. *Heredity* 83: 469–475.

**Robert, M. & Sorci, G.** 2001. The evolution of obligate inter-specific brood parasitism in birds. *Behavioral Ecology* 12: 128–133.

**Rothstein, S.I.** 1990. A model system for co–evolution: avian brood parasitism. *Annual Review of Ecology and Systematics* 21: 481–508.

## » Questions

1. Should manipulative parasites modify only one or several dimensions of their host's phenotype?

2. Can you imagine one or more mechanisms to explain why hosts do not differentiate their parasite's chicks under conditions not covered by Lotem's (1993) hypothesis (see Section 17.3.2.3)? In a few cases, it has been suggested that hosts may discriminate parasitic nestlings and desert the brood. What are the possible evolutionary consequences of this host defence?

Part Six

# Humans and Animals

Readers will certainly have noticed that we rarely used human examples in previous parts. This could give rise to the impression that behavioural ecology does not have anything important to say about human behaviour. Yet humans are biological organisms and must in one way or another be under similar selective forces as any other living organism. The information that shapes our phenotype, including our behaviour, is likely to be transmitted through similar mechanisms as in other organisms. This part thus focuses on some evolutionary issues that are relevant to human behaviour.

Studies developed here are currently building up into emerging domains of behavioural ecology, some in very recent years. Approaches are in essence interdisciplinary, whether with conservation biology, or human sciences including physiology and medicine on the one hand, and psychology, anthropology, neurosciences, and cognition on the other.

Chapter 18 treats the question of the specific input that behavioural ecology can bring to conservation biology. This is a field in behavioural ecology that we hope will develop into a mature domain in the near future.

Chapter 19 deals with the question of the role of evolutionary processes in shaping human behaviour. This will raise interesting issues about whether behavioural ecology may help in understanding our biological nature.

Chapter 20 deals with an emerging topic in behavioural ecology, that of the potential existence of cultural inheritance in a wide variety of animals. This chapter is linked to the review presented in Chapter 4. One of the main goals of this chapter is to provide a general non-human-centred and testable definition of culture in order to allow the study of animal culture. Another important goal is to show how important cultural processes can be in evolution in general, not only for humans and some primates but also for a wide array of animals.

Chapter 20 ends the book by underlining the major, and often underestimated, role of behaviour in evolutionary processes. Authors often claim that behaviour is an important part of adaptation, but few accept that it is a major actor of information inheritance, and thus of evolution (reviewed in Avital and Jambloka 2000). It is surprising to observe that it is mainly students of behaviour that seek genes that code for behavioural variance, rather than incorporating social influences in their reasoning. With cultural transmission, behaviour and learning become as important as DNA duplication for the study of biological evolution in general. Behaviour is thus at the origin of another system of information transmission across generations, cultural transmission. Our prediction is that the study of animal culture is likely to develop into a major field of behavioural ecology in the near future.

18

# 18

# Behavioural Ecology and Conservation

Anders P. Møller and Étienne Danchin

## 18.1 Introduction

The past 500 years have seen a level of extinction that is unprecedented in the geological history of the globe. This rate of extinction is now 100–1000 times larger than at any time in the planet's previous history (Lawton and May 1975; Stattersfield *et al.* 1998). Many, many more species are threatened with extinction due to the loss of pristine habitat, change of climate, and human activity; current estimates of the loss of species during this century extend to more than 50% of all species present. This conservation crisis has placed issues of conservation and biodiversity on the priority list of national and international meetings. The conservation of biodiversity today represents the absolute priority in environmental politics (Myers 1989; Wilson 1992; Lawton and May 1995). Many different sciences contribute to the study of biodiversity and conservation such as systematics, ecology, genetics, molecular biology, and economics, but it is only most recently that behavioural ecology has begun to address these issues (Clemmons and Buchholz 1997; Caro 1998; Gosling and Sutherland 2000).

In the first part of this chapter, we briefly illustrate the use of behavioural ecology in problems of conservation. This is done by considering how a good knowledge of sexual selection can provide important **information** for decision-makers in conservation. Sexual selection (see Chapter 11) may seem a far cry from what is needed as a tool to make viable conservation plans. However, most animals and plants are sexually reproducing organisms, and sexual selection affects **life history traits**. This approach is exemplified by using demographic models to investigate the effects of sexual selection on the risk of extinction. A second issue that is treated in the first part of the chapter is the question of lack of reproduction in small populations. Many small populations experience negative **density-dependence** in population growth, with the populations actually going extinct. This is the so-called **Allee effect** named after the US ecologist and zoologist Warder Clyde Allee (1885–1955), who first described it in the 1930s (Allee 1931). Why are these individuals not reproducing? That is the million dollar question for conservation biologists and managers of threatened populations, especially in zoos.

The second part of the chapter deals with how studies of behaviour can positively affect captive breeding and reintroduction of threatened species. When breeding and reintroduction programmes started, it was often assumed that it was sufficient to provide plenty of food and shelter for populations to increase in size. The lack of success of many captive breeding programmes has suggested that this issue is not as simple as first thought. Similarly, reintroductions were often just 'done' without any preparation or follow-up studies of the fate of individuals (see Sarrazin and Barbault 1996). Current reintroductions usually have a more thorough approach that attempt to adapt these to the needs of the species in question.

## 18.2 Sexual selection and conservation

Why do populations go extinct? There are many different answers to this question and this chapter is far too short to give an exhaustive account. This first part provides two answers and briefly discusses the relationship between **sexual selection** and conservation. In the following two parts, other answers will be explored.

**Demographic stochasticity** is the demographic parallel to **genetic drift**. It is easy to imagine a theoretical population where the average mortality rate is 23.5%, the sex ratio 1.0, and where only 30% of the population reproduces. Such mean values would apply well to an infinite, or at least a very large, population. However, any real population is composed of individuals that are either alive or dead, male or female, reproducer or non-reproducer. Demographic stochasticity arises from the fact that random realizations of demographic parameters in small populations may create deviations from our expectation based on mean values. For example, all individuals of a small population may for random reasons die in the next year although all these individuals have a probability of survival that differs from zero. Such a random event could cause the population to disappear for **stochastic** reasons.

Scientists generally consider that random events may cause deviations that can have important implications for small populations, typically when populations contain fewer than 50 individuals. However, theoretical approaches show that because of the effects of sexual selection this may be the case even for populations that are not so small. The second section deals with the Allee effect and why there is often no, or reduced, reproduction in small populations. Again, this relates directly to problems of demographic stochasticity in small populations.

### 18.2.1 Demographic stochasticity and sexual selection

Sexual selection is important for the immigration success and survival of small populations. As we saw in Chapter 11, sexual selection results in the production of extravagant secondary sexual characters that are costly to produce and maintain. Such traits evolve towards exaggerated forms until the costs in terms of viability are balanced by the benefits in terms of mating success (Andersson 1994). However, these costs are paid by all individuals in the population, while usually only a small fraction of the population benefits from the extravagant trait. If the viability cost of sexual selection were absent, it would make the whole population fitter. However, because the benefits of individuals are what drives the evolution of an exaggerated male trait, in the end, the average individual pays the fitness cost of sexual selection. Furthermore, because only a small proportion of males ends up contributing to reproduction the effective population size is smaller, which may have consequences for additional demographically stochastic effects. Thus, sexual selection can be an important force that increases the risks of extinction (McLain 1993). This has important consequences for population processes and therefore also conservation.

#### 18.2.1.1 *Sexual selection and introduction success of species*

Introductions of alien species to islands in the Pacific have provided a test case of the effects of sexual selection on the probability of extinction. Many European and Asian species of birds have been introduced to islands such as the Hawaiian Archipelago and Tahiti. The fate of these introduction events has been recorded and this allows the testing of the factors contributing to the establishment success of small populations. McLain *et al.* (1995) analysed introductions to Oahu in the Hawaiian Archipelago and Tahiti. The probability of successful introduction was almost twice as high for sexually monochromatic than dichromatic species of birds. Because sexual dichromatism presumably has arisen and is maintained by sexual selection, this provides evidence of an effect of sexual selection on introduction success. However, in McLain *et al.*'s (1995) study it was not possible to control statistically for the number of release sites and the number of individuals released. A subsequent analysis of birds introduced to New Zealand had much more detailed information on these parameters, but still found a significant effect of sexual dichromatism (Sorci *et al.* 1998). Because most introductions failed during the first year, it seems unlikely that inbreeding had played a role in causing this effect of sexual dichromatism. Why should sexual selection be important for introduction success?

#### 18.2.1.2 *Demographic stochasticity*

Demographic stochasticity arises from the fact that demographic parameters have average population values, but these values are realized at random for individuals of a population. For example, a male migratory bird might find itself in a region without other conspecifics. Imagine that another individual of the same species makes the same navigation error and ends up in the same area. That individual has a 50% probability of being female (if the sex ratio is even in the population of origin), but in this random process the individual is either a male or a female. The two individuals can only reproduce and perhaps establish a new population if they happen to be of different sex. This is clearly the outcome of a stochastic process.

This example illustrates well why demographic stochasticity plays a particularly important role in small populations. If the population consisted of 100 individuals, the probability of all of them being of the same sex would be very close to zero ($2/0.5^{100} = 10^{-30}$). Demographic stochasticity can be important even in populations with 50 individuals, and many species on the red list of endangered species have populations of that size.

#### 18.2.1.3 *Sex ratio and population viability*

Legendre *et al.* (1999) modelled the importance of sexual selection in the production of stochastic effects with demographic variation in the sex ratio. Obviously, other demographic parameters like survival, age to maturity, fecundity, and others could have similar effects that might exacerbate the effects of the sex ratio.

Legendre *et al.*'s (1999) modelling consisted of running the simulations very many times for a period of '100 years', using mean demographic parameters that were representative of the introduced species (a small passerine bird with high annual mortality and fecundity). In the demographic models using information on the number of individuals released in New Zealand it was possible to predict the introduction success for different species with a very high accuracy. The proportion of theoretical populations that had gone extinct in the '100 years' after their introduction corresponded well with the observed risk of extinction for a diverse assemblage of real species introduced to New Zealand over 100 years ago (Legendre *et al.* 1999).

#### 18.2.1.4 *Mating system and population viability*

When the animal's **mating system** (see Chapter 12), either strict monogamy or polygyny, was introduced into Legendre *et al.*'s (1999) simulation, demographic stochasticity had a greater impact on probability of extinction in monogamous species (Figure 18.1). This finding makes intuitive sense because more individuals will remain unmated for random reasons in a strictly monogamous species than in a polygynous one.

#### 18.2.1.5 *Differential parental investment and population viability*

Females may reduce their **parental investment** if mated to an unattractive partner. In fact, females may reproduce at an elevated rate when mated to an attractive male, but at a reduced rate when mated to an unattractive male; a phenomenon termed differential parental investment (Burley 1986). The most important finding in Legendre *et al.*'s (1999) simulations was obtained for a monogamous mating system with a 10% reduction in fecundity due to females being mated to unattractive males. This situation also mimics the situation where there is a general viability cost of sexual selection applied to all individuals in a population. When fecundity was reduced only by 10% on average, there was a dramatic increase in the population size necessary to maintain a viable population with a high probability (Figure 18.1). A literature review by Møller and Legendre (2001) showed numerous cases of reductions in fecundity much greater than 10%, suggesting that demographic stochasticity will play an important role even in populations with a size of several hundred individuals, when females invest differentially in reproduction. This result has important consequences for captive breeding programmes, but also for estimating the size of minimum viable populations.

Clearly, these results are in need of empirical testing preferentially using field experiments. Nevertheless, Legendre *et al.*'s (1999) results show that

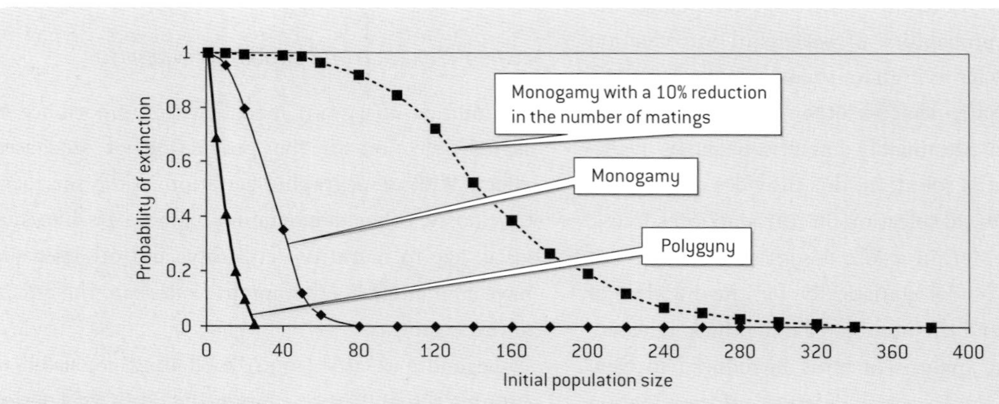


Figure 18.1 **Extinction probability and mating system**

Probability of extinction of a population of a small passerine in relation to initial population size and mating system based on simulations over '100 years'. A monogamous mating system is more sensitive to risk of extinction because of demographic stochasticity. Situations were the number of matings was reduced by 10% showed a much higher probability of extinction.

Adapted from Legendre *et al.* (1999).

sexual selection can produce demographic stochasticity effects, even in populations as large as several hundred individuals. Numerous species on the Red Lists of species in risk of extinction have global population sizes of that magnitude. Obviously, many other demographic parameters like survival, age at maturity, and fecundity can have similar effects (review in Møller 2003), increasing the effects of sex ratio shown here. These considerations will have important implications for the size of minimum viable populations.

#### 18.2.1.6 *Sexual selection and species turnover within a community*

Sexual selection may also affect the extinction probability of local populations. Indeed, factors affecting the balance between birth rate and mortality can have profound effects on population persistence. Sexual selection has been identified as one such factor because populations under strong sexual selection experience several costs such as increased predation and parasitism directed to highly ornamented males, or enhanced sensitivity to environmental and demographic stochasticity by these males. Thus local populations of strongly sexually-selected species should be more prone to local extinctions than species under weak sexual selection.

This hypothesis was tested by Paul Doherty and collaborators through analysis of the dynamics of natural bird communities at a continental scale over a period of 21 years (1975–1996; Doherty *et al.* 2003). They used the massive and impressive data set on presence/absence of bird species that were obtained from the North American Breeding Bird Survey, a continent-wide survey of breeding birds performed along survey routes by volunteers each year. To account for the intensity of sexual selection, they used sexual dichromatism that reveals highly sexually selected species, while sexually monochromatic species are under milder sexual selection. Previous studies using a comparative analysis had shown that sexual dichromatism was associated with higher mortality rate within passerines and waterfowl (Promislow *et al.* 1992, 1994). This suggests that dichromatic bird species are indeed more exposed to local extinctions. Doherty *et al.* (2003) used capture–mark–recapture statistical methods to differentiate between situations when a species was present but was missed by the observers during surveys from situations when the species had really disappeared from a given place.

The combination of such a monumental data set and the use of state of the art statistical methods allowed Doherty *et al.* (2003) to show that, as expected, dichromatic birds had on average a 23% higher local extinction rate than monochromatic species. They further showed that, despite higher local extinction probabilities, the number of dichromatic species did not decrease over the period covered by the study. This was because dichromatic species also had a higher local turnover rate than monochromatic species implying that the former species were more likely to recolonize patches of habitat. The combination of higher extinction and colonization rates in dichromatic species, thus results in relatively stable communities for both groups of species. The study suggests that bird communities function as 'metacommunities': independent communities that are linked by **dispersal** (see Chapter 10), with frequent local extinctions followed by colonization. Finally, the study showed that the intensity of the difference between dichromatic and monochromatic species varied spatially at the scale of the whole North American continent suggesting that local conditions may also influence the impact of sexual selection on population persistence.

### 18.2.2 The Allee effect

The Allee effect has been invoked as an explanation for the evolution of aggregation and sociality in animals and deals with the consequences of social aggregation for individuals within the aggregation. The consequences of aggregation are often beneficial to the individuals as, for example, when the per capita rate of desiccation decreases in aggregated woodlice. The Allee effect concerns a situation where a small population goes extinct because it has a low rate of reproduction that is insufficient to maintain the population. The effect is not necessarily restricted to small populations of the type that are usually considered in the conservation context. Even very abundant species such as the passenger pigeon (*Ectopistes migratorius*), which numbered perhaps in the hundreds of millions of individuals, went extinct over a relative short time span as a result of dramatic declines over its entire range. Although the cause of this extinction is still hotly debated, and factors such as human persecution and disease have been suggested, the Allee effect has been involved after a substantial reduction of the population of this highly social species (Blockstein and Tordoff 1985). Conservation biologists have become interested in the Allee effect because of its obvious relevance for the study of extinction risk in small populations (Lande 1987; Dennis 1989).

#### 18.2.2.1 *Multiple origins of the Allee effect*

The Allee effect is the product of several phenomena, and it is not necessarily the case that any one single factor will be responsible for the negative density-dependence that characterizes it. The mechanisms underlying Allee effects are not as clear as the simple description of the phenomenon. Examples of potential mechanisms include: (1) increased risk of predation at low population densities resulting in decreasing population size and eventually population extinction (Andrewartha and Birch 1954); (2) reduced mating efficiency due to the difficulty of finding a partner (Andrewartha and Birch 1954); and (3) reduced foraging efficiency when information about rewarding foraging sites must be obtained by parasitizing the collective efforts of an assemblage of individuals (Ward and Zahavi 1973; see Chapters 8, 14, and 15).

#### 18.2.2.2 *The recurrent problem of reproduction in zoos*

Lack of breeding in zoos is commonplace. Yet given the enormous literature on chronic lack of reproduction or reduced reproductive rates among in zoo animals, none of these effects may seem particularly likely to apply to zoos. For example, giant pandas (*Ailuropoda melanoleuca*) often refrain from reproduction in captivity and artificial insemination does not work well. The result is that reproduction then requires reliance on a costly airline transportation programme that takes pandas across the world to insure fertilization (Kleiman 1994). Similarly, mated female Hamadryas baboons (*Papio cynocephalus hamadryas*) may spend years as non-reproducers even when having the opportunity to copulate with more than one male (see Biquand *et al.* 1994). Such examples of Allee-like effects have been described for most animal taxa (see Fowler and Baker 1991), but the cause of this effect still remains unknown.

#### 18.2.2.3 *Sexual selection as a cause of certain Allee effects?*

Recently, Anders Møller and Stéphane Legendre suggested that sexual selection could be the cause of some of the reported Allee effects in the literature (Møller and Legendre 2001). They argued that if individuals within a population were unable to find a compatible or a preferred mate, then reproductive rates would be low or nil. This kind of Allee effect has been shown in numerous studies in which individuals were either allowed to choose a mate or forced to mate randomly. Mate choice considerably increases the reproductive output in threatened species such as Puerto Rican parrots (*Amazona vittata*; Brock and White 1992), California condors (*Gymnogyps californianus*; Cox *et al.* 1993), Mauritius kestrels (*Falco punctatus*; Jones *et al.* 1991), and whooping cranes (*Grus Americana*; Lewis 1990), compared with individuals that were given little or no choice. An important mechanism underlying this effect is differential parental investment (see section 18.2.1.5; Burley 1986). Females' investment in reproduction often depends on the relative quality of their mate. Experimental studies of birds and other organisms have shown that simply changing the phenotype of a male is sufficient to increase the reproductive output of a female by a factor two or more (review in Møller and Legendre 2001). This will have obvious implications for the Allee effect because small numbers of individuals would render the probability of mating with an attractive individual small, simply for stochastic reasons.

### a *The role of stochasticity associated with sexual selection*

To study the potential role of sexual selection as a generator of the Allee effect, Møller and Legendre (2001) investigated the relationship between population size and mating success using simulation models. In a truly demographic stochastic model with random variation in **sex ratio**, the proportion of individuals mating in a large population is 100%, 90% in a population of 50 individuals, and only 75% in a population of 10 individuals. When they added mating preferences to the model, things got worse. They modelled two different mating preference scenarios; a **directional preference** (i.e. a fixed mating preference for a given type of male) and a **compatibility preference** where individuals had to find a partner that differed from them in genotype. In the case of a directional preference using a 50% partner acceptance rate, mating success was as low as 55% for a population of 10 individuals. In the case of compatibility the mating success increases to 65% for populations of 10 individuals and up to 85% for populations of 50 individuals (Figure 18.2). Even when they removed the influence of chance realizations of the sex ratio by using a balanced sex ratio, as could be done say in a breeding programme, they still found that in small populations of 10 individuals the mating success would remain as low as 65%.

Thus, in small populations, random realizations of the sex ratio combined with directional or compatibility mating preferences can considerably reduce the mating success. In addition, adding the effects of differential parental investment will tend to increase the effects of small population size on population growth.

### b *Infanticide and conservation*

In a very general way most demographic models, even those in conservation biology, do not consider sex (Møller 2003). For example, males are harvested preferentially in many species (for example by hunting), potentially reducing population growth and introducing demographic stochastic effects.

*Infanticidal bears*

In 1995, Robert Wielgus and Fred Bunnell proposed an illustrative example for brown bears (*Ursus arctos*). Like the lions we discussed in Chapter 2, males of many primates and carnivores kill dependent

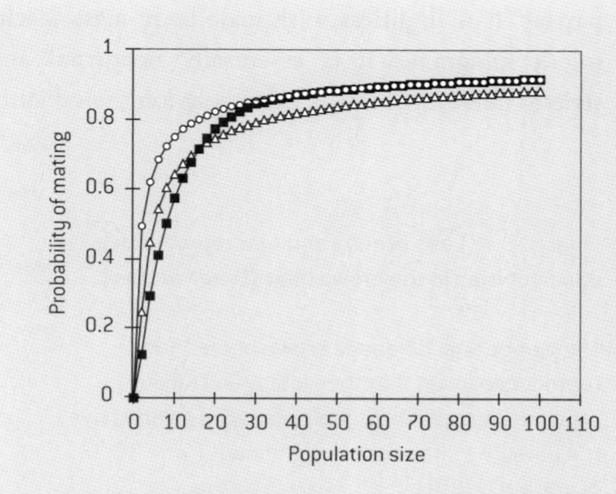


Figure 18.2 **Probability of mating and population size**

Probability of mating in a monogamous mating system as a function of population size when the sex ratio is 0.5 for three different scenarios: (1) chance realization of the number of males and females (empty circles); (2) same scenario but with a compatible mating preference involving two incompatible phenotypes A and B in equal proportions, with mating between compatible phenotypes being the only fertile ones (open triangles); (3) same scenario as the first one, but with a directional mating preference, each individual choosing its mate with probability 0.5 (filled squares). Results are based on numerical simulations. See text.

Adapted from Møller and Legendre (2001).

offspring when they replace another male that has disappeared from its territory (Hrdy 1979). The main evolutionary explanation for this behaviour is that females come into **oestrous** more rapidly when not suckling dependent offspring, and that these offspring have no relatedness to the replacement male and hence are of no fitness value to him. Wielgus and Bunnell (1994, 1995) studied two populations of brown bears, one in Canada, the other in the USA. In the Canadian population hunting differentially affects males, potentially causing an increase in male turnover and hence in the number of infanticidal males. In that population, cub survival was low. In the USA population where there is no hunting, male replacement is rare and cub survival is high. This correlational evidence indicates that the hunting of males, or perhaps any other difference between the two populations, was responsible for the reduction in cub survival.

Swenson *et al.* (1997) resolved the issue by using data from Scandinavia where only adult males are hunted. Cubs were almost exclusively lost during breeding in May–June (75% of 20 cubs) (Swenson *et al.* 1997). The disappearance of cubs resulted in a shortening of the mother's time to the next conception (Figure 18.3). Cub survival in an area where five male bears were killed was only 72%, whereas it was significantly higher at 98% in an area where bears were not hunted (Swenson *et al.* 1997). A similar difference in survival was seen 1.5 years after loss of the resident male, while survival was exactly the same in the two populations 2.5 years after male loss (Fig. 17.4). The fact that if no male bears were killed 1.5 years earlier, cub survival was high and almost identical (100% in one area and 98% in the other) suggests that resident males do not kill cubs. Computer simulations by Swenson *et al.* (1997) showed that reduced cub survival lowered the **population growth rate** ($\lambda$) from 1.18 to 1.14, and the net reproductive output was reduced by 30%.

In conclusion, it seems obvious now that gender and sexual selection can generate Allee effects in small populations that can negatively affect their maintenance. Ignorance of behavioural processes can cause managers responsible for guarding and maintaining viable populations to make decisions that eventually may accelerate the decline of small populations.

### 18.2.3 The role of deterministic processes: the case of the kakapo

Deterministic sex ratios as well as other demographic parameters can have important consequences for conservation. The lek-breeding New Zealand parrot, the flightless kakapo (*Strigops habroptilus*), is a good example of this (Figure 18.5).

#### 18.2.3.1 *An atypical parrot on the brink of extinction*

The kakapo has several characteristics atypical of parrots. It is flightless with male body mass reaching 4 kilograms. It is essentially nocturnal and strictly herbivorous. It is extremely long-lived, only

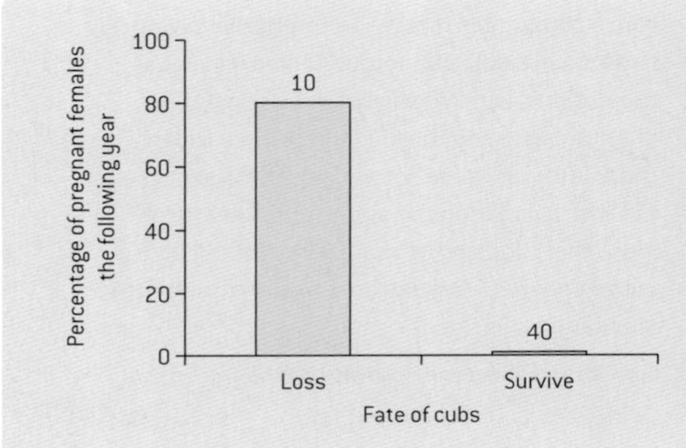


Figure 18.3 **Loss of cubs and new reproductive opportunities in the brown bear (*Ursus arctos*)**

The percentage of female brown bears that reproduced given that they lost their cubs the previous year or not. Sample sizes are given above the bars. The difference is significant ($\chi^2 = 32.37$, df $= 1$, $P < 0.0001$).

Drawn from data in Swenson *et al.* (1997).

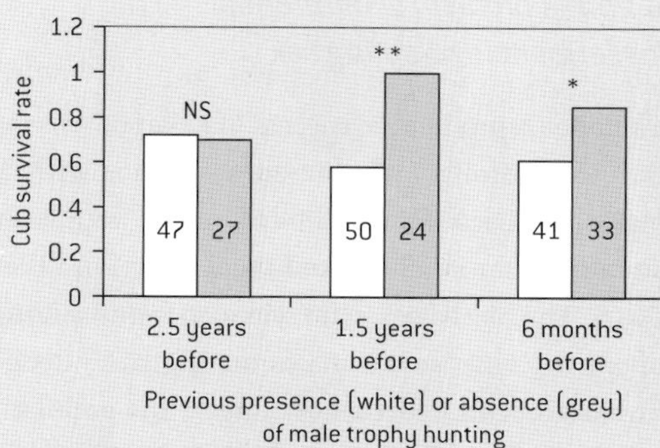


Figure 18.4 **Survival of brown bear cubs according to whether adult males were killed (white bars) or not killed (grey bars) in the area 0.5, 1.5, 2.5 years before the births of the cub**

Survival of brown bear cubs during their first years of life as a function of the interval since some of the resident males were killed. The data were obtained from a large 11,200 km$^2$ area from 1985 to 1995. In the north hunting zone, hunting was effective for only 4/11 years. The cubs were not individually marked, but the mothers carried a radio-transmitter collar allowing localization and identification. The hunting season was in autumn just before hibernation, implying that male movements and their effects on cub survival occurred during the following spring. Survival of cubs was reduced in regions with hunting (white bars all together; 0.72, $N=74$) compared to the region with no hunting (grey bars all together; 0.98, $N=50$, $P=0.0004$). Sample sizes are reported on the bars. The statistical significance of results: *, $P=0.016$; **, $P=0.0005$; NS, not significant. Other data showed that these results were not confounded by population density ($P=0.77$). Therefore, the survival of cubs was affected by hunting during at least 18 months after birth. After that interval there was no significant effect of hunting of males on cub survival.

Redrawn from Swenson *et al.* (1997).

Figure 18.5 **The kakapo (*Strigops habroptilus*) is a flightless parrot that lives only in New Zealand**

This female named 'Alice' is pictured in the nest at night while feeding her female offspring named 'Manu' aged 12 days. The two birds were still alive in 2002 on the island Codfish, New Zealand. The photo was taken in 1997 on the island Whenua and is courtesy of Don Merton, pioneer of conservation in New Zealand. Don Merton is famous for his leading of (among others) the Kakapo and Black robin (*Petroica traversi*) conservation projects.

reproducing when food is sufficiently abundant every 3–4 years. Finally, it forms **leks** (Chapter 12) and has an extreme sexual size dimorphism with adult males being 30–40% larger than females. The body mass of adults varies strongly among seasons and years (Clout *et al.* 2002).

Reproduction is synchronized with the periodic fruit set of podocarpal trees at intervals of 2–5 years. During years of reproduction males form exploded leks (i.e. display territories are not highly clumped so that leks span over large areas) at traditional sites where they display at night to attract females. The display consists in booming vocalizations that can be heard at distances of several kilometres. Each male maintains several connected 'bowls' (slight depressions dug in the ground). Males often fight at leks, sometimes until death. As is typical of lekking species, males do not participate in any reproductive activity other than copulation. Reaching reproductive condition seems to be more easy for males than for females since not all females come to breed during years of reproduction. The clutch of 1–4 eggs is placed in a natural cavity on the ground. Incubation lasts approximately 30 days, with the eggs left unguarded when the female feeds at night. The young stay in the nest for about 10 weeks and usually only one or two young fledge from the nest. All these characteristics render the kakapo particularly susceptible to predation by introduced mammals (apart from bats, New Zealand has no native mammals).

Outside the breeding season, kakapos are solitary and live in forests where they feed on seeds, leaves, shoots, and roots of numerous plant species. The species is now extinct in nature because of mammalian predation by introduced cats and stoats. The last individuals were captured and transferred to safe locations at the end of the 1970s and the beginning of the 1980s.

#### 18.2.3.2 *A desperate situation*

Before attempting to save the species a total of 82 survivors (22 females only) were translocated to predator-free small islands by 1982. On these islands adult survivorship is very high, with a value of 98%. Since then, the population decreased until 2001. At the beginning of 2002 the total population contained 62 individuals of which 21 were adult females. Several of these individuals were more than 25 years old, and certain females had not reproduced since 1982. The main cause of the decline was the very low reproductive success, with only 15 young being produced since 1977 (Clout *et al.* 2002).

Conservation efforts concentrated on increasing reproductive success. The prime means of doing so was to provide high quality food (nuts, apples, and sweet potatoes) distributed among feeding sites in the territories of females. For various reasons, not all females were provisioned each year. The provisioning programme initiated in 1989 had a positive effect on several parameters (Clout *et al.* 2002). For instance, provisioned females weighed on average 310 grams (15%) more than unfed females. While provisioned females that reproduced had higher reproductive success, provisioning itself had no effect on whether a female reproduced or not. Given these positive effects, provisioning was extended to the entire population.

#### 18.2.3.3 *A problem of differential investment in offspring sex*

The poor reproductive success of kakapos was mostly due to a strong male-biased sex ratio, with three males for each female in the adult population, a situation that is related to sexual selection (Trewick 1997). The biased sex ratio was also found among the offspring, with seven males among nine fledglings produced, which further endangers the population.

The Spanish scientist José Luis Tella (2001) recently emphasized that the extreme male biased sex ratio might be due to females that are in prime physical condition as a result of low population density, superior resource availability, and supplemental feeding year-round. We have mentioned in Chapter 13 some of the evolutionary reasons for such an association between sex ratio and female condition. Trivers and Willard (1973) suggest that females should adjust their brood sex ratio in response to resource availability to optimize their fitness. In lek-breeding species, males have, without doubt, a much greater variance in reproductive success than females, and sons are more costly to raise than daughters because they weigh 40% more than daughters. A considerable amount of evidence supports this prediction (see Chapter 13). Thus, mothers in prime condition are expected to produce sons that will also be in prime condition and therefore will be likely to produce more offspring. In contrast, it will be more advantageous for mothers in poor condition to produce less costly daughters because daughters' reproductive success is less variable than sons', a prediction that has been supported by several tests.

In the kakapo, sons are more costly to produce than daughters because they are bigger. Therefore, mothers should produce more sons when in good condition, but more daughters when in poor condition. Tella (2001) suggested that keeping females on a diet, or at least abandoning supplemental feeding, could skew the sex ratio of newborns in favour of females. This suggestion may be a last resort to save one of the most endangered and peculiar bird species in the world.

| | Offspring sex | | Sex ratio | |
|---|---|---|---|---|
| | Males | Females | | |
| Provisioned mothers | 13 | 5 | 2.6 | Seven times more daughters in unprovisioned mothers |
| Unprovisioned mothers | 4 | 11 | 0.36 | |

Table 18.1 **Sex ratio in relation to food provisioning of mother kakapos**

Data on reproduction from 1982 to 2002: six individuals produced on Stewart Island, 1977–1982 (before any food provisioning), 15 individuals produced after 1982 after translocation to predator-free islands, five unhatched eggs, and seven nestlings that died in the nest. Each mother was classified as either provisioned or unprovisioned. Young were sexed according to morphological criteria or using molecular techniques (embryos and dead nestlings). In total, 33 young from 18 broods were used.

Adapted from Clout and Merton (1998) and Clout *et al.* (2002).

Following this suggestion, Mick Clout and his collaborators (2002) analysed data on sex ratios of offspring produced by fed and non-fed mothers in the past. The prediction was borne out because the sex ratio of the offspring of fed mothers was 2.6 sons for each daughter, while it was 0.36 sons for each daughter among non-fed mothers (Table 18.1). Non-fed mothers thus produced seven times more daughters than fed mothers. Therefore, there was experimental evidence consistent with Tella's (2001) prediction.

#### 18.2.3.4 *A real size experiment on sex ratio*

Based on these data, the Department of Conservation in New Zealand, after having observed that food supplementation did not affect reproductive rates during years of high fruit set, decided in 2002 to feed females only after egg laying. This allowed for a possible bias of the sex ratio towards daughters owing to an absence of feeding before laying, but simultaneously allowed for an elevated reproductive success due to the effects of feeding after egg laying.

*An excess of daughters produced in a favourable year . . .*
Again, the results were consistent with the prediction (Table 18.2). Twenty of 21 females reproduced. They produced a total of 67 eggs of which 42 were fertile. Among the 26 hatchlings, 24 reached fledging and were sexed. Among these surviving offspring, 15 were females, or a sex ratio of 0.60 males for each female (Table 18.2).

| | Males | Females | Sex ratio |
|---|---|---|---|
| Production of young in 2002 with females only provisioned after egg laying | 9 | 15 | 0.60 |
| Sex ratio of embryos that did not produce an independent young | 12 | 6 | 2.00 |

Table 18.2 **Sex ratio of eggs laid during 2002 in the kakapo**

The data on young produced in 2002 were kindly provided by Mick N. Clout at the International Ornithologist Congress in Beijing 2002, and commented on in Sutherland (2002).

This result is particularly interesting because 2002 was one of the most favourable years for reproduction in the kakapo, giving rise to an expectation of a male-biased sex ratio, contrary to the actual observation. Therefore, this suggests that the sex ratio at independence will likely always be female biased. This implies that food supplementation *ad libitum* had caused mothers to reach abnormally favourable body condition. In those situations females produced almost exclusively sons.

*. . . despite an equal primary sex ratio*

The results were even more interesting when dead embryos were sexed (Table 18.2). Dead embryos in this particular year were statistically biased towards sons (sex ratio of 2; Table 18.2) such that the sex ratio was even at laying but in favour of females at fledging (sex ratio of 0.6 at fledging; Table 18.2). This suggests that an important component of a mother's control of the sex ratio of her brood occurs during the incubation of eggs and raising of the hatchlings.

#### 18.2.3.5 *The species is probably saved for now*

The decision to provision females only after they have laid resulted in doubling the number of surviving females in a single year. The result occurs after biologists had been attempting to resolve the chronic female deficit for over 50 years, a solution that was required to save the species. This opens encouraging perspectives for survival of the kakapo. The Department of Conservation in New Zealand has now put in place a rat extermination programme on several other islands to allow the translocation of the new kakapos that will be produced in coming years. Placing individuals on a greater number of islands could reduce the risk of extinction due to a stochastic event at a given site. Since the 1950s this is the first time that conservationists can speak of the future of this species with some optimism.

This example demonstrates how behavioural ecological reasoning in particular, and evolutionary reasoning in general, can sometimes help solve desperate conservation problems. Sometimes the solution to such conservation crises can be completely counter-intuitive as in the case of the kakapo. This is an important lesson that should be kept in mind.

### 18.2.4 Can good genes save populations?

The **good genes theory** of sexual selection is based in the hypothesis that extravagant sexual displays reliably reflect the ability of an individual to carry the burden of the display (see Chapter 11). This implies that males with more extravagant sexual displays, on average, are more viable because they carry genes that confer to the bearer a selective advantage in terms of general viability, or more specifically resistance to parasites (Hamilton and Zuk 1982). Thus, if sexual selection is disrupted because populations are small and/or propagated in captivity under the management of humans, then the genetic advantages of sexual selection may be weakened or completely disappear. There are plenty of studies showing that reproduction in these circumstances is reduced considerably (review in Møller and Legendre 2000), but the actual cause of this problem is less well known. Most cases of captive breeding with poor reproductive performance seem to be related to problems of direct phenotypic responses by females to the absence of attractive partners. Effects of good genes on population persistence may only be discernible on a longer time scale, although we cannot exclude the possibility that lack of sexual selection would already show up as a reduction in fertilization success, hatching success or survival of offspring.

Surprisingly, there are no experimental studies that have investigated the effects of sexual selection on the viability of small populations. A simple design would be to maintain small populations under two different mating regimes: one with random mating imposed by the experimenter and the second with free mate choice. The population would in this case be the individual replicate. Both types of population should be maintained under similar conditions, and the social and sexual environment should be similar outside the actual period of mating. An important example of the effects of sexual selection on population viability comes from a long-term study of artificial breeding of salmon (*Salmo salar*) in a Swedish river (Grahn *et al.* 1998). All fish in this river were captured and bred in captivity, and the breeding regime has for several decades consisted of a random breeding design imposed by humans. The fry are subsequently released in the river system, and they will eventually recruit to the same breeding population. This breeding regime has resulted in a reduction in population size, mainly caused by a reduction in the general viability of the fish. In particular, resistance to a viral disease has decreased dramatically during this regime. Recently, experiments have been conducted by letting the individual males with the most extreme secondary sexual characters contribute disproportionately to reproduction, and this has caused an increase in parasite resistance in the population. Similar experiments, but under more controlled conditions, may help reveal the relative importance of sexual selection in successful propagation of small populations.

## 18.3 Studies of behaviour in captive breeding and reintroduction

How can we improve the success of captive breeding and subsequent reintroduction into the wild? Ever since Noah's ark, animals in captivity have been bred in monogamous situations. Many zoological gardens still keep animals in single male–female pairs despite the fact that almost all fish, amphibians, reptiles, and mammals are not monogamous. Although most birds are socially monogamous, there is considerable evidence suggesting that females of most species regularly copulate with more than a single male, resulting in frequent **sperm competition** (see Chapters 11 and 12). Sperm competition is also prominent among other taxa. Hence the maintenance of animals in pairs arose as a consequence of particular moral standards of the people that kept these animals and often has had important consequences for the success of captive breeding programmes.

### 18.3.1 Sexual selection and captive breeding

Many captive breeding programmes are based on confinement of individuals or 'pairs' in individual cages. While such an approach may be justifiable for disease management, this is not necessarily the **optimal** solution for reproductive success. The importance of mate choice or male–male competition in determining reproductive success cannot be neglected (Andersson 1994). Although many captive programmes provide each female with a single male partner, such enforced social monogamy may be very different from the natural breeding system. While the behaviour and ecology of a given population is closely tied to its mating system, captive breeding under artificial social conditions may reduce or even prevent successful reproduction. A second issue that has been ubiquitous in most captive breeding including zoological gardens is the absence of female copulations with multiple partners. Whereas enforced sexual monogamy may be a remnant of 19th century human moral standards when many zoos were established, we have seen in Chapter 12 that there is plenty of current evidence suggesting that multiple mating and sperm competition is the rule rather than the exception (review in Birkhead and Møller 1998). The behaviour, physiology, and reproductive anatomy of numerous organisms seem adjusted to this aspect of reproduction. Maintaining captive animals under artificial conditions that prevent such processes from occurring may be a direct cause of low or even no reproductive success.

### 18.3.2 Imprinting on humans and particular habitats

A classical photograph from the history of ethology is Konrad Lorenz being followed by a brood of grey-lag goslings (*Anser anser*). The detection of **imprinting** was one of many results involved in giving Konrad Lorenz a share of the 1973 Nobel Prize in Physiology or Medicine (see Chapter 1). Imprinting is a learning phenomenon that occurs mostly during a relatively short period after birth or hatching when the animal learns to recognize its mother, conspecifics, and in some cases even the appearance of future sexual partners.

#### 18.3.2.1 *Imprinting on potential sexual partners*

Imprinting, however, was not taken seriously by breeders for a long time because numerous captive breeding experiments were initiated without the slightest concern about animals imprinting on humans. It was only later when these individuals became adults and potential reproducers that imprinting was considered a problem. Captive bred animals often seek the company of humans, and when human persecution is a main reason for the decline of the species, such affiliative tendencies are often associated with high risks of premature death after release. This problem has been recognized first in the captive breeding of condors and whooping cranes and now in many other animals, such that people involved with the young animals wear a special outfit bearing some details that resemble the species-specific appearance of the adult to avoid imprinting on people and facilitate species recognition in the future. Another, equally problematic, method of captive breeding consists of using individuals of another species as foster parents of the offspring of a threatened species. For example, the threatened lesser white-fronted goose (*Anser erythropus*) has been reared by grey-lag geese foster parents in a breeding programme. As a consequence of this, **hybridization** with grey-lag geese has arisen as a major problem that causes 'genetic pollution' of the original stock of lesser white-fronted geese. A second problem arising from this use of heterospecific foster parents is that geese that habitually follow conspecifics on migration to the wintering grounds now follow foster parents to the Iberian Peninsula rather than to their usual destination, Kazakhstan.

The same heterospecific fostering method was used in conservation programmes of the New Zealand black robin, the population of which was reduced to only a few males and a single fertile female. The programme removed the first clutch of a black robin nest and placed them in the nest of another passerine species, the Chatham Island warbler (*Gerygone albofrontata*), with similar ecology. In such conditions, black robin females produced a replacement clutch, which enabled conservationists to increase the productivity of the relict population. However, hatchlings had to be put back in their original nest to be reared by their genetic parents to avoid imprinting on the foster parents. This allowed boosting reproduction and the restoration of population growth. The population is now strong, comprising several hundred individuals.

#### 18.3.2.2 *Imprinting on habitats*

Early behavioural experience can also influence responses to particular habitats and important elements of the environment such as potential predators (Curio 1976). Several captive breeding programmes have improved their release success by allowing individuals to be released after adjustment to the novel habitat. Provisioning of food for some time after release will also allow released individuals to adapt gradually to finding natural food in the novel environment.

These two methods have been used in the successful reintroduction programme of the griffon vulture (*Gyps fulvus*) in France (Figure 18.6). Long-term studies of this reintroduction have led to the accumulation of information that will prove important for successful conservation. For example, the estimation of demographic parameters of individuals released as adults has shown that they pay a significant survival release cost during their first year in the wild (survival rate of 0.74 against 0.99 for the following years for adults born in captivity; Sarrazin *et al.* 1994) as well as a lifelong fecundity cost (Sarrazin *et al.* 1996). These reductions in demographic parameters occurred despite regular food provisioning. The effects perhaps arose as a consequence of captive raising that may have prevented the acquisition of certain key behaviour patterns required to survive. These important costs raise the question about the appropriate age for releasing individuals in a reintroduction programme. In the case of the population

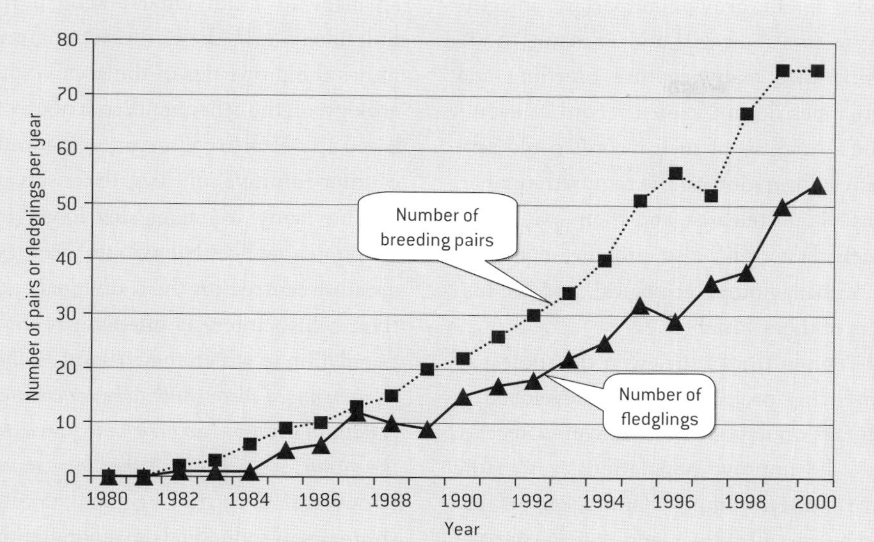


Figure 18.6 **Population size of griffon vultures (*Gyps fulvus*) reintroduced in the Cévennes, France**

Number of reproducing pairs (dotted line with squares) and number of young produced per year (line with triangles) in the population of vultures reintroduced to the Cévennes. The population consisted of 60 adults released 1981–1986. Since then, the population has increased to reach 80 reproducing pairs and approximately 300 individuals in 2000. That population is still growing though starting to level off.

Adapted from Sarrazin (1998).

of griffon vultures, Sarrazin and Legendre (2000) have shown that despite these costs, it remains preferable to release adult vultures rather than young because adults considerably increase the probability of success of a reintroduction.

Encounters with potential predators before release are commonly staged in many captive breeding programmes. This allows individuals bred in captivity to learn to recognize and avoid potential predators. The use of conspecific alarm calls in the presence of a particular stimulus is an efficient method of training an otherwise naive individual to recognize the danger associated with the stimulus (Curio *et al.* 1978). In fact, this method can induce naive birds to emit alarm calls to completely benign features such as an empty bottle if they have previously been exposed to the sound of species-specific alarm calling in the presence of a similar bottle (Curio *et al.* 1978).

### 18.3.3 What determines and limits reproductive rates?

**Social** and **sexual facilitation** are thought to be important factors that enhance reproduction. The Allee effect, which we saw earlier, describes this set of phenomena in general, although the exact causes of the Allee effect remain obscure. It seems likely that social and sexual factors may play important roles. However, experimental studies are urgently needed to test the relative importance of these factors in determining reproductive rates and hence introduction success.

The ultimate goal of captive breeding experiments is successful release into the wild. Unfortunately, very little is known about the factors that contribute to the causes of release failure. This is partly because the persons involved in conservation and release programmes do not often cooperate with population biologists, who might provide important insights into limiting factors (Sarrazin and Barbault 1996). It is surprising that many capture breeding and release programmes do not keep a record of events during releases. Analysis of such records could provide important information on the required number, the size of individual releases, and their spatial distribution. Such data could also provide important information on behavioural, ecological, and life history correlates of these variables.

Cade and Temple (1995) reviewed the success of 30 captive breeding projects on birds developed to enhance natural populations of endangered species. Whereas 43% did improve population sizes in the wild, 17% have ended in complete failure. Factors that contribute to successful release and management of threatened populations include limiting nest sites, alleviation of competition, predation and parasitism, supplemental feeding, and manipulation of breeding biology (such as removal of offspring in species with obligate fratricide). Although many of the reviewed studies proved to be successful, there was little evidence that could be used to assess what would have happened if the interventions had not occurred. This may seem trivial, but the only way to make scientific predictions would be to base projects on rigorous data obtained experimentally in other species.

A final alternative that has not really been adopted systematically concerns the investigation of the limiting factors of population growth in several model systems that are typical of species that are at risk of extinction. Extensive experiments on such model species could provide valuable information on the factors that determine the growth of small populations.

## 18.4 Conclusions

### 18.4.1 What does behavioural ecology contribute to conservation that other sciences do not provide?

Behavioural ecology is almost completely neglected by conservation biology, which has more easily used contributions from population genetics and demography that are based on means, frequencies, and (at best) population variances. However, populations are made up of individuals and the properties of these individuals and their interactions are what determines the properties of the population. Behavioural ecology, on the other hand, contributes an individual-based approach to conservation that can often provide counter-intuitive solutions to conservation problems.

The many examples developed in the current chapter show how behavioural ecology brings a perspective centred on the individuals and shows how their selfish interests often clash with those of the population to which they belong. '*Only characters that have managed successfully to be transmitted during events of natural selection can pretend to be selected at the level of the group . . . , even when that may result in extinction*' (Gouyon *et al.* 1997). In general, taking the different strategies of individuals in a population into account most often results in a profound change in understanding of the way in which populations are regulated. This is a major contribution of behavioural ecology to conservation biology.

The examples of the kakapo and the brown bear have illustrated the utility for conservation biology of searching for the evolutionary interests of different categories of individuals within a given population. Conservation action can only be proposed when such knowledge is available, sometimes potentially saving a species from certain extinction as in the case of the kakapo.

The example of the kakapo is particularly illuminating. Scientists had accumulated information on reproduction of the species for many decades. Although the biased sex ratio was known to limit population growth in the kakapo, the male-biased sex ratio was based on the calculation of average sex ratios among all offspring produced. It was only when scientists started to consider condition-dependence that patterns of sex ratio distortion started to emerge, finally allowing efficient conservation actions to be developed.

Another example concerns the process of mate choice, especially among species under risk of extinction that are kept in zoos. Often managers transport a single male to a female in the hope that this will result in reproductive success. In such situations females do not have much choice. Females may thus refuse to mate with the male for several different behavioural reasons. Even in the case of copulation, offspring may still have poor viability due to problems of incompatibility, poor immune defence, or differential parental investment. Many conservation efforts may be of no consequence if we do not rely on the important mechanisms of sexual selection that have evolved specifically to ensure the production of viable offspring. Ignorance of behavioural aspects of conservation may put overall conservation efforts at risk.

The major lesson to draw from these arguments is that we cannot necessarily apply the same treatment to all individuals. It is possible that in the future it would be better to apply different treatments to individuals that differ in quality. If behavioural ecologists do not become actively involved in conservation, we run the risk of seeing many species disappear despite considerable efforts from our societies.

### 18.4.2 Conservation biology provides opportunities for real experiments

Most of the time, conservation actions constitute real experiments which are based on a series of assumptions that need to be made as explicit as possible. A lack of efficiency of conservation efforts may indeed be because they are based on improper assumptions. In such cases conservationists need to modify their assumptions and hence the specific conservation actions. In fact, such underlying assumptions are rarely stated clearly, nor is the evidence on which they are based. Often such underlying assumptions are based on knowledge derived from related species, or they are entirely theory-based. It is the interaction between theory, assumptions, experimentation, and results that can ultimately allow maximum conservation efficiency. If a proper scientific approach is not adopted, this may result in conservation efforts actually increasing rather than decreasing the risk of extinction.

The case of the kakapo clearly illustrates this problem. Supplementary feeding of females rested on the untested assumption that limited access to high quality food was the factor limiting population growth. A second untested assumption underlying this approach was that food supplementation did not have any negative side effects, or if that was the case, such effects would be less important than the positive effects. Another lesson from the kakapo story is that despite biologists having known the problem of the biased sex ratio for decades, nobody had made a literature survey on that subject in an attempt to find a solution to the problem. The lesson to draw from this experience is that conservation biology, like all other sciences, must exploit knowledge and concepts from other areas of research.

Often conservation measures are adopted when the crisis is already there, and conservationists can therefore not afford to commit errors. We have even less right to persist in erroneous approaches. If the kakapo had not been a particularly long-lived species, it would have disappeared long ago despite many efforts to secure its future.

## » Further reading

> *This chapter only deals with a few ways in which behavioural ecology can contribute to conservation biology. Many other examples can be found in Clemmons and Buchholz (1997), Caro (1998), and Gosling and Sutherland (2000).*

**Caro, T.** 1998. *Behavioral Ecology and Conservation Biology*. Oxford University Press, New York.

**Clemmons, J.R. & Buchholz, R.** 1997. *Behavioral Approaches to Conservation in the Wild*. Cambridge University Press, Cambridge, UK.

**Gosling, L.M. & Sutherland, W.J.** 2000. *Behaviour and Conservation*. Cambridge University Press, Cambridge, UK.

## » Questions

1. Discuss the various mechanisms accounting for the Allee effect.

2. How would you design an experiment to test the effect of inbreeding and sexual selection on extinction risk?

3. Discuss the role of 'human nature' and culture in determining how natural systems are exploited.

4. What is the relative role of conservation projects in determining the natural resources of the future? Do not hesitate to use the knowledge acquired in different chapters of this book.

# 19

# 19

# Behavioural Ecology and Humans

Anders P. Møller and Étienne Danchin

## 19.1 Introduction

Some of Darwin's (1871, 1872) fundamental contributions were the rejection of any idea of a clear-cut separation between humans and animals, and the assertion, on the one hand, of a continuity of mental processes across living organisms, and, on the other hand, of the importance of **natural selection** in the **evolution** of human behaviour. Iconoclastic in its time (Browne 2002), this proposal by Darwin continues to feed a continuously renewed debate about human nature and its biological bases (Wilson 1978; Rose *et al.* 1984). As with other behavioural sciences, behavioural ecology contributes to this debate. Ever since the publication of *Sociobiology* (Wilson 1975), several approaches have led to apparently different conclusions about the relative importance of the history of mankind and that of the history of individuals (in the sense of their ontogenesis; Brown 1999; Sterelny 2001; see Laland and Brown (2002) for an account of the debate). Among those approaches are at one end, those that consider that the action of natural selection in the past played a major role in determining human contemporary behaviours (for example Cartwight 2000). At the other end, there are those that grant learning processes and **cultural transmission** a major role, seeing in the approach of the first only an outrageous reductionism (Rose and Rose 2000). The object of this chapter is obviously not to put an end to the debate. Our goal is not to oppose scientific approaches or to claim a moral superiority, but to show that behavioural ecology can and must add its vision to the other disciplines that try to elucidate the bases of human behaviour. We thus propose to illustrate some applied domains where, in line with other disciplines such as evolutionary psychology or evolutionary anthropology (both tending to be hardly distinguishable), behavioural ecology brings an original point of view that, at the very least, deserves to be considered.

## 19.2 What sets humans apart?

Several among us find it easier to accept continuity between animals and humans at the biological and physiological levels than at the behavioural level. Consider how efficiently medicine finds treatments for our diseases from work undertaken on animal models such as rats, pigs, and primates. Medical efficiency seems a tangible proof of this physiological continuity. However, continuity at the behavioural level is not easily accepted by all. Human beings are thus often set apart by philosophers and scientists of social and human sciences (see Lestel 1996). This distinction undoubtedly goes back a very long time (see Chapter 1). Obviously, humans differ from other animals in several important aspects, just as the cat is distinct from the dog. It would be false to deny these fundamental differences such as humans' hyper-developed culture, our reasoning abilities, and our disproportionate potential to transform our environment so as to influence our own future (Odling-Smee *et al.* 2003). We, like Darwin before us, claim that these distinctive features reveal quantitative differences rather than clear-cut discontinuities between animals and mankind.

*Culture?*

We will see in Chapter 20 that if the transmission and the accumulation of cultural traits are particularly marked in humans, they are not necessarily absent in other animal species (see Avital and Jablonka 2000). For example, the song of birds is generally transmitted in a way that is akin to cultural transmission (Catchpole and Slater 1995) and it results in the evolution of regional dialects that characterize individuals according to their geographical source. Several vertebrates, including great apes, show culturally transmitted traits, such as, for example, some traditions of food preparation (Lefebvre *et al.* 1997, 1998; Jablonka *et al.* 1998; Avital and Jablonka 2000; see Chapter 20). So, even if human culture seems overwhelming in terms of diversity and complexity compared with animal **cultures**, we will see in the following chapter that culture is unlikely to be unique to humans. The difference between humans and animals is in fact only a matter of degree in the importance of culture, and not a matter of the existence or absence of cultural processes.

Considering the importance of culture in human societies, one can also raise the question of the origin and maintenance of culture. Culture is not a free phenomenon that would be disconnected from the remainder of organisms' phenotype (Cavalli-Sforza and Feldman 1981; Boyd and Richerson 1985; Richerson and Boyd 1992). In other words, culture must in one way or another be influenced by the process of natural selection. Moreover, the effectiveness of cultural transmission depends on sensors and nervous tissues whose current state results from the processes of natural and sexual selection that operated in the past.

Relative to other animal species, cultural processes in humans must constitute a large part of the **information** that is transmitted between generations. In other words, the cultural component of transmittability (see Chapter 20) must be much larger in humans than in any other animal species. This suggests that the behavioural variation observed between human populations and individuals may have a weaker genetic component than in other animals.

*The relative importance of nature and nurture?*

Some researchers still insist on opposing the action of the environment to that of the genes, allotting some to nature (genetics) and others to nurture (experience). This dichotomy between nature and nurture is artificial and erroneous. A gene can produce a phenotype only by interacting with the environment. Chapters 5 and 6 are entirely devoted to describing the mechanisms involved in the interaction of these two components. Thus traits that are purely of one or the other origin cannot exist. To accept a biological continuity between animals and humans does not imply that human nature (and that of all organisms) are only determined by genes. It is rather a question of accepting that humans, like any organism, are the product of a complex interaction between their genetic heritage and the environment in which they developed and live. Studying human behaviour without taking into account their biological origins is quite as erroneous as studying the biology of behaviour without taking into account the environment in which it develops and expresses itself.

*Our environmental impact?*

Some consider humans as unique in that they can influence their own environment in a dramatic way and, by doing so, they directly influence the mode of selection to which they are subjected. However, the same is true for many organisms. For example, ants constitute a significant part of the biomass of many ecosystems, particularly in the tropics (Wilson 1990). They can have an important impact on the environment by transport, distribution and use of their resources. They belong to the most important ecological engineers, and their activities have a considerable impact on their own environment and that of all other organisms in their ecosystem (Wilson 1990). Theoretical models have tackled the question of a species' ability to create its own ecological niche, and have done so in relation to the presence of cultural processes (Laland *et al.* 2000; Odling-Smee *et al.* 2003).

*The size of our brain?*

It is also claimed that human beings differ from other organisms by their larger brain size, and consequently by their greater mental abilities. Humans are not unique when it comes to increased brain volume (see Figure 3.5) because several other organisms also have large brains compared with their body size. It is true for invertebrates in the case of cephalopods and for vertebrates of whales and dolphins. Likewise, hints of intelligence and the use of cognitive abilities in social interactions were highlighted in several primate species, in particular in great apes (Byrne and Whiten 1988). The level of cognitive abilities and their use in great apes seem to differ from humans in a quantitative rather than a qualitative way.

*Tool use?*

It was once claimed that tool use was unique to humans, but since the discovery of tool use in chimpanzees by Jane Goodall in the 1960s, evidence of tool use has accumulated not only in primates but also in a variety of mammals and birds (see Lefebvre *et al.* 2002; Chapter 20) at the very least.

### 19.2.1 A warning

It is so difficult to find a point of obvious discontinuity between humans and animals that there is no objective reason to consider that our species completely escapes the influence of our biological nature. It thus seems that we have to accept that human beings may be studied using an evolutionary approach, and more particularly in the context of behavioural ecology. Can humans and human nature be better understood by adopting an evolutionary prospect? Can biology and behavioural ecology in particular contribute to the understanding of human beings?

It goes without saying, however, that speaking about human behaviour raises all kinds of controversies. Yet we decided to gamble that taking such an approach in this chapter provides readers with the opportunity of deciding for themselves. The chapter should be taken first of all as an essay whose primary function is to elicit constructive thinking about the relevance of the central concepts of behavioural ecology in the sphere of human behaviour. In this kind of exercise, one is likely to shock, or to be shocked, by an apparently too objective, or perhaps too simple cause-and-effect vision of human behaviour. Of course, each point tackled in this chapter is open to discussion given that science always considers any knowledge as provisional. It seems to us, however, that the ability, to some extent at least, to explain human behaviour as complex as, say, the relative investment in offspring sex (see Chapter 13), illustrates the tremendous explanatory power of evolutionary reasoning, even when it is applied to humans for whom we would be tempted to think that innumerable layers of conscience, culture, and morals tend to cover up any trace of such processes.

Far from being exhaustive, this chapter tackles three different subjects that illustrate the potential contributions of a behavioural ecology approach to human behaviour. The first part relates to human sexual behaviour because this question is central for the understanding of sex roles and the determinants of human behaviour. The second part briefly tackles the question of an evolutionary human medicine, i.e. how considering humans as a product of evolution can help develop medical treatments. Lastly, we show how some work, through the consideration of our biological nature, can provide elements of answers to questions as difficult as perhaps child abuse or some forms of criminality.

## 19.3 Human sexual behaviour

Compared with other primates, humans are characterized by a period of child development and dependence that is long relative to the period during which the individual can reproduce. This characteristic must have a strong impact on human reproductive behaviour, and it is more useful to consider the mating system as a whole rather than only the mode of reproduction and **parental care**. Among primates, humans are characterized by a slightly larger male size, the prevalence of a **monogamist** mode of reproduction, with a tendency to **polygyny**, and an important role for paternal care. Although there are variations from these traits, they are found in many cultures (Deliège 1996; Cézilly 2006).

### 19.3.1 Human sexual behaviour

#### 19.3.1.1 *The role of ecological conditions*

Classically in animals, **sexual selection**, modes of reproduction, and modes of parental care have been considered in relation to ecological conditions, in that some conditions can allow the monopolization of several individuals of the opposite sex, and some conditions can increase the **potential reproductive rate** of a sex compared with the other (see Chapter 12). For example, a certain degree of polygyny is observed in almost all human cultures, whereas only three cultures show a strong rate of **polyandry** (Smith 1984; Deliège 1996). The influence of ecological conditions is rather obvious. Polygyny is particularly frequent in cultures marked by a very strong variance in the quantity of resources available to males, whereas the very rare cases of polyandry are observed in environments where resources are very limited and within societies whose economy rests largely on a division of labour (Levine 1988; Deliège 1996).

#### 19.3.1.2 *Sexual dimorphism in relation to the rate of polygyny*

Human sexual dimorphism is weak compared with other primates, but men are generally stronger and taller than women. This would indicate a stronger sexual selection (see Chapter 11) on men than on women. In fact, the variance in reproductive success is considerably larger in men. According to the *Guinness Book of World Records*, the man having had the greatest descent was the Sultan Moulay Ismail of Morocco with 888 children, whereas the record for a woman is established at 'only' 69 children over 27 pregnancies. The difference between the two records is of a factor of more than 10.

The patterns of sexual dimorphism among human cultures follow those of polygyny, with a stronger rate of polygyny in the societies showing the greatest size dimorphism (Figure 19.1). The mode of reproduction of our hominid ancestors is not precisely established. It has long been considered that prehistoric men presented a much greater sexual dimorphism than today, which was viewed as an indication of a polygynous mode of reproduction (Lewin 1999). One analysis, however, led to the conclusion that the sexual dimorphism of *Australopithecus afarensis* was similar to that of modern humans (Reno *et al.* 2003), thus tending to support the existence of a monogamous mode of reproduction in one of the most ancient hominids. However, this conclusion has recently been criticized, based on a re-analysis of the same data set (Plavcan *et al.* 2005). Caution is thus

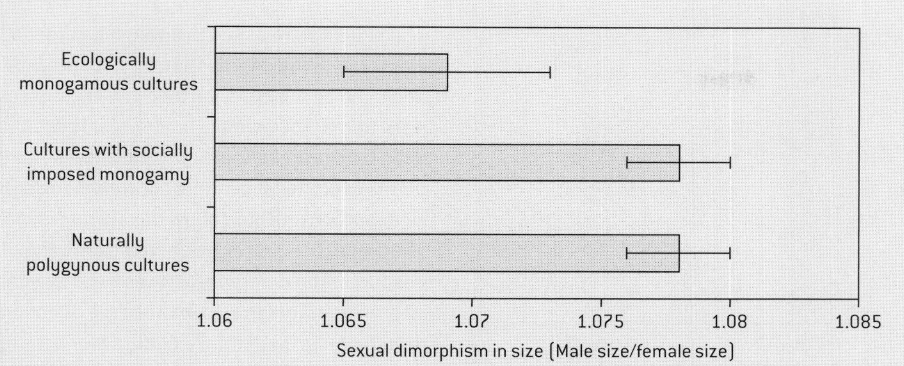


Figure 19.1 **Sexual dimorphism for body size and rate of polygyny in various current cultures**

Bars indicate the dimorphism in male relative to female body size observed on average (± standard error) in 46 polygynous cultures, 31 cultures with socially imposed monogamy, and 16 ecologically monogamous cultures.

After data from Alexander *et al.* (1979).

required, in particular because the fossil data examined often span rather broad time and space scales, which can introduce many biases into the analysis of sexual dimorphism.

#### 19.3.1.3 *Important paternal care*

Among mammals, humans are special due to the importance of intensive paternal care. In most mammals paternal care is absent, except for some monogamous or semi-polyandrous species (see Chapter 12). The adaptationist point of view considers that human paternal care is associated on the one hand with a strong certainty of paternity, and, on the other hand, with the prolonged dependence of offspring that makes it difficult for the mother alone to rear offspring successfully. In such a situation, men would have more maturing offspring by contributing to parental care than by seeking extra-pair mates.

### 19.3.2 Human sexual selection

We now will consider several aspects of human sexual behaviour by interpreting them in the light of sexual selection theory, a detailed description of which is given in Chapter 11. Four questions are addressed successively. Can one interpret human beauty in relation to the process of sexual selection? What are the roles of parasites and human diseases in sexual selection? How does developmental stability relate to sexual selection? Do odours and smell play a role in human sexual selection?

#### 19.3.2.1 *Human beauty and sexual selection*

##### a *Men prefer beauty, women prefer resources*

A vast study of criteria used in mate choice showed that, in a sample of 37 cultures from countries spread over all the continents, men and women rate and rank the 18 proposed criteria in very similar ways (coefficients of similarity greater than 0.92; Buss *et al.* 1990). However, there were differences, with men ranking beauty higher in the list of characteristics preferred in women, whereas women rank the resources of men higher in the list (Buss 1994; Figure 19.2).

##### b *Beauty, a concept with an evolutionary basis*

The very concept of beauty can have an evolutionary origin. Female traits qualified as beautiful are those

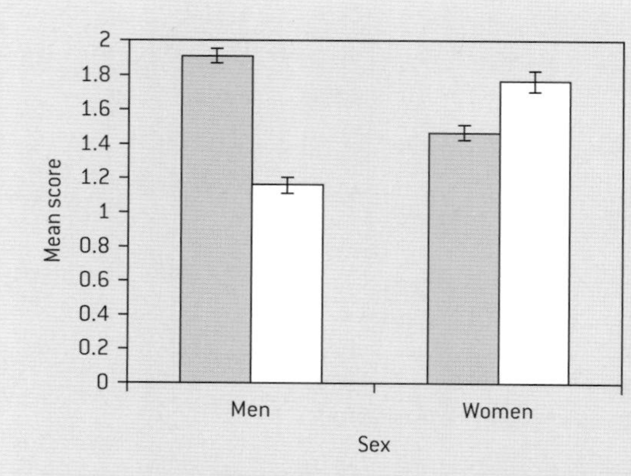


Figure 19.2 **Relative frequency of the attributes preferred in a mate for men and women belonging to 37 different cultures**

The people were asked to arrange by order of importance the criteria that significantly influence their choice of sexual partner. The criterion could vary from 0 (of no importance) to 3 (essential). The grey bars represent the average scores (± standard error) over the 37 cultures allotted to the criterion of beauty. White bars show the average scores over the same 37 cultures given to the criterion of resource. The difference between the evaluations of men and women was in the same direction and significant in 34 of the 37 cultures for beauty criterion, and in 36 of the 37 cultures for the criterion of resources.

After data from Buss (1989).

that attract men sexually. Many of them actually reveal age, good health, and fertility. This comes down to saying that men say they are attracted by women whose traits reveal a good capacity to bear and raise children which, from an evolutionary point of view, is completely sensible. On their side, women are attracted by male traits that will ensure the best possible development for their children, namely the resources they control. From an evolutionary point of view this is also sensible because, owing to the differential of investment in offspring (see Chapter 11), females reproductive output tends to be limited by their access to resources.

### C *Do beauty criteria have a genetic basis?*

Charles Darwin (1871) was the first to write a synthesis about the perceived criteria of human beauty from a biological point of view. Since then, data coming from many cultures have been collected and it appears that there is no single criterion of beauty common to all human cultures. This observation may seem to contradict the biological approach but it is not necessarily the case. This diversity can have an ecological explanation. Indeed, the study of mate preferences in more than thirty animal species as diverse as Coleoptera, fruit flies, and fish showed in each case the existence of genetic bases to female mate preferences. In short, related females are more similar in their preference criteria than two females taken randomly in a given population.

Another result of all these studies is that female mate preferences and male secondary sexual traits can diverge considerably among populations of the same species and this over a relatively short time span. For example, some populations of guppies, *Poecilia reticulata*, live in sympatry with predatory fish that use the orange colours of male's secondary sexual traits to locate their prey, whereas other populations of guppies live in rivers devoid of predators. Female guppies drawn from populations living in sympatry with predators show mate preferences for males presenting less orange while the reverse is true for females drawn from populations without predators (Houde 1997). Mate preferences can thus be shaped locally by selection pressures.

Note, however, that most of these studies, if not all, did not really ensure that the parents/offspring resemblance was not partly the result of cultural processes. Indeed, all the results presented above on the genetic support of female preferences were interpreted in terms of genetic **heritability** but could result from a purely cultural transmission of female preferences, with young females imitating the choices of their elder. We will discuss the important question of the genetic versus cultural basis of mate preference in details in Chapter 20. We will in particular review the evidence for what is called

**eavesdropping** and **mate choice copying**, two phenomena that could very well generate the transmission of cultural differences across generations.

In any case, if one extends this result to human populations, one can infer that mate preferences can also depend on local conditions. So it is perhaps not surprising that preferences for large reserves of body fat are prevalent in populations living in environments where slimming diseases like schistosomiasis, malaria, and others prevail (Buss 1994), whereas a thinner body is preferred in environments with few parasites (Singh 1993).

Darwin's argument for the lack of universal beauty criteria can appear in contradiction with the general success and multinational success of the beauty industry. However, it is clear that the current cultural globalization of Western beauty criteria through media such as cinema and television can perhaps explain a certain homogenization of beauty criteria across cultures. In any case, the human obsession with beauty in Western societies is not so different from the comparable phenomena in other societies. The fact that many women are assiduous users of cosmetic products and that several of these products are sold by praising their capacity to provide eternal youth nevertheless raises the possibility that our **choices** of aesthetic criteria are based on general principles. Any book on the use of cosmetics constitutes a handbook of know-how to accentuate the traits that announce health and fertility: a full and symmetrical face, a healthy skin, colours that reflect health. With the development of plastic surgery, these sought out and admired characteristics of human beauty can even be acquired in a more permanent way. It is thus not surprising that the purpose of most plastic surgery interventions is to correct asymmetries (see below) and to exaggerate traits considered as beautiful.

#### 19.3.2.2 *The perception of health and beauty in animals and humans*

Parasites and diseases play a fundamental role in human populations, and perhaps even more so than in most of the great apes because of the very high density that can be reached by human populations. Parasites exert an enormous selection pressure on their hosts by increasing their mortality and reducing their reproductive output (see Chapter 17). It has long been known that individuals differ in their susceptibility to parasites. These variations imply genetic differences in resistance. So the choice of healthy mates obviously provides an important selective advantage to the choosy individuals (Hamilton and Zuk 1982). Inter-sexual selection can, in relation to parasites, favour choosy individuals that avoid acquiring mates bearing contagious parasites that could infect them and their offspring, and instead prefer healthier mates that will be more effective parents, and perhaps produce offspring that are more resistant to parasites (Møller *et al.* 1999).

There are several lines of evidence in a large variety of organisms that secondary sexual traits reliably reflect the parasitic level of infection and the immune capacity of individuals (Møller *et al.* 1999). For instance, studies in a great diversity of plants and animals suggest that parasites exert costs that make their hosts asymmetrical and thus less attractive than unparasitized individuals (Møller 1996a). The same appears to be true in humans: men, of a great diversity of cultures, tend to rate beauty in women more than women do. If we accept that beauty is an assessment of a women's health then this statement is not so surprising. Moreover, the importance that men place on beauty seems to be highest in those cultures where diseases like malaria and other parasites of similar level of virulence have an important impact suggests that the capacity to discriminate against unhealthy parasitized mates is important in humans too (Gangestad and Buss 1993; Figure 19.3). However, this pattern of men placing much importance on beauty when parasites are common suffers from some important exceptions, California being a prime example. In that state youth, beauty and health are universally worshipped and yet there is no strong incidence of debilitating parasites. Such counter-examples can result from particular socio-economic situations that would be interesting to explore.

Over time, hosts have come to evolve acquired immune defences that allow them to reduce the negative effects of parasites. The immune system could thus play a fundamental role in sexual selection if

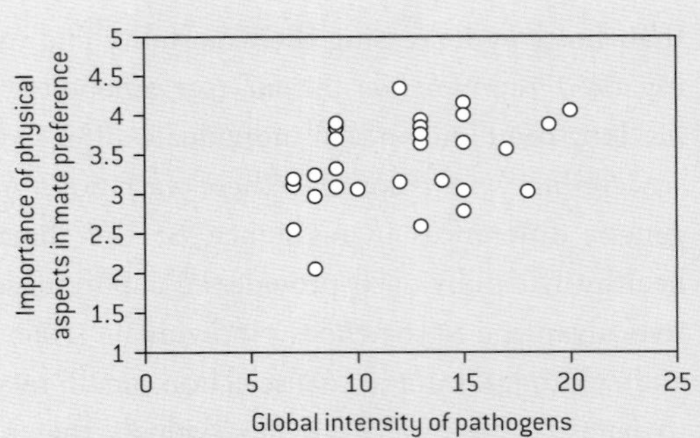


Figure 19.3 **Importance of physical beauty in mate choice in relation to the impact of serious diseases over 29 cultures all over the world**

*x*-axis: the diseases concerned were leshmaniasis, trypanosomiasis, malaria, schistosomiasis, lymphatic filariasis, Lyme disease and plague. Their presence was either null (coded 1), existing but without indication of a severe occurrence (coded 2), or with a severe occurrence (coded 3). The score of the *x*-axis is the sum of the scores for all diseases.

*y*-axis: Each subject was asked to allot a degree of importance (from 0 = without interest, to 3 = essential) to 18 indices likely to influence mate choice. The *y*-axis adds up the mean index by culture allotted to the beauty criterion by men and women. The relation is significant ($r = 0.38$, $P < 0.05$): the higher the parasitic risk, the more men and women value physical beauty, an indicator of good health.

After Gangestad and Buss (1993).

secondary sexual traits indeed reflect an individual's **immunocompetence** (Folstad and Karter 1992; see Chapter 11). We have seen in Chapter 6 how sexual **hormones** affect the development of secondary sexual traits and other sexual signals, like the behaviour and vocalizations of animals in general. The same is true in humans where, at puberty, the changes that occur in the shape of the face and the body proportions, along with the voice and behaviour, are driven by hormones. However, sex hormones have antagonistic effects on how the immune system operates, and only individuals that happen to be in very good physical condition are able to develop the most stringent secondary sexual traits without compromising their immune defences. It is thus probable that only high-quality individuals can afford high sex hormone levels without jeopardizing their immune defences.

#### 19.3.2.3 *Developmental stability and beauty*

*Symmetry as an external index of developmental stability*

Over the many upheavals occurring during development, incidents can arise that will tend to destabilize development. Mechanisms of control exist that prevent these incidents from affecting the phenotype. Developmental stability is the capacity of individuals to maintain a stable development of their phenotype in a given environment (Møller and Swaddle 1997). Ludwig (1932) proposed that fluctuating asymmetry provides an external measure of developmental instability. Fluctuating asymmetry is the asymmetry that can occur when symmetry constitutes a standard from which any deviation can randomly occur on either side. Most human traits like the ear length, finger length, and fist size show such a fluctuating asymmetry. Likewise, the cases that clearly depart from this standard, for example when the heart is on the right cavity of the body or the presence of an even number of fingers on a hand, also reflect such a developmental instability.

Symmetry being the optimal phenotype because it favours biological performance, any deviation from symmetry can be regarded as a sub-optimal solution to a design problem. It was probably difficult for our prehistoric ancestors to escape the lion (*Panthera leo*), but it was even more difficult to do so with two legs of unequal length. For example, the remainders of skeletons of prehistoric Indians showed that the individuals who died old had more symmetrical bones than those that died young (Ruff and Jones 1981). This result is particularly interesting because the continual reshaping of bones over life generally leads to an increase in the asymmetry of bones in older individuals. This result is thus **conservative**.

*A selection for symmetry . . .*

It is perhaps not surprising that symmetry turned out to be important in plants and animals (including humans) when they must fight to ensure their survival or to reach mating partners (Møller and Swaddle 1997). Continuous selection for symmetry starts as early as the stage of sperm and ovules inside the female in species with internal fertilization: only a small fraction of gametes indeed fulfil their function, and it is mainly those with a deviating phenotype that are disadvantaged (Møller 1996b). This negative selection against deviating gametes and eggs seems to constitute a very general process. For example, seed and fruit abortion is very common in plants. Experimental studies show that in fireweeds, *Epilobium angustifolium*, about three-quarters of all embryos abort during the very first cellular divisions because of developmental errors (Møller 1996b). Interestingly, the frequency of abortion is directly connected to the asymmetry both of the flowers from which pollen comes and of those that receive pollen. This implies that similar mechanisms of development are involved in the maintenance of even phenotypes in embryos and flowers.

*. . . in many animals, including humans*

The same kind of phenomenon was described in many organisms from invertebrates to vertebrates, including humans (Møller 1997). For example, a study of scorpions has shown that newborns must climb on their mother's backs during their first day. Some young born with deformities were unable to fulfil this task and were eaten by their mother. This behaviour makes evolutionary sense because malformed offspring, raised apart from their mother along with offspring having a normal phenotype, always performed poorly as adults (Møller 1997). Infanticide was and still is a common practice in many human societies and is mainly directed against children with deviating phenotypes (Daly and Wilson 1984). Even if it may seem shocking to say so, it is possible that the behaviour has an evolutionary origin if ancestral parents that were infanticidal had a higher **fitness**, hence produced more successfully maturing offspring over their lifetime than those parents that were not infanticidal, perhaps because of the energy, time and resources saved by such a behaviour.

Fluctuating asymmetry is also important for mating. Fluctuating asymmetry and inter-sexual selection are associated in organisms as different as plants, flies, grasshoppers, fish, birds, and mammals (see Møller and Thornhill 1998). For example, in the barn swallow (*Hirundo rustica*), the manipulation of tail streamers' symmetry showed that females strongly prefer males with symmetrical compared with males with asymmetrical streamers (Figure 19.4).

Reminiscent of results with swallows, women, it seems, also prefer men with symmetrical traits such as faces and bodies (see Grammer and Thornhill 1994; Thornhill and Gangestad 1994), and an individual's lifetime number of mates appears to be directly related to the symmetry of its skeleton (Figure 19.5). If we accept that body symmetry somehow is an indicator of an individual's potential fitness, then females seeking symmetrical males would end up mating with individuals that are better able to monopolize resources and transmit genes for greater developmental stability to their offspring.

#### 19.3.2.4 *Sexual attraction, odour, and smell*

Whereas studies of human sexual preferences involving vision can be interpreted as resulting primarily, if not exclusively, from cultural biases, it is less the case for attraction to odours because we are unaware of the kind of odours involved in sexual attraction given they are often at subliminal levels. It is often stated that human beings are less receptive to odours than other mammals, and it is true that our sense of smell does not compare with that of a dog or a mouse. However, studies of human sexual preferences classify odours on the same level of importance as visual cues, and even more so in women (Kohl and Francoeur 1995). It is noteworthy that women's

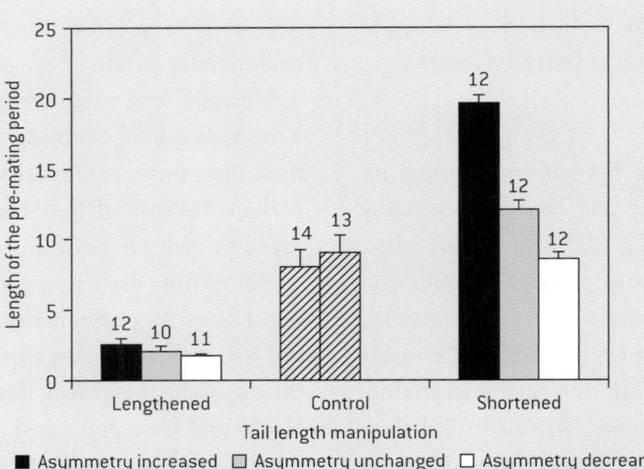


Figure 19.4 **Duration of the pre-mating period and independent manipulation of tail length and tail symmetry in barn swallows (*Hirundo rustica*)**

A total of 96 males were allotted randomly to 8 treatments.

Tail length was either: *shortened* by 20 mm, which corresponded to a reduction of 17.3% for a final length of 91 mm; *lengthened* by 20 mm leading to a 16.7% increase for an average final length of 126 mm; Control I: the outermost tail feathers were cut then re-glued resulting in a very small 1% reduction of final tail length of 105 mm; Control II: the individuals were captured but did not undergo any streamer manipulation with a tail length of 106 mm.

The other treatment consisted in manipulating the symmetry of the outermost tail feathers by either: *increasing the asymmetry* by 20 mm (up to a mean of 23 mm) without changing the average tail length (black bars); *reducing the asymmetry* to 0 mm (white bars); or keeping control males without changing the average natural asymmetry of 3 mm (grey bars).

In the two groups on the left and on the right, tail length was increased (on the left) or was decreased (on the right). In each one of these groups, asymmetry was also either decreased (white bars), increased (black bars) or left unchanged (grey bars). The two hatched bars in the centre give the average durations of mating periods of the control individuals for which neither tail length nor symmetry was manipulated. No difference in tail length before treatment was detectable. On the other hand tail length varied significantly between treatments after manipulation ($F_{7,88} = 35.15$, $P < 0.0001$). Durations are given in days, ± standard error. The interaction between manipulating tail length and tail symmetry was highly significant ($P < 0.001$), showing a joint effect of the two manipulations. The numbers above bars give sample sizes.

Redrawn from Møller (1992).

sensitivity to olfactive cues varies over the course of the menstrual cycle: the odours of androgen derivatives such as androstenone and androstenol, both associated with sweet, musky odours are preferred by women around the time of ovulation (Grammer 1993). The crucial importance of perfumes in human societies and in all the historical chronicles brings an indirect argument in favour of an important role of smell in human sexual relations.

### a *A story of T-shirts*

An important component of immunity in vertebrates involves what is called the major histocompatibility complex (MHC). This complex results from an aggregate of highly variable genes involved in resistance against a great number of parasites such as malaria and schistosoma in humans. Since the 1980s, studies undertaken on mice suggested that adults can

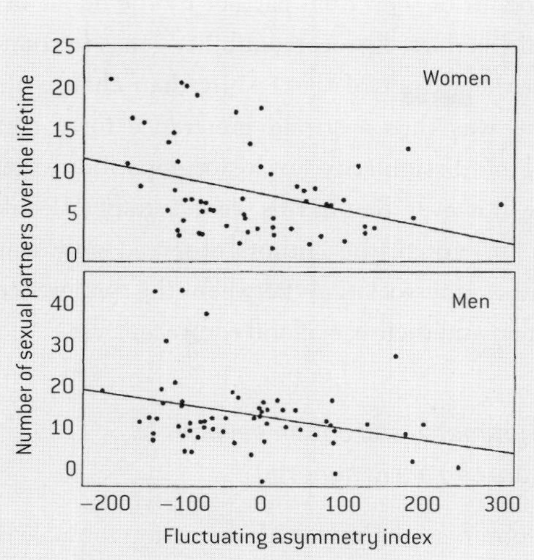


Figure 19.5 **Number of mates over a lifetime in relation to the degree of fluctuating asymmetry in men and women**

This analysis took age into account because it influences the number of mates over the lifetime. After taking this effect into account, the effect of fluctuating asymmetry was significant and had a negative slope both in men ($P < 0.02$) and women ($P < 0.01$).

Redrawn from Thornhill and Gangestad (1994).

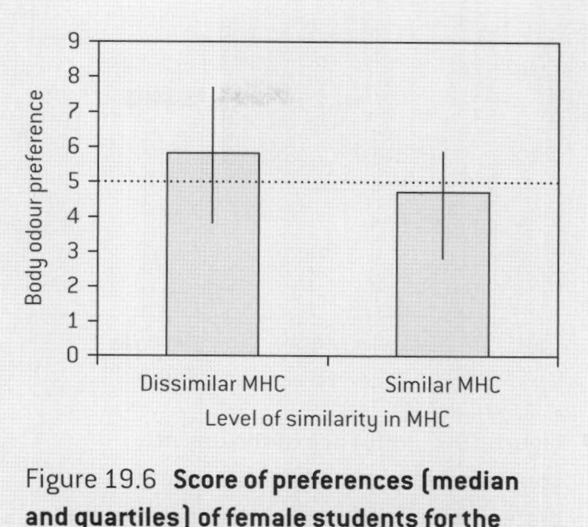

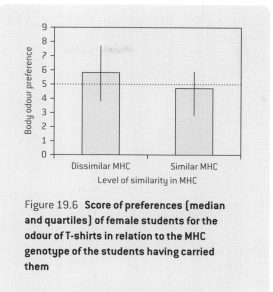

Figure 19.6 **Score of preferences (median and quartiles) of female students for the odour of T-shirts in relation to the MHC genotype of the students having carried them**

The bars represent the average scores (± standard deviation) given by the female students not taking oral contraceptives (number of boys = 38, $P = 0.04$). The horizontal dotted line represents a random score.

Redrawn from Wedekind *et al.* (1995).

distinguish the MHC genotypes of potential mates on the sole basis of their odours. Experiments in humans show discriminatory abilities quite as astonishing. In Switzerland, Claus Wedekind and his colleagues (1995) distributed T-shirts to male students with the instruction to wear them over a single night, to use neither perfume nor deodorant, not to smoke, to avoid eating garlic or to have other activities known to influence body odours. The T-shirts were then collected in individual food plastic bags (i.e. odourless, among other things) and given to assess to female students who were not informed of the identity of the wearers of the T-shirts. The MHC groups of the male and female students were then determined. The girls had stronger preference for male odours when the carrier had a dissimilar MHC genetic group (Figure 19.6).

The direction of this effect is important as it is known in medicine that spontaneous miscarriages and sterility of couples trying to have children very often involve partners whose MHC genetic groups are very similar. Moreover, this preference for those persons having a distinct resistance to parasites produces a greater genetic diversity for these MHC genes in their progeny, thus conferring them a greater resistance to parasites. What was more surprising in this study was that preferences of women for the T-shirts depended on whether they used oral contraceptives or not (Figure 19.7). The use of such contraceptives reverses women preferences: they now prefer the odours of T-shirts coming from men with histocompatibility genes similar to theirs. This reversion is perhaps because oral contraceptives affect hormonal levels by imitating the gestation period.

This preference for individuals carrying similar histocompatibility genes during the gestation period but dissimilar otherwise, could indicate that at the biological level women would prefer men unrelated

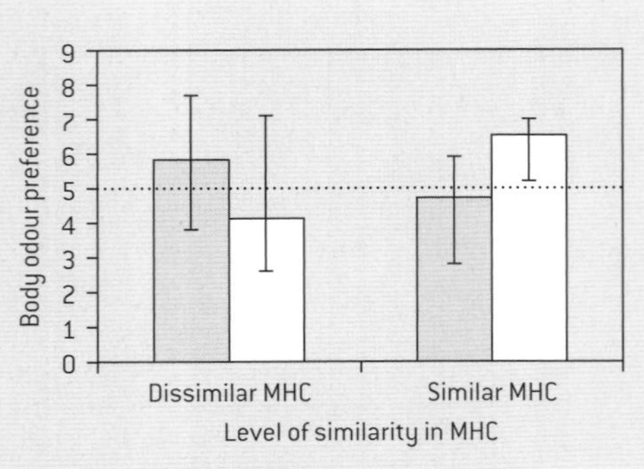


Figure 19.7 **Influence of the use of contraception on the female student preferences (median and quartiles) for the T-shirts' odour in relation to the genotype of the MHC of the male students having carried them**

The **grey bars** represent the average scores (± standard deviation) given by female students not taking oral contraceptives (number of males = 38, $P$ = 0.04). The **white bars** represent the scores given by female students taking oral contraceptives (number of males = 23, $P$ = 0.02). All statistical estimates are made with two-tailed tests. The horizontal dotted line represents a random score.

Redrawn from Wedekind *et al.* (1995).

to themselves when looking for the sire of their children, but related to themselves when it is a question of recruiting help to raise their children. This variable preference is not surprising in the context of mammalian biology where in most species females live in groups of related females (see Chapter 12). Young males are rejected from their family group at puberty. They will be able to reproduce only when managing to be accepted in one way or another in groups of females unrelated to them. These results have been supported in a later study that showed again a strong negative relationship between the pleasantness of a T-shirt odour and the degree of genetic similarity of the major complex of histocompatibility between the carrier of the T-shirt and the person smelling the odour (Wedekind and Füri 1997). Moreover, men and women who remembered the odour of their own partner by the odour of the T-shirt had less histocompatibility genes in common with the wearer of the T-shirt than chance alone. There was thus a strong preference for a certain level of dissimilarity in histocompatibility genes. Note, however, that in this second study (Wedekind and Füri 1997), the authors observed only a non-significant association between the preference of women and their use of oral contraceptives.

### b *Symmetry and odour: several factors can act in synergy*

The observation that female attraction for male body traits like odours changes over the menstrual cycle was replicated in another experiment performed by Steven Gangestad and Randy Thornhill (1998). In this experiment, the experimenters first measured the extent of their male subjects' body symmetry based on nine phenotypic traits such as, among others, ear length, wrist thickness and foot width. Women were asked to score the T-shirts worn by these men according to their preference for the odour. The women established a coherent ranking. Results show that the pleasant aspect of the preferred odour is at its maximum for the women who were ovulating or who were close to that period, i.e. when the probability of fertilization was high. Moreover, these women showed a marked preference for the odours of T-shirts worn by men whose bodies were scored as most symmetrical. The women who were far from their ovulation date had odour preferences that bore no association with the body symmetry of the men that had worn the T-shirt (Figure 19.8).

Thus, the closer a women is to her period of fertility, the more she seems to prefer the odours of symmetrical men. We also know, thanks to former work, that the ovulation period also corresponds to the moment when extra-pair sexual intercourse is most frequent. That result is more surprising if you consider that no clear pattern in change in the frequency of sexual relations within a stable couple has ever been associated with a women's menstrual cycle (Baker and Bellis 1995). The possible link between symmetry and the major histocompatibility

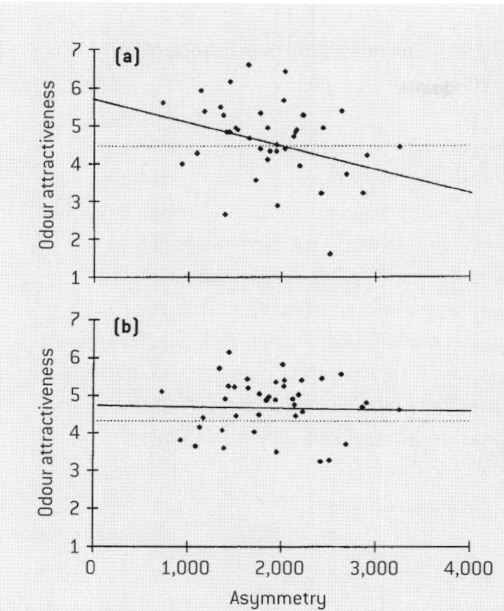


Figure 19.8 **Women's preference for the odour of symmetrical men (the negative slope indicates an increased attractiveness of symmetrical men) according to the date in the menstrual cycle**

**a.** Score of the average attractiveness of male body odours made by women at the time of the menstrual cycle when they have a high probability of fertility. The black line indicates the regression line estimated by the least squares method ($R=-0.31, P<0.03$). The dashed line indicates the average attractiveness of a clean and unknown T-shirt.

**b.** Score of the average attractiveness of male body odours made by women at the time of the menstrual cycle when they have only weak fertility. The black line indicates the regression line estimated by the least squares method ($R=-0.02$, non significant). The dashed line indicates the average attractiveness of a clean and unknown T-shirt.

Redrawn from Gangestad and Thornhill (1998).

complex at the origin of body odour variations, however, remains to be elucidated.

Women also produce odours that affect male preferences. Recent experiments led with men show that subjects are sensitive to the presence of copulines, fatty acids of vaginal origin. The subjects had to rank photographs without copulines. Then in a second ranking exercise they were presented with cotton soaked with copulines or not. The presence of copulines attenuated the differences in classification between the photographs of women previously ranked. These particular female odours thus tended to render the women also more attractive (Grammer *et al.* 1998). Such studies on the role of odours in human mating preferences are interesting because their perception is almost unconscious.

### 19.3.3 Sex ratio at birth in humans

*From the prediction of birth sex ratio in humans . . .*

As seen in Chapter 13, there are strong selective pressures maintaining the sex ratio of a population to a stable evolutionary equilibrium value of one male for one female. This is because every time the sex ratio moves away from this equilibrium, the parents who produce F1 progeny of the rarer sex are favoured because on average, these children will produce more progeny of the F2 generation than those of the more common sex. This reasoning thus led to the prediction that when at the scale of the population a sex becomes rare, an excess of birth of individuals of that sex must immediately occur.

*. . . to the testing of birth sex ratio in humans . . .*

This prediction can be tested in humans using data from birth patterns after the two world wars. These massive events indeed produced a higher mortality in men than in women, skewing the adult sex ratio in favour of women. After these wars, an excess birth of males was noted in countries that were involved in the conflicts but not in those that were not (Figures 19.9 and 19.10).

In humans, there is always a slight excess in the number of male births of about 51 to 52 males per 100 births. This slight bias is itself easily explainable from an evolutionary point of view because the mortality of young boys during the period of parental investment is higher than that of young girls (you can see this pattern in Figure 19.11; see also Chapter 13).

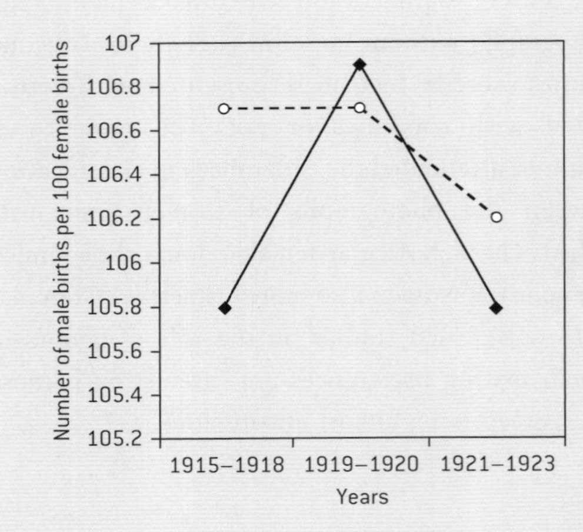


Figure 19.9 **Sex ratio at birth in Europe after World War I**

Number of male births for 100 female births in the countries involved (continuous line and black diamonds) and those not involved (dotted line and empty circles) in World War I in the years that followed the conflict. The data of the countries involved are extracted from the statistics of 12 European countries like France, Belgium, Germany, Italy, United Kingdom, and Bulgaria. Data for the countries not involved include seven other European countries like Switzerland, Finland, Norway, and Denmark.

Redrawn from Trivers (1985).

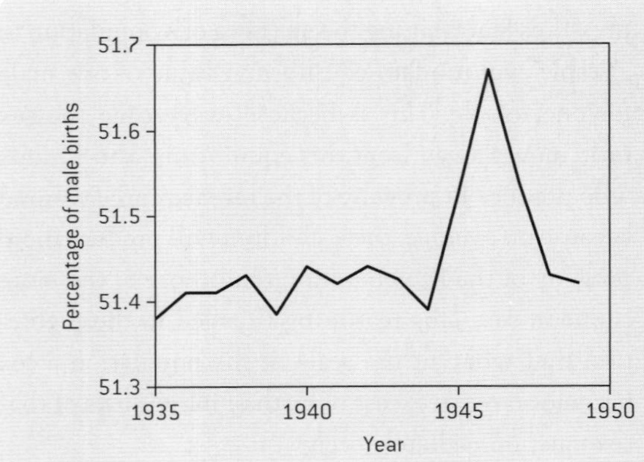


Figure 19.10 **Sex ratio at birth after World War II in five states of the United States of America from 1935 to 1949**

The choice of using data from the United States of America is justified by the fact that only men were involved in the conflict thus suggesting that the consequence back home was a biased sex ratio in favour of women.

After Trivers (1985).

*. . . An unusual situation in studies of human behaviour*

All in all, these results are interesting because the authors' approach is clear: they elaborated their predictions from a theory following from the mechanisms of evolution, and they tested these predictions on humans. Naturally, it is ethically impossible to conduct such experiments with humans. However the special circumstances afforded by the wars provided a natural experiment; both wars acted to decrease the sex ratio artificially, exactly as if an experimenter had chosen to do so. Moreover, in the case of World War I, the natural experiment even contains a true internal control provided by the countries not involved in the conflict. Lastly, World War II provides a replicate for World War I. So we can have more confidence in the prediction with humans given that the same result was obtained on two independent populations after two sex-ratio-reducing manipulations that differed both in their impact and the time at which they occurred. It is not common that all these conditions are met by a natural experiment, let alone one that involves humans.

*What is the proximate mechanism?*

The result with humans naturally raises questions about the underlying physiological mechanism by which a sex ratio can change in response to local conditions. One of the putative explanations in the human example described above is that the sperm cells that carry the Y chromosome are smaller and can therefore mature sooner that those carrying the X chromosome (see Section 13.9.1). If a separation caused by a traumatic event such as a war leads to more frequent sex in couples that reform after a long separation then it is possible that the slightly longer maturation time required for X sperm causes a reduction in their representation in subsequent ejaculations and hence a male-biased conception rate. This difference in the maturation time of the two types of sperm is actually used by doctors when they advise couples wishing to have a boy to have lots of sex, but advise those wanting to have a girl to have only one intercourse at the favourable time of the women's cycle.

## 19.4 Human medicine and behaviour

Medicine has major effects on our well-being, but also in our everyday life. It is at the very least surprising to note that even if medicine is rooted in biology, it is at the same time totally divorced from an evolutionary perspective (Short 1997). In most, if not all, countries of the world, most physicians never receive any training in evolution, and so it is not surprising that most doctors are totally unaware of the adaptive nature of some of our body's reactions. Take fever, for example. This elevation in body temperature is seen as a dysfunction of the body that must be corrected and so most doctors when faced with patients with fever prescribe drugs or physical actions whose objective is to lower body temperature. What if this rise in body temperature in the presence of infectious agents was a perfectly adaptive defence against a pathogen (Nesse and Williams 1997)? That would mean that by lowering body temperature of individuals physicians may actually be prolonging infections or making them worse. The ignorance of evolution by medicine is surprising given that a good share of the human phenotype has an evolutionary origin. Thus, our phenotype cannot be fully understood without some evolutionary thinking.

Darwinian medicine aims precisely to introduce evolution into the programmes of medicine (Nesse and Williams 1994, 1997). In this chapter we will consider only two aspects of medicine that can perhaps be better understood with an evolutionary perspective. These questions relate to the frequency of cancers of the reproductive tract and the sex differences in mortality.

### 19.4.1 Reproductive cancers in modern societies

In that humans live to increasingly older ages in Western countries, a certain number of causes of mortality that are rare in hunter–gatherer societies have become prevalent in ours. Mean longevity in Western countries is well beyond 70 years, for both men and women. Such an age must have been exceptional in hunter–gatherer peoples and thus probably also among our ancestors. The most common causes of death in the Western countries are coronary diseases, various cancers, and diabetes.

In particular, the prevalence of cancers is high, and the media classically tackle the problem of this disease under the angle of the development of a medical treatment that would cure it definitively. Such an attitude is rather naive because cancers, like other diseases, are in fact a heterogeneous series of illnesses characterized by specific symptoms. Some cancers have a genetic basis, others do not. Belonging to a certain family can increase the risks of some types of cancers by a factor of 20 or 30. Breeds of laboratory mice can be very liable to cancer because control mechanisms of cancer are lacking. Natural toxins, like nicotine or alcohol, radiations, and diets too rich in carbohydrates and lipids can all have carcinogenic effects, probably because the human body has never been selected to face such conditions.

Cancers often result from a reduction of regulation abilities of the organism during ageing. Cellular growth and proliferation occur throughout life, but become less effective with age. Organs' primary functions, and hence, oxygen and nutrient supply to their cells, also worsen during ageing. The progressive failure of regulation systems leads to an uncontrolled cellular division (Prescott and Flexner 1986). From an evolutionary point of view, senescence can be explained by the fact that genes underlying advantageous traits during an organism's youth will be selected for even if the expression of these genes cause major disorders during old age, well after the organism has ceased reproducing (Williams 1957; Hamilton 1968). So the genetical ability to control cancers during youth can be transformed into an inability to control them during old age.

#### 19.4.1.1 *Why are reproductive cancers so common in Western societies?*

A good example of a group of cancers that became very common recently in Western societies is female reproductive cancers: breast, uterus, and ovarian cancers. A recent synthesis indicates that this increase is associated with a particular reproduction pattern in the Western nations, a pattern that deviates substantially from the one that prevailed in our recent and older past (Eaton *et al.* 1994). It seems that the risk of contracting one of these cancers is directly related to a women's total lifetime number of menstrual cycles. This would imply, within Western societies, that a healthy woman having had her first periods precociously and a late menopause would be particularly likely to develop such a cancer. Why should this be so?

##### a *A correlation with the number of cycles in a lifetime*

Let us imagine the situation of prehistoric humans. A typical Stone Age woman was to have her puberty around 15 years of age or later, mainly because of difficult conditions and the incidence of parasitism. This woman would get pregnant relatively quickly and then would nurse her child for a period probably ranging from 2 to 4 years. Each weaning was to be followed by some menstrual cycles and then a new pregnancy. A similar return to a few menstrual cycles and then pregnancy would follow a miscarriage or an early mortality of the child. This alternation between pregnancy, lactation, and a few menstrual cycles would continue until death or menopause that would occur then around 45 years of age. On the whole, such a woman would be expected to have perhaps 150 menstrual cycles in her lifetime. If you compare that number to current Western societies, where women probably have 300–500 menstrual cycles (Eaton *et al.* 1994) in their lifetime, the difference is considerable.

Menstrual cycles are characterized by huge variations of plasma hormonal concentration having important effects on the state of mammary, ovarian, and uterine cells. Superficially, the organism's cells all share the same genes and so have a coefficient of relatedness close to unity. However, other factors predispose cells not to be the ideal partner within an efficient and functional organism as we would have expected on the basis of relatedness. Cells interact with their neighbours by producing costly signals indicating their quality, and a cell that emits few signals is often eliminated (Pagel 1993; Krakauer and Pagel 1996; Møller and Pagel 1998), the low level of signalling being interpreted as a sign of poor health. Conversely, cells signalling too much and thus requiring too many resources can be regarded as selfish and will also be eliminated because they can turn into cancerous cells (Pagel 1993; Krakauer and Pagel 1996; Møller and Pagel 1998).

##### b *A possible role of the hormonal changes over the cycle*

The behavioural and tissue responses of the female reproductive organs to hormonal changes are adaptive in that they increase the chances of a successful pregnancy. Yet such adaptations can also have costs such as an increased risk of escaping control of cellular growth and division. The costs related to these changes can be minimized for the periods during which menstrual cycles are interrupted. However, if such interruptions never or seldom occur, the capacity to control or minimize the risks resulting from such processes of cell propagation can strongly decrease. For example, studies of breast cancer show that high levels of oestrogens (hormones occurring in large quantities during the menstrual cycle) constitute an important risk factor. The large number of menstrual cycles in women of Western societies is in fact connected with a defective functioning of the organism, hence we cannot hope for the existence of adaptations to control for the increased risks of cancer that result from it. Moreover, such risks express themselves only late in life. They thus have only a very weak impact on fitness and are probably only slightly counter-selected. There is therefore no reason to expect that selection will support the emergence of a physiological resistance against this new situation.

Women can, however, reduce these risks to some extent by avoiding carcinogenic situations, like radiations or nicotine, and by having a diet without excess of fatty acids. Another way of decreasing the risks of reproductive cancers would be to use hormonal treatments. For example, oral contraceptives induce a hormonal state simulating gestation and indeed some contraceptives, like gestation, notably decrease the prevalence of ovarian and uterine cancers (but apparently not breast cancers; Eaton *et al.* 1994).

#### 19.4.1.2 *What lessons can be learned?*

This example illustrates how a simple evolutionary reasoning suggests novel types of explanation and more importantly, perhaps, new solutions for problems of public health. Evolutionary thinking may orient research in new directions and thus increase the speed with which we develop effective medical solutions to health problems. However, the menstrual cycle hypothesis obviously cannot apply to male reproductive cancers, which have also increased recently in Western societies. The medical and/or evolutionary reasons for such cancers must be sought elsewhere. Likewise, the difference in mortality related to coronary diseases between men and women in Western societies deserves to be tackled under an evolutionary angle.

We repeatedly exposed (in particular in Chapter 3) the risks associated with using a purely correlative approach to understand cause and effect. Because it is almost impossible to adopt an experimental approach for obvious ethical reasons with humans, most human-oriented research will be correlational. It is important therefore that readers appreciate the constant risk of misinterpreting the causes of the observed correlations.

### 19.4.2 Differences in mortality related to sex

Why do men die younger than women? We have seen in Chapter 11 that differences in mortality related to sex often result from the effects of sexual selection. In many species, the excess male mortality is the consequence of exaggerated behaviour expressed during the period of the life where males must compete among each other for access to mates. Thus, when the among-individual variance in the reproductive success increases for a given sex, the competition between individuals of that sex intensifies. This phenomenon is explained by the relative importance of potential benefits in terms of fitness. If the among-individual variance is large, there is a high risk of having only a low reproductive success. Individuals belonging to the competing sex are thus selected to take greater risks, simply because that increases the probability of obtaining a substantial reproductive benefit in return.

*A difference in mortality particularly present in young adults*

In this context, the study of human mortality patterns is particularly interesting. There is a significant excess of male mortality from puberty on, this difference fading beyond 60 years of age. The difference culminates between the ages of 20 and 25 years old, when young men's mortality exceeds that of young women's by a factor of higher than two (Daly and Wilson 1983; Figure 19.11). This difference in mortality is particularly striking when death is caused by accidents and violence. For example, the rate of homicides is much larger for men than for women. It is also known that characteristics of driver behaviour like speed, acceleration, and risk taking differs dramatically between sexes and age groups with a peak for 15- to 25-year-old men. The relative risk of male

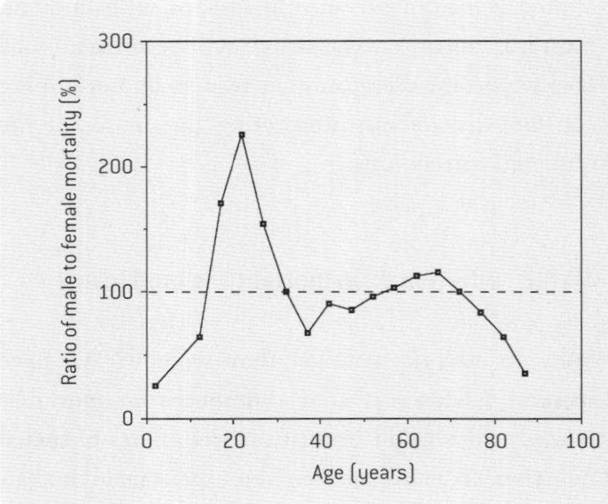


Figure 19.11 **Ratio between men's and women's mortality according to age**

It appears clearly that for almost all ages, the mortality of male individuals is higher than that of female individuals. In addition, the ratio is at its maximum in young adults, i.e. at the age when couples form. Data are from Canada, but similar results are also observed in other industrialized countries. The dashed line corresponds to equal mortalities.

Modified from Daly and Wilson (1983).

death is more than 2.4 times higher than that of women in this age group. This difference of a factor higher than two remains even after correcting for the number of kilometres driven (Daly and Wilson 1983). This differential mortality constitutes a clearly established fact today.

*What is the proximate mechanism?*

In terms of **proximate** mechanisms that can explain these facts, it appears that this pattern of mortality difference related to sex follows the pattern of variation of circulating testosterone, which could indicate the existence of a link between sexual selection and driver behaviour. According to this interpretation, the reasons for greater risk taking by men on the roads would be related to sexual selection: men would take more risks and hence expose others to more risk when they are in **competition** to obtain social status and mates. According to this interpretation, differences in 'constraints' related to sexual selection would have generated distinct mechanisms of decision-making towards risk that translate in a Western society into different behaviours and death rates between sexes.

Another external cause of death showing a clear difference between sexes relates to substance abuse. Young men in their twenties are more involved in the abuse of illicit substances than young women and pay the consequences in terms of increased risks of death. Again, an interpretation in terms of sexual selection can be suggested for these sex differences in risk taking. A natural 'experiment' testing this assertion recently occurred in the ex-communist countries of Eastern Europe. After the fall of communist governments, large quantities of resources were redistributed, some individuals building huge fortunes whereas most citizens became considerably poorer than before. In that women of all societies prefer men having many resources (Buss 1994), men rather than women are expected to try to build such fortunes in such periods of resource redistribution, by risky behaviours. In agreement with this reasoning, mean longevity in Russia decreased by 14 years for men and by 7 years for women between 1990 and 2000. A large part of this increased mortality is due to external causes in relation to physical violence, the abuse of dangerous substances, and other similar causes. Despite the multiplicity of causes for the decreased longevity of ex-USSR men, those that refer to differences in cognitive processes surrounding the assessment and weighting of risks deserve further attention.

*A need for replication*

This result is consistent with what an evolutionary reasoning would predict in such circumstances. It should be noted that in this study each country actually constitutes a replicate. The comparison with other countries can thus provide important complementary information. Many countries of the USSR block that remained under a communist and totalitarian regime, such as Belarus, Kazakhstan, or Uzbekistan for instance, did not show such a change. Albania and Bulgaria can also be compared. The confirmation by similar studies in equivalent situations of resource redistribution would constitute an excellent means of giving a little weight to such an interpretation.

## 19.5 Childhood and child abuse

The nature of parent–offspring interactions are tainted by the evolutionary process related to the costs and benefits that parents and offspring obtain from their behavioural decisions. Each party naturally seeks to maximize its own fitness. For example, the parental care allocated to a given offspring can provide benefits to the offspring but impose costs on the offspring to come later in the parent's lifetime (Trivers 1974; see Chapter 12).

### 19.5.1 Conflicts among offspring and with the parents

Full siblings have a coefficient of relatedness ($r$) of 0.5 with both their mother and father but because of the risks of extra-pair paternity, among other things, the $r$ shared with their brothers and sisters can be notably lower. If the cost of reproductive effort from the parents is measured in terms of future offspring produced by the parents then offspring will seek to obtain more resources from their parents than the parents are willing to provide. This is true because the parent shares the same $r$ with all present and future offspring whereas the siblings share a lower $r$. This asymmetry will be even stronger when the brothers and sisters are only half-siblings, that is when fathers are different for each of them as would be the case say for a serially monogamous female. This asymmetry in relatedness may generate intense **parent–offspring conflicts** (see Chapter 12, Section 12.6.3).

The food-begging behaviour often expressed by juveniles, such as vocalizations, movements, or intense gaping, could have evolved because they helped convince the parents to provide resources to the individuals that use such behaviours (see Götmark and Ahlström 1997; Kilner 1997). According to some authors, such behaviour would constitute honest signals of the offspring fitness, because the fitter individuals would grow the fastest and hence require more resources than slower growing less fit individuals. The parents would be selected to be sensitive to such **signals** (Saino *et al.* 2000; refer to Chapter 12 for a more exhaustive treatment of this question) because this way they can shunt limited resources to the fittest offspring. If this is so, then parental neglect of some offspring can be an adaptive response to information concerning the fitness of a juvenile either in relation to the current offspring or to those to come.

*Baby cries as an indicator of their health?*

Human babies also produce many signals involved in parent–child **communication**. These signals can be visual (e.g. colour) or vocal (e.g. number and pitch of cries). It seems that baby cries are reliable indicators of their health (Furlow 1997; Thornhill and Furlow 1998). Children in good health produce cries whose main frequency is around 300–600 Hz, although babies vary tremendously in the tone of their cries. Many diseases like diabetes, jaundice, asthma, and meningitis are expressed directly in the type of cries through a frequency reaching 1000–2000 Hz. It so happens that parents respond differently to cries of different frequency. The responses to high-frequency cries range from indifference to dislike and roughness.

Why is it that the shrill cry of babies (1000–2000 Hz) is perceived as irritating, even unbearable to some? Such negative responses of parents towards their babies may be problematic today because of the cultural significance of maternity and paternity. Perhaps in our close ancestors parental rejection of infants emitting signals of ill health provided parents with a selective advantage over those that did not discriminate by allowing them to avoid wasting efforts on young whose chances of becoming adult were very low. Parents neglecting such children might simply have had higher fitness.

*How can we use such a result?*

Of course, evolutionary reasoning cannot be used to decide today on the responsibility of parents that neglect their children. Clearly, parents do not usually neglect their children when they are very sick. Nonetheless, it is not because the reasoning appears cruel that it should not be made, because in it perhaps there is some hope of providing more effective means of preventing cases of child neglect. To deny the existence of such mechanisms if they did exist would be quite as reprehensible. We argue that it is more important to face the reality of our potential biological nature than to try to ignore it. The above results suggest that the study of human responses to baby signals can bring new light on health and social problems, in that they can indicate important hints to solve, *a priori* rather than *a posteriori*, psychological problems having apparently an ancient origin.

### 19.5.2 Kin selection and child abuse

The very fairy tales we read to children report many stories of cruel step-parents, often stepmothers. Canadian evolutionary psychologists Martin Daly and Margo Wilson (1999) argue that there may well be some degree of biological realism behind these stories. Recall that sharing genes by descent means sharing the same evolutionary interests (Chapters 2 and 15). Relatedness, as we saw earlier in the case of parent–offspring conflict can have played a role in patterns of child abuse.

*Two types of infanticide*

Infanticide is a rather widespread phenomenon among many animal groups from invertebrates to fish, birds and mammals (see Hausfater and Hrdy 1984). Two different types of infanticide are commonly observed. The first kind we explored in the previous section. It involves mothers that kill their offspring whose survival chances are low in order to concentrate their reproductive efforts on those that express the best signs of viability. The other type of infanticide involves males that often kill the still-dependent offspring of a previous male (see Chapter 12, Section 12.3.2.2).

*Infanticide in human societies*

Human societies are not exempt from infanticide, from the mythical accounts of the past to today with reports of infanticide in India and China (Daly and Wilson 1984), the two most populated nations in the world. For centuries, mothers or midwives from many nations have regularly disposed of presumably non-viable newborn babies. The abortion of non-viable embryos or those having malformations remains a common practice, in particular in Western societies equipped with modern technology. The fact that parents often do not grant the same value to a child according to his/her sex contributed to extending the practice of infanticide in some societies. For example, abortions are often practised differentially according to the sex of the fetus, or newborn girls are sometimes discarded in China following the governmental policy of limiting couples to one child. This practice reached such a point that many Chinese men are doomed to remain unmarried because of an insufficient number of women of an age compatible with theirs.

For some ecologists, the differential value granted to a child according to his/her sex in some human societies and the influence of this difference in value on the probability of practising infanticide is reminiscent of the adaptive adjustment of the sex ratio practised by females in other species. Infanticide initiated by men living with a spouse taking care of children from a former relation is also frequent among various human cultures (Daly and Wilson 1984, 1999). The phenomenon is particularly obvious in Western societies. Daly and Wilson (1988) studied cases of the ill-treatment of children, sometimes ending in infanticide, according to whether the children live with both of their biological parents, only one or neither, using statistical data collected in North America and Scandinavia. The frequency of ill-treatment of children by step-parents was 70 times

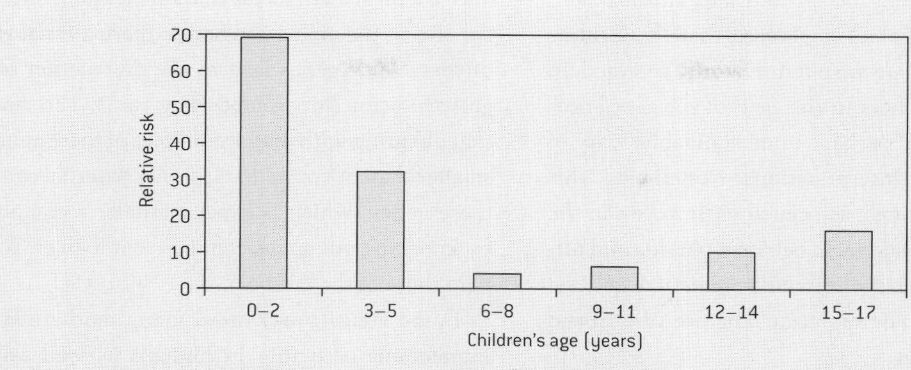


Figure 19.12 **Frequency of ill-treatment in recomposed and non-recomposed human families**

Frequency of ill-treatment of children living with their two biological parents relative to those living with a couple made up of one biological parent and a step parent, according to the age of the child. Values above 1 reveal higher risks in children living with a step parent. Data come from Canada between 1974 and 1983. These data are based on very large samples.

Redrawn from Daly and Wilson (1988).

higher in the absence of one of the biological parents, and even higher in the absence of both (Daly and Wilson 1988; Figure 19.12). This effect was independent of other factors likely to increase the risk of ill-treatment.

In addition, infanticides by one of the adults living under the same roof are committed significantly more often by the step than the biological parent (Figure 19.13). Also, the studies undertaken by Flinn (1988) within recomposed families in Trinidad revealed that step-parents treat the children of their spouse more severely than their own children, independently of the duration of the relation between the adult and the child.

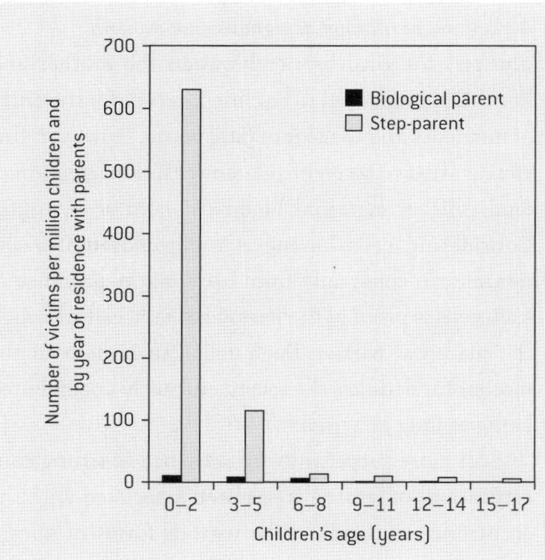


Figure 19.13 **Frequency of infanticides by biological or step-parents in relation to children age in humans**

Data come from Canada between 1974 and 1983. These data are based on very large samples.

Redrawn from Daly and Wilson (1988).

*Evolutionary reasoning and precautionary measures*

Again, it should be acknowledged that these interpretations of the data are not universally accepted. In particular, researchers in the field of social sciences are extremely, and perhaps understandably, wary of the evolutionary interpretations. Nonetheless this increase in infanticide associated with a drop in the level of genetic relatedness between parents and offspring is a typical evolutionary prediction and yet had not been seen or appreciated before Wilson and Daly pointed it out.

Naturally the results are correlational and cannot really provide strong support for a cause-and-effect relationship. Perhaps the proposed interpretation is incorrect and a different evolutionary interpretation could provide a better fit to the observations. However, we argue that it is better to ponder even a disturbing interpretation than to ignore it and perhaps avoid coming up with a solution to a universally acknowledged problem. In this context, some evolutionary considerations have started to be incorporated in adoption programmes in the United States.

### 19.5.3 *In utero* mother–child conflicts

The interactions between a mother and her offspring start early in the uterus. These interactions can be understood in the light of parent–offspring conflict we dealt with in Section 19.5.1 (see also Chapter 12) as well as **genomic imprinting**. Genomic imprinting relates to the situations where genes have a different action according to their parental origin (Peters and Beechey 2004). Indeed, the genes inherited from the mother are selected to avoid overexploitation of maternal resources because that would compromise the future reproductive success of the mother. This is not the case for genes inherited from the father. Those were not necessarily selected to exploit maternal resources only moderately. This is particularly true when faithfulness between mating partners is not very high. Studies in the mouse showed that some genes, when they are of paternal origin, produce a fetal growth factor that considerably increases the size of the embryo, whereas the maternal copies of these same genes lead to the destruction of this growth factor (Moore and Haig 1991). The result is that offspring with maternal copies of these genes are smaller than offspring having only paternal copies of these genes (which is experimentally accomplished by knocking out genes coming from a given parent, here those of the mother).

David Haig (1993) listed many mother–embryo interactions occurring in humans as well as their biochemical bases. Most of the fetal substances highlighted relate to the transport of resources from the mother towards the embryo, whereas the substances of maternal origin rather limit the export of maternal resources towards the embryo. For example, during the implant, the fetal cells invade the maternal endometer and transform the endometrial spiral arteries into vessels of lower resistance that are unable of vasoconstriction. This provides the fetus with direct access to maternal blood by disconnecting the volume of blood arriving at the placenta from maternal control. Moreover, the placenta thus becomes able to release hormones and other substances directly into maternal blood flow. Such mother–offspring relations during gestation resemble conflicting interactions more than a peaceful association.

*A placental barrier not as tight as once thought*

The physiological barrier between the mother and her child is supposed to be closed, except for the entry of nutrients and the exit of fetal waste. However, this view started to be reconsidered with the detection of fetal cells in maternal blood (Bianchi *et al.* 1996). Considering cells having a Y chromosome in the mother can come only from her son, the presence of such cells is proof of the passage of fetal cells through the placental barrier. Such cells can remain in the mother for decades; the record currently documented being as long as 27 years!

Still more surprising, the presence of strong concentrations of fetal cells has been associated with the occurrence of diseases like various forms of scleroderma and autoimmune diseases late in the mother's life. As we saw in the case of cancer, the efficiency of natural selection in eliminating such defects depends on the moment when the process occurs relative to an individual's reproductive life. A strong negative selection during old age has in fact very little effect on fitness, because the reproduction already took place. So fetal cells can invade the mother for the immediate benefit of the child, even if this can have strongly negative effects for the mother later in life. In addition, from an evolutionary point of view, this invasion of fetal cells into the maternal organism can also benefit the mother, in the sense that it increases her fitness given that it increases the number and/or the health of her offspring, even if that is done at the expense of her life expectancy beyond her active reproductive period. In fact, one cannot expect to find adaptations in the maternal organism preventing the expression of such consequences at the end of her lifetime.

*Acting preventively during pregnancies?*

If there is indeed a link between the incidence of these diseases occurring late in a women's life and her pregnancies, then to avoid such diseases it may be necessary to take action during or right after the pregnancies. The medical implications are obvious and are provided by an evolutionary approach.

## 19.6 Conclusions: why deal with humans at all?

We feel there are two reasons for adopting an evolutionary approach when dealing with humans: first, it can provide insight on the biological nature of our behaviours and attitudes; secondly, it can enhance our ability to cure disease. However, it is also important that we be aware of the important shortcomings in the application of an evolutionary approach to humans. We conclude by dealing with each of these issues.

### 19.6.1 Understanding the bases of human nature?

The facts considered in this chapter suggest that there can be some predictive and explanatory power to the application of the evolutionary approach to human behaviour. Yet, it is also important to acknowledge that the evolutionary approach does not exclude the social sciences from being equally efficient at predicting human behaviour. The aim of understanding something as complex as human nature may in fact call for the interactions from a diverse set of disciplines. However, we feel it would be wrong to exclude *a priori* any evolutionary explanation for current human behaviour.

### 19.6.2 Enhancing prevention and cure of our diseases?

Medicine, we feel, has much to gain from adopting an evolutionary perspective. Consider an obstetrician working on a mother and her child during fetal development. Were the physician to study the physiological interaction between mother and offspring on the assumption that these must be of a cooperative type, he or she would likely miss a whole set of fundamental processes that follow from parent–offspring conflict. Evolutionarily speaking, the mother's interest is to provide her child with all the resources it needs but not more because that would jeopardize her ability to produce further offspring. However, the child's evolutionary interest is to divert many more resources, well beyond the optimal provision quantities of its mother. This fundamental conflict, as we saw, can generate diseases in the mother several decades later. Admittedly, understanding the origin of these diseases does not allow one to develop an immediate cure, but it can conceivably allow for prevention. For example, the measurement of fetal cell concentration in the mother, followed by a suitable treatment if necessary after childbirth, should allow for decrease the risks incurred by the mother at old age. Naturally this is just one example of how evolutionary thinking can influence heath practices. Other examples involve the evolution of virulence in hospital strains of bacteria, the response to infections, the conditions for predicting pandemics, etc.

### 19.6.3 Human beings: a study model for behaviour?

Working on human behaviour opens a whole range of possibilities compared with other living species. Humans can be asked questions and provide subtle and precise answers. However, working on humans comes with a great deal of problems that are difficult to circumvent.

*The lack of replicates*

First of all, it is very important to have replicates of the cases studied. For obvious reasons, it can be difficult to obtain replicates with humans. Without replicates, one runs the risk of coming up with nice stories on the basis of a single observation. As seen in Chapter 3, scientific interpretation rests primarily on the consistency and, as much as possible, on the similarity of a response in repeated situations. It is often impossible to obtain such replication with humans, though we have seen several examples in this chapter where replicates existed.

*A primarily correlative approach*

For practical reasons, studies implying humans are often only correlative. Again, there is a risk of building nice stories based on the causal interpretation of a correlation. In the absence of experiments, the necessity of having carefully designed replicates is thus even more compelling. Furthermore, we have seen several examples in this chapter of quasi-experiments, and there are several ways to design real experiments in humans without raising any ethical problem.

However, it should be reiterated that most studies on humans reported here rest on a prediction-testing process: if we postulate that the evolutionary processes are also at play in our species, then such-or-such relation is predicted. Predictions are thus made *a priori* and not *a posterior*, as can be *ad hoc* interpretations.

*The importance of prejudice*

Individuals can very well have biases of various kinds that will purposely or not lead them to provide slightly (or not so slightly) distorted answers to the questions we ask them during interviews meant to provide behavioural data. Sometimes, one has means to detect this type of problem. For example, there are frequent investigations into the number of sexual partners declared by heterosexual men and women. This often leads to an impossible result: women on average acknowledge half the number of sexual partners men do. This difference persists even after correcting for a possible effect of a slightly female biased sex ratio. However, such a result is impossible, because sex requires one man and one woman. While the variance may differ between males and females, men and women must have the same average number of heterosexual partners. This result can only be explained by the fact that for some reason or another, interviewed men and women give a systematically biased answer, men increasing the number of their partners, and/or women underestimating this same number. Psychologists who often rely on interview techniques to obtain data have developed numerous methods of circumventing this type of difficulty.

*The place of culture*

As we have seen in this book, what is fundamental in evolution is the transmission of information from one generation to the next. It is this information that allows the development of the phenotype. We have seen in Chapters 4 and 5 how this information can take two extremely different forms: genetic information and cultural information. Both cultural and genetic transmissions lead to the transmission of differences across generations, a concept that is termed **transmittability** in Chapter 20.

As for animals, analyses in many human studies explain only a small part of the variance in behaviour: usually the coefficient of determination (coefficient

$r^2$) has only a relatively low value. In most of the examples provided in this chapter, only 10–20% of the variance in behaviour is actually explained by the proposed mechanisms. Thus a lot of the variance still remains to be explained.

So clearly humans pose several specific problems to studying them. Nonetheless, we feel it is important to understand the limitations that are imposed by these problems without sacrificing the potential knowledge that can be generated by applying an evolutionary approach to humans. The significant effects of the evolutionary processes found in the above examples despite the numerous layers of culture, morals, and social constraints acting on our behaviour are indicative of the utility of behavioural ecology to cast some light on human behaviour. As formulated by Dobzhansky: '*Nothing in biology makes sense except in the light of evolution*'.

## » Further reading

> *For an excellent introduction to the evolutionary approach of human behaviour:*
**Cartwright, J.** 2000. *Evolution and Human Behaviour*. MacMillan, London.

> *The influence of Darwinian thought on the analysis of human behaviour, and particularly on the bases of morals, is accessible by reading the essay by Robert Wright:*
**Wright, R.** 1994. *The Moral Animal: Evolutionary Psychology and Everyday Life*. Little Brown, London.

> *The importance and the originality of the Darwinian point of view in medicine are perfectly exposed in the following works:*
**Nesse, R.M. & Williams, G.C.** 1994. *Why We Get Sick: The New Science of Darwinian Medicine*. Times Books, New York.
**Stearns, S.C.** 1998. *Evolution in Health and Disease*. Oxford University Press, Oxford.

> *For a critical presentation on the excesses of the adaptationist approach of human behaviour:*
**Rose, H. & Rose, S.** 2000. *Alas, Poor Darwin*. Vintage, London.

## » Questions

1. Is the existence of culture a valid objection against the biological study of human behaviour?

2. Can morals have a biological basis?

3. Can the knowledge of the sexual selection process predict the phenomena of fashion in relation to the notion of beauty and clothing?

20

# 20

# Cultural Evolution

Étienne Danchin and Richard H. Wagner

## 20.1 Introduction

As we have seen in Chapter 4, life is essentially a matter of self-replication, passing information from one generation to the next. We also saw how genes are not merely pieces of DNA, but bits of **information** coded by DNA (see also Section 2.2.2). The general impression that often emanates from such a view is that this self-replicated information is exclusively genetic. However, in Chapter 4 we saw that non-genetically coded information can be transmitted from individual to individual in many circumstances. This final chapter uses the taxonomy of biological information presented in Chapter 4 to discuss the possibility of transmission of information across generations through social influences and then explores the evolutionary consequences of such a non-genetic system of transmission of parent–offspring resemblance.

### a *Salmon that transmit spawning site location across generations*

Chapter 4 presented several candidate non-genetic, self-replicating information systems. One example concerns information about the value of a given environment that is often transmitted from parent to young in species where some of the offspring return to their natal site. Atlantic salmon (*Salmo salar*) provide an excellent example of this where the young of both sexes return to the very same river where their parents had spawned: a self-replicating, non-genetically transmitted behavioural tradition. This transmission is mainly thought to involve early **imprinting** on the river of birth. In other words, it involves some simple form of **learning**.

### b *Intriguing behavioural variations among primate populations*

Several authors have documented intriguing behavioural differences among wild chimpanzee populations across their entire range in Africa (McGrew 1992; Whiten *et al.* 1999; van Schaik 2004; Whiten 2006). As they could not detect any genetic or ecological explanation to such variations, these authors claimed that such variations are of 'cultural' origin. Similarly, several authors studying mate copying (Brooks 1998; Dugatkin 1998; White and Galef 2000b), as well as others working on song dialect (Ficken and Popp 1995; Whitehead 1998; MacDougall-Shackleton and MacDougall-Shackleton 2001) interpreted their results as evidence for animal culture. However, how can we ascertain that certain variations in behaviour are transmitted culturally? And what do we mean exactly by **culture** when dealing with animals?

### c *Human examples*

Concerning humans, we intuitively know that much of our knowledge about the environment (whether physical, biological, or social), as well as many of our decisions, are deeply influenced by our social experience. For example, our food preferences are typically influenced by what we had learned is considered 'tasty'. Similarly, our sexual preferences are likely to be shaped by our experiences early in life.

An even better example is that of language. Our mother tongue is clearly learned during our infancy through our repeated interactions with other individuals who have already mastered a given language. Despite the fact that it is likely that we have genetic abilities to learn languages, it is highly unlikely that we have genetic predispositions to speak a specific language. This is demonstrated by babies adopted from a foreign country that rapidly learn to speak the language of their foster parents. Most of the time as adults these adopted individuals do not show particular capacities to learn the language of the country they were born. More generally, no reasonable person would argue that variance in human behaviour is only due to genetic variation. Clearly many human actions are shaped by social influences early in life. Such influences are often so profound that we tend to think that what we learned has a universal value until we are confronted with people from other cultures who may have habits that we find shocking. These considerations, plus the many animal examples presented in Chapter 4, strongly suggest that much human and animal behavioural variation has somehow been 'learned', and certainly not transmitted by genes alone.

### d *The fundamentals of non-genetic replication*

We know that the substrate of genetically transmitted information is DNA or RNA molecules and their mode of self replication has been the focus of much of the early advances in molecular biology. However, if we are to make any progress in exploring non-genetic replication we must first define what biological information is exactly and understand the various forms of information in biology. This was the goal of Chapter 4, which provides a taxonomy of biological information summarized in Figure 4.3. In short, a given individual can have three forms of **non-genetic information** about its environment. First, its parent may have transmitted information through **parental effects** that moulded its phenotype to current environmental conditions. Secondly, the individual can learn about his environment by interacting with it to gather **personal information**. Finally, it might acquire information by observing other individuals interacting with the environment, thus obtaining **social information**. This last form on non-genetic information is likely to play a major role in non-genetic replication because it allows individuals to **copy** or **imitate** others, leading to the transmission of behavioural patterns among individuals.

To explore non-genetic replication we also need to elucidate where such non-genetic information resides, what is its substrate, and how it replicates by passing from one organism to the other. We argue that the substrate of at least part of this non-genetic information must be provided by the neurons of an animal's central nervous system. The mechanism of self-replication, its passage from one individual to another can, for the sake of simplicity, be ascribed to a very general phenomenon: 'learning', and more specifically social learning[1], both of which have traditionally been studied almost exclusively by animal experimental psychologists but now should be more explicitly integrated into behavioural ecology (Figure 4.19).

Behaviour is to a large extent a matter of decision-making, and decision-making requires information and hence information gathering, memorizing, and processing. Consequently, behavioural ecology has been increasingly adopting an information-driven approach to the study of behaviour (as developed in Section 4.4) and has just begun to consider the evolutionary consequences of individuals having the ability to use information extracted from the behaviour of others.

#### e *Goal of this chapter*

The main goal of this chapter is to provide a definition of cultural evolution that can be tested empirically. For this purpose we will need to apply the methods of quantitative genetics to apportion variance of both genetic and non-genetic components of behaviour. In doing so, we will introduce the concept of transmittability as a generalization of inheritance. This will lead to a definition of culture based on a series of four **testable criteria** that need to be verified **simultaneously** to demonstrate that a given behavioural pattern is transmitted culturally. This definition will lead to a new and more general definition of natural selection and of evolution. The impetus for studying animal culture is to embrace a broader approach to understanding animal behaviour, which is produced both by genetic and non-genetic information. In effect, this broader approach attempts to reconcile the dilemma of nature versus nurture. Finally, we analyse some of the many potential consequences of the existence of animal culture for evolution. We consider how non-genetic, or learned, behaviours may influence biological evolution. This is stressed at the end of this chapter (Section 20.3) where we review evidence of the exciting possibility that non-genetic, or cultural, changes within subpopulations may lead to reproductive isolation and ultimately to speciation.

1 There are no fewer than 30 different definitions of social learning (Avital and Jablonka 2000). Here we use the phrase social learning in its broadest meaning to encompass any form of learning that involves social information.

## 20.2 An evolutionary definition of culture

### 20.2.1 Social influences and culture

In Chapter 4 we saw many examples of inadvertent social information (ISI) use. Implicitly, ISI use may lead to the transmission of behavioural patterns among individuals, a process that is akin to what we call 'culture' in humans. For instance, in some peculiar circumstances, an entire subpopulation of females may prefer males of a given phenotype, while another subpopulation nearby may have acquired a different preference. As we now know, the origin of such variation in preferences may at least partly be created by mate choice copying (Figure 4.13) rather than genetic differences among females of different subpopulations. We may be tempted to qualify differences in female preferences among populations as **cultural variation**.

This raises the question of what exactly is culture in the context of evolution? The goal of this section, therefore is to provide an evolutionary definition of culture.

### 20.2.2 Transmittability: a generalization of the concept of heritability

#### 20.2.2.1 *The origins*

Natural selection occurs when three conditions are fulfilled (Chapter 2): (1) there is **variation** in some trait; (2) that variation has **fitness implications**; and (3) that **variation is transmitted** to offspring. The last condition is usually reduced to saying that the trait must be heritable, and evolution through natural selection is thus said to be the change that results from selection acting on heritable variation. The concept of heritability is thus central to evolutionary sciences.

#### 20.2.2.2 *Can we account for all environmental influence when measuring heritability?*

Because heritability has a purely genetic definition (see Section 2.2.3), measurements of heritability reduce to estimating similarities in a given trait between parents and offspring when all environmental influences have been controlled. A classical way to do that with field data is by performing a regression of the trait of the offspring against that of their parents while accounting for environmental effects (Falconer 1981). The implicit assumption when measuring heritability is that we can account for all of the part ($V_E$) of phenotypic variance ($V_P$) that is due to environmental variation. Then, we can decompose the genetic part ($V_G$) of phenotypic variance as in Chapter 2, Section 2.2.3.1, and interpret all the variation that is transmitted to offspring as resulting from additive genetic variance ($V_{AG}$).

However, the accuracy of that reasoning rests on our capacity to account for all the variance that is due to the environment. A way to check whether this is correct is by decomposing $V_E$ in turn into its own components, an approach rarely taken relative to the classic decomposition of genetic variance.

#### 20.2.2.3 *Decomposing the environmental component of phenotypic variation*

Environmental variance (see Equation 20.1) may have four major components:

$$V_E = V_{DE} + V_{AE} + V_{AP} + V_{AS} \qquad (20.1)$$

##### a $V_{DE}$: *the direct influence of the environment on the phenotype*

$V_{DE}$ is the part of phenotypic variation that directly results from the effect of the environment while controlling for genetic differences. For instance, animals with the same genotype grow large if there is a lot of food, and small if food is lacking. This component is easy to handle when measuring heritability just by accounting for the conditions under which the measurement is performed in the field, or by providing organisms with the same controlled conditions in the laboratory. Evolutionary biologists often tend to equate $V_E$ with $V_{DE}$. However, several other components may be involved.

##### b $V_{AE}$: *genotype–environment correlation*

$V_{AE}$ is the part of $V_E$ that may result from the casual association between a given **genotype** and a particular environment. An example is territorial inheritance in animals, which is also a classic phenomenon in humans. Sons often inherit the territory of their fathers, with the consequence that some lineages may become associated with the same kind of habitat for a long time. In bowerbirds (family *Ptilonorhynchidae*) for instance, sons may inherit the bower of their fathers. Bowers are sophisticated structures built by males that function as attractors to females (Figure 20.1). Since there is within-population variation in bower characteristics (Diamond 1982a, b), one may find a highly significant father–son correlation and infer high heritability of bower characteristics. However, such a correlation may not indicate the existence of additive genetic variance in the capacity to build bowers. Thus, in species where males inherit their fathers' breeding territory, significant father–son regressions in the capacity to attract females do not necessarily reveal genetic variance.

Furthermore, male competitive ability may be influenced by the environment in which males grew up. Males born in high-quality territories may become good at acquiring high-quality territories for developmental reasons, which would lead to father–son correlations. Such processes may become particularly prevalent because of **niche construction** (Laland *et al.* 2000), which occurs when the activity of individuals modifies the habitat in which they live. Many human activities deeply modify the environment in which we live. However, similar although less intense processes occur in many other organisms, from beavers to bacteria.

The important point is that genotype–environment correlations may lead to parent–offspring resemblance that may not reveal the existence of **additive genetic variance**. In other words, by

Figure 20.1 **The bower of the great bowerbird (*Chlamydera nuchlis*) in Australia**

**a.** The bower seen from its main axis (note the various objects brought by the male to attract females).
**b.** A closer view shows various objects of human origin brought by the male.

Picture by Étienne Danchin.

ignoring that phenomenon, one may include part of $V_{AE}$ into $V_{AG}$. The best way to control for this potential problem is by performing **partial cross-fostering experiments** that allow the disentangling of the genetic and environmental components of phenotypic variance (reviewed in Merilä and Sheldon 2001).

### C $V_{PAP}$: *the role of parental effects*

$V_{PAP}$ stands for the part of $V_E$ that results from parental effects (see Section 4.2.3.1). $V_{PAP}$ may lead to significant parent–offspring regressions. In other words, it may have an **additive effect**[2] (Khombe *et al.* 1995). To some extent, genotype–environment correlations belong to such parental effects. **Partial cross-fostering experiments** can control for relatively late parental effects (Merilä and Sheldon 2001). However, parental effects may occur very early in life, sometimes even before fertilization. For instance, the capacity of female kittiwakes to deposit antibodies in their eggs against the spirochaete *Borrelia burgdorferi*, the agent of Lyme disease, may lead to significant parent–offspring regressions and high estimates of heritability for resistance to that parasite. Part of that resemblance may be due to additive genetic variance, but may also result from early parental effects because only mothers confronted with the *Borrelia* deposit antibodies against it in their eggs (Gasparini *et al.* 2001). Thus, ignoring parental influences may lead to the spurious inclusion of $V_{AP}$ in $V_{AG}$.

A problem with this component is that it is extremely difficult to control properly for such very early parental influences. Again, we are very likely to include part of $V_{AP}$ into $V_{AG}$.

2 The term additive is borrowed from the statistical terminology. Here it describes the fact that variations in parents are transmitted to offspring: tall parents have offspring that are on average taller than offspring of small parents. Thus additive genetic variance is the part of genetic variation that is transmitted to offspring and that leads to significant parent–offspring regressions (after controlling for every potential environmental effect). However, as we develop in this chapter, offspring may resemble their parents for many reasons other than genes.

### d $V_{AS}$: *the role of social influences*

In Equation 20.1, $V_{AS}$ stands for the part of $V_E$ that may result from social influences. Social influences involve some form of social learning, or learning from other individuals. Social influences are thus likely to result in the transmission of behavioural patterns across generations. In other words, $V_{AS}$ has an additive effect because it can lead to significant parent-offspring resemblance in behaviour.

In humans, $V_{AS}$ clearly includes cultural influences. Many of our preferences are culturally inherited from the observation of our elders or neighbours and are acquired during development. Our language is clearly the result of learning from our parents and neighbours. If we were to measure parent–offspring regressions in language among people from various countries we would obtain an estimate of the slope of the relationship that would be very close to unity. Indeed, do you know someone who does not speak the language of their parents? Nobody would argue that this resemblance is due to additive genetic variance for language. Thus, clearly this very strong parent–offspring resemblance does not result from $V_{AG}$, but rather from $V_{AS}$.

In Chapter 4 we saw that subtle social influences occur in various contexts in a wide variety of animals and possibly in plants. Thus, we are likely to include part of $V_{AS}$ in $V_{AG}$ if we do not control for social influences in our protocols. Partial cross-fostering experiments may allow controlling for part of $V_{AS}$. However, the question is why should we remove this important potential component of parent–offspring resemblance? This component is important because it can generate slopes of parent-offspring regression that are very close to unity. This suggests that some components of phenotypic variance that result from social influences may be transmitted completely. Such highly additive social components raise the question of their role in evolution, a question that is central to this chapter.

### e *Several components of environmental variance are additive*

In summary, several components of environmental effects can have additive effects that can lead to parent–offspring resemblance (note that this is why we put an 'A' in their subscript). In other words, the phrase 'the heredity of differences' that we used in Section 2.2.3.1 to define heritability in fact encompasses any additive components of phenotypic variance.

As we have seen, several methods were designed to try to remove some of these undesirable effects when measuring $V_{AG}$. This is undoubtedly necessary for estimating the portion of such resemblance that is due to genes. However, this raises methodological issues. The above considerations suggest that it is likely that none of those methods really control for every additive effect, particularly environmental influences early in life. In fact, very few studies have specifically controlled for the possibility of early parental and social effects (Danchin *et al.* 2004). Thus, even in carefully designed experiments, it is likely that some undesirable non-genetic effects would be captured in the measurements. Consequently, most estimates of heritability include some early parental and social effects.

### 20.2.2.4 *Why do we need the concept of heritability?*

To proceed, we first need to return to the origin of the concept of heritability. In fact, if we already had the concept of heredity, why did quantitative geneticists create the concept of hereditability?

### a *A concept derived from agricultural practices*

Historically, the origin derives from agricultural practices: people needed a concept to quantify the evolutionary potential of a given trait in a breed. Breeders of livestock to make milk aimed at having cows that made as much milk as possible every day. Our ancestors understood a long time ago that this could be achieved by allowing the cows that produce more milk to reproduce. In doing so, they assumed that this trait had the potential to evolve. This potential was later called heritability.

In more Darwinian terms, the fundamental reason for defining the concept of heritability was that evolution in a trait, whether through natural selection or genetic drift, can only occur if differences among individuals of one generation **are at least partly transmitted to the next generation**. In the absence of such transmission of trait variation, there is no way to see that trait evolve in time. As soon as a trait is transmitted it has the potential to evolve.

### b *Estimating the actual evolutionary potential of a population*

If our goal is to estimate **the part of phenotypic variation that is open to selection** and can thus lead to evolution, there is no reason to eliminate non-genetic additive components. Hence we may not be interested in measuring heritability ($V_{AG}/V_P$) only rather the total portion of phenotypic variance that is transmitted to offspring:

$$(V_{AG} + V_{AE} + V_{AP} + V_{AS} + V_{S*G})/V_P \quad (20.2)$$

where $V_{S*G}$ quantifies the fact that social influences probably interact with the genetic predispositions to be socially influenced, in other words with social learning capacities.

Note that for practical reasons scientists very often use imperfect estimates of heritability. This is the case in field situations where it is very difficult to control for every environmental effect. Hence, estimates almost certainly include part of the non-genetic components of Equation 20.2 and they are called 'broad sense heritabilities'. However, broad sense heritability clearly differs from transmittability as in the former case the goal is to reduce environmental effects as much as possible, whereas in the latter case the goal is to capture all of the additive effects, whether of genetic or environmental origin.

### c *From heritability to transmittability*

In origin the term 'heritability' thus encompassed any form of variation that is transmitted across generations and on which selection and evolution was possible. However, because of the discovery of the fantastic potential of genes, and because breeders of livestock were mainly interested in traits that are strongly influenced by genes, the definition of heritability was little by little reduced to its genetic component (additive genetic variance). However, for an evolutionary biologist, this does not need to be so.

The key point is that evolution can occur as soon as variation is transmitted, **whatever the mechanism of transmission of the variation**. It may, and often does, involve additive genetic variation, but it may also involve other ways of transmission of information across generations. **All the additive components of phenotypic variation are in fact open to evolution through natural selection or drift**. This implies that we should not eliminate additive environmental effects if we want to understand evolution in all its components. Today some authors tend to broaden the scope of heritability back to its original meaning, leading to unresolved debates with quantitative geneticists (see, for example, Avital and Jablonka 2000; Danchin *et al.* 2004). A way to avoid such needless arguments is to create a new term encompassing any form of information transmitted across generations. That term would thus measure the total evolutionary potential of a population, measured as phenotypic change. Equation 20.2 provides such an estimate without restricting it to its genetic component (Figure 20.2).

> The concept of heritability can thus be generalized into that of **transmittability which is the heredity of differences whatever the mechanism of transmission involved** (Figure 20.2). We will see in Section 20.2.7 that accounting for those various types of transmittable information may greatly influence the way evolution functions.

### d *To what extent does transmittability differ from heritability?*

The question remains of the extent to which transmittability differs significantly from heritability. Obviously the answer to this question will depend on the type of trait considered. For morphological traits it is plausible that the non-genetic additive components are negligible. However, this is unlikely because some studies showed that accounting for

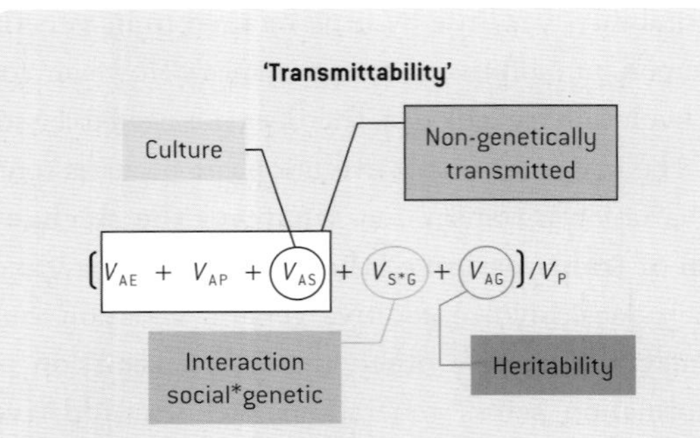


Figure 20.2 **Definition of transmittability**

Transmittability is the 'heredity of differences', the part of phenotypic variance that is transmitted to offspring, **whatever the mechanism that leads to transmission across generations**. Transmittability thus involves any variation that is susceptible of evolving; it is the evolutionary potential of a population. The term $V_{S*G}$ highlights the fact that social influences are not independent from the genetic components of learning with which social influences necessarily interact. Transmittability includes heritability as a fundamental component, but also other forms of transmission across generations, one of which is **cultural transmission**. Heredity is the fact that dogs do not make cats and vice versa; transmittability is the fact that big dogs (or cats) tend to make bigger dogs (or cats) than small dogs (or cats). Note that the equation corresponds to Equation 20.2. It thus deals with any within-population variation.

heritable **maternal effects** revealed a greater than threefold increase in the potential for evolution in some morphological traits relative to that predicted by heritability alone (McAdam *et al.* 2002). Furthermore, although it may be acceptable to neglect the non-genetic additive effects of equation 20.2 for some morphological traits, such an assumption is likely to be incorrect in most behavioural traits. We have seen in Chapter 4 the prominent role of social information and more specifically ISI in animal behaviour. In particular, we saw how information extracted from the performance of conspecifics may greatly influence individual decisions. This strongly suggests that the non-genetic additive components of behavioural variation cannot be neglected. The example of language in humans shows how important the social component of transmittability can be. If we were to measure the heritability of language properly, we would find zero heritability. Yet parent–offspring regressions can be highly significant, suggesting that the evolutionary potential of this behavioural trait is high, despite the absence of heritability.

### 20.2.3 From quantitative genetics to culture

Anthropologists and psychologists have traditionally viewed culture as an exclusively human characteristic (Boesch 2006). Hence most existing definitions attempt to capture what is unique in human culture. A consequence is that anthropocentric definitions have been introduced that are unlikely to be useful for studying animal evolution.

#### 20.2.3.1 *An evolutionary definition of culture*

An evolutionary definition can be derived from our above considerations and earlier definitions (Soltis *et al.* 1995; Dugatkin 1999; Witte and Noltemeier 2002; Henrich and McElreath 2003; Barnard 2004; Danchin *et al.* 2004; Mesoudi *et al.* 2004).

> Culture is the part ($V_{AS}$) of phenotypic variance that results from information transmitted across generations through social influences. It is the part of transmittability that results from social learning.
>
> Culture is thus the sum of information, the transmission of which between generations rests on **social influences**. **For a trait to be considered cultural, social influences must modify the phenotype lastingly, and, whenacquired in a situation, it should be generalized to similar situations** (Danchin et al. 2004).

Thus, a trait should verify **four conditions simultaneously to be considered as a cultural trait**:

1) The transmission of the trait must result from 'social learning', that is learning from others, which necessarily interact with genes coding for learning capacities.
2) Variation in information must be **transmitted across generations**.
3) The modification of the phenotype must be **generalized to any kind of similar situation**.
4) Social influences **must modify the phenotype of the individual lastingly** so that the resulting variation becomes part of $V_{AS}$.

We now detail the meaning of each of these conditions.

### a *The transmission of the trait must involve 'social learning', which is learning from others*

Learning is usually defined as **an adaptive change in behaviour through the effect of acquired information, which provides experience** (Dudai 1989; Avital and Jablonka 2000). Cognition scientists use more than 30 precise and often overlapping definitions of social learning, but the meaning of this expression can be broadened to encompass any form of learning that involves social information (Galef and Giraldeau 2001).

The term 'social learning' thus includes any form of learning from other individuals, such as social imprinting, and any form of **imitation**, **copying**, **teaching**, etc. (see Avital and Jablonka (2000) for an extensive review on these processes). This definition encompasses any situation in which information obtained by watching other individuals influences decision-making. The crucial point here is not in the precise mechanism of acquiring information (which leads to new patterns of behaviour), but rather in the processes that allow behavioural variation to be transmitted across generations (Avital and Jablonka 2000).

Mechanisms involved in social learning are influenced by genes. Individuals are naturally more or less predisposed to gather information from others and to learn socially. In humans, we know that there are important variations in learning capacities among individuals. Because some of this variance is likely to be genetic, the genetic components of the capacity to acquire and transmit cultural variation must necessarily interact with cultural transmission. Futhermore, an indivdual's learning capacity is also enhanced by the practice of learning. Thus, this interaction is likely to influence our measures of transmittability (Figure 20.2).

The growing evidence for the role of socially mediated learning processes in many fitness enhancing decisions is reviewed in Chapter 4. That review led to the conclusion that ISI plays a major role in social learning. For instance, hungry animals tend to monitor feeding conspecifics and prefer to go to the place where they saw others feeding most successfully (Section 4.3.1). Similarly, animals may prospect breeding areas of conspecifics to gather ISI about the breeding performance of others and choose their future breeding site accordingly (Section 4.3.2.1). This has been termed **habitat copying** because such a process leads new breeders to **copy** the breeding habitat choices of successful conspecifics. In all these processes, animals 'learn' about their environment by monitoring the performance of other individuals. They thus use ISI to learn from others.

### b *Variation in information must be transmitted across generations*

Unlike genetically acquired traits, cultural traits are often transmitted horizontally among individuals of the same generation. This is the case of fashion and fads in humans. However, for culture to be transmittable, cultural variation must also be transmitted vertically across generations. Thus it is learning from members of the previous generation that makes cultural variation transmittable. A consequence is that for a trait to be cultural, individuals of successive generations must interact. This is clearly the case in human societies. But this is also true of very many animal species.

According to this criterion, any non-genetic information that is transmitted across generations is not necessarily cultural. For instance, in the case of the Atlantic salmon that return to their birthplace so that lineages are attached to a specific river, we have seen that this leads to non-genetically transmitted behavioural traditions. However, this does not mean that such tradition is cultural because individuals from the previous generation are long dead when young salmon hatch. The transmission of this tradition thus cannot involve social learning. This example fixes the limit in the range of culture: as Figure 20.2 shows, cultural variation is only a part of the non-genetic component of transmittability.

### C *The modification of the phenotype must be generalized to any kind of similar situation*

The concept of **generalization** comes from the scientific literature of experimental psychology (reviewed in Avital and Jablonka 2000). Generalization can take various forms that are all linked to the animal and human capacity to make categories (of objects, or situations, etc.).

In the context of cultural transmission, this criterion specifies that for a trait to be cultural, the effect

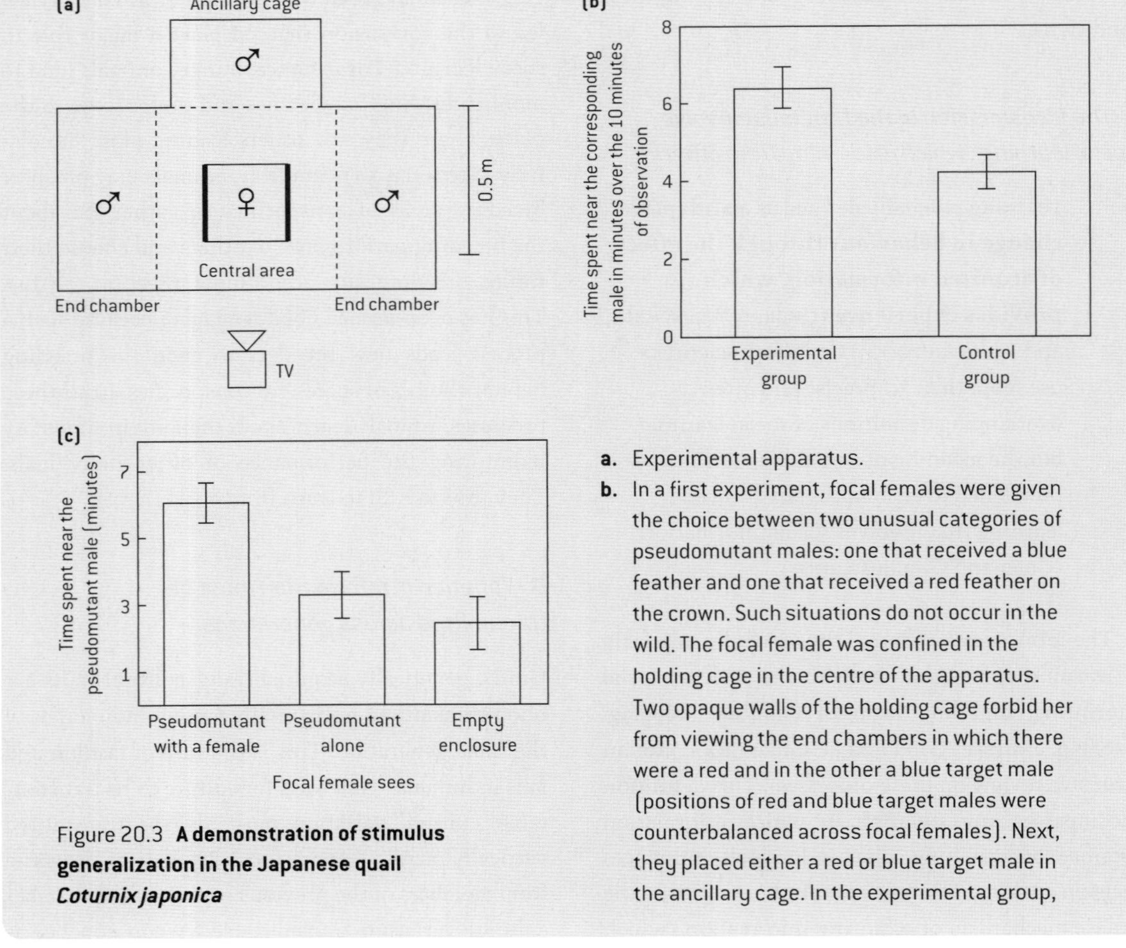


Figure 20.3 **A demonstration of stimulus generalization in the Japanese quail *Coturnix japonica***

**a.** Experimental apparatus.

**b.** In a first experiment, focal females were given the choice between two unusual categories of pseudomutant males: one that received a blue feather and one that received a red feather on the crown. Such situations do not occur in the wild. The focal female was confined in the holding cage in the centre of the apparatus. Two opaque walls of the holding cage forbid her from viewing the end chambers in which there were a red and in the other a blue target male (positions of red and blue target males were counterbalanced across focal females). Next, they placed either a red or blue target male in the ancillary cage. In the experimental group, of social influences must be transposed to any similar situation. For instance, in the context of breeding habitat choice (Chapter 9), habitat copying may lead to cultural transmission of habitat use. This would be the case if after observing that conspecifics breeding in habitat type A have higher reproductive success than those breeding in habitat type B, prospectors tend to systematically prefer to breed in type A habitat patches.

Another example is that of the transmission of female preference for a given male phenotype through mate copying (see Section 4.3.3.1). Most mate copying experiments show that providing a focal female with ISI suggesting that a given male $A_1$ of phenotype A is more attractive than a given male $B_1$ of phenotype B leads the focal female to prefer male $A_1$ over male $B_1$ in subsequent tests (see Figures 4.13 and 20.3). However, although this is indeed an impressive result, according to our definition, this does not demonstrate cultural transmission of a female preference for male $A_1$. For mate copying to lead to cultural transmission, such preference must be generalized to any male of the same phenotype as male A. Socially influenced focal

but not in the control group, a model female was placed in the ancillary cage with the pseudomutant male. After 10 minutes, they placed an opaque partition between the ancillary cage and the central area of the apparatus and raised the holding cage to permit the focal subject to move freely. During the next 10 minutes they recorded the time the focal female spent closer to each of the two end chambers of the apparatus, one containing a red and one a blue target male. Such tests of affiliative preferences of female quail have proven strong predictors of their actual mate choices (see Figure 20.7). They recorded each female's behaviour on videotape, and two observers independently scored the time that each of 10 focal females spent nearer each end chamber. Inter-observer repeatability was higher than 0.99. Focal females assigned to the experimental condition spent significantly longer nearer the end chamber holding a target male of the same colour as the one they saw copulating during the first phase, than did focal females assigned to the control condition (Student's *t* test: $P<0.01$). Bars represent the mean $\pm$ standard error of the number of minutes during the 10 minute second phase that focal female quail of the experimental and control groups spent nearer the target male that was of the same colour as the male that they saw copulating during the observation phase.

**c.** A second experiment, using the same apparatus aimed at determining whether generalized attraction to a class of males sharing a trait after seeing one of the members of that class mate would be found if the trait that the males shared was a rather subtle, naturally occurring one. Ten pseudomutant males were created by gluing white feathers taken from albino quails onto the crown of these males. Ten control males received the same treatment by gluing normal quail feathers on the crown. To ensure that the results of the experiment did not reflect a sampling error in assignment of males to pseudomutant and control conditions, the authors swapped males from the control and experimental groups by changing the feathers they had attached to males' crowns, so that each control male became a pseudomutant and each pseudomutant became a control. Again the experiment consisted of a demonstration and a test phase each lasting 10 minutes. As in the first experiment, during the demonstration the focal female could only see the male in the ancillary cage. That cage received either a pseudomutant male with a sexually active female (10 focal females), or a pseudomutant male alone (10 focal females), or no male and no female. During the 10 minute test phase, the female could choose between pseudomutant and a control male confined in the two end-chambers of the apparatus. Bars provide the mean $\pm$ standard error time spent near the pseudomutant male during that second phase. The treatment effect was significant ($F_{2,28}=7.43, P<0.003$).

Modified from White and Galef (2000b).

females should not only prefer mating with male $A_1$ over male $B_1$, but when given the choice between any pair of males of phenotypes A vs B, she must systematically prefer males of phenotype A. Hence if ISI revealed that the smaller male was more attractive, then the focal female should not only prefer that small male over that big male, but generally prefer any small male over big males. This has been shown only very rarely in mate copying studies. The most convincing such demonstration was provided by the group of Bennett G. Galef Jr in the Japanese quail (*Coturnix japonica*; Galef and White 2000; White and Galef 2000b). They showed that they could trigger sexual preferences for any 'pseudo-mutant' male with some white crown feathers (from albino quail) or even blue or red crown feathers (colours that are never observed in nature) just by providing ISI suggesting that such pseudomutant males are more attractive to females than males with a normal phenotype (Figure 20.3).

Similarly, (Godin *et al.* 2005) showed that female guppies (*Poecilia reticulata*), when socially influenced to prefer a male with a given colour pattern, generalize that preference to other males of a similar colour phenotype.

#### d *Durability of social influences*

**Finally, social influences must modify the phenotype of the individual so that the resulting variation becomes part of $V_{AS}$.** In other words, the effect of social influence must be retained for a sufficient proportion of the organism's lifespan so that the resulting variation may be transmitted to the next generation. If social influences last a relatively short period of time then, despite being interesting, such influences cannot lead to cultural traits as they are unlikely to be transmitted. This is an important characteristic as it is the only way for phenotypic variation to be transmitted across generations. Short-term effects of social influences are unlikely to lead to lineages of individuals sharing the same culturally inherited behavioural traits. This criterion has only been tested a few times (see for instance Witte and Noltemeier 2002).

This criterion was already underlined by Cecilia Heyes and Bennett G. Galef Jr (1996) who argued that culture also requires something that makes socially acquired behaviour resilient to change as a result of individual learning. For instance, something forces languages to stay more or less as they are learned. If socially acquired traits could be changed easily by personal experience traditions would not occur.

#### 20.2.3.2 *Some properties of culture*

A major characteristic of cultural variation is thus that it is transmittable. This is because culture is a by-product of learning from others and particularly from individuals of the previous generation. The example of language shows that learning makes cultural variation transmittable. Thus, to evolutionary biologists culture is just another means of producing non-genetic, yet transmittable, differences among populations. Furthermore, our four-point definition of culture makes it unlikely that cultural transmission occurs in species without overlapping generations because social interactions across generations most likely imply the meeting of individuals from different generations.

A recurrent confusion is the use of the word 'culture' for any kind of non-genetically transmitted information. However, Equation 20.2 and Figure 20.2 show that there are several other forms of transmittable variation that involve different environmental influences. As the above definition underlines, cultural traits are only the part of such variation that directly involves some social learning. For instance, parental effects (see Section 4.2.3.1) is another source of non-genetic yet transmittable information. Other potential processes include other forms of non-genetic information transferred across generations such as territorial inheritance and niche construction (equation 20.2).

One may also view cultural transmission as a particular form of parental effects. However, there are two major differences. First, as our definition states, culture is 'learned' from others through social information acquisitions, implying cognition, whereas parental effects are mainly mediated through physiological processes. Culture is derived from behaviour, which excludes parental effects and genes, although all three can interact. Secondly, while parental effects are by definition transmitted vertically from parents to offspring, cultural transmission has the very peculiar property of also being transmitted horizontally and obliquely (see Figure 20.4).

#### 20.2.3.3 *Cultural evolution*

A major characteristic of culture as defined above is that it fulfils the three conditions that enable natural selection. There may be variation; some cultural variants may be favoured in some conditions; and more importantly, learning from others makes cultural variation transmittable. This implies that cultural variation may undergo **cultural evolution** through the selection of cultural variants, just as genetic variation may undergo evolution through natural selection.

### 20.2.4 Is there convincing evidence of culturally evolved traits in animals?

The case of whether song dialects in whales and dolphins should be called culture gave rise to a fierce debate (Galef 1992; Whiten and Ham 1992; Rendell and Whitehead 2001). Most of the debate in fact came down to **agreeing on a clear and testable definition** of what we mean by culture. The definition presented in Section 20.2.3.1 may provide the long-sought conceptual tool necessary for the study of animal culture in an evolutionary perspective. In fact, no study has yet demonstrated that mate copying, and more generally any behavioural trait has simultaneously fulfilled our four criteria of a cultural trait. However, evidence has been provided that the four criteria are fulfilled **separately** in different species and for different behaviours. So far, the only situation that seems to fit our definition of culture lies in bird song dialects (see below). This means that despite the fact that we are on the verge of demonstrating the existence of animal culture, there is no definitive evidence for it yet (Laland and Janik 2006).

### 20.2.5 Inadvertent social information and culture

As we have seen in Chapter 4, ISI has been shown to be involved in many decisions, particularly in the learning processes that make behavioural variation transmittable. Many of the examples developed in Chapter 4 resemble cultural transmission. This section discusses how ISI use may interact with cultural transmission.

#### a *Inadvertent social information, mate choice copying, and cultural transmission*

A potential example is that of the transmission of female sexual preferences. Results of Figure 20.3 show that there is within-population variation in female preference in the quail. Figure 4.13 shows similar within-population variations in sailfin mollies (*Poecilia latipinna*). Because current models of sexual selection by female choice require transmittable variation in female mating preferences for sexual selection to operate, and because transmittability is commonly considered to be due to variation in genes only, researchers have assumed that such within-population variation was the result of some genetic variance and thus interpreted it as evidence of a heritable component in female preference (Wilcockson *et al.* 1995; Iyengar *et al.* 2002; Haesler and Seehausen 2005). However, to ascertain that genetic variation is involved, experiments need to control specifically for parental and social influences. The existence of mate choice copying demonstrates that such social influences may greatly affect sexual preferences to the point that socially influenced females may reverse their sexual preferences (see Figure 4.13). Obviously, part of mother–daughter resemblances results from social influences, so that genetic variation is unlikely be the only cause of variation in female preference.

Several authors have suggested that their demonstration of mate choice copying illustrates the existence of cultural transmission in animals (Dugatkin 1996; Brooks 1998; White and Galef 2000b). However, as implied by our definition of culture (Section 20.2.3.1), the classic protocol to demonstrate mate choice copying (see Section 4.3.3.1) falls short of fulfilling our four conditions of culture.

### b *Inadvertent social information and dialects as cultural processes*

Dialects in some song birds may be one of the most promising phenomena for examining culture in animals. There is important variation in male songs across populations of the same species. It is clearly demonstrated that most of that variation is learned from the songs of surrounding conspecific males and particularly from the father during development. This learning process leads to the transmission of slight variations in song types across generations. For the moment, only a few studies were specifically aimed at studying the nature of the information involved in song learning. However, we can imagine that young males may learn the song of winning males more readily than that of losers. Playback experiments during development involving different song types and mimicking territorial conflicts as in the **eavesdropping** experiment detailed in Section 4.3.3.2 would allow the testing of the impact of the singers' social performance on song transmission across generations. Indeed, the current knowledge about song learning in birds strongly suggests that this process verifies all the characteristics of cultural transmission as defined in section 20.2.5.b.

The other well-documented case of song dialects is that of cetaceans (see Whitehead and Rendell 2004) where variation in song type is mainly observed between rather than within groups. Furthermore, if dialects are cultural they should change over time, as was found in whales (see Chilton and Lein 1996). However, because of the difficulty of performing laboratory experiments, little is known about the actual mechanisms of transmission of such variation in whales.

### c *Inadvertent social information and tool use*

An early example of potential animal culture is tool making and use, which has been documented in many taxa such as apes and corvids. Tool use in apes is classically considered as a cultural trait. Chimpanzees learn from their parents how to use various tools (Boesch 1991; Lonsdorf *et al.* 2004). We will see in Section 20.2.8 that parents even seem to teach their offspring how to use such tools. However, such observations are so time-consuming that no one has tried to analyse whether performance is involved in such learning processes. It is very likely, however, that young individuals would not learn how to use a specific tool if tool use did not improve the performance of the tool user. This could be tested in the laboratory by manipulating the intensity of the benefit of a given tool, for instance by providing nuts of various sizes or nut-cracking tools of various qualities. If tool use creates ISI, young individuals should learn it faster when the benefits of tool use are greater. Similarly, one could manipulate the efficacy of various tools for the same purpose. The expectation would be that young observing their parents using the most efficient tools would learn faster than those whose parents were given inefficient tools. This would show that ISI is involved in the acquisition of these cultural traits.

The now classic bird example is that of the New Caledonian crow (*Corvus moneduloides*), which has been shown to manufacture various types of tools, often involving hooks (Hunt 1996; Hunt *et al.* 2001; Weir *et al.* 2002; Hunt and Gray 2003). Field studies have for instance demonstrated functional lateralization or 'handedness' in tool making and use (Hunt 2000; Hunt *et al.* 2001, 2006), a trait that was thought to be very specific to humans and to be linked to symbolic thought and language. Furthermore, laboratory experiments showed impressive capacities of tool making: a given individual, named Betty, was able to bend or unbend an aluminium strip according to the situation in order to function as a food retrieving tool (Weir and Kacelnik 2006). In these experiments there was no demonstrator. One could imagine experiments in which naive individuals have to learn what tool is necessary and how to make it, either on their own or by watching another individual performing the task. One would expect a naive bird that observed demonstrators to learn faster and better than solitary naive birds. We can also expect learning to be expedited as the benefit to the demonstrator increases.

#### d *Inadvertent social information and behavioural imprinting*

Another important source of additive social effects may be behavioural imprinting. Imprinting is defined as early learning that occurs at some well circumscribed periods in the animal's life and which has a persistent and long-term effect on its behaviour (Avital and Jablonka 2000). If young individuals are imprinted by the behaviour of their parents, they may grow up behaving like their parents, leading to parent–offspring resemblance. No research has examined effects of ISI on behavioural imprinting. However, one may expect that young individuals learn better from good than bad performers. Young male songbirds, for instance, may learn songs of neighbouring dominant males more readily than those of subordinate males. On the other hand, some studies suggested that when learning discrimination tasks, animals may learn better from a tutor that makes mistakes than from a perfect one (Templeton 1998). More generally, the natural tendency of young animals to imitate their parents sets the stage for the transmission of behavioural patterns across generations, leading to cultural transmission. It is thus no surprise to observe that social influences are widespread taxonomically and in terms of types of decisions (reviewed in Chapter 4).

#### e *Inadvertent social information as a conceptual tool for the study of animal culture*

In summary, just as genetics is the study of the mechanisms and outcomes of the transmission of genetic information across generations, culture is the study of the mechanisms and outcomes of the transmission of cultural information across generations. An important point is that the concepts of ISI and signalling, as well as those of learning, make the link between animal decision-making and the existence of transmittable cultural variation. They thus provide practical and conceptual tools for studying the existence and evolutionary impact of cultural processes (Danchin *et al.* 2004). This opens a new field of research for behavioural ecologists in which behaviour is a major actor of evolutionary processes.

### 20.2.6 More general definitions of natural selection and evolution

> All the above considerations led some authors to propose more general definitions of natural selection and evolution. Natural selection is the change that results from the selection of transmittable variation (compare with the definition provided in Sections 2.2.5.1 and 20.2.2.1), whereas 'evolution is the process by which the frequencies of **variants** change over time' (Bentley *et al.* 2004; compare with the definition provided in Chapter 2, Section 2.2.2a).

The term **variant** in this definition includes any additive component of phenotypic variance. Of course it includes genetic variance as a major component. However, it may also incorporate part of $V_{AE}$, $V_{AP}$, and $V_{AS}$, which are genetic–environmental correlations, parental effects, and social inheritance, respectively, and these effects may result from niche construction, territorial inheritance, and cultural transmission.

Interestingly, some researchers are transposing all the concepts of population genetics to the study of cultural variance (Cavalli-Sforza and Feldman 1981). For instance, R. Alexander Bentley from University College London and collaborators (Bentley *et al.* 2004) applied the neutral model of Kimura (1983) and Gillespie (1991) to three human cultural traits: first names, archaeological pottery, and applications for technology patents. The beauty of that approach is that it is possible to get very big data sets on human cultural variants, much bigger than any population geneticist could ever expect. For instance, for the first-name data set, they used first-name frequencies in a sample of 6.3 million North Americans (about one-fortieth of the US population) from the 1990 census. Impressively, the fit to the neutral model was as high as 0.990!

This kind of study opens a new avenue of research and shows the parallels that can be made between genetic and cultural variations, suggesting that they are both maintained by similar processes of selection.

### 20.2.7 Cultural and genetic transmission

An important question is that of how genetic and cultural transmission interact (Avital and Jablonka 2000). It is clear that there can be genes without culture, but there cannot be culture without genes. This sets a major asymmetry between the two processes. Culture is the by-product of genes, in particular those genes that are responsible for learning capacities. Thus, the dynamics of cultural and genetic variations must interact. We will return to this theoretical issue in Section 20.1.9. However, beforehand, it is important to understand how genetic and cultural transmissions differ, because such differences will play a major role in their interactions (Figure 20.4).

#### 20.2.7.1 *Genes and memes*

Individual organisms of a given generation (*n*) receive their gene pool at fertilization (Figure 20.4). Then organisms grow to an adult stage through developmental processes, one of which is 'learning' (taken in its broadest sense). Developmental processes are constantly influenced by the environment either through parental effects or through the organism's direct interaction with it. The latter interaction produces personal information. In parallel, organisms may learn about the environment through performance information. We have seen in Chapter 4 that not only **intentionally** transmitted information (true **communication**) is involved in social influences, but

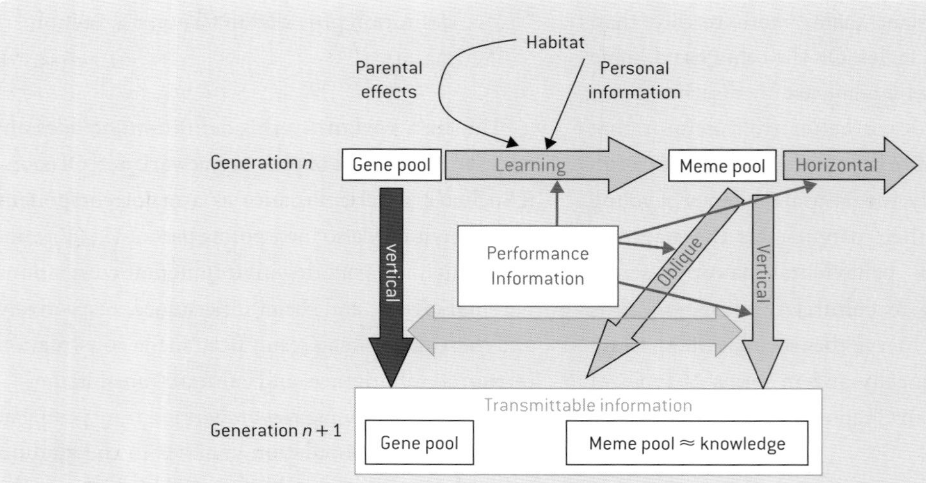


Figure 20.4 **Performance information and genetic and cultural evolution**

Social influences are involved in the learning processes that are represented by the four light-grey arrows. See text for further comments and Figure 4.3 for the definition of **performance information**. Note that the light-grey arrow entitled 'Learning' results from the joint effects of the light grey vertical, horizontal, and oblique arrows from generation $n-1$ to $n$. See text.

Modified from Danchin *et al.* (2004).

also the inadvertent component of performance information that has been called ISI (Danchin *et al.* 2004).

*The knowledge of individuals . . .*

The sum of these different forms of non-genetic information constitutes the **knowledge** the organism has about its environment (Figure 20.4). Here we call that knowledge the meme pool, using a concept introduced by Richard Dawkins in his book '*The Selfish Gene*' (Dawkins 1976). A meme is to culture what a gene is to the genotype. A meme is an abstract concept that is meant to represent a unit of cultural information. Through learning, organisms acquire memes the sum of which constitutes the individual's total knowledge. The part of that knowledge that is transmitted to other individuals is culture.

Individuals in the next generation, $(n + 1)$ inherit the genes and memes of their parents and other group members which together constitute the pool of transmittable information (Figure 20.4). Genes are acquired once at fertilization whereas memes can be acquired continually during an organism's lifetime.

*. . . is transmitted across generations . . .*

An important difference between genetic and cultural information is in how they are transmitted to the next generation. In eukaryotes, genetic information is only transmitted vertically (dark arrow in Figure 20.4), from parents to offspring. Vertical transmission of culture is also present (as for language in humans). However, culture is also transmitted **horizontally**, among individuals of the same generation (as in fashion in humans), and **obliquely** among non-kin individuals of different generations (as in teaching in humans). No such possibilities exist in genetic transmission. We have reviewed in Chapter 4 the evidence that performance information (ISI and signals) is likely to play a major role in the vertical, horizontal, and oblique transmission of information (central box and arrows pointing to light grey arrows in Figure 20.4).

*. . . which may impact evolution*

Horizontal and oblique transmissions of transmittable information profoundly change the rules of information transfer across generations. This has major consequences for the way evolution may function. Processes that are impossible with purely vertical transmission may become possible with cultural transmission. For instance, **group selection** is impossible with purely genetic transmission. However, one effect of horizontal cultural transmission is to homogenize individuals belonging to the same group while increasing the variance among groups, leading to the emergence of the cultural group as a new unit of selection. Several models have shown that this reasoning is biologically sound (Hochberg *et al.* 2003; Jansen and Van Baalen 2006). For instance, a model by Michael Hochberg *et al.* (2003) showed that social selection can lead to prezygotic reproductive isolation. More specifically, with altruistic and selfish acts, the evolution of social discrimination causes the congealing of phenotypically similar individuals into different, spatially distinct tribes. Furthermore, the reduced fitness of hybrids at tribal borders leads to the selection of mating preferences for mates of the same tribe on both sides of the border. This is the reinforcement process developed in Section 11.9.3. Such mating preferences then spread to the core areas of the respective tribes. This shows that unlike other resource competition models, social behaviour can generate reproductive isolation in an ecologically homogeneous environment.

The general message is that taking cultural transmission (or more generally non-genetic transmission) into account greatly widens the range of possible evolutionary outcomes.

#### 20.2.7.2 *The importance of gene–culture interactions*

The term $V_{G*E}$ in Equation 2.1 (Section 2.2.3) formalizes the genotype–environment interaction. This term is represented in Figure 20.4 by the large horizontal light grey arrow joining genetic and cultural transmission. For instance, cultural transmission must depend upon individual capacities of learning from others, and we know that there is tremendous within-population variation in learning capacities. Some individuals are excellent at copying; others are not particularly good or can even be very bad at it. It is interesting that such considerations led Susan Blackmore to suggest that the main distinctiveness of humans is in our incredible skills for imitation (Blackmore 1999). Such variations in imitation and copying capacities show how culture and genes may interact, and accordingly models of cultural evolution incorporate these types of interactions (Section 20.2.9).

### 20.2.8 Do animals teach?

Teaching may be an example of oblique transmission when individuals of one generation intentionally transmit information to unrelated individuals of younger generations. Teaching is common in humans, and is even the main goal of this book. However, the evidence for teaching in animals is very scant. Evidence comes from tool use in apes. For instance, Boesch (1991) observed a mother chimpanzee that actively demonstrated nut-cracking to her offspring. In contrast, in the context of tool use for termite fishing in chimpanzees, Elizabeth Lonsdorf *et al.* (2004) did not detect any significant teaching from mothers. They found that offspring observe their mother's termite fishing and learn little by little to do it by themselves, but mothers did not seem to behave differently in the presence or absence of their offspring. They found interesting differences in intensity and speed of learning between sexes. Females spent more time observing their mothers and learned significantly faster than their brothers.

#### a *What is teaching?*

> The definition of teaching comprises three criteria (Caro and Hauser 1992): (1) an individual A modifies its behaviour only in the presence of a naïve observer B; (2) individual A incurs some cost or at least does not derive any immediate benefit; and (3) A's behaviour results in B acquiring knowledge or skills more rapidly or efficiently than it would in the absence of A, or it would not have learned at all in the absence of A.

#### b *Meerkats as teachers*

Recently, J. Alex Thornton and Katherine McAuliffe (Thornton and McAuliffe 2006) found correlational evidence that meerkats (*Suricata suricatta*), a cooperatively breeding mammal, routinely engage in opportunity teaching (Figure 20.5). They performed experiments that demonstrate that **helpers** (i.e. individuals older than 3 months) teach younger individuals how to handle dangerous prey: scorpions of the genera *Parabuthus* and *Opistophthalamus*. The former possess neurotoxins potent enough to kill a human, whereas the latter have milder toxins but are more aggressive, defending themselves with large, powerful pincers. Pups are initially unable to find and handle their own prey. They begin to follow foragers at about 30 days of age and are provisioned by all group members in response to begging calls, until they reach nutritional independence at around 90 days of age.

Helpers typically kill or disable prey with rapid bites to the head or abdomen before provisioning pups. Scorpions are normally disabled by removing the stinger. Helpers adjust the frequency with which they kill or disable mobile prey according to pup age, gradually introducing pups to live prey. The proportion of highly mobile prey fed when dead or disabled decreases with pup age while the proportion of prey fed intact increases (Thornton and McAuliffe 2006). To show that helpers adapt their feeding behaviour to the age of the pups in the group, they performed a playback experiment. Helpers respond to playback experiments. Playbacks of begging calls of pups of the same age as those in the group (control group) or the opposite age extreme (experimental group) were broadcast to groups with foraging pups. The results were very convincing in showing that helpers did adapt their behaviour according to pup age (Figure 20.6). Helpers with young pups fed significantly more intact prey when the experimenter broadcast begging calls of older pups than when broadcasting begging calls of pups of the same age as

(a)

(b)

(c)

Figure 20.5 **The meerkat (*Suricata suricatta*), a highly social mammal inhabiting the Kalahari Desert in South Africa**

a. A sentinel adult meerkat with a six-week-old pup (photograph by Katherine McAuliffe).
b. Helper watching a pup eating a dead scorpion (photograph by Alex Thornton).
c. Alex Thornton weighing meerkats in the field while a standing adult on the left watches for predators (photograph by Hansjoerg Kunc).

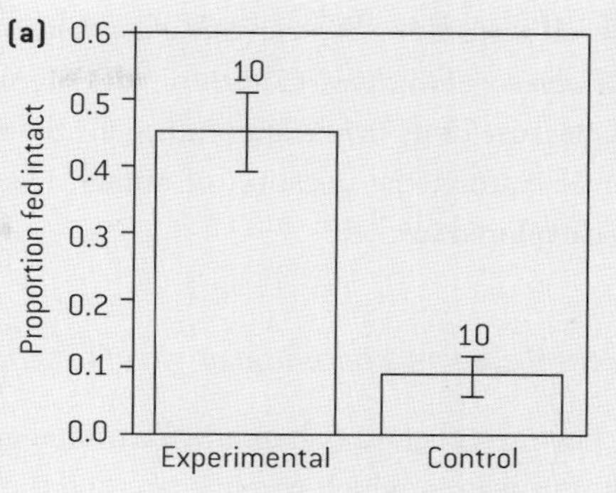


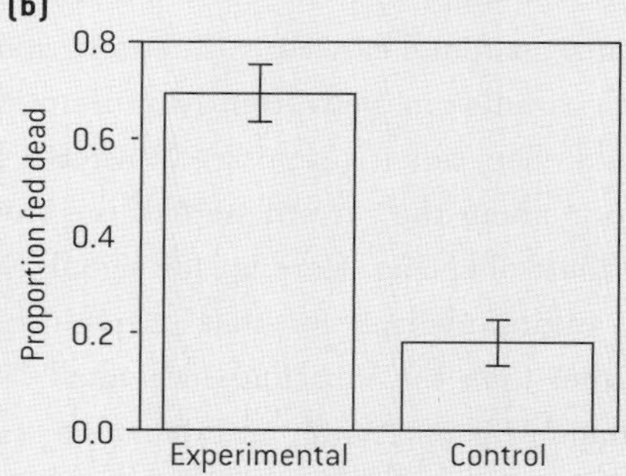


Figure 20.6 **Experimental evidence of teaching in meerkats**

a. Helpers in groups with young pups (28–37 days old) fed significantly more **intact** prey under experimental (voice of older pups) than control (voice of same age pups) playbacks (paired *t*-test, $P = 0.002$).
b. Helpers in groups with old pups (71–86 days old) fed significantly more **dead** prey under experimental (voice of younger pups) than control (voice of same age pups) playbacks (paired *t*-test, $P < 0.001$). Numbers above bars show sample sizes.

Modified from Thornton and McAuliffe (2006).

those of the group (Figure 20.6a) and vice versa (Figure 20.6b).

Thornton and McAuliffe (2006) produced many other results that show that helping behaviour in relation to pups verify the three criteria that define teaching. Thornton's results provide the strongest evidence so far of teaching in non-human animals. They conclude that '*The lack of evidence for teaching in species other than humans may reflect problems in producing unequivocal support for the occurrence of teaching, rather than the absence of teaching*' (Thornton and McAuliffe 2006). Because of its major importance for the study of animal culture, the question of animal teaching must be explored further.

#### c *Parental care as a potential origin of teaching*

As we have seen before (Chapter 4), parental effects constitute a way in which parents can tune the phenotype of their offspring to the current environment. In particular, parental care constitutes a period during which parents can inadvertently or intentionally (we would then call it teaching) transfer a lot of information about the current state of the environment to their offspring. Here, as for signals, intentionality means evolved for that purpose and no other, rather than the conscious individual's intention. Chapter 4 of Avital and Jablonka (2000) provides an extensive review on the importance of parental care as a process that enhances parent–offspring resemblance in animals. In this way, one could consider parental care as the platform for the origin of teaching. Avital and Jablonka (2000) also underline the existence of an 'information lacuna', or gap, in the study of parental care. Indeed most, if not all, of the literature on parental care ignores the potential evolutionary benefits of parent to offspring information transfers during parental care. This information lacuna might also explain the current relative paucity of evidence for animal teaching.

However, in the context of cultural evolution, this facet of parental care can be a powerful process of information transfer across generations. It is noticeable that none of the examples presented in Chapter 4 of the present book specifically involve information transfer from parent to offspring (i.e. vertical transmission). All these examples may equally concern horizontal, oblique, or vertical transmission (Figure 20.4). It is, however, likely that, historically, horizontal and oblique transmissions were by-products of selection for vertical transmission. These considerations suggest that a great deal remains to be discovered about social information of parental origin and animal decision-making in general.

### 20.2.9 The major theoretical approaches to the evolution of culture

Authors have long debated the potential for cultural evolution theoretically (Table 20.1). For reasons developed above, all models have analysed the relationship between genetic and cultural evolution. Conclusions differed according to the type of relationship that was assumed. The main difference among models is in the extent to which culture was considered as having its own **replicator** system (see Section 2.2.2.) that evolves more or less independently from genes, with the two systems influencing each other. We will describe the various approaches briefly, starting from those that allow little independence of culture, to models that consider memes as independent replicators. Complementary descriptions can be found in Barnard (2004) and in the original publications cited in Table 20.1.

#### a *Culturegens*

The renowned Harvard evolutionary biologist Edward O. Wilson, working with a physicist, Charles Lunsden, transposed models of co-evolution between hosts and parasites to the study the co-evolution of genes and memes (Lunsden and Wilson 1981). Their theory assumed that cultural traits show some degree of independence but are inextricably linked to genes (Table 20.1). Wilson stated this in saying that 'culture is on a leash' (Wilson 1978). They called units of cultural information 'culturegens'. The model transposes host and parasites models in which host and parasites may become gradually adapted to each other by increased resistance and decreased virulence.

#### b *Phenogenotype*

The second type of approach involves a family of models initiated by Cavalli-Sforza and Feldman (1981). Their approach views cultural traits as evolving more independently from genes, but still ultimately being selected through their effect on genetic fitness (Table 20.1). In these models, culture is still on a leash, but the leash is now much longer, allowing greater freedom of movement. According to these

| Model name | Assumptions of the model | Relationship between memes and genes | Application | Reference |
|---|---|---|---|---|
| Culturegens | Cultural traits can show some degree of independence, but are inextricably tied to genes that express them | 'Culture on a leash'. Culturegens are not replicators in their own rights. | The evolution of the shape of teddy bears | Lunsden and Wilson (1981) |
| Gene-culture, pheno-genotype | Cultural traits evolve more independently of genes, but are still ultimately selected through their effect on genetic fitness. Considers the fitness of a genotype/cultural amalgam (called phenogenotype) rather than a mutually co-evolved relationship between cultural traits and genotype | Cultural leash still on, but allows greater freedom of movement. Can lead to conflicts between cultural traits and genotype. Can drive a male trait to fixation if female preference for that trait reaches a critical frequency | From human language and altruism to mating preference and sex ratio | Cavalli-Sforza and Feldman (1981); Laland (1994); Feldman and Laland (1996) |
| Dual inheritance model | Individuals learn both through performance information (social learning) and personal information. These two forms of learning interact with genetically determined (i.e. direct) biases for basic characteristics of the traits available as well as with indirect biases such as conformism (the tendency to copy the most common trait). | Genes and cultural traits evolve independently. May lead to a runaway cultural selection for extreme traits in a manner akin to the Fisher runaway effect in mate choice (Chapter 11, Section 11.2.2). The leash is still on but less constraining. | Acquisition of human food preferences | Boyd and Richerson (1985, 1988) |
| Memes and Memetics | Genes and memes are fully independent replicators. Both genes and memes succeed in being copied. Good copying potential in genes arises from their effect on their bearer's fitness. Good meme copying potential arises from gaining the attention of prospective imitators. | Memes are replicators in their own right. There is no more leash. The relationship may even be inversed, with memes beginning to exert their own selection pressure on genes. Maximum potential for conflicts between replicators | Mainly human culture, language, brain size, etc. | Dawkins (1976); Blackmore (1999) and references therein |

Table 20.1 **The various models of cultural evolution: from culture on a leash to two independent replicators**

models, cultural traits can spread despite an enormous cost in genetic fitness. An empirical example is that of the Foré tribe of New Guinea, where rituals in honour of the dead involve cannibalism and the smearing of brain tissue from the dead over their bodies. This resulted in the spread of a neurodegenerative disease similar to Creutzfeld–Jakob disease.

### C *The dual inheritance model*

To remove Wilson's leash, Robert Boyd and Peter Richerson (1985) developed a model in which genes and cultural traits evolve independently (the dual inheritance model; Table 20.1). Boyd and Richerson (1985, 1988) used an **evolutionarily stable strategy** (ESS) approach to model the relative importance of social and personal learning in different environments. The outcomes suggest that the stable strategy for acquiring information depends on how difficult it is to learn accurately through personal information and the degree of temporal autocorrelation of the environment (see Figure 4.1). Although the dual inheritance model gives far greater play to cultural evolution than previous models, it does not assume the existence of independently replicating cultural units relative to genetic inheritance (Blackmore 1999).

### d Memes and memetics

The idea that cultural traits are driven by an independent evolutionary process based on a non-genetic replicator was provocatively introduced by Richard Dawkins (1976). He introduced the abstract concept of meme that is meant to represent a unit of cultural information that can be stored in the brain and transmitted by social learning. This concept gave rise to a full field of research called **memetics** (Blackmore 1999). The key assumption of memetics is that memes are fully independent replicators (Table 20.1). Whereas genes are replicators that dwell in cell nuclei and are passed by DNA duplication during reproduction, memes are replicators that dwell in the brain and are passed by imitation (Table 20.2).

Memes are mainly learned socially, a process that involves performance information. The main goal of memetics is to understand why some memes are copied much more successfully than others. Such questions may be answered by concentrating on the nature of the performance information that renders memes transmittable.

### e Conclusion

This brief overview (see Barnard 2004 for a more comprehensive overview) shows that all the theoretical approaches to the role of culture in evolution were mainly human-centred. However, most of the conclusions of these models probably hold for animal culture as well. This human bias may exist because most definitions of culture have aimed at capturing the originality of human culture, making them largely inapplicable to animals. Thus, evolutionary biologists have tended to overlook the possibility of such processes in animals. However, this chapter in conjunction with Chapter 4 has presented evidence that animal culture may be a significant component of the evolution of behaviour.

One consequence of this human bias is that we currently have more theories (Table 20.1) than facts because experiments are difficult in humans. Researchers have documented cultural variation in apes (see Section 20.3.1), but for practical reasons (Whiten 2006) virtually nothing is known about the mechanisms that produce such patterns. Results presented in Section 4.3 suggest that such questions can be studied in a wide variety of model animals. In Section 4.4 we showed how important it is to adopt an information-driven approach to study such consequences of animal behaviour. Furthermore, we have already suggested that the concepts of performance information, and particularly inadvertent social information, may provide the long-sought conceptual tool to experimentally study animal culture (Danchin *et al.* 2004, 2005).

| | Genetic | Culture |
|---|---|---|
| Information unit (replicator) | Gene | Meme |
| Information vector | DNA | Behaviour and central nervous system |
| Transmission mechanism | DNA duplication | Social learning: imitation, copying, social facilitation, imprinting, teaching |
| When transmitted | At reproduction, even if not expressed | During imitation, only if expressed |
| Mutation | Duplication errors, Pseudo-genes | Learning errors, innovation |
| Potential rate of innovation | Low | Possibly intermediate or high |
| Impact of most mutations | Deleterious | Probably deleterious |
| Heritability (genetic) | Yes (low) | No (very low) |
| Transmittability | Yes (higher than heritability) | Yes (moderate to very high) |

Table 20.2 **Similarities and differences between genes and memes**

## 20.3 Some implications of culture

This last section aims at producing a series of snapshots illustrating some of the evidence and implications of animal culture. In it, we will assume that the various traits we are talking about have been demonstrated to fulfil the four criteria that define cultural traits. This assumption is not problematic for many traits, such as mate copying because, despite the fact that for the moment no study has yet successfully and simultaneously tested the four criteria that define culture for this pattern in the same species, as we have discussed before, we are now on the verge of demonstrating that mate copying at least fits our definition and can thus be safely considered as a cultural trait in a first approximation.

In other instances, such as those developed in Section 20.3.5, the assumption that those traits are cultural might be more problematic. Indeed, as we have seen in the first section of this chapter, demonstrating that dialects differ among whale populations, does not necessarily demonstrate that these are transmitted culturally.

### 20.3.1 Patterns of cultural variation in apes (chimpanzees and orang-utans)

Cultural variation in animals as been documented in apes such as chimpanzees (see Lefebvre 1995; Whiten *et al.* 1999; van Schaik 2004; Whiten 2006). The patterns of behavioural traits in different populations throughout tropical Africa reveal surprising variation among populations. Some behavioural patterns are present and common in some populations but absent or very rare in others. Such variation appears to be relatively stable over time. Similar patterns of behavioural variation were described in orang-utans (van Schaik 2005). Although these patterns suggest the existence of cultural variation among populations, it would be difficult in such animal species to determine whether the described variation fulfils the four criteria of culture in animals (Section 20.2.3.1). Simple field experiments suggest that such variation is learned, which fulfils at least one of the criteria (Whiten 2006). Nonetheless, it is now necessary to decompose phenotypic variance into its various components to understand better the mechanisms underlying such variation (Danchin *et al.* 2004; Laland and Janik 2006). We will revisit primates in Section 20.3.5.

### 20.3.2 Bird song and dialects

We have briefly examined the existence of song dialects in birds and cetaceans (Sections 20.2.4 and 20.2.5). We have also seen how the dialects are mainly transmitted through the inter-generational learning of songs. Song dialects are therefore the by-products of learning from individuals of the previous generation and may provide a good example of animal cultural transmission. Several of the four conditions that are necessary to consider a trait as cultural are fulfilled: dialects vary among populations, they are learned socially across generations, and once individuals have learned a dialect they sing it for their entire life. The criterion of generalization is also likely to apply to female preference for a given dialect. Thus, the four criteria of culture have been demonstrated for song dialects independently in different species, suggesting that bird song dialect may provide a model system to test the existence of animal culture.

In South Pacific sperm whales (*Physeter macrocephalus*), dialect characteristics appear linked to foraging success (Whitehead and Rendell 2004). Furthermore, low diversities of mitochondrial DNA have been found in four species of matrilineal whales. Culture that seems to be an important part of the lives of those whales has been proposed as a possible explanation for this apparent anomaly (Whitehead 1998). The selection of matrilineally transmitted cultural traits, upon which neutral mitochondrial DNA alleles are supposed to 'hitchhike', has the potential to strongly reduce genetic variation. Thus, as in humans, culture may be an important evolutionary force in matrilineal whales. Moreover, in mountain white-crowned sparrows (*Zonotrichia leucophrys oriantha*), males singing a local dialect have higher paternity, which suggests female preference for local dialects. The consequence is that most of the population variation at microsatellite loci that could not be attributed to individuals was attributable to differences among, rather than within, dialect regions (MacDougall-Shackleton and MacDougall-Shackleton 2001). If the effect of dialects on female mating preferences is prevalent, then any two populations evolving different dialects after a sufficiently long separation may ultimately become unable to interbreed successfully (Grant and Grant 2002), creating a culturally induced first step toward speciation.

### 20.3.3 Performance information and animal aggregations

The study of the evolution of coloniality provides a good example of the insight that an information-driven approach of behaviour may cast on some evolutionary questions. In Chapter 9 we have seen how selection favours the evolution of sophisticated animal decision-making in the context of breeding habitat. We showed that according to the habitat copying hypothesis, animals gather and use ISI from the reproductive performance of other individuals in breeding habitat choice. This can lead to skewed distributions of animals over the habitat, with some habitat patches being overcrowded while others remain under exploited. In Chapter 14 we noted that real or additive aggregation may lead to the further aggregation of breeding animals in colonies. Danchin and Wagner (Danchin and Wagner 1997; Wagner *et al.* 2000) proposed that in a first step colonial breeding may not have evolved because of its possible benefits, but rather as the by-product of habitat selection and mate choice based on social information.

A consequence of habitat copying is that animals tend to aggregate in the same areas for generations, perhaps ignoring alternative suitable areas, because they have been occupied in the past. Thus, colony locations might be viewed as a socially transmitted trait resulting from a common knowledge about the habitat that is transmitted across generations by the use of ISI.

### 20.3.4 Fitness consequences of sexual preference

Chapter 4 reviewed the evidence that animals belonging to varied taxa use ISI in decision-making. However, we did not deal with the adaptive function of such strategies. In fact, there are very few experiments on the fitness consequences of performance information or ISI use. The series of experiments by Bennet G. Galef and colleagues for mate copying (as detailed in Section 4.3.3.1 and Figure 20.3) in the Japanese quail (*Coturnix japonica*) constitutes one of the exceptions. In that example, as in most other studies, female sexual preference was first measured as the time a female spent near a given male. However, such affiliative preference does not necessarily reflect the female's copulating and fertilization preferences. Females may affiliate with a male but prefer copulating with and having their ova fertilized by another male.

#### a *Affiliation preferences that reveal mating preferences . . .*

White and Galef (1999) tested the relationship between affiliative, mating, and fertilization preferences in an impressive series of experiments. They first allowed a focal female to choose between two target males for 10 minutes in a standard apparatus to test their affiliative preferences. They then allowed the same focal female to choose for 10 minutes between the same two target males, tethered at opposite ends of a straight alley as partners for mating. The focal female could run up and down the alley and mate with whichever male she wished. Results showed that the target male that a female spent more time close to in the affiliative choice test was also the male that she mated with most often during the 10 minute test (Figure 20.7).

#### b *. . . as well as cryptic female preferences . . .*

However, copulation does not always result in fertilization of a female's ova. The disjunction between copulation and fertilization is particularly salient in

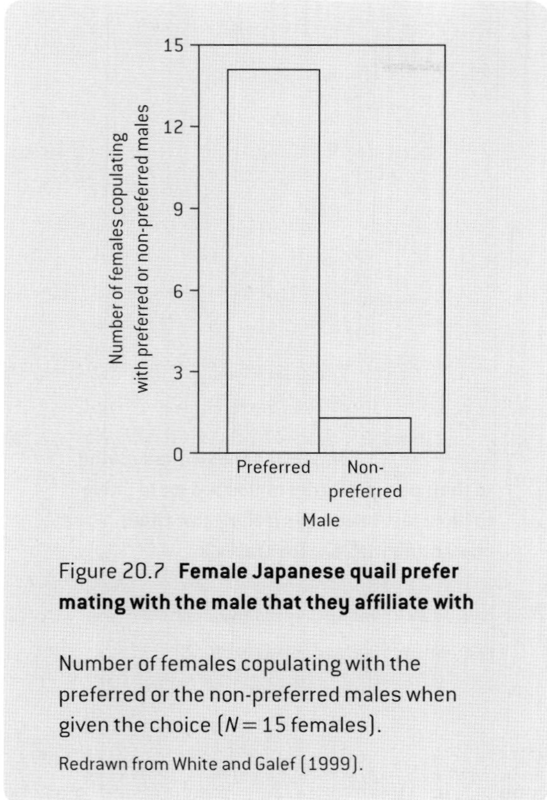


Figure 20.7 **Female Japanese quail prefer mating with the male that they affiliate with**

Number of females copulating with the preferred or the non-preferred males when given the choice ($N = 15$ females).

Redrawn from White and Galef (1999).

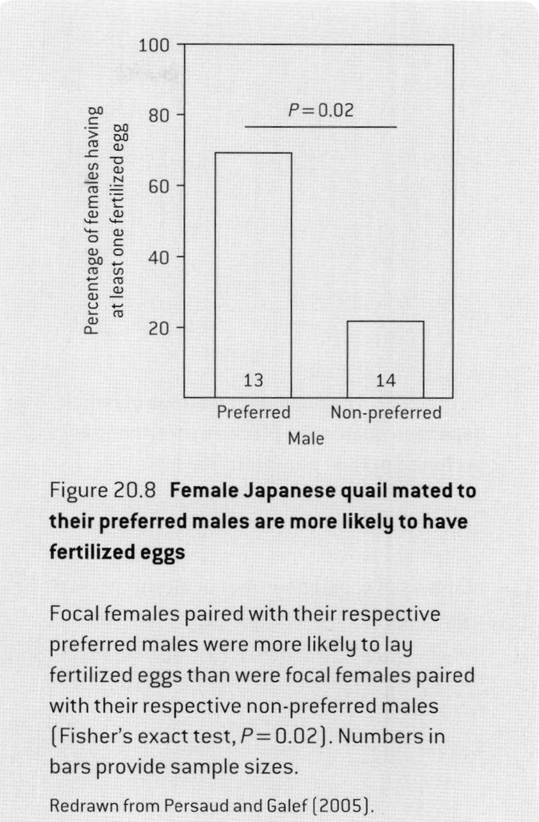


Figure 20.8 **Female Japanese quail mated to their preferred males are more likely to have fertilized eggs**

Focal females paired with their respective preferred males were more likely to lay fertilized eggs than were focal females paired with their respective non-preferred males (Fisher's exact test, $P = 0.02$). Numbers in bars provide sample sizes.

Redrawn from Persaud and Galef (2005).

birds, because some female birds, Japanese quail among them, can eject sperm from their reproductive tracts after they have been inseminated by a male (Pizzari and Birkhead 2000; Siva-Jothy 2000; Wagner *et al.* 2004). Consequently, insemination by a male may not lead to fertilization of a female's ova.

Persaud and Galef (2005) looked at effects of female affiliative preference on the fertilization success of males (Persaud and Galef 2005). In a two-step experiment, females were allowed to choose between two males for 10 minutes in the standard test of affiliative preference. Next, females were placed in an enclosure randomly with either their preferred male or their non-preferred male for 10 minutes, and allowed to mate. Then, females were put back in their home cage and all eggs that they laid for the next 10 days were collected. After 5 days of incubation, all eggs were opened to determine whether they had been fertilized. The percentage of females laying at least one fertilized egg was higher when they had been allowed to mate with their preferred than with their non-preferred males (Figure 20.8). This result was obtained despite the fact that females copulated equally often with their preferred and non-preferred males during the 10 minutes they were together (Figure 20.9).

### C . . . *that are influenced by inadvertent social information!*

As we have seen, a female's mating preferences can be strongly influenced by her observing one of the males mating with another female (for example Figure 20.3). Interestingly, similar results were obtained for mating preference and egg fertilization after such a mate copying experiment in the Japanese quail (Persaud and Galef 2005). In a classic mate-copying experiment, focal Japanese quail females that first preferred male A in the pre-test, then saw male B with a model female during the demonstration step, then affiliated preferentially with male B when

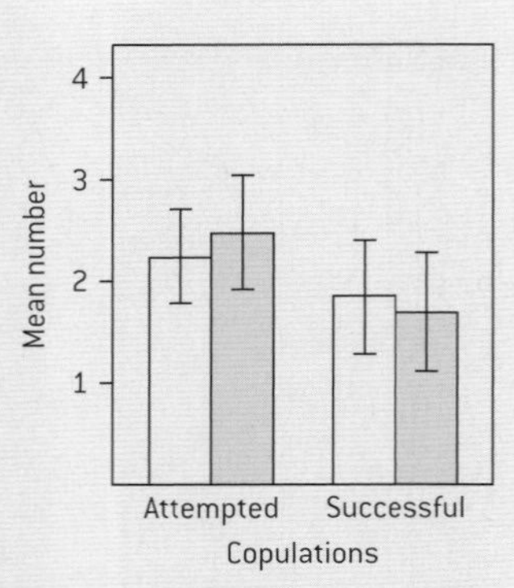


Figure 20.9 **The number of matings of female Japanese quail when placed with either their preferred or their non-preferred male**

Experimental protocol as in Figure 20.8. White bars provide mean ± standard error numbers of copulations for females that were placed with their preferred male. Dark bars are for females that were placed with their non-preferred male. Total sample size was 61 females.

Redrawn from Persaud and Galef (2005).

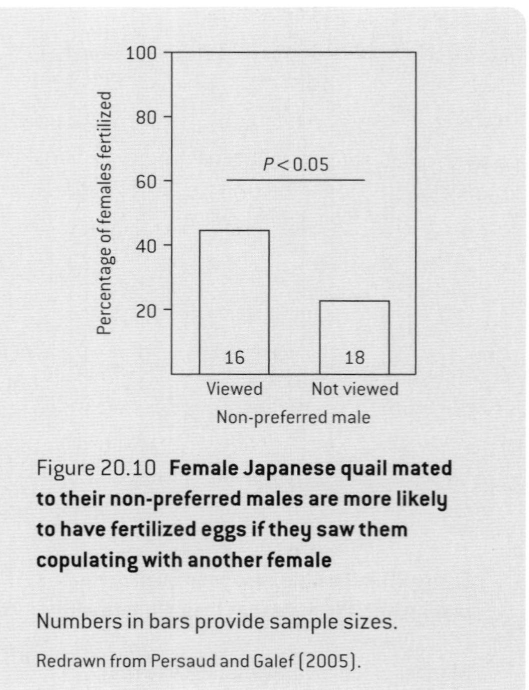


Figure 20.10 **Female Japanese quail mated to their non-preferred males are more likely to have fertilized eggs if they saw them copulating with another female**

Numbers in bars provide sample sizes.

Redrawn from Persaud and Galef (2005).

tested in a third phase. Furthermore, as Figure 20.8 shows, if half of the focal females are randomly allowed to copulate with A or B after the pre-test, more eggs are fertilized when they copulate with A than with B.

The surprising result is that if the same fertilization test is performed after the demonstration, this time more eggs are fertilized when the focal female is allowed to mate with the B male, the one that she did not prefer in the pre-test, but now prefers after seeing it copulating with another female (Figure 20.10). Again, this was true despite the fact that numbers of copulations with A and B males did not differ.

This series of experiments clearly demonstrates that mate copying not only influences female affiliative preferences, but also their tendency to mate with a given male and when she copulates, the tendency for her eggs to actually get fertilized in the end. These experiments were not designed to study the mechanisms by which females controlled fertilization. Females may simply eject all the sperm just after copulation. However, other mechanisms may be involved that would deserve attention. Another question is that of the benefits of such behaviour. As sperm ejection may be involved, benefits may be linked to those of sperm ejection (Pizzari and Birkhead 2000; Siva-Jothy 2000; Wagner *et al.* 2004).

Galef's group also demonstrated that social influences have the opposite effect on males: males tend to affiliate and copulate less with females that they saw copulating with another male (White and Galef 2000a). This difference is probably adaptive as males that would copulate with already-inseminated females would face high sperm competition. However, by roughly one day after copulating with a first male, the probability of a female's eggs being fertilized by a second male returns to baseline (Galef 2005). Thus, as expected, social influences on sexual preferences disappear in 48 hours in males but persist at least over two days in females (Galef 2005).

Whatever the proximate and ultimate mechanisms, this suggests that cultural influences can have dramatic consequences that may profoundly affect sexual selection, and thus evolutionary processes.

### 20.3.5 Stability and accumulation of transmittable cultural variation

The questions of the stability and accumulation of genetic variation are the two pillars of genetic evolution. Thus the questions of the stability and accumulation of cultural variation have been heavily debated.

#### a *Stability of cultural variation across generations*

Despite having been demonstrated only rarely, current evidence suggests that the transmission of a given pattern of behaviour can occur from one generation to the next. However, the question of the stability of that pattern over several generations remains open. Indeed, a fundamental difference between genes and memes is that genes are transmitted even if they are not expressed, whereas memes can only be transmitted through their expression in order to allow social learning. This implies that the absence of the expression of a culturally transmitted behavioural pattern in one generation will cause its disappearance in the lineage. This may greatly diminish the long-term inter-generational durability of cultural variation. Furthermore, imitation and copying are far more prone to errors of transmission than DNA duplication. Such errors might be assimilated to a form of mutation (Table 20.2). Such cultural mutations may occur randomly, in which case they might often be deleterious, or they might be oriented by current environmental conditions, in which case they might be adaptive. Hence, cultural variation is open to the addition of newly invented patterns of behaviour. That characteristic has played a major role in the evolution of human culture and has allowed the fantastic development of new technologies ranging from the discovery of how to master fire to the invention and general use of computers and the Internet.

#### b *Does animal cultural variation accumulate over generations?*

The example of the evolution of human culture underlines another important characteristic of evolutionary processes: for evolution to occur, variation needs to accumulate over generations. In the context of genetic evolution, the development of wings in birds likely implied the accumulation of several, if not many, mutations that transformed the forearm and the skin of the ancestral dinosaur into a functional wing. This accumulation process is fundamental for adaptation. Similarly, in the context of culture, tools as sophisticated as computers are the result of the accumulation of many discoveries over generations. To some extent, the origin of computers can be traced back to the origin of language and writing.

There is an important debate about whether this accumulation occurs in animal culture (Avital and Jablonka 2000). The best example of such accumulation is provided by the washing of potatoes in the macaque (*Macaca fascicularis*). Japanese researchers began a study of a population on the island of Koshima. To attract animals in open areas to study their behaviour, they put potatoes on the beach. Soon a young female named Imo started washing potatoes in a stream and later in the sea, thus removing remains of soil from the potatoes. The new habit spread to other monkeys (Kawai 1965). Imo and other young monkeys also bit the potatoes before dipping them in the salty water, thus also seasoning them.

Later, the researchers put wheat on the sandy shore and observed how macaques would deal with this new food source. They expected monkeys to collect wheat grain by grain to remove the sand. But the same female, Imo, found a way around the problem. She threw the mixture of wheat and sand in the water which allowed her to separate the heavy sand that sank from the lighter wheat that floated on the surface of the water. She thus used differences in density between sand and wheat in a process often used by humans, for instance by gold prospectors who use differences in density between the heavy gold and lighter sand.

Interestingly, the habit of coming to the beach to forage led to a series of social innovations. Young monkeys brought by their mothers to the beach started to play in the water, swimming, jumping, and diving. They also cooled in the water in summer. Such habits were then adopted by adults and beach activities became a popular activity. Another consequence occurred later: monkeys started to eat raw fish. This habit spread from peripheral hungry males. Apparently fish is not particularly tasty, but becomes valuable when there is nothing better to eat (Avital and Jablonka 2000).

The consequence is that a first habit of washing food in the sea generated an accumulation of new habits that led the population to spend more and more time on the beach, a habitat that is not natural for these monkeys. Monkeys thus acquired a new lifestyle. Although this example was artificially created by researchers bringing food to the beach, it clearly shows that cultural variation may also accumulate in animals. The fact that evidence of such processes is lacking does not demonstrate its absence in nature. Such a lack may have several reasons other than the absence of accumulation. A reason may simply be the lack of research on the topic. Furthermore, documenting the accumulation of cultural variation is probably extremely difficult in the field because accumulation can only occur over years. It is thus highly improbable that anyone can actually witness that process. In the Koshima macaque example, the artificial nature of the experiment created a brand new situation that probably accelerated the process, thus allowing researchers to document it fully in only one decade (for a more comprehensive description of this example see Avital and Jablonka (2000), pp. 97–100).

### 20.3.6 Culture and population genetics

#### a *Cultural transmission of fitness in humans and whales*

**The genetic consequence of the cultural transmission of any behaviour that has an effect on fitness has been termed cultural transmission of fitness** (CTF). A well studied example of CTF is the existence of among-family variation in the number of offspring that remain in their birth population (but not in those that leave it) as well as the transmission of that component of fitness across generations. We know for instance that several human cultures living in the same country may have very different average numbers of offspring. Such differences can be maintained across many generations and are almost certainly transmitted culturally. CTF has detectable effects on population genetic structure, and may thus influence the evolution of such populations. It has been documented in several human populations (see Heyer *et al.* 2005) and in a whale population (Whitehead 1998). It helps to explain the increased carrier frequency of inherited disorders that can be explained by none of the classical population genetics mechanisms such as founder effects, selection, or drift. In this context, CTF is viewed as a third mechanism explaining the observed variation in gene frequencies across generations.

A well-studied example is that of the human population that settled the Saguenay-Lac-St-Jean (SLSJ) area in north–central Québec, Canada (Austerlitz and Heyer 1998). That case study was the first in which demographic and genetic data were studied together to understand the impact of CTF. This population of about 300,000 is descended from a few families all from the same original region of Québec (about 5,000 settlers), who themselves had come from French settlers who had established themselves in Québec in the 17th century (about 12 generations ago). The SLSJ population is known for its increased carrier frequency of several inherited disorders, which cannot be explained either by founder effect or pure drift. Demographic analyses (Austerlitz and Heyer 1998) with pedigrees on that large database revealed that, unlike the assumptions made using classical population genetics theory, the effective family size (i.e. the number of children per family who reproduce in their native population) correlates from one generation to the next. This correlation measures the intensity of CTF. A simulation study showed that an inter-generational correlation of effective family size was necessary to explain the increased frequency of inherited disorders in that population (Austerlitz and Heyer 1998). Several such examples are known in human populations (reviewed in Heyer *et al.* 2005).

Cultural transmission of fitness was also suggested in whales. A comparative study showed that matrilineal whales, in which females spend most of their lives in groups of close relatives (a typical situation in social mammals), are much less polymorphic than non-matrilineal whales (Whitehead 1998). This was suggested to be a consequence of fertility transmission in matrilineal groups (Rosenbaum *et al.* 2002). Several hypotheses have been used to explain this transmission in matrilineal whales. It could be a consequence of cultural difference, because these species show transmission of vocal dialects (Whitehead and Rendell 2004); however, it could also be the effect of environmental heterogeneity because the environment is shared by a mother and her daughters in matrilineal groups (Tiedemann and Milinkovitch 1999).

Whereas in humans CTF likely involves some real cultural transmission of behaviour, in other species such as whales an alternative explanation may be the effect of a common environment. In the latter case, the term 'cultural' in the expression 'cultural transmission of fitness' could be incorrect because it would not necessarily result from social learning. In other words, these results in whales may reveal either the effect of the social ($V_{AS}$) or the genotype–environment correlation ($V_{AE}$) additive components of phenotypic variance (Equation 20.1), or both. As we have seen, because both terms influence variance additively, they are open to evolution through selection, but only the term $V_{AS}$ which results from social influences can be qualified as cultural.

### b *Dairying and the capacity to make use of lactose*

Another example of the co-evolution between genes and culture is that of the human capacity to make use of the milk sugar lactose in some populations that use cows as a source of energy. Fresh milk contains the sugar lactose, which can be broken down into its useful components, glucose and galactose, by an enzyme called lactase-I, which all mammals are able to synthesize (Durham 1991). Young mammals usually have high concentration of that enzyme which decreases dramatically at weaning (Avital and Jablonka 2000). Milk is thus digestible only to suckling babies. As a consequence, adults usually do not really benefit from drinking fresh milk. A similar situation is found in humans, but there are some exceptions. A high proportion of adults in North European populations and nomadic African pastoralists such as the Tutsi population in the Congo Basin possess genes that enable them to make use of the lactose throughout their adult life. In other populations, among which some depend on pastoralism, lactose absorbers are much less common.

Such distributions may be explained by historical and ecological factors. The domestication of cattle led to an increase in beef eating, but also about 4000–6000 years ago, to the consumption of milk products such as cheese. Milk processing eliminates most of the lactose so that processed milk products are much more digestible than fresh milk. However, the lifestyle of the nomadic pastoralists of the Congo Basin probably made it difficult to process milk, constraining them to depend on fresh milk as an important food source. Adults with the genetic ability to digest lactose thrived and reproduced better than others, and thus their lineage increased in frequency in the population, so that a high proportion of those populations can still extract energy from milk as adults.

However, the high frequency of lactose absorbers in the sedentary populations of northern Europe cannot be explained in that way. These populations can and do process milk so that they should not need to drink fresh milk. However, the lactose is not only an excellent energy source, but also acts as a vitamin D supplement facilitating the absorption of calcium. Deficiency of vitamin D prevents the absorption of calcium which leads to rickets, the crippling softening of the bones. Usually, the synthesis of vitamin D from precursor steroids is made possible by the action of sunlight on cell metabolism. In southern populations sunlight is not limiting, whereas it is limiting in northern countries. In consequence, people living in sun-limited countries that have the genetic ability to make use of lactose have an evolutionary advantage and their lineage increased in frequency across generations. However, although interesting, the validity of such interpretations needs to be supported by replication.

This example shows how cultural transmission, here the domestication of cattle, can interact with ecological constraints and influence the changes in gene frequencies across generations. In this example, the cultural trait appears to have influenced the selection of individuals in the concerned populations by favouring individuals able to make use of the lactose contained in fresh milk. To paraphrase the expression of Lunsden and Wilson (1981): in this example it is genes not memes that appear to be on a leash.

### 20.3.7 Culture and the evolution of ornaments and speciation

#### a *A puzzle: dull species among bright species*

In 2001 John Wiens published a rather puzzling review paper (Wiens 2001). He reported that in many animal taxa, sexually selected extravagant traits could be lost through evolutionary time. His argument was derived from phylogenetic analyses. In some species, males show obvious secondary sexual characters. Such male characters may differ greatly among species. They may consist of a long bright or dull tail, flashy colours, antlers in cervids or in some insects, etc. Such males with obvious sexually selected secondary sexual characters are often called 'bright males' as opposed to 'dull males' which lack such **ornaments**. The surprise was that the use of molecular phylogenies clearly showed that in some taxa that mainly comprised species with bright males there was sometimes one species with dull males. Often the dull species are placed in the phylogeny so that the only explanation was that the ancestral bright species lost its brightness.

Such a result is puzzling because a classical theory of exaggerated male ornaments involves the Fisher **runaway process** (see Chapter 11, Sections 11.2.2 and 11.5.2). According to this process, the selection for exaggerated male traits results from female selection on males, and once females have acquired a preference for a given male ornament, the ornament can only become increasingly exaggerated. The expression 'runaway process' clearly states that this is a one-way process that may even lead to a situation when that trait involves a high cost for the male that expresses it. Wiens' (2001) observation implicitly contradicted that common view: sometimes the Fisherian process may be reversed. This is quite surprising as it raises the question of how a male with reduced ornaments can be selected for in a population with females commonly preferring exaggerated male traits?

#### b *A reformulation of the Fisherian runaway process . . .*

The Fisherian runaway process states that when females mate with their preferred kind of males, their offspring inherit both the male's exaggerated trait and the female's preference for that trait. This has led to the classic, though incorrect, statement that 'genes that code for male traits become "linked" to genes that code for female preference' (see Chapter 5 for a detailed explanation of what underlines the term 'linked' in this expression). This is because most of the time, we envisage female preference and male characters to be exclusively determined by genes. However, we have seen that things are not as simple as that. In fact, females use variation in male characters to select their mate because that trait is condition-dependent and thus reveals performance (see Chapter 4). Thus, by definition, environmental factors also influence that character.

A more suitable way of describing the Fisherian runaway process (or other processes such as the handicap principle, Zahavi and Zahavi 1997) would then be that 'when females mate with preferred males, **the transmittable components of male traits become linked to the transmittable components of female preference**'. This formulation allows the accounting of all the components of transmittability (see Section 20.2.2.4). In other words, there is no need to reduce transmittability to its sole genetic components. This formulation thus allows us to account for the fact that female preferences are strongly influenced by performance information.

### C ... *can explain the observed reversions of ornaments*

With such a reformulation of the evolution of ornaments, it becomes possible to explain how they may sometimes be reversed. Imagine a population in which females prefer bright males. In such a population, on the long run, male conspicuousness is likely to become a handicap that honestly reveals male quality (Zahavi and Zahavi 1997) because ornaments can evolve to extravagance via female preference. Whether the ornament evolved by a Fisherian, Zahavian or other processes, empirical evidence (Wiens 2001) suggests that ornaments can decrease or vanish. Such a reversal might be envisioned for instance by the appearance of a new predator that is attracted to the most conspicuous males. This may lead to the predation of all ornamented mature males in an isolated subpopulation. In this subpopulation, young females will observe that older females mate only with unornamented males. If mate preferences are acquired through mate choice copying then in that population the young females will learn to prefer unornamented males. Now, if unfavourable conditions for ornamented males remain long enough, it may start an informational cascade leading to a population in which females prefer unornamented males, thus selecting for males without ornaments. After a long isolation, this may lead females of that subpopulation to ignore the ornamented males of the main population when they contact again.

In this scenario, pre-zygotic reproductive isolation is acquired through culturally transmitted traits (here female sexual preference). With this logic, the relative instability of cultural traits over many generations may increase biodiversity. This suggests how accounting for the non-genetic component of transmittable variation may explain results that would be difficult, if not impossible, to understand under purely genetic heredity.

This scenario raises many interesting questions. For instance, models to explain the evolution of male ornaments often assume that ornaments reveal male quality (note that this is not the case at the initiation of the Fisherian process). Thus, for that scenario to function, one probably needs to assume that at some time in the process the conditions that favoured ornamented males changed. Another possibility lies in the fact that phenogenotype models showed that cultural traits can spread despite an enormous cost in genetic fitness (Cavalli-Sforza and Feldman 1981; Table 20.1). Furthermore, existing models show that at least part of this scenario is plausible in that mate choice can lead to pre-zygotic reproductive isolation (Hochberg *et al.* 2003). Hochberg *et al.* (2003) concluded that mate choice can lead to socially mediated speciation in some circumstances. Very similar results were also obtained in the more general context of green beard effects by Jansen and Van Baalen (2006). These authors showed that the conditions under which the green beard effect can be selected for encompass the characteristics of culture, suggesting that culture may play the role of a green beard.

## 20.4 Conclusion

We saw in Chapter 4 that most social information is derived from condition-dependent traits that produce performance information. In turn, performance information, whether intentional (signals) or inadvertent (ISI), reveals interactions between individual animals and environmental quality. Chapter 4 also showed that the use of ISI is much more common and influential than previously recognized.

This chapter has presented the consequences of the widespread use of performance information. We first returned to the link between biological information and phenotypic variance. In doing so, we suggested that genetic variance is not the only source of additive effects because several components of environmental variance produce parent–offspring resemblance. The concept of transmittability thus encompasses any form of information, whether genetic or environmental, that leads to parent–offspring resemblance and which is thus open to evolution. This led us to propose **an evolutionary definition of culture that is rooted in quantitative genetics**. That definition is based on four testable criteria that need to be verified simultaneously to be able to identify culture in an evolutionary context.

That definition has several important implications. First, it views **social information as the vector of cultural transmission**. Thus, **concepts of biological information** (Figure 4.3) **provide conceptual tools** for studying animal culture. Secondly, Chapter 4's review on the ubiquity of social information use in animal decision-making suggests that **cultural transmission likely exists in a wide array of organisms**. Thirdly, that definition implies that **learning is to cultural variation and evolution what DNA duplication, repairing, mutation, and selection are to genetic variation and evolution**. Some researchers have even suggested that learning may generate a new replicator (defined in Chapter 2), called memes, which functions interactively with genes (Dawkins 1976; Blackmore 1999).

The second part of this chapter briefly presented some of the potential consequences of cultural effects in evolution. Such consequences include: (1) continental patterns of behavioural variance in apes as well as bird and whale dialects and consequences in terms of the traditional use of some parts of the environment; (2) fitness consequences of social influences in the context of sexual selection; (3) the stability of accumulation of cultural variation over generations; (4) detectable effect of cultural transmission of fitness on population genetic structure; and (5) potential consequences in terms of speciation. We saw that cultural transmission may potentially impact evolutionary dynamics.

## 20.5 The future

The major goal of this chapter in conjunction with Chapter 4 has been to emphasize the importance of information use in evolution in general and that of behaviour in particular. However, we have deliberately ignored forms of non-genetic yet transmittable information, other than culture, despite the fact that other processes such as parental effects, territorial inheritance, and niche construction probably also influence evolution (Laland *et al.* 2000). Incorporating all these factors may broaden the study of evolution by considering non-genetic information as an important component. Because behaviour is the raw material of culture, it is a natural topic for behavioural ecology.

At the end of this book we might want to re-examine the field of behavioural ecology in the light of all the previous chapters. Throughout this book, we have envisaged behaviour as a decision-making process that involves information gathering and processing. Thus, behavioural ecology naturally adopts an information-driven approach to behavioural evolution. Because behaviour is a property of individual organisms, behavioural ecology is a part of evolutionary science that mainly views the individual as a unit of selection. It is the study of how individuals' behaviour affects fitness. Behavioural ecology is thus an important domain of evolution. One of its great strengths is its interdisciplinary nature, which allows behavioural ecologists to integrate such diverse fields as ethology, population biology, physiology, genetics, cognition, psychology, and neuroscience. This integration is helping to satisfy Tinbergen's (1963) plea for a synthesis of multiple approaches to the study of behaviour and more generally to the study of any trait. Information use and cultural evolution are on the cusp of joining this field. The emergence of these topics may set the stage for the next new synthesis.

## » Further reading

> *For an extensive and outstanding review on animal traditions, their cultural origin, the underlying mechanisms and their implications:*

**Avital, E. & Jablonka, E.** 2000. *Animal Traditions – Behavioural Inheritance in Evolution.* Cambridge University Press, Cambridge, UK.

> *For a review on non genetic transmission:*

**Jablonka, E., Lamb, M.J. & Avital, E.** 1998. 'Lamarckian' mechanisms in Darwinian evolution. *Trends in Ecology and Evolution* 13: 206–210.

> *For a review on the links between ISI and cultural evolution:*

**Danchin, E., Giraldeau, L.A., Valone, T.J. & Wagner, R.H.** 2004. Public information: from nosy neighbors to cultural evolution. *Science* 305: 487–491.

**Danchin, E., Giraldeau, L.A., Valone, T.J. & Wagner, R.H.** (2005). Defining the concept of public information – response. *Science* 308: 355–356.

> *For the first modelling approaches to culture:*

**Cavalli-Sforza, L.L. & Feldman, M.W.** (1981). *Cultural Transmission and Evolution: A Quantitative Approach.* Princeton University Press, Princeton, NJ.

**Cavalli-Sforza, L.L. & Feldman, M.W.** (1983). Cultural versus genetic adaptation. *Proceedings of the National Academy of Sciences of the USA* 80: 4993–4996.

**Feldman, M.W. & Cavalli-Sforza, L.L.** 1984. Cultural and biological evolutionary processes: gene–culture disequilibrium. *Proceedings of the National Academy of Sciences of the USA* 81: 1604–1607.

**Boyd, R. & Richerson, P.J.** 1988. An evolutionary model of social learning: the effects of spatial and temporal variation. In *Social Learning Psychological and Biological Perspectives* (Zentall, T.R. & Galef, B.G.J. eds), pp. 29–48. Lawrence Erlbaum Associates, Hillsdale, New Jersey, Hove and London.

**Feldman, M.W. & Zhivotovsky, L.A.** 1992. Gene–culture coevolution: toward a general theory of vertical transmission. *Proceedings of the National Academy of Sciences of the USA* 89: 11935–11938.

**Laland, K.N.** 1994. Sexual selection with a culturally transmitted mating preference. *Theoretical Population Biology* 45: 1–15.

> *Students interested in memetic review:*

**Blackmore, S.** 1999. *The Meme Machine.* Oxford University Press, Oxford.

> *About the application of population genetic concepts to the study of cultural variation:*

**Bentley, R.A., Hahn, M.W. & Shennan, S.J.** 2004. Random drift and culture change. *Proceedings of the Royal Society of London Series B* 271: 1443–1450.

> *About the importance of imitation, copying etc. in the transmission of behaviour:*

**Dugatkin, L.A.** 1999. *The Imitation Factor. Evolution beyond the Gene.* The Free Press, New York.

> *For a more human science centred review on culture:*

**Barnard, C.** 2004. *Animal behaviour. Mechanism, Development, Function and Evolution.* Pearson, Prentice Hall.

**Bikhchandani, S., Hirshleifer, D. & Welch, I.** 1992. A theory of fads, fashion, custom, and cultural changes as informational cascades. *Journal of Political Economy* 100: 992–1026.

**Bikhchandani, S., Hirshleifer, D. & Welch, I.** 1998. Learning from the behavior of others: conformity, fads, and informational cascades. *Journal of Economic Perspectives* 12: 151–170.

> *For a recent series of papers on 'Social Intelligence: From Brain to Culture' (a special issue of the journal with 20 papers):*

volume 362, number 1480, 29 April 2007 of the *Philosophical Transactions of the Royal Society of London Series B* from page 485 to page 754.

## » Questions

1. Try to imagine the various ways genetic and cultural evolution may interact. Can you imagine situations in which these two systems of information transfer across generations may be in conflict?

2. Imagine a protocol aiming at estimating the various components of transmittability. To what extent can one tell early parental effects apart from purely genetic effects?

3. Dialects are often regarded as 'by-products of learning from others'. On the other hand, as we have developed in this chapter, animal dialects are regarded as one of the behavioural traits that are most likely to reveal culture. To what extent are these two statements are contradictory? In other words, does the expression 'by-product of social learning' provides a good definition of culture?

4. There are many human centred definitions of culture. To what extent does the definition of animal culture presented here encompass the peculiar situation of human culture?

5. About the role of culture in speciation. Imagine experiments that would allow one to demonstrate that the prezygotic separation of two closely related species living in sympatry is essentially of cultural origin?

# Glossary

Each entry of the glossary is in bold and appears in a different typeface. When the definition makes use of words also described in the glossary, those words also appear in this format. When necessary, reference to the chapter where more explicit information can be found is provided.

## A

**Adaptation**: **In evolution**, can mean one of two things: A phenotypic trait that has become fixed or stable within a **population** through the process of **natural selection** (having an adaptation); or the gradual process of adjustment of a species' characteristics to the conditions of its **environment** under the effect of natural selection (the process of adaptation). **In physiology**, the gradual adjustment in physiological parameters that occurs shortly after an individual is exposed to a modified environment. For example, the metabolic rate of poikiloterms will decline when placed in a cold environment but may slowly adapt and increase to higher levels after a few hours of exposure to the lower temperature.

**Aggregation economy**: Situation where the fitness of individuals increases with density, at least initially. Fitness is maximized at some intermediate density. See also **Allee effect**, and Chapter 8.

**Aggregation**: Distribution of individuals in space or time that is different from predictions of the **ideal free distribution**, and hence does not result solely from the variation in the spatial or temporal distribution of **resources**. See **habitat mediated aggregation, real aggregation**, and Chapter 14.

**Allee effect**: Proposed by Warder Clyde Allee from the United States to depict situations where at some initial density the presence of conspecifics is beneficial.

**Allometry**: The fact that most phenotypic traits are related to the size of the bearers. Allometry might result from direct effects of physics and do not necessarily reveal **adaptation**. Divergence from allometry is more likely to reveal **adaptation**.

**Allopatric divergence**: Hypothesis suggesting that traits that lead to isolation diverge randomly within **populations** that are geographically separated either through the accumulation of mutations, contrasting selective pressures or **genetic drift**. When the two populations meet again for one reason or another mate recognition traits may have differed to a point where members of either population can no longer breed together. See **sympatric divergence** and Chapter 11.

**Altricial**: Concerns young that have long periods of parental dependency within the nest, as opposed to **precocial** young, which rapidly leave a nest.

**Altruism**: A **behaviour** that is a priori paradoxical from an evolutionary point of view because it reduces the **fitness** of the actor while increasing the fitness of one or more conspecific recipients. A behaviour that is apparently altruistic at the phenotypic level can be interpreted as selfish at the level of the genes supporting the behaviour. Alternatively, an altruistic behaviour can be maintained if it is part of direct or **indirect reciprocity** among protagonists (reciprocal altruism) See **cooperation, reciprocity**, and Chapters 2 and 15.

**Amplexus**: See **mate guarding**.

**Amplifier**: Morphological or **behaviour**al trait that makes it easier for another individual to assess one's physical condition, as opposed to **attenuator**. Chapter 16.

**Analogy**: Similarity of traits born by distinct species as a result of the action of a similar selective pressure applied to structures of differing origin (phenomenon of **convergent evolution**). The similarity of body shapes of fish and cetaceans is an example of an analogy. An equivalent term is homoplasy.

**Anisogamy**: An asymmetry in the size of gametes among the sexes such that females produce fewer large usually less mobile gametes and males produce a greater number of smaller more mobile gametes. Chapter 11.

**Antagonistic sexual conflict**: Arises in the context of **polygynous** situations where one mating partner is favoured to enhance its own reproductive success (in the context of mate competition), even if it negatively impacts the future reproductive success of its partner. This is because the partner subject to polygyny may only gain high **fitness** from the current mating, and has much less of a genetic stake (i.e., progeny) in future matings by the partner.

**Aposematism**: The association between some **prey** species bright visual **signal**s and their bad or foul taste. The colouration is thought to help predators learn to avoid brightly coloured and hence potentially toxic prey. Many species within a **community** of prey can converge towards similar looking warning signals (Mullerian **mimicry**). Other prey species in the same community can exploit the situation and adopt similar looking warning colouration while being totally edible (Batesian **mimicry**). See **mimicry** and Chapter 16.

**Armament**: Any offensive or defensive trait that has evolved by **intra-sexual selection**, as opposed to an **ornament**. See Chapter 11.

**Arms race**: antagonistic **co-evolution** characterized by the mutual escalation of investments in attack capacity in one species and defence capacity in the other. In reference to the nuclear arms race between the United Sates and the Soviet Union during the Cold War.

**Assumption**: An implicit condition under which a model or **hypothesis** is correctly applied. Should not be confused with the hypothesis. Unlike a postulate an assumption can be questioned.

**Atavism**: The fact that ancestral states that seem to have disappeared during evolution are likely to reappear later in the species evolution.

**Attenuator**: Morphological or behavioural trait that makes assessment of one's physical condition by another individual more difficult, as opposed to **amplifier**. Chapter 16.

**Audience**: One or more individuals that are witness to a social interaction without participating in it. Chapter 4.

**Auto-communication (also self communication)**: Use by the same individual of variations in emitted and received versions of its own **signal** in order to gain **information** about ambient conditions. The same individual is both the emitter and the **receiver** (e.g. echolocation and electrolocation). Chapter 16.

## B

**Badge**: Trait that reveals the social status of the individual bearing it.

**Bayesian**: In reference to Bayes' theorem that allows for the combination of prior and **current information** following a sampling event to update the probabilities associated with the values of alternatives.

**Behaviour**: There are several potential definitions (1) **Motor definition**: the motor components that determines an organism' position, movement, sound, electric field and odour emission; (2) **Functional definition**: the way organisms adjust their state to environmental changes; (3) **Biological nature**: a development which constantly modifies the phenotype. Behaviour is thus a **decision**-making process—that naturally implies **information gathering** and processing. Chapters 2, 4, and 20.

**Behavioural epistasis**: The fact that **sender**s and **receivers** interact during communication and this interaction generates a fitness effect that couples the selective fate of **sender** and receiver loci, even if the loci are unlinked. See **epistasis, genetic epistasis, physiological epistasis, fitness epistasis**, and Chapter 5.

**Behavioural syndrome**: Suites of behaviours that are correlated across situations. Also called personality.

**Benefit**: Any gain expressed either as a currency of **fitness** or fitness itself that is the consequence of a given **behaviour** or strategy. Benefits can be direct when they are expressed in the individual performing the behaviour, or indirect when they are expressed in following generations. Chapter 2.

**Bluff**: A **signal** whose intensity is exaggerated such that it does not convey honest **information**. See **deceit**, **cheating**, and Chapter 16.

**Breeding dispersal**: See **dispersal** and Chapter 10.

**Brood parasitism**: A reproductive **strategy** in which females lay eggs in the nests of non-related individuals in order to exploit their **parental investment**. Could be intra- or inter-specific. Chapter 17.

C

**Carrying capacity**: The maximum **population** size that can be supported by an **environment**. It is often denoted by the variable $k$ in population dynamic models.

**Central place foraging**: Refers to foraging **behaviour** that is organized as a series of return trips between a **resource patch** and a central place (nest, burrow, etc.). The resource is loaded and transported to the central place instead of being exploited at the point where it is captured. The model describing this behaviour is a variant of the marginal value theorem. Chapters 7 and 14.

**Chase-away sexual selection**: A type of **inter-sexual selection** characterized by an ongoing **co-evolution** in the form of an **arms race** between male **ornaments** that are preferred by females because of their sensory bias. Females are in turn selected to resist such **sensory exploitation**. At the onset a mutant male bears a trait that females prefer by virtue of some pre-existing sensory bias. The chances that the trait happens to signal male genetic quality are slim. So females that prefer such males are not choosing the best males thus creating a selective pressure favouring females that show resistance towards the trait. Increasing resistance to the trait by females then creates a selective pressure favouring exaggeration of the male ornament to overcome resistance. Chapter 11.

**Cheating**: See **deceit**.

**Choice**: Refers to selection by an individual among a series of alternative options available at a given time. This choice does not necessarily imply conscious **decision**; however, non-random choice among alternatives implies the use of some **information**. Chapters 7–9.

**Choosiness**: Tendency for a given individual to select its sexual partners on the basis of specific phenotypic attributes. Chapter 11.

**Choosy**: Is said to the sex that chooses its mates. Generally females choose male partners. Chapter 11.

**Co-evolution**: The joint evolution of two types (e.g. species, sexes within a species) through their mutual influence on one another. By extension, we may consider that two traits within the same species could have co-evolved if a change in one of the traits systematically leads to a change in the **optimal** value of the other.

**Cognitive ability**: An organism's aptitude at perceiving, acquiring, memorizing and using **information** extracted from the characteristics of its **environment**. Chapters 4 and 20.

**Coloniality**: This term has two meanings. It may first be used to describe species that reproduce on territories that contain no other **resource** but the nest and that form **aggregations** in space. This form of coloniality implies that individuals must leave the territory to gather food. It is thus an aggregate of breeding individuals that only cooperate within pairs. Examples are seabird or seal colonies. Chapter 14. Secondly, the term colony is also used to describe aggregates of cooperative individuals such as in eusocial Hymenoptera, termites or slime moulds. Chapter 15.

**Commodity**: Any one of the **resource**s required for a given activity as well as any of the factors (climatic conditions, protection against predators or against weather) that is likely to influence the success of that activity. Chapter 14.

**Commodity selection**: The approach that envisages the choice of all the commodities necessary to breed that lead animals to aggregate as a by-product. See **commodity**, **coloniality**, and Chapter 14.

**Communication**: Broadly speaking: use by an individual (**receiver**) of **information** that has been emitted (voluntarily or not) by another individual (**sender**). Strictly, (true communication), emission of a **signal** whose function is to influence the **behaviour** of a **receiver**. Communication is often pictured as an interaction between only two individuals. However, when an **audience** is present we are dealing with a **communication network**. Chapter 16.

**Communication network**: See **communication**. Chapter 4.

**Community**: An assemblage of **populations** of different species that coexist in the same place and time within an ecosystem.

**Comparative approach**: A procedure that consists of establishing the adaptive nature of a trait by looking for correlations among species between a given ecological variable and some trait. The correlation analysis must control for phylogenetic associations among the species in order to be statistically valid. Chapter 3.

**Compatibility preference**: Mate preference based on the individual's compatibility with the chooser. Chapter 12.

**Competition**: See **interference competition**, **scramble competition**, **inter-sexual selection**, **intra-sexual selection**, **local mate competition**, **local resource competition**.

**Condition-dependent handicap**: A **handicap** trait for which the intensity of expression depends on the bearer's condition. Weak males would develop only weak expressions of the condition-dependent handicap trait. Also called epistatic handicap. See **handicap principle** and Chapter 11.

**Conflict**: Interaction within which the interests of protagonists differ.

**Confusion effect**: Confusion resulting from the simultaneous use of diverse and unpredictable trajectories by a group of individuals. The confusion effect is argued to reduce a **predator**'s ability of capturing social **prey**. Chapter 14.

**Conservative**: We say that a statistical test is conservative when the measurement biases of the required parameters go against the prediction under test.

**Convergence**: See **analogy**.

**Cooperation**: Joint action that is mutually beneficial in terms of **fitness** to all protagonists. See **altruism** (if the action imposes a **cost** to the actor), **reciprocity** (if the altruistic actions are exchanged among individuals), and **mutualism** (when both actor and recipients **benefit** from the cooperation). Chapter 15.

**Cooperative breeding**: Refers to a particular social organization within which non-reproducing individuals (often called **helpers**) assist the reproduction of other members of the group, most notably by providing them with **parental care**. Chapter 15.

**Copying**: See **imitation**. Chapters 4 and 20.

**Cost**: Any decline in **fitness** (often measured in terms of **currency of fitness**) linked to a **behaviour** or a trait. (Equivalent term: disadvantage).

**Cost function**: Function that an individual should minimize through its choices to maximize its **fitness**. The adaptationist approach considers that the difference between **objective** and cost functions should be small. Chapter 7.

**Cost of reproduction**: The suite of **costs** directly associated with reproduction. These costs can be measured as a decline in survival or residual reproductive potential.

**Crossing over**: A process during meiosis whereby chromosomes of a diploid cell exchange fragments of DNA. This process insures genetic recombination.

**Cryptic female choice**: The capacity of females to control, by various internal (hence cryptic) mechanisms the fertilization probabilities of their eggs by sperm from different males. Chapter 11.

**Cue**: Any detectable fact providing **information** on the state of an individual or **environment** but that has not been designed for that purpose by **natural selection** in the context of **communication**. (as opposed to **signal**). Chapter 4.

**Cultural evolution**: See **cultural selection**.

**Cultural selection**: Cultural variations can give rise to **natural selection** because (1) there exists variation among cultures, (2) these variations can be subjected to differential selective pressures, and (3) the cultural variants are **transmittable**. We can then speak of **cultural evolution**. To evolutionary biologists, culture is another means of generating transmittable differences that can be subjected to **selection**. See **culture**, and Chapter 20.

**Cultural transmission**: The fact that cultural variation is transmitted across generations. Chapter 20.

**Culture**: For an evolutionary biologist it is the part of phenotypic variance that is transmitted across generations through social influences. It is thus the suite of a group's or a population's **information** whose transmission across generations depends on social interactions (**imprinting**, **imitation**, **copying**, **learning**, and **teaching**). A trait can be considered cultural if it meets four testable conditions simultaneously: (1) its transmission results from **social learning**, (2) its variation is transmitted across generations, (3) the resulting modification of the phenotype is generalized to any similar situation, and (4) is durable. Culture is therefore the collection of non-genetic yet **transmittable** information that participates to phenotypic variance. See **transmittability**, **heritability**, **cultural transmission**, **cultural selection**, and Chapter 20.

**Currency of fitness**: A component of **fitness** that is used to compare the selective value of alternative courses of action. The currency must be directly linked to **fitness**. Chapter 7.

**Current information**: Any **information** obtained directly from a sample. Drawing an ace of hearts from a complete deck of cards provides current information about the remaining cards in the deck.

## D

**Deceit**: Production of non-reliable **information** in order to manipulate the **behaviour** of other individuals. Occurs in various domains: **social relationships**, when one of the players uses a **strategy** whose aim is to lure a partner in order to obtain a **selfish** gain; **communication**, when the **information** emitted does not correspond to reality either because it involves (1) lying (providing false **information** conveyed to a **receiver** faced with incompatible alternatives such (e.g. emitting a false alarm)), (2) concealing information (keeping information from receivers (have I found food or not?)), (3) **Bluffing**, or (4) **attenuators**. Chapter 16.

**Decision**: The non-random process leading to the adoption of one of the alternative options available to an individual in a **choice** situation. Decision does not necessarily imply a conscious process; however, it implies the use of **information** concerning the alternatives. Decision theory explores the cognitive factors that can influence an individual's decision.

**Demographic stochasticity**: Random fluctuation of the level (or demographic parameters) of a population. Chapters 5 and 18.

**Density dependence**: When phenotypic **fitness** varies with **population** density. The relationship can be positive, negative of bell-shaped: fitness first increases with density, peaks and beyond some threshold declines with increasing population density. The latter case is known as the **Allee effect**. Reverse density-dependence refers to situations where a reduction in population density has a deleterious effect on **population growth rate**.

**Dichromatic species**: Is said of species where male and female individuals are of different colouration. By extension applies to species where males and females differ by one (or many) morphological trait(s) (length and shape of feathers, hair, etc.). As opposed to **monochromatic species** where the sexes are of similar appearance.

Can more generally be part of sexual dimorphism versus sexual monomorphism.

**Dilution**: Reduction of the probability of falling victim to a predator by virtue of the presence of alternative potential victims. The probability of being the victim within a group of $n$ individuals is $1/n$. Chapter 14.

**Direct or indirect benefit**: **In sexual selection**. Direct: **benefit** in terms of viability of the progeny that comes from the quality of the mate or parent. Indirect: benefit that comes from the genetic quality of the mate that is **heritable** and will be effective only in the next generation. **In kin selection**. Direct: Part of **inclusive fitness** that consists only of the individual's own reproductive output without any consideration of its social **environment**. Indirect: the contribution of an individual's social environment to its own **inclusive fitness**. Chapters 11 and 12.

**Directional preference**: During mate **choice**, a systematic preference in one direction for a given type of variant among the many available (e.g. preference for the largest, tallest, oldest, etc.).

**Directional selection**: a type of selection that favours only one of the extreme values of the trait. It has no specific effect on trait variance. See **diversifying selection**, **stabilizing selection** and Chapter 2.

**Disinformation**: See **deceit**.

**Dispersal**: The movement of individuals from their place of birth to their first place of reproduction (natal dispersal) or between successive reproductions (breeding dispersal). Chapter 10.

**Dispersion economy**: A case where in the absence of **benefits**, **costs** increase with increasing conspecific density. **Fitness** is maximized when the density is minimized. Chapter 8.

**Disposition**: Potential of accomplishing an act without necessarily doing it. See **performance**.

**Diversifying selection**: A type of selection that favours extreme phenotypes and so tends to increase the variance of a trait's distribution in a **population**. See **directional selection**, **stabilizing selection** and Chapter 2.

## E

**Eavesdropping**: In the context of **communication**, the **behaviour** of a **receiver** that extracts **information** from the **signals** or **cues** resulting from an interaction in which it does not take part. Chapter 4.

**Economic approach**: Approach that consists of analysing the adaptive value of a trait through a **cost benefit** analysis in terms of its fitness. Most commonly in behavioural ecology these involve **optimization** and **game theory**.

**Effective population size**: The number of individuals in a **population** with efficient reproduction.

**Emancipation**: Developmental phase of the young characterized by the end of **parental care**.

**Emergence**: See **emergent property**.

**Emergent property**: Property of a whole that cannot be deduced from knowledge of its individual components.

**Emitter**: See **sender**.

**Empiricism**: An approach that considers knowledge as arising from observation and **experimentation** as opposed to the theoretical approach.

**Encounter rate**: In foraging theory it is the number of encountered prey of a given type divided by the time spent in search. Chapter 7.

**Energy budget rule**: In the context of **foraging** theory this rule states that animals should be **risk-prone** when at the end of some given **time horizon** they expect an energetic deficit but **risk-averse** when instead at the end of the time horizon they expect an energetic surplus. Chapter 7.

**Environment**: The term environment comprises all the ecological components that surround an individual animal. It thus includes the **habitat** as well as the social component of an individuals' ecology. See **habitat**, **habitat suitability**, and **habitat quality**.

**Epistasis**: Qualifies the fact that the expression of a trait involves the interactions among many genes (or as noted below many interacting physiological or behavioural pathways). See **pleiotropy**, **physiological epistasis**, **behavioural epistasis**, **fitness epistasis**, and Chapter 5.

**Eusociality**: A type of social organization that is common in social hymenoptera and termites. It is characterized by the existence of sterile casts that assist in the reproduction of other members of the social group. Chapter 15.

**Evolutionarily stable strategy (ESS)**: A **strategy** that, once it has arisen in a **population**, cannot be replaced by any other alternative and hence corresponds to **evolutionary stability**. See **game theory**. Chapter 2.

**Evolutionary stability**: State of a **population** that can no longer evolve. Any perturbation away from this point gives rise to selective forces that move it back towards that point.

**Excludability**: The fact that some **information** may remain **private** (excludable) or may be detected by other individuals (non-excludable), and then become part of **public information**. The concept of excludability *vs* non-excludability entirely overlap those of private versus public information. See **information**, **public information**, **private information**, **personal information**, and Chapters 4 and 14.

**Experiment**: Practical test involving the manipulation of a single component in order to examine its influence in a phenomenon. All experiments have a control that serves as a reference against which to compare the effect of the manipulation. Unlike a correlation approach, experimentation allows one to conclude that an observed effect is the consequence of the manipulated component.

**Experimental design**: It is a plan that sets out the organization of an **experiment** or the collection of field data as well as its statistical analysis.

**Exploitation**: **With respect to resources**: The act of using a resource. **With respect to others**: an act that takes advantage of the effort invested by another. It can refer to a social relationship where one individual takes advantage of the efforts of others such as in **producer–scrounger** relationships. Chapter 8. **With respect to communication**: The use of **cues** or **signals** by an outsider at the expense of the actors. See **eavesdropping** and Chapter 16.

**Exploitation function**: Function depicting the cumulative gain in **resource** within a **patch** as a function of the time spent in the patch exploiting it. Chapter 7.

**Extended phenotype**: Manifestation of an organism's phenotype outside its own body. Hence termite mounds, bird nests and spider webs are all extended phenotypes. In a host-parasite relationship, the parasite's genotype is often capable of extending its expression into the host's phenotype through various physiological alterations.

**Extra-pair copulation (EPC)**: In socially **monogamous** species, the mating of an individual of one sex with any other individual of the opposite sex that is not its social partner. EPCs may translate into extra-pair paternity (EPP). Chapters 11 and 12.

**Extra-pair paternity (EPP)**: See **extra-pair copulation.**

## F

**Feedback**: Refers to the consequence of an action on its cause. The effect can be either positive or negative.

**Fitness** (or phenotypic fitness, or individual fitness): The average capacity of a phenotype of producing mature offspring relative to other individuals in the same population at the same time. See also **Genotypic fitness**. Chapter 2.

**Fitness epistasis**: The fact that nonlinearity among traits and **fitness** may be so extreme that they create alternative optima on a fitness landscape. See **diversifying selection**, **epistasis**, **physiological epistasis** and **genetic epistasis**. Chapter 5.

**Fitness return**: Increase in phenotypic **fitness** as a consequence of investing in some activity, also called pay-off.

**Fixed handicap**: Case of the **handicap principle** where the full expression of a handicap trait is coded by a single allele. All individuals bearing this allele will express the full handicap trait. Weaker individuals are more penalized for expressing the trait than stronger ones. Selection acts to increase the fraction of individuals exhibiting the highest viability among those bearing the handicap but not within those without the handicap. By choosing to mate with individuals bearing a handicap individuals often obtain '**good genes**' for their offspring. There is therefore no correlation between the degree of expression of a handicap trait and an individual's quality. See also **handicap principle**. Chapter 11.

**Foraging**: The suite of activities involved in the exploitation of a **resource**. The term is mostly used for food resources but can be used for other resources such as nesting material, mates, etc. Chapters 7 and 8.

**Frequency-dependence**: Characterizes a **strategy** whose payoffs depend on the proportion of strategists in the **population**. Frequency-dependence is positive when payoffs increase with frequency and negative when payoffs decline with increasing frequency. Chapter 2.

## G

**Game theory**: A theoretical approach developed in economics to analyse the best course of action in a social interaction. In behavioural ecology evolutionary game theory can be distinguished from **simple optimality** in two ways: (1) the solution is based on **evolutionary stability** rather than **fitness** maximization; (2) the gains of a **strategy** depend on the frequency of the strategy in the population. See **optimization** and **evolutionarily stable strategy**.

**Generation time**: Average time that separates the birth of a given generation to that of the following generation. Can be estimated as the average age of females at the time they give birth. See **Life cycle**. Chapter 5.

**Genetic drift**: Change in the frequency of genes due to stochastic effects related to sampling error. Chapter 2.

**Genetic epistasis**: When a gene interacts with another gene, perhaps shutting off that gene's expression or amplifying its gene products in a highly nonlinear way. See **epistasis**, **physiological**, **behavioural**, and **fitness epistasis**. Chapter 5.

**Genetic information:** The information transmitted across generation by genes. See **information**, **transmittability**, and Chapters 4 and 20.

**Genetic monogamy**: A **mating system** in which a single male and female are genetically involved in the production of offspring in one or more reproductive episodes. This mating system implies absolute and total sexual fidelity of the mates. Chapter 12.

**Genome**: The collection of genetic **information** carried by an individual. Thus, apart from identical twins, two individuals are extremely unlikely to have the same genome. In a more restricted sense also the collection of genes of any functional unit (e.g. mitochondrial genome, genome of a polyploid individual that is inherited from an ancestral species, etc.). See **genotype** and Chapter 2.

**Genomic imprinting**: Reflects a situation in which genetic transmission of traits to progeny is expressed in a parent-specific pattern, where one parent's genetic contribution is silenced while the other parent's contribution is intact. Chapter 5.

**Genotype**: As defined by population geneticists, the allelic composition of the gene locus or loci studied in an individual. See **genome** and Chapter 2.

**Genotypic fitness**: Absolute or relative measure of a genotype's success within a given population estimated from the difference in its frequency between two generation. See **fitness**, **genotype**, and Chapter 2.

**Good genes hypothesis**: A group of theories according to which females can choose males according to their genetic quality. See **handicap** and Chapter 11.

## H

**Habitat**: A type of place that provides more or less favourable characteristics for an organisms' activity (reproduction, foraging, mating, shelter, etc.). Usually the concept of habitat does not include the social component of the **environment**. See **environment** and Chapter 9.

**Habitat copying**: A **strategy** of breeding habitat choice in which **prospecting** individuals use the breeding performance of conspecifics as a source of **information** about breeding habitat quality, and thus copy the choice of their **successful** conspecifics. By extension, the term habitat copying also encompasses the fact that such a habitat **choice** strategy leads to **aggregated** nesting distributions and hence to **colonial** breeding. See **inadvertent social information**, **performance information**, **coloniality**, and Chapters 9 and 14.

**Habitat matching effect**: The fact that a disperser with a particular **phenotype** chooses a **patch** whose quality matches the quality of its own phenotype, i.e. where the **disperser** will have a high **fitness**. See **silver spoon effect** and Chapter 5.

**Habitat-mediated aggregation**: In highly heterogeneous **environment**, even if animals are distributed in an **ideal free way**, they may appear aggregated because most individuals are in the few best **patches**. However, as such distribution simply mirrors the variation of **habitat quality** this is not **real aggregation**. See **aggregation**, **real aggregation**, and Chapter 14.

**Habitat patch**: See **habitat**.

**Habitat quality**: See **habitat**, **habitat suitability** and Chapter 9.

**Habitat suitability**: Describes the quality of an environment made up of its **intrinsic quality** (the **fitness** returns afforded by the habitat in the absence of any other companions) and the realized quality that takes into account the effects of the social environment which can increase (**Allee effects**) or decreased (competition) expected **fitness**. See **habitat quality**, **environment**, and **ideal free distribution**.

**Handicap principle**: A principle meant to explain in a context of **communication** the evolution of extravagant traits that are *a priori* costly to the bearer (such as the peacock's tail). The full expression of the trait can only occur in perfectly healthy and fit individuals that are the only ones capable of bearing the trait's full **cost**. The traits are thus **honest** signals of individual condition. The high costs guarantee the honesty of the **signal**. See also **condition-dependent handicap**, **fixed handicap** and Chapters 11 and 16.

**Handicap**: See **handicap principle**.

**Harem**: A particular case of **polygyny** where a male instead of defending a territory monopolizes a group of females that he defends against intrusion attempts by other males. Males and their harems are often associated with a nomadic lifestyle. Chapter 12.

**Helper**: Non-reproducing individuals (or individuals whose reproduction failed) that contribute to the reproduction of a pair of conspecifics by helping them raise their young. The term applies mostly to **cooperatively breeding** vertebrates (birds and mammals). Chapter 15.

**Heritability**: It is the genetic component of the variation that is **transmitted** to offspring and that is thus open to evolution through **natural selection** and **genetic drift**. It is the part of the evolutionary potential of a population that involves genetic variation. Measuring heritability implies controlling for any environmental influences. It is, however, almost impossible to remove all non-genetic additive effects, so that heritability often remains so difficult to measure that evolutionary biologists talk of 'broad sense heritability' to specify that true heritability values are impossible to measure properly. See **transmittability**, **natural selection**, **evolution**, **cultural transmission**, and Chapters 2 and 20.

**Hidden lek (hypothesis)**: The hypothesis that the discrepancy between the social and genetic mating systems of many monogamous species can lead to the **real aggregation** of nesting territories. In some species such as razorbills (*Alca torda*) **extra-pair copulations (EPCs)** occur mainly in mating arenas that are separated from the nesting area. Such arenas function in ways that are very similar to **leks**. However, in most species the nesting and display territories are confounded and the lek is in fact hidden behind **social**

**monogamy**. In such species, the mechanisms that explain the aggregation of display territories in leks can explain the aggregation of nesting territories in **colonies**. See **coloniality** and Chapters 12 and 14.

**Homogamy**: Describes within a **population** mates whose **phenotype**s resemble each other more (positive homogamy) or less (negative homogamy) than would be expected from random assortment. Chapter 2.

**Homology**: Refers to the type of resemblance between traits of two or more species that can be attributed to phylogeny. For instance, bird wings, fish pectoral fins and tetrapod forelegs are homologous structures (opposed to analogous). See **analogy**, **homoplasy**, and **convergence**.

**Homoplasy**: see **analogy**.

**Honest**: Qualifies a **signal** that provides reliable **information** concerning an individual's state, intentions, status or quality. See **deceit**, **handicap principle**, and Chapter 16.

**Horizontal transmission**: As opposed to **vertical transmission**. **In cultural transmission**: a **behaviour** transmitted among individuals of the same generation. **In parasitology**: parasite transmission from host to host by: (1) contact between two individual hosts, (2) consumption of an infected individual, (3) an insect vector, or (4) the emission of parasite (sexually or asexually produced) propagules outside the host and that can enter a new host. Chapter 20.

**Hormone**: A chemical substance secreted into the bloodstream by an endocrine gland which, once in the bloodstream, has an effect on target cells. The action of the hormone on a target cell implies the existence of a specific hormone receptor. Chapter 6.

**Hybridization**: A cross between two distinct species.

**Hypothesis**: Proposition that results from observation or logical deduction whose validity is submitted to an **experiment**. The hypothesis rests on a certain number of **assumptions**.

**Hypothetico-deductive approach**: A scientific approach that consists of deducing a **hypothesis** from first principles or observations and then subjecting its predictions to **empirical** testing. Lack of congruence between observations and predictions lead to rejection of the deductive logic used to generate the hypothesis. Chapter 3.

**Ideal despotic distribution**: See **ideal free distribution**.

**Ideal free distribution**: A model of the equilibrium distribution of individuals of a **population** over areas that differ in **habitat suitability**. The baseline model assumes that all individuals are equal competitors, have perfect **information** concerning the quality of alternatives (ideal individuals) and are capable of moving to each **habitat** without **cost** (free individuals). At equilibrium all individuals obtain the same mean **fitness** whichever habitat they exploit. There are different variations of the model. For instance, the **ideal despotic distribution** applies to individuals of different competitive abilities in which one or more despots impose entry costs to the habitat patches they exploit. Chapters 8, 9, 10, and 14.

**Identical by descent**: Two genes can be identical for a number of historical reasons. They are identical by descent when they are copies of the same recent ancestral gene (e.g. two brothers have on average 50% chance of sharing genes identical by descent because these genes are copies of their mother's or father's genes). The extensive genetic similarity reported, say between chimpanzees and humans, is not from recent ancestors and so cannot be considered to be identical by descent. Chapter 2, 13, and 15.

**Idiosyncrasy**: Patterns of response that differ between individuals but are repeatable within individuals or clones. By extension, patterns that are taxon-specific.

**Imitation**: Action that reproduces the **behaviour** or the **choice** (**habitat**, mate) of another individual. Can be part of **learning**. Also called **copying**. See Chapters 4 and 20.

**Immunocompetence**: The collection of capacities of an individual's immune system that allows it to resist pathogens. In the absence of further precision it is a generic term that covers all components of the immune system.

**Imprinting**: Often called behavioural imprinting. A form of **learning** that is generally limited to a sensitive period in which an individual acquires a more or less irreversible representation of either its parent (filial imprinting) or sexual partner (sexual imprinting). Imprinting is similar in some respects to song learning, which is also limited in time to sensitive periods. Chapter 20.

**Inadvertent social information**: A form of **social information** based on **social cues** produced inadvertently by individuals engaged in efficient performance of their activities. See **social information**, **information**, **cue**, **signal**, Figure 4.3 and Chapter 4.

**Inbreeding**: Reproduction among directly related individuals as opposed to outbreeding.

**Inclusive fitness**: Refers to the **fitness** of an individual taking into account the consequences of its actions on both its own reproduction and that of individuals to whom it is genetically related. It can be measured by the number of mature offspring produced directly increased or decreased by effects from its social **environment** (direct component), added to the positive and negative effects of its own **behaviour** on its social environment (indirect component) weighted by the coefficient of genetic relatedness with members of its social environment. Chapter 2.

**Indirect reciprocity**: An extension of the social **prestige** theory to situations where there are no differences in quality among the members of a population. The reciprocity is indirect inasmuch as individuals that are the target of an altruistic act are not necessarily the ones that will be altruistic towards their benefactor. See **prestige**, **reciprocity**, and Chapter 15.

**Information**: In its broadest sense any cause of phenotypic variation ($V_P$). In a more behaviourally oriented perspective, information is genes and detectable facts that reduce uncertainty, potentially allowing a more adaptive response. The term 'detectable' means anything that organisms can sense, which includes chemical recognition in plants and microorganisms, i.e. also organisms without brains. Non-genetic information is thus information that can be extracted from detectable facts. The value of information resides in its ability to increase an individual's **fitness** when faced with alternative courses of action. See **personal information**, **social information**, **cue**, **signals**, and Chapter 4.

**Information gathering**: The process of collecting **information**. This expression does not necessarily imply some active information seeking or transfer. See **information**, **excludability**, and Chapter 14.

**Information sharing**: When **information** is shared within a group then all its members always have the same information at all times. See **information**, **information gathering**, **inadvertent social information**, **commodity selection**, and Chapter 14.

**Information transfer**: See **information gathering**.

**Information value**: The extent to which **information** allows accurate prediction of future events or states. See **information**.

**Intensity of selection**: Within a population it is the difference in the mean values of a trait before and after selection, divided by the standard deviation of the trait's distribution prior to selection. Chapter 2.

**Intentionality**: Concerns the **adaptive** function of behaviour: intent means that that **behaviour** has evolved for that purpose and no other and not the performer's intent. Thus, intentionality implies that the behaviour provides a **fitness benefit**, directly or indirectly to the performer. Signals are intentional patterns as they are the product of a history of selection for them to convey information. See **communication** and Chapter 16.

**Interest**: In reference to **benefits** in terms of **fitness**.

**Interference competition**: Also called contest competition. Reduction in **resource** use resulting solely from the antagonistic interactions among competitors. See **scramble competition**.

**Inter-sexual selection**: Selection of a trait that is advantageous in the indirect competition by members of one sex for access to gametes of the other sex. The competition is indirect because selection depends on the **choices** that members of one sex make among individuals of the other sex bearing the trait. See **ornament** and Chapter 11.

**Intra-sexual selection**: Selection in favour of traits that are advantageous in the direct competition among individuals of one sex to gain access to the gametes of the other sex. This competition can occur by **interference** (see **armament**) or by exploitation. Chapter 11.

**Intrinsic habitat quality**: (or realized or simply habitat quality). The intrinsic quality of a **habitat** that results from its physical and biological characteristics. Habitat quality does not comprise the social component of the **environment**. Habitat quality for a given species is measured by the **fitness returns** afforded by the habitat in the absence of any other companions. See **habitat**, **habitat suitability**, **ideal free distribution** and Chapter 9.

**Intrinsic rate of increase**: Usually denoted by the symbol $r$. The intrinsic capacity of a given population (or a **phenotype** within a population) to grow in number over time. The intrinsic rate of increase is linked to the **population growth rate** ($\lambda$) as $r = \log(\lambda)$. Chapter 5.

**Intrinsic value or basic suitability**: In **ideal free distribution** refers to the value of a habitat before it is colonized by any single individual.

**Isogamy**: The production of gametes of similar size by both sexes. Chapter 11.

**Iteroparous**: A species whose individuals reproduce repeatedly during their lifetimes. As opposed to **semelparous**.

**Kin selection**: A type of selection that operates when the consequences of **behaviour** affect individuals with whom one is genetically related by descent. See also **inclusive fitness** and Chapter 2.

L

**Learning**: Process through which an individual's **phenotype** (including **behaviour**) is modified through the effect of acquired **information** whose memory is stored in the nervous system. This learning can be the result of **imitation**, **copying**, **imprinting**, or **teaching**. See also **culture**, **social learning**, and Chapter 20.

**Lek paradox**: The theory behind the origin of a **lek** is that it offers a place where females can select the highest quality mates among a collection of males that vary in genetic quality. Yet, if a male's mating success depends on its genetic quality, **good genes** will increase in the male population over generations until they become fixed. At that point all males have the same genetic quality and hence the selective pressures to maintain a lek will wane. The solution to this paradox is to be sought in the mechanisms responsible for maintaining genetic heterogeneity among males. (Also referred to as the **good genes** paradox.) Chapter 11.

**Lek**: A spatial **aggregation** of male courtship territories (generally these territories contain no other **resources** of interest to females). The word refers both the place as well as the corresponding **mating system**. Chapter 12.

**Lie**: see **deceit**.

**Life cycle**: The description of the way organisms are born, grow, reproduce and die. See **life history strategy**, **life history trait**, and Chapter 5.

**Life history strategy**: The set of parameters that describe the life cycle of a species or a group of individuals in a population. Each parameter is a **life history trait**. See **life cycle**, **life history trait**, and Chapter 5.

**Life history trait (or parameter)**: A component of a species' or a **phenotype**'s bio-demographic characteristics such as age at first reproduction, survival, fecundity, etc. See **life cycle**, **life history strategy**, and Chapter 5.

**Lifetime reproductive success (LRS)**: The number of descendants produced by an individual over its lifetime.

**Limiting resource (or factor)**: The specific **resource** whose abundance prevents a **population** from growing further. See **carrying capacity**.

**Linkage disequilibrium**: In a population the non-random association of alleles from different loci.

**Local mate competition**: Competition among related individuals for access to mates. More commonly it involves competition among brothers for access to females.

**Local mate enhancement**: **Cooperation** between brothers for access to females.

**Local resource competition**: Competition among related individuals for access to **resources**. More commonly it involves sisters competing for access to resources required for reproduction.

**Local resource enhancement**: **Cooperation** among related individuals for gaining access to **resources**.

**Long-lasting interaction**: Any long-lasting interaction between individuals of different species. Parasitism, mutualism, and **symbiosis** are examples of long-lasting interactions. Chapter 17.

## M

**Major evolutionary transition**: A stage in the history of life corresponding to the emergence of a higher level of complexity. For instance, the transition from unicellular to multicultural forms of life is one of the major evolutionary transitions.

**Manipulation**: Change in a host's **phenotype** induced by its parasite and that leads to an increase in the parasite's **fitness** at the expense of the host's. Chapter 17.

**Mate choice copying** (or simply mate copying): When a naive individuals chooses the same mate that it witnessed being chosen by previous individuals. See also **public information** and Chapters 4 and 20.

**Mate guarding**: Tendency for individuals of one sex (usually males) of maintaining close proximity to their reproductive partner, especially during periods of fertility. In cases of pre-copulatory mate guarding or **amplexus**, males tend to remain close or attached to females up to copulation and hence guarding ceases. In cases of post-copulatory mate guarding the male stays with the female for some time after copulation which reduces a female's ability to re-mate and hence limits the risk of **sperm competition**.

**Maternal effect**: The influence of the maternal **phenotype** on the offspring's phenotype that is independent from genetic similarity. The expression **parental effect** is used to encompass both maternal and paternal effects. Chapters 4 and 5.

**Mating arena**: Refers to a place where males and females meet to mate. See also **lek**, **hidden lek**, and Chapter 14.

**Mating system**: Refers to the way in which individuals of a species have access to their reproductive partners, the number of reproductive partners with whom they interact with within a reproductive season, the length of the association between mates and the relative involvement of the sexes in **parental care**. Examples are **monogamy**, **polygyny**, **polyandry**, **lek**, and **promiscuity**. Chapter 12.

**Meme**: A meme is to **culture** what a gene is to the **genotype**. A meme is an abstract concept that is meant to represent a unit of cultural information. Organisms acquire memes through **learning** the sum of which constitutes the individual's total knowledge. The part of that knowledge that is transmitted to other individuals is **culture**. See also **culture** and Chapter 20.

**Metapopulation**: Collection of **populations** that are more or less connected by **dispersal** of which some are subject to recurrent extinction and recolonization by propagules from other populations of the metapopulation. See **source–sink** and Chapter 10.

**Migration**: Back and forth movement of animals from their reproductive and wintering grounds. In population genetics refers to the movement of genes among populations within a **metapopulation**. Chapter 10.

**Mimicry**: Evolutionary convergence towards greater resemblance between (1) an individual and the substrate on which it lives or (2) an individual and members of another species. **Aggressive mimicry**: resemblance between a **prey** and a dangerous species that can frighten **predators**. **Mullerian mimicry**: converging morphological resemblance of a community of phylogenetically distinct venomous or toxic species. See **aposematism**. **Batesian mimicry**: when a non-toxic or venomous species evolves resemblance towards a toxic or venomous species with whom it is sympatric. Chapter 16.

**Monochromatic species**: See **dichromatic species**.

**Monogamy**: See **genetic monogamy** and **social monogamy**.

**Motivation**: Inclination to exhibit a **behaviour** under the influence of internal mechanisms (physiology, **hormones**, internal clock, etc.) and external releasers (climatic event, light intensity, presence of food or a mate). Motivation essentially varies according to the fitness consequences of the concerned behaviour. For instance, a hungry individual has a higher motivation to forage than a satiated one.

**Mutualism**: Type of **long-lasting interaction** that affords **benefits** to all protagonists involved. See **symbiosis** and Chapter 17.

## N

**Nash equilibrium**: Within **game theory** an equilibrium situation characterized by a state where any unilateral change of **strategy** by a player reduces its gains. Chapter 3.

**Natural selection**: Selection that occurs over generations among phenotypic variants on the basis of their survival and ability to produce fecund descendants. Natural selection requires three conditions: (1) variation in a trait, (2) concomitant variation in the trait's phenotypic fitness, and (3) **heritability** (or more generally **transmittability**) of the trait. Natural selection leads to **adaptations**. There are two subcategories of natural selection: **utilitarian selection** and **sexual selection**. See also **transmittability**.

**Negative genetic correlation**: Reflects a genetic **trade-off** between two traits that both affect **fitness** in a positive way, but the traits are negatively related to each other. Chapter 5.

**Neo-Darwinism**: Conceptual synthesis started in the middle of the 20th century that incorporated modern genetics, development, palaeontology and systematics within Darwin's original formulation of his theory. It is also referred to as the New Synthesis.

**Non-genetic information**: Any form of **information** that is not acquired through genes. See Figure 4.3 and Chapter 4.

**Norm of reaction**: Or reaction norm. A collection of phenotypes that can be obtained from a single **genotype** in a host of **environments**. The extent of a norm of reaction illustrates **phenotypic plasticity**. Chapter 5.

## O

**Objective function**: The function that is maximized through an individual's **choice**.

**Oestrus**: Female mammals' period of receptivity and fertility.

**Ontogeny**: The collection of processes involved in the development, maturation, and organization of an organism including its **behaviour** from conception to death.

**Operational sex ratio**: The effective **sex ratio** at the time of mating, which is the number of males divided by the number of females that are actually available for mating at a given time. Chapters 11 and 12.

**Optimal**: Qualifies a behavioural option that is the most desirable and hence expected by virtue of its maximization of **fitness**.

**Optimization**: An approach that consist of assuming that **phenotypic** traits tend, under the effect of **natural selection**, to take on values that maximize **fitness**. This approach generates predictions that can then be tested by **experiment**. See **simple optimality** and **game theory**.

**Ornament**: A secondary sexual character that has evolved by **inter-sexual selection**. As opposed to an **armament**. Chapter 11.

P

**Paedomorphism**: The fact that some species have lost the adult stage, with reproduction occurring in individuals that retained characteristics of juveniles. Also called neoteny. Chapter 5.

**Panmixy**: Qualifies a population in which matings are equally likely among any possible pair of adults of different sexes. Opposed to **viscous** populations in which there is some spatial structure within the population so that matings are more likely with close neighbours.

**Paradox of group size**: When animals are free to join groups in which they can obtain **fitness benefits**, group size will increase as long at it offers benefits over solitary living. Groups therefore grow until they offer no greater fitness than solitary living. See **Allee effect**.

**Parasite**: Species whose existence depends on its association with another species that suffers in return a reduction in viability. The greater part of the parasite's life cycle occurs in close relationship with a host that provides it with shelter and nourishment. We distinguish endoparasites that live inside their host from ectoparasites that live on the surface of their host's body. See **long-lasting interaction** and Chapter 17.

**Parasitoid**: Type of parasitic insect whose larval stage develops inside a host using it as a **resource**. Emergence usually leads to host death.

**Parental care**: Any parental **behaviour** whose function is to increase its offspring's' **fitness**. Parental care includes nest building, production of nutritious substances in the egg, care for the eggs and young, both inside and outside the parent's body, feeding the young before and after birth, and any assistance provided to the young once they have acquired nutritional independence. See **parental investment** and Chapter 12.

**Parental effect**: See **maternal effect**.

**Parental expense**: The share of parental **resources** (in the form of time or energy) that is invested in care devoted to offspring. Relative **parental care** refers to the fraction of resources allocated to parental care.

**Parental investment**: It is defined from the consequences of **parental care** on the **fitness** of the parents. The expense can affect their survival and reproduction from the short- to the long-term. It consists of any **parental expense** that **benefits** the immediate progeny at the expense of the parents' future reproduction (Trivers, R.L. 1972. Parental investment and sexual selection. In *Sexual selection and the descent of man*, 1871–1971 (B. Campbell ed.), pp. 136–179, Aldine Transaction, Chicago, IL). The investment can be measured as **costs** of reproduction. Also called **parental care** or parental expenditure. See **mating systems** and Chapter 12.

**Parent–offspring conflict**: Conflict involving the differing Darwinian interests of parents and their offspring. Chapter 12.

**Partial preference**: In the context of the **prey** model of **optimal** foraging theory it describes the instance where the same prey type is sometimes rejected and sometimes attacked when encountered. Chapter 7.

**Patch**: An homogenous **resource** containing area (or part of **habitat**) separated from others by areas containing little or no resources (or suitable habitat). A patchy distribution implies a heterogeneous distribution where resources are located within patches. See **environment**.

**Pattern**: The overall appearance (or distribution) of a phenomenon. A pattern is generated by processes but it is difficult perhaps even impossible to infer the process from the pattern given that several different processes can produce the same pattern.

**Performance**: Realized level of an activity. Ultimately measured in terms of **fitness** components. See **disposition**. Chapter 4.

**Performance information**: The **information** that can be derived from the performance of other individuals having similar ecological requirements. Encompasses both signals and cues that being both condition dependent reveal performance. See **Inadvertent social information, social information, personal information, social cue, signal, communication**, Figure 4.3, and Chapter 4.

**Personal information**: The **information** obtained by an individual from its own interaction with the environment. See **private information, social information, inadvertent social information, public information**, Figure 4.3, and Chapter 4.

**Phenotype**: The set of characteristics of an organism that results from the interaction between its **genotype** and the **environment** within which it has developed. As all individuals differ by at least one character two individuals cannot share the same phenotype. Therefore, in the narrow sense the term phenotype refers to a subset of the characteristics of an organism, usually the one under study. See **extended phenotype, genotype**, and Chapter 2.

**Phenotypic engineering**: **Phenotypic** manipulations allowing us to evaluate the **utility** of traits by modifying them and comparing the **performance** of modified individuals with those individuals that were not manipulated. This approach demonstrates the current **utility** of a trait. Chapters 5 and 6.

**Phenotypic fitness**: See **fitness**.

**Phenotypic plasticity**: Capacity of a given genotype of producing different **phenotypes** according to the **environment** in which it is expressed. See **norm of reaction**, and Chapters 2 and 5.

**Pheromone**: Molecules secreted outside the body by exocrine glands that facilitate **communication** among individuals. Those detected between species are allochemical that comprise allomones and allohormones that **benefit** the **sender** and kairomones that benefit the **receiver**.

**Physiological epistasis**: The fact that the action of a given **hormone** depends on levels of other hormones in a **sender** and **receiver** fashion involving hormones. See **epistasis, genetic epistasis, behavioural epistasis, fitness epistasis**, and Chapter 5.

**Pigment**: A molecule that, by absorbing certain wavelengths of the visible spectrum while transmitting others is responsible for an organism's colouration. Chapter 16.

**Playback experiment**: A type of experiment where animals are exposed to audio recordings of auditory signals. Could perhaps be extended to video playback situations.

**Pleiotropy**: The capacity of a single gene to influence different functionally unrelated phenotypic traits. Chapter 5.

**Polyandry**: **Mating system** consisting of the association of one female with several males during the same mating season. In most cases the males provide a large part of the **parental care**. Chapter 12.

**Polygenic**: A phenotypic trait whose expression depends on several genes.

**Polygyny**: **Mating system**: characterized by the association of a single male with several females while each female reproduces only with a single male. During the mating season the male may associate with several females simultaneously (simultaneous polygyny) or sequentially (sequential polygyny). In polygynous systems the larger part of **parental care** is generally afforded by females. Chapter 12.

**Population**: A collection of individuals that are more likely to reproduce with each other than with any other member of their species.

**Population growth rate**: Usually denoted by the symbol $\lambda$. See **intrinsic rate of increase**. Chapter 5.

**Potential reproductive rate**: The maximum number of independent offspring that a parent can produce per unit time. This variable directly determines which sex becomes limiting and is thus selecting the other. Chapter 11.

**Precocial**: See **altricial**.

**Predator**: A functional name given to any organism that consumes others whether animal (carnivore) or plant (herbivore).

**Predictability**: A quantity is predictable over time (or space) if its measurement at one instant (place) allows a prediction, within a certain margin of error, of its value $n$ time steps later (or at distance $n$). Predictability is measured by the coefficient of auto-correlation that varies from –1 to 1.

**Prediction**: Result expected in specific circumstances from a **hypothesis** or theory. A prediction is valuable inasmuch as it can be tested, that is one can show it to be wrong, and it can allow a distinction among competing hypotheses (or theories).

**Prestige**: According to social prestige theory (an extension of the **handicap principle**) an act of **altruism**, because of its intrinsic **cost** to the actor, can be used as an **honest signal** of an individual's quality. In this way individuals can acquire social prestige by the use of altruistic **behaviour** which in turn increases the likelihood they will themselves **benefit** from other individuals' altruism in the future. See **indirect reciprocity** and Chapter 15.

**Prevalence**: Ratio of the number of cases observed of a given phenomenon to the whole population of events. The prevalence of a parasite in a **population** corresponds to the proportion of infected individuals.

**Prey**: Organism (animal or plant) that is eaten in part or whole by another (the **predator**).

**Principle of optimality**: the principle that states that the observed traits (including **behaviour**) result from a history of selection and are thus **optimal**. See **adaptation**. Chapter 3.

**Prior information**: **Information** about an environment that precedes any sampling. For instance, knowing that all complete decks of cards contain four aces is prior information.

**Private information**: **Information** that an animal has that is not available to others. By opposition to **public information**. The taste of a given food items is private **information**. Sometimes called **excludable information**. See **information**, Figure 4.3 and Chapters 4 and 14.

**Process**: A series of phenomena that are linked by a causal chain. See **pattern**.

**Producer–scrounger game**: A game involving two alternative and mutually exclusive resource acquisition strategies: **producer** that searches for the resource, and **scrounger** that searches for producers that have discovered resources. Chapter 8.

**Profitability**: In foraging theory refers to the ratio of a **prey** item's energetic content to its handling time. Chapter 7.

**Promiscuity**: (also called polygynandry) A **mating system** for which within the same mating season each male and female has many mates. Chapter 12.

**Prospecting**: An action that consists in visiting sites without directly exploiting their main **resource**. For instance, a prospector can be present at a potential reproductive site without reproducing there at that time.

**Proximate**: That which concerns immediate (physiological, neurological, motivational) causes of **behaviour**. As opposed to **ultimate**.

**Public information**: this term has two different meanings in the literature (1) Usual meaning: **Information** concerning the quality of a **resource** that is obtained by witnessing the performance (i.e. the **behaviour** and success in an activity) of other individuals. According to this definition, the act of foraging, the number of **prey** items captured at a patch by another individual, or the number and health of chicks in a nest can be public information. (2) **In this book** the expression public information depicts any **information** that is non-private, and we use the expression **inadvertent social information** to designate what is usually called public information (see Figure 4.3). In this meaning public information is not necessarily social, it also comprises physical characteristics of the environment, such as the height and orientation of a cliff. Sometimes called non-excludable information. See **information, social information, cue, signals**. Chapters 4 and 14.

## Q

**Quasi-social or para-social**: Evolution of group living by quasi-social pathway implies parents retaining young to form a **society** whereas the para-social pathway involves separate individuals that form societies having independently chosen to live together.

R

**Real aggregation**: Any discrepancy from the **ideal free distribution**. See **aggregation**, **habitat mediated aggregation** and Chapter 14.

**Receiver**: In intentional or true **communication**, the individual toward which a signal is targeted. Also called receptor. See **signal**, **communication**, and Chapter 16.

**Receptive period**: Period during which an individual is physiologically ready to accept a mating (often called time in). Inversely, a period of non-receptivity or refractory period designates the period where the animal is not physiologically or behaviourally ready to accept a mating (time out). Chapter 11.

**Receptor**: See **receiver**.

**Reciprocity**: Situations of **cooperation** in which altruism is directed towards individuals likely to reciprocate in the future. Implies some level of individual recognition. See **cooperation**, **prestige**, **indirect reciprocity**. Chapter 15.

**Reinforcement**: **In an evolutionary context**: following **allopatric divergence** between two **populations** the reduced viability of hybrids leads to a reinforcement of the reproductive characters that accentuate the differences between members of either population in the contact zone. During this reinforcement there is still some genetic exchange between populations because they are not completely isolated at the contact zone. See **reproductive character displacement** and **allopatric** and **sympatric** divergence. **In a learning context**: refers to an event that affects the likelihood that a given **behaviour** will be repeated. The event is a positive reinforcement if it increases the likelihood of repetition and a negative reinforcement if it reduces it.

**Replication**: **In statistics**: Repetition of an experimental treatment on a new group of subjects or of a correlation analysis using a new set of data. Other term often used: replicate. **In biology**: process by which a **replicator** (gene) makes a copy of itself.

**Replicator**: The smallest entity capable of making copies of itself. Usually refers to **genes**. Chapter 2.

**Reproductive bottleneck**: Period during which only a restricted number of individuals within a population contribute to reproduction resulting in reduced genetic diversity.

**Reproductive character displacement**: Following **allopatric divergence** of two **populations** the lower viability of hybrids can lead to the displacement of reproductive characters that amplifies the differences between the two populations in the contact zone. The mechanisms of character displacement occur only once the populations are completely isolated, that is when separate species are involved. The displacement reduces the chances of wasting gametes in infertile mating, reduces the chances of species interfering with each other by making the reproductive signals more efficient. See **allopatric** and **sympatric divergence**, **reinforcement process**, and Chapter 11.

**Reproductive effort**: The quantity of **resources** and energy that an individual invests in producing viable offspring. Also called reproductive investment. Chapter 12.

**Resource**: A biotic or abiotic element that is available in limited quantities in a given **environment** whose exploitation contributes to maintaining or increasing one's phenotypic **fitness**. Also used to describe an individual's total capacity to invest in some vital activity: resources invested in reproduction, or in defence, etc.

**Response to selection**: The change in allelic frequencies within a **population** as the result of **selection**.

**Revealing handicap**: Case related to the **handicap principle** where all individuals develop the **handicap** trait whatever their intrinsic quality. Because a handicap trait requires maintenance, only those individuals in excellent condition will have the time and energy required to keep the full expression of the **ornament**. Sick or weaker animals will have more difficulty maintaining the revealing handicap. Because an individual's propensity of falling ill depends on its genetic quality, individuals that choose their partner in terms of the quality of their ornamentation will obtain indirect **benefits** in the way of **good genes** for their progeny. Chapter 11.

**Risk-averse**: From **foraging** theory, refers to an individual's avoidance of alternatives (**prey** item or **patch**) having a more uncertain outcome. It is important not to confuse this technical usage of 'risk' with its usual meaning of danger or hazard such as the hazard (or risk) of predation. See **risk-prone** and Chapter 7.

**Risk-prone**: Qualifies an individual that prefers alternatives offering the greatest uncertainty. See **risk-averse** and Chapter 7.

**Runaway process**: **Co-evolution** of a phenotypic trait in one sex and its preference in the other. Given that descendants inherit both the trait and the preference, the positive association between the **heritable** (or **transmittable**) component of the trait and the preference results in a positive **feedback** loop that leads to traits whose form is exaggerated beyond its strictly utilitarian value. See **sexual selection** and Chapter 11.

## S

**Scramble competition**: Also called competition by exploitation. Reduction in **resource** use resulting solely from the removal of the resource by competing individuals. The reduction cannot be due to interactions among competitors. See **interference competition**.

**Scrounger**: One of two strategies in the producer scrounger game where the producer's behaviour consists of making a resource available while the scrounger strategy consists of usurping the resource made available by the producer. The payoffs to scrounger are negatively-frequency dependent.

**Selection**: This term can have two meanings. The most common is in the sense of **natural selection**. The second is equivalent to **choice**, as in habitat selection. In this book we tried to avoid using this term in the latter meaning and instead used the term **choice**.

**Selfish herd**: A group that results from each individual's attempt to place itself behind a companion to shield itself from a potential predator. Chapter 14.

**Selfishness**: A **behaviour** that **benefits** only the actor (as opposed to **cooperation** or **altruism**).

**Semelparous**: A species whose individuals reproduce only once in their lifetime, generally massively. Chapter 5. See **iteroparous**.

**Sender**: In **intentional** or true **communication**, the individual that emits a **signal**. Also called **emitter**, or **transmitter**. Chapter 16.

**Sensory drive**: Constraints on the efficiency of signals imposed by characteristics of the physical or social **environment**. For example, except for special circumstances, a green **signal** is unlikely to evolve in an environment dominated by green colour. Chapter 16.

**Sensory exploitation**: Principle used to explain the evolution of a **signal** as a result of a pre-existing sensory bias in the **receivers**. See also **chase-away sexual selection** and Chapter 16.

**Sex allocation**: Also sometimes referred to as sex investment ratio. Relative proportion of parental **resource**s invested in sons and daughters. Chapter 13.

**Sex ratio**: Numerical ratio of the number of males to females. Can be defined at the level of a **population** or a single individual's offspring. Generally measured as the number of males divided by the number of females. Chapter 16.

**Sexual facilitation**: Increase in sexual and reproductive activity as a result of the presence of others.

**Sexual selection**: Selection among variants that occurs according to their ability at securing access to the gametes of the other sex. There are two subcategories of sexual selection, **inter-sexual** selection and **intra-sexual** selection that lead to secondary sexual characters known as **ornaments** or **armament, respectively**. Chapter 11.

**Signal**: Any trait or **behaviour** whose adaptive function is to alter the behaviour of one or more **receivers**. Chapters 4 and 16.

**Silver spoon effect**: The fact that **dispersers** in good condition are selecting the best **habitats** while those in poor conditions are confined to poor habitats. See **habitat matching effect** and Chapter 5.

**Simple optimality**: A formal approach that consists of identifying the behavioural option that corresponds to maximal **fitness** gains. It is 'simple' when the calculation of the **benefits** of a given option does not require taking into account the frequency of the options used in the population. See **game theory**.

**Social attraction**: The tendency of being attracted by the presence of conspecifics (e.g. while choosing a habitat). Chapter 4.

**Social cues**: detectable facts that are unintentionally produced by organisms. Although 'social' typically refers to interactions among individuals of the same species, we view social cues as also encompassing heterospecific interactions. See **information**, Figure 4.3 and Chapter 4.

**Social facilitation**: Increase in a given activity resulting from the presence of other individuals engaged in the same activity.

**Social genetic information: Information** that can be extracted from components of phenotypic variance that are rather robust to environmental variation. May also provide valuable information about an individual, particularly in the context of mate choice. Example: eye colour in humans. Animals might also use such environmentally robust phenotypic information to assess the contents of another individual's genotype, something useful in mating and in estimating genetic kinship. See Figure 4.3 and Chapter 4.

**Social information: Information** obtained from conspecifics or other individuals sharing some ecological requirements. See **information**, **personal** and **public information**, **cue** and **signals**, Figure 4.3 and Chapter 4.

**Social learning**: There are more that 30 different definitions of social learning. In this book we mean any form of **learning** resulting from **social information**. See **learning**, **imprinting**, and Chapter 20.

**Social monogamy: Mating system** that consists in the association of a single male and female in the course of one or several reproductive episodes sometimes involving some level of **cooperation** in **parental care**. Does not necessarily imply sexual fidelity between mates. Chapter 12.

**Social systems**: Social systems comprise all interactions among juveniles, males, and females. While juveniles of species may not interact with adults during the reproductive season, they may disperse to avoid adults or remain philopatric. Levels of genetic similarity, which generate social **trade-offs**, also structure social systems.

**Sociality**: Tendency for organisms of living within groups of conspecifics. Chapters 14 and 15.

**Society**: The association of individuals of a given species that exhibits a particular structure and a certain degree of coordination among its members.

**Sociobiology**: An area of behavioural ecology that studies the biological bases of social **behaviour** using an adaptationist approach.

**Source–sink**: A **metapopulation** composed of some **populations** (sources) that produce an excess number of individuals (**intrinsic growth rate** greater than 1) while others (sinks) are declining (an **intrinsic growth rate** less than 1). Sinks are maintained by the arrival of individuals from the sources. Chapter 10.

**Sperm competition**: Any competition among spermatozoa from different males to fertilize the ova of the same female. Chapter 11.

**Stabilizing selection**: a type of selection that favours intermediate values of the trait. It diminishes the trait variance. See **directional selection**, **diversifying selection**, and Chapter 2.

**State: In communication**: a state refers to a more or less permanent **signal** such as colour or odour.

**In stochastic dynamic modelling**: a state refers to an internal factor (satiation, level of energy or sperm reserves, etc.) that affects an individual's **decision**.

**Stochasticity**: Usually denotes the fact that some events occur randomly. For instance, the expression **demographic stochasticity** denotes the fact that even though demographic parameters are estimated as the mean survival or fecundity at the scale of the population, they qualify processes that are not continuous: a given individual can either survive or die during the winter, or produce 1, 2, or 3 offspring this year, not 1.78 offspring. The average probability of dying may be 0.25, in the adult population, however individuals either die (**fitness** = 0) or survive (fitness = 1). The result is that in some populations (particularly small ones) all individuals may die in a given year, 'just by chance', in spite of the fact their average annual survival rate is 0.75. Demographic stochasticity only becomes negligible in populations of quasi-infinite size. See **genetic drift**.

**Strategy**: A structured suite of **behaviours** or **decision** rules that have arisen by **natural selection**.

**Supergene**: Two or more genes whose fates are inextricably linked by common mechanisms of physiological or endocrine regulation or behavioural mechanisms such as in **communication** between a **sender** and **receiver**. Notice that endocrine cascades consist of sender–receiver systems that are intrinsic to the organism or a property of their physiology. Behavioural communication consists of sender-receiver systems that are extrinsic to the organism. This definition of supergene unifies concepts of the physiological control of **behaviours** (intrinsic endocrine regulation involving sender and receiver **hormone** molecules) and **information** flow used in behaviours (extrinsic signals involving sender and receivers such as in the male–female or the male–male interactions of **sexual selection**). Chapter 5.

**Symbiosis**: A specialized form of **mutualism** involving highly developed relationships between partners that could not survive without each other.

**Sympatric divergence**: According to this hypothesis traits involved in the reproductive isolation of two subspecies continue to evolve in secondary contact zones once **allopatric divergence** has occurred. The individuals of a sub-species that either resemble most the individuals of the other or fail to discriminate between the two, risk mating with individuals of the other type. If these hybrid matings are less viable, **selection** will favour individuals that are better able to discriminate between the two sub-species. See **reinforcement process, reproductive character displacement**, and Chapter 11.

T

**Teaching**: The definition of teaching comprises three criteria: (1) an individual A modifies its **behaviour** only in the presence of a naive observer B, (2) A incurs some cost or at least does not derive any immediate benefit, and (3) A's behaviour results in B acquiring knowledge or skills more rapidly or efficiently than it would in the absence of A, or that it would not have learned at all. Chapter 20.

**Teleonomy**: Interpretation of the apparent goal of a **behaviour** or trait as resulting from **natural selection** (as opposed to reference to final causes or teleology).

**Time horizon**: A bounded interval during which an animal can perform a given activity. The duration of a given time horizon depends on the context. It can vary from a few hours for a diurnal forager that must acquire some quantity of energy before nightfall, up to several weeks for an individual searching for mates at the start of the breeding season.

**Trade-off**: Adaptive compromise between two characters (or more) that cannot be optimized simultaneously. Chapter 5.

**Transmittability**: The heredity of differences whatever the mechanism of transmission involved. Transmittability includes **heritability** as a fundamental component, but also other forms of transmission across generations, one of which is **cultural transmission**. It is the evolutionary potential of a **population**. Heredity is the fact that dogs do not make cats and vice versa; transmittability is the fact that big dogs (or cats) tend to make bigger dogs (or cats) than small dogs (or cats). See Figure 20.2 and Chapter 20.

**Transmitter**: See **sender**.

**Travel time**: In **foraging** theory, the time spent searching for **patches**. In **central place foraging**, the time spent going from patch to central place and back again including any mandatory time required in the central place.

**Tropism**: **In plants**: growth that is oriented by the influence of an outside stimulus. **In animals**: orientation movements.

**Two-armed bandit**: A name given to a choice situation that is analogous to gamblers playing a two-lever slot machine. The animal must decide which of two uncertain alternatives is most likely to provide a reward. Chapter 7.

## U

**Ultimate**: In reference to adaptive or evolutionary explanations of a trait or a **behaviour**. As opposed to **proximate**.

**Utilitarian selection**: Process of selection that chooses among traits in relation to their contribution to survival and fecundity (irrespective of their access to mates).

**Utility**: In microeconomics the **choice** of consumers is assumed to be governed by an attempt to maximize some abstract value referred to as utility. Consumers are assumed to be rational, that is they assign a different utility value to various alternatives and these utility rankings remain consistent across **choice** situations. In behavioural ecology utility is substituted by **phenotypic fitness**.

## V

**Vertical transmission**: As opposed to **horizontal transmission**. In **cultural transmission**: transmission of a **behaviour** across generations. See **culture**. **In parasitology:** transmission of a parasite from parents to offspring.

**Viability**: The capacity of a **population**, a **strategy** or a **phenotype** of persisting over time. Viability studies are particularly important in conservation biology. Chapter 18.

**Viable population**: Minimum size of a **population** that insures its lasting existence. See **viability** and Chapter 18.

**Vicarious sampling**: When one individual obtains **information** through the sampling **behaviour** of another individual. Leads to the acquisition of **social information**. See **information, public** versus **private information**, and Chapter 4.

**Viscosity**: Applies to an environment to describe the difficulty with which individuals can move. Chapter 15.

# References

## Chapter 1


A

**Allee, W.C.** 1931. *Animal Aggregations. A Study in General Sociology*. University of Chicago Press, Chicago.

**Allee, W.C.** 1933. *Animal Life and Social Growth*. Williams & Wilkins, Baltimore, Maryland.

**Allee, W.C.** 1938. *The Social Life of Animals*. Norton, New York.

**Ashmole, N.P.** 1963. The regulation of numbers of tropical oceanic birds. *Ibis* 103: 458–473.

B

**Baerends, G.P.** 1976. The functional organization of behaviour. *Animal Behaviour* 24: 726–738.

**Baerends, G.P., Drent, R.H., Glas, P. & Groenewold, H.** 1970. An ethological analysis of incubation behaviour in the herring gull. *Behaviour Suppl.* 17: 135–235.

**Barash, D.P.** 1982. *Sociobiology and Behavior*, 2nd edition. Hodder & Stoughton, London.

**Barlow, G.W.** 1968. Ethological units of behavior. In *The Central Nervous System of Fish Behavior* (Ingle, D. ed.), pp. 217–232. University of Chicago Press, Chicago.

**Barlow, G.W.** 1989. Has sociobiology killed ethology or revitalized it? In *Perspectives in Ethology*, vol. 8 (Bateson, P.P.G. & Klopfer, P.H. eds), pp. 1–45. Plenum Press, New York.

**Beach, F.A.** 1955. The descent of instinct. *Psychological Review* 62: 401–410.

**Birkhead, T.R. & Møller, A.P.** 1992. *Sperm Competition in Birds. Evolutionary Causes and Consequences*. Academic Press, London.

**Birkhead, T.R. & Møller, A.P.** 1998. *Sperm Competition and Sexual Selection*. Academic Press, London.

**Bowler, P.J.** 2003. *Evolution. The History of an Idea*. University of California Press, Berkeley.

**Brain, P.F.** 1989. Ethology and experimental psychology: from confrontation to partnership. In *Ethoexperimental Approaches to the Study of Behavior* (Blanchard, R.J., Brain, P.F., Blanchard, D.C. & Parmigiani, S. eds), pp. 18–27, NATO ASI Series D, vol. 48. Kluwer Academic Publishers, Dordrecht.

**Brown, J.L.** 1964. The evolution of diversity in avian territorial systems. *Wilson Bulletin* 76: 160–169.

**Browne, J.** 2002. *Charles Darwin. The Power of Place*. Alfred Knopf, New York.

**Buican, D.** 1989. *L'Evolution and les Evolutionnismes*. PUF, 'Que sais-je?', Paris.

C

**Cartwight, J.** 2000. *Evolution and Human Behaviour*. MacMillan Press, London.

**Charnov, E.L.** 1976. Optimal foraging: the marginal value theorem. *Theoretical Population Biology* 9: 129–136.

**Chatelin, Y.** 2001. *Audubon. Peintre, Naturaliste, Aventurier*. Editions France-Empire, Paris.

**Chatfield, C. & Lemon, R.E.** 1970. Analysing sequences of behavioural events. *Theoretical Biology* 29: 427–445.

**Collias, N.E.** 1991. The role of American zoologists and behavioural ecologists in the development of animal sociology, 1934–1964. *Animal Behaviour* 41: 613–631.

**Corsi, P.** 2001. *Lamarck. Genèse and Enjeux du Transformisme 1770–1830*. CNRS Editions, Paris.

**Craig, W.** 1908. The voices of pigeons as a means of social control. *American Journal of Sociology* 14: 86–100.

**Crook, J.H.** 1964. The evolution of social organization and visual communication in the weaver birds (Ploceinae). *Behaviour* (Supplement) 10: 1–178.

D

**Darwin, C.** 1859. *On the Origin of Species by Means of Natural Selection*. John Murray, London.

**Darwin, C.** 1871. *The Descent of Man and Selection in Relation to Sex*. John Murray, London.

**Darwin, C.** 1872. *The Expression of Emotions in Man and Animals*. John Murray, London.

**Dawkins, R.** 1976. *The Selfish Gene*. Oxford University Press, Oxford.

**Dawkins, R.** 1989. *The Blind Watchmaker*. Oxford University Press, Oxford.

**Dawkins, R. & Dawkins, M.** 1973. Decisions and the uncertainty of behaviour. *Behaviour* 45: 83–103.

**Delius, J.D.** 1969. Stochastic analysis of the maintenance behaviour of skylarks. *Behaviour* 33: 137–178.

**Dewsbury, D.A.** 1989. A brief history of the study of animal behavior in North America. In *Perspectives in Ethology*, vol. 8 (Bateson, P.P.G. & Klopfer, P.H. eds), pp. 85–122. Plenum Press, New York.

**Dewsbury, D.A.** 1999. The proximate and ultimate: past, present, and future. *Behavioural Processes* 46: 189–199.

**Dunlap, P.J.** 1919. Are there any instincts? *Journal of Abnormal Psychology* 14: 35–50.

**Durant, J.R.** 1986. The making of ethology: The Association for the Study of Animal Behaviour, 1936–1986. *Animal Behaviour* 34: 1601–1616.

E

**Emlen, J.M.** 1966. The role of time and energy in food preferences. *American Naturalist* 100: 611–617.

**Espinas, A.** 1876. *Des Sociétés Animales*. Baillière, Paris.

F

**Fabre, J.H.** 1989. *Souvenirs Entomologiques. Etudes sur l'Instinct and les Mœurs des Insectes*. Editions Robert Laffont, Paris.

**Fentress, J.C. & Stilwell, F.P.** 1973. Grammar of a movement sequence in inbred mice. *Nature* 244: 52–53.

**Flourens, P.** 1842. *Recherches Expérimentales sur les Propriétés and les Fonctions du Système Nerveux dans les Animaux Vertébrés*. Baillière, Paris.

**von Frisch, K.** 1955. *Vie and Mœurs des Abeilles*. Albin Michel, Paris.

**Futuyma, D.J.** 1998. *Evolutionary Biology*, 3rd edition. Sinauer, Sunderland, Massachussets.

G

**Gayon, J.** 1998. *Darwinism's Struggle for Survival. Heredity and the Hypothesis of Natural Selection*. Cambridge University Press, Cambridge.

**Gervet, J.** 1980. Où en est l'étude du comportement? Ou dix thèses sur l'éthologie. *Revue des Questions Scientifiques* 151: 305–334.

**Giard, A.** 1904. *Controverses transformistes*. C. Naud, Paris.

**Gould, S.J.** 1974. Darwin's dilemma. *Natural History* 83: 16–22.

**Gould, S.J. & Lewontin, R.** 1979. The spandrels of San Marco and the Panglossian paradigm: a critique of the adaptationnist programme. *Proceedings of the Royal Society of London Series B* 205: 581–598.

**Gross, M.R.** 1994. The evolution of behavioral ecology. *Trends in Ecology and Evolution* 9: 358–360.

H

**Hachet-Souplet, P.** 1928. *Les Sociétés Animales*. Alphonse Lemerre, Paris.

**Hamilton, W.D.** 1964a. The genetical evolution of social behaviour, I. *Journal of Theoretical Biology* 7: 1–16.

**Hamilton, W.D.** 1964b. The genetical evolution of social behaviour, II. *Journal of Theoretical Biology* 7: 17–52.

**Hamilton, W.D.** 1971. Geometry for the selfish herd. *Journal of Theoretical Biology* 31: 295–311.

**Hebb, D.O.** 1953. Heredity and environment in mammalian behaviour. *British Journal of Animal Behaviour* 1: 43–47.

**Heinroth, O.** 1911. Beiträge zur Biologie, namentlich Ethologie und Psychologie der Anatiden. *Verhandl. 5 International Ornithologische Kongreß*: 589–702.

**Hess, E.H.** 1962. Ethology: an approach toward the complete analysis of behavior. In *New Directions in Psychology* (Brown, R., Galanter, E., Hess, E.H. & Mandler, G. eds), Holt, Rinehart and Winston, New York.

**Hughes, C.** 1998. Integrating molecular techniques with field methods in studies of social behavior: a revolution results. *Ecology* 79: 383–399.

**Huxley, J.S.** 1914. The courtship habits of the great crested grebe (*Podiceps cristatus*). *Proceedings of the Zoological Society of London* 1914: 491–562.

**Huxley, J.S.** 1942. *Evolution: The Modern Synthesis*. Allen & Unwin, London.

J

**Jarman, P.J.** 1974. The social organization of antelopes in relation to their ecology. *Behaviour* 58: 215–267.

**Jaynes, J.** 1969. The historical origins of 'ethology' and 'comparative psychology'. *Animal Behaviour* 17: 601–606.

K

**Kaye, H.L.** 1986. *The Social Meaning of Modern Biology*. Yale University Press, New Haven.

**Kennedy, J.S.** 1954. Is modern ethology subjective? *British Journal of Animal Behaviour* 2: 12–19.

**Kennedy, J.S.** 1992. *The New Anthropomorphism*. Cambridge University Press, Cambridge.

**Klopfer, P.H.** 1962. *Behavioral Aspects of Ecology*. Prentice-Hall, Englewood Cliffs, New Jersey.

**Köhler, W.** 1925. *L'Intelligence des Singes Supérieurs*. CEPL, Paris.

**Kortlandt, A.** 1940. Wehselwirkiung Zwischen Instinkten. *Archives Néerlandaises de Zoologie* 4: 443–520.

**Krebs, J.R.** 1985. Sociobiology ten years on. *New Scientist* 1476: 40–43.

**Krebs, J.R. & Davies, N.B.** 1978. *Behavioural Ecology, An Evolutionary Approach*. Blackwell Scientific Publications, Oxford.

**Krebs, J.R. & Davies, N.B.** 1984. *Behavioural Ecology An Evolutionary Approach* 2nd edition. Sinauer Associates, Sunderlans, ÉUA.

**Krebs, J.R. & Davies, N.B.** 1991. *Behavioural Ecology. An Evolutionary Approach* 3rd edition. Blackwell, Oxford.

**Krebs, J.R. & Davies, N.B.** 1997. *Behavioural Ecology An Evolutionary Approach* 4th edition. Blackwell Science, Oxford, R.U.

**Kruuk, H.** 2003. *Niko's Nature. A Life of Niko Tinbergen and his Science of Animal Behaviour*. Oxford University Press, Oxford.

**Kuo, Z.Y.** 1924. A psychology without heredity. *Psychological Review* 31: 427–451.

L

**Laland, K.N. & Brown, G.R.** 2002. *Sense and Nonsense. Evolutionary perspectives on human behaviour*. Oxford University Press, Oxford.

**Lamarck, J.-B. P.A. de Monet de.** 1809. *Philosophie Zoologique*. Dentu, Paris.

**Lehrman, D.S.** 1953. A critique of Konrad Lorenz's theory of instinctive behaviour. *Quarterly Review of Biology* 28: 337–363.

**Leroy, C.-G.** 1802. *Lettres Philosophiques sur l'Intelligence and la Perfectibilité des Animaux avec quelques lettres sur l'homme*. Imprimerie de Valade, Paris.

**Little, T.J., Hultmark, D. & Read, A.F.** 2005. Invertebrate immunity and the limits of mechanistic immunology. *Nature Immunology* 6: 651–654.

**Lorenz, K.** 1935. Der Kumpan in der Welt des Vogels. *Journal für Ornithologie* 83: 137–213, 289–413.

**Lorenz, K.** 1950. The comparative method in studying innate behaviour patterns. *Symposium of the Society for Experimental Biology* 4: 221–268.

**Lorenz, K.** 1958. The evolution of behavior. *Scientific American* 199: 67–78.

**Lorenz, K.Z.** 1965. *Evolution and Modification of Behavior*. University of Chicago Press, Chicago.

**Maddison, W.P. & D.R. Maddison.** 2003. *Mesquite: A Modular System for Evolutionary Analysis* version 1.0. http: //mesquiteproject.org.

**Mangel, M. & Clark, C.W.** 1988. *Dynamic Modelling in Behavioral Ecology*. Princeton University Press, Princeton.

**Martins, E.P.** 1996. *Phylogenies and the Comparative Method in Animal Behaviour*. Oxford University Press, Oxford.

**Martins, E.P.** 2000. Adaptation and the comparative method. *Trends in Ecology and Evolution* 15: 296–299.

**Matthews, R.** 2000. Storks deliver babies ($p$ = 0.008). *Teaching Statistics* 22: 36–38.

**Maynard Smith, J.** 1982. *Evolution and the Theory of Games*. Cambridge University Press, Cambridge.

**Maynard Smith, J.** 1984. Game theory and the evolution of behaviour. *Behavior and Brain Sciences* 7: 95–125.

**McCracken, K.G. & Sheldon, F.H.** 1998. Molecular and osteological heron phylogenies: source of incongruence. *Auk* 115: 127–141.

**McFarland, D.J. & Houston, A.I.** 1981. *Quantitative Ethology. The State Space Approach*. Pitman, London.

**Metz, K.J. & Weatherhead, P.J.** 1992. Seeing red: uncovering coverable badges in red-winged blackbirds. *Animal Behaviour* 43: 223–229.

**Møller, A.P.** 1988. Female choice selects for male sexual tail ornaments in the monogamous swallow. *Nature* 332: 640–642.

**Møller, A.P.** 1990. Male tail length and female mate choice in the monogamous swallow *Hirundo rustica*. *Animal Behaviour* 39: 458–465.

**Møller, A.P.** 1994. *Sexual Selection and the Barn Swallow*. Oxford University Press, Oxford.

**Moore, A.J. & Boake, C.R.B.** 1994. Optimality and evolutionary genetics: complementary procedures for evolutionary analysis in behavioural ecology. *Trends in Ecology and Evolution* 9: 69–72.

**Morand, S. & Poulin, R.** 1998. Density, body mass and parasite species richness of terrestrial mammals. *Evolutionary Ecology* 12: 717–727.

**Mousseaux, T.A. & Roff, D.A.** 1987. Natural selection and the heritability of fitness components. *Heredity* 59: 181–197.

**O**

**Oster, G.F. & Wilson, E.O.** 1978. *Caste and Ecology in the Social Insects*. Princeton University Press, Princeton. (ED 702)

**P**

**Pagel, M.** 1994. Detecting correlated evolution on phylogenies: a general method for the comparative analysis of discrete characters. *Proceedings of the Royal Society of London Series B* 255: 37–45.

**Pagel, M.** 1997. Inferring evolutionary processes from phylogenies. *Zoologica Scripta* 26: 331–348.

**Peek, F.W.** 1972. An experimental study of the territorial function of vocal and visual display in the male red-winged blackbird (*Agelaius phoenicus*). *Animal Behaviour* 20: 112–118.

**R**

**Riechert, S.E. & Hedrick, A.V.** 1990. Levels of predation and genetically based anti-predator behaviour in the spider *Agelenopsis aperta*. *Animal Behaviour* 40: 679–687.

**Roff, D.A.** 1997. *Evolutionary Quantitaive Genetics*. Chapman and Hall, New York.

**Rohlf, J.F.** 2001. Comparative methods for the analysis of continuous variables: geometric interpretations. *Evolution* 55: 2143–2160.

**Rolland, C., Danchin, E. & de Fraipont, M.** 1998. The evolution of coloniality in birds in relation to food, habitat, predation, and life-history traits: a comparative analysis. *The American Naturalist* 151: 514–529.

**Røskaft, E. & Rohwer, S.** 1987. An experimental study of the function of the red epaulettes and the black body colour of red-winged blackbirds. *Animal Behaviour* 35: 1070–1077.

**Rothenbuhler, W.** 1964. Behaviour genetics of nest cleaning in honey bees. *American Zoologist* 4: 111–123.

**S**

**Samuelson, P.A.** 1965. *Foundations of Economic Analysis*. Harvard University Press, Cambridge, Massachusetts.

**Sarrazin, F. & Barbault, R.** 1996. Reintroduction: challenges and lessons for basic ecology. *Trends in Ecology and Evolution* 11: 474–478.

**Sibley, C.G. & Ahlquist, J.E.** 1987. Avian phylogeny reconstructed from comparisons of the genetic material, DNA. Pp. 95–121 in *Molecules and Morphology in Evolution: Conflict or Compromise* (Patterson, C. ed.). Cambridge University Press, Cambridge.

**Sinervo, B. & Basolo, A.L.** 1996. Testing adaptation using phenotypic manipulation. Pp. 149–185 in *Adaptation* (Rose, M.R. & Lauder, G.V. eds). Academic Press, New York.

**Smith, D.G.** 1972. The role of epaulets in the red-winged blackbird (*Agelaius phoniceus*) social system. *Behaviour* 41: 251–268.

**Smith, H.G. & Montgomerie, R.** 1991. Sexual selection and the tail ornaments of North American barn swallows. *Behavioral Ecology and Sociobiology* 28: 195–201.

**Sokolowski, M.B.** 1980. Foraging strategies of *Drosophila melanogaster*: a chromosomal analysis. *Behavior Genetics* 10: 291–302.

**Sokolowski, M.B., Kent, C. & Wong, J.** 1984. *Drosophila* larval foraging behaviour: Developmental stages. *Animal Behaviour* 32: 645–651.

**Sokolowski, M.B., Pereira, H.S. & Hughes, K.** 1997. Evolution of foraging behavior in *Drosophila* by density-dependent selection. *Proceedings of the National Academy of Sciences of the USA* 94: 7373–7377.

**Stephens, D.W. & Krebs, J.R.** 1986. *Foraging Theory*. Princeton University Press, Princeton.

**T**

**Tatar, M.** 2000. Transgenic organisms in evolutionary ecology. *Trends in Ecology and Evolution* 15: 207–211.

**Tinbergen, N., Broekhuysen, G.J., Feekes, F., Houghton, J.C.W., Kruuk, H. & Szulc, E.** 1962a. Egg shell removal by the black-headed gull *Larus ridibundus*, L.; a behaviour component of camouflage. *Behaviour* 19: 74–117.

**Tinbergen, N., Kruuk, H., Paillette, M. & Stamm, R.** 1962b. How do black-headed gulls distinguish between eggs and egg-shells? *British Birds* 55: 120–129.

**Torres-Villa, L.M., Rodriguez-Molina, M.C., Gragera, J. & Bielza-Lino, P.** 2001. Polyandry in Lepidoptera: a heritable trait in *Spodoptera exigua* Hübner. *Heredity* 86: 177–183.

W

**Weibel, E.R., Taylor, C.R. & Bolis, L.** 1998. *Principles of Animal Design. The Optimization and Symmorphosis Debate*. Cambridge University Press, Cambridge.

**Wolf, J.B.** 2001. Integrating biotechnology and the behavioral sciences. *Trends in Ecology and Evolution* 16: 117–119.

**Wood-Gush, D.G.M.** 1960. A study of sex drive of two strains of cockerel through three generations. *Animal Behaviour* 8: 43–53.

Z

**Zangerl, R.** 1948. The methods of comparative anatomy and its contribution to the study of evolution. *Evolution* 2: 351–374.

## Chapter 4

A

**Aragón, P., Clobert, J. & Massot, M.** 2006. Individual dispersal status influences space use of conspecific residents in the common lizard, *Lacerta vivipara*. *Behavioral Ecology and Sociobiology* 60: 430–438.

**Avital, E. & Jablonka, E.** 2000. *Animal Traditions. Behavioural Inheritance in Evolution*. Cambridge University Press, Cambridge.

B

**Barta, Z. & Giraldeau, L.A.** 2001. Breeding colonies as information centers: a reappraisal of information-based hypotheses using the producer–scrounger game. *Behavioral Ecology* 12: 121–127.

**Bikhchandani, S., Hirshleifer, D. & Welch, I.** 1998. Learning from the Behavior of Others: Conformity, Fads, and Informational Cascades. *Journal of Economic Perspectives* 12: 151–170.

**Blomqvist, D., Andersson, M., Küpper, C., Cuthill, I.C., Kis, J., Lanctot, R.B., Sandercock, B.K., Székely, T., Wallander, J. & Kempenaers, B.** 2002. Genetic similarity between mates and extra-pair parentage in three species of shorebirds. *Nature* 419: 613–615.

**Both, C., Dingemanse, N.J., Drent, P.J. & Tinbergen, J.M.** 2005. Pairs of extreme avian personalities have highest reproductive success. *Journal of Animal Ecology* 74: 667–674.

**Boulinier, T. & Danchin, E.** 1997. The use of conspecific reproductive success for breeding patch selection in territorial migratory species. *Evolutionary Ecology* 11: 505–517.

**Boulinier, T., Danchin, E., Monnat, J.Y., Doutrelant, C. & Cadiou, B.** 1996. Timing of prospecting and the value of information in a colonial breeding bird. *Journal of Avian Biology* 27: 252–256.

**Boyd, R. & Richerson, P.J.** 1988. An evolutionary model of social learning: the effects of spatial and temporal variation. In *Social Learning Psychological and Biological Perspectives* (Zentall, T.R. & Galef, B.G.J. eds), pp. 29–48. Lawrence Erlbaum Associates, Hillsdale, New Jersey, Hove and London.

**Bradbury, J.W. & Vehrencamp, S.L.** 2000. Economic models of animal communication. *Animal Behaviour*, 59: 259–268.

C

**Cadiou, B., Monnat, J.Y. & Danchin, E.** 1994. Prospecting in the kittiwake, *Rissa tridactyla*: different behavioural patterns and the role of squatting in recruitment. *Animal Behaviour*, 47: 847–856.

**Cheverud, J.M. & Moore, A.J.** 1994. Quantitative genetics and the role of the environement provided by relatives in behavioral ecology. In *Quantitative Genetic Studies of Behavioral Evolution* (Boake, C.R.B. ed.), pp. 67–100. University of Chicago Press, Chicago.

**Chivers, D.P. & Smith, R.J.F.** 1998. Chemical alarm signaling in aquatic predator–prey systems: a review and prospectus. *Ecoscience* 5: 338–352.

**Clobert, J., Danchin, E., Dhondt, A.A. & Nichols, J.D.** 2001. *Dispersal*. Oxford University Press, New York.

**Coolen, I., Dangles, O. & Casas, J.** 2005a. Social learning in noncolonial insects? *Current Biology* 15: 1931–1935.

**Coolen, I., van Bergen, Y., Day, R.L. & Laland, K.N.** 2003. Species differences in adaptive use of public information in sticklebacks. *Proceedings of the Royal Society of London Series B* 270: 2413–2419.

**Coolen, I., Ward, A.J.W., Hart, P.J.B. & Laland, K.N.** 2005b. Foraging nine-spined sticklebacks prefer to relay on public information over simple social cues. *Behavioral Ecology* 16: 865–870.

D

**Dall, S.R.X., Giraldeau, L.-A., Olsson, O., McNamara, J.M. & Stephens, D.W.** 2005. Information and its use by animals in evolutionary ecology. *Trends in Ecology and Evolution* 20: 187–193.

**Danchin, E., Boulinier, T. and Massot, M.** 1998. Conspecific reproductive success and breeding habitat selection: implications for the evolution of coloniality. *Ecology* 79: 2415–2428.

**Danchin, E., Heg, D. & Doligez, B.** 2001. Public information and breeding habitat selection. In *Dispersal* (Clobert, J., Danchin, E., Dhondt, A.A. & Nichols, J.D. eds), pp. 243–258. Oxford University Press, Oxford and New York.

**Danchin, E., Giraldeau, L.A., Valone, T.J. & Wagner, R.H.** 2004. Public information: from nosy neighbors to cultural evolution. *Science* 305: 487–491.

**Danchin, E., Giraldeau, L.A., Valone, T.J. & Wagner, R.H.** 2005. Defining the concept of public information—response. *Science* 308: 355–356.

**Deutsch, J.C. & Nefdt, R.J.C.** 1992. Olfactory cues influence female choice in two lek-breeding antelopes. *Nature, London* 356: 596–598.

**Dingemanse, N.J. & de Goede, P.** 2004. The relation between dominance and exploratory behavior is context-dependent in wild great tits. *Behavioral Ecology* 15: 1023–1030.

**Dingemanse, N.J., Both, C., Drent, P.J. & Tinbergen, J.M.** 2004. Fitness consequences of avian personalities in a fluctuating environment. *Proceedings of the Royal Society of London Series B* 271: 847–852.

**Dingemanse, N.J., Both, C., van Noordwijk, A.J., Rutten, A.L. & Drent, P.J.** 2003. Natal dispersal and personalities in great tits (*Parus major*). *Proceedings of the Biological Society of London Series B* 270: 741–747.

**Doligez, B., Danchin, E. & Clobert, J.** 2002. Public information and breeding habitat selection in a wild bird population. *Science* 297: 1168–1170.

**Doligez, B., Danchin, E., Clobert, J. & Gustafsson, L.** 1999. The use of conspecific reproductive success for breeding habitat selection in a non-colonial, hole-nesting species, the collared flycatcher. *Journal of Animal Ecology* 68: 1193–1206.

**Doligez, B., Cadet, C., Danchin, E. & Boulinier, T.** 2003. When to use public information for breeding habitat selection? The role of environmental predictability and density dependence. *Animal Behaviour* 66: 973–988.

**Dolman, C., Templeton, J. & Lefebvre, L.** 1996. Mode of foraging competition is related to tutor preference in *Zenaida aurita*. *Journal of Comparative Psychology* 110: 45–54.

**Doutrelant, C. & McGregor, P.K.** 2000. Eavesdropping and mate choice in female fighting fish. *Behaviour* 137: 1655–1669.

**Doutrelant, C., McGregor, P.K. & Oliveira, R.F.** 2001. The effect of an audience on intra-sexual communication in male Siamese fighting fish, *Betta splendens*. *Behavioral Ecology* 283–286.

E

**Elgar, M.A.** 1986. House sparrows establish foraging flocks by giving chirrup calls if the resources are divisible. *Animal Behaviour* 34: 169–174.

**Emery, N.J. & Clayton, N.S.** 2001. Effects of experience and social context on prospective caching strategy by scrub jays. *Nature* 414: 443–446.

F

**Feldman, M.W. & Laland, K.N.** 1996. Gene-culture coevolutionary theory. *Trends in Ecology and Evolution* 11: 453–457.

**Foerster, K., Delhey, K., Johnsen, A., Lifjeld, J.T. & Kempenaers, B.** 2003. Females increase offspring heterozygosity and fitness through extra-pair matings. *Nature* 425: 714–717.

**Freeman-Gallant, C.R., Meguerdichian, M., Wheelwright, N.T. & Sollecito, S.V.** 2003. Social pairing and female mating fidelity predicted by restriction fragment length polymorphism similarity at the major histocompatibility complex in a songbird. *Molecular Ecology* 12: 3077–3083.

G

**Galef, B.G.J. & White, D.J.** 2000. Evidence of social effects on mate choice in vertebrates. *Behavioural Processes* 51: 167–175.

**Galef, B.G. & Giraldeau, L.A.** 2001. Social influences on foraging in vertebrates: causal mechanisms and adaptive functions. *Animal Behaviour* 51: 3–15.

**Galef, B.G., Kennett, D.J. & Wigmore, S.W.** 1984. Transfer of information concerning distant foods in rats: A robust phenomenon. *Animal Learning and Behavior* 12: 292–296.

**Gasparini, J., McCoy, K.D., Haussy, C., Tveraa, T. & Boulinier, T.** 2001. Induced maternal response to the Lyme disease spirochaete *Borrelia burgdorferi sensu lato* in a colonial seabird, the kittiwake, *Rissa tridactyla*. *Proceedings of the Royal Society of London Series B* 268: 647–650.

**Gasparini, J., McCoy, K., Staszewski, V., Haussy, C. & Boulinier, T.** 2006. Dynamics of anti-*borrelia* antibodies in Black-legged Kittiwakes (*Rissa tridactyla*) chicks suggest a maternal educational effect. *Canadian Journal of Zoology* 84: 623–627.

**Gilddon, C.J. & Gouyon, P.H.** 1989. The units of selection. *Trends in Ecology and Evolution* 4: 204–208.

**Giraldeau, L.-A.** 1997. The ecology of information use. In (Krebs, J. R. and Davies, N.B. eds.) *Behavioural Ecology An Evolutionary Approach*: Blackwell Scientific Publications, 42–68.

**Giraldeau, L.A., Valone, T.J. & Templeton, J.J.** 2002. Potential disadvantages of using socially acquired information. *Philosophical Transactions of the Royal Society Series B* 357: 1559–1566.

H

**Hardin, G.** 1968. The tragedy of the commons. *Science* 162: 1243–1248.

J

**Johnstone, R.A.** 2001. Eavesdropping and animal conflict. *Proceedings of the National Academy of Science of the USA* 98: 9177–9180.

K

**Karban, R. & Maron, J.** 2002. The fitness consequences of interspecific eavesdropping between plants. *Ecology* 83: 1209–1213.

L

**Lefebvre, L., Templeton, J., Brown, K. & Koelle, M.** 1997. Carib grackles imitate conspecific and Zenaida dove tutors. *Behaviour* 134: 1003–1017.

**Lotem, A., Wagner, R.H. & Balshine-Earn, S.** 1999. The overlooked signaling component of nonsignaling behavior. *Behavioral Ecology* 10: 209–212.

M

**Massot, M. & Clobert, J.** 2000. Processes at the origin of similarities in dispersal behaviour among siblings. *Journal of Evolutionary Biology* 13: 707–719.

**McGregor, P.K. & Peake, T.M.** 2000. Communication networks: social environments for receiving and signalling behaviour. *Acta Ethologica* 2: 71–81.

**Mennill, D.J., Ratcliffe, L.M. & Boag, P.T.** 2002. Female eavesdropping on male song contest in songbirds. *Science* 296: 873.

**Méry, F. & Kawecki, T.J.** 2004. The effect of learning on experimental evolution of resource preference in *Drosophila melanogaster*. *Evolution* 58: 757–767.

**Milinski, M., Semmann, D. & Krambeck, H.-J.** 2002. Reputation helps solve the tragedy of the commons. *Nature* 415: 424–426.

N

**Nocera, J.J., Forbes, G.J. & Giraldeau, L.A.** 2006. Inadvertent social information in breeding site selection of natal dispersing birds. *Proceedings of the Royal Society of London Series B* 273: 349–355.

**Nowak, M.A. & Sigmund, K.** 1998. Evolution of indirect reciprocity by image scoring. *Nature* 393: 573–577.

O

**Oliveira, R.F., Lopes, M., Carneiro, L.A. & Canario, A.V.M.** 2001. Watching fights raises fish hormone levels. *Nature* 409: 475.

**Otter, K., McGregor, P.K., Terry, A.M.R., Burford, F.R.L., Peake, T.M. & Dabelsteen, T.** 1999. Do female great tits (*Parus major*) assess males by eavesdropping? A field study using interactive song playback. *Proceedings of the Royal Society of London Series B* 266: 1305–1309.

P

**Parejo, D. & Avilés, J.M.** 2007. Do avian brood parasites eavesdrop on heterospecific sexual signals revealing host quality? A review of the evidence. *Animal Cognition* 10: 81–88.

**Parejo, D., White, J.F., Clobert, J., Dreiss, A.N. & Danchin, E.** 2007. Blue tits use fledging quantity and quality as public information in breeding choice habitat. *Ecology* 88: 2373–2382.

**Peake, T.M.** 2005. Eavesdropping in communication networks. In *Animal Communication Networks* (McGregor, P.K. ed.), pp. 13–37. Cambridge University Press, Cambridge.

**Peake, T.M., Terry, A.M.R., McGregor, P.K. & Dabelsteen, T.** 2001. Male great tits eavesdrop on simulated male-to-male vocal interactions. *Proceedings of the Royal Society of London Series B* 268: 1183–1187.

**Pöysä, H.** 2006. Public information and conspecific nest parasitism in goldeneyes: targeting safe nests by parasites. *Behavioral Ecology* 17: 1–7.

S

**Semmann, D., Krambeck, H.-J. & Milinski, M.** 2003. Volunteering leads to rock–paper–scissors dynamics in a public goods game. *Nature* 425: 390–393.

**Shannon, C.E.** 1948. A mathematical theory of communication. *The Bell System Technical Journal*, 27: 379–423 and 623–656.

**Smith, J.W., Benkman, C.W. & Coffey, K.** 1999. The use and misuse of public information by red crossbils. *Behavioral Ecology* 10: 54–62.

**Stephens, D.W.** 1991. Change, regularity and value in the evolution of animal learning. *Behavioral Ecology* 2: 77–89.

**Stoddard, P.K.** 1988. The 'Bugs' call of the cliff swallow: a rare food signal in a colonially nesting bird species. *Condor* 90: 714–715.

**Switzer, P.V.** 1997. Past reproductive success affects future habitat selection. *Behavioral Ecology and Sociobiology* 40: 307–312.

T

**Taylor, J.B.** 1998. *Economics*. Houghton Mifflin Company, Boston and New York.

**Templeton, J.J. & Giraldeau, L.A.** 1995. Patch assessment in foraging flocks of European starlings: evidence for the use of public information. *Behavioral Ecology* 6: 65–72.

**Tinbergen, N.** 1963. On aims and methods of ethology. *Zietschrift für Tierpsychologie* 20: 410–433.

**Tinbergen, J.M. & Drent, R.H.** 1980. The starling as a successful forager. In *Bird Problems in Aagriculture* Wright (Inglis, E.N., Fear, J.R., C.J. eds), pp. 83–101. British Crop Protection Council, Croydon, UK.

V

**Valone, T.J.** 1989. Group foraging, public information & patch estimation. *Oikos* 56: 357–363.

**Valone, T.J. & Giraldeau, L.A.** 1993. Patch estimation by group foragers: what information is used? *Animal Behaviour* 45: 721–728.

**Valone, T.J. & Templeton, J.J.** 2002. Public information for the assessment of quality: a widespread social phenomenon. *Philosophical Transactions of the Royal Society Series B* 357: 1549–1557.

W

**Ward, P. & Zahavi, A.** 1973. The importance of certain assemblages of birds as information centres for food finding. *Ibis* 115: 517–534.

**Wedekind, C. & Milinski, M.** 2000. Cooperation through image scoring in humans. *Science* 288: 850–852.

**Wedekind, C., Seebeck, T., Bettens, F. & Paepke, A.J.** 1995. MHC-dependent mate preferences in humans. *Proceedings of the Royal Society of London Series B* 260: 245–249.

**Whitfield, J.** 2002. Nosy neighbours. *Nature* 419: 242–243.

**Witte, K. & Noltemeier, B.** 2002. The role of information in mate-choice copying in female sailfin mollies (*Poecilia latipinna*). *Behavioral Ecology and Sociobiology* 52: 194–202.

Z

**Zahavi, A.** 1995. Altruism as a handicap—the limitation of kin selection and reciprocity. *Journal of Avian Biology* 26: 1–3.

**Zahavi, A. & Zahavi, A.** 1997. *The Handicap Principle: A Missing Piece of Darwin's Puzzle*. Oxford University Press, Oxford.

## Chapter 5

A

**Alexander, R.** 1974. The evolution of social behavior. *Annual Review of Ecology and Systematics* 5: 325–383.

**Axelrod, R., Hammond, R.A. & Grafen, A.** 2004. Altruism via kin-selection strategies that rely on arbitrary tags with which they coevolve. *Evolution* 58(8): 1833–1838.

**Andrewartha, H.G. & Birch, L.C.** 1954. *The Distribution and Abundance of Animals*. University of Chicago Press, Chicago.

**Arnqvist, G. & Rowe, L.** 2002. Antagonistic coevolution between the sexes in a group of insects. *Nature* 415 (6873): 787–789.

B

**Bateman, A.J.** 1948. Intrasexual selection in *Drosophila*. *Heredity* 2: 349–368.

**Berrigan, D., Purvis, A., Harvey, P.H. & Charnov, E.L.** 1993. Phylogenetic contrasts and the evolution of mammalian life histories. *Evolutionary Ecology* 7: 270–278.

**Bleay, C. & Sinervo, B.** 2007. Discrete genetic variation in mate choice and a condition dependent preference function in the side blotched lizard: Implications for the formation and maintenance of co-adapted gene complexes. *Behavioural Ecology* 18: 304–310.

**Bradshaw, W.E., Haggerty, B.P. & Holzapfel, C.M.** 2005. Epistasis Underlying a Fitness Trait Within a Natural Population of the Pitcher-Plant Mosquito, *Wyeomyia smithii*. *Genetics* 169: 485–488.

**Brantley, R.K., Wingfield, J.C. & Bass, A.H.** 1993. Sex steroid levels in *Porichthys notatus*, a fish with alternative reproductive tactics, and a review of the hormonal bases for male dimorphism among teleost fishes. *Hormones and Behavior* 27: 332–347.

**Brooks, D.R. & McLennan, D.A.** 1991. *Phylogeny, Ecology, and Behavior*. University of Chicago Press, Chicago.

C

**Calder, W.A. III** 1984. *Size, Function and Life History*. Harvard University Press, Cambridge, Massachusetts.

**Calsbeek, R. & Sinervo, B.** 2002a. The ontogeny of territoriality during maturation. *Oecologia* 132: 468–477.

**Calsbeek, R. & Sinervo, B.** 2002b. Uncoupling direct and indirect components of female choice in the wild. *Proceedings of the National Academy of Sciences of the USA* 99: 14897–14902.

**Calsbeek, R. & Sinervo, B.** 2004. Within-clutch variation in offspring sex determined by differences in sire body size: cryptic mate choice in the wild. *Journal of Evolutionary Biology* 17: 464–470.

**Calsbeek, R., Alonzo, S.H., Zamudio, K. & Sinervo, B.** 2002. Sexual selection and alternative mating behaviours generate demographic stochasticity in small populations. *Proceedings of the Royal Society of London Series B* 269: 157–164.

**Carson, H.L.** 1975. Genetics of speciation at diploid level. *American Naturalist* 109: 83–92.

**Caswell, H.** 2001. *Matrix Population Models*. Sinauer, Sunderland, Massachusetts, USA.

**Chadwick, C.S.** 1950. Observations on the behavior of the larvae of the common American newt during metamorphosis. *The American Midland Naturalist* 43: 392–398.

**Charmantier, A., Perrins, C., McCleery, R.H. & Sheldon, B.C.** 2006. Quantitative genetics of age at reproduction in wild swans: Support for antagonistic pleiotropy models of senescence. *Proceedings of the National Academy of Sciences of the USA* 103: 6587–6592.

**Charnov, E.L.** 1993. *Life History Invariants: Some Explorations of Symmetry in Evolutionary Ecology*. Oxford University Press, Oxford.

**Charnov, E.L. & Berrigan, D.** 1991. Dimensionless numbers and the assembly rules for life histories. *Philosophical Transactions of the Royal Society of London Series B* 332: 41–48.

**Chevrud, J.M.** 1984. Quantitative genetics and developmental constraints on evolution by selection. *Journal of Theoretical Biology* 110: 155–71

**Clobert, J., Garland Jr, T. & Barbault, R.** 1998. The evolution of demographic tactics in lizards: a test of some hypotheses concerning life history evolution. *Journal of Evolutionary Biology* 11: 329–364.

**Clobert, J. Oppliger, A., Sorci, G., Ernande, B., Swallow, J. & Garland, T.** 2000. Endurance at birth, activity rate, and parasitism in the common lizard. *Journal of Functional Ecology* 14: 675–684.

**Clobert, J., Perrins, C.M., McCleery, R.H. & Gosler, A.** 1988. Survival rate in the Great Tit *Parus major* in relation to sex, age, and status. *Journal of Animal Ecology* 57: 287–306.

**Comendant, T., Sinervo, B., Svensson, E.I. & Wingfield, J.** 2003. Social competition, corticosterone and survival in female lizard morphs. *Journal of Evolutionary Biology* 16: 948–955.

**Costa, D. & Sinervo, B.** 2004. Field physiology: physiological insights from animals in nature. *Annual Review of Physiology* 66: 209–238

**Cote, J. & Clobert, J.** 2007. Social personalities influence natal dispersal in a lizard. *Proceedings of the Royal Society of London Series B* 274: 383–390.

D

**Dawkins, R.** 1976. *The Selfish Gene*. Oxford University Press, Oxford.

**Denver, R.J.** 2000. Evolution of the corticotrophin-releasing hormone signaling and its role in stress-induced developmental plasticity. *American Zoologist* 40: 995–996.

**DeWitt, T.J.** 1998. Costs and limits of phenotypic plasticity: tests with predator-induced morphology and life history in a freshwater snail. *Journal of Evolutionary Biology* 11: 465–480.

**DeWitt, T.J., Sih, A. & Wilson, D.S.** 1998. Costs and limits of phenotypic plasticity. *Trends in Ecology and Evolution* 13: 77–81.

**Dieckmann, U.** 1997. Can adaptive dynamics invade? *Trends in Ecology and Evolution* 12: 128–131.

**Dufty, A.M., Clobert, J. & Moller, A.P.** 2002. Hormones, developmental plasticity and adaptation. *Trends in Ecology and Evolution* 17: 190–196.

E

**Emlen, S. & Oring, L.** 1977. Ecology, sexual selection, and the evolution of mating systems. *Science* 197: 215–222.

**Enquist, M.** 1985. Communication during aggressive interactions with particular reference to choice of behavior. *Animal Behavior* 33: 1151–1161.

**Enquist, M. & Leimar, O.** 1983. Evolution of fighting behaviour: decision rules and assessment of relative strength. *Journal of Theoretical Biology* 102: 387–410.

**Enquist, B.J., West, G.B. & Brown, J.H.** 1998. Allometric scaling of plant energetics and population density. *Nature* 395: 163–165.

**Ernande, B. & Dieckmann, U.** 2004. The evolution of pnenotypic plasticity in spatially-structured environments: implications of intraspecific competition, plasticity costs and environmental charactersitics. *Journal of Evolutionary Biology* 17: 613–628.

**Ernande, B., Boudry, P., Clobert, J. & Haure, J.** 2004. Plasticity in resource allocation based life history traits in the Pacific oyster, *Crassostrea gigas*. I. Spatial variation in food abundance. *Journal of Evolutionary Biology* 17: 342–356.

F

**Ferrière, R. & Clobert, J.** 1992. Evolutionary stable age at first reproduction in a density-dependent model. *Journal of Theoretical Biology* 157: 253–267.

**Fisher, R.A.** 1930. *The Genetical Theory of Natural Selection*. Clarendon Press, London.

**Foster, K.R., Fortunato, A., Strassmann, J.E. & Queller, D.C.** 2002. The costs and benefits of being a chimera. *Proceedings of the Royal Society of London Series B* 269: 2357–2362.

**Freedman, L.P. & Luisi, B.F.** 1993. On the mechanism of DNA-binding by nuclear hormone receptors: a structural and functional perspective. *Journal of Cellular Biochemistry* 51: 140–150.

G

**Gaillard, J.-M., Pontier, D. Allainé, D., Lebreton, J.-D., Trouvilliez, J. & Clobert, J.** 1989. An analysis of demographic tactics in birds and mammals. *Oikos* 56: 59–76.

**Gavrilets, S. & Hayashi, T.I.** 2005. Speciation and sexual conflict. *Evolutionary Ecology* 19: 167–198

**Gavrilets, S. & Hayashi, T.I.** 2006. The dynamics of two- and three-way sexual conflicts over mating. *Philosophical Transactions of the Royal Society B* 361: 345–354.

**Gibson, J.R., Chippindale, A.K. & Rice, W.R.** 2002. The X chromosome is a hot spot for sexually antagonistic fitness variation. *Proceedings of the Royal Society of London Series B* 269: 499–505.

**Goodnight, C.J.** 1995. Epistasis and the increase in additive genetic variance: Implications for phase 1 of Wright's shifting-balance process. *Evolution* 49: 502–511.

**Gwynne, D.T.** 1981. Sexual difference theory: Mormon crickets show role reversal in mate choice. *Science* 213: 779–780.

**Gwynne, D.T. & Simmons, L.W.** 1990. Experimental reversal of courtship roles in an insect. *Nature* 346: 171–174.

H

**Haig, D. & Westoby, M.** 1989. Parent-specific gene expression and the triploid endosperm. *American Naturalist* 134: 147–155.

**Hamilton, W.D.** 1964. The evolution of social behavior I and II. *Journal of Theoretical Biology* 7: 1–52.

**Hamilton, W.D. & May, R.M.** 1977. Dispersal in stable habitats. *Nature* 269: 578–581

**Hanken, J. & Wake, D.B.** 1993. Miniaturization of body size: organismal consequence and evolutionary significance. *Annual Review of Ecology and Systematics* 21: 501–519.

**Hayes, T.B.** 1997a. Hormonal mechanisms as potential constraints on evolution: examples from the Anura. *American Zoologist* 37: 482–490.

**Hayes, T.B.** 1997b. Steroids as potential modulators of thyroid hormone activity in anuran metamorphosis. *American Zoologist* 37: 185–194.

**Holland, B. & Rice, W.R.** 1999. Experimental removal of sexual selection reverses intersexual antagonistic coevolution and removes a reproductive load. *Proceedings of the National Academy of Sciences of the USA* 96: 5083–5088.

**Houde, A.E.** 1994. Effect of artificial selection on male colour patterns and mating preference of female guppies. *Proceedings of the Royal Society of London Series B* 256: 125–130.

**Huey, R.B., Hertz, P.E. & Sinervo, B.** 2003. Behavioral drive versus behavioural inertia in evolution: a null model approach. *American Naturalist* 161: 357–366.

J

**Jansen, V.A. & van Baalen, M.** 2006. Altruism through beard chromodynamics. *Nature* 440: 663–666.

**Julliard, R., Leirs, H., Stenseth, N.C., Yoccoz, N.G., Prevot-Julliard, A.C., Verhagen, R. & Verheyen, W.** 1999. Survival-variation within and between functional categories of the African multimammate rat. *Journal of Animal Ecology* 68: 550–561.

K

**Keller, L. & Ross, K.G.** 1998. Selfish genes: a greenbeard in the red fire ant. *Nature* 374: 573–575.

**Kelly, J.K.** 2000. Epistasis, linkage and balancing selection. In *Epistasis and Evolutionary Process* (Wolf, J.B., Brodie III, E.D. & Wade, M.J. eds), pp. 146–157. Oxford University. Press, New York.

**Kingsolver, J.G., Hoekstra, H.E., Hoekstra, J.M., Berrigan, D., Vignieri, S.N., Hill, C.E., Hoang, A., Gibert, P. & Beerli, P.** 2001. The strength of phenotypic selection in natural populations. *American Naturalist* 157: 245–261.

**Kokko, H.** 2001. Fisherian and 'good genes' benefits of mate choice: how (not) to distinguish between them. *Ecology Letters* 4: 322–326.

**Krebs, J.R. & Davies, N.B.** 1987. *An Introduction to Behavioral Ecology*. Blackwell Scientific, Oxford.

L

**Lande, R.** 1981. Models of speciation by sexual selection on polygenic traits. *Proceedings of the National Academy of Sciences USA* 78: 3721–3725.

**Le Galliard, J.-F., Fitze, P., Ferrière, R. & Clobert, J.** 2005. Sex ratio bias, male aggression, and population collapse in lizards. *Proceedings of the National Academy of Sciences USA* 102: 18231–18236.

**Legendre, S. & Clobert, J.** 1995. ULM, software for conservation and evolutionary biologists. *Journal of Applied Statistics* 22: 817–834.

**Levins, R.** 1962a. Theory of fitness in a heterogeneous environment. I. The fitness set and adaptive function. *American Naturalist* 96: 361–373.

**Levins, R.** 1962b. Theory of fitness in a heterogeneous environment. II. Developmental flexibility and niche selection. *American Naturalist* 97: 74–90.

**Linke, R., Roth, G. & Rottluff, B.** 1993. Comparative studies on the eye morphology of lungless salamanders, family Plethodontidae, and the effect of miniaturization. *Journal of Morphology* 189: 131–143.

**Lively, C.M.** 1986. Canalization versus developmental conversion in a spatially variable environment. *American Naturalist* 128: 561–572.

**Luetenegger, W.** 1979. Evolution of litter size in primates. *American Naturalist* 114: 525–531.

**Lynch, M. & Walsh, B.** 1998. *Genetics and Analysis of Quantitative Traits*. Sinauer Associates, Sunderland, Massachusetts.

M

**MacArthur, R.H. & Wilson, E.O.** 1967. *The Theory of Island Biogeography*. Princeton University Press, Princeton, NJ.

**Madsen, T., Shine, R., Loman, A. & Hakansson, T.** 1992. Why do female adders copulate so frequently? *Nature* 355: 440–442.

**Marler, C.A. & Moore, M.C.** 1991. Energetics of aggression: Supplemental feeding compensates for testosterone-induced costs of aggression in male mountain spiny lizards (*Sceloporus jarrovi*). *Animal Behaviour* 42: 209–219.

**May, R.M.** 1976. Simple mathematical models with very complicated dynamics. *Nature* 261: 459–467.

**Maynard Smith, J.** 1982. *Evolution and Theory of Games*. Cambridge University Press, Cambridge, Massachusetts.

**McMahon, T.** 1973. Size and shape in biology. *Science* 179: 1201–1204.

**Mueller, R.L., Macey, J.R., Jaekel, M., Wake, D.B. & Boore, J.L.** 2004. Morphological homoplasy, life history evolution, and historical biogeography of plethodontid salamanders inferred from complete mitochondrial genomes. *Proceedings of the National Academy of Sciences of the USA* 38: 13820–13825.

N

**Newman, R.A.** 1994. Genetic variation for phenotypic plasticity in the larval life history of spadefoot toads (*Scaphiopus couchii*). *Evolution* 48: 1773–1785.

O

**Orians, G.H.** 1969. On the evolution of mating systems in birds and mammals. *American Naturalist* 103: 589–603.

P

**Peters, R.H.** 1983. *The Ecological Implications of Body Size*. Cambridge University Press, Cambridge.

**Pfennig, D.W.** 1992. Proximate and functional causes of polyphenism in an anuran tadpole. *Functional Ecologist* 6: 167–174.

**Pfennig, D.W., Reeve, H.K. & Sherman, P.W.** 1993. Kin recognition and cannibalism in spadefoot toad tadpoles. *Animal Behavior* 46: 67–84.

**Phillips, P.C. & Arnold, S.J.** 1989. Visualizing multivariate selection. *Evolution* 43: 1209–1222.

**Pianka, E.R.** 1970. On *r* and *K* selection. *American Naturalist* 100: 592–597.

**Pope, P.H.** 1928. The life history of *Triturus viridescens*: some further notes. *Copeia* 168: 61–73.

Q

**Queller, D.C., Ponte, E., Bozzaro, S. & Strassmann, J.E.** 2003. Single-gene greenbeard effects in the social amoeba *Dictyostelium discoideum*. *Science* 299: 105–106.

R

**Rankin, M.A.** 1991. Endocrine effects on migration. *American Zoologist* 31: 217–230.

**Reynolds, J.D., Goodwin, N.B. & Freckleton, R.P.** 2002. Evolutionary transitions in parental care and live-bearing in vertebrates. *Proceedings of the Royal Society of London Series B* 357: 269–281.

**Reznick, D., Bryant, M. & Holmes, D.** 2006. The evolution of senescence and post-reproductive lifespan in guppies (*Poecilia reticulata*). *Plos Biology* 4: 136–143.

**Reznick, D., Nunny, L. & Tessier, A.** 2000. Big houses, big cars, superfleas, and the cost of reproduction. *Trends in Ecology and Evolution* 15: 421–425.

**Rice, W.R.** 1984. Sex chromosomes and the evolution of sexual dimorphism. *Evolution* 38: 735–742.

**Rice, W.R.** 2000. Dangerous liaisons. *Proceedings of the National Academy of Sciences of the USA* 97: 12953–12955.

**Rice, W.R. & Chippindale, A.K.** 2001. Intersexual ontogenetic conflict. *Journal of Evolutionary Biology* 14: 685–693.

**Riska, B., Atchley, W.R. & Rutledge, J.J.** 1986. A genetic analysis of targeted growth rate in mice. *Genetics* 107: 79–101.

**Roff, D.A.** 1992. *The Evolution of Life Histories*. Chapman & Hall, New York.

**Roff, D.A. & Fairbairn, D.J.** 2006. The evolution of trade-offs: where are we? *Journal of Evolutionary Biology* 20: 433–447.

**Rolland, C., Danchin, E. & de Fraipont, M.** 1998. The evolution of coloniality in birds in relation to food, habitat, predation, and life-history traits: a comparative analysis. *American Naturalist* 151: 514–529.

**Rose, M. & Charlesworth, B.** 1980a. A test of evolutionary theories of senescence. *Nature* 287: 141–142.

**Rose, M. & Charlesworth, B.** 1980b. Genetics of life history in *Drosophila melanogaster*. II. Exploratory selection experiments. *Genetics* 97: 187–196.

**Ross, K.G., Vargo, E.L. & Keller, L.** 1996. Simple genetic basis for important social traits in the fire ant *Solenopsis invicta*. *Evolution* 50: 2387–2399.

**Rousset, F.** 2004. *Genetic Structure and Selection in Subdivided Populations*. Princeton University Press, Princeton, NJ.

**Rowe, L., Arnqvist, G., Sih, A. & Kruppa, J.J.** 1994. Sexual conflict and the evolutionary ecology of mating patterns: water striders as a model system. *Trends in Ecology and Evolution* 9: 289–293.

**Ryan, M.** 1997. Sexual selection and mate choice. In *Behavioral Ecology: An Evolutionary Approach* (Krebs, J.H. & Davies, N.B. eds), pp. 179–202. Blackwell, Oxford.

S

**Sanchez, R., Nguyen, D. & Rocha, W.** 2002. Diversity in the mechanisms of gene regulation by estrogen receptors. *Bioessays* 24: 244–254.

**Sawada, K., Ukena, K., Kikuyama, S. & Tsutsui, K.** 2002. Identification of cDNA encoding a novel amphibian growth hormone-releasing peptide and localization of its transcript. *Journal of Endocrinology* 174: 395–402.

**Schraden, C. & Anzenburger, G.** 1999. Prolactin, the hormone of paternity. *News in the Physiological Sciences* 14: 223–231.

**Semlistch, R.D.** 1988. Intraspecific heterochrony and life history: decoupling somatic and sexual development in a facultatively paedomorphic salamander. *Proceedings of the National Academy of Sciences of the USA* 95: 5643–5648.

**Semlitsch, R.D., Scott, D.E. & Pechmann, J.H.K.** 1988. Time and size at metamorphosis related to adult fitness in *Ambystoma talpoideum*. *Ecology* 69: 184–192.

**Sessions, S.K. & Larson, A.** 1987. Developmental correlates of genome evolution in plethodontid salamanders and their implications for genome evolution. *Evolution* 41: 1239–1251.

**Shaffer, H.B., Austin, C.C. & Huey, R.B.** 1991. The consequences of metamorphosis on salamander (*Ambystoma*) locomoter performance. *Physiological Zoology* 64: 212–231.

**Shine, R. & Charnov, E.L.** 1992. Patterns of survivorship, growth and maturation in lizards and snakes. *American Naturalist* 139: 1257–1269.

**Shuster, S.M. & Wade, M.J.** 2003. *Mating Systems and Strategies*. New Jersey, Princeton University Press.

**Sinervo, B.** 1990. The evolution of maternal investment in lizards: an experimental and comparative analysis of egg size and its effects on offspring performance. *Evolution* 44: 279–294.

**Sinervo, B.** 1999. Mechanistic analysis of natural selection and a refinement of lacks and Williams principles. *American Naturalist* 154: S26–S42.

**Sinervo, B.** 2000. Adaptation, natural selection, and optimal life history allocation. In: *Adaptive Genetic Variation in the Wild* (Mousseau, T.A. Sinervo, B. & Endler, J.A. eds), pp. 41–64. Oxford University Press, New York.

**Sinervo, B.** 2001. Runaway social games, genetic cycles driven by alternative male and female strategies, and the origin of morphs. In *Macroevolutionary and Microevolutionary Process* (Hendry, A. & Kinnison, M. eds). *Genetica* 112: 417–434.

**Sinervo, B. & Basolo, A.** 1996. Measuring adaptation with phenotypic manipulations. In *Adaptation* (Rose, M.R. & Lauder, G.V. eds), pp. 149–185. Academic Press, New York.

**Sinervo, B. & Calsbeek, R.** 2003. Ontogenetic conflict and morphotypic selection on physiology, life history, and adaptive sex allocation. In *Selection and Evolution of Performance in Nature* (Kingsolver, J. & Huey, R.B. eds). *Integrative and Comparative Biology* 43: 419–430.

**Sinervo, B. & Calsbeek, R.** 2006. The developmental and physiological causes and consequences of frequency dependent selection in the wild. *Annual Review of Ecology and Systematics* 37: 581–610.

**Sinervo, B. & Clobert, J.** 2003. Morphs, dispersal, genetic similarity and the evolution of cooperation. *Science* 300: 1949–1951.

**Sinervo, B. & Huey, R.B.** 1990. Allometric engineering: an experimental test of the causes of interpopulational differences in locomotor performance. *Science* 248: 1106–1109.

**Sinervo, B. & Licht, P.** 1991. Proximate constraints on the evolution of egg size, egg number and total clutch mass in lizards. *Science* 252: 1300–1302.

**Sinervo, B. & Svensson, E.I.** 2002. Correlational selection and the evolution of genomic architecture. *Heredity* 89: 329–338.

**Sinervo, B., Bleay, C. & Adamopoulou, C.** 2001. Social causes of selection and the resolution of a heritable throat color polymorphism in a lizard. *Evolution* 55: 2040–2052.

**Sinervo, B., Svensson, E.I. & Comendant, T.** 2000b. Density cycles and an offspring quantity and quality game driven by natural selection. *Nature* 406: 985–988.

**Sinervo, B., Calsbeek, R., Comendant, T., Both, C., Adamopoulou, C. & Clobert, J.** 2006b. Genetic and maternal determinants of effective dispersal: the effect of sire genotype and size at birth in side-blotched lizards. *American Naturalist* 168: 88–99.

**Sinervo, B., Chaine, A., Clobert, J., Calsbeek, R., McAdam, A., Hazard, L., Lancaster, L., Alonzo, S.H., Corrigan, G. & Hochberg, M.** 2006a. Self-recognition, color signals and cycles of greenbeard mutualism and transient altruism. *Proceedings of the National Academy of Sciences of the USA* 103: 7372–7377.

**Sinervo, B., Huey, R.B., Doughty, P. & Zamudio, K.** 1992. Allometric engineering: a causal analysis of natural selection on offspring size. *Science* 258: 1927–1930.

**Sinervo, B., Miles, D.B., Frankino, W.A., Klukowski, M. & DeNardo, D.F.** 2000a. Testosterone, endurance, and Darwinian fitness: natural and sexual selection on the physiological bases of alternative male behaviors in side-blotched lizards. *Hormones and Behavior* 38: 222–233.

**Stearns, S.C.** 1983. The effect of size and phylogeny on patterns of covariation among life history traits in the mammals. *Oikos* 41: 173–187.

**Stearns, S.C.** 1984. The effect of size and phylogeny on patterns of covariation in the life history traits of lizards and snakes. *American Naturalist* 123: 56–72.

**Stearns, S.C.** 1992. *The Evolution of Life Histories*. Oxford University Press, Oxford.

**Svensson, E.I., Sinervo, B. & Comendant, T.** 2001a. Condition, genotype-by-environment interaction, and correlational selection in lizard life-history morphs. *Evolution* 55: 2053–2069.

**Svensson, E.I., Sinervo, B. & Comendant, T.** 2001b. Density-dependent competition and selection on immune function in genetic lizard morphs. *Proceedings of the National Academy of Sciences of the USA* 98: 12561–12565 23: 2001.

**Svensson, E.I., Sinervo, B. & Comendant, T.** 2002. Mechanistic and experimental analysis of condition and reproduction in a polymorphic lizard. *Journal of Evolutionary Biology* 15: 1034–1047.

**Svensson, E.I., Abbott, J. & Hardling, R.** 2005. Female polymorphism, frequency dependence, and rapid evolutionary dynamics in natural populations. *American Naturalist* 165: 567–576.

T

**Thompson, J.N.** 2005. *The Geographic Mosaic of Coevolution*. 443 pages. University of Chicago Press, IL.

**Thompson, J.N. & Pellmyr, O.** 1992. Mutualism with pollinating seed parasites amid co-pollinators – constraints on specialization. *Ecology* 73: 1780–1791.

**Trivers, R.L.** 1972. Parental investment and sexual selection. In *Sexual Selection and Descent of Man*. (Campbell, B. ed.), pp. 136–179. Aldine, Chicago, IL.

V

**van Rhijn, J.G. & Vodegel, R.** 1980. Being honest about ones intentions: an evolutionary stable strategy. *Journal of Theoretical Biology* 85: 623–641.

**Vercken, E., Massot, M. Sinervo, B. & Clobert, J.** 2007. Colour polymorphism and alternative reproductive strategies in females of the common lizard *Lacerta vivipara*. *Journal of Evolutionary Biology* 20: 221–232.

**Voss, S.R. & Shaffer, H.B.** 1997. Adaptive evolution via a major gene effect: paedomorphosis in the Mexican axolotl. *Proceedings of the National Academy of Sciences of the USA* 94: 14185–14189.

W

**Wachtel, S., Demas, S., Tiersch, T., Pechan, P. & Shapiro, D.** 1991. Bkm satellite DNA and Zfy in the coral-reef fish *Anthias squamipinnis*. *Genome* 34: 612–617.

**Waddington, C.H.** 1966. *Principles of Development and Differentiation*. MacMillan Pub. Co., New York, New York.

**Wade, M.J.** 2002. A gene's eye view of epistasis, fitness, and speciation. *Journal of Evolutionary Biology* 15: 337–346.

**Wake, D.B.** 1991. Homoplasy, the result of natural selection or of design limitations. *American Naturalist* 138: 543–567.

**Wake, D.B. & Larson, A.** 1987. Multidimensional analysis of an evolving lineage. *Science* 238: 42–48.

**Wallace, B.** 1975. Hard and soft selection revisited. *Evolution* 29: 465–473.

**West, G.B., Brown, J.H. & Enquist, B.J.** 1997. A general model for the origin of allometric scaling laws in biology. *Science* 276: 122–126.

**West, S.A., Griffin, A.S. & Gardner, A.** 2007. Social semantics: altruism, cooperation, mutualism, strong reciprocity and group selection. *Journal of Evolutionary Biology*, 20: 415–432.

**Western, D.** 1979. Size, life-history and ecology in mammals. *African Journal of Ecology* 17: 185–204.

**Whitlock, M.C., Phillips, P.C., Moore, F.B.G. & Tonsor, S.J.** 1995. Multiple fitness peaks and epistasis. *Annual Review of Ecology and Systematics* 26: 601–629.

**Wiens, J.J., Bonett, R.M. & Chippindale, P.T.** 2005. Ontogeny discombobulates phylogeny: paedomorphosis and higher level salamander relationships. *Systematic Biology* 54: 91–110.

**Wiley, R.H.** 2000. Sexual selection and mate choice: trade-offs for males and females. In *Vertebrate Mating Systems* (Appolonio, M., Festa-Bianchet, M. & Mainardi, D. eds) (Zichichi, A. series editor), pp. 8–48. Proceedings of the 14th Course of the International School of Ethology, World Scientific Publishing Co, Singapore.

**Williams, P.D., Day, T., Fletcher, Q. & Rowe, D.** 2006. The shaping of senescence in the wild. *Trends in Ecology and Evolution* 21: 458–463.

**Wright, S.** 1969. *Evolution and the Genetics of Populations*, vol. 2. Chicago University Press, Chicago, IL.

Z

**Zajac, J.D. & Chilco, P.J.** 1995. Transcriptional control and the regulation of endocrine genes. *Clinical and Experimental Pharmacology* 22: 935–943.

**Zamudio, K. & Sinervo, B.** 2000. Polygyny, mate-guarding, and posthumous fertilization as alternative male mating strategies. *Proceedings of the National Academy of Sciences of the USA* 97: 14427–14432.

**Zera, A.J. & Harshman, L.G.** 2001. The physiology of life history trade-offs in animals. *Annual Review of Ecology and Systematics* 32: 95–126.

**Zoltán, B. & Szép, T.** 1994. Frequency-dependent selection on information-transfer strategies at breeding colonies: a simulation study. *Behavioral Ecology* 6: 308–310.

## Chapter 6


A

**Abbott, D.H., Keverne, E.B., Bercovitch, F.B., Shively, C.A., Mendoza, S.P., Saltzman, W., Snowdon, C.T., Ziegler, T.E., Banjevic, M., Garland Jr, T. & Sapolsky, R.M.** 2003. Are subordinates always stressed? A comparative analysis of rank differences in cortisol levels among primates. *Hormones and Behavior* 43: 677–682.

**Able, K.P.** 1999. *Gatherings of Angels: Migrating Birds and Their Ecology*. Cornell University Press, Ithaca, New York.

**Ader, R.** 2000. On the development of psychoneuroimmunology. *European Journal of Pharmacology* 405: 167–176.

**Adkins, E.K.** 1976. Embryonic exposure to an antiestrogen masculinizes behavior of female quail. *Physiology and Behavior* 17: 357–359.

**Adkins-Regan, E.** 1987. Sexual differentiation in birds. *Trends in Neuroscience* 10: 517–522.

**Adkins-Regan, E.** 2005. Activity dependent brain plasticity: does singing increase the volume of a song system nucleus? Theoretical comment on Sartor and Ball (2005). *Behavioral Neuroscience* 119: 346–348.

**Agrawal, A.A.** 2001. Phenotypic plasticity in the interactions and evolution of species. *Science* 294: 321–326.

**Ahima, R.S. & Flier, J.S.** 2000. Leptin. *Annual Review of Physiology* 62: 413–437.

**Ahima, R.S., Dushay, J., Flier, S.N., Prabakaran, D. & Flier, J.S.** 1997. Leptin accelerates the onset of puberty in normal female mice. *Journal of Clinical Investigation* 99: 391–395.

**Akana, S.F., Strack, A.M., Hanson, E.S., Horsley, C.J., Milligan, E.D., Bhatnagar, S. & Dallman, M.F.** 1999. Interactions among chronic cold, corticosterone and puberty on energy intake and deposition. *Stress* 3: 131–146.

**Andersson, M.** 1994. *Sexual Selection. Monographs in Behavior and Ecology*. Princeton Unviversity Press.

**Andreassen, H.P. & Ims, R.A.** 1990. Responses of female grey-backed voles *Clethrionomys rufocanus* to malnutrition: a combined laboratory and field experiment. *Oikos* 59: 107–114.

**Anisman, H., Zaharia, M.D., Meaney, M.J. & Merali, Z.** 1998. Do early life events permanently alter behavioral and hormonal responses to stressors? *International Journal of Developmental Neuroscience* 16: 149–164.

**Arendash, G.W. & Gorski, R.A.** 1983. Effects of discrete lesions of the sexually dimorphic nucleus of the preoptic area or other medial preoptic regions on the sexual behavior of male rats. *Brain Research Bulletin* 10: 147–154.

**Arnold, A.P.** 1975. The effects of castration and androgen replacement on song, courtship and aggression in zebra finches (*Poephilia guttata*). *Journal of Experimental Zoology* 191: 309–326.

**Arnold, A.P.** 1996. Genetically triggered sexual differentiation of brain and behavior. *Hormones and Behavior* 30: 495–505.

**Astheimer, L.B., Buttemer, W.A. & Wingfield, J.C.** 2000. Corticosterone treatment has no effect on reproductive hormones or aggressive behavior in free-living male tree sparrows, *Spizella arborea*. *Hormones and Behavior* 37: 31–39.

B

**Bairlein, F.** 1990. Nutrition and food selection in migratory birds. In *Bird Migration: Physiology and Ecophysiology* (Gwinner, E. ed.), pp. 198–213. Springer-Verlag, Berlin.

**Balaban, E.** 2005. Brain switching: studying evolutionary behavioral changes in the context of individual brain development. *International Journal of Developmental Biology* 49: 117–124.

**Ball, G.F.** 1991. Endocrine mechanisms and the evolution of avian parental care. In *Acta XX Congr. Int. Ornithol.*, pp. 984–991.

**Balthazart, J. & Ball, G.F.** 1998. New insights into the regulation and function of brain estrogen synthase (aromatase). *Trends in Neuroscience* 21: 243–249.

**Balthazart, J. & Foidart, A.** 1993. Neural bases of behavioral sex differences in the quail. In *The Development of Sex Differences and Similarities in Behavior* (Haug, M. ed.), pp. 51–75. Kluwer Academic, Amsterdam.

**Balthazart, J., Baillien, M. & Ball, G.F.** 2006. Rapid control of brain aromatase activity by glutamatergic inputs. *Endocrinology* 147: 359–366.

**Barrett, J., Abbott, D.H. & George, L.M.** 1993. Sensory cues and the suppression of reproduction in subordinate female marmoset monkeys, *Callithrix jacchus*. *Journal of Reproduction and Fertility* 97: 301–310.

**Bass, A.H.** 1995. Alternative life history strategies and dimorphic males in an acoustic communication system. In: *Proceedings of the Fifth International Symposium on the Reproductive Physiology of Fish*, pp. 258–260. Austin, TX.

**Bass, A.H.** 1996. Shaping brain sexuality. *American Scientist* 84: 352–363.

**Bass, A.H. & Groberb, M.S.** 2001. Social and neural modulation of sexual plasticity in teleost fish. *Brain, Behavior and Evolution* 57: 293–300.

**Bass, A.H., Horvth, B.J. & Brothers, E.B.** 1996. Non-sequential developmental trajectories lead to dimorphic vocal circuitry for males with alternative reproductive tactics. *Journal of Neurobiology* 30: 493–504.

**Baulieu, E.E.** 1998. Neurosteroids: a novel function of the brain. *Psychoneuroendocrinology* 23: 963–987.

**Beatty, W.W.** 1979. Gonadal hormones and sex differences in nonreproductive behaviors in rodents: organizational and activational influences. *Hormones and Behavior* 12: 112–163.

**Beletsky, L.D., Gori, D.F., Freeman, S. & Wingfield, J.C.** 1995. Testosterone and polygyny in birds. *Current*

*Ornithology* (Powers, D.M. ed.), pp. 1–41. Plenum, New York.

**Belthoff, J.R. & Dufty, A.M., Jr.** 1995. Activity levels and the dispersal of western screech-owls, *Otus kennicottii*. *Animal Behaviour* 50: 558–561.

**Belthoff, J.R. & Dufty, A.M., Jr.** 1998. Corticosterone, body condition and locomotor activity: a model for dispersal in screech owls. *Animal Behaviour* 55: 405–415.

**Berdanier, C.D.** 1989. Role of glucocorticoids in the regulation of lipogenesis. *FASEB Journal* 3: 2179–2183.

**Bernard, D.J. & Ball, G.F.** 1997. Photoperiodic condition modulates the effects of testosterone on song control nuclei volumes in male European starlings. *General and Comparative Endocrinology* 105: 276–283.

**Bernard, D.J., Wilson, F.E. & Ball, G.F.** 1997. Testis-dependent and -independent effects of photoperiod on volumes of song control nuclei in American tree sparrows (*Spizella arborea*). *Brain Research* 760: 163–169.

**Berthold, P.** 1990. Genetics of migration. In *Bird Migration: Physiology and Ecophysiology* (Gwinner, E. ed.), pp. 269–280. Springer-Verlag, Berlin.

**Berven, K.A.** 1981. Mate choice in the wood frog, *Rana sylvatica*. *Evolution* 35: 707–722.

**Bishop, C.M., Butler, P.J. & Atkinson, N.M.** 1995. The effect of elevated levels of thyroxine on the aerobic capacity of locomotor muscles of the tufted duck, *Aythya fuligula*. *Journal of Comparative Physiology B* 164: 618–621.

**Blem, C.R.** 1990. Avian energy storage. In Powers, D.M. (ed.), *Current Ornithology*, vol. 7, pp. 59–113. Plenum, New York.

**Bondrup-Nielsen, S.** 1992. Emigration of meadow voles, *Microtus pennsylvanicus*: the effect of sex ratio. *Oikos* 65: 358–360.

**Boorse, G.C. & R.J. Denver.** 2003. Endocrine mechanisms underlying plasticity in metamorphic timing in spadefoot toads. *Integrative and Comparative Biology* 43: 646–657.

**Boswell, T., P.J. Sharp, M.R. Hall & A.R. Goldsmith.** 1995. Migratory fat deposition in European quail: a role for prolactin? *Journal of Endocrinology* 146: 71–79.

**Bottjer, S.W., J.N. Schoonmaker & A.P. Arnold.** 1986. Auditory and hormonal stimulation interact to produce neural growth in adult canaries. *Journal of Neurobiology* 17: 605–612.

**Bowden, R.M., M.A. Ewert & C.E. Nelson.** 2000. Environmental sex determination in a reptile varies seasonally and with yolk hormones. *Proceedings of the Royal Society of London Series B* 267: 1745–1749.

**Bowden, R.M., Ewert, M.A., Lipar, J.L. & Nelson, C.E.** 2001. Concentrations of steroid hormones in layers and biopsies of chelonian egg yolks. *General and Comparative Endocrinology* 121: 95–103.

**Brantley, R.K. & Bass, A.H.** 1994. Alternative male spawning tactics and acoustic signals in the plainfin midshipman fish, *Porichthys notatus* Girard (Teleostei, Batrachoididae). *Ethology* 96: 213–223.

**Brantley, R.K., Wingfield, J.C. & Bass, A.H.** 1993. Sex steroid levels in *Porichthys notatus*, a fish with alternative reproductive tactics and a review of the hormonal bases for male dimorphism among teleost fish. *Hormones and Behavior* 27: 332–347.

**Brenowitz, E.A.** 2004. Plasticity of the adult avian song control system. *Annals of the New York Academy of Science* 1016: 560–585.

**Breuner, C.W. & Orchinik, M.** 2000. Downstream from corticosterone: seasonality of binding globulins, receptors and behavior in the avian stress response. In *Avian Endocrinology* (Dawson, A. & Chaturvedi, C.M. eds), pp. 1–12. Narosa Publishing House, New Delhi.

**Breuner, C.W., Orchinik, M., Hahn, T.P., Meddle, S.L., Moore, I.T., Owen-Ashley, N.T., Sperry, T.S. & Wingfield, J.C.** 2003. Differential mechanisms for regulation of the stress response across latitudinal gradients. *American Journal of Physiology* 285: R594–600.

**Bridges, R.S.** 1996 Biochemical basis of parental behavior in the rat. In: *Advances in the Study of Behavior* (Rosenblatt, J.S. & Snowden, C.T. eds), vol. 25, pp. 215–242. Academic Press, Orlando, FL.

**Bridges, R.S. & Mann, P.E.** 1994. Prolactin–brain interactions in the induction of material maternal behavior in rats. *Psychoneuroendocrinology* 19: 611–622.

**Bridges, R.S., Robertson, M.C., Shiu, R.P.C., Sturgis, J.D., Henriquez, B.M. & Mann, P.E.** 1997. Central lactogenic regulation of maternal behavior in rats: steroid dependence, hormone specificity and behavioral potencies of rat prolactin and rat placental lactogen I. *Endocrinology* 138: 756–763.

**Bronson, F.H.** 1989. *Mammalian Reproductive Biology*. University of Chicago Press.

**Buntin, J.D.** 1989. Time course and response specificity of prolactin-induced hyperphagia in ring doves. *Physiology and Behavior* 45: 903–909.

**Buntin, J.D., Advis, J.P., Ottinger, M.A., Lea, R.W. & Sharp, P.J.** 1999. An analysis of physiological mechanisms underlying the antigonadotropic action of intracranial prolactin in ring doves. *General and Comparative Endocrinology* 114: 97–107.

**Buntin, J.D., el Halawani, M.E., Ottinger, M.A., Fan, Y. & Fivizzani, A.J.** 1998. An analysis of sex and breeding stage differences in prolactin binding activity in brain and hypothalamic GnRH concentration in Wilson's phalarope, a sex role-reversed species. *General and Comparative Endocrinology* 109: 119–132.

**Burke, W.H. & Henry, M.H.** 1999. Gonadal development and growth of chickens and turkeys hatched from eggs injected with aromatase inhibitor. *Poultry Science* 78: 1019–1033.

**Burmeister, S. & Wilczynski, W.** 2000. Social signals influence hormones independently of calling behavior in the treefrog (*Hyla cinerea*). *Hormones and Behavior* 38: 201–209.

**Butler, C.G.** 1954. The method and importance of the recognition by a colony of honey bees (*A. mellifera*) of the presence of its queen. *Transactions of the Royal Entomological Society of London* 105: 11–29.

**Butler, P.J. & Woakes, A.J.** 2001. Seasonal hypothermia in a large migrating bird: saving energy for fat deposition? *Journal of Experimental Biology* 204: 1361–1367.

C

**Cardinali D.P., Cutrera, R.A. & Esquifino, A.I.** 2000. Psychoimmune neuroendocrine integrative mechanisms revisited. *Biological Signals and Receptors* 9: 215–230.

**Caro, S.P., Lambrechts, M.M. & Balthazart, J.** 2005. Early seasonal development of brain song control nuclei in male blue tits. *Neuroscience Letters* 386: 139–144.

**Cheng, M.F.** 1986. Female cooing promotes ovarian development in ring doves. *Physiology and Behavior* 37: 371–374.

**Cheng, M.F., Peng, J.P. & Johnson, P.** 1998. Hypothalamic neurons preferentially respond to female nest coo stimulation: demonstration of direct acoustic stimulation of luteinizing hormone release. *Journal of Neuroscience* 18: 5477–5489.

**Cheng, M.F., Desiderio, C., Havens, M. & Johnson, A.** 1988. Behavioral stimulation of ovarian growth. *Hormones and Behavior* 22: 388–401.

**Cherel, Y., Robin, J.-P., Walch, O., Karmann, H., Netchitailo, P. & le Maho, Y.** 1988. Fasting in king penguin. I. Hormonal and metabolic changes during breeding. *American Journal of Physiology* 254: R170–R177.

**Choi, S.M., Yoo, S.D. & Lee, B.M.** 2004. Toxicological characteristics of endocrine-disrupting chemicals: developmental toxicity, carcinogenicity and mutagenicity. *Journal of Toxicology and Environmental Health Part B: Critical Reviews* 7: 1–32.

**Christensen L.W., Nance, D.M. & Gorski, R.A.** 1977. Effects of hypothalamic and preoptic lesions on reproductive behavior in male rats. *Brain Research Bulletin* 2: 137–141.

**Christian, J.J., Lloyd, J.A. & Davis, D.E.** 1965. The role of endocrines in the self-regulation of mammalian populations. *Recent Progress in Hormone Research* 21: 501–578.

**Clobert, J., Danchin, E., Dhondt, A.A. & Nichols, J.D.** 2001. *Dispersal*. Oxford University Press.

**Clode, D., Birks, J.D.S. & Macdonald, D.W.** 2000. The influence of risk and vulnerability on predator mobbing by terns (*Sterna* spp.) and gulls (*Larus* spp.) *Journal of Zoology* 252: 53–59.

**Cole, C.J.** 1984. Unisexual lizards. *Scientific American* 250: 94–100.

**Considine R.V., Sinha, M.K, Heiman, M.L., Kriauciunas, A., Stephens, T.W., Nyce, M.R., Ohannesian, J.P., Marco, C.C., McKee, L.J., Bauer, T.L.** *et al.* 1996. Serum immunoreactive-leptin concentrations in normal-weight and obese humans. *New England Journal of Medicine* 334: 292–295.

**Cote, J., Clobert, J., Meylan, S. & Fitze, P.S.** 2006. Experimental enhancement of corticosterone levels positively affects subsequent male survival. *Hormones and Behavior* 49: 320–327.

**Crespi, E.J. & Denver, R.J.** 2005. Role of stress hormones in food intake regulation in anuran amphibians throughout the life cycle. *Comparative Biochemistry and Physiology A* 141: 381–390.

**Crews, D.** 1987. Diversity and evolution of behavioral controlling mechanisms. In *The Psychobiology of Reproductive Behavior* (Crews, D. ed.), pp. 88–119. Prentice Hall, Englewood Cliffs, NJ.

**Crews, D.** 1997. Species diversity and the evolution of behavioral controlling mechanisms. *Annals of the New York Academy of Sciences* 807: 1–21.

**Crews, D.** 2003. Sex determination: where environment and genetics meet. *Evolution and Development* 5: 50–55.

**Crews, D., Bull, J.J. & Billy, A.J.** 1988. Sex determination and sexual differentiation in reptiles. In *Handbook of Sexology*, vol. 6: *The Pharmacology and Endocrinology of Sexual Function* (Sitsen, J.M.A. ed.), pp. 98–121. Elsevier, New York.

**Crews, D., Grassman, M. & Lindzey, J.** 1986. Behavioral facilitation of reproduction in sexual and unisexual whiptail lizards. *Proceedings of the National Academy of Sciences of the USA* 83: 9547–9550.

**Cushing, B.S. & Kramer, K.M.** 2005. Mechanisms underlying epigenetic effects of early social experience: the role of neuropeptides and steroids. *Neuroscience and Biobehavioral Reviews* 29: 1089–1105.

D

**Dallman, M.F.** 2005. Fast glucocorticoid actions on brain: back to the future. *Frontiers in Neuroendocrinology* 26: 103–108.

**Danchin, E.** 1980. Étude immunocytologique du complexe neuro-endocrine hypothalamo-hypophysaire au cours du développement chez le macaque (*Macaca fascicularis*) and le porc (*Sus scrofa*). Thesis, Université Pierre and Marie Curie, 20 November.

**Danchin, E. & Dang, D.C.** 1981. La différentiation & le fonctionnement de l'axe hypothalamo-hypophysaire chez le foetus de deux espèces de mammifères: le porc domestique (*Sus scrofa*) and le macaque crabier (Macaca fascicularis), primate non humain. *Cahiers d'Anthropologie* 1981: 43–106.

**Danchin, E., Dang, D.C., Dubois, M.P.** 1981. An immunocytological study of the adult crab-eating macaque (*Macaca fascicularis*) pituitary and its cytological differentiation during fetal life. *Reproduction Nutrition Development* 21: 441–454.

**Danchin, E., Dang, D.C. & Dubois, M.P.** 1982. Immunocytological study of the chronology of pituitary cytogenesis in the domestic pig (*Sus scrofa*) with special reference to the functioning of the hypothalamo–pituitary–gonadal axis. *Reproduction Nutrition Development*. 22: 135–151.

**Davidson, J.M.** 1966. Activation of the male rat's sexual behavior by intracerebral implantation of androgen. *Endocrinology* 79: 783–784.

**Dawkins, R.** 1982. *The Extended Phenotype*. Oxford University Press.

**de Fraipont, M., Clobert, J., John-Alder, H. & Meylan, S.** 2000. Increased pre-natal maternal corticosterone promotes philopatry of offspring in common lizards *Lacerta vivipara*. *Journal of Animal Ecology* 69: 404–413.

**De Vries, G.J. & Simerly, R.B.** 2002. Anatomy, development and function of sexually dimorphic neural circuits in the mammalian brain. *Hormones, Brain, and Behavior* 4: 137–191.

**DeMartini, E.E.** 1988. Spawning success of the male plainfin midshipman. I. Influences of male body size and area of spawning site. *Journal of Experimental Marine Biology and Ecology* 121: 177–192.

**Demir, E. & Dickson, B.J.** 2005. Fruitless splicing specifies male courtship behavior in *Drosophila*. *Cell* 121: 785–794.

**DeNardo, D.F. & Sinervo, B.** 1994. Effects of steroid hormone interaction on activity and home-range size of male lizards. *Hormones and Behavior* 28: 273–287.

**Denver, R.J.** 1997. Environmental stress as a developmental cue: corticotrophin-releasing hormone is a proximate mediator of adaptive phenotypic plasticity in amphibian metamorphosis. *Hormones and Behavior* 31: 169–179.

**Denver, R.J.** 1998. Hormonal correlates of environmentally induced metamorphosis in the Western spadefoot toad, *Scaphiopus hammondii*. *General and Comparative Endocrinology* 110: 326–336.

**Denver, R.J.** 1999. Evolution of the corticotrophin-releasing hormone signaling system and its role in stress-induced phenotypic plasticity. *Annals of the New York Academy of Sciences* 897: 46–53.

**Desvages, G. & Pieau, C.** 1992. Aromatase activity in gonads of turtle embryos as a function of the incubation temperature of eggs. *Journal of Steroid Biochemistry and Molecular Biology* 41: 851–853.

**Desvages, G., Girondot, M. & Pieau, C.** 1993. Sensitive stages for the effects of temperature on gonadal aromatase activity in embryos of the marine turtle *Dermochelys coriacea*. *General and Comparative Endocrinology* 92: 54–61.

**Deviche, P.** 1995. Androgen regulation of avian premigratory hyperphagia and fattening: from eco-physiology to neuroendocrinology. *American Zoologist*. 35: 234–245.

**Dietz, M.W., Piersma, T. & Dekinga, A.** 1999. Body-building without power training: endogenously regulated pectoral muscle hypertrophy in confined shorebirds. *Journal of Experimental Biology* 202: 2831–2837.

**Dingle, H.** 1996. *Migration: The Biology of Life on the Move*. Oxford University Press.

**Doutrelant, C., Blondel, J., Perret, P. & Lambrechts, M.M.** 2000. Blue tit song repertoire size, male quality and interspecific competition. *Journal of Avian Biology* 31: 360–366.

**Du Toit, L., Bennett, N.C., Katz, A.A., Kalló, I. & Coen, C.W.** 2006. Relations between social status and the gonadotrophin-releasing hormone system in females of two cooperatively breeding species of African mole-rats, *Cryptomys hottentotus hottentotus* and *Cryptomys hottentotus pretoriae*: neuroanatomical and neuroendocrinological studies. *Journal of Comparative Neurology* 494: 303–313.

**Dudai, Y.** 1988. Neurogenetic dissection of learning and short-term memory in *Drosophila*. *Annual Review of Neuroscience* 11: 537–563.

**Dufty, A.M., Jr.** 1989. Testosterone and survival: a cost of aggressiveness? *Hormones and Behavior* 23: 185–193.

**Dufty, A.M., Jr. & Belthoff, J.R.** 2001. Proximate mechanisms of natal dispersal: The role of body condition and hormones. In *Dispersal* (Clobert, J., Danchin, E., Dhondt, A.A. & Nichols, J.D. eds), pp. 217–229. Oxford University Press.

**Dufty, A.M., Jr. & Wingfield, J.C.** 1986a. Temporal patterns of circulating LH and steroid hormones in a brood parasite, the brown-headed cowbird, *Molothrus ater*. I. Males. *Journal of Zoology* 208: 191–203.

**Dufty, A.M., Jr. & Wingfield, J.C.** 1986b. The influence of social cues on the reproductive endocrinology of male brown-headed cowbirds: field and laboratory studies. *Hormones and Behavior* 20: 222–234.

**Dufty, A.M., Jr. & Wingfield, J.C.** 1990. Endocrine response of captive male brown-headed cowbirds to intrasexual cues. *Condor* 92: 613–620.

**Dufty, A.M., Jr., Goldsmith, A.R. & Wingfield, J.C.** 1987. Prolactin secretion in a brood parasite, the brown-headed cowbird, *Molothrus ater*. *Journal of Zoology* 212: 669–675.

**Dunlap, K.D. & Wingfield, J.C.** 1995. External and internal influences on indices of physiological stress. I. Seasonal and population variation in adrenocortical secretion of free-living lizards, *Sceloporus occidentalis*. *Journal of Experimental Zoology* 271: 36–46.

F

**Farner, D.S. & Follett, B.K.** 1979. Reproductive periodicity in birds. In *Hormones and Evolution* (Barrington, E.J.W. ed.), pp. 829–872. Academic Press, New York.

**Farner, D.S. & Wingfield, J.C.** 1980. Reproductive endocrinology of birds. *Annual Review of Physiology* 42: 457–472.

**Faulkes, C.G., Abbott, D.H. & Jarvis, J.U.** 1990. Social suppression of ovarian cyclicity in captive and wild colonies of naked mole-rats, *Heterocephalus glaber*. *Journal of Reproduction and Fertility* 88: 559–568.

**Foran, C.M. & Bass, A.H.** 1999. Preoptic GnRH and AVT: axes for sexual plasticity in teleost fish. *General and Comparative Endocrinology* 116: 141–152.

**Francis, C.M., Anthony, E.L.P., Brunton, J.A. & Kunz, T.H.** 1994. Lactation in male fruit bats. *Nature* 367: 691–692.

**Francis, D.D. & Meaney, M.J.** 1999. Maternal care and the development of stress responses. *Current Opinions in Neurobiology* 9: 128–134.

**Friedman, J.M. & Halaas, J.L.** 1998. Leptin and the regulation of body weight in mammals. *Nature* 395: 763–770.

**Fuchs, A.-R. & Dawood, M.Y.** 1980. Oxytocin release and uterine activation during parturition in rabbits. *Endocrinology* 107: 1117–1126.

**Fugger, H.N., Cunningham, S.G., Rissman, E.F. & Foster, T.C.** 1998. Sex differences in the activational effect of ERα on spatial learning. *Hormones and Behavior* 34: 163–170.

G

**Gasparini, J.,, McCoy, K.D., Haussy, C., Tveraa, T. & Boulinier, T.** 2001. Induced maternal response to the Lyme disease *spirochaete Borrelia burgdorferi senus lato* in a colonial seabird, the kittiwake, *Rissa tridactyla. Proceedings of the Royal Society of London Series B* 268: 647–650.

**George, F.W., Johnson, L. & Wilson, J.D.** 1989. The effect of a 5 alpha-reductase inhibitor on androgen physiology in the immature male rat. *Endocrinology* 125: 2434–2438.

**Gil, D., Leboucher, D.G., Lacroix, A., Cue, R. & Kreutzer, M.** 2004. Female canaries produce eggs with greater amounts of testosterone when exposed to preferred male song. *Hormones and Behavior* 45: 64–70.

**Gil, D., Graves, J., Hazon, N. & Wells, A.** 1999. Male attractiveness and differential testosterone investment in zebra finch eggs. *Science* 286: 126–128.

**Gingrich, J.A. & Hen, R.** 2000. The broken mouse: the role of development, plasticity and environment in the interpretation of phenotypic changes in knockout mice. *Current Opinions in Neurobiology* 10: 146–152.

**Girondot, M., Zaborski, P., Servan, J. & Pieau, C.** 1994. Genetic contribution to sex determination in turtles with environmental sex determination. *Genetics Research* 63: 117–127.

**Godfrey, K.M.** 2002. The role of the placenta in fetal programming – a review. *Placenta* 23 Suppl. A1: S20–S27.

**Godwin J. & Crews, D.** 1999. Hormonal regulation of progesterone receptor mRNA expression in the hypothalamus of whiptail lizards: regional and species differences. *Journal of Neurobiology* 39: 287–293.

**Goldsmith, A.R.** 1982. Plasma concentrations of prolactin during incubation and parental feeding throughout repeated breeding cycles in canaries (*Serinus canarius*). *Journal of Endocrinology* 94: 51–59.

**Goldsmith, A.R.** 1983. Prolactin in avian reproductive cycles. In: *Hormones and Behaviour in Higher Vertebrates* (Balthazart, J., Pröve, E. & Gilles, R. eds), pp. 375–387. Springer-Verlag, Berlin/Heidelberg.

**Goodfellow, P.N. & Lovell-Badge, R.** 1993. *SRY* and sex determination in mammals. *Annual Review of Genetics* 27: 71–92.

**Gorski, R.A., Gordon, J.H., Harlan, R.E., Jacobson, C.D., Shryne, J.E. & Southam, A.M.** 1980. Evidence for the existence of a sexually dimorphic nucleus in the preoptic area of the rat. *Journal of Comparative Neurology* 193: 529–539.

**Götz, J., Streffer, J.R., David, D., Schild, A., Hoerndli, F., Pennanen, L., Kurosinski, P. & Chen, F.** 2004. Transgenic animal models of Alzheimer's disease and related disorders: histopathology, behavior and therapy. *Molecular Psychiatry* 9: 664–683.

**Gray, P. & Brooks, P.J.** 1984. Effect of lesion location within the medial preoptic-anterior hypothalamic continuum on maternal and male sexual behaviors in female rats. *Behavioral Neuroscience* 98: 703–711.

**Gréco, B., Blasberg, M.E., Kosinski, E.C. & Blaustein, J.D.** 2003. Response to ERalphaα-IR and ERbetaβ-IR cells in the forebrain of female rats to mating stimuli. *Hormones and Behavior* 43: 444–453.

**Greenwood, P.J.** 1980. Mating systems, philopatry and dispersal in birds and mammals. *Animal Behaviour* 28: 1140–1162.

**Groothuis, T.G.G., Müller, W., von Engelhardt, N., Carere, C. & Eising, C.** 2005. Maternal hormones as a tool to adjust offspring phenotype in avian species. *Neuroscience and Biobehavioral Reviews* 29: 329–352.

**Gubbay, J., Collignon, J., Koopman, P., Capel, B., Economou, A., Müsterberg, A., Vivian, N., Goodfellow, P. & Lovell-Badge, R.** 1990. A gene mapping to the sex-determining region of the mouse Y chromosome is a member of a novel family of embryonically expressed genes. *Nature* 346: 245–250.

**Gubernick, D.J. & T. Teferi.** 2000. Adaptive significance of male parental care in a monogamous mammal. *Proceedings of the Royal Society of London Series B* 267: 147–150.

**Gutiérrez-Ospina, G., Jiménez-Trejo, F.J., Favila, R., Moreno-Mendoza, N.A., Rojas, L.G., Barrios, F.A., Díaz-Cintra, S. & Merchant-Larios, H.** 1999. Acetylcholinesterase-positive innervation is present at undifferentiated stages of the sea turtle *Lepidochelys olivacea* embryo gonads: Implications for temperature-dependent sex determination. *Journal of Comparative Neurology* 410: 90–98.

**Gwinner, E.** 1996. Circannual clocks in avian reproduction and migration. *Ibis* 138: 47–63.

**Gwinner, E. & Wiltschko, W.** 1980. Circannual changes in migratory orientation of the garden warbler, *Sylvia borin. Behavioral Ecology and Sociology* 7: 73–78.

H

**Haddad, J.J., Saadé, N.E. & Safieh-Garabedian, B.** 2002. Cytokines and neuro–immune–endocrine interactions: a role for the hypothalamic–pituitary–adrenal revolving axis. *Journal of Neuroimmunology* 133: 1–19.

**Hadley, M.E.** 1996. *Endocrinology*, 4th edition. Prentice Hall, Englewood Cliffs, NJ.

**Hamilton, K.S., King, A.P., Sengelaub, D.R. & West, M.J.** 1998. Visual and song nuclei correlate with courtship skills in brown-headed cowbirds. *Animal Behaviour* 56: 973–982.

**Haussmann M.F., Carroll, J.A., Weesner, G.D., Daniels, M.J., Matteri, R.L. & Lay, D.C., Jr.** 2000. Administration of ACTH to restrained, pregnant sows alters their pigs' hypothalamic–pituitary–adrenal (HPA) axis. *Journal of Animal Science* 78: 2399–2411.

**Heath, J.A.** 1997. Corticosterone levels during nest departure of juvenile American kestrels. *Condor* 99: 806–811.

**Heath, J.A. & Dufty, A.M., Jr.** 1998. Body condition and the adrenal stress response in captive American kestrel juveniles. *Physiological Zoology* 71: 67–73.

**Hennessey, A.C., Wallen, K. & Edwards, D.A.** 1986. Preoptic lesions increase the display of lordosis by male rats. *Brain Research* 370: 21–28.

**Hews D.K. & Moore, M.C.** 1996. A critical period for the organization of alternative male phenotypes of tree lizards by exogenous testosterone? *Physiology and Behavior* 60: 425–429.

**Hews, D.K., Knapp, R. & Moore, M.C.** 1994. Early exposure to androgens affects adult expression of alternative male types in tree lizards. *Hormones and Behavior* 28: 96–115.

**Hews D.K., Thompson, C.W., Moore, I.T. & Moore, M.C.** 1997. Population frequencies of alternative male phenotypes in tree lizards: geographic variation and common-garden rearing studies. *Behavioral Ecology and Sociobiology* 41: 371–380.

**Hintz, J.V.** 2000. The hormonal regulation of premigratory fat deposition and winter fattening in red-winged blackbirds. *Comparative Biochemistry and Physiology* 125A: 239–249.

**Hofer, H. & East, M.L.** 1998. Biological conservation and stress. In *Advances in the Study of Behavior* (Møller, A.P., Milinski, M. & Slater, P.J.B. eds), vol. 27, pp. 405–525. Academic Press, New York.

**Holberton, R.L.** 1999. Changes in patterns of corticosterone secretion concurrent with migratory fattening in a Neotropical migratory bird. *General and Comparative Endocrinology* 116: 49–58.

**Holberton, R.L. & Dufty, A.M., Jr.** 2005. *Hormone Patterns and Variation in Life History Strategies of Migratory and Non-migratory Birds in Birds of Two Worlds: The Ecology and Evolution of Migratory Birds* (Marra, P. & Goldberg, R., eds), Johns Hopkins Press, Baltimore.

**Holberton, R.L., Marra, P.P. & Moore, F.L.** 1999. Endocrine aspects of physiological condition, weather and habitat quality in landbird migrants during the non-breeding period. In *Proceedings of the 22nd International Ornithological Congress* (Adams, N.J. & Slotow, R.H. eds), pp. 847–866. BirdLife South Africa, Johannesburg.

**Holberton, R.L., Parrish, J.D. & Wingfield, J.C.** 1996. Modulation of the adrenocortical stress response in neotropical migrants during autumn migration. *Auk* 113: 558–564.

**Holekamp, K.E. & Sherman, P.W.** 1989. Why male ground squirrels disperse. *American Scientist* 77: 232–239.

**Holekamp, K.E., Smale, L., Simpson, H.B. & Holekamp, N.A.** 1984. Hormonal influences on natal dispersal in free-living Belding's ground squirrels (*Spermophilus beldingi*). *Hormones and Behavior* 18: 465–483.

**Honkaniemi, J., Kononen, J., Kainu, T., Pyykonen, I. & Pelto-Huikko M., M.** 1994. Induction of multiple immediate early genes in rat hypothalamic paraventricular nucleus after stress. *Brain Research* 25: 234–241.

**Hsu, Y., Earley, R.L. & Wolf, L.L.** 2006. Modulation of aggressive behaviour by fighting experience: mechanisms and contest outcomes. *Biological Reviews* 81: 33–74.

**Huck, U.W. & Banks, E.M.** 1984. Social olfaction in male brown lemmings (*Lemmus sibiricus = trimucronatus*) and collared lemmings (*Dicrostonyx groenlandicus*): I. Discrimination of species, sex, and estrous condition. *Journal of Comparative Psychology* 98: 54–59.

**Hunt, K.E., Hahn, T.P & Wingfield, J.C.** 1999. Endocrine influences on parental care during a short breeding season: testosterone and male parental care in Lapland longspurs (*Calcarius lapponicus*). *Behavioral Ecology and Sociobiology* 45: 360–369.

I

**Ims, R.A.** 1989. Kinship and origin effects on dispersal and space sharing in *Clethrionomys rufocanus*. *Ecology* 70: 607–616.

**Ims, R.A.** 1990. Determinants of natal dispersal and space use in grey-sided voles, *Clethrionomys rufocanus*: a combined field and laboratory experiment. *Oikos* 57: 106–113.

J

**Jacobs, J.D. & Wingfield, J.C.** 2000. Endocrine control of life-cycle stages: a constraint on response to the environment? *Condor* 102: 35–51.

**Jakubowski, M. & Terkel, J.** 1986. Female reproductive function and sexually dimorphic prolactin secretion in rats with lesions in the medial preoptic-anterior hypothalamic continuum. *Neuroendocrinology* 43: 696–705.

**Jenni, L., Jenni-Eiermann, S., Spina, F. & Schwabl, H.** 2000. Regulation of protein breakdown and adrenocortical response to stress in birds during migratory flight. *American Journal of Physiology* 278: R1182–1189.

**Jennings, D.H., Painter, D.L. and Moore, M.C.** 2004. Role of the adrenal gland in early post-hatching differentiation of alternative male phenotypes in the tree lizard (*Urosaurus ornatus*). *General and Comparative Endocrinology* 135: 81–89.

**Jennings, D., Moore, M.C., Knapp, R.K., Matthews, L. & Orchinik, M.** 2000. Plasma steroid-binding globulin mediation of differences in stress reactivity in alternative male phenotypes in tree lizards, *Urosaurus ornatus*. *General and Comparative Endocrinology* 120: 289–299.

**Jin, H. & Clayton, D.F.** 1997. Localized changes in immediate–early gene regulation during sensory and motor learning in zebra finches. *Neuron* 19: 1049–1059.

**Jones, J.S. & Wynne-Edwards, K.E.** 2000. Paternal hamsters mechanically assist the delivery, consume amniotic fluid and placenta, remove fetal membranes and provide parental care during the birth process. *Hormones and Behavior* 37: 116–125.

K

**Kanai, Y., Hiramatsu, R., Matoba, S. & Kidokoro, T.** 2005. From SRY to SOX9: mammalian testis differentiation. *Journal of Biochemistry (Tokyo)* 138: 13–19.

**Ketterson, E.D. & Nolan, V., Jr.** 1999. Adaptation, exaptation, and constraint: a hormonal perspective. *American Naturalist* Suppl. 154: S4–S25.

**Ketterson, E.D., Nolan, V., Jr., Wolf, L., Ziegunfus, C., Dufty, A.M., Jr., Ball, G.F. & Johnsen, T.S.** 1991. Testosterone and avian life histories: the effect of experimentally elevated testosterone on corticosterone and body mass in dark-eyed juncos. *Hormones and Behavior* 25: 489–503.

**Keverne, E.B. & Kendrick, K.M.** 1994. Maternal behaviour in sheep and its neuroendocrine regulation. *Acta Paediatrica* Suppl. 397: 47–56.

**Knackstedt, M.K., Hamelmann, E. & Arck, P.C.** 2005. Mothers in stress: Consequences for the offspring. *American Journal of Reproductive Immunology* 54: 63–69.

**Knapp, R.** 2003. Endocrine mediation of vertebrate male alternative reproductive tactics: The next generation of studies. *Integrative and Comparative Biology* 43: 658–668.

**Knapp, R., Wingfield, J.C. & Bass, A.H.** 1999. Steroid hormones and paternal care in the plainfin midshipman fish (*Porichthys notatus*). *Hormones and Behavior* 35: 81–89.

**Knapp, R., Hews, D.K., Thompson, C.W., Ray, L.E. & Moore, M.C.** 2003. Environmental and endocrine correlates of tactic switching by nonterritorial male tree lizards (*Urosaurus ornatus*). *Hormones and Behavior* 43: 83–92.

**Kobayashi, A. & Behringer, R.R.** 2003. Developmental genetics of the female reproductive tract in mammals. *Nature Reviews Genetics* 4: 969–980.

**Koch, K.A., Wingfield, J.C. & Buntin, J.D.** 2004. Prolactin-induced parental hyperphagia in ring doves: are glucocorticoids involved? *Hormones and Behavior* 46: 498–505.

**Koopman, P., Gubbay, J., Vivian, N., Goodfellow, P. & Lovell-Badge, R.** 1991. Male development of chromosomally female mice transgenic for *Sry*. *Nature* 351: 117–121.

**Korte, S.M., Koolhaas, J.M., Wingfield, J.C. & McEwen, B.S.** 2005. The Darwinian concept of stress: benefits of allostasis and costs of allostatic load and the trade-offs in health and disease. *Neuroscience and Biobehavioral Reviews* 29: 3–38.

**Kroodsma, D.S. & Byers, B.** 1991. The function(s) of bird song. *American Zoologist* 31: 318–328.

L

**Lachlan, R.F. & Slater, P.J.B.** 1999. The maintenance of vocal learning by gene-culture interaction: the cultural trap hypothesis. *Proceedings of the Royal Society of London Series B* 266: 701–706.

**Lambin, X.** 1994. Litter sex ratio does not determine natal dispersal tendency in female Townsend's voles. *Oikos* 69: 353–356.

**Landys, M.M., Ramenofsky, M., Guglielmo, C.G. & Wingfield, J.C.** 2004. The low-affinity glucocorticoid receptor regulates feeding and lipid breakdown in the migratory Gambel's white-crowned sparrow *Zonotrichia leucophrys gambelii*. *Journal of Experimental Biology* 207: 143–154.

**Landys-Ciannelli, M.M., Ramenofsky, M., Piersma, T., Jukema, J., Castricum Ringing Group & Wingfield, J.C.** 2002. Baseline and stress-induced plasma corticosterone during long-distance migration in the bar-tailed godwit, *Limosa lapponica*. *Physiological and Biochemical Zoology* 75: 101–110.

**Lee, M.K., Borchelt, D.R., Wong, P.C., Sisodia, S.S. & Price, D.L.** 1996. Transgenic models of neurodegenerative diseases. *Current Opinions in Neurobiology* 6: 651–660.

**Lehrman, D.S.** 1965. Interaction between internal and external environments in the regulation of the reproductive cycle of the ring dove. In *Sex and Behavior* (Beach, F.A. ed.), pp. 335–380. Wiley, New York.

**Léna, J.-P., Clobert, J., de Fraipont, M., Lecomte, J. & Guyot, G.** 1998. The relative influence of density and kinship on dispersal in the common lizard. *Behavioral Ecology* 9: 500–507.

**Levin, R.N. & Johnston, R.E.** 1986. Social mediation of puberty: an adaptive female strategy? *Behavioral and Neural Biology* 46: 308–324.

**Li, D. & Sánchez, E.R.** 2005. Glucocorticoid receptor and heat shock factor 1: novel mechanism of reciprocal regulation. *Vitamins and Hormones* 71: 239–262.

**Lipar, J.L., Ketterson, E.D., Nolan, V., Jr. & Casto, J.M.** 1999. Egg yolk layers vary in the concentration of steroid hormones in two avian species. *General and Comparative Endocrinology* 115: 220–227.

**Liu, Y.-C., Salamone, J.D. & Sachs, B.D.** 1997b. Lesions in medial preoptic area and bed nucleus of stria terminalis: differential effects on copulatory behavior and noncontact erection in male rats. *Journal of Neuroscience* 17: 5245–5253.

**Liu, D., Tannenbaum, B., Caldji, C., Francis, D., Freedman, A., Sharma, S., Pearson, D., Plotsky, P.M. & Meaney, M.J.** 1997a. Maternal care, hippocampal glucocorticoid receptor gene expression and hypothalamic–pituitary–adrenal responses to stress. *Science* 277: 1659–1662.

**Long, J.A. & Holberton, R.L.** 2004. Corticosterone secretion, energetic condition and a test of the migration modulation hypothesis in the hermit thrush. *Auk* 121: 1094–1102.

**Lordi, B., Protais, P., Mellier, D. & Caston, J.** 1997. Acute stress in pregnant rats: effects on growth rate, learning and memory capabilities of the offspring. *Physiology and Behavior* 62: 1087–1092.

M **MacDougall-Shackleton, S.A. & Ball, G.F.** 1999. Comparative studies of sex differences in the song-control system of songbirds. *Trends in Neuroscience* 22: 432–436.

**Maguire, E.A., Gadian, D.G., Johnsrude, I.S., Good, C.D., Ashburner, J., Frackowiak, R.S. & Frith, C.D.** 2000. Navigation-related structural change in the hippocampi of taxi drivers. *Proceedings of the National Academy of Sciences of the USA* 97: 4398–4403.

**Majzoub, J.A., McGregor, J.A., Lockwood, C.J., Smith, R., Taggart, M.S. & Schulkin, J.** 1999. A central theory of preterm and term labor: putative role for corticotrophin-releasing hormone. *American Journal of Obstetrics and Gynecology* 180: S232–S241.

**Margulis, S.W., Saltzman, W. & Abbott, D.H.** 1995. Behavioral and hormonal changes in female naked mole-rats (*Heterocephalus glaber*) following removal of the breeding female from a colony. *Hormones and Behavior* 29: 227–247.

**Marler, P., Peters, S., Ball, G.F., Dufty, A.M., Jr. & Wingfield, J.C.** 1988. The role of sex steroids in the acquisition and production of birdsong. *Nature* 336: 770–772.

**Marra, P.P. & Holberton, R.L.** 1998. Corticosterone levels as indicators of habitat quality: Effects of habitat segregation in a migratory bird during the non-breeding season. *Oecologia* 116: 284–292.

**Marra, P.P., Hobson, K.A. & Holmes, R.T.** 1998. Linking winter and summer events in a migratory bird by using stable-carbon isotopes. *Science* 282: 1884–1886.

**Marshall Graves, J.A. & Shetty, S.** 2001. Sex from W to Z: evolution of vertebrate sex chromosomes and sex determining genes. *Journal of Experimental Zoology* 290: 449–462.

**Massot, M. & Clobert, J.** 1995. Influence of maternal food availability on offspring dispersal. *Behavioral Ecology and Sociobiology* 37: 413–418.

**McCormick, C.M., Smythe, J.W., Sharma, S. & Meaney, M.J.** 1995. Sex-specific effects of prenatal stress on hypothalamic–pituitary–adrenal responses to stress and brain glucocorticoid receptor density in adult rats. *Brain Research and Developmental Brain Research* 84: 55–61.

**McCormick, J.A., Lyons, V., Jacobson, M.D., Noble, J., Diorio, J., Nyirenda, M., Weaver, S., Ester, W., Yau, J.L.W., Meaney, M.J., Seckl, J.R. & Chapman, K.E.** 2003. 5′-Heterogeneity of glucocorticoid receptor messenger RNA is tissue specific: Differential regulation of variant transcripts by early-life events. *Molecular Endocrinology* 14: 506–517.

**McEwen, B.S.** 1998. Protective and damaging effects of stress mediators. *New England Journal of Medicine* 338: 171–179.

**McEwen, B.S.** 1999. Stress and hippocampal plasticity. *Annual Review of Neuroscience* 22: 105–122.

**McEwen, B.S. & Wingfield, J.C.** 2003. The concept of allostasis in biology and biomedicine. *Hormones and Behavior* 43: 2–15.

**McEwen, B.S., Brinton, R.E. & Sapolsky, R.M.** 1988. Glucocorticoid receptors and behavior: implications for the stress response. *Advances in Experimental Medicine and Biology* 245: 35–45.

**McKibben, J.R. & Bass, A.H.** 1998. Behavioral assessment of acoustical parameters relevant to signal recognition and preference in a vocal fish. *Journal of the Acoustic Society of America* 104: 3520–3533.

**McLean, M., Bisits, A., Davies, J., Woods, R., Lowry, P. & Smith, R.** 1995. A placental clock controlling the length of human pregnancy. *Nature Medicine* 1: 460–463.

**McWilliams, S.R. & Karasov, W.H.** 2001. Phenotypic flexibility in digestive system structure and function in migratory birds and its ecological significance. *Comparative Biochemistry and Physiology A* 128: 577–591.

**Meaney, M.J. & Szyf, M.** 2005. Environmental programming of stress responses through DNA methylation: life at the interface between a dynamic environment and a fixed genome. *Dialogues in Clinical Neuroscience* 7: 103–123.

**Meaney, M.J., Viau, V., Bhatnagar, S., Betito, K., Iny, L.J., O'Donnell, D. & Mitchell, J.B.** 1991. Cellular mechanisms underlying the development and expression of individual differences in the hypothalamic–pituitary–adrenal stress response. *Journal of Steroid Biochemistry and Molecular Biology* 39: 265–274.

**Meek, S.B. & Robertson, R.J.** 1994. Effects of male removal on the behaviour and reproductive success of female eastern bluebirds *Sialia sialis*. *Ibis* 136: 305–312.

**Meier, A.H. & Farner, D.S.** 1964. A possible endocrine basis for premigratory fattening in the white-crowned sparrow, *Zonotrichia leucophrys gambelii* (Nuttall). *General and Comparative Endocrinology* 4: 584–595.

**Meier, A.H. & Martin, D.D.** 1971. Temporal synergism of corticosterone and prolactin controlling fat storage in the white-throated sparrow, *Zonotrichia albicollis*. *General and Comparative Endocrinology* 17: 311–318.

**Mello, C.V., Vicario, D.S. & Clayton, D.F.** 1992. Song presentation induces gene expression in the songbird forebrain. *Proceedings of the National Academy of Sciences of the USA* 89: 6818–6822.

**Meylan, S. & Clobert, J.** 2003. Is corticosterone-mediated phenotype development adaptive? Maternal corticosterone treatment enhances survival in male lizards. *Hormones and Behavior* 48: 44–52.

**Meylan, S., Clobert, J. & de Fraipont, M.** 2001. Maternal stress and juvenile dispersal in the common lizard. In *Proceedings of the Annual Meeting, Soc. Int. Comp. Biol. Chicago, IL.* [Abstract.]

**Meylan, S., Dufty, A.M. & Clobert, J.** 2003. The effect of transdermal corticosterone application on plasma corticosterone levels in pregnant *Lacerta vivipara*. *Comparative Biochemistry and Physiology A* 134: 497–503.

**Miller, D., Summers, J. & Silber, S.** 2004. Environmental versus genetic sex determination: a possible factor in dinosaur extinction? *Fertility and Sterility* 81: 954–964.

**Moore, F.L. & Evans, S.J.** 1999. Steroid hormones use non-genomic mechanisms to control brain functions and behaviors: a review of evidence. *Brain, Behavior and Evolution* 54: 41–50.

**Moore, M.C.** 1991. Application of organization–activation theory to alternative male reproductive strategies: a review. *Hormones and Behavior* 25: 154–179.

**Moore, M.C. & Crews, D.** 1986. Sex steroid hormones in natural populations of a sexual whiptail lizard *Cnemidophorus inornatus*, a direct evolutionary ancestor of a unisexual parthenogen. *General and Comparative Endocrinology* 63: 424–430.

**Moore, M.C., Hews, D.K. & Knapp, R.** 1998. Hormonal control and evolution of alternative male phenotypes: generalizations of models for sexual differentiation. *American Zoologist* 38: 133–151.

**Moore, M.C., Thompson, C.W. & Marler, C.A.** 1991. Reciprocal changes in corticosterone and testosterone levels following acute and chronic handling stress in the tree lizard, *Urosaurus ornatus*. *General and Comparative Endocrinology* 81: 217–226.

**Moreno-Mendoza, N., Harley, V.R. & Merchant-Larios, H.** 2001. Temperature regulates SOX9 expression in cultured gonads of *Lepidochelys olivacea*, a species with temperature sex determination. *Developmental Biology* 229: 319–326.

N

**Nair, N.G., Pant, K. & Chandola-Saklani, A.** 1994. Environmental and hormonal control of vernal migration in red-headed bunting (*Emberiza bruniceps*). *Journal of Biosciences* 19: 453–466.

**Nelson, R.J.** 1997. The use of genetic 'knock-out' mice in behavioral endocrinology research. *Hormones and Behavior* 31: 188–196.

**Nelson, R.J.** 2000 *An Introduction To Behavioral Endocrinology*. Sinauer, Sunderland, Massachusetts.

**Nelson, R.J. & Demas, G.E.** 1996. Seasonal changes in immune function. *Quarterly Review of Biology* 71: 511–548.

**Nicholls, T.J., Goldsmith, A.R. & Dawson, A.** 1988. Photorefractoriness in birds and comparison with mammals. *Physiological Reviews* 68: 133–176.

**Nichols, R. & Bondrup-Nielsen, S.** 1995. The effect of a single dose of testosterone propionate on activity & natal dispersal in the meadow vole, *Microtus pennsylvanicus*. *Annales Zoologici Fennici* 32: 209–215.

**Nizielski, S.E., Lechner, P.S., Croniger, C.M., Wang, N.D., Darlington, G.J. & Hanson, R.W.** 1996. Animal models for studying the genetic basis of metabolic regulation. *Journal of Nutrition* 126: 2697–2708.

**Norman, A.W. & Litwack, G.** 1987. *Hormones*. Academic Press, Orlando, FL, USA.

**Nottebohm, F. & Arnold, A.P.** 1976. Sexual dimorphism invocal control areas of the song bird brain. *Science* 194: 211–213.

**Nowicki, S., Searcy, W. & Peters, S.** 2002. Brain development, song learning and mate choice in birds: a review and experimental test of the 'nutritional stress hypothesis'. *Journal of Comparative Physiology A: Sensory, Neural and Behavioral Physiology* 188: 1003–1114.

**Nunes, S. & Holekamp, K.E.** 1996. Mass and fat influence the timing of natal dispersal in Belding's ground squirrels. *Journal of Mammalogy* 77: 807–817.

**Nunes, S., Co-Diem, T.H., Garrett, P.J., Mueke, E.-M., Smale, L. & Holekamp, K.E.** 1998. Body fat and time of year interact to mediate dispersal behaviour in ground squirrels. *Animal Behaviour* 55: 605–614.

O

**O'Connor, T.G., Ben-Shlomo, Y., Heron, J., Golding, J., Adams, D. & Glover, V.** 2005. Prenatal anxiety predicts individual differences in cortisol in pre-adolescent children. *Biological Psychiatry* 58: 211–217.

**O'Connell, M.E., Reboulleau, C., Feder, H.H. & Silver, R.** 1981b. Social interactions and androgen levels in birds. I. Female characteristics associated with increased plasma androgen levels in the male ring dove (*Streptopelia risoria*). *General and Comparative Endocrinology* 44: 454–463.

**O'Connell, M.E., Silver, R., Feder, H.H. & Reboulleau, C.** 1981a. Social interactions and androgen levels in birds. II. Social factors associated with a decline in plasma androgen levels in male ring doves (*Streptopelia risoria*). *General and Comparative Endocrinology* 44: 464–469.

**Ogawa, S., Chester, A.E., Hewitt, S.C., Walker, V.R., Gustafsson, J.-Å., Smithies, O., Korach, K.S. & Pfaff, D.W.** 2000. Abolition of male sexual behaviors in mice lacking estrogen receptors α and β (αβERKO). *Proceedings of the National Academy of Sciences of the USA* 97: 14737–14741.

**Oppliger, A., Clobert, J., Lecomte,J., Lorenzon, P., Boudjemadi, K. & John-Alder, H.B.** 1998. Environmental stress increases the prevalence and intensity of blood parasite infection in the common lizard *Lacerta vivipara*. *Ecology Letters* 1: 129–138.

**Orchinik, M. & McEwen, B.S.** 1995. Rapid actions in the brain: a critique of genomic and non-genomic mechanisms. In *Genomic and Non-Genomic Effects of Aldosterone* (Wehling, M. ed.), pp. 77–108. CRC Press, Boca Raton, FL.

**Orchinik, M., Murray, T.F. & Moore, F.L.** 1991. A corticosteroid receptor in neuronal membranes. *Science* 252: 1848–1851.

**O'Reilly, K.M. & Wingfield, J.C.** 1995. Spring and autumn migration in Arctic shorebirds: Same distance, different strategies. *American Zoologist* 35: 222–233.

**Oring, L.W., Fivizzani, A.J., Colwell, M.A. & el Halawani, M.E.** 1988. Hormonal changes associated with natural and manipulated incubation in the sex-role reversed Wilson's phalarope. *General and Comparative Endocrinology* 72: 247–256.

**Oring, L.W., Fivizzani, A.J., El Halawani, M.E. & Goldsmith, A.** 1986. Seasonal changes in prolactin and luteinizing hormone in the polyandrous spotted sandpiper, *Actitis macularia*. *General and Comparative Endocrinology* 62: 394–403.

P

**Panzica, G.C., Castagna, C., Viglietti-Panzica, C., Russo, C., Tlemçani, O. & Balthazart, J.** 1998. Organizational effects of estrogens on brain vasotocin and sexual behavior in quail. *Journal of Neurobiology* 37: 684–699.

**Panzica, G.C., Viglietti-Panzica, C., Calacagni, M., Anselmetti, G.C., Schumacher, M. & Balthazart, J.** 1987. Sexual differentiation and hormonal control of the sexually dimorphic medial preoptic nucleus in the quail. *Brain Research* 416: 59–68.

**Park, S.Y. & Jameson, J.L.** 2005. Minireview: transcriptional regulation of gonadal development and differentiation. *Endocrinology* 146: 1035–1042.

**Pelleymounter, M.A., Cullen, M.J., Baker, M.B., Hecht, R., Winters, D., Boone, T. & Collins, F.** 1995. Effects of the obese gene product on body weight regulation in ob/ob mice. *Science* 269: 475–543.

**Perfito, N., Meddle, S.L., Tramontin, A.D., Sharp, P.J. & Wingfield, J.C.** 2005. Seasonal gonadal recrudescence in song sparrows: response to temperature cues. *General and Comparative Endocrinology* 143: 121–128.

**Phoenix, C.H., Goy, R.W., Gerall, A.A. & Young, W.C.** 1959. Organizing action of prenatally administered testosterone propionate on the tissues mediating mating behavior in the female guinea pig. *Endocrinology* 65: 369–382.

**Pieau, C.** 1996. Temperature variation and sex determination in reptiles. *Bioessays* 18: 19–26.

**Piersma, T.** 1998. Phenotypic flexibility during migration: optimization of organ size contingent on the risks and rewards of fueling and flight? *Journal of Avian Biology* 29: 511–520.

**Piersma, T., Gudmundsson, G.A. & Lilliendahl, K.** 1999. Rapid changes in the size of different functional organ and muscle groups during refueling in a long-distance migrating shorebird. *Physiological and Biochemical Zoology* 72: 405–415.

**Piersma, T., Reneerkens, J. & Ramenofsky, M.** 2000. Baseline corticosterone peaks in shorebirds with maximal energy stores for migration: a general preparatory mechanism for rapid behavioral and metabolic transitions? *General and Comparative Endocrinology* 120: 118–126.

**Pike, I.L.** 2005. Maternal stress and fetal responses: evolutionary perspectives on preterm delivery. *American Journal of Human Biology* 17: 55–65.

**Pulido, F., Berthold, P., Mohr, G. & Querner, U.** 2001. Heritability of the timing of autumn migration in a natural bird population. *Proceedings of the Royal Society Series B* 268: 953–959.

**Pusey, A.E.** 1987. Sex-biased dispersal and inbreeding avoidance in birds and mammals. *Trends in Ecology and Evolution* 2: 295–299.

R

**Raisman, G. & Field, P.M.** 1973. Sexual dimorphism in the neutrophil of the preoptic area of the rat and its dependence on neonatal androgen. *Brain Research* 54: 1–29.

**Ramenofsky, M.** 1990. Fat storage and fat metabolism in relation to migration. In *Bird Migration: Physiology and Ecophysiology* (Gwinner, E. ed.), pp. 214–231. Springer-Verlag, Berlin.

**Rand, M.S. & Crews, D.** 1994. The bisexual brain: sex behavior differences and sex differences in parthenogenetic and sexual lizards. *Brain Research* 663: 163–167.

**Raouf, S.A., Parker, P.G., Ketterson, E.D., Nolan, V., Jr. & Ziegenfus, C.** 1997. Testosterone affects reproductive success by influencing extra-pair fertilizations in male dark-eyed juncos (Aves: *Junco hyemalis*). *Proceedings of the Royal Society of London Series B* 264: 1599–1603.

**Raynaud, A. & Pieau, C.** 1985. Embryonic development of the genital system. In *Biology of the Reptilia*, vol. 15(B) (Gans, C. ed.), pp. 149–300. Wiley and Sons, New York.

**Raynaud, A. & Raynaud, J.** 1961. L'activité sécrétoire précoce des glandes endocrines de l'embryon d'orvet (*Anguis fragilis*). *Comptes Rendus Hebdomadaires des Séances de l'Académie des Sciences, Paris* 253: 2254–2256.

**Reiner, A., Perkel, D.J., Bruce, L.L.** *et al.* 2004. Revised nomenclature for avian telencephalon and some related brainstem nuclei. *Journal of Comparative Neurology* 473: 377–414.

**Rey, R. & Picard, J.Y.** 1998. Embryology and endocrinology of genital development. *Baillieres Clinical Endocrinology and Metabolism* 12: 17–33.

**Rhen, T. & Crews, D.** 1999. Embryonic temperature and gonadal sex organize male-typical sexual and aggressive behavior in a lizard with temperature-dependent sex determination. *Endocrinology* 140: 4501–4508.

**Rhen, T. & Crews, D.** 2000. Organization and activation of sexual and agonistic behavior in the leopard gecko, *Eublepharis macularius*. *Neuroendocrinology* 71: 252–261.

**Richard-Mercier, N., Dorizzi, M., Desvages, G., Girondot, M. & Pieau, C.** 1995. Endocrine sex reversal of gonads by the aromatase inhibitor Letrozole (CGS 20267) in *Emys orbicularis*, a turtle with temperature-dependent sex determination. *General and Comparative Endocrinology* 100: 314–326.

**Richardson, R.D., Boswell, T., Raffety, B.D., Seeley, R.J., Wingfield, J.C. & Woods, S.C.** 1995. NPY increases food intake in white-crowned sparrows: effect in short and long photoperiods. *American Journal of Physiology* 268: R1418–R1422.

**Rissman, E.F.** 1996. Behavioral regulation of gonadotropin-releasing hormone. *Biology of Reproduction* 54: 413–419.

**Robin, J.-P., Boucontet, L., Chillet, P. & Groscolas, R.** 1998. Behavioral changes in fasting emperor penguins: evidence for a 'refeeding signal' linked to a metabolic shift. *American Journal of Physiology* 274: R746–R753.

**Romero, L.M. & Remage-Healey, L.** 2000. Daily and seasonal variation in response to stress in captive starlings (*Sturnus vulgaris*): corticosterone. *General and Comparative Endocrinology* 119: 52–59.

**Romero, L.M., Ramenofsky, M. & Wingfield, J.C.** 1997. Season and migration alters the corticosterone response to capture and handling in an Arctic migrant, the white-crowned sparrow (*Zonotrichia leucophrys gambelii*). *Comparative Biochemistry and Physiology* 116C: 171–177.

**Romero, R.D.** 2003. Puberty: a period of both organizational and activational effects of steroid hormones on neurobehavioural development. *Journal of Neuroendocrinology* 15: 1185–1192.

**Ronce, O., Clobert, J. & Massot, M.** 1998. Natal dispersal and senescence. *Proceedings of the National Academy of Sciences of the USA* 95: 600–605.

**Rosenblatt, J.S., Siegel, H.I. & Mayer, A.D.** 1979. Blood levels of progesterone, estradiol and prolactin in pregnant rats. *Advances in the Study of Behavior* 10: 225–311.

**Ryan, B.C. & Vandenbergh, J.G.** 2002. Intrauterine position effects. *Neuroscience and Biobehavioral Reviews* 6: 665–678.

**Ryffel, B.** 1996. Gene knockout mice as investigative tools in pathophysiology. *International Journal of Experimental Pathology* 77: 125–141.

S

**Sachs, B.D. & Meisel, R.** 1988. The physiology of male sexual behavior. In *The Physiology of Male Sexual Behavior* (Knobil, E. & Neill, J. eds), pp. 1393–1485. Raven Press, New York.

**Sachser, N.** 1998. Of domestic and wild guinea pigs: studies in sociophysiology, domestication, and social evolution. *Naturwissenschaften* 85: 307–317.

**Salame-Mendez, A., Herrera-Munoz, J., Moreno-Mendoza, N. & Merchant-Larios, H.** 1998. Response of diencephalon but not the gonad to female-promoting temperature with elevated estradiol levels in the sea turtle *Lepidochelys olivacea*. *Journal of Experimental Zoology* 280: 304–313.

**Saldanha, C.J., Schlinger, B.A. & Clayton, N.S.** 2000. Rapid effects of corticosterone on cache recovery in mountain chickadees (*Parus gambeli*). *Hormones and Behavior* 37: 109–115.

**Sapolsky, R.M.** 1982. The endocrine stress-response and social status in the wild baboon. *Hormones and Behavior* 16: 279–292.

**Sapolsky, R.M.** 1992. Neuroendocrinology of the stress-response. In *Behavioral Endocrinology* (Becker, J.B., Breedlove, S.M. & Crews, D. eds), pp. 287–324. MIT Press, Cambridge, MA.

**Sapolsky, R.M.** 1996. Why stress is bad for your brain. *Science* 273: 749–750.

**Sapolsky, R.M., Romero, L.M. & Munck, A.U.** 2000. How do glucocorticoids influence stress responses? Integrating permissive, suppressive, stimulatory and preparative actions. *Endocrine Reviews* 21: 55–89.

**Sartor, J.J. & Ball, G.F.** 2005. Social suppression of song is associated with a reduction in volume of a song-control nucleus in European starlings (*Sturnus vulgaris*). *Behavioral Neuroscience* 119: 233–244.

**Sauer, J.R., Hines, J.E., Gough, G., Thomas, I. & Peterjohn, B.G.** 1997. *The North American Breeding Bird Survey: Results and Analysis*. Version 96.4. Patuxent Wildlife Research Center, Laurel, MD.

**Schlinger, B.A.** 1997. The activity and expression of aromatase in songbirds. *Brain Research Bulletin* 44: 359–364.

**Schlinger, B.A.** 1998. Sexual differentiation of avian brain and behavior: current views on gonadal hormone-dependent and independent mechanisms. *Annual Review of Physiology* 60: 407–429.

**Schneider, M.L., Roughton, E.C., Koehler, A.J. & Lubach, G.R.** 1999. Growth and development following prenatal stress exposure in primates: an examination of ontogenetic vulnerability. *Child Development* 70: 263–274.

**Schum, J.E. & Wynne-Edwards, K.E.** 2005. Estradiol and progesterone in paternal and non-paternal hamsters (*Phodopus*) becoming fathers: conflict with hypothesized roles. *Hormones and Behavior* 47: 410–418.

**Schwabl, H.** 1993. Yolk is a source of maternal testosterone for developing birds. *Proceedings of the National Academy of Sciences of the USA* 90: 11446–11450.

**Schwabl, H.** 1995. Individual variation of the acute adrenocortical response to stress in the white-throated sparrow. *Zoology* 99: 113–120.

**Schwabl, H.** 1997. The contents of maternal testosterone in house sparrow *Passer domesticus* eggs vary with breeding conditions. *Naturwissenschaften* 84: 406–408.

**Schwabl, H., Bairlein, F. & Gwinner, E.** 1991. Basal and stress-induced corticosterone levels of garden warblers, *Sylvia borin*, during migration. *Journal of Comparative Physiology B* 161: 576–580.

**Schwabl, H., Schwabl-Benzinger, I., Goldsmith, A.R. & Farner, D.S.** 1988. Effects of ovariectomy on long-day-induced premigratory fat deposition, plasma levels of luteinizing hormone and prolactin, and molt in white-crowned sparrows, *Zonotrichia leucophrys gambelii*. *General and Comparative Endocrinology* 71: 398–405.

**Schwilch, R., Grattarola, A., Spina, F. & Jenni, L.** 2002. Protein loss during long-distance migratory flight in passerine birds: adaptation and constraint. *Journal of Experimental Biology* 205: 687–695.

**Searcy, W.A.** 1984. Song repertoire size and female preferences in song sparrows. *Behavioral Ecology and Sociobiology* 14: 281–286.

**Seiler, H.W., Gahr, M., Goldsmith, A.R. & Guttinger, H.R.** 1992. Prolactin and gonadal steroids during the reproductive cycle of the Bengalese finch (*Lonchura striata* var. *domestica*, Estrildidae), a nonseasonal breeder with biparental care. *General and Comparative Endocrinology* 88: 83–90.

**Sharp, P.J. & Blache, D.** 2003. A neuroendocrine model for prolactin as the key mediator of seasonal breeding in birds under long- and short-day photoperiods. *Canadian Journal of Physiology and Pharmacology* 81: 350–358.

**Shaw, B.K. & Kennedy, G.G.** 2002. Evidence for species differences in the pattern of androgen receptor distribution in relation to species differences in an androgen-dependent behavior. *Journal of Neurobiology* 52: 203–220.

**Silverin, B.** 1997. The stress response and autumn dispersal behaviour in willow tits. *Animal Behaviour* 53: 451–459.

**Silverin, B. & Goldsmith, A.R.** 1983. Reproductive endocrinology of free living pied flycatchers (*Ficedula hypoleuca*): Prolactin and FSH secretion in relation to incubation and clutch size. *Journal of Zoology* 200: 119–130.

**Silverin, B. & Goldsmith, A.R.** 1990. Plasma prolactin concentrations in breeding pied flycatchers (*Ficedula hypoleuca*) with an experimentally prolonged brooding period. *Hormones and Behavior* 24: 104–113.

**Silverin, B. & Wingfield, J.C.** 1982. Patterns of breeding behavior and plasma levels of hormones in a free-living population of pied flycatchers *Ficedula hypoleuca*. *Journal of Zoology* 198: 117–129.

**Silverin, B., Arvidsson, B. & Wingfield, J.C.** 1997. The adrenocortical responses to stress in breeding willow warblers *Phylloscopus trochilus* in Sweden: effects of latitude and gender. *Functional Ecology* 11: 376–384.

**Sims, C.G. & Holberton, R.L.** 2000. Development of the corticosterone stress response in young Northern Mockingbirds (*Mimus polyglottos*). *General and Comparative Endocrinology* 119: 193–201.

**Sinervo, B., Miles, D.B., Frankino, W.A., Klukowski, M. & DeNardo, D.F.** 2000. Testosterone, endurance and Darwinian fitness: natural and sexual selection on the physiological bases of alternative male behaviors in side-blotched lizards. *Hormones and Behavior* 38: 222–233.

**Sisk, C.L. & Foster, D.L.** 2004. The neural basis of puberty and adolescence. *Nature Neuroscience* 7: 1040–1047.

**Sisk, C.L., Schulz, K.M. & Zehr, J.L.** 2003. Puberty: a finishing school for male social behavior. *Annals of the New York Academy of Science* 1007: 189–198.

**Smith, M.A., Kim, S.-Y., van Oers, H.J. & Levine, S.** 1997. Maternal deprivation and stress induce immediate early genes in the infant rat brain. *Endocrinology* 138: 4622–4628.

**Soma, K.K., Tramontin, A. & Wingfield, J.C.** 2000. Oestrogen regulates male aggression in the non-breeding season. *Proceedings of the Royal Society of London Series B* 267: 1089–1096.

**Soma, K.K., Sinchak, K., Lakhter, A.D., Schlinger, B.A. & Micevych, P.E.** 2005. Neurosteroids and female reproduction: estrogen increases 3β-HSD mRNA and activity in rat hypothalamus. *Endocrinology* 146: 4386–4390.

**Son, G.H., Geum, D., Chung, S., Kim, E.J., Jo, J.H., Kim, C.M., Lee, K.H., Kim, H., Choi, S., Kim, H.T., Lee, C.-J. & Kim, K.** 2006. Maternal stress produces learning deficits associated with impairment of NMDA receptor-mediated synaptic plasticity. *Journal of Neuroscience* 26: 3309–3318.

**Sorci, G. & Clobert, J.** 1995. Effects of maternal parasite load on offspring life-history traits in the common lizard (*Lacerta vivipara*) increases sprint speed and philopatry in female offspring of the common lizard. *Journal of Evolutionary Biology* 8: 711–723.

**Sorci, G., Massot, M. & Clobert, J.** 1994. Maternal parasite load predicts offspring sprint speed in the philopatric sex. *American Naturalist* 144: 153–164.

**Specker, J.L. & Kishida, M.** 2000. Mouthbrooding in the black-chinned tilapia, *Sarotherodon melanotheron* (Pisces: Cichlidae): the presence of eggs reduces androgen and estradiol levels during paternal and maternal parental behavior. *Hormones and Behavior* 38: 44–51.

**Stamatakis, A., Mantelas, A., Papaioannou, A., Pondiki, S., Fameli, M. & Stylianopoulou, F.** 2006. Effect of neonatal handling on serotonin 1A sub-type receptors in the rat hippocampus. *Neuroscience* 140: 1–11.

**Stearns, S.** 1992. *The Evolution of Life Histories*. Oxford University Press, New York.

**Stenseth, N.C. & Lidicker, W.Z., Jr.** 1992. *Animal Dispersal: Small Mammals as a Model*. Chapman and Hall, London.

**Stern, J.M.** 1996. Trigeminal lesions and maternal behavior in Norway rats: II. Disruption of parturition. *Physiology and Behaviour* 60: 187–190.

**Storey, A.E., Walsh, C.J., Quinton, R.L. & Wynne-Edwards, K.E.** 2000. Hormonal correlates of paternal responsiveness in new and expectant fathers. *Evolution and Human Behavior* 21: 79–95.

**Straub, R.H., Cutolo, M., Zietz, B. & Scholmerich, J.** 2001. The process of aging changes the interplay of the immune, endocrine and nervous systems. *Mechanisms of Ageing and Development* 122: 1591–1611.

T

**ten Cate, C., Lea, R.W., Ballintijn, M.R. & Sharp, P.J.** 1993. Brood size affects behavior, interclutch interval, LH levels and weight in ring dove (*Streptopelia risoria*) breeding pairs. *Hormones and Behavior* 27: 539–550.

**Totzke, U., Hübinger, A. & Bairlein, F.** 1997. A role for pancreatic hormones in the regulation of autumnal fat deposition of the garden warbler (*Sylvia borin*)? *General and Comparative Endocrinology* 107: 166–171.

**Tramontin, A.D., Wingfield, J.C. & Brenowitz, E.A.** 1999. Contributions of social cues and photoperiod to seasonal plasticity in the adult avian song control system. *Journal of Neuroscience* 19: 476–483.

V

**Vaillant, S., Dorizzi, M., Pieau, C. & Richard-Mercier, N.** 2001. Sex reversal and aromatase in chicken. *Journal of Experimental Zoology* 290: 727–740.

**Valverde, R.A., Owens, D.W., MacKenzie, D.S. & Amoss, M.S.** 1999. Basal and stress-induced corticosterone levels in olive ridley sea turtles (*Lepidochelys olivacea*) in relation to their mass nesting behavior. *Journal of Experimental Zoology* 284: 652–662.

**van der Steen, W.J.** 1998. Bias in behaviour genetics: an ecological perspective. *Acta Biotheoretica* 46: 369–377.

**van Oers, H.J.J., de Kloet, E.R., Li, C. & Levine, S.** 1998. The ontogeny of glucocorticoid negative feedback: influence of maternal deprivation. *Endocrinology* 139: 2838–2846.

**Velthuis, H.H.W.** 1976. Egg laying, aggression and dominance in bees. In *Proceedings of the XV International Congress of Entomology, Washington DC*, pp. 436–449.

**Vigier, B., Forest, M.G., Eychenne, B., Bézard, J., Garrigou, O., Robel, P. & Josso, N.** 1989. Anti-Müllerian hormone produces endocrine sex reversal of fetal ovaries. *Proceedings of the National Academy of Sciences of the USA* 86: 3684–3688.

**Virgin, C.E., Jr. & Sapolsky, R.M.** 1997. Styles of male social behavior and their endocrine correlates among low-ranking baboons. *American Journal of Primatology* 42: 25–39.

**Visser, J.A. & Themmen, A.P.N.** 2005. Anti-Müllerian hormone and folliculogenesis. *Molecular and Cellular Endocrinology* 234: 81–86.

**Vleck, C.M., Bucher, T.L., Reed, W.L. & Kristmundsdottir, A.Y.** 1999. Changes in reproductive hormones and body mass through the reproductive cycle in the Adélie penguin (*Pygoscelis adeliae*), with associated data on courting-only individuals (Adams, N. & Slotow, R. eds). *Proceedings of the 22nd International Ornithological Congress, Durban, University of Natal*, pp. 1210–1223. University of Natal, Durban.

**Vleck, C.M., Ross, L.L., Vleck, D. & Bucher, T.L.** 2000. Prolactin and parental behavior in Adelie penguins: effects of absence from nest, incubation length and nest failure. *Hormones and Behavior* 38: 149–158.

**vom Saal, F.S.** 1984. The intrauterine position phenomenon: effects on physiology, aggressive behavior and population dynamics in house mice. *Progress in Clinical Biological Research* 169: 135–179.

**von Engelhardt, N., Carere, C., Dijkstra, C. & Groothuis, T.G.G.** 2006. Sex-specific effects of yolk testosterone on survival, begging and growth of zebra finches. *Proceedings of the Royal Society of London Series B* 273: 65–70.

W

**Wadhwa, P.D., Porto, M., Garite, T.J., Chicz-DeMet, A. & Sandman, C.A.** 1998. Maternal corticotrophin-releasing hormone levels in the early third trimester predict length of gestation in human pregnancy. *American Journal of Obstetrics and Gynecology* 179: 1079–1085.

**Walker, B.G., Boersma, P.D. & Wingfield, J.C.** 2005. Field endocrinology and conservation biology. *Integrative and Comparative Biology* 45: 12–18.

**Wang, M.-H. & vom Saal, F.S.** 2000. Maternal age and traits in offspring. *Nature* 407: 469–470.

**Wang, Q. & Buntin, J.D.** 1999. The roles of stimuli from young, previous breeding experience, and prolactin in regulating parental behavior in ring doves (*Streptopelia risoria*). *Hormones and Behavior* 35: 241–532.

**Wendelaar Bonga, S.E.** 1997. The stress response in fish. *Physiological Reviews* 77: 591–625.

**West, M.J. & King, A.P.** 1988. Visual displays affect the development of male song in the cowbird. *Nature* 334: 244–246.

**Whitney, O. & Johnson, F.** 2005. Motor-induced transcription but sensory-regulated translation of ZENK in socially interactive songbirds. *Journal of Neurobiology* 65: 251–259.

**Wibbels, T. & Crews, D.** 1994. Putative aromatase inhibitor induces male sex determination in a female unisexual lizard and in a turtle with temperature-dependent sex determination. *Journal of Endocrinology* 141: 295–299.

**Wiberg, U.H. & Günther, E.** 1985. Female wood lemmings with the mutant X*-chromosome carry the H-Y transplantation antigen. *Immunogenetics* 21: 91–96.

**Wikelski, M., Hau, M. & Wingfield, J.C.** 1999. Social instability increases plasma testosterone in a year-round territorial Neotropical bird. *Proceedings of the Royal Society of London Series B* 266: 551–556.

**Williams, C.L. & Meck, W.H.** 1991. The organizational effects of gonadal steroids on sexually dimorphic spatial ability. *Psychoneuroendocrinology* 16: 155–176.

**Williams, T.D.** 1999. Parental and first generation effects of exogenous 17β-estradiol on reproductive performance of female zebra finches (*Taeniopygia guttata*). *Hormones and Behavior* 35: 135–143.

**Wingfield, J.C.** 1984a. Environmental and endocrine control of reproduction in the song sparrow, *Melospiza melodia*. I. Temporal organization of the breeding cycle. *General and Comparative Endocrinology* 56: 406–416.

**Wingfield, J.C.** 1984b. Androgens and mating systems: testosterone-induced polygyny in normally monogamous birds. *Auk* 101: 665–671.

**Wingfield, J.C. & Hahn, T.P.** 1994. Testosterone and territorial behaviour in sedentary and migratory sparrows. *Animal Behaviour* 47: 77–89.

**Wingfield, J.C., Hegner, R.E., Dufty, A.M., Jr. & Ball, G.F.** 1990a. The 'challenge hypothesis': theoretical implications for patterns of testosterone secretion, mating systems and breeding strategies. *American Naturalist* 136: 829–846.

**Wingfield, J.C. & Ramenofsky, M.** 1999. Hormones and the behavioral ecology of stress. In *Stress Physiology in Animals* (Balm, P.H.M. ed.), pp. 1–51. Sheffield Academic Press, Sheffield UK.

**Wingfield, J.C. & Silverin, B.** 1986. Effects of corticosterone on territorial behavior of free-living male song sparrows *Melospiza melodia*. *Hormones and Behavior* 20: 405–417.

**Wingfield, J.C., O'Reilly, K.M. & Astheimer, L.B.** 1995. Ecological bases of the modulation of adrenocortical responses to stress in Arctic birds. *American Zoologist* 35: 285–294.

**Wingfield, J.C., Schwabl, H. & Mattocks, P.W., Jr.** 1990b. Endocrine mechanisms of migration. In *Bird Migration: Physiology and Ecophysiology* (Gwinner, E. ed), pp. 232–256. Springer-Verlag, Berlin.

**Wingfield, J.C., Vleck, C.M. & Moore, M.C.** 1992. Seasonal changes in the adrenocortical response to stress in birds of the Sonoran Desert. *Journal of Experimental Zoology* 264: 419–428.

**Wingfield, J.C., Maney, D.L., Breuner, C.W., Jacobs, D., Lynn, S., Ramenofsky, M. & Richardson, R.D.** 1998. Ecological bases of hormone–behavior interactions: the 'emergency life history stage'. *American Zoologist* 38: 191–206.

**Wynne-Edwards, K.E.** 2001. Hormonal changes in mammalian fathers. *Hormones and Behavior* 40: 139–145.

**Wynne-Edwards, K.E. & Reburn, C.J.** 2000. Behavioral endocrinology of mammalian fatherhood. *Trends in Ecology and Evolution* 15: 464–468.

Y

**Young, L.J., Lim, M.M., Gingrich, B. & Insel, T.R.** 2001. Cellular mechanisms of social attachment. *Hormones and Behavior* 40: 133–138.

Z

**Zera, A.J. & Denno, R.F.** 1997. Physiology and ecology of dispersal polymorphism in insects. *Annual Review of Entomology* 42: 207–230.

## Chapter 7

B

**Barnard, C.J. & Brown, C.J.A.** 1985. Risk-sensitive foraging in common shrews (*Sorex araneus* L.) *Behavioral Ecology and Sociobiology* 16: 161–164.

**Begon, M., Harper, J.L. & Townsend, C.R.** 1990. *Ecology: Individuals, Populations and Communities*, 2nd edition. Blackwell Scientific Publications, Boston, Massachusetts.

C

**Charnov, E.** 1976. Optimal foraging, the marginal value theorem. *Theoretical Population Biology* 9: 129–136.

E

**Elner, R.W. & Hughes, R.N.** 1978. Energy maximization in the diet of the shore crab, *Carcinus maenas*. *Journal of Animal Ecology* 47: 103–116.

G

**Getty, T. & Krebs, J.R.** 1985. Lagging partial prefences for cryptic prey: a signal detection analysis of great tit foraging. *American Naturalist* 125: 39–60.

**Giraldeau, L.-A.** 1997. The ecology of information use. In *Behavioural Ecology: An Evolutionary Approach*, 4th edition (Krebs, J.R. & Davies, N.B. eds), pp. 42–68. Blackwell Scientific Publications.

**Giraldeau, L.-A., Kramer, D.L., Deslandes, I. & Lair, H.** 1994. The effect of competitors and distance on central place foraging in eastern chipmunks, *Tamias striatus*. *Animal Behaviour* 47: 621–632.

**Giraldeau, L.-A. & Caraco, T.** 2000. *Social Foraging Theory*. Princeton University Press, Princeton.

**Giraldeau, L.-A. & Kramer, D.L.** 1982. The marginal value theorem: a quantitative test using load size variation in a central place forager the eastern chipmunk, *Tamias striatus*. *Animal Behaviour* 30: 1036–1042.

K

**Kacelnik, A.** 1984. Central place foraging in starlings (*Sturnus vulgaris*). I. Patch residence time. *Journal of Animal Ecology* 53: 283–299.

**Kramer, D.L. & Nowell, W.** 1980. Central place foraging in the eastern chipmunk *Tamias striatus*. *Animal Behaviour* 28: 772–778.

**Krebs, J.R., Erichsen, J.T., Webber, M.I. & Charnov, E.L.** 1977. Optimal prey-selection by the great tit (*Parus major*). *Animal Behaviour* 25: 30–38.

**Krebs, J.R. & Davies, N.B.** 1978. *Behavioural Ecology: An Evolutionary Approach*. Sinauer Associates, Sunderland, Massachusetts.

**Krebs, J.R. & Davies, N.B.** 1987. *An Introduction to Behavioural Ecology*, 2nd edition. Sinauer Associates, Sunderland, Massachusetts.

L

**Lair, H., Kramer, D.L. & Giraldeau, L.-A.** 1994. Interference competition in central place foragers: the effect of imposed waiting on patch use decisions of eastern chipmunks. *Behavioral Ecology* 5: 237–244.

**Lima, S.** 1984. Downy woodpecker foraging behaviour: efficient sampling in simple stochastic environments. *Ecology* 65: 166–174.

M

**MacArthur, R.H. & Pianka, E.R.** 1966. On optimal use of a patchy environment. *American Naturalist* 100: 603–609.

O

**Orians, G.H. & Pearson, N.E.** 1979. On the theory of central place foraging. In *Analysis of Ecological Systems* (Horn, D.J., Mitchell, R.D. & Stairs, G.R. eds). Ohio State University Press, Columbus.

P

**Parker, G.A.** 1978. Searching for mates. In *Behavioural Ecology: An Evolutionary Approach*

(Krebs, J.R. & Davies, N.B. eds). Sinauer Associates, Sunderland, Massachusetts.

**Parker, G.A. & Stuart, R.A.** 1976. Animal behaviour as a strategy optimizer: evolution of resource assessment strategies and optimal emigration thresholds. *American Naturalist* 110: 1055–1076.

R

**Real, L. & Caraco, T.** 1986. Risk and foraging in stochastic environments. *Annual Review of Ecology and Systematics* 17: 371–390.

S

**Stephens, D.W. & Krebs, J.R.** 1986. *Foraging Theory.* Princeton University Press, Princeton.

Y

**Ydenberg, R.C., Giraldeau, L.-A. & Kramer, D.L.** 1986. Interference competition, payoff asymmetries and the social relationships of central place foragers. *Theoretical Population Biology* 30, 26–44.

## Chapter 8

B

**Baird, R.N. & Dill, L.M.** 1996. Ecological and social determinants of group size in transient killer whales. *Behavioral Ecology* 7: 408–416.

**Barnard, C.J. & Sibly, R.M.** 1981. Producers and scroungers: a general model and its application to captive flocks of house sparrows. *Animal Behaviour* 29: 543–555.

**Beauchamp, G. & Giraldeau, L.-A.** 1997. Patch exploitation in a producer–scrounger system: test of a hypothesis using flocks of spice finches (*Lonchura punctulata*). *Behavioral Ecology* 8: 54–59.

**Begon, M., Harper, J. & Townsend, C.R.** 1990. *Ecology: Individuals, Populations and Communities*, 2nd edition. Blackwell Scientific Publications, Boston.

**Bertram, B.C.R.** 1978. Living in groups: predators and prey. In *Behavioural Ecology: An Evolutionary Approach* (Krebs, J.R. & Davies, N.B. eds), pp. 64–96. Sinauer Associates, Sunderland, Massachusetts.

C

**Caraco, T. & Pulliam, R.** 1984. Sociality and survivorship in animals exposed to predation. In *A New Ecology: Novel Approaches to Interactive Systems* (Price, P.W., Slobodchikoff, C.N. & Gaud, W.S. eds), pp. 179–309. Wiley Interscience, New York.

**Clark, C.C. & Mangel, M.** 1984. Foraging and flocking strategies: information in an uncertain environment. *American Naturalist* 123: 626–641.

**Coolen, I., Giraldeau, L.-A. & Lavoie, M.** 2001. Head position as an indication of producer scrounger tactics in a ground-feeding bird. *Animal Behaviour* 61: 895–903.

D

**Davis, M.D.** 1970. *Game Theory: A Nontechnical Introduction.* Basic Books, New York.

F

**Fretwell, S.D.** 1972. *Populations in a Seasonal Environment.* Princeton University Press, Princeton.

**Fretwell, S.D. & Lucas, H.L.** 1970. On territorial behaviour and other factors influencing habitat distribution in birds. *Acta Biotheoretica* 19: 16–36.

G

**Gillis, D.M. & Kramer, D.L.** 1987. Ideal interference distributions: population density and patch use by zebrafish. *Animal Behaviour* 35: 1875–1882.

**Giraldeau, L.-A. & Caraco, T.** 2000. *Social Foraging Theory.* Princeton University Press, Princeton.

**Giraldeau, L.-A. & Livoreil, B.** 1998. *Game theory and social foraging.* In *Game Theory and Animal Behavior* (Dugatkin, L.A. & Reeve, H.K. eds), pp. 16–37. Oxford University Press, New York.

**Godin, J.-G. & Keenleyside, M.H.A.** 1984. Foraging on patchily distributed prey by a cichlid fish (Teleostei, Cichlidae): a test of the ideal free distribution theory. *Animal Behaviour* 32: 120–131.

H

**Hamilton, W.D.** 1971. Geometry for the selfish herd. *Journal of Theoretical Biology* 31: 295–311.

**Heinsohn, R. & Packer, C.** 1995. Complex cooperative strategies in group-territorial African lions. *Science* 269: 1260–1262.

**Heller, R.** 1980. On optimal diet in a patchy environment. *Theoretical Population Biology* 17: 201–214.

K

**Kennedy, M. & Gray, R.D.** 1993. Can ecological theory predict the distribution of foraging animals? A critical evaluation of experiments on the ideal free distribution. *Oikos* 68: 158–166.

M

**Maynard Smith, J.** 1982. *Evolution and the Theory of Games.* Cambridge University Press, Cambridge.

**Milinki, M.** 1979. An evolutionarily stable feding strategy in sticklebacks. *Zietschrift für Tierpsychologie* 51: 36–40.

**Mitchell, W.A.** 1990. On optimal control theory of diet selection: the effects of resource depletion and exploitative competition. *Oikos* 58: 16–24.

P

**Parker, G.A.** 1978. Searching for mates. In *Behavioural Ecology: An Evolutionary Approach* (Krebs, J.R. & Davies, N.B. eds), pp. 214–244. Sinauer Associates, Sunderland, Massachusetts.

**Pulliam, R.H.** 1973. On the advantages of flocking. *Journal of Theoretical Biology* 38, 419–422.

**Pulliam, R.H. & Caraco, T.** 1985. Living in groups: is there an optimal group size? *Behavioural Ecology: An Evolutionary Approach*, 2nd edition (Krebs, J.R. & Davies, N.B. eds), pp. 122–147. Sinauer Associates, Sunderland, Massachusetts.

S

**Sibly, R.M.** 1983. Optimal group size is unstable. *Animal Behaviour* 31: 947–948.

**Sigmund, K.** 1993. *Games of Life.* Penguin Books, London.

**Sjerps, M. & Haccou, P.** 1994. Effects of competition on optimal patch leaving: a war of attrition. *Theoretical Population Biology* 46: 300–318.

**Smith, R.L. & Smith, T.M.** 1998. *Elements of Ecology*, 4th edition. Benjamin Cummings, Menlo Park, California.

**Sutherland, W.J.** 1983. Aggregation and the 'ideal free' distribution. *Journal of Animal Ecology* 52: 821–828.

**Sutherland, W.J. & Parker, G.A.** 1985. Distribution of unequal competitors. In *Behavioural Ecology* (Sibly,

R.M. & Smith, R.H. eds), pp. 255–274. Blackwell Scientific Publications, Oxford.

T

**Templeton, J.J. & Giraldeau, L.-A.** 1996. Vicarious sampling: the use of personal and public information by starlings foraging in a simple patchy environment. *Behavioral Ecology and Sociobiology* 38: 105–113.

**Tregenza, T.** 1995. Building on the ideal free distribution. *Advances in Ecological Research* 26: 253–307.

V

**Valone, T.** 1989. Group foraging, public information and patch estimation. *Oikos* 56: 357–363.

**Vollrath, F.** 1982. Colony foundation in a social spider. *Zietschrift für Tierpsychologie* 60: 313–324.

## Chapter 9

A

**Arlt, D. & Pärt, T.** 2007. Nonideal breeding habitat selection: a mismatch between preference and fitness. *Ecology* 88: 792–801.

**Arthur, S.M., Manly, B.F.J., McDonald, L.L. & Garner, G.W.** 1996. Assessing habitat selection when availability changes. *Ecology* 77: 215–227.

B

**Battin, J.** 2004. When good animals love bad habitats: ecological traps and the conservation of animal populations. *Conservation Biology* 18: 1482–1491.

**Bernstein, C., Kacelnik, A. & Krebs, J.R.** (1988). Individual decisions and the distribution of predators in a patchy environment. *Journal of Animal Ecology* 57: 1007–1026.

**Bernstein, C., Krebs, J.R. & Kacelnik, A.** 1991. Distribution of birds amongst habitat: theory and relevance to conservation. In Perrins, C.M., Lebreton, J.-D. & Hirons, G.J.M. (eds), *Bird Population Studies*, pp. 317–345. Oxford: Oxford University Press.

**Blums, P., Nichols, J.D., Hines, J.E. & Mednis, A.** 2002. Sources of variation in survival and breeding site fidelity in three species of European ducks. *Journal of Animal Ecology* 71: 438–450.

**Boulinier, T. & Danchin, E.** 1997. The use of conspecific reproductive success for breeding patch selection in territorial migratory species. *Evolutionary Ecology* 11: 505–517.

**Boulinier, T. & Lemel, J.-Y.** 1996. Spatial and temporal variations of factors affecting breeding habitat quality in colonial birds: some consequences for dispersal and habitat selection. *Acta Oecologica* 17: 531–552.

**Boulinier, T., Danchin, E., Monnat, J.-Y., Doutrelant, C. & Cadiou, B.** 1996. Timing of prospecting and the value of information in a colonial breeding bird. *Journal of Avian Biology* 27: 252–256.

**Boulinier, T., Yoccoz, N.G., McCoy, K.D., Erikstad, K.E. & Tveraa, T.** 2002. Testing the effect of conspecific reproductive success on dispersal and recruitment decisions in a colonial bird: design issues. *Journal of Applied Statistics* 29: 509–520.

**Brown, C.R. & Brown, M.B.** 1996. *Coloniality in the Cliff Swallow. The Effect of Group Size on Social Behavior.* University of Chicago Press.

**Brown, C.R., Bomberger Brown, M. & Danchin, E.** 2000. Breeding habitat selection in cliff swallows: the effect of conspecific reproductive success on colony choice. *Journal of Animal Ecology* 69: 133–142.

C

**Cadiou, B., Monnat, J.Y. & Danchin, E.** 1994. Prospecting in the kittiwake, *Rissa tridactyla*: different behavioural patterns and the role of squatting in recruitment. *Animal Behaviour* 47: 847–856.

**Clark, R.G. & Shutler, D.** 1999. Avian habitat selection: pattern from process in nest-site use by ducks. *Ecology* 80: 272–287.

**Clobert, J., Danchin, E., Dhondt, A. & Nichols, J.D.** 2001. *Dispersal.* Oxford University Press, Oxford.

**Combes, C.** 2001. *Parasitism: The Ecology and Evolution of Intimate Interactions*, 1st edition. University of Chicago Press, Chicago.

D

**Danchin, E., Boulinier, T. & Massot, M.** 1998. Habitat selection based on conspecific reproductive success: implications for the evolution of coloniality. *Ecology* 79: 2415–2428.

**Danchin, E., Heg, D. & Doligez, B.** 2001. Public information and breeding habitat selection. In *Dispersal* (Clobert, J., Danchin, É., Dhondt, A.A. & Nichols, J. eds), pp. 243–258. Oxford University Press, Oxford.

**Danchin, E., Cadiou, B., Monnat, J.Y. & Rodriguez Estrella, R.** 1991. Recruitment in long-lived birds: conceptual framework and behavioural mechanisms. In *Acta XXth Congressus Internationalis Ornithologicus*, pp. 1641–1656. Hutcheson, Bowman and Stewart, Wellington.

**Danchin, E., Giraldeau, L.-A., Valone, T.J. & Wagner, R.H.** 2004. Public information: from nosy neighbors to cultural evolution. *Science* 305: 487–491.

**Delibes, M., Ferreras, P. & Gaona, P.** 2001. Attractive sinks, or how individual behavioural decisions determine source–sink dynamics. *Ecology Letters* 4: 401–403.

**Doligez, B., Danchin, E. & Clobert, J.** 2002. Public information and breeding habitat selection in a wild bird population. *Science* 297: 1168–1170.

**Doligez, B., Pärt, T. & Danchin, E.** 2004a. Prospecting in the collared flycatcher: gathering public information for breeding habitat selection? *Animal Behaviour* 67: 457–466.

**Doligez, B., Cadet, C., Danchin, E. & Boulinier, T.** 2003. When to use public information for breeding habitat selection? The role of environmental predictability and density dependence. *Animal Behaviour* 66: 973–988.

**Doligez, B., Danchin, E., Clobert, J. & Gustafsson, L.** 1999. The use of conspecific reproductive success for breeding habitat selection in a non-colonial, hole-nesting species, the collared flycatcher. *Journal of Animal Ecology* 68: 1193–1206.

**Doligez, B., Pärt, T., Danchin, E., Clobert, J. & Gustafsson, L.** 2004b. Availability and use of public information and conspecific density for settlement decisions in the collared flycatcher. *Journal of Animal Ecology* 73: 75–87.

E

**Ens, B.J., Weissing, F.J. & Drent, R.H.** 1995. The despotic distribution and deferred maturity – 2 sides of the same coin. *American Naturalist* 146: 625–650.

**Erwin, R.M., Nichols, J.D., Eyler, T.B., Stotts, D.B. & Truitt, B.R.** 1998. Modeling colony-site dynamics: a case study of gull-billed terns (*Sterna nilotica*) in coastal Virginia. *Auk* 115: 970–978.

F

**Forbes, L.S. & Kaiser, G.W.** 1994. Habitat choice in breeding seabirds: when to cross the information barrier. *Oikos* 70: 377–383.

**Frederiksen, M. & Bregnballe, T.** 2001. Conspecific reproductive success affects age of recruitment in a great cormorant, *Phalacrocorax carbo sinensis*, colony. *Proceedings of the Royal Society of London Series B* 268: 1519–1526.

**Fretwell, S.D. & Lucas, H.L. Jr.** 1970. On territorial behaviour and other factors influencing habitat distribution in birds. *Acta Biotheoretica* 19: 16–36.

G

**Giraldeau, L.-A.** 1997. The ecology of information use. In Krebs, J.R., Klopfer, P.H. & Davies, N.B. (eds), *Behavioural Ecology: An Evolutionary Approach*, pp. 42–68. Sinauer Associates, Sunderland, Massachusetts.

**Giraldeau, L.A., Valone, T.J. & Templeton, J.J.** 2002. Potential disadvantages of using socially-acquired information. *Philosophical Transactions of the Royal Society of London Series B* 357: 1559–1566.

H

**Haas, C.A.** 1998. Effects of prior nesting success on site fidelity and breeding dispersal: an experimental approach. *Auk* 115: 929–936.

**Hakkarainen, H., Ilmonen, P., Koivunen, V. & Korpimäki, E.** 2001. Experimental increase of predation risk induces breeding dispersal of Tengmalm's owl. *Oecologia* 126: 355–359.

**Heg, D.** 1999. *Life history decisions in oystercatchers*. Thesis, University of Groningen.

J

**Johnson, D.H.** 1980. The comparison of usage and availability measurements for evaluating resource preference. *Ecology* 61: 65–71.

**Jones, J.** 2001. Habitat selection studies in avian ecology: A critical review. *Auk* 118: 557–562.

K

**Klopfer, P.H. & Ganzhorn, J.U.** 1985. Habitat selection: behavioral aspects. In *Habiat Selection in Birds* (Cody, M.L. ed.), pp. 435–453. Academic Press, San Diego.

**Kokko, H. & Ekman, J.** 2002. Delayed dispersal as a route to breeding: territorial inheritance, safe havens & ecological constraints. *American Naturalist* 160: 468–484.

**Kokko, H. & Lundberg, P.** 2001. Dispersal, migration and offspring retention in saturated habitats. *American Naturalist* 157: 188–202.

**Kokko, H. & Sutherland, W.J.** 1998. Optimal floating and queuing strategies: consequences for density dependence and habitat loss. *American Naturalist* 152: 354–366.

**Kokko, H. & Sutherland, W.J.** 2001. Ecological traps in changing environments: ecological and evolutionary consequences of a behaviourally mediated Allee effect. *Evolutionary Ecology Research* 3: 537–551.

**Kress, S.W.** 1998. Applying research for effective management: case studies in seabird restauration. In *Avian Conservation* (Marzluff, J.M. & Sallabanks, R. eds), pp. 141–154. Island Press, Washington, DC.

L

**Lima, S.L. & Zollner, P.A.** 1996. Towards a behavioral ecology of ecological landscapes. *Trends in Ecology and Evolution* 11: 131–135.

M

**Manuwald, D.A.** 1974. Effects of territoriality on breeding in a population of Cassin's auklet. *Ecology* 55: 1399–1406.

**Martin, T.E.** 1992. Nest predation and nest sites: new perspectives on old patterns. *Bioscience* 43: 523–532.

**Meadows, P.S. & Campbell, J.I.** 1972. Habitat selection by aquatic invertebrates. In *Advances in Marine Biology* (Russell, F.S. & Yonge, M. eds), pp. 271–382. Academic Press, London and New York.

**Milinski, M. & Parker, G.A.** 1991. Competition for resources. In *Behavioural Ecology. An Evolutionary Approach* (Krebs, J.R. & Davis, N.B. eds), pp. 137–168. Blackwell, Oxford.

**Monnat, J.Y., Danchin, E. & Rodriguez Estrella, R.** 1990. Évaluation de la qualité du milieu dans le cadre de la prospection et du recrutement: le squatterisme chez la Mouette tridactyle. *Comptes Rendus Hebdomadaires des Séances de l'Académie des Sciences, Paris Série 3* 311: 391–396.

N

**Nocera, J.J., Forbes, G.J. & Giraldeau, L.-A.** 2006. Inadvertent social information in breeding site selection of natal dispersing birds. *Proceedings of the Royal Society of London Series B* 273: 349–355.

O

**Orians, G.H. & Wittenberger, J.F.** 1991. Spatial and temporal scales in habitat selection. *American Naturalist* 137: S29–S49.

**Oro, D. & Ruxton, G.-D.** 2001. The formation and growth of seabird colonies: Audouin's gull as a case study. *Journal of Animal Ecology* 70: 527–535.

P

**Pärt, T. & Doligez, B.** 2003. Gathering public information for habitat selection: prospecting birds cue on parental activity. *Proceedings of the Royal Society of London Series B* 270: 1809–1813.

**Petit, L.J. & Petit, D.R.** 1996. Factors governing habitat selection by prothonotary warblers: field tests of the Fretwell–Lucas models. *Ecological Monographs* 66: 367–387.

**Pierotti, R. & Annett, C.A.** 1991. Diet choice in the herring gull: constraints imposed by reproductive and ecological factors. *Ecology* 72: 319–328.

**Pulliam, H.R.** 1988. Sources, sinks and population regulation. *American Naturalist* 132: 652–661.

**Pulliam, H.R.** 2000. On the relationship between niche and distribution. *Ecology Letters* 3: 349–361.

**Pulliam, H.R. & Danielson, B.J.** 1991. Sources, sinks and habitat selection: a landscape perspective on population dynamics. *American Naturalist* 137: S50–S66.

R

**Ray, C., Gilpin, M. & Smith, A.T.** 1991. The effect of conspecific attraction on metapopulation dynamics. *Biological Journal of the Linnean Society* 42: 123–134.

**Reed, J.M. & Dobson, A.P.** 1993. Behavioural constraints and conservation biology: conspecific attraction and recruitment. *Trends in Ecology and Evolution* 8: 253–256.

**Reed, J.M. & Oring, L.W.** 1992. Reconnaissance for future breeding sites by spotted sandpipers. *Behavioral Ecology* 3: 310–317.

**Reed, J.M., Boulinier. T., Danchin, E. & Oring, L.** 1999. Informed dispersal: prospecting by birds for breeding sites. *Current Ornithology* 15: 189–259.

**Rodenhouse, N.L., Sherry, T.W. & Holmes, R.T.** 1997. Site dependent regulation of population size: a new synthesis. *Ecology* 78: 2025–2042.

S

**Safran, R.J.** 2004. Adaptive site selection rules and variation in group size of barn swallows: individual decisions predict population patterns. *American Naturalist* 164: 121–131.

**Sarrazin, F., Bagnolini, C., Pinna, J.L. & Danchin, E.** 1996 Breeding biology during establishment of a reintroduced Griffon vulture (*Gyps fulvus*) population. *Ibis* 138: 315–325.

**Schjørring, S.** 2002. The evolution of informed dispersal: inherent versus acquired information. *Evolutionary Ecology Research* 4: 227–238.

**Schjørring, S., Gregersen, J. & Bregnballe, T.** 1999. Prospecting enhances breeding success of first-time breeders in the great cormorant, *Phalacrocorax carbo sinensis*. *Animal Behaviour* 57: 647–654.

**Schuck-Paim, C. & Alonso, W.J.** 2001. Deciding where to settle: conspecific attraction and web-site selection in the orb-web spider *Nephilengys cruentata*. *Animal Behaviour* 62: 1007–1012.

**Serrano, D., Tella, J.L. & Forero, M.G.** 2001. Factors affecting breeding dispersal in the facultatively colonial lesser kestrel: individual experience vs. conspecific cues. *Journal of Animal Ecology* 70: 568–578.

**Smith, A.T. & Peacock, M.M.** 1992. Conspecific attraction and the determination of metapopulation coloniozation rates. *Conservation Biology* 4: 320–323.

**Stamps, J.A.** 1988. Conspecific attraction and aggregation in territorial species. *American Naturalist* 131: 329–347.

**Stamps, J.A.** 1991. The effects of conspecifics on habitat selection in territorial species. *Behavioral Ecology and Sociobiology* 28: 29–36.

**Stamps, J.A.** 2001. Habitat selection by dispersers: integrating proximate and ultimate approaches. In *Dispersal* (Clobert, J., Danchin, E., Dhondt, A.A., Nichols, J. eds), pp. 230–242. Oxford University Press, Oxford.

**Stearns, S.C.** 1992. *The Evolution of Life History Strategies.* Oxford University Press, Oxford.

**Stephens, D.W.** 1989. Variance and the value of information. *American Naturalist* 134: 128–140.

**Suryan, R.M. & Irons, D.B.** 2001. Colony and population dynamics of black-legged kittiwakes in a heterogeneous environment. *Auk* 118: 636–649.

**Sutherland, W.D.** 1996. *From Individual Behaviour to Population Ecology.* Oxford University Press, Oxford.

**Switzer, P.V.** 1993. Site fidelity in predictable and unpredictable habitats. *Evolutionary Ecology* 7: 533–555.

**Switzer, P.V.** 1997. Past reproductive success affects future habitat selection. *Behavioral Ecology and Sociobiology* 40: 307–312.

T

**Templeton, J.J. & Giraldeau, L.-A.** 1996. Vicarious sampling: the use of personal and public information by starlings foraging in a simple patchy environment. *Behavioral Ecology and Sociobiology* 38: 105–113.

U

**Ueta, M.** 2001. Azure-winged magpies avoid nest predation by breeding synchronously with Japanese lesser sparrowhawks. *Animal Behaviour* 61: 1007–1012.

V

**Valone, T.J.** 1989. Group foraging, public information and patch estimation. *Oïkos* 56: 357–363.

**Valone, T.J. & Templeton, J.J.** 2002. Public information for the assessment of quality: a widespread phenomenon. *Philosophical Transactions of the Royal Society of London Series B* 357: 1549–1557.

**Van Horne, B.** 1983. Density as a misleading indicator of habitat quality. *Journal of Wildlife Management* 47: 893–901.

**van Teeffelen, A.J.A. & Ovaskainen, O.** 2007. Can the cause of aggregation be inferred from species distributions? *Oikos* 116: 4–16.

**Veen, J.** 1977. Functional and causal aspects of nest distribution in colonies of the sandwich tern (*Sterna sandvicensis* Lath.). *Behaviour Supplement* 20: 1–193.

W

**Wagner, R.H. & Danchin, E.** 2003. Conspecific copying: a general mechanism of social aggregation. *Animal Behaviour* 65: 405–408.

**Wagner, R.H., Danchin, E., Boulinier, T. & Helfenstein, F.** 2000. Colonies as byproducts of commodity selection. *Behavioral Ecology* 11: 572–573.

**Ward, M.P.** 2005. Habitat selection by dispersing yellow-headed blackbirds: evidence of prospecting and the use of public information. *Oecologia* 145: 650–657.

**Ward, S.A.** 1987. Optimal habitat selection in time-limited dispersers. *American Naturalist* 129: 568–579.

**Wiens, J.A.** 1985. Habitat selection in variable environments: shrub–steppe birds. In *Habitat Selection in Birds* (Cody, M.L. ed.), pp. 227–251. Academic Press, San Diego.

## Chapter 10

A

**Anderson, P.K.** 1989. Dispersal in rodents: a resident fitness hypothesis. *American Society of Mammologists Special Publication* 9: 1–141.

**Arcese, P.** 1989a. Intrasexual competition, mating system and natal dispersal in song sparrows. *Animal Behaviour* 38: 958–979.

**Arcese, P.** 1989b. Territory acquisition and loss in male song sparrows. *Animal Behaviour* 37: 45–55.

B

**Belthoff, J.R. & Dufty, A.M. Jr.** 1998. Corticosterone, body condition, and locomotor activity: a model for natal dispersal. *Animal Behaviour* 54: 405–15.

**Bengtsson, B.O.** 1978. Avoding inbreeding: at what cost? *Journal of Theoretical Biology* 73: 439–444.

**Bengtsson, G., Hedlund, K. & Rundgren, S.** 1994. Food- and density-dependent dispersal: evidence from a soil collembolan. *Journal of Animal Ecology* 63: 513–520.

**Berthold, P. & Pulido, F.** 1994. Heritability of migratory activity in a natural bird population. *Proceedings of the Royal Society of London Series B* 257: 311–315.

**Blaustein, A.R. & Waldman, B.** 1992. Kin recognition in anuran amphibians. *Animal Behaviour* 44: 207–221.

**Boonstra, R. & Krebs, C.J.** 1977. A fencing experiment on a population of *Microtus townsendii*. *Canadian Journal of Zoology* 55: 1166–1175.

**Both, C., Dingemanse, N.J., Drent, P.J. & Tinbergen, J.M.** 2005. Pairs of extreme avian personalities have highest reproductive success. *Journal of Animal Ecology* 74: 667–674.

C

**Cadet, C., Ferrière, R., Metz, J.A.J. & van Baalen, M.** 2003. The evolution of dispersal under demographic stochasticity. *American Naturalist* 162: 427–441.

**Clobert, J., Ims, R. & Rousset, F.** 2004. Causes, mechanisms and consequences of dispersal. In *Metapopulation Biology* (Hanski, I. and Gaggiotti, O.E. eds), pp. 307–355. Academic Press.

**Cote, J. & Clobert, J.** 2007. Social personalities influence natal dispersal in a lizard. *Proceedings of the Royal Society of London Series B* 274: 383–390.

D

**Dawkins, R.** 1976. *The Selfish Gene*. Oxford University Press, Oxford.

**de Fraipont, M., Clobert, J., John-Alder, H. & Meylan, S.** 2000. Increased prenatal maternal corticosterone promotes philopatry of offspring in common lizards *Lacerta vivipara*. *Journal of Animal Ecology* 69: 404–413.

**de Fraipont, M., Clobert, J., John-Alder, H. & Melan, S.** (2000). Pre-natal stress increases offspring philopatry. *Journal of Animal Ecology* 69: 404–413.

**Dingemanse, N.J. & Réale, D.** 2005. Natural selection and animal personality. *Behaviour* 142: 1165–1190.

**Dingemanse, N.J., Both, C., Drent, P.J. & Tinbergen, J.M.** 2004a. Fitness consequences of avian personalities in a fluctuating environment. *Proceedings of the Royal Society of London Series B* 271: 847–852.

**Dingemanse, N.J., Both, C., van Noordwijk, A.J., Rutten, A.L. & Drent, P.J.** 2004b. Natal dispersal and personalities in great tits (*Parus major*). *Proceedings of the Royal Society of London Series B* 270: 741–747.

**Doncaster, C.P., Clobert, J., Doligez, B., Gustafsson, L. & Danchin, E.** 1997. Balanced dispersal between spatially varying local populations: an alternative to the source–sink model. *American Naturalist* 150: 425–445.

**Doums, C., Cabrera, H. & Peeters, C.** 2002. Population genetic structure and male-biased dispersal in the queenless ant *Diacamma cyaneiventre*. *Molecular Ecology* 11: 2251–2264.

**Dufty, A.M. Jr., Clobert, J. & Møller, A.P.** 2002. Hormones, developmental plasticity, and adaptation. *Trends in Ecology and Evolution* 17: 190–196.

E

**Ens, B.J., Weissing, F.J. & Drent, R.H.** 1995. The despotic distribution and deferred maturity: two sides of the same coin. *American Naturalist* 146: 625–650.

F

**Fletcher, D.J.C. & Michener, C.D.** (eds). 1987. *Kin Recognition in Animals*. Wiley, Chichester.

**Fretwell, S.D. & Lucas, H.L.** 1970. On territorial behavior and other factors influencing habitat distribution in birds. I. Theoretical developments. *Acta Biotheoretica* 19: 16–36.

G

**Gandon, S.** 1999. Kin competition, the cost of inbreeding and the evolution of dispersal. *Journal of Theoretical Biology* 200: 345–360.

**Gandon, S. & Michalakis, Y.** 2001. Multiple causes of the evolution of dispersal. In *Dispersal* (Clobert, J., Danchin, E., Dhondt, A.A. & Nichols, J.D. eds), pp. 155–167. Oxford University Press, New York.

**Greenwood, P.J.** 1980. Mating systems, philopatry and dispersal in birds and mammals. *Animal Behaviour* 28: 1140–1162.

**Greenwood, P.J. & Harvey, P.H.** 1982. The natal and breeding dispersal of birds. *Annual Review of Ecology and Systematics* 13: 1–21.

**Gundersen, G., Andreassen, H.P. & Ims, R.A.** 2002. Individual and population level determinants of immigration success on local habitat patches: an experimental approach. *Ecology Letters* 5: 294–301.

H

**Haag, C.R., Hottinger, J.W., Riek, M. & Ebert, D.** 2002. Strong inbreeding depression in a *Daphnia* metapopulation. *Evolution* 56: 518–526.

**Hamilton, W.D. & May, R.M.** 1977. Dispersal in stable habitats. *Nature* 269: 5578–581.

**Hanski, I. & Thomas, C.D.** 1994. Metapopulation dynamics and conservation: a spatially explicit model applied to butterflies. *Biological Conservation* 68: 167–180.

**Haugen, T.O., Winfield, I.J., Vøllestad, L.A., Fletcher, J.M., James. J.B. & Stenseth, N.Chr.** 2006. The ideal free pike: 50 years of fitness-maximizing dispersal in Windermere. *Proceedings of the Royal Society of London Series B* 273: 2917–2924

**Heg, D.** 1999. Thesis, Life history decisions in oystercatchers. University of Groningen.

**Hepper, P.G.** (ed). 1991. *Kin Recognition*. Cambridge University Press, Cambridge, UK.

**Herzig, A.L.** 1995. Effects of population density on long-distance dispersal in the goldenrod beetle *Trirhabda virgata*. *Ecology* 76: 2044–2054.

**Holt, R.D. & Barfield, M.** 2001. On the relationship between the ideal free distribution and the evolution of dispersal. In *Dispersal* (Clobert, J., Danchin, E., Dhondt, A.A. & Nichols, J.D. eds), pp. 83–95. Oxford University Press, New York.

I

**Imbert, E.** 1999. The effects of achene dimorphism on the dispersal in time and sapce in *Crepis sancta* (*Asteraceae*). *Canadian Journal of Botany* 77: 508–513.

**Ims, R.A.** 1989. Kinship and origin effects on dispersal and space sharing in *Cletthrionomys rufocanus*. *Ecology* 70: 607–616.

**Ims, R.A.** 1990. Determinants of natal dispersal and space use in grey-sided voles, *Clethrionomys rufocanus*: a combined field and laboratory experiment. *Oikos* 57: 106–113.

**Ims, R.A. & Hjermann, D.Ø.** 2001. Condition-dependent dispersal. In *Dispersal* (Clobert, J., Danchin, E., Dhondt, A.A. & Nichols, J.D. eds), pp. 203–216. Oxford University Press, New York.

K

**Keller, L.F., Arcese, P., Smith, J.N.M., Hochachka, W.M. & Stearns, S.C.** 1994. Selection against inbred song sparrows during a natural population bottleneck. *Nature* 372: 356–7.

**Komdeur, J.** 1996. Influence of helping and breeding experience on reproductive performance in the Seychelles warbler: a translocation experiment. *Behavioral Ecology* 7: 326–33.

**Krebs, C.J., Keller, B.L. & Tamarin, R.H.** 1969. *Microtus* population biology: demographic changes in fluctuating populations of *Microtus ochrogaster* and *M. pennsylvanicus* in southern Indiana. *Ecology* 50: 587–607.

L

**Lambin, X.** 1994. Natal philopatry, competition for ressources, and inbreeding avoidance in Townsend's voles (*Microtus townsendii*). *Ecology* 75: 224–235.

**Lambin, X. & Yoccoz, N.G.** 1998. The impact of population kin-structure on nestling survival in Townsend's voles, *Microtus townsendii*. *Journal of Animal Ecology* 67: 1–16.

**Lambin, X., Aars, J. & Piertney, S.B.** 2001. Dispersal, intraspecific competition, kin competition and kin facilitation. In *Dispersal* (Clobert, J., Danchin, E., Dhondt, A.A. & Nichols, J.D. eds), pp. 110–122. Oxford University Press, New York.

**Le Gaillard, J.-F., Ferrière, R. & Clobert, J.** 2003. Mother–offspring interactions affect natal dispersal in a lizard. *Proceedings of the Royal Society of London Series B* 270: 1163–1169.

**Le Galliard, J.-F., Fitze, P., Ferrière, R. & Clobert, J.** 2005. Sex ratio bias, male aggression, and population collapse in lizards. *Proceedings of National Academy of Sciences of the USA* 102: 18231–18236.

**Lecomte, J. & Clobert, J.** 1996. Dispersal and connectivity of the common lizard *Lacerta vivipara*: an experimental approach. *Acta Oecologica* 17: 585–598.

**Lecomte, J., Boudjemadi, K., Sarrazin, F. & Clobert, J.** 2004. Connectivity and homogenisation of population sizes: an experimental approach in *Lacerta vivipara*. *Journal of Animal Ecology* 73: 179–189.

**Lefranc, A.** 2001. Etude des facteurs déterminant les comportements de dispersion & de sélection d'habitat chez *Drosophila melanogaster*. Thesis, Université Pierre & Marie Curie, Paris France.

**Lemel, J.Y., Belichon, S., Clobert, J. & Hochberg, M.E.** 1997. The evolution of dispersal in a two-patch system: some consequences of differences between migrants and residents. *Evolutionary Ecology* 11: 613–629.

**Léna, J.-P., Clobert, J., de Fraipont, M., Lecomte, J. & Guyot, G.** 1998. The relative influence of density and kinship on dispersal in the common lizard. *Behavioral Ecology* 9: 500–507.

**Leturque, H. & Rousset, F.** 2002. Dispersal, kin competition, and the ideal free distribution in a spatially heterogeneous population. *Theoretical Population Biology* 62: 169–180.

**Levins, R. & MacArthur, R.** 1966. The maintenance of genetic polymorphism in a spatially heterogeneous environment: variations on a thème by Howard Levene. *American Naturalist* 100: 585–589.

**Linsenmair, K.E.** 1987. Kin recognition in subsocial arthropods, in particular in the desert isopod *Hemilepistus reaumuri*. In *Kin Recognition in Animals* (Fletcher, D.J.C. & Michener, C.D. eds), pp. 21–208. Wiley, Chichester.

M

**Massot, M., Clobert, J., Lorenzon, P. & Rossi, J.M.** 2002. Condition dependent dispersal and ontogeny of the dispersal behavior: an experimental approach. *Journal of Animal Ecology* 71: 235–261.

**McNeil, J.N., Cusson, M., Delisle, J., Orchard, I. & Tobe, S.S.** 1995. Physiological integration of migration in Lepidoptera. In *Insect Migration: Tracking Resources through Space and Time* (Drake, V.A. and Gatehouse, A.G. eds), pp 279–302. Cambridge University Press, Cambridge.

**McPeek, M.A. & Holt, R.D.** 1992. The evolution of dispersal in spatially and temporally varying environments. *Americcan Naturalist* 140: 1010–1027.

**Meylan, S., Belliure, J. & de Fraipont, M.** 2002. Stress and body condition as prenatal and postnatal determinants of dispersal in the common lizard (*Lacerta vivipara*). *Hormones and Behavior* 42: 319–326.

**Meylan, S., De Fraipont, M. & Clobert, J.** 2004. Maternal size and stress and offspring philopatry: an experimental study in the common lizard (*Lacerta vivipara*). *Ecoscience* 11: 123–129.

**Motro, U.** 1991. Avoiding inbreeding and sibling competition: the evolution of sexual dimorphism for dispersal. *American Naturalist* 137: 108–115.

**Murren, C.J., Julliard, R., Schlichting, C.D. & Clobert, J.** 2001. Dispersal, individual phenotype, and phenotypic plasticity. In *Dispersal* (Clobert, J., Danchin, E., Dhondt, A.A. & Nichols, J.D. eds), pp. 261–282. Oxford University Press, New York.

O

**O'Riain, M.J. & Braude, S.** 2001. Inbreeding versus outbreeding in captive and wild populations of naked mole-rats. In *Dispersal* (Clobert, J., Danchin, E., Dhondt, A.A. & Nichols, J.D. eds), pp. 143–154. Oxford University Press, New York.

P

**Peacock, M.M. & Ray, C.** 2001. Dispersal in pikas (*Ochotona princeps*): combining genetic and demographic approaches to eveal spatial and temporal patterns. In *Dispersal* (Clobert, J., Danchin, E., Dhondt, A.A. & Nichols, J.D. eds), pp. 43–56. Oxford University Press, New York.

**Peeters, C. & Ito, F.** 2001. Colony dispersal and the evolution of queen morphology in social hymenoptera. *Annual Review of Entomology* 46: 601–630.

**Perrin, N. & Goudet, J.** 2001. Inbreeding, kinship, and the evolution of natal dispersal. In *Dispersal* (Clobert, J., Danchin, E., Dhondt, A.A. & Nichols, J.D. eds), pp. 123–142. Oxford University Press, New York.

**Perrin, N. & Mazalov, V.** 2000. Local competition, inbreeding, and the evolution of sex-biased dispersal. *American Naturalist* 155: 116–127.

**Pusey, A.E.** 1987. Sex biased dispersal and inbreeding avoidance in birds and mammals. *Trends in Ecology and Evolution* 2: 295–299.

R

**Roff, D.A. & Fairbairn, D.J.** 2001. The genetic basis of dispersal and migration, and its consequences for the evolution of correlated traits. In *Dispersal* (Clobert, J., Danchin, E., Dhondt, A.A. & Nichols, J.D. eds), pp. 191–202. Oxford University Press, New York.

**Ronce, O. & I. Olivieri** 2004. Life history evolution in metapopulations. In *Metapopulation Biology* (Hanski, I. & Gaggiotti, O.E. eds), pp. 227–257. Academic Press, London.

**Ronce, O., Clobert, J. & Massot, M.** 1998. Natal dispersal and senescence. *Proceedings of the National Academy of Science of the USA* 95: 600–605.

**Ronce, O., Olivieri, I., Clobert, J. & Danchin, E.** 2001. Perspectives on the study of dispersal evolution. In *Dispersal* (Clobert, J., Danchin, E., Dhondt, A.A. & Nichols, J.D. eds), pp. 340–357. Oxford University Press, New York.

**Ross, K.G.** 2001. How to measure dispersal: the genetic approach. The example of the fire ants. In *Dispersal* (Clobert, J., Danchin, E., Dhondt, A.A. & Nichols, J.D. eds). Oxford University Press, New York.

S

**Saccheri, I., Kuussaari, M., Kankare, M., Vikman, P., Fortelius, W. & Hanski, I.** 1998. Inbreeding and extinction in a butterfly metapopulation. *Nature* 392: 491–494.

**Silverin, B.** 1997. The stress response and autumn dispersal behaviour in willow tits. *Animal Behaviour* 53: 451–459.

**Sinervo, B. & Clobert, J.** 2003. Morphs, dispersal behavior, genetic similarity, and the evolution of cooperation. *Science* 300: 1949–1951.

**Sinervo, B. & Lively, C.M.** 1996. The rock–paper–scissors game and the evolution of alternative male reproductive strategies. *Nature* 380: 240–243.

**Sinervo, B., Bleay, C. & Adamopoulou, C.** 2001. Social causes of correlational selection and the resolution of a heritable throat color polymorphism in a lizard. *Evolution* 55: 2040–2052.

**Sinervo, B., Calsbeeck, R. & Clobert, J.** (in press). Genetic and maternal determinants of dispersal in color morphs of side-blotched lizards. *Evolution.*

**Sinervo, B., Calsbeek, R., Comendant, T., Both, C., Adamopoulou, C. & Clobert, J.** 2006 Genetic and maternal determinants of effective dispersal: the effect of sire genotype and size at birth in side-blotched lizards. *American Naturalist* 168: 88–99.

**Stamps, J.A.** 1987. Conspecifics as cues to territory quality: a preference of juvenile lizard (*Anolis aeneus*) for previously used territories. *American Naturalist* 129: 629–642.

**Stamps, J.A.** 2006. The silver sponn effect and habitat selection by dispersers. *Ecology Letters* 9: 1179–1185.

**Stenseth, N.C. & Lidicker, W.Z. Jr.** (eds). 1992. *Animal Dispersal: Small Mammals as a Model.* Chapman & Hall, London.

T

**Travis, J.M.J., Murrell, D.J. & Dytham, C.** 1999. The evolution of density-dependent dispersal. *Proceedings of the Royal Society of London Series B* 266: 1837–1842.

**Trefilov, A., Berard, J., Krawczak, M. & Schmidtke, J.** 2000. Natal dispersal in rhesus macaques is related to transporter gene promoter variation. *Behavior Genetics* 30: 295–301.

V

**Van Dijken, F.R. & Scharloo, W.** 1980. Divergent selection for locomotor activity in *Drosophila melanogaster*. *Behavioral Genetics* 9: 543–553.

**Van Valen, L.** 1971. Group selection and the evolution of dispersal. *Evolution* 25: 591–598.

W

**Weisser, W.W.** 2001. The effect of predation on dispersal. In *Dispersal* (Clobert, J., Danchin, E., Dhondt, A.A. & Nichols, J.D. eds), pp. 180–188. Oxford University Press, New York.

**Werner, D.I., Baker, E.M., Gonzalez, E.C. & Sosa, I.R.** 1987. Kinship recognition and grouping in hatchling green iguanas. *Behavioral Ecology and Sociobiology* 21: 83–89.

**Woiwood, I.P., Reynolds, D.R. & Thomas, C.D.** (eds). 2001. *Insect Movement: Mechanisms and Consequences*. CAB Publications, Wallingford, UK.

**Wolff, J.O.** 1992. Parents suppress reproduction and stimulate dispersal in opposite-sex juvenile white-footed mice. *Nature* 359: 409–10.

**Woolfenden, G.E. & Fitzpatrick, J.W.** 1990. Florida scrub jay after 19 years of study. In *Cooperative Breeding in Birds* (ed. Stacey, P.B. and Koenig, W.D.), pp. 241–66. Cambridge University Press, New York.

**Wright, S.** 1932. The roles of mutation, inbreeding, crossbreeding and selection in evolution. *Proceedings of the Sixth International Congress of Genetics*, vol. 1, pp. 356–366.

Z

**Zamudio, K. & Sinervo, B.** 2000. Polygyny, mate-guarding, and posthumous fertilization as alternative male mating strategies. *Proceedings of the National Academy of Science of the USA* 97: 14427–14432.

## Chapter 11

A

**Ahearn, J.N. & Templeton, A.R.** 1989. Interspecific hybrids of *Drosophila heteroneura* and *D. silvestris* courtship success. *Evolution* 43: 347–361.

**Alatalo, R.V., Lundberg, A. & Glynn, C.** 1986. Female pied flycatchers choose territory quality and not male characteristics. *Nature* 323: 738–753.

**Alcock, J. & Pyle, D.W.** 1979. The complex courtship behaviour of *Physiofora demendata* (F.) (Diptera: Otitidae). *Zeitscrift für Tierpsychologie* 49: 352–362.

**Alexander, R.D.** 1975. Natural selection and specialized chorusing behavior in acoustical insects. In *Insects, Science and Society*. (Pimentel, D. ed.). Academic Press, New York.

**Amos, W., Worthington Wilmer, J., Fulard, K., Burg, T.M., Croxall, J.P., Bloch, D. & Coulson, T.** 2001. The influence of parental relatedness on reproductive success. *Proceedings of the Royal Society of London Series B* 268: 2021–2027.

**Andersson, M.** 1982. Sexual selection, natural selection and quality advertisement. *Biological Journal of the Linnean Society* 17: 375–393.

**Andersson, M.** 1986. Evolution of condition-dependent sex ornaments and mating preferences: sexual selection based on viability differences. *Evolution* 40: 804–816.

**Andersson, M.B.** 1994. *Sexual Selection. Monographs in Behavior and Ecology*. Princeton Unviversity Press, Princeton.

**Arak, A.** 1988. Female mate selection in the natterjack toad: active choice or passive attraction? *Behavioural Ecology and Sociobiology* 22: 317–327.

**Arnold, S.J. & Wade, M.J.** 1984a. On the measurement of natural and sexual selection: theory. *Evolution* 38: 709–719.

**Arnold, S.J. & Wade, M.J.** 1984b. On the measurement of natural and sexual selection: applications. *Evolution* 38: 709–719.

**Arnqvist, G. & Nilsson, T.** 2000. The evolution of polyandry: multiple matings and female fitness in insects. *Animal Behaviour* 60: 145–164.

**Arnqvist, G., Edvardsson, M., Friberg, U. & Nilsson, T.** 2000. Sexual conflict promotes speciation in insects. *Proceedings of the National Academy of Sciences* 97: 10460–10464.

**Avital, E. & Jablonka, E.** 2000. *Animal Traditions. Behavioural Inheritance in Evolution*. Cambridge University Press, Cambridge.

B

**Badyaev, A.V., Whittingham, L.A. & Hill, G.E.** 2001. The evolution of sexual dimorphism in the house finch. III. Developmental basis. *Evolution* 55: 176–189.

**Baker, R.H., Ashwell, R.I.S., Richards, T.A., Fowler, K., Chapman, T. & Pomiankowski, A.** 2001. Effects of multiple mating and male eye span on female reproductive output in the stalk-eyed fly *Cytodiopsis dalmanni*. *Behavioral Ecology* 12: 732–739.

**Baker, R.R. & Bellis, M.A.** 1988. 'Kamikaze' sperm in mammals? *Animal Behaviour* 36: 936–939.

**Baker, R.R. & Bellis, M.A.** 1995. *Human Sperm Competition*. Chapman and Hall. London.

**Bakker, T.C.M.** 1986. Aggressiveness in sticklebacks (*Gasterosteus aculeatus* L.): a behaviour–genetic study. *Behaviour* 98: 1–44.

**Bakker, T.C.M.** 1990. Genetic variation in female mating preferences. *Netherlands Journal of Zoology* 40: 617–642.

**Bakker, T.C.M.** 1993. Positive genetic correlation between female preference and preferred male ornament in sticklebacks. *Nature* 363: 255–257.

**Bakker, T.C.M. & Pomiankowski, A.** 1995. The genetic basis of female mate preference. *Journal of Evolutionary Biology* 8: 129–171.

**Basolo, A.** 1990. Female preference predates the evolution of the sword in the swordtail fish. *Science* 250: 808–810.

**Basolo, A.** 1995a. A further examination of a pre-existing bias favouring a sword in the genus *Xiphophorus*. *Animal Behaviour* 50: 365–375.

**Basolo, A.** 1995b. Phylogenetic evidence for the role of a pre-existing bias in sexual selection. *Proceedings of the Royal Society Series B* 259: 307–311.

**Beani, L. & Dessi-Fulgheri, F.** 1995. Mate choice in the grey partridge, *Perdix perdix*: role of physical and behavioural male traits. *Animal Behaviour* 49: 347–356.

**Bell, G.** 1978. The handicap principle in sexual selection. *Evolution* 32: 872–885.

**Bell, P.D.** 1979. Acoustic attraction of herons by crickets. *Journal of the New York Entomological Society* 87: 126–127.

**Berglund, A., Bisazza, A. & Pilastro, A.** 1996. Armaments and ornaments: an evolutionary explanation of traits of dual utility. *Biological Journal of the Linnean Society* 58: 385–399.

**Birkhead, T.R.** 1998. Cryptic female choice: criteria for establishing female sperm choice. *Evolution* 52: 1212–1218.

**Birkhead, T.R.** 2000. Defining and demonstrating postcopulatory female choice – again. *Evolution* 54: 1057–1060.

**Birkhead, T.R. & Møller, A.P.** 1992. *Sperm Competition in Birds. Evolutionary Causes and Consequences*. Academic Press, London.

**Birkhead, T.R. & Møller, A.P.** 1998. *Sperm Competition and Sexual Selection*. Academic Press, Cambridge, UK.

**Bisazza, A. & Marin, G.** 1991. Male size and female mate choice in the eastern mosquitofish (*Gambisia holbrooki*, Poeciliidae). *Copeia* 1991: 730–735.

**Bisazza, A., Marconato, A. & Marin, G.** 1989a. Male competition and female choice in *Padogobius martensi* (Pisces, Gobiidae). *Animal Behaviour* 38: 406–413.

**Bisazza, A., Marconato, A. & Marin, G.** 1989b. Male preference in the mosquitofish *Gambusia holbrooki*. *Ethology* 83: 335–343.

**Blomqvist, D., Andersson, M., Küpper, C., Cuthill, I.C., Kis, J., Lanctot, R.B., Sandercock, B.K., Székely, T., Wallander, J. & Kempenaers, B.** 2002. Genetic simlarity between mates and extra-pair parentage in three species of shorebirds. *Nature* 419: 613–615.

**Blows, M.W.** 1999. Evolution of the genetic covariance between male and female components of mate recognition: an experimental test. *Proceedings of the Royal Society of London Series B* 266: 2169–2174.

**Borgia, G.** 1979. Sexual selection and the evolution of mating systems. In: *Sexual selection and reproductive competition in insects.* (Blum, M.S. & Blum, N.A. eds): 19–80. Academic press, New York.

**Borgia, G.** 1981. Mate selection in the fly *Scatophaga stercoraria*: females choice in a male controlled system. *Animal Behaviour* 29: 71–80.

**Bradbury, J.W. & Gibson, R.M.** 1983. Leks and mate choice. In *Mate Choice* (Bateson, P. ed.), pp. 109–138. Cambridge University Press, Cambridge.

**Briceno, R.D. & Eberhard, W.G.** 1995. The functional morphology of male cerci and associated characters in 13 species of tropical earwigs (Dermaptera: Forficulidae, Labiidae, Carcinophoridae, Pygidicranidae). *Smithsonian Contributions to Zoology* 555: 1–63.

**Brodie, E., Moore, A. & Janzen, F.** 1995. Visualizing and quantifying natural selection. *Trends in Ecology and Evolution* 10: 313–318.

**Brooks, R.** 1998. The importance of mate copying and cultural inheritance of mating preferences. *Trends in Ecology and Evolution* 13: 45–46.

**Brooks, R.** 2000. Negative genetic correlation between male sexual attractiveness and survival. *Nature* 406: 67–70.

**Brown, C.R. & Laland, K.N.** 2003. Social learning in fishes: a review. *Fish and Fisheries* 4: 280–288.

**Burns, K.J.** 1998. A phylogenetic perspective on the evolution of sexual dichromatism in tanagers (Thraupidae): the role of female versus male plumage. *Evolution* 52: 1219–1224.

**Buskirk, R.E.C., Frohlich, C. & Ross, K.G.** 1984. The natural selection of sexual cannibalism. *American Naturalist* 123: 612–625.

**Butlin, R.K., Woodhatch, C.W. & Hewitt, G.M.** 1987. Male spermatophore investment increases female fecundity in a grasshopper. *Evolution* 41: 221–225.

**Byers, J.A., Moodie, J.D. & Hall, N.** 1994. Pronghorn females choose vigorous mates. *Animal Behaviour* 47: 33–43.

C

**Calder, W.A. III.** 1984. *Size, Function and Life History*. Harvard University Press, Harvard.

**Candolin, U. & Reynolds, J.D.** 2001. Sexual signaling in the European bitterling: females learn the truth by direct inspection of the resource. *Behavioral Ecology* 12: 407–411.

**Carson, H.L., Kaneshiro, K.Y. & Val, F.C.** 1989. Natural hybridization between the sympatric hawaiian species *Drosophila silvestris* and *Drosophila heteroneura*. *Evolution* 43: 190–203.

**Chapman, T., Hutchings, J., Partridge, L.** 1993. No reduction in the cost of mating for *Drosophila melanogaster* females mating with spermless males. *Proceedings of the Royal Society Series B* 253: 211–217.

**Chapman, T., Lindsay, F., Liddle, F., Kalb, J.M., Wolfner, M.F. & Partridge, L.** 1995. Cost of mating *Drosophila melanogaster* females is mediated by male accessory gland products. *Nature* 373: 241–244.

**Civetta, A. & Clark, A.G.** 2000. Correlated effects of sperm competition and postmating female mortality. *Proceedings of the National Academy of Sciences of the USA* 97: 13162–13165.

**Clayton, D.H.** 1991. The influence of parasites on host sexual selection. *Parasitology Today* 7: 329–334.

**Clutton-Brock, T.H. & McComb, K.** 1993. Experimental tests of copying and mate choice in fallow deer (*Dama dama*). *Behavioral Ecology* 4: 191–193.

**Clutton-Brock, T.H. & Parker, G.A.** 1992. Potential reproductive rates and the operation of sexual selection. *Quarterly Review of Biology* 67: 437–456.

**Clutton-Brock, T.H. & Vincent, A.C.J.** 1991. Sexual selection and the potential reproductive rates of males and females. *Nature* 351: 58–60.

**Clutton-Brock, T.H., Guiness, F.E. & Albon, S.D.** 1982. *Red Deer. Behavior and Ecology of Two Sexes.* University of Chicago Press, Chicago.

**Cohen, J.A.** 1984. Sexual selection and the psychophysics of female choice. *Zeitchrift für Tierpsychologie* 64: 1–8.

**Cohen, L.B. & Dearborn, D.C.** 2004. Great frigatebirds, *Fregata minor*, choose mates that are genetically similar. *Animal Behaviour* 68: 1229–1236.

**Colegrave, N., Kotiaho, J. & Tomkins, J.** 2002. Reconciling models of female choice for good genes and genetic compatability. *Evolutionary Ecology Research* 4: 911–917.

**Collins, S.C.** 1993. Is there only one type of male handicap? *Proceedings of the Royal Society of London Sreries B* 252: 193–197.

**Córdoba-Aguilar, A.** 1999. Male copulatory sensory stimulation induces female ejection of rival sperm in a damsefly. *Procedings of the Royal Society of London Series B* 266: 779–784.

**Córdoba-Aguilar, A.** 2005. Possible coevolution of male and female genital form and function in a calopterygid damselfly. *Journal of Evolutionary Biology* 18: 132–137.

**Coté, I.M. & Hunte, W.** 1989. Male and female mate choice in the redlip blenny: why bigger is better. *Animal Behaviour* 38: 78–88.

**Cox, C.R. & LeBoeuf, B.J.** 1977. Female incitation of male competition: a mechanism in sexual selection. *American Naturalist* 111: 317–335.

**Coyne, J.A. & Orr, H.A.** 1989. Patterns of speciation in *Drosophila*. *Evolution* 43: 362–381.

**Cronly-Dillon, J. & Sharma, S.C.** 1968. Effect of season and sex on the photopic spectral sensitivity of the three-spined stickleback. *Journal of Experimental Biology* 49: 679–687.

D

**Dahlgren, J.** 1990. Females choose vigilant males: an experiment with the monogamous grey partridge, *Perdrix perdrix*. *Animal Behaviour* 39: 646–651.

**Darwin, C.** 1859. *On the Origin of Species by Means of Natural Selection*. Murray, London.

**Darwin, C.** 1871. *The Descent of Man and Selection in Relation to Sex*. Murray, London.

**Darwin, C.** 1874. *The Descent of Man and Selection in Relation to Sex*, 2nd edition. Murray, London.

**Davies, N.B.** 1983. Polyandry, cloaca-pecking and sperm competition in dunnocks. *Nature* 302: 334–336.

**Davies, N.B.** 1991. Mating systems. In *Behavioural Ecology. An Evolutionary Approach*, 3rd edition (Krebs, J.R. & Davies, N.B. eds), pp. 263–299. Blackwell, Oxford.

**Deutsch, C.J., Haley, M.P. & LeBoeuf, B.J.** 1990. Reproductive effort of male northern elephant seals: estimate from mass loss. *Canadian Journal of Zoology* 68: 2580–2593.

**Dewsbury, D.A.** 1982. Ejaculate cost and male choice. *American Naturalist* 119: 601–610.

**Dobzhansky, T.** 1940. Speciation as a stage in evolutionary divergence. *American Naturalist* 74: 312–332.

**Doutrelant, C. & McGregor, P.K.** 2000. Eavesdropping and mate choice in female fighting fish. *Behaviour* 137: 1655–1669.

**Doutrelant, C., McGregor, P.K. & Oliveira, R.F.** 2001. The effect of an audience on intra-sexual communication in male Siamese fighting fish, *Betta splendens*. *Behavioral Ecology*, 283–286.

**Doutrelant, C., Leitao, A., Otter, K. & Lambreshts, M.M.** 2000. Effect of blue tit song syntax on great tit territorial responsiveness – an experimental test of the character shift hypothesis. *Behavioral Ecology and Sociobiology* 48: 119–124.

**Duffy, D.L., Bentley, G.E., Drazen, D.L. & Ball, G.F.** 2000. Effects of testosterone on cell-mediated and humoral immunity in non-breeding adult European starlings. *Behavioral Ecology* 11: 654–662.

**Dugatkin, L.A.** 1992. Sexual selection and imitation: females copy the mate choice of others. *American Naturalist* 139: 1384–1389.

**Dugatkin, L.A.** 1996a. Copying and mate choice. In *Social Learning in Animals: The Roots of Culture*, pp. 85–106. Academic Press, San Diego.

**Dugatkin, L.A.** 1996b. The interface between culturally-based preferences and genetic preferences: female mate choice in *Poecilia reticulata*. *Proceedings of the National Academy of Sciences of the USA* 93: 2770–2773.

**Dugatkin, L.A.** 1998. Genes, copying and female mate choice: shifting thresholds. *Behavioral Ecology* 9: 323–327.

**Dugatkin, L.A. & Godin, J.G.J.** 1992. Reversal of female mate choice by copying in the guppy (*Poecilia reticulata*). *Proceedings of the Royal Society of London Series B* 249: 179–184.

**Dugatkin, L.A. & Godin, J.G.J.** 1993. Female mate copying in the guppy (*Poecilia reticulata*): age-dependent effects. *Behavioral Ecology* 4: 289–292.

E

**Eberhard, W.G.** 2000. Criteria for demonstrating postcopulatory female choice. *Evolution* 54: 1047–1050.

**Eberhard, W.G.** 1993. Evaluating models of sexual selection: genitalia as a test case. *American Naturalist* 142: 564–571.

**Eberhard, W.G.** 1996. *Female Control: Sexual Selection by Cryptic Female Choice*. Princeton University Press, Princeton.

**Eberhard, W.G.** 2000. Criteria for demonstrating postcopulatory female choice. *Evolution* 54: 1047–1050.

**Eberhard, W.G. & Gutierrez, E.E.** 1991. Male dimorphisms in beetles and earwigs and the question of developmental constraints. *Evolution* 45: 18–28.

**Elgar, M.A.** 1992. Sexual cannibalism in spiders and other invertebrates. In *Cannibalism: Ecology and Evolution among Diverse Taxa* (Elgar, M.A. & Crespi, B.J. eds), pp. 128–155. Oxford University Press, Oxford.

**Elgar, M.A. & Nash, D.R.** 1988. Sexual cannibalism in the garden spider *Araneus diadematus*. *Animal Behaviour* 36: 1511–1517.

**Elwood, R.W. & Dick, J.T.A.** 1990. The amorous *Gammarus*: the relationship between precopula duration and size-assortative mating in *G. pulex*. *Animal Behaviour* 39: 828–833.

**Emlen, S.T. & Oring, L.W.** 1977. Ecology, sexual selection and the evolution of mating systems. *Science* 197: 215–223.

**Endler, J.A.** 1978. A predator's view of animal colour patterns. *Evolutionary Biology* 11: 319–364.

**Endler, J.A.** 1980. Natural selection on color patterrns in *Poecilia reticulata*. *Evolution* 34: 76–91.

**Endler, J.A.** 1983. Natural and sexual selection in poeciliid fishes. *Environmental Biology of Fishes* 9: 173–190.

**Endler, J.A.** 1986. *Natural Selection in the Wild*. Princeton Unveristy Press, Princeton.

**Endler, J.A.** 1987. Predation, light intensity and courtship behaviour in *Poecilia reticulata* (Pisces: Poeciliidae). *Animal Behaviour* 35: 1376–1385.

**Endler, J.A. & Basolo, A.L.** 1998. Sensory ecology, receiver biases and sexual selection. *Trends in Ecology and Evolution* 13: 415–420.

**Enquist, M. & Arak, A.** 1993. Selection of exaggerated male traits by female aesthetic senses. *Nature* 361: 446–448.

**Eshel, I.** 1978. On the handicap principle – a critical defence. *Journal of Theoretical Biology* 70: 245–250.

**Etges, W.J.** 1996. Sexual selection operating in a wild population of *Drosophila robusta*. *Evolution* 50: 2095–2100.

F

**Faivre, B., Préault, M., Salvadori, F., Théry, M., Gaillard, M. & Cézilly, F.** 2003. Bill colour and immunocomptence in the European Blackbird. *Animal Behaviour* 65: 1125–1131.

**Faivre, B., Préault, M., Théry, M., Secondi, J., Patris, B. & Cézilly, F.** 2001. Pairing pattern, morphological characters and individual quality in an urban population of blackbirds *Turdus merula*. *Animal Behaviour* 61: 969–974.

**Farr, J.A. & Travis, J.** 1986. Fertility advertisement by female sailfin mollies, *Poecilia latipinna* (Pisces: Poeciliidae). *Copeia* 1986: 467–472.

**Fisher, D.O. & Lara, M.C.** 1999. Effects of body size and home range on access to mates and paternity in male bridled nailtail wallabies. *Animal Behaviour* 58: 121–130.

**Fisher, R.A.** 1915. The evolution of sexual preferences. *Eugenics Review* 7: 184–192.

**Fisher, R.A.** 1930. *The Genetical Theory of Natural Selection*. Clarendon Press, Oxford.

**Fisher, R.A.** 1958. *The Genetical Theory of Natural Selection*. Dover: New York.

**Foerster, K., Delhey, K., Johnsen, A., Lifjeld, J.T. & Kempanaers, B.** 2003. Females increase offspring heterozygosity and fitness through extra-pair matings. *Nature* 425: 714–717.

**Folstad, I. & Karter, A.J.** 1992. Parasites, bright males and the immunocompetence handicap. *American Naturalist* 139: 603–622.

**Forslund, P.** 2000. Male–male competition and large size mating advantage in European earwigs, *Forficula auricularia*. *Animal Behaviour* 59: 753–762.

**Freeberg, T.M.** 2000. Culture and courtship in vertebrates: a review of social learning and transmission of courtship systems and mating patterns. *Behavioural Processes* 51: 177–192.

**Freeman-Gallant, C.R., Meguerdichian, M., Wheelwright, N.T. & Sollecito, S.V.** 2003. Social pairing and female mating fidelity predicted by restriction fragment length polymorphism similarity at the major histocompatibility complex in a songbird. *Molecular Ecology* 12: 3077–3083.

G

**Galef, B.G. & White, D.J.** 1998. Mate choice copying in Japanese quail, *Coturnix coturnix japonica*. *Animal Behaviour* 55: 545–552.

**Galef, B.G. & White, D.J.** 2000. Evidence of social effects on mate choice in vertebrates. *Behavioural Processes* 51: 167–175.

**Gerhardt, H.C.** 1982. Sound pattern recognition in some North American treefrogs (Anura: Hylidae): implications for mate choice. *American Zoologist* 22: 581–595.

**Gerhardt, H.C.** 1994. Reproductive character displacement of female mate choice in the grey treefrog, *Hyla chrysoscelis*. *Animal Behaviour* 47: 959–969.

**Ghiselin, M.T.** 1974. *The Economy of Nature and the Evolution of Sex*. University of California Press, Berkeley.

**Gibson, R.M., Bradbury, J.W. & Verhencamp, S.** 1991. Mate choice in lekking sage grouse revisited: The roles of vocal display, female site fidelity and copying. *Behavioral Ecology and Sociobiology* 2: 165–180.

**Godin, J.-G.J., Herdman, E.J.E. & Dugatkin, L.A.** 2005. Social influences on female mate choice in the guppy, *Poecilia reticulata*: generalized and repeatable trait-copying behaviour. *Animal Behaviour* 69: 999–1005.

**Gomendio, M., Harcourt, A.H. & Roldán, E.R.S.** 1998. Sperm competition in mammals. In *Sperm Competition and Sexual Selection* (Birkhead, T.R. & Møller, A.P. eds), pp. 667–756. Academic Press, San Diego.

**Gomulkiewicz, R.S. & Hastings, A.** 1990. Ploidy and evolution by sexual selection: A comparison of haploid and diploid female choice models near fixation equilibria. *Evolution* 44: 757–770.

**Gonzalez, G., Sorci, G. & De Lope, F.** 1999. Seasonal variation in the relationship between cellular immune response and badge size in male house sparrows (*Passer domesticus*). *Behavioral Ecology and Sociobiology* 46: 117–122.

**Gould, S.J.** 1974. The origin and function of 'bizarre' structures: antler size and skull size in the 'Irish elk', *Megaloceros giganteus*. *Evolution* 28: 191–201.

**Gouyon, P.H., Henry, J.P. & Arnould, J.** 1997. *Les Avatars du Gène. La Théorie Néodarwinnienne de l'Évolution*. Belin, Paris.

**Grafen, A.** 1990a. Sexual selection unhandicapped by the Fisher Process. *Journal of Theoretical Biology* 144: 473–516.

**Grafen, A.** 1990b. Biological signals as handicaps. *Journal of Theoretical Biology* 144: 517–546.

**Grafen, A. & Ridley, M.** 1983. A model of mate guarding. *Journal of Theoretical Biology* 102: 549–567.

**Graves, H.B., Hable, C.P. & Jenkins, T.H.** 1985. Sexual selection in *Gallus*: effects of morphology and dominance on female spatial behavior. *Behavioural Processes* 11: 189–197.

**Gray, D.A. & Cade, W.H.** 1999a. Correlated-response-to-selection experiments designed to test for a genetic correlation between female preferences and male traits yield biased results. *Animal Behaviour* 58: 1325–1327.

**Gray, D.A. & Cade, W.H.** 1999b. Quantitative genetics of sexual selection in the field cricket, *Gryllus integer*. *Evolution* 53: 848–854.

**Gross, M.R.** 1994. The evolution of behavioural ecology. *Trends in Ecology and Evolution* 9: 358–360.

**Gwynne, D.T.** 1981. Sexual difference theory: Mormon crickets show role reversal in mate choice. *Science* 213: 779–780.

**Gwynne, D.T.** 1984. Coutship feeding increases female reproductive success in bushcrickets. *Nature* 307: 361–363.

H

**Hamilton, W.D. & Zuk, M.** 1982. Heritable true fitness and bright birds: a role for parasites? *Science* 218: 384–387.

**Hansson, B. & Westerberg, L.** 2002. On the corelation between heterozygosity and fitness in natural populations. *Molecular Ecology* 11: 2467–2474.

**Hansson, B., Jack, L., Christians, J.K., Pemberton, J.M., Åkesson, M., Westerdahl, H., Bensch, S. & Hasselquist, D.** 2007. No evidence for inbreeding avoidance in a great reed warbler population. *Behavioral Ecology* 18: 157–164.

**Harvey, P.H. & Bradbury, J.W.** 1991. Sexual selection. In *Behavioural Ecology: An Evolutionary Approach* (Krebs, J.R. & Davies, N.B. eds), pp. 203–233. Oxford University Press, Oxford.

**Harvey, P.H. & Pagel, M.D.** 1991. *The Comparative Method in Evolutionary Biology*. Oxford: Oxford University Press.

**Hasselquist, D., Marsh, J.A., Sherman, P.W. & Wingfield, J.C.** 1999. Is avian humoral immunocompetence suppressed by testosterone? *Behavioural Ecology and Sociobiology* 45: 167–175.

**Hasson, O.** 1989. Amplifiers and the handicap principle in sexual selection: a different emphasis. *Proceedings of the Royal Society of London Series B* 235: 383–406.

**Hasson, O.** 1991. Sexual displays as amplifiers: practical examples with an emphasis on feather decorations. *Behavioral Ecology* 2: 189–197.

**Hayashi, F. & Tsuchiya, K.** 2005. Functional association between female sperm storage organs and male sperm removal organs in calopterygid damselflies. *Entomological Science* 8: 245–252.

**Helfenstein, F., Wagner, R.H. & Danchin, E.** 2003. Sexual conflict over sperm ejection in monogamous pairs of kittiwakes *Rissa tridactyla*. *Behavioral Ecology and Sociobiology* 54: 370–376.

**Heywood, J.S.** 1989. Sexual selection by the handicap principle. *Evolution* 43: 1387–1397.

**Hill, G.E.** 1991. Plumage coloration is a sexually selected indicator of male quality. *Nature* 350: 337–339.

**Hoelzer, G.A.** 1989. The good parent process of sexual selection. *Animal Behaviour* 38: 1067–1078.

**Hoffmann, A.A.** 1988. Heritable variation for territorial success in two *Drosophila melanogaster* populations. *Animal Behaviour* 36: 1180–1189.

**Hoglund, J. & Alatalo, R.V.** 1995. *Leks*. Princeton University Press, Princeton.

**Holland, B. & Rice, W.R.** 1998. Chase-away sexual selection: anatonistic seduction versus resistance. *Evolution* 52: 1–7.

**Holland, B. & Rice, W.R.** 1999. Experimental removal of sexual selection reverses intersexual antagonistic coevolution and removes a reproductive load. *Proceedings of the National Academy of Sciences of the USA* 96: 5083–5088.

**Hosken, D.J., Garner, T.W.J. & Ward, P.I.** 2001. Sexual conflict selects for male and female reproductive characters. *Current Biology* 11: 489–493.

**Houde, A.E.** 1994. Effect of artificial selection on male colour pattern on mating preference of female guppies. *Proceedings of the Royal Society of London Series B* 256: 125–130.

**Houde, A.E.** 1997. *Sex, Color and Mate Choice in Guppies*. Princeton University Press, Princeton.

**Houde, A.E. & Endler, J.A.** 1990. Correlated evolution of female mating preference and male color patterns in the guppy *Poecilia reticulata*. *Science* 248: 1405–1408.

**Howard, R.D.** 1978. The influence of male-defended oviposition sites on early embryo mortality in bullfrogs. *Ecology* 59: 789–798.

**Hunt, J., Brooks, R. & Jennions, M.D.** 2005. Female mate choice as a condition-dependent life-history trait. *American Naturalist* 166: 79–92.

**Hurst, L.D. & Peck, J.R.** 1996. Recent advances in understanding of the evolution and maintenance of sex. *Trends in Ecology and Evolution* 11: 46–52.

**Huxley, J.S.** 1938a. Darwin's theory of sexual selection and the data subsumed by it, in the light of current research. *American Naturalist* 72: 416–433.

**Huxley, J.S.** 1938b. The present standing of the theory of sexual selection. In *Evolution: Essays on Aspects of Evolutionary Biology* (de Beer, G.R. ed.). Clarendon Press, Oxford.

**Huxley, J.S.** 1942. *Evolution the Modern Synthesis*. Allen and Unwin, London.

I

**Iwasa, Y. & Pomiankowski, A.** 1994. The evolution of mate preferences for multiple sexual ornaments. *Evolution* 48: 853–867.

**Iwasa, Y., Pomiankowski, A. & Nee, S.** 1991. The evolution of costly mate preferences II. The 'handicap' principle. *Evolution* 45: 1431–1442.

J

**Jamieson, I.** 1995. Do female fish prefer to spawn in nests with eggs for reasons of mate choice copying or egg survival? *American Naturalist* 145: 824–832.

**Jennions, M.D., Møller, A.P. & Petrie, M.** 2001. Sexually selected traits and adult survival: a meta-analysis. *Quarterly Review of Biology* 76: 3–36.

**Johnson, S.G.** 1991. Effects of predation, parasites and phylogeny on the evolution of bright colorations in North American male passerines. *Evolutionary Ecology* 5: 52–62.

**Johnstone, R.A.** 1995. Sexual selection, honest advertisement and the handicap principle: reviewing the evidence. *Biological Review* 70: 1–65.

**Johnstone, R.A.** 1996. Multiple displays in animal communication: 'backup signals' and 'multiple messages'. *Philosophical Transactions of the Royal Society of London Series B* 351: 329–338.

**Johnstone, R.A. & Norris, K.** 1993. Badges of status and the cost of aggression. *Behavioural Ecology and Sociobiology* 32: 127–134.

**Jones, G.P.** 1981. Spawning-site choice by female *Pseudolabrus celidotus* (Pisces: Labridae) and its influence on the mating success. *Behavioural Ecology and Sociobiology* 7: 107–112.

**Jones, I.L. & Hunter, F.M.** 1998. Heterospecific mating preferences for a feather ornament in least auklets. *Behavioural Ecology* 9: 187–192.

K

**Karubian, J. & Swaddle, J.P.** 2001. Selection on females can create 'larger males'. *Proceedings of the Royal Society of London Series B* 268: 725–728.

**Kempenaers, B., Foerster, K., Questiau, B., Robertson, B.C. & Vermeirssen, E.L.M.** 2000. Distinguishing between female sperm choice versus male sperm competition: a comment on Birkhead. *Evolution* 54: 1050–1052.

**Kempenaers, B., Verheyen, G.R., Van den Broeck, M., Burke, T., Van Broeckhoven, C. & Dhondt, A.** 1992. Extra-pair paternity results from female preference for high-quality males in the blue tit. *Nature* 357: 494–496.

**Keyser, A.J. & Hill, G.E.** 2000. Structurally based plumage coloration is an honest signal of male quality in male blue grosbeaks. *Behavioral Ecology* 11: 202–209.

**Kirkpatrick, M.** 1982. Sexual selection and the evolution of female choice. *Evolution* 36: 1–12.

**Kirkpatrick, M.** 1986. The handicap mechanism of sexual selection does not work. *American Naturalist* 127: 222–240.

**Kirkpatrick, M.** 1987. The evolutionary forces acting on female mating preferences in polyginous animals. In *Sexual Selection: Testing the Alternatives* (Bradbury, J.W. & Andersson, M.B. eds), pp. 67–82. John Wiley & Sons, Chichester.

**Kirkpatrick, M.** 1996. Good genes and direct selection in the evolution of mating preferences. *Evolution* 50: 2125–2140.

**Kirkpatrick, M. & Barton, N.H.** 1997. The strength of indirect selection on female mating preferences. *Proceedings of the National Academy of Sciences of the USA* 94: 1282–1286.

**Kirkpatrick, M. & Dugatkin, L.A.** 1994. Sexual selection and the evolutionary effects of copying mate choice. *Behavioural Ecology and Sociobiology* 34: 443–449.

**Kirkpatrick, M. & Ryan, M.J.** 1991. The evolution of mating preferences and the paradox of the lek. *Nature* 350: 33–38.

**Klump, G.M. & Gerhardt, H.C.** 1987. Use of non-arbitrary acoustic criteria in mate choice by female gray treefrog. *Nature* 326: 286–288.

**Kodric-Brown, A. & Brown, J.H.** 1984. Truth in advertising: the kinds of traits favored by sexual selection. *American Naturalist* 124: 309–323.

**Kokko, H. & Monaghan, P.** 2001. Predicting the direction of sexual selection. *Ecology Letters* 4: 159–165.

**Kokko, H. & Ots, I.** 2006. When not to avoid inbreeding. *Evolution* 60: 467–475.

**Kokko, H., Brooks, R., McNamara, J.M. & Houston, A.I.** 2002. The sexual selection continuum. *Proceedings of the Royal Society of London Series B* 269: 1331–1340.

L

**Laland, K.** 1994. Sexual selection with a culturally transmitted mating preference. *Theoretical Population Biology* 45: 1–15.

**Lampert, K.P., Bernal, X.E., Rand, A.S., Mueller, U.G. & Ryan, M.J.** 2006. No evidence for female mate choice based on genetic similarity in the tungara frog, *Physalaemus pustulosus*. *Behavioral Ecology and Sociobiology* 59: 796–804.

**Lande, R.** 1979. Quantitative genetics of multivariate evolution applied to brain–body size allometry. *Evolution* 33: 402–416.

**Lande, R.** 1981. Models of speciation by sexual selection on polygenic traits. *Proceedings of the National Academy of Science of the USA* 78: 3721–3725.

**Lande, R.** 1982. Rapid origin of sexual isolation and character divergence in a cline. *Evolution* 36: 213–223.

**Lande, R. & Arnold,** 1983. The measurement of selection on correlated characters. *Evolution* 37: 1210–1226.

**LeBoeuf, B.J.** 1974. Male–male competition and reproductive success in elephant seals. *American Zoologist* 14: 163–176.

**LeBoeuf, B.J. & Reiter, J.** 1988. Lifetime reproductive success in northern elephant seals. In *Reproductive Success* (Clutton-Brock, T.H. ed.), pp. 344–362. University of Chicago Press, Chicago.

**Legrand, R.S. & Morse, D.H.** 2000. Factors driving extreme sexual size dimorphism of a sit-and-wait predator under low density. *Biological Journal of the Linnean Society* 71: 643–664.

**Linville, S.U., Breitwisch, R. & Schilling, A.** 1998. Plumage brightness as an indicator of parental care in northern cardinals. *Animal Behaviour* 55: 119–127.

**Liske, E. & Davis, W.J.** 1987. Courtship and mating behaviour of the Chinese praying mantis *Tenodera aridfolia sinensis*. *Animal Behaviour* 35: 1524–1538.

**Loehle, C.** 1997. The pathogen transmission avoidance theory of sexual selection. *Ecological Modelling* 103: 231–250.

M

**Madsen, T., Shine, R. Loman, J. & Hakansson, T.** 1993. Determinants of mating success in male adders, *Vipera berus*. *Animal Behaviour* 45: 491–499.

**Manly, B.F.J.** 1985. *The Statistics of Natural Selection*. Chapman and Hall, London.

**Markow, T.A.** 1988. *Drosophila* males provide a material contribution to offspring sired by other males. *Functional Ecology* 2: 77–79.

**Marler, P.** 1956. Behaviour of the chaffinch *Fringilla coelebs*. *Behaviour* (Suppl.) 5: 1–84.

**Maynard Smith, J.** 1976. Sexual selection and the handicap principle. *Journal of Theoretical Biology* 57: 239–242.

**Maynard Smith, J.** 1982. *Evolution and the Theory of Games*. Cambridge University Press, Cambridge, UK.

**Maynard Smith, J.** 1985. Mini review: sexual selection, handicaps and true fitness. *Journal of Theoretical Biology* 57: 239–242.

**Maynard Smith, J.** 1991. Honest signalling: the Philip Sidney game. *Animal Behaviour* 42: 1034–1035.

**Maynard Smith, J. & Brown, R.L.W.** 1986. Competition and body size. *Theoretical Population Biology* 30: 166–179.

**Mayr, E.** 1963. *Animal Species and Evolution*. Harvard University Press, Cambridge, Massachusetts.

**Mays, H.L. & Hill, G.E.** 2004. Choosing mates: good genes versus genes that are a good fit. *Trends in Ecology and Evolution* 19: 554–559.

**McLain, D.K.** 1998. Non-genetic benefits of mate choice: fecundity enhancement and sexy sons. *Animal Behaviour* 55: 1191–1201.

**McLain, D.K. & Boromisa, R.D.** 1987. Male choice, fighting ability, assortative mating and the intensity of sexual selection in the milkweed longhorn beetle *Tetraopes tetraophtalmus* (Coleoptera, Cerambycidae). *Behavioural Ecology and Sociobiology* 20: 239–246.

**Mead, L.S. & Arnold, S.J.** 2004. Quantitative genetic models of sexual selection. *Trends in Ecology and Evolution* 19: 264–271.

**Mennill, D.J.** *et al.* 2002. Female eavesdropping on male song contest in songbirds. *Science* 296: 873.

**Møller, A.P.** 1994. *Sexual Selection and the Barn Swallow*. Oxford University Press, Oxford.

**Møller, A.P. & Alatalo, R.V.** 1999. Good-genes effects in sexual selection. *Proceedings of the Royal Society of London Series B* 266: 85–91.

**Møller, A.P. & Jennions, M.D.** 2001. How important are direct benefits of sexual selection? *Naturwissenschaften* 88: 401–415.

**Møller, A.P., Biard, C., Blount, J.D., Houston, D.C., Ninni, P., Saino, N. & Surai, P.F.** 2000. Carotenoid-dependent signals: indicators of foraging efficiency, immunocompetence or detoxification ability? *Poultry and Avian Biology Reviews* 11: 137–159.

**Moore, A.J.** 1990. The inheritance of social dominance, mating behaviour and attractiveness to mates in male *Nauphoeta cinerea*. *Animal Behaviour* 39: 388–397.

**Moore, A.J. & Moore, P.J.** 1999. Balancing sexual selection through opposing mate choice and male competition. *Proceedings of the Royal Society of London Series B* 266: 711–716.

**Moore, A.J. & Wilson, P.** 1993. The evolution of sexually dimorphic earwig forceps: social interactions among adults of the toothed earwig, *Vostox apicedentatus*. *Behavioral Ecology* 4: 40–48.

**Moritz, R.F.A. & Hillesheim, E.** 1985. Inheritance of dominance in honeybees (*Apis mellifera capensis* Esch.). *Behavioral Ecology and Sociobiology* 17: 87–89.

**Morris, M.R., Wagner, W.E. & Ryan, M.J.** 1996. A negative correlation between trait and mate preference in *Xiphophorus pygmaeus*. *Animal Behaviour* 52: 1193–1203.

N

**Neff, B. & Pitcher, T.E.** 2005. Genetic quality and sexual selection: an integrated framework for good genes and compatible genes. *Molecular Ecology* 14: 19–38.

**Nisbet, I.C.T.** 1973. Courtship feeding, egg size and breeding success in common terns. *Nature* 241: 141–142.

**Norris, K.J.** 1990. Female choice and the quality of parental care in the great tit *Parus major*. *Behavioural Ecology and Sociobiology* 27: 275–281.

**Norris, K.J.** 1993. Heritable variation in a plumage indicator of viability in male great tits *Parus major*. *Nature* 362: 537–539.

**Norris, K. & Evans, M.R.** 2000. Ecological immunology: life-history trade-offs and immune defense in birds. *Behavioral Ecology* 11: 19–26.

O

**O'Riain, M.J., Bennett, N.C., Britherton, P.N.M., McIlrath, G. & Clutton-Brock, T.** 2000. Reproductive suppression and inbreeding avoidance in wild populations of co-operatively breeding meercats (*Suricata suricatta*). *Behavioral Ecology and Sociobiology* 48: 471–477.

**O'Donald, P.** 1962. The theory of sexual selection. *Heredity* 17: 541–552.

**O'Donald, P.** 1967. A general model of sexual selection and natural selection. *Heredity* 22: 499–518.

**O'Donald, P.** 1980. *Genetic Models of Sexual Selection*. Cambridge University Press, Cambridge, UK.

**O'Donald, P.** 1983. Sexual selection by female choice. In *Mate Choice* (Bateson, P. ed.), pp. 53–66. Cambridge University Press, Cambridge.

**Olsson, M.** 1993. Male preference for large females and assortative mating for body size in the sand lizard (*Lacerta agilis*). *Behavioural Ecology and Sociobiology* 31: 337–341.

**Otte, D.** 1989. Speciation in Hawaiian crickets. In *Speciation and its Consequences* (Otte, D. & Endler, J.A. eds), pp. 482–526. Sinauer, Sunderland, Massachusetts.

**Owens, I.P.F. & Wilson, K.** 1999. Immunocompetence: a neglected life history trait or conspicuous red herring? *Trends in Ecology and Evolution* 14: 170–172.

P

**Pagel, M.** 1994. Detecting correlated evolution on phylogenies: a general method for the comparative analysis of discrete characters. *Proceedings of the Royal Society of London Series B*, 255: 37–45.

**Pagel, M.** 1997. Inferring evolutionary processes from phylogenies. *Zoologica Scripta* 26: 331–348.

**Palokangas, P., Korpimäki, E., Hakkarainen, H., Huhta, E., Tolonen, P. & Alatalo, R.V.** 1994. Female kestrels gain reproductive success by choosing brightly ornamented males. *Animal Behaviour* 47: 443–448.

**Parker, G.A.** 1970. Sperm competition and its evolutionary consequences in insects. *Biological Review* 45: 525–567.

**Parker, G.A.** 1974. Courtship persistence and female-guarding as male time investment strategies. *Behaviour* 58: 157–184.

**Parker, G.A.** 1979. Sexual selection and sexual conflict. In *Sexual Selection and Reproductive Competition in Insects* (Blum, M.S. & Blum, N.A. eds), pp. 123–166. Academic Press, New York.

**Parker, G.A.** 1983a. Arms races in evolution – an ESS to the opponent–independent costs game. *Journal of Theoretical Biology* 101: 619–648.

**Parker, G.A.** 1983b. Mate quality and mating decisions. In *Mate Choice* (Bateson, P. ed.), pp. 141–164. Cambridge University Press, Cambridge.

**Parker, G.A., Baker, R.R. & Smith, W.G.F.** 1972. The origin and evolution of gamete dimorphism and the male–female phenomenon. *Journal of Theoretical Biology* 36: 529–533.

**Petrie, M.** 1994. Improved growth and survival of offsprings of peacocks with more elaborate trains. *Nature* 371: 598–599.

**Pinto, J.D.** 1980. Behavior and taxonomy of the *Epicauta maculata* group (Coleoptera: Meloidae). *University of California Publications on Entomology* 89: 1–111.

**Pitnick, S. & Brown, W.D.** 2000. Criteria for demonstrating female sperm choice. *Evolution* 54: 1052–1056.



**Pitnick, S. & Brown, W.D.** 2000. Criteria for demonstrating female sperm choice. *Evolution*, 54: 1052–1056.



**Pitnick, S., Brown, W.D. & Miller, G.T.** 2001a. Evolution of female remating behaviour following experimental removal of sexual selection. *Proceedings of the Royal Society of London Series B* 1467: 557–563.

**Pitnick, S., Miller, G.T., Reagan, J. & Holland, B.** 2001b. Male's evolutionary responses to experimental removal of sexual selection. *Proceedings of the Royal Society of London Series B* 1471: 1071–1080.

**Pizzari, T. & Birkhead, T.R.** 2000. Female feral fowl eject sperm of subdominant males. *Nature* 405: 787–789.

**Plaistow, S., Bollache, L. & Cézilly, F.** 2003. Energetically costly pre-copulatory mate-guarding in the amphipod *Gammarus pulex*: causes and consequences. *Animal Behaviour* 65: 683–691.

**Podos, J.** 2001. Correlated evolution of morphology and vocal signal structure in Darwin's finches. *Nature* 409: 185–188.

**Pomiankowski, A.** 1987a. Sexual selection: the handicap principle does work – sometimes. *Proceedings of the Royal Society of London Series B* 231: 123–145.

**Pomiankowski, A.** 1987b. The costs of choice in sexual selection. *Journal of Theoretical Biology* 128: 195–218.

**Pomiankowski, A.** 1988. The evolution of female mate preferences for male genetic quality. *Oxford Surveys in Evolutionary Biology* 5: 136–184.

**Pomiankowski, A. & Møller, A.P.** 1995. A resolution of the lek paradox. *Proceedings of the Royal Society of London Series B* 260: 21–29.

**Pomiankowski, A., Iwasa, Y. & Nee, S.** 1991. The evolution of costly mate preferences. I. Fisher and biased mutation. *Evolution* 45: 1422–1430.

**Poole, J.** 1989. Mate guarding, reproductive success and female choice in African elephants. *Animal Behaviour* 37: 842–849.

**Price, T., Schluter, D. & Heckman, N.E.** 1993. Sexual selection when the female directly benefits. *Biological Journal of the Linnean Society* 48: 187–211.

**Pruett-Jones, S.** 1992. Independent versus non-independent mate choice: do females copy each other? *American Naturalist* 140: 1000–1009.

**Pruett-Jones, S., Pruett-Jones, M.A. & Jones, H.I.** 1991. Parasites and sexual selection in a New Guinea avifauna. *Current Ornithology* 8: 213–245.

Q

**Qvarnström, A., Brommer, J.E. & Gustafsson, L.** 2006. Testing the genetics underlying the co-evolution of mate choice and ornament in the wild. *Nature* 441: 84–86.

R

**Radesäter, T. & Halldorsdottir, H.** 1993. Two male types of the common earwig: male–male competition and mating success. *Ethology* 95: 89–96.

**Read, A.F.** 1987. Comparative evidence supports the Hamilton–Zuk hypothesis on parasites and sexual selection. *Nature* 328: 68–70.

**Read, A.F.** 1991. Passerine polygyny: a role for parasites. *American Naturalist* 138: 434–459.

**Read, A.F. & Harvey, P.H.** 1989. Reassessment of comparative evidence for Hamilton and Zuk theory on the evolution of secondary sexual characters. *Nature* 339: 618–620.

**Rice, W.R.** 1996. Sexually antagonistic male adaptation triggered by experimental arrest of female evolution. *Nature* 361: 232–234.

**Ridley, M.** 1983. *The Explanation of Organic Diversity: The Comparative Method and Adaptations for Mating.* Clarendon Press, Oxford.

**Ritchie, M.G.** 1992. Setbacks in the search for mate-preference genes. *Trends in Ecology and Evolution* 7: 328–329.

**Roberts, S.C. & Gosling, L.M.** 2003. Genetic similarity and quality interact in mate choice decisions by female mice. *Nature Genetics* 35: 103–106.

**Roberts, S.C., Hale, M.L. & Petrie, M.** 2006. Corelations between heterozygosity and measures of genetic similarity: implications for understanding mate choice. *Journal of Evolutionary Biology* 19: 558–569.

**Robinson, B.W. & Doyle, R.W.** 1985. Trade-off between male reproduction (amplexus) and growth in the amphipod *Gammarus lawrencianus. Biological Bulletin* 168: 482–488.

**Rodd, F.H., Hughes, K.A., Grether, G.F. & Baril, C.T.** 2002. A possible origin for mate preference: are male guppies mimicking fruit? *Proceedings of the Royal Society of London Series B* 269: 475–481.

**Roeder, K.D.** 1935. An experimental analysis of the sexual behavior of the praying mantis (*Mantis religiosa*). *Biological Bulletin* 69: 203–220.

**Rosenthal, G.G. & Evans, C.S.** 1998. Female preference for swords in *Xiphophorus helleri* reflects a bias for large apparent size. *Proceedings of the National Academy of Sciences of the USA* 95: 4431–4436.

**Rowe, L.V., Evans, M.R. & Buchanan, K.L.** 2001. The function and evolution of the tail streamer in hirundines. *Behavioral Ecology* 12: 157–163.

**Rowe, L. & Houle, D.** 1996. The lek paradox and the capture of genetic variance by condition-dependent traits. *Proceedings of the Royal Society of London Series B* 263: 1415–1421.

**Ryan, M.J.** 1990. Sexual selection, sensory systems and sensory exploitation. *Oxford Surveys in Evolutionary Biology* 7: 157–195.

**Ryan, M.J.** 1997. Sexual selection and mate choice. In *Behavioural Ecology. An Evolutionary Approach* (Krebs, J.R. & Davies, N.B. eds), pp. 179–202. Blackwell, Oxford.

**Ryan, M.J.** 2001. Food, song and speciation. *Nature* 409: 139–140.

**Ryan, M.J. & Rand, A.S.** 1990. The sensory basis of sexual selection for complex calls in the tungara frog, *Physalaemus Pustolosus* (sexual selection for sensory exploitation). *Evolution* 44: 305–314.

**Ryan, M.J. & Wagner, W.E.** 1987. Asymetries in mating preferences between species: female swordtails prefer heterospecific males. *Science* 236: 595–597.

**Ryan, M.J., Fox, J.H., Wilczynski, W. & Rand, A.S.** 1990. Sexual selection for sensory exploitation in the frog *Physalaemus pustulosus. Nature* 343: 66–67.

S

**Saino, N. & Møller, A.P.** 1996. Sexual ornamentation and imunocompetence in the barn swallow. *Behavioral Ecology* 7: 227–232.

**Saino, N., Bolzern, A.M. & Møller, A.P.** 1997a. Immunocompetence, ornamentation and viability of male barn swallows (*Hirundo rustica*). *Proceedings of the National Academy of Sciences of the USA* 94: 549–552.

**Saino, N., Galeotti, P., Sacchi, R. & Møller, A.P.** 1997b. Song and immunological condition in male barn swallows (*Hirundo rustica*). *Behavioral Ecolology* 8: 364–371.

**Saino, N., Stradi, R., Ninni, P. & Møller, A.P.** 1999. Carotenoid plasma concentration, immune profile and plumage ornamentation of male barn swallows (*Hirundo rustica*). *American Naturalist* 154: 441–448.

**Sasvári, L.** 1986. Reproductive effort of widowed birds. *Journal of Animal Ecology* 55: 553–564.

**Savalli, U.D. & Fox, C.W.** 1998. Sexual selection and the fitness consequences of male body size in the seed beetle *Stator limbatus*? *Animal Behaviour* 55: 473–483.

**Shaw, K.** 1995. Phylogenetic tests of the sensory exploitation model of sexual selection. *Trends in Ecology and Evolution* 10: 117–120.

**Shuster, S.M. & Wade, M.J.** 1991. Female copying and sexual selection in a marine isopod crustacean, *Paracerceis sculpta. Animal Behaviour* 42: 1071–1078.

**Shuster, S.M. & Wade, M.J.** 2003. *Mating Systems and Strategies.* Princeton University Press, Princeton.

**Sikkel, P.C.** 1989. Egg presence and developmental stage influence spawning-site choice by female garibaldi. *Animal Behaviour* 38: 447–456.

**Siller, S.** 1998. The epistatic handicap principle does work. *Journal of Theoretical Biology* 191: 141–161.

**Simmons, L.W.** 1991. Female choice and the relatedness of mates in the field cricket, *Gryllus bimaculatus* (de Greer). *Animal Behaviour* 41: 493–501.

**Singer, M.C.** 1982. Sexual selection for small size in male butterflies. *American Naturalist* 119: 440–443.

**Siva-Jothy, M.T.** 1995. 'Immunocompetence': conspicuous by its absence. *Trends in Ecology and Evolution* 10: 205–206.

**Smith, C., Barber, I., Wootton, R. & Chittka, L.** 2004. A receiver bias in the origin of the three-spined stickleback mate choice. *Proceedings of the Royal Society of London Series B* 271: 949–955.

**Smith, R.L.** (ed.) 1984. *Sperm Competition and the Evolution of Animal Mating Systems.* Academic Press, Orlando.

**Stålhandske, P.** 2001. Nuptial gift in the spider *Pisaura mirabilis* maintained by sexual selection. *Behavioral Ecology* 12: 691–697.

**Stow, A.J. & Sunnucks P.** 2004. Inbreeding avoidance in Cunningham's skinks (*Egernia cunninghami*) in natural and fragmented habitat. *Molecular Ecology* 13: 443–447.

**Swaddle, J.P., Cathey, M.G. & Correll, M.** 2005. Socially transmitted mate preferences in a monogamous bird:

a non genetic mechanism of sexual selection. *Proceedings of the Royal Society of London Series B* 272: 1053–1058.

T

**Taigen, T.L. & Wells, K.D.** 1985. Energetics of vocalization by an anuran amphibian (*Hyla versicolor*). *Journal of Comparative Physiology* 155: 163–170.

**Tarvin, K.A., Webster, M.S., Tuttle, E.M. & Pruett-Jones, S.** 2005. Genetic similarity of social mates predicts the level of extrapair paternity in splendid fairy-wrens. *Animal Behaviour* 70: 945–955.

**Tasker, C.R. & Mills, J.A.** 1981. A functional analysis of courtship-feeding in the red-billed gull (*Larus novaehollandiae*). *Behaviour* 77: 222–241.

**Taylor, P.D. & Williams, G.C.** 1982. The lek paradox is not resolved. *Theoretical Population Biology* 22: 392–409.

**Thornhill, R.** 1983. Cryptic female choice and its applications in the scorpionfly *Harpobittacus nigriceps. American Naturalist* 122: 765–788.

**Thornhill, R.** 1984. Alternative female choice tactics in the scorpionfly *Harpobittacus nigriceps* (Mecoptera) and their implications. *American Zoologist* 24: 367–383.

**Thornhill, R.** 1988. The jungle fowl hen's cackle incites male competition. *Verhandlungen Deutschen Zoologisches Gesellshaft* 81: 145–154.

**Thornhill, R. & Alcock, J.** 1983. *The Evolution of Insect Mating Systems*. Harvard University Press, Cambridge, MA.

**Tomkins, J.L. & Simmons, L.W.** 1996. Dimorphism and fluctuating asymetry in the forceps of male earwigs. *Journal of Evolutionary Biology* 9: 753–770.

**Tomlinson, I.P.M.** 1988. Diploid models of the handicap principle. *Heredity* 60: 283–293.

**Trail, P.W.** 1985. Courtship disruption modifies mate choice in a lek-breeding bird. *Science* 227: 778–780.

**Trivers, R.** 1972. Parental investment and sexual selection. In *Sexual Selection and the Descent of Man 1871–1971* (Campbell, B. ed.), pp. 139–179. Aldine Press, Chicago.

V

**Vahed, K.** 1998. The function of nuptial feeding in insects: a review of empirical studies. *Biological Review* 73: 43–78.

**Verhemcamp, S.L., Bradbury, J.W. & Gibson, R.M.** 1989. The energetic cost of display in male sage grouse. *Animal Behaviour* 38: 885–896.

**Verhulst, S., Dieleman, S.J. & Parmentier, H.K.** 1999. A trade off between immunocompetence and sexual ornamentation in domestic fowl. *Proceedings of the National Academy of Sciences of the USA* 96: 4478–4481.

**Viken, Å., Fleming, I.A. & Rosenquist, G.** 2006. Premating avoidance of inbreeding absent in female guppies (*Poecilia reticulata*). *Ethology* 112: 716–723.

W

**Waage, J.K.** 1979. Reproductive character displacement in *Calopteryx*. *Evolution* 33: 104–116.

**Wagner, R.H. Helfenstein, F. & Danchin, E.** 2004. Female choice of young sperm in a genetically monogamous bird. *Proceedings of the Royal Society of London Series B (Supplement)*, doi: 10.1098/rsbl.2003.0142.

**Wallace, A.R.** 1889. *Darwinism*, 3rd edition. Macmillan, London.

**Wallace, A.R.** 1891. *Natural Selection and Tropical Nature*. Macmillan, London.

**Ward, P.I.** 1983. Advantages and disadvantages of large size for male *Gammarus pulex* (Crustacea: Amphipoda). *Behavioral Ecology and Sociobiology* 14: 69–76.

**Watson, P.J.** 1990. Female-enhanced male competition determines the first mate and the principal sire in the spider *Linyphia litigiosa* (Linyphiidae). *Behavioral Ecology and Sociobiology* 26: 77–90.

**Weary, D.M., Guilford, T.C. & Weisman, R.G.** 1993. A product of discriminative learning may lead to female preferences for elaborate males. *Evolution* 47: 333–336.

**Weatherhead, P.J., Bennett, G.F. & Shutler, D.** 1991. Sexual selection and parasites in wood-warblers. *Auk* 108: 147–152.

**Wedekind, C.** 1994. Mate choice and maternal selection for specific parasite resistances before, during and after fertilization. *Philosophical Transactions of the Royal Society of London Series B* 346: 303–311.

**Wedell, N.** 1994. Variation in nuptial gift quality in bush crickets (Orthoptera: Tettigoniidea). *Behavioural Ecology* 5: 418–425.

**Wedell, N. & Tregenza, T.** 1999. Successful fathers sire successful sons. *Evolution* 53: 620–625.

**West, M.J. & King, A.P.** 1980. Enriching cowbird song by social deprivation. *Journal of Comparative and Physiological Psychology* 94: 263–270.

**West, M.J., King, A.P. & Eastzer, D.H.** 1981. Validating the female bioassay of cowbird song: relating differences in song potency to mating success. *Animal Behaviour* 29: 490–501.

**White, D.J. & Galef, B.G.** 1999 Mate choice copying and conspecific cueing in Japanese quail *Coturnix coturnix japonica*. *Animal Behaviour* 57: 465–473.

**White, D.J. & Galef, B.G.** 2000. Culture in quail: social influences on mate choice in female *Coturnix japonica*. *Animal Behaviour* 59: 975–979.

**Wicklund, C. & Fagerström, T.** 1977. Why do males emerge before females? A hypothesis to explain the incidence of protandrie in butterflies. *Oecologia* 31: 153–158.

**Wiehn, J.** 1997. Plumage characteristics as an indicator of male parental quality in the American kestrel. *Journal of Avian Biology* 28: 47–55.

**Wiens, J.J.** 2001. Widespread loss of sexually selected traits: how the peacock lost its spots. *Trends in Ecology and Evolution* 16: 517–523.

**Wiggins, D.A. & Morris, R.D.** 1986. Criteria for female choice of mates: courtship feeding and parental care in the common tern. *American Naturalist* 128: 126–129.

**Wilkinson, G.S. & Reillo, P.R.** 1994. Female choice response to artificial selection on an exagerated male trait in a stark-eyed fly. *Proceedings of the Royal Society of London Series B* 255: 1–6.

**Williams, G.C.** 1966. *Adaptation and Natural Selection: A Critique of Some Current Evolutionary Thought.* Princeton University Pres, Princeton, NJ.

**Witte, K. & Noltemeier, B.** 2002. The role of information in mate choice copying in female sailfin mollies (*Poecilia latipinna*). *Behavioral Ecology and Sociobiology* 52: 194–202.

**Witte, K. & Ryan, M.J.** 2002. Mate choice copying in the sailfin molly, *Poecilia latipinna*, in the wild. *Animal Behaviour* 63: 943–949.

**Wolfner, M.F.** 1997. Tokens of love: functions and regulation of *Drosophila* male accessory gland products. *Insect Biochemistry and Molecular Biology* 27: 179–192.

Z

**Zahavi, A.** 1975. Mate selection – a selection for a handicap. *Journal of Theoretical Biology* 53: 205–214.

**Zahavi, A.** 1977. The cost of honesty (further remarks on the handicap principle). *Journal of Theoretical Biology* 67: 603–605.

**Zeh, D.W. & Zeh, J.A.** 1988. Condition-dependent sex ornaments and field tests of sexual-selection theory. *American Naturalist* 132: 454–459.

**Zuk, M. & Johnsen, T.S.** 1998. Seasonal changes in the relationship between ornamentation and immune response in red jungle fowl. *Proceedings of the Royal Society of London Series B* 265: 1631–1635.

**Zuk, M., Johnsen, T.S. & MacLarty, T.,** 1995. Endocrine–immune interactions, ornaments and mate choice in red jungle fowl. *Proceedings of the Royal Society of London Series B* 260: 205–210.

## Chapter 12

A

**Alatalo, R.V., Carlson, A., Lundberg, A. & Ulfstrand, S.** 1981. The conflict between male polygamy & female monogamy: the case of the pied flycatcher *Ficedula hypoleuca*. *American Naturalist* 117: 285–291.

**Alexander, R.D.** 1974. The evolution of social behavior. *Annual Reviews of Ecology and Systematics* 5: 325–383.

**Andersson, M.** 1994. *Sexual Selection.* Princeton University Press, Princeton.

**Andersson, M.** 2005. Evolution of classical polyandry: three steps to female emancipation. *Ethology* 111: 1–23.

**Arcese, P.** 1989. Intrasexual competition and the mating system in primarily monogamous birds: the case of the song sparrow. *Animal Behaviour* 38: 96–111.

**Arroyo, B.E., De Cornulier, T. & Bretagnolle, V.** 2002. Parental investment and parent–offspring conflicts during the postfledging period in Montagu's harriers. *Animal Behaviour* 63: 235–244.

B

**Baker, A.J., Bales, K.L. & Dietz, J.M.** 2002. Mating systems and group dynamics in lion tamarins. In *Lion Tamarins: Biology and Conservation* (Kleinman, D.G. & Rylands, A.B. eds), pp. 188–212. Smithonian Institution Press, Washington, DC.

**Baker, A.J., Dietz, J.M. & Kleinman, D.G.** 1993. Behavioural evidence for monopolization of paternity in multi-male groups of golden lion tamarins. *Animal Behaviour* 46: 1091–1101.

**Barlow, G.W.** 2000. *The Cichlid Fishes. Nature's Grand Experiment in Evolution.* Perseus Publishing, Cambridge, Massachusetts.

**Bart, J. & Tornes, A.** 1989. Importance of monogamous male birds in determining reproductive success: evidence from house wrens and a review of male-removal experiments. *Behavioral Ecology and Sociobiology* 24: 109–116.

**Baur, B.** 1994. Multiple paternity and individual variation in sperm precedence in the simultaneously hermaphroditic land snail *Arianta arbustorum. Behavioral Ecology and Sociobiology* 35: 413–421.

**Baur, B.** 1998. Sperm competition in molluscs. In *Sperm Competition and Sexual Selection* (Birkhead, T.R. and Møller, A.P.), pp. 253–305. Academic Press, San Diego.

**Baylis, J.R.** 1981. The evolution of parental care in fishes, with reference to Darwin's rule of male and sexual selection. *Environmental Biology of Fish* 6: 223–251.

**Beauchamp, G. & Kacelnick, A.** 1990. On the fitness functions relating parental care to reproductive values. *Journal of Theoretical Biology* 146: 513–522.

**Beck, C.W.** 1998. Mode of fertilization and parental care in anurans. *Animal Behaviour* 55: 439–449.

**Beehler, B.M.** 1983. Lek behaviour of the lesser bird of paradise. *Auk* 100: 992–995.

**Beehler, B.M. & Foster, M.S.** 1988. Hotshots, hotspots and female preference in the organization of lek mating systems. *American Naturalist* 131: 203–219.

**Bennett, P.M. & Owens, I.P.F.** 2002. *Evolutionary Ecology of Birds: Life Histories, Mating Systems and Extinction.* Oxford University Press, Oxford.

**Berglund, A. & Rosenquist, G.** 2003. Sex role reversal in pipefish. *Advances in the Study of Animal Behaviour* 32: 131–167.

**Berglund, A., Rosenquist, G. & Robinson-Wolrath, S.** 2006. Food or sex – males and females in a sex role reversed pipefish have different interests. *Behavioral Ecology and Sociobiology* 60: 281–287.

**Berglund, A., Rosenquist, G. & Svensson, I.** 1986. Reversed sex roles and parental energy investment in zygotes of two pipefish (Syngnathidae) species. *Marine Ecology Progress Series* 29: 209–215.

**Berglund, A., Rosenquist, G. & Svensson, I.** 1989. Reproductive success of females limited by males in two pipefish species. *American Naturalist* 133: 506–516.

**Bertram, B.C.R.** 1975. Social factors influencing reproduction in wild lions. *Journal of Zoology* 177: 462–482.

**Birkhead, T.R. & Møller, A.P.** 1992. *Sperm Competition in Birds. Evolutionary Causes and Consequences.* Academic Press, London.

**Black, J.M.** (ed.). 1996. *Partnerships in Birds. The Study of Monogamy.* Oxford University Press, Oxford.

**Borgerhoff Mulder, M.** 1990. Kipsigis women prefer wealthy men: evidence for female choice in mammals? *Behavioral Ecology and Sociobiology* 27: 255–264.

**Bourlière, F.** 1967. *The Natural History of Mammals*, 3rd edition. Alfred A. Knopf, New York.

**Bradbury, J.W.** 1981. The evolution of leks. In *Natural Selection and Social Behavior: Research and Theory* (Alexander, R.D. & Twinkle, D.W. eds), pp. 138–169. Chiron, New York.

**Bradbury, J.W.** 1985. Contrasts between insects and vertebrates in the evolution of male display, female choice & lek mating. In *Experimental Ecology and Sociobiology* (Hölldobler, B. & Lindauer, M. eds), pp. 273–289. Gustav Fischer Verlag, New York.

**Bradbury, J.W. & Gibson, R.M.** 1983. Leks and mate choice. In *Mate Choice* (Bateson, P. ed.), pp. 109–138. Cambridge University Press, Cambridge.

**Bull, C.M.** 2000. Monogamy in lizards. *Behavioural Processes* 51: 7–20.

**Burley, N.** 1988. The differential allocation hypothesis: an experimental test. *American Naturalist* 132: 611–628.

**Butchart, S.H.M.** 2000. Population structure and breeding system of the sex-role reversed, polyandrous bronze-winged jacana, *Metopidius indicus*. *Ibis* 142: 93–102.

**Butchart, S.H.M., Seddon, N. & Ekstrom, J.M.M.** 1999. Polyandry and competition for territories in bronze-winged jacanas. *Journal of Animal Ecology* 68: 928–939.

C

**Cam, E., Link, W.A., Cooch, E.G., Monnat, J.Y. & Danchin, E.** 2002. Individual covariation in life-history traits: seeing the tree despite the forest. *American Naturalist* 159: 96–105.

**Carlson, A.A. & Isbell, L.A.** 2001. Causes and consequences of single-male and multimale mating in free-ranging patas monkeys, *Erythrocebus patas*. *Animal Behaviour* 62: 1047–1058.

**Carter, C.S., Lederhendler, I.I. & Kirkpatrick, B.** (eds). 1999. *The Integrative Neurobiology of Affiliation.* MIT Press, Cambridge, Massachusetts.

**Cartwright, J.** 2000. *Evolution and Human Behaviour.* MacMillan Press, London.

**Catry, P., Ratcliffe, N. & Furness, R.W.** 1997. Partnerships and mechanisms of divorce in the great skua. *Animal Behaviour* 54: 1475–1482.

**Cézilly, F.** 1993. Nest desertion in the greater flamingo *Phoenicopterus ruber roseus*. *Animal Behaviour* 45: 1038–1040.

**Cézilly, F. & Johnson, A.R.** 1995. Re-mating between and within seasons in the greater flamingo, *Phoenicopterus ruber roseus*. *Ibis* 139: 543–546.

**Cézilly, F. & Nager, R.G.** 1996. Age and breeding performance in monogamous birds: the influence of pair stability. *Trends in Ecology and Evolution* 11: 27.

**Cézilly, F., Dubois, F. & Pagel, M.** 2000a. Is mate fidelity related to site fidelity? A comparative analysis in ciconiiforms. *Animal Behaviour* 59: 1143–1152.

**Cézilly, F., Tourenq, C. & Johnson, A.R.** 1994. Variation in parental care with offspring age in the greater flamingo. *Condor* 96: 809–812.

**Cézilly, F. Préault, M., Dubois, F., Faivre, B. & Patris, B.** 2000b. Pair-bonding in birds and the active role of females: a critical review of the empirical evidence. *Behavioural Processes* 51: 83–92.

**Chase, I.D.** 1980. Cooperative and noncooperative behaviour in animals. *American Naturalist* 115: 827–857.

**Choe, J.C. & Crespi, B.J.** (eds) 1997. *The Evolution of Mating Systems in Insects and Arachnids.* Cambridge University Press, Cambridge.

**Chuang-Dobbs, H.C., Webster, M.S. & Holmes, R.T.** 2001. Paternity and parental care in the black-throated blue warbler, *Dendroica caerulescens*. *Animal Behaviour* 62: 83–92.

**Clark, A.B. & Lee, W.H.** 1998. Red-winged blackbird females fail to increase feeding in response to begging call playbacks. *Animal Behaviour* 56: 563–570.

**Clark, M.M., Moghaddas, M. & Galef, G. Jr.** 2002. Age at first mating affects parental effort and fecundity of female Mongolian gerbils. *Animal Behaviour* 63: 1129–1134.

**Clutton-Brock, T.H.** 1984. Reproductive effort and terminal investment in iteroparous animals. *American Naturalist* 123: 212–229.

**Clutton-Brock, T.H.** 1989. Mammalian mating systems. *Proceedings of the Royal Society of London Series B* 236: 339–372.

**Clutton-Brock, T.H.** 1991. *The Evolution of Parental Care.* Princeton University Press, Princeton.

**Clutton-Brock, T.H., Guinness, F.E. & Albon, S.D.** 1982. *Red Deer: The Behaviour and Ecology of Two Sexes.* University of Chicago Press, Chicago.

**Clutton-Brock, T.H., Price, O.F. & MacColl, A.D.C.** 1992. Mate retention, harassment, and the evolution of ungulate leks. *Behavioral Ecology* 3: 234–242.

**Clutton-Brock, T.H., Green, D., Hiraiwa-Hasegawa, M. & Albon, S.D.** 1988. Passing the buck: ressource defence, lekking and mate choice in fallow deer. *Behavioral Ecology and Sociobiology* 23: 281–296.

**Cotton, P.A., Kacelnik, A. & Wright, J.** 1996. Chick begging as a signal: are nestlings honest? *Behavioral Ecology* 7: 178–182.

**Coulson, J.C.** 1966. The influence of the pair-bond and age on the breeding biology of the kittiwake gull *Rissa tridactyla*. *Journal of Animal Ecology* 35: 269–279.

**Cronin, E.W. & Sherman, P.W.** 1977. A resource-based mating system: the orange-rumped honey guide. *Living Bird* 15: 5–32.

**Cunningham, E.J.A. & Russel, A.F.** 2000. Egg investment is influenced by male attractiveness in the mallard. *Nature* 404: 74–76.

D

**Dale, S., Rinden, H. & Slagsvold, T.** 1992. Competition for male restricts mate search of female pied flycatchers. *Behavioral Ecology and Sociobiology* 30: 165–176.

**Davies, N. B.** 1991. Mating systems. In *Behavioural Ecology. An Evolutionary Approach*, 2nd edition (Krebs, J.R. & Davies, N.B. eds), pp. 263–294. Blackwell, Oxford.

**Davies, N.B.** 1992. *Dunnock Behaviour and Social Evolution*. Oxford University Press, Oxford.

**Davison, G.W.H.** 1985. Avian spurs. *Journal of Zoology* 206: 353–366.

**del Hoyo, J., Eliott, A. & Sargatal, J.** 1994. *Handbook of the Birds of the World*, vol. 2. Lynx Edicions, Barcelona.

**DeLay, L.S., Faaborg, J., Naranjo, J., Paz, S.M., de Vries, T. & Parker, P.G.** 1996. Parental care in the cooperatively polyandrous Galapagos hawk. *Condor* 98: 300–311.

**Dhondt, A.A. & Adriaensen, F.** 1994. Causes and effects of divorce in the blue tit *Parus caeruleus* L. *Journal of Animal Ecology* 63: 979–987.

**Dixon, A., Ross, D., O'Malley, S.L. & Burke, T.** 1994. Paternal investment inversely related to degree of extra-pair paternity in the reed bunting. *Nature* 371: 698–700.

**Dixson, A.F.** 1998. *Primate Sexuality. Comparative Studies of the Prosimians, Monkeys, Apes, and Human Beings*. Oxford University Press, Oxford.

**Drent, R.H. & Daan, S.** 1980. The prudent parent: energetic adjustments in avian breeding. *Ardea* 68: 225–252.

**Drummond, H.** 2001. A revaluation of the role of food in broodmate aggression. *Animal Behaviour* 61: 517–526.

**Drummond, H., Rodriguez, C., Vallarino, A., Vaderrabano, C., Rogel, G. & Tobon, E.** 2003. Desperado siblings: uncontrollably aggressive junior chicks. *Behavioral Ecology and Sociobiology* 53: 287–296.

**Dubois, F. & Cézilly, F.** 2002. Breeding success and mate retention in birds: a meta-analysis. *Behavioral Ecology and Sociobiology* 52: 357–364.

**Dubois, F., Cézilly, F. & Pagel, M.** 1998. Mate fidelity and coloniality in waterbirds: a comparative analysis. *Oecologia* 116: 433–440.

**Dunbar, R.I.M.** 1988. *Primate Social Systems*. Chapman & Hall, London.

**Dunbar, R.I.M.** 1995. The mating system of callitrichid primates: I. Conditions for the coevolution of pair bonding and twinning. *Animal Behaviour* 50: 1057–1070.

**Dunbar, R.I.M. & Dunbar, P.** 1980. The pairbond in klipspringer. *Animal Behaviour* 28: 219–229.

**Dunn, P.O. & Robertson, R.J.** 1992. Geographic variation in the importance of male parental care and mating systems in tree swallows. *Behavioral Ecology* 3: 291–299.

E

**Emlen, S.T.** 1993. Ethics and experimentation: hard choices for the field ornithologist. *Auk* 110: 406–409.

**Emlen, S.T. & Oring, L.W.** 1977. Ecology, sexual selection and the evolution of animal mating systems. *Science* 197: 215–223.

**Emlen, S.T. & Wrege, P.H.** 2004a. Size dimorphism, intrasexual competition, and sexual selection in wattled jacana (*Jacana jacana*), a sex-role reversed shorebird in panama. *Auk* 121: 391–403.

**Emlen, S.T. & Wrege, P.H.** 2004b. Division of labour in parental care behaviour of a sex-reversed shorebird, the wattled jacana. *Animal Behaviour* 68: 847–855.

**Emlen, S.T., Demond, N.J. & Emlen, D.J.** 1989. Experimental induction of infanticide in female wattled jacanas. *Auk* 106: 1–7.

**Ens, B.J., Choudhury, S. & Black, J.M.** 1996. Mate fidelity and divorce in monogamous birds. In *Partnerships in Birds. The Study of Monogamy* (Black, J.M. ed.), pp. 344–395. Oxford University Press, Oxford.

**Ens, B.J., Safriel, U.N. & Harris, M.P.** 1993. Divorce in the long-lived and monogamous oystercatcher, *Haematophagus ostralegus*. Incompatibility or choosing a better option? *Animal Behaviour* 45, 1199–1217.

**Erckmann, W.J.** 1983. The evolution of polyandry in shorebirds: an evaluation of hypotheses. In *Social Behavior of Female Vertebrates* (Wasser, S.K. ed.), pp. 113–168. Academic Press, New York.

**Ewald, P.W. & Rohwer, S.** 1982. Effects of supplemental feeding on timing of breeding, clutch size and polygamy in red-winged blackbirds, *Agelaius phoenicus*. *Journal of Animal Ecology* 51: 429–450.

F

**Faaborg, J., Parker, P.G., DeLay, L.S., de Vries, T., Bednarz, J.C., Paz, S.M., Nranjo, J. & Waite, T.A.** 1995. Confirmation of cooperative polyandry in the Galapagos hawk (*Buteo galapagoensis*). *Behavioral Ecology and Sociogbiology* 36: 83–90.

**Fincke, O.M., Waage, J.K. & Koenig, W.D.** 1997. Natural and sexual selection components of odonate mating patterns. In *Mating Systems in Insects and Arachnids* (Choe, J.C. & Crespi, B.J. eds), pp. 58–74. Cambridge University Press, Cambridge.

**Forslund, P. & Pärt, T.** 1995. Age and reproduction in birds – hypotheses and tests. *Trends in Ecology and Evolution* 10: 374–378.

**French, J.A. & Inglette, B.J.** 1989. Female–female aggression and male indifference in response to unfamiliar intruders in lion tamarins. *Animal Behaviour* 37: 487–497.

**Fricke** 1975. Evolution of social systems through site attachment in fish. *Zeitschrift für Tierpsychologie* 39: 206–210.

G

**Gibbons, D.W. & Pain, D.** 1992. The influence of river flow rate on the breeding behaviour of *Calopteryx* damselflies. *Journal of Animal Ecology* 61: 283–289.

**Gillette, J.R., Jaeger, R.G. & Peterson, M.G.** 2000. Social monogamy in a territorial salamander. *Animal Behaviour* 59: 1241–1250.

**Godfray, H.C.J.** 1995. Evolutionary theory of parent–offspring conflict. *Nature* 376: 133–138.

**Gomendio, M. & Roldan, E.R.S.** 1993. Mechanisms of sperm competition: linking physiology and behavioural ecology. *Trends in Ecology and Evolution* 8: 95–100.

**Gomendio, M., Harcourt, A.H. & Roldan, E.R.S.** 1998. Sperm competition in mammals. In *Sperm Competition and Sexual Selection* (Birkhead, T.R. & Møller, A.P. eds), pp. 665–751. Academic Press, San Diego.

**Goodall, J.** 1986. *The Chimpanzees of Gombe: Patterns of Behavior.* Belknap Press, Harvard.

**Gould, J.L. & Gould, C.G.** 1989. *Sexual Selection.* Scientific American Library, Freeman & Co., New York.

**Gowaty, P.A.** 1996. Battles of the sexes and the origins of monogamy. In *Partnerships in Birds. The Study of Monogamy* (Black, J.M. ed.) pp. 21–52. Oxford University Press, Oxford.

**Granadeiro, J.P., Bolton, M., Silva, M.C., Nunes, M. & Furness, R.W.** 2000. Responses of breeding Cory's shearwater *Calonectris diomedea* to experimental manipulation of chick condition. *Behavioral Ecology* 11: 274–281.

**Green, D.J.** 2002. Pair bond duration influences paternal provisioning and the primary sex ratio of brown thornbill broods. *Animal Behaviour* 64: 791–800.

**Greenfield, M.D.** 1997. Sexual selection in defense polygyny: lessons from territorial grasshoppers. In *Mating Systems in Insects and Arachnids* (Choe, J.C. & Crespi, B. J. eds), pp. 75–88. Cambridge University Press, Cambridge.

**Griffith, B., Owens, I.P.F. & Thuman, K.A.** 2002. Extra pair paternity in birds: a review of interspecific variation and adaptive function. *Molecular Ecology* 11: 2195–2212.

**Grinnel, J. & McComb, K.** 1996. Maternal grouping as a defense against infanticide by males: evidence from field playback experiments on African lions. *Behavioral Ecology* 7: 55–59.

**Gross, M.R. & Sargent, R.C.** 1985. The evolution of male and female parental care in fishes. *American Zoologist* 25: 807–822.

H

**Hadfield, M.G. & Switzer-Dunlap, M.** 1984. Opisthobranchs. In *The Mollusca*, vol. 7, *Reproduction* (Tompa, A.S., Verdonk, N.H. & van den Biggelaar, J.A.M. eds), pp. 209–350. Academic Press, London.

**Hamilton, W.D.** 1964a. The genetical evolution of social behaviour I. *Journal of Theoretical Biology* 7: 1–16.

**Hamilton, W.D.** 1964b. The genetical evolution of social behaviour II. *Journal of Theoretical Biology* 7: 17–52.

**Harfenist, A. & Ydenberg, R.C.** 1995. Parental provisioning and predation risk in rhinoceros auklets (*Cerorhinca monocerata*): effects on nestling growth and fledging. *Behavioral Ecology* 6: 82–86.

**Harris, M.P.** 1980. Breeding performance of puffins *Fratercula arctica* in relation to hatching date and growth. *Ibis* 127: 243–250.

**Heeb, P., Schwander, T. & Faoro, S.** 2003. Nestling detectability affects parental feeding preferences in a cavity nesting bird. *Animal Behaviour* 66: 637–642.

**Höglund, J. & Alatalo, R.V.** 1995. *Leks.* Princeton University Press, Princeton.

**Höglund, J., Alatalo, R.V., Lundberg, A., Rintamäki, P.T. & Lindell, J.** 1999. Microsatellite markers reveal the potential for kin selection on black grouse leks. *Proceedings of the Royal Society of London Series B* 266: 813–816.

**Hunt, J. & Simmons, L.W.** 2002. Behavioural dynamics of biparental care in the dung beetle *Ontophagus taurus. Animal Behaviour* 64: 65–75.

I

**Iacovides, S. & Evans, R.M.** 1998. Begging as graded signals of need for food in young ring-billed gulls. *Animal Behaviour* 56: 79–85.

J

**Jehl, J.R. Jr. & Murray, B.G. Jr.** 1986. The evolution of normal and reverse sexual size dimorphism in shorebirds and other birds. In *Current Ornithology*, vol. 3 (Johnston, R.F. ed.), pp. 1–86. Plenum Press, New York.

**Jenni, D.A. & Collier, G.** 1972. Polyandry in the American jacana (*Jacana spinosa*). *Auk* 89: 743–765.

**Jiguet, F., Arroyo, B. & Bretagnolle, V.** 2000. Lek mating systems: a case study in the little bustard *Tetrax tetrax. Behavioural Processes* 51: 63–82.

**Johnston, V. & Ryder, J.P.** 1987. Divorce in larids: a review. *Colonial Waterbirds* 10: 16–26.

**Johnstone, R.A.** 1996. Begging signals and parent–offspring conflict: do parents always win? *Proceedings of the Royal Society of London Series B* 263: 1677–1681.

**Jones, A.G., Rosenquist, G., Berglund, A. & Avise, J.C.** 1999. The genetic mating system of a sex-role reversed pipefish (*Syngnathus typhle*): a molecular inquiry. *Behavioral Ecology and Sociobiology* 46: 357–365.

**Jones, A.G., Walker, D. & Avise, J.C. Jones, A.G., Walker, D. & Avise, J.C.** 2001. Genetic evidence for extreme polyandry and extraordinary sex-role reversal in a pipefish. *Proceedings of the Royal Society of London Series B* 268: 2531–2535.

**Jones, D.N., Dekker, R.W.R.J. & Roselaar, C.S.** 1995. *The Megapodes.* Oxford University Press, Oxford.

**Jones, T.M. & Quinnell, R.J.** 2002. Testing predictions for the evolution of lekking in the sandfly, *Lutzomyia longipalis. Animal Behaviour* 63: 605–612.

**Jourdie, V., Moureau, B., Bennett, A.T.D. & Heeb, P.** 2004. Ultraviolet reflectance by the skin of nestlings. *Nature* 431: 262.

K

**Kempenaers, B. & Sheldon, B.C.** 1997. Studying paternity and paternal care: pitfalls and problems. *Animal Behaviour* 53: 4223–427.

**Kempenaers, B., Lanctot, R.B. & Robertson, R.J.** 1998. Certainty of paternity and paternal investment in eastern bluebirds and tree swallows. *Animal Behaviour* 55: 845–860.

**Kilner, R.** 1995. When do canary parents respond to nestling signals of need? *Proceedings of the Royal Society of London Series B* 260: 343–348.

**Kilner, R. & Johnstone, R.A.** 1997. Begging the question: are offsrping solicitation behaviours signals of need? *Trends in Ecology and Evolution* 12: 11–15.

**Kleinman, D.G.** 1977. Monogamy in mammals. *Quarterly Review of Biology* 52: 39–69.

**Kleinman, D.G. & Malcom, J.R.** 1981. The evolution of male parental investment in mammals. In *Parental Care in Mammals* (Gubernick, D.J. & Klopfer, P.H. eds), pp. 347–387. Plenum Press, New York.

**Klemperer, H.G.** 1983. The evolution of parental behaviour in Scarabaeinae (Coleoptera, Scarabaeidae): an experimental approach. *Ecological Entomology* 8: 49–59.

**Knowlton, N.** 1979. Reproductive synchrony, parental investment, and the evolutionary dynamics of sexual selection. *Animal Behaviour* 27: 1022–1033.

**Kokko, H.** 1999. Cuckoldry and the stability of biparental care. *Ecology Letters* 2: 247–255.

**Kokko, H. & Lindström, J.** 1996. Kin selection and the evolution of leks: whose success do young males maximize? *Proceedings of the Royal Society of London Series B* 263: 919–923.

**Kölliker, M. & Richner, H.** 2001. Parent–offspring conflict and the genetics of offspring solicitation and parental response. *Animal Behaviour* 62: 395–407.

**Komers, P.E.** 1996. Obligate monogamy without parental care in Kirk's dikdik. *Animal Behaviour* 51: 131–140.

L

**Lack, D.** 1968. *Ecological Adaptations for Breeding in Birds*. Methuen, London.

**Lazarus, J.** 1989. The logic of mate desertion. *Animal Behaviour* 39: 672–684.

**Lazarus, J. & Inglis, I.R.** 1986. Shared and unshared parental investment, parent–offspring conflict and brood size. *Animal Behaviour* 34: 1791–1804.

**Le Boeuf, B.J.** 1974. Male–male competition and reproductive success in elephant seals. *American Zoologist* 14: 163–176.

**Le Boeuf, B.J.** 1978. Social behaviour in some marine and terrestrial carnivores. In *Contrasts in Behavior* (Reese, E.S. & Lighter, F.J. eds), pp. 251–279. Wiley, New York.

**Leonard, M.L. & Horn, A.G.** 1996. Provisioning rules in tree swallows. *Behavioral Ecology and Sociobiology* 38: 341–347.

**Lifjeld, J.T., Anthonisen, K., Blomquist, D., Johnsen, A., Krokene, C. & Rigstad, K.** 1998. Studying the influence of paternity on parental effort: a comment on Kempenaers and Sheldon. *Animal Behaviour* 55: 235–238.

**Lightbody, J.P. & Weatherhead, P.J.** 1988. Female settling patterns and polygyny: tests of a neutral mate-choice hypothesis. *American Naturalist* 132: 20–33.

**Ligon, J.D.** 1999. *The Evolution of Avian Breeding Systems*. Oxford University Press, Oxford.

**Lindburg, D.G.** 1983. Mating behaviour and estrus in the Indian rhesus monkey. In *Perpsectives in Primate Biology* (Seth, P.K. ed.), pp. 45–61. Today and Tomorrow, New Delhi.

**Linden, M. & Møller, A.P.** 1989. Cost of reproduction and covariation of life history traits in birds. *Trends in Ecology and Evolution* 4: 367–371.

**Lotem, A.** 1998. Differences in begging behaviour between barn swallows, *Hirundo rustica*, nestlings. *Animal Behaviour* 55: 809–818.

**Lozano, G.A. & Lemon, R.E.** 1996. Male plumage, paternal care and reproductive success in yellow warblers, *Dendroica petechia*. *Animal Behaviour* 51: 265–272.

**Lyon, B.E., Eadie, J.M. & Hamilton, L.D.** 1994. Parental choice selects for ornamental plumage in American coot chicks. *Nature* 371: 240–243.

M

**Markman, S., Yom-Tov, Y. & Wright, J.** 1995. Male parental care in the orange-tufted sunbird: behavioural adjustments in provisioning and nest guarding effort. *Animal Behaviour* 50: 655–669.

**Markman, S., Yom-Tov, Y. & Wright, J.** 1996. The effect of male removal on female parental care in the orange-tufted sunbird. *Animal Behaviour* 52: 437–444.

**Marlowe, F.** 2000. Paternal investment and the human mating system. *Behavioural Processes* 51: 45–61.

**Martin, T.L.F. & Wright, J.** 1993. Cost of reproduction and allocation of food between parent and young in the swift (*Apus apus*). *Behavioral Ecology* 4: 213–223.

**Mathews, L.M.** 2002. Territorial cooperation and social monogamy: factors affecting intersexual behaviours in pair-living snapping shrimp. *Animal Behaviour* 63: 767–777.

**Matsumoto, K. & Yanagisawa, Y.** 2001. Monogamy and sex role reversal in the pipefish *Corythoicthys haematopterus*. *Animal Behaviour* 61: 163–170.

**Mauck, R.A. & Grubb, Jr. T.C.** 1995. Petrel parents shunt all experimentally increased reproductive costs to their offspring. *Animal Behaviour* 49: 999–1008.

**Maynard Smith, J.** 1977. Parental investment: a propsective analysis. *Animal Behaviour* 25: 1–9.

**Mazuc, J., Chastel, O. & Sorci, G.** 2003. No evidence for differential maternal allocation to offspring in the house sparrow (*Passer domesticus*). *Behavioral Ecology* 14: 340–346.

**McVey, M.E.** 1988. The opportunity for sexual selection in a territorial dragonfly *Erythemis simplicicollis*. In *Reproductive Success: Studies of Individual Variation in*

*Contrasting Breeding Systems* (Clutton-Brock, T.H. ed.), pp. 44–58. University of Chicago Press, Chicago.

**Ménard, N., Scheffrahn, W., Vallet, D, Zidane, C. & Reber, C.** 1992. Application of blood protein electrophoresis and DNA fingerprinting to the analysis of paternity and social characteristics of wild barbary macaques. In *Paternity in Primates: Genetic Tests and Theories*, pp. 155–174. Karger, Bâle.

**Mock, D.W. & Parker, G.A.** 1997. *The Evolution of Sibling Rivalry*. Oxford University Press, Oxford.

**Møller, A.P.** 2000. Male parental care, female reproductive success and extra-pair paternity. *Behavioral Ecology* 11: 161–168.

**Møller, A.P. & Thornhill, R.** 1998. Male parental care, differential parental investment by females and sexual selection. *Animal Behaviour* 55: 1507–1515.

O

**Orell, M., Rytokönen, S. & Koivula, K.** 1994. Causes of divorce in the monogamous willow tit, *Parus montanus*, and consequences for reproductive success. *Animal Behaviour* 48: 1143–1154.

**Orians, G.H.** 1969. On the evolution of mating systems in birds and mammals. *American Naturalist* 103: 589–603.

**Oring, L.W.** 1982. Avian mating systems. In *Avian Biology*, vol. 6 (Farner, S., King, J.R. & Parkes, C. eds), pp. 1–92. Academic Press, New York.

**Oring, L.W.** 1986. Avian polyandry. In *Current Ornithology*, vol. 3 (Johnston, R.F. ed.), pp. 309–351. Plenum Press, New York.

**Otter, K. & Ratcliffe, L.** 1996. Female initiated divorce in a monogamous songbird abandoning mate for males of higher quality. *Proceedings of the Royal Society of London Series B* 263: 351–354.

**Owen-Smith, N.** 1977. On territoriality in ungulates and an evolutionary model. *Quarterly Review of Biology* 52: 1–38.

P

**Packer, C., Herbst, L., Pusey, A.E., Bygott, J.D., Hanby, J.P., Cairns, S.J. & Borgerhoff Mulder, J.** 1988. Reproductive success in lions. In *Reproductive Success: Studies of Individual Variation in Contrasting Breeding Systems* (Clutton-Brock, T.H. ed.), pp. 363–383. University of Chicago Press, Chicago.

**Parker, G.A.** 1985. Models of parent–offspring conflict. V. Effects of the behaviour of two parents. *Animal Behaviour* 33: 519–533.

**Parker, G.A. & Macnair, M.R.** 1978. Models of parent–offspring conflict. I. Monogamy. *Animal Behaviour* 26: 97–110.

**Parker, G.A. & Macnair, M.R.** 1979. Models of parent–offspring conflict. IV. Suppression: evolutionary retaliation by the parent. *Animal Behaviour* 27: 1210–1235.

**Parker, G.A. & Mock, D.W.** 1987. Parent–offspring conflict over clutch size. *Evolutionary Ecology* 1: 161–174.

**Petrie, M., Krupa, A. & Burke, T.** 1999. Peacoks lek with relatives even in the absence of social and environmental cues. *Nature* 401: 155–157.

**Pleszczynska, W.K.** 1978. Microgeographic prediction of polygyny in the lark bunting. *Science* 201: 935–937.

**Price, K.** 1998. Benefits of begging for yellow-headed blackbird nestlings. *Animal Behaviour* 56: 571–577.

**Price, K., Harvey, H. & Ydenberg, R.** 1996. Begging tactics of nestling yellow-headed blackbirds, *Xanthocephalus xanthocephalus*, in relation to need. *Animal Behaviour* 51: 421–435.

**Pugesek, B.H.** 1995. Chick growth in the California gull: reproductive effort and parental experience hypothesis. *Animal Behaviour* 49: 641–647.

**Pugesek, B.H. & Diem, K.L.** 1990. The relationship between reproduction and survival in known-aged California gulls. *Ecology* 71: 811–817.

Q

**Quillfeldt, P.** 2002. Begging in the absence of sibling competition in Wilson's storm petrels, *Oceanites oceanicus*. *Animal Behaviour* 64: 579–587.

R

**Real, L.** 1990. Search theory and mate choice. I. Models of single sex discrimination. *American Naturalist* 136: 376–404.

**Redondo, T. & Castro, F.** 1992. Signalling of nutritional need by magpie nestlings. *Ethology* 92: 193–204.

**Reichard, U.** 1995. Extra-pair copulations in a monogamous gibbon (*Hylobated lar*). *Ethology* 100: 99–112.

**Reichard, U.H.** 2003. Monogamy: past and present. In *Monogamy. Mating Strategies and Partneships in Birds, Humans and Other Mammals* (Reichard, U.H. & Boesch, C. eds), pp. 3–25. Cambridge University Press, Cambridge.

**Reynolds, J.D.** 1996. Animal breeding systems. *Trends in Ecology and Evolution* 11: 68–72.

**Robel, R.J. & Ballard, W.B.** 1974. Lek social organization and reproductive success in the greater prairie chicken. *American Zoologist* 14: 121–128.

**Roberts, R.L., Williams, J.R., Wang, A.K. & Carter, C.S.** 1998. Cooperative breeding and monogamy in prairie voles: influence of the sire and geographical variation. *Animal Behaviour* 55: 1131–1140.

**Robertson, D.R. & Hoffman, S.G.** 1977. The roles of female mate choice and predation in the mating system of some tropical labroid fishes. *Zeitchtrift für Tierpsychologie* 45: 298–320.

**Rodriguez-Gironés, M.A.** 1999. Sibling competition stabilizes signalling resolution of models of parent–offspring conflict. *Proceedings of the Royal Society of London Series B* 266: 2399–2402.

**Rollinson, D., Kane, R.A. & Lines, J.R.L.** 1989. An analysis of fertilization in *Bulinus cernicus* (Gastropoda: Planorbidae). *Journal of Zoology* 217: 295–310.

**Rosenblatt, J.S. & Snowdown, C.T.** 1996. *Parental Care: Evolution, Mechansims, and Adaptive Significance.* Academic Press, San Diego.

**Royle, N.J., Hartley, I.R. & Parker, G.A.** 2002. Begging for control: when are offspring sollicitation behaviours honnest? *Trends in Ecology and Evolution* 17: 434–440.

**Rubenstein, D.I.** 1986. Ecology and sociality in horses and zebras. In *Ecological Aspects of Social Evolution* (Rubenstein, D.I. & Wrangham, R.W. eds), pp. 282–302. Princeton University Press, Princeton.

**Rudman, W.B.** 1981. Further studies on the anatomy and ecology of opisthobranch molluscs feeding on the scleractinian coral *Porites*. *Biological Journal of the Linnean Society* 71: 373–412.

**Runcie, M.J.** 2000. Biparental care and obligate monogamy in the rock-haunting possum, *Petropseudes dahli*, from tropical Australia *Animal Behaviour* 59: 1001–1008.

**Ruusila, V. & Pöysä, H.** 1998. Shared and unshared parental investment in the precocial goldeneye (Aves: Anatidae). *Animal Behaviour* 55: 307–132.

S

**Saether, B.E., Andersen, R. & Pedersen, H.C.** 1993. Regulation of parental effort in a long-lived seabird: an experimental manipulation of the cost of reproduction in the Antarctic petrel *Thalassoica antarctica*. *Behavioral Ecology and Sociobiology* 33: 147–150.

**Sæther, S.A.** 2002. Kin selection, female preferences and the evolution of leks: direct benefits may explain kin structuring. *Animal Behaviour* 63: 1017–1019.

**Saino, N. & Møller, A.P.** 1995. Testosterone-induced depression of male parental behaviour in the barn swallow: female compensation & effects on seasonal fitness. *Behavioral Ecology and Sociobiology* 36: 151–157.

**Saino, N., Stradi, R., Ninni, P., Pini, E. & Møller, A.P.** 2000. Better red than dead: carotenoid-based mouth coloration reveals infection in barn swallow nestlings. *Proceedings of the Royal Society of London Series B* 267: 57–61.

**Schamel, D., Tracy, D.M. & Lank, D.B.** 2004. Male mate choice, male availability and egg production as limitations on polyandry in the red-necked phalarope. *Animal Behaviour* 67: 847–853.

**Schawgmeyer, P.L., Mock, D.W. & Parker, G.A.** 2002. Parental care in house sparrows: negotiation or sealed bid? *Behavioral Ecology* 13: 713–721.

**Schoener, T.W. & Schoener, A.** 1982. Intraspecific variation in home range size in some *Anolis* lizards. *Ecology* 63: 809–823.

**Searcy, W.A. & Yasukawa, K.** 1989. Alternative models of territorial polygyny in birds. *American Naturalist* 134: 323–343.

**Seyfarth, R.M.** 1978. Social relationships among adult male and female baboons, I: behaviour during sexual consortship. *Behaviour* 64: 204–226.

**Sheldon, B.C.** 2000. Differential allocation: tests, mechanisms and implications. *Trends in Ecology and Evolution* 15: 397–402.

**Shelly, T.E. & Whittier, T.S.** 1997. Lek behavior of insects. In *The Evolution of Mating Systems in Insects and Arachnids* (Choe, J.C. & Crespi, B.J. eds), pp. 273–293. Cambrdige University Press, Cambridge.

**Shelly, T.E., Greenfield, M.D. & Downum, K.R.** 1987. Variation in host plant quality: influences on the mating system of a desert grasshopper. *Animal Behaviour* 35: 1200–1209.

**Sherman, K.J.** 1983. The adaptive significance of post-copulatory mate guarding in a dragonfly *Pachydiplax longipennis*. *Animal Behaviour* 35: 1200–1209.

**Shorey, L., Piertney, S., Stone, J. & Höglund, J.** 2000. Fine-scale genetic structuring on *Manacus manacus* leks. *Nature* 408: 352–353.

**Shuster, S.M. & Wade, M.J.** 2003. *Mating Systems and Strategies*. Princeton University Press, Princeton.

**Siva-Jothy, M.T., Gibbons, D.W. & Pain, D.** 1995. Female oviposition-site preference and egg hatching success in the damselfly *Calopteryx spledens xanthostoma*. *Behavioral Ecology and Sociobiology* 37: 39–44.

**Slagsvold, T., Amundsen, T., Dale, S. & Lampe, H.** 1992. Female–female aggression explains polyterritoriality in male pied flycatchers. *Animal Behaviour* 43: 397–407.

**Smith, J.L.D., McDougal, C. & Sunquist, M.E.** 1987. Female land tenure system in tigers. In *Tigers of the World* (Tilson, R.L. & Seal, U.S. eds), pp. 97–109. Noyes Publications, Park Ridge.

**Smith, R.L.** 1980. Evolution of exclusive post-copulatory parental care in the insect. *Florida Entomologist* 63: 65–78.

**Stamps, J., Metcalf, R.A. & Krishnan, V.V.** 1978. A genetic analysis of parent–offspring conflict. *Behvioral Ecology and Sociobiology* 3: 369–392.

**Stenmark, G.T., Slagsvold, T. & Lifjeld, T.** 1988. Polygyny in the pied flycatcher, *Ficedula hypoleuca*: a test of the deception hypothesis. *Animal Behaviour* 36: 1646–1657.

**Stillman, R.A., Clutton-Brock, T.H. & Sutherland, W.J.** 1993. Black holes, mate retention, and the evolution of ungulate leks. *Behavioral Ecology* 4: 1–6.

**Strohm, E. & Marliani, A.** 2002. The cost of parental care: prey hunting in a digger wasp. *Behavioral Ecology* 13: 52–58.

**Sullivan, M.S.** 1994. Mate choice as an information gathering process under time constraint: implications for behaviour and signal design. *Animal Behaviour* 47: 141–151.

**Sunquist, F. & Sunquist, M.** 1988. *Tiger Moon*. University of Chicago Press, Chicago.

**Szekely, T. & Reynolds, J.D.** 1995. Evolutionary transitions in parental care in shorebirds. *Proceedings of the Royal Society of London Series B* 262: 57–64.

T

**Tavecchia, G., Pradel, R., Boy, V., Johnson, A.R. & Cézilly, F.** 2001. Sex- and age-related variation in survival probability and cost of first reproduction in greater flamingos. *Ecology* 82: 165–174.

**Terborgh, J. & Goldizen, A.W.** 1985. On the mating system of the cooperatively-breeding saddle-back tamarin (*Saguinus fuscicollis*). *Behavioral Ecology and Sociobiology* 16: 293–299.

**Théry, M.** 1992. The evolution of leks through female choice: differential clustering and space utilization in six sympatric manakins. *Behavioral Ecology and Sociobiology* 30: 227–237.

**Thirgood, S.J., Langbein, J. & Putnam, R.J.** 1999. Intraspecific variation in ungulate mating strategies: the case of the flexible fallow deer. *Advances in the Study of Behavior* 28: 333–361.

**Tinbergen, N.** 1953. *Social Behaviour in Animals*. T.J. Press, Padstow, Cornwall.

**Tokuda, H. & Seno, H.** 1994. Some mathematical considerations on the parent–offspring conflict phenomenon. *Journal of Theoretical Biology* 170: 145–157.

**Trivers, R.L.** 1972. Parental investment and sexual selection. In *Sexual Selection and the Descent of Man* (Campbell, B. ed.), pp. 136–179. Aldine, Chicago.

**Trivers, R.L.** 1974. Parent–offspring conflict. *American Zoologist* 11: 249–264.

**Trumbo, S.T.** 1992. Monogamy to communal breeding: exploitation of a variable resource base in burying beetles (*Nicrophorus*). *Ecological Entomology* 17: 289–298.

**Trumbo, S.T. & Eggert, A.-K.** 1994. Beyond monogamy: territory quality influences sexual advertisement in male burying beetles. *Animal Behaviour* 48: 1043–1047.

**Tutin, C.E.G.** 1979. Mating patterns and reproductive strategies in a community of wild chimpanzees (*Pan troglodytes schweinfurthii*). *Behavioral Ecology and Sociobiology* 6: 29–38.

V

**van de Pol, M., Heg, D., Bruinzeel, L.W., Kuijper, B. & Verhulst, S.** 2006. Experimental evidence for a causal effect of pair-bond duration on reproductive performance in oystercatchers. *Behavioral Ecology* 17: 982–991.

**Van Noordwijk, M.A.** 1985. Sexual behaviour of Sumatran long-tailed macaques (*Macaca fascicularis*). *Zeitschtrift für Tierpsychologie* 70: 277–296.

**Van Schaik, C.P. & Dunbar, R.I.M.** 1990. The evolution of monogamy in large primates: a new hypothesis and some crucial tests. *Behaviour* 115: 30–62.

**Veasey, J.S., Houston, D.C. & Metcalfe, N.B.** 2000. Flight muscle atrophy and predation risk in breeding birds. *Functional Ecology* 14: 115–121.

**Veiga, J.P.** 1992. Why are house sparrows predominantly monogamous: a test of hypotheses. *Animal Behaviour* 43: 361–370.

**Verner, J. & Wilson, M.F.** 1966. The influence of habitats on mating systems of North American passerine birds. *Ecology* 47: 143–147.

**Vincent, A.C.J. & Sadler, L.** 1995. Faithful pair bonds in wild seahorses, *Hippocampus whitei*. *Animal Behaviour* 50: 1557–1569.

W

**Wagner, R.H.** 1998. Hidden leks: sexual selection and the clustering of avian territories. In *Extra-Pair Mating Tactics in Birds* (Parker, P.G. & Burley, N. eds), pp. 123–145. Washington, DC: Ornithological Monographs, American Ornithologists' Union.

**Wagner, R.H., Shug, M.D. & Morton, E.S.** 1996. Confidence of paternity and parental effort by purple martins. *Animal Behaviour* 52: 123–132.

**Warham, J.** 1990. *The Petrels. Their Ecology and Breeding Systems*. Academic Press, London.

**Watanuki, Y.** 1986. Moonlight avoidance behavior in Leach's storm petrels as a defense against slaty-backed gulls. *Auk* 103: 14–22.

**Weatherhead, P.J. & Robertson, R.J.** 1979. Offspring quality and the polygyny threshold: 'the sexy son hypothesis'. *American Naturalist* 113: 201–208.

**Weatherhead, P.J., Montgomerie, R., Gibbs, H.L. & Boag, P.T.** 1995. The cost of extra-pair fertilizations to female red-winged blackbirds. *Proceedings of the Royal Society of London Series B* 258: 315–320.

**Westneat, D.F. & Sargent, R.C.** 1996. Sex and parenting: the effects of sexual conflict and parentage on parental strategies. *Trends in Ecology and Evolution*, 11: 87–91.

**Westneat, D. F. & Sherman, P.W.** 1993. Parentage and the evolution of parental behavior. *Behavioral Ecology* 4: 66–77.

**Whillans, K.V. & Falls, J.B.** 1990. Effects of male removal on parental care of female white-throated sparrows, *Zonotrichia albicolis*. *Animal Behaviour* 39: 869–878.

**Wickler, W. & Seibt, U.** 1983. Monogamy: an ambiguous concept. In *Mate Choice* (Bateson, P. ed.), pp. 33–50. Cambridge University Press, Cambridge.

**Wiley, R.H.** 1991. Lekking in birds and mammals: behavioral and evolutionary issues. *Advances in the Study of Behavior* 20: 201–291.

**Williams, G.C.** 1966. *Adaptation and Sexual Selection*. Princeton University Press, Princeton.

**Wilson, A.B., Ahnesjö, I., Vincent, A.C.J. & Meyer, A.** 2003. The dynamics of male brooding, mating patterns, and sex roles in pipefishes and seahorses (family Syngnathidae). *Evolution* 57: 1374–1386.

**Winkler, D.W.** 1987. A general model for parental care. *American Naturalist* 130: 526–543.

**Wolf, L. Ketterson, E.D. & Nolan, V. Jr.** 1990. Behavioural responses of female dark-eyed juncos to the experimental removal of their mates: implications for the evolution of parental care. *Animal Behaviour* 39: 125–134.

**Wright & Cuthill** 1990. Biparental care: short-term manipulation of partner contribution and brood size in the starling *Sturnus vulgaris*. *Behavioral Ecology* 1: 116–124.

**Wright, J.** 1998. Paternity and paternal care. In *Sperm Competition and Sexual Selection* (Birkhead, T.R. & Møller, A.P. eds), pp. 117–145. Academic Press, San Diego.

Y

**Yezerinac, S.M., Weatherhead, P.J. & Boag, P.T.** 1996. Cuckoldry and lack of parentage-dependent paternal care in yellow warblers: a cost–benefit approach. *Animal Behaviour* 52: 821–832.

**Yom-Tov, Y.** 2001. An updated list and some comments on the occurrence of intraspecific nest parasitism in birds. *Ibis* 143: 133–143.

Z

**Zabel, C.J. & Taggart, S.J.** 1989. Shift in red fox, *Vulpes vulpes*, mating system associated with El Niño in the Bering Sea. *Animal Behaviour* 38: 830–838.

## Chapter 13

B

**Badyaev, A.V., Hill, G.E., Beck, M.L., Dervan, A.A., Duckworth, R.A., McGraw, K.J., Nolan, P.M. & Whittingham, L.A.** 2002. Sex-biased hatching order and adaptive population divergence in a passerine bird. *Science* 295: 316–318.

**Boomsma, J.J.** 1989. Sex-investment ratios in ants: has female bias been systematically overestimated? *American Naturalist* 133: 517–532.

**Boomsma, J.J.** 1991. Adaptive colony sex ratios in primitively eusocial bees. *Trends in Ecology and Evolution* 6: 92–95.

**Boomsma, J.J.** 1996. Split sex ratios and queen–male conflict over sperm allocation. *Proceedings of the Royal Society of London Series B* 263: 697–704.

**Boomsma, J.J. & Grafen, A.** 1990. Intraspecific variation in ant sex ratios and the Trivers–Hare hypothesis. *Evolution* 44: 1026–1034.

**Boomsma, J.J. & Grafen, A.** 1991. Colony-level sex ratio selection in the eusocial Hymenoptera. *Journal of Evolutionary Biology* 4: 383–407.

**Bourke, A.F.G.** 2005. Genetics, relatedness and social behaviour in insect societies. In *Insect Evolutionary Ecology* (Fellowes M.D.E., Holloway, G.J. & Rolff, J. eds), pp. 1–30. CABI Publishing, Wallingford, UK.

**Bourke, A.F.G. & Franks, N.R.** 1995. *Social Evolution in Ants. Monographs in Behavior and Ecology*. Princeton University Press, Princeton, NJ.

**Brown, G.R. & Silk, J.B.** 2002. Reconsidering the null hypothesis: is maternal rank associated with birth sex ratios in primate groups? *Proceedings of the National Academy of Science of the USA* 99: 11252–11255.

**Bull, J.J.** 1983. *Evolution of Sex Determining Mechanisms*. Benjamin/Cummings, Menlo Park, California.

**Bull, J.J. & Charnov, E.L.** 1988. How fundamental are Fisherian sex ratios? *Oxford Surveys in Evolutionary Biology* 5: 96–135.

C

**Chan, G.L. & Bourke, A.F.G.** 1994. Split sex ratios in a multiple-queen ant population. *Proceedings of the Royal Society of London Series B* 258: 261–266.

**Chapuisat, M. & Keller, L.** 1999. Testing kin selection with sex allocation data in eusocial Hymenoptera. *Heredity* 82: 473–478.

**Chapuisat, M., Sundström, L. & Keller, L.** 1997. Sex ratio regulation: the economics of fratricide in ants. *Proceedings of the Royal Society of London Series B* 264: 1255–1260.

**Charnov, E.L.** 1979. The genetical evolution of patterns of sexuality: Darwinian fitness. *American Naturalist* 113: 460–480.

**Charnov, E.L., Los-den Hartogh, R.L., Jones, W.T. & Van den Assem, J.** 1981. Sex ratio evolution in a variable environment. *Nature* 289: 27–33.

**Clark, A.B.** 1978. Sex ratio and local resource competition in a prosimian primate. *Science* 201: 163–165.

**Clutton-Brock, T.H.** 1991. *The Evolution of Parental Care. Monographs in Behavior and Ecology*. Princeton University Press, Princeton, NJ.

**Clutton-Brock, T.H. & Iason, G.R.** 1986. Sex ratio variation in mammals. *Quarterly Review of Biology* 61: 339–374.

**Clutton-Brock, T.H., Albon, S.D. & Guinness, F.E.** 1984. Maternal dominance, breeding success and birth sex-ratios in red deer. *Nature* 308: 358–360.

**Crozier, R.H. & Pamilo, P.** 1996. *Evolution of Social Insect Colonies: Sex Allocation and Kin Selection*. Oxford Series in Ecology and Evolution. Oxford: Oxford University Press.

D

**Darwin, C.** 1871. *The Descent of Man and Selection in Relation to Sex*. John Murray, London.

**de Menten, L., Fournier, D., Brent, C., Passera, L., Vargo, E.L. & Aron, S.** 2005. Dual mechanism of queen influence over sex ratio in the ant *Pheidole pallidula*. *Behavioral Ecology and Sociobiology* 58: 527–533.

**Deslippe, R.J. & Savolainen, R.** 1995. Sex investment in a social insect: the proximate role of food. *Ecology* 76: 375–382.

**Dreiss, A., Richard, M., Moyen, F., White, J., Moller, A.P. & Danchin, E.** 2006. Sex ratio and male sexual characters in a population of blue tits, *Parus caeruleus*. *Behavioral Ecology* 17: 13–19.

E

**Edwards, A.W.F.** 1998. Natural selection and the sex ratio: Fisher's sources. *American Naturalist* 151: 564–569.

**Edwards, A.W.F.** 2000. Carl Dusing (1884) on the regulation of the sex-ratio. *Theoretical Population Biology* 58: 255–257.

**Ellegren, H., Gustafsson, L. & Sheldon, B.C.** 1996. Sex ratio adjustment in relation to paternal attractiveness in a wild bird population. *Proceedings of the National Academy of Science of the USA* 93: 11723–11728.

**Evans, J.D.** 1995. Relatedness threshold for the production of female sexuals in colonies of a polygynous ant, *Myrmica tahoensis*, as revealed by microsatellite DNA analysis. *Proceedings of the National Academy of Science of the USA* 92: 6514–6517.

F

**Fisher, R.A.** 1930. *The Genetical Theory of Natural Selection*. Clarendon Press, Oxford.

**Frank, S.A.** 1990. Sex allocation theory for birds and mammals. *Annual Review of Ecology and Systematics* 21: 13–55.

**Frank, S.A.** 1998. *Foundations of Social Evolution. Monographs in Behavior and Ecology*. Princeton: Princeton University Press.

**Freedberg, S. & Wade, M.J.** 2001. Cultural inheritance as a mechanism for population sex-ratio bias in reptiles. *Evolution* 55: 1049–1055.

G

**Gosling, L.M.** 1986. Selective abortion of entire litters in the coypu: adaptive control of offspring production in relation to quality and sex. *American Naturalist* 127: 772–795.

**Griffin, A.S., Sheldon, B.C. & West, S.A.** 2005. Cooperative breeders adjust offspring sex ratios to produce helpful helpers. *American Naturalist* 166: 628–632.

**Griffith, S.C., Ornborg, J., Russell, A.F., Andersson, S. & Sheldon, B.C.** 2003. Correlations between ultraviolet coloration, overwinter survival and offspring sex ratio in the blue tit. *Journal of Evolutionary Biology* 16: 1045–1054.

H

**Hamilton, W.D.** 1967. Extraordinary sex ratios. *Science* 156: 477–488.

**Hammond, R.L., Bruford, M.W. & Bourke, A.F.G.** 2002. Ant workers selfishly bias sex ratios by manipulating female development. *Proceedings of the Royal Society of London Series B* 269: 173–178.

**Hastings, M.D., Queller, D.C., Eischen, F. & Strassmann, J.E.** 1998. Kin selection, relatedness, and worker control of reproduction in a large-colony epiponine wasp, *Brachygastra mellifica*. *Behavioral Ecology* 9: 573–581.

**Helms, K.R.** 1999. Colony sex ratios, conflict between queens and workers, and apparent queen control in the ant *Pheidole desertorum*. *Evolution* 53: 1470–1478.

**Helms, K.R., Fewell, J.H. & Rissing, S.W.** 2000. Sex ratio determination by queens and workers in the ant *Pheidole desertorum*. *Animal Behaviour* 59: 523–527.

**Henshaw, M.T., Strassmann, J.E. & Queller, D.C.** 2000. The independent origin of a queen number bottleneck that promotes cooperation in the African swarm-founding wasp, *Polybioides tabidus*. *Behavioral Ecology and Sociobiology* 48: 478–483.

**Herre, E.A.** 1985. Sex ratio adjustment in fig wasps. *Science* 228: 896–898.

**Herre, E.A.** 1987. Optimality, plasticity and selective regime in fig wasp sex ratio. *Nature* 329: 627–629.

**Hewison, A.J.M. & Gaillard, J.M.** 1999. Successful sons or advantaged daughters? The Trivers–Willard model and sex-biased maternal investment in ungulates. *Trends in Ecology and Evolution* 14: 229–234.

J

**Jaenike, J.** 2001. Sex chromosome meiotic drive. *Annual Review of Ecology and Systematics* 32: 25–49.

**Johnson, C.N.** 1988. Dispersal and the sex ratio at birth in primates. *Nature* 332: 726–728.

K

**Keller, L. & Chapuisat, M.** 1999. Cooperation among selfish individuals in insect societies. *BioScience* 49: 899–909.

**Komdeur, J.** 1998. Long-term fitness benefits of egg sex modification by the Seychelles warbler. *Ecology Letters* 1: 56–62.

**Komdeur, J., Daan, S., Tinbergen, J. & Mateman, C.** 1997. Extreme adaptive modification in sex ratio of the Seychelles warbler's eggs. *Nature* 385: 522–525.

**Krackow, S.** 1995. Potential mechanisms for sex ratio adjustment in mammals and birds. *Biological Review of the Cambridge Philosophical Society* 70: 225–241.

**Kruuk, L.E.B., Clutton-Brock, T.H., Albon, S.D., Pemberton, J.M. & Guinness, F.E.** 1999. Population density affects sex ratio variation in red deer. *Nature* 399: 459–461.

L

**Leech, D.I., Hartley, I.R., Stewart, I.R.K., Griffith, S.C. & Burke, T.** 2001. No effect of parental quality or extrapair paternity on brood sex ratio in the blue tit (*Parus caeruleus*). *Behavioral Ecology* 12: 674–680.

**Liautard, C., Brown, W.D., Helms, K.R. & Keller, L.** (2003) Temporal and spatial variations of gyne production in the ant *Formica exsecta*. *Oecologia* 136: 558–564.

M

**Maynard Smith, J.** 1982. *Evolution and the Theory of Games*. Cambridge University Press, Cambridge.

**Mueller, U.G.** 1991. Haplodipoidy and the evolution of facultative sex ratios in a primitively eusocial bee. *Science* 254: 442–444.

N

**Nur, U., Werren, J.H., Eickbush, D.G., Burke, W.D. & Eickbush, T.H.** 1988. A selfish B-chromosome that enhances its transmission by eliminating the paternal genome. *Science* 240: 512–514.

P

**Packer, C. & Pusey, A.E.** 1987. Intrasexual cooperation and the sex ratio in African lions. *American Naturalist* 130: 636–642.

**Packer, P. & Owen, R.E.** 1994. Relatedness and sex ratio in a primitively eusocial halictine bee. *Behavioral Ecology and Sociobiology* 34: 1–10.

**Packer, C., Collins, D.A. & Eberly, L.E.** 2000. Problems with primate sex ratios. *Philosophical Transactions of the Royal Society of London Series B* 355: 1627–1635.

**Pamilo, P.** 1991. Evolution of colony characteristics in social insects. I. Sex allocation. *American Naturalist* 137: 83–107.

**Passera, L., Aron, S., Vargo, E.L. & Keller, L.** 2001. Queen control of sex ratio in fire ants. *Science* 293: 1308–1310.

**Petrie, M., Schwabl, H., Brande-Lavridsen, N. & Burke, T.** 2001. Sex differences in avian yolk hormone levels. *Nature* 412: 498.

**Pike, T.W. & Petrie, M.** 2005. Offspring sex ratio is related to paternal train elaboration and yolk corticosterone in peafowl. *Biology Letters* 1: 204–207.

Q

**Queller, D.C., Strassmann, J.E., Solis, C.R., Hughes, C.R. & DeLoach, D.M.** 1993. A selfish strategy of social insect workers that promotes social cohesion. *Nature* 365: 639–641.

R

**Ratnieks, F.L.W. & Keller, L.** 1998. Queen control of egg fertilization in the honey bee. *Behavioral Ecology and Sociobiology* 44: 57–61.

**Ratnieks, F.L.W., Foster, K.R. & Wenseleers, T.** 2006. Conflict resolution in insect societies. *Annual Review of Entomology* 51: 581–608.

**Rosenheim, J.A., Nonacs, P. & Mangel, M.** 1996. Sex ratios and multifaceted parental investment. *American Naturalist* 148: 501–535.

**Rosivall, B., Torok, J., Hasselquist, D. & Bensch, S.** 2004. Brood sex ratio adjustment in collared flycatchers (*Ficedula albicollis*): results differ between populations. *Behavioral Ecology and Sociobiology* 56: 346–351.

**Rosset, H. & Chapuisat, M.** 2006. Sex allocation conflict in ants: when the queen rules. *Current Biology* 16: 328–331.

S

**Schwarz, M.P.** 1988. Local resource enhancement and sex ratios in a primitively social bee. *Nature* 331: 346–348.

**Shaw, R.F. & Mohler, J.D.** 1953. The selective advantage of the sex ratio. *American Naturalist* 87: 337–342.

**Sheldon, B.C. & West, S.A.** 2004. Maternal dominance, maternal condition, and offspring sex ratio in ungulate mammals. *American Naturalist* 163: 40–54.

**Sheldon, B.C., Andersson, S., Griffith, S.C., Ornborg, J. & Sendecka, J.** 1999. Ultraviolet colour variation influences blue tit sex ratios. *Nature* 402: 874–877.

**Shuker, D.M. & West, S.A.** (2004) Information constraints and the precision of adaptation: Sex ratio manipulation in wasps. *Proceedings of the National Academy of Science of the USA* 101: 10363–10367.

**Silk, J.B., Willoughby, E. & Brown, G.R.** 2005. Maternal rank and local resource competition do not predict birth sex ratios in wild baboons. *Proceedings of the Royal Society of London Series B* 272: 859–864.

**Starks, P.T. & Poe, E.S.** 1997. 'Male-stuffing' in wasp societies. *Nature* 389: 450.

**Sundström, L.** 1994. Sex ratio bias, relatedness asymmetry and queen mating frequency in ants. *Nature* 367: 266–268.

**Sundström, L., Chapuisat, M. & Keller, L.** 1996. Conditional manipulation of sex ratios by ant workers: a test of kin selection theory. *Science* 274: 993–995.

**Svensson, E. & Nilsson, J.A.** 1996. Mate quality affects offspring sex ratio in blue tits. *Proceedings of the Royal Society of London Series B* 263: 357–361.

T

**Tella, J.L.** 2001. Sex-ratio theory in conservation biology. *Trends in Ecology and Evolution* 16: 76–77.

**Trivers, R.L.** 1985. *Social Evolution*. Benjamin/Cummings Publishing Company, Menlo Park, California.

**Trivers, R.L. & Hare, H.** 1976. Haplodiploidy and the evolution of the social insects. *Science* 191: 249–263.

**Trivers, R.L. & Willard, D.E.** 1973. Natural selection of parental ability to vary the sex ratio of offspring. *Science* 179: 90–92.

W

**Walin, L. & Seppä, P.** 2001. Resource allocation in the red ant *Myrmica ruginodis* – an interplay of genetics and ecology. *Journal of Evolutionary Biology* 14: 694–707.

**Werren, J.H.** 1997. Biology of *Wolbachia*. *Annual Review of Entomology* 42: 587–609.

**Werren, J.H. & Beukeboom, L.W.** 1998. Sex determination, sex ratios, and genetic conflict. *Annual Review of Ecology and Systematics* 29: 233–261.

**West, S.A. & Sheldon, B.C.** 2002. Constraints in the evolution of sex ratio adjustment. *Science* 295: 1685–1688.

**West, S.A., Shuker, D.M. & Sheldon, B.C.** 2005. Sex-ratio adjustment when relatives interact: a test of constraints on adaptation. *Evolution* 59: 1211–1228.

## Chapter 14

A

**Alatalo, R.V., Hoglund, J., Lundburg, A. & Sutherland, W.J.** 1992. Evolution of black grouse leks – female preferences benefit males in larger leks. *Behavioral Ecology* 3: 53–59.

**Alexander, R.D.** 1974. The evolution of social behavior. *Annual Reviews of Ecology and Systematics* 5: 325–383.

B

**Baldi, R., Campagna, C., Pedraza, S. & Le Boeuf, B.J.** 1996. Social effects of space availability on the breeding behaviour of elephant seals in Patagonia. *Animal Behaviour* 51: 717–724.

**Barta, Z. & Giraldeau, L.-A.** 2001. Breeding colonies as information centres: a re-appraisal of information-based hypotheses using the producer–scrounger game. *Behavioral Ecology* 12: 121–127.

**Beehler, B.M. & Foster, M.S.** 1988. Hotshots, hotspots and female preference in the organization of lek mating systems. *American Naturalist* 131: 203–219.

**Bertram, B.C.R.** 1978. Living in groups: predators and prey. In *Behavioural Ecology: An Evolutionary Approach* (Krebs, J.R. & Davies, N.B. eds). Blackwell, Oxford.

**Bonenfant, M. & Kramer, D.L.** 1996. The influence of distance to burrow on flight initiation distance in the woodchuck, *Marmota monax*. *Behavioural Ecology* 7: 299–303.

**Boulinier, T. & Danchin, E.** 1997. The use of conspecific reproductive success for breeding patch selection in territorial migratory species. *Evolutionary Ecology* 11: 505–517.

**Bradbury, J.W.** 1981. The evolution of leks. In *Natural Selection and Social Behavior: Research and Theory* (Alexander, R.D. & Twinkle, D.W. eds), pp. 138–169. Chiron, New York.

**Brown, C.R. & Bomberger Brown, M.B.** 1986. Ectoparasitism as a cost of coloniality in cliff swallows (*Hirundo pyrrhonota*). *Ecology* 67: 1206–1218.

**Brown, C.R. & Bomberger Brown, M.** 1987. Group-living in cliff swallows as an advantage in avoiding predators. *Behavioral Ecology and Sociobiology* 21: 97–107.

**Brown, C.R. & Bomberger Brown, M.** 1996. *Coloniality in the Cliff Swallow. The Effect of Group Size on Social Behavior*. University of Chicago Press, Chicago.

**Brown, C.R., Bomberger Brown, M. & Danchin, É.** 2000. Breeding habitat selection in cliff swallows: the effect of conspecific reproductive success on colony choice. *Journal of Animal Ecology* 69: 133–142.

**Brown, C.R., Bomberger Brown, M. & Ives, A.R.** 1992. Nest placement relative to food and its influence on the evolution of avian coloniality. *American Naturalist* 139: 205–217.

**Brown, C.R., Stutchbury, B.J. & Walsh, P.D.** 1990. Choice of Colony Size in Birds. *Trends in Ecology and Evolution* 5: 398–403.

C

**Cairns, D.K.** 1992. Population regulation of seabird colonies. In *Current Ornithology*, volume 9 (Power, D.M. ed.), pp. 37–61. Plenum Press, New York.

**Caraco, T. & Pulliam, R.H.** 1984. Sociality and survivorship in animals exposed to predation. In *A New Ecology: Novel Approaches to Interactive Systems* (Price, P.W., Sloboschikoff, C.N. & Gaud, W.S. eds), pp. 279–309. Wiley Interscience, New York.

**Clark, C.W. & Mangel, M.** 1984. Foraging and flocking strategies: Information in an uncertain environment. *American Naturalist* 123: 626–641.

**Collias, N.E. & Colias, E.C.** 1969. Size of breeding colony related to attraction of mates in a tropical passerine bird. *Ecology* 50: 481–488.

**Coulson, J.C.** 1986. A new hypothesis for the adaptive significance of colonial breeding in the Kittiwake *Rissa tridactyla* and other seabirds. In *Proceedings of the XVIII International Ornithological Congress, Moscow*, pp. 892–899.

D

**Danchin, E. & Richner, H.** 2001. Viable and unviable hypotheses for the evolution of raven roosts. *Animal Behaviour* 61: F7–F11.

**Danchin, E. & Wagner, R.H.** 1997. The evolution of coloniality: the emergence of new perspectives. *Trends in Ecology and Evolution* 12: 342–347.

**Danchin, E., Wagner, R. & Boulinier, T.** 1998a. The evolution of coloniality: does commodity selection explain it all? Reply. *Trends in Ecology and Evolution* 13: 76.

**Danchin, E., Boulinier, T. & Massot, M.** 1998b. Conspecific reproductive success and breeding habitat selection: implications for the study of coloniality. *Ecology* 79: 2415–2428.

**Darling, F.F.** 1938. *Bird Flocks and the Breeding Cycle. A Contribution to the Study of Avian Sociality.* Cambridge University Press, Cambridge.

**Doligez, B., Cadet, C., Danchin, E. & Boulinier, T.** 2003. When to use public information for breeding habitat selection? The role of environmental predictability and density dependence. *Animal Behaviour* 66: 973–988.

**Draulans, D.** 1988. The importance of heronries for mate attraction. *Ardea* 76: 187–192.

**Dybas, H.S. & Lloyd, M.** 1974. The habitats of 17 year periodical cicadas (Homoptera: Cicadidae: *Magicicada* spp.). *Ecological Monographs* 44: 279–324.

E

**Elgar, M.A.** 1986. House sparrows establish foraging flocks by giving chirrup calls if the resources are divisible. *Animal Behaviour* 34: 169–174.

**Emlen, S.T. & Oring, L.W.** 1977. Ecology, sexual selection and the evolution of animal mating systems. *Science* 197: 215–223.

**Endler, J.A.** 1995. Multiple-trait coevolution and environmental gradients in guppies. *Trends in Ecology and Evolution* 10: 22–29.

F

**Ferrière, R., Cazelles, B., Cézilly, F. & Desportes, J.P.** 1996. Predictability and chaos in bird vigilant behaviour. *Animal Behaviour* 52: 457–472.

**Fitzgibbon, C.D.** 1989. A cost to individuals with reduced vigilance in groups of Thompson's gazelles hunted by cheetahs. *Animal Behaviour* 37: 508–510.

**Foster, W.A. & Treherne, J.E.** 1981. Evidence for the dilution effect in the selfish herd from fish predation on a marine insect. *Nature* 295: 466–467.

G

**Giraldeau, L.-A. & T. Caraco.** 2000. *Social Foraging Theory*. Princeton, Princeton University Press.

**Giraldeau, L.A., Valone, T.J. & Templeton, J.J.** 2002. Potential disadvantages of using socially acquired information. *Philosophical Transactions of the Royal Society of London Series B* 357: 1559–1566.

**Gochfeld, M.** 1980. Mechanisms and adaptative value of reproductive synchrony in colonial seabirds. In *Behavior of Marine Animals. Current Perspectives in Research* (Burger, J., Olla, B.L. & Winn H.E. eds), vol. 4, *Marine Birds*, pp. 207–270. Plenum Press, New York and London.

**Gowaty, P.A. & N. Buschhaus.** 1998. Ultimate causation of aggressive and forced copulation in birds: female resistence, the CODE hypothesis, and social monogamy. *American Zoologist* 38: 207–225.

H

**Hamilton W.D.** 1971. Geometry for the selfish herd. *Journal of Theoretical Biology* 31: 295–311.

**Heinrich, B.** (1990) *Ravens in Winter*. Barrie and Jenkins, London.

**Heinrich, B., Marzluff, J.M. & Marzluff, C.** 1993. Common ravens are attracted by appeasement calls of food discoverers when attracted. *Auk* 110: 247–254.

**Höglund, J. & Alatalo, R.V.** 1995. *Leks*. Princeton University Press, Princeton.

**Hoi, H.** 1997. Assessment of the quality of copulation partners in the monogamous bearded tit. *Animal Behaviour* 53: 277–286.

**Hoi, H. & Hoi-Leitner, M.** 1997. An alternative route to coloniality in the bearded tit: females pursue extra-pair fertilizations. *Behavioral Ecology* 8: 113–119.

**Hoogland, J.L. & Sherman, P.W.** 1976. Advantages and disadvantages of bank swallow (*Riparia riparia*) coloniality. *Ecological Monographs* 46: 33–58.

**Horn, H.S.** 1968. The adaptative significance of colonial nesting in the Brewer's blackbird (*Euphages cyanocephalus*). *Ecology* 49: 682–694.

**Horner, J.R.** 1982. Evidence of colonial nesting and 'site fidelity' among ornithischian dinosaurs. *Nature* 297: 675–676.

K

**Kenward, R.E.** 1978. Hawks and doves: factors affecting success and selection in goshawk attacks on wood-pigeons. *Journal of Animal Ecology* 47: 449–460.

**Kramer, D.L.** 1985. Are colonies supraoptimal groups? *Animal Behaviour* 33: 1031–1032.

**Krieber, M. & Barrette, C.** 1984. Aggregation behaviour of harbour seals at Forillon National Park, Canada. *Journal of Animal Ecology* 53: 913–928.

L

**Lachmann, M., Sella, G. & Jablonka, E.** 2000. On advantages of information sharing. *Proceedings of the Royal Society of London Series B* 267: 1287–1293.

**Lima, S.L.** 1995. Back to the basics of anti-predatory vigilance: the group size effect. *Animal Behaviour* 49: 11–20.

M

**Marzluff, J.M. & Heinrich, B.** 1991. Foraging by common ravens in the presence and absence of territory holders, an experimental analysis of social foraging. *Animal Behaviour* 42: 755–770.

**Marzluff, J.M. & Heinrich, B.** 2001. Raven roosts are still information centers. *Animal Behaviour* 61: F14–F15.

**Marzluff, J.M., Heinrich, B. & Marzluff, C.S.** 1996. Raven roosts are mobile information centres. *Animal Behaviour* 51: 89–103.

**McKinney, F. & Evarts, S.** 1997. Sexual coercion in waterfowl and other birds. In *Avian Reproductive Tactics: Female and Male Perspectives* (Parker, P.G. & Burley, N.T. eds), pp. 163–195. Ornithological monograph No. 49, American Ornithologists' Union, Washington, DC.

**McKinney, F., Cheng, K.M. & Bruggers, D.J.** 1984. Sperm competition in apparently monogamous birds. In *Sperm Competition and the Evolution of Animal Mating Systems* (Smith, R.L. ed.), pp. 523–545. Academic Press, Orlando.

**Meadows, P.S. & Campbell, J.I.** 1972. Habitat selection by aquatic invertebrates. In *Advances in Marine Biology*, (Russell, F.S. & Yonge, M. eds), vol. 10, pp. 271–382. Academic Press, London and New York.

**Mock, D.W.** 2001. Comments on Danchin and Richner's 'Viable and unviable hypotheses for the evolution of raven roosts'. *Animal Behaviour* 61: F12–F13.

**Mock, D.W., Lamey, T.C. & Thompson D.B.A.** 1988. Falsifiability and the information centre hypothesis. *Ornis Scandinavica* 19: 231–248.

**Møller, A.P.** 1987. Advantages and disadvantages of coloniality in the swallow, *Hirundo rustica*. *Animal Behaviour* 35: 819–832.

**Møller, A.P. & Birkhead, T.R.** 1993. Cuckoldry and sociality: a comparative study of birds. *American Naturalist* 142: 118–140.

**Moratalla, J.J. & Powell, J.E.** 1994. Dinosaur nesting patterns. In *Dinosaur Eggs and Babies* (Carpenter, K., Hirsch, K.F. & Horner, J.R. eds), pp. 37–46, Cambridge University Press.

**Morton, E.S.** 1987. Variation in mate-guarding intensity by male purple martins. *Behaviour* 101: 211–224.

**Morton, E.S. & Derrickson, K.** 1990. The biological significance of age-specific return schedules in breeding purple martins. *Condor* 92: 1040–1050.

**Morton, E.S., Forman, L. & Braun, M.** 1990. Extrapair fertilization and the evolution of colonial breeding in purple martins. *Auk* 107, 275–283.

N

**Neill, S.R. & Cullen, J.M.** 1974. Experiments on whether schooling by their prey affects the hunting behaviour of cephalopods and fish predators. *Journal of Zoology* 172: 549–569.

O

**Olsthoorn, J.C.M., Nelson, J.B. & Hasson, O.** 1990. The availability of breeding sites for some British seabirds. *Bird Study* 37: 145–164.

P

**Parejo, D., White, J., Clobert, J., Dreiss, A. & Danchin, É.** 2007. Blue tits use fledgling quantity and quality as public information in breeding site choice: an experimental study of habitat copying. *Ecology* 88: 2373–2382.

**Post, W.** 1994. Are female boat-tailed grackle colonies neutral assemblages? *Behavioral Ecology and Sociobiology* 35: 401–407.

**Pruett-Jones, S.G.** 1988. Lekking versus solitary display: temporal variations in dispersion in the buff-beasted sandpiper. *Animal Behaviour* 36: 1740–1752.

**Pulliam, H.R., Pyke, G.H. & Caraco, T.** 1982. The scanning behavior of the juncos: a game-theoretical approach. *Journal of Theoretical Biology* 95: 89–103.

R

**Ramsay, S.M., Otter, K. & Ratcliffe, L.M.** 1999. Nest-site selection by female black-capped chickadees: settlement based on conspecific attraction? *Auk* 116: 604–617.

**Richner, H. & Danchin, E.** 2001. On the importance of slight nuances in evolutionary scenario. *Animal Behaviour* 61: F17–F18.

**Richner, H. & Heeb, P.** 1995. Is the information center hypothesis a flop? *Advances in the Study of Behavior* 24: 1–45.

**Richner, H. & Heeb, P.** 1996. Communal life: honest signaling and the recruitment center hypothesis. *Behavioral Ecology* 7: 115–118.

**Rodgers, J.A.** 1987. On the antipredator advantages of coloniality: a word of caution. *Wilson Bulletin* 99: 269–271.

**Rolland, C., Danchin, E & de Fraipont, M.** 1998. The evolution of coloniality in birds in relation to food, habitat, predation, and life-history traits: a comparative analysis. *American Naturalist* 151: 514–529.

S

**Seghers, B.H.** 1974. Schooling behaviour in the guppy *Poecilia reticulata*: an evolutionary response to predators. *Evolution* 28: 486–489.

**Shapiro, A. & Dworkin, M.** (eds) 1997. *Bacteria as Multicellular Organisms.* Oxford University Press, London.

**Shields, W.M., Crook, J.R., Hebblethwaite, M.L. & Wiles-Ehmann, S.S.** 1988. Ideal free coloniality in the swallows. In *The Ecology of Social Behavior* (Slobodchikoff, C.N. ed.), pp. 189–228.

**Sibly, R.M.** 1983. Optimal group size is unstable. *Animal Behaviour* 31: 947–948.

**Simon, C.** 1979. Debut of the seventeen-year-old cicada. *Natural History* 88: 38–45.

**Smith, S.M.** 1988. Extra-pair copulations in blacked capped chickadees: the role of the female. *Behaviour* 107: 15–23.

**Stoddart, P.K.** 1988. The 'Bugs' call of the cliff swallow: a rare food signal in a colonially nesting bird species. *Condor* 90: 714–715.

**Stutchbury, B.J.** 1988. Evidence that bank swallow colonies do not function as information centers. *Condor* 90: 953–955.

**Stutchbury, B.J.** 1991. The adaptive significance of male subadult plumage in purple martins: plumage dyeing experiments. *Behavioural Ecology and Sociobiology* 29: 297–306.

T

**Tarof, S.A., Ratcliffe, L.M., Kasumovic, M.M. & Boag, P.T.** 2005. Are least flycatcher (*Epidonax* minimus) clusters hidden leks? *Behavioral Ecology* 16: 207–217.

**Terhune, J.M. & Brillant, S.W.** 1996. Harbour seal vigilance decreases over time since haul out. *Animal Behaviour* 51: 757–763.

**Theraulaz, G., Bonabeau, E. & Deneubourg, J.L.** 1998. Response threshold reinforcement and division of labour in insect societies. *Proceedings of the Royal Society of London Series B* 265: 327–332.

V

**Varela, S.A.M., Danchin, É. & Wagner, R.H.** 2007. Does predation select for or against avian coloniality? A comparative analysis. *Journal of Evolutionary Biology.* 20: 1490–1503.

**Veen, J.** 1977. Functional and causal aspects of nest distribution in colonies of the sandwich tern (*Sterna sandvicensis* Lath.). *Behaviour Supplement* 20: 1–193.

W

**Wagner, R.H.** 1991. The use of extra-pair copulations for mate appraisal by razorbills. *Behavioral Ecology* 2: 198–203.

**Wagner, R.H.** 1992a. Behavioural and habitat-related aspects of sperm competiton in razorbills. *Behaviour* 123: 1–26.

**Wagner, R.H.** 1992b. Extra-pair copulations in a lek: the secondary mating system of monogamous razorbills. *Behavioral Ecology and Sociobiology* 31: 63–71.

**Wagner, R.H.** 1993. The pursuit of extra-pair copulations by female birds: a new hypothesis of colony formation. *Journal of Theoretical Biology* 163: 333–346.

**Wagner, R.H.** 1998. Hidden leks: sexual selection and the clumping of avian territories. In *Extra-Pair Mating Tactics in Birds.* (Parker, P.G. & Burley, N. eds), pp. 123–145. Ornithological Monographs, American Ornithologists' Union, Washington, D.C.

**Wagner, R.H.** 1999. Sexual selection and colony formation. In *Proceedings of the 22nd International Ornithological Congress* (Adams, N. & Slotow, R. eds), pp. 1304–1313. University of Natal Press, Durban.

**Wagner, R.H. & Danchin, E.** 2003. Conspecific copying: a general mechanism of social aggregation. *Animal Behaviour* 65: 405–408.

**Wagner, R.H., Schug, M.D. & Morton, E.S.** 1996. Condition-dependent control of paternity by female purple martins: implications for coloniality. *Behavioral Ecology and Sociobiology* 38: 379–389.

**Wagner, R.H., Danchin, E., Boulinier, T. & Helfenstein, F.** 2000. Colonies as byproducts of commodity selection. *Behavioral Ecology* 11: 572–573.

**Ward, P. & Zahavi, A.** 1973. The importance of certain assemblages of birds as 'information centres' for food finding. *Ibis* 115: 517–534.

**Westneat, D.F. & Sherman, P.W.** 1997. Density and extra-pair fertilization in birds: a comparative analysis. *Behavioral Ecology and Sociobiology* 41: 205–215.

**Westneat, D.F., Sherman, P.W. & Morton, M.L.** 1990. The ecology and evolution of extra-pair copulations in birds. In *Current Ornithology*, vol. 7 (Power, D.M. ed.), pp. 331–369. Plenum, New York.

**Wittenberger, J.F. & Hunt, G.L.** 1985. The adaptive significance of coloniality in birds. In *Avian Biology*, vol. VIII (Farner, D.S., King, J.R. & Parkes, K.C. eds), pp. 1–78. Academic Press, New York.

**Wright, J., Stone, R.E. & Brown, N.** 2003. Communal roosts as structured information centres in the raven, *Corvus corax. Journal of Animal Ecology* 72: 1003–1014.

Y

**Ydenberg, R.C. & Dill, L.M.** 1986. The economics of fleeing from predators. *Advances in the Study of Behavior* 16: 229–249.

Z

**Zahavi, A.** 1986. Some further comments on the gathering of birds. In *Proceedings of the XVIII International Ornithological Congress, Moscow*, pp. 919–920.

## Chapter 15

A

**Abbot, P., Withgott, J.H. & Moran, N.A.** 2001. Genetic conflict and conditional altruism in social aphid colonies. *Proceedings of the National Academy of Sciences of the USA* 98: 12068–12071.

**Agrawal, A.F.** 2001. Kin recognition and the evolution of altruism. *Proceedings of the Royal Society of London Series B* 268: 1099–1104.

**Alexander, R.D.** 1986. Ostracism and indirect reciprocity: the reproductive significance of humor. *Ethology and Sociobiology* 7: 253–270.

**Arnold, K.E. & Owens, I.P.F.** 1998. Cooperative breeding in birds: a comparative test of the life-history hypothesis.

*Proceedings of the Royal Society of London Series B* 265: 739–745.

**Arnold, K.E. & Owens, I.P.F.** 1999. Cooperative breeding in birds: the role of ecology. *Behavioral Ecology* 10: 465–471.

**Axelrod, R. & Hamilton, W.D.** 1981. The evolution of cooperation. *Science* 211: 1390–1396.

**B**

**Barron, A.B., Oldroyd, B.P. & Ratnieks, F.L.W.** 2001. Worker reproduction in honey-bees (*Apis*) and the anarchic syndrome: a review. *Behavioral Ecology and Sociobiology* 50: 199–208.

**Bell, G.** 1985. The origin and early evolution of germ cells as illustrated by the Volvocales. In *The Origin and Evolution of Sex* (Halvorson, H. & Mornoy, A. eds), pp. 221–256. Allan R. Liss.

**Bennett, N.C. & Faulkes, C.G.** 2000. *African Mole-Rats: Ecology and Eusociality*. Cambridge University Press.

**Benton, T.G. & Foster, W.A.** 1992. Altruistic housekeeping in a social aphid. *Proceedings of the Royal Society of London Series B* 247: 199–202.

**Bernasconi, G. & Strassmann, J.E.** 1999. Cooperation among unrelated individuals: the ant foundress case. *Trends in Ecology and Evolution* 14: 477–482.

**Blaustein, A.R., Bekoff, M., Byers, J.A. & Daniels, T.J.** 1991. Kin recognition in vertebrates: what do we really know about adaptive value? *Animal Behaviour* 41: 1079–1083.

**Boland, C.R.J., Heinsohn, R.H. & Cockburn, A.** 1997. Deception by helpers in cooperatively breeding white-winged choughs and its experimental manipulation. *Behavioral Ecology and Sociobiology* 41: 251–256.

**Brown, J.L.** 1987. *Helping and Communal Breeding in Birds: Ecology and Evolution*. Princeton University Press.

**Brown, J.S., Sanderson, M.J. & Michod, R.E.** 1982. Evolution of social behavior by reciprocation. *Journal of Theoretical Biology* 99: 319–339.

**Buss, L.W.** 1987. *The Evolution of Individuality*. Princeton University Press.

**C**

**Carlin, N.F. & Frumhoff, P.C.** 1990. Nepotism in the honey bee. *Nature* 346: 706–707.

**Carlisle, T.R. & Zahavi, A.** 1986. Helping at the nest, allofeeding and social status in immature Arabian babblers. *Behavioral Ecology and Sociobiology* 18: 339–351.

**Choe, J.C. & Crespi, B.J.** 1997. *The Evolution of Social Behavior in Insects and Arachnids*. Cambridge University Press.

**Clarke, F.M. & Faulkes, C.G.** 1999. Kin discrimination and female mate choice in the naked mole-rat *Heterocephalus glaber*. *Proceedings of the Royal Society of London Series B* 266: 1995–2002.

**Clarke, M.F.** 1984. Cooperative breeding by the Australian bell miner *Manorina melanophrys* Latham: a test of kin selection theory. *Behavioral Ecology and Sociobiology* 14: 137–146.

**Clutton-Brock, T.H.** 1998. Reproductive skew, concessions and limited control. *Trends in Ecology and Evolution* 13: 288–292.

**Clutton-Brock, T.H.** 2002. Breeding together: kin selection and mutualism in cooperative vertebrates. *Science* 296: 69–72.

**Clutton-Brock, T.H. & Parker, G.A.** 1995. Punishment in animal societies. *Nature* 373: 209–216.

**Clutton-Brock, T.H., Brotherton, P.N.M., Russell, A.F., O'Riain, M.J., Gaynor, D., Kansky, R., Griffin, A., Manser, M., Sharpe, L., McIlrath, G.M., Small, T., Moss, A. & Monfort, S.** 2001. Cooperation, control, and concession in meerkats groups. *Science* 291: 478–481.

**Clutton-Brock, T.H., Gaynor, D., Kansky, R., MacColl, A.D.C., McIlrath, G., Chadwick, P., Brotherton, P.N.M., O'Riain, J.M., Manser, M. & Skinner, J.D.** 1998. Costs of cooperative behaviour in suricates (*Suricata suricatta*). *Proceedings of the Royal Society of London Series B* 265: 185–90.

**Clutton-Brock, T.H., Gaynor, D., McIlrath, G.M., Maccoll, A.D.C., Kansky, R., Chadwick, P., Manser, M., Skinner, J.D. & Brotherton, P.N.M.** 1999. Predation, group size and mortality in a cooperative mongoose, *Suricata suricatta*. *Journal of Animal Ecology* 68: 672–683.

**Cockburn, A.** 1991. *An Introduction to Evolutionary Ecology*. Blackwell Science.

**Cockburn, A.** 1998. Evolution of helping behavior in cooperatively breeding birds. *Annual Review of Ecology and Systematics* 29: 141–177.

**Connor, R.C.** 1995. Altruism among non-relatives: alternatives to the 'Prisoner's Dilemma'. *Trends in Ecology and Evolution* 10: 84–86.

**Courchamp, F., Clutton-Brock, T.H. & Grenfell, B.** 1999. Inverse density dependence and the Allee effect. *Trends in Ecology and Evolution* 14: 405–410.

**Creel, S.R., Creel, N., Wildt, D.E. & Monfort, S.L.** 1992. Behavioural and endocrine mechanisms of reproductive suppression in Serengeti dwarf mongooses. *Animal Behaviour* 43: 231–245.

**Crespi, B.J.** 1992. Eusociality in the Australian gall thrips. *Nature* 359: 724–726.

**Crespi, B.J.** 1996. Comparative analysis of the origins and losses of eusociality: causal mosaics and historical uniqueness. In *Phylogenies and the Comparative Method in Animal Behaviour* (Martins, E. ed), pp. 253–287. Oxford University Press.

**Crespi, B.J.** 2001. The evolution of social behavior in microorganisms. *Trends in Ecology and Evolution* 16: 178–183.

**Crespi, B.J. & Choe, J.C.** 1997. Explanation and evolution of social systems. In *The Evolution Of Social Behavior in Insects and Arachnids* (Choe, J.C. & Crespi, B.J. eds), pp. 499–524. Cambridge University Press.

**Crespi, B.J. & Yanega, D.** 1995. The definition of eusociality. *Behavioral Ecology* 6: 109–115.

**Curry, R.L.** 1988. Influence of kinship on helping behavior in Galapagos mockingbird. *Behavioral Ecology and Sociobiology* 22: 141–152.

**Czaran, T. & Szathmary, E.** 2000. Coexistence of replicators in prebiotic evolution. In *The Geometry of Ecological Interactions: Simplifying Spatial Complexities* (Dieckmann, U., Law, R. & Metz, J.A.J. eds), pp. 116–134. Cambridge University Press.

D

**Danforth, B.N. & Eickwort, G.C.** 1997. The evolution of social behavior in the auglochorine sweat bees (Hymenoptera: Halictidae) based on a phylogenetic analysis of the genera. In *The Evolution of Social Behavior in Insects and Arachnids* (Choe, J.C. & Crespi, B.J. eds), pp. 270–291. Cambridge University Press.

**Darwin, C.** 1859. *The Origin of Species*. John Murray, London.

**Dawkins, R.** 1976. *The Selfish Gene*. Oxford University Press.

**Day, T. & Taylor, P.D.** 1998. Unifying genetic and game theoretic models of kin selection for continuous traits. *Journal of Theoretical Biology* 194: 391–407.

**Downs, S.G. & Ratnieks, F.L.W.** 2000. Adaptive shifts in honey bee (*Apis mellifera* L.) guarding behavior support predictions of the acceptance threshold model. *Behavioral Ecology* 11: 326–333.

**Duffy, J.E.** 1996. Eusociality in a coral-reef shrimp. *Nature* 381: 512–514.

**Duffy, J.E., Morrison, C.L. & Rios, R.** 2000. Multiple origins of eusociality among sponge dwelling shrimps (*Synalpheus*). *Evolution* 54: 503–516.

**Dugatkin, L.A.** 1997. *Cooperation among Animals*. Oxford University Press.

**Dugatkin, L.A. & Wilson, D.S.** 1991. ROVER: a strategy for exploiting cooperators in a patchy environment. *American Naturalist* 138: 687–701.

E

**Edwards, S.V. & Naeem, S.** 1993. The phylogenetic component of cooperative breeding in perching birds. *American Naturalist* 141: 754–789.

**Eickwort, G.C., Eickwort, J.M., Gordon, J. & Eickwort, M.A.** 1996. Solitary behavior in a high-altitude population of the social sweat bee *Halictus rubicundus* (Hymenoptera: Halictidae). *Behavioral Ecology and Sociobiology* 38: 227–233.

**Emlen, S.T.** 1982. The evolution of helping. I. An ecological constraints model. *American Naturalist* 119: 29–39.

**Emlen, S.T.** 1996. Reproductive sharing in different kinds of kin associations. *American Naturalist* 148: 756–763.

**Emlen, S.T.** 1997. Predicting family dynamics in social vertebrates. In *Behavioural Ecology: An Evolutionary Approach* (Krebs, J.R. & Davies, N.B. eds), pp. 228–253. Blackwell Science.

**Emlen, S.T. & Wrege, P.H.** 1988. The role of kinship in helping decisions among white-fronted bee-eaters. *Behavioral Ecology and Sociobiology* 23: 305–315.

**Emlen, S.T. & Wrege, P.H.** 1994. Gender, status and family fortunes in the white-fronted bee-eater. *Nature* 367: 129–132.

**Endler, J.A.** 1986. *Natural Selection in the Wild*. Princeton University Press.

**Ennis, H.L., Nao, D.N., Pukatzki, S.U. & Kessin, R.H.** 2000. *Dictyostelium amoebae* lacking an F-box protein form spores rather than stalk in chimeras wild type. *Proceedings of the National Academy of Sciences of the USA* 97: 3292–3297.

**Enquist, M. & Leimar, O.** 1993. The evolution of cooperation in mobile organisms. *Animal Behaviour* 45: 747–757.

**Eshel, I. & Cavalli-Sforza, L.L.** 1982. Assortment of encounters and evolution of cooperativeness. *Proceedings of the National Academy of Sciences of the USA* 79: 1331–1335.

F

**Faulkes, C.G. & Bennett, N.C.** 2001. Family values: group dynamics and social control of reproduction in African mole-rats. *Trends in Ecology and Evolution* 16: 184–190.

**Ferrière, R.** 1998. Help and you shall be helped. *Nature* 393: 517–518.

**Ferrière, R. & Michod, R.** 1995. Invading wave of cooperation in a spatial iterated prisoner's dilemma. *Proceedings of the Royal Society of London Series B* 259: 77–83.

**Ferrière, R. & Michod, R.** 1996. The evolution of cooperation in spatially heterogeneous populations. *American Naturalist* 147: 692–717.

**Ferrière, R., Bronstein, J.L., Rinaldi, S., Law, R. & Gauduchon, M.** 2002. Cheating and the evolutionary stability of mutualisms. *Proceedings of the Royal Society of London Series B* 269: 773–780.

**Frank, S.A.** 1998. *Foundations of Social Evolution*. Princeton University Press.

**Frank, S.A.** 2003. Repression of competition and the evolution of cooperation. *Evolution* 57: 693–705.

**Frumhoff, P.C. & Schneider, S.** 1987. The social consequences of honey bee polyandry: the effects of kinship on worker interactions within colonies. *Animal Behaviour* 55: 255–262.

G

**Gamboa, G.J., Reeve, H.K., Ferguson, I.D. & Wacker, T.L.** 1986. Nestmate recognition in social wasps: the origin and acquisition of recognition odours. *Animal Behaviour* 34: 685–695.

**Getz, W.M.** 1991. The honey bee as a model of kin recognition systems. In *Kin Recognition* (Hepper, P.G. ed.), pp. 358–412. Cambridge University Press.

**Grafen, A.** 1990. Do animals really recognize kin? *Animal Behaviour* 39: 42–54.

**Griffin, A.S. & West, S.A.** 2002. Kin selection: fact and fiction. *Trends in Ecology and Evolution* 17: 15–21.

**Grosberg, R.K. & Hart, M.W.** 2000. Mate selection and the evolution of highly polymorphic self/non self recognition genes. *Science* 289: 2111–2114.

**Grosberg, R.K. & Quinn, J.F.** 1986. The genetic control and consequences of kin recognition by the larvae of a colonial marine invertebrate. *Nature* 322: 457–459.

H

**Hamilton, W.D.** 1964a. The genetical evolution of social behaviour I. *Journal of Theoretical Biology* 7: 1–16.

**Hamilton, W.D.** 1964b. The genetical evolution of social behaviour II. *Journal of Theoretical Biology* 7: 17–52.

**Hamilton, W.D.** 1972. Altruism and related phenomena, mainly in social insects. *Annual Review of Ecology and Systematics* 3: 193–232.

**Hamilton, W.D.** 1995. *Narrow Roads of Gene Lands*. Freeman.

**Hamilton, W.D. & May, R.M.** 1977. Dispersal in stable habitats. *Nature* 269: 578–581.

**Hardin, G.** 1968. The tragedy of the commons. *Science* 162: 1243.

**Hart, B.L. & Hart, L.A.** 1992. Reciprocal allogrooming in impala, *Aepyceros melampus*. *Animal Behaviour* 44: 1073–1083.

**Hatchwell, B.J., Ross, D.J., Fowlie, M.K. & McGowan, A.** 2001. Kin discrimination in cooperatively breeding long-tailed tits. *Proceedings of the Royal Society of London Series B* 268: 885–890.

**Hauert, C., De Monte, S., Hofbauer, J. & Sigmund, K.** 2002. Volunteering as Red Queen mechanism for cooperation in public goods games. *Science* 296: 1129–1132.

**Heinsohn, R. & Cockburn, A.** 1994. Helping is costly to young birds in cooperatively breeding white-winged choughs. *Proceedings of the Royal Society of London Series B* 256: 293–298.

**Heinsohn, R.G.** 1991. Slow learning of foraging skills and extended parental care in cooperatively breeding white-winged choughs. *American Naturalist* 137: 864–881.

**Heinsohn, R.G. & Legge, S.** 1999. The cost of helping. *Trends in Ecology and Evolution* 14: 53–57.

**Heinze, J., Holldobler, B. & Peeters, C.** 1994. Conflict and cooperation in ant societies. *Naturewissenschaften* 81: 489–497.

**Hepper, P.G.** 1991. *Kin Recognition*. Cambridge University Press.

**Holmes, W.G.** 1986. Kin recognition by phenotype matching in female Belding's ground squirrels. *Animal Behaviour* 34: 38–47.

**Houston, A.I.** 1993. Mobility limits cooperation. *Trends in Ecology and Evolution* 8: 194–196.

**Hutson, V.C.L. & Vickers, G.T.** 1995. The spatial struggle of Tit-for-Tat and Defect. *Philosophical Transactions of the Royal Society of London Series B* 348: 393–404.

J

**Jaisson, P.** 1991. Kinship and fellowship in ants and social wasps. In *Kin Recognition* (Hepper, P.G. ed.), pp. 60–93. Cambridge University Press.

**Jarvis, J.U.M.** 1981. Eusociality in mammal cooperative breeding in naked mole rat *Heterocephalus glaber* colonies. *Science* 212: 571–573.

**Jarvis, J.U.M., Bennett, N.C. & Spinks, A.C.** 1998. Food availability and foraging by wild colonies of Damaraland mole-rats (*Cryptomis damarensis*): implications for sociality. *Oecologia* 113: 290–298.

**Jarvis, J.U.M., O'Riain, M.J., Bennett, N.C. & Sherman, P.W.** 1994. Mammalian eusociality: a family affair. *Trends in Ecology and Evolution* 9: 47–51.

**Johnstone, R.A. & Cant, M.A.** 1999. Reproductive skew and the threat of eviction: a new perspective. *Proceedings of the Royal Society of London Series B* 266: 275–279.

K

**Keane, B., Creel, S.R. & Waser, P.M.** 1990. No evidence of inbreeding avoidance or inbreeding depression in a social carnivore. *Behavioral Ecology* 7: 480–489.

**Keane, B., Waser, P.M., Creel, S.R., Creel, N., Elliot, L.F. & Minchella, D.J.** 1994. Subordinate reproduction in dwarf mongooses. *Animal Behaviour* 47: 65–75.

**Kéfi, S., Bonnet, O. & Danchin, É.** 2007. Accumulated gains in a prisoner's dilemma: which game is performed by the players? *Animal Behaviour*, in press.

**Keller, L.** 1995. Social life: the paradox of multiple-queen colonies. *Trends in Ecology and Evolution* 10: 355–360.

**Keller, L.** 1997. Indiscriminate altruism: unduly nice parents and siblings. *Trends in Ecology and Evolution* 12: 99–103.

**Keller, L. & Chapuisat, M.** 1999. Cooperation among selfish individuals in insect societies. *Bioscience* 49: 899–909.

**Keller, L. & Nonacs, P.** 1993. The role of queen pheromones in social insects: queen control or queen signal? *Animal Behaviour* 45: 787–794.

**Keller, L. & Reeve, H.K.** 1994. Partitioning of reproduction in animal societies. *Trends in Ecology and Evolution* 9: 98–102.

**Keller, L. & Ross, K.G.** 1998. Selfish genes: a green beard in the red fire ant. *Nature* 394: 573–575.

**Kitchen, D.M. & Packer, C.** 1999. Complexity in vertebrate societies. In *Levels of Selection in Evolution* (Keller, L. ed.), pp. 176–196. Princeton University Press.

**Kokko, H., Johnstone, R.A. & Clutton-Brock, T.H.** 2001. The evolution of cooperative breeding through group augmentation. *Proceedings of the Royal Society of London Series B* 268: 187–196.

**Komdeur, J.** 1992. Importance of habitat saturation and territory quality for evolution of cooperative breeding in the Seychelles warbler. *Nature* 358: 493–495.

**Komdeur, J.** 1996. Influence of helping and breeding experience on reproductive performance in the Seychelles warbler: a translocation experiment. *Behavioral Ecology* 7: 326–333.

**Komdeur, J. & Hatchwell, B.J.** 1999. Kin recognition: function and mechanism in avian societies. *Trends in Ecology and Evolution* 14: 237–241.

L

**Lacey, E.A. & Sherman, P.W.** 1991. Social organization of naked mole-rat colonies: evidence for division of labor. In *The Biology of the Naked Mole-Rat*

(Sherman, P.W., Jarvis, J.U.M. & Alexander, R.D. eds), pp. 267–301. Princeton University Press.

**Lacy, R.C. & Sherman, P.W.** 1983. Kin recognition by phenotype matching. *American Naturalist* 121: 489–512.

**Le Galliard, J.-F., Ferrière, R. & Dieckmann, U.** 2003. The adaptive dynamics of altruism in spatially heterogeneous populations. *Evolution* 57: 1–17.

**Le Galliard, J.-F., Ferrière, R. & Dieckmann, U.** 2005. Adaptive evolution of social traits: origin, history, and correlation patterns of altruism and mobility. *American Naturalist* 165: 206–224.

**Ligon, J.D. & Ligon, S.H.** 1978. Communal breeding in green woodhoopoes as a case for reciprocity. *Nature* 276: 496–498.

**Lindgren, K. & Nordahl, M.G.** 1994. Evolutionary dynamics of spatial games. *Physica D* 75: 292–309.

M

**Margulis, L. & Dorion, S.** 2002. *Acquiring Genomes: A Theory of the Origins of Species*. Perseus Book Group.

**Martin, S.J., Beekman, M., Wossler, T.C. & Ratnieks, F.L.W.** 2002. Parasitic Cape honeybee workers, *Apis mellifera capensis*, evade policing. *Nature* 415: 163–165.

**Marzluff, J.M. & Balda, R.P.** 1990. Pinyon jays: making the best of a bad job by helping. In *Cooperative Breeding in Birds: Long-Term Studies of Ecology and Behavior* (Stacey, P.B. & Koenig, W.D. eds), pp. 199–237. Cambridge University Press.

**Maynard Smith, J. & Szathmary, E.** 1995. *The Major Transitions in Evolution*. Oxford University Press.

**Michod, R.E.** 1999. *Darwinian Dynamics: Evolutionary Transitions in Fitness and Individuality*. Princeton University Press.

**Michor, F. & Nowak, M.A.** 2002. Evolution: the good, the bad and the lonely. *Nature* 419: 677.

**Milinski, M. & Wedekind, C.** 1998. Working memory constrains human cooperation in the Prisoner's Dilemma. *Proceedings of the National Academy of Sciences of the USA* 95: 13755–13758.

**Mitteldorf, J. & Wilson, D.S.** 2000. Population viscosity and the evolution of altruism. *Journal of Theoretical Biology* 204: 481–496.

**Monnin, T. & Peeters, C.** 1999. Dominance hierarchy and reproductive conflicts among subordinates in a monogynous queenless ant. *Behavioral Ecology* 10: 323–332.

**Monnin, T., Ratnieks, F.L.W., Jones, G.R. & Beard, R.** 2002. Pretender punishment induced by chemical signalling in a queenless ant. *Nature* 419: 61–65.

**Moritz, R.F.A., Kryger, P. & Allsopp, M.H.** 1996. Competition for royalty in bees. *Nature* 384: 31.

**Mumme, R.L.** 1992. Do helpers increase reproductive success? An experimental analysis in the Florida scrub jay. *Behavioral Ecology and Sociobiology* 31: 319–328.

N

**Noe, R. & Hammerstein, P.** 1995. Biological markets. *Trends in Ecology and Evolution* 10: 336–339.

**Nowak, M.A. & May, R.M.** 1992. Evolutionary games and spatial chaos. *Nature* 359: 826–829.

**Nowak, M. & Sigmund, K.** 1993. A strategy of win–stay, lose–shift that outperforms in the Prisoner's Dilemma game. *Nature* 364: 56–58.

**Nowak, M.A. & Sigmund, K.** 1998. Evolution of indirect reciprocity by image scoring. *Nature* 393: 573–577.

**Nowak, M. & Sigmund, K.** 2005. Evolution of indirect reciprocity. *Nature* 437: 1291–1298.

O

**Olroyd, B.P., Smolenski, A.J., Cornuet, J.-M. & Crozier, R.H.** 1994. Anarchy in the beehieve. *Nature* 371: 749.

**O'Riain, M.J. & Jarvis, J.U.M.** 1997. Colony member recognition and xenophobia in the naked mole-rat. *Animal Behaviour* 53: 487–498.

**Owens, D.D. & Owens, M.J.** 1984. Helping behaviour in brown hyenas. *Nature* 296: 740–742.

P

**Page, R.E.J., Robinson, G.E. & Fondrk, M.K.** 1989. Genetic specialists, kin recognition and nepotism in honey-bee colonies. *Nature* 338: 576–579.

**Pamilo, P.** 1991. Evolution of colony characteristics in social insects. II. Number of reproductive individuals. *American Naturalist* 138: 412–433.

**Peeters, C.** 1997. Morphologically 'primitive' ants: comparative review of social characters, and the importance of queen–worker dimorphism. In *The Evolution of Social Behavior in Insects and Arachnids* (Choe, J.C. & Crespi, B.J. eds). Cambridge University Press.

**Perrin, N. & Lehmann, L.** 2001. Is sociality driven by the costs of dispersal or the benefits of philopatry? A role for kin-discrimination mechanisms. *American Naturalist* 158: 471–483.

**Petrie, M., Krupa, A. & Burke, T.** 1999. Peacok leks with relatives even in the absence of social and environmental cues. *Nature* 401: 155–157.

**Pfennig, D.W., Reeve, H.K. & Sherman, P.W.** 1993. Kin recognition and cannibalism in spadefoot toad tadpoles. *Animal Behaviour* 46: 87–94.

**Pfennig, D.W., Gamboa, G.J., Reeve, H.K., Shellman Reeve, J. & Ferguson, I.D.** 1983. The mechanism of nestmate discrimination in social wasps (*Polistes*, Hymenoptera: Vespidae). *Behavioral Ecology and Sociobiology* 13: 299–305.

Q

**Queller, D.C.** 1992. Does population viscosity promote kin selection? *Trends in Ecology and Evolution* 7: 322–324.

**Queller, D.C.** 1994. Genetic relatedness in viscous populations. *Evolutionary Ecology* 8: 70–73.

**Queller, D.C. & Strassmann, J.E.** 1998. Kin selection and social insects. *Bioscience* 48: 165–175.

R

**Rabenold, K.N.** 1990. *Campylorhynchus* wrens: the ecology of delayed dispersal and cooperation in the Venezuelan savanna. In *Cooperative Breeding in Birds: Long-Term Studies of Ecology and Behavior* (Stacey, P.B.

& Koenig, W.D. eds), pp. 159–196. Cambridge University Press.

**Rabenold, P.P., Rabenold, K.N., Piper, W.H., Haydock, J. & Zack, S.W.** 1990. Shared paternity revealed by genetic analysis in cooperatively breeding tropical wrens. *Nature* 348: 538–540.

**Ratnieks, F.L.W. & Visscher, P.K.** 1989. Worker policing in the honeybee. *Nature* 342: 796–797.

**Ratnieks, F.L.W.** 1988. Reproductive harmony via mutual policing by workers in eusocial Hymenoptera. *American Naturalist* 112: 217–236.

**Reeve, H.K. & Keller, L.** 1997. Reproductive bribing and policing as mechanisms for the suppression of within-group selfishness. *American Naturalist* 150: S42–S58.

**Reeve, H.K. & Ratnieks, F.L.W.** 1993. Queen–queen conflict in polygynous societies: mutual tolerance and reproductive skew. In *Queen Number and Sociality in Insects* (Keller, L. ed), pp. 45–85. Oxford University Press.

**Reeve, H.K., Emlen, S.T. & Keller, L.** 1998. Reproductive sharing in animal societies: reproductive incentives or incomplete control by dominant breeders? *Behavioral Ecology* 9: 267–278.

**Reeve, K. & Sherman, P.W.** 1991. Intracolonial aggression and nepotism by the breeding female naked mole-rat. In *The Biology of Naked Mole-Rat* (Sherman, P.W., Jarvis, J.U.M. & Alexander, R.D. eds), pp. 337–357. Princeton University Press.

**Reyer, H.-U.** 1984. Investment and relatedness: a cost/benefit analysis of breeding and helping in the pied kingfisher (*Ceryle rudis*). *Animal Behaviour* 32: 1163–1178.

**Riolo, R.L., Cohen, M.D. & Axelrod, R.** 2001. Evolution of cooperation without reciprocity. *Nature* 414: 441–443.

**Roberts, G.** 1998. Competitive altruism: from reciprocity to the handicap principle. *Proceedings of the Royal Society of London Series B* 265: 427–431.

**Roberts, G. & Sherratt, T.N.** 1998. Development of cooperative relationships through increasing investment. *Nature* 394: 175–179.

**Rolland, C., Danchin, E. & de Fraipont, M.** 1998. The evolution of coloniality in birds in relation to food, habitat, predation, and life-history traits: a comparative analysis. *American Naturalist* 151: 514–529.

**Röseler, P.F.** 1991. Social and reproductive dominance among ants. *Naturewissenschaften* 78: 114–120.

**Ross, K.G. & Keller, L.** 1998. Genetic control of social organization in an ant. *Proceedings of the National Academy of Sciences of the USA* 95: 14232–14237.

**Rothenbuhler, N.** 1964. Behavior genetics of nest cleaning in honey bees. IV. Responses of F1 and backcross generations to disease-killed brood. *American Zoologist* 4: 111–123.

S

**Shellman-Reeve, J.S.** 1997. The spectrum of eusociality in termites. In *The Evolution Of Social Behavior in Insects and Arachnids* (Choe, J.C. & Crespi, B.J. eds), pp. 52–93. Cambridge University Press.

**Sherley, G.H.** 1990. Cooperative breeding in rifleman (*Acanthisitta chloris*): benefits to parents, offspring and helpers. *Behaviour* 112: 1–22.

**Sherman, P.W.** 1981. Kinship, demography, and Belding's ground squirrel nepotism. *Behavioral Ecology and Sociobiology* 8: 251–259.

**Sherman, P.W., Lacey, E.A., Reeve, H.K. & Keller, L.** 1995. The eusociality continuum. *Behavioral Ecology* 6: 102–108.

**Sigmund, K. & Nowak, M.A.** 1999. Evolutionary game theory. *Current Biology* 9: R503–R505.

**Spinks, A.C., Jarvis, J.U.M. & Bennett, N.C.** 2000. Comparative patterns of philopatry and dispersal in two common mole-rat populations: implications for the evolution of mole-rat sociality. *Journal of Animal Ecology* 69: 224–234.

**Stacey, P.B. & Ligon, J.D.** 1991. The benefits-of-philopatry hypothesis for the evolution of cooperative breeding: variation in territory quality and group size effects. *American Naturalist* 137: 831–846.

**Stephens, D.W., McLinn, C.M. & Stevens, J.R.** 2002. Discounting and reciprocity in an iterated Prisoner's Dilemma. *Science* 298: 2216–2218.

**Stern, D.L. & Foster, W.A.** 1997. The evolution of sociality in aphids: a clone's-eye view. In *The Evolution of Social Behavior in Insects and Arachnids* (Choe, J.C. & Crespi, B.J. eds), pp. 150–165. Cambridge University Press.

**Strassmann, J.E., Zhu, Y. & Queller, D.C.** 2000. Altruism and social cheating in the social amoeba *Dictyostelium discoideum*. *Nature* 408: 965–967.

**Szathmary, E. & Demeter, L.** 1987. Group selection of early replicators and the origin of life. *Journal of Theoretical Biology* 128: 463–486.

T

**Taylor, P.D.** 1992a. Altruism in viscous populationsan inclusive fitness approach. *Evolutionary Ecology* 6: 352–356.

**Taylor, P.D.** 1992b. Inclusive fitness in a homogeneous environment. *Proceedings of the Royal Society of London Series B* 249: 299–302.

**Taylor, P.D. & Frank, S.A.** 1996. How to make a kin selection model? *Journal of Theoretical Biology* 180: 27–37.

**Thorne, B.L.** 1997. Evolution of eusociality in termites. *Annual Review of Ecology and Systematics* 28: 27–54.

**Turner, P.E. & Chao, L.** 1999. Prisoner's dilemma in a RNA virus. *Nature* 398: 441–443.

V

**Varela, S.A.M., Danchin, É. & Wagner, R.H.** 2007. Does predation select for or against avian coloniality? A comparative analysis. *Journal of Evolutionary Biology* 20: 1490–1503.

**Vehrencamp, S.L.** 1983. Optimal degree of skew in cooperative societies. *American Zoologist* 23: 327–335.

**Vehrencamp, S.L., Koford, R.R. & Bowen, B.S.** 1988. The effect of breeding-unit size on fitness components in groove-billed anis. In *Reproductive Success: Studies on Individual Variation in Contrasting Breeding Systems* (Clutton-Brock, T.H. ed.), pp. 291–304. University of Chicago Press.

**Velicer, G.J., Kroops, L. & Lenski, R.E.** 2000. Developmental cheating in the social bacterium *Myxococcus xanthus*. *Nature* 404: 598–600.

**Velicer, G.J., Kroos, L. & Lenski, R.E.** 1998. Loss of social behaviors in *Myxococcus xanthus* during evolution in an unstructured habitat. *Proceedings of the National Academy of Sciences of the USA* 95: 12376–12380.

W

**Waldman, B.** 1988. The ecology of kin recognition. *Annual Review of Ecology and Systematics* 19: 543–571.

**Wcislo, W.T.** 1997. Are behavioral classification blinders to sudying natural variation? In *The Evolution of Social Behavior in Insects and Arachnids* (Choe, J.C. & Crespi, B.J. eds), pp. 8–13. Cambridge University Press.

**Wcislo, W.T. & Danforth, B.N.** 1997. Secondarily solitary: the evolutionary loss of social behavior. *Trends in Ecology and Evolution* 12: 468–474.

**Wedekind, C. & Milinski, M.** 1996. Human cooperation in the simultaneous and the alternating Prisoner's Dilemma: Pavlov versus Generous Tit-for-Tat. *Proceedings of the National Academy of Sciences of the USA* 93: 2686–2689.

**West, S.A., Pen, I. & Griffin, A.S.** 2002. Cooperation and competition between relatives. *Science* 296: 72–75.

**West, S.A., Murray, M.G., Machado, C.A., Griffin, A.S. & Herre, E.A.** 2001. Testing Hamilton's rule with competition between relatives. *Nature* 409: 510–513.

**Wilson, D.S.** 1997. Altruism and organism: disentangling the themes of multilevel selection theory. *American Naturalist* 150: S122–S134.

**Wilson, D.S., Pollock, G.B. & Dugatkin, L.A.** 1992. Can altruism evolve in purely viscous populations? *Evolutionary Ecology* 6: 331–341.

**Wilson, E.O.** 1971. *The Insects Societies*. Harvard University Press.

**Wilson, E.O.** 1975. *Sociobiology: The New Synthesis*. Harvard University Press.

**Wright, J.** 1997. Helping-at-the-nest in Arabian babblers: signalling social status or sensible investment in chicks? *Animal Behaviour* 54: 1439–1448.

**Wright, J.** 1999. Altruism as a signal: Zahavi's alternative to kin selection and reciprocity. *Journal of Avian Biology* 30: 108–115.

Z

**Zahavi, A.** 1990. Arabian babblers: the quest for status in cooperative breeder. In *Cooperative Breeding in Birds: Long Term Studies of Ecology and Behavior* (Stacey, P.B. & Koenig, W.D. eds), pp. 105–130. Cambridge University Press.

**Zahavi, A.** 1995. Altruism as a handicap – the limitations of kin selection and reciprocity. *Journal of Avian Biology* 26: 1–3.

**Zahavi, A. & Zahavi, A.** 1997. *The Handicap Principle: A Missing Piece of Darwin's Puzzle*. Oxford University Press.

## Chapter 16

A

**Andersson, S., Ömborg, J. & Andersson, M.** 1998. Ultraviolet sexual dimorphism and assortative mating in blue tits. *Proceedings of the Royal Society of London Series B* 263: 445–450.

**Aubin, T. & Jouventin, P.** 1998. Cocktail-party effect in king penguin colonies. *Proceedings of the Royal Society of London Series B* 265: 1665–1673.

**Aubin, T., Jouventin, P. & Hildebrand, C.** 2000. Penguins use the two-voice system to recognize each other. *Proceedings of the Royal Society of London Series B* 267: 1081–1087.

B

**Bazzaz, F.A. & Pickett, S.T.A.** 1980. Physiological ecology of succession: a comparative review. *Annual Review of Ecology and Systematics* 11: 287–310.

**Berthier, S.** 2007. *Iridescences: The Physical Color of Insects*. Paris, Springer.

**Boncoraglio, G. & Saino, N.** 2007. Habitat structure and the evolution of bird song: a meta-analysis of the evidence for the acoustic adaptation hypothesis. *Functional Ecology* 21: 134–142.

**Bradbury, J.W. & Vehrencamp, S.L.** 1998. *Principles of Animal Communication*. Sinauer Associates, Sunderland, Massachusetts.

**Bradbury, J.W. & Vehrencamp, S.L.** 2000. Economic models of animal communication. *Animal Behaviour* 59: 259–268.

**Brémond, J.-C., Aubin, T., Nyamsi, R.M. & Robisson, P.** 1990. Le chant du manchot empereur (*Aptenodytes forsteri*): recherche des paramètres utilisables pour la reconnaissance individuelle. *Comptes Rendus de l'Académie des Sciences de Paris Serie III* 311: 31–35.

**Bullock, T.H. & Heiligenberg, W.** 1986. *Electroreception*. John Wiley, New York.

C

**Cade, W.** 1975. Acoustically orienting parasitoids: fly phonotaxis to cricket song. *Science* 190: 1312–1313.

**Cairns, D.K.** 1986. Plumage colour in pursuit-diving seabirds: why do penguins wear tuxedos? *Bird Behaviour* 6: 58–65.

**Candolin, U.** 2000. Increased signalling effort when survival prospects decrease: male–male competition ensures honesty. *Animal Behaviour* 60: 417–422.

**Caro, T.M.** 1994. Ungulate predator behaviour. preliminary and comparative data from African bovids. *Behaviour* 128: 189–228.

**Caro, T.M.** 2005. *Antipredator Defenses in Birds and Mammals*. University of Chicago Press, Chicago and London.

**Catchpole, C.K. & Slater, P.J.B.** 1995. *Bird Song: Biological Themes and Variations.* Cambridge University Press.

**Cheney, D.L. & Seyfarth, R.M.** 1985. Social and non-social knowledge in vervet monkeys. *Philosophical Transactions of the Royal Society of London Series B* 308: 187–201.

**Cheney, D.L. & Seyfarth, R.M.** 1990. *How Monkeys See the World.* University of Chicago Press, Chicago.

**Clutton-Brock, T.H. & Albon, S.D.** 1979. The roaring of red deer and the evolution of honest advertisement. *Behaviour* 69: 145–170.

**Clutton-Brock, T.H., Albon, S.D., Gibson, R.M. & Guinness, F.E.** 1979. The logical stag: adaptive aspects of fighting in red deer. *Animal Behaviour* 27: 211–225.

**Cott, H.B.** 1940. *Adaptive Coloration in Animals.* London, Methuen.

**Craig, C.L., Weber, R.S. & Bernard, G.D.** 1996. Evolution of predator–prey systems: spider foraging plasticity in response to the visual ecology of prey. *American Naturalist* 147: 205–229.

**Curio, E.** 1978. The adaptive significance of avian mobbing. *Zeitschrift für Tierpsychologie* 48: 175–183.

D

**Davies, N.B. & Halliday, T.R.** 1978. Deep croaks and fighting assessment in toads *Bufo bufo. Nature* 275: 683–685.

**Dodson, G.N.** 1997. Resource defense mating system in antlered flies, *Phytalmia* spp. (Diptera: Tephritidae). *Annals of the Entomological Society of America* 90: 80–88.

**Doutrelant, C. & McGregor, P.K.** 2000. Eavesdropping and mate choice in female fighting fish. *Behaviour* 137: 1655–1669.

**Doutrelant, C., McGregor, P.K. & Oliveira, R.F.** 2001. The effect of an audience on intrasexual communication in male siamese fighting fish, *Betta splendens. Behavioral Ecology* 12: 283–286.

E

**Elgar, M.A.** 1986. House sparrows establish foraging flocks by giving chirrup calls if the resources are divisible. *Animal Behaviour* 34: 169–174.

**Endler, J.A.** 1986. *Natural Selection in the Wild.* Princeton University Press, Princeton, NJ.

**Endler, J.A.** 1987. Predation, light intensity, and courtship behaviour in *Poecilia reticulata. Animal Behaviour* 35: 1376–1385.

**Endler, J.A.** 1990. On the measurement and classification of colour in studies of animal colour patterns. *Biological Journal of the Linnean Society* 41: 315–352.

**Endler, J.A.** 1991. Variation in the appearence of guppy color patterns to guppies and their predators under visual conditions. *Vision Research* 31: 587–608.

**Endler, J.A.** 1992. Signals, signal conditions and the direction of evolution. *American Naturalist* 139: 5125–5153.

**Endler, J.A.** 1993. The color of light in forests and its implications. *Ecological Monographs* 63: 1–27.

**Endler, J.A. & Basolo, A.L.** 1998. Sensory ecology, receiver biases and sexual selection. *Trends in Ecology and Evolution* 13: 415–420.

**Endler, J.A. & Théry, M.** 1996. Interacting effects of lek placement, display behavior, ambient light, and color patterns in three neotropical forest-dwelling birds. *American Naturalist* 148: 421–452.

**Enquist, M.** 1985. Communication during aggressive interactions with particular reference to variation in choice of behaviour. *Animal Behaviour* 33: 1152–1611.

**Evans, C.S., Evans, L. & Marler, P.** 1993. On the meaning of alarm calls: functional reference in an avian vocal system. *Animal Behaviour* 46: 23–38.

F

**Feener, D.H., Jacobs, L.F. & Schmidt, J.O.** 1996. Specialized parasitoid attracted to a pheromone of ants. *Animal Behaviour* 51: 61–66.

**Fleishman, L.J., Loew, E.R. & Leal, M.** 1993. Ultraviolet vision in lizards. *Nature* 365: 397.

**Fleishman, L.J., Bowman, M., Saunders, D., Miller, W.E., Rury, M.J. & Loew, E.R.** 1997. The visual ecology of Puerto Rican anoline lizards: habitat light and spectral sensitivity. *Journal of Comparative Physiology A* 181: 446–460.

G

**Gates, D.M.** 1965. Energy, plants, and ecology. *Ecology* 46: 1–13.

**Gautier, J.-P. & Gautier, A.** 1977. Communication in old world monkeys. In *How Animals Communicate* (Sebeok, T.A. ed.), pp. 890–964. Indiana University Press, Bloomington.

**Gomez, D. & Théry, M.** 2004. Influence of ambient light on the evolution of colour signals: comparative analysis of a Neotropical rainforest bird community. *Ecology Letters* 7: 279–284.

**Gomez, G. & Théry, M.** 2007. Simultaneous crypsis and conspicuousness in color patterns: comparative analysis of a neotropical rainforest bird community. *American Naturalist* 169: S42–S61.

**Grafen, A.** 1990a. Biological signals as handicaps. *Journal of Theoretical Biology* 144: 517–546.

**Grafen, A.** 1990b. Sexual selection unhandicapped by the Fisher process. *Animal Behaviour* 39: 42–54.

**Green, S. & Marler, P.M.** 1979. The analysis of animal communication. In *Handbook of Behavioral Neurobiology*, vol. 3, *Social Behavior and Communication* (Marler, P. & Vandebergh, J.G. ed.), pp. 73–158. Plenum Press, New York.

H

**Hailman, J.P.** 1977. *Optical Signals, Animal Communication and Light.* Indiana University Press, Bloomington, Indiana.

**Harvey, P.H. & Pagel, M.D.** 1991. *The Comparative Method in Evolutionary Biology.* Oxford University Press, Oxford.

**Hasson, O.** 1994. Cheating signals. *Journal of Theoretical Biology* 167: 223–238.

**Hill, G.E. & McGraw, K.J.** 2006. *Bird Coloration*, vol. 1, *Mechanisms and Measurements.* Harvard University Press.

**Hill, G.E. & McGraw, K.J.** 2006. *Bird Coloration*, vol. 2. *Function and Evolution*. Harvard University Press.

**Hölldobler, B. & Wilson, E.O.** 1990. *The Ants*. Belknap Press of Harvard University Press, Cambridge, Massachusetts.

**Hunt, S., Bennett, A.T.D., Cuthill, I.C. & Griffiths, R.** 1998. Blue tits are ultraviolet tits. *Proceedings of the Royal Society of London Series B* 265: 451–455.

**Hunt, S., Cuthill, I.C., Bennett, A.T.D. & Griffiths, R.** 1999. Preferences for ultraviolet partners in the blue tit. *Animal Behaviour* 58: 809–815.

I

**Ichikawa, T.** 1976. Mutual communication by substrate vibrations in the mating behavior of planthoppers (Homoptera, Delphacidae). *Applied Entomology and Zoology* 11: 8–21.

J

**Johnstone, R.A.** 1997. The evolution of animal signals. In *Behavioural Ecology: An evolutionary approach*, 4th edition (Krebs, J.R. & Davies, N.B. eds), pp. 155–178. Blackwell Scientific Publications, Oxford.

**Johnstone, R.A.** 1998. Game theory and communication. In *Game Theory And Animal Behavior* (Dugatkin, L.A. & Reeve, H.K. eds), pp. 94–117. Oxford University Press, New York.

K

**Kelber, A., Vorobyev, M. & Osorio, D.** 2003. Animal colour vision – behavioural tests and physiological concepts. *Biological Reviews* 78: 81–118.

**Klump, G.M. & Shalter, M.D.** 1984. Acoustic behavior of birds and mammals in the predator context. I. Factors affecting the structure of alarm signals. II. The functional significance and evolution of alarm signals. *Zeitschrift für Tierpsychologie* 66: 189–226.

**Kroodsma, D.E & Miller, E.H.** 1996. *Ecology and Evolution of Acoustic Communication in Birds*. Comstock Publishing Associates, Cornell University Press, Ithaca.

L

**Le Boeuf, B.J.** 1974. Male–male competition and reproductive success in elephant seals. *American Zoologist* 14: 163–176.

**Lloyd, J.E.** 1965. Aggressive mimicry in *Photuris*: firefly femmes fatales. *Science* 149: 653–654.

**Lloyd, J.E.** 1975. Aggressive mimicry in *Photuris* fireflies: signal repertoires by femmes fatales. *Science* 197: 452–453.

**Lythgoe, J.N.** 1979. *The Ecology of Vision*. Oxford, Clarendon Press.

M

**Mangel, M. & Clark, C.W.** 1988. *Dynamic Modelling in Behavioral Ecology*. Princeton, Princeton University Press.

**Marchetti, K.** 1993. Dark habitats and bright birds illustrate the role of the environment in species divergence. *Nature* 362: 149–152.

**Marler, P.** 1977. The evolution of communication. In *How Animals Communicate* (Sebeok, T.A. ed.), pp. 45–70. Indiana University Press, Bloomington.

**Marler, P., Dufty, A. & Pickert, R.** 1986a. Vocal communication in the domestic chicken: I. Does a sender communicate information about the quality of a food referent to a receiver? *Animal Behaviour* 34: 188–193.

**Marler, P., Dufty, A. & Pickert, R.** 1986b. Vocal communication in the domestic chicken: II. Is a sender sensitive to the presence and nature of a receiver? *Animal Behaviour* 34: 194–198.

**Marten, K. & Marler, P.** 1977. Sound transmission and its significance for animal vocalizations. I. Temperate habitats. *Behavioral Ecology and Sociobiology* 2: 271–290.

**Martin, R.D. & Ross, C.F.** 2005. The evolutionary and ecological context of primate vision. In *The Primate Visual System – A Comparative Approach* (Kremers, J. ed.), pp. 1–36. Wiley, Chichester, UK.

**Martins, E.P.** 1996. *Phylogenies and the Comparative Method in Animal Behavior*. Oxford University Press, New York.

**Mathevon, N., Aubin, T. & Brémond, J.-C.** 1997. Propagation of bird acoustic signals: comparative study in starling and blackbird distress calls. *Comptes Rendus Hebdomadaires des Séances de l'Académie des Sciences, Paris Serie III* 320: 869–876.

**Mathevon, N., Aubin, T. & Dabalsteen, T.** 1996. Song degradation during propagation: importance of song post for the wren *Troglodytes troglodytes*. *Ethology* 102: 397–412.

**Maynard Smith, J. & Harper, D.** 2003. *Animal Signals*. Oxford University Press.

**McGregor, P.K. & Peake, T.M.** 2000. Communication networks: social environments for receiving and signalling behaviour. *Acta Ethologica* 2: 71–81.

**Møller, A.P.** 1988. False alarm calls as a means of resource usurpation in the great tit *Parus major*. *Ethology* 79: 25–30.

**Munn, C.A.** 1986. Birds that 'cry wolf'. *Nature* 319: 143–145.

N

**Naguib, M. & Todt, D.** 1997. Effects of dyadic interactions on other conspecific receivers in nightingales. *Animal Behaviour* 54: 1535–1543.

**Nelson, B.S.** 2000. Avian dependence on sound pressure level as an auditory distance cue. *Animal Behaviour* 59: 57–67.

O

**Oliveira, R.F., McGregor, P.K. & Latruffe, C.** 1998. Know thine enemy: fighting fish gather information from observing conspecific interactions. *Proceedings of the Royal Society of London Series B* 265: 1045–1049.

**Otter, K., McGregor, P.K., Terry, A.M.R., Burford, F.R.L., Peake, T.M. & Dabalsteen, T.** 1999. Do female great tits (*Parus major*) assess males by eavesdropping? A field study using interactive song playback. *Proceedings of the Royal Society of London Series B* 266: 1305–1310.

P

**Payne, R. & Webb, D.** 1971. Orientation by means of long range acoustic signalling in baleen whales. *Annals of the New York Academy of Sciences* 188: 110–141.

**Pomiankowski, A.** 1987. Sexual selection: the handicap principle does work sometimes. *Proceedings of the Royal Society of London Series B* 231: 123–145.

R

**Rees, J.-F., de Vergifosse, B., Noiset, O., Dubuisson, M., Janssens, B. & Thompson, E.M.** 1998. The origin of marine bioluminescence: turning oxygen defence mechanisms into deep-sea communication tools. *Journal of Experimental Biology* 201: 1211–1221.

**Richards, D.G. & Wiley, R.H.** 1980. Reverberations and amplitude fluctuations in the propagation of sound in a forest: implications for animal communication. *American Naturalist* 115: 381–399.

**Rodd, F.H., Hughes, K.A., Grether, G.F & Baril, C.T.** 2002. A possible non-sexual origin of mate preference: are male guppies mimicking fruit? *Proceedings of the Royal Society of London Series B* 269: 475–481.

**Ruxton, G.D., Sherratt, T.N. & Speed, M.P.** 2004. *Avoiding Attack, The Evolutionary Ecology of Crypsis, Warning Signals and Mimicry.* Oxford University Press, Oxford.

**Ryan, M.J.** 1990. Sexual selection, sensory systems, and sensory exploitation. *Oxford Surveys in Evolutionary Biology* 5: 156–195.

**Ryan, M.J.** 1998. Sexual selection, receiver biases, and the evolution of sex differences. *Science* 281: 1999–2003.

**Ryan, M.J. & Brenowitz, E.A.** 1985. The role of body size, phylogeny, and ambient noise in the evolution of bird song. *American Naturalist* 126: 87–100.

**Ryan, M.J., Tuttle, M.D. & Rand, A.S.** 1982. Bat predation and sexual advertisement in a neotropical anuran. *American Naturalist* 119: 136–139.

S

**Shettleworth, S.J.** 1998. *Cognition, Evolution, and Behavior.* Princeton University Press, New York.

T

**Théry, M.** 1987. Influence des caractéristiques lumineuses sur la localisation des sites traditionnels, parade & baignade des manakins (Passériformes: Pipridae). *Comptes Rendus Hebdomadaires des Séances de l'Académie des Sciences, Paris Serie III* 304: 19–24.

**Théry, M.** 1990a. Display repertoire and social organization of the white-fronted and white-throated manakins. *Wilson Bulletin* 102: 123–130.

**Théry, M.** 1990b. Influence de la lumière sur le choix de l'habitat & le comportement sexuel des Pipridae (Aves: Passériformes) en Guyane française. *Revue d'Ecologie (Terre & Vie)* 45: 215–236.

**Théry, M.** 2001. Forest light and its influence on habitat selection. *Plant Ecology* 153: 251–261.

**Théry, M. & Casas, J.** 2002. Predator and prey views of spider camouflage. *Nature* 415: 133.

**Théry M. & Endler J.A.** 2001. Habitat selection, ambient light and colour patterns in some lek-displaying birds. In *Nouragues: Dynamics and Plant–Animal Interactions in a Neotropical Rainforest* (Bongers, F., Charles-Dominique, P., Forget, P.-M. & Théry, M. ed.). Kluwer, Dordrecht.

**Théry, M. & Vehrencamp, S.L.** 1995. Light patterns as cues for mate choice in the lekking white-throated manakin (*Corapipo gutturalis*). *Auk* 112: 133–145.

**Thompson, T.J., Winn, H.E. & Perkins, P.J.** 1979. *Mysticete Sounds. Behavior of Marine Animals: Current Perspectives in Research.* H.E. Winn & B.L. Olla. Plenum Press, New York.

**Tinbergen, N.** 1951. *The Study of Instinct.* Oxford University Press, Oxford.

V

**Viitala, J., Korpimaki, E., Palokangas, P. & Koivula, M.** 1995. Attraction of kestrels to vole scent marks visible in ultraviolet light. *Nature* 373: 425–427.

**von Frish, K.** 1967. *The Dance Language and Orientation of Bees.* Harvard University Press, Cambridge, Massachusetts.

W

**Whitham, T.G.** 1979. Territorial defense in a gall aphid. *Nature* 279: 324–325.

**Wiley, R.H.** 1991. Associations of song properties with habitats for territorial oscine birds of eastern North America. *American Naturalist* 138: 973–993.

**Wilkinson, G.S. & Dodson, G.N.** 1997. Function and evolution of antlers and eye stalks in flies. In *Mating Systems in Insects and Arachnids* (Choe, J.C. & Crespi, B.J. ed.), pp 310–328. Cambridge University Press, Cambridge, Massachusetts.

**Wilkinson, G.S. & Reillo, P.R.** 1994. Female choice response to artificial selection on an exaggerated male trait in a stalk-eyed fly. *Proceedings of the Royal Society of London Series B* 255: 1–6.

**Winn, H.E. & Winn, L.K.** 1978. The song of the humpback whale, *Megaptera novenglia*, in the West Indies. *Marine Biology* 47: 97–114.

Y

**Yokoyama, S. & Yokoyama, R.** 1996. Adaptive evolution of photoreceptors and visual pigments in vertebrates. *Annual Review of Ecology and Systematics* 27: 543–567.

Z

**Zahavi, A.** 1975. Mate selection a selection for a handicap. *Journal of Theoretical Biology* 53: 205–214.

**Zahavi, A.** 1977. The cost of honesty (further remarks on the handicap principle). *Journal of Theoretical Biology* 67: 603–605.

**Zahavi, A.** 1987. The theory of signalling detection and some of its implications. In *International Symposium of Biological Evolution* (Delfino, V.P. ed.), pp. 305–327. Adriatica Editrice, Bari, Italy.

## Chapter 17

A

**Adamo, S.A.** 2002. Modulating the modulators: parasites, neuromodulators and host behavioral change. *Brain, Behavior and Evolution* 60: 370–377.

**Akino, T., Knapp, J.J., Thomas, J.A. & Elmes, G.W.** 1999. Chemical mimicry and host specificity in the butterfly *Maculinea rebeli*, a social parasite of *Myrmica* ant colonies. *Proceedings Royal Society of London B* 266: 1419–1426.

**Andersson, M.** 2001. Relatedness and brood parasitism. *American Naturalist* 158: 599–614.

**Aragon, S., Møller, A.P., Soler, J.J. & Soler, M.** 1999. Molecular phylogeny of cuckoos supports a polyphyletic origin of brood parasitism. *Journal of Evolutionary Biology* 12: 495–506.

**Arnal, C.** 2000. *Ecologie comportementale de la symbiose poisson nettoyeur/poisson client: motivations et honnêteté.* Thèse de doctorat, Université de Perpignan, Perpignan.

**Arnal, C. & Côté, I.** 1998. Interactions between cleaning gobies and territorial damselfish on coral reefs. *Animal Behaviour* 55: 1429–1442.

**Arnal, C. & Morand, S.** 2000. Importance of ectoparasites and mucus in cleaning interactions in the Mediterranean cleaner wrasse, *Symphodus melanocercus. Marine Biology* 138: 777–784.

B

**Bakyono, E.** 1988. *Contribution à l'éco-éthologie des indicateurs (Indicatoridae: Aves) de la savane herbeuse de Nazinga.* Mémoire de Diplôme Supérieur d'Etude et de Recherche, Université de Bourgogne, Dijon.

**Begon, M., Harper, J.L. & Townsend, C.R.** 1996. *Ecology*, 3rd edition, Blackwell, London.

**Berdoy, M., Webster, J.P. & Macdonald, D.W.** 2000. Fatal attraction in rats infected with *Toxoplasma gondii. Proceedings of the Royal Society of London Series B* 267: 1591–1594.

**Bethel, W.M. & Holmes, J.C.** 1973. Altered evasive behavior and responses to light in amphipods harboring acanthocephalan cystacanths. *Journal of Parasitology* 59: 945–956.

**Bethel, W.M. & Holmes, J.C.** 1977. Increased vulnerability of amphipods owing to altered behavior induced by larval acanthocephalan. *Canadian Journal of Zoology* 55: 110–115.

**Bollache, L., Gambade, G. & Cézilly, F.** 2001. The effects of two acanthocephalan parasites, *Pomphorhynchus laevis* and *Polymorphus minutus*, on pairing success in male *Gammarus pulex* (Crustacea: Amphipoda). *Behavioural Ecology and Sociobiology* 49: 296–303.

**Bollache, L., Rigaud, T. & Cézilly, F.** 2002. Effects of two acanthocephalan parasites on the fecundity and pairing status of female *Gammarus pulex* (Crustacea: Amphipoda). *Journal of Invertebrate Pathology* 79(2): 102–110.

**Boucher, D.H., James, S. & Keeler, K.H.** 1982. The ecology of mutualism. *Annual Review of Ecology and Systematics* 13: 315–347.

**Bouchon, D., Rigaud, T. & Juchault, P.** 1998. Evidence for widespread *Wolbachia* infection in isopod crustaceans: molecular identification & host feminisation. *Proceedings of the Royal Society of London Series B* 265: 1081–1090.

**Bron, J.E. & Treasurer, J.W.** 1992. Sea lice (Caligidae) on wrasse (Labridae) from selected British wild and salmon-farm source. *Journal of the Marine Biological Association of the United Kingdom* 72: 645–650.

**Bronstein, J.L.** 1994. Conditional outcomes in mutualistic interactions. *Trends in Ecology and Evolution* 9: 214–217.

**Brooke, M. de L. & Davies, N.B.** 1988. Egg mimicry by cuckoos *Cuculus canorus* in relation to discrimination by hosts. *Nature* 335: 630–632.

**Brooke, M. de L., Davies, N.B. & Noble, D.G.** 1998. Rapid decline of host defences in response to reduced cuckoo parasitism: behavioural flexibility of reed warblers in a changing world. *Proceedings of the Royal Society of London Series B* 265: 1277–1282.

**Brooker, M.G. & Brooker, L.C.** 1996. Acceptance by the splendid fairy-wren of parasitism by the Horsfield's bronze-cuckoo: further evidence for evolutionary equilibrium in brood parasitism. *Behavioural Ecology* 7: 395–407.

**Brown, C.R. & Brown, M.B.** 1988. A new form of reproductive parasitism in cliff swallows. *Nature* 331: 66–68.

**Bshary, R. & Würth, M.** 2001. Cleaner fish *Labroides dimidiatus* manipulate client reef fish by providing tactile stimulation. *Proceedings of the Royal Society of London Series B* 268: 1495–1501.

C

**Cézilly, F., Grégoire, A. & Bertin, A.** 2000. Conflict between co-occurring manipulative parasites? An experimental study of the joint influence of two acanthocephalan parasites on the behaviour of *Gammarus pulex. Parasitology* 120: 625–630.

**Clark, L.** 1990. Starlings as herbalists: countering parasites and pathogens. *Parasitology Today* 6: 358–360.

**Clark, L. & Mason, J.R.** 1985. Use of nest material as insecticidial and antipathogenic agents by the European starling. *Oecologia* 67: 169–176.

**Cleveland, L.R.** 1926. Symbiosis among animals with special reference to termites and their intestinal flagellates. *Quarterly Review of Biology* 1: 51–64.

**Combes, C.** 1995. *Interactions Durables. Ecologie and Evolution du Parasitisme.* Masson, Paris.

**Combes, C.** 2001. *Parasitism. The Ecology and Evolution of Intimate Interactions.* Chicago University Press, Chicago.

**Connor, R.C.** 1995. The benefits of mutualism: a conceptual framework. *Biological Reviews of the Cambridge Philosophical Society* 70: 427–457.

**Côté, I.M.** 2000. Evolution and ecology of cleaning symbioses in the sea. *Oceanography and Marine Biology: an Annual Review* 38: 311–355.

**Côté, I.M. & Poulin, R.** 1995. Parasitism and group size in social animals: a meta-analysis. *Behavioral Ecology* 6: 159–165.

**Côté, I.M., Arnal, C. & Reynolds, J.D.** 1998. Variation in posing behaviour among fish species visiting cleaning stations. *Journal of Fish Biology* 53 (suppl. A): 256–266.

**Cruz, A. & Wiley, J.W.** 1989. The decline of an adaptation in the absence of a presumed selection pressure. *Evolution* 43: 55–62.

**Cushman, J.H. & Beattie, A.J.** 1991. Mutualisms: assessing the benefits to hosts and visitors. *Trends in Ecology and Evolution* 6: 193–195.

D

**Davies, N.B.** 2000. *Cuckoos, Cowbirds and Other Cheats*. T. & A. D. Poyser, London.

**Davies, N.B., Brooke, M. de L. & Kacelnik, A.** 1996. Recognition errors and probability of parasitism determine whether reed warblers should accept or reject mimetic cuckoo eggs. *Proceedings of the Royal Society of London Series B* 263: 925–931.

**Davies, N.B., Kilner, R.M. & Noble, D.G.** 1998. Nestling cuckoos *Cuculus canorus* exploit hosts with begging calls that mimic a brood. *Proceedings of the Royal Society of London Series B* 265: 673–678.

**Dawkins, R.** 1982. *The Extended Phenotype*. Oxford University Press, Oxford.

**Diamond, A.W. & Place, A.R.** 1988. Wax digestion in black-throated honeyguides. *Ibis* 130: 558–561.

**Dickman, C.R.** 1992. Commensal and mutualistic interactions among terrestrial vertebrates. *Trends in Ecology and Evolution* 7: 194–197.

**Dunn, A.M., Hatcher, M.J., Terry, R.S. & Tofts, C.** 1995. Evolutionary ecology of vertically transmitted parasites: transovarial transmission of a microsporidian sex-ratio distorter in *Gammarus duebeni*. *Parasitology* 111: S91–S109.

E

**Eberhard, W.G.** 2000. Spider manipulation by a wasp larva. *Nature* 406: 255–256.

**Eibl-Eibesfeldt, I.** 1955. Über symbiosen, Parasitismus und andere besondere zwischenartliche Beziehungen tropischer Meerefische. *Zeitschrift für Tierpsychologie* 12: 203–219.

F

**Feder, H.M.** 1966. Cleaning symbiosis in the marine environment. In *Symbiosis* (Henry, S.M. ed.), pp. 327–380. Academic Press, New York.

**Francini-Filho, R.B., Moura, R.L. & Sazima, I.** 2000. Cleaning by the wrasse *Thalassoma noronhanum*, with two records of predation by its grouper client *Cephalopholis fulva*. *Journal of Fish Biology* 56: 802–809.

**Freeland, W.J.** 1976. Pathogens and the evolution of primate sociality. *Biotropica* 8: 12–24.

**Freeland, W.J.** 1977. Blood-sucking flies and primate polyspecific associations. *Nature* 269: 801–802.

**Friedmann, H. & Kern, J.** 1956. The problem of cerophagy or wax-eating in the honey-guides. *Quarterly Review of Biology* 31: 19–30.

G

**Gorlick, D.L.** 1980. Ingestion of host fish surface mucus by the Hawaiian cleaning wrasse, *Labroides phtirophagus* (Labridae), and its effect on host species preference. *Copeia* 1980: 863–868.

**Gorlick, D.L., Atkins, P.D. & Losey, G.S.** 1978. Cleaning stations as water holes, garbage dumps and sites for the evolution of reciprocal altruism. *American Naturalist* 112: 341–353.

**Grim, T.** 2007. Experimental evidence for chick discrimination without recognition in a brood parasite host. *Proceedings of the Royal Society of London Series B* 274: 373–381.

**Grim, T., Kleven, O. & Mikulica, O.** 2003. Nestling discrimination without recognition: a possible defence mechanism for hosts towards cuckoo parasitism? *Proceedings of the Royal Society of London Series B (Suppl.)* 270: S73–S75.

**Grutter, A.S.** 1996. Experimental demonstration of no effect by the cleaner wrasse *Labroides dimidiatus* (Cuvier & Valenciennes) on the host fish *Pomacentrus moluccensis* (Bleeker). *Journal of Experimental Marine Biology and Ecology* 196: 285–298.

**Grutter, A.S.** 1997. Spatiotemporal variation and feeding selectivity in the diet of the cleaner fish *Labroides dimidiatus*. *Copeia* 1997: 346–355.

**Grutter, A.S.** 1999. Cleaner fish really do clean. *Nature* 398: 672–673.

**Grutter, A.S.** 2001. Parasite infection rather than tactile stimulation is the proximate cause of cleaning behaviour in fish. *Proceedings of the Royal Society of London Series B* 268: 1361–1365.

**Grutter, A.S. & Hendrikz, J.** 1999. Diurnal variation in the abundance of juvenile parasitic gnathiid isopod coral reef fish: implications for parasite-cleaner fish interactions. *Coral Reefs* 18: 187–191.

H

**Hamilton, W.D.** 1971. Geometry for the selfish herd. *Journal of Theoretical Biology* 31: 295–311.

**Hamilton, W.J. & Orians, G.H.** 1965. Evolution of brood parasitism in altricial birds. *Condor* 67: 361–382.

**Hart, B.L., Hart, L.A. & Mooring, M.S.** 1990. Differential foraging of oxpeckers on impala in comparison with sympatric antelope species. *African Journal of Ecology* 28: 240–249.

**Hechtel, L.J., Johnson, C.L. & Juliano, S.A.** 1993. Modification of antipredator behavior of *Caecidotea intermedius* by its parasite *Acanthocephalus dirus*. *Ecology* 74: 710–713.

**Helluy, S.** 1983. Un mode de favorisation de la transmission parasitaire: la manipulation du comportement de l'hôte intermédiaire. *Revue d'Ecologie (Terre and Vie)* 38: 211–223.

**Helluy, S.** 1984. Relations hôtes–parasites du trématode *Microphallus papillorobustus* (Rankin 1940). III. Facteurs impliqués dans les modifications du comportement des *Gammarus* hôtes intermédiaires & tests de prédation. *Annales de Parasitologie Humaine and Comparée* 59: 41–56.

**Helluy, S. & Holmes, J.C.** 1990. Serotonin, octopamine and the clinging behavior induced by the parasite *Polymorphus paradoxus* (Acanthocephala) in *Gammarus lacustris* (Crustacea). *Canadian Journal of Zoology* 68: 1214–1220.

**Hobson, E.S.** 1971. Cleaning symbioses among California inshore fishes. *Fishery Bulletin* 69: 491–523.

**Hoese, H.D.** 1966. Ectoparasitism by juvenile sea catfish, *Galeichtys felis*. *Copeia* 1966: 880–881.

**Hölldobler, B. & Wilson, E.O.** 1990. *The Ants*. Belknap Press, Harvard.

**Hurd, H.** 2001. Host fecundity reduction: a strategy for damage limitation? *Trends in Parasitology* 17: 363–368.

**Hurst, G.D.D. & Majerus, M.E.N.** 1993. Why do maternally inherited microorganisms kill males? *Heredity* 71: 81–95.

I

**Isack, H.A. & Reyer, H.-U.** 1988. Honeyguides and honey gatherers: interspecific communication in a symbiotic relationship. *Science* 243: 1343–1346.

J

**Jiggins, F.M., Hurst, G.D.D. & Majerus, M.E.N.** 1998. Sex ratio distortion in *Acraea encedon* (Lepidoptera: Nymphalidae) is caused by a male-killing bacterium. *Heredity* 81: 87–91.

**Jiggins, F.M., Hurst, G.D.D. & Majerus, M.E.N.** 2000. Sex-ratio-distorting *Wolbachia* causes sex-role reversal in its butterfly host. *Proceedings of the Royal Society of London Series B* 267: 69–73.

**Johnsgard, P.A.** 1997. *The Avian Brood Parasites: Deception at the Nest*. Oxford University Press, Oxford.

K

**Kavaliers, M., Colwell, D.D. & Choleris, E.** 1999. Parasites and behavior: an ethopharmacology analysis and biomedical implications. *Neuroscience and Biobehavioral Research* 23: 1037–1045.

**Keddy, P.A.** 1989. *Competition*. Chapman and Hall, London.

**Kelly, C.** 1987. A model to explore the spread of mimicry and rejection in hypothetical populations of cuckoos and their hosts. *Journal of Theoretical Biology* 125: 283–299.

**Kilner, R.M., Noble, D.G. & Davies, N.B.** 1999. Signals of need in parent–offspring communication and their exploitation by the common cuckoo. *Nature* 397: 667–672.

**Koenig, W.D.** 1997. Host preferences and behaviour of oxpeckers: co-existence of similar species in a fragmented landscape. *Evolutionary Ecology* 11: 91–104.

L

**Lafuma, L., Lambrechts, M. & Raymond, M.** 2001. Aromatic plants in bird nests as a protection against blood-sucking flying insects? *Behavioural Processes* 56: 113–120.

**Lahti, D.C.** 2005. Evolution of bird eggs in the absence of cuckoo parasitism. *Proceedings of the Naional Academy of Sciences of the USA* 102: 18057–18062.

**Lahti, D.C.** 2006. Persistence of egg recognition in the absence of cuckoo brood parasitism: pattern and mechanism. *Evolution* 60: 157–168.

**Langmore, N.E., Hunt, S. & Kilner, R.M.** 2003. Escalation of a coevolutionary arms race through host rejection of brood parasitic young. *Nature* 422: 157–160.

**Lemaire, P. & Maigret, J.** 1987. Importance relative des différents stimuli dans le comportement de nettoyage de *Labroides dimidiatus* (Cuv. & Val., 1839). *Annales de l'Institut Océanographique* 63: 69–84.

**Loehle, C.** 1995. Social barriers to pathogen transmission in wild animal populations. *Ecology* 76: 326–335.

**Losey, G.S.** 1979. Fish cleaning symbiosis: proximate causes of host behaviour. *Animal Behaviour* 27: 669–685.

**Losey, G.S.** 1987. Cleaning symbiosis. *Symbiosis* 4: 229–258.

**Losey, G.S. & Margules, L.** 1974. Cleaning symbiosis provides a positive reinforcer for fish. *Science* 1984: 179–180.

**Lotem, A.** 1993. Learning to recognize nestlings is maladaptive for cuckoo *Cuculus canorus* hosts. *Nature* 362: 743–745.

**Lotem, A. & Nakamura, H.** 1998. Evolutionary equilibria in avian brood parasitism. In *Parasitic Birds and their Hosts* (Rothstein, S.I. & Robinson, S.K. eds), pp. 223–235. Oxford University Press, Oxford.

**Lotem, A., Nakamura, H. & Zahavi, A.** 1992. Rejection of cuckoo eggs in relation to host age: a possible evolutionary equilibrium. *Behavioral Ecology* 3: 128–132.

**Lyon, B.E.** 1998. Optimal clutch size and conspecific brood parasitism. *Nature* 392: 380–383.

M

**Major, P.F.** 1973. Scale feeding behavior of the leatherjacket, *Scombroides layson* and two species of the genus *Oligoplites* (Pisces: Carangidae). *Copeia* 1973: 151–154.

**Marchetti, K.** 1992. Costs to host defence and the persistence of parasitic cuckoos. *Proceedings of the Royal Society of London Series B* 248: 41–45.

**Maynard, B.J., DeMartini, L. & Wright, W.G.** 1996. *Gammarus lacustris* harboring *Polymorphus paradoxus* show altered patterns of serotonin-like immunoreactivity. *Journal of Parasitology* 82: 663–666.

**Maynard, B.J., Wellnitz, T.A., Zanini, N., Wright, W.G. & Dezfuli, B.S.** 1998. Parasite-altered behavior in a crustacean intermediate host: field and laboratory studies. *Journal of Parasitology* 84: 1102–1106.

**Mills, S.C. & Reynolds, J.D.** 2002 Host species preferences by bitterling, *Rhodeus sericeus*, spawing in freshwater mussels and consequences for offspring survival. *Animal Behaviour* 63: 1029–1036.

**Moore, J.** 1984. Altered behavioral responses in intermediate hosts – an acanthocephalan parasite strategy. *American Naturalist* 123: 572–577.

**Moore, J.** 2002. *Parasites and the Behaviour of Animals*. Oxford University Press, Oxford.

**Moore, J. & Gotelli, N.J.** 1996. Evolutionary patterns of altered host behavior and susceptibility in parasitized hosts. *Evolution* 50: 807–819.

**Moreau de Saint-Méry, M.L.E.** 1797. *Description Topographique, Physique, Civile, Politique et Historique de la Partie Française de l'Isle de Saint-Domingue*. Dupont, Paris.

**Moreau, J., Bertin, A., Cauber, Y. & Rigaud, T.** 2001. Sexual selection in an isopod with *Wolbachia*-induced sex reversal: males prefer real females. *Journal of Evolutionary Biology* 14: 388–394.

**Mouritsen, K.N. & Poulin, R.** 2003. Parasite-induced trophic facilitation exploited by a non-host predator: a manipulator's nightmare. *International Journal for Parasitology* 33: 1043–1050.

**Murray, B.G.** 1971. The ecological consequences of interspecific territorial behavior in birds. *Ecology* 52: 414–423.

O

**Olivier, R.C.D. & Laurie, W.A.** 1974. Birds associating with hippopotami. *Auk* 91: 169–170.

**Olson, V.A. & Owens, I.P.F.** 1998. Costly sexual signals: are carotenoids rare, risky or required? *Trends in Ecology and Evolution* 13: 510–514.

P

**Perrot-Minnot, M.-J., Kaldonski, N. & Cézilly, F.** 2007. Increased susceptibility to predation and altered anti-predator behaviour in an acanthocephalan-infected host. *International Journal for Parasitology* 37: 645–651.

**Petit, C., Hossaert-McKey, M., Perret, P., Blondel, J. & Lambrechts, M.** 2002. Blue tits use selected plants and olfaction to maintain an aromatic environment for nestlings. *Ecology Letters* 5: 585–589.

**Plaistow, S.J., Bollache, L. & Cézilly, F.** 2003. Energetically costly pre-copulatory mate-guarding in the amphipod *Gammarus pulex*: causes and consequences. *Animal Behaviour* 65(4): 683–691.

**Post, W. & Wiley, J.W.** 1977. Reproductive interactions of the shiny cowbird and the yellow-shouldered blackbird. *Condor* 79: 176–184.

**Potts, G.W.** 1973. The ethology of *Labroides dimidiatus* (Cuv. & Val.) (Labridae; Pisces) on Aldabra. *Animal Behaviour* 21: 250–291.

**Poulin, R.** 1995. 'Adaptive' changes in the behaviour of parasitized animals: a critical review. *International Journal for Parasitology* 25: 1371–1383.

**Poulin, R.** 1998. *Evolutionary Ecology of Parasites. From Individuals to Communities.* Chapman and Hall, London.

R

**Randall, J.E.** 1958. A review of the labrid fish genus *Labroides*, with description of two new species and notes on ecology. *Pacific Science* 12: 327–347.

**Reynolds, J.D., Debuse, V.J. & Aldridge, D.C.** 1997. Host specialisation in an unusual symbiosis: European bitterlings spawning in freshwater mussels. *Oikos* 78: 539–545.

**Rigaud, T.** 1997. Inherited microorganisms and sex determination of arthropod hosts. In *Influential Passengers. Inherited Microorganisms and Arthropod Reproduction* (O'Neill, S.L., Hoffmann, A.A. & Werren, J.H. eds), pp. 81–101. Oxford University Press, Oxford.

**Rigaud, T., Moreau, J. & Juchault, P.** 1999. *Wolbachia* infection in the terrestrial isopod *Oniscus asellus*: sex ratio distortion and effect on fecundity. *Heredity* 83: 469–475.

**Robert, M. & Sorci, G.** 1999. Rapid increase of host defence against brood parasites in a recently parasitized area: the case of village weavers in Hispaniola. *Proceedings of the Royal Society of London Series B* 266: 941–946.

**Robert, M. & Sorci, G.** 2001. The evolution of obligate interspecific brood parasitism in birds. *Behavioral Ecology* 12: 128–133.

**Robert, M., Sorci, G., Møller, A.P., Hochberg, M.E., Pomiankowski, A. & Pagel M.** 1999. Retaliatory cuckoos and the evolution of host resistance to brood parasites. *Animal Behaviour* 58: 817–824.

**Robinson, B.W. & Doyle, R.W.** 1985. Trade-off between male reproduction (amplexus) and growth in the amphipod *Gammarus lawrencianus*. *Biological Bulletin* 168: 482–488.

**Roff,** 1992. *The Evolution of Life Histories*. Chapman & Hall, London.

**Rohde, K.** 1980. Comparative studies on microhabitat utilization by ectoparasites of some marine fishes from the North Sea and Papua New Guinea. *Zoologischer Anzeiger* 204: 27–63.

**Rohwer, S. & Spaw, C.D.** 1988. Evolutionary lag versus bill-size constraints: a comparative study of the acceptance of cowbird eggs by old hosts. *Evolutionary Ecology* 2: 27–36.

**Rothstein, S.I.** 1975. Evolutionary rates and host defenses against avian brood parasitism. *American Naturalist* 109: 161–176.

**Rothstein, S.I.** 1990. A model system for coevolution: avian brood parasitism. *Annual Review of Ecology and Systematics* 21: 481–508.

**Rothstein, S.I.** 1993. An experimental test of the Hamilton–Orians hypothesis for the origin of avian brood parasitism. *Condor* 95: 1000–1005.

**Rothstein, S.I.** 1994. Brood parasitism and the Hamilton–Orians hypothesis revisited. *Condor* 96: 1117–1118.

S

**Sato, T.** 1986. A brood parasitic catfish of mouthbrooding cichlid fishes in Lake Tanganyika. *Nature* 323: 58–59.

**Sazima, I. & Moura, R.L.** 2000. Shark (*Carcharinus perezi*), cleaned by the Goby (*Elacatinus randalli*), at Fernando de Noronha Archipelago, western South Atlantic. *Copeia* 1: 297–299.

**Slagsvold, T.** 1998. On the origin and rarity of interspecific nest parasitism in birds. *American Naturalist* 152: 264–272.

**Smith, C., Reynolds, J.D., Sutherland, W.J. & Jurajda, P.** 2000. Adaptive host choice and avoidance of superparasitism in the spawning decisions of bitterling (*Rhodeus sericeus*). *Behavioral Ecology and Sociobiology* 48: 29–35.

**Smith, D.C.** 1992. The symbiotic condition. *Symbiosis* 14: 3–15.

**Smith Trail, D.R.** 1980. Behavioral interactions between parasites and hosts: host suicide and the evolution of complex life cycles. *American Naturalist* 116: 77–91.

**Svensden, Y.S. & Bogwald, J.** 1997. Influence of artificial wound and non-intact mucus layer on mortality of Atlantic salmon (*Salmo salar* L.) following a bath challenge with *Vibrio anguillarum* and *Aeronomas salmonicidae*. *Fish and Shellfish Immunology* 7: 317–325.

T

**Tain, L., Perrot-Minnot, M.-J. & Cézilly, F.** 2006. Altered host behaviour and brain serotonergic activity caused by acanthocephalans: evidence for specificity. *Proceedings of the Royal Society of London Series B* 273: 3039–3045.

**Takasu, F.** 1998. Why do all host species not show defense against avian brood parasitism: evolutionary lag or equilibrium? *American Naturalist* 151: 193–205.

**Thomas, J.A. & Elmes, G.W.** 1998. Higher productivity at the cost of host-specificity when *Maculinea* butterfly larvae exploit ant colonies through trophallaxis rather than by predation. *Ecological Entomology* 23: 457–464.

**Thomas, J.A., Elmes, G.W., Wardlaw, J.C. & Woyciechowski, M.** 1989. Host specificity among *Maculinea* butterfly in *Myrmica* ant nests. *Oecologia* 79: 452–457.

**Thompson, S.N. & Kavaliers, M.** 1994. Physiological bases for parasite-induced alterations of host behaviour. *Parasitology* 109: S119–S138.

U

**Urdal, K., Tierney, J.F. & Jakobsen, P.J.** 1995. The tapeworm *Schistocephalus solidus* alters the activity and response, but not the predation susceptibility of infected copepods. *Journal of Parasitology* 81: 330–333.

V

**Victoria, J.K.** 1972. Clutch characteristics and egg discrimination ability of the African village weaverbird *Ploceus cucullatus*. *Ibis* 114: 367–376.

W

**Wald, E.R., Dashefsky, B., Byers, C., Guerra, N. & Taylor, F.** 1988. Frequency and severity of infections in day care. *Journal of Pediatry* 112: 540–546.

**Webster, J.P.** 2001. Rats, cats, people and parasites: the impact of latent toxoplasmosis on behaviour. *Microbes and Infection* 3: 1037–1045.

**Webster, J.P., Gowtage-Sequeira, S., Berdoy, M. & Hurd, H.** 2000. Predation of beetles (*Tenebrio molitor*) infected with tapeworms (*Hymenolepis diminuata*): a note of caution for the Manipulation Hypothesis. *Parasitology* 120: 313–318.

**Weeks, P.** 1999. Interactions between red-billed oxpeckers, *Buphagus erythrorhynchus*, and domestic cattle, *Bos taurus*, in Zimbabwe. *Animal Behaviour* 58: 1253–1259.

**Weeks, P.** 2000. Red-billed oxpeckers: vampires or tickbirds? *Behavioral Ecology* 11: 154–160.

**Werren, J.H. & O'Neill, S.L.** 1997. The evolution of heritable symbionts. In *Influential Passengers. Inherited Microorganisms and Arthropod Reproduction* (O'Neill, S.L., Hoffmann, A.A. & Werren, J.H. eds), pp. 1–41. Oxford University Press, Oxford.

**Wilson, E.O.** 1971. *The Insect Societies*. Belknap Press of Harvard University Press, Cambridge, Massachusetts.

Y

**Yezerinac, S.M. & Dufour, K.W.** 1994. On testing the Hamilton–Orians hypothesis on the origin of brood parasitism. *Condor* 96: 1115–1116.

Z

**Zohar, A.S. & Holmes, J.C.** 1998. Pairing success of male *Gammarus lacustris* infected by two acanthocephalans: a comparative study. *Behavioral Ecology* 9: 206–211.

## Chapter 18

A

**Allee, W.C.** 1931. *Animal Aggregations*. University of Chicago Press, Chicago.

**Andersson, M.** 1994. *Sexual Selection*. Princeton University Press, Princeton.

**Andrewartha, H.G. & Birch, L.C.** 1954. *The Distribution and Abundance of Animals*. University of Chicago Press, Chicago.

B

**Biquand, S., Boug, A., Biquand-Guyot, V. & Gautier, J.P.** 1994. Management of commensal baboons in Saudi Arabia. *Reviews in Ecology* 49: 213–222.

**Birkhead, T.R. & Møller, A.P.** 1998. *Sperm Competition and Sexual Selection*. Academic Press, London, U.K.

**Blockstein, D.E. & Tordoff, H.B.** 1985. Gone forever – a contemporary look at the extinction of the passenger pigeon. *American Birds* 39: 845–851.

**Brock, M.K. & White, B.N.** 1992. Application of DNA fingerprinting to the recovery program of the endangered Puerto Rican parrot. *Proceedings of the National Academy of Science of the USA* 89: 11121–11125.

**Burley, N.** 1986. Sexual selection for aesthetic traits in species with biparental care. *American Naturalist* 127: 415–445.

C

**Cade, C.J. & Temple, S.A.** 1995. Management of threatened bird specis – evaluation of the hands-on approach. *Ibis* 137: S161–172.

**Caro, T.** (ed.) 1998. *Behavioral Ecology and Conservation Biology*. Oxford University Press, New York.

**Clemmons, J.R. & Buchholz, R.** 1997. *Behavioral Approaches to Conservation in the Wild*. Cambridge University Press, Cambridge, UK.

**Clout, M.N. & Merton, D.V.** 1998. Saving the kakapo: the conservation of the world's most peculiar parrot. *Bird Conservation International* 8: 281–296.

**Clout, M.N., Elliott, G.P. & Robertson, B.C.** 2002. Effects of supplementary feeding on the offspring sex ratio of kakapo: a dilemma for the conservation of a polyginous parrot. *Biological Conservation* 107: 13–18.

**Cox, C.R., Goldsmith, V.I. & Engelhardt, H.R.** 1993. Pair formation in California condors. *American Zoologist* 33: 126–138.

**Curio, E.** 1976. *The Ethology of Predation*. Springer-Verlag, Berlin.

**Curio, E., Ernst, U. & Vieth, W.** 1978. Cultural transmission of enemy recognition: one function of mobbing. *Science* 202: 899–901.

D

**Dennis, B.** 1989. Allee-effects: population growth, critical density, and the chance of extinction. *Natural Resource Modeling* 3: 481–538.

**Doherty, P.F.J., Sorci, G., Royle, J.A., Hines, J.E., Nichols, J.D. & Boulinier, T.** 2003. Sexual selection affects local extinction and turnover in bird communities. *Proceedings of the National Academy of Sciences of the USA* 100: 5858–5862.

F

**Fowler, C.W. & Baker, J.D.** 1991. A review of animal population dynamics at extremely reduced population levels. *Report of the International Whaling Commission* 41: 545–554.

G

**Gosling, M. & Sutherland, W.J.** (eds). 2000. *Behaviour and Conservation*. Cambridge University Press, Cambridge, UK.

**Gouyon, P.H., Henry, J.P. & Arnould, J.** 1997. *Les Avatars du Gène. La Théorie Néodarwinnienne de l'Évolution*. Belin, Paris. Collection 'Regards sur la science'.

**Grahn, M., Langefors, Å. & von Schantz, T.** 1998. The importance of mate choice in improving viability in captive populations. In *Behavioral Ecology and Conservation Biology* (Caro, T. ed.), pp. 341–363. Oxford University Press, New York.

H

**Hamilton, W.D. & Zuk, M.** 1982. Heritable true fitness and bright birds: a role for parasites? *Science* 218: 384–387.

**Hrdy, S.B.** 1979. Infanticide among animals: a review, classification, and examination of the implications for the reproductive strategies of females. *Ethology and Sociobiology* 1: 13–40.

J

**Jones, C.G., Heck, W., Lewis, R.E., Mungroo, Y., Slade, G. & Cade, T.** 1991. The restoration of the Mauritius kestrel *Falco punctatus* population. *Ibis* 137 (Supplement 1): S173–S180.

K

**Kleiman, D.G.** 1994. Animal behavior studies and zoo propagation programs. *Zoo Biology* 13: 411–412.

L

**Lande, R.** 1987. Extinction thresholds in demographic models of territorial populations. *American Naturalist* 130: 624–635.

**Lawton, J.H. & May, R.M.** (eds). 1995. *Assessing Extinction Rates*. Oxford University Press, Oxford.

**Legendre, S., Clobert, J., Møller, A.P. & Sorci, G.** 1999. Demographic stochasticity and social mating system in the process of extinction of small populations: The case of passerines introduced to New Zealand. *American Naturalist* 153: 449–463.

**Lewis, J.C.** 1990. Captive propagation in the recovery of the whooping crane. *Endangered Species Update* 8: 46–48.

M

**McLain, D.K.** 1993. Cope's rule, sexual selection, and the loss of ecological plasticity. *Oikos* 68: 490–500.

**McLain, D.K., Moulton, M.P. & Redfern, T.P.** 1995. Sexual selection and the risk of extinction of introduced birds on oceanic islands. *Oikos* 74: 27–34.

**Møller, A.P.** 2003. Sexual selection and extinction: Why sex matters and why asexual models are insufficient. *Annales Zoologici Fennici* 40: 221–230.

**Møller, A.P. & Legendre, S.** 2001. Allee effect, sexual selection and demographic stochasticity. *Oikos* 92: 27–34.

**Myers, N.** 1989. A major extinction spasm: predictable and inevitable? In *Conservation for the Twenty-First Century* (Western, D. & Pearl, M. eds), pp.42–49. Oxford University Press, Oxford.

P

**Promislow, D.E.L., Montgomerie, R. & Thomas, T.E.** 1992. Mortality costs of sexual dimorphism in birds. *Proceedings of the Royal Society of London Series B* 250: 143–150.

**Promislow, D., Montgomerie, R. & Martin, T.E.** 1994. Sexual selection and survival in North American waterfowl. *Evolution* 48: 2045–2050.

S

**Sarrazin, F.** 1998. Modelling establishment of a reintroduced population of Griffon vultures *Gyps fulvus* in southern France. In *Holartic Birds of Prey* (Chancellor, R.D., Meyburg, B.U. & Ferrero J.J. eds), pp. 405–416.

**Sarrazin, F. & Barbault, R.** 1996. Reintroduction: challenges and lessons for basic ecology. *Trends in Ecology and Evolution* 11: 474–478.

**Sarrazin, F. & Legendre, S.** 2000. Demographic approach to realising adults versus young in reintroductions. *Conservation Biology* 14: 488–500.

**Sarrazin, F., Bagnolini, C., Pinna, J.L. & Danchin, E.** 1996. Breeding biology during establishment of a reintroduced Griffon vulture *Gyps fulvus* population. *Ibis* 138: 315–325.

**Sarrazin, F., Bagnolini, C., Pinna, J.L., Danchin, E. & Clobert, J.** 1994. High survival estimates of Griffon vultures (*Gyps fulvus fulvus*) in a reintroduced population. *Auk* 111: 853–862.

**Sorci, G., Møller, A.P. & Clobert, J.** 1998. Plumage dichromatism of birds predicts introduction success in New Zealand. *Journal of Animal Ecology* 67: 263–269.

**Stattersfield, A.J., Crosby, M.J., Long, A.J. & Wege, D.C.** 1998. *Endemic Bird Areas of the World*. BirdLife International, Cambridge.

**Sutherland, W.J.** 2002. Science, sex and the kakapo. *Nature* 419: 265–266.

**Swenson, J.E., Sandegren, F., Söderberg, A., Bjärvall, A., Franzén, R. & Wabakken, P.** 1997. Infanticide caused by hunting of male bears. *Nature* 386: 450–451.

T

**Tella, J.L.** 2001. Sex-ratio theory in conservation biology. *Trends in Ecology and Evolution* 16: 76–77.

**Trewick, S.A.** 1997. On the skewed sex ratio of the kakapo *Strigops habroptilus*: sexual and natural selection in opposition? *Ibis* 139: 652–663.

**Trivers, R.L. & Willard, D.E.** 1973. Natural selection of parental ability to vary the sex ratio of offspring. *Science* 179: 90–92.

W

**Ward, P. & Zahavi, A.** 1973. The importance of certain assemblages of birds as 'information centers' for food finding. *Ibis* 115: 517–534.

**Wielgus, R.B. & Bunnell, F.L.** 1994. Sexual segregation and female grizzly bear avoidance of males. *Journal of Wildlife Management* 58: 405–413.

**Wielgus, R.B. & Bunnell, F.L.** 1995. Tests of hypotheses for sexual segregation in grizzly bears. *Journal of Wildlife Management* 59: 552–560.

**Wilson, E.O.** 1992. *The Diversity of Life*. Harvard University Press, Cambridge, Massachusetts.

## Chapter 19

A

**Alexander, R.D., Hoogland, J., Howard, R., Noonan, K. & Sherman, P.** 1979. Sexual dimorphisms and breeding systems in pinnipeds, ungulates, primates, and humans. In *Evolutionary Biology and Human Social Behavior* (Chagnon, N. & Irons, W. eds), pp. 402–435. Duxbury Press, North Scituate.

**Avital, E. & Jablonka, E.** 2000. *Animal Traditions. Behavioural Inheritance in Evolution*. Cambridge University Press, Cambridge, UK.

B

**Baker, R.R. & Bellis, M.** 1995. *Human Sperm Competition*. Chapman and Hall, London.

**Bianchi, D.W., Zickwolf, G.K., Weil, G.J., Sylvester, S. & DeMaria, M.A.** 1996. Male fetal progenitor cells persist in maternal blood for as long as 27 years postpartum. *Proceedings of the National Academy of Science of the USA* 93: 705–708.

**Boyd, R. & Richerson, P.J.** 1985. *Culture and the Evolutionary Process*. Chicago University Press, Chicago.

**Brown, A.** 1999. *The Darwin Wars. The Scientific Battle for the Soul of Man*. Simon and Schuster, London.

**Browne, J.** 2002. *Charles Darwin: The Power of Place*. Knopf, New York.

**Buss, D.M.** 1989. Sex differences in human mate preferences: evolutionary hypotheses tested in 37 cultures. *Behavioral and Brain Sciences* 12: 1–49.

**Buss, D.M.** 1994. *The Evolution of Desire: Strategies of Human Mating*. Basic Books, New York.

**Buss, D.M., Abbott, M., Angleitner, A. *et al.*** 1990. International preferences in selecting mates. A study of 37 cultures. *Journal of Cross-Cultural Psychology* 21: 5–47.

**Byrne, R. & Whiten, A.** 1988. *Machiavellian Intelligence*. Clarendon Press, Oxford.

C

**Cartwight, J.** 2000. *Evolution and Human Behaviour*. MacMillan, London.

**Catchpole, C.K. & Slater, P.J.B.** 1995. *Bird Song*. Cambridge University Press, Cambridge.

**Cavalli-Sforza, L.L. & Feldman, M.W.** 1981. *Cultural Transmission and Evolution: A Quantitative Approach*. Princeton University Press, Princeton.

**Cézilly, F.** 2006. *Le Paradoxe de l'Hippocampe*. Buchert Chastel, Paris.

D

**Daly, M. & Wilson, M.** 1983. *Sex, Evolution, and Behavior*, 2nd edition. Willard Grant, Boston.

**Daly, M. & Wilson, M.** 1984. A sociobiological analysis of human infanticide. In *Infanticide: Comparative and Evolutionary Perspectives* (Hausfater, G. & Hrdy, S.B. eds), pp. 487–502. Aldine, New York.

**Daly, M. & Wilson, M.** 1988. *Homicide*. Aldine de Gruyter, Hawthorne.

**Daly, M. & Wilson, M.** 1999. *The Truth about Cinderella: A Darwinian View of Parental Love*. Weidenfeld and Nicholson, London. (French translation: *La Vérité sur Cendrillon: Un Point de Vue Darwinien sur l'amour Parental*. Cassini, Paris, 2002.)

**Darwin, C.** 1871. *The Descent of Man, and Selection in Relation to Sex*. John Murray, London.

**Darwin, C.** 1872. *The Expression of Emotions in Man and Animals*. John Murray, London.

**Deliège, R.** 1996. *Anthropologie de la Parenté*. Armand Colin, Paris.

E

**Eaton, W.B., Pike, M.S., Short, R.V., Lee, N.C., Trussell, J., Hatcher, R.A., Wood, J.W., Worthman, C.M., Blurton Jones, N.G., Konner, M.J., Hill, K.R., Bailey, R. & Hurtado, A.M.** 1994. Women's reproductive cancers in evolutionary perspective. *Quarterly Review of Biology* 69: 353–367.

F

**Flinn, M.** 1988. Mate guarding in a Caribbean village. *Ethology and Sociobiology* 9: 335–369.

**Folstad, I. & Karter, A.J.** 1992. Parasites, bright males, and the immunocompetence handicap. *Amercian Naturalist* 139: 602–622.

**Furlow, F.B.** 1997. Human neonatal cry quality as an honest signal of fitness. *Evolution and Human Behavior* 18: 175–193.

G

**Gangestad, S.W. & Buss, D.M.** 1993. Pathogen prevalence and human mate preference. *Ethology and Sociobiology* 14: 89–96.

**Gangestad, S.W. & Thornhill, R.** 1998. Menstrual cycle variation in women's preferences for the scent of symmetrical men. *Proceedings of the Royal Society of London Series B* 265: 927–933.

**Götmark, F. & Ahlström, M.** 1997. Parental preference for red mouth of chicks in a songbird. *Proceedings of the Royal Society of London Series B* 264: 959–962.

**Grammer, K.** 1993. 5-α-androst-16en-3α-on: a male pheromone? A brief report. *Ethology and Sociobiology* 14: 201–214.

**Grammer, K. & Thornhill, R.** 1994. Human (*Homo sapiens*) facial attractiveness and sexual selection: The role of symmetry and averageness. *Journal of Comparative Psychology* 108: 233–242.

**Grammer, K., Jütte, A. & Fischmann, B.** 1998. Der Kampf der Geschlechter und der Krieg der Signale. In *Liebe, Lust und Leidenschaft. Sexualität im Spiegel der Wissenschaft* (Kanitscheider, B. ed.), pp. 9–35. Hirzel, Stuttgart.

H

**Haig, D.** 1993. Genetic conflicts in human pregnancy. *Quarterly Review of Biology* 68: 495–532.

**Hamilton, W.D.** 1968. The moulding of senescence by natural selection. *Journal of Theoretical Biology* 12: 12–45.

**Hamilton, W.D. & Zuk, M.** 1982. Heritable true fitness and bright birds: a role for parasites? *Science* 341: 289–290.

**Hausfater, G. & Hrdy, S.B.** (eds) 1984. *Infanticide: Comparative and Evolutionary Perspectives.* Aldine, New York.

**Houde, A.E.** 1997. *Sex, Color and Mate Choice in Guppies.* Princeton University Press, Princeton.

J

**Jablonka, E., Lamb, M.J. & Avital, E.** 1998. 'Lamarckian' mechanisms in Darwinian evolution. *Trends in Ecology and Evolution* 13: 206–210.

K

**Kilner, R. 1997.** Mouth colour is a reliable signal of need in begging canary nestlings. *Proceedings of the Royal Society of London Series B* 264: 963–968.

**Kohl, J.V. & Francoeur, R.T.** 1995. *The Scent of Eros.* Continuum, New York.

**Krakauer, D.C. & Pagel, M.** 1996. Selection by somatic signals: the advertisement of phenotypic state through costly intercellular signals. *Philosophical Transactions of the Royal Society of London Series B* 351: 647–658.

L

**Laland, K.N. & Brown, G.R.** 2002. *Sense and Nonsense.* Oxford University Press: New York.

**Laland, K.N., Odling-Smee, J. & Feldman, M.W.** 2000. Niche construction, biological evolution and cultural changes. *Behavioral and Brain Sciences*, 23: 131–175.

**Lefebvre, L., Gaxiola, A., Dawson, S.** *et al.* 1998. Feeding innovations and forebrain size in Australasian birds. *Behaviour* 135: 1077–1097.

**Lefebvre, L., Nicolakakis, N. & Boire, D.** 2002. Tools and brains in birds. *Behaviour* 139: 939–973.

**Lefebvre, L. Whittle, P., Lascaris, E. & Finkelstein, A.** 1997. Feeding innovations and forebrain size in birds. *Animal Behaviour* 53: 549–560.

**Lestel, D.** 1996. *L'Animalité. Essai sur le Statut de l'Humain.* Hatier, Paris.

**Levine, N.** 1988. *The Dynamics of Polyandry: Kinship, Domesticity and Population on the Tibetan Border.* Chicago University Press, Chicago.

**Lewin, R.** 1999. *Human Evolution. An Illustrated Introduction.* Blackwell, Oxford.

**Ludwig, W.** 1932. *Das Rechts-Links Problem im Tierreich und beim Menschen.* Springer-Verlag, Berlin.

M

**Møller, A.P.** 1992. Female swallow preference for symmetrical male sexual ornaments. *Nature* 357: 238–240.

**Møller, A.P.** 1996a. Parasitism and developmental stability of hosts: a review. *Oikos* 77: 189–196.

**Møller, A.P.** 1996b. Developmental stability of flowers, embryo abortion, and developmental stability of plants. *Proceedings of the Royal Society of London Series B* 263: 53–56.

**Møller, A.P.** 1997. Developmental stability and developmental selection against developmentally unstable offspring. *Journal of Theoretical Biology* 185: 415–422.

**Møller, A.P. & Pagel, M.** 1998. Developmental stability and signalling among cells. *Journal of Theoretical Biology* 193: 497–506.

**Møller, A.P. & Swaddle, J.P.** 1997. *Asymmetry, Developmental Stability, and Evolution.* Oxford University Press, Oxford.

**Møller, A.P. & Thornhill, R.** 1998. Developmental stability and sexual selection: a meta-analysis. *American Naturalist* 151: 174–192.

**Møller, A.P., Christe, P. & Lux, E.** 1999. Parasite-mediated sexual selection: Effects of parasites and host immune function. *Quarterly Review of Biology* 74: 3–20.

**Moore, D. & Haig, D.** 1991. Genomic imprinting in mammalian development: a parental tug-of-war. *Trends in Genetics* 7: 45–49.

N

**Nesse, R.M. & Williams, G.C.** 1994. *Why We Get Sick: The New Science of Darwinian Medicine.* Times Books, New York.

**Nesse, R.M. & Williams, G.C.** 1997. Evolutionary biology in the medical curriculum – what every physician should know. *BioScience* 47: 664–666.

O

**Odling-Smee, F.J., Laland K.N. & Feldman M.W.** 2003. *Niche Construction: The Neglected Process in Evolution.* Monographs in Population Biology 37. Princeton University Press, Princeton, NJ.

P

**Pagel, M.** 1993. Honest signalling among gametes. *Nature* 363: 539–541.

**Peters, J. & Beechey, C.** 2004. Identification and characterisation of imprinted genes in the mouse. *Briefings in Functional Genomics and Proteomics* 2: 320–333.

**Plavcan, J.M., Lokwood, C.A., Kimbel, W.H., Lague, M.R. & Harmon, E.H.** 2005. Sexual dimorphism in *Autralopithecus afarensis* revisited: How strong is the case for a human-like pattern of dimorphism? *Journal of Human Evolution* 48: 313–320.

**Prescott, D.M. & Flexner, A.S.** 1986. *The Misguided Cell*, 2nd edition. Sinauer, Sunderland.

R

**Reno, P.L., Meindl, R.S., McCollum, M. & Lovejoy, C.O.** 2003. Sexual dimorphism in *Australopithecus afarensis* was similar to that of modern humans. *Proceedings of the National Academy of Sciences USA* 100: 9404–9409.

**Richerson, P.J. & Boyd, R.** 1992. Cultural inheritance and evolutionary ecology. In *Evolutionary Ecology and Human Behaviour* (Smith, E.A. & Winterhalder, B. eds), pp. 61–92. Aldine de Gruyter, Chicago.

**Rose, H. & Rose, S.** 2000. *Alas, Poor Darwin.* Vintage, London.

**Rose, S., Lewontin, R. & Kamin, L.** 1984. *Not in Our Genes.* Harmondsworth, Penguin, London.

**Ruff, C.B. & Jones, H.H.** 1981. Bilateral asymmetry in cortical bone of the humerus and tibia: sex and age factors. *Human Biology* 53: 69–86.

S

**Saino, N., Ninni, P., Calza, S., Martinelli, R., De Bernardi, F. & Møller, A.P.** 2000. Better red than dead: carotenoid-based gape coloration reveals health status in barn swallow nestlings. *Proceedings of the Royal Society of London Series B* 267: 57–61.

**Short, R.R.** 1997. Darwin, have I failed you? *Trends in Ecology and Evolution* 9: 275.

**Singh, D.** 1993. Adaptive significance of female physical attractiveness: role of waist-to-hip ratio. *Journal of Personality and Social Psychology* 59: 1191–1201.

**Smith, R.L** 1984. Human sperm competition. In *Sperm Competition and the Evolution of Animal Mating Systems* (Smith, R.L. ed.), pp. 601–659. Academic Press, Orlando.

**Stearns, S.C.** (ed.). 1998. *Evolution in Health and Disease.* Oxford University Press, Oxford.

**Sterelny, K.** 2001. *Dawkins vs. Gould. Survival of the Fittest.* Icon Books, Cambridge.

T

**Thornhill, R. & Furlow, F.B.** 1998. Stress and human reproductive behavior: attractiveness, women's sexual development, postpartum depression, and baby's cry. *Advances in the Study of Behavior* 27: 319–369.

**Thornhill, R. & Gangestad, S.W.** 1994. Human fluctuating asymmetry and sexual behavior. *Psychological Science* 5: 297–302.

**Trivers, R.L.** 1974. Parent–offspring conflict. *American Zoologist* 14: 249–264.

**Trivers, R.L.** 1985. *Social Evolution.* Benjamin/Cummings, Menlo Park, California.

W

**Wedekind, C. & Füri, S.** 1997. Body odour preferences in men and women: do they aim for specific MHC combinations or simply heterozygosity? *Proceedings of the Royal Society of London Series B* 264: 1471–1479.

**Wedekind, C., Seebeck, T., Bettens, F. & Paepke, A.J.** 1995. MHC-dependent mate preferences in humans. *Proceedings of the Royal Society of London Series B* 260: 245–249.

**Williams, G.C.** 1957. Pleiotropy, natural selection, and the evolution of senescence. *Evolution* 11: 398–411.

**Wilson, E.O.** 1975. *Sociobiology: The New Synthesis.* Harvard University Press, Cambridge, Massachusetts.

**Wilson, E.O.** 1978. *On Human Nature.* Harvard University Press, Cambridge, Massachusetts.

**Wilson, E.O.** 1990. *Success and Dominance in Ecosystems: The Case of the Social Insects.* Ecology Institute, Oldendorf.

**Wright, R.** 1994. *The Moral Animal: Evolutionary Psychology and Everyday Life.* Little Brown, London. (French translation: *L'Animal Moral*, Editions Michalon, Paris, 1995.)

## Chapter 20

A

**Austerlitz, F. & Heyer, E.** 1998. Social transmission of reproductive behavior increases frequency of inherited disorders in a young expanding population. *Proceedings of the National Academy of Sciences of the USA* 95: 15140–15144.

**Avital, E. & Jablonka, E.** 2000. *Animal Traditions. Behavioural Inheritance in Evolution.* Cambridge University Press, Cambridge, UK.

B

**Barnard, C.** 2004. *Animal Behaviour. Mechanism, Development, Function and Evolution.* Pearson, Prentice Hall.

**Bentley, R.A., Hahn, M.W. & Shennan, S.J.** 2004. Random drift and culture change. *Proceedings of the Royal Society of London B* 271: 1443–1450.

**Blackmore, S.** 1999. *The Meme Machine.* Oxford University Press, Oxford.

**Boesch, C.** 1991. Teaching among wild chimpanzees. *Animal behaviour* 41: 530–532.

**Boesch, C.** 2006. Culture in evolution: towards an integration of chimpanzee and human culture. In *Explaining Culture Scientifically* (Brown, M.B. ed.). University of Washington Press, Washington.

**Boyd, R. & Richerson, P.J.** 1985. *Culture and the Evolutionary Process.* Chicago University Press, Chicago.

**Boyd, R. & Richerson, P.J.** 1988. An evolutionary model of social learning: the effects of spatial and temporal variation. In *Social Learning Psychological and Biological Perspectives* (Zentall, T.R. & Galef, B.G.J. eds), pp. 29–48. Lawrence Erlbaum Associates, Hillsdale, New Jersey, Hove and London.

**Brooks, R.** 1998. The importance of mate copying and cultural inheritance of mating preferences. *Trends in Ecology and Evolution* 13: 45–46.

C

**Caro, T.M. & Hauser, M.D.** 1992. Is there teaching in nonhuman animals. *Quarterly Review of Biology* 67: 151–174.

**Cavalli-Sforza, L.L. & Feldman, M.W.** 1981. *Cultural Transmission And Evolution: A Quantitative Approach.* Princeton University Press, Princeton, NJ.

**Chilton, G. & Lein, M.R.** 1996. Long-term changes in songs and song dialect boundaries of puget sound white-crowned sparrows. *Condor* 98: 567–580.

D

**Danchin, E. & Wagner, R.H.** 1997. The evolution of coloniality: the emergence of new perspectives. *Trends in Ecology and Evolution* 12: 342–347.

**Danchin, E., Giraldeau, L.A., Valone, T.J. & Wagner, R.H.** 2004. Public information: from nosy neighbors to cultural evolution. *Science* 305: 487–491.

**Danchin, E., Giraldeau, L.A., Valone, T.J. & Wagner, R.H.** 2005. Defining the concept of public information – response. *Science* 308: 355–356.

**Dawkins, R.** 1976. *The Selfish Gene.* Oxford University Press, Oxford.

**Diamond, J.M.** 1982a. Evolution of bowerbirds bowers – animal origins of the aesthetic sense. *Nature* 297: 99–102.

**Diamond, J.M.** 1982b. Rediscovery of the yellow-fronted gardener bowerbird. *Science* 216: 431–434.

**Dudai, Y.** 1989. *The Neurobiology of Memory*. Oxford University Press, Oxford.

**Dugatkin, L.A.** 1996. Interface between culturally based preferences and genetic preferences: Female mate choice in *Poecilia reticulata. Proceedings of the National Academy of Sciences of the USA* 93: 2770–2773.

**Dugatkin, L.A.** 1998. Interface between culturally based preferences and genetic preferences: Female mate choice in *Poecilia reticulata. Proceedings of the National Academy of Sciences of the USA* 93: 2770–2773.

**Dugatkin, L.A.** 1999. *The Imitation Factor. Evolution beyond the Gene*. New York: The Free Press.

**Durham, W.H.** 1991. *Co-evolution: Genes, Culture and Human Diversity*. Stanford University Press, Stanford.

F

**Falconer, D.S.** 1981. *Introduction to Quantitative Genetics*. Longman, New York.

**Feldman, M.W. & Laland, K.N.** 1996. Gene–culture coevolutionary theory. *Trends in Ecology and Evolution* 11: 453–457.

**Ficken, M.S. & Popp, J.W.** 1995. Long-term persistence of a culturally transmitted vocalization of the black-capped chickadee. *Animal Behaviour* 50: 683–693.

G

**Galef, B.G.** 2005. Social influences on the reproductive behavior of Japanese quail. In *Saint Andrews International Conference on Animal Social Learning*, pp. 3. Saint Andrews.

**Galef, B.G.J.** 1992. The question of animal culture. *Human Nature* 3: 157–178.

**Galef, B.G. Jr & Giraldeau L.-A.** 2001. Social influences on foraging in vertebrates: causal mechanisms and adaptive functions. *Animal Behaviour* 61: 3–15.

**Galef, B.G. Jr. & White, D.J.** 2000. Evidence of social effects on mate choice in vertebrates. *Behavioural Processes* 51: 167–175.

**Gasparini, J., McCoy, K.D., Haussy, C., Tveraa, T. & Boulinier, T.** 2001. Induced maternal response to the Lyme disease spirochaete *Borrelia burgdorferi* senus lato in a colonial seabird, the kittiwake, *Rissa tridactyla*. *Proceedings of the Royal Society of London Series B*, 268: 647–650.

**Gillespie, J.H.** 1991. *The Causes of Molecular Evolution*. Oxford University Press.

**Godin, J.-G.J., Herdman, E.J.E. & Dugatkin, L.A.** 2005. Social influences on female mate choice in the guppy, *Poecilia reticulata*: generalized and repeatable trait-copying behaviour. *Animal Behaviour* 69: 999–1005.

**Grant, B.R. & Grant, P.R.** 2002. Simulating secondary contact in allopatric speciation: an empirical test of premating isolation. *Biological Journal of the Linnean Society* 76: 545–556.

H

**Haesler, M.P. & Seehausen, O.** 2005. Inheritance of female mating preference in a sympatric sibling species pair of Lake Victoria cichlids: implications for speciation. *Proceedings of the Royal Society of London Series B* 272: 237–245.

**Henrich, J. & McElreath, R.** 2003. The evolution of cultural evolution. *Evolutionary Anthropology* 12: 123–135.

**Heyer, E., Sibert, A. & Austerlitz, F.** 2005. Cultural transmission of fitness: genes take the fast lane. *Trends in Genetics* 21: 234–239.

**Heyes, C.M. & Galef, B.G.J.** 1996. *Social Learning and Imitation: The Roots of Culture*. Academic Press, New York.

**Hochberg, M.E., Sinervo, B. & Brown, S.P.** 2003. Socially mediated speciation. *Evolution* 57: 154–158.

**Hunt, G.L. & Gray, R.D.** 2003. Diversification and cumulative evolution in New Caledonian crow tool manufacture. *Proceedings of the Royal Society of London Series B* 270: 867–874.

**Hunt, G.R.** 1996. Manufacture and use of hook-tools by New Caledonian crows. *Nature* 379: 249–251.

**Hunt, G.R.** 2000. Human like, population-level specialization in the manufacture of pandanus tools by New Caledonian crows *Corvus moneduloides*. *Proceedings of the Royal Society of London Series B* 267: 403–413.

**Hunt, G.R., Corballis, M.C. & Gray, R.D.** 2001. Laterality in tool manufacture by crows. *Nature* 414: 707.

**Hunt, G.R., Corballis, M.C. & Gray, R.D.** 2006. Design complexity and strength of laterality are correlated in New Caledonian crows pandanus tool manufacture. *Proceedings of the Royal Society of London Series B* 273: 1127–1133.

I

**Iyengar, V.K., Reeve, H.K. & Eisner, T.** 2002. Paternal inheritance of a female moth's mating preference. *Nature* 419: 830–832.

J

**Jansen, V.A.A. & Van Baalen, M.** 2006. Altruism through beard chromodynamics. *Nature* 440: 663–666.

K

**Kawai, M.** 1965. Newly acquired pre-cultural behaviour of the natural troop of Japanese monkeys on Koshima Islet. *Primates* 6: 1–30.

**Khombe, C.T., Hayes, J.F., Cue, R.I. & Wade, K.M.** 1995. Estimation of direct additive and maternal additive genetic-effects for weaning weight in Mashona cattle of Zimbabwe using individual animal-model. *Animal Science* 60: 41–48.

**Kimura, M.** 1983. *The Neutral Theory of Molecular Evolution*. Cambridge University Press.

L

**Laland, K.N.** 1994. Sexual selection with a culturally transmitted mating preference. *Theoretical Population Biology* 45: 1–15.

**Laland, K.N. & Janik, V.M.** 2006. The animal culture debate. *Trends in Ecology and Evolution* 21: 542–547.

**Laland, K.N., Odling-Smee, J. & Feldman, M.W.** 2000. Niche construction, biological evolution and cultural changes. *Behavioral and Brain Sciences* 23: 131–175.

**Lefebvre, L.** 1995. Culturally-transmitted feeding behaviour in primates: evidence for accelerating learning rates. *Primates* 36: 227–239.

**Lonsdorf, E.V., Eberly, L.E. & Ousey, A.E.** 2004. Sex differences in learning in chimpanzees. *Nature* 428: 715–716.

**Lotem, A., Wagner, R.H. & Balshine-Earn, S.** 1999. The overlooked signaling component of nonsignaling behavior. *Behavioral Ecology* 10: 209-212.

**Lunsden, C.J. & Wilson, E.O.** 1981. *Genes, Mind and Culture.* Harward University Press, Cambridge, Massachusetts.

M

**MacDougall-Shackleton, E.A. & MacDougall-Shackleton, S.A.** 2001. Cultural and genetic evolution in mountain white-crowned sparrows: song dialects are associated with population structure. *Evolution* 55: 2568–2575.

**McAdam, A.G., Boutin, S., Reale, D. & Berteaux, D.** 2002. Maternal effects and the potential for evolution in a natural population of animals. *Evolution* 56: 846–851.

**McGrew, W.C.** 1992. *Chimpanzee Material Culture: Implications for Human Evolution.* Cambridge University Press, Cambridge, UK.

**Merilä, J. & Sheldon, B.C.** 2001. Avian quantitative genetics. In *Current Ornithology*, vol. 16 (Val Nolan, J. & Thompson, C.F. eds) pp. 179–255. Kluver Academic/ Plenum Press, New York.

**Mesoudi, A., Whiten, A. & Laland, K.N.** 2004. Is human cultural evolution Darwinian? Evidence reviewed from the perspective of *The Origin of Species. Evolution* 58: 1–11.

P

**Persaud, K.N. & Galef, B.G.** 2005. Eggs of a female Japanese quail are more likely to be fertilized by a male that she prefers. *Journal of Comparative Physiology* 119: 251–256.

**Pizzari, T. & Birkhead, T.R.** 2000. Female feral fowl eject sperm of subdominant males. *Nature* 405: 787–789.

R

**Rendell, L. & Whitehead, H.** 2001. Culture in whales and dolphins. *Behavioral and Brain Sciences* 24: 309–382.

**Rosenbaum, H.C., Weinrich, M.T., Stoleson, S.A., Gibbs, J.P., Baker, C.S. & DeSalle, R.** 2002. The effect of differential reproductive success on population genetic structure: correlations of life history with matrilines in humpback whales of the Gulf of Maine. *Journal of Heredity* 93: 389–399.

S

**Siva-Jothy, M.T.** 2000. The young sperm gambit. *Ecology Letters* 3: 172–174.

**Soltis, J., Boyd, R. & Richerson, P.J.** 1995. Can group-functional behaviors evolve by cultural group selection? *Current Anthropology* 36: 473–494.

T

**Templeton, J.J.** 1998. Learning from others' mistakes: a paradox revisited. *Animal Behaviour* 55: 79–85.

**Thornton, A. & McAuliffe, K.** 2006. Teaching in wild meerkats. *Science* 313: 227–229.

**Tiedemann, R. & Milinkovitch, M.C.** 1999. Culture and genetic evolution in whales. *Science* 284: 2055.

**Tinbergen, N.** 1963. On aims and methods of ethology. *Zietschrift für Tierpsychologie* 20: 410–433.

V

**van Schaik, C.** 2004. Behavioural diversity in chimpanzees and bonobos. *Journal of Human Evolution* 46: 517–518.

**van Schaik, C.** 2005. Can we recognize innovations in the field? A test with Bornean orangutans. In *Saint Andrews International Conference on Animal Social Learning*, pp. 33. Saint Andrews.

W

**Wagner, R.H., Helfenstein, F. & Danchin, E.** 2004. Female choice of young sperm in a genetically monogamous bird. *Proceedings of the Royal Society of London Series B (Supplement)* 271: s134–s137.

**Wagner, R.H., Danchin, E., Boulinier, T. & Helfenstein, F.** 2000. Colonies as byproducts of commodity selection. *Behavioral Ecology* 11: 572–573.

**Weir, A.A.S. & Kacelnik, A.** 2006. A New Caledonian crow (*Corvus moneduloides*) creatively re-designs tools by bending or unbending aluminium strips. *Animal Cognition* 9: 317–334.

**Weir, A.A.S., Chapell, J. & Kacelnik, A.** 2002. Shaping of hooks in New Caledonian crows. *Science* 297: 981.

**White, B.N. & Galef, B.G.** 1999. Affiliative preferences are stable and predict mate choices in both sexes of Japanese quail, *Coturnix japonica. Animal Behaviour* 58: 865–871.

**White, B.N. & Galef, B.G.** 2000a. Differences between the sexes in direction and duration of response to seeing a potential sex partner mate with another. *Animal Behaviour* 59: 1235–1240.

**White, D.J. & Galef, B.G.** 2000b. Culture in quail: social influences on mate choices of female *Coturnix coturnix. Animal Behaviour* 59: 975–979.

**Whitehead, H.** 1998. Cultural selection and genetic diversity in matrilineal whales. *Science* 282: 1708–1711.

**Whitehead, H. & Rendell, L.** 2004. Movements, habitat use and feeding success of cultural clans of South Pacific sperm whales. *Journal of Animal Ecology* 73: 190–196.

**Whiten, A.** 2006. The second inheritance system of chimpanzees and humans. *Nature* 437: 52–55.

**Whiten, A. & Ham, R.** 1992. On the nature and evolution of imitation in the animal kingdom: reappraisal of a century of research. *Advances in the Study of Behaviour* 21: 239–283.

**Whiten, A., Goodall, J., McGrew, W.C., Nishida, T., Reynoldsk, V., Sugiyama, Y., Tutin, C.E.G., Wrangham, R.W. & Boesch, C.** 1999. Culture in chimpanzees. *Nature* 399: 682–685.

**Wiens, J.J.** 2001. Widespread loss of sexually selected traits: how the peacock lost its spots. *Trends in Ecology and Evolution* 16: 517–523.

**Wilcockson, R.W., Crean, C.S. & Day, T.H.** 1995. Heritability of a sexually selected character expressed in both sexes. *Nature* 374: 158 – 159.

**Wilson, E.O.** 1978. *On Human Nature*. Harvard University Press, Cambridge, Massachusetts.

**Witte, K. & Noltemeier, B.** 2002. The role of information in mate-choice copying in female sailfin mollies *Poecilia latipinna*. *Behavioral Ecology and Sociobiology* 52: 194–202.

Z

**Zahavi, A.** 1975. Mate selection: a selection for a handicap. *Journal of Theoretical Biology* 53: 205-214.

**Zahavi, A. & Zahavi, A.** 1997. *The Handicap Principle: A Missing Piece of Darwin's Puzzle*. Oxford University Press.

# Index of Species

**Note: page numbers in *italics* refer to Figures and Tables.**

## A



## B

## C

## D

# General Index

**Note: page numbers in *italics* refer to Figures and Tables, whilst those in bold refer to Glossary entries.**

## B

## C

## D

## F

## I

## Q

T